AF478585

MEDICAL APPLICATIONS OF LIPOSOMES

About the cover

Artistic impression of three different liposomes: small unilamellar vesicle with membrane associated Amphotericin B (top), small sterically stabilized unilamellar vesicle with encapsulated doxorubicin (middle) and sterically stabilized vesicle containing DNA and antibody fragments as targeting ligands on the far end of polymer chains (bottom).

Artist: Alenka Dvorzak Lasic
Technique: Mixed media
Fall 1997

MEDICAL APPLICATIONS OF LIPOSOMES

EDITED BY

D.D. LASIC

Liposome Consultations, 7512 Birkdale Drive, Newark, California, U.S.A.

D. PAPAHADJOPOULOS

California Pacific Medical Center Research Institute, Liposome Research Laboratory, San Francisco, California, U.S.A.

1998

ELSEVIER

AMSTERDAM–LAUSANNE–NEW YORK–OXFORD–SHANNON–SINGAPORE–TOKYO

ELSEVIER SCIENCE B.V.
Sara Burgerhartstraat 25
P.O. Box 211, 1000 AE Amsterdam, The Netherlands

Library of Congress Cataloging-in-Publication Data

Medical applications of liposomes / edited by D.D. Lasic, D.
Papahadjopoulos.
p. cm.
Includes index.
ISBN 0-444-82917-2 (alk. paper)
1. Liposomes--Therapeutic use. 2. Drug carriers. 3. Drug
targeting. I. Lasic, D. D. II. Papahadjopoulos, Demetrios.
[DNLM: 1. Liposomes--therapeutic use. 2. Drug Delivery Systems.
3. Neoplasms--drug therapy. 4. Communicable Diseases--drug therapy.
5. Vaccination. 6. Gene Therapy. 7. Clinical Trials. QU 93 M4895
1998]
RS201.L55M43 1998
615'.7--dc21
DNLM/DLC
for Library of Congress 98-22644
CIP

ISBN: 0-444-82917-2

∞ The paper used in this publication meets the requirements of ANSI/NISO Z39.48-1992 (Permanence of Paper).

Printed in The Netherlands.

Dedication

This book is dedicated to the following people:

To our wives: *Alenka Dvorzak Lasic* for her artistic contribution to this volume by designing the cover, and for her patience with my (D.L.) liposome pre-occupations and *Brigitte Papahadjopoulos-Sternberg* whose scientific liveliness and productivity has encouraged me (D.P.) to delay my retirement.

To our mentor and friend: *Alec Bangham*, without whom liposomes would be very different, if they were at all known, whose scientific diligence and enthusiasm kept us alive during the dark periods.

To one of the unsung heroes of Liposome Research: *Keelung Hong*, who joined my (D.P.) laboratory at UCSF twenty years ago, who was the primary inventor on several new liposome methods that were developed during this time, who has been the hands-on mentor for many of my post-doctoral fellows, and who continues to be a vital figure in my new laboratory at California Pacific Medical Center Research Institute, applying his wide knowledge in the field of liposomes with dedication, honesty and candor.

Finally, to the group of distinguished post-doctoral fellows who were trained in my (D.P.) lab, and having left the nest, have managed to become masters on their own, and enrich the liposome field with their own contributions: *Rudi Juliano*, *Frank Szoka*, *Jan Wilschut*, *Nejat Düzgüneş*, *Timothy Heath*, *Frank Martin*, *Robert Debs*, *Alberto Gabizon*, *David Daleke*, *Robert Straubinger*, *Kyung-Dall Lee*, *Paul Meers*, *S.K. (Ken) Huang*, *and Dmitri Kirpotin*.

Foreword

It is with great pleasure and excitement that we have undertaken the task of editing this book at this time of development in the liposome field. As we observe the progress of liposome research from today's prespective we see the often used description "from bench to bedside" as a full realization of our dreams and no longer just as as a hopeful statement.

The development started with liposomes as research tools to understand the function of biological membranes, and progressed into the area of cell biology and medicine as a carrier system for delivery of (macro)molecules at the cellular and organismal level. The synergy between membrane biophysics, cell biology and medicine has propelled liposomes to emerge as a leading drug delivery system, with several pharmaceutical products already approved for clinical use.

Liposomes are synthetic analogues of natural membranes. They are composed of polar lipids, which are molecules essential for the appearance of life on earth and its evolution. The special physico-chemical characteristics of polar lipids, such as their peculiar solubility, self-aggregation and membrane forming properties, as well as their phase behavior with their thermodynamic and kinetics effects, define the properties of liposomes. Their utility in biological systems derives from their biocompatibility, colloidal character and encapsulating properties. As a model membrane system, liposomes have helped unravel the mechanism of many cell membrane functions. As a carrier system for drugs and other macromolecules, they hold great promise for tissue- and cell-specific delivery of a variety of pharmaceuticals and biotechnology products.

The development in this volume of liposomes as a drug delivery system had an up and down path since the introduction of the concept in the late 1960s. While academic research of liposomes as a model membrane system did always flourish, as the exponential growth of papers can testify, the application of these findings to medicinally useful products went through several crises. After initial optimism in the 1970s and early 1980s, a period of severe scepticism ensued at the end of 1980s and beginning of 1990s, culminating in a moderate but real optimism in mid 1990s, after a successful launch of the first products in the United States and Europe.

In this collection of papers, we have attempted to gather the most promising ideas, approaches, applications and commercial developments. Because of the rather overwhelming response of the invited contributors, we feel that we have succeeded in our goal, which was to present an up-to-date collection of the present status of the field. This includes broad areas such as anti-cancer chemotherapy, immune stimulation and infectious diseases. In several areas for which we did

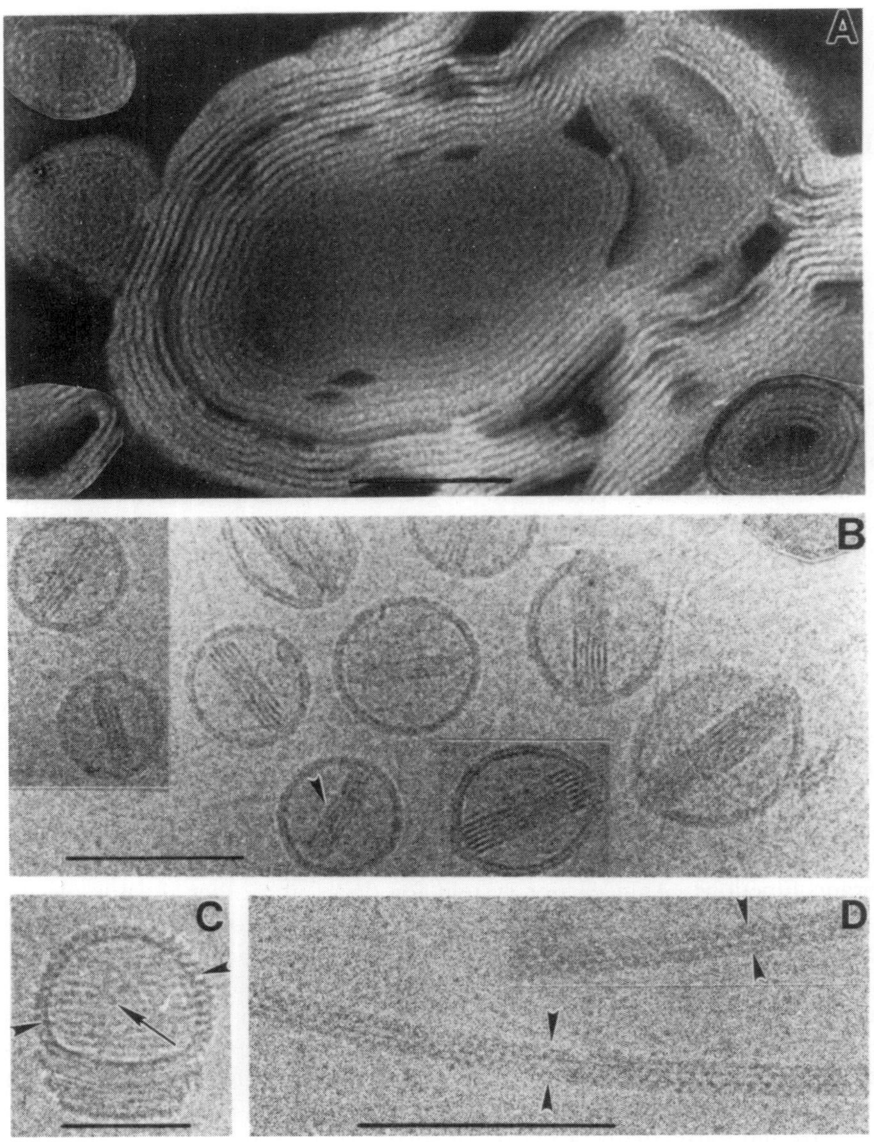

Liposomes – Past, Present, Future: Electron micrographs of various polar lipid structures. (a) *Multilamellar liposomes* of egg yolk phosphatidylcholine in water, negativelly stained with potassium phosphotungstate: from Bangham and Horne, J. Mol. Biol. 8, 660–668, 1964. This was the first published micrograph of a liposome. (b) *Unilamellar liposomes* encapsulating doxorubicin sulfate microcrystallites, as seen by cryo-electron microscopy; courtesy of Peter Frederik, Limburg University, Maastricht, The Netherlands. This type of liposome has been approved by FDA for the treatment of Kaposi's sarcoma (Chapter 8.2). (c) and (d) *cationic lipid – DNA complex* as seen by cryo-electron microscopy; courtesy of Brigitte Sternberg and Christopher Böttcher, Chapter 5.4. (c) a stacked DNA-lipid bilayer intercalated lamellar structure and (d) an elongated fibrilar structure with DNA surrounded by a lipid bilayer (Chapters 5.2 and 5.4). On all images the bar indicates 100 nm.

not request special chapters, we have briefly reviewed the current state in short introductions, before the collection of papers on this subject.

Currently, the major areas of progress are in delivery of anti-fungal agents by conventional liposomes or lipid-based carriers, and systemic anticancer therapy using long-circulating liposomes. The future applications as characterized by the direction of present day research is in specific targeting and delivery of informational molecules, such as DNA plasmids (genes), antisense oligonucleotides or ribozymes. Other future developments may be in topical delivery, vaccination and in diagnostics. Actually, the latter field is much more developed than most of the liposome scientists are aware of, but is not covered in this volume.

Although this book concentrates only on medical applications, it should be emphasized that liposome research today flourishes in numerous scientific disciplines, from mathematics (topology of two dimensional surfaces floating in a three dimensional space), theoretical physics (shapes of vesicles, phase segregation within membranes), colloid science (stability of colloids, interface phenomena), chemistry (catalysis, energy conversion, artificial photosynthesis, analytical assays, separations, organic synthesis), biochemistry (function of membrane proteins, signalling), biology (cellular functions, such as membrane transport, exo-, endocytons, membrane fusion), molecular biology (gene expression and function), pharmacology (action of drugs), as well as medicine (study of immune system, diagnostics and therapeutics).

With a large number of books focusing on liposomes already published (for a list of books and reviews see page 6, chapter 1.1), one may wonder about the rationale for another book on this subject. We feel that the field has now reached a point of maturity, and we have attempted to capture that spirit in this book. Thus, we have included chapters ranging from basic research to clinical findings, by the best people in the each area of interest. We may not agree with the conclusions of some of the contributors but we have included a variety of controversies, which represent the dynamic tension within a fast moving field.

In the relatively short time since their initial description three decades ago, liposomes, once a physicochemical curiosity in a few laboratories, have become a fact of every day life.

The electron micrograph collage "Past, Present, Future", shown opposite, illustrates the development of liposome research better than any words of ours can do.

The Editors
San Francisco, Spring 1998

Contents

SECTION III. Infectious diseases

SECTION IV. Cancer therapy

SECTION V. Gene therapy

SECTION VI. Other applications

SECTION VII. Industrial manufacturing and pre-clinical testing

Lasic and Papahadjopoulos (eds.), Medical Applications of Liposomes

CHAPTER 1.1

General introduction

Danilo D. Lasic[a] and Demetrios Papahadjopoulos[b]

[a]*Liposome Consultations, 7512 Birkdale Drive, Newark, CA 94560, U.S.A.*
[b]*Department of Cellular and Molecular Pharmacology, University of California, San Francisco, and California Pacific Medical Center Research Institute, San Francisco, CA 94115, U.S.A.*

The control over spatial and temporal distribution of drug molecules after systemic or localized administration represents the major challenge in drug delivery. While pharmacokinetics can be determined to some extent by the rate of drug introduction into the body, the spatial drug profile in various organs or its biodistribution, is much more difficult to control. Typically, for systemic applications, pharmacokinetics can be influenced by mechanical devices, and biodistribution mostly by drug carrier systems.

Chemotherapy and especially systemic administration of drugs is plagued by inefficient drug delivery to the desired site and toxic side effects, because there is practically no control over biodistribution of systemically administered drugs. Dosing and the use of mechanical devices, such as minipumps, microreservoirs with controlled leakage, skin patches, supositories, etc. can change mostly pharmacokinetics, i.e., temporal concentration profiles of the drug in the body but in general does not influence biodistribution to any important extent. For instance, infusion pumps or slow infusion can avoid peak levels of a drug administered as a bolus and provide sustained drug levels, but lack control of the disposition of the medicament. In contrast, particulate drug carriers can substantially influence not only pharmacokinetics but also biodistribution of the drug.[1] Typical particulate drug carriers are micelles, mixed micelles, emulsions, nano- and microparticles, and liposomes, which in the last twenty years, are emerging as a leading particulate drug delivery system.

Liposomes are colloidal particles in which a lipid bilayer membrane, composed from self-assembled lipid molecules encapsulates part of the aqueous phase in which they are dispersed.[2–4] They are characterized by their lipid composition, particle size distribution, number of lamellae, and inner/outer aqueous phases, all of which dictate their stability and interaction characteristics. Morphologically, we distinguish between large (L), small (S), uni (U), oligo (O) and multilamellar (ML) vesicles (V), as shown in Figure 1. Other combinations include also multivesicular liposomes in which smaller liposomes are entrapped randomly in larger vesicles where bilayers may form three dimensional network of chambers resulting in larger spherical structures resembling miniaturized foam. The commonly used

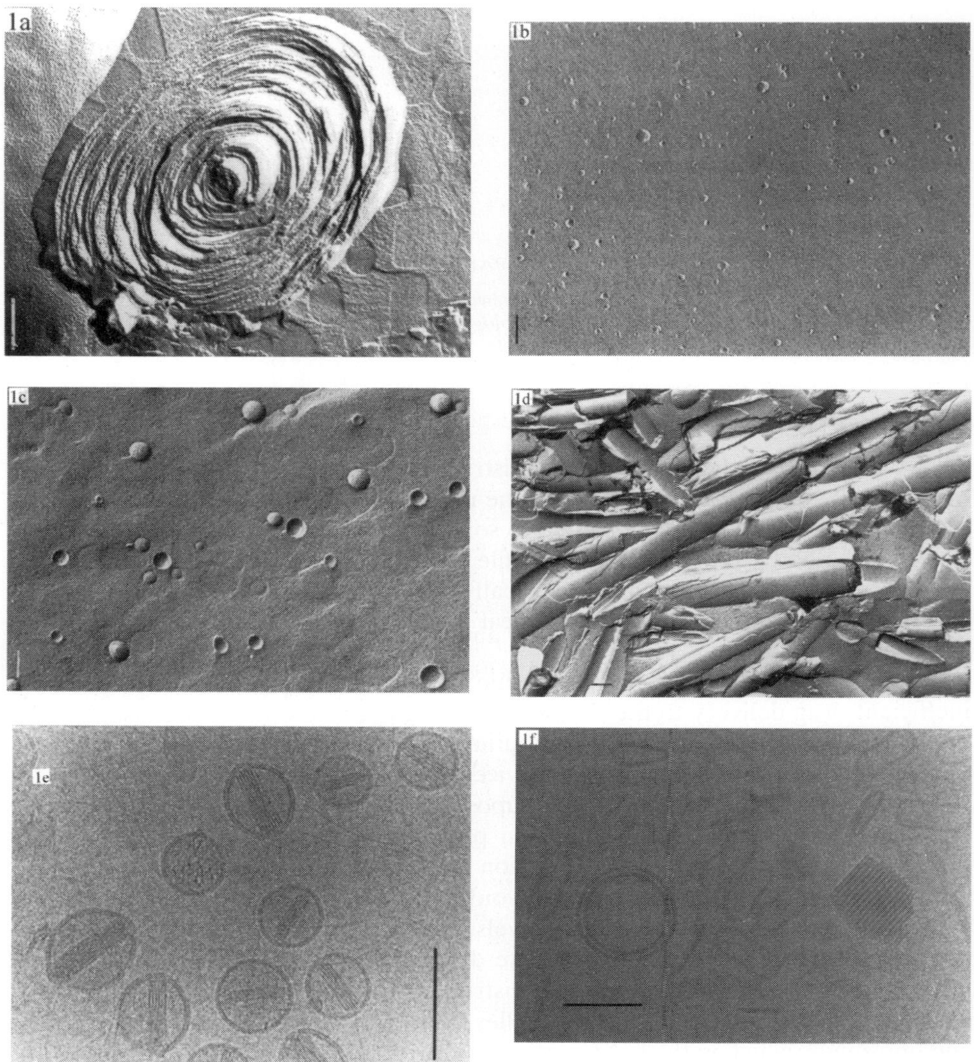

Fig. 1. Various types of liposomes according to a morphological classification. Small (SUV) typically means below 80–100 nm and giant is typically used for liposomes larger than 1 μm. Recent systemic applications often find optimal size between 100 and 150 (or 200) nm. These size ranges were historically counted as large while many recent references refer to them as small. Perhaps we should call them medium sized vesicles. Morphologically, they are mostly unilamellar.
(a) MLV, multilamellar vesicle;
(b) SUV, small unilamellar vesicle;
(c) LUV, large unilamellar vesicles;
(d) cochleate cylinders;
(e) SUV, with encapsulated doxorubicin precipitate (Doxil);
(f) cationic liposome-DNA complex (intercalated lamellar phase).
Bars indicate 100 mm. Freeze fracture electron microscopy: Courtesy of W.J. Vail (a, b, c) and B. Sternberg (d). Cryoelectron micrographs (e, f) are courtesy of P. Frederik (Limburg University, Maastricht).

liposomes are relatively stable systems. This is because they are kinetically trapped systems, not necessarily at thermodynamic equilibrium. This point will be discussed below more extensively.

With respect to interaction properties, we distinguish[5] between conventional liposomes which are characterized by a nonspecific reactivity with the milieu; sterically stabilized liposomes, which are relatively inert and therefore nonreactive to the environment; and polymorphic liposomes, which are very reactive towards specific agents. These latter include proton sensitive liposomes which aggregate, destabilize and/or fuse upon lowering of pH, or cationic liposomes, which upon interaction with nucleic acids change their structure. These are examples of liposomes which change their membrane permeability, phase and/or integrity upon relatively small stimulus from the surroundings. While liposome morphology is mostly dependent on the preparation procedure and to a lesser extent on their composition, the latter one is largely responsible for their functionality. The fact that liposome morphology is a function of preparation procedure is another indication of the kinetic contribution to the liposome characteristics and properties.

Several types of liposomes with numerous variations in lipid composition are used in drug delivery for chemotherapy of many diseases in animal models. For systemic administration, however, mostly small (and medium sized, 100–200 nm) unilamellar vesicles are used. Smaller size increases blood circulation times, increases the volume of biodistribution and allows extravasation through blood vessels, while, on the other hand, reducing the amount of encapsulated contents per mass of lipid. Therefore, for various systemic applications, the optimal size may vary depending on the target tissue. Preparation and manufacturing procedures for liposomes of various morphologies have been reviewed often and interested readers can find more information below (see list of books at end of this chapter) as well as in Section 7 of this volume.

The physico-chemical properties of liposomes as well as the fundamental physical and chemical concepts which underly their structure, stability and interaction characteristics have also been established and were reviewed elsewhere.[6,7] Briefly, liposome properties have been studied through some physically measurable quantities, such as order parameter of the bilayer, phase transitions and mechanical properties of the bilayer which are related to the bending and stretching elastic modulus, and surface properties, which can be explained by the Poisson-Boltzmann treatment in the case of electrostatic stabilization and with scaling concepts in the case of steric stabilization. This approach has enabled theoretical understanding of the observed phenomena as well as rational construction of liposomes with improved stability or specifically designed interaction properties. For a first approximation, liposome properties are based on their composition (including lipid degradation products), which defines their membrane mechanics (stretching elasticity, which is inversely proportional to membrane permeability) and surface properties, which define most of the interaction characteristics. While these concepts are already reasonably well understood, the physico-chemical explanations of liposome-anchored ligand—receptor interactions are only still emerging.

Compared to other delivery systems, the advantages of liposomes range from manufacturing and physico-chemical to biological reasons. These include biocompatibility, biodegradability, as well as relatively low toxicity and immunogenicity. With respect to liposome formulation issues, their advantages are relative ease of preparation and tayloring of their properties, as well as relatively accessible raw materials. From physico-chemical point of view the advantage of liposomes is the fact that they are not at thermodynamic equilibrium but represent a kinetically trapped system. In contrast to common belief, this makes an important difference: while systems at thermodynamic equilibrium are quickly affected by a change in the environment, kinetically trapped systems, such as liposomes, are not. For instance, liposomes are stable upon dilution while thermodynamically stable systems, such as micelles or microemulsions are not. They simply disintegrate or aggregate. Therefore, liposomes preserve their size, shape as well as encapsulated contents much better than micelles or microemulsions. Lipid bilayers also represent a strong support for attachment of various other molecules and ligands with specific function. Actually, if we take into account the small thickness of lipid membrane, we can see that mechanical properties of very cohesive bilayers, such as those composed of DSPC and cholesterol, are approaching Young modulus of van der Waals solids, such as polyethylene.

Some ten or fifteen years ago, sceptics were forecasting that liposomes cannot become a successful delivery system, because of several weaknesses often referred to in the pharmaceutical industry, as the triple S: scale up, stability and sterility. In most cases, these problems were successfully solved, as stated above, and nowadays it is possible to achieve reproducible preparations of over 100 liter quantities of well defined liposomes which can be stable for years either in liquid, frozen or freeze dried form. Smaller liposomes are sterilized by sterile 0.2 μm filtration, while larger liposomes have to be manufactured in aseptic conditions. While these conclusions seem to be quite general, in practice, however, each liposome–drug system has to be carefully optimized. In addition to the formulation issues, chemical stability of each particular drug may require thorough assessment of a series of buffers, freeze drying procedures, or storage conditions. In parallel, liposome specifications and corresponding quality control have already been established, including raw materials. Several companies are currently manufacturing and supplying apparatuses for liposome production as well as equipment for their characterization. The most important of those include extrusion, homogenization, detergent dialysis, particle size analysis, and zeta potential. For reviews, see Refs. 8–10.

Colloidal and chemical stability of liposomes were successfully solved. The former one by incorporation of charged and mostly polymer-bearing lipids in the membrane, and the latter one by the proper selection of lipids, addition of antioxidants, optimizing pH and adding metal chelators. For instance, if fluid membranes are desired without the presence of the relatively unstable double bonds, unsaturated lipids can be replaced by saturated ones with shorter and/or mixed hydrocarbon chains. Biological instability of liposomes and their short cirulation times in blood were improved drastically by coating liposome surface with inert hydrophilic polymers, such as polyethylene glycol, and other glycolipids.[11,12]

In addition to stable liposomes, efficient drug encapsulation is also an important parameter for many applications. In this respect, several "remote loading" techniques have been introduced in which preformed liposomes are filled with drug molecules added to the external solution. Molecules such as weak bases and possibly weak acids, as well as some permeable chelated metals, can be concentrated into the liposome interior space in response to specific transmembrane gradients[13–15]. The retention of some drug molecules in the liposome interior can be further enhanced by precipitating the encapsulated drug (Figure 1e). In general, however, the problem of drug encapsulation is more difficult than the literature admits. While this problem was rather neglected in the first twenty years of liposome research, now more and more groups are working on better encapsulation methods. The obvious solution for hydrophilic agents is "brute force", i.e., working at high lipid concentrations, thus obtaining closely packed spherical liposomes that encapsulate most (≤66%) of the aqueous space. Membrane embedded (hydrophobic) or electrostatically bound molecules can be rather strongly associated with liposomes during preparations. However, soon after injection or application, the equilibrium is shifted due to dilution and interactions with proteins or other substances, and in many cases the drug is released from the liposomes. Thus, the administration of hydrophobically or electrostatically associated drug molecules in liposomes may involve only a short-lived complex, which provides only temporary solubilization of the drug. In the case of stably encapsulated molecules, liposomes can not only help to dissolve some drugs and condense nucleic acids, but in addition, then can change their pharmacokinetics and biodistribution and therefore facilitate their uptake by specific tissues and internalization by target cells.

Because of the existance of such a variety of different types of liposomes, they can be used for carrying a wide spectrum of drugs. Liposomes can be formulated as a solution, dry powder, aerosol, cream or lotion, and therefore practically all conventional administration routes can be employed. While a wide variety of potent drugs that cannot be administered orally have been used in a liposomal form in numerous disease models, the majority of the successful applications are in cancer, in parasitic infections and in stimulation of the immune system. In the latter cases, the potential of the system has not been fully exploited yet[16]. The next important group of applications may involve infectious diseases and inflammation. A third tier of applications are vaccination as well as delivery of nucleic acids and other molecules with informational sequences. Some additional, less established applications will be represented in Section 6.

The aim of this selection of articles is to cover all these areas with an up-to-date review by researchers whose work in many cases has established the field or at least it had a strong influence in its development.

References

1. Matsumura Y, Maeda H. A new concept for macromolecular therapeutics in cancer chemotherapy; mechanism of tumoritropic accumulation of proteins and the antitumor activity of Smanes. Cancer Res 1986;46:6387–6392.

2. Bangham AD, Standish, MM, Watkins JC. Diffusion of univalent ions across the lamellae of swollen phospolipids. J Mol Biol 1965;13:238–252.
3. Papahadjopoulos D (ed.). Liposomes and their use in biology and medicine. Ann NY Acad Sci 1978;408:1–412.
4. Lasic DD. Liposomes: from physics to applications. Amsterdam: Elsevier, 1993.
5. Lasic DD, Papahadjopoulos D. Liposomes revisited. Science 1995:267:1267–1276.
6. Barenholz Y, Crommelin DJA: Liposomes as pharmaceutical dosage forms. In: Swarbick J, Boylan JC, eds. Encyclopedia of Pharmaceutical Technology, Vol 9, New York: M. Dekker, 1994;1–39.
7. Lasic DD, Needham D. Stealth liposomes: a prototypical biomaterial. Chem Rev 1995;95:2601–2628.
8. Szoka F, Papahadjopoulos D. Comparative properties and methods of preparation of lipid vesicles (liposomes). Ann Rev Biophys Bioeng 1980;9:476–508.
9. Woodle MC, Papahadjopoulos D. Liposome preparation and size characterization. Meth Enzy 1989;171:193–217.
10. Cullis PR, Hope MJ (eds.). Special issue: liposomes. Chem Phys Lipids 1986;40:2–4.
11. Woodle MC, Lasic DD. Sterically stabilized liposomes. Biochim Biophys Acta 1992;113:171–199.
12. Allen TM, Papahadjopoulos D. Sterically stabilized (stealth) liposomes: pharmacokinetic and therapeutic advantages. In: Gregoriadis G. ed. Liposome Technology, 2nd edn, Vol III. Ch 5, Boca Raton, Florida: CRC Press, 1992;59–72.
13. Mauk MR, Gamble RC. Preparation of lipid vesicles containing high levels of entrapped radioactive cations. Anal Biochem 1979;302–312.
14. Mayer LD, Bally MB, Hope MJ, Cullis PR. Techniques for encapsulating bioactive agents into liposomes. Chem Phys Lip 1986;40:333–345.
15. Haran G, Cohen R, Bar LK, Barenholz Y. Transmembrane ammonium sulfate gradients in liposomes produce efficient and stable entrapment of amphiphilic weak bases. Biochem Biophys Acta 1993;1151:201–215.
16. van Rooijen N, Sanders A. Liposome-mediated depletion of macrophages: mechanism of action, preparation of liposomes and applications. J Immunol Meth 1994;174:83–93.

List of major books and reviews in liposomology (June 1997)

1. Papahadjopoulos D (ed). Liposomes and their uses in biology and medicine, Vol 308: Annl N Y Acad Sci, 1978.
2. Tom BH, Six HR. Liposomes and immunobiology. Amsterdam: Elsevier/North Holland, 1980.
3. Gregoriadis G, Allison AC (eds). Liposomes in Biological Systems. John Wiley & Sons, Chichester, New York, 1980.
4. Knight CG (ed). Liposomes: from physical structure to therapeutic application. Amsterdam: Elsevier/North Holland, Biomedical Press, 1981.
5. Nicolau C, Paraf A (eds). Liposomes, Drugs, and Immunocompetent Cell Functions. Academic Press, London, New York, 1981.
6. Bangham AD (ed). Liposome letters, Academic Press, 1983.
7. Gregoriadis G (ed). Liposome technology, Vols 1, 2, 3, Boca Raton, FL: CRC Press, 1984.
8. Yagi K (ed). Medical applications of liposomes, Special issue. Tokyo: Scientific Societies Press, 1986.
9. Cullis PR, Hope MJ (ed). Liposomes, Special Issue. Chem Phys Lipids 1986;40(2–4):87.
10. Schmidt KH (ed). Liposomes as drug carriers. Stuttgart: George Thiem Verlag, 1986.
11. Ostro M (ed). Liposomes: from biophysics to therapeutics. New York: Marcel Dekker, 1987.
12. Machy P, Leserman L. Liposomes in Cell Biology and Pharmacology. John Libbey and Co., London, 1987.
13. Gregoriadis G, (ed). Liposomes as drug carriers. New York: Wiley, 1988.
14. Lichtenberg D, Barenholz Y. Liposomes: preparation, characterization and preservation. In: Glick D, ed. Methods in Biochemical Analysis, Vol 33, New York: Wiley, 1988.
15. Lopez-Berestein G, Fidler IJ (eds). Liposomes in the Therapy of Infectious Diseases and Cancer. New York: Alan R. Liss, 1989.
16. New RRC (ed). Liposomes—A practical approach. Oxford: IRL Press, 1990.
17. Marsh D, Phill D. Handbook of lipid bilayers. Boca Raton, FL: CRC Press, 1990.
18. Szoka FC. Liposomal drug delivery: Current status and future prospects. In: Wilschut J, Hoekstra D, eds. Membrane fusion. New York: Plenum, 1991.

19. Braun-Falco O, Korting HC, Maibach HI (eds.). Liposome dermatics. Berlin: Springer Verlag, 1992.
20. Gregoriadis G (ed). Liposome technology, 2nd edn, Vols 1, 2, 3. Boca Raton, FL: CRC Press, 1993.
21. Lasic DD. Liposomes: from physics to applications. Amsterdam: Elsevier, 1993.
22. Barenholz Y (ed). Special issue: Quality Control of Liposomes. Chem Phys Lipids 1993;64.
23. Cevc G (ed). Phospholipid Handbook. Marcel Dekker, 1994.
24. Barenholz Y, Crommelin D. Liposomes as pharmaceutical dosage forms. In: Swarbrick J, Boylan JCC, eds. Encyclopedia of pharmaceutical technology. New York: Marcel Dekker, 1994;9.
25. Philippot JR, Schuber F (ed). Liposomes as tools in basic research and industry. Boca Raton, FL: CRC Press, 1995.
26. Lasic DD, Martin FJ (ed). Stealth liposomes. Boca Raton, FL: CRC Press, 1995.
27. Nicolau C, Alving CR (eds). Special issue: Festschrift for D. Papahadjopoulos: Liposomes from art to science. J Liposome Res 1995;5(4).
28. Rossof M (ed). Vesicles. New York: M. Dekker, 1996.
29. Lasic DD, Barenholz Y (eds). A Handbook of Nonmedical Applications of Liposomes, Vol 1: From theory to basic science. Boca Raton, FL: CRC Press, 1996.
30. Barenholz Y, Lasic DD (eds). A Handbook of Nonmedical Applications of Liposomes, Vol 2: Models for biological phenomena. Boca Raton, FL: CRC Press, 1996.
31. Barenholz Y, Lasic DD (eds.). A Handbook of Nonmedical Application of Liposomes, Vol 3: From design to microreactors. Boca Raton, FL: CRC Press, 1996.
32. Lasic DD, Barenholz Y (eds). A Handbook of Nonmedical Application of Liposomes, Vol 4: From gene delivery and diagnostics to ecology. Boca Raton, FL: CRC Press, 1996.
33. Lasic DD. Liposomes in Gene Delivery. Boca Raton, FL: CRC Press, 1997.

Lasic and Papahadjopoulos (eds.), Medical Applications of Liposomes

CHAPTER 1.2

Liposome research in drug delivery and targeting: thoughts of an early participant

Gregory Gregoriadis

Centre for Drug Delivery Research, The School of Pharmacy, University of London, 29–39 Brunswick Square, London WC1N 1AX, England

Rational research in drug delivery began in the 1950's with the use of polyclonal antitumour antibodies for the targeting of cytostatic drugs to experimental tumours.[1] This work, although intellectually attractive, was unremarkable in terms of results. Equally attractive as a concept were attempts in the 1960's to deliver enzymes and other proteins via microcapsules made of nylon and suchlike materials.[2] Over the years, drug delivery and targeting evolved to produce advanced system versions of considerable sophistication and wider scope (e.g., humanized mouse monoclonal antibodies, an array of receptor-specific ligands and a multitude of designer polymers and microspheres). Effective targeting of drugs in the treatment or prevention of disease via the great range of delivery systems currently available is the aim of numerous research groups worldwide. It covers therapies such as those for cancer, microbial infections, hormone and enzyme deficiencies and gene malfunction, as well as vaccines. Central to the success of these systems have been on the one hand intimate knowledge of their structure and physical properties, and on the other, ways by which such properties influence the behaviour of systems within the biological milieu. However obvious it may now seem, the significance of the relationship between these two aspects in the design of pharmacologically optimal constructs was fully realized and systematically explored with the advent of liposomes.

When liposomes were first observed and their semipermeable nature described by Bangham and colleagues[3,4] in the early 1960's, biophysicists, cell biologists and biochemists were presented with a unique system for the study of natural membranes and their properties. The structural versatility of the liposome also allowed for its manipulation (by the liposome school that formed in the succeeding years) through the addition of lipid soluble agents into the bilayers or of water soluble agents in the aqueous compartments, both of which helped in the construction of minimalistic versions of cells and the study of some of the latter's functions. Work on this aspect of liposomology has contributed to such diverse areas of experimental cell biology as reconstituted pumps, membrane fusion and antigen presentation.[5] However, it took several years from the early to the late 60's before the system was looked upon as a candidate drug transporter, an evolutionary side jump so to speak. Had liposomes been discovered in today's climate of unbridled exploitation of any system that may show the slightest promise in drug delivery or targeting, their adoption as a drug carrier would have been instant, making its discoverers rich in the process! This scenario, however, puts the cart before the horse for, I would argue, it is the nature of events, experiences and acquired knowledge associated with liposome research that helped to drive the field to where it stands today.

Progress in liposome research on drug delivery and targeting has been often seen retrospectively in the context of decades marked by significant milestones, obviously with some overlaps in between. Thus, the 1970's are noted for the initial knowledge on the system's interaction with the biological milieu and, on the basis of results from such interaction, the proposition of a variety of applications in therapeutics.[6,7] Following a certain "disillusionment" in the hearts and minds of those who would, in another age, demand no less than a Jumbo jet from the Wright brothers the day after their first flight, the 1980's became known as a period of reflection, consolidation of previous in vivo work, imaginative solutions to problems encountered in the interaction of liposomes with blood and tissues, significant advances in the application of the system in a number of therapies, and last but not least, breakthroughs in drug entrapment.[8,9] Above all, the founding of three liposome-based companies in the USA ensured a systematic transition of some of the ideas (abundantly floating on both sides of the Atlantic) to realistic goals, backed with appropriate large-scale technology. The present decade is not over yet but it is clearly the decade of approved injectable products, new horizons and long-term optimism.[10–12] Below I discuss early key developments which I believe helped to shape the future of our field. The audience I have in mind includes those liposomologists who, now in their twenties and thirties, are entrapped in a labyrinth of a myriad of publications and reviews of varying elegance, clarity, insight, or even bias.

As alluded to earlier, the state of infancy in drug delivery at the time liposomes were discovered, and the interests of individuals engaged with the system, did not encourage thoughts on its alternative, drug carrier facet with which we all are now familiar. My own involvement with it, described in detail elsewhere,[13] was accidental in that I happened in 1969 to come across an advertisement for a research post with Brenda Ryman on enzyme delivery to the hepatic parenchymal cells. Having previously participated in the discovery of the hepatic galactose receptor,[14] the opportunity that I had been seeking to apply galactose-terminating ligands as such or in association with particulate systems in drug or enzyme targeting to the liver was not to be missed. On my arrival in England in the summer of 1970, it turned out that the candidate system for enzyme delivery was one called "liposomes" (a suggestion to that effect was also put forward independently by Gerry Weissmann's group in 1969).[15] Because of my familiarity with work on the fate of macromolecules (glycoproteins) in vivo at the cellular and subcellular level,[14] facts about the clearance of intravenously injected liposomes and entrapped markers, their distribution in tissues and the (endocytic) mechanism of uptake within intracellular organelles were easy to ascertain. Published in 1972, our data[16,17] supported the use of liposomes in enzyme replacement therapy, a notion that was successfully tested soon afterwards in a model of lysosomal storage disease.[18] Having a head start in this area, and being sufficiently aware of the need for specific drug action in a plethora of therapies, further exploration of the system's potential uses extended these to cancer[19,20] and anti-microbial therapy,[19] and established the concept[21] of vesicle targeting with surface bound antibodies and other cell specific ligands (e.g., asialoglycoproteins). Equally exciting but perhaps more significant in its implications, was the finding[22,23] (together with Tony Allison) that liposomes are capable of potentiating immune responses to entrapped antigens. Seen from today's perspective, these quarter of a century old papers could appear to some naive (or perhaps courageous) in their claims and vulnerable in their assertive optimism. It is an arresting thought however, that these innocent flights of fancy have now ended up wrapped in red ribbons, on the desks of hard-nosed lawyers, eagle-eyed patent attorneys and worried CEOs, or hidden in the "high"s and "low"s numbers of the NASDAQ stock list. Progress in liposome research in the 1970's accelerated and also branched into additional avenues by the entry into the race of a number of formidable individuals (notably the co-editor of this book Demetrios Papahadjopoulos) whose influence in the direction of the field, finally culminating in life-saving products,[10] has been seminal.[24–26] The Introduction to the book and some of its chapters will no doubt outline their achievements far more efficiently and eloquently than I could ever do.

It has been often stated that a major disadvantage of the liposomal carrier is its early interception by the fixed macrophages of the liver and spleen. Yet, participation of the reticuloendothelial system (RES) in vesicle uptake is the basis of the mode of action of several of the licenced liposome-based products,[10] including the recent vaccine against hepatitis A and, as our data suggest elsewhere in this book (Chapter 2.4), would explain the apparent success of liposome-mediated DNA vaccination.[27] It is nonetheless true that a significant delay of RES involvement would extend the circulation time of liposomes, thus enabling them to deliver their drug load to alternative sites and, as a result, enlarge the spectrum of the system's possibilities in therapy. The way with which the challenge of long-lived liposomes was met, one of the better paradigms of rational carrier design, has been discussed elsewhere in detail.[28] Briefly, it was based on the use of neutral small unilamellar vesicles shown[29] in 1975 to exhibit longer circulation times than larger liposomes. Addition of excess cholesterol[30] in the bilayers and substitution of unsaturated phospholipids with the high-melting distearoyl phosphatidylcholine[31] or with sphingomyelin[32,33] resulted in vesicles of greatly improved stability on exposure to plasma high density lipoproteins, a known agent of vesicle destabilization in vivo.[28,34] It turned out that the greater the bilayer stability (in terms of drug retention by the vesicles), the longer the vesicle half-life in the circulation.[28,31] A formulation of stable, long lived liposomes is now marketed by one of the liposome companies as a carrier of cytostatic drugs in the treatment of cancer.[10] An important subsequent innovation extended the half-life of such stable vesicles even further and also rendered it dose-independent. Thus, several groups[35–38] were able to show simultaneously that coating of the liposomal surface with the hydrophilic polyethylene glycol, which interferes with the adsorption of blood opsonins on the vesicles,[38] curtails their recognition by the RES. One of these formulations has recently become the flagship product of another liposome-based company for the treatment of certain cancers.[10]

Such is the versatile nature of the liposome that its manipulation to versions tailored for specific functions can only be limited by the imagination of the practitioner. Encouraging new developments[10,11] in this respect include applications in tumour targeting,[39] gene[40] and antisense[41] therapy, genetic vaccination,[27] immunomodulation,[42,43] lung therapeutics,[44] and cyclodextrin-controlled drug release in situ.[45] Not surprisingly, the culture of liposomes and its promises has already penetrated non-medical areas[12] including bioreactors, catalysis, cosmetics and ecology. The reader will have noted my optimism for the future of our system and might like me, have considered the difficulty of inventing an alternative one of similar attributes and potential, or the impossibility of disciplining every drug for erratic or dangerous behaviour in vivo through molecular modelling. Were either of these two little pigs to fly, and liposomes became the Titanic of drug delivery systems, I would rather go down with it than jump ship. But of course it ain't going to happen!

References

1. O'Neil GJ. The use of antibodies as drug carriers. In: Gregoriadis G, ed. Drug carriers in biology and medicine. London: Academic Press, 1979;23–41.
2. Chang TMS. Artificial cells as drug carriers in biology and medicine. In: Gregoriadis G, ed. Drug carriers in biology and medicine. London: Academic Press, 1979;271–285.
3. Bangham AD, Standish MM, Weismann G. Diffusion of univalent ions across the lamellae of swollen phospholipids. J Mol Biol 1965;13:238–252.
4. Papahadjopoulos D, Bangham AD. Biophysical properties of phospholipids. II. Permeability of phosphatidylserine liquid crystals to univalent ions. Biochim Biophys Acta 1966;126:185–188.
5. Bangham AD (ed). Liposome letters. London: Academic Press, 1983.
6. Gregoriadis G. The carrier potential of liposomes. New Engl J Med 1976;295:704–710 and 765–770.
7. Papahadjopoulos D (ed). Liposomes and their uses in biology and medicine. New York Academy of Sciences, 1978;308:1–462.
8. Gregoriadis G (ed). Liposomes as drug carriers: recent trends and progress. Chichester: John Wiley, 1988.
9. Gregoriadis G (ed). Liposome technology, vols I–III. Boca Raton: CRC Press, 1993.
10. Gregoriadis G. Engineering targeted liposomes: progress and problems. Trends in biotechnology 1995;5:635–639.
11. Puisieux F, Couvreur P, Delattre J, Devissagnet J-P (eds). Liposomes, new systems and new trends in their applications. Paris: Editions de Santé, 1995.
12. Lasic DD, Barenholz Y (eds). Non-medical applications of liposomes, vols I–IV. Boca Raton: CRC Press, 1996.
13. Gregoriadis G. . . . Twinkling guide stars to throngs of acolytes desirous of your membrane semi-barriers precursors of bion, potential drug carriers . . . . J Liposome Research 1995;5;329–346.
14. Gregoriadis G, Morell AG, Sternlieb I, Scheinberg IH. Catabolism of desialylated ceruloplasmin in the liver. J Biol Chem 1970;245:5833–5837.
15. Sessa G, Weismann G. Formation of artificial lysosome in vitro. J Clin Invest 1969;48:76a.
16. Gregoriadis G, Ryman BE. Fate of protein-containing liposomes injected into rats. An approach to the treatment of storage diseases. Eur J Biochem 1972;24:485–491.
17. Gregoriadis G, Ryman BE. Lysosomal localization of β-fructofuranosidase-containing liposomes injected into rats. Biochem J 1972;129:123–133.
18. Gregoriadis G, Buckland RA. Enzyme-containing liposomes alleviate a model for storage disease. Nature (London) 1973;2:170–172.

19. Gregoriadis G. Drug entrapment in liposomes. FEBS Lett 1973;36:292–296.
20. Gregoriadis G, Swain CP, Wills EJ, Tavill AS. Drug-carrier potential of liposomes in cancer chemotherapy. Lancet 1974;I:1313–1316.
21. Gregoriadis G, Neerunjun DE. Homing of liposomes to target cells. Biochem Biophys Res Comm 1975;65:537–544.
22. Allison AC, Gregoriadis G. Liposomes as immunological adjuvants. Nature (London) 1974; 252:252.
23. Gregoriadis G, Allison AC. Entrapment of proteins in liposomes prevents allergic reactions in preimmunised mice. FEBS Lett 1974;45:71–74.
24. Gregoriadis G. Demetrios Papahadjopoulos: An encounter of the Greek kind. J Liposome Res 1995;5:635–639.
25. Alving CR. Liposomes as carriers of antigens and adjuvants. Immunol Methods 1991;140:1–13.
26. Nicolau C, Paraf A (eds). Liposomes, drugs and immunocompetent cell functions. London: Academic Press, 1981.
27. Gregoriadis G, Saffie R, de Souza B. Liposome-mediated DNA vaccination. FEBS Lett 1997;402:107–110.
28. Gregoriadis G. Fate of liposomes in vivo and its control: A historical perspective. In: Lasic L, Martin F, eds. Stealth liposomes. Boca Raton: CRC Press Inc, 1995;7–12.
29. Juliano R, Stamp D. Effects of particle size and charge on the clearance of liposomes and liposome-encapsulated drugs. Biochem Biophys Res Comm 1975;63:651–658.
30. Kirby C, Clarke J, Gregoriadis G. Effect of the cholesterol content of small unilamellar liposomes on their stability in vivo and in vitro. Biochem J, 1980;186:591–598.
31. Senior J, Gregoriadis G. Is half-life of circulating small unilamellar liposomes determined by changes in their permeability? FEBS Lett, 1982;145:109–114.
32. Hwang KJ, Luke KFS, Baumier PL. Hepatic uptake and degradation of unilamellar sphingomyelin/cholesterol liposomes: A kinetic study. Proc Acad Sci USA 1980;77:4030–4034.
33. Gregoriadis G, Senior J. The phospholipid component of small unilamellar liposomes controls the rate of clearance of entrapped solutes from the circulation. FEBS Lett 1980;119:43–46.
34. Scherphof G, Roerdink DDDG, Waite M, Parks J. Disintegration of phosphatidylcholine liposomes in plasma as a result of interactions with high density lipoproteins. Biochim Biophys Acta 1978;542:296–307.
35. Blume G, Cevc G. Liposomes for the sustained drug release in vivo. Biochim Biophys Acta 1990;1029:91–97.
36. Klibanov AL, Marnyama K, Torchilin VP, Huang L. Amphipathic polyethyleneglycols effectively prolong the circulation time of liposomes. FEBS Lett 1990;268:235–237.
37. Papahadjopoulos D, Allen T, Gabizon A, Mayhew E, Matthay K, Huang K, Lee SK, Woodle MC, Lasic DD, Redemann C, Martin FJ. Sterically stabilized liposomes: improvements in pharmacokinetics, tissue disposition and anti-tumour therapeutic efficacy. Proc Natl Acad Sci USA 1991;88:11460–11464.
38. Senior JH, Delgado C, Fisher D, Tilcock C, Gregoriadis G. Influence of surface hydrophilicity of liposomes on their interaction with plasma proteins and clearance from the circulation: Studies with polyethylene glycol-coated vesicles. Biochim Biophys Acta 1991;1062:77–82.
39. Lasic DD, Papahadjopoulos D. Liposomes revisited. Science 1995;267:1275–1276.
40. Legendre J-Y, Szoka Jr FC. Liposomes for gene therapy. In: Puisieux F, Couvreur P, Delattre J and Devissagnet J-P, eds. Liposomes, new systems and new trends in their applications. Paris: Editions de Santé, 667–692.
41. Zelphati Oster, Wagner E, Leserman L. Synthesis and anti-HIV activity of thiocholesterol-coupled phosphodiester oligonucleotides incorporated into immunoliposomes. Antiviral Res 1994;25:13–25.
42. Barratt G, Morin C, Schuber F. Liposomal immunomodulators. In: Puisieux F, Couvreur P, Delattre J and Devissagnet J-P, eds. Liposomes, new systems and new trends in their applications. Paris: Editions de Santé, 461–506.
43. Gürsel M, Gregoriadis G. Interleukin-15 acts as an immunological co-adjuvant for liposomal antigen in vivo. Immunology Letters 1997;55:161–165.
44. Denizot B, Proust J-E, Tchoreloff P-C, Gulik A. Liposomes for the pulmonary route. In: Puisieux F, Couvreur P, Delattre J, Devissagnet J-P, eds. Liposomes, new systems and new trends in their applications. Editions de Santé, Paris, pp 574–614.
45. McCormack B, Gregoriadis G. Comparative studies of the fate of free and liposome-entrapped hydroxypropyl-β-cyclodextrin/drug complexes after intravenous injection into rats: Implications in drug delivery. Biochim. Biophys. Acta 1996;1291:237–244.

Lasic and Papahadjopoulos (eds.), Medical Applications of Liposomes

CHAPTER 2.1

Class I presentation of liposomal antigens

Mangala Rao and Carl R. Alving

Department of Membrane Biochemistry, Walter Reed Army Institute of Research, Washington, DC 20307-5100, USA

Overview

I. Introduction

The current paradigm for induction of an immune response against a protein antigen invokes the interaction of the antigen with an antigen presenting cell (APC) that partially degrades the antigen and channels peptides into the appropriate families of immunological responses.[1,2] Based on interactions with different types of major histocompatibility complex (MHC) molecules (class I or class II) on the APC, protein antigens can be processed and presented either by MHC class I or class II pathways. MHC class I and class II molecules are highly polymorphic membrane proteins, that bind and transport peptide fragments of intact proteins to the surface of APCs. The peptide-MHC complex then interacts with either $CD8^+$ or $CD4^+$ T lymphocytes. It is generally believed that endogenous proteins of a cell are presented via the MHC class I pathway, whereas, exogenous (extracellular) antigens are presented via the MHC class II pathway. In most cells, exogenous antigens cannot be presented by class I molecules because of the inability of antigens to gain access to the cytosol. Therefore, most soluble antigens are poor at priming MHC class I-restricted cytotoxic T lymphocyte (CTL) responses unless they are artificially introduced into the cytoplasm by osmotic loading,[3] covalently or noncovalently associated with lipid carriers,[4–6] conjugated to latex beads,[7,8] or encapsulated in liposomes.[9–15] For the purpose of this review we will focus on recent advances in understanding the role of liposome-encapsulated antigens in the class I presentation pathway.

II. Cellular fate of liposomal antigens

One of the original and most well-known observations in the field of liposome research is that when liposomes are injected intravenously, they are ingested by splenic macrophages and Kupffer cells in the liver. A comprehensive report on this aspect has been presented by Woodle and Lasic[16] and refs. therein. The best characterized mechanism of liposomal uptake is phagocytosis.[16–20] Studies have shown that binding of liposomes occurs as a result of specific opsonization followed by uptake. The liposomes are then degraded in lysosomal vacuoles.[16] Despite extensive studies, the mechanisms of liposome uptake leading to their localization in endosomal and lysosomal compartments are still poorly understood.

Macrophages are thought to be the predominant APCs responsible for processing and presentation of particulate antigens, including liposomal antigens.[21–25] The interaction with phagocytes probably is a general property for antigen presentation of particulate antigens, inasmuch as ovalbumin coupled to latex beads is also presented by macrophages.[24]

The initial studies demonstrating that phagocytes could present exogenous antigen on class I molecules were performed in vitro and a key question has been whether phagocytes are also operative as APCs in vivo. The antigen presenting function of the phagocyte has been demonstrated in vivo in studies showing the presentation of ovalbumin by class I molecules present on the surface of macrophages following injection of soluble ovalbumin into mice.[25] Other studies have analyzed the effect of depleting phagocytic function in vivo by injection of silica, carageenan or liposomes cytotoxic for macrophages. This inhibits the presentation of several forms of particulate antigens, including antigen encapsulated in liposomes.[26]

Recently, dendritic cells (DC) have also been shown to be involved in the induction of CTL responses to liposome-encapsulated antigens.[26,27] Nair et al.,[26] demonstrate that a potent primary CTL response against a soluble protein can be achieved by delivering antigen in pH-sensitive liposomes to DCs either in vivo or in vitro. However, if macrophages are depleted in vivo by the drug dichloromethylene diphosphate encapsulated in liposomes prior to antigen exposure, DCs are ineffective as APCs for CTL induction. These studies indicate a role for macrophages in enhancing the antigen presenting function of DCs.

II.1. Phagocytosis vs. Endocytosis

The ability to internalize antigens by phagocytosis or by endocytosis is a common feature of these APCs and may be important to the pathway of presentation. In studies on the ingestion of liposomes by cells the terms endocytosis and phagocytosis have often been used interchangeably. It has occurred to us that this looseness of nomenclature has resulted in poor differentiation between the processes of pinocytosis, macropinocytosis, endocytosis, and phagocytosis. These different processes can be distinguished by the size of the ingested particle and by the presence or absence of receptor mechanisms. In general, in our view phagocytosis can be considered as being restricted to relatively large particles (probably those greater than 0.2 μm) while pinocytosis and endocytosis are restricted to soluble proteins and small particles (probably those less than 0.2 μm). In addition, phagocytosis may be associated with the presence of specific receptors on phagocytic cells, such as the complement receptor or Fc receptor.[28] In contrast, macropinocytosis, a process that can cause nonphagocytic ingestion of soluble molecules or particles greater than 0.2 μm is not associated with any receptor activity.[29]

Because of the particulate nature of liposomes, the question arises whether nonphagocytic cells can serve as APCs for induction of CTLs by liposomal antigens, particularly whether endocytosis can cause the same intracellular cytoplasmic delivery that is required for induction of CTLs. This naturally raises the question whether small liposomes that presumably only undergo endocytosis rather than phagocytosis can actually enter the cytoplasm of cells. The ability of small anionic or cationic liposomes to deliver diphtheria toxin fragment A to the cytoplasm of nonphagocytic HeLa cells was demonstrated as determined by killing activity of fragment A in the cells despite the absence of fragment B that is responsible for binding of the toxin to the cells.[30] Based on this it must be presumed that small liposomes can be delivered to the cytoplasm of cells by the process of endocytosis or pinocytosis. However, it is widely believed that endocytosis (as opposed to phagocytosis) is not an efficient route by which liposomes could enter nonphagocytic cells or by which large labile molecules could gain access to the cytoplasm.[31] These different routes of uptake could result in either the partial or complete degradation of the antigen and consequently the antigen has the potential to enter different processing compartments such as the endosomes, lysosomes or the cytosol.[32]

Although it has often been stated that liposomes are lysosomotropic agents that efficiently deliver substances to endosomes and lysosomes, it has been demonstrated by Venna et al.,[18] by immunogold electron microscopy that after phagocytosis of liposomes containing a recombinant malaria antigen, epitopes derived from the liposome-encapsulated antigen can enter the cytoplasm of bone marrow-derived macrophages in large amounts (Figure 1). This observation of cytosolic delivery of liposomal antigen was also confirmed with a completely different antigen encapsulated in liposomes.[33] The unique observation by us[18] and by Zhou et al.,[33] that liposomal antigens can spill from endosomal vesicles into the cytoplasm raises the question of the ultimate fate of the intracytoplasmic liposomal antigen. The cytoplasmic liposomal antigens might thus gain access to the endoplasmic reticulum or to the Golgi apparatus, major cellular organelles that contain MHC class I molecules.

II.2. MHC Class I Pathway

In the classical pathway for presentation of intracellular (endogenous) antigens by MHC class I molecules on APCs, the endogenous proteins are hydrolyzed into peptides in the cytosol by proteasomes and then delivered to the endoplasmic reticulum (ER) by transporters associated with antigen processing (TAPs).[34,35] A

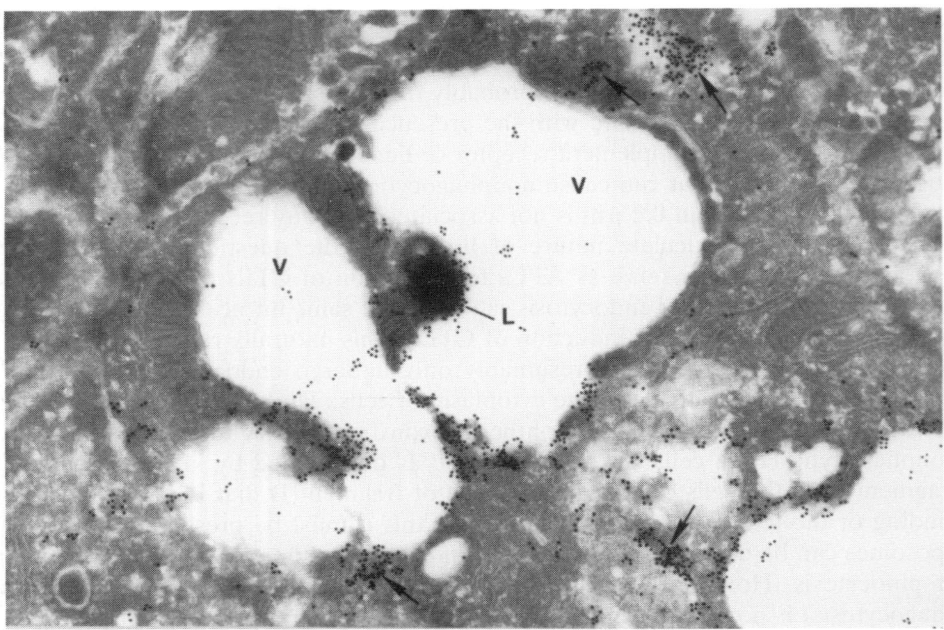

Fig. 1. Immunogold electron microscopy of cultured murine bone marrow-derived macrophages after phagocytosis of liposomes containing malaria antigen ($R32NS1_{81}$). Murine bone marrow-derived macrophages were fixed 6 h after incubation with liposome-encapsulated recombinant malaria antigen, $R32NS1_{81}$, and processed for electron microscopy. R32NS1 was detected by a monoclonal antibody (Pf 1B2.2) specific to the antigen, followed by treatment with gold labelled secondary antibody. The sections were stained with 2% uranyl acetate in 50% methanol, contrasted with Reynold's lead citrate, carbon coated in a vacuum evaporator, and examined with a JEOL 100 CX electron microscope. V, vacuole; L, liposome containing antigen. Four arrows indicate examples of locations of cytoplasmic antigen. From Verma et al.[18]

wide array of peptides are transported by TAP proteins. The transported peptides bind to nascent MHC class I molecules that generates stable trimeric MHC-I heavy chain-β2-microglobulin-peptide complexes which are then transported to the plasma membrane for recognition and activation of $CD8^+$ CTLs.[36–39]

As illustrated in Figure 2, cytoplasmic liposomal peptides derived from degraded liposomal antigen could easily be expected to participate in this process, either through interaction with the peptide transporter or through direct transfer of liposomal lipid-protein or peptide complexes to the Golgi. The peptide could then associate with the MHC class I molecules and undergo vesicular transport to the surface of the cells for presentation and induction of CTLs.[40] The involvement of the Golgi complex in the MHC class I pathway for presentation of intracellular antigens are incomplete in that the studies addressed only the trafficking pattern of the class I molecules and not that of the processed peptide.[41,42] The class I presentation of exogenous ovalbumin coupled to latex beads was inhibited by a mutation that disrupts TAP[8] and also by brefeldin A,[8,43] whose major function is

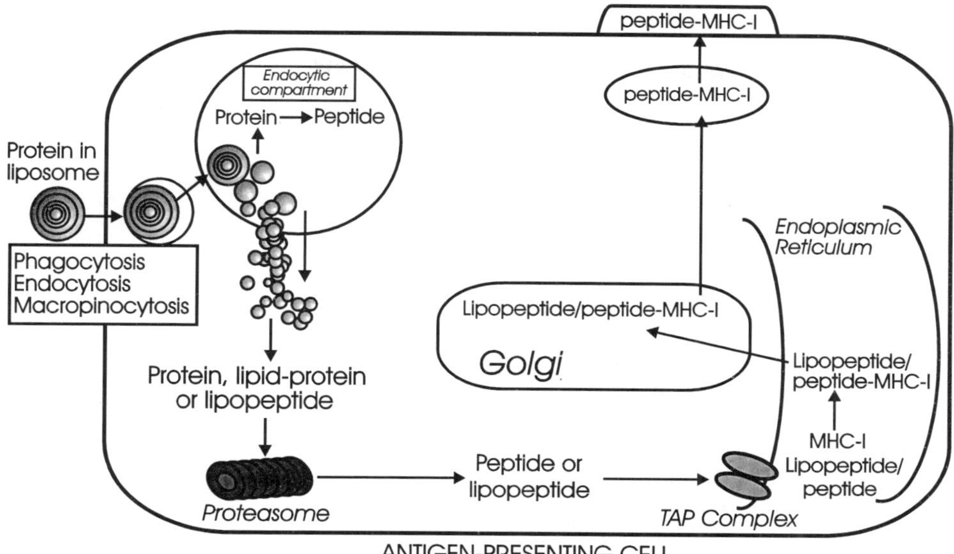

Fig. 2. Presentation of liposome-encapsulated antigen via the major histocompatibility complex (MHC) class I pathway by an antigen presenting cell. Liposome-encapsulated antigens are internalized either by phagocytosis, endocytosis or macropinocytosis by antigen presenting cells. Liposomal antigens enter the phagocytic or endocytic or pinocytotic vacuoles and are then released into the cytoplasmic compartment. Peptides or lipopeptides are then generated by the proteasome complex and transported into the endoplasmic reticulum (ER) by TAP proteins. The peptide complexes with the newly synthesized MHC class I molecule in the ER, and is transported to the Golgi complex and then to the plasma membrane.

to inhibit exocytosis of proteins out of the ER. These data indicate that peptides derived from exogenous antigens must initially be present in the cytosol and then transported into the ER. A similar situation might exist for liposome-encapsulated antigens. Using bone marrow-derived macrophages as the APCs, we are currently studying the mechanisms by which liposome-encapsulated antigens enter the MHC class I pathway. The pathways shown in Figure 2 therefore provide a theoretical basis for predicting that liposomal antigenic peptides containing CTL epitopes would be expected to interact with class I molecules in the induction of CTLs.

III. Induction of CTLs by liposomal antigens

Numerous studies have described class I presentation and induction of CTLs by liposomal antigens both in vivo and in vitro (Table I). Our laboratory has developed a liposome formulation that we refer to as Walter Reed Liposomes (WRAIR)[57] that has been shown to serve as an effective vehicle for delivery of proteins or peptides to antigen presenting cells (APCs) for presentation via the MHC class I pathway.[11–13,58] In addition, to DMPC, DMPG, and CHOL, the

Table 1

Induction of cytotoxic T lymphocytes by liposome-encapsulated antigen

Antigen	Liposome composition	Reference
In vitro studies		
MHC antigens (H-2 in mice)	Egg PC/CHOL (70:30 w/w)	Hale[44]
MHC antigens (H-2 in mice)	Egg PC/CHOL (70:30 w/w)	Hale & McGhee[45]
Human colon tumor antigens LS174T colon tumor cells	PC/CHOL/PA (7:2:1)	Raphael and Tom[46,47]
Ovalbumin	DOPC/DOPS (4:1) and DOPE-PHC (4:1)	Harding et al.[48]
Murine hemoglobin		
Bovine ribonuclease		
Hen egg lysozyme		
Ovalbumin	DOPE/DOSG (1:1) and DOPC/PS/CHOL (5:2:3)	Reddy et al.[49]
Ovalbumin	DOPE/DOSG (1:1) and DOPC/PS/CHOL (5:2:3)	Zhou et al.[50]
In vivo studies		
Haemagglutinin neuraminidase	MDP and MDP/CHOL (1:1 w/w)	Nerome et al.[51]
Ovalbumin β-galactosidase	DOPE/DOSG (1:1) and DOPC/PS/CHOL (5:2:3)	Reddy et al.[10]
Ovalbumin	DOPE/DOSG DOPE/DPSG DOPE/CHEMS DOPC/DOPS (all 4:1) ±50 μg lipid A DPPC/DPPG/CHOL (9:1:8) ±50 ug lipid A	Collins et al.[52]
Ovalbumin	PC/lysoPC/CHOL (6.9:0.1:3 neutral) PC/lysoPC/SA/CHOL (6.9:0.1:1:2 positive) PC/lysoPC/DCP/CHOL (6.9:0.1:1:2 negative)	Lopes and Chain[9]
Multiple antigen peptide system (MAPS) from GP120 of HIV-1 (B2M-P3C)	Egg PC/CHOL/SA (7.5:1:0.25 w/w)	Defoort et al.[53]
Glycoprotein B from HSV	Cationic lipids (dioleoyloxypropyl-trimethyl-ammonium methyl sulfate DOTAP)	Walker et al.[54]
Ovalbumin	Commercially available DOTAP	Chen et al.[55]
Repeatless *P. falciparum* CS protein	DMPC/DMPG/CHOL/lipid A (0.9:0.1:0.75:0.026)	White et al.[12]
HIV-1 gp120 protein-derived peptide (P18, CG-P18)	DMPC/DMPG/CHOL/lipid A (0.9:0.1:0.75:0.026)	White et al.[12]
P. falciparum CS protein-derived peptide (Myr-367–390)	DMPC/DMPG/CHOL/lipid A (0.9:0.1:0.75:0.026)	Alving et al.[13]
SIV gag protein-derived peptide (p11C)	PS/CHOL (9:1)-envelope glycoproteins and lipids of Sendi virus	Miller et al.[14]
SIV gag protein-derived peptide (M90–07A)	DMPC/DMPG/CHOL/lipid A (0.9:0.1:0.75:0.1)	Yasutomi et al.[56]

Abbreviations used: CHEMS cholesterol hemisuccinate; CHOL cholesterol; CS circumsporozoite; DCP dicetyl phosphate; DMPC dimyristoyl phosphatidylcholine; DMPG dimyristoyl phosphatidylglycerol; DOPC dioleoyl phosphatidylcholine; DOPE dioleoyl phosphatidylethanolamine; DOPS dioleoyl phosphatidylserine; DOSG 12-dioleoyl-sn-3succinylglycerol; DPPC dipalmitoyl phosphatidylcholine; DPPG dipalmitoyl phosphatidylglycerol; DPSG dipalmitoyl succinylglycerol; lysoPC lysophosphatidylcholine; MDP muramyl dipeptide; MHC major histocompatibility antigen gene complex; PA phosphatidic acid; PC phosphatidylcholine; PHC palmitoyl homocysteine; PS phosphatidylserine; SA stearylamine.

liposomes also contain either lipid A or its monophosphoryl derivative, MPL, as an intrinsic liposomal adjuvant. Liposomal lipid A appears to be extremely important and is often a requirement as an adjuvant for induction of CTLs against liposomal antigens.[11,12] Using either encapsulated or surface-bound HIV peptides, it has also been shown that antigenic surface expression of liposomal peptide is not required for induction of CTLs.[12,13] We have achieved success in using liposomes containing lipid A as a vehicle for inducing MHC class I restricted CTLs specific for the 367–390 sequence present in RLF, a recombinant malaria protein, derived from *Plasmodium falciparum*.[11,59] Furthermore, we have demonstrated that immunization of mice with myristoylated 367–390 peptide encapsulated in liposomes containing lipid A also elicits excellent CTLs.[13]

Other investigators have developed pH-sensitive liposomes to deliver antigens to the cytoplasmic compartment of APCs.[60–62] This type of liposome is mainly composed of dioleoylphosphatidylethanolamine (DOPE) and an acidic amphiphile. In acidic vesicles, such as in endosomes, the amphiphile is protonated leading to destabilization of liposomes and release of the entrapped antigens and fusion with each other or with cellular membranes.[61,63] Initially, this approach had been used to induce CTL in vivo.[10,49,52] Later it was shown that pH-insensitive liposomes could also deliver antigen to the cytosol and sensitize APCs for MHC class I presentation, although with a lower efficiency than the pH-sensitive liposomes.[33,48] It should be pointed out that although WRAIR liposomes are not pH-sensitive, they are delivered in large amounts to the cytoplasm of macrophages. Consequently, we have had very good success in inducing CTL responses against a wide variety of liposome-encapsulated antigens ranging from recombinant proteins to synthetic peptides.[11–13]

IV. Conclusion

Liposomes have been widely used as carriers of protein or peptide antigens. Experimental vaccines against viral and parasitic diseases such as HIV and malaria have been developed by using the Walter Reed liposomes and the liposomal vaccines have been shown to be safe and highly immunogenic in several human trials. A detailed understanding of the intracellular trafficking patterns of liposomal antigens and liposomal lipids is essential for developing effective vaccines. Although it is well established that one of the ways of inducing CTLs against exogenous protein antigens is to use liposomes as the delivery system, there are many unresolved issues regarding processing of exogenous liposome-encapsulated antigens for MHC class I presentation and priming of CTL responses. Perhaps the most important question is how does the antigen enter the cytoplasmic compartment from the endosomes? Are there specific transport proteins to facilitate this entry into the cytosol? What exactly is the role played by liposomal lipids in this process? Additionally, do liposomal lipids also get transported to the ER/Golgi complex and finally how are trimeric MHC-I-β_2-microglobulin complexes transported to the cell membrane?

One of the limitations of conventional protein-based vaccines is frequently their failure to prime CTL responses. Nonreplicating antigens which normally do not induce CTL responses because of their inability to access the MHC class I pathway can do so if they are encapsulated in liposomes. Because liposome-formulated vaccines have the potential to stimulate both antibody as well as T cell responses simultaneously by gaining entry to both the conventional MHC class I and class II pathways this approach could prove to be extremely useful in designing vaccine strategies.

References

1. Braciale TJ, Morrison LA, Sweetser MT, Sambrook J, Gething, MJ, Braciale VL. Antigen presentation pathways to class I and class II MHC-restricted T lymphocytes. Immunol Rev 1987;98:95–114.
2. Germain RN, Margoulies DH. The biochemistry and cell biology of antigen processing and presentation. Ann Rev Immunol 1993;11:403–450.
3. Moore MW, Carbone FR, Bevan MJ. Introduction of soluble protein into the class I pathway of antigen processing and presentation. Cell 1988;54:777–785.
4. Deres K, Schild H, Wiesmuller KH, Jung G, Rammensee HG. In vivo priming of virus-specific cytotoxic T lymphocytes with synthetic lipopeptide vaccine. Nature 1989;342:561–564.
5. Schild H, Norda M, Deres K, Falk K, Rotzschke O, Wiesmuller KH, Jung G, Rammensee HG. Fine specificity of cytotoxic T lymphocytes primed in vivo either with virus or synthetic lipopeptide vaccine or primed in vitro with peptide. J Exp Med 1991;174:1665–1668.
6. Martinon F, Gras-Masse H, Boutillon C, Chirat F, Deprez B, Guillet J-G, Gomard E, Tartar A, Levy JP. Immunization of mice with lipopeptides bypasses the prerequisite for adjuvant. Immune response of BALB/c mice to human immunodeficiency virus envelope glycoprotein. J Immunol 1992;149:3416–3422.
7. Harding CV, Song R. Phagocytic processing of exogenous particulate antigens by macrophages for presentation by class I MHC molecules. J Immunol 1994;153:4925–4933.
8. Kovacsovics-Bankowski M, Rock KL. A phagosome-to-cytosol pathway for exogenous antigens presented on MHC class I molecules. Science 1995;267:243–246.
9. Lopes LM, Chain BM. Liposome-mediated delivery stimulates a class I-restricted cytotoxic T cell response to soluble antigen. Eur J Immunol 1992;22:287–290.
10. Reddy R, Zhou F, Nair S, Huang L, Rouse BT. In vivo cytotoxic T lymphocyte induction with soluble proteins administered in liposomes. J Immunol 1992;148:1585–1589.
11. White K, Krzych U, Gordon DM, Porter TR, Richards RL, Alving CR, Deal CD, Hollingdale M, Silverman C, Sylvester DR, Ballou WR, Gross M. Induction of cytolytic and antibody responses using *Plasmodium falciparum* repeatless circumsporozoite protein encapsulated in liposomes. Vaccine 1993;11:1341–1346.
12. White WI, Cassatt DR, Madsen J, Burke SJ, Woods RM, Wassef NM, Alving CR, Koenig S. Antibody and cytotoxic T-lymphocyte responses to a single liposome-associated peptide antigen. Vaccine 1995;13:1111–1122.
13. Alving CR, Koulchin V, Glenn GM, Rao M. Liposomes as carriers of peptide antigens: Induction of antibodies and cytotoxic T lymphocytes to conjugated and unconjugated peptides. Immunol Rev 1995;145:5–31.
14. Miller MD, Gould-Fogerite S, Shen L, Woods RM, Koenig S, Mannino RJ, Letvin NL. Vaccination of rhesus monkeys with synthetic peptide in a fusogenic proteoliposome elicits simian immunodeficiency virus-specific $CD8^+$ cytotoxic T lymphocytes. J Exp Med 1992;176:1739–1744.
15. Alving CR, Wassef NM. Cytotoxic T lymphocytes induced by liposomal antigens: Mechanisms of immunological presentation. AIDS Res and Human Retrovir 1994;10:S91–S94.
16. Woodle MC, Lasic DD. Sterically stabilized liposomes. Biochim Biophys Acta 1992;1113:171–199.
17. Dijkstra J, Van Galen M, Scherphof G. Effects of (dihydro) cytochalasin B, colchicine, monensin and trifluoperazine on uptake and processing of liposomes by Kuffer cells in culture. Biochim Biophys Acta 1988;845:34–42.
18. Verma JN, Wassef NM, Wirtz RA, Atkinson, CT, Aikawa M, Loomis LD, Alving CR. Phago-

cytosis of liposomes by macrophages: intracellular fate of liposomal malaria antigen. Biochim Biophys Acta 1991;1066:229–238.

19. Segal AW, Willis EJ, Richmond JE, Slavin G, Black CDV, Gregoriadis G. Morphological observations on the cellular and subcellular destination of intravenously administered liposomes. Br J Exp Pathol 1974;55:320–327.
20. Alving CR. Liposomes as carriers of antigens and adjuvants. J Immunol Meth 1991;140:1–13.
21. Su D, Van Rooijen N. The role of macrophages in the immunoadjuvant action of liposomes: effects of elimination of splenic macrophages on the immune response against intravenously injected liposome-associated albumin antigen. Immunol 1989;66:466–470.
22. Verma JN, Rao M, Amselem S, Krzych U, Alving CR, Green SJ, Wassef NM. Adjuvant effects of liposomes containing lipid A: enhancement of liposomal antigen presentation and recruitment of macrophages. Infect Immun 1992;60:2438–2444.
23. Rao M, Wassef NM, Alving CR, Krzych U. Intracellular processing of liposome-encapsulated antigens by macrophages depends upon the antigen. Infect Immun 1995;63, 2396–2402.
24. Rock KL, Rothstein L, Gamble S, Fleischacker C. Characterization of antigen-presenting cells that present exogenous antigens in association with class I molecules. J Immunol 1993;150:438–446.
25. Grant EP, Rock KL. MHC class I-restricted presentation of exogenous antigen by thymic antigen-presenting cells in vitro and in vivo. J Immunol 1992;148:13–18.
26. Nair S, Buiting AM, Rouse RJ, Van Rooijen N, Huang L, Rouse BT. Role of macrophages and dendritic cells in primary cytotoxic T lymphocyte responses. Int Immunol 1995;7:679–688.
27. Nair S, Zhou F, Reddy R, Huang L, Rouse BT. Soluble proteins delivered to dendritic cells via pH-sensitive liposomes induce primary cytotoxic T lymphocyte responses in vitro. J Exp Med 1992;175:609–612.
28. Allen LA, Aderem A. Mechanisms of phagocytosis. Curr Opin Immunol 1996;8:36–40.
29. Swanson JA, Watts C. Macropinocytosis. Trends Cell Biol 1995;5:424–427.
30. Norrie DH, Pietrowski RA, Stephen J. Screening the efficiency of intracytoplasmic delivery of materials to HeLa cells by liposomes. Anal Biochem 1982;127:276–281.
31. Straubinger RM, Hong K, Friend DS, Papahadjopoulos D. Endocytosis of liposomes and intracellular fate of encapsulated molecules: encounter with a low pH compartment after internalization in coated vesicles. Cell 1983;32:1069–1079.
32. Rock KL. A new foreign policy: MHC class I molecules monitor the outside world. Immunol Today 1996;17:131–137.
33. Zhou F, Watkins SC, Huang L. Characterization and kinetics of MHC class I-restricted presentation of a soluble antigen delivered by liposomes. Immunobiol 1994;190:35–52.
34. Attaya M, Jameson S, Martinez CK, Hermel E, Aldrich C, Forman J, Lindahl, KF, Bevan M, Monaco, JJ. Ham-2 corrects the class I antigen-processing defect in RMA-S cells. Nature (London) 1992;355:647–649.
35. Spies T, DeMars R. Restored expression of major histocompatibility class I molecules by gene transfer of a putative peptide transporter. Nature (London) 1991;351:323–324.
36. Roelse J, Gromme M, Momburg F, Hammerling G, Neefjes JJ. Trimming of TAP-translocated peptides in the endoplasmic reticulum and in the cytosol during recycling. J Exp Med 1994;180,1591–1597.
37. Bergeron JJ, Brenner MB, Thomas DY, Williams DB. Calnexin: a membrane-bound chaperone of the endoplasmic reticulum. Trends Biochem Sci 1994;9:124–128.
38. Jondal M, Schirmbeck R, Reimann J. MHC class I-restricted CTL responses to exogenous antigens. Immunity 1996;5:295–302.
39. Townsend ARM, Bodmer HC. Antigen recognition by class I-restricted T lymphocytes. Ann Rev Immunol 1989;7:601–624.
40. Alving C. Liposomal Vaccines: Clinical status and immunological presentation for humoral and cellular immunity. Combined vaccines and simultaneous administration Annals of New York Academy of Sciences 1995;754:143–152.
41. Neefjes JJ, Stollorz V, Peters PJ, Geuze H, Ploegh HL. The biosynthetic pathway of MHC class II but not class I molecules intersects the endocytic route. Cell 1990;61:171–183.
42. Peters PJ, Neefjes JJ, Oorschot V, Ploegh HL, Geuze H. Segregation of MHC class II molecules from MHC class I molecules in the Golgi complex for transport to lysosomal compartments. Nature 1991;349:669–676.
43. Reise Sousa C, Germain RN. Major histocompatibility complex class I presentation of peptides derived from soluble exogenous antigen by a subset of cells engaged in phagocytosis. J Exp Med 1995;182:841–852.

44. Hale AH. H-2 antigens incorporated into phospholipid vesicles elicit specific allogeneic cytotoxic T lymphocytes. Cell Immunol 1980;55:328–341.
45. Hale AH, McGee MP. A study of the inability of subcellular fractions to elicit primary anti-H-2 cytotoxic T lymphocytes. Cell Immunol 1981;58:277–285.
46. Raphael L, Tom BH. In vitro induction of primary and secondary xenoimmune responses by liposomes containing human colon tumor cell antigens. Cell Immunol 1982;71:224–240.
47. Raphael L, Tom BH. Liposome facilitated xenogeneic approach for studying human colon cancer immunity: carrier and adjuvant effect of liposomes. Clin Exp Immunol 1984;55:1–13.
48. Harding CV, Collins DS, Kanagawa O, Unanue ER. Liposome-encapsulated antigens engender lysosomal processing for class II MHC presentation and cytosolic processing for class I presentation. J Immunol 1991;147:2860–2863.
49. Reddy R, Zhou F, Huang L, Carbone F, Bevan M, Rouse BT. pH sensitive liposomes provide an efficient means of sensitizing target cells to class I restricted CTL recognition of a soluble protein. J Immunol Meth 1991;141:157–163.
50. Zhou F, Rouse BT, Huang L. An improved method of loading pH-sensitive liposomes with soluble proteins for class I restricted antigen presentation. J Immunol Meth 1991;145:143–152.
51. Nerome, K, Yoshioka Y, Ishida M, Okuma K, Oka T, Kataoka T, Inoue A, Oya A. Development of a new type of influenza subunit vaccine made by muramyldipeptide-liposome: enhancement of humoral and cellular immune responses. Vaccine 1990;8:503–509.
52. Collins DS, Findlay K, Harding CV. Processing of exogenous liposome-encapsulated antigens in vivo generates class I MHC-restricted T cell responses. J Immunol 1992;148:3336–3341.
53. Defoort J-P, Nardelli B, Huang W, Tam JP. A rational design of synthetic peptide vaccine with a built-in adjuvant. Int J Pep Prot Res 1992;40:214–221.
54. Walker C, Selby M, Erickson A, Cataldo D, Valensi J, Van Nest G. Cationic lipids direct a viral glycoprotein into the class I major histocompatibility complex antigen presentation pathway. Proc Natl Acad Sci USA 1992;89:7915–7918.
55. Chen W, Carbone FR, McCluskey J. Electroporation and commercial liposomes efficiently deliver soluble protein into the MHC class I presentation pathway. J Immunol Meth 1993;160:49–57.
56. Yasutomi Y, Alving CR, Wassef NM, Conrad P, Conley AJ, Emini EA, Madsen J, Woods R, Koenig S, Letvin NL. Combined modality immunization for elicitation of SIV mac gag-specific CTL. Vaccine 1994.
57. Vogel FR, Powell MF. Section on Walter Reed Liposomes in A Compendium of Vaccine Adjuvants. In: Powell MF, Newman MJ, eds. Vaccine Design: The Subunit and Adjuvant Approach. New York: Plenum Press, 1995;226–227.
58. Wassef NM, Alving CR, Richards, RL. Liposomes as carriers for vaccines. Immuno Methods 1994;4:217–222.
59. Heppner DG, Gordon DM, Gross M, Wellde B, Leitner W, Krzych U, Schneider I, Wirtz RA, Richards RL, Trofa A, Hall T, Sadoff JC, Boerger P, Alving CR, Sylvester DR, Porter TG, Ballou WR. Safety, immunogenicity and efficacy of *Plasmodium falciparum* repeatless circumsporozoite protein vaccine encapsulated in liposomes. J Infect Dis 1996;174:361–366.
60. Connor J, Huang L. Cytoplasmic delivery of a fluorescent dye by pH-sensitive immunoliposomes. J Cell Biol 1985;101:582–58.
61. Duzgunes N, Straubiger RM, Baldwin PA, Friend DS, Papahadjopoulos D. Proton-induced fusion of oleic acid-phosphatidylethanolamine liposomes. Biochem 1985;24:3091–3098.
62. Ellens H, Bentz J, Szoka FC. pH-induced destabilization of phosphatidylethanolamine-containing liposomes. Role of bilayer contact. Biochem 1984;23:1532–1538.
63. Connor J, Yatvin MB, Huang L. pH-sensitive liposomes: Acid induced liposome fusion. Proc Natl Acad Sci USA 1984;81:1715–1718.

Lasic and Papahadjopoulos (eds.), Medical Applications of Liposomes
Elsevier Science B.V.

CHAPTER 2.2

Major histocompatibility complex class II molecules, liposomes and antigen presentation

LEE LESERMAN AND NICOLAS BAROIS

Centre d'Immunologie INSERM-CNRS de Marseille-Luminy, Parc Scientifique et Technologique de Luminy, Case 906, 13288 Marseille CEDEX 9, France

Overview

I. Introduction: Context and purpose of this review

The success of liposomes as immunological vectors was originally attributed to their capacity to sequester antigens and release them slowly,[1,2] but it is now clear that this is not the only reason. Liposomes also resemble or may be made to resemble the structures that the immune system evolved to protect us against, viruses, bacteria and other microorganisms. The success of vaccination with liposomes also depends on this resemblance, promoting liposome uptake by several cell types called antigen presenting cells (APC) and conversion of liposome-associated antigen by those cells into structures that stimulate T lymphocytes.[3] APC include macrophages, B cells and dendritic cells. These cells are dispersed in tissues and difficult to study in situ. We will review recent immunological and cell biological results on the mechanism of antigen recognition and presentation which are primarily derived from in vitro experiments. Despite the fact that in vitro results may not necessarily predict in vivo consequences, we feel that understanding the mechanisms generating forms of antigen that stimulate T lymphocytes will help us to improve the capture and intracellular processing of liposomes and related antigen carriers. This chapter emphasizes studies oriented towards certain T cells called helper T cells, which are principally involved in the induction of immune responses. The role of another set of T lymphocytes, cytotoxic T cells, which are cells implicated primarily in the effector phase of the immune response by killing infected cells is presented in the chapter of Rao and Alving. The use of liposomes for vaccine development is discussed in other chapters in this book and has been the subject of excellent recent reviews.[4–7] Readers interested in more detailed explanations and specific references are encouraged to consult recent immunology textbooks.[8,9]

II. Overview of antigen recognition by B and T lymphocytes

Lymphocytes are divided into B and T cells. B cells express receptors for antigen on their surface and synthesize and secrete antibody, which is a soluble form of the receptor, as a result of antigen stimulation. The structure of the antigen-binding site of this receptor varies from B cell to B cell as a consequence of genetic recombination and somatic mutation. Antibodies specific for given viruses or bacteria can block infection by activating complement and mediating their lysis or by neutralizing molecules on the surface of the microrganisms used to bind and to enter cells. The antigen receptor of B cells and the secreted antibodies recognize antigen in intact form independent of other structures. These antigens may be protein, carbohydrate or lipid. Since the estimated number of different antigens which may potentially be recognized varies widely (from 10^4 to $>10^{10}$ different possible antigens) and may exceed the 10^9 lymphocytes available in a mouse,[10] it is possible that one or only a few specific B cells will be stimulated by any given antigen. These few cells will necessarily produce limited quantities of antibody, so the production of sufficient antibody to provide protection requires amplification of the antigen-specific B cells. In some instances antigenic stimulus alone is sufficient to drive proliferation and antibody production by B cells in the absence of T cells. This was shown twenty years ago by Kinsky's group using rigid liposomes with antigenic hapten determinants expressed at high density and has been shown recently to be valid for certain bacteria or viruses expressing repetitive antigenic determinants.[11,12] This response occurs when a large number of B cell antigen receptors are engaged. The response is transient and in the absence of T cell participation does not generate "memory" which provides enhanced responses following subsequent contact with the same antigen. When antigen is less dense on microorganisms or expressed on the surface of less rigid liposomes, B cells cannot respond alone but require additional signals provided by T cells. The purpose of vaccination is to stimulate specific T cells, in order to increase their number and to induce this memory.

Each T cell also recognizes only one antigenic specificity or a limited number of closely related molecular mimics of that specificity. In contrast to B cells, T cells do not secrete their antigen receptors but each T cell receptor also recognizes a particular antigen determinant expressed at the surface of APC in conjunction with two types of proteins called major histocompatibility class I and class II molecules.[13,14] These molecules have a similar three-dimensional structure, but are made up of different subunits. The genes encoding these molecules are located in a region called the major histocompatibility complex (MHC), on chromosome 6 in humans and chromosome 17 in the mouse. In contrast to B cells, most T cells recognize only protein antigens. These are not perceived in their native conformation but as short peptide fragments derived from the antigen that are tightly associated with the MHC class I or II molecules. T cells may be broadly divided into cytotoxic T and helper T cells. Practically all the cells of the body express or may be induced to express class I molecules. Cytotoxic T cells recognize a molecular complex which includes MHC class I molecules and a peptide sequence derived from the particular antigen for which that T cell is specific. In general, cytotoxic T cells recognize antigens synthesized by target cells. Since this includes viral proteins in the case of virus-infected cells, these cells are important in defense against viral infection. Cytotoxic T cells may be identified because they bear CD8 cell surface molecules.

T helper cells bear on their surface CD4 molecules and recognize molecular complexes formed between 27MHC class II molecules and antigen-derived peptides for which they are specific. In general, these antigens are concentrated in compartments in APC accessible from the exterior of the cells, called endosomes. Antigen taken up from microorganisms in the extracellular environment or from intra-cellular parasites, such as *Leishmania* and *Mycobacteria tuberculosis* are processed by APC and presented as peptides derived from those organisms associated with class II molecules. Cells expressing class II molecules are primarily macrophages, B cells and dendritic cells. The stimulation of T helper cells by peptide plus class II increases their number and augments their capacity to "help" macrophages, B cells and cytotoxic T cells function. APC tend to concentrate in lymph nodes and other lymphoid tissues, where they encounter T cells that pass through these tissues.[15] Contact between T cells and APC is transient unless the T cell antigen receptor is engaged by a sufficient number of appropriate class II-peptide complexes, which initiates the process of T cell proliferation. In addition to class II-peptide complexes, other molecules expressed on APC, called co-stimulatory molecules are necessary for the productive activation of naive T cells. In the absence of their expression stimulation of the T cells' receptors for class II plus peptide may result in tolerance, rather than in induction of immunity.[16] In response to full stimulation T cells will proliferate and secrete molecules called cytokines, which are growth factors for proliferation and differentiation. This secretion may be directed toward those cells with which the T helper cells are in contact, rendering the effect of these factors relatively cell-specific.[17] T helper cells have been further divided into those which predominantly help cytotoxic T cells, or macrophages in inflammatory responses (Th1) and those which predominantly help B cells (Th2).[18] These two types of helper cells secrete different classes of cytokines and are regulated by factors secreted by APC.

The importance of CD4-bearing T helper cells has been underscored by the demonstration that destruction of these cells by infection with the HIV-1 virus causes the Acquired Immune Deficiency Syndrome. Macrophages require activation by T helper cells in order to kill bacteria and other intracellular parasites. B cells depend on T help for proliferation and production of antibody to most antigens. In humans, activated T cells also express class II, and most cytotoxic T cells depend on T helper cells in order to become fully competent in recognizing and killing cells that have become infected by viruses. Thus, impairment of the function of CD4 T cells has repercussions throughout the immune system and eventually leaves the individual susceptible to death from intercurrent infections which do not occur, or which are easily controlled, in normal individuals.

III. MHC class II molecules and role of liposomes in revealing their antigen presentation function

Techniques for the culture and expansion of CD4-bearing T cells of known antigen specificity developed in the early 1980's revealed that T cells could not be directly stimulated by antigen alone, but required participation of APC expressing class II molecules. T cell recognition of antigen could be measured by the induction of T cell proliferation or production of interleukins. T cells specific for a bacterial antigen could be stimulated by macrophages pre-incubated with the bacteria, but not by bacteria or macrophages alone. Antigen presentation by macrophages to T cells was shown to be dependent on proteolysis of the bacteria, because it could be blocked by inhibitors of proteolysis or the intracellular acidification necessary for proteolytic enzymes to act optimally.[19,20] These same requirements were observed for protein antigens, which had to be processed into peptides.[21]

Other experiments established that the cellular expression of class II molecules and the presence of peptides derived from antigen was sufficient for the stimulation of the relevant T helper cells. Fibroblasts, which do not express class II molecules could not stimulate T cells specific for a given combination of antigen and class II molecules but would do so if transfected with genes encoding those class II molecules, provided that peptides from the antigen were also present. The same class II-bearing fibroblasts were less potent in stimulating T cells if the intact protein antigen, from which the peptide was derived, was used.[22,23] Since fibroblasts are not phagocytic, the function of macrophages was confirmed to be related to uptake and degradation of the antigen into peptide fragments, as had been indicated by the pharmacologic inhibition of proteolysis of antigens they ingested.

To unravel the function of class II molecules, Watts and McConnell solubilized cell membranes in detergent, isolated class II molecules from them by affinity chromatography and reconstituted them into planar membranes. These were shown to be able to bind particular peptides derived from antigens, defining class II proteins as peptide-binding molecules. The same class II molecules lacked the capacity to bind the intact protein from which peptides were derived. This observation permitted testing whether class II molecules plus peptides, independent of other components of APC, would be sufficient to stimulate antigen-specific cloned T cells. Class II molecules reconstituted in planar bilayers stimulated mature ovalbumin-specific T cells in the presence of peptide fragments derived from ovalbumin. The same peptides associated with class II reconstituted in liposomes were not perceived.[24–26] This may reflect the relatively poor peptide loading of class II molecules that could be achieved in this experimental system. In some experiments T cells could be stimulated by liposomes bearing class II molecules plus high concentrations of covalently-coupled antigen or in the presence of other membrane components,[27–30] but these concentrations of antigen plus class II may be difficult to achieve on the surface of APC, in which a given peptide is bound to only a minority of class II molecules. Further, most T cells used in these experiments have been maintained in culture for long periods and are more easily stimulated than naive T cells. Thus, if administered in vivo these liposomes would probably not be perceived directly by T cells but would require uptake and presentation by APC. Nevertheless, we see from these experiments that expression of peptide-class II complexes at the surface of a cell or a bilayer is necessary and sufficient for stimulation of mature T cells of known specificity. Over the last few years a considerable amount of information has been obtained on the mechanism of assembly and intracellular routing of class II molecules, the mechanism of degradation of antigen into peptides and the manner in which these molecules find each other in APC.

IV. Biosynthesis of class II and associated molecules

Class II molecules are transmembrane proteins formed by the non-covalent association of 33 kD and 29 kD polypeptide chains called α and β chains, respectively. The structure of class II molecules determined by X-ray crystallography shows that they have a "cavity" sufficiently large to bind peptides.[31,32] Sensitive techniques now permit purification of class II molecules from cells and the determination of the sequences of the antigen-derived peptides associated with them. The size of peptides stably associated with class II molecules varies from 13 to 24 residues,[33] depending on the class II molecule. Many allelic forms of the MHC genes coexist and MHC molecules are highly polymorphic within the population, though the MHC molecules in a given individual are the same on all cells which express them. Correlations had been observed for many years between the expression of given MHC class II molecules in individuals and the ability to respond to certain antigenic peptide determinants. It is now known that class II binding of peptide is much less specific than antibody binding to antigen. Each class II molecule has a preferred set of peptides to which it binds.[33] This "degenerate binding" permits many peptides derived from antigenic proteins to be presented and reduces the risk that mutation of a gene encoding a protein in an infectious organism results in the failure of antigenic peptide presentation.

Two accessory molecules, the invariant chain (Ii) and HLA-DM (in humans; H-2M in mice), bear an important role in the formation of MHC class II molecules. Ii is required during class II assembly and transport while HLA-DM acts at the final stage of class II peptide loading by catalyzing the exchange of Ii-derived peptide for antigen peptides within the groove of MHC II. Ii molecules have several functions. The α, β and Ii chains are synthesized in the endoplasmic reticulum (ER). Ii initially acts as a "chaperone" which associates with and assists newly synthesized proteins during this folding process. Spleen cells from mice lacking Ii have a low surface expression of $\alpha\beta$ dimers because these molecules are recognized as being misfolded and are retained in the ER.[34] In contrast, correctly folded complexes can leave the ER and reach the Golgi apparatus. Secondly, Ii directs the transport of $\alpha\beta$ complexes from the Golgi apparatus to compartments called endosomes in which antigen is concentrated and processed.[35,36] Ii contains in its cytoplasmic domain two amino acid sequences responsible for targeting these molecules to endosomal compartments.[37,38] Transport of class II molecules from endocytic compartments to the cell surface requires the dissociation of Ii by proteolytic cleavage and the binding of antigen-derived peptides.[35,39,40] The third role of Ii is to prevent the peptide binding site of a majority of class II molecules from being occupied with peptides derived from endogenous proteins synthesized in the APC.[41–43] The region of Ii necessary for interaction and interference with the class II peptide binding site is called CLIP, for class II-associated invariant chain peptide.[44] Peptides derived from CLIP have been eluted from class II molecules[45] and X-ray crystallography has shown that CLIP occupies the same site as antigen-derived peptides.[46]

In the endocytic compartments, proteases degrade Ii sequentially into peptide fragments associated with class II molecules by CLIP.[47–50] These peptides can be observed by inhibition of Ii degradation with protease inhibitors which block binding of antigen derived peptides with class II molecules as well as their expression at the cell surface.[48,51–53] The unique Ii gene codes for two different Ii forms due to alternative splicing. One of these molecules has an additional cysteine-rich domain of 64 amino acids[54] and inhibits cysteine proteases present in the endocytic compartments.[55] Inhibition of Ii degradation permits MHC class II molecules to be retained for a longer time in the intracellular compartments[39] and also limits antigen degradation to peptides of a size adequate for the binding site of class II.[56] In conclusion, Ii serves important functions for the MHC class II peptide loading: folding and export of the $\alpha\beta$ dimers in the ER, blockade of class II binding sites for peptides present in the ER, transport and retention of class II molecules in endocytic compartments for antigen peptide binding and finally, in conjunction with HLA-DM, regulation of peptide binding in these compartments.

After complete degradation of Ii, CLIP is removed from the peptide binding site and replaced by an antigen peptide.[35] HLA-DM acts in this final maturation step by favoring peptide exchange on the class II binding site. HLA-DM never reaches the cell surface and is accumulated in an intracellular compartment with characteristics of lysosomes.[57–59] Sanderson et al., have reported that HLA-DM accumulated in specialized compartments containing MHC class II molecules.[58,60]

In mice lacking H2-M molecules, class II molecules are primarily loaded with CLIP, affecting antigen binding and presentation.[61–63] Transfection of HLA-DM deficient cells with HLA-DM genes restored the capacity of association between peptides and class II molecules and restored antigen presentation.[57,64,65] Several in vitro studies have shown that HLA-DM increased CLIP release from class II and also increased binding of exogenous antigen-derived peptides.[66,67] Recently, two studies have reported that HLA-DM could remove peptides lacking the consensus amino acid sequence needed for stable association with class II molecules and exchange unstable peptides for peptides which are more stably associated with class II.[68,69] By doing so, HLA-DM may affect the repertoire of class II-restricted presentation. Finally, peptide-bound MHC class II molecules leave endocytic compartments to go to the cell surface by a poorly elucidated pathway.[70]

V. Intracellular transport of class II molecules in relation to endocytosis and degradation of proteins

The presentation pathway described above is called the biosynthetic-coupled pathway. It affects newly synthesized class II molecules and depends on the presence of Ii. A second presentation pathway depends on the recycling of surface class II molecules and is independent of expression of Ii. This alternative pathway, first described in B lymphocytes,[71,72] relies on the recycling of class II molecules in a compartment in which the acidic pH is sufficient to promote the association with newly generated antigen peptides.[73] When APC are treated with an inhibitor of protein synthesis or with an inhibitor of ER-Golgi transport the biosynthesis-coupled presentation pathway is affected while the alternative presentation pathway is not.[74,75] Moreover, these two pathways may present different antigens or different peptides from the same antigen.[76,77] Different pathways could result in the selection of different class II-associated sets of peptides in distinct endocytic compartments. Class II molecules associated with peptides derived from chicken lysozyme are differentially recognized by specific T cells when lysozyme is delivered into cells from pH sensitive liposomes which release their contents in early endosomes, or from pH insensitive liposomes, which release their contents in late endosomes.[78] Newly synthesized and recycled class II molecules are transported through different compartments of the endocytic pathway subjected to specific acidification and maturation processes. These compartments can be differentiated by microscopy or by physical separation techniques discriminating their internal pH, density, morphology, the presence of characteristic proteins, or the presence of internalized proteins or receptors.[79]

The first compartment in which internalized antigens are found corresponds to early endosomes. This compartment also contains class II/Ii complexes and is characterized by pH of 6 to 6.5, a light density on sucrose gradients following cellular fractionation and the presence of the transferrin receptor.[80] Internalized antigens start to be degraded in the early endosomes, though their content of proteases is very low.[80] Ii begins to be degraded in the same compartment. Degradation first releases peptides present in accessible loops of the proteins, as well as on Ii associated with class II molecules. As shown by several groups, peptides can associate with recycled class II molecules.[71,72,81] Alternatively, HLA-DM can exchange newly generated peptides for pre-associated ones.[68,69] Recent results indicate that HLA-DM could release $\alpha\beta$ dimers from Ii fragments and allow them to bind peptides.[82] Other antigen processing steps occur in late endosomes. This compartment has pH of 5.5 to 6, a medium density and is defined by the presence of the mannose 6-phosphate receptor and other characteristic proteins.[83,84] It also has a higher content of proteolytic enzymes and exchanges solute and membrane material with lysosomes. This latter compartment is distinguished by an acidic pH of 4.5 to 5.5, a high density and sometimes a particular multivesicular structure.[79] Lysosomes contain proteolytic enzymes in abundance[47] and specific lysosomal membrane glycoproteins.[85] Endocytosed antigens reach lysosomes[86] for final degradation into amino acids, in some instances releasing internal peptides[87] which can gain access to newly synthesized class II molecules. It is also plausible that recycled class II molecules could also reach lysosomes and once loaded with peptides, return to the cell surface. Finally, it has recently been shown that peptide-class II complexes can leave lysosomes directly and reach the cell surface without passing through the endosomes.[88]

Increasing evidence shows some antigen delivered to endocytic vesicles can leave the endosomal system and enter into the cytoplasm. This is well documented for particulate antigens, including liposomes.[89–94] Entry into the cytoplasm allows these antigens to be presented in association with class I molecules, eliciting T cytotoxic cell responses (see Rao and Alving in this book). There is also increasing evidence that antigen synthesized within cells may become associated with class II molecules.[95] This is especially clear for peptides derived from transmembrane molecules normally expressed at the cell surface. These may be accessible to the recycling class II presentation discussed above but also to the biosynthetic-coupled pathway of presentation. Since many encapsulated viruses synthesize capsular proteins in excess, this mechanism permits peptides from these antigens to meet the class II association pathway, allowing identification of infected cells by T cells.

VI. Peptide loading compartments

In addition to the early and late endosomes and lysosomes, which are found in all cell types, several groups have described specialized compartments for peptide-loading in class II-expressing cells. The first observation of a compartment enriched in MHC class II molecules was provided in human B lymphoblastoid cells using electron microscopy.[79] A multivesicular compartment was identified which contains a high concentration of class II molecules, lysosomal proteases and antigen but lacked markers of early and late endosomes, such as transferrin receptors and mannose 6-phosphate receptors, respectively. The compartment, with characteristics of lysosomes, has been called MIIC for MHC class II compartment and its existence seems to be dependent on the expression of class II molecules.[96] HLA-DM is also present in a variety of MIIC[58] and is more concentrated in this compartment than in the other compartments of the endocytic pathway.[97]

To determine whether MIIC was effectively a peptide-loading compartment, several studies were performed using subcellular fractionation techniques (density gradient centrifugation or organelle electrophoresis) and techniques allowing the detection of the presence of peptide-loaded class II molecules. Harding and Geuze have shown that, in macrophages, MIIC was found in the high density fractions with a multivesicular morphology. It was distinct from lysosomes and contained peptide-associated class II.[98] Subsequent studies on murine B cell lymphoma and on human B lymphoblastoid cells have confirmed by fractionation and radiolabeling that complexes were formed in a dense compartment.[99,100] Another technique of subcellular fractionation, density gradient electrophoresis, was also used to study the peptide-loading compartment in a human melanoma cell line and showed that Class II molecules entered in this compartment with invariant chain which is rapidly degraded.[101] Ii degradation is likely to occur in this compartment or before, since in the majority of studies Ii has not been detected. Using the technique of free flow electrophoresis, Amigorena et al., have characterized in murine B cell lymphoma cells a compartment in which newly synthesized class II molecules are transiently accumulated before reaching the cell surface. The compartment, called CIIV for class II-containing vesicles, contains the transferrin receptor and has a low density.[102] CIIV are closer to early endosomes than lysosomes and contain detectable amounts of Ii. Internalized antigens have rapid access to CIIV and can bind to $\alpha\beta$ dimers after Ii degradation. Thus, specialized compartments exist in APC in which class II-antigen complexes may be concentrated.

VII. Role of antigen receptors

In addition to having specialized intracellular compartments for the meeting between antigen-derived peptides and the class II molecules to which they bind, APC have mechanisms to increase acquisition of antigen. This is particularly well established in the case of B cells. A B cell with receptors specific for a given antigen can bind that antigen and concentrate to levels hundreds or thousands of times greater than its concentration by irrelevant B cells.[103,104] These receptors mediate antigen internalization into endocytic vesicles, initiating a process which can generate a sufficiently high level of class II plus antigen-derived peptides to stimulate specific T cells. In the absence of a relevant receptor, insufficient antigen is acquired by non-specific endocytic mechanisms to generate these levels of class II-antigen peptide complexes. This process ensures that only antigen-specific B cells receive help from T cells recognizing class II peptides derived from the antigen. At the same time antigen binds to surface immunoglobulin it initiates activation signals which may alter antigen processing, such as by regulating proteases or intracellular trafficking of class II molecules.[105] These signals also augment the ability of B cells to interact with T cells by inducing their expression of co-stimulatory molecules.

Because of the inefficient nature of antigen uptake by B cells in the absence of binding to surface antibody, it was originally thought that liposomes and other particulate antigens could not be taken up by these cells.[106] However, it is now established that liposomes[107,108] and other particles bearing ligands capable of binding to specific surface immunoglobulin[109] are taken up hundreds or thousands of times more efficiently by antigen-specific B cells than liposomes lacking the ligand, or the same liposomes in the presence of irrelevant B cells. Antigens

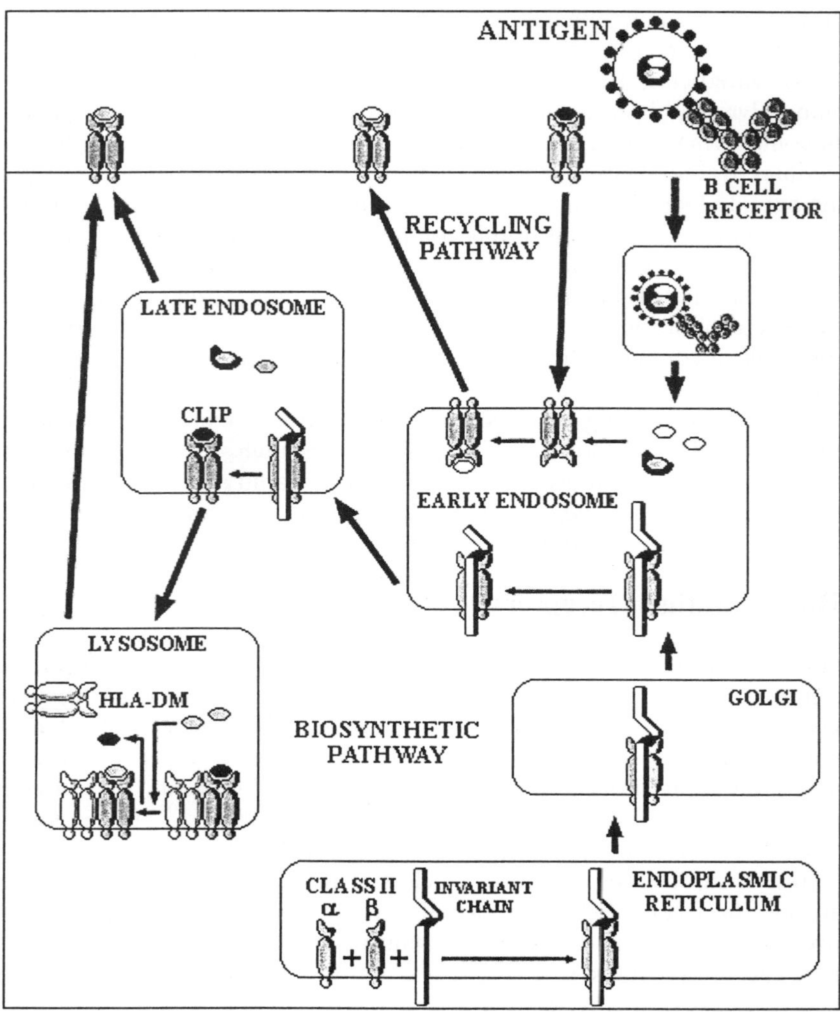

Fig. 1. Some of the intracellular compartments implicated in antigen internalization and peptide association with class II molecules. Antigen (in this instance a liposome, bacteria or virus) is taken up by receptor-mediated endocytosis by surface antibody or B cells or via other receptors on dendritic cells or macrophages. After partial degradation, peptide fragments derived from any of the component proteins of the antigen can associate with recycling empty class II molecules and the new peptide-class II complex re-expressed at the cell surface. Alternatively the peptide may become associated with newly synthesized class II molecules in deeper compartments. This process is catalyzed by HLA-DM and the peptide is exchanged for CLIP, which is a part of the invariant chain. The MIIC and CIIV compartments are not shown here, as their relation to the better studied intracellular compartments is not well defined.

encapsulated in these liposomes may be efficiently presented to T cells. This mimics the role of B cells in immunity. B cells take up microorganisms which express surface antigen for which they are specific and may present any peptides derived from that microorganism which will associate with class II molecules. These B cells can thus be helped by T cells specific for any of these peptides to make antibody able to bind to the surface of those microorganisms and mediate their lysis or neutralization. In summary, antigen receptors serve both to augment the concentration of class II-antigen peptide complexes and to promote upregulation of co-stimulatory molecules, increasing the probability of activating the relevant specific T cell when it is encountered.

VIII. The role of macrophages and dendritic cells

B cells can thus identify antigen with great specificity, but B cells specific for a given antigen are rare, so that immune responses are normally initiated by other APC, such as macrophages and dendritic cells. Macrophages and dendritic cells are heterogeneous populations scattered throughout the body and undoubtedly have multiple functions. Another population of cells called follicular dendritic cells exists in lymphoid organs;[110] these probably have a separate origin, do not endocytose antigen and will not be discussed here. Macrophages and dendritic cells are not rare, but unlike B cells, these cells do not express antigen-specific receptors, though they are more active in endocytosis than B cells and they have been shown to present peptides from antigens taken up "non-specifically" in this way. The concentration of antigen in body fluids is likely to be low relative to the concentration of self protein but dendritic cells and macrophages may "home" to antigen by following chemotactic signals released by injured tissue or bacteria.

Part of the response to microorganisms depends on what is called "innate" immunity. Innate immunity permits rapid responses to infection with pre-formed effector molecules and cells lacking true antigen receptors, such as neutrophils and macrophages. This evolved before "acquired" immunity, which depends on the relatively slow expansion of rare, antigen-specific lymphocytes. Innate immunity permits recognition of conserved characteristics of viruses and bacteria. These include expression of certain lipids and carbohydrates which may be discriminated from normal self proteins.[111] LPS-binding protein, mannose binding lectin and C-reactive protein are opsonins present in body fluids which can bind to microorganisms.[112] Cellular receptors exist for these "pattern recognition" molecules. These receptors on APC permit antigen identification, capture and presentation to T cells. A particularly interesting example of molecules implicated in both innate and acquired immunity, is complement, a cascade of plasma proteins which acts at multiple stages in the immune response.[113] It catalyzes the lysis of bacteria and viruses to which antibody has bound but can also bind directly to microorganisms in the absence of antibody. Complement which undergoes proteolytic changes after binding to substrate may be taken up by cellular receptors for complement fragments. Complement receptors are found on APC. APC consequently bind microorganisms to which complement has bound more avidly than in the absence of complement. Even when antigen-specific receptors exist, as on B cells, the additional presence of complement receptors on those cells results in more efficient stimulation by their specific antigens when complement components have bound to the antigen than in the absence of complement.[114] Liposomes may also bind complement, to an extent which may be increased[115,116] or decreased,[117] depending on their size and lipid composition.

Other systems permitting antigen recognition are known. Many APC including macrophages and dendritic cells express receptors for the conserved regions of antibody molecules; these are called antibody Fc receptors.[118] These receptors have low affinity for antibody in solution but when antibodies bind to many different antigens on the surface of a microorganism the array of multiple antibody molecules may interact simultaneously with several Fc receptors on the same cell. This multimeric binding is stable and results in antigen uptake by these APC. This uptake process is hundreds or thousands of times more efficient than the uptake of antigen by non-specific endocytosis in the absence of Fc receptors.[119] Costimulatory molecules are even more richly expressed on dendritic cells and macrophages than on B cells[120] and it is also possible that signals initiated by stimulation of the Fc receptor have a role in induction of co-stimulatory molecules. Experiments in our laboratory have demonstrated that antigen-containing liposomes made with a dinitrophenyl determinant coupled to a fraction of the phospholipids in their membranes are efficiently taken up by dendritic cells in the presence of anti-dinitrophenyl antibody, that is, by the Fc receptor, but not in the absence of the antibody. Antigens within the liposomes are efficiently presented to T cells (L. Leserman et al., manuscript in preparation). Fc receptors are present on B cells, but do not mediate antigen uptake. Antibody may thus act both as an inducer as well as effector molecule of the immune response, increasing antigen uptake and presentation by permitting non-B cell APC to "identify" antigen by means of antibody secreted by specific B cells, as well as by mediating neutralization and lysis of microorganisms.

Fc receptors have also been shown to increase the uptake of liposomes administered in vivo.[121] In fact, as shown by Alving and others, pre-formed antibodies to lipid components of liposomes can be found in the serum of many animals. Since complement binding to antigen is markedly enhanced by antibody binding, the action of Fc and complement receptors may be synergistic. This means that a significant component of the "non-specific" uptake of liposomes may be in part mediated specifically by antibody and complement. Many other serum proteins are known to bind to liposomes based on charge or other membrane characteristics[122,123] and it is reasonable to assume that receptors for these proteins exist on APC.

Both macrophages and dendritic cells may be obtained from blood monocytes, depending on the culture conditions and growth factors present.[124] Macrophages occur in most multicellular organisms and have evolved many functions in innate immunity, while dendritic cells have appeared more recently in evolutionary time and apparently interact only with T cells. Dendritic cells are the last identified and thus least studied APC but nevertheless have been shown to be a major cell type involved in antigen presentation.[125–128] Both macrophages and dendritic cells are known to increase their expression of co-stimulatory molecules after contact with conserved bacterial products, such as lipopolysaccharide (LPS). The ability to increase homing of APC and to induce upregulation of co-stimulatory molecules probably explain the adjuvant activities of LPS and its constituents, such as lipid A, and of other adjuvants associated with liposomes.[129,130] Even though recently developed culture systems have permitted us to appreciate the importance of dendritic cells in antigen presentation, it is difficult to distinguish immature dendritic cells from macrophages in vivo. Intravenous injection of liposomes containing certain cytotoxic drugs has been shown to induce immunodeficiency, presumably by blocking antigen presentation and the target cell for this action was though to be macrophages, which are killed by this treatment.[131,132] However, it has recently been reported that this treatment also eliminates immature dendritic cells.[133] Since these cells are much more potent than macrophages in vitro the true importance of macrophages versus dendritic cells as APC in vivo remain to be elucidated.

IX. Questions and directions for future research

This overview describes some of the mechanisms of antigen presentation that explain why liposomes are effective vehicles for immunization. They can be easily recognized by APC by virtue of their opsonization by proteins present in the blood, or their uptake may be increased by the expression on their surface of antigens recognized by surface antibody of B cells. Their size, which may range from that of viruses to bacteria, permits engagement of a sufficient number of cellular receptors to mediate efficient uptake and signaling. They can sequester antigen for delivery into endocytic vesicles, permitting association with class II molecules. They can carry potent adjuvants to upregulate expression of class II and costimulatory molecules, maximizing chances of stimulating relevant T cells. They lack the mechanisms which various viruses and bacteria have evolved to avoid recognition by subverting numerous steps in antigen presentation.[134] Nevertheless, there are gaps in our understanding of antigen presentation that need to be filled in order to design liposomes improved for this purpose. In particular, in vitro assays measure short term events in the absence of cell migration. The rules which determine the outcome of antigen administration in vivo over longer periods are largely unknown. Our ignorance is particularly profound in the area of tumor immunity. The slow growth of tumors may not be accompanied by many of the "danger" stimuli associated with innate immunity induced by infection that normally alert the immune system to the presence of potential harm and facilitate antigen presentation and costimulation.[135] Immunization protocols using tumor-derived peptides to stimulate immunity to tumor antigens have in at least one instance promoted tumor growth, possibly by engaging T cell receptors for antigen in the absence of costimulation and inducing tolerance.[136] Increased appreciation of the role of costimulation has led to the development of gene therapy protocols to induce cytokine expression by tumor cells, in order to increase attractiveness

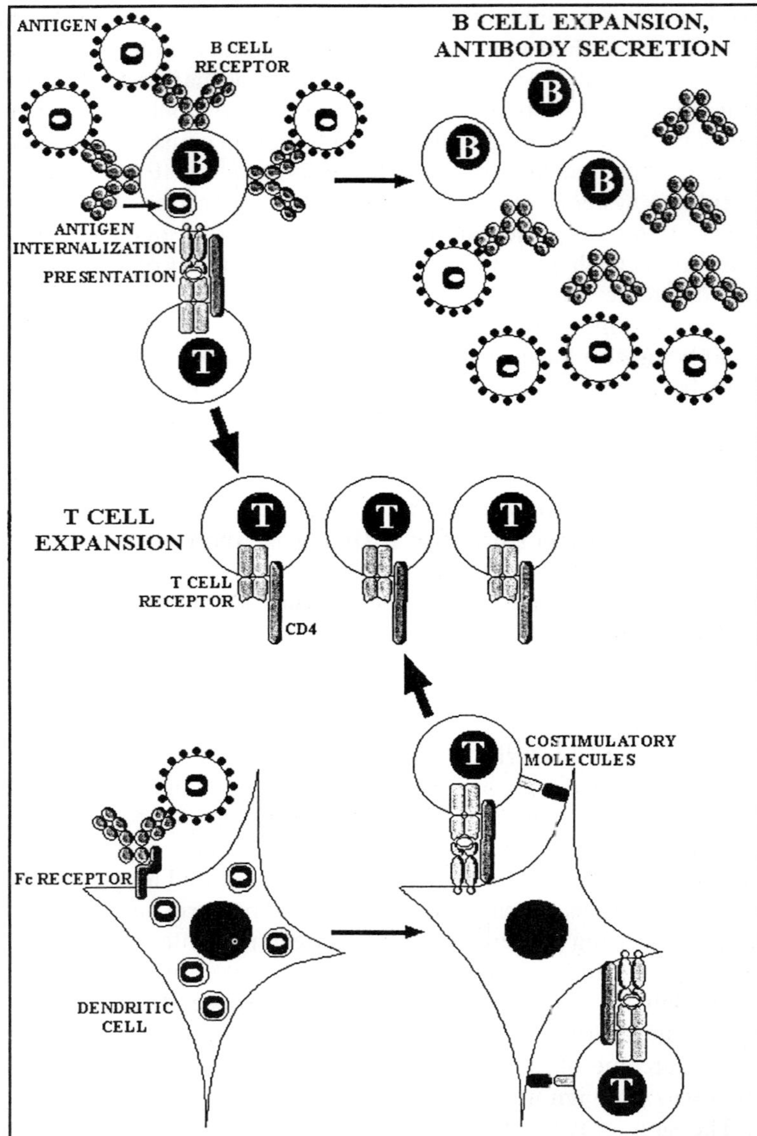

Fig. 2. Overview of antigen presentation by B cells or dendritic cells. Antigen taken up by receptor-mediated endocytosis by surface antibody of B cells (top) or by one of several receptors of dendritic cells, such as the immunoglobulin Fc receptor (bottom) is internalized, degraded into fragments and associated with MHC class II molecules in one of several intracellular compartments. Surface expression of class II molecules plus peptide will stimulate relevant T cells through their T cell receptors for antigen. These cells also express CD4 surface molecules. These cells will expand and will stimulate proliferation and antibody production by the B cells which present the antigen. Dendritic cells are terminally differentiated and do not divide.

of these tissues for dendritic cells and to upregulate costimulatory molecules. However, it is important to remember that cytokines may also be growth factors for tumors.[137] In addition, antigen presentation need not necessarily lead to the proliferation of relevant T cells. APC have also been reported to downregulate responses by inhibiting other APC or killing or preventing proliferation of T cells in certain experimental conditions.[138–141] While this is a handicap for immunization protocols, the possibility of using APC to induce tolerance may have important applications in organ transplantation.[142] All that is required now is for us to understand these processes sufficiently well that we can achieve the desired outcome.

Acknowledgments

We thank Jean Davoust, Patrick Machy and Anne-Marie Schmitt-Verhulst for helpful discussion and criticism of the manuscript. N.B. was supported by a fellowship from le Ministère de l'Enseignement Supérieure et de la Recherche. Studies from the authors' laboratories were supported by grants from the Association pour la Recherche sur le Cancer, the Ligue National Française contre le Cancer and the European Community, and by institutional grants from the Centre National de la Recherche Scientifique and the Institut National de la Santé et de la Recherche Médicale.

References

1. Allison A, Gregoriadis G. Liposomes as immunological adjuvants. Nature 1974;252:252.
2. Gregoriadis G, Allison AC. Entrapment of proteins in liposomes prevents allergic reactions in pre-immunized mice. FEBS Lett 1974;45:71–74.
3. Garcon NMJ, Six HM. Universal vaccine carrier. Liposomes that provide T-dependent help to weak antigens. J Immunol 1991;146:3697–3702.
4. Gregoriadis G. Immunological adjuvants: a role for liposomes. Immunol Today 1990;11:89–96.
5. Alving C. Immunologic aspects of liposomes: presentation and processing of liposomal protein and phospholipid antigens. Biochim Biophys Acta 1992;1113:307–322.
6. Alving C, Koulchin V, Glenn G, Rao M. Liposomes as carriers of peptide antigens: induction of antibodies and cytotoxic T lymphocytes to conjugated and unconjugated peptides. Immunol Rev 1995;145:5–31.
7. Wassef N, Alving C, Richards R. Liposomes as carriers for vaccines. Immunomethods 1994;4:217–222.
8. Lichtman A, Pober J, Abbas A. Cellular and Molecular Immunology, 2nd edn. W.B. Saunders, 1994.
9. Janeway C, Travers P. Immunobiology. The immune system in health and disease, 3rd edn. New York: Garland Publishing, Inc, 1997.
10. Bachmann MF, Kündig TM, Kalberer CP, Hengartner H, Zinkernagel RM. How many specific B cells are needed to protect against a virus? J Immunol 1994;152:4235–4241.
11. Yasuda T. Immunogenicity of liposomal model membranes in mice. Dependence on phospholipid composition. Proc Natl Acad Sci USA 1977;74:1234–1236.
12. Bachmann M, Zinkernagel R. The influence of virus structure on antibody responses and virus serotype formation. Immunol Today 1996;17:553–558.
13. Unanue ER. Antigen-presenting function of the macrophage. Annu Rev Immunol 1984;2:395–429.
14. Townsend ARM, Bodmer H. Antigen recognition by class I MHC-restricted T lymphocytes. Annu Rev Immunol 1989;7:601–624.
15. Gretz JE, Kaldjian EP, Anderson AO, Shaw S. Sophisticated strategies for information encounter

in the lymph node. The reticular network as a conduit of soluble information and a highway for cell traffic. J Immunol 1996;157:495–499.
16. Schwartz RH. Costimulation of T lymphocytes: The role of CD28:CTLA-4, and B7/BB1 in interleukin-2 production and in immunotherapy. Cell 1992;71:1065–1068.
17. Kupfer A, Mosmann T, Kupfer H. Polarized expression of cytokines in cell conjugates of helper T cells and splenic B cells. Proc Natl Acad Sci USA 1991;88:775–779.
18. Mosmann TR, Coffman RL. Heterogeneity of cytokine secretion patterns and functions of helper T cells. Adv Immunol 1989;46:111–148.
19. Ziegler K, Unanue ER. Identification of a macrophage antigen-processing event required for I-region-restricted antigen presentation to T lymphocytes. J Immunol 1981;127:1869–1875.
20. Ziegler HK, Unanue ER. Decrease in macrophage antigen catabolism caused by ammonia and chloroquine is associated with inhibition of antigen presentation to T cells. Proc Natl Acad Sci. 1982;79:175–178.
21. Allen PM, Strydom DJ, Unanue ER. Processing of lysozyme by macrophages: Identification of the determinants recognized by two T-cell hybridomas. Proc Natl Acad Sci USA 1984;81:2489–2493.
22. Shastri N, Malissen B, Hood L. Ia-transfected L-cell fibroblasts present a lysozyme peptide but not the native protein to lysozyme-specific T cells. Proc Natl Acad Sci USA 1985;82:5885–5889.
23. Malissen B, Shastri N, Pierres M, Hood L. Cotransfer of the Ed α and Ad β genes into L cells results in the surface expression of a functional mixed-isotype Ia molecule. Proc Natl Acad Sci USA 1986;83:3958–3962.
24. Watts T, Gariepy J, Schoolnik G, McConnell H. T-cell activation by peptide antigen: effect of peptide sequence and method of antigen presentation. Proc Natl Acad Sci USA 1985;82:5480–5484.
25. Watts T, Brian A, Kappler J, Marrack P, McConnell H. Antigen presentation by supported planar membranes containing affinity-purified I-Ad. Proc Natl Acad Sci USA 1984;81:7564–7568.
26. McConnell H, Watts T, Weis R, Brian A. Supported planar membranes in studies of cell-cell recognition in the immune system. Biochim Biophys Acta 1986;864:95–106.
27. Walden P, Nagy Z, Klein J. Induction of regulatory T-lymphocyte responses by liposomes carrying major histocompatibility complex molecules and foreign antigen. Nature 1985;315:327–329.
28. Walden P, Nagy Z, Klein J. Antigen presentation by liposomes: inhibition with antibodies. Eur J Immunol 1986;16:717–720.
29. Bakouche O, Gerlier D. Enhancement of immunogenicity of tumour virus antigen by liposomes: the effect of lipid composition. Immunology 1986;58:507–513.
30. Walden P, Nagy Z, Klein J. Major histocompatibility complex-restricted and unrestricted activation of helper T cell lines by liposome-bound antigens. J Mol Cell Immunol 1986;2:191–197.
31. Brown JH, Jardetzky TS, Gorga JC, Stern LJ, Urban RG, Strominger JL, Wiley DC. The three-dimensional structure of the human class II histocompatibility antigen HLA-DR1. Nature 1993;364:33–39.
32. Stern LJ, Brown JH, Jardetzky TS, Gorga JC, Urban RG, Strominger JL, Wiley DC. Crystal structure of the human class II MHC protein HLA-DR1 complexed with an influenza virus peptide. Nature 1994;368:215–221.
33. Chicz RM, Urban RG, Lane WS, Gorga JC, Stern LJ, Vignali DAA, Strominger JL. Predominant naturally processed peptides bound to HLA-DR1 are derived from MHC-related molecules and are heterogeneous in size. Nature 1992;358:764–758.
34. Elliott EA, Drake JR, Amigorena S, Elsemore J, Webster P, Mellman I, Flavell RA. The invariant chain is required for intracellular transport and function of major histocompatibility complex class II molecules. J Exp Med 1994;179:681–694.
35. Roche PA, Cresswell P. Proteolysis of the class II-associated invariant chain generates a peptide binding site in intracellular HLA-DR molecules. Proc Natl Acad Sci USA 1991;88:3150–3154.
36. Arneson LS, Miller J. Efficient endosomal localization of major histocompatibility complex class II-invariant chain complexes requires multimerization of the invariant chain targeting sequence. J Cell Biol 1995;129:1217–1228.
37. Pieters J, Bakke O, Dobberstein B. The MHC class II-associated invariant chain contains two endosomal targeting signals within its cytoplasmic tail. J Cell Sci 1993;106:831–846.
38. Odorizzi CG, Trowbridge IS, Xue L, Hopkins CR, Davis CD, Collawn JF. Sorting signals in the MHC class II invariant chain cytoplasmic tail and transmembrane region determine trafficking to an endocytic processing compartment. J Cell Biol 1994;126:317–330.

39. Loss GE Jr, Sant AJ. Invariant chain retains MHC class II molecules in the endocytic pathway. J Immunol 1993;150:3187–3197.
40. Germain RN, Rinker AG Jr. Peptide binding inhibits protein aggregation of invariant-chain free class II dimers and promotes surface expression of occupied molecules. Adv Cancer Res 1993;363:725–728.
41. Roche PA, Cresswell P. Invariant chain association with HLA-DR molecules inhibits immunogenic peptide binding. Nature 1990;345:615–618.
42. Teyton L, O'Sullivan D, Dickson PW, Lotteau V, Sette A, Fink P, Peterson PA. Invariant chain distinguishes between the exogenous and endogenous antigen presentation pathways. Nature 1990;348:39–44.
43. Long EO, LaVaute T, Pinet V, Jaraquemada D. Invariant chain prevents the HLA-DR-restricted presentation of a cytosolic peptide. J Immunol 1994;153:1487–1494.
44. Romagnoli P, Germain RN. The CLIP region of invariant chain plays a critical role in regulating major histocompatibility complex class II folding, transport, and peptide occupancy. J Exp Med 1994;180:1107–1113.
45. Freisewinkel IM, Schenck K, Koch N. The segment of invariant chain that is critical for association with major histocompatibility complex class II molecules contains the sequence of a peptide eluted from class II polypeptides. Proc Natl Acad Sci USA 1993;90:9703–9706.
46. Ghosh P, Amaya M, Mellins E, Wiley DC. The structure of an intermediate in class II MHC maturation: CLIP bound to HLA-DR3. Nature 1995;378:457–462.
47. Bond JS, Butler PE. Intracellular proteases. Annu Rev Biochem 1987;56:333–364.
48. Maric MA, Taylor MD, Blum JS. Endosomal aspartic proteinases are required for invariant-chain processing. Proc Natl Acad Sci USA 1994;91:2171–2175.
49. Xu M, Capraro GA, Daibata M, Reyes VE, Humphreys RE. Cathepsin B cleavage and release of invariant chain from MHC class II molecules follow a staged pattern. Mol Immunol 1994;31:723–731.
50. Riese RJ, Wolf PR, Brömme D, Natkin LR, Villadangos JA, Ploegh HL, Chapman HA. Essential role for cathepsin S in MHC class II-associated invariant chain processing and peptide loading. Immunity 1996;4:357–366.
51. Neefjes JJ, Ploegh HL. Inhibition of endosomal proteolytic activity by leupeptin blocks surface expression of MHC class II molecules and their conversion to SDS resistant $\alpha\beta$ heterodimers in endosomes. EMBO J 1992;11:411–416.
52. Nguyen QV, Knapp W, Humphreys RE. Inhibition by leupeptin and antipain of the intracellular proteolysis of Ii. Hum Immunol 1989;24:153–163.
53. Demotz S, Danieli C, Wallny H-J, Majdic O. Inhibition of peptide binding to DR molecules by a leupeptin-induced invariant chain fragment. Mol Immunol 1994;31:885–893.
54. Koch N, Lauer W, Habicht J, Dobberstein B. Primary structure of the gene for the murine Ia antigen-associated invariant chain (Ii). An alternatively spliced exon encodes a cysteine-rich domain highly homologous to a repetitive sequence of thyroglobulin. EMBO J 1987;6:1677–1683.
55. Fineschi B, Sakaguchi K, Appella E, Miller J. The proteolytic environment involved in MHC class II-restricted antigen presentation can be modulated by the p41 form of invariant chain. J Immunol 1996;157:3211–3215.
56. Rodriguez GM, Diment S. Destructive proteolysis by cysteine proteases in antigen presentation of ovalbumin. Eur J Immunol 1995;25:1823–1827.
57. Denzin LK, Robbins NF, Carboy-Newcomb C, Cresswell P. Assembly and intracellular transport of HLA-DM and correction of the class II antigen-processing defect in T2 cells. Immunity 1994;1:595–606.
58. Sanderson F, Kleijmeer MJ, Kelly A, Verwoerd D, Tulp A, Neefjes JJ, Geuze HJ, Trowsdale J. Accumulation of HLA-DM, a regulator of antigen presentation, in MHC class II compartments. Science 1994;266:1566–1569.
59. Schäfer W, Stroh A, Berghöfer S, Seiler J, Vey M, Kruse M-L, Kern HF, Klenk H-D, Garten W. Two independent targeting signals in the cytoplasmic domain determine *trans*-Golgi network localization and endosomal trafficking of the proprotein convertase furin. EMBO J 1995;14:2424–2435.
60. Sanderson F, Thomas C, Neefjes J, Trowsdale J. Association between HLA-DM and HLA-DR in vivo. Immunity 1996;4:87–96.
61. Fung-Leung W-P, Surh CD, Lijedahl M, Pang J, Leturcq D, Peterson PA, Webb SR, Karlsson L. Antigen presentation and T cell development in H2-M-deficient mice. Science 1996;271:1278–1291.
62. Miyazaki T, Wolf P, Tourne S, Waltzinger C, Dierich A, Barois N, Ploegh H, Benoist C, Mathis

D. Mice lacking H2-M complexes, enigmatic elements of the MHC class II peptide-loading pathway. Cell 1996;84:531–541.
63. Martin WD, Hicks GG, Mendiratta SK, Leva HI, Ruley HE, Van Kaer L. H2-M mutant mice are defective in the peptide loading of class II molecules, antigen presentation, and T cell repertoire selection. Cell 1996;84:543–550.
64. Fling SP, Arp B, Pious D. HLA-DMA and -DMB genes are both required for MHC class II/peptide complex formation in antigen-presenting cells. Nature 1994;368:554–558.
65. Morris P, Shaman J, Attaya M, Amaya M, Goodman S, Bergman C, Monaco JJ, Mellins E. An essential role for HLA-DM in antigen presentation by class II major histocompatibility molecules. Nature 1994;368:551–554.
66. Sherman MA, Weber DA, Jensen PE. DM enhances peptide binding to class II MHC by release of invariant chain-derived peptide. Immunity 1995;3:197–205.
67. Denzin LK, Cresswell P. HLA-DM induces CLIP dissociation from MHC class II $\alpha\beta$ dimers and facilitates peptide loading. Cell 1995;82:155–165.
68. Katz JF, Stebbins C, Appella E, Sant AJ. Invariant chain and DM edit self-peptide presentation by major histocompatibility complex (MHC) class II molecules. J Exp Med 1996;184:1747–1753.
69. Van Ham SM, Grüneberg U, Malcherek G, Bröker I, Melms A, Trowsdale J. Human histocompatibility leukocyte antigen (HLA)-DM edits peptides presented by HLA-DR according to their ligand binding motifs. J Exp Med 1996;184:2019–2024.
70. Germain RN, Margulies DH. The biochemistry and cell biology of antigen processing and presentation. Annu Rev Immunol 1993;11:403–450.
71. Harding CV, Roof RW, Unanue ER. Turnover of Ia-peptide complexes is facilitated in viable antigen-presenting cells: Biosynthetic turnover of Ia vs. peptide exchange. Proc Natl Acad Sci USA 1989;86:4230–4234.
72. Salamero J, Humbert M, Cosson P, Davoust J. Mouse B lymphocyte specific endocytosis and recycling of MHC class II molecules. EMBO J 1990;9:3489–3496.
73. Marsh EW, Dalke DP, Pierce SK. Biochemical evidence for the rapid assembly and disassembly of processed antigen-major histocompatibility complex class II complexes in acidic vesicles of B cells. J Exp Med 1992;175:425–436.
74. St.-Pierre Y, Watts TH. MHC class II-restricted presentation of native protein antigen by B cells is inhibitable by cycloheximide and brefeldin A. J Immunol 1990;145:812–818.
75. Nadimi F, Moreno J, Momburg F, Heuser A, Fuchs S, Adorini L, Hämmerling GJ. Antigen presentation of hen egg-white lysozyme but not of ribonuclease A is augmented by the major histocompatibility complex class II-associated invariant chain. Eur J Immunol 1991;21:1255–1263.
76. Momburg F, Fuchs S, Drexler J, Busch R, Post M, Hämmerling GJ, Adorini L. Epitope-specific enhancement of antigen presentation by invariant chain. J Exp Med 1993;178:1453–1458.
77. Vidard L, Rock KL, Benacerraf B. Diversity in MHC class II ovalbumin T cell epitopes generated by distinct proteases. J Immunol 1992;149:498–504.
78. Harding CV, Collins DS, Slot JW, Geuze HJ, Unanue ER. Liposome-encapsulated antigens are processed in lysosomes, recycled and presented to T cells. Cell 1991;64:393–401.
79. Peters PJ, Neefjes JJ, Oorschot V, Ploegh HL, Geuze HJ. Segregation of MHC class II molecules from MHC class I molecules in the Golgi complex for transport to lysosomal compartments. Nature 1991;349:669–672.
80. Daro E, van der Sluijs P, Galli T, Mellman I. Rab4 and cellubrevin define different early endosome populations on the pathway of transferrin receptor recycling. Proc Natl Acad Sci USA 1996;93:9559–9564.
81. Pinet V, Vergelli M, Martin R, Bakke O, Long EO. Antigen presentation mediated by recycling of surface HLA-DR molecules. Nature 1995;375:603–606.
82. Stebbins CC, Peterson ME, Suh WM, Sant AJ. DM-mediated release of a naturally occurring invariant chain degradation intermediate from MHC class II molecules. J Immunol 1996;157:4892–4898.
83. Méresse S, Gorvel J-P, Chavrier P. The rab7 GTPase resides on a vesicular compartment connected to lysosomes. J Cell Sci 1995;108:3349–3358.
84. Chavrier P, Parton RG, Hauri HP, Simons K, Zerial M. Localization of low molecular weight GTP binding proteins to exocytic and endocytic compartments. Cell 1990;62:317–329.
85. Granger BL, Green SA, Gabel CA, Howe CL, Mellman I, Helenius A. Characterization and cloning of lgp 110, a lysosomal membrane glycoprotein from mouse and rat cells. J Biol Chem 1990;265:12036–12043.
86. Mitchell RN, Barnes KA, Grupp SA, Sanchez M, Misulovin Z, Nussenzweig MC, Abbas AK.

Intracellular targeting of antigens internalized by membrane immunoglobulin in B lymphocytes. J Exp Med 1995;181:1705–1714.

87. Collins DS, Unanue ER, Harding CV. Reduction of disulfide bonds within lysosomes is a key step in antigen processing. J Immunol 1991;147:4054–4059.
88. Wubbolts R, Fernandez-Borja M, Oomen L, Verwoerd D, Janssen H, Calafat J, Tulp A, Dusseljee S, Neefjes J. Direct vesicular transport of MHC class II molecules from lysosomal structures to the cell surface. J Cell Biol 1996;135:611–622.
89. Harding CV, Collins DS, Kanagawa O, Unanue ER. Liposome-encapsulated antigens engender lysosomal processing for MHC class II presentation and cytosolic processing for Class I presentation. J Immunol 1991;147:2860–2863.
90. Zhou F, Rouse BT, Huang L. Induction of cytotoxic T lymphocytes in vivo with protein antigen entrapped in membranous vesicles. J Immunol 1992;149:1599–1604.
91. Kovacsovics-Bankowski M, Clark K, Benacerraf B, Rock KL. Efficient major histocompatibility complex class I presentation of exogenous antigen upon phagocytosis by macrophages. Proc Natl Acad Sci USA 1993;90:4942–4946.
92. Kovacsovics-Bankowski M, Rock KL. A phagosome-to-cytosol pathway for exogenous antigens presented on MHC class I molecules. Science 1995;267:243–246.
93. White W, Cassatt D, Madsen J, Burke S, Woods R, Wassef N, Alving C, Koenig S. Antibody and cytotoxic T-lymphocyte responses to a single liposome-associated peptide antigen. Vaccine 1995;13:1111–1122.
94. York IA, Rock KL. Antigen processing and presentation by the class I major histocompatibility complex. Annu Rev Immunol 1996;14:369–396.
95. Rudensky A, Janeway CJ. Studies on naturally processed peptides associated with MHC class II molecules. Chem Immunol 1993;57:134–151.
96. Calafat J, Nijenhuis M, Janssen H, Tulp A, Dusseljee S, Wubbolts R, Neefjes J. Major histocompatibility complex class II molecules induce the formation of endocytic MIIC-like structures. J Cell Biol 1994;126:967–978.
97. Schafer PH, Green JM, Malapati S, Gu L, Pierce SK. HLA-DM is present in one-fifth the amount of HLA-DR in the class II peptide-loading compartment where it associates with leupeptin (LIP)-HLA-DR complexes. J Immunol 1996;157:5487–5495.
98. Harding CV, Geuze HJ. Immunogenic peptides bind to class II MHC molecules in an early lysosomal compartment. J Immunol 1993;151:3988–3998.
99. Qiu Y, Xu X, Wandinger-Ness A, Dalke DP, Pierce SK. Separation of subcellular compartments containing distinct functional forms of MHC class II. J Cell Biol 1994;125:595–605.
100. West MA, Lucocq JM, Watts C. Antigen processing and class II MHC peptide-loading compartments in human B-lymphoblastoid cells. Nature 1994;369:147–151.
101. Tulp A, Verwoerd D, Dobberstein B, Ploegh HL, Pieters J. Isolation and characterization of the intracellular MHC class II compartment. Nature 1994;369:120–126.
102. Amigorena S, Drake JR, Webster P, Mellman I. Transient accumulation of new class II MHC molecules in a novel endocytic compartment in B lymphocytes. Nature 1994;369:113–120.
103. Chesnut RW, Grey HM. Antigen presentation by B cells and its significance in T-B interactions. Adv Immunol 1986;39:51–94.
104. Lanzavecchia A. Antigen uptake and accumulation in antigen-specific B cells. Immunol Rev 1987.
105. Barois N, Fourquet F, Davoust J. Selective modulation of the MHC class II antigen presentation pathway following B cell receptor ligation and PKC activation. J Biol Chem 1997;272:3641–3647.
106. Dal Monte P, Szoka FJ. Effect of liposome encapsulation on antigen presentation in vitro. Comparison of presentation by peritoneal macrophages and B cell tumors. J Immunol 1989;142:1437–1443.
107. Grivel J, Crook K, Leserman L. Endocytosis and presentation of liposome-associated antigens by B cells. Immunomethods 1994;4:223–8.
108. Gerlier D, Trescol-Biemont M, Varior-Krishnan G, Naniche D, Fugier-Vivier I, Rabourdin-Combe C. Efficient major histocompatibility complex class II-restricted presentation of measles virus relies on hemagglutinin-mediated targeting to its cellular receptor human CD46 expressed by murine B cells. J Exp Med 1994;179:353–358.
109. Vidard L, Kovacsovics-Bankowski M, Kraeft S-K, Chen LB, Benacerraf B, Rock KL. Analysis of MHC class II presentation of particulate antigens by B lymphocytes. J Immunol 1996;156:2809–2818.
110. Szakal AK, Kapasi ZF, Haley ST, Tew JG. A theory of follicular dendritic cell origin. Curr Top Microbiol Immunol 1995;201:1–13.

111. Janeway C. Approaching the asymptote: evolution and revolution in immunology. Cold Sprng Harb Symp Quant Biol 1989;54:1–13.
112. Sastry K, Ezekowitz R. Collectins: pattern recognition molecules involved in first line host defense. Curr Opin Immunol 1993;5:59–66.
113. Morgan B. Physiology and pathophysiology of complement: progress and trends. Crit Rev Clin Lab Sci 1995;32:265–298.
114. Dempsey P, Allison M, Akkaraju S, Goodnow C, Fearon D. C3d of complement as a molecular adjuvant: bridging innate and acquired immunity. Science 1996;271:348–350.
115. Marjan J, Xie Z, Devine D. Liposome-induced activation of the classical complement pathway does not require immunoglobulin. Biochim Biophys Acta 1994;1192:35–44.
116. Devine D, Wong K, Serrano K, Chonn A, Cullis P. Liposome-complement interactions in rat serum: implications for liposome survival studies. Biochim Biophys Acta 1994;1191:43–51.
117. Wassef N, Alving C. Complement-dependent phagocytosis of liposomes. Chem Phys Lipids 1993;64:239–248.
118. Ravetch JV, Kinet JP. Fc receptors. Ann Rev Immunol 1991;9:457.
119. Sallusto F, Cella M, Danieli C, Lanzavecchia A. Dendritic cells use macropinocytosis and the mannose receptor to concentrate macromolecules in the major histocompatibility complex Class II compartment: downregulation by cytokines and bacterial products. J Exp Med 1995;182:389–400.
120. Cassell DJ, Schwartz RH. A quantitative analysis of antigen-presenting cell function: Activated B cells stimulate naive CD4 T cells but are inferior to dendritic cells in providing co-stimulation. J Exp Med 1994;180:1829–1840.
121. Aragnol D, Leserman L. Immune clearance of liposomes inhibited by an anti-Fc receptor antibody in vivo. Proc Natl Acad Sci USA 1986;83:2699–2703.
122. Bonte F, Juliano R. Interactions of liposomes with serum proteins. Chem Phys Lipids 1986;40:359–372.
123. Chonn A, Semple S, Cullis P. Association of blood proteins with large unilamellar liposomes in vivo. Relation to circulation lifetimes. J Biol Chem 1992;267:18759–18765.
124. Caux C, Vanbervliet B, Massacrier C, Dezutter-Dambuyant C, De Saint-Vis B, Jacquet C, Yonead K, Imamura S, Schmitt D, Banchereau J. $CD34^+$ hematopoietic progenitors from human cord blood differentiate along two independent dendritic cell pathways in response to GM-CSF + TNFα. J Exp Med 1996;184:695–706.
125. Steinman R. The dendritic cell system and its role in immunogenicity. Ann Rev Immunol 1991;9:271–296.
126. Ibrahim M, Chain B, Katz D. The injured cell: the role of the dendritic cell system as a sentinel receptor pathway. Immunol Today 1995;16:181–186.
127. Austyn J. New insights into the mobilization and phagocytic activity of dendritic cells. J Exp Med 1996;183:1287–1292.
128. Pieters J, Gieseler R, Thiele B, Steinbach F. Dendritic cells: from ontogenic orphans to myelomonocytic descendants. Immunol Today 1996;17:273–278.
129. Alving CR, Richards RL. Liposomes containing lipid A: a potent nontoxic adjuvant for a human malaria sporozoite vaccine. Immunol letters 1990;25:275–280.
130. Ullrich S, Fidler I. Liposomes containing muramyl tripeptide phosphatidylethanolamine (MTP-PE) are excellent adjuvants for induction of an immune response to protein and tumor antigens. J Leukoc Biol 1992;52:489–494.
131. Su D, van Rooijen N. The role of macrophages in the immunoadjuvant action of liposomes: effects of elimination of splenic macrophages on the immune response against intravenously injected liposome-associated albumin antigen. Immunology 1989;66:466–470.
132. Claassen I, Van Rooijen N, Claassen E. A new method for removal of mononuclear phagocytes from heterogeneous cell populations in vitro, using the liposome-mediated macrophage 'suicide' technique. J Immunol Methods 1990;134:153–161.
133. Leenen P, Voerman J, Radosevic K, van Rooijen N, van Ewijk W. Spleen dendritic cells: heterogeneity and in vivo phagocytic activity. In: Ricciardi-Castagnoli P, ed. 4th International Symposium on Dendritic Cells in Fundamental and Clinical Immunology. Venice: Ricerca Scientifica ed Educazione Permanente, 1996.
134. Zinkernagel R. Immunology taught by viruses. Science 1996;271:173–178.
135. Matzinger P. Tolerance, danger, and the extended family. Annu Rev Immunol 1994;12:991–1045.
136. Toes R, Blom R, Offringa R, Kast W, Melief C. Enhanced tumor outgrowth after peptide

vaccination. Functional deletion of tumor-specific CTL induced by peptide vaccination can lead to the inability to reject tumors. J Immunol 1996;156:3911–3918.

137. Han D, Pottin-Clemenceau C, Imro M, Scudeletti M, Doucet C, Puppo F, Brouty-Boye D, Vedrenne J, Sahraoui Y, Brailly H, Poggi A, Jasmin C, Azzarone B, Indiveri F. IL2 triggers a tumor progression process in a melanoma cell line MELP derived from a patient whose metastasis increased in size during IL2/INFα biotherapy. Oncogene 1996;12:1015–1023.
138. Holt P, Oliver J, Bilyk N, McMenamin C, McMenamin P, Kraal G, Thepen T. Downregulation of the antigen presenting cell function(s) of pulmonary dendritic cells in vivo by resident alveolar macrophages. J Exp Med 1993;177:397–407.
139. Suss G, Shortman K. A subclass of dendritic cells kills CD4 T cells via Fas/Fas-ligand-induced apoptosis. J Exp Med 1996;183:1789–1796.
140. Kronin V, Winkel K, Suss G, Kelso A, Heath W, Kirberg J, von BH, Shortman K. A subclass of dendritic cells regulates the response of naive CD8 T cells by limiting their IL-2 production. J Immunol 1996;157:3819–3827.
141. Hölsberg P, Batra V, Dressel A, Hafler D. Induction of anergy in CD8 T cells by B cell presentation of antigen. J Immunol 1996;157:5269–5276.
142. Coulombe M, Yang H, Guerder S, Flavell R, Lafferty K, Gill R. Tissue immunogenicity: the role of MHC antigen and the lymphocyte co-stimulator B7-1. J Immunol 1996;157:4790–4795.

Lasic and Papahadjopoulos (eds.), Medical Applications of Liposomes

CHAPTER 2.3

Systemic activation of macrophages by liposomes containing synthetic immunomodulators for treatment of metastatic disease

LAURA L. WORTH[a], ISAIAH J. FIDLER[b] AND EUGENIE S. KLEINERMAN[c]

[a]*Department of Pediatrics, Box 87;* [b,c]*Department of Cell Biology, Box 173, M.D. Anderson Cancer Center, 1515 Holcombe Boulevard, Houston, TX 77030, USA*

Overview

I. Introduction

Although the ability to diagnose and treat cancer has improved dramatically over the past 15–20 years, treatment of metastases continues to be a challenge. In many tumor types,[1] including osteosarcoma, micrometastases already can be present at the time of diagnosis. These metastases can contain cells that are different from those in the primary tumor and from each other, resulting in variable responses to chemotherapy and radiotherapy.[1,2] Searches for new agents or treatment modalities have sought to circumvent the heterogeneity that results in resistance. Activation of the immune system is a treatment modality that attempts to address this problem, as tumor cells share the property of being susceptible to destruction by tumoricidal macrophages irrespective of their sensitivity to chemotherapy.[3,4] Macrophages are an active component of the immune response that are responsible for identifying and phagocytosing microorganisms, foreign material, and cellular debris, as well as for recognition of self versus altered self, e.g., neoplastic cells. When activated, macrophages will selectively kill tumor cells but not normal cells.[5,6]

The mechanism by which macrophages recognize and destroy tumor cells is not clear; however, it does require direct macrophage-tumor cell contact.[5–9] This recognition appears to be independent of transplantation antigens, cell-specific antigens, tumor-specific antigens, cell cycle time, or phenotypes associated with transformation.[6] Resistance to macrophage-mediated tumor lysis is rare.[6] Attempts have been made to select tumor cells that were resistant to macrophage-mediated cell killing. Eight cell lines were incubated with syngeneic tumoricidal macrophages, and the cells that survived the incubation were expanded and re-exposed to tumoricidal macrophages.[10] After six sequential interactions, the treated cells were as susceptible to tumor lysis as were the parent cells.

If macrophages could be activated to their tumoricidal state, they would make an ideal modality for treatment of metastatic disease that is resistant to other forms of therapy. Accumulating evidence indicates that agents that activate macrophages may offer a new therapeutic approach to the world of cancer treatment.

II. In vivo activation of macrophages

II.1. Microorganisms and their products

Macrophages can be activated to become tumoricidal by two major pathways in vivo. The first is by interaction with microorganisms or their products, such as bacterial cell wall components or endotoxins. However, the systemic administration of microorganisms or their products in attempts to activate macrophages was shown to induce major side effects, including granuloma formation and allergic reactions.[11,12] Therefore, further attempts at activation of macrophages in vivo were not persued until the discovery of muramyl dipeptide (MDP). MDP is the minimal structural unit of *Mycobacterium*[13–15] that can stimulate the immune response[16] without inducing granuloma formation or allergic reactions. MDP is a water-soluble, low-molecular weight (MW = 495) synthetic moiety of *N*-acetylmuramyl-L-alanyl-D-isoglutamine that when administered intravenously in mice is cleared in the urine within 60 minutes. Although muramyl derivatives are able to activate macrophage tumoricidal function in vitro, they are rapidly cleared from the circulation after systemic administration[17,18] before they are able to activate macrophages.[19,20]

II.2. Cytokines

Macrophages can also be activated in vitro following interaction with various cytokines or macrophage-activating factors[21–23] such as interferon-γ.[24–26] However, systemic administration of cytokines showed a similar deficiency in the ability to activate tumoridical activity of macrophages. Under in vitro conditions, human monocytes require an 8-hour incubation with cytokines to become activated.[27] The half-life of the systemically administered cytokines was very brief[28] and consequently failed to activate the tumoricidal properties of the macrophages.

II.3. Liposomes containing immunomodulators

Because macrophages are phagocytic, the problem of rapid clearance of systemically administered cytokines and other immunomodulators can be overcome by encapsulating the activating agents into mutilamellar vesicles (MLV) or liposomes composed of phospholipids. A lipophilic MDP derivative, *N*-acetyl muramyl-L-alanyl-D-isoglutamyl-L-alanyl-2-(1′,2′ dipalmitoyl)-sn-glycero-3′-phosphorylethylamide (MTP-PE) has been synthesized[29] that can be incorporated into the lipid bilayer of liposomes (L-MTP-PE) in an attempt to delay rapid clearance from the body. When these agents are administered systemically, the distribution and retention of liposomes are affected by their size, composition, and surface charge.[3] The majority of systemically administered liposomes are phagocytosed by the cells of the reticuloendothelial system, mainly in the liver, spleen, and to a lesser extent, the lungs.[30,31]

The lung is a major site of metastatic disease in many pediatric and adult cancers. Therefore, our goal was to identify liposomes that could deliver MTP-PE to the pulmonary macrophages. Large liposomes (>0.1 μm) are retained more efficiently in the lung than smaller, otherwise identical liposomes. When negatively charged lipids (phosphotidylserine) are added to liposomes containing neutral phospholipid (phosphotidylcholine) in a ratio of 7:3 (PC:PS), they are more rapidly phagocytosed and efficiently arrested in the lungs.[30,31] Macrophages that reside in the lung parenchyma are not those activated by systemic administration of L-MTP-PE; instead, monocytes in the lung capillaries engulf liposomes and then migrate into the alveoli.[32] The intravenous injection of MLV containing MTP-PE (L-MTP-PE) resulted in the activation of mouse alveolar macrophages to the tumoricidal state. However, twice weekly injections were needed to maintain this level of tumoricidal activity.

Macrophage activation was found to be from the direct interaction of the L-MTP-PE with macrophages and not via an interaction with T cells causing the release of cytokines that subsequently activated the macrophages. T cells do not phagocytize liposomes containing MTP-PE.[33] Furthermore, we have been unable to demonstrate production of any cytokines when isolated T cells are incubated in vitro with L-MTP-PE. In addition, macrophages from mice with impaired T cell function, mice exposed to UV radiation, thymectomized adult mice exposed to X-rays, and athymic nude mice were all rendered tumoricidal by the systemic administration of liposome-encapsulated MTP-PE but not by control liposome preparations.[33]

III. In vitro activation of human monocytes by liposomes containing MTP-PE

Activated monocyte tumoricidal function has been correlated with the secretion of cytokines,[19] specifically interleukin-1 (IL-1),[34] tumor necrosis factor (TNF),[35] IL-6, IL-8, and monocyte chemotactic and activating factor (MCAF).[36,37] Furthermore, the anti-tumor effect of L-MTP-PE has been correlated with its ability to stimulate cytokine production. As shown in Figure 1, when peripheral monocytes from normal donors were incubated with L-MTP-PE, monocyte IL-1α, IL-1β, IL-6, IL-8, TNF and MCAF mRNA levels were rapidly up-regulated.[36,37] The expression of MCAF was short-lived, returning to baseline within 4 hours; whereas, increased expression of IL-1α, IL-1β, IL-6, IL-8 mRNA persisted for 72 hours. TNF mRNA levels remained elevated for 24 hours.[37] Nuclear run-on studies indicated that the increased levels of cytokine mRNA are due to increased transcription and not to the stabilization of message already present.[36,37] This increase in mRNA levels resulted in increased protein production. TNF protein levels peaked at 8 hours and persisted for 72 hours. IL-6 and IL-8 protein levels first became evident by 2 and 4 hours, respectively, and both persisted for 72 hours. IL-1 levels peaked at 8 hours, but IL-1α and IL-1β proteins remained intracellular, requiring a second stimulus (IFN-γ) before secretion took place.[37]

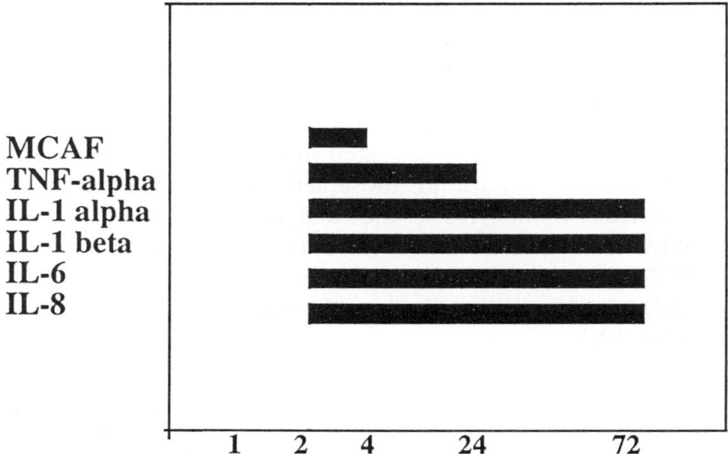

Fig. 1. Upregulated cytokine mRNA expression following liposome-encapsulated muramyl tripeptide phosphatidylethanolamine (L-MTP-PE). Time course in hours of increased cytokine mRNA stimulated by L-MTP-PE. Bars indicate range of upregulation of each cytokine in monocytes treated with L-MTP-PE (1.67 μg/ml) compared with expression in control monocytes. MCAF, monocyte chemotactic and activating factor; TNF, tumor necrosis factor; IL, interleukin.

MCAF protein was detectable by 4 hours.[36,37] In addition to the up-regulation of cytokine genes, L-MTP-PE increased the cell surface expression of specific adhesion molecules, including ICAM-1 and the integrins αL, α5, and β1. The αL integrin was crucial in the monocytes' ability to kill tumor cells following stimulation with L-MTP-PE.[38] These data indicate that L-MTP-PE activates monocyte tumoricidal function through up-regulation of both cytokine genes and cell surface adhesion molecules and the subsequent secretion of the specific cytokines.

IV. Preclinical in vivo studies using liposomal MTP-PE in the adjuvant setting

IV.1. Murine studies

The B16/BL6 mouse melanoma model was first used to study the effect of L-MTP-PE on lung metastases. When the B16/BL6 melanoma cells were injected into the footpad of a C57BL/6 mouse, the tumor metastasized to the lymph nodes and lung in more than 90% of the mice.[39] Four to 5 weeks after injection when the tumor in the footpad measured 10–15 mm in diameter, metastases were visible in the lymph nodes and could be visualized in the lung with a dissecting microscope. No gross lung nodules were detected at this time. In one study,[40] the midfemur of the tumor bearing leg, including the popliteal lymph node, was amputated. Treatment with L-MTP-PE (5 μmol lipid, 2.5 μg MDP) was initiated twice weekly for 4 weeks as this injection schedule was necessary to maintain pulmonary macrophage activation. Seventy percent of the animals survived for more than 190 days. These mice were considered disease free, because in the natural course of the disease, death occurs within 40 days. Although 30% of animals treated with L-MTP-PE had progressive disease and died, they did show a response to therapy, with a decrease in the median number of metastases relative to control animals and those treated with empty liposomes. Subsequent studies[41] looked at the timing of initiating therapy and survival. When initiation of liposomal MTP-PE was delayed after amputation of the primary tumor and a larger metastatic tumor burden was present, fewer mice survived. These studies suggested that L-MTP-PE therapy would only be effective when used in a setting of microscopic disease.

IV.2. Canine studies

Liposomal MTP-PE also eradicated lung metastases in dogs with spontaneous (autochthonous) osteosarcoma. Most dogs, like humans, have micrometastases to the lung at the time of diagnosis.[42] Surgical resection of the primary tumor produces a median survival of 3–6 months. In a double-blinded, placebo-controlled study,[43] 14 dogs were injected with liposomal MTP-PE (2 mg/m^2 MTP-PE; 500 mg liposomes) and 13 dogs with empty liposomes twice weekly for 8 weeks after the amputation of the primary tumor. The median survival for the dogs treated with empty liposomes was 77 days, which was the same as surgery without additional therapy. In contrast, the dogs treated with L-MTP-PE had a median survival time of 222 days, with four of these dogs alive and disease free 4 years after their initial surgery. The treatment did not produce toxic side effects and was well tolerated.[43] Because the major limitation for eradication of metastases by macrophages is tumor burden,[44] it has been hypothesized that in some dogs, the tumor burden exceeded the levels that can be destroyed by macrophages.[43] These studies indicated that L-MTP-PE therapy can indeed be an effective therapy for the eradication of microscopic lung metastases in an adjuvant or neoadjuvant setting.

V. Clinical trials with L-MTP-PE

V.1. Phase I trial

The tissue distribution, toxicity, maximally tolerated dose, and optimal biological dose of L-MTP-PE has been studied using a stable reproducible preparation of L-MTP-PE produced by CIBA-GEIGY (Basel, Switzerland).[44,45] L-MTP-PE was administered intravenously over one hour twice weekly on an out patient basis for 9 weeks to 28 patients with metastatic solid tumors unresponsive to all other therapies. The dosages ranged from 0.05–12 mg/m^2. Four of the 28 patients received ^{99m}Tc-labeled liposomes containing MTP-PE. The labeled liposomes were taken up predominantly in the spleen, liver, nasopharynx, thyroid, and, to a lesser extent, lung. Two patients showed uptake of label in their pulmonary metastases. The side effects were fever (70%), chills (80%), fatigue (60%), malaise, and myalgias. Four patients required dose reduction of in L-MTP-PE because of severe malaise and recurrent fever >38.8°C. Toxic reactions were not cumulative, however. The maximally tolerated dose was 6 mg/m^2. Because the optimal biological dose of biological response modifiers is frequently not the maximally tolerated dose, indicators of the immune response were also analyzed. The immune parameters that were analyzed included plasma cytokine levels, lymphocyte surface markers, acute phase reactants (fibrinogen, β_2-microblobulin, ceruloplasmin, and C-reactive protein). At doses of 2 mg/m^2, increases in acute phase reactants were seen with increases in IL-1β, white blood cell and granulocyte count as well as decreases in cholesterol. The tumoricidal properties of monocytes were also studied using a monocyte cytotoxicity assay (MTA). Essentially no tumoricidal activity was seen when patients received doses of <0.5 mg/m^2. At doses between 0.5 and 2.0 mg/m^2, monocytes exhibited elevated tumoricial activity at 24 and 72 hours following the infusion. At doses greater than 2.0 mg/m^2, monocytes only exhibited tumoricidal activity 24 hours following L-MTP-PE administration. Therefore, although the maximally tolerated dose was 6.0 mg/m^2, the dose providing the optimal biological activity was determined to be 2.0 mg/m^2, and this dose was used in all subsequent studies.

V.2. Pilot study in stage III and IV melanoma

Eighteen stage III or IV melanoma patients with resectable lymph node, skin, lung, or subcutaneous tissue were treated with L-MTP-PE.[46] Patients received 2 mg/m^2 L-MTP-PE twice a week for 4 weeks before the surgical excision of all disease. Therapy was then resumed for an additional 8 weeks at the twice weekly schedule, followed by once a week therapy for 12 weeks. Plasma cytokine, neopterin, and monocyte cytotoxicity activity were measured at various times during the first week of therapy. In addition, tumor-infiltrating lymphocyte (TIL) proliferation and TIL cytotoxicity, as well as tumor cell proliferation assays, were performed on the surgically resected specimens. The only predictors of patient response were the monocyte tumoricidal assay and the tumor proliferation assay. Three of the 18 patients had progressive disease after 4 weeks of preoperative therapy, and the melanoma cells removed from these patients proliferated rapidly in vitro. Twelve of the remaining 15 patients had tumor cells that did not proliferate under ideal culture conditions. Interestingly, one patient had tumor cells from the initial resection that did not proliferate well; however, the cells surgically resected at the time of relapse did grow well in vitro. Patients whose tumor cells showed positive in vitro growth had a median progression-free interval of 4.5 months compared with 8.25 months for those with a negative tumor cell proliferation assay response ($P = 0.056$). These data suggest that the ability of tumor cells to proliferate in vitro correlates with clinical outcome. The other parameter that correlated with clinical response was the MTA which did not increase during the first week of therapy in the three patients who had progressive disease. The median progression-free interval for the MTA-positive group (nine patients) was greater than 24 months compared with 4.25 months for the MTA negative-group (nine patients; $P < 0.05$). Four of the 18 patients remain disease-free at 52+, 59+, 69+, and 73+ months. Two additional patients were disease-free for 19 and 28 months. Six of the 18 patients are alive 52, 59, 64, 69, 71, and 73 months. Based on these findings, the use of L-MTP-PE in metastatic melanoma warrants further investigation.

V.3. Phase II trial in relapsed osteosarcoma

Based on our preclinical animal studies, the setting in which L-MTP-PE would be expected to have its optimal effect is with minimal residual disease as opposed to bulk tumor.[4] To assess the effect of L-MTP-PE in a phase II clinical trial, we chose relapsed osteosarcoma patients who had been rendered disease free of any visible or palpable tumor. Ordinarily, 80% of these patients will relapse within 1 year,[47] and chemotherapy has played little role in prolonging the disease-free interval. The study design specifically looked at patients who presented with pulmonary metastases that did not respond to chemotherapy or patients who developed pulmonary disease while receiving adjuvant chemotherapy. All visible and palpable tumor was removed, and L-MTP-PE therapy was administered intravenously twice a week for 12 weeks. The side effects were the same as identified by the phase I trial. Seven of the first 12 patients relapsed within 6 weeks of completing liposomal therapy. Metastases that had been surgically excised from these patients showed a unique histological picture: peripheral fibrosis surrounding the tumor with infiltration of inflammatory cells (histiocytes and macrophages).[48] In constrast, pulmonary nodules removed following treatment with chemotherapeutic agents showed central necrosis with viable tumor at the periphery and no evidence of an inflammatory response. This pattern is reminiscent of the pattern seen with pulmonary tuberculosis where the lesion is walled off and necrosis proceeds from the periphery inward. Fibrotic tissue gradually replaces the tubercle.[49] This process is slow, and active bacilli can persist for months. As a result of this finding, the treatment phase of the study was extended to 24 weeks. The median disease-free survival of relapsed osteosarcoma patients following surgery and salvage chemotherapy is 4.5 months, with 85% relapse by 1 year (historical control group). The 12 patients treated with L-MTP-PE for only 12 weeks had a median disease-free survival of 6.8 months.[50] However, the 16 patients treated for 24 weeks had a median disease-free interval of 9 months, which significantly exceeded that of our control group ($P < 0.003$). Seventy-five percent of the patients were still disease free at 6 months compared with 43% of the control group.[50] As observed earlier, with increasing time after the cessation of L-MTP-PE therapy, more patients relapsed. While the numbers are small, they suggest that a more prolonged therapy (9–12 months) may be even more beneficial to patients and result in longer disease-free intervals.

VI. Interaction of chemotherapy and liposomal MTP-PE in vitro and in vivo

It is unlikely that L-MTP-PE can serve as a single modality in treating metastatic disease. If the ratio of macrophages to tumor cells required for optimal macrophage-mediated tumoricidal activity in vivo is similar to that operating in vitro, then tumor burden exceeding 10^9 cells would be too large for the number of macrophages in the body. As an adjuvant against microscopic disease, we envision L-MTP-PE as an additional therapy with chemotherapy and surgical regimens. Once the major tumor burden is eliminated by surgery, the activated macrophages can perhaps destroy the residual tumor cells not eradicated by chemotherapy.

We believe it is important to combine L-MTP-PE with chemotherapy early in the treatment course of osteosarcoma in an attempt to cure the 30 to 40% of those patients who appear to harbor drug-resistant cells. Prior to designing such a study, we felt it was imperative to evaluate two factors: (1) Does L-MTP-PE increase the toxic side effects of chemotherapeutic agents, and (2) does chemotherapy inhibit the ability of L-MTP-PE to activate cytotoxic-mediated toxicity?

VI.1. Murine Studies

Prior to initiating therapy with L-MTP-PE and chemotherapy in the clinical situation, we assessed in mice if L-MTP-PE produced additional toxic side effects or inhibited the anti-tumor effects of various chemotherapeutic agents. Mice were treated with either L-MTP-PE alone, chemotherapeutic agents alone (doxorubicin, 10 mg/kg on days 0 and 7; cisplatin, 10 mg/kg on days 0 and 7; or ifosfamide, 2.5 mg on days 0, 3 and 5), or received simultaneous administration of one chemotherapeutic agent plus L-MTP-PE.[51] Total leukocyte counts and hematocrits levels decreased with all chemotherapeutic agents. The leukocyte differential was also altered. Therapy with one chemotherapeutic agent plus liposomal MTP-PE did not increase the myelosuppression normally seen at 2 weeks. Indeed, L-MTP-PE prevented the myelosuppression caused by doxorubicin. Additionally, the anti-tumor effect of combination therapy was also analyzed. Mice were inoculated with syngeneic splenic, renal, or cutaneous tumor cells prior to therapy with doxorubicin or ifosfamide with or without L-MTP-PE (days 11, 14, 18, 21, 25, and 28). L-MTP-PE did not interfere with the anti-tumor effects of the chemotherapeutic agents.

VI.2. Effect of chemotherapy administration on L-MTP-PE's activity in vitro

Human peripheral blood monocytes were assessed for their ability to respond to L-MTP-PE in vitro after treatment with chemotherapy. Peripheral blood monocytes were incubated with 5–500 ng/ml of adriamycin. After 3 hours, the drug was washed out, and the cells were treated with L-MTP-PE. Cultured cell supernatants showed secretion of IL-6, IL-8, and TNFα and increases in intracellular IL-1. The secretion of these cytokines was not increased when cells were incubated with adriamycin. As was seen in cells treated with only L-MTP-PE, cells treated sequentially with adriamycin and L-MTP-PE showed increased mRNA expression of IL-1α, IL-1β, TNFα, IL-6, and IL-8, suggesting that treatment with adriamycin did not suppress the ability of human monocytes to respond to L-MTP-PE.[52]

To assess the effect of chemotherapy in vivo, peripheral blood monocytes were isolated from patients before and after treatment with intraarterial cisplatin (150 mg/m^2), methotrexate (12 gm/m^2), cytoxan (120 mg/m^2), adriamycin (75 mg/m^2), or co-administration of cytoxin and adriamycin. These monocytes were then incubated with L-MTP-PE or empty liposomes in vitro, and tumoricidal activity was quantified.[53] The tumoricidal properties of monocytes isolated from patients with osteosarcoma after single agent chemotherapy could be activated to levels equal to or greater than those expressed by normal control donors. However, when adriamycin and cytoxan were administered together on the same day, profound suppression in monocyte activation was observed. The suppressed function returned to normal by 3 weeks.[53] These studies suggested that single agent chemotherapy did not interfere with L-MTP-PE's ability to activate important parameters associated with monocyte tumoricidal function. However, combination chemotherapy may indeed have a deterimental effect on cells, inhibiting immune stimulation by L-MTP-PE.

VI.3. Phase IIb trial with combination ifosfamide and L-MTP-PE therapy

Although the preclinical data suggested that co-administration of chemotherapy with L-MTP-PE did not increase the toxic side effects of the chemotherapeutic agent or decrease the ability of L-MTP-PE to stimulate the immune response, a small clinical trial[54] was undertaken in relapsed osteosarcoma patients to verify this theory. Ifosfamide was chosen as the chemotherapeutic agent because it has been shown to be effective against relapsed disease.[55] At that time, the adjuvant chemotherapy at our institution did not include ifosfamide as a front-line agent. Relapsed patients could therefore receive a potentially active drug not previously used in their therapy. The goals of this study were to (1) determine the tolerability of L-MTP-PE given in combination with ifosfamide, (2) assess whether L-MTP-PE increases the toxic side effects of ifosfamide, and (3) determine whether ifosfamide alters the in vivo response to L-MTP-PE. Patients were entered on one of two strata. Stratum A patients were rendered disease free after which they received ifosfamide, (IFX; 1.8 mg/m^2) and mesna (1.8 mg/m^2) daily for 5 days every 3 weeks for eight cycles. L-MTP-PE was administered twice a week for 12 weeks, then once a week for an additional 12 weeks. Stratum B patients received three cycles of chemotherapy (one cycle consisting of 5 days of IFX and 3 weeks of twice weekly L-MTP-PE), prior to surgical resection of lung metastases. Patients then received an additional five courses of chemotherapy postoperatively. This stratum allowed for the histological evaluation of tumor responsiveness to therapy. Cytokines were measured, and peripheral blood monocytes were assayed for their tumoricidal activity to assess the effect of IFX on L-MTP-PE's immunostimulatory capacity.

This study showed that L-MTP-PE did not increase the toxic side effects of IFX. No delays in IFX administration were necessary due to neutropenia. Similarly, the toxic side effects of L-MTP-PE were not increased by IFX. In addition, the immune response to L-MTP-PE and IFX (as measured by increases in C-reactive protein, IL-6, IL-8, TNFα, neopterin, and monocyte tumoricidal activity) was not different from the responses seen in patients who received L-MTP-PE alone. Surgically resected pulmonary metastases from the stratum B patients showed both the effects of chemotherapy (dead, amorphorous, acellular osteoid with cell drop out) and a finding indicative of the effect of L-MTP-PE (peripheral fibrosis and infiltration with inflammatory cells). Taken together, these results further substantiated the conclusion that IFX does not ablate the effect of L-MTP-PE[56] and strengthens the rationale for combination therapy in a newly diagnosed population of patients with osteosarcoma.

VII. Phase III trial

In summary, we have demonstrated that monocytes from osteosarcoma patients can be rendered cytotoxic to tumor cells by both in vitro incubation with L-MTP-PE and following the intravenous administration of this agent.[52,56] L-MTP-PE can be given safely to both adults and children with minimal side effects.[44,45,50,56] The whole body distribution of ^{99m}Tc-labeled liposomes containing MTP-PE confirms that the agent is taken up by the lungs.[45] Biological activity in osteosarcoma patients is revealed by the elevations in plasma levels of several cytokines plus stimulation of monocyte-mediated cytotoxicity following L-MTP-PE infusion and by histological changes in the pulmonary lesions.[48,50,56] Ifosfamide therapy given in combination with L-MTP-PE does not suppress the immune response, as judged by both plasma cytokine levels and tumor histology. Finally, L-MTP-PE has been shown to be effective as a single agent against relapsed osteosarcoma.[50] Based on these findings, it was concluded that the addition of L-MTP-PE to the postoperative adjuvant chemotherapy regimen may indeed improve the 2-year disease-free survival of osteosarcoma patients; a survival rate that has been stagnant for the past 10 years. Our hypothesis is that L-MTP-PE will activate the patient's pulmonary macrophages to destroy residual tumor cells that were not eliminated by chemotherapy.

The Children's Cancer Group and Pediatric Oncology Group currently are conducting a phase III clinical trial evaluating the role of L-MTP-PE in newly diagnosed osteosarcoma patients. The trial addresses two main questions: (1) Is there benefit to adding ifosfamide to an effective chemotherapeutic regimen (adriamycin, cytoxan, methotrexate) and (2) Does the addition of L-MTP-PE to chemotherapy improve the disease free survival in patients with osteosarcoma? Patients are randomized to the two different arms (chemotherapy ± ifosfamide and also ± L-MTP-PE) at the onset of the study. After receiving chemotherapy for 10 weeks, patients undergo resection of the primary tumor and any metastatic disease. Postoperatively, they continue with chemotherapy alone or combined with L-MTP-PE. The L-MTP-PE is administered at 2 mg/m^2 intravenously twice a week for 12 weeks, followed by weekly for 24 weeks.

Although the phase IIb trial showed that ifosfamide did not suppress the patients' immune responses to L-MTP-PE in vivo,[56] the co-administration of adriamycin and cytoxan profoundly suppressed the patients' monocyte activation.[56] In this study IFX/adriamycin and adriamycin/cisplatin are co-administrated in two cycles. At various strategic time points throughout this protocol, blood samples are collected to assess monocyte tumoricidal activity and plasma cytokines (IL-1α, IL-1β, TNF, IL-6, IL-8, IFNγ, IL-2).

In addition to the two main therapy questions being asked, another two questions are also being addressed. Is one chemotherapeutic regimen more compatible with L-MTP-PE's ability to stimulate the immune system, and does this also influence the patient's outcome?

VIII. Future directions

We anticipate that one of the conclusions from the phase III clinical trial evaluating the role of L-MTP-PE in newly diagnosed osteosarcoma patients will be an increased disease-free survival in patients receiving L-MTP-PE. At the present time, L-MTP-PE is administered intravenously *only*. Although therapy with L-MTP-PE requires only a short stay (1–2 h) in the outpatient clinic, twice weekly clinic visits are expensive and time consuming. If oral administration were found to be at least as effective as when intravenously given, medical costs would be reduced substantially. In addition, considerably less time would be lost from school and work for patients and their families. This is especially true during the weeks when chemotherapy is not being administered. During these times, the patient could take the oral L-MTP-PE at home.

Recent studies[57] in BALB/c mice demonstrate that L-MTP-PE administered orally is rapidly absorbed in the intestines of mice, reaching the systemic circulation in 4 hours. Alveolar macrophages harvested 24 hours after a single oral dose of L-MTP-PE were tumoricidal, and peritoneal macrophages were also activated and produced TNFα and IL-6. Furthermore, daily oral therapy was effective in inhibiting the development of experimental renal cell carcinoma lung metastases.

We are, therefore, currently investigating the activity of orally administered L-MTP-PE in a metastatic human osteosarcoma nude mouse model. Optimal dosage and dosing schedules are being compared against that of intravenously administered L-MTP-PE. We see the role of L-MTP-PE as expanding beyond the osteosarcoma and melanoma patient populations, promising new therapy for any tumor that metastasizes to the lung.

References

1. Fidler IJ, and Balch CM. The biology of cancer metastasis and implications for therapy. Curr Probl Surg 1987;24:129–209.
2. Fidler IJ, Poste G. The cellular heterogeneity of malignant neoplasms: implications for adjuvant chemotherapy. Semin Oncol 1985;12:207–221.
3. Fidler IJ. Targeting of immunomodulators to mononuclear phagocytes for therapy of cancer. Advanced Drug Delivery Review 1988;2:69–106.
4. Fidler IJ, Poste G. Macrophage-mediated destruction of malignant tumor cells and strategies for therapy of metastatic disease. Springer Seminars Immunopathology 1982;5:161–174.
5. Fidler IJ, Kleinerman ES. Lymphokine-activated human blood monocytes destroy tumor cells but not normal cells under cocultivation conditions. J Clin Oncol 1984;2:937–943.
6. Fidler IJ, Schroit AJ. Recognition and destruction of neoplastic cells by activated macrophages: discrimination of altered self. Biochim Biophys Acta 1988;948:151–173.
7. Hibbs JB Jr. Discrimination between neoplastic and non-neoplastic cells in vitro by activated macrophages. J Natl Cancer Inst 1974;53:1487–1492.
8. Bucana C, Hoyer LC, Hobbs B, Breesman S, McDaniel M, Hanna MG Jr. Morphological evidence for the translocation of lysosomal organelles from cytotoxic macrophages into the cytoplasm of tumor target cells. Cancer Res 1976;36:4444–4458.
9. Bucana CD, Hoyer LC, Schroit AJ, Kleinerman ES, Fidler IJ. Ultrastructural studies of the interaction between liposome-activated human blood monocytes and allogeneic tumor cells in vitro. Am J Pathol 1983;112,101–111.
10. Fogler WE, Fidler IJ. Nonselective destruction of murine neoplastic cells by syngeneic tumoricidal macrophages. Cancer Res 1985;45:14–18.
11. Allison AC. On the role of mononuclear phagocytes in immunity against viruses. Prog Med Virol 1974;18:15–31.
12. Allison AC. Mode of action of immunological adjuvants. J Reticuloendothelial Society 1979;26:619–630.
13. Lederer E. Synthetic immunostimulants derived from bacterial cell wall. J Med Chem 1980;23:819–825.
14. Chedid L, Audibert F, Johnson AG. Biological activities of muramyl dipeptide, a synthetic glycopeptide analogous to bacterial immunoregulating agents. Progress Allergy 1978;25:63–105.
15. Ellouz F, Adam A, Ciorbaru R, Lederer E. Minimal structural requirements for adjuvant activity of bacterial peptidoglycan derivatives. Biochem Biophys Res Commun 1974;59:1317–1325.

16. Fogler WE, Fidler IJ. Modulation of the immune response by muramyl dipeptide. In: Chirigos MA, Fenichel RL, eds. Immune Modulation Agents and Their Mechanisms, New York: Marcel Dekker, 1984;499–512.
17. Parant M, Parant F, Chedid L, Yapo A, Petit JF, Lederer E. Fate of the synthetic immunoadjuvant, muramyl dipeptide (^{14}C-labelled) in the mouse. Int J Immunopharmacol 1979;1:35–41.
18. Fogler WE, Wade R, Brundish DE, Fidler IJ. Distribution and fate of free and liposome-encapsulated [^{3}H] nor-muramyldipeptide and [^{3}H] muramyl-tripeptide phosphotidylethenolamine in mice. J Immunol 1985;135:1372–1377.
19. Fidler IJ, Barnes Z, Fogler WE, Kirsh R, Bugelski P, Poste G. Involvement of macrophages in eradication of established metastases following intravenous injection of liposomes containing macrophage activators. Cancer Res 1982;42:496–501.
20. Fidler IJ, Fogler WE, Tarcsay L, Schumann G, Braun DG, Schroit AJ. Systemic activation of macrophages and treatment of cancer metastases by liposomes containing hydrophilic or lipophilic muramyl dipeptide. In: Haddon JW et al., eds. Advances in Immunopharmacology, Vol 2. Oxford: Pergamon Press, 1983;253–352.
21. Fidler IJ. The MAF dilemma. Lymphokine Res 1984;3:51–54.
22. Kleinerman ES, Schroit AJ, Fogler WE, Fidler IJ. Tumoricidal activity of human monocytes activated in vitro by free and liposome-encapsulated human lymphokines. J Clin Invest 1983;72:304–315.
23. Kleinerman ES, Schroit AJ, Fogler WE, Fidler IJ. Tumoricidal activity of human monocytes activated in vitro by free and liposome encapsulated human lymphokines. J Clin Invest 1983;72:304–315.
24. Saika I, Fidler IJ. Synergistic activation by recombinant mouse gamma-interferon and muramyl dipeptide of tumoricidal properties in mouse macrophages. J Immunol 1985;135:684–688.
25. Saiki I, Sone S, Fogler WE, Kleinerman ES, Lopez-Berestein G, Fidler IJ. Synergism between human recombinant gamma-interferon and muramyl dipeptide encapsulated in liposomes for activation of antitumor properties in human blood monocytes. Cancer Res 1985;45:6188–6193.
26. Fidler IJ, Fogler WE, Kleinerman ES, Saiki I. Abrogation of species specificity for activation of tumoricidal properties in macrophages in recombinant mouse or human gamma-interferon encapsulated in liposomes. J Immunol 1985;135:4289–4296.
27. Kleinerman ES, Fogler WE, Fidler IJ. Intracellular activation of human and rodent macrophages by human lymphokines encapsulated in liposomes. J Leuko Biol 1985;37:571–574.
28. Poste G, Kirsh R, Fogler WE, Fidler IJ. Activation of tumoricidal properties in mouse macrophages by lymphokines encapsulated in liposomes. Cancer Res 1979;39:881–892.
29. Gisler RH, Dietrich FM, Bachang G, Brownhill A, Schumann G, Staber FB, Tarcsay L, Wachsmuth ED, Dukor P. New developments in drugs enhancing the immune response: activation of lymphocytes and accessory cells by muramyl peptides. In: Turk JL, Danker D, eds. Immune Responsiveness. Macmillan: London, 1979;113–160.
30. Fidler IJ, Raz A, Fogler WE, Kirsh R, Bugalski P, Poste G. Design of liposomes to improve delivery of macrophage-augmenting agents to alveolar macrophages. Cancer Res 1980;40:4460–4466.
31. Schroit AJ, Fidler IJ. Effects of liposome structure and lipid composition on the activation of the tumoricidal properties by liposomes containing muramyl dipeptide. Cancer Res 1982;42:161–167.
32. Poste G, Bucana C, Raz A, Bugelski P, Kirsh R, Fidler IJ. Analysis of the fate of systemically administered liposomes and implications for their use in drug delivery. Cancer Res 1982;42:1412–1422.
33. Fidler IJ. The in situ induction of tumoricidal activity in alveolar macrophages by liposomes containing muramyl dipeptide is a thymus-independent process. J Immunol 1981;127:1719–1920.
34. Lovett D, Koran B, Hadam M, Resch K, Gemsa D. Macrophage toxicity: Interleukin-1 as a mediator of tumor cytostasis. J Immunol 1996;136:340–347.
35. Urban JL, Shephard HM, Rothstein JL, Sugarman BJ, Schreiber H. Tumor necrosis factor: A potent effector molecule for tumor cell killing by activated macrophages. Proc Natl Acad Sci USA 1986;83:5233–5237.
36. Asano T, Matsushima K, Kleinerman ES. Liposome-encapsulated muramyl tripeptide up-regulates monocyte chemotactic and activating factor gene expression in human monocytes at the transcriptional and posttranscriptional levels. Cancer Immunol Immunother 1994;38:16–22.
37. Maeda M, Knowles RD, Kleinerman ES. Muramyl tripeptide phosphatidylethanolamine encapsulated in liposomes stimulates monocyte production of tumor necrosis factor and interleukin-1 in vitro. Cancer Communications 1991;3:313–321.
38. Asano T, McIntyre BW, Bednarczyk JL, Wygant JN, Kleinerman ES. Liposomal muramyl tripep-

tide upregulates adhesion molecules on the surface of human monocytes. Oncol Res 1995;7:253–257.
39. Fidler IJ. Therapy of spontaneous metastases by intravenous injection of liposomes containing lymphokines. Science 1980;208:1469–1471.
40. Fidler IJ, Sone S, Fogler WE, Barnes W. Eradication of spontaneous metastases and activation of alveolar macrophages by intravenous injection of liposomes containing muramyl dipeptide. Proc Natl Acad Sci USA 1981;78:1680–1684.
41. Fidler IJ. Optimization and limitations of systemic treatment of murine melanoma metastases with lipsomes containing muramyl tripeptide phosphotidylethanolamine. Cancer Immunol Immunother 1986;21:169–173.
42. Brodey RS, Abt DA. Results of surgical treatment in 65 dogs with osteosarcoma. J Am Vet Med Assoc 1976;168:1032–1035.
43. MacEwen EG, Kurzman ID, Rosenthal RC, Smith BW, Manley PA, Roush JK, Howard PE. Therapy for osteosarcoma in dogs with intravenous injection of liposome-encapsulated muramyl tripeptide. J Natl Cancer Inst 1989;81:935–938.
44. Creaven PJ, Cowen JW, Brenner DE, Dadey BM, Han T, Huben R, Karakousis C, Frost H, LeSher D, Hanagan J, Andrejcio K, Cushman MK. Initial clinical trial of macrophage activator muramyl tripeptide-phosphatidyl-ethanolamine encapsulated in liposomes in patients with advanced cancer. J Biological Response Modifiers 1990;9:492–498.
45. Murray JL, Kleinerman ES, Cunningham JE, Tatom JR, Andrejcio K, Lepe-Zuniga J, Lamki LM, Rosenblum MG, Frost H, Gutterman JU, Fidler IJ, Krakoff IH. Phase I trial of liposomal muramyl tripeptide phosphatidylethanolamine in cancer patients. J Clin Oncol 1989;7:1915–1925.
46. Fujimaki W, Itoh K, An T, Gano JB, Ross MI, Mansfield PF, Balch CM, Augustus LB, Kartevitch DD, Johnston D, Fidler IJ, Kleinerman ES. Cytokine production and immune cell activation in melanoma patients treated with liposomal muramyl tripeptide (CGP 19835A) Cancer Biother 1993;8:307–318.
47. Goorin AM, Schuster JJ, Baker A, Horowitz ME, Meyer WH, Link MP. Changing pattern of pulmonary metastases with adjuvant chemotherapy in patients with osteosarcoma: results from the multiinstitutional osteosarcoma study. J Clin Oncol 1991;9:600–605.
48. Kleinerman ES, Raymond AK, Bucana CD, Jaffe N, Harris MB, Krakoff IH, Benjamin R, Fidler IJ. Unique histological changes in lung metastases of osteosarcoma patients following therapy with liposomal muramyl tripeptide (CGP 19835A lipid). Cancer Immunol Immunother 1992;34:211–220.
49. Dannenberg AM Jr, Thomasshefski JF Jr. Pathogenesis of pulmonary tuberculosis. In: Fishman AP, ed. Pulmonary Diseases and Disorders, 2nd edn. New York: McGraw Hill, 1988;18–21.
50. Kleinerman ES, Gano JB, Johnston DA, Benjamin RS, Jaffe N. Efficacy of liposomal muramyl tripeptide (CGP 19835A) in the treatment of relapsed osteosarcoma. Am J Clin Oncol 1995;18:93–99.
51. Killion JJ, Kleinerman ES, Wilson MR, Tanaka M, Fidler IJ. Sequential therapy with chemotherapeutic drugs and liposome-encapsulated muramyl tripeptide: determination of potential interactions between these agents. Oncol Res 1992;4:413–418.
52. Asano T, Fujimaki W, McWatters A, An T, Matsushima K, Kleinerman ES. Effect of adriamycin on lipsomal muramyl tripeptide's ability to up-regulate monocyte cytokine expression. Cancer Immunol Immunother 1993;37:408–411.
53. Kleinerman ES, Snyder JS, Jaffe N. Influence of chemotherapy administration on monocyte activation by liposomal muramyl tripeptide phosphatidylethanolamine in children with osteosarcoma. J Clin Oncol 1991;9:259–267.
54. Kleinerman ES, Meyer PA, Raymond AK, Gano JB, Jia S-F, Jaffe N. Combination therapy with ifosfamide and liposome-encapsulated muramyl tripeptide: tolerability, toxicity, and immune stimulation. J Immunother 1995;17:181–193.
55. Harris MB, Cantor A, Goorin A, Ayala A, Link MP. Response to ifosfamide in patients with osteosarcoma: a comparison of results in newly diagnosed patients versus those with recurrent disease after adjuvant chemotherapy. Proceedings Society Clinical Oncology 1991;10:315.
56. Kleinerman ES, Jia S-F, Griffin J, Seibel NL, Benjamin RS, Jaffe N. Phase II study of liposomal muramyl tripeptide in osteosarcoma: the cytokine cascade and monocyte activation following administration. J Clin Oncol 1992;10:1310–1316.
57. Tanguay S, Bucana CD, Wilson MR, Fidler IJ, von Eschenbach AC, Killion JJ. In vivo modulation of macrophage tumoricidal activity by oral administration of the liposome-encapsulated macrophage activator CGP 19835A. Cancer Res 1994;54:5882–5888.

Lasic and Papahadjopoulos (eds.), Medical Applications of Liposomes

CHAPTER 2.4

DNA vaccination: A role for liposomes

Gregory Gregoriadis, Brenda McCormack, Yvonne Perrie and Roghieh Saffie

Centre for Drug Delivery Research, The School of Pharmacy, University of London, 29–39 Brunswick Square, London WC1N 1AX, UK

Overview

I. Introduction

DNA immunization is one of the more exciting spin offs of gene therapy. It arose from the unexpected observation[1] that injected (usually into skeletal muscle) purified plasmid DNA containing encoding sequences for a protein immunogen and regulatory elements necessary for their expression, transfects cells whereupon the protein produced induces humoural and cell-mediated immunity. Experiments in animals immunized with DNA encoding sequences corresponding to such diverse immunogens as those from influenza,[1] human immunodeficiency,[2] hepatitis B[3] and herpes simplex[4] viruses, mycobacterium leprosy[5] and the malaria parasite,[6] have shown that conferred immunity, often associated with cytotoxic T lymphocyte response, is protective. Immunity follows DNA uptake by skeletal muscle fibres, leading to the expression and extracellular release of the antigen[7,8] and its subsequent interaction with antigen presenting cells (APC).

A major advantage[9] of DNA-based immunization over conventional subunit vaccines is that it mimics, at least to some extent, immune responses seen on viral infections. For instance, antigen presentation can occur through the major histocompatibility complex (MHC) class I pathway, leading to the induction of cytotoxic T lymphocytes. Other features of DNA immunization include expression of the gene over a prolonged period of time thus eliminating the need of booster injections, avoidance of virulence associated with live or attenuated vaccines, resistance to interference by any pre-existing immune response as might occur with protein or peptide immunogens, and reduced costs. On the other hand, the reliance of the approach on the uptake of DNA by muscle cells may be a disadvantage: although such cells can express class I MHC molecules, they are not likely to possess the accessory molecules that are responsible for T cell activation.[9] Moreover, only a minor fraction of muscle cells participates in DNA uptake. This and exposure of the naked DNA to nucleases in the interstitial fluid necessitates the injection of relatively large quantities of material (typically up to 300 μg or more[9]), often into regenerating muscle (previously treated with appropriate agents), in order to enhance immunity.[2,5,10]

We have recently proposed (see below) that APC may be a preferred alternative to muscle cells as targets for DNA uptake and expression. To that end, administration of antigen-encoding plasmid DNA entrapped in liposomes would circumvent the need of muscle involvement and facilitate instead its uptake by APC. APC infiltrating the site of injection or in the lymphatics are known[11,12] to take up liposomes avidly and have, indeed, been implicated[13] in the liposomal immunoadjuvant activity. At the same time, liposomes would protect[14] their DNA content from nuclease attack. Moreover, transfection of APC with liposome-entrapped DNA could be promoted by the judicial choice of vesicle surface charge, size and lipid composition. In this respect, a method developed recently in this laboratory allows for the quantitative entrapment of plasmid DNAs into neutral, anionic and cationic liposomes which were shown to transfect cells in vitro[14] with varying efficiency. More importantly, liposomes containing one of the plasmid DNAs (encoding the S region of the hepatitis B surface antigen) were found[15] much more effective in inducing humoural and cell mediated immunity against the encoded antigen in mice than either naked DNA or DNA complexed with preformed similar (cationic) liposomes.

II. Incorporation of DNA into liposomes

A wide range of molecules (e.g., small drugs,[16] drug-cyclodextrin complexes,[17] peptides,[18] interleukins,[19] antigens[20] and other proteins[21]) can be quantitatively entrapped into the aqueous phase of liposomes by the dehydration rehydration procedure[16] in the absence of sonication, detergents or organic solvents. The procedure consists of mixing "empty" liposomes, preferably small unilamellar vesicles (produced by sonication, microfluidization or extrusion) with a solution of the drug destined for entrapment, dehydration of the mixture by freeze-drying, and subsequent controlled[16] rehydration of the formed powder. This leads to the formation of multilamellar[22] dehydration-rehydration vesicles (DRV) containing up to 80% or more of the original solute. Significantly, microfluidization of the drug-containing DRV in the presence of non-entrapped solute generates smaller vesicles (down from about 800 nm to 100 nm average mean diameter, depending on the number of microfluidization cycles) with much of the originally entrapped solute still contained by the vesicles.[14,21,23]

Table 1

Incorporation of plasmid DNA into liposomes

Liposomes	Incorporated plasmid DNA (% of used)				
	pGL2	pRc/CMV HBS	pRSVGH	pCMV4.65	pCMV4.EGFP
PC, DOPE[a]	44.2	55.4	45.6	28.6	
PC, DOPE[b]	12.1		11.3		
PC, DOPE, PS[a]	57.3				
PC, DOPE, PS[b]	12.6				
PC, DOPE, PG[a]			53.5		
PC, DOPE, PG[b]			10.2		
PC, DOPE, SA[a]	74.8				
PC, DOPE, SA[b]	48.3				
PC, DOPE, BisHOP[a]	69.3				
PC, DOPE, DOTMA[a]	86.8				
PC, DOPE, DC-Chol[a]		87.1	76.9		
PC, DOPE, DC-Chol[b]			77.2		
PC, DOPE, DOTAP[a]		80.1	79.8	52.7	71.9
PC, DOPE, DOTAP[b]		88.6	80.6	67.7	
PC, DOPE, DODAP[a]			57.4		
PC, DOPE, DODAP[b]			64.8		

^{35}S-labelled plasmid DNA (10–500 μg) was incorporated ([a]) into or mixed ([b]) with neutral (PC, DOPE), anionic (PC, DOPE, PS or PG) or cationic (PC, DOPE, SA, BisHOP, DOTMA, DC-Chol, DOTAP or DODAP) DRV. Incorporation values for the different amounts of DNA used for each of the DRV formulations did not differ significantly and were therefore pooled (values shown are means of values obtained from 3–5 experiments). PC (16 μmoles) was used in molar ratios of 1:0.5 (neutral) and 1:0.5:0.25 anionic and cationic DRV). PC, egg phosphatidylcholine; DOPE, dioleoyl phosphatidylcholine; PS, phosphatidylserine; PG, phosphatidylglycerol; SA, stearylamine; BisHOP, 1,2-bis (hexadecylcycloxy)-3-trimethylaminopropane; DOTMA, *N*[1-(2,3-dioleoyloxy)propyl]-*N*,*N*,*N*, triethylammonium; DC-Chol, 3β-(*N*,*N*-dimethylaminoethane) carbonyl cholesterol; DOTAP, 1,2-dioleoyloxy-3(trimethylammonium) propane; DODAP, 1,2-dioleoyl-3-trimethylammonium propane.

Plasmid DNA used encoded luciferase (pGL2), hepatitis B surface antigen (S region) (pRc/CMV HBS), human growth hormone (pRSVGH), mycobacterium leprosy protein (pCMV 4.65) and "fluorescent green protein" (pCMV 4.EGFP).

In recent work with DRVs incorporating a number of plasmid DNAs (see Table 1) incorporation values were considerable and, in view of the anionic nature of the solute, dependent on whether or not a cationic lipid was one of the liposomal lipid components. Table 1 shows values of 29–55% for neutral and 53–57% of the amount used for anionic DRV. There was no apparent relationship between amount of DNA used (10–500 μg) and values of incorporation in liposomes of compositions and lipid mass shown in Table 1. It appears that much of the DNA was incorporated within the vesicles (rather than being adsorbed to their surface) as incubation of preformed DRV with free DNA resulted in only a modest proportion (10–13%) of it being recovered with the DRV. As expected, incorporation values of DNA in cationic DRV were greater (57–87%). However, in contrast with the neutral or anionic DRV, after incubation of preformed cationic DRV with free DNA, as much as 48–81% of the material used was recovered with the DRV, presumably as vesicle surface bound.

Our results have also shown[14] that most of the DNA incorporated in neutral (45–72%), anionic (58–69%) or cationic (68–86% of total) is resistant to degradation by deoxyribonuclease (DNase), indicating significant true entrapment: retention of DNA adsorbed to the surface of neutral or anionic DRV after exposure to the enzyme was much lower (17–18%). However, with DNA complexed with preformed cationic (e.g., SA) liposomes, a substantial proportion (41–58%) of it was not available for digestion by DNase.[14] This could be attributed[24] to the condensed state attained by DNA on complexation with cationic vesicles. Thus, the extent of DNA incorporation by the present procedure within cationic DRV (as opposed to DNA bound to their surface) is difficult to estimate accurately. Data on liposomal DNA digestion after exposure to DNase were confirmed[14] in experiments where samples of free or liposomal DNA were exposed to DNase and then subjected to agarose gel electrophoresis: whereas free DNA was completely digested, DNA incorporated in cationic DRV was largely protected. There was less protection for DNA incorporated in neutral or anionic DRV.

III. Transfection with liposomal DNA in vitro

Exposure of Cos-7 cells to liposomal pGL2 plasmid DNA (expressing the luciferase reporter gene) revealed significant levels of luciferase activity with each of the formulations tested.[14] However, values obtained with the cationic PC/DOPE/DOTMA and PC/DOPE/SA DRV (as described in Table 1 and legend) were approximately 10-fold higher than those seen with neutral or anionic DRV. As transfection efficiency with liposomes may be related to vesicle size, Cos-7 cells were exposed[14] to pGL2 incorporated into DRV microfluidized to a size of about 210–380 nm diameter. (Extensive microfluidization of DRV to a size of about 100–200 nm resulted in DNA damage as revealed by agarose gel electrophoresis[14]). Results[14] showed a ten-fold further improvement in transfection efficiency for the microfluidized cationic PC/DOPE/DOTMA DRV. Nonetheless, even this latter DRV formulation of DNA was 10–15-fold less efficient than LipofectAMINE® complexed with pGL2.[14]

The considerable in vitro transfection efficiency of DNA incorporated into DRV[14] or complexed with a variety of cationic vesicles[24,25] has been attributed,[24] at least in part, to their augmented association with cells as a result of electrostatic interaction between the positive and negative surface charges of the vesicles and cells respectively. However, as uptake of liposomes by certain populations of cells (e.g., those of the reticuloendothelial system) in injected animals is avid and largely independant of surface charge,[13] the presence of a cationic lipid in the liposomal structure may not be essential for transfection to occur in vivo in such cells. This is supported by experiments where plasmid DNA constructs incorporated in neutral or anionic liposomes were expressed in injected rodents.[26–28] Moreover, it is well known from early liposome electrophoresis studies[29] that the possitive surface charge of cationic liposomes is masked by plasma proteins which impose a net negative charge on the surface of the vesicles. On the other hand, it is also apparent from a number of studies with DNA complexed with cationic vesicles (or indeed other cationic agents)[30] that transfection in vivo is greatly facilitated by the judicial choice of cationic lipids[24,30–34] and, more recently, by vesicle size[35] and by replacing DOPE with cholesterol.[35,53] Although the mechanism of action for cationic lipids in improving transfection is not clear at present, it is probably[24] related to the cationic charge-induced condensed state of the complexed DNA.

IV. Transfection with liposomal DNA in vivo: immunization experiments

Having established[14] that plasmid DNA incorporated into DRV liposomes is capable of transfecting cells in vitro, immunization experiments were carried out[15] using a variety of DRV plasmid DNA formulations. Mice (Balb/c) were immunized with a single or several successive intramuscular (hind leg) injections of 50 μl 0.15 M sodium phosphate buffer (pH 7.4) supplemented with 0.9% NaCl (PBS) containing 1–10 μg naked, liposome-entrapped or complexed (to preformed cationic DRVs) pRc/CMV HBS (see legend to Figure 1 for details). pRc/CMV HBS, expressing sequences coding for the S (small) protein of the hepatitis B virus surface antigen (HBsAg, subtype ayw), was cloned by Dr R. Whalen using pRc/CMV as vector backbone. Animals were bled (tail vein) at time intervals after the first injection and sera tested[18] for anti-HBsAg (S region) IgG_1, IgG_{2a} and IgG_{2b} by the enzyme-linked immunoadsorbent assay (ELISA), using the same antigen to coat the plates. In some experiments, spleens from intact (control) and immunised mice were assayed[36] for endogeneous IFN-γ and IL-4. Significance levels for IgG titers and cytokine values were determined by the Student's test for unpaired observations. Figure 1 and legend show that in animals injected repeatedly during a 37 day period with 5 or 10 μg of pRc/CMV HBS entrapped in cationic DOTAP, DC-Chol or SA liposomes, antibody (IgG_1) responses against the encoded antigen were up to at least 100-fold greater at all times tested than those seen in mice immunized with naked DNA. IgG_{2a} and IgG_{2b} responses for the liposomal plasmid DNA were also greater, but to a lesser extent (about 10-fold).

As Lipofectin® complexed with antigen-encoding plasmid DNA did not appear to augment antibody responses to the antigen in a single previous study,[37] we compared entrapped (charged and uncharged liposomes) and complexed (with cationic liposomes) pRc/CMV HBS in terms of immune responses to the encoded antigen using the same protocol of immunization as in Figure 1. Results in Figure 2 show that DNA entrapped in cationic (DOTAP) liposomes produced greater (over 80-fold) IgG_1 responses than complexed DNA (10 μg dose; 28 days). DNA (10 μg) entrapped in uncharged liposomes was also found capable of transfection albeit with reduced efficiency (Figure 2). Animals from the experiment in Figure 2 were also tested for T cells responses. To that end, levels of IFN-γ and IL-4 in their spleens were measured respectively as indicators of Th1 and Th2 subset T cell activation. Figure 3 indicates that activation of Th1 and Th2 subsets was greater with liposome-entrapped DNA than with complexed or naked DNA. It thus appears that immunization with liposomal plasmid DNA induces both humoral and cell-mediated immunity. Work is in progress to establish whether

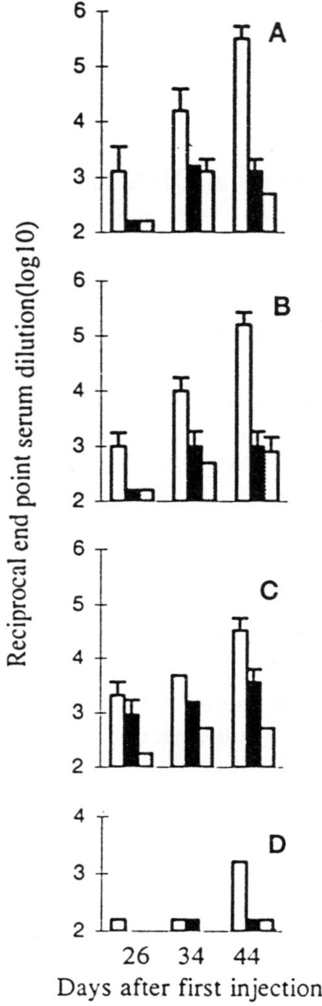

Fig. 1. Comparison of immune responses in mice injected with plasmid DNA as such or entrapped in different cationic liposomes. Balb/c mice in groups of four were injected intramuscularly on days 0, 10, 20, 27 and 37 with 5 μg of pRc/CMV HBS entrapped in cationic liposomes composed of PC, DOPE and DOTAP (A), DC-Chol (B) or SA(C), or in the naked form (D). Animals were bled 7, 15, 26, 34 and 44 days after the first injection and sera tested by ELISA for IgG_1 (white bars), IgG_{2a} (black bars) or IgG_{2b} (dotted bars) responses against the encoded hepatitis B surface antigen (HBsAg; S region, ayw subtype). Values are means ±SD of $\log_{10}$ of reciprocal end point serum dilutions required for OD to reach readings of about 0.200. Similar values (all groups) were obtained in mice injected as above with 10 μg DNA in a separate experiment (results not shown). Sera from untreated mice gave $\log_{10}$ values of less than 2.0. Immune responses were mounted by all mice injected with liposomal DNA but became measurable only at 26 days. Differences in $\log_{10}$ values (all IgG subclasses at all time intervals) in mice immunized with liposomal DNA and mice immunized with naked DNA were statistically significant ($P < 0.0001$–0.002) (Reproduced with permission from ref 15).

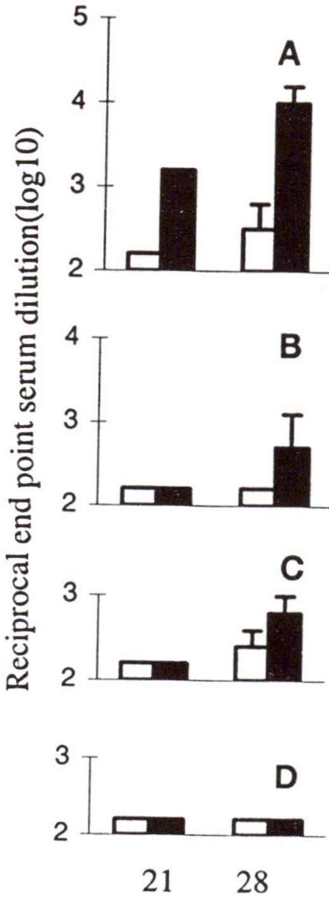

Fig. 2. Comparison of immune responses in mice injected with complexed or liposome-entrapped plasmid DNA. Balb/c mice in groups of four were injected intramuscularly on days 0, 7, 14, 21 and 28 with 1 (white bars) or 10 μg (black bars) of pRc/CMV HBS entrapped in cationic liposomes composed of PC, DOPE and DOTAP (A), uncharged liposomes composed of PC and DOPE (B), complexed with similar preformed cationic DOTAP liposomes (C) or in naked form (D). Sera from animals bled at 7, 14, 21 and 28 days after the first injection were analysed for anti-HBsAg IgG_1. Immune responses were mounted by all mice injected with liposomal DNA but became measurable only at 21–28 days. For other details see legend to Figure 1. Differences in $\log_{10}$ values (10 μg dose, 21 and 28 days) between mice immunized with cationic liposomal DNA and mice immunized with neutral liposomal, complexed and naked DNA were statistically significant ($P < 0.0001$–0.0032). (Reproduced with permission from ref 15).

liposome-entrapped antigen encoding plasmid DNA is also capable of inducing a cytotoxic T lymphocyte response.

In most published studies[4,9,30,38–43] on naked DNA vaccination, protocols of multiple injections have been employed. However, a single dose is also known to produce a humoural response to the encoded antigen.[7,44] Under the present conditions of single immunization with much lower doses of pRc/CMV HBS (2 and 10 μg) than those (up to 300 μg or more) normally used in naked DNA vaccination,[7] anti-HBsAg IgG_1 responses for naked and complexed DNA were barely detectable even after ten weeks (Figure 4). On the other hand, there was a pronounced response (IgG_1) for DNA entrapped in cationic (peaking at 5–7

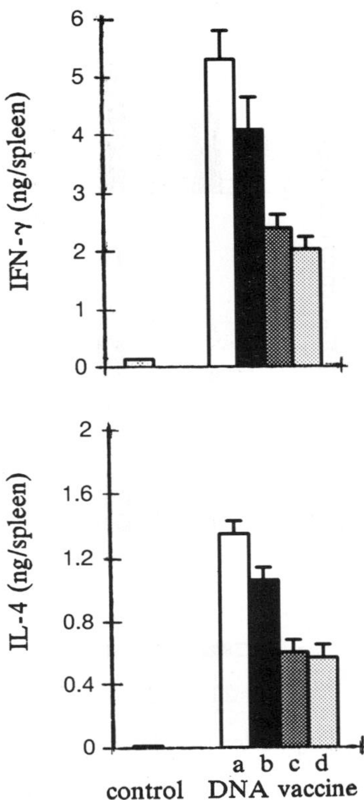

Fig. 3. Cytokine levels in the spleens of mice immunized with naked, complexed or liposome-entrapped plasmid DNA. Mice were immunized as in Figure 2 with pRc/CMV HBS entrapped into either cationic (a) or uncharged liposomes (b), complexed with cationic liposomes (c), or in naked form (d). "Control" represents cytokine levels in normal unimmunized mice. Three weeks after the final injection, mice were killed and their spleens subjected to cytokine analysis. Each bar respresents the mean ±SE of a group of 4 mice. Cytokine values in mice immunized with cationic liposomes were significantly higher than those in the other groups ($p < 0.001$–0.05). (Reproduced with permission from ref. 15).

weeks) and a delayed but significant response for DNA entrapped in neutral or anionic liposomes (Figure 4 and legend).

V. Possible mechanisms of liposomal DNA vaccination

Several, possibly concurrent, pathways leading to Th1 and Th2 immunity following naked DNA vaccination have been suggested.[9,39] They include secretion of the antigen by the transfected muscle cells and its subsequent processing and eventual presentation by resident Langerhans or infiltrating APC; release of antigen from transfected muscle cells following their death via a cytotoxic T cell response; and, possibly, transfection of both muscle cells and resident APC leading to simulta-

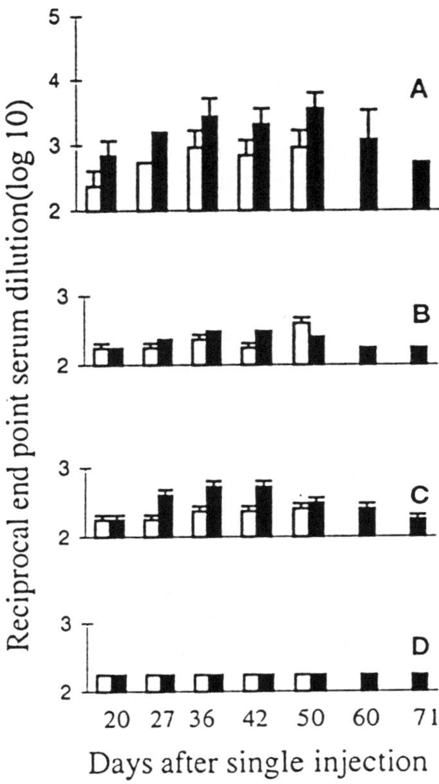

Fig. 4. Immune responses in mice after a single injection of plasmid DNA. Balb/c mice in groups of four were injected once intramuscularly with 2 (white bars) or 10 μg (black bars) of pRc/CMV HBS entrapped in cationic liposomes composed of PC, DOPE and DOTAP (A), uncharged liposomes composed of PC and DOPE (B), complexed with preformed similar DOTAP liposomes (C) or in naked form (D). Anti-HBsAg IgG_1 responses were analysed in sera obtained at time intervals after injection. Immune responses were mounted by all mice injected with liposomal DNA but became measurable only at 20–27 days. For other details see legend to Figure 1. Differences in $\log_{10}$ values (both doses; all time intervals) between mice immunized with cationic liposomal DNA and mice immunized with naked DNA were statistically significant ($P < 0.0001$–0.002). In a fifth group of four mice immunized once as above with 10 μg pRc/CMV HBS entrapped in anionic liposomes composed of PC, DOPE and PS, IgG_1 immune responses values ($\log_{10}$) were 2.25 ± 0.0 and 2.73 ± 0.0 at 21 and 29 days respectively. (Reproduced with permission from ref. 15).

neous activation of the T cell subsets. However, on the basis of present knowledge[13,45,46] on the fate of liposomes in vivo, there is no evidence of significant vesicle uptake by skeletal muscle cells after local injection. On the other hand, it has long been established[11,12,47,48] that liposomes enter the lymphatic system to localize in the lymph nodes and, as already mentioned, such fate has been implicated in the ability of liposomes to act as immunological adjuvants.[13] Although cationic (DNA-containing) liposomes could conceivably bind to the negatively charged surface of muscle cells and be taken up by them, proteins in the interstitial fluid would render[29] the liposomal surface negatively charged and thus prevent such binding.

It is much more likely that cationic (as well as neutral or anionic) liposomes are phagocytosed by resident or infiltrating APC, or by APC in the lymphatics. As phagocytosis of DNA complexes may be facilitated[49] by the condensed, supercoiled state of the DNA that is attained[24,50,51] on interaction with cationic lipids, the presence[14] of about 25% of the total DNA content of cationic DRV on their surface may contribute to this process. However, phagocytosis of DNA incorporated in neutral or anionic liposomes could not be so much less extensive to justify their considerably reduced ability (Figures 2 and 4) to promote immune responses. It is thus probable that the key ingredient of the DNA DRV formulations that is responsible for enhancing immune responses (presumably through enhanced transfection), is the cationic lipid. Moreover, judging from data in Figure 1 showing little differences in immune responses obtained with three different cationic lipids (DOTAP, DC-Chol and SA), the structural identity of the cationic lipid does not appear to contribute significantly to transfection efficiency under the present conditions. As both naked DNA (especially in large doses)[9] and DNA incorporated in uncharged or anionic liposomes[15] result in the elicitation of immune responses against the encoded antigen, it is legitimate to suggest that, following endocytosis, some of the DNA escapes the endocytic valuoles prior to their fusion with lysosomes, to gain entry into the cytoplasm for eventual episomal transfection and presentation of the encoded antigen. It is probably at the stage of intracellular trafficking of DNA (spanning its putative escape from the endosomes and access to the nucleus) that a cationic agent plays a significant but as yet unravelled role. In this respect, it has been recently suggested[52] that endocytosed cationic lipid-DNA complexes destabilise the endosomal membrane whereupon, through a lateral diffusion of anionic lipids from the cytoplasm-facing monolayer of endosomes, DNA is displaced from the complex and released into the cytoplasm. However, the question as to why liposome "entrapped" DNA in the present work is much more efficient in promoting immune responses to the encoded antigen than complexed DNA still remains. Experiments are in progress to establish the significance of the gel liquid crystalline transition temperature of phospholipids and of the presence or absence of a variety of lipids (including DOPE and cholesterol) in liposomes in terms of DNA transfection as monitored by measurements of immunity in injected mice. It appears, however, that regardless of the mechanisms involved, vaccination with liposome-entrapped DNA is more effective than with naked DNA.

VI. Conclusions

A variety of plasmid DNAs can be quantitatively incorporated by the dehydration-rehydration method into multilamellar liposomes composed of PC and DOPE alone or supplemented with anionic or cationic lipids. Much of the incorporated DNA appears to be entrapped in the vesicles rather than externally bound. In vitro studies indicate that DNA entrapped in such liposomes is capable of transfecting cells, with vesicles bearing a cationic charge being the most effective. This was confirmed in DNA immunization experiments with intramuscularly injected Balb/c mice where plasmid DNA (in relatively small amounts) entrapped in cationic liposomes was much more effective in inducing both humoural and cell-mediated immunity to the encoded antigen than naked DNA or DNA entrapped in uncharged and anionic or complexed with cationic vesicles. Recent data (unpublished) from this laboratory with DNA (pRc/CMV HBS) vaccination indicate that (a) DNA in cationic liposomes is also effective in inducing immunity when given by alternative routes (e.g., intravenous, subcutaneous or intraperitoneal), supporting our notion that in DNA vaccination skeletal muscle cell involvement is not essential; (b) outbred T/O mice immunized by a variety of routes with pRc/CMV HBS in cationic liposomes produced IgG_1 responses to the encoded antigen and exhibited splenic IFN-γ and IL-4 levels that were similar to or even greater than those seen in Balb/c mice. This finding suggests that immunization with liposomal DNA is not MHC restricted and could have important implications in human and veterinary immunization programmes.

Acknowledgements

We thank Dr Steven Hart for advice and help with the gel electrophoresis and in vitro transfection studies, Dr Brian de Souza for useful discussions and help with the assay of splenic cytokines, Dr Ricardo Tascon for providing the pCMV 4.65 and pCMV 4.EGFP and Mrs Concha Perring for excellent secretarial assistance.

References

1. Ulmer JB, Donnelly JJ, Parker SE, Rhodes GH, Felgner PL, Dwarki VJ, Gromkowski SH, Deck RR, DeWitt CM, Friedman A, Hawe LA, Leander, KR, Martinez D, Perry HC, Shiver JW, Montgomery DL, Liu MA. Heterologous protection against influenza by injection of DNA encoding a viral protein. Science 1993;259:1745–1749.
2. Wang BK, Ugen E, Srikanton V, Agcoliganyan MG, Dang K, Refaeli Y, Sato AI, Boyer S, Williams WV. Gene inoculation generates immune responses against human immunodeficiency type 1. Proc Natl Acad Sci USA 1993;90:4156–4160.
3. Davis HL, Michel M-L, Mancini M, Schleef M, Whalen RG. Direct gene transfer in skeletal muscle: Plasmid DNA-based immunization against the hepatitis B virus surface antigen. Vaccine 1994;12:1503–1509.
4. Manickan E, Rouse Richard JD, Yu Z, Wire WS, Rouse BT. Genetic immunization against herpes simplex virus: Protection is mediated by CD4+ T lymphocytes. Journal of Immunology 1995;155:259–265.
5. Tascon RE, Colston MJ, Ragno S, Stavropoulos E, Gregory D, Lowrie DB. Vaccination against tuberculosis by DNA injection. Nature Medicine 1996;2:888–892.
6. Mor G, Yamshchikov G, Sedegah M, Takeno M, Wang R, Houghten RA, Hoffman S, Klinman M. Induction of neonatal tolerance by plasmid DNA vaccination of mice. The Journal of Clinical Investigation 1996;98:2700–2705.
7. Davis HL, Demeneix BA, Quantin B, Coulombe J, Whalen RG. Plasmid DNA is superior to viral vectors for direct gene transfer in adult mouse skeletal muscle. Hum Gene Ther 1993;4:733–740.
8. Xiang ZQ, Spitalnik S, Tran M, Wunner WH, Cheng J, Ertl HCJ. Vaccination with a plasmid vector carrying the rabies virus glycoprotein gene induces protective immunity against rabies virus. Virology 1994;199:132–140.

9. Whalen RG, Davis HL. DNA-mediated immunization and the energetic immune response to hepatitis B surface antigen. Clinical Immunology and Immunopathology 1995;75:1–12.
10. Vitadello M, Schiaffmo MV, Picard A, Scarpa M, Schiaffno S. Gene-transfer in regenerating muscle. Hum Gen Ther 1994;5:11–18.
11. Tumer A, Kirby C, Senior J, Gregoriadis G. Fate of cholesterol-rich unilamellar liposomes containing ^{111}Inlabelled bleomycin after subcutaneous injection into rats. Biochim Biophys Acta 1983;760,119–125.
12. Velinova M, Read N, Kirby C, Gregoriadis G. Morphological observations on the fate of liposomes in the regional lymph nodes after footpad injection into rats. Biochim Biophys Acta 1996;1299: 207–215.
13. Gregoriadis G. Immunological adjuvants. A role for liposomes. Immunology Today 1990;11:89–97.
14. Gregoriadis G, Saffie R, Hart SL. High yield incorporation of plasmid DNA within liposomes: Effect on DNA integrity and transfection efficiency. J Drug Targeting 1996;3:469–475.
15. Gregoriadis G, Saffie R, de Souza B. Liposome-mediated DNA vaccination. FEBS Lett 1997;402:107–110.
16. Kirby C, Gregoriadis G. Dehydration-rehydration vesicles (DRV): A new method for high yield drug entrapment in liposomes. Biotechnology 1984;2:979–984.
17. McCormack B, Gregoriadis G. Comparative studies of the fate of free and liposome-entrapped hydroxypropyl-β-cyclodextrin/drug complexes after intravenous injection into rats: Implications in drug delivery 1996;1291:237–244.
18. Gregoriadis G, Wang Z, Barenholz Y, Francis MJ. Liposome-entrapped T-cell peptide provides help for a co-entrapped B-cell peptide to overcome genetic restriction in mice and induce immunological memory. Immunology 1993;80:535–540.
19. Gürsel M, Gregoriadis G. Interleukin-15 acts as an immunological co-adjuvant for liposomal antigen in vivo. Immunology Letters 1997;55:161–165.
20. Gregoriadis G, Davis D, Davies A. Liposomes as immunological adjuvants: Antigen incorporation studies. Vaccine 1987;5:143–149.
21. Skalko N, Bouwstra J, Spies F, Gregoriadis G. The effect of microfluidization of protein-coated liposomes on protein distribution on the surface of generated small vesicles. Biochim Biophys Acta 1996;1301:249–254.
22. Gregoriadis G, Garcon N, da Silva H, Sternberg B. Coupling of ligands to liposomes independently of solute entrapment: Observations on the formed vesicles. Biochim Biophys Acta 1993;1147:185–193.
23. Gregoriadis G, da Silva H, Florence AT. A procedure for the efficient entrapment of drugs in dehydration-rehydration liposomes (DRV). Int J Pharmaceutics 1990;65:235–242.
24. Legendre J-Y, Szoka FC Jr. Liposomes in gene therapy. In: Puisieux F et al., eds. Liposomes, New Systems and New Trends in their Applications. Paris: Editions de Santé, 1995;667–692.
25. Felgner PL. Gene therapeutics. Nature 1991;349:351–352.
26. Nicolau C, LePape C, Soriano P, Fargette F, Juhel M-F. In vivo expression of rat insulin after intravenous administration of the liposome-entrapped gene for rat insulin I. Proc Natl Acad Sci USA 1983;80:1068–1072.
27. Baru M, Axelrod JH, Nur I. Liposome-encapsulated DNA-mediated gene transfer and synthesis of human factor IX in mice. Gene 1995;161:143–150.
28. Alino SF, Crespo J, Bobadilla M, Lejarreta M, Blaya C, Crespo A. Expression of human α_1 antitrypsin in mouse after in vivo gene transfer to hepatocytes by small liposomes. Biochem Biophys Res Comm 1994;204:1023–1030.
29. Black CDV, Gregoriadis G. Interaction of liposomes with blood plasma proteins. Biochem Soc Trans 1976;4:253–256.
30. Gregoriadis G, McCormack B (eds). Targeting of Drugs: Strategies for Oligonucleotide and Gene Delivery in Therapy. New York: Plenum Press, 1996.
31. Thierry AR, Lunardi-Iskandar Y, Bryant JL, Rabinovich P, Gallo RC, Mahan LC. Systemic gene therapy: Biodistribution and long-term expression of a transgene in mice. Proc Natl Acad Sci USA 1995;92:9742–9746.
32. Zhu N, Liggit D, Liu Y, Debs R. Systemic gene expression after intravenous DNA delivery into adult mice. Science 1993;261:209–211.
33. Nabel GJ, Nabel EG, Yang Z-Y, Fox BA, Plautz GE, Gao X, Huang L, Shu S, Gordon D, Chang AE. Direct gene transfer with DNA-liposome complexes in melanoma: expression, biologic activity, and lack of toxicity in humans. Proc Natl Acad Sci USA 1993;90:1130–1138.
34. Solodin I, Brown C, Bruno M, Chow C, Jang E-H, Debs R, Heath T. High efficiency in vivo

gene delivery with a novel series of amphilic imadazolinium compounds. Biochemistry 1995;34:13537–13544.
35. Liu Y, Mounkes LC, Liggit HD, Brown CS, Solodin I, Heath TD, Debs RJ. Factors influencing the efficiency of cationic liposome-mediated intravenous gene delivery. Nature Biotechnology 1997;15:167–173.
36. De Souza JB, Ling IT, Ogun SA, Holder AA, Playfair JHL. Cytokines and antibody subclass associated with protective immunity against blood-stage malaria in mice vaccinated with the C terminus of merozoite surface protein 1 plus a novel adjuvant. Infect Immun 1996;64:3532–3536.
37. Sedegah M, Hedstrom R, Hobart P, Hoffman SL. Protection against malaria by immunization with circumsporozoite protein plasmid DNA. Proc Nat Acad Sci USA 1994;91:9866–9870.
38. Ulmer JB, Deck RR, DeWitt CM, Friedman A, Donnelly JJ, Liu MA. Protective immunity by intramuscular injection of low doses of influenza-virus DNA vaccines. Vaccine 1994;12:1541–1544.
39. Xiang ZQ, Spitalnik S, Tran M, Wunner WH, Cheng J, Ertl HCJ. Vaccination with a plasmid vector carrying the rabies virus glycoprotein gene induces protective immunity against rabies virus. Virology 1994;199:132–140.
40. Cox JM, Zamb TJ, Babiuk LA. Bovine herpes virus 1: Immune responses in mice and cattle injected with plasmid DNA. J Virol 1993;67:5664–5667.
41. Fynan EF, Webster RG, Fuller DH, Haynes JR, Santoro JC, Robinson HL. DNA vaccine: protective immunization by parenteral mucosal and gene-gun inoculations. Proc Natl Acad Sci USA 1993;90:11478–11482.
42. Montgomery DL, Shiver JW, Leander KR, Perry HC, Friedman A, Martinez D, Ulmer JB, Donnelly JJ, Liu MA. Heterologous and homologous protection against influenza A by DNA vaccination: optimization of DNA vectors. DNA Cell Biol 1993;12:777–783.
43. Yokoyama M, Zhang J, Whitton JL. DNA immunization confers protection against lethal lymphocytic choriomeningitis virus infection. J Virol 1995;69:5664–5667.
44. Raz E, Carson DA, Parker SE, Par TB, Abai AM, Aichinger G, Gromkowski SH, Singh M, Lew D, Yankauckas MA, Baird SM, Rhodes GH. Intradermal gene immunization: the possible role of DNA uptake in the induction of cellular immunity to viruses. Proc Natl Acad Sci USA 1994;91:9519–9523.
45. Papahadjopoulos D. Optimal liposomal drug action: From serendipity to targeting. In: Gregoriadis G, ed. Liposome Technology, vol III. Boca Raton: CRC Press Inc, 1993;1–14.
46. Gregoriadis G. Engineering targeted liposomes: Progress and problems. Trends in Biotechnology 1995;13:527–537.
47. Segal AW, Gregoriadis G, Black CDV. Liposomes as vehicles for the local release of drugs. Clin Sci Mol Med 1975;49:99–106.
48. Ryman, BE, Jewkes RF, Jeasingh K, Osborne MP, Patel HM, Richardson VJ, Tattersall MHN, Tyrrell DA. Potential applications of liposomes to therapy. Ann NY Acad Sci 1978;308:281–307.
49. Wagner E, Cohen M, Foisner R, Birnstiel ML. Transferrin polycation-DNA complexes: The effect of polycation on the structure of the complex and DNA delivery to cells. Proc Natl Acad Sci USA 1991;88:4255–4259.
50. Sternberg B, Sorgi FL, Huang L. New structures in complex formation between DNA and cationic liposomes visualized by freeze-fracture electron microscopy. FEBS Lett 1994;356:361–366.
51. Sternberg B. Morphology of cationic liposome/DNA complexes in relation to their chemical composition. J Liposome Research 1996;6:515–533.
52. Szoka FC, Xu Y, Zelphati O. How are nucleic acids released in cells from cationic lipid-nucleic acid complexes? J Liposome Research 1996;6:567–587.
53. Hong K, Zheng W, Baker A, Papahadjopoulos D. FEBS Lett 1997;400:233–237.

Lasic and Papahadjopoulos (eds.), Medical Applications of Liposomes

CHAPTER 2.5

"Virosomes", a new liposome-like vaccine delivery system

Reinhard Glück and Alfred Wegmann

Swiss Serum and Vaccine Institute, P.O. Box CH 3001, Berne, Switzerland

Overview

I. Introduction

Whole, intact proteins from pathogens have been shown to be effective immunogens. The surface glycoproteins of many pathogenic organisms contain regions that are utilized by the immune system of the infected host to mount a protective response against infection.[1] These regions include neutralizing B cell epitopes (recognized by B lymphocytes that manufacture and secrete antibodies that inhibit pathogen infectivity), T helper cell epitopes (regions which are needed to activate specific T helper cells), and cytotoxic T cell epitopes. Combinations of these epitopes are used by an infected host to rise an immunological defense against pathogen challenge.

Many pathogens, however, have evolved mechanisms for subverting the immune system. The surface glycoproteins of some pathogens, e.g., human immunodeficiency virus (HIV) and malaria, contain regions termed dominant B cell epitopes, which are strong stimulators of antibody production. The antibodies reactive against these regions, however, do not interfere with the infectivity of the pathogen. Indeed, in some cases they enhance pathogen infectivity. In addition, while eliciting a strong response to themselves, dominant B cell epitopes can reduce or suppress the immune response to more weakly immunogenic neutralizing B cell epitopes. Other, less well-defined regions, can also inhibit or suppress the activity of helper and cytolytic T cells. Therefore, an immunogen most capable of inducing an immune response protective against pathogen challenge would be one in which small, synthetic peptides, representing regions of pathogen proteins that can elicit beneficial (i.e., protective) activities are included, while the regions that are subversive to the immune response are excluded.

In the 23 years since Allison and Gregoriadis[2] demonstrated that liposomes could enhance the antibody response to diphtheria toxoid, the role of liposomes as adjuvants in stimulating an immune response has received a great deal of attention. Since that time liposomes have been shown to be effective adjuvants for a large number of protein antigens.[2,3,4] Liposomes are being investigated in the design of subunit vaccines for viral diseases and are able to efficiently present the surface glycoproteins of many enveloped viruses for stimulation of a protective immune response. The viral glycoproteins are anchored in the liposomal bilayer via a transmembrane segment and assume a conformation analogous to their native conformation in the viral envelope.[5]

While proteins reconstituted into phospholipid bilayers have been shown to be effective antigens, the ability of liposomes to potentiate the immune response to peptides have been sometimes less impressive, often requiring that the liposome-peptide complex be emulsified in Freund's adjuvant in order to elicit antibody production.[6,7] This may be due to the fact that the peptides have been associated with the liposomes either by encapsulation in the internal aqueous space of the liposome, or by adsorbing the peptide to the liposomal surface.

This chapter is intended to review current advances in virosome design and immunological effects of different vaccine antigens in man.

II. History of liposomes

The formation of artificial lipid vesicles by allowing phospholipids to swell in aqueous media was first described by Bangham et al.[8] Comprised of concentric bilayers that alternate with aqueous compartments, these vesicles (liposomes) entrapped ions in the aqueous phase between the lipid bilayers and had permeability properties resembling those of biological membranes. They were initially used as models of lipid bilayers[9] to study ion transport across cell membranes, and those early experiments set the stage for a whole series of studies in membrane biophysics.[10] Since their introduction the use of liposomes as a research tool has undergone an impressive evolution. Liposomes have become the preferred system for studying the reconstitution of membrane transport proteins and enzymes, the mode of action of ionophoric peptides and a variety of anaesthetics and other drugs.[11] Thus liposomes have played an essential role in developing our current understanding of the structure and function of biological membranes in areas such as membrane fusion,[12,13,14,15] antigen-antibody interactions,[16,17] the complement system,[18] blood coagulation[19,20] and arteriosclerosis.[21,22] During the early 1970s, liposome research went beyond basic membrane physics and into the area of therapeutic application, as a vector system for altering the tissue disposition of various macromolecules in vitro,[23] and for introducing foreign macromolecules into cells in vitro.[21] With these two new developments, liposome research bridges three different fields: biophysics, cell biology and medicine. Their potentials as magic bullets caught the imagination of many researchers but yielded little success. However, in recent years there has been a greater appreciation of the limitation of liposomes as a drug delivery and targeting system and of the enormous difficulties of in vivo administration. The early unrealistic expectations produced disappointments although an alert scientific interest in liposomes has survived. In the past 25 years, investigators have contributed more than 20,000 scientific articles on liposomes in fields as diverse as gene transfer and mutation, and more than 200 patents have been issued covering their formation, structure, manufacture and use.[24]

To date the most successful examples of liposomal pharmaceutical products are those with doxorubicin (an antineoplastic drug),[25,26] amphotericin (an antifungal drug)[27,28] and vaccines [hepatitis A vaccine[29]]. Liposomal amphotericin and liposomal hepatitis A vaccine[30] are the first formulations of liposomes to become licensed for clinical use, while doxorubicin encapsulated in sterically stabilized liposomes became the first product on sale in USA.[31]

III. General properties of liposomes

Liposomes are vesicular bilayer structures generally composed of phospholipids and cholesterol.[32,33,34,35] The first report of liposomes as immunological adjuvants was made 20 years ago,[2] and since then, numerous studies showing the adjuvant action of liposomes have been published. Applications for liposomes include haptens,[36] hepatitis B-derived polypeptides,[37] subunit antigens from the influenza virus,[38,39] adenovirus type 5 hexon,[40] allergens,[41] and polysaccharide-protein conjugates,[42] to name a few. Several laboratories have studied liposomes made with detergent-extracted envelope glycoproteins from HIV-1,[43] and synthetic peptide carrying a CTL epitope from the simian immunodeficiency virus gag protein.[44] As testament to the increased interest in liposome technology, 353 patents on liposomes and related topics were issued in the United States between 1975 and October, 1990,[45] the most in 1989–1990. Universities and individuals hold the largest share of patents (27%) filed during the 15-year period.

Based on particle size and mode of preparation, liposomes can be separated into various types (see Table 1). A detailed description of nomenclature and formulation processes has been given by Szoka and Papahadjopoulos.[46,47] Although the mechanism(s) responsible for enhanced immunogenicity by liposomes have not been defined, prolonged retention of antigen at the site of administration along with increased antigen delivery to macrophages are thought to be responsible for this phenomenon.[40,48] Immune enhancement by liposomes can be affected by liposome type, size, bilayer composition, lamellarity, net surface charge,[49,50] and the mode of a physical association between antigen and liposome.[51] Methods have been developed for the chemical coupling of antibodies,[52,53,54,55,56,57,58] antigens,[59,60,61,62,63] and sugars[64,65] to liposomes. In addition, several groups have

Table 1

Nomenclature and description of various liposomes (See: Szoka and Papahadjopoulos 1980 (Lit. 46) for details)

Liposome type	Abbreviation	Size (nm)	Mode of preparation
Small unilamellar vesicles	SUV	25–50	Sonication
		30–100	Ethanol injection
		100–200	Detergent dialysis
Large unilamellar vesicles	LUV	150–250	Ether infusion
		200–1000	Calcium-induced fusion
LUV/reverse phase evaporation	REV	100–1000	Reverse phase evaporation
Large unilamellar vesicles by extrusion	LUVET	100–1000	Membrane extrusion
Multilamellar vesicles	MLV	25–100	French Press extrusion
		400–5000	Rotoevaporation
Freeze and thaw multilamellar vesicles	FT-MLV	100–500	Freeze and thaw MLV
Stable plurilamellar vesicles	SPLV	100–1000	Ether infusion/osmotic buffers

shown that phospholipid composition can affect the antibody IgG subclass response to viral antigens.[66,67] Whatever the final liposome formulation, it seems clear that immune enhancement is linked to the physicochemical properties inherent in the *combined* antigen-liposome formulation.

The past 10 years has brought forth a plethora of information on novel vaccine adjuvants. Acronyms such as ISCOMs, SAF-1, MF-59, QS-21, MPL, RIBI, MDP, MAP, Pam3Cys, Pluronic L121, and MLV are as commonplace in vaccine literature as is CFA. In animal models, each of these adjuvants and/or immunomodulators has demonstrated potentiation of antigen-specific immune responses. The degree of immune enhancement as well as the effects on humoral or cell-mediated immune response were clearly dependent on the antigen-adjuvant formulation.

Key aspects influencing the development of new and effective adjuvants for human vaccines are the safety, purity, and physicochemical characteristics of the final adjuvanted formulation. As reviewed herein, liposomes or virosomes have been safely used clinically as vaccine adjuvants for parasitic and viral antigens. Thus, clinical-scale liposome manufacturing processes and quality-control release tests establishing purity, safety, and stability of liposomal vaccine have been confirmed. Such successes continue to support the rationale for incorporating liposome-like adjuvants into the design of new human vaccines.

IV. The virosomal vaccine approach

The most significant impediment to the use of synthetic peptides as vaccines has been that they are only weakly or non-immunogenic when injected by themselves into animals.[3,68] This property has necessitated the use of carriers, usually large, highly "immunogenic" proteins, to which the peptides are covalently coupled. These carriers, although helpful in producing an initial antibody response, have no relationship to the pathogen against which the vaccine is designed and therefore do not elicit pathogen-specific T cell help. Therefore, when an individual who has been vaccinated with a peptide-carrier complex is challenged with the pathogen a primary rather than a secondary (faster, stronger, higher affinity) response results. Also, booster immunizations often lead to a stronger antibody response to the carrier and a diminishing one to the peptide. In addition, these peptide-carrier complexes must usually be combined with other adjuvants (for example, Freund's) to enhance the response to the peptide. These adjuvants frequently induce undesirable side effects which make them unacceptable for use in humans.[3,4,6,68,69,70,71,72,73,74,75,76]

It has been hypothesized that anchorage of a peptide in the liposomal bilayer might mimic the normal presentation of antigen on an infectious agent (i.e., multivalent and projecting outward from an anchor on the surface of the cell) and thereby potentiate the immune response to the peptide. To test this hypothesis, peptides were covalently linked to a phospholipid, providing a hydrophobic anchorage into the phospholipid bilayer.

It has been found that when molecules capable of stimulating T helper cells (either viral envelope proteins or peptides representing defined Th cell epitopes) are integrated into the same phospholipid matrix as a B cell epitope, a highly efficient immunogen is produced.[77,78] Sequences which are not recognized by T helper cells do not elicit antibody responses even when formulated into peptide-phospholipid complexes.[79]

Current concepts regarding the mechanisms through which peptide epitopes are presented to CD8+, MHC Class I-restricted cytotoxic T lymphocytes indicate that a crucial aspect of this process is the capacity to introduce antigen into the cytoplasm (but not endosomes) of antigen presenting cells.[80] This explains, at least in part, the success of live-attenuated and live-vector vaccines for stimulating cell mediated immune responses.

In order to obtain a similar mechanism, methods have been introduced for integrating lipid-linked peptides (membrane proteins) into the lipid bilayer of large, mainly unilamellar liposomes.[81] For example, glycoproteins of influenza and parainfluenza type I (Sendai) viruses maintain their activities of receptor binding and receptor induced endocytosis when reconstituted into protein lipid vesicles (virosomes).[82,83] In addition, water soluble materials can be encapsulated within the aqueous interior of such vesicles at high efficiency. It could even be shown that these vesicles act as effective delivery vehicles for drugs, proteins, and DNA. Using a liposome based system they were employed to achieve the first stable gene transfer in animals.[82,84]

Virosomes proved to be also highly effective immunogens in mice, rabbits and monkeys.[44,85] This included the ability to stimulate strong CD8+ cytotoxic T cell responses (CTL) to lipid bilayer-integrated glycoproteins or lipid linked peptides, as well as to encapsulated peptides, proteins and formalin-fixed whole viruses.[44,85,79]

V. Immunopotentiating reconstituted influenza virosomes (IRIV)

Twenty years after the discovery of the immunological adjuvant properties of liposomes[2] and the ensuing multitude of related animal immunization studies,[34] liposomes as adjuvants have come of age[86,29] with the first liposome-based vaccine against hepatitis A being licensed for use in humans. Vaccines based on novasomes (non phospholipid biodegradable, pausilamellar vesicles formed from single-chain amphiphiles, with or without other lipids) have also been licensed for the immunization of fowl against Newcastle disease virus and avian rheovirus.

As mentioned above, the way in which liposomes induce immune responses to antigen is not clear, but has been attributed to a depot effect (slow release of antigen and the ability of vesicles and the associated antigen to migrate to regional lymph nodes following local injection.) In the case of liposomes, further improvement of adjuvanticity has been achieved by the use of co-adjuvants such as lipopolysaccharides, positively or negatively charged lipids, interleukin 2 and by ligand-mediated targeting to antigen-presenting cells.[87]

The approach adapted for the immunopotentiating reconstituted influenza virosome (IRIV) vaccines is of particular interest, as it combines several components that are known to contribute to immunostimulation and that are at the same time harmless.

IRIV are spherical, unilamellar vesicles with a mean diameter of ~150 nm. They show short surface projections of 10–15 nm (Figure 1). IRIVs are prepared by detergent removal of influenza surface glycoproteins and a mixture of natural and synthetic phospholipids containing 70% egg yolk phosphatidylcholine (EYPC), 20% synthetic phosphatidylethanolamine (PE) and 10% envelope phospholipids originating from H1N1 influenza virus (A/Singapore/6/86)[29] (Figure 2).

EYPC is known to be well tolerated in man and is an important constituent in commercial solutions for i.v. applications in undernourished persons. EYPC has been used in nearly all liposomal preparations which were produced for the enhancement of immune responses. PE was chosen for two reasons: First it is known that hepatitis A virus (HAV) attachment to host cells occurs via binding to PE regions of the cell membrane.[88] Furthermore, it has been shown that liposomes containing PE are able to directly stimulate B cells to produce antibodies without any T cell determinant being present.[89] There were several reasons for including influenza virus envelope glycoproteins: The hemagglutinin (HA) plays a key role in the mode of action of the IRIVs. HA is the major antigen of influenza virus, containing epitopes on both HA1 and HA2 polypeptides, and is responsible for the fusion of the virus with the endosomal membrane.[14,90] The HA1 globular head groups contain the sialic acid site for HA and it is therefore expected that the IRIVs bind to such receptors of antigen presenting cells (e.g., macrophages, lymphocytes) initiating a successful immune response. The entry of influenza viruses into cells occurs through HA-receptor mediated endocytosis.[91] It is likely

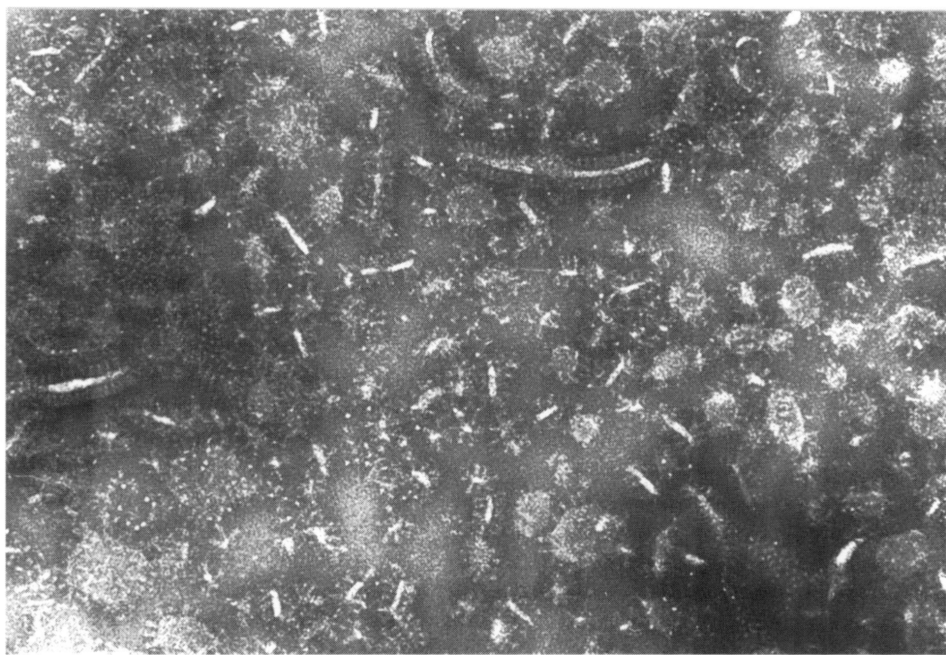

Fig. 1. Transmission electron micrograph of IRIV–HAV vaccine (×100,000). The electron photomicrograph of the IRIV–HAV vaccine shows spherical unilamellar vesicles with a mean diamter of ~150 nm.

that this mechanism also functions with the IRIV particles. The HA2 subunit of HA mediates the fusion of viral and endosomal membranes, which is required in order to initiate infection of cells. At the low pH of the host cell endosome (~pH 5), a conformational change occurs in the HA that is a prerequisite for fusion to occur. Fusion activity tests have shown that there was no difference of activity between influenza virus and IRIV (Figure 3). It is expected that this mediates the rapid release of the transported antigen into the membranes of the target cells.[92]

Further immunopotentiating effects have recently been described for the influenza virus hemagglutinin: Studies provide evidence for an alternative stimulation of peritoneal B lymphocytes by HA, a so called B cell "superstimulatory" antigen.[93] This finding implies that the B cell superstimulatory influenza virus glycoprotein has been evolutionarily adapted to activate not only conventional B2 cells, but in addition a B cell subset that represents a major weapon in the first line of defense against invading microorganisms. The great potency of B1 cells to build up an immediate immune response against microbial antigen is paralleled by its increased susceptibility to cross react with "third party" antigen. This phenomenon has been further investigated by showing that this new example of B cell stimulation by multivalent type-2 antigen (e.g., HA) seems to be mediated by a phospha-

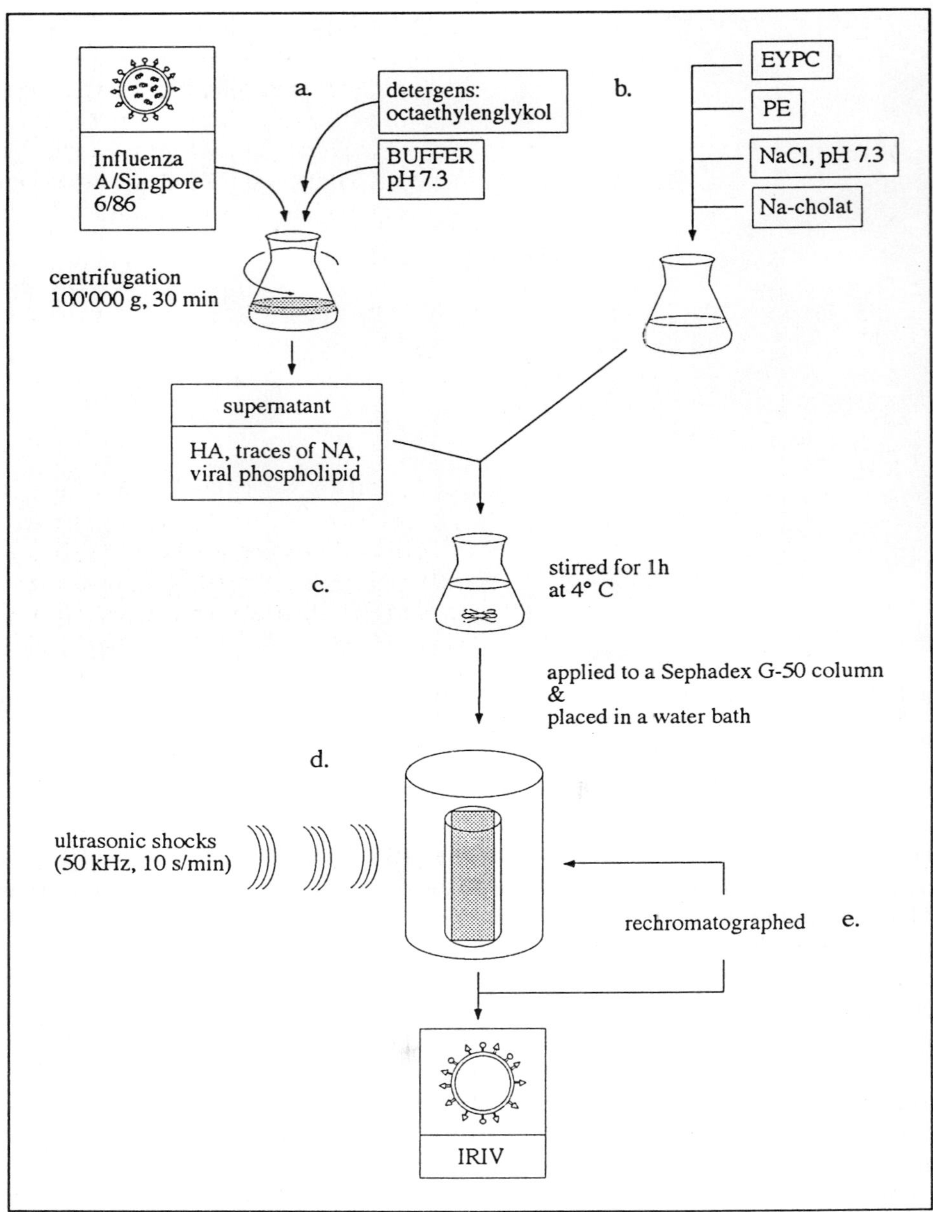

Fig. 2. Preparation of IRIVs.

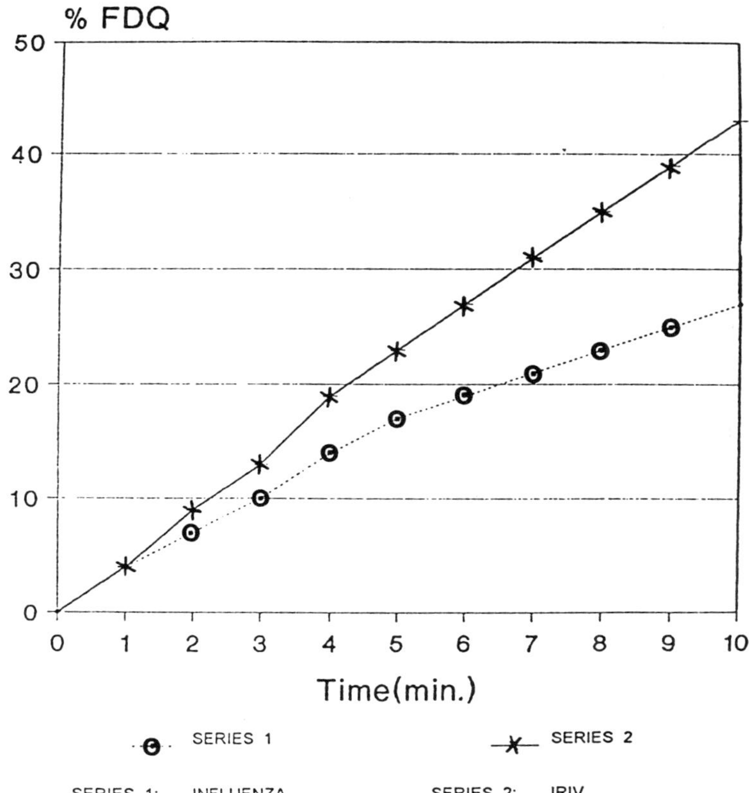

Fig. 3. Comparison of fusion rates between influenza virus and IRIV. In a fusion test described by Lüscher and Glück[13] IRIVs were compared with live influenza virus in fusion activity with model membranes. The figure shows the kinetics of fluorescence de-quenching with DOPC-cholesterol liposomes. The increase in fluorescence is expressed in % FDQ, calculated according to Lüscher and Glück.

tidyl-inositol—and Ca^{2+}-independent signaling pathway.[94] In addition it has been reported that B cell superstimulatory influenza virus (H2-subtype) induced B cell proliferation by a PKC-activating, Ca^{2+}-independent mechanism.

Furthermore, influenza A virus has been described as a model system for the study of viral antigen presentation to cytotoxic T cells (CTL).[95] In the clinical part of this chapter the potent effect of IRIV designed influenza vaccine on the cellular immune system is mentioned.

The second influenza glycoprotein exposed on the IRIV surface, the enzyme neuraminidase (NA), is a tetramer composed of four equal, spherical subunits that are hydrophobically embedded in the membrane by a central stalk. The entire enzymatic activity takes place in the region of the head. NA catalyzes the cleavage of *N*-acetylneuraminic acid (sialic acid) from bound sugar residues.[96] In the mucus, this process leads to a decrease in viscosity and allows the influenza virus easier access to epithelial cells. In the area of the cell membrane, the same process leads to destruction of the HA receptor. The consequence of this is, first, that newly formed virus particles do not adhere to the host cell membrane after budding, and second, that aggregation of the viruses is prevented. NA therefore allows the influenza virus to retain its mobility. In terms of the IRIV, these characteristics of NA can, in theory, be utilized in that, after coupling with HA, IRIVs not taken up by phagocytosis could be cleaved off again and would therefore not be lost. Also, the reduction in viscosity of the mucus could be useful in connection with the development of a nasal IRIV vaccine.

Recently, a chimeric influenza virus has been constructed that expressed the highly conserved amino acid sequence ELDKWA of gp41 of human immunodeficiency virus type 1 (HIV-1).[97] Muster et al. could demonstrate that intranasal immunization of mice with this chimeric virus was also able to induce a humoral immune response at the mucosal level. They concluded that influenza virus can be used to efficiently induce antibodies against antigens from foreign pathogens by mucosal immunization.[97]

The excellent characteristics of IRIVs as adjuvants have been demonstrated in several systems. IRIVs were first utilized in the manufacture of a hepatitis A vaccine. This contains formalin-inactivated and highly purified hepatitis A viruses (HAV) of strain RG–SB, cultured on human diploid cells, which are electrostatically coupled to the IRIV vesicle.[29] The surface spikes (HA and NA) of three currently circulating influenza strains were jointly inserted in the vesicle membrane of the IRIVs and successfully tested clinically. A combined hepatitis A–hepatitis B vaccine was also produced, based on IRIVs. The highly purified, inactivated hepatitis A virions and the hepatitis B surface antigens (HBsAg) genetically engineered in yeast were together covalently coupled to the surface of the IRIV.[98] Finally, combination vaccines were developed, for example a combined diphtheria-tetanus-hepatitis A vaccine. For this, the diphtheria toxoid, the alpha-tetanus toxoid, the beta-tetanus toxoid, and the inactivated hepatitis A virion were covalently bound via cross-linker molecules to the IRIV surface. A "supercombined" vaccine based on IRIV was developed too, containing covalently bound HAV, Hbs, diphtheria, alpha- and beta-tetanus as well as HA and NA from three different influenza strains.[99,100]

VI. Preclinical evaluation of virosomal vaccines

The persian philosopher-physician Avicenna (980–1037) is reported to have observed: "The experimentation must be done with the human body, for testing a drug on a lion or a horse might not prove anything about its effect on man." (Crombie, A.C. Avicenna, scientist and philosopher, Wicken G.M., Ed. London: Luzac & Company, 1982: 89.) Despite the scientific advances that have been made, Avicenna's words strike a cord of truth, even today. Although useful information is frequently obtained by studies in animals, animal models can only approximate the pathophysiology of an immune response to disease in humans. Furthermore, experimental animals, such as nonhuman primates whose disease states and immune mechanisms could closely resemble those of humans are frequently too rare and its experimental use often creates an ethical problem. It is not surprising, therefore, that the determination of the effect of a vaccine in humans must ultimately rest on the actual administration of that vaccine to humans.

A number of studies have been conducted to evaluate the capacity of virosomal vaccines to enhance antibody responses in animals: Gregoriadis et al.[101] described an experiment with a virosomal hepatitis B vaccine containing a B cell epitope peptide from the S region of the hepatitis B surface antigen (HBsAg) and a Th cell epitope from the pre-S1 region exposed on dehydration–rehydration vesicles (DRV) which were inoculated into mice.[101] Comparative results obtained with these preparations and with vaccines containing only the soluble antigens without vesicles showed that the Th cell epitope provided help for the pre-S1 peptide only when the two peptides were coadministered in the same vesicle. This helper effect was found to correlate with the ability of S peptide (co-entrapped with the pre-S1) to stimulate T cell proliferation in vitro. There was no IgG1 response against pre-S1 peptide in mice immunized with a mixture of the free peptides or a mixture of separately entrapped peptides. A helper effect, albeit much weaker, was also observed in mice immunized with the two peptides emulsified in incomplete Freund's adjuvant. It was concluded that hepatitis B virosomes could serve not only as an immunological adjuvant but also as a carrier for Th- and B-cell epitopes, thus eliminating the need for covalent linkage to a carrier protein.

In another experiment Ben Almeida et al.[67] showed the superiority of influenza virosomes over aqueous A/Sichuan influenza surface preparations in unprimed mice and mice primed by prior infection with a H1N1 subtype of influenza virus.

Influenza virosomes were also used for intranasal immunization in mice:[5] A vaccine was constructed composed of glycoproteins from the envelopes of either influenza or Sendai virus embedded in a lipid bilayer (liposomes). Intranasal immunization with these virosomes induced an adequate systemic immune response and a superior local IgA response. These animals were specifically protected from a virus challenge.

In the course of development of a virosomal hepatitis A vaccine, several preparations containing different compositions of liposomes and influenza virosomes (IRIV) have been tested in small animals.

The RG–SB strain hepatitis A virus was cultivated in MRC-5 cells and, after harvesting, subjected to a high-level purification procedure:
After extraction from the MRC-5 cells, the virus was separated from the cell debris by centrifugation (30 min. 2500 rpm). The virus was then ultrafiltered, lipids removed with *n*-heptane, ultracentrifuged in the saccharose gradient and, after dialysis against EDTA borate buffer, inactivated with 0.55% formalin solution (3 days, 37°C/9 days, 22°C). After inactivation, the virus was purified from unbound formalin by ultracentrifugation and then coupled to influenza virosomes (IRIV) or liposomes.

In order to test whether the chemical phospholipid composition of the IRIV vesicle and influenza glycoprotein spiking has an effect on the immunogenicity of the vaccine preparation, the following virosomes and liposomes were produced:

(A) virosomes consisting of influenza HA and NA, influenza envelope phospholipids and egg phosphatidylcholine (PC);
(B) virosomes consisting of influenza HA and NA, influenza envelope phospholipids, egg PC and phosphatidylethanolimine (PE);
(C) virosomes consisting of influenza HA and NA, influenza envelope phospholipids, egg PC, PE and succinimidyl-4-(*p*-male-imidophenyl) butyrate (SMPB) (crosslinker molecule for covalent binding of the HAV antigen);
(D) liposomes consisting only of egg PC and PE;
(E) liposomes consisting only of egg PC, PE and SMPB.

In all preparations, the inactivated hepatitis A virus (HAV) particles were adsorbed or bound to the surface. Five mice per group were inoculated i.m. with the antigen preparation (each group received the HAV antigen from one FB pool).

The immune response (HAV antibodies) was compared in 5 mice (female, NMRI-strain) per group: All mice received one 0.5 mL dose (i.m.) on day 0 and day 7. Blood was taken on day 28.

The evaluation of the results (Figure 4) revealed that all of the mice that were inoculated with an HAV-IRIV or HAV-liposome preparation had positive HAV antibody titres on day 28. However, there were differences in the level of the titres in the individual groups. The highest antibody titres were achieved in groups B and C. In comparison, the antibody titres in group A were lower. Analysis of the preparations had shown that the percentage of HAV particles associated with the IRIVs was lower than in B and C. The lowest HAV antibody titres were achieved with the HAV liposome preparations D and E. This result shows that influenza IRIVs containing PE have the strongest immunostimulatory effect when tested in animals (mouse). PE increases HAV adsorption to the vesicle and the influenza spikes probably promote immunostimulation.

Several antigens have been coupled to the IRIV showing high immunogenicity in mice. In one experiment the SPf(66)n malaria antigen was coupled to the IRIV using a crosslinking PE-GMBS molecule (unpublished results). 14 mice were previously primed with a H1N1 influenza antigen simulating a situation similar to that in a human population and vaccinated then according to the following schedule: 0.1 ml of vaccine were administered intramusculary into the hind leg on days 1 and 29. Five mice received 75 μg of SPf66 antigen, five mice 10 μg and 4 mice 1 μg each. A second series of similar groups was vaccinated with the exception that the mice were not preprimed with influenza antigen. After 10 days the mice were bled, and the probes were tested by ELISA for seroconversion using known methods. As shown in Table 2 all mice developed anti-SPf66 antibodies after an IRIV-SPf66 -dose of 75 μg or 10 μg. Even in the group vaccinated with 1 μg of vaccine dose 50% of the animals developed anti SPf66 antibodies. It was interesting to note that the mice which were not primed with influenza antigen responded to a much lesser extent. The experiments show that previous influenza immunization additionally enhance antibody response after vaccination with IRIV-adsorbed antigens. Since all human beings can be considered to be primed with influenza

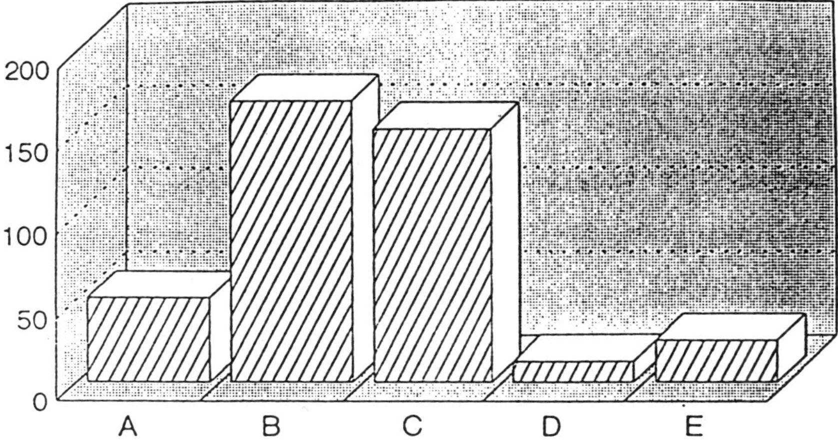

Fig. 4. HAV immunization in mice: HAV–IRIV versus HAV liposomes. The following HAV-vesicle preparations were tested in mice (5 animals per group): Equal amounts of inactivated HAV antigen were adsorbed on:

(A) Virosomes consisting of influenza HA and NA, influenza envelope phospholipids and egg PC;
(B) Virosomes consisting of influenza HA and NA, influenza envelope phospholipids, egg PC and PE;
(C) Virosomes consisting of influenza HA and NA, influenza envelope phospholipids, egg PC, PE and succinimidyl-4-(*p*-male-imidophenyl) butyrate (SMPB) (crosslinker molecule for covalent binding of the HAV antigen);
(D) Liposomes consisting only of egg PC and PE;
(E) Liposomes consisting only of egg PC, PE and SMPB.

The bars indicate the GMTs of HAV antibodies.

virus this experiment shows the positive effect of previous contact with influenza on the immunization with IRIV coupled antigens. Furthermore, analysis of antibody subtypes revealed that the IRIV coupled with the malaria antigen induced antibody titres of IgG_1, IgG_2 and IgG_3 whereas the alum vaccine used as a control raised only an IgG_1 humoral response. In addition the IRIV vaccine also raised CD4 and CD8 cellular immune response while the conventional alum vaccine raised neither CD4 nor CD8 cells specific for the malarial antigen.

In the case of IRIV-based vaccines, the common safety and efficacy tests, as recommended by WHO, the European Pharmacopoeia, or the FDA, were performed. Since no special guidelines exist for liposomal vaccines, the so-called

Table 2

Immune response of mice primed and not primed with influenza after IRIV-SPf66 malaria antigen application on day 1 and 29

Preimmunization	SPf66-dose (Virosome-coupled)	% Mice which IgG response	Strong IgG response	Week IgG response	No IgG response
Influenza	75 μg	100%	4/5	1/5	—
Influenza	10 μg	100%	4/5	1/5	—
Influenza	1 μg	50%	2/4	—	2/4
None	75 μg	80%	1/5	3/5	1/5
None	10 μg	20%	—	1/5	4/5
None	1 μg	0%	—	—	4/4

Table 3

Anti-phospholipid antibody response following immunisation of human volunteers

	GM ELISA titre (range)*		
	Phosphatidylethanolamine	Phosphatidylcholine	Cardiolipin
Pre-immunisation (95% CL)	97 (19–669) (37.6–157.2)	127 (31–724) (59.1–194.8)	109 (20–763) (34.5–183.2)
Post-immunisation (95% CL)	81 (13–677) (19.9–142.6)	122 (32–809) (49.2–195.3)	99 (22–655) (45.6–153.3)
No ⩾4-fold rise	0/30	0/30	0/30

*Data represent results from 30 subjects immunized with the virosome formulation.
CL = confidence limits.

"case by case (creative)" tests were established. For instance, extended immunological studies in animals had to show that the phospholipids employed did not provoke any immunological responses[102] or any pathological reactions. Another test series should demonstrate the physiological and chemical consistency of the IRIV structure over a certain period. In vitro tests and in vivo studies in animals had to demonstrate the mode of action of IRIV-based vaccines in the immune system. Specific fusion tests with "nude" liposomes should show the intact biological activity of the IRIV vaccine.[13] Finally, an important question was the stability of such a novel product: During a period of 2 years the vaccine had to be consistent in biophysical, biochemical, and immunological characteristics, a difficult goal to achieve with liposomal products.

Although PC and PE are naturally occurring mammalian cell phospholipids, when complexed with certain antigens in the form of liposomes, there was a possibility that anti-PC and anti-PE antibodies would be engendered. Therefore, Glück et al.[30] analyzed 30 pre-immunization and post-immunization sera for antibodies to PC, PE, and cardiolipin (Table 3). There was rise in GMT to any of the phospholipids and no subject showed a significant (⩾4-fold) rise in titre.

VII. Clinical evaluation of virosomal hepatitis A vaccine

Clinical data of virosomal hepatitis A vaccine were first presented by Glück et al.[29] It was also the first IRIV vaccine to be investigated in a clinical trial. Nowadays, this aluminium-free hepatitis A vaccine is widely used mainly in travel medicine. To evaluate the advantage of the IRIV hepatitis A vaccine 120 HAV-seronegative (titre <10 IU/L) healthy adults were enrolled. The subjects were randomly divided into 3 groups of 40 volunteers to receive either fluid, alum adsorbed, or IRIV vaccine. The groups were well matched in regard to age and sex. The vaccines were administered intramuscularly into the deltoid region. Volunteers were observed for 30 min after vaccination for immediate-type reactions. Each volunteer was asked to record all adverse reactions on a report sheet for the 4 days after immunization. Serum samples for HAV antibody determinations were taken at the time of immunization and on days 14, 28, 180, and 1 year later. The adverse reactions noted in connection with the injection of the 3 different vaccines are shown in Table 3.

Pain at the injection site was the most frequently reported complaint with all the vaccines (Table 4). Such discomfort was classified as moderate by one vaccinee (2.5%) who received the fluid formulation, 9 (23%) who were immunized with the alum-adsorbed vaccine, and one (2.5%) who received the IRIV preparation. Severe pain was reported by one subject who received the alum-adsorbed vaccine. All other subjects who reported a "painful" reaction graded it as mild. Immunization with the alum-adsorbed vaccine was associated with a significantly ($p < 0.01$) higher incidence of both pain and swelling/induration compared with either the fluid or IRIV formulations. No systemic reactions attributable to vaccination were noted.

The HAV antibody response engendered at various times postvaccination is shown in Table 5. At 2 weeks, immunization with the fluid vaccine yielded a geometric mean titre (GMT) of 16 IU/L with 30% of the subjects seroconverting (≥20 IU/L). Although the alum-adsorbed vaccine induced both a moderately higher GMT (21 IU/L) and seroconversion rate (44%), neither was significantly greater than that obtained with the fluid vaccine. In contrast, the IRIV vaccine formulation elicited a far more vigorous antibody response. The GMT of 140 IU/L was significantly ($p < 0.0001$) higher compared with either of the two other

Table 4

Adverse reactions associated with immunization

		Local reactions (%)			Systemic reactions		
Vaccine	N	Pain	Swelling/ Induration	Redness	Fever	Headache	Malaise
Fluid	40	42*	0′	0	0	0	0
$Al(OH)_3$-adsorbed	40	88+	23″	0	0	0	0
IRIV	40	25&	5§	0	0	0	0

+ versus * or &, $p < 0.01$. ″ versus ′ or §, $p < 0.01$.

Table 5

Immunogenicity of fluid, $Al(OH)_3$-adsorbed, and IRIV-adjuvanted hepatitis A vaccines

		Geometric mean titre (range)[IU/L]				
Vaccine formulation	N	Day 0	Day 14	Day 28	Day 180	Day 352
Fluid	40	<10	16 (<10–100)[A]	388 (100– > 1.000)[D]	211 (14–1.043)[G]	39 (<1–133)[J]
$Al(OH)_3$-adsorbed	40	<10	21 (10–100)[B]	871 (100–1.000)[E]	535 (18–1.758)[H]	57 (12–211)[K]
IRIV	40	<10	140 (25–300)[C]	831 (100– > 1.000)[F]	1.499 (130–3.819)[I]	655 (59–2.112)[L]

Subjects received a single dose of vaccine on day 0.
C vs A or B, $p < 0.0001$; E or F vs. D, $p < 0.001$;
E vs. F, $p > 0.05$; I vs. G or H, $p < 0.001$;
H vs. G, $p > 0.05$; L vs. J or K, $p < 0.001$;
J vs K, $p > 0.05$

vaccines. All but one vaccinee possessed ⩾100 IU/L. Of greater importance was the fact that all vaccinees seroconverted by day 14 versus <50% for the other vaccine formulations ($p < 0.005$).

By 4 weeks, all subjects in the three groups had seroconverted and the HAV antibody GMT had increased substantially. The GMT obtained for the groups that received either the alum-adsorbed (871 IU/L) or the IRIV formulation (831 IU/L) were comparable, with both being significantly ($p < 0.001$) higher than that obtained with the fluid vaccine (388 IU/L).

By 6 months time, the GMTs for the groups receiving either the fluid or alum-adsorbed vaccines had declined by ~30% from the peak levels seen on day 28 and were not significantly different ($p > 0.05$). In contrast, the GMT of the IRIV vaccine recipients nearly doubled and was significantly ($p < 0.001$) higher than in the other two groups. The seroconversion rates were 80% and 95% for the recipients of the fluid and alum-adsorbed groups, respectively. In contrast, all subjects in the IRIV group still maintained >20 IU/L with >65% having >1.000 IU/L.

By 1 year postimmunization, as many as 7 of 14 (50%) of the fluid vaccine group and 4 of 10 (40%) of the alum-adsorbed vaccine group possessed <20 IU/L. All 22 subjects available for follow-up who had received the IRIV vaccine had >20 IU/L, with 6 maintaining levels >1.000 IU/L ($p < 0.01$). The GMT for the IRIV vaccine recipients was >10-fold higher than for the other two vaccine groups.

These findings indicate that the IRIV formulated vaccine for use in humans is highly immunogenic and has a very low reactogenic potential.

In 1994, Loutan et al.[103] as well as Frösner et al.[104] reported on long-term observations and booster vaccinations at 1 year after the one dose basic immunization, respectively. In the first of the two clinical investigations,[103] 119 healthy adult HAV-negative volunteers were involved, of whom 104 could be controlled at days 14 and 28, 78 at month 8, 94 at month 12 and 71 at month 13 (i.e., 1 month after a booster injection). In the second trial,[104] 99 healthy, adult HAV-negative volunteers participated at day 0, 94 at days 14–19, 99 at days 21–47, 91 after 1 year and 88 after about 13 months (i.e., 1 month after a booster injection).

In the first trial, the IRIV vaccine was considered as being tolerated well. No alteration of liver function or of other biological tests was observed. There was no long-term adverse reaction. In most subjects reporting some reaction, symptoms did not last more than the initial 24 hours. Only 1 volunteer (1%) had a temperature above 38°C on the day of immunization. Overall, 10% of vaccinees reported some general or local significant reaction (temperature above 38°C or moderate symptoms). At this time, 112 of 114 (98%) were satisfied that the immunization was well tolerated and 99% accepted a second injection. After the second injection at 12 months, significant adverse reactions (general or local) were reported only by 3% of the volunteers. 1 volunteer complained of moderate headache and another had some inflammation at the injection site.

In the second trial involving 99 volunteers, 6 probands experienced mild and transient local symptoms: pain in 5 cases; induration, swelling, and redness in 1 case each. 13 persons complained about moderate general symptoms such as

Table 6

Seroconversion rate (%) after 2 doses of IRIV hepatitis A vaccine given at day 0 and at 1 year thereafter

	Day 14	Day 28	Month 8	1 Year	1 Year + 1 Month
Seroconversion					
Loutan et al.	98	100	100	100	100
Frösner et al.	95	100	N.D.	100	100

headache ($n = 8$), nausea and vomitus, and (in one case) loss of appetite, dizziness, diarrhea and salt taste.

The antibody titres obtained after the first vaccination and after the booster injection are jointly reported in Tables 6 and 7.

In both clinical investigations the high immunogenicity of the IRIV vaccine was substantially confirmed. The numeric differences between the antibody titres of the two series is the result of the potency of the test kit used in either series and there is a good correlation between the two series. Thus, these do not express a true difference between the respective immune responses. Both series clearly demonstrate an unusually rapid development of hepatitis A antibodies since a seroconversion rate of ⩾95% is attained after two weeks only and a 100% already after 1 month. A protective antibody titre is maintained over 1 year after the single basic immunization. The booster injection performed after 1 year elicited a ~20-fold increase of the GMT which evoked a memory (T-) cell mediated immune response.

Similar results were obtained during an observation period of 1 year by Poovoravan et al.[105] who vaccinated 61 seronegative students during an outbreak of hepatitis A within the Chulalongkorn Hospital in Bangkok. The serological controls as shown in Table 8 confirmed the results obtained by others.

Most side reactions were classified as mild and transient, lasting for 1 day or less. The most frequent complaints associated with vaccination were pain and swelling at the injection site (16.4% and 13%, respectively), malaise (10%), and headache (7.6%). Low-grade fever and redness at the injection site were reported

Table 7

Hepatitis A serum antibody titre (GMT IU/L) after one injection of IRIV hepatitis A vaccine given at day 0 and at 1 year thereafter

	Day 14	Day 28	Month 8	1 Year	1 Year + 1 Month
GMT (range)					
Loutan et al.*	544	1393	821	770	17,928
(Lit. 103)	(9–5,110)	(209–15,513)	(74–11,516)	(28–9,885)	(3,215–122,432)
Frösner et al.**	51	138	(N.D.)	124	2,684
(Lit. 104)	(0–738)	(14–1,298)		(10–1,134)	(564–9,000)

*Reagent: Boehringer Ingelheim.
**Reagent: RG-SB Berna.

Table 8

Magnitude and duration of the immune response (IU/L) following a single dose of virosmal hepatitis A vaccine.

	Day 0	1 Month	3 Months	6 Months	12 Months
Seroconversion					
%	–	100	100	100	100
GMT (range)	<20	441	510	575	394
	–	(51–10,000)	(80–10,000)	(66–4,220)	(62–1,799)

by 2.5% of vaccinees. Liver function tests performed 30 days after vaccination were within the normal range.

In a rather large study Holzer et al.[106] compared the immunogenicity of IRIV bound hepatitis A antigen with that of an $Al(OH)_3$ adjuvanted commercial vaccine. For this purpose 201 adult volunteers were vaccinated with one dose of the IRIV vaccine and 97 adult volunteers were vaccinated with two doses of the alum adjuvanted commercial vaccine. The results are documented in Table 9.

The results clearly demonstrated that with the IRIV hepatitis A vaccine seroconversion rates and antibody titres can be obtained by a single vaccination and by a booster injection which are comparable to those obtained by immunization with an aluminium adjuvanted vaccine given as double vaccination containing an estimated HAV antigen content more then twice as high as in the IRIV formulation. Furthermore, local reactions with the IRIV vaccine were substantially less frequent than with the alum preparation (Table 10).

Since the immune response to parenteral antigens is reduced in individuals who have been splenectomized, Simmen et al.[107] vaccinated 26 such patients with the IRIV hepatitis A vaccine. Thus, 26 patients (23 men and 3 women, mean age 34.9 ± 9.7 (25–65) years) with anti-hepatitis A virus antibody titres <20 IU/L (maximally 17 IU/L) were given a single dose of 0.5 mL of the IRIV hepatitis A vaccine 1 to 14 (mean 9.4) years after splenectomy. Immediately after the immunisation and 14 and 28 days afterwards the anti-HAV titres were determined. A titre rise to >20 IU/L was counted as seroconversion. At the first and last titre measurement immunoglobulins, neopterin and beta-microglobulin levels were also measured as additional markers. The seroconversion rate was 69.2% (18/26) after 14 days (geometric mean titre 39 IU/L; range 3–185) and rose to 88.5% (23/26)

Table 9

Immunogenicity of hepatitis A IRIV vaccine and $Al(OH)_3$ adjuvanted vaccine. A booster injection was given after 12 months

	N	Day 14	Day 28	Month 12	Month 13
Seroconversion %					
IRIV (500 RU)	201	90.6	98.0	93.9	100
$Al(OH)_3$ (2 × 720 EU)*	97	87.6	97.9	94.1	100

*2 doses of 720 EU at Day 0.

Table 10

Frequency of local and other adverse effects during the three days following initial and booster vaccination

	IRIV–HAV	AL–HAV	P*
Initial vaccination**			
	201	95	
Local adverse effects			
Pain	25 (12.4%)	60 (63.2%)	<0.0001
Induration	14 (7.0%)	16 (16.8%)	0.02
Swelling	5 (2.5%)	9 (9.5%)	0.02
Other adverse effects			
Headache	41 (20.4%)	14 (14.7%)	0.3
Malaise	23 (11.4%)	17 (17.9%)	0.18
Nausea	17 (8.5%)	9 (9.5%)	0.9
Fever	12 (6.0%)	3 (3.2%)	0.5
Booster vaccination[+]			
N	147	67	
Local adverse effects			
Pain	38 (25.8%)	24 (35.8%)	0.18
Induration	21 (14.3%)	9 (13.4%)	0.9
Swelling	13 (8.8%)	7 (10.5%)	0.9
Other adverse effects			
Headache	18 (12.2%)	7 (10.5%)	0.9
Malaise	6 (4.1%)	5 (7.5%)	0.5
Nausea	5 (3.4%)	3 (4.5%)	0.9
Fever	2 (1.4%)	1 (1.5%)	0.9

*Chi-square test with continuity correction or Fisher's exact test.
**One injection with IRIV-HAV, two injections with Al-HAV.
[+]One injection in both groups.

after 28 days (geometric mean titre 74 IU/L; range 3–614). Seroconversion occurred in the three nonresponders after a second dose of the vaccine. All measurements of the immunological markers were within normal limits.

It was concluded that a single dose of the vaccine conferred adequate protection in most cases who had undergone splenectomy.

A clinical trial with 80 student volunteers[108] was intended to elucidate whether the intramuscular administration of immunoglobulin (10 mL) simultaneously with the hepatitis A vaccination would unduly inhibit the immune response to the active immunization. The results of this study are shown in Table 11. While hepatitis A antibody titres measured 14 and 28 days after the vaccination were virtually equal in the two groups, the simultaneous administration of immunoglobulin seemed to slightly affect the seroconversion rate and the serum antibody levels as observed after 3 months. Thus, it can be assumed that active/passive immunization confers a high protection for persons who can not await the response to the active immunization and need the immediate protection conferred by the administration of immunoglobulin. In case of a prolonged stay in an endemic area or when returning to such an area a booster injection (for example after 1 year) is advisable.

Table 11

Seroconversion rate and antibody titres after active immunization against hepatitits A by IRIV vaccine without and with simultaneous administration of immunoglobulin (IgG).

	Day 14	Day 28	Day 90
Seroconversion (%)			
IRIV	93	95	98
IRIV + IgG	98	100	85
GMT (IU/L)			
IRIV	63	315	130
IRIV + IgG	43	253	92

It could be expected that vaccination against hepatitis A with the IRIV vaccine could possibly also elicit the development of antibodies directed against the hemagglutinins incorporated in the liposomal membranes. In a clinical trial comprising 64 adult volunteers, F. Ambrosch, Vienna (unpublished) clearly demonstrated that this is not the case. When taking serum samples at days 0 and 28, it was found that the hemagglutinin antibody titre remained unchanged in 56 out of 64 probands (87.5%). In 4 cases (6.25%) the hemagglutinin antibody titre increased. However, with one exception the titre increased only by a dilution factor of 2 or less. In 4 cases (6.25%) the titre decreased by a dilution factor of 2. The lack of hemagglutinin antibody titre rise can be explained by the fact that the influenza antigen content of an IRIV adjuvanted dose is considerably lower than in classical influenza vaccines. On the contrary, the hepatitis A antibody titre was not affected by the non-significant changes occurring in the hemagglutinin antibody titre.

Another question which could be raised is whether the IRIV could induce anti-phospholipid antibodies. This question was answered by Glück and Wälti.[109] The anti-PE and anti-PC antibody response was studied in 100 volunteers (students, who had received two doses of the liposomal hepatitis A vaccine with a year between doses.[110] Furthermore, a group of 61 nursing home residents was studied for anti-PE, anti-PC as well as anti-CL antibodies after vaccination with a new liposomal (trivalent) influenza vaccine.[30] Two groups each of ca. 30 volunteers received either a commercial whole virus- or a subunit vaccine. The anti-phospholipid antibodies were determined by EIA. No subject in the hepatitis A group showed a significant (⩾4 fold) rise in anti-PE or anti-PC antibody titre over baseline following immunization. Furthermore, vaccination did not engender a significant ($p > 0.05$) rise in the geometric mean anti-PC, anti-PE or anti-CL titres. In the influenza group, no significant ($p > 0.05$) rise in geometric mean anti-PC or anti-CL titres was observed following immunization with the liposomal vaccine or with any of the two commercial control vaccines (whole virus or subunit) (see also section "Preclinical Evaluation of Virosomal Vaccines").

In a further study, Ambrosch et al.[111] compared the seroconversion rate and the antibody titres as determined with the neutralization test (NT) with the results obtained by using the enzyme immuno assay (EIA) in order to evaluate the protective efficacy of the antibodies usually found by the measurements with the

Table 12

Seroconversion rate and geometric mean titres (GMT) after a first vaccination on day 0 and after a booster injection one year later with the IRIV hepatitis A vaccine. The antibody determinations were performed by enzyme immuno assay (EIA) in 112 cases and additionally by neutralization test (NT) in 25 cases

	Day 0	Day 14	Day 28	Month 6	Month 12	Month 13
Seroconversion (%)						
EIA	–	97	99	99	99	100
NT	–	84	96	100*	96*	100
GMT (1 U/L)	<10	109	176	179	175	1884
EIA						
NT	<10	20.6	29.4	36.0*	36.8*	N.D.

*N = 24.

EIA. Thus, 112 adult seronegative volunteers were vaccinated on day 0 and boosted after 1 year with one dose each time of the IRIV hepatitis A vaccine. The 25 sera with the lowest antibody response as determined with the EIA were additionally examined with the NT. The results obtained with both techniques are demonstrated in Table 12. It was remarkable that protective neutralizing antibodies were already measurable in 84% of these low responder volunteers after 14 days and in 96% after 28 days. 96% of the probands still showed protective antibodies after 12 months and 100% after the booster injection. It was concluded that the antibodies elicited by the IRIV vaccine are protective and have a high affinity to hepatitis A virus.

Laboratory controls of the stability of the IRIV hepatitis A vaccine revealed a decrease of the in vitro detectable hepatitis A antigen content after prolonged storage. However, after addition of a detergent, the original antigen content became detectable again by EIA. It was concluded that the hepatitis A antigens may have become internalized within the lipid bilayer membrane. The question whether the IRIV vaccine containing internalized hepatitis A antigens as a result of long lasting storage is still suitable for vaccination, was affirmatively answered by a clinical study performed by Wegmann et al.[112] In this trial the immunogenicity of a single injection of a vaccine lot stored at 4°C for 32 months was compared by parallel antibody determination with that of a single injection of a freshly prepared lot of the same antigen content. The results shown in Table 13 demonstrate that the duration of storage (at 4°C) had no influence on the immunogenicity of the vaccine whatsoever.

VIII. Clinical evaluation of two virosomal influenza vaccines

An early clinical trial with a virosomal influenza vaccine containing different amounts of hemagglutinin (HA) and neuraminidase (NA) as well as various concentrations of muramyl dipeptide (MDP) as an adjuvant (Table 14) was performed by Kaji et al.[113] In this study, three strains, A/Yamagata/120/86 (H1N1), A/Fukuoka/C29/86 (H3N2) and B/Nagasaki/1/87, provided by the National Institute

Table 13

Hepatitis A antibody titre 28 days after the injection of one dose of IRIV hepatitis A vaccine, freshly produced (A) or stored at 4°C for 32 months (B), respectively

	A	B
N	29	10
Seroconversion (%)	100	100
GMT (IU/L)	226	237
Range	40–610	134–506

Table 14

Compositions of vaccines tested

Vaccine	Influenza vaccine protein antigen (μg/mL)	B30-MDP (μg/mL)	Number of vaccinees
E (5–50)	70	100	10
A (10–100)	140	200	10
B (KD-2)	180	—	10
C (5–150)	70	300	9
D (10–200)	140	400	10
F (10–200)	140	400	10
G (10–200A)	230	670	10
H (10–300A)	230	1000	8

of Health, Tokyo, were used to prepare the vaccines. 77 volunteers were divided into eight groups, and each individual was injected subcutaneously once, or twice at a 4 week interval, in their upper arm with one of seven different MDP-virosome vaccine preparations or with HA vaccine as control.

Local reactions caused by MDP vaccine were more frequent than those caused by HA vaccine, especially reddening and swelling which were seen more frequently 1–4 days after vaccination. Reddening tended to increase in size depending on the MDP concentration. With regard to pain and the other reactions, no relation with MDP concentration was observed. More MDP-vaccinees complained of pain at the site of injection or of reddening, but no case was severe and the local reactions disappeared in about 5 days after injection.

Hemagglutination inhibition (HI) tests were performed with sera taken before and 4 weeks after vaccination. As shown in Table 15, MDP vaccine stimulated production of a larger amount of HI antibody than did HA vaccine. Particularly in group G, given only a single injection, the average antibody titre 4 weeks after injection increased 8.7-fold with strain A/Yamagata/120/86, and 15.6-fold with strain A/Fukuoka/C29/85. It was considered that sufficient antibody to protect against infection was produced. As for strain B/Nagasaki/1/89, antibody production by MDP vaccine and HA vaccine was poor.

In a further clinical investigation the safety and immunogenicity of a trivalent virosomal influenza vaccine was compared with the effects of a commercial whole-

Table 15

Changes in average H1 titre (2″) of each vaccine group

Vaccine	A/Yamagata/120/86 (H1N1)			A/Fukuoka/C29/85 (H3N2)			B/Nagasaki/1/87		
	Before vaccination	After vaccination	Rate of increase	Before vaccination	After vaccination	Rate of increase	Before vaccination	After vaccination	Rate of increase
E	7.5	8.6	1.1	6.7	7.6	0.9	5.5	6.2	0.7
A	7.3	9.1	1.8	7.0	8.3	1.3	6.1	6.8	0.7
B	7.3	9.0	1.7	6.6	8.5	1.9	5.5	6.3	0.8
C	7.1	9.3	2.2	6.5	8.2	1.7	6.1	7.5	1.4
D	7.5	9.1	1.6	7.9	9.2	1.3	6.6	7.1	0.5
F	7.5	9.7	2.2	5.9	8.8	2.9	5.9	7.3	1.4
G	7.3	10.3	3.0	6.0	10.0	4.0	6.6	8.0	1.4
H	6.8	9.3	2.5	7.3	9.8	2.5	6.6	8.1	1.5

virus vaccine and of a subunit vaccine in elderly people.[30] Elderly and ill people are at increased risk of complications associated with influenza.[114,115] Influenza outbreaks and epidemics cause an increase in hospital admissions and deaths among the elderly.[116] This has led many public health authorities to recommend routine immunization of these high-risk individuals.

The formulation of the virosomal vaccine included the purified hemagglutinins (HA) of the influenza virus strains H1N1(A) Singapore 6/86, H3N2A Beijing 353/89, and B/Yamagata 16/88. HA extracted from all 3 strains of influenza virus could be incorporated into the PC liposome bilayer yielding unilamellar virosomes with an average diameter of 150 nm. The final vaccine, which contained 15 μg of HA per viral strain was sterile, nontoxic, and nonpyrogenic. The intact virion vaccine was Inflexal Berna (Swiss Serum and Vaccine Institute, Berne, Switzerland); the subunit vaccine Influvac-92 (Methocompany, Weesp, Netherlands). These vaccines, produced for the 1992/1993 season, contained the three influenza virus strains described above. 126 elderly people (⩾63 years of age) were recruited (Table 16). Mean age was 78, and 70% of the participants were female. 12% had been vaccinated against influenza during the previous year.

There were no reports of serious local or serious systemic reactions. The immune responses are shown in Table 17. All 3 vaccines caused a significant ($p < 0.01$) rise in geometric mean antibody titre (GMT) over baseline against all 3 viral strains. The highest GMTs were seen in the group which received virosomal vaccine. Interestingly, the baseline GMT was approximately 10-fold higher for the H3N2 strain as compared with the other 2 strains. The number of people immunized mounting a ⩾4-fold rise in titre was significantly ($p = 0.039–0.0016$) higher in the virosome group against all 3 vaccine components than in the 2 other groups. Overall, the subunit vaccine was the least immunogenic.

The immunogenicity of the 3 vaccines was compared by determining the proportion of those immunized who had protective levels of antibody following immunization—the best indicator of vaccine efficacy (Table 18). Relatively few people (<10%) had protective baseline antibody titres (⩾40) against either H1N1 or B/Yamagata strains, although >50% had against the H3N2 strain. Following immunization with the virosome formulation, the number of individuals who achieved protective levels of antibody against the H1N1 and B/Yamagata strains was significantly ($p = 0.022–0.007$ and $0.035–0.0017$, respectively) higher than in the groups immunized with the other 2 vaccines. No differences were seen for the

Table 16

Characteristics of study groups

Vaccine	No	Male/Female	Mean age (Range)	No. vaccinated against influenza in past 24 months (%)
Whole virus	32	9:23	78 (64–88)	1 (3)
Virosome	63	22:41	78 (63–102)	10 (16)
Subunit	31	7:41	79 (63–102)	4 (13)

Table 17

Immune response following vaccination with various vaccine preparations

Vaccine	Geometric mean anti-HA titre (range)						No ≥4-fold rise/total (%)		
	H1N1 Singapore		H3N2 Beijing		B/Yamagata		H1N1 Singapore	H3N2 Beijing	B/Yamagata
	Pre	Post	Pre	Post	Pre	Post			
Whole virus	2.9	29.1*	25.2	130*	3.4	16.3*	19/32 (59) ⇒ $p = 0.009$	19/32 (59) ⇒ $p = 0.039$	14/32 (44) ⇒ $p = 0.031$
	(0–80)	(0–320)	(0–160)	(10–640)	(0–20)	(0–640)			
Virosome	2.4	44.2*	20.6	142*	2.4	17.5*	52/63 (83) ⇒ $p = 0.0019$	50/63 (79) ⇒ $p = 0.0033$	42/63 (67) ⇒ $p = 0.0016$
	(0–80)	(0–640)	(0–320)	(20–2560)	(0–40)	(0–320)			
Subunit	1.8	14*	26.4	121*	2.9	11.4*	14/31 (45)	16/32 (52)	10/31 (32)
	(0–40)	(0–320)	(0–160)	(10–5120)	(0–40)	(0–160)			

*Denotes values which are significant ($p < 0.01$) as compared with baseline values.

Table 18

Protective antibody titres in study subjects before and after vaccination. Protective anti-HA antibody titre: ≥40

Vaccine	H1N1 Singapore		H3N2 Beijing		B/Yamagata	
	Pre	Post	Pre	Post	Pre	Post
Whole virus *n* (%)	3/32 (9)	17/32 (53) ⇒ $p = 0.022$	18/32 (56)	27/32 (84)	0/32 (0) ⇒ $p = 0.035$	12/32 (38)
Virosome *n* (%)	5/63 (8)	48/63 (76) ⇒ $p = 0.007$	32/63 (51)	59/63 (94)	4/63 (6) ⇒ $p = 0.0017$	38/63 (60)
Subunit *n* (%)	2/31 (6)	15/31 (48)	16/31 (52)	29/31 (94)	1/31 (3)	8/31 (26)

Table 19

Production of protective anti-HA antibody levels following vaccination in subjects with non-protective baseline antibody titres

Vaccine	No. of subjects with protective levels of antibody after immunization (%)*		
	H1N1 Singapore	H3N2 Beijing	B/Yamagata
Whole virus n (%)	12/29 (41) $\Rightarrow p = 0.0049$	9/14 (64)	12/32 (39) $\Rightarrow p = 0{,}08$
Virosome n (%)	42/58 (72) $\Rightarrow p = 0.0049$	27/31 (87)	31/57 (54) $\Rightarrow p = 0.006$
Subunit n (%)	12/29 (41)	12/14 (86)	7/30 (23)

*Protective anti-HA antibody titre: $\geqslant 40$.

H3N2 strain, most likely due to the high percentage of individuals who presented protective baseline antibody levels.

The 3 vaccines were also evaluated for their ability to raise anti-HA antibody titres to protective levels in susceptible individuals (baseline titre of <40) (Table 19). Following immunization with the virosomal vaccine, 72% of susceptible individuals attained protective titres against the H1N1 strain. This was significantly ($p = 0.0049$) higher than the 41% observed for either the whole virus or subunit vaccine. No differences among the vaccines were observed for the H3N2 strain. However, the virosomal formulation induced a significantly ($p = 0.006$) higher response rate to the B/Yamagata component when compared to the subunit, but not to the whole virus vaccine ($p = 0.08$). Overall, the IRIV vaccine clearly showed superior immunogenicity specially in the elderly who lacked protective serum antibody levels.

Immunity to influenza can be conferred by both the humoral and cellular arms of the immune system.[117] Antigens delivered as part of virosomes are more effective at stimulating a cell-mediated immune response.[118] Preliminary findings indicate that a virosome vaccine can elicit a cell-mediated response directed to the influenza HA (unpublished observations), which may result in a higher level of protection against disease.

The above findings demonstrate the safety and superior immunogenicity of a trivalent IRIV influenza vaccine in elderly home residents. Additional studies are planned to investigate this vaccine in other high-risk groups such as debilitated patients and infants.[119,120] The presently described vaccine may play an important part in preventing morbidity and mortality associated with influenza.

IX. Clinical evaluation of an IRIV based vaccine combining 7 viral and bacterial antigens

As demonstrated in Sections IV and V, IRIVs have turned out as efficient adjuvant in inactivated hepatitis A as well as in inactivated influenza A and B virus vaccines with a very low reactogenic potential. Due to the size and structure of IRIVs it is theoretically possible to attach covalently an additional number of different immunogenic proteins and to develop multivalent vaccines. A clinical study on the efficacy and reactogenicity of different antigen combinations including hepatitis A, hepatitis B, tetanus and diphtheria toxoid as well as 3 strains of influenza type A and B was published by Mengiardi et al.[98] The reactogenicity and immunogenicity of the combined vaccines based on IRIV were compared with that of the corresponding single vaccines applied simultaneously at different sites. For this purpose all antigens were bound to the IRIV surface by PE whose free amino group allows a covalent coupling. First *N*-succinimidyl 4-(*p*-maleimidophenyl) butyrate was coupled to PE. The generated *n*-4-(*p*-maleimidophenyl) butyrate PE could bind the antigens via a sulfid bridge of cystein. In contrast to the hepatitis A-IRIV single vaccine all antigens had to be activated for the combination IRIV vaccine by a reduction method previously described:[52] free cystein groups of the antigens were obtained by treatment with 40 mM 1,4-dithio-DL-threitol. IRIV vesicles were separated from unconjugated antigens using ultrafiltration or sucrose density ultracentrifugation. About 30–50% of the reduced antigens were associated with the vesicles.

The formula of the combined IRIV vaccine was as follows: Hepatitis A antigen 0.5 μg, HBsAg 10 μg, diptheria toxoid 1 Lf, tetanus toxoid 10 Lf, influenza antigen A/Singapore (6/86)-like (H1N1) 15 μg, A/Beijing (353/89)-like (H3N2) 15 μg and B/Yamagata (16/88)-like 15 μg. The vaccines used for separate injection had a formula according to the combined vaccine with the exception of the HBsAg vaccine which yielded an aluminium adjuvant.

The results of the hepatitis A antibody determinations in the serum of adult volunteers who received the combined IRIV vaccine and of the probands who received the monovaccines are presented in Table 20. There was a clear inhibition of the development of hepatitis A antibodies in the combined vaccine group on day 28 since the seroconversion rate was only 65% against 100% in the monovaccine group, the GMT being about 25% of the value found in the monovaccine group ($p = 0.03$).

The HBs antibody titres did not reveal any significant primary immunization effect. This indicates that a single injection of IRIV-HBsAg must be followed by successive vaccination(s) in order to obtain protective HBs antibody titres as is the case when conventional HBsAg vaccines are injected.

The results of the diphtheria toxoid antibody titre determinations after vaccination with the combined IRIV vaccine and with the monovaccines are shown in Table 21.

On day 0 all but two probands had titres above the protective level of 0.1 IU/mL and an antibody titre rise was found which corresponded to a booster response. On day 28, the antibody titre in the combined vaccine group was twice as high as in the monovaccine group, the difference being statistically significant ($p = 0.024$). However, on days 90 and 180 the difference was no more substantial and statistically not significant.

The results of the tetanus antibody titre determinations in the combined IRIV vaccine group and in the monovaccine group are compared in Table 22. All

Table 20

Hepatitis A antibody titres on days 0/28/90/180. Group A ($N = 23$) received the combined vaccine on an IRIV base, group B ($N = 23$) received each vaccine as a monovaccine (HAV = IRIV based) in separate injections

	Day 0		Day 28		Day 90		Day 180	
	A. Comb.	B. Mono.	A. Comb.	B. Mono.	A. Comb.	B. Mono.	A. Comb.	B. Mono.
Seroconversion rate %			65	100	83	100	91	100
GMT (IU/L)	6.6	6.0	46.2	166.5	57.3	74.8	71.7	53.7
*p**	0.31		0.03		0.27		0.80	

*Mann-Whitney-Wilcoxon test.

Table 21

Diphtheria toxoid antibody titres on days 0/28/90/180. Group A ($N = 23$) received the IRIV based combined vaccine, group B ($N = 23$) received each vaccine as a monovaccine (diphtheria toxoid = alum based) in separate injections.

	Day 0		Day 28		Day 90		Day 180	
	A. Comb.	B. Mono.	A. Comb.	B. Mono.	A. Comb.	B. Mono.	A. Comb.	B. Mono.
GMT (IU/mL)	0.48	0.43	1.64	0.76	4.57	3.78	3.78	3.5
p*	0.78		0.024		0.054		0.16	

*Mann-Whitney-Wilcoxon test.

Table 22

Tetanus toxoid antibody titres on days 0/28/90/180. Group A ($N = 23$) received the IRIV based combined vaccine, group B ($N = 23$) received each vaccine as a monovaccine (tetanus toxoid = alum based) in separate injections

	Day 0		Day 28		Day 90		Day 180	
	A. Comb.	B. Mono.	A. Comb.	B. Mono.	A. Comb.	B. Mono.	A. Comb.	B. Mono.
GMT (IU/mL)	2.6	4.3	16.1	13.1	30.4	22.7	23.4	20.3
p*	0.06		0.45		0.08		0.72	

*Mann-Whitney-Wilcoxon test.

Table 23

Influenza antibody titres on days 0 and 28. Group A ($N = 23$) received the IRIV based combined vaccine while group B ($N = 23$) received single injections with monovaccines

	Day 0		Day 28	
	A. Comb.	B. Mono.	A. Comb.	B. Mono.
Influenza A Singapore H1N1				
Seroconversion rate %			86	95
GMT	12	10	300	226
p*	0.87		0.27	
Influenza A Beijing H3N2				
Seroconversion rate %			90	86
GMT	34	31	525	411
p*	0.71		0.40	
Influenza B Yamagata				
Seroconversion rate %			90	82
GMT	10	8	94	165
p*	0.58		0.11	

*Mann-Whitney-Wilcoxon test.

subjects had protective antibody titres above 0.1 IU/mL already before vaccination. The GMT as measured before vaccination was almost twice as high in the monovaccine group as in the combined vaccine group ($p = 0.06$). On days 28, 90 and 180 the tetanus antibody titres of the combined vaccine group were slightly but statistically not significantly higher than in the monovaccine group.

In view of the annual strain shifts and of the long-lasting antibody mediated protection against influenza the observation of the antibody titre changes 28 days after vaccination was sufficient for comparing the two vaccines used in the study. Influenza seroconversion was defined according to the Working Group for Biotechnology and Pharmacy of the EU as the increase of a titre of <10 to $\geqslant 40$ or of a titre of >10 to a $\geqslant 4$-fold value. As may be seen in Table 23, the seroconversion rates of 86%, 90% and 90% obtained with the combined vaccine and of 95%, 86% and 82% with the monovalent vaccines for the three influenza strains appear to be comparable and very satisfactory in both groups. No relevant or statistically significant strain-specific antibody response differences were found between the GMT of the two groups of vaccinees.

In order to investigate a possible *epitope suppression*, 15 volunteers were vaccinated with a mixture of HAV-IRIV's, diphtheria toxoid IRIV's and tetanus toxoid IRIV's. Hepatitis A antibodies were determined on days 0/28/90/180. The results of these determinations were compared with the corresponding results obtained when hepatitis A and B, diphtheria and tetanus toxoid, Hbs and influenza A + B IRIV bound monoantigens had been injected separately (on the base of an early study it was taken for given that HBsAg is not interfering after day 14 while influenza virus antigen is an integral part of IRIV).

The results of the hepatitis A antibody determination showed an inhibition of antibody production on day 28 but no more later. This inhibition did not exceed that observed after the injection of the combined vaccine. This clearly demonstrated that inhibition of antibody formation as observed before was *not due to epitope suppression*. As expected, the development of diphtheria and tetanus toxoid antibodies remained unaffected by the new formula.

In order to investigate a possible *antigenic suppression*, 15 volunteers were vaccinated with hepatitis A virus antigen (RG–SB) combined with a diphtheria and tetanus toxoid amount of 50% of that previously used.

The induction of hepatitis A antibodies by the IRIV formula with reduced diphtheria and tetanus toxoid amounts was highly improved and equaled the levels obtained with monovalent vaccine injections. This observation indicates that the impairment of the hepatitis A antibody production as observed before was due to an *antigenic suppression by the higher amounts of toxoid antigens*. The diphtheria and the tetanus toxoid antibody production was not impaired when the amount of toxoids reduced by 50% was injected.

Thus, by coupling of up to 7 vaccinal antigens on the same IRIV satisfactory results could be obtained in adult volunteers. Nevertheless, it was found that the immunogenicity of the HAV antigen was significantly reduced. Interestingly, this was also the case when a mixture of IRIV containing the HAV antigen and of IRIV containing the diphtheria and tetanus toxoid antigens were injected. The suppression of the production of hepatitis A antibody was shown to be an *antigenic suppression* preventable by halving the amounts of diphtheria and tetanus toxoid antigens without impairing the immune response to the latter.

The IRIV based vaccine combinations were found to be well tolerated. This is in good agreement with observations published by Glück[121] who found a substantial reduction of local and general symptoms induced by a multivalent IRIV vaccine (diphtheria and tetanus toxoid, hepatitis A, hepatitis B) as compared with the side reactions due to a vaccine combining the analogous antigens of commercial vaccines.

X. Summary

Liposomes are vesicular structures limited by a bilayer membrane generally composed of phospholipids and cholesterol. It was found that viral glycoproteins maintain their capacity of receptor binding and receptor induced endocytosis when reconstituted into protein lipid vesicles (virosomes). Virosomes proved to be highly effective immunogens in mice, rabbits and monkeys. This included the ability to stimulate strong CD8+ cytotoxic T-cell responses to lipid bilayer integrated glycoproteins or lipid linked peptides, as well as to encapsulated peptides, proteins and formalin-fixed whole viruses.

The immunopotentiating reconstituted influenza virosome (IRIV) vaccines combine several components known to promote immunostimulatory processes. IRIV's are prepared by detergent removed influenza surface glycoproteins and a mixture of natural and synthetic phospholipids. The influenza virus glycoproteins incorporated into the virosomal bilayer membrane are thought to facilitate the binding to the receptors of the antigen presenting cells, the endocytosis and the fusion with the endosomal membrane thus mediating the rapid release of the transported antigen into the membranes of the target Th-cells. Other immunostimulatory effects of the hemagglutinins and neuraminidase incorporated into the IRIV membrane are also discussed.

Preclinical experiments with a virosomal hepatitis B vaccine containing a B-cell epitope from the S region as well as a Th-cell epitope from the pre-S1 region showed that the pre-S1 Th-cell epitope was only efficient when the two peptide antigens were placed on the same vesicle. Other preclinical experiments performed in mice with virosomal influenza vaccines revealed immunogenic superiority when compared with the effect of an aqueous vaccine and, when applied by intranasal administration, the capacity to induce an excellent systemic and local immunity which provided protection from virus challenge. Extended investigations performed in mice with hepatitis A IRIV vaccine demonstrated that the incorporation of influenza hemagglutinins and neuraminidase as well as kephalin is indispensable for obtaining an optimal immune response. Further experiments in mice showed that priming with H1N1 influenza antigen had an enhancing effect on the immune response to the vaccination of the animals with an IRIV-SPf(66)n malaria antigen. This confirmed the importance of incorporating influenza virus antigen into the IRIV vaccines.

A test series with adult human volunteers failed to demonstrate any induction of antibodies against the phospholipids contained in the IRIV's.

Another vaccination trial showed that an aqueous solution of hepatitis A IRIV vaccine stored for more than 2 years had not lost any of its immunogenic activity.

While up to now most investigations on virosomal antigen preparations remained restricted to in vitro studies or animal experiments, a series of IRIV based vaccines have yet undergone a large variety of clinical trials.

First, it could clearly be established that an IRIV vaccine yielding formalin-inactivated hepatitis A virions bound to the bilayer surface was not only more immunogenic than a fluid vaccine with the same antigen content but that after the injection of an unique vaccine dose the hepatitis A antibodies appeared much earlier (i.e., on day 14 versus day 28) at a protective level and also persisted at a tenfold higher level after one year when compared with the antibody levels induced by a doubled hepatitis A antigen dose adjuvanted by aluminium hydroxide. Thus, large clinical trials showed that with the IRIV vaccine seroconversion rates of 91–100% can be obtained already on day 14 and of 98–100% on day 28, and protective serum antibody titres were found in 94–100% one year after a single vaccination. A booster injection performed after one year induced a 20- to 30-fold antibody titre increase which evokes a memory T-cell mediated response. Local adverse reactions, i.e., pain, induration or swelling, were highly reduced when the IRIV vaccine was used instead of aluminium vaccine. It was also demonstrated that the simultaneous administration of immunoglobulin would not substantially impair the vaccine induced antibody production. In a further study it was demonstrated that the hepatitis A antibody titres induced by the IRIV vaccine could be determined as easily by neutralization test (NT) as by the most usual enzyme immuno-assay (EIA) suggesting the antibodies to confer protection and to have a high affinity to hepatitis A virus.

A trivalent IRIV influenza vaccine yielding the appropriate H1N1, H3N2 and B strain assayed in a community of home inmates aged 63 to 102 years (average 78 years) partially showed a significantly higher immunogenic potential specially in the elderly who before vaccination lacked protective antibody levels when the seroconversion rates induced by a whole virus or a subunit vaccine were compared.

In view of the multitude of vaccinations required already in early life and later on, investigations with an IRIV vaccine combining 5 viral and 2 toxoid antigens were undertaken with student volunteers. The combined IRIV vaccine contained formalin-inactivated hepatitis A virions (500 RIA units), hepatitis B surface antigen (HBsAg; 10 μg), diphtheria toxoid (1 Lf), tetanus toxoid (10 Lf) and 3 influenza virus antigens (H1N1, H3N2, B; 15 μg each). The antibody titres induced were compared with those elicited by the corresponding monovaccines injected simultaneously at different sites. It resulted that the hepatitis A antibody production was significantly delayed in the combined vaccine group. Hepatitis B antibodies remained absent in either group thus confirming the necessity of repeated doses for providing protection against the disease. Toxoid as well as influenza antibody levels all showed to be unaffected by the IRIV antigen combination. A complementary vaccination series with a vaccine containing a *mixture* of hepatitis A virion IRIV's, diphtheria toxoid IRIV's and tetanus toxoid IRIV's failed to indicate any impairment of the immune response. By this, *an epitope suppression* seemed to be excluded. However, when volunteers were immunized with an IRIV vaccine combining the hepatitis A virions with diphtheria and tetanus toxoid amounts reduced by 50%, the immune response to all 3 antigens proved to be rapid and complete. This observation indicated that the reduction of the hepatitis A antibody formation observed before was due to an *antigenic suppression* by the higher yield of toxoid antigens. Thus, it is obviously possible to develop fully effective IRIV based combined vaccines which could play a major role in future vaccination activities.

References

1. Zanetti M, Sercarz E, Salk J. The immunology of a new generation of vaccines. Immunol Today 1987;8:18–23.
2. Allison AC, Gregoriadis G. Liposomes as immunological adjuvants. Nature 1974;252:252–255.
3. Morein B, Simons K. Subunit vaccines against enveloped viruses: virosomes, micelles and other protein complexes. Vaccine 1985;3:83–93.
4. Boudreault A, Thibodeau L. Mouse response to influenza immunosomes. Vaccine 1985;3:231–234.
5. El Guink N, Kris R, Goodman-Snitkoff GW, Small PA, Mannino RJ. Intranasal immunization with proteoliposomes protects against influenza. Vaccine 1989;7:147–151.

6. Neurath AR, Kent SBH, Strick N. Antibodies to hepatitis B surface antigen (HBsAG) elicited by immunization with a synthetic peptide covalently linked to liposomes. J Gen Virol 1984;65:1009–1014.
7. Lifshitz R, Gitler C, Moses E. Liposomes as immunological adjuvants in eliciting antibodies specific to the synthetic polypeptide poly(LTyr. LGlu)—poly(DLAla)-poly(LLys) with high frequency of site associated idiotypic determinants. Eur J Immunol 1981;11:398–404.
8. Bangham AD, Standish MM, Watkins JC. Diffusion of univalent ions across the lamellae of swollen phospholipids. J Mol Biol 1965;13:238–252.
9. Bangham AD. Membrane models with phospholipids. Prog Biophys Mol Biol 1968;18:29–37.
10. Bangham AD, Hill MW, Miller NGA. Preparation and use of liposomes as models of biological membranes. In: ED Korn, ed. Methods in membrane biology, New York: Plenum Press, 1974;1–23.
11. Brotherus JR, Griffith OH, Brotherus MO. Lipid-protein multiple binding equilibria. Biochemistry 1981;20:5261–5267.
12. Duzzqunes N, Papahadiopoulos D. Ionotropic effects on phospholipid membranes: Calcium/magnesium specificity in binding, fluidity and fusion. In: RC Aloia, ed. Membrane fluidity in biology New York: Academic Press, 1983;187–197.
13. Lüscher-Mattli M, Glück R. Dextran sulfate inhibits the fusion of influenza virus with model membranes, and suppresses influenza virus replication in vivo. Antivir Res 1990;14:39–50.
14. Tsurudome M, Glück R, Graf R, Falchetto R, Schaller U, Brunner J. Lipid interactions of the hemagglutinin HA2NH2-terminal segment during influenza virus-induced membrane fusion. J Biol Chemistry 1992;267/28:20225–20232.
15. Lüscher-Mattli M, Glück R, Kempf C, Zanoni-Grassi M. A comparative study of the effect of dextran sulfate on the fusion and the in vitro replication of influenza A and B, Semliki Forest, vesicular stomatitis, rabies, Sendai and mumps virus. Arch Virol 1993;130:317–326.
16. Honegger JL, Isakron PC, Kinsky SC. Murine immunogenicity of *N*-substituted phosphatidylethanolamine derivatives in liposomes: Response to the hapten phosphocholine. J Immunol 1980;124:669–675.
17. Hafeman DG, Lewis TJ, McConnel HM. Triggering of the macrophage and neutrophil respiratory burst by antibody bound to a spin-label phospholipid hapten in model lipid bilayer membranes. Biochemistry 1980;19:5387–5394.
18. Müller-Eberhard HJ. Molecular organization and function of the complement system. Ann Rev Biochem 1988;57:321–347.
19. Bangham AD. A correlation between surface charge and coagulant action of phospholipids. Nature 1961;192:1197–1198.
20. Papahadjopoulos D, Hougie C, Hanacham DJ. Influence of surface charge of phospholipids on their clot-promoting activity. Proc Soc Exp Biol Res 1962;111:412–416.
21. Papahadjopoulos D. Cholesterol and cell membrane functions: A hypothesis concerning the etiology of arteriosclerosis. J Theoret Biol 1974;43:329–337.
22. Small DM, Shipley GG. Physical-chemical basis of the lipid deposition in arteriosclerosis. Science 1974;185:222–229.
23. Gregoriadis G, Leathwood PD. Enzyme entrapment in liposomes. FEBS Lett 1971;14:95–99.
24. Dean T. Liposomes in allergy and immunology. Clin Exp Allergy 1993;23:557–563.
25. Treat J, Greenspan A, Trost D. Antitumour activity of liposome-encapsulated doxorubicin in advanced breast cancer: Phase II study. J Nat Cancer Inst 1990;82:1706–1710.
26. Wälti E, Glück R. Virosomes: A new specific drug delivery system for cancer therapy. J Molec Recogn 1993;4/6(1):21.
27. Lopez-Berestein G. Treatment of systemic fungal infections with liposomal amphotericin B. In: G Lopez-Berestein, IJ Fidler, eds. Liposomes in the Therapy of Infection Diseases and Cancer, New York: Alan R Liss, 1989;317–327.
28. Davidson RN, Croft SL, Scott A. Liposomal amphotericin B in drug resistant visceral leishmaniasis. The Lancet 1991;337:1061–1062.
29. Glück R, Mischler R, Brantschen S, Just M, Althaus B, Cryz SJ Jr. Immunopotentiating reconstituted influenza virosome (IRIV) vaccine delivery system for immunization against hepatitis A. J Clin Invest 1992;90:2491–2495.
30. Glück R, Mischler R, Finkel B, Que JU, Scarpa B, Cryz SJ. Immunogenicity of new virosome influenza vaccine in the elderly people. Lancet 1994;344:160–163.
31. Lasic D. Doxorubicin in sterically stabilized liposomes. Nature 1996;380:561–562.
32 Tom RH. An overview: Liposomes and immunobiology—macrophages, liposomes, and tailored

immunity. In: Tom BH, Six HR, eds. Liposomes and Immunobiology, Amsterdam:Elsevier North-Holland, 1980;3–22.
33. Alving CR. Liposomes as carriers for vaccines. In: Ostro JM, ed. Liposomes from Biophysics to Therapeutics, New York: Dekker, 1987;195–218.
34. Gregoriadis G. Immunological adjuvants: A role for liposomes. Immunol Today 1990;11:89–97.
35. Gregoriadis G. Overview of liposomes. J Antimicrob Chemother 1991;28(Suppl B):39–48.
36. Uemura K-I, Claflin JL, Davie JM, Kinsky SC. Immune response to liposomal model membranes: Restricted IgM and IgG anti-dinitrophenyl antibodies produced in guineapigs. J Immunol 1975;114:958–961.
37. Sanchez Y, Ionescu-Matiu I, Dreesman GR, Kramp W, Six HR, Hollinger FB, Melnick JL. Humoral and cellular immunity to hepatitis B virus-derived antigens: Comparative activity of Freund complete adjuvant, alum and liposomes. Infect Immun 1980;30:728–733.
38. Tan L, Loyter A, Gregoriadis G. Incorporation of reconstituted influenza virus envelopes into liposomes: Studies of the immune response in mice. Biochem Soc Trans 1989;17:129–130.
39. Burkhanov SA, Mazhul LA, Torchilin VP, Ageyeva ON, Saatov TS, Andzhanaridze OG. Protective action of influenza-virus surface antigens incorporated in liposomes in various methods of immunization. Vop Virusol 1988;33:151–153.
40. Kramp WJ, Six HR, Kasel JA. Postimmunization clearance of liposome entrapped adenovirus type 5 hexon. Proc Soc Exp Biol Med 1982;169:135–139.
41. McWilliam AS, Stewart GA. Production of multilamellar, small unilamellar and reverse-phase liposomes containing house dust mite allergens. J Immunol Methods 1989;121:53–69.
42. Bruyere T, Wachsmann D, Klein J-P, Scholler M, Frank RM. Local response in rat to liposome-associated Streptococcus mutans polysaccharide-protein conjugate. Vaccine 1987;5:39–42.
43. Thibodeau L, Chagnon M, Flamand L, Oth D, Lachapelle L, Tremblay C, Montagnier L. Role of liposomes in the presentation of HIV envelope glycoprotein and the immune response in mice. CR Acad Sci 1989;309:741–747.
44. Miller MD, Gould-Fogerite S, Shen L, Woods RM, Koenig S, Mannino RJ, Letvin NL. Vaccination of rhesus monkeys with synthetic peptide in a fusogenic proteoliposome elicits simian immunodeficiency virus-specific CD8+ cytotoxic T lymphocytes. J Exp Med 1992;176:1739–1744.
45. Grove CF, Jensen M. Liposome-related U.S. patents. In: Gregoriadis G, ed. Liposome Technology, Vol II, 2nd edn. Boca Raton: CRC Press, 1993;487–500.
46. Szoka F, Papahadjopoulos D. Comparative properties and methods of preparation of liposomes. Annu Rev Biophys Bioeng 1980;9:467–508.
47. New RRC. Preparation of liposomes. In: New RRC, ed. Liposomes: A Practical Approach, Oxford University Press, London, 1990;33–104.
48. Verma JN, Wassef NM, Wirtz RA, Atkinson CT, Aikawa M, Loomis LD, Alving CR. Phagocytosis of liposomes by macrophages: Intracellular fate of liposomal malaria antigen. Biochim Biophys Acta 1991;1066:229–238.
49. Raz A, Bucana C, Fogler WE, Poste G, Fidler IJ. Biochemical, morphological, and ultrastructural studies on the uptake of liposomes by murine macrophages. Can J Res 1981;41:487–494.
50. Schroit AJ, Fidler IJ. Effects of liposome structure and lipis composition on the activation of the tumoricidal properties of macrophages by liposomes containing muramyl dipeptide. Can J Res 1982;42:161–167.
51. Therien H-M, Shahum E. Importance of physical association between antigen and liposomes in liposomes adjuvanticity. Immunol Lett 1989;22:253–258.
52. Martin FJ, Papahadjopoulos D. Irreversible coupling of immunoglobulin fragments to preformed vesicles. J Biol Chem 1982;257:286–288.
53. Shen D-F, Huang A, Huang L. An improved method for covalent attachment of antibody to liposomes. Biochim Biophys Acta 1982;689:31–37.
54. Huang A, Ysao YS Kennel SJ, Huang L. Characterization of antibody covalently coupled to liposomes. Biochim Biophys Acta 1982;716:140–150.
55. Leserman LD, Machy P, Barbet J. Covalent coupling of monoclonal antibodies and protein A to liposomes: Specific interaction with cells in vitro and in vivo. In: Gregoriadis G, ed. Liposome Technology, Vol III, Boca Raton: CRC Press, 1983;29–40.
56. Derksen JTP, Scherphof GL. An improved method for the covalent coupling of proteins to liposomes. Biochim Biophys Acta 1985;814:151–155.
57. Chua M-M, Fan S-T, Karush F. Attachment of immunoglobulin to liposomal membrane via protein carbohydrate. Biochim Biophys Acta 1984;800:291–300.
58. Bogdanov AA, Klibanov AL, Torchilin VP. Protein immobilization on the surface of liposomes

via carbodiimide activation in the presence of *N*-hydroxysulfosuccinimide. Fed Eur Biochem Soc 1988;231(2):381–384.
59. Torchilin VP, Goldmacher VS, Smirnov VN. Covalent studies on covalent and noncovalent immobilization of protein molecules on the surface of liposomes. Biochem Biophys Res Commun 1978;85:983–990.
60. Van Rooijen N, van Nieuwmegen R. Liposomes in immunology: Evidence that their adjuvant effect results from surface exposition of the antigen. Cell Immunol 1980;149:402–407.
61. Gregoriadis G, Davis D, Davis A. Liposomes as immunological adjuvants: Antigen incorporation studies. Vaccine 1987;5:145–151.
62. Shahum E, Therien H-M. Comparative effect on humoral response of four different proteins covalently linked on BSA-containing liposomes. Cell Immunol 1989;123:36–43.
63. DalMonte PR, Szoka FC. Antigen presentation by B cells and macrophages of cytochrome e and its antigenic fragment when conjugated to the surface of liposomes. Vaccine 1989;7:401–408.
64. Weissig V, Lasch J, Gregoriadis G. Covalent coupling of sugars to liposomes. Biochim Biophys Acta 1989;1003:54–57.
65. Garçon N, Gregoriadis G, Taylor M, Summerfield J. Mannose-mediated targeted immunoadjuvant action of liposomes. Immunology 1988;64:743–745.
66. Balkovic ES, Florack JA, Six HR. Immunoglobulin subclass antibody responses of mice to influenza virus antigens given in different forms. Antiviral Res 1987;8:151–160.
67. Ben Ahmeida ETS, Jennings R, Tan L, Gregoriadis G, Potter CW. The subclass IgG response of mice to influenza surface proteins formulated into liposomes. Antiviral Res 1993;21:217–231.
68. Muller GM, Shapira M, Arnon R. Anti-influenza response achieved by immunization with a synthetic conjugate. Proc Natl Acad Sci USA 1982;79:569–573.
69. Lerner RA, Green N, Alexander H, Liu F-T, Sutcliffe JG, Shinnick TM. Chemically synthesized peptides predicted from the nucleotide sequence of the hepatitis B virus genome elicit antibodies reactive with the native envelope protein of Dane particles. Proc Natl Acad Sci USA 1981;78:3403–4307.
70. Bhatnagar PK, Papas E, Blum HE, Milich DR, Nitecki D, Karels MJ, Vyas GN. Immune response to synthetic peptide analogues of hepatitis B surface antigen specific for the A determinant. Proc Natl Acad Sci USA 1982;79;4400–4404.
71. Bittle JL, Houghton RA, Alexander H, Shinnik TM, Sutcliffe JG, Lerner RA. et al. Protection against foot and mouth disease by immunization with a chemically synthesized peptide predicted from the viral nucleotide sequence. Nature 1982;298:30–33.
72. DiMarchi R, Brooke G, Gale C, Cracknell V, Doel T, Mowat N. Protection of cattle against foot-and-mouth disease by a synthetic peptide. Science 1986;232:639–641.
73. Kennedy RC, Henkel RD, Pauletti D, Allan JS, Lee TH, Essex M, Dressman GR. Antiserum to a synthetic peptide recognizes the HTLV-III envelope glycoprotein. Science 1986;231:1556–1559.
74. Miller LH, Howard RD, Carter R, Good MD, Nussenzweig V, Nussenzweig RS. Research toward malaria vaccines. Science 1986;234:1349–1356.
75. Caryanniotis G, Barber BH. Adjuvant-free IgG responses induced with antigen coupled to antibodies against class II MHC. Nature 1987;327:59–61.
76. Francis MJ, Fry CM, Rowlands DJ, Brown F, Bittle JL, Houghten RA, Lerner RA. Immunological priming with synthetic peptides of foot-and-mouth disease virus. J Gen Virol 1985;66:2347–2354.
77. Goodman-Snitkoff G, Eisele LE, Heimer EP, Felix AM, Andersen TT, Fuerst TR, Mannino RJ. Defining minimal requirements for antibody production to peptide antigens. Vaccine 1990;8:257–262.
78. Goodman-Snitkoff G, Good MR, Berzofsky JA, Mannino RJ. Role of intrastructural/intermolecular help in immunization with peptide-phospholipid complexes. J Immunol 1991;147:410–415.
79. Mannino RJ, Gould-Fogerite S. Lipid matrix-based vaccines for mucosal and systemic immunization. In: Powell MF, Newman MJ, eds. Vaccine Design: The Subunit and Adjuvant approach. New York: Plenum Press, 1995;363.
80. Allen PM. Antigen processing at the molecular level. Immunol Today 1987;8:270–273.
81. Gould-Fogerite S, Mannino RJ. Rotary dialysis: Its application to the production of large liposomes and large proteoliposomes (protein-lipid vesicles) with high encapsulation efficiency and efficient reconstitution of membrane proteins. Anal Biochem 1985;148:15.
82. Gould-Fogerite S, Mazurkiewicz J, Lehman J, Raska K Jr, Voelkerding K, Mannino RJ. Chimerasome-mediated gene transfer in vitro and in vivo. Gene 1989;84:429–438.
83. Gould-Fogerite S, Mazurkiewicz JE, Bhisitkul D, Mannino RJ. The reconstitution of biologically

active glycoproteins into large liposomes: Use as a delivery vehicle to animal cells. In: Kim CH, Diwan J, Tedeschi H, Salerno JJ, eds. Advances in Membrane Biochemistry and Bioenergetics. New York: Plenum Publishing Corp, 1987;569.

84. Mannino RJ, Gould-Fogerite S. Liposome Mediated Gene Transfer. Biotechniques 1988;6:682–690.
85. Gould-Fogerite S, Edghill-Smith Y, Kheiri M, Wang Z, Das K, Feketeova E, Canki M, Mannino RJ. Lipid matrix-based subunit vaccines: A structure-function approach to oral and parenteral immunization. AIDS Res and Hum Retrovirus 1994;10:99.
86. Glück R. Combined Vaccines—The European Contribution. Biologicals 1994;22:347–351.
87. Gregoriadis G. Engineering liposomes for drug delivery: progress and problems. TIBTECH 1995;13:527–537.
88. Seganti L, Superti F, Orsini N, Gabrielli R, Divizia M, Panà A. Membrane lipid components interacting with hepatitis A virus. Microbiologica 1989;12:225–230.
89. Garçon NM, Six HR. Universal Vaccine Carrier. Liposomes that provide T-dependent help to weak antigen. J Immunol 1991;146/11:3697–3702.
90. Durrer P, Galli C, Hoenke S, Corti C, Glück R, Vorherr T, Brunner J. H^+-induced membrane insertion of influenza virus hemagglutinin involves the HA2 amino-terminal fusion peptide but not the coiled coil region. J Biol Chem 1996;271/23:13417–13421.
91. Matlin KS, Reggio H, Helenius A, Simmons K. Infections entry pathway of influenza virus in a canine kidney cell line. J Cell Biol 1981;91:601–613.
92. Glück R. Liposomal Presentation of Antigens for Human Vaccines. In: Powell MR, Newman MJ, eds. Vaccine Design. New York: Plenum Press and London, 1995a;325–345.
93. Rott O, Charreire J, Mignon-Godefroy K, Cash E. B Cell Superstimulatory Influenza Virus Activates Peritoneal B Cells. J Immunology 1995a;155:134–142.
94. Rott O, Charreire J, Semichon M, Bismuth G, Cash E. B Cell Superstimulatory Influenza Virus (H2-Subtype) Induces B Cell Proliferation by a PKC-Activating, Ca^{2+}-Independent Mechansim. J Immunology 1995b;154:2092–2103.
95. Parker EE, Gould KG. Influenza A virus—a model for viral antigen presentation to cytotoxic T lymphocytes. Virology 1996;7:61–73.
96. Air GM, Laver WG. The neuraminidase of influenza virus. Proteins 1989;6:341–356.
97. Muster T, Ferko B, Klima A, Purtscher M, Trkola A, Schulz P, Grassauer K, Engelhardt OG, Garcia-Sastre A, Palese P, Katinger H. Mucosal model of immunization against human immunodeficiency virus Type 1 with a chimeric influenza virus. J Virology 1995;69/11:6678–6686.
98. Mengiardi B, Berger R, Just M, Glück R. Virosomes as carriers for combined vaccines. Vaccine 1995;13:1306–1315.
99. Glück R. Towards a universal IRIV traveller vaccine. Abstract No. 259. In: Third Conference on International Travel Medicine. Paris, France 1993.
100. Just M. Experience with IRIV as carrier for hepatitis A and other antigens. Abstract No. 277. In: Third Conference on International Travel Medicine, Paris, France 1993.
101. Gregoriadis G, Wang Z, Barenholz Y, Francis MJ. Liposome-entrapped T-cell peptides help for a co-entrapped B-cell peptide to overcome genetic restriction in mice and induce immunological memory. Immunology 1993;80:535–540.
102. Alving CR. Immunologic aspects of liposomes. Presentation and processing of liposomal protein and phospholipid antigens. Biochim Biophys Acta 1992;1113:307–322
103. Loutan L, Bovier P, Althaus B, Glück R. Inactivated virosome hepatitis A vaccine. Lancet 1994;343:322–324.
104. Frösner GG, Dathe O, Althaus B, Glück R, Erhardt F, Eisenburg J, Phillip J, Heun M, Fröhlich C. Hohe Immunogenität und geringe Nebenwirkungen eines neuartigen Liposomen-Impfstoffes gegen Hepatitis A. In: Maass G, Stück B, eds. Virushepatitis A–E, Diagnose, Therapie, Prophylaxe, Kongressberichte des Deutschen Grünen Kreuzes, KilianVerlag Marburg, BRD, 1994;222–224.
105. Poovorawan Y, Theamboonlers A, Chumdermpadetsuk S, Glück R, Cryz SJ. Safety, immunogenicity, and kinetics of the immune response to a single dose of virosome-formulated hepatitis A vaccine in Thais. Vaccine 1995;13/10:891–893.
106. Holzer BR, Hatz C, Schmidt-Sissolak D, Glück R, Althaus B, Egger M. Immunogenicity and adverse effects of inactivated virosome versus alum-adsorbed hepatitis A vaccine: a randomized controlled trial. Vaccine 1996;14:982–986.
107. Simmen HP, Robustelli L, Althaus B, Glück R, Trentz O, Vogt M. Immunantwort einer Einzeldosis eines neuartigen Hepatitis-A-Impfstoffes nach Splenektomie. Dtsch. med. Wschr. 1996;121:295–298.

108. Berger R, Just M, Althaus B. Time course of hepatitis A antibody production after active, passive and active/passive immunization: the results are highly dependent of the antibody test system used. J Virol Methods 1993;43:287–298.
109. Glück R, Wälti E. Are anti-phospholipid antibodies to be expected after proteoliposomal hepatitis A vaccination? J Liposome Res 1996;6(2):415–439.
110. Cryz SJ, Que JU, Glück R. A virosome vaccine antigen delivery system does not stimulate an antiphospholipid antibody response in humans. Vaccine 1996;14:1381–1383.
111. Ambrosch F, Wiedemann G, Jonas S, Althaus B, Finkel B, Glück R, Herzog Ch. Immunogenicity and protectivity of a new liposomal hepatitis A vaccine. In: IX Triennial International Symposium on Viral Hepatitis and Liver Disease. Rome, 1996.
112. Wegmann A, Zellmeyer M, Glück R, Finkel B, Flückiger A, Berger R, Just M. Immunogenität und Stabilität eines aluminiumfreien liposomalen Hepatitis-A-Imppfstoffes (Epaxal Berna[R]). Schweiz. med. Wschr. 1994;124:2053–2056.
113. Kaji M, Kaji Y, Kaji M, Ohkuma K, Honda T, Oka T, Sakoh M, Nakamura S, Kurachi K, Sentoku M. Phase 1 clinical tests of influenza MDP-virosome vaccine (KD-5382;. Vaccine 1992;10:663–667.
114. Couch RB, Kasel JA, Glezen WP, Cate TR, Six HR, Taber CH. Influenza: its control in persons and populations. J Infect Dis 1986;153:431–440.
115. Kohn RP. Cause of death in very old people. JAMA 1982;247:2793–2797.
116. Perrotta DM, Decker M, Glezen WP. Acute respiratory disease. Hospitalization as a measure of impact of epidemic influenza. Am J Epidemiol 1985;122:468–476.
117. McMichael AF, Gotch FM, Noble GR, Beare PS. Cytotoxic T-cell immunity to influenza. N Engl J Med 1983;309:13–17.
118. Loh D, Ross AH, Hale AH, Baltimore D, Eisen HN. Synthetic phospholipid vesicles containing a purified viral antigen and cell membrane proteins stimulate the development of cytotoxic T lymphocytes. J Exp Med 1979;150:1067–1075.
119. Groothuis JR, Levin MJ, Rabalais GP, Meiklejohn G, Lauer BA. Immunization of high-risk infants younger than 18 months of age with split-product influenza vaccine. Pediatrics 1991;87:823–828.
120. Glezen WP, Paredes A, Taber LH. Influenza in children: relationship to other respiratory agents. JAMA 1980;243:1345–1349.
121. Glück R. Liposomal hepatitis A vaccine and liposomal multiantigen combination vaccines. J Liposome Res 1995b;5:467–479.

Lasic and Papahadjopoulos (eds.), Medical Applications of Liposomes

CHAPTER 2.6

Liposomes and virosomes as immunoadjuvant and antigen-carrier systems in vaccine formulations

TOOS DAEMEN, AALZEN DE HAAN, ANNEMARIE ARKEMA AND JAN WILSCHUT

Department of Physiological Chemistry, Groningen-Utrecht Institute for Drug Exploration (GUIDE), University of Groningen, Ant. Deusinglaan 1, 9713 AV Groningen, The Netherlands

Overview

I. Introduction and scope

Current vaccination procedures against infectious diseases employ either live-attenuated or killed whole pathogens, or (recombinant) subunit vaccines. Live, replicating, vaccines have the advantage that they closely mimick the actual infection and therefore induce a broad and physiologically relevant immune response, involving both a humoral immune response (antibody production) and cell-mediated immunity (cytotoxic T lymphocytes). However, despite the superior immunity induced by live vaccines, there is an increasing concern about the adverse side effects that may occur as a result of vaccination with replicating pathogen preparations. On the other hand, killed nonreplicating vaccines, including both whole pathogen preparations and subunit vaccines, in general induce a humoral response only. Furthermore, apart from the fact that these vaccines do not normally induce cytotoxic T lymphocyte (CTL) activity, killed vaccines, subunit preparations in particular, are usually much less immunogenic than the pathogen they are derived from.

One of the reasons for the fact that nonreplicating vaccines are usually less efficacious than live vaccines relates to the way in which the antigens in either case are processed and presented to the immune system. However, new insights in these antigen processing routes suggest that it may be possible to introduce nonreplicating antigens into presentation pathways which are usually available for live vaccines only. This would require the use of proper antigen delivery systems and, thus, the nature of the resulting immune response may be modulated. Specifically, it would appear to be feasible now to deliver protein antigens in such a way that not only helper T-cells and the humoral arm of the immune response is activated, but also the cytotoxic T-cell arm, without the risk involved in the use of live pathogens. Since specific targeting and delivery as well as the display of antigens on the surface of professional antigen-presenting cells is a key issue in the design and development of new-generation vaccines for induction of both humoral and cell-mediated immunity, in Section II we will first briefly discuss current views on antigen processing pathways. In addition, we will indicate how antigen delivery systems, liposomes and virosomes in particular, may act to modulate antigen processing and thus the nature of the induced immune response.

Another aspect of vaccine design, especially with regard to viral infections that are transmitted via the airways, orally, or by sexual intercourse, is that not only a systemic response but also a strong mucosal immune response should be induced to provide a first line of defense against the incoming pathogen. However, induction of mucosal immunity requires local administration of the antigen, and local immunization is usually not very efficacious, unless, again, a live attenuated pathogen is used. In addition, local administration of nonreplicating antigens in many cases results in a state of immunological tolerance. This implies that powerful immunoadjuvants are needed that have the capacity to stimulate induction of mucosal immunity and to break tolerance. We have recently provided evidence that liposomes may act as a mucosal immunoadjuvant system when administered to the respiratory tract. This mucosal adjuvant activity is distinct from the well-documented function of liposomes as an antigen-carrier system. For example, it does not require the association of the antigen with the liposomes. In Section III, we present an overview of our own work on the mucosal immunoadjuvant activity of liposomes, involving the induction of strong secretory IgA (S-IgA) responses against influenza and measles virus. The mechanism by which liposomes may enhance the immune response, independent of their carrier function, is discussed.

Virosomes represent a unique system for presentation of antigens to the immune system. First, virosomes closely resemble the envelope of the virus they are derived from and therefore constitute an antigen-presentation form superior to isolated surface antigens. In addition, properly assembled virosomes retain the membrane fusion activity of the native virus and, therefore, virosomes may be used to deliver encapsulated, unrelated, antigens to the cytosol of antigen-presenting cells. In this respect, virosomes differ from conventional liposomes which will target enclosed antigens primarily to the phagolysosomal system of macrophages. We have recently exploited both aspects of virosomes, derived from influenza virus, to induce (i) enhanced influenza-specific antibody responses, and (ii) CTL activity against a virosome-encapsulated antigenic peptide. In Section IV, we give a summary of these data.

Finally, Section V presents a number of conclusions and perspectives on the potential application of liposomes or virosomes in new-generation vaccines.

II. Background

II.1. Pathways of antigen processing and presentation to the immune system

Almost all cells present peptides derived from endogenously synthesized proteins to the immune system by means of Major Histocompatibility Class I (MHC-1) molecules. In general, these peptides are derived from autologous proteins and, due to self-tolerance, they are ignored by the immune system.[1] However, peptides derived from, for example, endogenously synthesized viral antigens or oncogenic proteins, are generally recognized as "foreign" and therefore elicit an immune reaction. Presentation of such foreign peptides in the context of MHC-1 molecules may then result in the induction of a specialized population of T lymphocytes that is able to kill the "affected" cells. For presentation, endogenous antigens are processed to peptides in the cytosol by proteasome-mediated degradation of the antigens (see Figure 1A). Subsequently, these peptides are transported into the lumen of the endoplasmic reticulum (ER) by socalled transporters for antigen presentation (TAP). Within the ER, peptides associate with MHC-I molecules and are then transported, via the Golgi system, to the plasma membrane, where the MHC-I/peptide complex is presented to $CD8^+$ T-cells.[2] "Foreign" MHC-I/peptide complexes can thus activate $CD8^+$ T cells that subsequently mature to cytotoxic T lymphocytes (CTL).[3]

MHC-I molecules are expressed on essentially all cells in the body and, therefore, in principle, all cells, once affected by a viral infection or transformed to a tumorigenic state, constitute targets for attack by CTLs. On the other hand, not all cells, even though they may abundantly express MHC-I molecules, have the capacity to prime $CD8^+$ T-cells for proliferation to mature CTLs. The reason is that CTL activation requires not only interaction of the MHC-I/peptide complex on the presenting cell with the T-cell receptor (TCR) on the lymphocyte, but also interaction of socalled costimulatory molecules, such as B7-1, on the presenting cell with CD28 molecules on the T cell. In general, only a specialized population of socalled antigen-presenting cells (APCs) carry these costimulatory molecules and thus, have the capacity to prime $CD8^+$ T cells. Specifically, dendritic cells are likely to play a major role in antigen presentation to T-cells and induction of CTL activity. On the other hand, the lack of costimulatory molecules on, for example, tumor cells might explain why these cells, despite the fact that they may express

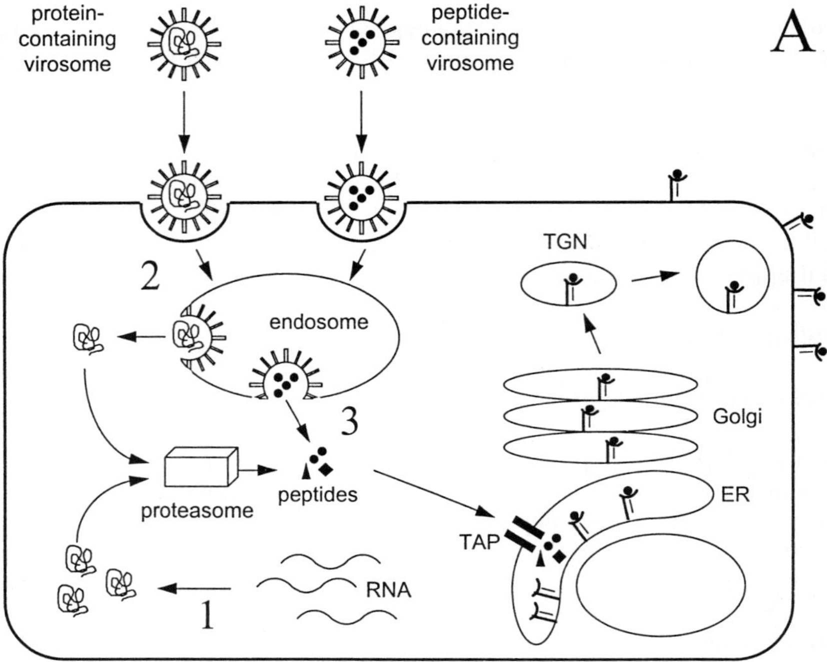

Fig. 1. Pathways of antigen processing and presentation to the immune system. Panel A depicts the MHC class I presentation pathway. Antigens synthesized de novo by the antigen-presenting cell (APC) are degraded to peptides by the proteasome complex in the cell cytosol (1). These antigens may be of e.g., viral origin in the case of infected APCs. The peptides which are generated are transported to the lumen of the endoplasmic reticulum (ER) by "transporters of antigenic peptides" (TAP), where they combine with MHC class I molecules. Subsequently, the complexes of the class I molecules and peptides are transported via the Golgi apparatus and the trans-Golgi network (TGN) to the plasma membrane where they may interact with $CD8^+$ T lymphocytes. Clearly, a fusogenic virosome containing proteins (2) or peptides (3) may deliver exogenous antigenic material to the conventional class I presentation pathway.

"foreign" antigens, by themselves constitute poor antigen-presenting cells for CTL activation.

The immune system can also recognize antigens that are not synthesized by the organism, but taken up from the environment. These exogenous antigens (e.g., bacteria, viruses, or bacterial/viral products) are internalized by specialized APCs through phagocytosis or endocytosis and processed within the endosomal/lysosomal cell compartment. In this compartment specific antigenic peptides derived from the protein antigen associate with MHC class II (MHC-II) molecules (see Figure 1B). These complexes are then transported to the plasma membrane and presented to helper T (Th) cells, carrying the CD4 marker.[4,5] These $CD4^+$ helper cells are stimulated and assist in the induction of antibody responses, mediated by B cells, and/or in the generation of specific CTL activity. While MHC-I mole-

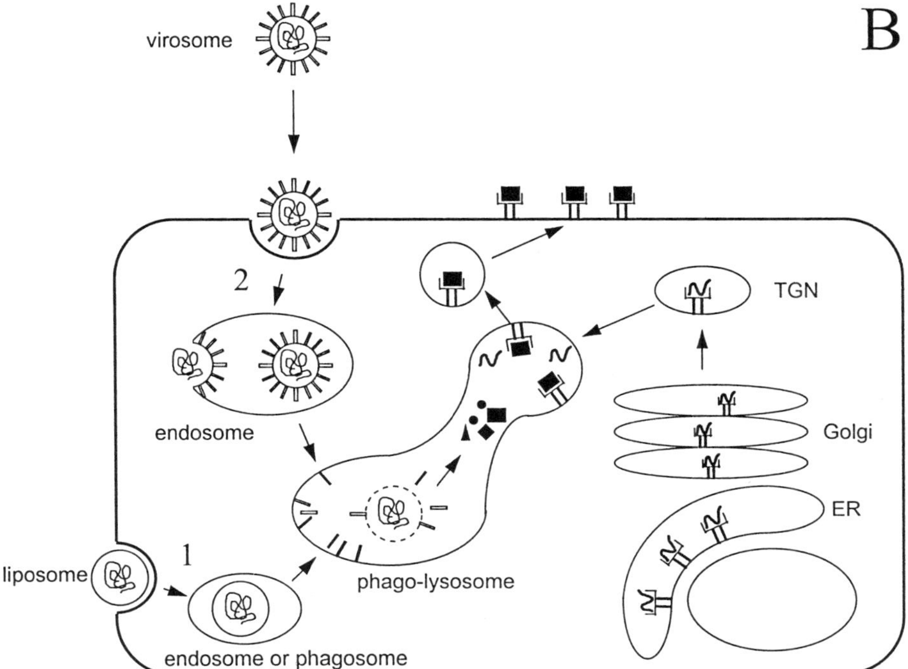

Fig. 1 (Cont.). Pathways of antigen processing and presentation to the immune system. Panel B depicts the MHC class II presentation pathway. In this case, exogenous antigens are internalized by the APC through phagocytosis or endocytosis. These antigens may be antigen-containing liposomes (1) or virosomes (2), but also protein antigens, viruses or bacteria. Invariably, the antigenic material ends up in the phago-lysosomal system of the cells, even in the case of fusogenic virosomes, where after the fusion process the viral spike proteins remain within the endosomal-lysosomal pathway and, in addition, a fraction of the virosomes will not fuse with the endosomal membrane. Within phagolysosomes, peptides are generated which combine with MHC class II molecules, synthesized and assembled in the ER. These class II molecules contain the invariant chain during transit through the Golgi and TGN; the invariant chain is replaced for peptides in phagolysosomes and the complexes of class II molecules and peptides are transported to the plasma membrane where they may interact with $CD4^+$ T lymphocytes.

cules are found on virtually all cell types, MHC-II molecules are constitutively expressed only by APCs, such as macrophages, dendritic cells, and B-cells.

The different pathways of processing and presentation of endogenous and exogenous antigens, as outlined above, imply that killed, nonreplicating, vaccines in principle will only induce the humoral arm of the immune system by activating $CD4^+$ Th cells. Since exogenous antigens are unable to enter the cytosol of APCs, such killed vaccines, in the absence of a specific cytosolic delivery system, will not normally activate $CD8^+$ T cells. Clearly, the way exogenous antigens are processed by APCs forms the basis for the current interest in antigen-carrier systems, as in principle these delivery systems may modulate the presentation process and thus activate the CTL arm of the immune response, which would otherwise remain

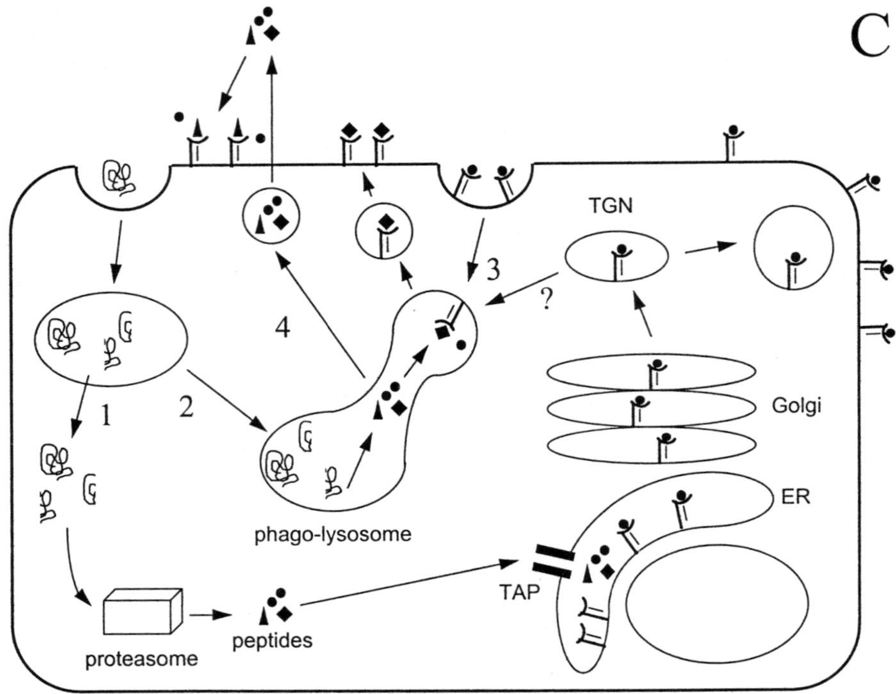

Fig. 1 (Cont.). Pathways of antigen processing and presentation to the immune system. Panel C shows a number of recently described "alternative" routes for class I MHC-restricted presentation of exogenous antigens. First, a subpopulation of macrophages and/or dendritic cells may have the capacity to transport protein antigens of fragments from phagosomes or endosomes to the cell cytosol by an as yet unidentified mechanism. Subsequently, peptides may be generated and processed in a TAP-dependent manner as in the class I presentation route (1). Another possibility involves delivery of the proteins to the phagolysosomes (2), where they are degraded to peptides essentially as in the conventional class II presentation pathway. Here, the peptides combine, in a TAP-independent manner, with class I molecules derived from the plasma membrane through endocytosis or directly from the TGN (3), although this latter option is still speculative. Finally, peptides may be transported from phagolysosomes to the extracellular medium where they may exchange with peptides on class I molecules at the cell surface (4).

unaccessible. Any delivery system that would deposit protein antigens directly into the cytosol of APCs is expected to access the class I presentation pathway. As discussed more extensively below, liposomes and particularly fusogenic virosomes may be able to deliver antigens to APCs such that these antigens are processed in the class I presentation route (see Figure 1A).

In addition, there is increasing evidence that one or more "alternative processing pathways" for exogenous antigens exist. Recent studies have shown that exogenous, particulate, antigens can be presented in the context of MHC-I molecules by a subpopulation of macrophages and probably also dendritic cells[6–9] (see Figure 1C). Although these alternative pathways have not been completely unraveled yet, there is evidence to suggest that peptides derived from exogenous antigens may be loaded onto MHC-I molecules within the endosomal or a related compartment in a TAP-independent manner. For example, class I molecules may be internalized from the plasma membrane into the endosomal pathway and subsequently combine with peptides derived from exogenous proteins.[10] As a variation on this theme, it is also possible that peptides generated within the endosomal/lysosomal system are released into the medium (peptide "regurgitation") to combine with MHC-I molecules on the surface of the same cell or to be further processed by other cells. Alternatively, MHC-I molecules may reach a subset of endosomal/lysosomal vacuoles directly, to be loaded with peptides derived from exogenous proteins for subsequent display on the cell surface.[11–15] Yet another possibility is that, within these specialized APCs, exogenous protein antigens or antigenic fragments may be released, in an unconventional manner, from the endosomal/lysosomal system to the cytosol. In this latter case, peptides would enter the conventional class-I presentation pathway, involving proteasome-catalyzed degradation of the proteins or protein fragments and TAP-mediated transport of the generated peptides to the ER lumen.[16,17] Physiologically, alternative class I processing pathways would allow induction of $CD8^+$ T-cell responses to intracellular pathogens that reside within the vacuolar system. Importantly, within the context of the discussion in this chapter, alternative processing may be involved in the immune response to immunization with certain liposomal antigen formulations, as discussed in more detail below.

II.2. Immunoadjuvant activity of liposomes

Liposomes are artificial lipid vesicles that represent a promising carrier system for delivery of drugs, biologically-active molecules and antigens. Lipid composition, liposome size and specific targeting devices can be chosen such that specific cells and or tissues are reached. Generally, the lipids used for liposome preparation are non-toxic, biodegradable and non-immunogenic. Water-soluble antigens can be encapsulated within the aqueous compartment of the liposomes, while hydrophobic antigens can be inserted within the liposomal membrane. Alternatively, antigens can be covalently coupled to liposomal membranes. For immunopotentiation, immunomodulating drugs such as muramyl dipeptide, lipopolysaccharide and lipid A can be co-incorporated or co-inserted.

After the first report on liposomes as immunological adjuvants,[18] a large variety of antigens has been evaluated in liposomal vaccine formulations (reviewed by, for example, Alving,[19] Gregoriadis[20] and Van Rooijen[21]), such as bacterial polysaccharides,[22] Streptococcus mutant carbohydrate antigen,[23–25] influenza subunit antigen,[26–28] tetanus toxoid,[29] filamentous hemagglutinin and detoxified pertussis toxin of Bortella pertussis,[30] measles virus,[31] polio virus peptides[32] and cholera toxin.[33] In many cases, a substantial stimulation of the systemic antibody responses has been observed. Incorporation or coadministration of immunomodulators (e.g., lipid A or muramyl dipeptide, MDP) and adjuvants (e.g., cholera toxin or its B subunit) generally further enhanced the response.[24,34–37]

With respect to the mechanism of the immunoadjuvant action of liposomes, clearly, liposomes may act as a depot for slow release and sustained presentation of antigen to the immune system. However, this is not the main basis for the immune-stimulatory activity of liposomes. The facts that after i.v. injection, liposomes are avidly phagocytosed by macrophages and that macrophages can present antigenic peptides in the context of MHC class II molecules to Th cells, thereby stimulating a humoral immune response, strongly suggest that macrophages play a key role in the immune-stimulatory action of liposomes (see Figure 1B). Indeed, the general view is that liposomes act as a carrier system for efficient delivery of antigens to macrophages, and perhaps also other APCs such as dendritic cells, where the carrier vesicles are degraded to release the enclosed antigens within the phagolysosomal system for degradation and conventional presentation of peptides in the context of class-II molecules.[38–40] Thus, by liposome encapsulation or association, a T-cell-independent antigen may become a T-cell-dependent antigen, with a concomitant enhancement and modulation of the induced antibody response. A specific property of liposomes in this context is their ability to combine different antigens within one vesicular entity, resulting in targeting of all liposome-associated and/or -enclosed antigens to the same cell. For example, a powerful helper antigens may be encapsulated within the lumen of a liposome carrying one or more weakly immunogenic, or T-independent, unrelated antigenic epitopes or haptens at its surface, such that the surface antigens become strongly immunogenic in a T-cell-dependent fashion. Accordingly, it has been demonstrated that antigens exclusively coupled to the surface of liposomes tend to induce an IgM response mainly, whereas antigens encapsulated within liposomes, presumably through the obligatory involvement of the Th-cell arm of the immune response, preferentially induce IgG antibodies.[21,41] In this regard, it is puzzling how antigens which are completely enclosed within the aqueous lumen of liposomes interact with B-cell surface Ig receptors in the initial process of B-cell selection and activation. Probably, under in vivo conditions some of the liposomes will break down and release the enclosed antigen. Alternatively, it is possible that liposomes are preprocessed by macrophages to release large antigenic fragments while preserving the nature of the (conformational) epitopes of the antigen.[21,42]

An important question relates to the issue as to whether liposomes are capable of delivering antigens, via either the conventional or alternative processing pathways, to the MHC class I presentation route. As indicated above, in many instances it would be highly desirable to tailor nonreplicating vaccine formulations in such a way that they induce not only a humoral, but also a cellular, i.e., CTL, response. According to the conventional processing pathways, for a successful induction of a class I MHC-restricted CTL response, antigen should gain access to the cytosol of APCs. Since liposomes are generally degraded within the endosomal system, encapsulated materials are not expected to enter the cytoplasm (see Figure 1A,B). However, antigen-containing acid-sensitive or pH-sensitive liposomes, e.g., liposomes that may destabilize the endosomal or lysosomal membranes, might be able to deliver their contents to the cytosol (see Figure 1A). Reddy et al.,[43] demonstrated that acid-sensitive liposomes, containing ovalbumin (OVA), were able to sensitize target cells for recognition by class I MHC-restricted OVA-specific CTL.

OVA-containing acid-resistant liposomes or native OVA failed to sensitize the target cells. The authors furthermore demonstrated that CTLs could be primed in vitro by interaction of antigen-containing acid-sensitive liposomes with dendritic cells but not with macrophages.[44] In contrast to the in vitro results, Collins et al.,[45] demonstrated that, in vivo, immunization with OVA encapsulated in acid-resistant as well as acid-sensitive liposomes generated class I MHC-restricted T-cell responses. Thus, it would appear as though the pathway for class I processing of liposomal antigens by dendritic cells is accessed in vitro by acid-sensitive liposomes only, whereas in vivo antigens from both acid-sensitive and acid-resistant liposomes may enter class I presentation. Subsequently, Nair et al.,[46] demonstrated that in vivo macrophages play an important role in enhancing the antigen-presenting function of dendritic cells. In this study dendritic cells and macrophages, isolated from mice injected with acid-sensitive liposomal OVA, were tested for their antigen-presenting capacity to induce CTLs; dendritic cells were always superior to macrophages. However, if the macrophages were depleted before antigen administration, the isolated dendritic cells were ineffective in antigen presentation.

Thus far, the exact mechanism by which at least some liposomal antigens may enter the MHC class I pathway is unknown. Is antigen preprocessed in vivo by macrophages to allow for efficient presentation by dendritic cells and/or is antigen released from the macrophage into the cytosol with both acid-sensitive and acid-resistant liposomes? Although most of the above studies were performed using OVA as an antigen, several studies indicated that CTL responses can also be induced, both in vitro and in vivo, against other proteins encapsulated in acid-resistant liposomes, such as a *Plasmodium falciparum* circumsporozoite protein[47] or tetrapeptide[48] or a peptide derived from the V3 loop of HIV-1 gp 120.[49] By immuno-gold electron microscopy it was shown that liposomal antigenic epitopes derived from the circumsporozoite of *Plasmodium falciparum* can actually spill from endosomes into the cytoplasm of cultured cells. Subsequently, the antigen would be delivered to the ER to associate with class I molecules. The latter observation would suggest involvement of one of the "alternative class I presentation routes", as proposed by Rock and coworkers[16,17] (see Figure 1C). However, at the same time, many liposomal antigens have failed to induce a class I MHC-restricted CTL response. It is possible that the answer to the question as to whether conventional acid-resistant liposomes have the capacity to mediate class I MHC-restricted CTL responses to enclosed antigens will ultimately depend on the nature of the antigen.

II.3. Virosomes as an antigen-carrier system

Virosomes are vesicular particles reconstituted from viral envelopes.[50–52] They can be prepared in various ways, generally involving detergent-mediated disassembly of viral membranes, followed by separation of the viral capsid containing the genetic material from the dissolved membrane components, and, after optional addition of excess lipids, final removal of the detergent from the membrane components to induce reassembly of membranous vesicles carrying the viral surface proteins. By virtue of the fact that reconstituted viral envelopes closely mimick the outer surface of the virus they are derived from, virosomes represent a very useful system for presentation of antigens to induce antibody responses against the native virus.[53–57] Virosomes may also be used to incorporate other unrelated antigens in the virosomal membrane. For example, Glück and coworkers[58–60] have incorporated hepatitis A virions (HAV) in influenza-derived virosomes and observed a strong stimulation of the HAV-specific antibody response.

Properly assembled virosomes retain the membrane fusion activity of the virus they are derived from.[50–52] This property makes virosomes a highly suited system for direct delivery of other unrelated antigens into the cytosol of APCs. Dependent on the origin of the viral membrane, virosomes can either fuse directly with the plasma membrane of APCs (e.g., virosomes derived from Sendai virus) or intracellularly with endosomal membranes (e.g., virosomes derived from influenza or Semliki Forest virus). In the latter case, the virosomes enter the cell by receptor-mediated endocytosis where, due to the local mildly acidic condition, the virosomal membrane fuses with the endosomal membrane. Similar to liposomes, the size of the virosomes and the method of preparation allow efficient encapsulation of hydrophilic, hydrophobic and amphipathic molecules. But, again, the major advantage of virosomes over liposomes is their fusogenic capacity. Encapsulated hydrophilic material can be introduced in the cytoplasm of cells, while hydrophobic molecules inserted/entrapped in the virosomal membrane can be inserted into cellular membranes. For example, we have demonstrated that encapsulation of the A chain of diphteria toxin (DTA) in fusogenic influenza virosomes induces efficient delivery of the DTA to the cell cytosol resulting in a complete inhibition of the cellular protein synthesis.[61] On the other hand, incubation of cells with free DTA, empty virosomes or fusion-inactivated DTA-containing virosomes had no effect. Furthermore, we have shown, that lipopolysaccharide (LPS) incorporated in the virosomal membrane has the capacity to stimulate lymphocytes after fusion-mediated insertion of the LPS molecule into the B-cell membranes.[62] Thus, as a result of their fusogenic capacity, virosomes are able to deliver incorporated molecules to different sites of the cell. With respect to the use of virosomes as antigen-carriers, virosomes, in contrast to liposomes, will be able to directly deliver antigenic material into the cytosol of cells, which may result in antigen presentation in the context of MHC class I molecules, thereby stimulating CTL induction (see Figure 1A).

Since fusion-mediated delivery of encapsulated antigens to the cytosol of APCs for induction of class I MHC-restricted CTL responses and fusion-mediated insertion of membrane-associated virosomal antigens into the endosomal membrane of APCs for stimulation of antibody responses represent novel approaches which have not been addressed by others before, in Section IV of this review we will briefly summarize our recent data in this area.

III. Liposomes as a mucosal immunoadjuvant system

III.1. Intranasal administration of influenza subunit antigen; induction of systemic IgG and secretory IgA

As indicated above, an important aspect in which many current vaccines require improvement is at the level of the induction of mucosal immunity in the respiratory tract, gut and/or the urogenital tract, to provide a first barrier against entry of pathogens via these epithelial surfaces. Conventional (subunit) vaccines do not normally induce a substantial mucosal secretory IgA (S-IgA) response. Yet, for example in the case of influenza, it has been demonstrated by Renegar and Small[63,64] that S-IgA plays a crucial role in the protection against infection of the upper respiratory tract. The authors showed that an HA-specific monoclonal polymeric IgA antibody was specifically transported into nasal secretions upon i.v. administration to mice, protecting the mice against nasal infection.[63] In addition, it was shown that i.n. inoculation of immune mice with an antiserum against the alpha-chain of IgA abrogated nasal protection against influenza infection.[64] The importance of the induction of S-IgA is further emphazised by the fact that S-IgA antibodies are more cross-reactive against various influenza virus subtypes than serum IgG antibodies, which would imply that mucosal S-IgA is likely to mediate a broader and perhaps longer-lasting protection against influenza infections.

We have recently investigated the immune-stimulatory activity of liposomes in an influenza subunit vaccine formulation administered i.n. to mice.[65,66] The original rationale for these studies involved the use of liposomes as an antigen-carrier system for efficient targeting of the viral subunit antigen to APCs, in line with the generally acknowledged concept of the immunoadjuvant activity of liposomes as discussed under II.2. However, as shown below, it turned out that the observed immune stimulation was not due to antigen delivery, but rather the result of a bona fide mucosal adjuvant activity of the liposomes.

Figure 2 demonstrates that i.n. administered liposomal influenza virus subunit antigen, consisting mainly of the isolated envelope glycoprotein of the virus hemagglutinin (HA), induces not only high levels of serum IgG but, more importantly, also a strong pulmonary S-IgA response directed against HA.[26] Mice were immunized under light ether anaesthesia resulting in antigen deposition throughout the respiratory tract, including the lungs.[67] The onset of the IgG response occurred about 7 days after the immunization and high titers of serum IgG were still present after 21 weeks (not shown). Similarly, S-IgA levels in the lung determined 21 weeks postimmunization were of the same magnitude as the S-IgA titers 2 weeks postimmunization. The serum IgG and the pulmonary IgA responses were similar to the corresponding antibody levels in mice recovered from a prior influenza infection (Figure 2, bars D). Moreover, the serum IgG response to i.n. immunization with the liposomal vaccine was comparable or slightly superior to the response after intramuscular (i.m.) administration of the subunit antigen alone, the current procedure for human flu vaccination.

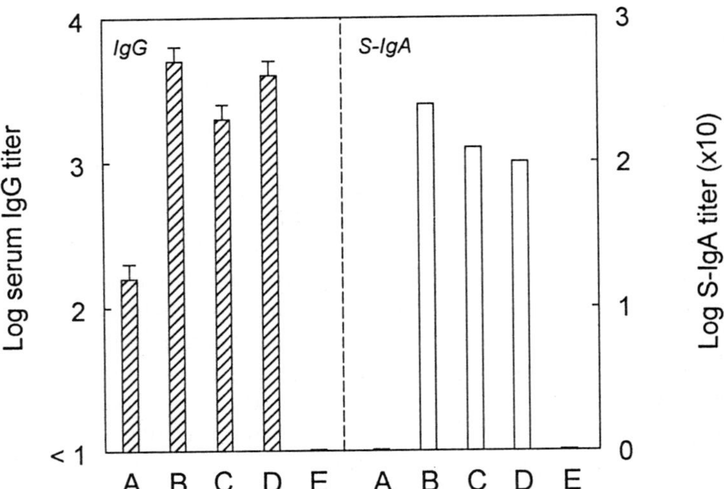

Fig. 2. Serum IgG and respiratory IgA responses upon i.n. immunization of mice with free or liposomal influenza virus subunit antigen or upon infection. Groups of mice were immunized i.n. with 1 μg of viral hemaggglutinin (HA) per dose in a free form (A), 1 μg of HA incorporated in liposomes (1.7 mg of lipid per dose) (B), 1 μg of HA admixed with liposomes (1.7 mg of lipid per dose) (C) or mice were given an aerosolized sublethal dose of influenza virus (strain A/PR/8/34) (D). Liposomes consisted of egg-phosphatidylcholine, cholesterol and dicetylphosphate in a molar ratio of 4:5:1. Blood samples and lung washes were taken 4 weeks post-immunization. Blood samples were assayed for antigen-specific IgG by ELISA (crossed bars). S-IgA titers in lung washings (open bars) were determined, after pooling and concentration of the lavage fluid, also by ELISA. Titers are given as geometric mean titers (GMT ± s.e.m). (Data adapted from Refs 26 and 73).

III.2. Role of alveolar macrophages

As discussed above, in general, the immune-stimulatory activity of liposomes has been ascribed to an increased uptake of liposome-associated antigen by macrophages, thereby facilitating MHC-II presentation of antigenic peptides to Th cells. Therefore, in our initial studies the viral subunit antigen was incorporated in the liposomes by preparation of the vesicles in the presence of the antigen. Surprisingly, when the subunit material was not incorporated, but simply mixed with preformed liposomes, immune stimulation was still observed (Figure 2, bars C). Moreover, liposomes administered up to 48 h before administration of the subunit antigen also significantly enhanced the immune response (Table 1).[26] These observations indicate that the liposomes under the conditions of these experiments do not function as antigen carriers, delivering the antigen to APCs such as macrophages.

Yet, it is likely that macrophages, alveolar macrophages (AM) in particular, are involved in the observed mucosal immune-stimulatory activity of the liposomes. First, immune stimulation is observed only with liposomes containing negatively charged lipids, such as dicetylphosphate (DCP), phosphatidylserine (PS) (not shown) or phosphatidylglycerol (PG) but not with zwitterionic lipids

Table 1

Effect of liposome pretreatment on serum IgG and lung S-IgA responses to influenza subunit antigen

Antigen[a]	Time of pretreatment with liposomes[b]	Serum IgG ($\log_{10}$ titer)[c]	S-IgA in lung wash ($\log_{10}$ titer × 10)[d]
HA	–	<1	<0.3
HA + liposomes	–	3.2 ± 0.2	2.1
HA + liposomes	6 hr	2.7 ± 0.1	1.2
HA + liposomes	16 hr	3.3 ± 0.1	1.5
HA + liposomes	24 hr	3.1 ± 0.1	1.2
HA + liposomes	48 hr	3.0 ± 0.2	1.8

[a]Groups of five mice were immunized i.n. with influenza virus subunit antigen (5 μg HA) with or without negatively charged liposomes (1.7 mg of lipid).
[b]Liposomes were given at different time points before immunization or simultaneously with the antigen.
[c]Blood samples were taken 4 weeks post-immunization; antibody titers were determined by ELISA and are presented as geometric mean titres ± s.e.m.
[d]Lung washings were taken 4 weeks post-immunization; S-IgA levels were determined in pooled and concentrated lung washings. (Data adapted from Ref. 26).

Table 2

Effect of liposomal surface charge on serum IgG and lung S-IgA responses to influenza subunit antigen

Antigen[a]	Liposome composition	Serum IgG ($\log_{10}$ titer)[b]	S-IgA in lung wash ($\log_{10}$ titer × 10)[c]
HA	–	2.0 ± 0.4	<0.3
HA + liposomes	PC:Chol:DCP (4:5:1)	3.5 ± 0.1	2.1
HA + liposomes	PC:Chol (1:1)	2.4 ± 0.1	<0.3
HA + liposomes	PC:Chol:DOPG (4:5:1)	3.4 ± 0.1	2.4

[a]Groups of five mice were immunized i.n. with influenza virus subunit antigen (0.5 μg HA) with or without liposomes (0.6 mg of lipid) consisting of the indicated lipids. (PC, egg-yolk phosphatidylcholine; Chol, cholesterol; DCP, dicetylphosphate; POPG, palmitoyl-oleoyl-phosphatidylglycerol).
[b]Blood samples were taken 4 weeks post-immunization; antibody titers were determined by ELISA and are presented as geometric mean titres ± s.e.m.
[c]Lung washings were taken 4 weeks post-immunization; S-IgA levels were determined in pooled and concentrated lung washings. (Data adapted from Ref. 26).

(Table 2).[26] It is known that a negative surface charge facilitates uptake of liposomes by macrophages.[68,69] The nature of the negative charge does not influence the immune-stimulatory effect of the liposomes. Second, the liposomes exert their mucosal adjuvant activity primarily at the level of the lungs. Immune stimulation is observed only when antigen and liposomes are given i.n. in a relatively large volume to anesthetized mice, while the same dose of antigen given i.n. in a small volume to unanesthetized mice does not induce an S-IgA response.[27] As shown by Yetter et al.,[67] the former administration procedure results in deposition of the preparation throughout the respiratory tract whereas a small volume, given to unanesthetized mice, remains confined to the nasal cavity. The requirement for deposition of antigen and liposomes in the lower respiratory tract is consistent with a role for AM in the observed immune stimulation by the liposomes.

Several mechanisms could explain our observations. First, liposomes may be taken up by alveolar macrophages (AM), thereby inhibiting AM-dependent immune suppression.[70] It is becoming increasingly clear that AM are not primarily involved in antigen presentation; rather, AM appear to suppress the immune response in the lung by uptake and elimination of antigenic material without concomitant presentation of antigenic peptides to immune-competent cells.[71] It is possible that negatively charged liposomes are taken up by AM and, thus, suppress their further phagocytic activity and elimination of coadministered antigen. Thus, the antigen would remain available for dendritic cells, resulting in more efficient antigen presentation. Second, liposomes may activate the AM, thereby upregulating MHC class II expression and antigen presentation. This latter mechanism, however, seems less plausible since in vivo depletion of AM by the procedure developed by Van Rooijen[72] facilitated an enhanced systemic and local antibody response against influenza subunit antigen deposited in the lower respiratory tract.[69] Therefore, our currrent hypothesis for the mucosal immunoadjuvant activity of liposomes is that the liposomes are taken up by an AM population that under normal conditions suppress immune responses against inhaled material. Uptake of liposomes by these macrophages may then result in downregulation of the immune suppression and coadministered or subsequently administered antigen may then, due to the inhibition of the suppressive state, be presented efficiently by dendritic cells or by an "activated" or recruited different population of macrophages.

III.3. Induction of protective immunity against influenza

In order to assess the protective immunity conferred by the liposomal vaccine preparation, mice were challenged with infectious homologous or heterologous influenza A virus.[73] Immunity induced by i.n. immunization with liposome-supplemented subunit antigen was equal or even slightly superior to that induced by i.m. immunization with subunit antigen alone (Figure 3). In addition, immunization with the subunit antigen derived from an A/Taiwan strain also conferred complete protection against a A/PR/8/34 challenge, while after i.m. immunization with the subunit antigen only partial protection against a heterologous challenge was conferred.

It can be concluded that i.n. immunization with liposome-supplemented influenza virus subunit antigens provides a promising and potentially superior alternative to conventional i.m. influenza vaccination.

III.4. Migration of IgA-producing B-cells to distant mucosal tissues

The S-IgA response is initiated in the Mucosa-Associated Lymphoid Tissue (MALT). Upon stimulation, IgA-committed B cells migrate to the local lymph nodes and the systemic circulation and, subsequently, IgA plasma cells home to

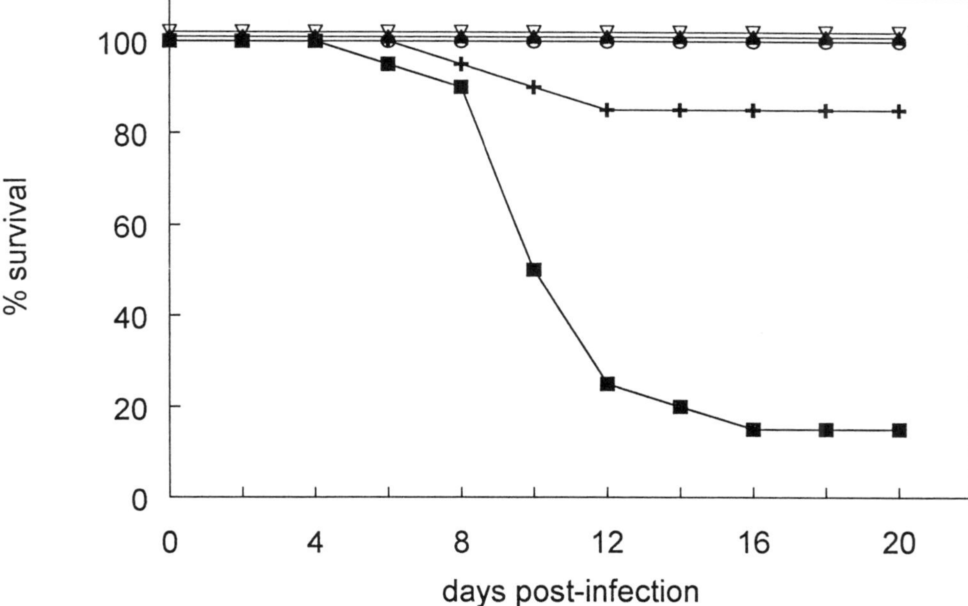

Fig. 3. Protection of mice against a lethal influenza virus infection after i.n. immunization with liposomal subunit antigen or s.c. immunization with subunit antigen. Groups of 20 mice were not immunized (■) or immunized i.n. with 5 μg of HA supplemented with liposomes (▽, ▲) or s.c. with 5 μg of HA alone (○, +). The subunit antigen was derived from A/PR8 (H1N1) virus (○, ▲) or from A/Taiwan (an H1N1 variant; ▽, +) virus. A booster immunization was given at week 4. Immunized mice and untreated control mice were given a lethal dose of aerosolized A/PR8 virus 2 weeks after the booster immunization. Survival was scored over a 3-week period. (Data adapted from Ref. 73).

the mucosal sites where the antigen was first encountered. Interestingly, however, these IgA plasma cells also home to distant mucosal sites.[74] This feature is very interesting since, for example, this would imply that i.n. immunization could induce an S-IgA response in the female urogenital tract to prevent sexually transmitted diseases. Apart from being convenient to the vaccinee, such a strategy would be of major interest, since the induction of an urogenital mucosal S-IgA response following local administration of antigen has been proven to be very difficult. Studies in our laboratory have indicated that liposomes co-administered with influenza subunit antigen to the lower respiratory tract, indeed have the capacity to induce an antigen-specific S-IgA response in the murine female urogenital tract.[27] Figure 4 shows, that significant levels of influenza-specific S-IgA could be generated, not only in the lung, but also in vaginal secretions by i.n. immunization with the liposome-supplemented antigen in anesthetized mice (immunization of the total respiratory tract). After i.n. immunization of unanesthetized mice (nasal deposition only) or after oral immunization, no lung and vaginal S-IgA could be detected, again suggesting a role for alveolar macrophages in the initiation of the mucosal S-IgA response as discussed above. These data demonstrate that

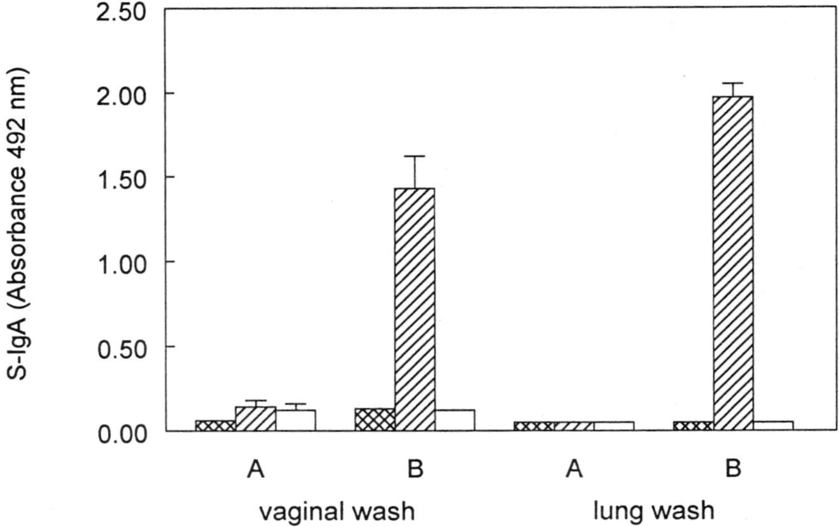

Fig. 4. Induction of s-IgA in vaginal and pulmonary secretions following i.n. immunization of mice with a liposome-containing influenza subunit vaccine. Groups of mice were immunized i.n. while awake (bars A) or i.n. while under anesthesia using pentobarbital (bars B) with 5 μg of HA mixed with liposomes (1 mg of lipid per dose; hatched bars). Control groups received the same amount of liposomes without antigen (crossed bars) or antigen alone (open bars). Washes were taken 4 weeks post-immunization and screened for antigen-specific s-IgA by ELISA. Bars represent mean absorbance values (± s.e.m.) for undiluted pooled vaginal washes and lung washes. (Data adapted from Ref. 27).

upon stimulation with liposomes, the lymphoid tissue associated with the lung can effectively be primed for an S-IgA response that subsequently disseminates throughout the common mucosal immune system. Recently, we have observed a similar dissemination of IgA-producing B-cells after priming of the nasal mucosa with influenza antigen supplemented with (nontoxic variants of) the *Escherichia coli* heat-labile enterotoxin as a mucosal immunoadjuvant.[75,76]

Obviously, further studies using antigens more relevant to infections of the urogenital tract are needed. Nonetheless, this study clearly demonstrates the potential of liposomes as an adjuvant system in vaccines for the induction of mucosal immunity, including immunity against infections in the urogenital tract.

III.5. Antigens other than the influenza subunit: whole inactivated measles virus

Our studies have demonstrated that the mucosal immunoadjuvant activity of liposomes is not restricted to the influenza virus subunit antigen.[31] Figure 5 shows that i.n. immunization with inactivated whole measles virus supplemented with liposomes, stimulated the serum IgG response relative to the response to the

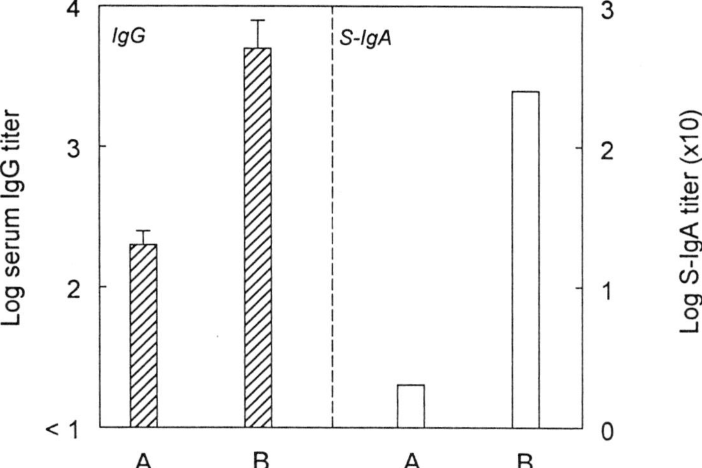

Fig. 5. Serum IgG and s-IgA responses in the respiratory tract of mice immunized i.n. with inactivated measles virus alone or mixed with liposomes. Mice were immunized i.n. with inactivated measles virus (5 μg of protein) (bars A) or inactivated measles virus (5 μg of protein) supplemented with liposomes as in Figure 1 (1 mg of liposomal lipid) (bars B). Blood samples and lung washes were taken 4 weeks post-immunization. Blood samples were assayed for antigen-specific IgG (crossed bars). S-IgA titers in lung washings (open bars) were determined after pooling and concentration of the lavage fluids. Titers are given as geometric mean titers (GMT ± s.e.m.) (Data adapted from Ref. 31).

inactivated virus alone. But perhaps more importantly, while the inactivated virus did not induce an S-IgA response in the lungs the liposomal vaccine did. In this respect, it is important to note that it is likely that mucosal antibodies provide a first line of defense against invading airborne virus, while a serum IgG response is required for prevention of a systemic infection.

Although the current measles vaccines, live attenuated virus formulations, are very effective and confer adequate long-term protection,[77–79] there is a clear need for the development of an inactivated vaccine. First, an inactivated vaccine has the advantage of being more stable, which is of particular importance in developing countries. Second, inactivated whole virus or subunit vaccines are effective in the presence of maternal antibodies while replication of the virus, required for a live vaccine, is sensitive to maternal antibodies. However, inactivated measles vaccines are usually less immunogenic, do not always induce relevant virus-neutralizing antibodies, and the antibody levels drop rapidly relative to antibody titers elicited by a viral infection. A liposomal measles vaccine formulation seems promising since, in mice, not only high levels of mucosal S-IgA and serum IgG were obtained, but also the antibody responses persisted for at least 6 months after i.n. administration.

IV. Fusogenic virosomes as an antigen-carrier system

IV.1. Stimulation of antibody responses against influenza

The above experiments have demonstrated the immunoadjuvant activity of liposomes admixed with an influenza virus subunit preparation. This subunit preparation consists of aggregates of HA mainly. Previous studies have shown that influenza viral antigens, reconstituted in virosomes, may also serve as efficacious vaccines for induction of protective humoral immune responses.[53–57] Virosomes closely mimick the virus they are derived from and, therefore, represent an optimal vehicle for antigen presentation to the immune system. Furthermore, properly assembled virosomes retain the membrane fusion activity of the native virus. In the case of influenza virosomes this implies that the HA, which is both the membrane fusion protein and the major surface antigen of the virus, will be introduced into the endosomal membrane of APCs, after uptake of the virosomes through receptor-mediated endocytosis.

We have evaluated the effect of the membrane fusion activity of influenza virosomes, in the i.n. liposome-supplemented formulation described above, on the induction of systemic IgG and mucosal S-IgA responses in mice. Figure 6 (upper panel) shows that i.n. administration of virosomes alone induces high levels of serum IgG. Supplementation of liposomes to the i.n. virosomal vaccine significantly stimulated serum IgG responses. Serum IgG levels induced by the liposome-supplemented virosome vaccine are superior to the IgG levels observed after intramuscular (i.m.) administration of virosomes. In contrast to i.n. administration of virosomes alone and i.m. administration of virosomes, i.n. administration of the liposome-supplemented virosome vaccine induced a mucosal S-IgA response in the respiratory tract.

To investigate the role of the fusion activity of the virosomes in the induction of the antibody responses, we compared serum IgG and S-IgA responses induced by fusogenic (untreated) and fusion-inactivated virosomes. The fusogenic capacity of virosomes was irreversibly inactivated by a short low-pH preincubation of the virosomes which induces an irreversible conformational change in the HA. Fusion-inactivation is readily demonstrated by the inability of the treated virosomes to fuse with (and lyse) red blood cells (hemolysis assay; data not shown). This treatment does not destroy B-cell epitopes of the HA, since HA-specific antisera obtained from previous immunization experiments, using the same HA type, reacted normally with fusion-inactivated virosome preparations (data not shown). Figure 6 (lower panel) shows that inhibition of the fusogenic capacity of the virosomes significantly reduces its immunogenicity. This is demonstrated by significantly reduced serum IgG titers observed after both i.n. and i.m. administration of the fusion-inactivated virosomes (compare upper and lower panels, Figure 6). Supplementation of liposomes to the i.n. administered fusion-inactivated virosomes again stimulated serum IgG responses and induced a S-IgA response in the respiratory tract, albeit to a lower final extent than the antibody levels induced by liposome-supplemented fusogenic virosomes. In a sense, the immunogenicity

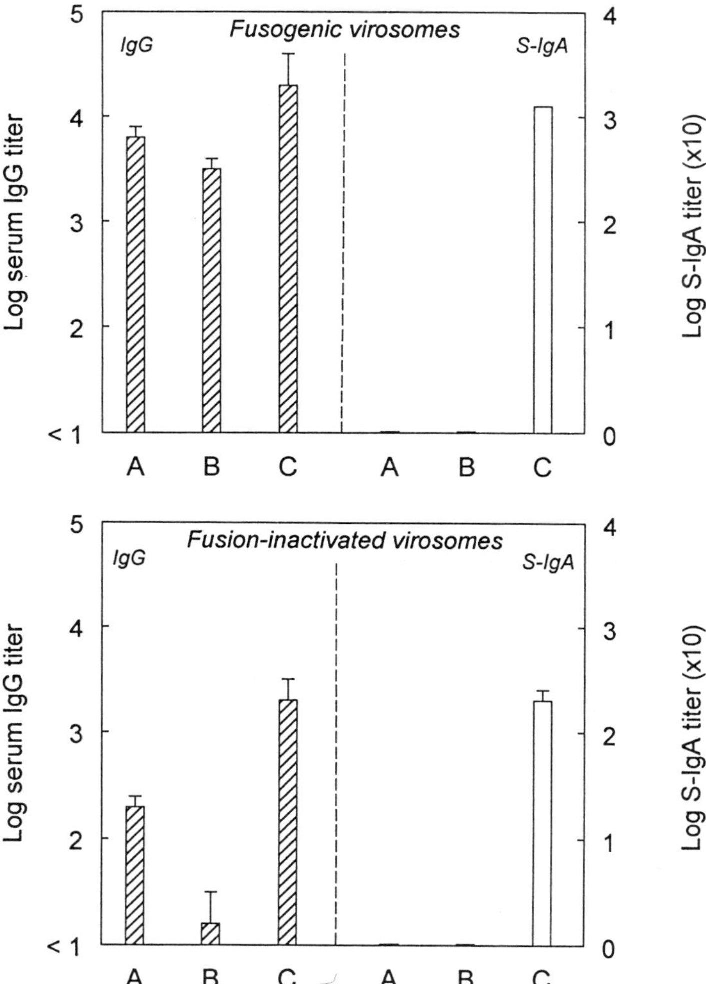

Fig. 6. Serum IgG and lung secretory IgA responses in mice immunized with virosomes alone or virosomes supplemented with liposomes. Groups of 5 BALB/c mice were immunized i.m. with a single dose of virosomes (prepared from NIB26 influenza virus) containing 5 μg of HA (bars A) or i.n. with NIB26 virosomes (5 μg of HA) alone (bars B), or i.n. with NIB26 virosomes supplemented with negatively charged liposomes (bars C). Virosomes were left untreated ("fusogenic virosomes"; upper panel) or were fusion-inactivated ("fusion-inactivated virosomes"; lower panel). Liposomes consisted of PC, cholesterol and DCP in a molar ratio of 4:5:1. The liposomal dose was 1 mg of total lipid. Blood samples and lung washes were taken 4 weeks post-immunization. Titers (hatched bars, serum IgG; open bars, lung s-IgA) are given as geometric mean titres (GMT ± s.e.m.).

of fusion-inactivated virosomes is comparable to that of subunit antigen, as presented in Figure 2.

In conclusion, these data demonstrate that influenza virosomes are efficacious in inducing high serum IgG titers upon i.n. or i.m. administration to mice. In this respect, the fusogenic capacity of virosomes clearly contributes to their potential to induce a humoral immune response, possibly via improved interaction with APCs and fusion of the antigen "into" the endosomal membrane of the APC. Supplemented with liposomes in the i.n. liposomal adjuvant formulation, virosomes induce, in addition to serum antibodies, a potent mucosal S-IgA response in the respiratory tract.

IV.2. Cytosolic delivery of a peptide antigen and induction of CTL activity

CTLs play an important role in the control of viral infections. As discussed in Section II, the induction of class I MHC-restricted CTL activity is optimally achieved by synthesis of viral antigens within APCs, for example after immunization with live attenuated virus. Despite the superior immunity induced with live vaccines there is a certain risk of causing disease. Therefore, vaccine delivery systems are sought that will enable introduction of nonreplicating antigen into the class I presentation pathway. Because of their membrane fusion activity, virosomes would appear to be ideally suited for cytosolic delivery of encapsulated antigens (see also Figure 1A). In addition, since not all virosomes taken up by an APC will fuse with the endosomal membrane, a fraction of the virosomes will be degraded within the endosomes leading to presentation of antigenic peptides in the context of class II molecules. These antigenic peptides will induce a Th response and a humoral immune response. An additional advantage of influenza-derived virosomes would be, that the immune response is likely to be further enhanced by the strong helper activity of the hemagglutinin (HA), which even after fusion of the virosomes with the endosomal membrane would remain confined to the lumen of the endosomal/lysosomal system.

Several studies have demonstrated the induction of antigen-specific CTLs after immunization with synthetic peptide preparations. The peptides were either emulsified in Freunds adjuvants,[80,81] or associated with ISCOMs[82] or liposomes.[83] The mechanism by which these peptides, in either preparation, are routed into the class-I presentation pathway is unclear. As indicated above, the current paradigm of antigen presentation is that endogenously produced antigens are presented in association with MHC-I molecules. The alternative class I presentation route for exogenous proteins implies that the antigen or antigenic peptides are introduced in the cytoplasm and routed into the endoplasmic reticulum or the peptides are regurgitated and bind to empty MHC-I molecules.[84]

We have recently demonstrated that encapsulation in fusogenic influenza virosomes of a synthetic 9-mer peptide, derived from the influenza nucleoprotein and corresponding to a major CTL epitope in Balb/c mice, indeed mediates the induction of a potent CTL response against influenza-infected target cells upon i.m. administration of the virosomes to mice (manuscript in preparation). Spleen cells

Table 3

Induction of influenza virus-specific CTL activity by immunization of mice with an NP-peptide encapsulated in fusogenic virosomes

	% cytotoxicity[b]	
Immunization[a]	Virus-infected targets[c]	Noninfected targets[d]
NP-virosomes (i.m.)	50.8	5.7
Influenza virus (i.p.)	59.6	11.8
NP-peptide + IFA (s.c.)	4.3	0.5
NP-peptide + IFA (i.m.)	2.5	0.5

[a]Mice were immunized twice, with a two-week interval, with 0.5 μg of NP-peptide incorporated in influenza-derived virosomes (i.m. administration), or 100 μg of the peptide was admixed with Incomplete Freunds Adjuvant (IFA) (s.c. or i.m. administration). As a positive control, mice were inoculated once with 100 HAU of infectious influenza virus (i.p.).
[b]Spleen cells from immunized mice were restimulated for 5 days with influenza virus infected naive spleen cells. Cytotoxicity was determined by a standard ^{51}Cr-release assay using P815 mastocytoma cells as targets. Effector-to-target cell ratio was 30:1.
[c]For specific cytotoxicity, P815 cells were infected with influenza virus.
[d]For nonspecific release, noninfected P815 cells were used.

from immunized mice were isolated and restimulated for 5 days in vitro with virus-infected syngeneic spleen cells from a naive mouse and subsequently CTL activity was determined against virally infected or uninfected target cells. Table 3 shows that the CTL activity induced by immunization with the peptide-containing virosomes ("NP-virosomes") was as high as that induced in mice after a prior infection with influenza virus. Additional experiments showed that administration of a similar dose of liposome-encapsulated NP-peptide or NP-peptide incorporated in fusion-inactivated virosomes did not induce a CTL response (not shown).

Since fusion-inactivated virosomes containing the NP-peptide fail to induce a detectable CTL response, we interpret these results to indicate that the peptide encapsulated in fusion-active virosomes is efficiently delivered to the cytosol of APCs in vivo, to be further processed in the conventional class I presentation pathway involving TAP-mediated transport to the ER, association with MHC class I molecules and final display of the MHC-I/peptide complexes on the cell surface. It is likely that the powerful helper activity of the virosomal HA is involved in additional stimulation of the response. However, the key event would appear to be the delivery of the virosome-encapsulated peptide to the cytosol of APCs. This then opens new opportunities for virosome-mediated delivery of entire protein antigens rather than small peptides. Clearly, one of the drawbacks of peptide vaccination strategies relates to the problem of MHC restriction due to the MHC (or HLA) polymorphism in the human population. Delivery of protein antigens to the cytosol of APCs would circumvent this problem. Therefore, fusion-active virosomes represent a promising antigen delivery system for induction of class I MHC-restricted CTL activity with nonreplicating viral and perhaps also tumor antigens.

V. Conclusions and perspectives

It is clear that a viral or bacterial infection provides optimal immunological protection against reinfection. Accordingly, vaccination with live attenuated pathogens is usually more efficacious than vaccination with nonreplicating vaccines, such as killed whole pathogens, subunit or recombinant preparations or synthetic peptide vaccines. These nonreplicating vaccines are generally deficient in induction of mucosal immunity and CTL activity. Nonetheless, primarily because of safety considerations, there is a trend toward the use of pure well-defined recombinant or synthetic vaccines, despite the price that has to be paid in terms of reduced efficacy. Indeed, well-defined vaccines will require the application of powerful adjuvants and antigen-delivery systems in order to stimulate the magnitude of the induced immune response and to broaden the scope of the response to include mucosal and cell-mediated immunity.

In this review, we have attempted to indicate how liposomes and fusogenic virosomes may act to stimulate and modulate immune responses to admixed or encapsulated protein antigens. It is clear that, in general, liposome-associated or -encapsulated antigens, upon parenteral administration, are processed efficiently by macrophages, resulting in class II MHC-restricted peptide presentation, activation of $CD4^+$ Th cells and stimulation of systemic antibody responses. As discussed in Section II, there is also evidence to indicate that at least certain liposome-associated antigens may be processed to enter the MHC class I presentation pathway, and thus to prime $CD8^+$ T-cells for maturation to CTLs. Whether or not this processing involves one of the recently described "alternative routes" for class I presentation remains to be established. Once the nature of the processing of liposomal antigens is better understood, it may be possible to further improve liposomal vaccine formulations for more efficient induction of CTL activity.

We have presented in this review an unprecedented aspect of liposomal immunoadjuvant activity, unrelated to the generally accepted antigen-delivery function of liposomes. Clearly, negatively charged liposomes when administered to the lower respiratory tract of mice exhibit mucosal immunoadjuvant activity. This adjuvant activity does not require association of the antigen to the liposomes and is even seen when the liposomes and the antigen are administered separately in time. It is likely that the mucosal immunoadjuvant activity of liposomes is related to a temporary interruption of the immune-suppressive action of the alveolar macrophages, which normally moderate the response to inhaled antigenic material. An important aspect of the induced S-IgA response is that it appeared to disseminate througout the common mucosal immune system, including the urogenital mucosa. This, in principle, opens the possibility to design intranasal vaccines against sexually transmitted infectious diseases.

By virtue of their potent membrane fusion activity, reconstituted viral envelopes (virosomes) would appear to be ideally suited for delivery of protein antigens to the cytosol of APCs, and thus for introduction of antigenic peptides into the class I presentation pathway. In this review, we have presented a preliminary proof-of-principle in this regard. Potent CTL activity could be induced by immunization of mice with an antigenic peptide encapsulated in influenza-derived virosomes. The action of the virosomes is likely to involve both delivery of the enclosed antigen to the cytosol of APCs and the powerful helper activity of the virosomal hemagglutinin, but the cytosolic delivery of the antigen seems crucial as fusion-inactivated virosomes were ineffective. It is interesting that to note that virosome-mediated delivery of protein antigens would circumvent problems associated with MHC restriction and HLA polymorphism in the human population, since the APC would select its "own" peptides. A promising potential application of fusogenic virosomes involves the development of strategies for CTL induction in cancer immunotherapy through virosome-mediated delivery of tumor antigens. Fusogenic virosomes derived from influenza virus also appeared to be very efficacious in induction of an antibody response against the viral hemagglutinin, including production of mucosal S-IgA in the respiratory tract, after local administration of the virosomes supplemented with liposomes as a mucosal immunoadjuvant.

In conclusion, lipid vesicle systems, fusogenic vesicles in particular, represent promising vehicles for efficient delivery of antigens to specific subcellular compartments of antigen-presenting cells. In addition, liposomes exhibit potent mucosal immunoadjuvant activity in the respiratory tract. Thus, liposomal and virosomal systems may stimulate, modulate and broaden the immune response to associated or admixed protein antigens. This, in principle, provides the opportunity to induce systemic and mucosal antibody responses as well as cell-mediated immunity with safe, well-defined, nonreplicating, antigens.

Acknowledgements

We wish to thank all those who have contributed to our work discussed in this chapter, particularly Drs. Parker Small Jr., Bradley Bender, and Kathryn Renegar of The University of Florida in Gainesville, Drs. Guus van Scharrenburg, Ruud Brands, Bram Palache, Harm Geerligs and Piet Huchshorn from Solvay Duphar in Weesp (The Netherlands), Dr. Nico van Rooijen at the Free University in Amsterdam, and Drs. Jochum Prop, Greetje Groen, and Chris Tomee at The University of Groningen. We wish to thank Drs. Etienne Agsteribbe, Pieter Schoen and Anke Huckriede from our own department for stimulating discussions on mucosal immunity, virosomes and antigen presentation pathways, respectively. We gratefully acknowledge Dr. Sebo Withoff for generating the antigen presentation cartoons of Figure 1. We also acknowledge the financial and material support of The Netherlands Organization for Scientific Research (NWO) under the auspices of the Chemical and Technology Foundations (SON/STW), Solvay Duphar BV (Weesp, The Netherlands), and Inex Pharmaceuticals Corporation (Vancouver, BC, Canada).

References

1. Kappler JW, Roehm N, Marrack P. T cell tolerance by clonal elimination in the thymus. Cell 1987;49:273–280.
2. Cox JC, Yewdell JW, Eisenlohr LC, Johnson PR, Bennink JR. Antigen presentation requires transport of MHC class I molecules from endoplasmic reticulum. Science 1990;247:715–718.
3. Townsend ARM, Bodmer H. Antigen recognition by class I restricted lymphocytes. Ann Rev Immunol 1989;7:601–624.
4. Bevan MJ. Class discrimination in the world of immunology. Nature 1987;325:192–194.
5. Germain RN. The ins and outs of antigen processing and presentation. Nature 1986;322:687–689,
6. Reimann J, Bohm W, Schirmbeck R. Alternative processing pathways for MHC class I-restricted epitope presentation to $CD8^+$ cytotoxic T lymphocytes. Biol Chem Hoppe-Seyler 1994;375:731–736.
7. Bevan M. Antigen presentation to cytotoxic T lymphocytes in vivo. J Exp Med 1995;182:639–641.
8. Harding CV. Phagocytic processing of antigens for presentation by MHC molecules. Trends in Cell Biol 1995;5:105–109.
9. Rock KL. A new foreign policy: MHC class I molecules monitor the outside world. Immunol Today 1996;17:131–137.
10. Schirmbeck R, Reimann J. 'Empty' L^d molecules capture peptides from endocytosed hepatitis B surface antigen particles for major histocompatibility complex class I-restricted presentation. Eur J Immunol 1996;26:2812–2822.
11. Pfeifer JD, Wick MJ, Roberts RL, Findlay K, Normark SJ, Harding CV. Phagocytic processing of bacterial antigens for class I MHC presentation to T cells. Nature 1993;361:359–362.
12. Harding CV, Song R. Phagocytic processing of exogenous particulate antigens by macrophages for presentation by class I MHC molecules. J Immunol 1994;153:4925–4933.
13. Bachmann MF, Lutz MB, Layton GT, Harris SJ, Fehr T, Rescigno M, Ricciardi-Castagnoli P. Dendritic cells process exogenous viral proteins and virus-like particles for class I presentation to $CD8^+$ cytotoxic T lymphocytes. Eur J Immunol 1996;26:2595–2600.
14. Harding CV, Song R, Griffin J, France J, Wick MJ, Pfeifer JD, Geuze HJ. Processing of bacterial antigens for presentation to class I and II MHC-restricted T lymphocytes. Infect Agents and Dis 1995;4:1–12.
15. Bachmann MF, Oxenius A, Pircher H, Hengartner H, Ashton-Richardt PA, Tonegawa S, Zinkernagel RM. TAP1-independent loading of class I molecules by exogenous viral proteins. Eur J Immunol 1995;25:1739–1743.
16. Kovacsovics-Bankowski M, Rock KL. A phagosome-to-cytosol pathway for exogenous antigens presented on MHC Class I molecules. Science 1995;267:243–246.
17. Shen ZH, Reznikoff G, Dranoff G, Rock KL. Cloned dendritic cells can present exogenous antigens on both MHC class I and class II molecules. J Immunol 1997;158:2723–2730.
18. Allison AC, Gregoriadis G. Liposomes as immunological adjuvants. Nature 1974;252:252–253.
19. Alving CR. Immunologic aspects of liposomes: presentation and processing of liposomal protein and phospholipid antigens. Biochim Biophys Acta 1992;1113:307–322.
20. Gregoriadis G. Immunological adjuvants: a role for liposomes. Immunol Today 1990;11:89–97.
21. Van Rooijen N. Liposome mediated immunopotentiation and immunomodulation. In: Gregoriadis G et al., eds. Vaccines: New-Generation Immunological Adjuvants. New York: Plenum Press, 1995;15–24.
22. Abraham E. Intranasal immunization with bacterial polysaccharide containing liposomes enhances antigen-specific pulmonary secretory antibody responses. Vaccine 1992;10:461–468.
23. Childers NK, Zhang SS, Michalek SM. Oral immunization with dehydrated liposomes containing Streptococcus mutans glucosyltransferase (GTF) in humans. Adv Exp Med Biol 1995;371B:1481–1484.
24. Wachsmann D, Klein JP, Scholler M, Ogier J, Ackermans F, Frank RM. Serum and salivary antibody responses in rats orally immunized with Streptococcus mutans carbohydrate protein conjugate associated with liposomes. Infect Immun 1986;52:408–413.
25. Michalek SM, Childers NK, Katz J, Dertzbaugh M, Zhang S, Russell MW, Macrina FL, Jackson S, Mestecky J. Liposomes and conjugate vaccines for antigen delivery and induction of mucosal immune responses. Adv Exp Med Biol 1992;327:191–198.
26. De Haan A, Geerligs HJ, Huchshorn JP, Van Scharrenburg GJM, Palache AM, Wilschut J. Mucosal immunoadjuvant activity of liposomes: induction of systemic IgG and secretory IgA

responses in mice by intranasal immunization with an influenza subunit vaccine and coadministered liposomes. Vaccine 1995;13:155–162.
27. De Haan A, Renegar KB, Small Jr PA, Wilschut J. Induction of a secretory IgA response in the murine female urogenital tract by immunization of the lungs with liposome-supplemented viral subunit antigen. Vaccine 1995;13:613–616.
28. Ben Ahmeida ET, Potter CW, Gregoriadis G, Adithan C, Jennings R. IgG subclass response and protection against challenge following immunisation of mice with various influenza A vaccines. J Med Microbiol 1994;40:261–269.
29. Davis D, Gregoriadis G. Primary response to liposomal tetanus toxoid in mice: the effect of mediators. Immunology 1989;68:277–282.
30. Guzman CA, Molinari G, Fountain MW, Rohde M, Timmis KN, Walker MJ. Antibody responses in the serum and respiratory tract of mice following oral vaccination with liposomes coated with filamentous hemagglutinin and pertussis toxoid. Infect Immun 1993;61:573–579.
31. De Haan A, Tomee JFC, Huchshorn JP, Wilschut J. Liposomes as an immunoadjuvant system for stimulation of mucosal and systemic antibody responses against inactivated measles virus administered intranasally to mice. Vaccine 1995;13:1320–1324.
32. Tan L, Weissig V, Gregoriadis G. Comparison of the immune response against polio peptides covalently-surface-linked to and internally-entrapped in liposomes. Asian Pac J Immunol 1991;9:25–30.
33. Pierce NF, Sacci JB, Alving CR, Richardson EC. Enhancement by lipid A of mucosal immunogenicity of liposome-associated cholera toxin. Rev Infect Dis 1984;6:563–566.
34. Zhou F, Kraehenbuhl JP, Neutra MR. Mucosal IgA response to rectally administered antigen formulated in IgA-coated liposomes. Vaccine 1995;13:637–644.
35. Harokopakis E, Childers NK, Michalek SM, Zhang SS, Tomasi M. Conjugation of cholera toxin or its B subunit to liposomes for targeted delivery of antigens. J Immunol Methods 1995;185:31–42.
36. Vadolas J, Davies JK, Wright PJ, Strugnell RA. Intranasal immunization with liposomes induces strong mucosal immune responses in mice. Eur J Immunol 1995;25:969–975.
37. Alving CR, Verma JN, Rao M, Krzych U, Amselem S, Green SM, Wassef NM. Liposomes containing lipid A as a potent non-toxic adjuvant. Res Immunol 1992;143:197–198.
38. Gregoriadis G. Liposomes as immunological adjuvants: approaches to immunopotentiation including ligand-mediated targeting to macrophages. Res Immunol 1992;143:178–185.
39. Szoka Jr FC. The macrophage as the principal antigen-presenting cell for liposome-encapsulated antigens. Res Immunol 1992;143:186–187.
40. Su D, Van Rooijen N. The role of macrophages in the immunoadjuvant action of liposomes: effects of elimination of splenic macrophages on the immune response against intravenously injected liposome associated albumin antigen. Immunology 1989;66:466–470.
41. Van Rooijen N. Liposomes as carrier and immunoadjuvant of vaccine antigens. Adv Biothechnol Processess 1990;13:255–279,
42. Van Rooijen N. Macrophages as accessory cells in the "in vivo" humoral immune response: From processing of particulate antigens to regulation by suppression. Sem Immunol 1992;4:237–245.
43. Reddy R, Zhou F, Huang L, Carbone F, Bevan M, Rouse BT. pH sensitive liposomes provide an efficient means of sensitizing target cells to class I restricted CTL recognition of a soluble protein. J Immunol Methods 1991;141:157–163.
44. Nair S, Zhou F, Reddy R, Huang L, Rouse BT. Soluble proteins delivered to dendritic cells via pH-sensitive liposomes induce primary cytotoxic T lymphocyte responses in vitro. J Exp Med 1992;175:609–612.
45. Collins DS, Findlay K, Harding CV. Processing of exogenous liposome-encapsulated antigens in vivo generates class I MHC-restricted T cell responses. J Immunol 1992;148:3336–3341.
46. Nair S, Buiting AM, Rouse RJ, Van Rooijen N, Huang L, Rouse BT. Role of macrophages and dendritic cells in primary cytotoxic T lymphocyte responses. Int Immunol 1995;7:679–688.
47. White K, Krych U, Gordon DM, Porter TG, Richards RL, Alving CR, Deal CD, Hollingdale M, Silverman C, Sylvester DR et al. Induction of cytolytic and antibody responses using *Plasmodium falciparum* repeatless circumsporozoite protein. Vaccine 1993;11:1341–1346.
48. Alving CR, Wassef NM. Cytotoxic T lymphocytes induced by liposomal antigens: mechanisms of immunological presentation. AIDS Res. Hum. Retroviruses 1994;10:S91–94.
49. White WI, Cassatt DR, Madsen J, Burke SJ, Woods RM, Wassef NM, Alving CR, Koenig S. Antibody and cytotoxic T-lymphocyte responses to a single liposome-associated peptide antigen. Vaccine 1995;13:1111–1122.
50. Wilschut J, Bron R. The influenza virus hemagglutinin: membrane fusion activity in intact virions

and reconstituted virosomes. In: Bentz J, ed. Viral Fusion Mechanisms. Boca Raton, FL: CRC Press, 1993;133–161.
51. Bron R, Ortiz A, Dijkstra J, Stegmann, Wilschut J. Preparation, properties, and applications of reconstituted influenza virus envelopes (Virosomes). Methods in Enzymol 1993;220:313–331.
52. Stegmann T, Morselt HWM, Booy FP, Van Breemen JFL, Scherphof G, Wilschut J. Functional reconstitution of influenza virus envelopes. EMBO J 1987;6:2651–2658.
53. Almeida JD, Edwards DC, Brands CM, Heath TD. Formation of virosomes from influenza subunits and liposomes. The Lancet 1975;2:899–901.
54. Glück R, Mischler R, Finkel B, Que JU, Scarpa B, Cryz SJ. Immunogenicity of new virosome influenza vaccine in elderly people. Lancet 1994;344:160–163.
55. Ando S, Tsuge H, Mayumi T. Preparation of influenza virosome vaccine with muramyldipeptide derivative B30-MDP. J Microencaps 1997;14:79–90.
56. Boudreault A, Thibodeau L. Mouse response to influenza immunosomes. Vaccine 1985;3:231–234.
57. El Guink N, Kris RM, Goodman-Snitkoff G, Small Jr PA, Mannino RJ. Intranasal immunization with proteoliposomes protects against influenza. Vaccine 1989;7:147–151.
58. Glück R, Mischler R, Bratschen S, Just M, Althaus B, Cryz SJ. Immunopotentiating reconstituted influenza virosome vaccine delivery system for immunization against hepatitis A. J Clin Invest 1992;90:2491–2495.
59. Loutan L, Bovier P, Althaus B, Glück R. Inactivated virosome hepatitis A vaccine. Lancet 1994;343:322–324.
60. Mengiardi B, Berger R, Just M, Glück R. Virosomes as carriers for combined vaccines. Vaccine 1995;13:1306–1315.
61. Bron R, Ortiz A, Wilschut J. Cellular cytoplasmic delivery of a polypeptide toxin by reconstituted influenza virus envelopes (virosomes). Biochemistry 1994;33:9110–9117.
62. Dijkstra J, Bron R, Wilschut J, De Haan A, Ryan JL. Activation of murine lymphocytes by polysaccharide incorporated in fusogenic, reconstituted influenza virus envelopes (virosomes). J Immunol 1996;157:1028–1036.
63. Renegar KB, Small Jr PA. Passive transfer of local immunity to influenza virus infection by IgA antibody. J Immunol 1991;146:1972–1978.
64. Renegar KB, Small Jr PA. Immunoglobulin A mediation of murine nasal anti-influenza virus immunity. J Virol 1191;65:2146–2148.
65. De Haan A, Wilschut J. Liposomes and antiviral mucosal immunity. In: Shek PN, ed. Liposomes in Biomedical Applications. Chur, Switzerland: Harwood Acad Publishers, 1995;69–83.
66. Wilschut J, De Haan A, Geerligs HJ, Huchshorn JP, Van Scharrenburg GJM, Palache AM, Renegar KB, Small Jr PA. Liposomes as a mucosal adjuvant system: an intranasal liposomal influenza subunit vaccine and the role of IgA in nasal anti-influenza immunity. J Lip Res 1994;4:301–314.
67. Yetter RA, Lehrer S, Ramphal R, Small Jr PA. Outcome of influenza infection: effect of site of initial infection and heterotypic immunity. Infect Immun 1980;29:654–662.
68. Daemen T, Regts J, Scherphof GL. Liposomal phosphatidylserine inhibits tumor cytotoxicity of liver macrophages induced by muramyl dipeptide and lipopolysaccharide. Biochim Biophys Acta 1996;1285:219–228.
69. De Haan A, Groen G, Prop J, Van Rooijen N, Wilschut J. Mucosal immunoadjuvant activity of liposomes: role of alveolar macrophages. Immunology 1996;89:488–493.
70. Van Rooijen N. Macrophages as accessory cells in the in vivo humoral immune response: from processing of particulate antigens to regulation by suppression. Sem Immunol 1992;4:237–245.
71. Thepen T, Van Rooijen N, Kraal G. Alveolar macrophage elimination in vivo is associated with an increase in pulmonary immune responses in mice. J Exp Med 1989;170,499–509.
72. Van Rooijen N. The liposome-mediated macrophage "suicide" technique. J Immunol Meth 1989;124:1–6.
73. De Haan A, Van Scharrenburg GJM, Masihi KN, Bender BS, Small Jr PA, Wilschut J. Protective immunity induced by liposome-supplemented influenza virus subunit antigen administered intranasally to mice. Vaccine Res 1995;4:207–216.
74. McGhee JR, Mestecky J, Dertzbaugh MT, Eldridge JH, Hirasawa M, Kiyono H. The mucosal immune system: from fundamental concepts to vaccine development. Vaccine 1992;10:75–88.
75. De Haan L, Verweij WR, Holtrop M, Agsteribbe E, Wilschut J. Mucosal immunogenicity of *Escherichia coli* heat-labile enterotoxin: role of the A subunit. Vaccine 1996;14:620–626.
76. De Haan L, Verweij WR, Feil IK, Lijnema TH, Hol WGJ, Agsteribbe E, Wilschut J. Mutants of

the *Escherichia coli* heat-labile enterotoxin with reduced ADP-ribosylation or no activity retain the immunogenic properties of the native holotoxin. Infect Immun 1996;64:5413–5416.
77. Norrby E, Oxman MN. Measles virus. In: Fields BN, Knipe DM, eds. Virology. New York: Raven Press, 1990;1013–1044.
78. Osterhaus ADME, De Vries P. Vaccination against acute respiratory virus infections and measles in man. Immunobiol 1992;184:180–192.
79. Osterhaus ADME, De Vries P, Van Binnendijk RS. Measles vaccines:novel generations and new strategies. J Infect Dis 1994;170:S42–S55.
80. Schulz M, Zinkernagel RM, Hengartner H. Peptide-induced antiviral protection by cytotoxic T cells. Proc Natl Acad Sci USA 1991;88:991–993.
81. Kast WM, Roux L, Curren J, Blom HJ, Voordouw AC, Meloen RH. Protection against lethal Sendai virus infection by in vivo priming of virus-specific cytotoxic T lymphocytes with a free synthetic peptide. Proc Natl Acad Sci 1991;88:2283–2287.
82. Takahashi H, Takeshitta T, Morein B, Putney S, Germain RN, Berzofsky JA. Induction of $CD8^+$ cytotoxic cells by immunization with purified HIV-1 envelope protein in ISCOMs. Nature 1990;344:873–875.
83. Collins DS, Findlay K, Harding CV. Processing of exogenous liposome-encapsulated antigens in vivo generates class I MHC-restricted T cell responses. J Immunol 1992;148:3336–3341.
84. Yewdell JW, Bennink JR, Hosaka Y. Cells process exogenous proteins for recognition by cytotoxic T lymphocytes. Science 1988;239:637–640.

Lasic and Papahadjopoulos (eds.), Medical Applications of Liposomes
Elsevier Science B.V.

CHAPTER 2.7

Theoretical basis for development of liposomes as carriers of vaccines

CARL R. ALVING

Department of Membrane Biochemistry, Walter Reed Army Institute of Research, Washington, DC 20307–5100, USA

Overview

I. Introduction

The fundamental reasons for the success of vaccine strategies are not always obvious, particularly at an early stage of development of the vaccine. Are antibodies important? Cytotoxic T lymphocytes? Mucosal Immunity? What are the roles of adjuvants? Cytokines? Accessory molecules? Antigen presenting cells? What are the predictive correlates for protection? Are certain animal models important or misleading? As the list is long and the strategies complex, especially when the disease, such as HIV, attacks the immune system itself, the struggle to develop a universal basis for rational vaccine development is a considerable challenge. In the liposomal approach to vaccine development that my colleagues and I have undertaken we have focused on cellular mechanisms that are important in the immune response. In our program, interactions of liposomes containing antigens and adjuvants have been examined with antigen presenting cells, particularly phagocytic antigen presenting cells. The purpose of this review is to summarize some of these and certain other theoretical issues that may have a substantial impact on vaccine development.

Numerous previous reviews of liposomes as carriers of antigens and adjuvants, including listings of antigens and adjuvants that have been utilized in experimental animals and in humans, have been published.[1–5]

II. Role of phagocytic cells

II.1. In vivo studies

It is now clear that the expression of the antigen-derived peptide on the surface of the antigen presenting cell (APC) in conjuction with the appropriate major histocomptibility gene complex (MHC) product leads to the presentation of the peptide to T lymphocytes and to the generation of the immune response.[6] The presentation of the liposomal antigen to T lymphocytes, whether encapsulated or surface bound, is thought to occur in the context of this system.

For a variety of reasons it is believed that phagocytic cells, such as macrophages or dendritic cells, serve as the predominant APCs for liposomes.[7,8] Based on intravenous (IV) injection of "suicide" liposomes containing a cytotoxic chemical that depletes splenic and hepatic macrophages it was determined that the immune response against a liposome-encapsulated antigen that was subsequently also injected IV was strongly suppressed.[8–10] In contrast to liposomal antigen, subsequent studies with soluble antigen that was locally coadministered with suicide liposomes suggested that killing of macrophages in several organs (such as the lung or intraperitoneal cavity) actually enhanced the immune response against soluble (nonliposomal) antigens delivered to the same tissues.[8,11] Thus it was concluded that phagocytic cells have an enhancing effect for inducing immune responses against particulate antigens (such as liposomal antigens) and a suppressive effect against nonparticulate soluble antigens.

Further support for the in vivo enhancing role of macrophages was suggested by studies in which liposomes containing antigen were preincubated with splenic cells in vitro followed by adoptive transfer of the cells into animals. A secondary immune response in this system was blocked by depletion of macrophages in the splenic population with nonliposomal leucine methyl ester.[12] Other evidence for beneficial effects of macrophages was provided by the ability of peritoneal macrophages, but not B cell tumors, to present liposome-encapsulated antigen to antigen-specific T cells in vitro.[13]

II.2. Cell culture studies: Are phagocytic cells always necessary for processing and presentation of liposomal antigen?

Uptake of liposomal antigen by phagocytic cells results in processing of the antigen in endosomes for entry of the antigen into the MHC class II pathway that eventually leads to induction of antibodies. Inhibition of uptake and processing of liposomes, for example by fixation of cultured macrophages with glutaraldehyde, prevents presentation by the macrophages of a liposome-bound protein antigen to antigen specific T lymphocytes.[14] In contrast, a liposome-bound peptide (derived from the same antigen) that did not require processing was presented to the antigen-specific lymphocytes, albeit inefficiently, either by glutaraldehyde-fixed macrophages or by B cells.[14]

Further in vitro experimental proof that confirmed that processing and presentation of a liposomal antigen could be carried out by macrophages was demonstrated by utilizing a system comprising: (a) liposomes having encapsulated conalbumin; (b) bone marrow-derived macrophages (BMs) that served as APCs; and (c) a conalbumin-specific T cell clone that proliferated in the presence of syngeneic APCs that were expressing conalbumin peptides on their cell surfaces.[15,16] Proliferation of the conalbumin-specific T cells occurred after exposure of the lymphocytes to BMs that had previously ingested liposomes. When the BMs were exposed to inhibitors of phagocytosis and intracellular processing, including chloroquine, NH_4Cl, leupeptin, brefeldin A, monensin, antimycin A, NaF, and cycloheximide, or when BMs were treated with gluteraldehyde, proliferation of the conalbumin-specific T cells was abolished, thus demonstrating that phagocytosis and specific intracellular processing were both necessary for presentation of the liposomal antigen by macrophages.

In contrast to the above, it has been demonstrated with at least one liposomal protein antigen that presentation does not invariably require intracellular processing of the antigen. Proof that this pathway could be provided by macrophages in the absence of phagocytosis was demonstrated with a system comprising: (a) liposomes having an encapsulated recombinant malaria antigen (R32NS1) containing repeating NANP peptide sequences derived from the circumsporozoite protein of *Plasmodium falciparum*; (b) bone marrow-derived macrophages (BMs) that served as APCs; and (c) a NANP-specific T cell clone that proliferated in the presence of syngeneic APCs that were presenting NANP.[16] When the BMs were exposed to inhibitors of phagocytosis and intracellular processing, including chloroquine, NH_4Cl, leupeptin, brefeldin A, monensin, antimycin A, NaF, and cycloheximide or when BMs were treated with gluteraldehyde, proliferation still occurred normally, thus indicating that presentation of this particular antigen could occur directly at the surface of the BMs without undergoing internalization and intracellular processing. Presentation occurred in a MHC-restricted manner, thus indicating that NANP epitopes on the surface of the liposomes must have interacted with the MHC molecules at the surface of the BMs through an unknown mechanism and that the NANP on the recombinant protein was appropriately complexed with the MHC class II molecules leading to presentation to T helper lymphocytes.

The conclusion from the above contrasting studies with liposomal conalbumin and liposomal R32NS1 was that the requirement for intracellular processing and presentation of liposomal antigen for presentation of derived peptide in conjunction with MHC class II molecules depends on the antigen itself. In those instances where multiple repeats of a short peptide (such as NANP) are present, the surface exposure of the antigen on the liposomes might be more important than the encapsulated antigen. Thus with certain antigens it might be possible to obtain an immune response with an APC such as a B lymphocyte that does not exhibit phagocytic activity. This confirms the previous conclusions of Dal Monte and Szoka that was based on research in which they found that gluteraldehyde-fixed macrophages did not process protein antigen coupled to the surface of liposomes but that a surface-bound peptide that did not require processing was presented either by fixed macrophages or B lymphocytes.[14]

III. Effect of surface expression of antigen vs. encapsulation

Some years ago a controversy arose as to the possible role that the physical location of the antigen within the liposome might play in the induction of an immune response. Arguments have been made for and against the requirement of surface exposure of liposomal antigen in order to achieve antigen presentation (reviewed in references 17 and 18). This is an important issue because it could have important implications with respect to antigen selection and manufacturing procedures for vaccines. For example, it was found that a surface bound antigen was presented much more efficiently by macrophages when compared with the same antigen encapsulated in liposomes.[14] The above in vitro studies with R32NS1 suggested that at least in some instances antigen expressed on the surface of liposomes might be directly presented at the surface of the APC.

An indirect theoretical argument for the possible importance of liposomal surface expression has been made with a peptide antigen.[5] In response to a need to produce a monoclonal antibody to the active site of acetylcholinesterase (AChE), a 25 amino acid peptide consisting of the amino acids surrounding the active site serine was synthesized and encapsulated without conjugation within liposomes containing lipid A. The liposomal formulation was highly effective for producing murine monoclonal antibodies that proved to be useful for probing the active site of AChE.[19] The antigenic epitopes conferring specificity on the monoclonal antibodies were examined by synthesizing overlapping octapeptides directly on the heads of polystyrene pins, and enzyme-linked immunosorbent assays were performed on the pins using the monoclonal antibodies. To the surprise of the investigators, the monoclonal antibodies recognized epitopes at opposite ends of the 25 amino acid peptide (Figure 1C), thus suggesting that the peptide was actually folded and the antibodies were recognizing a conformational specificity.

Upon molecular modeling of the above 25 amino acid peptide, an interesting folding pattern was noticed that was compatible with the above folding hypothesis for conformational specificity. The molecule has two interesting features (Figure 1A). First, it is folded such that the F and H residues located at positions 6 and 18, respectively, on the peptide (reading from the N terminus), both of which are present in the antigenic epitopes recognized by the monoclonal antibodies, are located adjacent to each other on the top of the model. Second, widely separated K and R residues, representing two of the three positively charged residues, which are present at opposite ends of the peptide at positions 1 and 25, respectively,

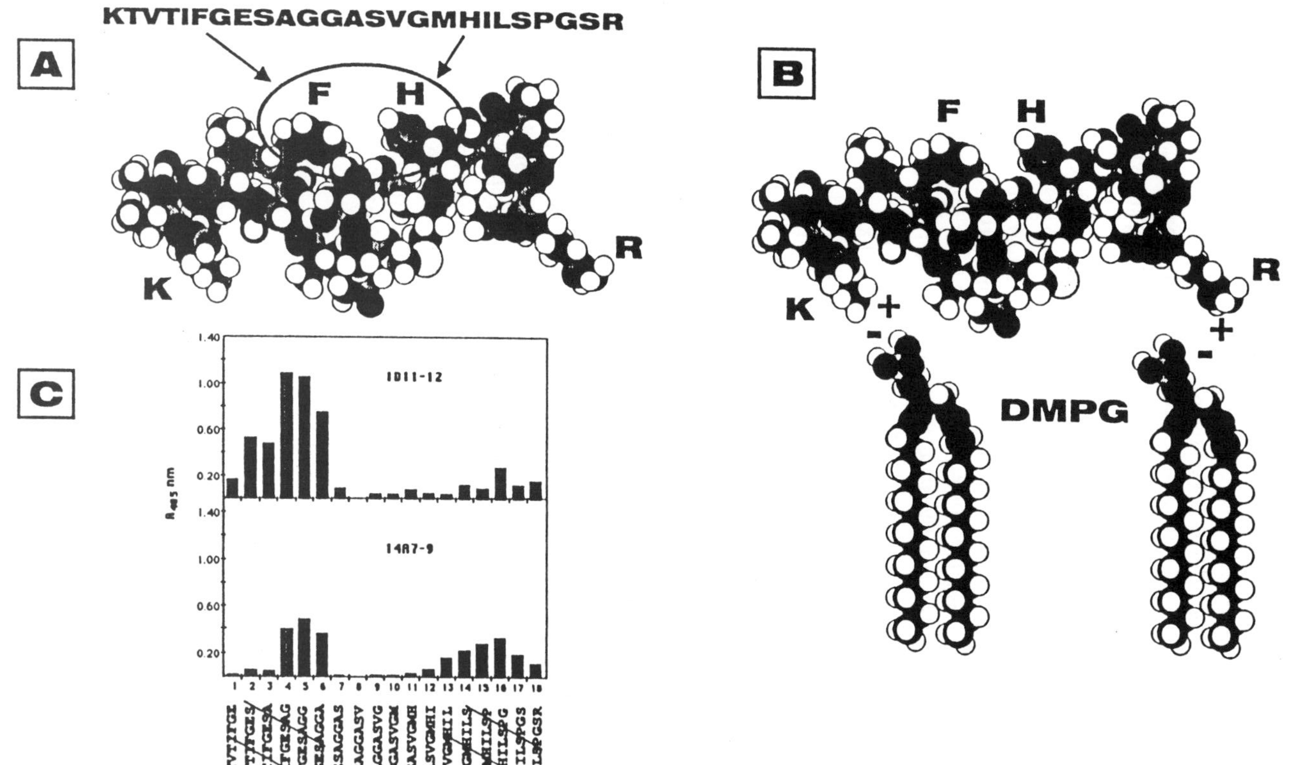

Fig. 1. Molecular modeling of antigenic epitopes of AChE active site peptide.[5] (A) Peptide models were built (Alchemy III, Windows version, Tripos Associates, Inc., St. Louis, MO) with consecutive addition of single amino acids from the N terminus with energy minimization (gradient cutoff at 0.01, and charge and PI-interactions included) at each step. (B) Dimyristoyl phosphatidylglycerol (DMPG) was built with HyperChem, version 3.0 for Windows, Hypercube, Inc., Waterloo, Ont., Canada. The minimized peptide model was exported as a sybylmol2 file to the HyperChem program, and the image was exported on Windows clipboard to CorelDraw (Ottawa, Ont., Canada). (C) Binding of two mAbs to overlapping octapeptides as tested by ELISA by Ogert et al.[19]). See text for further details.

both project down and are on the other side (the bottom) of the model structure when compared to the F and H residues on the top (Figure 1A). It was therefore hypothesized that the physical location of the antigenic peptide was on the outer surface of the liposomes and that it was oriented such that the K and R residues were anchored by electrostatic bonds to the phosphatidylglycerol in the liposomes thus projecting the conformational antigenic epitopes to a greater distance away from the surface where they then interacted with T or B lymphocytes to induce an immune response (Figure 1B).

Direct demonstration of the critical role of liposomal surface exposure of a small peptide antigen was also achieved in vivo.[5,20] A 15 amino acid peptide, P18, derived from the tip of the V3 loop of the envelope gp120 of the IIIB strain of HIV-1 was synthesized. During the synthesis a palmitoyl moiety was added either to the N (PA-CGP18) or carboxy (CGP18-PA) terminus. Upon incorporation into the lipid bilayer of liposomes the PA-CGP18, but not the CGP18-PA, was detected by fluorescence-activated cell sorting (FACS) analysis of the the liposomes with a monoclonal antibody directed against the CGP18 (Figure 2).[20] Molecular modeling of the constructs suggested that the epitope, including P, R, and F, recognized by the monoclonal antibodies in the FACS could have been cryptic in liposomes. As illustrated in the Figure 2 the reactive epitope of CGP18-PA has shifted clockwise and has been brought closer to the hydrophobic region, a conformation that may have prevented binding of the mAbs in the FACS analysis due to steric hindrance to binding of the antibody by the adjacent phospholipid headgroups.[5] Upon injection of the liposomes into mice, the PA-CGP18, but not the CGP18-PA, induced antibodies to the peptide. In contrast, both constructs induced murine CTLs. It was concluded that the surface expression of the PA-CGP18, as detected by FACS analysis, was critical for induction of the antibodies in vivo. Although the mechanism for this has not yet been completely worked out, I would speculate that although both types of liposomes, containing either PA-CGP18 or CGP18-PA, stimulated antigen presentation by macrophages, subsequent proliferation of B cells was induced by PA-CGP18 liposomes but not by CGP18-PA liposomes. Thus, surface expression of liposomal antigenic epitopes is not required for antigen presentation by macrophages, but may be required for subsequent B cell proliferation for production of antibodies.

IV. Role of liposomal lipid composition

IV.1. Effect of antigen composition

Experience has proven that generalizations based on scientific studies with a single antigen or adjuvant are risky when it comes to advocacy of a given liposome composition as being optimal for vaccines. In addition to scientific issues, such advocacy also represents a potential pitfall in protecting the intellectual property value of a vaccine in the process of documenting the metes and bounds of a liposomal formulation as an invention for commercial development. For example,

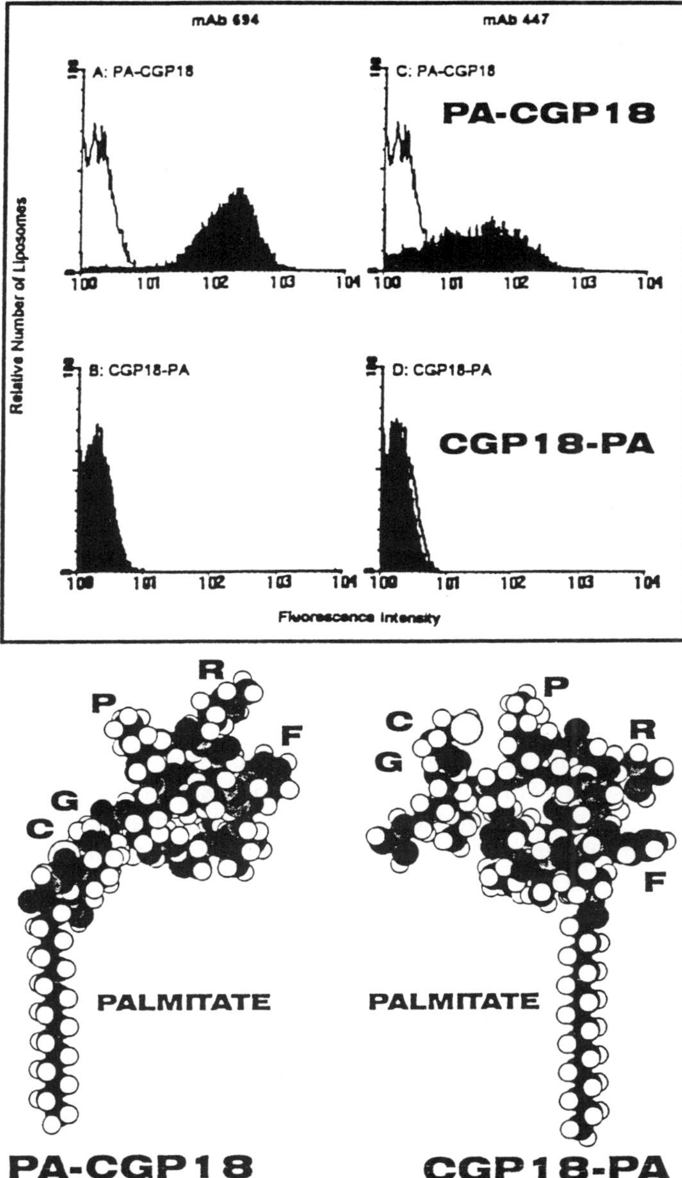

Fig. 2. Molecular modeling of antigenic epitopes of PA-CGP18 and CGP18-PA.[5] See legend to Figure 1 for the computer modeling technique. The FACS analysis, performed with mAb 694 and mAb 447, illustrates the relative number of liposomes containing peptide (shaded areas) compared to staining by secondary antibody alone (unshaded areas), is derived from White et al.[20]

as described both in a U.S. patent[21] and also in the scientific literature,[22] early studies with liposome-encapsulated diphtheria toxoid or tetanus toxoid suggested that the use of negatively charged liposomes effectively induced a higher immune response to diphtheria toxoid in mice, but that the use of positively charged liposomes led to an immune response that was equivalent to, or weaker, than that obtained with the use of either diphtheria toxoid alone. However, it was subsequently found that when other antigens were used, positively charged and negatively charged liposomes were equally effective as adjuvants.[23,24] It is apparent that whether a particular liposomal lipid formulation can act as an adjuvant for a liposomal antigen is a property that varies from antigen to antigen.

IV.2. Walter Reed liposomes

In the course of our human vaccine research we have concentrated on a particular liposomal composition that we refer to as Walter Reed liposomes,[25,26] a hand shaken multilamellar vesicle formulation that contains dimyristoyl phosphatidylcholine (DMPC), dimyristoyl phosphatidylglycerol (DMPG), cholesterol (CHOL), and lipid A. We have used monophosphoryl lipid A (MPLA; also known as MPL™) in the Walter Reed liposomes, but detoxification of lipid A by incorporation of the molecule into liposomes would even allow the use of native diphosphoryl lipid A instead of monophosphoryl lipid A. The molar ratios of DMPC/DMPG/CHOL are 9/1/7.5, and MPLA is present in different concentrations. Depending on the particular human vaccine formulation we have adjusted the individual dose of MPLA to range from approximately 100 μg to 2,200 μg. Although MPLA is somewhat reduced in toxicity when compared to native diphosphoryl lipid A (it had approximately 40-fold reduced pyrogenicity in our experience[27]), the remaining pyrogenicity and toxicity of MPLA which renders it highly reactogenic in humans,[28] is essentially completely eliminated in humans by incorporation into the liposomes.[29] The liposomes containing antigen and lipid A may then be adsorbed on aluminum hydroxide, although aluminum salts can cause destabilization of liposomes and such adsorption is not always thought to be beneficial.[30]

The Walter Reed liposomes have a high level of potency for induction of antibodies in humans.[29] When compared with other adjuvant formulations, such as alum or Ribi Detox®, smaller doses of liposome-encapsulated antigen induced higher antibody titres.[31] The liposome formulation, despite the presence of high doses of MPLA, was also significantly less reactogenic than antigen formulated with Detox®. From the standpoint of stability, the Walter Reed liposomes also have an extremely long shelf life. A recent analysis of a three-year old lot of Walter Reed liposomes containing an encapsulated malaria antigen that had been used in a human clinical trial demonstrated no detectable leakage. When compared with numerous other adjuvants the Walter Reed liposomes remain a prime candidate for application in a malaria vaccine.[31,32]

V. Routes of immunization

V.1. Liposomes as depots: Stealth is not required and may not be desired

At the present time all successful commercial vaccines are administered either by intracutaneous, subcutaneous (SC), intramuscular (IM), or oral routes (PO). The classical pharmaceutical approach for development of vaccines thus contrasts greatly with the classical IV injection drug delivery strategy that has been so successful in the commercial advancement of liposomes having a long circulation time as drug carriers. The tactic of camouflaging liposomes, for example by coating them with polyethylene glycol to prevent opsonization, thereby avoiding endocytic uptake or uptake by phagocytic cells (the so-called steric stabilization, or stealth, strategy),[33–35] while desirable for maximizing delivery of intravenously-injected liposomes to tumors or sites of infection, presumably would have the detrimental effect (from a vaccine standpoint) of promoting the avoidance of local interactions of subcutaneously- or intramuscularly-injected liposomes with the immune system.[36,37]

It has been determined that IM injection of Walter Reed liposomes results in a long-lasting depot effect at the site of injection, an effect in which inflammatory cells slowly ingest the liposomes, resulting in slow distribution of fluorescent liposomal lipid into the local lymphatic circulation.[38] There was virtually no detectable delivery of such liposomes to the spleen. In contrast, according to Allen et al.,[36] stealthy liposomes that are extremely small are readily transported from a subcutaneous injection site into the blood. Local inflammations and granulomas induced by IM or SC nonstealthy liposomal vaccines, particularly liposomes containing adjuvants such as endotoxin or cytokines, would be expected to encourage enhanced uptake of liposomes by phagocytic cells and interactions of liposomes with lymphatic cells at the site of injection. Suppression of such local interactions with inflammatory cells, for example by the use of so-called "Stealth"™ liposomes[33] that are not readily phagocytized would be expected to have a detrimental effect on the immune response. I believe that it might be theoretically more useful to facilitate, rather than discourage, the interactions of locally injected liposomes with inflammatory cells—an antistealth strategy.

From the above one might expect that liposomal vaccines would act at least partly through a depot effect at the site of injection. Although I believe that this is true, it has also been determined that the liposomes are not acting simply as semipermeable devices for slow release of free antigen. This conclusion is based on a study in which a mixture consisting of two types of liposomes, liposomes containing antigen but lacking lipid A and liposomes lacking antigen but containing lipid A, had a markedly reduced ability to generate antibodies in mice when compared with a single formulation of liposomes containing both antigen and lipid A in the same liposomes. The above experiment suggests that the liposome is viewed by the immune system as a single type of integrated particle containing both antigen and adjuvant, and that liposomes reside in a depot to which antigen presenting cells migrate for the purpose of ingesting and processing liposomes and liposomal antigen. Based on this it may be predicted that liposomes that are highly leaky to antigen would function much less well as a vaccine formulation than more stable liposomes that remain intact for long periods of time at the site of injection. However, methods have been described for efficient encapsulation of granulocyte stimulating factor within liposomes leading to slow release of granulocyte stimulating factor for adjuvant activity.[39]

V.2. Influence of rapid uptake of liposomes

Rapidity of uptake of liposomes by macrophages, a variable factor that would obviously be influenced by the depot effect of liposomes and representing as it does a dynamic mechanism for clearance of liposomes as drug carriers, has been extensively studied, for example by Fortin and Thérien.[40] Based on the presumed more rapid uptake of mannosylated liposomes compared to the slower uptake of nonmannosylated liposomes after intraperitoneal injection, it was concluded that rapid uptake may not optimize the immune response when excess liposomal antigen is available.[40,41] In contrast, rapid uptake of liposomes by phagocytic cells might optimize the immune response at low doses of liposomal antigen that would otherwise give a signal of sub-threshold intensity over a longer period of time.[40,42] Although not directly considered by the above authors, it would seem to me that this same principle of rapid vs. slow delivery of antigen could apply to IM depot vs IV rapid delivery of liposomal vaccines.

V.3. Targeting strategy

Although the IV administration of liposomal antigen is a highly effective route of administration, it has been claimed that targeting liposomes preferentially to splenic macrophages rather than to hepatic Kupffer cells may be advantageous for generating an immune response in mice.[43] The targeting was achieved by utilizing large liposomes containing egg phosphatidylcholine, cholesterol and ganglioside G_{M1}. It may or may not be significant that liposomes containing G_{M1} were previously reported to have "stealth" properties in mice[44,45] but not in rats,[46] and that G_{M1} has a suppressive effect on complement-dependent phagocytosis of liposomes by murine macrophages in vitro[47,48] but enhances the uptake of liposomes by Kupffer cells in rats due to increased complement activation.[46] The enhanced complement-dependent uptake in the rat compared to mouse may have been due to the presence of anti-G_{M1} antibodies in rat but not mouse plasma.[49]

V.4. Could there be a role for an intravenous liposomal vaccine?

Evidence from our laboratory and from others have suggested that complement activation, either via the classical or alternative pathway, may be one of the major opsonizing systems that are responsible for rapid removal of liposomes from the circulation.[50–54] A literature that goes back more than twenty years indicates rather convincingly that complement (C) activation also plays an important role in the immune response against certain T-cell dependent antigens.[55–58] In the case of large particulate antigens (such as any antigen that might be associated with large liposomes), the enhancing effects of C presumably are due to interactions of antigen-C complexes with complement receptors on phagocytic antigen presenting cells such as macrophages or dendritic cells.[54,59] Complement receptor 1 (CR1, CD35)-deficient and complement receptor 2 (CR2, CD21)-deficient mice have a markedly reduced primary and secondary antibody response against certain T-dependent antigens.[57]

Leaky liposomes, if they were to be used as a vaccine formulation, would probably be much more effective when injected IV or IP rather than IM, due to rapid uptake by macrophages in the liver, spleen, or peritoneal cavity after IV or IP administration. Indeed, liposomes that presumably were leaky because they were comprised of egg lecithin, cholesterol, and phosphatidic acid, did appear to induce high primary antibody titres to liposome-encapsulated diphtheria toxoid upon IV injection.[22] Upon SC or IM injection with the same antigen strong but somewhat lower titres were observed than than those obtained with the IV primary responses; however, the secondary antibody responses against the same liposomal antigen were not greater after SC or IM injection than those obtained by the free antigen alone.[22] In contrast, although the data were not shown, with liposomes that lacked cholesterol and were therefore presumably even more leaky, the primary immune response to IV liposomal albumin was reported to have been very low, lower than after the SC or IM routes of injection.[23] Perhaps in the latter case the liposomes had lost essentially all of the encapsulated antigen because of extreme leakiness of the liposomes.

In our own experience with IV injection, in which the antigen (cholera toxin) was either encapsulated or strongly bound to the ganglioside G_{M1} receptor on the surface of the liposomes, the immune responses, as detected by RIA of serum antibodies with purified cholera toxin as capture antigen, were extremely strong.[60,61] Regardless of the relative efficacy of IV vs SC or IM route of administration, because of the perceived logistical difficulties and costs that would occur with mass immunization of humans by the IV route, IV vaccines have never been seriously considered to date. However, for certain extremely difficult but very important vaccines (such as a vaccine against HIV), or for vaccine immunotherapy of cancer, the IV route, if it were the only recourse, could be considered.

In addition to the above, I should like to propose a possible role of IV injection of liposomes for vaccines, namely for efficient delivery of liposomes to the spleen for induction of cytotoxic T lymphocytes (CTLs). The subject of induction of splenic CTLs by liposomal antigen has been extensively studied in mice and class I presentation of liposomal antigens is discussed in detail in a separate chapter in this volume.[62] Because the depot effect of intramuscularly-injected liposomes may result in negligible or nonexistent delivery of liposomes to the spleen,[38] the only efficient method of gaining access to the spleen with large liposomes would be through IV injection.[34] Immunization by the IV route is used widely for induction of CTLs in mice.[63,64]

It has been established that a liposomal recombinant protein antigen (RLF), derived from the circumsporozoite antigen of *Plasmodium falciparum*, that contains both murine and human CTL epitopes, efficiently stimulates splenic CTLs after IP injection in mice[65] but is inefficient as a vaccine for stimulating circulating CTLs after IM injection in humans.[66] I would like to propose the following possible reasons for this common observation that CTL results are often efficient in mice but apparently inefficient in primates, and particularly in humans. First it should be noted that murine CTLs are almost always experimentally derived only from the spleens of sacrificed animals, but in humans only the blood is examined for CTLs. If the spleens are the actual sources of the CTLs in each species, it is likely that mouse spleen is much more accessible than human spleen for interaction with vaccine antigen. There are several possible reasons for this. (A) After IM injection the volume of vaccine per muscle fiber injected into a mouse muscle is generally large relative to the volume of vaccine per muscle fiber injected into a human muscle, thus potentially allowing escape of large amounts of vaccine that might travel out of the mouse muscle to the spleen. (B) The mouse is often injected IV or IP, again permitting delivery of antigen to the spleen, while humans are almost always injected either IM or SC. (C) After IM or SC injection of vaccine resulting in a depot of vaccine at the site of injection in humans, the actual transit distance to the spleen might be huge (perhaps approximating a meter), while the transit distance from the site of depot injection in a mouse is relatively small (perhaps approximating a centimeter).

The above analysis suggests that the geometric and physical differences between humans and mice might explain the apparent frequent inability to detect circulating CTLs derived from spleen after depot injection of vaccine into a distal IM or SC site in humans. We are currently engaged in several studies, including a human immunotherapeutic cancer vaccine trial, to determine the relative benefits of IV vs IM injection of liposomes for efficient induction of circulating CTLs.

It should be pointed out that the ability of a liposome-adjuvanted antigen to induce CTLs after IM injection in humans could also be dependent in part on the antigen itself. As evidence of this, a liposomal influenza A virus vaccine that did not enhance the induction of antibodies over the nonliposomal vaccine in elderly individuals after IM injection did significantly enhance the anti-influenza virus CTL activity.[67] The continuation of this latter study presumably will provide valuable information regarding the actual benefits (or lack of benefits) of consistent induction of CTLs against a viral infection in the elderly.

V.5. Delivery of liposomes to mucosal sites

Numerous laboratories have investigated the possibility of immunization by the oral route, pursuing the admirable dream of developing a potent and effective synthetic mucosal vaccine, a vision that has not yet translated to a widely used product.[68–70] The experience of my colleagues and myself has led us to believe that oral immunization with liposomes could occur, particularly with a gut-associated antigen such as an enteric bacterial toxin as an antigen, but we also believe that a considerable amount of additional development would be required for practical realization of a useful product.

The feasibility of utilizing the Walter Reed liposomes for immunizing orally with an enteric toxin was suggested by a set of experiments carried out in collaboration with Nathaniel Pierce who was then at Johns Hopkins University.[71,72] In those studies a murine model was utilized in which the animals were first immunized orally (enterically primed) either with liposomal cholera toxin (CT) or free CT, and the immune response was then boosted in each case by a second oral immunization with free CT. The immunogenicity was quantified by measuring the number of antitoxin (IgA)-containing plasma cells in jejunal lamina propria by use of a fluorescent antibody technique. The CT was either mixed with preformed empty liposomes, attached to the outer surface of liposomes by binding to the G_{M1} receptor (a process that totally eliminated toxicity of CT), or encapsulated CT. Mixing the CT antigen with empty liposomes had no effect at all on enteric priming when compared to CT alone; but when the CT was bound to G_{M1} there was a sharp drop in the amount of enteric priming induced by the liposomal CT. In contrast, as shown in Figure 3, when the liposomes containing G_{M1} also contained lipid A (the Walter Reed liposomes) there was a 5.7-fold increase in enteric priming by G_{M1}-bound CT when compared to liposomes lacking lipid A. Based on this it appears that liposomal lipid A can serve as a potent adjuvant for an orally administered liposomal antigen.

Cholera toxin, or the cholera toxin subunit B (CTB), comprise a different class of molecules that have also been proposed as mucosal adjuvants.[73,74] As shown

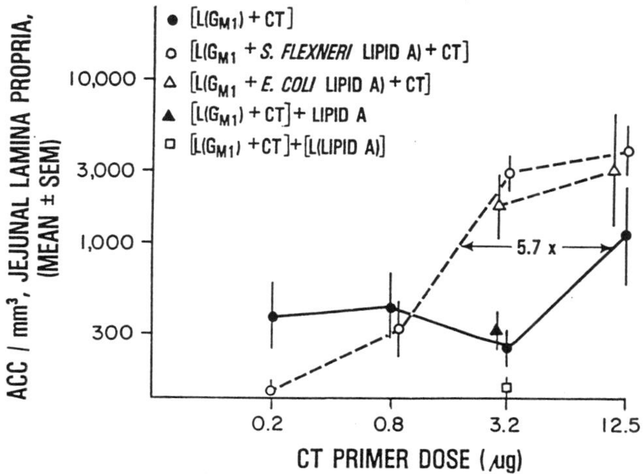

Fig. 3. Effect of liposomal lipid A on enteric priming after direct intraduodenal inoculation of CT bound to G_{M1} on the surface of liposomes.[71] After boosting with free CT the number of antitoxin-containing plasma cells (ACC) were enumerated in jejunal lamina propria by use of a fluorescent antibody technique. Results for each group of five to seven identically immunized rats are expressed as the geometric mean ACC/mm^3 in jejunal lamina propria ±SE. See text and primary reference for further details.

by the above studies,[71,72] all of the adjuvant activity of CT is lost if the CT is bound to G_{M1}. Conjugation of CT or CTB to liposomes lacking G_{M1} has been proposed as a strategy for targeting liposomes to Peyer's patch for adjuvant activity.[75] Even without conjugation CT and CTB serve as adjuvants for intranasal or rectal immunization with liposomal antigens.[76,77]

Oral immunization with liposome-encapsulated glutathione S-transferase from *Schistosoma mansoni*, an antigen that is known to induce protective immunity in various models, resulted in both a mucosal and systemic immune response to the enzyme in mice.[78] Specific IgA was detected in gut and specific IgG1 and IgG2b were detected in serum. Protective immunity was achieved by challenge, thus raising the possibility of developing a protective oral vaccine against schistosomiasis.

One of the most interesting recent developments in the use of liposomes as carriers for mucosal immunity is the apparent efficacy of liposomes administered intranasally for induction of both systemic immunity and pulmonic mucosal immunity.[79–84] Immunization via the IN route actually results in a secretory IgA response against influenza virus in the murine female urogenital tract.[80,83] Protective efficacy against influenza virus or against a plant toxin, ricin, was observed against experimental challenges with influenza virus[80,81] or with 5 LD_{99} doses of ricin toxin,[84] respectively. An unusual aspect of the IN immunization route is that the liposomes reportedly have adjuvant activity for induction of serum IgG antibodies as well as secretory IgA antibodies when the antigen is simply coadministered with the liposomes as opposed to being encapsulated within the liposomes.[80–83] Adjuvant effects of the coadministered "empty" liposomes could be accomplished even when the liposomes were administered as long as 48 hours prior to IN immunization.[81] Although the mechanism of this nonspecific effect of coadministered liposomes is not clear, it is possible that the immunostimulatory effect of the liposomes is due to a reversal, caused by liposome treatment, of an inhibitory effect of alveolar macrophages on the immune response.[85] Depletion of alveolar macrophages facilitated the antibody response against influenza virus, and it was speculated that the alveolar macrophages exerted a suppressive effect that was reversed by liposomes.[85]

VI. Rational selection of a liposomal vaccine formulation

The selection of a liposome formulation may be based on many different underlying commercial needs, among the most prominent of which are manufacturing issues. Manufacturing costs, and particularly the ability to scale up the manufacturing to high volumes, are necessarily important for many formulations, but may be less important for a low-volume, but high profit margin vaccine such as an immunotherapeutic vaccine against cancer. In the manufacture of liposomes, sterile filtration is not a viable option for formulations containing large particles (generally for those greater than 0.2 μm). Therefore, for large liposomes, such as hand-shaken liposomes, sterile manufacture is required. However, there are substantial drawbacks to the use of small liposomes: the techniques of sizing down, often consisting of complex procedures that may require harsh treatments such as high shear forces, may be detrimental to the protein antigen; the capture volume, which decreases with the cube of the diameter, may be quite low; and procedures for washing the liposomes free of unencapsulated antigen or drug may be complicated by the small size of the liposomes.

It is likely that methods of attaching proteins or peptides to the outer surface of liposomes would be highly effective and useful both because of the continuing close connection of the antigen with the liposomes, and because of the possibility of formulating a lyophilized formulation. This might be particularly beneficial for vaccines that must be applied in third world countries where vaccines are so widely needed and where storage conditions may not always be optimal. This also has the theoretical advantage, discussed earlier, that surface expression of epitopes may be a more efficient mechanism for stimulation of B lymphocytes for induction of antibodies.

Easy application, including oral or nasal application, an area of intense interest but as yet not yet sufficiently developed for liposome vaccines, remains a highly desirable direction for rational vaccine development. Liposomes are similar to other particles that have been proposed for mucosal vaccines both in the advantages and disadvantages. The case has not yet been settled either in favor or against such a strategy for synthetic particulate vaccines.

The Swiss Serum Institute has pioneered, under the leadership of Reinhard Glück, the first commercially licensed liposomal vaccine, registered first in Switzerland under the trade name EPAXAL BERNA in 1994.[86,87] One of the most interesting and potentially useful vaccine strategies, liposomal combination vaccines containing two or more antigens, has also been proposed by the Swiss Serum Institute. Combination or "super combination" vaccines (simultaneously containing more than three antigens),[83–85] although having understandably attractive virtues, also have potential drawbacks due to the possibility of antigenic competition. In our experience an extremely strong antigen that also had adjuvant properties (cholera toxin) exhibited an immunodominant role that apparently prevented the immune response against a coencapsulated weak malaria recombinant construct (SPf 66).[89] The immunodominance associated with the strong antigen was not strictly due to coencapsulation in the liposomes, but rather due to coimmunization. This was demonstrated by the observation that immunodominance that still occurred when the cholera toxin and SPf66 that were present in separate liposomes were mixed together during the immunization procedure.[89] Obviously antigenic competition does not occur with every antigen, but the possibility must always be considered with complex antigen combinations. I anticipate that novel new liposomal multiantigen combination vaccines will provide an interesting and broad additional commercial applicability to the future pharmaceutical development of liposomes.

References

1. Gregoriadis G. Immunological adjuvants: A role for liposomes. Immunol Today 1990;11:89–97.
2. Alving CR. Liposomes as carriers of antigens and adjuvants. J Immunol Meth 1991;140:1–13.

3. Phillips NC. Liposomal carriers for the treatment of acquired immune deficiency syndromes. Bull Inst Pasteur 1992;90:205–230.
4. Alving, CR. Lipopolysaccharide, lipid A, and liposomes containing lipid A as immunologic adjuvants. Immunobiology 1993;187:430–446.
5. Alving CR, Koulchin V, Glenn GM, Rao M. Liposomes as carriers of peptide antigens: Induction of antibodies and cytotoxic T lymphocytes to conjugated and unconjugated peptides. Immunol Rev 1995;145:5–31.
6. Germain RN, Margoulies DH. The biochemistry and cell biology of antigen processing and presentation. Ann Rev Immunol 1993;11:403–450.
7. Alving, CR. Immunologic aspects of liposomes: Presentation and processing of liposomal protein and phospholipid antigens. Biochim Biophys Acta (Rev Biomembranes) 1992;1113:307–322.
8. van Rooijen N. Macrophages as accessory cells in the in vivo humoral immune response: from processing of particulate antigens to regulation by suppression. Sem Immunol 1992;4:237–245.
9. Shek PN, Lukovich S. The role of macrophages in promoting the antibody response mediated by liposome-associated protein antigens. Immunol Lett 1982;5:305–309.
10. Su·D, van Rooijen N. The role of macrophages in the immunoadjuvant action of liposomes: effects of elimination of splenic macrophages on the immune response against intravenously injected liposome-associated albumin antigen. Immunology 1989;66:466–470.
11. van Rooijen N. Immunoadjuvant activities of liposomes: Two different macrophage-mediated mechanisms. Vaccine 1993;11:1170.
12. Bakouche O, Gerlier D. Presentation of an MuLV-related tumour antigen in liposomes as a potent tertiary immunogen after adoptive transfer. Immunology 1986;57:219–223.
13. Dal Monte PR, Szoka Jr FC. Effect of liposome encapsulation on antigen presentation in vitro. J Immunol 1989;142:1437–1443.
14. Dal Monte PR, Szoka Jr FC. Antigen presentation by B cells and macrophages of cytochrome *c* and its antigenic fragment when conjugated to the surface of liposomes. Vaccine 1989;7:401–408.
15. Verma JN, Rao M, Amselem S, Krzych U, Alving CR, Green SJ, Wassef NM. Adjuvant effects of liposomes containing lipid A: Enhancement of liposomal antigen presentation and recruitment of macrophages. Infect Immun 1992;60:2438–2444.
16. Rao M, Wassef NM, Alving CR, Krzych U. Intracellular processing of liposome-encapsulated antigens by macrophages depends on the antigen. Infect Immun 1995;63:2396–2402.
17. van Rooijen N, Su D. Immunoadjuvant action of liposomes: Mechanisms. In: Gregoriadis G, Allison AC, Poste G, eds. Immunological Adjuvants and Vaccines. New York: Plenum Press, 1989;95–106.
18. Alving CR. Liposomes as carriers for vaccines. In: Ostro MJ, ed. Liposomes: Biophysics to Therapeutics. New York: Marcel Dekker Inc., 1987;195–218.
19. Ogert RA, Richardson EC, Gentry MK, Abrahmson S, Alving CR, Taylor PT, Doctor, BP. Studies on the topography of catalytic site of acetylcholinesterase using polyclonal and monoclonal antibodies. J Neurochem 1990;55:756–763.
20. White WI, Cassatt DR, Madsen J, Burke SJ, Woods RM, Wassef NM, Alving CR, Koenig S. Induction of both antibody and cytotoxic T lymphocyte responses to a liposome-associated HIV-1 peptide. Vaccine 1995;13:1111–1122.
21. Allison AC, Gregoriadis G. Immunological preparations. US Patent No. 4,053,585, Issued October 11, 1977.
22. Allison AC, Gregoriadis G. Liposomes as immunological adjuvants. Nature 1974;252:252.
23. Heath TD, Edwards DC, Ryman BE. The adjuvant properties of liposomes. Biochem Soc Trans 1976;4:129–133.
24. van Rooijen N, van Nieuwmegen R. Liposomes in immunology: multilamellar phosphatidylcholine liposomes as a simple, biodegradable and harmless adjuvant without any immunogenic activity of its own. Immunol Commun 1980;9:243–256.
25. Vogel FR, Powell MF. Section on Walter Reed Liposomes, in A Compendium of Vaccine Adjuvants. In: Powell MF, Newman MJ, eds. "Vaccine Design: The Subunit and Adjuvant Approach". New York: Plenum Press, 1995;226–227.
26. Wassef, NM, Alving CR, Richards, RL. Liposomes as carriers for vaccines. Immuno Methods 1994;4:217–222.
27. Richards RL, Swartz Jr GM, Schultz C, Hayre MD, Ward GS, Ballou WR, Chulay JD, Hockmeyer WT, Berman SL, Alving CR. Immunogenicity of liposomal malaria sporozoite antigen in monkeys: Adjuvant effects of aluminum hydroxide and nonpyrogenic liposomal lipid A. Vaccine 1989;7:506–512.

28. Vosika GJ, Barr C, Gilbertson D. Phase-I study of intravenous modified lipid A. Cancer Immunol Immunother 1984;18:107–112.
29. Fries LF, Gordon DM, Richards RL, Egan JE, Hollingdale MR, Gross M, Silverman C, Alving CR. Liposomal malaria vaccine in humans: A safe and potent adjuvant strategy. Proc Natl Acad Sci USA, 1992;89:358–362.
30. Muderhwa JM, Wassef NM, Spitler LE, Alving CR. Effects of aluminum adjuvant compounds, Tweens and Spans on the stability of liposome permeability. Vaccine Res 1996;5:1–13.
31. Gordon DM. Use of novel adjuvants and delivery systems to improve the humoral and cellular immune response to malaria vaccine candidate antigens. Vaccine 1993;11:591–593.
32. Hui GSN. Liposomes, muramyl dipeptide derivatives, and nontoxic lipid A derivatives as adjuvants for human malaria vaccines. Am J Trop Med Hyg 1994;50(Suppl 4):41–51.
33. Allen TM. Stealth™ liposomes: avoiding reticuloendothelial uptake. In: Lopez-Berestein G, Fidler I, eds. "Liposomes in the Therapy of Infectious Diseases and Cancer". New York: Alan R. Liss, 1989;405–415.
34. Papahadjopoulos D, Allen TM, Gabizon A, Mayhew E, Matthay K, Huang SK, Lee K-D, Woodle MC, Lasic DD, Redemann C, Martin FJ. Sterically stabilized liposomes: improvements in pharmacokinetics and antitumor therapeutic efficacy. Proc Natl Acad Sci USA 1991;88:11460–11464.
35. Woodle MC, Lasic DD. Sterically stabilized liposomes. Biochim Biophys Acta 1992;1113:171–199.
36. Allen TM, Hansen CB, Guo LSS. Subcutaneous administration of liposomes: a comparison with the intravenous and intraperitoneal routes of injection. Biochim Biophys Acta 1993;1150:9–16.
37. Allen TM, Hansen CB, Peliowski A. Subcutaneous administration of sterically stabilized (Stealth) liposomes is an effective sustained release system for 1-β-D-arabinofuranosylcytosine. Drug Del 1993;1:55–60.
38. Richards RL, Hailey JR, Egan JE, Gordon DM, Johnson AJ, Alving CR, Wassef NM. Intramuscular fate and local effects of liposomes as carriers for drugs and vaccines. In: Shek PN, ed. "Liposomes in Biomedical Applications". Amsterdam: Harwood Academic Publishers, 1995;167–178.
39. Meyer J, Whitcomb L, Collins D. Efficient encapsulation of proteins within liposomes for slow release in vivo. Biochem Biophys Res Commun 1994;199:433–438.
40. Fortin A, Thérien H-M. Mechanism of liposome adjuvanticity: an in vivo approach. Immunobiol 1993;188:316–322.
41. Latif N, Bachhawat BK. The effect of surface sugars on liposomes in immunopotentiation. Immunol Lett 1984;8:75–78.
42. Garcon N, Gregoriadis G, Taylor M, Summerfield J. Mannose-mediated targeted immunoadjuvant action of liposomes. Immunology 1988;64,743–745.
43. Liu D, Wada A, Huang L. Potentiation of the humoral response of intravenous antigen by splenotropic liposomes. Immunol Lett 1992;31:177–182.
44. Gabizon A, Papahadjopoulos D. Liposome formulations with prolonged circulation time in blood and enhanced uptake by tumors. Proc Natl Acad Sci USA 1988;85:6949–6953.
45. Allen TM, Hansen C, Rutledge J. Liposomes with prolonged circulation times: factors affecting uptake by reticuloendothelial and other tissues. Biochim Biophys Acta 1989;981:27–35.
46. Liu D, Liu F, Song YK. Monosialoganglioside GM_1 shortens the blood circulation time of liposomes in rats. Pharmaceut Res 1995;12:508–512.
47. Wassef NM, Matyas GR, Alving CR. Complement-dependent phagocytosis of liposomes by macrophages: Suppressive effects of "stealth" lipids. Biochem Biophys Res Comm, 1991;176:866–874.
48. Alving CR, Wassef NM. Complement-dependent phagocytosis of liposomes: Suppression by "stealth" lipids. J Liposome Res 1992;2:383–395.
49. Liu D, Song YK, Liu F. Antibody dependent, complement mediated liver uptake of liposomes containing GM_1. Pharmaceut Res 1995;12:1775–1780.
50. Roerdink F, Wassef NM, Richardson EC, Alving CR. Phagocytosis of liposomes opsonized by complement: Effects of negatively charged lipids. Biochim Biophys Acta, 1983;734:33–39.
51. Chonn A, Semple SC, Cullis PR. Separation of large unilamellar liposomes from blood components by a spin column procedure: towards identifying plasma proteins which mediate liposome clearance in vivo. Biochim Biophys Acta 1991;1070:215–222.
52. Szebeni J, Wassef NM, Spielberg H, Rudolph AS, Alving CR. Complement activation in rats by liposomes and liposome-encapsulated hemoglobin. Biochem Biophys Res Comm 1994;205:255–263.
53. Szebeni J, Wassef NM, Rudolph AS, Alving CR. Complement activation in human serum by

liposome-encapsulated hemoglobin: The role of natural anti-phospholipid antibodies. Biochim Biophys Acta (Biomembranes) 1996;1285:127–130.

54. Szebeni J, Wassef NM, Hartman KR, Rudolph AS, Alving, CR. Complement activation in vitro by the red blood cell substitute, liposome-encapsulated hemoglobin: Mechanism of activation and inhibition by soluble complement receptor type 1. Transfusion 1997;37:150–159.
55. Cooper PD. Vaccine adjuvants based on gamma inulin. In: Powell MF, Newman MJ, eds. "Vaccine Design: The Subunit and Adjuvant Approach". New York: Plenum Press, 1995;559–580.
56. Pepys MM. Role of complement in induction of antibody production in vivo. Effect of cobra factor and other C3-reactive agents on thymus-dependent and thymus-independent antibody responses. J Exp Med 1974;140:126–145.
57. Molina H, Holers VM, Li B, Fang Y, Mariathasan S, Goellner J, Strauss-Schoenberger J, Karr RW, Chaplin DD. Markedly impaired humoral immune response in mice deficient in complement receptors 1 and 2. Proc Natl Acad Sci USA 1996;93:3357–3361.
58. Fischer MB, Ma M, Goerg S, Zhou X, Xia J, Finco O, Han S, Kelsoe G, Howard RG, Rothstein TL, Kremmer E, Rosen FS, Carroll MC. Regulation of the B cell response to T-dependent antigens by classical pathway complement. J Immunol 1996;157:549–556.
59. Fearon DT, Wong WW. Complement-ligand interactions that mediate biological responses. Annu Rev Immunol 1983;1:243–271.
60. Alving CR, Banerji B, Clements J, Richards RL. Adjuvanticity of lipid A and lipid A fractions in liposomes, In: Tom BH, Six HR, eds. Liposomes and Immunobiology. New York: Elsevier/North-Holland, 1980;67–78.
61. Alving CR, Richards RL, Moss J, Alving LI, Clements JD, Shiba T, Kotani S, Wirtz RA, Hockmeyer WT. Effectiveness of liposomes as potential carriers of vaccines. Applications to cholera toxin and human malaria sporozoite antigen. Vaccine 1986;4:166–172.
62. Rao M, Alving CR. Class I presentation of liposomal antigens. 1998;[PRESENT VOLUME]. 15–24.
63. Lopes LM, Chain BM. Liposome-mediated delivery stimulates a class I-restricted cytotoxic T cell response to soluble antigen. Eur J Immunol 1992;22:287–290.
64. Reddy R, Zhou F, Nair S, Huang L, Rouse BT. In vivo cytotoxic T lymphocyte induction with soluble proteins administered in liposomes. J Immunol 1992;148:1585–1589.
65. White K, Krzych U, Gordon DM, Porter TG, Richards RL, Alving CR, Deal CD, Hollingdale M, Silverman C, Sylvester DR, Ballou WR, Gross M. Induction of cytolytic and antibody responses using *P. falciparum* repeatless circumsporozoite protein encapsulated in liposomes. Vaccine 1993;11:1341–1346.
66. Heppner DG, Gordon DM, Gross M, Wellde B, Leitner W, Krzych U, Schneider I, Wirtz RA, Richards RL, Trofa A, Hall T, Sadoff JC, Boerger P, Alving CR, Sylvester DR, Porter TG, Ballou WR. Safety, immunogenicity and efficacy of *Plasmodium falciparum* repeatless circumsporozoite protein vaccine encapsulated in liposomes. J Infect Dis 1996;174:361–366.
67. Powers DC, Manning MC, Hanscome PJ, Pietrobon PJF. Cytotoxic T lymphocyte responses to a liposome-adjuvanted influenza A virus vaccine in the elderly. J Infect Dis 1995;172:1103–1107.
68. McGhee JR, Mestecky, J. In defense of mucosal surfaces. Development of novel vaccines for IgA responses protective at the portals of entry of microbial pathogens. Infect Dis Clin N Amer 1990;4:315–341.
69. Gilligan CA, Po ALW, Oral vaccines: Design and delivery, Int J Pharmaceut 1991;75:1–24.
70. Alving CR. Liposomes as vehicles for vaccines: Induction of humoral, Cellular, and Mucosal Immunity. In: Iglewski B, Vaughan M, Tu AT, Moss J, eds. "Handbook of Natural Toxins, Volume 8: Microbial Toxins". New York: Marcel Dekker, Inc., 1995;47–58.
71. Pierce NF, Sacci Jr JB, Alving CR, Richardson EC. Lipid A enhances mucosal immunogenicity of liposome-associated cholera toxin. Rev Inf Dis 1984;6:563–566.
72. Pierce NF, Alving CR, Richardson EC, Sacci Jr JB. Enhancement of specific mucosal antibody response by locally administered adjuvants. In: Kuwahara S, Pierce NF, eds. Advances in Research on Cholera and Related Diarrheas Vol 2. KTK Scientific Publishers, 1985;163–170.
73. Elson CO, Ealding W. Cholera toxin feeding did not induce oral tolerance in mice and abrogated oral tolerance to an unrelated protein antigen. J Immunol 1984;133:2892–2897.
74. Lycke N, Holmgren J. Strong adjuvant properties of cholera toxin on gut mucosal immune responses to orally presented antigens. Immunology 1986;59:301–308.
75. Harokopakis E, Childers NK, Michalek SM, Zhang SS, Tomasi M. Conjugation of cholera toxin or its B subunit to liposomes for targeted delivery of antigens. J Immunol Meth 1995;185:31–42.
76. Vadolas J, Davies JK, Wright PJ, Strugnell RA. Intranasal immunization with liposomes induces strong mucosal immune responses in mice. Eur J Immunol 1995;25:969–975.

77. Zhou F, Kraehenbuhl J-P, Neutra MR. Mucosal IgA response to rectally administered antigen formulated in IgA-coated liposomes. Vaccine 1995;13:637–644.
78. Ivanoff N, Phillips N, Schacht A-M, Heydari C, Capron A, Riveau G. Mucosal vaccination against schistosomiasis using liposome-associated Sm 28 kDa glutathione S-transferase. Vaccine 1996;14:1123–1131.
79. Aramaki Y, Fujii Y, Yachi K, Kikuchi H, Tsuchiya S. Activation of systemic and mucosal immune response following nasal administration of liposomes. Vaccine 1994;12:1241–1245.
80. Wilschut J, de Haan A, Geerligs HJ, Huchshorn JP, van Scharrenburg GJM, Palache AM, Renegar KB, Small Jr PA. Liposomes as a mucosal adjuvant system: an intranasal liposomal influenza subunit vaccine and the role of IgA in nasal anti-influenza immunity. J Liposome Res 1994;4:301–314.
81. de Haan A, Geerligs HJ, Huchshorn JP, van Scharrenburg GJM, Palache AM, Wilschut J. Mucosal immunoadjuvant activity of liposomes: induction of systemic IgG and secretory IgA responses in mice by intranasal immunization with an influenza subunit vaccine and coadministered liposomes. Vaccine 1995;13:155–162.
82. de Haan A, Tomee JFC, Huchshorn JP, Wilschut J. Liposomes as an immunoadjuvant system for stimulation of mucosal and systemic antibody responses against inactivated measles virus administered intranasally to mice. Vaccine 1995;13:1320–1324.
83. de Haan A, Renegar KB, Small Jr PA, Wilschut J. Induction of a secretory IgA response in the murine female urogenital tract by immunization of the lungs with liposome-supplemented viral subunit antigen. Vaccine 1995;13:613–616.
84. Matyas GR, Alving CR. Protective prophylactic immunity against intranasal ricin challenge induced by liposomal ricin A subunit. Vaccine Res 1996;5:163–172.
85. de Haan A, Groen G, Prop J, van Rooijen N, Wilschut J. Mucosal immunoadjuvant activity of liposomes: role of alveolar macrophages. Immunology 1996;89:488–493.
86. Glück R. Liposomal hepatitis A vaccine and liposomal multiantigen combination vaccines. J Liposome Res 1995;5:467–479.
87. Mengiardi B, Berger R, Just M, Glück R. Virosomes as carriers for combined vaccines. Vaccine 1995;13:1306–1315.
88. Glück R. Liposomal presentation of antigens for human vaccines. In: Powell MF, Newman MJ, eds. Vaccine Design: The Subunit and Adjuvant Approach. New York: Plenum Press, 1995;325–345.
89. Glenn GM, Rao M, Richards RL, Matyas GR, Alving CR. Murine IgG subclass antibodies to antigens incorporated in liposomes containing lipid A. Immunol Lett 1995;47:73–78.

Lasic and Papahadjopoulos (eds.), Medical Applications of Liposomes

CHAPTER 3.1

The development of liposomal amphotericin B: An historical perspective

KISHOR M. WASAN[a] AND GABRIEL LOPEZ-BERESTEIN[b]

[a]*Division of Pharmaceutics and Biopharmaceutics, Faculty of Pharmaceutical Sciences, The University of British Columbia, Vancouver, BC, Canada*
[b]*Section of Immunobiology and Drug Carriers, Department of Bioimmunotherapy, Division of Medicine, The University of Texas M.D. Anderson Cancer Center, Houston, Texas 77030, USA*

Overview

Abstract

In the past twenty years the increase in life-threatening systemic fungal infections, particularly in cancer, diabetic, and immunocompromised patients is alarming. Amphotericin B (AmB) has remained the most effective and widely used agent in the treatment of these infections, however, its use has been limited by dose-dependent kidney toxicity. In the early 1980s a number of promising lipid-based AmB formulations were developed. Our laboratory in the last 13 years has developed and investigated one of these liposomal AmB formulations.

I. Introduction

Over the last twenty years the frequency of life-threatening fungal infections have increased dramatically, particularly in immunocompromised patients.[1–4] Several factors have contributed to this rise: improved recognition and diagnosis of fungal infections; prolonged survival of patients with defects in their host defense mechanisms, including patients with cancer, organ transplant recipients, diabetics, and patients with AIDS; more invasive surgical procedures; the use of prosthetic devices and indwelling catheters; increased administration of parenteral nutrition; and the use of peritoneal dialysis and hemodialysis.[5–7] In these patients invasive fungal infections may account for as many as 30% of deaths.[8]

Despite the development of a number of new antifungal agents, amphotericin B (AmB), a polyene macrolide antibiotic, remains the gold standard agent in the treatment of systemic fungal infections.[9] AmB, produced by *Streptomyces*, has the broadest spectrum of activity of any clinically useful antifungal compound.[10–12] AmB interacts with ergosterol in the plasma membrane, causing membrane disruption, increased permeability, leakage of vital intracellular constituents, and eventual cell death.[13,14] Recent evidence suggests that AmB can cause oxidative damage, which may contribute to its fungicidal activity.[10] AmB has a higher affinity for the fungal sterol, ergosterol, than its mammalian counterpart, cholesterol, and is thus less toxic to mammalian cells.[15] Since the clinical use of AmB has been limited by its renal toxic effects,[16–18] an important question is how to best direct the drug specifically to the fungus and keep it away from sites of toxicity. One strategy is to use a vehicle preparation other than the commercially available preparation of AmB, which is a mixture of AmB, a detergent sodium deoxycholate, and a buffer that forms a micellar colloidal dispersion.

II. The early years

For many years, liposomes, originally described by Bangham and coworkers in the mid 1960s,[19,20] were used as models for biological membranes. In the early 1980s several research groups developed new AmB formulations by incorporating the drug into liposomes. New and coworkers initially investigated the antileshmanial activity of AmB and other antifungal agents entrapped in liposomes.[21] Six months later Graybill and coworkers published the first extensive paper investigating the treatment of murine cryptococcosis with liposome-associated AmB.[22] In this study BALB/c mice were challenged with *Cryptococcus neoformans* and given liposome-associated AmB or AmB-deoxycholate intravenously. They found mice that were treated with liposome-associated AmB survived longer and had lower tissue counts of cryptococci than mice treated with AmB-deoxycholate or untreated control mice. They concluded that the reduced acute toxicity of liposome-associated AmB permitted much larger doses of AmB to be given than were possible with AmB-deoxycholate.

In the early 1980s our laboratory incorporated AmB into liposomes consisting of dimyristoyl phosphatidylcholine (DMPC) and dimyristoyl phosphatidylglycerol (DMPG) in a lipid-to-drug weight ratio of 12:1.[23–30] We have shown, in both experimental and clinical studies, that this formulation has less toxicity than conventional AmB which allows increased doses to be given.

III. Animal studies

The effectiveness of free AmB (Fungizone®, Bristol-Myers Squibb, Nutley, NJ, USA; consisting of AmB and sodium deoxycholate) and that of L-AmB were tested against experimentally induced systemic candidiasis in mice.[23,31] Mice were inoculated intravenously with a strain of *Candida albicans* isolated from a patient with systemic candidiasis. Two days after the inoculation, a severe infection was detected in the liver, spleen, and kidneys. All treatments were administered intravenously starting 2 days after the inoculation with *Candida albicans*. Empty liposomes did not affect mouse survival. Multiple doses of free AmB at its maximum tolerated dose enhanced survival of mice significantly; however, a similar regimen using L-AmB was superior to one with free AmB and led to prolonged survival (greater than 60 days) and to a 60% cure rate (no histopathologic or microbiologic evidence of infection). L-AmB injected as a single cumulative dose corresponding to the total dose of the other regimens led to a statistically significant enhancement in survival compared with the result obtained with free AmB. These data demonstrate that L-AmB is far more active than free AmB in this model. The findings in experimentally induced candidiasis were later confirmed by others.[32,33] AmB incorporated into liposomes was also shown to be effective against experimentally induced histoplasmosis[34] and induced leishmaniasis in hamsters and nonhuman primates.[35] L-AmB was shown to be from 331 to 750 times more active than meglumine antimonate and from 2 to 5 times more active than free AmB in hamsters infected with *Leishmania donovani*. In squirrel monkeys infected with *Leishmania donovani*, L-AmB led to a 99% suppression of amastigotes in the liver. In two murine models of cutaneous *leishmaniasis* no significant decrease in tissue parasite density was observed when treated with liposome-intercalated AmpB seven days following inoculation with *Leishmania tropica*.[36] However, a case of visceral leishmaniasis in mice unresponsive to several courses of treatment with standard antifungal agents, was successfully cured by a 21 day course (50 mg/day) of liposomal amphotericin (AmBisome, Vestar, Inc.).[37] Furthermore, AmBisome appears to be effective following multiple dose-therapy to mice infected with Leishmania infantum.[38]

These experiments with L-AmB suggest that the therapeutic index is increased due to the better tolerance of high AmB dosages, but that the efficacy of a given dose is similar or even slightly decreased with L-AmB. However, Leishmaniasis appears to be an exception since L-AmB efficacy is increased at low unitary dosage.

IV. Clinical studies

Early clinical trials with L-AmB in patients with systemic mycoses refractory to free AmB and other antifungal agents were conducted at The University of Texas M.D. Anderson Cancer Center between 1983 and 1989.[30,39,40] Though the dosage was modified according to each patient's tolerance, the standard regimen was 2 mg/kg body weight of AmB daily for 3 days, which, when well tolerated (no fever, chills, or changes in kidney function), was increased by 1 mg/kg every fourth dose until 5 mg/kg was reached. Then 5 mg/kg of AmB was infused once daily for 3 days until the patient had received a total of 75 mg AmB/kg. The maximal single dose administered in those studies was 6 mg/kg. It is important to point out that the single maximum tolerated dose in mice ranged from 16 to 20 mg AmB/kg body weight and the active dose was 1 mg/kg so a therapeutic blood concentration of 10 to 20 μg/ml was maintained in the animal.

All patients tolerated L-AmB well; mild fever and chills occurred in only a few. Potassium supplements were required in most patients, particularly in those who received doses greater than 2 mg/kg body weight of L-AmB. Clinical improvement was observed in most patients during the first week of treatment, and no long-term renal, hepatic, or central nervous system toxicity's were observed.

An additional study included 46 cancer patients who had developed a variety of systemic fungal infections and were treated with L-AmB.[40] Twenty-one of these patients had disseminated candidiasis, 19 had aspergillosis, and the rest had a variety of other fungal infections. Forty patients failed to respond to conventional AmB therapy, and 6 were given L-AmB because therapy with conventional AmB drug severe side effects (e.g., nausea, vomiting, and hypokalemia). Twenty-four patients had a complete response, and 22 patients had none. No short- or long-term toxicity's were observed. Acute side effects associated with conventional AmB therapy (e.g., fever, chills, and potassium loss) were infrequent and milder in patients given L-AmB than commonly observed in patients given conventional AmB. No chronic renal, hematological, or central nervous system side effects were observed following therapy with L-AmB.

L-AmB is effective and less toxic than free AmB in the treatment of fungal infections caused by *Candida albicans* and *Aspergillus niger*; even in patients with neutropenia. The administration of L-AmB allowed for antileukemic treatment despite the presence of an active fungal infection and chemotherapy-induced neutropenia which usually compromises treatment of the fungal infection. L-AmB therapy is easier than AmB therapy, in part because of lower fluid volumes and shorter intravenous infusion times that enable patients to continue antifungal treatment as outpatients.

Recently, it has been observed two cases of visceral *leishmaniasis* in patients infected with HIV that L-AmpB has effective in curing these patients without toxicity.[41] Fusai and coworkers have also reported L-AmpB to be an effective anti-leishmania agent resulting in complete remission of polyresistant visceral *leishmaniasis* following L-AmpB therapy.[42] In addition, a multi-center trial has further shown the effectiveness of AmBisome in the treatment of 31 patients with visceral leishmaniasis.[43]

V. Mechanistic studies

V.1. Phagocyte transport of L-AmB

Liposomes are avidly taken up by phagocytes in the circulation and in tissues. We and others[44–46] have shown that liposomes are distributed in animals and man in organs rich in mononuclear phagocyte system cells. We previously observed that liposome incorporation enhanced the delivery of AmB to *Candida* infected organs in mice.[27] A potential exists, therefore, that monocytes and macrophages in peripheral blood may take up the drug-laden liposomes and transport them to the infected sites.

The in vitro uptake of AmB and L-AmB by murine peritoneal macrophages was studied.[47] Resident peritoneal macrophages were incubated at several time intervals with L-AmB or AmB in RPMI 1640 supplemented with 10% fetal calf serum. After each time interval, the supernatants were discarded, and the monolayer was washed three times with warm phosphate-buffered saline. The cells were then lysed, and radioactivity in the cell lysates was measured. Maximal uptake of L-AmB was observed after 8 hours, with a gradual decrease from 24 to 72 hours. The macrophage's liposome uptake capacity was maximal at 8 hours with a lipid saturation capacity of 100 μg lipid per million macrophages. No uptake was observed for AmB. However, recovery of the drug in organs rich in mononuclear phagocytes does not necessarily mean that the drug is taken up by phagocytes in vivo.

In the setting of patients with neutropenia, phagocytic transport is less likely to play a major role. Furthermore, since resident macrophages are less sensitive to cytotoxic agents than are other cell types, we believe that in the neutropenic setting a different transport of AmB may exist.

V.2. Lipoprotein transport of L-AmB

Serum lipoproteins have been hypothesized to influence the pharmacokinetics, tissue distribution, and pharmacological activity of AmB in rats.[48] In man, large volume of distribution of AmB appears to be a result of the drug's high accumulation in the kidney, liver, and lung tissues.[49] Injection of drug-free liposomes (DMPC:DMPG 7:3 w/w ratio) into the human circulation has resulted in a large volume of distribution and a long terminal half-life.[43] When AmB was injected intravenously into mice, only 15% of the original dose could be accounted for, 10% in the lung and 5% in the liver.[50] Furthermore, pharmacokinetics studies in humans have shown AmB to have a long terminal half-life ($t_{1/2b}$ = 15 days), and a very short distribution half-life ($t_{1/2a}$).[49] Intravenous injection of AmB into animals has resulted in slow or sustained release of the drug and altered tissue kinetics and distribution.[51] It has been suggested that the unusual pharmacokinetics of AmB may be a result of the slow release of the drug from a tissue or organ site due to the high-affinity binding of it to cholesterol in serum lipoproteins or cell membranes.[52–55]

Brajtburg and coworkers examined the interactions of AmB with human serum lipoproteins in vitro in an attempt to understand these interactions and how they might affect the pharmacological behavior of AmB. Their studies showed AmB to be equally associated with high-density lipoprotein (HDL) and low-density lipoprotein (LDL) fractions after 1 hour of incubation at 25°C.[52] Furthermore, AmB injected in LDL to rabbits was toxic: 70% of the rabbits died from a non toxic dose of AmB (1.0 mg/kg), which implies that LDL association would increase AmB toxicity.[54]

The results we obtained demonstrated that changes in temperature and liposomal lipid composition affect the distribution of AmB in serum lipoproteins.[56] Human serum obtained from healthy volunteers was incubated with known concentrations of AmB or different liposomal formulations of AmB (1 to 100 μg/ml) at 37°C for various time intervals (5, 10, 20, 30, 45, and 60 minutes). At the end of each time interval serum was removed and separated into HDL and LDL fractions by affinity chromatography. AmB in each lipoprotein fraction was quantified by high pressure liquid chromatography, and lipoprotein content was assessed. Equal distribution of AmB was found in human serum lipoprotein fractions following one hour incubation at 25°C. In contrast at 37°C, over 90% of the concentration of AmB was found in the HDL fraction following 1 hour incubation. AmB incorporated into liposomes composed of DMPC and DMPG showed a HDL:LDL ratio of 9:1. Liposomes composed of DMPG alone showed a HDL:LDL ratio of 1:1. Liposomes composed of DMPC or DMPC and stearlyamine (SA) showed a HDL:LDL ratio of 6:4. Studies were subsequently conducted where human serum was incubated with L-AmB (DMPC:DMPG 7:3 wt/wt ratio with a DMPG:AmB 4:1 M ratio) for 60 minutes at 37°C, serum separated into its lipoprotein fractions, and DMPG and AmB quantified by HPLC. Ninety percent of the drug and 80% of the lipid was found in the HDL fraction in a 3:1 M ratio (DMPG:AmB), while in a 6:1 M ratio (DMPG:AmB) in the LDL fraction. These experiments further suggested that AmB and DMPG may co-transfer as an intact drug-lipid complex to serum lipoproteins.

The modification of the distribution of AmB to serum lipoproteins at 37°C may be related to the transition temperature of lipoproteins, which is between 27°C and 34°C.[57,58] At the transition temperature, cholesteryl esters within the lipoprotein core exist as an isotropic solution, while below this temperature, they form disordered smectic liquid crystals.[57,58] The core of HDL becomes more ordered at the higher temperature, thus making it easier for the AmB molecule to associate with it. This hypothesis is based on the assumption that AmB is incorporated into the lipophilic core of these lipoproteins.

DMPG as an anionic exogenous phospholipid may distribute into HDL as opposed to LDL and be partially responsible for the concurrent transport of AmB to HDL. Since HDL and LDL are not found in an equimolar ratio in human serum, but at an LDL:HDL ratio of 6:1,[56] the data suggest that some mechanism besides random collision must drive this drug-liposome complex towards HDL rather than LDL. When L-AmB (4:1 molar ratio DMPG:AmB) was incubated for 1 hour at 37°C in human serum, AmB and DMPG seemed to co-transfer to the serum lipoproteins.[56] These observations suggest that phospholipids with a negative charge may be responsible for the altered AmB-lipoprotein distribution patterns. Furthermore, the DMPG:AmB mole ratio found in the lipoprotein fractions is similar to the initial mole ratio of the liposomes prior to incubation, which suggests that the drug-lipid association remains intact as it travels to HDL. Barwicz and coworkers have suggested that AmB association with LDL and very-low-density lipoproteins (VLDL) may be responsible for AmB nephrotoxicity in vivo and that hindering this complex formation results in a decrease in AmB nephrotoxicity.[59]

The rate at which AmB appears in the HDL fraction increases when AmB is incorporated in a liposome composed of DMPC and DMPG. Morton and Zilversmit have also demonstrated that a highly purified lipid transfer protein (LTP) facilitates the transfer of CE, triglyceride, and phosphatidylcholine between lipoprotein classes.[60,61] They demonstrated that LTP interacts with HDL, LDL, and VLDL; with the HDL-LTP interaction being the most likely to occur.[60,61] This interaction is attributed to HDL's ability to attract negatively charged particles (as the negative charge density increases on the surface of the lipoprotein, the lipoprotein-LTP interaction is more prominent) as well as to LTP's tendency to associate with negatively charged particles.[62] This interaction between HDL and LTP appears to be reversible.[60,61] In addition, work by Surewicz and coworkers has suggested the formation of thermally stable complexes between anionic phospholipids such as DMPG and apolipoprotein AI, one of the predominant protein components associated with HDL.[63] We have observed that when Fungizone® (AmB and sodium deoxycholate) and sodium deoxycholate were incubated for 120 minutes at 37°C in delipidated human serum containing 0.64 μg total protein/ml of LTP, CE transfer from HDL to LDL was not impaired (Figure 1A).[64] However, L-AmB at all concentrations greater than 10 μg/ml of AmB significantly decreased CE transfer compared with that of controls.[64] Since AmB interacts with cholesterol and CE[65] and the transfer of CE between lipoproteins is regulated by LTP,[60,61]

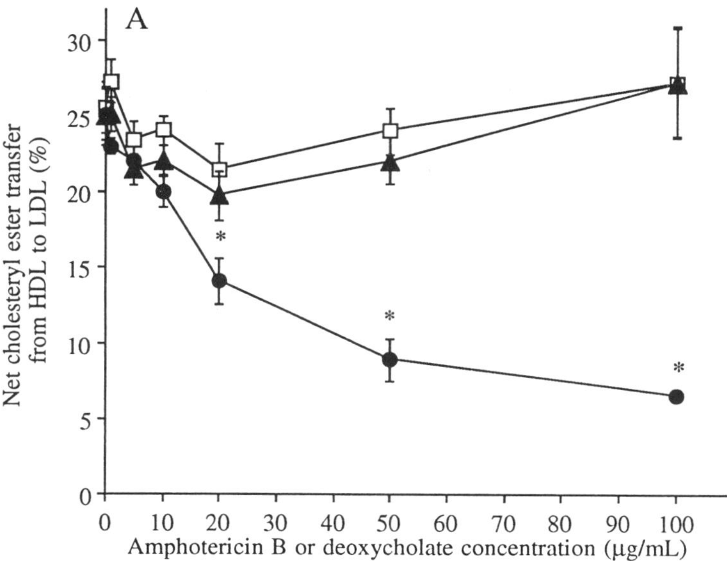

Fig. 1A. Effect of amphotericin B (AmB+ sodium deoxycholate) [▲], sodium deoxycholate [□], and liposomal amphotericin B (L-AmB) [●] on the cholesteryl ester transfer from high-density lipoproteins (HDL) to low-density lipoproteins (LDL). *$p < 0.05$ vs. amphotericin B (AmB+ deoxycholate) and deooxycholate alone. Data expressed as mean ± standard deviation (number of individual experiments = 6).

Table 1

Effect of lipid transfer protein (LTP) on the distribution of AmB and LAmB into serum lipoproteins after 60 minutes incubation in pooled human serum

Lipoprotein fraction	AmB[a]		LAmB[a]	
	No LTO added %	LTP added[b] %	No LTO added %	LTP added[b] %
HDL	74 ± 0.5	48.6 ± 4.9*	92.0 ± 5.0	88.0 ± 5.4
LDL	22 ± 5.5	45.6 ± 4.4*	Not detected	9.0 ± 7.8

Abbreviations: AmB, amphotericin B; LAmB, liposomal amphotericin B; HDL, high-density lipoproteins; LDL, low-density lipoproteins; [a]percent of initial AmB; [b]0.64 μg/ml LTP added; $*p < 0.05$ vs. AmB with no LTP added; mean ± standard deviation ($n = 6$). Modified from Ref. 64.

we conducted studies to determine the influence of LTP on the distribution of AmB between HDL and LDL and the influence of liposomal-lipid composition on LTP-regulated transfer of CE from HDL to LDL.

Our results demonstrated that the presence of LTP facilitates the transfer of AmB (incubated as Fungizone) between HDL and LDL. The addition of LTP resulted in increased distribution of AmB to the LDL fraction (Table 1). Furthermore, the presence of Fungizone or sodium deoxycholate alone did not reduce the CE transfer activity of LTP (Figure 1A).[64] These observations suggest that the redistribution of AmB from HDL to LDL may be regulated by LTP. Furthermore, previous investigators have suggested that AmB interacts with CE and cholesterol[65] upon incubation in human serum, thus supporting the hypothesis that it is AmB-associated CE which is being transferred from HDL to LDL by LTP.

AmB association with HDL increases when AmB is incorporated in liposomes composed of DMPC, DMPG, and SA.[56] However, we found that the addition of LTP facilitated only a minimal transfer of AmB from HDL into LDL when AmB was incorporated into liposomes composed of DMPC and SA (data not shown) or DMPC and DMPG (Table 1).[64] Furthermore, the presence of empty or AmB-containing DMPC/SA or DMPC/DMPG liposomes decreased LTP-regulated transfer of CE from HDL to LDL and therefore preventing the transfer of AmB from HDL to LDL (Figure 1B).[64] These observations may be explained in part by the influence of lipid surface charge on lipid transfer among lipoproteins. Billheimer and Gaylor observed the decrease of CE transfer between HDL and LDL in the presence of both DMPC and DMPG liposomes.[66] Those investigators found that phosphatidylglycerol increases CE exchange between HDL and the liposome in the presence of LTP but not in the absence of LTP. The presence of unsaturated acyl chains in the phospholipid enhances exchange. However, neutral phospholipids, such as sphingomyelin, drastically decrease cholesterol exchange with the liposome.

The presence of DMPC, DMPG, and SA phospholipids in liposomes results in the reduction of LTP-mediated transfer of CE from HDL to LDL; since AmB

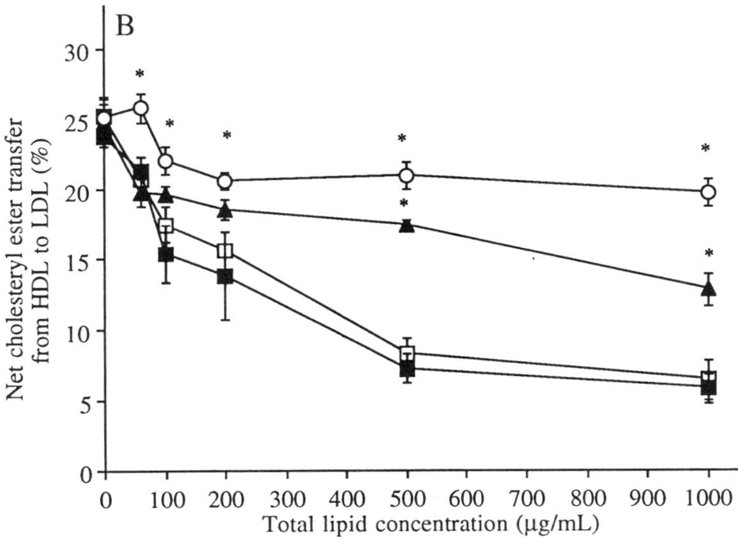

Fig. 1B. Effect of liposomal-lipid composition on the cholesteryl ester transfer from high-density lipoproteins (HDL) to low-density (LDL). Dimyristoyl phosphatidylcholine (DMPC) [○], dimyristoyl phosphatidylglycerol (DMPG) [▲], DMPC:DMPG 7:3 wt/wt [□], or DMPC:stearylamine (SA) 7:1 wt/wt [■] liposomes were incubated in human plasma for 60 minutes at 37°C. $*p < 0.05$ vs. DMPC:SA liposomes. Data expressed as mean ± standard deviation (number of individual experiments = 6).

interacts with CE, this finding may explain in part the lower distribution of AmB into LDL when AmB is incorporated into these liposomes.

V.3. Pharmacological implications of the AmB-lipoprotein complex

Preliminary investigations by others have suggested that the renal toxicity of AmB can be influenced by liposomal-phospholipid surface charge, phospholipid acyl chain length, chain saturation, and the liposomal-lipid/AmB ratio.[67,68] For example, AmB-containing liposomes composed of phospholipids with unsaturated acyl chains are as toxic as AmB to mammalian cells; however, those composed of phospholipids with saturated acyl chains are less toxic.[67,69] Previous studies have demonstrated a decrease of AmB cytotoxicity when the drug is delivered in the form of L-AmB to LLC PK1 cells (a pig kidney epithelial cell line)[67,70] and to primary cultures of rabbit proximal tubule cells.[71] To date, the mechanisms that result in the decreased renal cytotoxic effects of L-AmB are not fully understood. Krause and Juliano have suggested that the decreased toxicity of L-AmB compared with AmB is related to a selective transfer of the drug from liposomes to fungal but not mammalian cell membranes.[70] This selective toxicity shown towards the fungal membrane is probably regulated by physical characteristics of the donor and of the target membrane.[67] Brajtburg and coworkers demonstrated that AmB is highly bound to plasma lipoproteins[52] and that AmB-induced cytotoxic effects

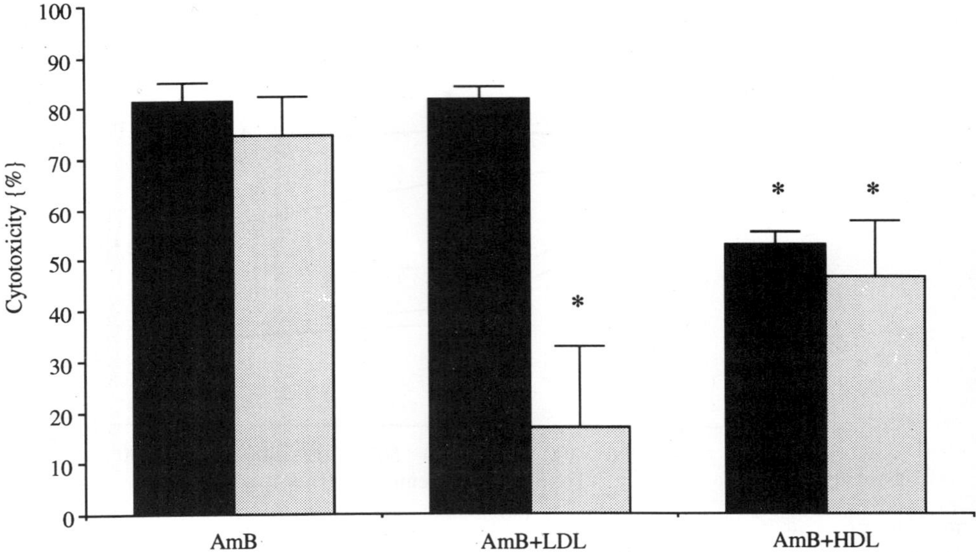

Fig. 2. Influence of reduced low-density lipoprotein (LDL) receptor number on high-density lipoprotein (HDL)- and LDL-associated amphotericin B (AmB) toxicity to LLC PK1 renal cells. Control cells [■] and cells with reduced LDL receptor expression [□] number of individual experiments = 3; mean ± standard deviation; $*p < 0.05$ vs. AmB.

on mammalian red blood cells, but not *Candida albicans* cells decreased in the presence of either HDL or LDL.[72] Previously, we demonstrated that HDL-associated AmB and HDL-associated L-AmB are less toxic to LLC PK1 renal cells than are AmB or LDL-associated AmB (Figure 2).[73] The reduced toxicity of HDL-associated AmB may be explained by the low level of expression of HDL receptors in LLC PK1 cells.[73] The sustained toxicity observed with AmB alone in trypsinized cells may be related to a direct membrane effect. However, when AmB is associated with LDL, the toxicity is maintained, which suggests that both direct membrane- and non-membrane-related toxicity may occur. Furthermore, a study to determine if a relationship existed between serum lipoprotein cholesterol concentration and the severity of AmB-induced renal toxicity in patients suggested that patients with higher serum LDL-cholesterol concentrations are more susceptible to AmB-induced renal toxicity (Table 2).[74]

HDL and LDL associated-AmB were equally toxic to fungal cells, which suggests that the presence of lipoproteins does not alter the antifungal activity of AmB. Such effects may be related to the liberation of monomeric AmB associated with lipoproteins or L-AmB, by fungal[68,69] or endothelial derived phospholipases.[68,69] The low concentrations of unbound and water-soluble monomeric AmB present in L-AmB[75–77] may be sufficient for fungal toxicity but not adequate for forming AmB aggregates that are toxic to mammalian cells.[69,70] AmB complexed

Table 2

The serum low density lipoprotein (LDL)-cholesterol concentration, amount of amphotericin B (AmB) associated with plasma LDL, serum creatinine concentrations, and cumulative amphotericin B dose following 10 days of therapy administered in patients with an anticipated or confirmed fungal infection

Treatment groups	LDL-Cholesterol* mg/dl	LDL-Associated amphotericin B %$^{+}$	Serum creatinine$^{\#}$ $^{++}$	Cumulative amphotericin B dose mg
A. Total AmB dose >180 mg				
	95.4 ± 12.2	51.4 ± 4.2	46.8 ± 11.0	202.1 ± 70
	65.0 ± 3.6**	30.8 ± 6.1**	−0.5 ± 19.2**	196.3 ± 11.0
B. Total AmB dose <180mg				
	92.4 ± 17.4	46.6 ± 13.9	−9.8 ± 8.8**	79.0 ± 14.6**

*LDL-cholesterol concentration prior to amphotericin B therapy (Note LDL-cholesterol level does not significantly change during amphotericin B therapy); $^{+}$Percent of total amphotericin B serum concentration after final dose; $^{++}$Percent change of serum creatinine concentration from baseline after final amphotericin B dose; $^{\#}$Serum creatinine levels are an indirect measure of kidney function; Data expressed as mean ± standard deviation ($n = 5$ each group) **$p < 0.05$ versus total amphotericin dose >180 mg group A. Modified from Ref. 74.

with lipid is less toxic than the self associated form of AmB in medium, but the monomeric form of AmB interacts with fungal cells membrane and is non toxic against mammalian cells membrane as shown by Bolard et al.[76]

Differences in the pharmacokinetics and tissue distribution of free AmB were demonstrated in healthy in comparison to hyperlipidemic rats induced with diabetes. In contrast, the pharmacokinetics and tissue distribution of L-AmB were unchanged in diabetic rats which suggests an independence of this delivery mechanism from the diabetic disease state and endogenous triglyceride and cholesterol levels.[48] However, a limitation of this study was that we could not determine if changes in the pharmacokinetics and tissue distribution of AmB was a direct result of the plasma hyperlipidemia or other diabetic-inflicted physiologic alterations (e.g., blood flow, liver metabolism, renal metabolism). Recent work by Wasan and Conklin suggest that following administration of a single intravenous dose, AmB and L-AmB appear to be less effective in killing *C. albicans* isolates in hypercholesterolemic diabetic than in normocholesterolemic nondiabetic rats, while they were found to improve the renal functions of rats in both treatment groups.[78]

To determine if the pharmacokinetics and tissue distribution of AmB and L-AmB were altered in plasma dyslipidemia (hypercholesterolemia) independent of other physiologic alterations, rats were administered a continuous infusion of Intralipid. Intralipid is a fatty acid/triglyceride emulsion administered intravenously as a nutritional supplement in debilitated patients. We found that in rats administered a continuous infusion of Intralipid for 5 days resulted in an increase in total serum cholesterol and HDL cholesterol concentrations without altering LDL cholesterol or total serum triglyceride concentrations.[79]

The influence of 5% Intralipid and 0.45% normal-saline infusions on the concentration in serum and distribution in tissue of AmB (Fungizone® consisting of amphotericin B and sodium deoxycholate) and L-AmB in rats were compared.[80] In animals receiving a continuous Intralipid infusion, concentrations of AmB in kidneys and lungs were significantly higher, but the concentration of AmB in serum was significantly lower in animals administered AmB versus those given L-AmB. In animals receiving a continuous normal-saline infusion concentrations of AmB in kidneys and the spleen were significantly higher, but the concentration of AmB in serum was significantly lower in animals administered AmB versus those given L-AmB. These results suggest that the increased total serum cholesterol and high-density lipoprotein cholesterol during the Intralipid infusion decreased the clearance of AmB from the bloodstream and decreased the L-AmB concentration in the kidney and lung.

VI. Recent development of other lipid-based amphotericin B products

Two other AmB lipid-based formulations are also being prepared on a large scale and available for clinical use. AmB colloidal dispersion (ABCD; Amphocil®; SEQUUS Pharmaceuticals, Menlo Park, CA, USA) is a stable complex of AmB and cholesteryl sulfate in a 1:1 molar ratio.[81] ABCD has equivalent antifungal activity but decreased toxicity than does the commercially available form of AmB, AmB plus deoxycholate (Fungizone®, Bristol-Myers Squibb, Nutley, NJ, USA).[81] In vitro studies have shown that the drug-lipid complex does not hemolyze erythrocytes and binds less to plasma lipoproteins than does the conventional form of AmB.[82,83] Studies in healthy volunteers indicated that drug disposition of ABCD was similar to that of Fungizone®. Acute side effects after ABCD administration were comparable with those of AmB but occurred at doses with 1.5 mg/kg/day compared to 0.5–0.75 mg/kg/day with the conventional preparation. The renal toxicity of ABCD is believed to be reduced because the AmB is bound as a cholesterol complex, so less "free" drug is available to interact with renal tubules.[83]

AmBisome (Nexstar, Boulder, CO, USA)[82,84] is supplied as a lyophilized powder, which must be reconstituted before intravenous infusion. It is the only liposomal AmB preparation currently licensed in the United Kingdom. This formulation consist of hydrogenated soy phosphatidylcholine, cholesterol, distearoyl phosphatidylglycerol, alpha-tocopherol, sucrose, and disodium succinate hexahydrate. A starting dose of 1.0 mg/kg/day has been recommended, increasing to 3.0 mg/kg/day, although doses up to 5.0 mg/kg/day have been used in compassionate studies, where exposure to the conventional AmB preparation led to unacceptable toxicity.[84–87] The highest concentrations of AmBisome are found in the liver and spleen; however, concentrations in the lung and kidney are highly inconsistent.

Since this review was written, a third lipid-based formulation (ABELCET®; The Liposome Co., Princeton, NJ) has been approved for clinical use, and is described in detail in Chapter 8.3.

VII. The future of liposomal amphotericin B

Fungal infections are on the rise worldwide, particularly as the population of immunocompromised patients continues to grow. By itself, AmB is an effective antifungal agent, though it is highly toxic, particularly to the kidneys. The goal of these lipid formulations of the AmB are to transport the drug through the body without exposing it to sensitive organs and tissues and then to deliver it in concentrated doses to the target site. To an certain extent all three of these formulations accomplish this goal. The maximum tolerable dose of AmB is about 1 mg/kg/day. However, these lipid formulations allow physicians to go up to 5 times the dose of AmB without increasing infusion-related toxicity's. All three lipid formulations of AmB demonstrate improved efficacy, primarily because of the higher administered dose, and reduced kidney toxicity, compared to AmB. As such, the future of L-AmB is bright and it is apparent that these lipid-based products will replace AmpB as the mainstays in the treatment of systemic fungal infections.

Acknowledgments

Doctor Papahadjopoulos asked me to address the historical development of liposomal-Amphotericin B. This was a wonderful request and at the same time a sort of awareness call. This study spans now almost two decades, that's something. There were ups and downs in its development, but the persistence and causation of several groups that this was a good idea certainly prevailed. Kish and I will provide here our perspective, how it was seen from the beginning and where we think it is headed. Kish and I would like to recognize everyone involved but it will be a long one, we decided that recognition comes from the papers cited. However, our work was a team effort: Rudy Juliano, Kapil and Reeta Mehta, Roy Hopfer, Leela Kasi, Tom Haynie, Evan Hersh, Eli Anaissie and Victor Fainstein were part of this unbelievably wonderful group of friends to work with. As the work progressed, there were those individuals that advanced these efforts: George Mackaness, Richard Sykes, Marc Ostro, Bob Lenk and others.

Gabriel Lopez-Berestein

References

1. Anaissie EJ. Opportunistic mycoses in the immunocompromised host: experience at a cancer center and review. Clin Infect Dis 1992;14(Suppl 1):43–53.
2. Pfaller MA, Wenzel R. The impact of changing epidemiology of fungal infections in the 1990s. Eur J Clin Microbiol Infect Dis 1992;11:287–291.
3. Richardson MD. Opportunistic and pathogenic fungi. J Antimicrobial Chemother 1991;28(Suppl A):1–11.
4. Walsh TJ. Invasive fungal infections: problems and challenges in developing new antifungal compounds. In: Sutcliffe J, Georgopapadakou NH, eds. Emerging targets in antibacterial and antifungal chemotherapy. New York: Chapman & Hall, 1992;349–373.
5. Beck-Sague CM, Jarvis WR. Secular trends in the epidemiology of nosocomial fungal infections in the United States, 1980–1990. J Infect Dis 1993;167:1247–1251.

6. Denning DW. Epidemiology and pathogenesis of systemic fungal infections in the immunocompromised host. J Antimicrob Chemother 1991;28(Suppl B):1–6.
7. Diamond RD. The growing problem of mycoses in patients infected with the human immunodeficiency virus. Rev Infect Dis 1991;13:480–486.
8. Bodey GP. Fungal infection and fever of unknown origin in neutropenic patients. Am J Med 1986;80:112–119.
9. Meyer RD. Current role of therapy with amphotericin B. Clin Infect Dis 1992;14:s154–s160.
10. Brajtburg J, Powderly WG, Kobayashi GS, Medoff G. Amphotericin B: current understanding of mechanisms of action. Antimicrob Agents Chemother 1990;34:183–188.
11. Gallis HA, Drew RH, Pickard WW. Amphotericin B: 30 years of clinical experience. Rev Infect Dis 1990;12:308–329.
12. Gallis HA. Amphotericin B: A commentary on its role as an antifungal agent and as a comparative agent in clinical trials. Clin Infect Dis 1996;22:s145–s147.
13. Bolard J. How do the polyene macrolide antibiotics affect the cellular membrane properties? Biochim Biophys Acta 1986;864:257–304.
14. Georgopapadakou NH, Walsh TJ. Antifungal agents: Chemotherapeutic targets and immunologic strategies. Antimicrob Agents Chemother 1996;40:279–291.
15. Warnock DW. Amphotericin B: an introduction. J Antimicrob Chemother 1991;28:27–38.
16. Chabot GG, Pazdur R, Valeriote FA, Baker LH. Pharmacokinetics and toxicity of continuous infusion of amphotericin B in cancer patients. J Pharm Sci 1989;78:307–310.
17. Tolins JP, Raij L. Adverse effect of amphotericin B administration on renal hemodynamics in the rat: neurohumoral mechanisms and influence of calcium channel blocker. J Pharmacol Exp Ther 1988;245:594–599.
18. Gardner ML, Godley P, Wasan SM. Sodium loading treatment of amphotericin B-induced nephrotoxicity. DICP 1990;24:940–945.
19. Bangham AD, Horne RW. Negative staining of phospholipids and their structural modification by surface-active agents as observed in the electron microscope. J Mol Biol 1964;8:660–668.
20. Bangham AD, Standish MM, Watkins JC. Diffusion of univalent ions across the lamellae of swollen phospholipids. J Mol Biol 1965;13:238–252.
21. New RR, Chance ML, Heath S. Antileishmanial activity of amphotericin and other antifungal agents entrapped in liposomes. J Antimicrob Chemother 1981;8:371–381.
22. Graybill JR, Craven PC, Taylor RL et al. Treatment of murine cryptococcosis with liposome-associated amphotericin B. J Infect Dis 1982;145:748–752.
23. Lopez-Berestein, G, Mehta, R, Hopfer, RL, Mills K, Kasi L, Mehta K, Fainstein V, Luna M, Hersh EM, Juliano RL. Treatment and prophylaxis of disseminated infection due to Candida albicans in mice with liposome-encapsulated amphotericin B. J Infect Dis 1983;147:939–945.
24. Lopez-Berestein G, Hopfer RL, Mehta R et al. Prophylaxis of Candida albicans infection in neutropenic mice with liposome-encapsulated amphotericin B. Antimicrob Agents Chemother 1984;25:366–377.
25. Hopfer RL, Mills K, Mehta R, et al. In vitro antifungal activities of amphotericin B and liposome-encapsulated amphotericin B. Antimicrob Agents Chemother 1984;25:387–389.
26. Mehta R, Lopez-Berestein G, Hopfer R et al. Liposomal amphotericin B is toxic to fungal cells but not to mammalian cells. Biochimica et Biophysica Acta 1984;770:230–234.
27. Lopez-Berestein G, Rosenblum MG, Mehta R. Altered tissue distribution of amphotericin B by liposomal encapsulation: comparison of normal mice to mice infected with Candida albicans. Cancer Drug Delivery 1984;1:199–205.
28. Lopez-Berestein G, Hopfer RL, Mehta R et al. Liposome-encapsulated amphotericin B for treatment of disseminated candidiasis in neutropenic mice. J Infect Dis 1984;150:278–283.
29. Lopez-Berestein G, McQueen T, Mehta K. Protective effect of liposomal-amphotericin B against C. albicans infection in mice. Cancer Drug Delivery 1985;2:183–189.
30. Lopez-Berestein G, Fainstein V, Hopfer R et al. Liposomal amphotericin B for the treatment of fungal infections in patients with cancer; a preliminary study. J Infect Dis 1985;151:704–710.
31. Wiebe VJ, De Gregorio MW. Liposome encapsulated amphotericin B: a promising new treatment for disseminated fungal infections. Rev Infect Dis 1988;10:1097–1101.
32. Tremblay C, Barza M, Fiore C, Szoka F. Efficacy of liposome-intercalated amphotericin B in the treatment of systemic candidiasis in mice. Antimicrob Agents Chemother 1984;26:170–173.
33. Tremblay C, Baraza M, Szoka F et al. Reduced toxicity of liposome-associated amphotericin B injected intravitreally in rabbits. Invest Opthal Vis Sci 1985;26:711–718.
34. Taylor RL, Williams DM, Craven PC et al. Amphotericin B in liposomes: novel therapy of histoplasmosis. Am Rev Respir Dis 1982;125:610–616.
35. Berman JD, Hanson WL, Chapman WL et al. Antileishmanial activity of liposome-encapsulated amphotericin B in hamsters and monkeys. Antimicrobial Agents Chemother 1986;30:847–51.

36. Ponosian CB, Barza M, Szoka F, Wyler DJ. Treatment of experimental cutaneous leishmaniasis with liposome-intercalated amphotericin B. Antimicrob Agents Chemother 1984;25:655–656.
37. Croft SL, Davidson RN, Thornton EA. Liposomal amphotericin B in the treatment of visceral leishmaniasis. J Antimicrob Chemother 1991;28:111–118.
38. Gradoni L, Davidson RN, Orsini S, Betto P. Activity of liposomal amphotericin B (AmBisome) against *Leishmania infantum* and tissue distribution in mice. J Drug Targeting 1993;1:311–316.
39. Lopez-Berestein G, Bodey GP, Frankel LS, Mehta K. Treatment of hepatosplenic candidiasis with liposomal-amphotericin B. J Clin Oncol 1987;5:310–317.
40. Lopez-Berestein G, Bodey GP, Fainstein V et al. Treatment of systemic fungal Infections with liposomal amphotericin B. Arch Intern Med 1989;149:2533–2538.
41. Torre-Cisneros J, Villanueva JL, Kindelan JM, Jurado R, Sanchez-Guijo P. Successful treatment of antimony-resistant visceral leishmaniasis with liposomal amphotericin B in patients infected with human immunodeficiency virus. Clin Infect Dis 1993;17:625–627.
42. Fusai T, Durand R, Boulard Y, Paul M, Bories C, Rivollet D, Houin R, Deniau M. Importance of drug carriers in the treatment of visceral leishmaniasis. Medecine Tropicale 1995;55:73–78.
43. Davidson RN, Di Martino L, Gradoni L, Giacchino R, Russo R, Gaeta GB, Pempinello R, Scott S, Raimondi F, Cascio A et al. Liposomal amphotericin B (AmBisome) in Mediterranean visceral leishmaniasis: a multi-centre trial. Quart J Med 1994;87:75–81.
44. Lopez-Berestein G, Kasi L, Rosenblum MG et al. Clinical pharmacology of 99mTc-labeled liposomes in patients with cancer. Cancer Res 1984;44:375–378.
45. Kasi LP, Lopez-Berestein G, Mehta K et al. Distribution and pharmacology of intravenous ^{99m}Tc-labeled multilamellar liposomes in rats and mice. Int J Nucl Med Biol 1984;11:35–37.
46. Perez-Soler R, Lopez-Berestein G, Kasi L et al. Distribution of technetium-99m-labeled multilamellar liposomes in patients with Hodgkin's disease. J Nucl Med 1985;26:743–747.
47. Mehta RT, McQueen TJ, Keyhani A, Lopez-Berestein G. Phagocyte transport as mechanism for enhanced therapeutic activity of liposomal amphotericin B. Exp Chemother 1994;40:256–262.
48. Wasan KM, Vadiei K, Lopez-Berestein G, Luke DR. Pharmacokinetics, tissue distribution, and toxicity of free and liposomal amphotericin B in diabetic rats. J Infect Dis 1990;161:562–566.
49. Atkinson AJ, Bennett JE. Amphotericin B pharmacokinetics in humans. Antimicrobial Agents Chemother 1978;13:271–278.
50. Lopez-Berestein G. Liposomes as carriers of antifungal drugs. Annals of the New York Academy of Sciences 1988;544:590–597.
51. Wasan KM, Lopez-Berestein G. Targeted Liposomes in fungi: Modifying the therapeutic index of amphotericin B by its incorporation into negatively charged liposomes. J Liposome Res 1995;5:883–903.
52. Brajtburg J, Elberg S, Bolard J, Medoff G. Interaction of plasma proteins and lipoproteins with amphotericin B. J Infect Dis 1984;149:986–992.
53. Andreoli TE. The anatomy of amphotericin B-cholesterol pores in lipid bilayer membranes. Kidney Int 1973;4:337–345.
54. Koldin MH, Kobayashi GS, Brajtburg J, Medoff G. Effects of elevation of serum cholesterol and administration of amphotericin B complexed to lipoproteins on amphotericin B-induced toxicity to rabbits. Antimicrobial Agents Chemother 1985;28:144–145.
55. Christansen KJ, Bernard EM, Gold JWM, Armstrong D. Distribution and activity of amphotericin B in humans. J Infect Dis 1985;152:762–765.
56. Wasan KM, Brazeau GA, Keyhani A, Hayman AC, Lopez-Berestein G. Role of liposome composition and temperature on the distribution of amphotericin B in serum lipoproteins. Antimicrobial Agents Chemother 1993;37:246–250.
57. Babiak J, Rudel LL. Lipoproteins and atherosclerosis. Bailliere's Clin Endocrinol Metab 1987;1:515–521.
58. Cushley RJ, Treleaven WD, Parmar YI et al. Surface diffusion in human serum lipoproteins. Biochem Biophys Res Commun 1987;146:1139–1145.
59. Barwicz J, Gareau R, Audet A, et al. Inhibition of the interaction between lipoproteins and amphotericin B by some delivery systems. Biochem Biophys Res Commun 1991;181:722–726.
60. Morton RE, Zilversmit DB. Purification and characterization of lipid transfer protein(s) from human lipoprotein-deficient plasma. J Lipid Res 1982;23:1058–1067.
61. Morton RE, Zilversmit DB. Inter-relationship of lipids transferred by the lipid-transfer protein Isolated from human lipoprotein-deficient plasma. J Biol Chem 1983;258:11751–11757.
62. Pattnaik NM, Zilversmit DB. Interaction of cholesteryl ester exchange protein with human plasma lipoproteins and phospholipid vesicles. J Biol Chem 1979;254:2782–2786.
63. Surewicz WK, Epand RM, Pownall HJ et al. Human apolipoprotein A-I forms thermally stable

complexes with anionic but not with zwitterionic phospholipids. J Biol Chem 1986;261:16191–16197.

64. Wasan KM, Morton RE, Rosenblum MG, Lopez-Berestein G. Association of amphotericin B with high density lipoproteins is responsible for the decreased toxicity of liposomal amphotericin B: Role of lipid transfer protein. J Pharm Sci 1994;83:1006–1010.
65. Bolard J, Seigneuret M, Boudet G. Interaction between phospholipid bilayer membranes and the polyene antibiotic amphotericin B. Lipid state and cholesterol content dependence. Biochim Biophys Acta 1980;599:280–293.
66. Billheimer JT, Gaylor JL. Effect of lipid composition on the transfer of sterols mediated by non-specific lipid transfer protein (sterol carrier protein 2). Biochim Biophys Acta 1990;1046(2):136–143.
67. Juliano RL, Grant CWM, Barber KR, Kalp MA. Mechanism of the selective toxicity of amphotericin B incorporated into liposomes. Mol Pharmacol 1987;31:1–11.
68. Janoff AS, Boni LT, Popescu MC et al. Unusual lipid structures selectively reduce the toxicity of amphotericin B. Proc Natl Acad Sci USA 1988;85:6122–6126.
69. Perkins WR, Minchey SR, Boni LT et al. Amphotericin B-phospholipid interactions responsible for reduced mammalian cell toxicity. Biochim Biophys Acta 1992;1107:271–282.
70. Krause HJ, Juliano RL. Interactions of liposome-Incorporated amphotericin B with kidney epithelial cells. Mol Pharmacol 1988;34:286–297.
71. Joly V, Line SJ, Carbon C, Yeni P. Interactions of free and liposomal amphotericin B with renal proximal tubular cells in primary culture. J Pharmacol Exp Ther 1990;255:17–22.
72. Brajtburg J, Elberg S, Kobayashi GS, Medoff G. Effects of serum lipoproteins on damage to erythrocytes and Candida albicans cells by polyene antibiotics. J Infect Dis 1986;153:623–626.
73. Wasan KM, Rosenblum MG, Cheung L, Lopez-Berestein G. Influence of lipoproteins on renal cytotoxicity and antifungal activity of amphotericin B. Antimicrobial Agents Chemother 1994;38:223–227.
74. Wasan KM, Conklin JS. Enhanced Amphotericin B Nephrotoxicity in Intensive Care Patients with Elevated Low-Density Lipoprotein Cholestero. Clin Infect Dis 1997;24:78–80.
75. Jullien S, Vertut-Croquin AJ, Brajtburg J, Bolard J. Circular dichroism for the determination of amphotericin B binding to liposomes. Anal Biochem 1988;1972:197–202.
76. Bolard J, Legrand J, Heitz F, Cybulska, B. One-sided action of amphotericin B on cholesterol-containing membranes is determined by its self association in the medium. Biochem 1991;30:5707–5715.
77. Jullien S, Brajtburg J, Bolard J. Affinity of amphotericin B for phosphatidylcholine vesicles as a determinant of the in vitro cellular toxicity of liposomal preparations. Biochim Biophys Acta 1990;1021:39–45.
78. Wasan KM, Conklin JS. Evaluation of renal toxicity and antifungal acitivity of free and liposomal amphotericin B following a single intravenous dose to diabetic rats with systemic candidiasis. Antimicrob Agents Chemother 1996;40:1806–1810.
79. Wasan KM, Grossie Jr VB, Lopez-Berestein G. Effects of Intralipid infusion on rat serum lipoproteins. Lab Animals 1994;28:138–142.
80. Wasan KM, Grossie Jr VB, Lopez-Berestein G. Concentrations in serum and tissue distribution of free and liposomal amphotericin B in rats on continuous Intralipid infusion. Antimicrobial Agents Chemother 1994;38:2224–2226.
81. Lasic DD. Mixed micelles in drug delivery. Nature 1992;355:279–280.
82. Gates C, Pinney RJ. Amphotericin B and its delivery by liposomal and lipid formulations. J Clin Pharmacy Ther 1993;18:147–153.
83. Saunders SW, Buchi KN, Goddard MS et al. Single-dose pharmacokinetics and tolerance of cholesterol sulphate complex of amphotericin B administered to healthy volunteers. Antimicrob Agents Chemother 1991;35:1029–1034.
84. Chopra R, Blair S, Strang J et al. Liposomal amphotericin B (AmBisome) in the treatment of fungal infections in neutropenic patients. J Antimicrob Agents 1991;28:93–104.
85. Ringden O, Meunier F, Tollemar J et al. Efficacy of amphotericin B (AmBisome) in the treatment of invasive fungal infections in immunocompromised patients. J Antimicrob Agents 1991;28:73–82.
86. Meunier F, Prentice HG, Ringden O. Liposomal amphotericin B (AmBisome): safety data from a phase II/III clinical trial. J Antimicrob Agents 1991;28:83–91.
87. Sculier JP, Coune A, Meunier F, Brassinne C, Laduron C, Hollaert C, Collette N, Heymans C, Klastersky J. Pilot study of Amphoterium B entrapped in sonicated liposomes in cancer patients with fungal infections. Eur J Cancer Clin Oncology 1998;24:527–538.

Lasic and Papahadjopoulos (eds.), Medical Applications of Liposomes

CHAPTER 3.2

Long-circulating liposomes containing antibacterial and antifungal agents

IRMA A.J.M. BAKKER-WOUDENBERG, ELS W.M. VAN ETTEN

Department of Medical Microbiology and Infectious Diseases, Erasmus University Rotterdam The Netherlands

Overview

I. Introduction

In clinical practice infectious complications caused by bacteria, fungi, viruses and parasites frequently occur. The incidence of severe infections is related to (recent) developments in clinical medicine, such as new therapeutic modalities, an increased use of prosthetic and other medical devices, frequent diagnostic and therapeutic intervention, and an increasing number of immunocompromised patients (malignancies, transplantations). As a consequence a growing number of patients are prone to severe (nosocomial) infections that are often difficult to treat. These infections remark a major cause of morbidity and mortality in these patients. Failure of antibiotic treatment occurs despite the availability of potent antibiotics. Intensification of antibiotic treatment is needed and should meet various requirements.

1. Antibiotic treatment failure may be related to moderate antibiotic susceptibility of the microorganism. In those cases high antibiotic concentrations in the infected tissues are needed, to prevent dissemination of the infection. Targeting of antibiotic to the site of infection should be effected. Application of liposomes to

achieve site-specific drug delivery resulting in high and prolonged antibiotic concentrations in the infected tissues may be of great value. In this respect passive targeting is important as the site of infection, particularly in immunocompromised patients, is often not known. To achieve this purpose long circulating sterically-stabilized liposomes may be an important tool. They remain in the vascular compartment for prolonged periods of time without the requirements of high lipid dose or rigid nature of the lipid bilayers.[1,2] It is speculated that these liposomes extravasate in areas of inflammation as result of locally increased vascular permeability and endothelial leakage, both developing at the infected site during the progression of the infection. Supporting evidence for the role of sterically stabilized liposomes in the treatment of severe lung infection in animals is obtained and described in this chapter.

2. Antibiotic treatment failure may be related to the intracellular location of the microorganism. Relevant in this respect are disseminated intracellular mycobacterial infections. Moreover mycobacteria grow very slowly in their intracellular location. To kill the intracellular mycobacteria in the "dormant state" long term treatment with high doses of antibiotic is required. Such treatment schedules facilitate the development of antibiotic resistance. In those cases application of liposomes aims for achieving high intracellular concentrations of antibiotic in the infected cells again with the purpose of site-specific drug delivery. Therefore it is important that these liposomes show relatively long circulation half-lives and reduced hepatosplenic uptake, in order to be successful as carriers in disseminated infections. Supporting evidence for the role of long circulating liposomes in the treatment of mycobacterial infections is obtained. Important in this respect are amikacin-containing long circulating liposomes: MiKasome®. Preclinical data as well as phase I clinical data are described in this chapter.

3. Antibiotic treatment failure may be related to insufficient availability of antibiotic due to extremely low half-life in blood. In those cases liposomes as carriers of antibiotics may be used as microreservoir of antibiotic during circulation. Again relatively long circulating liposomes are needed for this purpose. Experimental evidence in the field of antibacterial agents to support the application of liposomes in this way, is not yet available.

4. Antibiotic treatment failure may be related to toxic side effects of antibiotic, and as a result insufficient clinical efficacy because toxicity limits the dose of antibiotic. The rationale to apply liposomes as carriers of antibiotic in this respect is to achieve site-avoidance drug delivery. In this area extensive studies have been performed with amphotericin B (AMB) encapsulated in liposomes or bound to other lipid carriers. The reduction of toxicity of the AMB-lipid formulations is thought to result from a reduced affinity of AMB to cholesterol in the human cell membrane, compared to the lipids of the carrier and the ergosterol in the fungus membrane. As a result relatively high doses of AMB-lipid formulations are tolerated, and an increase of the therapeutic index of AMB is achieved. Animal

studies using various AMB-lipid formulations showed promising results, and were followed by clinical studies.[3,4] Three industrially prepared AMB-lipid formulations are now available (see Sections VII and VIII). Their efficacy is clearly demonstrated in compassionate use studies in immunocompromised patients with life-threatening fungal infections and a variety of underlying diseases who show intolerance to or failure on conventional AMB. Various clinical trials comparing the efficacy of the individual AMB-lipid formulations with conventional AMB in patients with fungal infections are ongoing.

One of the industrially prepared AMB-lipid formulations is AmBisome®. These AMB-liposomes have a small particle size and a rigid bilayer, and their blood residence time is primarily dependent on the lipid dose administered. At the clinically effective dose half-life in blood is about 32 h in man.[5] The role of prolonged blood residence time of AMB-liposomes was investigated in animal studies, and appeared to be of importance for therapeutic efficacy.[6] Sterically stabilized AMB-containing liposomes were prepared in our laboratory, and show long circulation half-life without the constraints of high lipid dose, small particle size, or rigid nature of the bilayer. Supporting evidence for the role of sterically stabilized liposomes containing AMB in the treatment of fungal infections in animals is obtained[6] and described in this chapter.

II. Antibacterial agents in long circulating liposomes

II.1. Sterically stabilized liposomes containing gentamicin or ceftazidime

In a rat model of leftsided pneumonia caused by *Klebsiella pneumoniae* (fatal infection within 5 days) the behaviour of sterically stabilized liposomes composed of PEG-DSPE : PHEPC : Chol (molar ratio, 0.15 : 1 : 1.85) with a mean particle size of 80 nm was investigated.[7] The circulation half-life for these liposomes in blood was about 20 h. The liposomes showed relatively low hepatosplenic uptake. After intravenous administration these liposomes are passively targeted towards the infected lung tissue. In the rats with severe left lung infection the localization of these liposomes in the infected left lung tissue was up to 10-fold higher compared to liposome localization in the left lung of uninfected rats, and was strongly correlated with the severity of infection. Up to 9% of the liposome dose was recovered from the infected lung tissue. Compared to the infected left lung, in the uninfected right lung of the infected rats the localization of liposomes was not different compared to that in uninfected rats. The efficacy of gentamicin or ceftazidime encapsulated in these liposomes was investigated in this experimental pneumonia model.[8] At a single-dose treatment schedule started at 24 h after bacterial inoculation a superior therapeutic efficacy of the liposome-encapsulated antibiotic was observed compared to the effects of free antibiotic in terms of increased survival of the infected rats, as well as increased bacterial killing in the infected lung tissue. In vitro the antibiotic-containing liposomes did not show bactericidal activity. Therefore it is concluded that after localization of the sterically stabilized liposomes at the site of infection, release of encapsulated antibiotic occurs. During the long circulation in blood the antibiotic-containing liposomes are relatively stable.

II.2. MiKasome®, amikacin-containing liposomes

A 55 nm unilamellar liposomal preparation of amikacin, MiKasome® consisting of HSPC:Chol:DSPG (molar ratio, 2:1:0.1) was developed (NeXstar Pharmaceuticals Inc., San Dimas, CA), and pharmacokinetics and toxicity was investigated in animal models. In addition, therapeutic efficacy studies were performed in animals particularly in experimental infections caused by *Mycobacterium avium* complex. Aminoglycosides such as amikacin are essential antibiotics in the treatment of serious infections. They are a valuable alternative or addition to treatment of Mycobacterium infections, particularly *M. avium* and multidrug-resistant strains of *Mycobacterium tuberculosis*, an emerging problem in the HIV infected population. However the efficacy of aminoglycosides in these infections may be limited by their inability to achieve high intracellular concentrations and their potential nephrotoxic and ototoxic side effects. Liposomal encapsulation of aminoglycosides could offer the possibility of overcoming these limitations. Important to investigate is whether slow release of aminoglycosides from the liposomes occurs resulting in sustained low concentrations in the blood, and hence an increased nephrotoxicity as is also observed when continuous infusion of low doses of aminoglycosides is used. Therefore toxicity studies are of high priority.

Data on biodistribution, in rats at relatively high dosage show that MiKasome® can more effectively deliver amikacin to liver, spleen, lung and kidney than free drug treatment.[9] In beige mice infected with *M. avium* complex MiKasome® appeared to be more efficacious in clearing *M. avium* complex from the liver and spleen.[10] Mice were treated thrice weekly for 1, 3, 5 or 7 weeks beginning 5 days after infection. MiKasome® was well tolerated and resulted in significantly increased killing of *M. avium* in the liver and spleen. The animal data suggest that MiKasome® might be a suitable candidate for treating human *M. avium* complex infections.

First clinical studies were performed to evaluate the renal safety of MiKasome®, and were reported in 1994.[11] In healthy symptom free HIV seropositive volunteers MiKasome® was administered in escalating single doses ranging from 1 to 20 mg/kg of amikacin intravenously. In spite of high total and free amikacin peaks (being 325 μg/ml and 32 μg/ml, respectively at the dosage of 20 mg/kg), and prolonged elimination half lives (being 40 h at the dosage of 20 mg/kg) single doses of MiKasome® caused no significant renal toxicity based on serum creatinine. Subclinical renal alterations at the largest doses (15 and 20 mg/kg) were observed in terms of phospholipid urea and human IAP excretion. In view of the pharmacokinetic profile and the low toxicity studies are continued with multiple dose administration. MiKasome® is now in phase I clinical trials in the U.S. (none of the data available for publication yet).

In more recent animal studies the efficacy of MiKasome® was investigated in a model of intraperitoneal infection caused by *Klebsiella pneumoniae* in immunosuppressed mice, resulting in sepsis. In prophylactic treatment MiKasome® showed improved efficacy in terms of survival of animals and bacterial killing in blood, liver and spleen.[12] An increased therapeutic efficacy was also observed: even in the lung an increased bacterial killing was observed following administration of MiKasome® compared to free amikacin.[13]

III. Antifungal agents in long circulating liposomes

III.1. AmBisome®, amphotericin B-containing liposomes

AmBisome® shows prolonged blood residence time at therapeutically effective doses, the elimination half-life being about 32 h in man.[5] Preclinical data on AmBisome® in a number of animal models have recently been reviewed.[3,4] From the animal studies it can be concluded that, depending on the model of fungal infection, the immune status of the host, and the parameters for efficacy used, the antifungal activity of AmBisome® is either somewhat less or equal to that of conventional AMB at equivalent dosages. However, using AmBisome® much higher dosages are tolerated, and these high doses result in improved antifungal efficacy, even in severe infection in immunocompromised animals. The clinical data on AmBisome® will be presented elsewhere in this book. Fundamental studies on the mechanism of action of AmBisome® show that intact AmBisome® can reach the site of infection; at the site of infection direct interaction between AmBisome® and the fungal cell may occur or AMB may be released from AmBisome® in the close vicinity of the fungus.[14] Prolonged blood residence probably allows the localization of intact liposomes at sites of infection outside MPS-tissues. For AmBisome®, HSPC:DSPG:Chol (molar ratio, 1:0.4:0.5) with a mean particle size of 80 nm, the blood residence time is dependent on the lipid dose administered.[15–18] To achieve prolonged circulation of AMB liposomes without the constraint of high lipid dose, sterically stabilized AMB-containing liposomes were prepared at our laboratory.

III.2. Sterically stabilized liposomes containing amphotericin B

Two different formulations of AMB in PEG-grafted liposomes have been studied. Liposome preparations consisted of PEG-DSPE : HSPC : Chol : DSPG : AMB (molar ratio 0.29 : 2 : 1 : 0.8 : 0.4), further refered to as PEG/DSPG-AMB, and PEG-DSPE : HSPC : Chol : AMB (molar ratio, 0.21 : 1.79 : 1 : 0.32), further refered to as PEG-AMB. The two different preparations showed a large difference in toxicity in uninfected mice. PEG/DSPG-AMB was as toxic as conventional AMB, whereas the liposomal formulation PEG-AMB greatly reduced the toxicity of AMB.[5,19] The in vitro antifungal activity of PEG-AMB during 6 h exposure of *Candida albicans* was similar to that of conventional AMB.[5] These data show that it is possible to reduce the toxicity of AMB by lipid formulation without reducing its intrinsic antifungal activity. For AmBisome® the reduction of AMB's toxicity following liposomal encapsulation is associated with a substantial reduction of antifungal activity.[5] Prolonged blood residence time of PEG-AMB was demonstrated in mice, the elimination half-life being approximately 20 h. For a dose range of 5–85 μmol lipid/kg blood circulation time was dosage independent.[5,19] In the same strain of mice the elimination half-life of AmBisome® was about 8 h at a lipid dose of 70 μmol lipid/kg.[18] Therapeutic efficacy of PEG-AMB was studied in two different animal models of invasive fungal infections. Clinically relevant issues including persistent leukopenia and dissemination of infection were addressed. It was shown in our animal model of severe invasive *C. albicans* infection in persistently leukopenic mice that only treatment with PEG-AMB, given as a single dose, resulted in decreasing numbers of viable *Candida albicans* in the kidney within a short period of time after *Candida* infection. This effect could not be achieved with AmBisome® at an equivalent dose, even when administered repeatedly.[20,21] In our model of *Aspergillus fumigatus* one-sided pulmonary infection in persistently leukopenic rats[22] it was demonstrated that survival of the animals was significantly prolonged after only a single dose of PEG-AMB.[21] Similar therapeutic efficacy in this infection model was reported for AmBisome®, when administered repeatedly.[22] In conclusion, the PEG-AMB formulation shows three characteristics that are expected to be important for improved antifungal efficacy: low toxicity, high intrinsic antifungal activity, and prolonged circulation time of intact AMB-containing liposomes in blood.

IV. Potential for the future

The earliest therapeutic application of liposomal antimicrobial therapy was for the treatment of the protozoal infection leishmaniasis in experimental animals,[23,24] and published in 1978. In these experiments classical liposomes were used for delivery of antimicrobial agents to the infected Kupffer cells harboring the *Leishmania* intracellularly. Today, administration of liposomal amphotericin B results in safe and effective treatment of multidrug-resistant visceral leishmaniasis. The clinical experience is described by Lopez-Berestein et al., in this book. Both site-specific delivery and site-avoidance delivery of amphotericin B is contributing to the increase in therapeutic index in the treatment of visceral leishmaniasis.

The characteristic of site-avoidance drug delivery obtained with lipid formulations of amphotericin B is clearly manifested in the increased therapeutic index observed in patients with severe fungal infections; the data summarized also by Lopez-Berestein et al. (Chapter 3.1). It should be emphasized that high dosages of amphotericin B-lipid formulations are needed for treatment to be effective, for the reason that lipid formulation of amphotericin B results in reduction of toxicity, however intrinsic antifungal activity is also reduced. In sterically stabilized liposomes containing amphotericin B the antifungal activity is fully retained. This amphotericin B-lipid formulation may be of great value and needs thorough investigation.

With respect to bacterial infections numerous studies were published demonstrating the superior efficacy of antibiotic when administered in the liposome-encapsulated form in a variety of models of intracellular infections in liver and spleen caused by a variety of intracellular pathogens.[25] In most studies "classical" liposomes were applied. The applicability of classical liposomes for achieving delivery of antibiotics to infections localized outside the liver and spleen is limited. These infections are of high clinical relevance. An example is disseminated infection caused by *Mycobacterium* species. In this respect MiKasome® show great promise as these amikacin-containing liposomes show a relatively long circulation half-life. The development of MPS-avoiding sterically stabilized liposomes characterized by long blood circulation time opens new ways to achieve improved delivery of antimicrobial agent in extracellular infections outside the liver and spleen. Particularly as the sterically stabilized liposomes show prolonged blood circulation without the constraint of high lipid dose, small particle size, or rigid nature of the bilayer. In addition, in these liposomes variation in the lipid composition does not affect the prolonged circulation properties. This provides the opportunity to influence antibiotic release from liposomes at the site of infection. This is important in view of the difference in pharmacodynamics of different classes of antibiotics (see Chapters 4.1, 4.3 and 4.4).

References

1. Woodle MC, Newman MS, Cohen JA. Sterically stabilized liposomes: physical and biological properties. J Drug Targeting 1994;2:397–403.
2. Marjan MJ, Allen TM. Long circulating liposomes: past, present and future. Biotechnol Adv 1996;14:151–175.
3. Leenders ACAP, De Marie S. The use of lipid formulations of amphotericin B for systemic fungal infections. Leukemia 1996;10:1570–1575.
4. Hiemenz JW, Walsh TJ. Lipid formulations of amphotericin B: recent progress and future directions. Clin Infect Dis 1996;22(Suppl 2):S133–144.
5. NeXstar Pharmaceuticals, Inc. AmBisome® Liposomal Amphotericin B, Product Monograph, 1994.
6. Van Etten EWM, Ten Kate MT, Stearne-Cullen LET, Bakker-Woudenberg IAJM. Amphotericin B liposomes with prolonged circulation in blood: in vitro antifungal activity, toxicity, and efficacy in systemic candidiasis in leukopenic mice. Antimicrob Agents Chemother 1995;39:1954–1958.
7. Bakker-Woudenberg IAJM, Lokerse AF, Ten Kate MT, Mouton JW, Woodle MC, Storm G. Liposomes with prolonged blood circulation and selective localization in *Klebsiella pneumoniae* infected lung tissue. J Infect Dis 1993;168:164–171.
8. Bakker-Woudenberg IAJM, Ten Kate MT, Stearne-Cullen LET, Woodle MC. Efficacy of gentamicin or ceftazidime entrapped in liposomes with prolonged blood circulation and enhanced localization in *Klebsiella pneumoniae* infected lung tissue. J Infect Dis 1995;171:938–947.
9. Proffitt RT, Grayson JB, Chiang SM, Coulter DM, Satorius AL, Petersen EA. Biodistribution and therapeutic efficacy of liposomal amikacin. The Third Liposome Research Days Conference, 19–22 June, 1994, Vancouver, British Columbia, Canada (Abstract A-18).
10. Petersen EA, Grayson JB, Hersh EM, Dorr RT, Chiang SM, Oka M, Proffitt RT. Liposomal amikacin: improved treatment of *Mycobacterium avium* complex infection in the beige mouse model. J Antimicrob Chemother 1996;38:819–828.
11. Eestermans GH, Van Laethem Y, Hermans P, Ross ME, Nuyts GD, Clumeck N. A single dose pharmacokinetic and tolerance assessment of liposomal amikacin in HIV seropositive patients. Conference on Liposomes in Biomedical Research, 5–8 October, 1994, Berlin, Germany.
12. Eng E, Satorius A, Proffitt RT, Adler-Moore JP. Prophylaxis of *Klebsiella pneumoniae* sepsis by MiKasome®, a liposomal formulation of amikacin. 96th General Meeting of the American Society for Microbiology, 19–23 May, 1996, New Orleans, La (Abstract A-7).
13. Eng E, McAndrews B, Satorius A, Proffitt RT, Adler-Moore J. Comparative efficacy of amikacin and liposomal amikacin (MiKasome) in the treatment of *Klebsiella pneumoniae* infection in mice.

97th General Meeting of the American Society for Microbiology, 4–8 May, 1997, Miami Beach, Fla.

14. Adler-Moore JP, Proffitt RT. Development, characterization, efficacy and mode of action of AmBisome®, a unilammelar formulation of amphotericin B. J Liposome Res 1993;3:429–450.
15. Francis P, Lee JW, Hoffman A, Peter J, Francesconi A, Bacher J, Shelhamer J, Pizzo PA, Walsh TJ. Efficacy of unilamellar liposomal amphothericin B in treatment of pulmonary aspergillosis in persistently granulocytopenic rabbits: The potential role of bronchoalveolar D-mannitol and serum galactomannan as markers of infection. J Infect Dis 1994;169:356–368.
16. Gondal JA, Swartz RP, Rahman A. Therapeutic evaluation of free and liposome-encapsulated amphotericin B in the treatment of systemic candidiasis in mice. Antimicrob Agents Chemother 1989;33:1544–1548.
17. Proffitt RT, Satorius A, Chiang SM, Sullivan L, Adler-Moore JP. Pharmacology and toxicology of a liposomal formulation of amphotericin B (AmBisome) in rodents. J Antimicrob Chemother 1991;28(Suppl B):49–61.
18. Van Etten EWM, Otte-Lambillion M, Van Vianen W, Ten Kate MT, Bakker-Woudenberg IAJM. Biodistribution of liposomal amphotericin B (AmBisome) versus amphotericin B-desoxycholate (Fungizone) in immunocompetent uninfected mice as well as in leucopenic mice infected with *Candida albicans*. J Antimicrob Chemother 1995;35:509–519.
19. Van Etten EWM, Van Vianen W, Tijhuis RHG, Storm G, Bakker-Woudenberg IAJM. Sterically stabilized amphotericin B-liposomes: toxicity and biodistribution in mice. J Control Release 1995;37:123–129.
20. Van Etten EWM, Snijders SV, Verbrugh HA, Bakker-Woudenberg IAJM. Efficacy of pegylated long-circulating amphotericin B-liposomes versus AmBisome® in the treatment of systemic candidiasis in leukopenic mice in relation to the severity of infection (submitted).
21. Van Etten EWM, Stearne-Cullen LET, Snijders SV, Verbrugh HA, Bakker-Woudenberg IAJM. Efficacy of pegylated long-circulating amphotericin B-liposomes in the treatment of pulmonary aspergillosis in leukopenic rats (submitted).
22. Leenders ACAP, de Marie S, Ten Kate MT, Bakker-Woudenberg IAJM, Verbrugh HA. Liposomal amphotericin B (AmBisome®) reduces dissemination of infection as compared to amphotericin B deoxycholate (Fungizone®) in a newly developed animal model of one-sided pulmonary aspergillosis. J Antimicrob Chemother 1996;38:215–225.
23. Alving CR, Steck EA, Chapman WL et al. Therapy of leishmaniasis: superior efficacies of liposome-encapsulated drugs. Proc Natl Acad Sci USA 1978;75:2959–2963.
24. New RRC, Chance ML, Thomas SC, Peters W. Antileishmanial activity of antimonials entrapped in liposomes. Nature 1978;272:55–56.
25. Bakker-Woudenberg IAJM. Liposomes in the treatment of parasitic, viral, fungal and bacterial infections. J Liposome Res 1995;5:169–191.

Lasic and Papahadjopoulos (eds.), Medical Applications of Liposomes

CHAPTER 3.3

Treatment of human immunodeficiency virus, *Mycobacterium avium* and *Mycobacterium tuberculosis* infections by liposome-encapsulated drugs*

Nejat Düzgüneş

Department of Microbiology, School of Dentistry, University of the Pacific, 2155 Webster Street, San Francisco, CA 94115, USA

Overview

I. Therapy of human immunodeficiency virus type 1 (HIV-1) infection

The onset of the acquired immunodeficiency syndrome (AIDS) epidemic in the 1980s, and the identification of the etiologic agent of the disease as human immunodeficiency virus type 1 (HIV-1),[1] has lead to the development of numerous drugs that target virus-specific processes.[2–4] These drugs include inhibitors of the viral

*This chapter is dedicated to the memories of my father Professor Orhan Düzgüneş, and my father-in-law Professor John Flasher.

reverse transcriptase and protease, antisense oligonucleotides complementary to the mRNA for various viral proteins, and therapeutic genes. One of the major problems in the therapy of AIDS has been the development of HIV-1 strains resistant to the drugs that are being used in patients, for example, the reverse transcriptase inhibitor 3′-azido-3′-deoxythymidine (AZT), and the protease inhibitor ritonavir.[5–9] The emergence of these strains is attributed to the rapid rate of mutation of the virus. Nevertheless, studies utilizing HIV-infected cells have shown that it is possible to block the emergence of resistant strains by using "knocking out" concentrations of drugs which are not toxic.[10] Thus, one of the reasons for the development of drug resistance may be the inability to achieve inhibitory, as well as "knocking out," concentrations of the drugs at the sites of HIV-1 infection. Another major problem associated with the use of the currently available anti-HIV agents is toxicity to the host. For example, AZT causes malaise, nausea, vomiting, anemia, neutropenia, and myopathy.[11–15] Adverse reactions caused by dideoxyinosine (ddI) include pancreatitis, peripheral neuropathy, and diarrhea,[14] and that caused by dideoxycytidine (ddC) include peripheral neuropathy, pancreatitis, esophageal ulcers and cardiomyopathy.[11,14] An additional problem in the oral administration of drugs to some AIDS patients is the malabsorption syndrome,[16,17] which limits the bioavailability of the drugs.

Liposomes may be useful in the therapy of HIV infection in several ways: (i) Targeting liposomes containing anti-HIV drugs to cells and tissues infected with HIV-1 may enhance the efficacy and reduce the toxicity of the drugs. The lymph node localization of certain types of liposomes administered either intravenously or subcutaneously may provide a particular advantage, since recent studies have shown HIV to be rapidly replicating in these tissues.[18–20] (ii) Water-insoluble drugs such as protease inhibitors may be solubilized in the membrane phase of liposomes and delivered intravenously or subcutaneously. This formulation may be beneficial in delivering potent protease inhibitors which have not been developed further for clinical use because of their low oral bioavailability. The ability to deliver such drugs in liposomes would increase the type of available protease inhibitors which may be useful in combatting the emergence of drug-resistance. (iii) Large molecular weight drugs such as antisense oligonucleotides may be delivered more effectively to the cytoplasm of infected cells by encapsulation in appropriate liposomes. Delivery of oligonucleotides in liposome-encapsulated form may also provide protection against nucleases. (iv) Therapeutic genes such as those expressing HIV-specific ribozymes may be targeted to stem/progenitor cells or infected lymphocytes and macrophages by complexation with liposomes containing cationic lipids.

II. Therapy of *Mycobacterium avium* and *Mycobacterium tuberculosis* infections

Mycobacterium avium complex causes the most common bacterial opportunistic infection in AIDS. It invades macrophages in various tissues, including the lungs, liver, spleen, bone marrow and the gastrointestinal tract, and is also found in

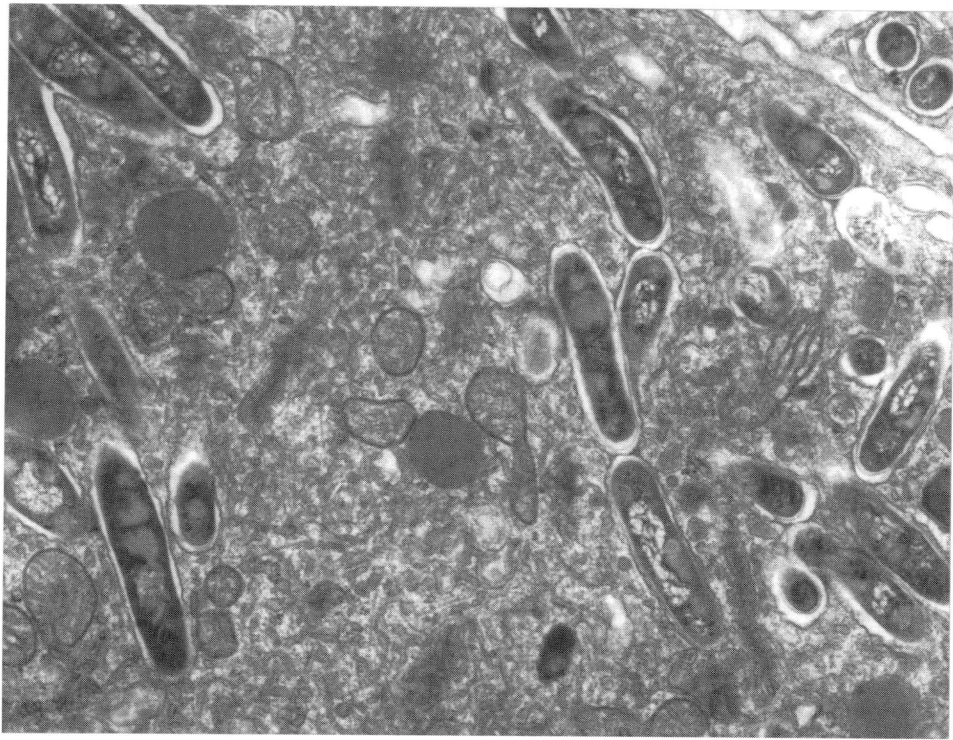

Fig. 1. *Mycobacterium avium* complex inside a human macrophage. Electron micrograph of a human monocyte-derived macrophage infected with *Mycobacterium avium* complex (strain 101), 24 hours after the addition of bacteria. Magnification: ×26,000. (Courtesy of Barbara Plowman and Diana Flasher).

blood[21–25] (Figure 1). AIDS patients with *M. avium* complex disease have a reduced survival rate compared to AIDS patients without MAC.[26] Many *M. avium* complex strains are resistant to conventional antimycobacterial drugs.[21,27–30] Although *M. avium* complex infection appears late in AIDS when the CD4-positive cell counts are low, treatment of the infection increases modestly the survival time of the patients.[31,32] Several recent clinical trials utilizing multiple drug therapy have shown that the blood levels of the microorganism can be reduced, with improved clinical sysmptoms.[33] Nevertheless, a significant percentage of patients (25–69%, depending on the drug combination and dose used in the different studies) could not complete the treatment due to adverse reactions. Single-agent therapy with clarithromycin or azithromycin has been shown to be very effective in reducing bacteremia, but results in the development of significant drug-resistance.[34–37] Another problem in the treatment of patients with disseminated *M. avium* complex is that serum levels of oral antimycobacterial drugs have been found to be below the expected range in these patients, possibly because of impaired drug absorption.[17,38]

The recent upsurge in the incidence in tuberculosis, particularly in AIDS patients, and the emergence of multidrug-resistant *Mycobacterium tuberculosis* strains, has focused much attention to the development of effective treatments for this disease.[39,40] New molecular targets in the microorganism are being identified for the development of new drugs. However, the effective delivery of currently available antibiotics to tissues and cells infected with *M. tuberculosis* is essential. An important problem in the therapy of tuberculosis is patient compliance with the therapeutic regimen, which is necessarily prolonged over many months. Another complication in AIDS patients infected by this microorganism is the enhancement of HIV-1 replication by *M. tuberculosis* by transcriptional activation.[41]

Liposomes are naturally targeted to macrophages of the reticuloendothelial system.[42,43] Antibiotics can therefore be targeted to cells infected with *M. avium* complex or *M. tuberculosis* by delivery in liposomes. The use of liposome encapsulated antibiotics for the therapy of mycobacterial infections may have several advantages: (i) The dose and the frequency of administration of antibiotics necessary to achieve a particular therapeutic effect may be lowered compared to the free drug, thereby reducing toxic side effects. (ii) The decrease in the necessary dose of the antibiotic may result in reduced drug interactions; this may be particularly significant for AIDS patients who have to take many different types of drugs. (iii) Liposomal delivery may be important in the delivery of antibiotics which may not be efficiently absorbed in AIDS patients due to gastrointestinal disorders. (iv) Liposomes with prolonged circulation in blood may be useful in providing a delivery system for hydrophobic antibiotics which normally deposit in tissues and do not remain in the circulation. (v) Novel drugs developed against new molecular targets in these microorganisms may be delivered effectively in liposome-encapsulated form before highly orally bioavailable forms of the drugs are developed.

III. Treatment of *Mycobacterium tuberculosis* infections by liposome-encapsulated antibiotics

The use of liposome-encapsulated antibiotics in the treatment of various intracellular infections has been reviewed previously.[44–47] The first studies on liposome-encapsulated antibiotics for the treatment of mycobacterial infections were performed in *Mycobacterium tuberculosis*-infected mice. Streptomycin sulfate encapsulated in phosphatidylcholine liposomes prepared by detergent dialysis (size range: 0.04–0.08 μm in diameter), and administered intravenously at a dose of 50 mg/kg on days 4, 7 and 10 after infection, significantly reduced the colony forming units (CFU) in the spleen, while free streptomycin did not cause a statistically significant change in the CFU.[48] A slight reduction of CFU in the lungs was also noted. The survival of the infected animals increased from about 12 days in untreated controls to almost 20 days in those treated with liposomal streptomycin; free streptomycin at the same dose increased survival to about 16 days.

A subsequent study utilized rifampicin and isoniazid encapsulated in multilamellar liposomes composed of phosphatidylcholine : cholesterol : cardiolipin (7 : 2 : 1) and in the size range 0.1–0.3 μm.[49] The liposomes were administered intravenously twice a week with seven 12 mg/kg doses of isoniazid and rifampicin each, into mice infected with the H37Rv strain of *Mycobacterium tuberculosis* 14 days previously. The group treated with the liposomal antimycobacterials showed the lowest degree of infection in the spleen, compared to the free drugs, empty liposome controls and untreated controls.[50] The treatments improved the percentage of surviving animals. However, there was no difference in the survival between the animals treated with free or liposomal drugs. One concern with the liposomes used in this study is the method of determination of encapsulated isonizid. Encapsulation was assessed by measuring the amount of ^{3}H-glucosamine as an analog of isoniazid.[49] In preliminary studies performed in our laboratory, we were unable to show stable encapsulation of isoniazid in phosphatidylcholine:-phosphatidylglycerol:cholesterol liposomes, as measured by an HPLC procedure (D. Flasher, M.V. Reddy, P. Gangadharam & N. Düzgüneş, unpublished data).

Orozco et al.,[51] have reported that intravenous administration of liposomes resulted in considerably higher levels of accumulation of the encapsulated radioactive marker, 99m-Tc-(Sn)-diethylene triamine pantaacetic acid (DTPA), in the liver, spleen and lungs of normal mice compared to animals heavily infected with *M. tuberculosis*. We should indicate that in this report it appears that the figures pertaining to the normal and infected mice were interchanged; thus, perusal of the data as presented would lead to the opposite interpretation.

Rifampin encapsulated in the membrane phase of sonicated egg phosphatidylcholine liposomes in the size range 0.03–0.07 μm, and delivered intravenously to *M. tuberculosis*-infected mice at twice weekly doses of 10 mg/kg, was more effective than free rifampin.[52] The liposomal formulation reduced the CFU in the lungs by almost 3 log units compared to controls, and 1.7 log units compared to free rifampicin. It also reduced the bacterial CFU levels in the liver and spleen by two orders of magnitude. This antimycobacterial effect was enhanced when the tetrapeptide macrophage activator tuftsin was coupled to the liposomes. Twice weekly treatments with these preparations were significantly more effective than daily treatments.

We examined the effect of streptomycin encapsulated in conventional or sterically stabilized liposomes on *M. tuberculosis* infection in C57BL/6 mice (M.V. Reddy, D. Flasher, N. Düzgüneş & P. Gangadharam, unpublished data). Liposomal streptomycin was administered intravenously or subcutaneously at a dose of 15 mg/kg twice a week for 2 weeks, while free streptomycin was administered subcutaneously at a dose of 150 mg/kg 5 days a week for 4 weeks. All the untreated control animals died within 3 weeks. Intravenously administered liposomal streptomycin as well as free streptomycin were effective in preventing mortality, while the drug encapsulated in sterically stabilized liposomes administered subcutaneously reduced mortality in only 1/3 of the animals. These experiments indicate that liposomal streptomycin was as effective as a 50-fold higher total dose of the free antibiotic. They also point to the possibility that subcutaneously administered sterically stabilized liposomes could be used for the therapy of tuberculosis, if increasing the dose and duration of treatment resulted in enhanced survival.

Drug resistant strains of *M. tuberculosis* are a serious public health concern. We examined the effect of free and liposome-encapsulated of sparfloxacin on two multi-drug resistant strains of *M. tuberculosis* (MTB 2219 and MTB 2227) inside the murine macrophage-like cell line J774. In this cell culture system both free sparfloxacin and the drug encapsulated in multilamellar phosphatidylglycerol: phosphatiylcholine: cholesterol (1:1:1) liposomes inhibited the growth of both strains.[53]

IV. Therapy of *M. avium* complex infections by liposome-encapsulated antibiotics

The first study on the effect of liposome-encapsulated antibiotics for the treatment of *M. avium* complex utilized negatively charged multilamellar liposomes composed of phosphatidylcholine: cholesterol: dipalmitoylphosphatidic acid (1:1:3), and encapsulating the aminoglycoside amikacin.[54] The efficacy of the liposomes was tested in human macrophages infected with the microorganism. Free amikacin in the range 10–30 μg/ml had no statistically significant effect on the colony forming units (CFU) of mycobacteria. In contrast, 20 μg/ml liposomal amikacin reduced the CFU by 92% within 2–4 days of treatment.

Similar results were obtained when mouse peritoneal macrophages were treated with 20 μg/ml amikacin encapsulated in large unilamellar phosphatidylglycerol: cholesterol (2:1) liposomes prepared by reverse phase evaporation and subsequent extrusion through polycarbonate membranes of 0.2 μm pore diameter.[55,56] Liposomal amikacin was more effective than the free drug in all three experimental protocols employed in this study, where the liposomes were added either 48 hours prior to infection, during infection or 48 hours following infection.

The beige mouse has been established as a useful model of *M. avium* complex disease.[57,58] Liposome-encapsulated amikacin was shown to be highly effective against *M. avium* complex in this animal model.[55,59] Amikacin was encapsulated in large unilamellar phosphatidylglycerol: phosphatidylcholine: cholesterol (1:1:1) liposomes prepared by reverse phase evaporation and extruded through 0.4 μm pore-diameter filters. Treatment of infected mice with weekly intravenous injections of only 5 mg/kg liposomal amikacin was shown to be effective in inhibiting the growth of the organism in the liver, spleen and kidneys by 3 orders of magnitude compared to untreated controls. Intravenous administration of the same dose of free amikacin was not effective. At the early stages of the treatment, amikacin in liposomes was more effective in the liver and spleen than amikacin given intramuscularly at a 60-fold higher total dose. At the 2 week time point, the reduction in CFU in the spleen per unit dose of liposome-encapsulated antibiotic was 1,280-fold greater than that of free intramuscular amikacin.[59] In the liver, the reduction per unit dose was 390-fold greater. In a separate experiment, the administration of 10 mg of amikacin per kg encapsulated in multilamellar liposomes (2–3 μm in diameter) was more effective in the liver and spleen than 5 or 10 mg amikacin per kg in unilamellar liposomes of 0.2 μm diameter. Liposomal amikacin had only a slight effect on the CFU in the lungs.

Cynamon et al.[60] increased the dose of amikacin encapsulated in plurilamellar phosphatidylcholine liposomes to 40 mg/kg and the frequency of administration to daily injections, and found a significant decrease in the number of viable *M. avium* in the spleen, liver and lungs. Twice weekly injections of 110 mg liposomal amikacin per kg body weight were more effective, reducing the CFU by about 3 log units in the liver and spleen, and by about 2 log units in the lungs. In comparison with the study described above, it appears that increasing the dose and frequency of administration of liposomal amikacin increased its efficacy in the lungs.

Bermudez et al.[61] also used a relatively high dose of liposomal amikacin (50 mg/kg) and a frequency of administration of every other day. The antibiotics were encapsulated in oligolamellar liposomes composed of partly hydrogenated phosphatidylcholine : phosphatidylglycerol : cholesterol : α-tocopherol at an unspecified ratio. At the end of the experiment, the *M. avium* counts were reduced by 92.7% in the liver, 92.8% in the spleen, and 98.5% in the blood. Similar results were obtained with gentamicin. Equivalent doses of free aminoglycosides did not result in significant reductions in CFU. Even a 10 mg/kg dose of liposomal amikacin resulted in a 95.7% reduction in the CFU in the liver, 69.7% in the spleen and 89.1% in the blood. Based on the urinary excretion of amikacin, these investigators concluded that liposome-encapsulated amikacin served as a sustained release system although the liposomes were cleared by the liver and spleen.

Gentamicin encapsulated in plurilamellar phosphatidylcholine liposomes in the size range 1.2–9.6 μm, and administered daily at a dose of 20 mg/kg was more effective than free gentamicin in the livers and spleens of beige mice infected with *M. avium* complex.[62] Prophylaxis studies involving the injection of free or liposomal gentamicin one day before infection, followed by daily treatments, indicated that the liposome-encapsulated antibiotic was more effective in these organs as well as in the lungs. Increasing the dose of gentamicin to 40 or 60 mg/kg had only a slight effect on the levels of infection in all three organs. Twice weekly administration of 20 mg/kg liposomal gentamicin was also more effective than the free antibiotic in the spleen and liver, decreasing the CFU by 2.5 log units.

Intraperitoneal injection of liposome-encapsulated kanamycin once a week for 8 weeks at doses ranging from 50–200 μg/mouse (2.5–10 mg/kg for a 20 g animal) caused significant reductions in the number of viable *Mycobacterium intracellulare* organisms in the lungs, liver, spleen and kidneys compared to untreated controls or free kanamycin.[63] Liposomal kanamycin was bactericidal in the liver, but only bacteriostatic in the spleen and kidneys. In the lungs, the treatment retarded the growth of the microorganism. The multilamellar liposomes used in this study consisted of phosphatidylcholine : dicetylphosphate : cholesterol (7 : 2 : 1), and had an average diameter of about 5 μm.

We examined the effect of streptomycin in large unilamellar liposomes extruded through filters of 0.2 μm pore diameter or multilamellar liposomes composed of phosphatidylglycerol : phosphatidylcholine : cholesterol (1 : 9 : 5), against *M. avium* complex in beige mice. Encapsulated streptomycin administered intravenously at a weekly dose of 15 mg/kg reduced the CFU in the liver and spleen by an extent similar to that obtained with a 50–100-fold higher total dose of free streptomycin given intramuscularly.[64,65] The enhanced effect of liposomal streptomycin was also observed in experiments involving *M. avium* complex-infected murine perioneal macrophages[66] or human monocyte/macrophages.[67]

Streptomycin encapsulated in sterically stabilized liposomes with prolonged circulation time was also highly effective against *M. avium* complex infection in beige mice.[68] These experiments employed two liposome types with prolonged circulation, composed of poly(ethylene glycol)-distearoylphosphatidylethanolamine : distearoylphosphatidylcholine : cholesterol (1:9:6.7) or plant phosphatidylinositol : distearoylphosphatidylcholine : cholesterol (1:9:6.7), and conventional liposomes composed of phosphatidylglycerol : phosphatidylcholine : cholesterol (1:9:6.7). The liposomes were administered twice weekly for 2 weeks, and the CFU were determined 2 weeks later. All the liposome preparations were bactericidal to *M. avium* complex in the spleen, compared to the level of infection before the initiation of treatment.[68] In the liver, phosphatidylinositol- and phosphatidylglycerol-containing liposomes encapsulating streptomycin were bactericidal. Significantly, conventional liposomes and sterically stabilized liposomes containing poly(ethylene glycol)-distearoylphosphatidylethanolamine reduced the level of infection in the lungs by more than 3 orders of magnitude, compared to untreated controls. Previous studies utilizing weekly injections of relatively low doses of amikacin or streptomycin encapsulated in conventional liposomes were rather ineffective in the lungs.[59,64,65]

The fluoroquinolone antibiotic ciprofloxacin encapsulated in multilamellar phosphatidylglycerol : phosphatidylcholine : cholesterol (1:9:5) liposomes was considerably more effective than the free drug against *M. avium* complex in human macrophages derived from monocytes.[67] Similar observations were made with differentiated monocytic THP-1 cells infected with *M. avium* complex.[69] The role of liposome composition in the antimycobacterial effect of ciprofloxain was examined by using liposomes into which the antibiotic was "remote-loaded" via pH- and potential-gradients.[70] The efficacy of the liposomal ciprofloxacin against *M. avium* complex inside J774 macrophages, measured as the decrease in the IC_{50}, was enhanced as the mole fraction of the negatively charged component in the membrane (distearoylphosphatidylglycerol) was increased. In contrast, the fluoroquinolone sparfloxacin encapsulated in the membrane phase of multilamellar phosphatidylglycerol : phosphatidylcholine : cholesterol (1:1:1) liposomes was only as effective as the free drug against *M. avium* complex-infected J774 macrophages.[71] Ofloxacin encapsulated in multilamellar phosphatidylcholine : dicetylphosphate : cholesterol (7:2:1) liposomes was shown to be more effective in reducing the CFU of *M. avium* complex in human monocyte/macrophages.[72] However, in this study the microorganism did not exhibit any growth inside untreated control macrophages over the course of the experiments.

The macrolide antibiotics azithromycin and clarithromycin have been identified recently as highly effective antibiotics against *M. avium* complex.[34,73–76] Azithromycin encapsulated in the membrane phase of small unilamellar liposomes prepared by sonication was shown to be much more effective than the free antibiotic against *M. avium* complex inside J774 macrophages, as long as the liposomes contained a high mole fraction of distearoylphosphatidylglycerol.[70] The most effective liposome composition in this system was distearoylphosphatidylglycerol:cholesterol (2:1), which reduced the IC_{50} from 18 μM for the free drug to 0.44 μM. Although Oh et al.[70] could observe a reduction in the IC_{50} of azithromycin even with distearoylphosphatidylglycerol:distearoylphosphatidylcholine:cholesterol (1:1:1) liposomes, clarithromycin encapsulated in the membrane phase of similar egg phosphatidylglycerol:egg phosphatidylcholine:cholesterol (1:1:1) liposomes at 30 mol% of the phospholipids showed an efficacy equivalent to the free antibiotic at low concentrations (1–2 μg/ml) against *M. avium* complex in human macrophages (I.I. Salem & N. Düzgüneş, unpublished data). At a higher concentration (4 μg/ml), however, liposomal clarithromycin was significantly more effective than the free drug. Similar observations were made by Onyeji et al.[72] using liposomes encapsulating clarithromycin in the aqueous phase.

Since these macrolides are known to accumulate in cells and tissues,[77] the utility of liposomal delivery vehicles for these antibiotics may be questioned. Mor et al.[78] have noted, however, that it is not possible to anticipate the enhancement of the activity of antibiotics against bacteria residing in macrophages from the intracellular accumulation of these molecules. It is likely that these antibiotics accumulate in lysosomes[77] which may be disconnected from the phagosomes harboring mycobacteria.[79]

The second line antituberculosis drug capreomycin was shown to have a slight (approximately 0.5 log reduction in CFU), but significant, effect against *M. avium* complex in the liver, spleen, blood and lungs of infected beige mice following intravenous administration inside multilamellar dipalmitoylphosphatidylcholine liposomes, while the free drug was ineffective.[80]

Liposome-encapsulated clofazimine had reduced toxicity compared to the free antibiotic both in macrophage cultures and in vivo, with the maximum tolerated dose in mice increasing by a factor of eight.[81] The liposomes used in this study were multilamellar and composed of dimyristoylphosphatidylcholine:dimyristoylphosphatidylglycerol (7:3), with clofazimine being incorporated in the membrane phase at a drug:lipid ratio of 1:10. The liposomes were injected intravenously twice a week, with a total of 6 injections) into beige mice with an established infection (28 days following inoculation of *M. avium* complex organisms). At 10 mg/kg, liposomal clofazimine reduced the CFU in the liver and kidneys to significantly lower levels than free clofazimine.[81] Liposome encapsulation permitted the administration of higher doses of clofazimine than was possible with the free drug. Thus, at 50 mg/kg, liposomal clofazimine reduced the CFU by 4 logs in the liver and 5 logs in the spleen, compared to untreated animals.

Resorcinomycin A, a recently discovered antibiotic with antimycobacterial activity, was encapsulated in the membrane phase of multilamellar dimyristoylphosphatidylcholine:plant phosphatidylinositol (9:1) liposomes, and tested against *M. avium* complex in murine peritoneal macrophages.[82] Liposome-encapsulated resorcinomycin A was more effective than the free antibiotic throughout the concentration range studied (3–50 μg/ml) in inhibiting the growth of intracellular mycobacteria.

The studies outlined above indicate clearly that liposome-encapsulated antibiotics have significant potential for the treatment of *M. avium* complex and *M. tuberculosis* in humans. Liposome-encapsulated gentamicin (TLC G-65) has been tested in AIDS patients with *M. avium* complex bacteremia, using twice weekly injections for a four-week period in the dose range 1.7–5.1 mk/kg.[83] The CFU in the blood were reduced by more than 75% in all patients. Transient renal insufficiency developed in one of 21 patients, and no other complications were observed. Phase II clinical trials with liposomal amikacin (MiKasome) are in progress.[84] The identification of the optimal liposomal antibiotics, their combinations, liposome compositions and schedule of administration await further studies in this area.

V. Liposome-encapsulated HIV reverse transcriptase inhibitors

Inhibitors of the HIV-1 reverse transcriptase were the first series of drugs that were approved for use against AIDS. Since macrophages have been identified as one of the major reservoirs of HIV-1 in infected individuals[85,86] (Figure 2), targeting of reverse transcriptase inhibitors to these cells in vivo may increase the efficacy of the inhibitors. Szebeni et al.[87] encapsulated 2′,3′-dideoxycytidine triphosphate (ddCTP) in liposomes composed of egg phosphatidylcholine : phosphatidylserine : cholesterol (37 : 18 : 45) and extruded through polycarbonate membranes of 0.4 μm pore diameter. The triphosphate derivative was chosen because of its higher retention time in liposomes compared to ddC. Monocyte-derived macrophages were infected with HIV-1_{BaL} for one day and then treated continuously. Liposome-encapsulated ddCTP was as effective as free ddCTP, inhibiting virus production by more than 95% at or above 62.5 nM. It was proposed that the ddCTP leaked from the liposomes in lysosomes and dephosphorylated to ddC, which then entered the cytoplasm. Although liposome-mediated delivery did not provide an advantage in cultured macrophages, the authors suggested that delivery in liposomes may increase the therapeutic index in vivo, considering the short circulation half-life, neurotoxicity and mucocutaneous side-effects of ddC.[87]

Zelphati et al.,[88] investigated the antiviral activity of 2′,3′-dideoxyuridine triphosphate (ddUTP) or 2′,3′-dideoxyuridine monophosphate (ddUMP) encapsulated in dipalmitoylphosphatidylcholine : cholesterol (64 : 35) liposomes. The liposomes, containing 1 mol% dipalmitoylphosphatidylethanolamine derivatized with *N*-succinimidyl-3-(2-pyridyldithio)propionate for protein coupling and extruded through 0.08 μm pore-diameter filters, were first coupled to Protein A to mediate association to the Fc region of antibodies. They were then incubated with HIV-1_{BRU}-infected T-lymphoblastoid CEM cells in the presence of antibodies to cell surface CD7 or HLA-class I molecules to "target" the liposomes to the cells. At a ddUTP concentration of 2 μM, liposomes targeted to HLA-class I molecules inhibited virus production by 95%, measured by reverse transcriptase activity, while those targeted to CD7 inhibited virus production by 67 percent.[88] The

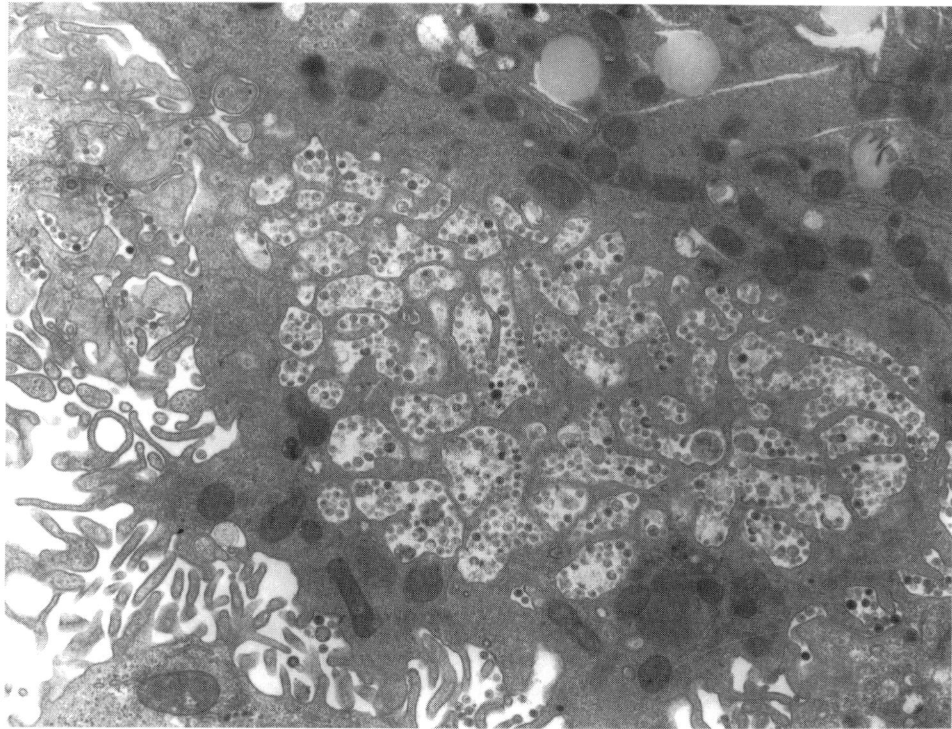

Fig. 2. HIV-1 inside a human macrophage-like cell. Electron micrograph of differentiated chronically infected THP-1/HIV-1_{IIIB} cells, showing large amounts of virions in intracellular vacuoles. Magnification: ×18,200. (The micrograph was reproduced from Konopka et al.,[135] with permission, and kindly provided by Barbara Plowman).

presence of a nonspecific antibody did not mediate any antiviral activity. Similar results were obtained with ddUMP. Free ddUMP or ddUTP were ineffective in this system even at concentrations of 100 μM. ddU itself was not used by these investigators both because it is inactive as an antiviral agent (presumably since it is phosphorylated poorly in human cells), and because it rapidly leaked out of liposomes.

Although the two studies cited above reported that the dideoxynucleosides ddC and ddU diffused out of the liposomes used in their studies, Makabi-Panzu et al.,[89] were able to encapsulate sufficient amounts of ddC in dipalmitoylphosphatidylcholine : dicetylphosphate : cholesterol (4 : 1 : 5) liposomes to determine its antiviral activity and biodistribution. These liposomes were extruded sequentially through 1 and 0.4 μm pore-diameter polycarbonate membranes. At 10 nM, liposome-encapsulated ddC was more effective than free ddC in inhibiting virus production by promonocytic U937 cells acutely infected with HIV-1_{IIIB}. When injected intravenously, liposomal ddC accumulated in the liver and spleen within 1 h, while the levels achieved with the free drug were much lower. Liposome-encapsulated ddC did not localize in the brain, lungs or bone marrow. Intraperitoneal injection also resulted in preferential accumulation of the drug in the reticuloendothelial system, particularly 3 h after injection. The authors suggested that ddC therapy directed to the macrophages of the reticuloendothelial system may have a therapeutic advantage, since it would prevent the dissemination of HIV to other cells and tissues.

In this respect, we should note that liver macrophages (Kupffer cells) from 3 of 7 AIDS patients examined were found to be infected with HIV-1.[90] In SIV-infected rhesus macaques, which is one of the best animal models of AIDS, Kupffer cells were found to be heavily infected with the virus.[91]

Dideoxycytidine encapsulated in multilamellar liposomes composed of dioleoylphosphatidylcholine : dipalmitoylphosphatidylglycerol : cholesterol : triolein at a mole ratio of 9.3 : 2.1 : 15 : 1.8, and injected into the cerebrospinal fluid was found to have a considerably longer half-life than free ddC. The half-life of free ddC was 1.1 h, while that of liposome-encapsulated ddC was 23 h, suggesting that anti-HIV agents that do not cross the blood-brain barrier could be delivered to the central nervous system in liposomes as a slow-releasing depot system.[92]

In contrast to the results obtained with ddC, liposome-encapsulated 2′,3′-dideoxyinosine (ddI) was less effective against HIV-1 replication in U937 cells compared to the free drug, consistent with the lower uptake of the liposomal drug by these cells.[93] In this study, ddI was encapsulated in distearoylphosphatidylcholine : distearoylphosphatidylglycerol (10 : 3) liposomes extruded through polycarbonate membranes of 0.2 μm pore diameter. The average diameter of the liposomes was assessed to be 0.18 μm by quasi-elastic light scattering. The uptake of liposomal ddI into the murine macrophage-like cell line RAW264.7 was also lower than that of free ddI, contrary to the findings with liposomal ddC.[89] These observations stress the necessity to design and utilize liposomes that not only carry the antiviral agent to the sites of infection, but also deliver them intracellularly at an effective concentration. Following intravenous injection, the plasma and spleen levels of the liposomal ddI was dramatically higher than that of the free drug (Table 1). The apparent systemic clearance of the liposomal drug (calculated as the ratio of the dose to the area under the plasma concentration-time curve) was determined to be 120 times lower than that of the free drug,[93] indicating that less frequent administration of liposomal ddI, compared to the free antiviral, may be sufficient to achieve therapeutic levels.

Sufficiently high drug concentrations were observed in lymph nodes for about 3 h when ddI was administered in liposomes, compared with only 30 min for the free drug.[94] For these experiments ddI was encapsulated in liposomes of the same composition as above, but extruded sequentially through membranes of 1 μm and 0.1 μm pore size, producing liposomes with an average diameter of 0.11 μm. Liposomes of smaller size (extruded through 0.05 μm filters, with an average diameter of 0.08 μm) did not retain the drug in serum or buffer as efficiently as the larger liposomes, and the plasma half-life of ddI encapsulated in these liposomes was more than 3-fold shorter than that of ddI in the larger liposomes.

Table 1

The ratios of the areas under the curve for liposome-encapsulated and free foscarnet and 2′,3′-dideoxyinosine in various tissues following intravenous administration

Tissues	Ratio of areas under the curve (liposomal drug/free drug)	
	foscarnet[a]	ddI[b]
Lymph nodes	8.1	ND[c]
Liver	52.1	3.7
Lungs	39.8	1.8
Spleen	1495.3	20.6
Brain	13.2	1.4

[a]Data from Dusserre et al.[101]
[b]Data from Désormeaux et al.[93]
[c]Not determined

Substantial amounts of liposomal lipids localized in lymph nodes over a 24 h period. This observation indicates that strategies to increase the retention of ddI inside liposomes could enhance and prolong the delivery of the drug to lymph nodes as well as to other tissues. Harvie et al.,[94] also reported that the levels of drug and lipid in cervical or mesenteric lymph nodes following subcutaneous administration of liposomes were similar to or lower than that following intravenous administration. Subcutaneous administration resulted in much lower levels of drug in the liver and spleen.

Phillips et al.,[95] encapsulated AZT in multilamellar liposomes composed of dipalmitoylphosphatidylcholine : dimyristoylphosphatidylglycerol (10 : 1). Although the drug was retained in liposomes upon storage at 4°C, it leaked out at 37°C with a half-life of less than 4 h. The use of distearoylphosphatidylcholine : dimyristoylphosphatidylglycerol liposomes in a subsequent study enhanced considerably the retention of AZT at 37°C.[96] Liposome-encapsulated AZT administered intravenously to mice localized in the liver, spleen and lungs, and the levels in the kidneys and bone marrow were reduced significantly compared to free AZT. The plasma levels of AZT injected intravenously dropped to less than 0.1% of the initial dose within 2 h, while that of liposomal AZT was about 1%. Free AZT at 0.4–10 mg/kg caused a significant reduction in bone marrow cellularity and in leukocyte and erythrocyte counts; within this dose range liposomal AZT caused no toxicity in the bone marrow.[95,96] AZT treatment of mice infected with LP-BM5 leukemia retrovirus, which results in murine AIDS similar to the HIV-induced disease in humans, delayed the development of reverse transcriptase activity in plasma (a measure of retrovirus levels), but had no effect on retrovirus-induced depression of $L3T4^+$ helper T cells. Liposome-encapsulated AZT, however, prevented the elevation of plasma reverse transcriptase levels and maintained the normal helper T cell numbers.[96]

Several laboratories have investigated the anti-HIV-1 activity of lipophilic derivatives of antiretroviral agents inserted in liposomes.[97–100] Hostetler et al.[100] utilized dioleoylphosphatidyl-ddC or dipalmitoylphosphatidyl-AZT incorporated into sonicated liposomes composed of dioleoylphosphatidylcholine :

Table 2

The ratios of the areas under the curve for lipid-coupled and free 3′-azido-3′-deoxythymidine (AZT) and 2′,3′-dideoxycytidine (ddC) in various tissues following intraperitoneal administration

Tissues	Ratio of areas under the curve (lipid-coupled drug/free drug)	
	AZT[a]	ddC[a]
Lymph nodes	5.2	3.8
Liver	32.6	50.1
Lungs	3.4	4.3
Spleen	65.8	107.3
Plasma	34.3	14.2

[a]Data from Hostetler et al.[99]

dioleoylphosphatidylglycerol : cholesterol : drug at a molar ratio of 5 : 1 : 3 : 1. While liposome-associated ddC was as active as free ddC in HIV-1-infected CD4-expressing HeLa cells (HT4–6C cells), liposomal AZT was less active than the free drug. The tissue levels of the liposome-associated drugs administered intraperitoneally were considerably higher than that of the free drugs (Table 2). In Rauscher leukemia virus-infected mice, liposome-associated AZT reduced the spleen weight and reverse transcriptase activity to levels that were comparable to that obtained with an equivalent dose of free AZT provided in the drinking water.[99]

Because of its slow phosphorylation by cellular thymidine kinase, 3′-deoxythymidine (3dT) is a weakly active HIV-1 inhibitor. However, its lipidic derivative 3dT diphosphate dimyristoylglycerol, incorporated into liposomes of a similar composition as the above study, was found to reduce the IC_{50} to 1.6 μM, from 29 μM for 3dT tested in HIV-1-infected CEM cells.[98]

Lipophilic dinucleoside phosphate derivatives of AZT (N^4-hexadecyl-dC-AZT and N^4-palmitoyl-dC-AZT) had IC_{50} values of 50 nM compared to 5 nM for free AZT, in HIV-1-infected H9 cells.[100] However, in Rauscher leukemia virus-infected mice, 380–1140 mg/kg intraperitoneal free AZT inhibited splenomegaly by 10–30%, while treatment with an equivalent dose of the derivatives resulted in 37–94% inhibition. Intravenous injection of AZT was ineffective in this system, while liposome-associated AZT inhibited splenomegaly by 48%.

Phosphonoformate, or foscarnet, is a non-nucleoside inhibitor of the HIV reverse transcriptase and has been shown to reduce viral p24 levels in patients with AIDS.[13] Foscarnet encapsulated in liposomes composed of dipalmitoylphosphatidylcholine : dipalmitoylphosphatidylglycerol (10 : 3) was shown to accumulate more efficiently in RAW 264.7 cells than the free drug, and to be slightly more effective against HIV-1_{IIIB} replication in U937 cells compared to free foscarnet.[101] The liposomes used in these experiments were extruded through 0.2 μm pore-diameter membranes and had an average diameter of 0.17 μm. The systemic clearance of liposomal foscarnet was 77 fold lower than than that of the free drug.

The acyclic nucleoside phosphonate 9-(2-(phosphonylmethoxy)ethyladenine (PMEA) is an inhibitor of reverse transcription via chain termination and inhibits HIV-1 replication in macrophages and lymphocytes.[102] Two of the drawbacks of

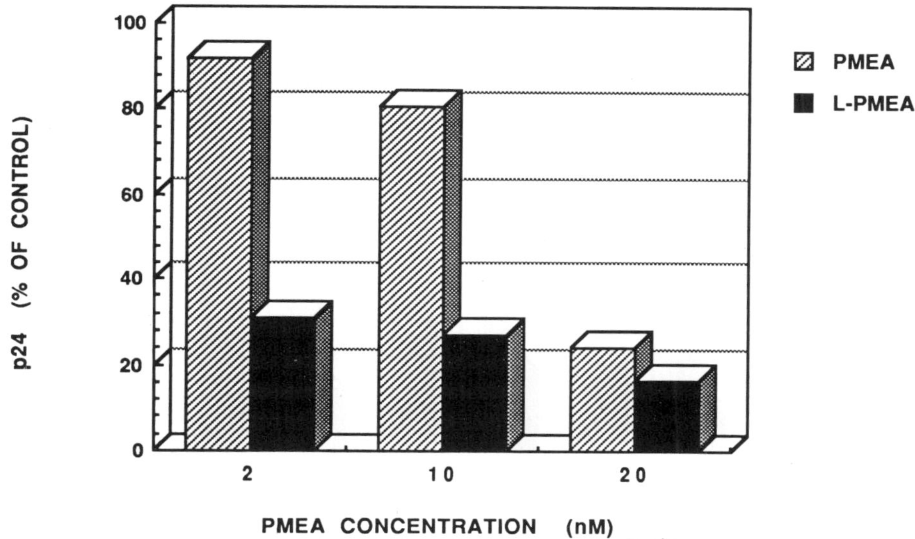

Fig. 3. Treatment of HIV-1_{BaL}-infected monocyte/macrophages with 9-(2-(phosphonylmethoxy)ethyl-adenine in free form (PMEA) or encapsulated in pH-sensitive liposomes (L-PMEA). The macrophages were treated for 8 days following infection, and viral p24 in culture supernatants were determined on day 15 after infection. The results are given as the percentage of untreated control cells (E. Pretzer, S. Simões, E. De Clercq and N. Düzgüneş, unpublished data).

PMEA and related inhibitors are their slow cellular uptake by an endocytosis-like process and their poor oral bioavailability.[103] When PMEA was delivered to HIV-infected macrophages in pH-sensitive liposomes, the antiviral effect of the drug, as measured by the inhibition of virus production, was enhanced (Figure 3). The EC_{50} of the liposome-encapsulated PMEA was about 10-fold lower than that of the free antiviral (E. Pretzer, S. Simões, E. De Clercq and N. Düzgüneş, unpublished data).

VI. Enhanced effect of a liposome-encapsulated HIV protease inhibitor against HIV infection of macrophages

To reduce the viral load in infected individuals it is important to inhibit virus production by cells in which the proviral DNA has been integrated in the genome, as well as to prevent the reverse transcription of HIV-1 RNA that has entered host cells. The HIV-1 protease is crucial to the cleavage of the viral Gag-Pol precursor polyprotein, whose components are essential for the generation of infectious virions.[104–106] A number of inhibitors have been developed that have a high specificity for the viral protease over cellular proteases.[107–112] Since macrophages are recognized to be a major reservoir of HIV-1 in infected individuals,[85,90,113,114] effective delivery of protease inhibitors to macrophages is likely to reduce the viral burden and reduce the risk of virus transmission to T cells. The compound

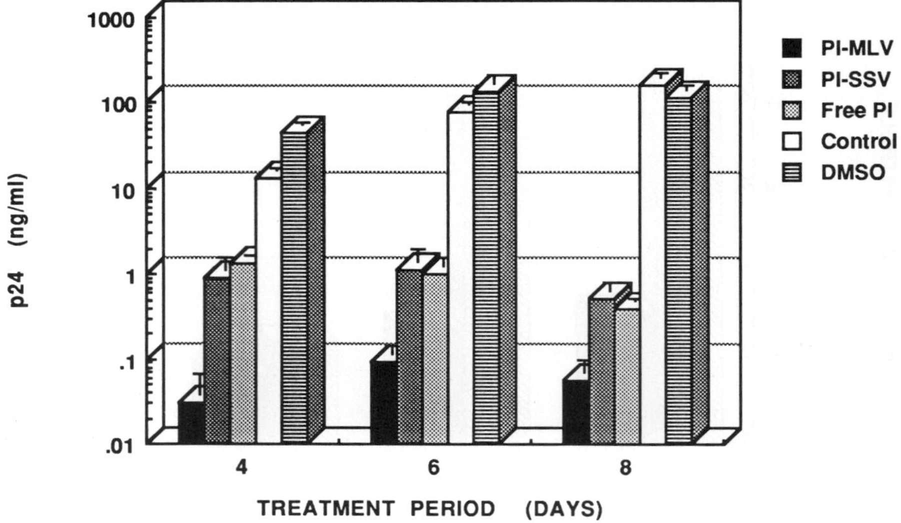

Fig. 4. Effect of the HIV-1 protease inhibitor L-689,502 on virus production by macrophages infected with HIV-1_{BaL}. The drug was added in free form (Free PI) or encapsulated in multilamellar (PI-MLV) or sterically stabilized (PI-SSV) liposomes. The effect of dimethylsulfoxide (DMSO), used to solubilize the free protease inhibitor, is also shown (Data from Pretzer et al.[117]).

L-689,502 was shown to inhibit the spread of the virus from infected monocyte-derived macrophages.[115] Our laboratory has investigated the effect of free and liposome-encapsulated L-689,502 on virus production by macrophages infected with a monocytotropic HIV-1 strain.[116,117] Treatment of the cells continuously with 100 nM L-689,502 encapsulated in sterically stabilized liposomes (polyethylene glycol-distearoyl phosphatidylethanolamine : partially hydrogenated egg phosphatidlycholine : cholesterol (0.15 : 1.85 : 1)) or the free inhibitor reduced viral p24 production by 10–100 fold compared to untreated controls (Figure 4), depending on the day following initial infection. In contrast, virus production in macrophages treated with the inhibitor encapsulated in multilamellar liposomes composed of egg phosphatidylcholine : egg phosphatidylglycerol : cholesterol (1 : 1 : 1) was about 1/10 the level of the other treatments. The differences between the treated and control wells increased during the treatment period, as the p24 levels of the treated wells remained relatively steady throughout the experiment, while the untreated control levels increased. This observation indicates that liposome-mediated administration of protease inhibitors can be more effective than the free drug in reducing the viral burden in infected macrophages.

The acyl chain region of the liposome bilayer constitutes a matrix in which hydrophobic protease inhibitors can be embedded for intravenous or subcutaneous delivery. Whether protease inhibitors are retained in liposomes in plasma to a sufficient degree to be transported to infected macrophages is not known. Sterically stabilized unilamellar liposomes containing protease inhibitors were not as effective as the multilamellar liposomes in infected macrophages, but were as potent as the free drug in reducing virus production. It is likely that the protease inhibitor leaks out more readily from the membrane of unilamellar sterically stabilized liposomes than from multilamellar liposomes. Since sterically stabilized liposomes can localize in lymph nodes, and hence may confer a special advantage in the delivery of protease inhibitors to tissues where the highest level of virus replication takes place in infected individuals, it may be necessary to develop prodrugs that are either lipid-associated or stably encapsulatable in the aqueous interior of the liposomes. Another potential advantage of liposome-mediated delivery of protease inhibitors would be to overcome the inhibitory activity of serum proteins on these drugs.[117] In the presence of 80% fetal bovine serum, the IC_{75} of the inhibitor KNI-272 was increased by a factor of 25–100.[118] It is likely that liposomes, and particularly sterically stabilized liposomes, would protect the protease inhibitors from binding by serum proteins until they reached relevant target tissues.

It is of interest to note that the protease inhibitor saquinavir loaded into nanoparticles composed of polyhexylcyanoacrylate also showed a superior antiviral effect compared to the free drug in monocyte/macrophages. While the free drug was ineffective in the range 0.1–1 nM, the nanoparticle formulation had significant anti-HIV-1 activity even at 0.1 nM, reducing the IC_{50} to 0.4 nM from 4.2 nM for the free drug.[119] Bender et al.[119] suggested that the use of nanoparticles as a carrier system for antiviral agents could improve their delivery to the mononuclear phagocyte system, overcome pharmakokinetic problems and enhance their anti-HIV activity.

VII. pH-Sensitive liposomes for the delivery of antisense oligonucleotides to HIV-infected macrophages

Antisense oligodeoxynucleotides are short segments of DNA or modified DNA that are complementary to specific sequences of target RNA. They inhibit the function of the target RNA by interfering with RNA transport, splicing or translation.[120] The oligodeoxynucleotide-RNA hybrid may also be a substrate for ribonuclease H, which selectively hydrolyzes the target RNA. Antisense oligodeoxynucleotides have been tested widely against HIV in vitro and shown to inhibit even chronic HIV infection.[121–126] Several antisense oligodeoxynucleotides against HIV are in phase I or II clinical trials.[120,127] Some of the advantages of antisense oligodeoxynucleotides in HIV therapy are that highly conserved target sequences can be chosen, and that a longer period of time may be required for the virus to develop resistance to antisense oligonucleotides treatment than to other antiviral drugs such as AZT.[125] The main hurdles in the therapeutic development of oligonucleotides are sequence-nonspecific interactions, sensitivity to nucleases, and low intracellular delivery.[120,122,128] The use of liposomes for the transport of antisense oligonucleotides as a method to overcome some of these drawbacks was evaluated in several laboratories. Liposome encapsulation protected oligo-dN from nuclease digestion and lead to substantially improved cellular accumulation and intracytoplasmic localization.[129,130]

Encapsulation in liposomes targeted to major histocompatibility complex HLA-B and C molecules enabled anti-*rev* antisense phosphorothioate oligonucleotides to inhibit chronic HIV infection in CEM cells in a sequence-specific manner.[131] The small unilamellar liposomes used in this study were coupled to protein A which then interacted with antibodies in solution or bound to the surface of target cells. Oligonucleotides in free form were not effective in this chronic infection model.

pH-sensitive liposomes were shown previously to deliver highly charged molecules or macromolecules into cultured cells.[132–134] We encapsulted a 28-mer antisense phosphorothioate oligodeoxynucleotide against *rev*[121] in pH-sensitive oleic acid/dioleoylphosphatidylcholine (3:7) liposomes, as well as in conventional liposomes, and examined its effect on viral replication in chronically infected differentiated THP-1/HIV-1_{IIIB} cells previously developed in our laboratory.[135] Although these cells produce very high quantities of HIV-1, a relatively low concentration of antisense oligodeoxynucleotides encapsulated in pH-sensitive liposomes reduced virus production to 52% of untreated controls 4 days after the end of the treatment period. Oligodeoxynucleotides encapsulated in non-pH-sensitive liposomes reduced the p24 levels to 76% of the controls, while free oligodeoxynucleotides reduced the p24 to only 88% of controls (E. Pretzer, D. Flasher & N. Düzgünes, unpublished data).

The enhanced effect of similar pH-sensitive liposomes was also observed in the case of antisense oligodeoxynucleotides against Friend murine leukemia virus.[136] Interestingly, Ropert et al.,[137] found that pH sensitive liposomes were taken up preferentially by virus-infected cells in which the virus budding process was intact. They suggested that the perturbation of the cell membrane during virus budding, and the associated enhancement of pinocytotic activity, may facilitate the uptake of particulate drug carriers. In this respect, it is worth noting that cationic liposomes are preferentially toxic to HIV-1-infected cells.[138] It is likely that the continuous production (budding) of the virus and expression of viral envelope proteins can alter the susceptibility of the cell membrane of HIV-infected cells to interaction with cationic liposomes.

Utilizing pH-sensitive liposomes composed of cholesterylhemisuccinate:dioleoylphosphatidylethanolamine (4:6), a composition similar to that described originally by Ellens et al.,[139] we investigated the antiviral effect of two antisense oligodeoxynucleotides in human monocyte-derived macrophages infected with HIV-1. The 28-mer anti-*rev* oligodeoxynucleotide in pH-sensitive liposomes inhibited HIV-1 production in a dose dependent manner. This formulation, incubated with the cells for 8 days, protected the cells from the cytopathic effect of HIV-1 infection as measured after 18 days from the beginning of the expriment, while the free oligodeoxynucleotide was ineffective in this respect.[140] A 15-mer antisense oligodeoxynucleotide against the Rev-responsive element of HIV-1 incubated in free form with infected macrophages was found to have no effect on virus production. In contrast, the oligonucleotide encapsulated in pH-sensitive liposomes reduced virus production by 91% at a dose of 3 μM.

We have shown recently that pH-sensitive liposomes can be sterically stabilized and still deliver their contents to macrophage-like cells as effectively as normal pH-sensitive liposomes.[141] Both the 28-mer and 15-mer antisense oligonucleotides encapsulated in sterically stabilized liposomes could inhibit HIV-1 replication in macrophages at a level comparable to that obtained with pH-sensitive liposomes.[140]

Delivery of oligodeoxynucleotides in liposomes may have several advantages: (i) Liposomes can protect oligodeoxynucleotides from degradation by nucleases in the extracellular milieu. (ii) The intracellular delivery of oligodeoxynucleotides by liposomes may be more efficient than that of the free compound. (iii) Liposomes may provide a targeting vehicle for the delivery of oligodeoxynucleotides to specific cells, in order to enhance the specificity of the drug.[142,143] (iv) Liposomes with prolonged circulation may provide a reservoir of antisense oligodeoxynucleotides, and small sterically stabilized liposomes may be able deliver their contents to lymph nodes. The lymphatic localization of subcutaneously injected liposomes may provide an additional route for targeting oligodeoxynucleotides to HIV-infected cells in lymph nodes.

VIII. Use of liposomes for gene therapy of HIV infection

Gene delivery via complexation of plasmids with cationic liposomes is being used in vitro and in vivo as an alternative to viral vectors.[144–148] The use of liposomes for gene delivery presents several advantages and disadvantages when compared to viral vectors. The advantages are that liposomes can be targeted to specific cells or tissues, they can carry large pieces of DNA, potentially up to chromosome size, they are not immunogenic, they are safe relative to viral vectors, and large scale production of liposomes is relatively easy compared to viruses.[147–149] The disadvantages of liposomal vectors include the use of unnatural cationic lipids that can be toxic, limited efficiency of delivery and gene expression, relatively large particle size, and potentially adverse interactions with biological milieu rich in negatively charged macromolecules.[138,149–152]

Gene therapy of HIV infection can be achieved by the efficient expression of various therapeutic genes, including antisense RNAs, ribozymes, RNA decoys, mutant HIV regulatory proteins, and toxins.[153,154] HIV-regulated expression of the diphtheria toxin A fragment gene (HIV-DT-A) is a potential gene therapy approach to AIDS. The diphtheria toxin A fragment coding sequence has been linked to *cis*-acting control elements unique to HIV, resulting in the expression of the toxin in a manner which is highly dependent on *trans*-activation by the HIV regulatory proteins Tat and Rev.[155] Cationic liposomes were used to transfect HIV-DT-A (pTHA43) or the HIV-regulated luciferase gene (pLUCA43) into HIV-infected or uninfected HeLa cells. The liposome compositions were either 2′3′-dioleyloxy-*N*[2(sperminecarboxamido)ethyl]-*N*,*N*-dimethyl-1-propanaminium trifluoroacetate : dioleoylphosphatidylethanolamine (3 : 1), or 1′,2′-dimyristoyloxy-propyl-3-dimethylhydroxyethylammonium bromide : dioleoylphosphatidylethanolamine (1 : 1). The HIV-regulated luciferase gene was expressed at a one thousand-fold higher level in chronically infected HeLa/LAV than in uninfected HeLa cells, while the extent of expression of Rous sarcoma virus (RSV)-regulated luciferase (pRSVLUC) was the same in both cell lines.[152] Co-transfection of HeLa cells with HIV-DT-A and the proviral HIV clone, HXBΔBgl, resulted in complete inhibition of virus production. Thus, when both the virus and DT-A genes were delivered into the same cells by cationic liposomes, HIV-DT-A was highly effective in inhibiting virus production. In contrast, the delivery of HIV-DT-A to chronically infected HeLa/LAV or HeLa/IIIB cells did not have a specific effect on virus production, since treatment of cells with control plasmids also reduced virus production. This reduction was ascribed to the cytotoxicity caused by the reagents. Studies on the efficiency of transfection, as measured by the percentage of cells expressing β-galactosidase, indicated that cationic liposome-mediated delivery of HIV-DT-A was too inefficient in this cell system to inhibit virus production.[152] Thus, the successful use cationic liposomes for the delivery of therapeutic genes in vivo will require the enhancement of their transfection efficiency.

The Rev protein of HIV-1 controls gene expression by binding to a Rev-responsive element in the viral mRNA. A segment of the Rev-responsive element binds Rev with high affinity, and is termed the Rev-binding element. An RNA decoy consisting of the Rev-binding element was shown to inhibit HIV-1 replication in T cells.[156] A Rev-binding aptamer gene[157] was inserted into the pTZU6+27 plasmid, and transfected into HeLa cells, together with the HIV proviral plasmid HXBΔBgl, using transferrin-associated cationic liposomes as a vector.[158] The production of viral p24 was inhibited specifically by the aptamer gene, the extent of inhibition depending on the ratio of the aptamer to viral genes transfected.[159] At a 1:1 ratio, the inhibition was 30% of the vector control plasmid, while at at an 11:1 ratio, the inhibition was 70%.

IX. Liposome targeting to HIV-1-infected cells

HIV-1 is produced by budding from the plasma membranes of actively infected cells, which necessarily express the viral envelope glycoprotein gp120/gp41 on their surface.[1,160] The surface protein gp120 is known to bind to the cell membrane antigen CD4.[1,161] Recombinant soluble CD4, the ectodomain of CD4, has a high affinity for gp120.[162] Incorporation of a recombinant transmembrane CD4 into liposomes was shown by fluorescence microscopy to result in targeting of the liposomes to HIV-infected cells.[163] Liposomes coupled to recombinant soluble CD4 could also be targeted specifically to HIV-infected cells, as shown by flow cytometry. CD4-coupled liposomes associated specifically with chronically HIV-1-infected H9/HTLV-IIIB or THP-1/HIV-1_{IIIB} cells, but not with uninfected H9 or THP-1 cells[164] (Figure 5).

The chimeric molecule CD4-immunoadhesin that combines the V1 and V2 domains of CD4 with the constant region of IgG, could also be used as a ligand to target liposomes to H9/HTLV-IIIB cells. In this case, liposomes were covalently coupled to Protein A, which binds the Fc region of the CD4-immunoadhesin. Significant association of Protein A-coupled liposomes with infected cells was observed in the presence of CD4-IgG, while control liposomes with or without

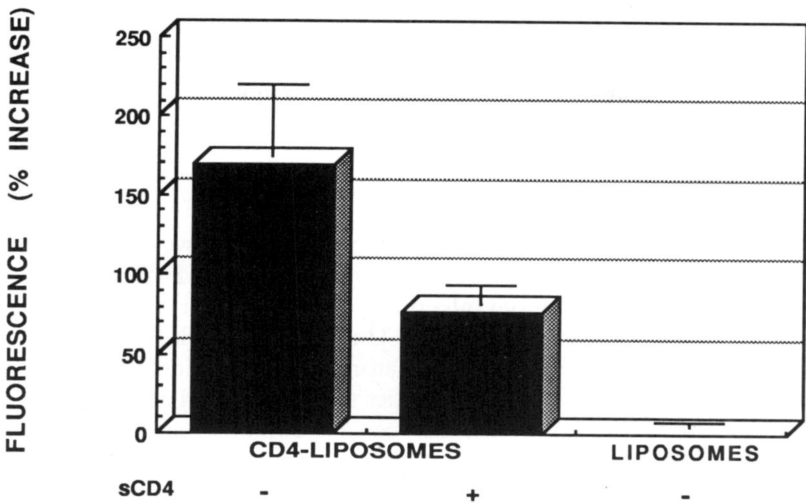

Fig. 5. Association of CD4-coupled and control liposomes with chronically infected H9/HTLV-IIIB cells, measured by flow cytometry. Rhodamine-phosphatidylethanolamine-labelled liposomes were incubated with the cells for 1 h at 37°C. Cell association is expressed as the % fluorescence increase relative to the autofluorescence of the cells. The effect of the presence of soluble CD4 on cell association is also shown (Data from Flasher et al.[164]).

CD4-immunoadhesin, or Protein A-liposomes without CD4 immunoadhesin, did not bind significantly to the cells.[164]

These studies indicate that antiviral agents, including protease inhibitors, antisense oligonucleotides or therapeutic genes, can be targeted to cells producing virions, thereby increasing the efficiency of the agents and possibly reducing toxicity to uninfected cells. HIV-infected cells in the lymph nodes appear to be actively producing virions throughout the clinically latent stages of AIDS.[18,20] Since sterically stabilized liposomes with prolonged circulation can extravasate into tissues including lymph nodes,[165,166] CD4-mediated targeting of these liposomes to infected cells is likely to facilitate their uptake once they arrive in the lymph nodes. The intracellular delivery of large drug molecules such as antisense oligonucleotides may require the use of pH-sensitive liposomes. It would therefore be of interest to determine whether sterically stabilized pH-sensitive liposomes[141] can be targeted to HIV-infected cells via CD4 coupled to reactive poly(ethylene glycol)-phospholipid derivatives.[167,168]

Although soluble CD4 was shown to be very effective in inhibiting the infectivity of laboratory isolates,[162,169–172] later studies indicated that high concentrations of the reagent are required to inhibit primary HIV-1 isolates.[173] Nevertheless, soluble CD4-toxin conjugates were demonstrated to be equally, if not more, cytotoxic to cells infected with primary HIV isolates as to cells infected with laboratory adapted strains.[174] The affinities of soluble CD4 for gp120 generated from primary isolates and laboratory strains were similar.[175–177] These observations indicate that soluble

CD4 may have greater potential as a means of targeting liposomes to HIV-1-infected cells than as an inhibitor of HIV-1 infectivity. However, it is also possible that CD4 coupled to liposomes in the proper orientation may be able to interact more efficiently with the gp120 molecules on virions of primary isolates.

X. Concluding remarks

The studies outlined in this chapter show the potential as well as the limitations of liposome-encapsulated drugs for the therapy of HIV-1 and Mycobacterial infections. In the case of Mycobacterial infections, the therapeutic advantage of certain liposomal antibiotics is clearly established. Unpublished work from our laboratory performed in collaboration with the laboratory of P. Gangadharam has indicated that not all liposome-encapsulated antibiotics show enhanced efficacy in vivo. This appears to be particularly the case for certain antibiotics encapsulated in the membrane phase of liposomes. However, this cannot be generalized, since rifampin and clofazimine embedded in the membrane phase of liposomes have superior efficacy against *M. tuberculosis* and *M. avium*, respectively. At this point, it is important to establish the optimal combinations of liposomal antibiotics, combinations of free and liposome-encapsulated drugs, and the frequency of administration that will reduce the CFU in all the major organs by 2–3 orders of magnitude compared to the initial level of infection in vivo. It is also essential to explore the ability of sterically stabilized liposomes to carry antibiotics to deep tissue macrophages. Finally, methods must be developed to stably encapsulate novel antibiotics that become available, and the liposomal formulations must be compared with the free antibiotics in animal models.

The delivery of certain anti-HIV-1 drugs in liposomes may be advantageous, particularly for potent compounds with low oral bioavailability. Prototype protease inhibitors which were not developed further because of poor oral bioavailability or very low aqueous solubility and which may be effective against emerging drug-resistant strains would become useful by delivery in liposomes. Problems such as the the high cost of the synthesis of protease inhibitors, the large oral doses necessary to achieve therapeutic levels in the circulation, the gastrointestinal side-effects of the drugs and poor absorption in patients with gastrointestinal problems, as well the advantages of less frequent administration of liposomal inhibitors, their high bioavailability and ability to localize in lymph nodes, should be weighed against the disadvantage of having to inject the liposomal formulation intravenously or subcutaneously.[117] The delivery of antisense oligo-dN in liposomes, specifically in sterically stabilized pH-sensitive liposomes, may alleviate many problems asociated with the administration of free oligo-dN. Although not the most effective drug against HIV-1, antisense oligo-dN have the advantage that drug resistance is much less likely to arise compared to reverse transcriptase or protease inhibitors. In addition, potentially unlimited types of antisense oligo-dN against any mutant strains that may arise can be synthesized readily without having to design novel drugs. Since gene therapy approaches to the treatment of AIDS hold great promise,[153,154] the development of liposomal vectors for the efficient and effective delivery of therapeutic genes is imperative, particularly since viral vectors are immunogenic. Such vectors will be important for both in vitro transfection to be used for ex vivo gene therapy, and in vivo gene delivery for gene therapy in vivo.

Acknowledgements

The work from my laboratory described in this chapter was supported by grants AI32399, AI35231, and AI33833 from the National Institute of Allergy and Infectious Diseases, and grants from the University of California Universitywide AIDS Research Program, Liposome Technology Inc., and Chiron Corporation. Additional support was provided by the University of the Pacific School of Dentistry.

I thank the members of my laboratory, Krystyna Konopka, Diana Flasher, Elizabeth Pretzer, Vladimir Slepushkin, Isam Salem, Gerhard Steffan and Sérgio Simões for their invaluable contributions. I also thank Pattisapu Gangadharam and members of his laboratory for their collaboration on the therapy of Mycobacterial infections, and Barbara Plowman for the electron micrographs.

References

1. Levy JA. Pathogenesis of human immunodeficiency virus infection. Microbiol Rev 1993;57:183–289.
2. Hirsch MS, D'Aquila RT. Therapy for human immunodeficiency virus infection. N Engl J Med 1993;328:1686–1695.
3. Johnston MI, Hoth DF. Present status and future prospects for HIV therapies. Science 1993;260:1286–1293.
4. Richman DD. HIV therapeutics. Science 1996;272:1886–1889.
5. Mitsuya H, Yarchoan R, Kageyama S, Broder S. Targeted therapy of human immunodeficiency virus-related disease. FASEB J 1991;5:2369–2381.
6. Richman DD. Resistance of clinical isolates of human immunodeficiency virus to antiretroviral agents. Antimicrob Agents Chemother 1993;37:1207–1213.
7. Otto MJ, Garber S, Winslow DL, Reid CD, Aldrich P, Jadhav PK, Patterson CE, Hodge CN, Cheng Y-SE. In vitro isolation and identification of human immunodeficiency virus (HIV) variants with reduced sensitivity to C-2 symmetrical inhibitors of HIV type 1 protease. Proc Natl Acad Sci USA 1993;90:7543–7547.
8. Wei X, Ghosh SK, Taylor ME, Johnson VA, Emini EA, Deutsch P, Lifson JD, Bonhoeffer S, Nowak MA, Hahn BH, Saag MS, Shaw GM. Viral dynamics in human immunodeficiency virus type 1 infection. Nature 1995;373:117–122.
9. Ho DD, Neumann AU, Perelson AS, Chen W, Leonard JM, Markowitz M. Rapid turnover of plasma virions and CD4 lymphocytes in HIV-1 infection. Nature 1995;373:123–126.
10. Balzarini J, Karlsson A, Pérez-Pérez M-J, Camarasa M-J, De Clercq E. Knocking-out concentrations of HIV-1-specific inhibitors completely suppress HIV-1 infection and prevent the emergence of drug-resistant virus. Virology 1993;196:576–585.
11. Hirsch MS. Chemotherapy of human immunodeficiency virus infections: Current practice and future prospects. J Infect Dis 1990;161:845–857.
12. Yarchoan R, Pluda JM, Perno C-F, Mitsuya H, Broder S. Anti-retroviral therapy of human immunodeficiency virus infection: Current strategies and challenges for the future. Blood 1991;78:859–884.
13. Richman DD. Antiviral therapy of HIV infection. Annu Rev Med 1991;42:69–90.
14. Neuzil KM. Pharmacologic therapy for human immunodeficiency virus infection: A review. Am J Med Sci 1994;307:368–372.
15. Cupler EJ, Danon MJ, Jay C, Hench K, Ropka M, Dalakas MC. Early features of zidovudine-

associated myopathy: histopathological findings and clinical correlations. J Infect Dis 1995;90:1–6.

16. Berning SE, Huitt GA, Iseman MD, Peloquin CA. Malabsorption of antituberculosis medications by a patient with AIDS. N Engl J Med 1992;327:1817–1818.
17. Peloquin CA, MacPhee AA, Berning SE. Malabsorption of antimycobacterial medications. N Engl J Med 1993;329:1122–1123.
18. Pantaleo G, Graziosi C, Demarest JF, Butini L, Montroni M, Fox CH, Orenstein JM, Kotler DP, Fauci AS. HIV infection is active and progressive in lymphoid tissue during the clinically latent stage of disease. Nature 1993;362:355–358.
19. Pantaleo G, Fauci AS. New concepts in the immunopathogenesis of HIV infection. Annu Rev Immunol 1995;13:487–512.
20. Embretson J, Zupancic M, Ribas JL, Burke A, Racz P, Tenner-Racz K, Haase AT. Massive covert infection of helper T lymphocytes and macrophages by HIV during the incubation period of AIDS. Nature 1993;362:359–362.
21. Young LS, Inderlied CB, Berlin OG, Gottlieb MS. Mycobacterial infections in AIDS patients, with an emphasis on the *Mycobacterium avium* complex. Rev Infect Dis 1986;8:1024–1033.
22. Zakowski P, Fligiel S, Berlin GW, Johnson L. Disseminated *Mycobacterium avium-intracellulare* infection in homosexual men dying of acquired immunodeficiency. J Am Med Assoc 1982;248:2980–2982.
23. Iseman MD, Corpe RF, O'Brien RJ, Rosenzweig DY, Wolinsky E. Disease due to *Mycobacterium avium-intracellulare*. Chest 1985;87:139S-149S.
24. Horsburgh CR Jr. *Mycobacterium avium* complex infection in the acquired immunodeficiency syndrome. N Engl J Med 1991;324:1332–1338.
25. Eng RHK, Bishburg E, Smith SM, Mangia A. Diagnosis of Mycobacterium bacteremia in patients with acquired immunodeficiency syndrome by direct examination of blood films. J Clin Microbiol 1989;27:768–769.
26. Benson C. Disseminated *Mycobacterium avium* complex disease in patients with AIDS. AIDS Res Hum Retroviruses 1994;10:913–916.
27. Hawkins CC, Gold JWM, Whimbey E, Kiehn TE, Brannon P, Cammarata R, Brown AE, Armstrong D. *Mycobacterium avium* complex infections in patients with the acquired immunodeficiency syndrome. Ann Intern Med 1986;105:184–188.
28. Horsburgh CR, Mason III UG, Farhi DC, Iseman MD. Disseminated infection with *Mycobacterium avium-intracellulare*. A report of 13 cases and a review of the literature. Medicine (Baltimore) 1985;64:36–48.
29. Bermudez LE, Young LS. New drugs for the therapy of mycobacterial infections. Curr Opin Infect Dis 1995;8:428–437.
30. Hoy J, Mijch A, Sandland M, Grayson L, Lucas R, Dwyer B. Quadruple-drug therapy for *Mycobacterium avium-intracellulare* bacteremia in AIDS patients. J Infect Dis 1990;161:801–805.
31. Horsburgh CR Jr, Havlik JA, Ellis DA, Kennedy E, Fann SA, Dubois RE, Thompson SE. Survival of patients with acquired immune deficiency syndrome and disseminated *Mycobacterium avium* complex infection with and without antimycobacterial chemotherapy. Am Rev Respir Dis 1991;144:557–559.
32. Chin DP, Reingold AL, Stone EN, Vittinghoff E, Horsburgh CR Jr, Simon EM, Yajko DM, Hadley WK, Ostroff SM, Hopewell PC. The impact of *Mycobacterium avium* complex bacteremia and its treatment on survival of AIDS patients—A prospective study. J Infect Dis 1994;170:578–584.
33. Benson CA, Ellner JJ. *Mycobacterium avium* complex infection and AIDS: Advances in theory and practice. Clin Infect Dis 1993;17:7–20.
34. Dautzenberg B, Truffot C, Legris S, Meyohas M-C, Berlie HC, Mercat A, Chevret S, Grosset J. Activity of clarithromycin against *Mycobacterium avium* infection in patients with the acquired immune deficiency syndrome: A controlled clinical trial. Am Rev Respir Dis 1991;144:564–569.
35. Dautzenberg B, Saint Marc T, Meyohas MC, Eliaszewitch M, Haniez F, Rogues AM, De Wit S, Cotte L, Chauvin JP, Grosset J. Clarithromycin and other antimicrobial agents in the treatment of disseminated *Mycobacterium avium* infections in patients with acquired immunodeficiency syndrome. Arch Intern Med 1993;153:368–372.
36. Young LS, Wiviott L, Wu M, Kolonoski P, Bolan R, Inderlied CB. Azithromycin for treatment of *Mycobacterium avium-intracellulare* complex infection in patients with AIDS. Lancet 1991;338:1107–1109.
37. Chaisson RE, Benson CA, Dube MP, Heifets LB, Korvick JA, Elkin S, Smith T, Craft JC, Sattler FR, AIDS Clin Trials Grp Protoc 157 Std Tm. Clarithromycin therapy for bacteremic

Mycobacterium avium complex disease. A randomized, double-blind, dose-ranging study in patients with AIDS. Ann Intern Med 1994;121:905–911.
38. Gordon SM, Horsburgh CR Jr, Peloquin CA, Havlik JA Jr, Metchock B, Heifets L, McGowan JE Jr, Thompson SE III. Low serum levels of oral antimycobacterial agents in patients with disseminated *Mycobacterium avium* complex disease. J Infect Dis 1993;168:1559–1562.
39. Bloom BR and Murray CJL. Tuberculosis: Commentary on a reemergent killer. Science 1992;257:1055–1064;
40. Collins FM. Tuberculosis: The return of an old enemy. Crit Rev Microbiol 1993;19:1–16.
41. Zhang Y, Nakata K, Weiden M, Rom WN. *Mycobacterium tuberculosis* enhances human immunodeficiency virus-1 replication by transcriptional activation at the long terminal repeat. J Clin Invest 1995;95:2324–2331.
42. Poste G. Liposome targeting in vivo: Problems and opportunities. Biol Cell 1983;47:19–38.
43. Scherphof GL. In vivo behavior of liposomes: Interactions with the mononuclear phagocyte system and implications for drug targeting. In: Juliano RL, ed. Handbook of Experimental Pharmacology, Vol. 100. Berlin: Springer-Verlag, 1991;285–327.
44. Popescu MC, Swensen CE, Ginsberg RS. Liposome-mediated treatment of viral, bacterial and protozoal infections. In: Ostro MJ, ed. Liposomes: From Biophysics to Therapeutics. New York: Marcel Dekker, 1987;219–251.
45. Karlowsky JA, Zhanel GG. Concepts on the use of liposomal antimicrobial agents: applications for aminoglycosides. Clin Infect Dis 1992;15:654–667.
46. Bakker-Woudenberg IAJM, Lokerse AF, Ten Kate MT, Melissen PMB, Van Vianen W, Van Etten EWM. Liposomes as carriers of antimicrobial agents or immunomodulatory agents in the treatment of infections. Eur J Clin Microbiol Infect Dis 12 Suppl. 1993;1:61–67.
47. Bermudez LE. Use of liposome preparation to treat mycobacterial infections. Immunobiology 1994;191:578–583.
48. Vladimirsky MA, Ladigina GA. Antibacterial activity of liposome-entrapped streptomycin in mice infected with *Mycobacterium tuberculosis*. Biomed Pharmacother 1982;36:375–377.
49. Wasserman M, Beltrán RM, Quintana FO, Mendoza PM, Orozco LC, Rodriguez G. A simple technique for entrapping rifampicin and isoniazid into liposomes. Tubercle 1986;67:83–90.
50. Orozco LC, Quintana FO, Beltrán RM, de Moreno I, Wasserman M, Rodriguez G. The use of rifampicin and isoniazid entrapped in liposomes for the treatment of murine tuberculosis. Tubercle 1986;67:91–97.
51. Orozco LC, Forero M, Wasserman M, Ahumada JJ. Distribution of liposomes in tuberculous mice. Tubercle 1990;71:209–214.
52. Agarwal A, Kandpal H, Gupta HP, Singh NB, Gupta CM. Tuftsin-bearing liposomes as rifampin vehicles in treatment of tuberculosis in mice. Antimicrob Agents Chemother 1994;38:588–593.
53. Düzgüneş N, Reddy MV, Flasher D, Luna-Herrera J, Gangadharam PRJ. Sparfloxacin therapy of multidrug-resistant *Mycobacterium tuberculosis* infection in macrophages. J Dent Res 1996;75:108 (Abstract).
54. Bermudez LEM, Wu M, Young LS. Intracellular killing of *Mycobacterium avium* complex by rifapentine and liposome-encapsulated amikacin. J Infect Dis 1987;156:510–513.
55. Gangadharam PRJ, Perumal VK, Kesavalu L, Debs RJ, Goldstein JA, Düzgüneş N. Comparative activities of free and liposome encapsulated amikacin against *Mycobacterium avium* complex (MAC). In: Lopez-Berestein G, Fidler IJ, eds. Liposomes in the Therapy of Infectious Diseases and Cancer. New York: Alan R. Liss, 1989;177–190.
56. Kesavalu L, Goldstein JA, Debs RJ, Düzgüneş N, Gangadharam PRJ. Differential effects of free and liposome encapsulated amikacin on the growth of *Mycobacterium avium* complex in mouse peritoneal macrophages. Tubercle 1990;71:215–218.
57. Gangadharam PRJ, Perumal VK, Farhi DC, LaBrecque J. The beige mouse model for *Mycobacterium avium* complex (MAC) disease: Optimal conditions for the host and parasite. Tubercle 1989;70:257–271.
58. Gangadharam PRJ. Beige mouse model for *Mycobacterium avium* complex disease. Antimicrob Agents Chemother 1995;39:1647–1654.
59. Düzgüneş N, Perumal VK, Kesavalu L, Goldstein JA, Debs RJ, Gangadharam PRJ. Enhanced effect of liposome-encapsulated amikacin on *Mycobacterium avium–M. intracellulare* complex in beige mice. Antimicrob Agents Chemother 1988;32:1404–1411.
60. Cynamon MH, Swenson CE, Palmer GS, Ginsberg RS. Liposome-encapsulated-amikacin therapy of *Mycobacterium avium* complex infection in beige mice. Antimicrob Agents Chemother 1989;33:1179–1183.
61. Bermudez LE, Yau-Young AO, Lin J-P, Cogger J, Young LS. Treatment of disseminated

Mycobacterium avium complex infection of beige mice with liposome-encapsulated aminoglycosides. J Infect Dis 1990;161:1262–1268.

62. Klemens SP, Cynamon MH, Swenson CE, Ginsberg RS. Liposome-encapsulated-gentamicin therapy of *Mycobacterium avium* complex infection in beige mice. Antimicrob Agents Chemother 1990;34:967–970.
63. Tomioka H, Saito H, Sato K, Yoneyama T. Therapeutic efficacy of liposome-encapsulated kanamycin against *Mycobacterium intracellulare* infection induced in mice. Am Rev Respir Dis 1991;144:575–579.
64. Düzgüneş N, Ashtekar DR, Flasher DL, Ghori N, Debs RJ, Friend DS, Gangadharam PRJ. Treatment of *Mycobacterium avium-intracellulare* complex infection in beige mice with free and liposome-encapsulated streptomycin: Role of liposome type and duration of treatment. J Infect Dis 1991;164:143–151.
65. Gangadharam PRJ, Ashtekar DA, Ghori N, Goldstein JA, Debs RJ, Düzgüneş N. Chemotherapeutic potential of free and liposome-encapsulated streptomycin against experimental *Mycobacterium avium* complex infections in beige mice. J Antimicrob Chemother 1991;28:425–435.
66. Ashtekar D, Düzgüneş N, Gangadharam PRJ. Activity of free and liposome-encapsulated streptomycin against *Mycobacterium avium* complex (MAC) inside peritoneal macrophages. J Antimicrob Chemother 1991;28:615–617.
67. Majumdar S, Flasher D, Friend DS, Nassos P, Yajko D, Hadley WK, Düzgüneş N. Efficacies of liposome-encapsulated streptomycin and ciprofloxacin against *Mycobacterium avium–M. intracellulare* complex infections in human peripheral blood monocyte/macrophages. Antimicrob Agents Chemother 1992;36:2808–2815.
68. Gangadharam PRJ, Ashtekar DR, Flasher D, Düzgüneş N. Therapy of *Mycobacterium avium* complex infections in beige mice with streptomycin encapsulated in sterically stabilized liposomes. Antimicrob Agents Chemother 1995;39:725–730.
69. Düzgüneş N, Flasher D, Reddy MV, Luna-Herrera J, Salem II, Gangadharam PRJ. Liposome-encapsulated sparfloxacin, paromomycin, ciprofloxacin and clarithromycin for the therapy of *Mycobacterium avium* complex and *Mycobacterium tuberculosis* in macrophages. 35th Intersci Conf Antimicrob Agents Chemother 1995;86 (Abstract).
70. Oh YK, Nix DE, Straubinger RM. Formulation and efficacy of liposome-encapsulated antibiotics for therapy of intracellular *Mycobacterium avium* infection. Antimicrob Agents Chemother 1995;39:2104–2111.
71. Düzgüneş N, Flasher D, Reddy MV, Luna-Herrera J, Gangadharam PRJ. Treatment of intracellular *Mycobacterium avium* complex infection by free and liposome-encapsulated sparfloxacin. Antimicrob Agents Chemother 1996;40:2618–2621.
72. Onyeji CO, Nightingale CH, Nicolau DP, Quintiliani R. Efficacies of liposome-encapsulated clarithromycin and ofloxacin against *Mycobacterium avium–M. intracellulare* complex in human macrophages. Antimicrob Agents Chemother 1994;38:523–527.
73. Dautzenberg B, Piperno D, Diot P, Truffot-Pernot C, Chauvin J-P, Clarithromycin Study Group of France. Clarithromycin in the treatment of *Mycobacterium avium* lung infections in patients without AIDS. Chest 1995;107:1035–1040.
74. Ives DV, Davis RB, Currier JS. Impact of clarithromycin and azithromycin on patterns of treatment and survival among AIDS patients with disseminated *Mycobacterium avium* complex. AIDS 1995;9:261–266.
75. Masur H. Recommendations on prophylaxis and therapy for disseminated *Mycobacterium avium* complex disease in patients infected with the human immunodeficiency virus. N Engl J Med 1993;329:898–903.
76. Cynamon MH, Klemens SP, Grossi MA. Comparative activities of azithromycin and clarithromycin against *Mycobacterium avium* infection in beige mice. Antimicrob Agents Chemother 1994;38:1452–1454.
77. Schlossberg D. Azithromycin and clarithromycin. Med Clin North Am 1995;79:803–815.
78. Mor N, Vanderkolk J, Mezo N, Heifets L. Effects of clarithromycin and rifabutin alone and in combination on intracellular and extracellular replication of *Mycobacterium avium*. Antimicrob Agents Chemother 1994;38:2738–2742.
79. Xu S, Cooper A, Sturgill-Koszycki S, Van Heyningen T, Chatterjee D, Orme I, Allen P, Russell DG. Intracellular trafficking in *Mycobacterium tuberculosis* and *Mycobacterium avium*-infected macrophages. J Immunol 1994;153:2568–2578.
80. Le Conte P, Le Gallou F, Potel G, Struillou L, Baron D, Drugeon HB. Pharmacokinetics, toxicity, and efficacy of liposomal capreomycin in disseminated *Mycobacterium avium* beige mouse model. Antimicrob Agents Chemother 1994;38:2695–2701.

81. Mehta RT. Liposome encapsulation of clofazimine reduces toxicity in vitro and in vivo and improves therapeutic efficacy in the beige mouse model of disseminated *Mycobacterium avium* complex infection. Antimicrob Agents Chemother 1996;40:1893–1902.
82. Gomez-Flores R, Hsia R, Tamez-Guerra R, Mehta RT. Enhanced intramacrophage activity of resorcinomycin A against *Mycobacterium avium–Mycobacterium intracellulare* complex after liposome encapsulation. Antimicrob Agents Chemother 40:2545–2549.1996;
83. Nightingale SD, Saletan SL, Swenson CE, Lawrence AJ, Watson DA, Pilkiewicz FG, Silverman EG, Cal SX. Liposome-encapsulated gentamicin treatment of *Mycobacterium avium–Mycobacterium intracellulare* complex bacteremia in AIDS patients. Antimicrob Agents Chemother 1993;37:1869–1872.
84. Gregoriadis G. Engineering liposomes for drug delivery: progress and problems. Trends Biotechnol 1995;13:527–537.
85. Meltzer MS, Skillman DR, Gomatos PJ, Kalter DC, Gendelman HE. Role of mononuclear phagocytes in the pathogenesis of human immunodeficiency virus infection. Annu Rev Immunol 1990;8:169–194.
86. Crowe SM, Kornbluth RS. Overview of HIV interactions with macrophages and dendritic cells: The other infection in AIDS. J Leukocyte Biol 1994;56:215–217.
87. Szebeni J, Wahl SM, Betageri GV, Wahl LM, Gartner S, Popovic M, Parker RJ, Black CDV, Weinstein JN. Inhibition of HIV-1 in monocyte/macrophage cultures by 2′,3′-dideoxycytidine-5′-triphosphate, free and in liposomes. AIDS Res Hum Retroviruses 1990;6:691–702.
88. Zelphati O, Degols G, Loughrey H, Leserman L, Pompon A, Puech F, Maggio A-F, Imbach J-L, Gosselin G. Inhibition of HIV-1 replication in cultured cells with phosphorylated dideoxyuridine derivatives encapsulated in immunoliposomes. Antiviral Res 1993;21,181–195.
89. Makabi-Panzu B, Lessard C, Perron S, Désormeaux A, Tremblay M, Poulin L, Beauchamp D, Bergeron MG. Comparison of cellular accumulation, tissue distribution, and anti-HIV activity of free and liposomal 2′,3′-dideoxycytidine. AIDS Res Hum Retroviruses 1994;10:1463–1470.
90. Hufert FT, Schmitz J, Schreiber M, Schmitz H, Rácz P, von Laer DD. Human Kupffer cells infected with HIV-1 in vivo. J Acquir Immune Defic Syndr 1993;6:772–777.
91. Persidsky Y, Steffan A-M, Gendrault J-L, Hurtrel B, Berger S, Royer C, Stutte H-J, Muchmore E, Aubertin A-M, Kirn A. Permissiveness of Kupffer cells for simian immunodeficiency virus (SIV) and morphological changes in the liver of rhesus monkeys at different periods of SIV infection. Hepatology 1995;21:1215–1225.
92. Kim S, Scheerer S, Geyer MA, Howell SB. Direct cerebrospinal fluid delivery of an antiretroviral agent using multivesicular liposomes. J Infect Dis 1990;162:750–752.
93. Désormeaux A, Harvie P, Perron S, Makabi-Panzu B, Beauchamp D, Tremblay M, Poulin L, Bergeron MG. Antiviral efficacy, intracellular uptake and pharmacokinetics of free and liposome-encapsulated 2′,3′-dideoxyinosine. AIDS 8:1545–1553.1994;
94. Harvie P, Désormeaux A, Gagné N, Tremblay M, Poulin L, Beauchamp D, Bergeron MG. Lymphoid tissues targeting of liposome-encapsulated 2′,3′-dideoxyinosine. AIDS 1995;9:701–707.
95. Phillips NC, Skamene E, Tsoukas C. Liposomal encapsulation of 3′-azido-3′-deoxythymidine (AZT) results in decreased bone marrow toxicity and enhanced activity against murine AIDS-induced immunosuppression. J Acquir Immune Defic Syndr 1991;4:959–966.
96. Phillips NC, Tsoukas C. Liposomal encapsulation of azidothymidine results in decreased hematopoietic toxicity and enhanced activity against murine acquired immunodeficiency syndrome. Blood 1992;79:1137–1143.
97. Hostetler KY, Stuhmiller LM, Lenting HBM, Van den Bosch H, Richman DD. Synthesis and antiretroviral activity of phospholipid analogs of azidothymidine and other antiviral nucleosides. J Biol Chem 1990;265:6112–6117.
98. Hostetler KY, Richman DD, Carson DA, Stuhmiller LM, Van Wijk GMT, Van den Bosch H. Greatly enhanced inhibition of human immunodeficiency virus type 1 replication in CEM and HT4-6C cells by 3′-deoxythymidine diphosphate dimyristoylglycerol, a lipid prodrug of 3′-deoxythymidine. Antimicrob Agents Chemother 1992;36:2025–2029.
99. Hostetler KY, Richman DD, Sridhar CN, Felgner PL, Felgner J, Ricci J, Selleseth DW, Ellis MN. Phosphatidylazidothymidine and phosphatidyl-ddC: Assessment of uptake in mouse lymphoid tissues and antiviral activities in human immunodeficiency virus-infected cells and in Rauscher leukemia virus-infected mice. Antimicrob Agents Chemother 1994;38:2792–2797.
100. Schwendener RA, Gowland P, Horber DH, Zahner R, Schertler A, Schott H. New lipophilic alkyl/acyl dinucleoside phosphates as derivatives of 3′-azido-3′-deoxythymidine: Inhibition of

HIV-1 replication in vitro and antiviral activity against Rauscher leukemia virus infected mice with delayed treatment regimens. Antiviral Res 1994;24:79–93.

101. Dusserre N, Lessard C, Paquette N, Perron S, Poulin L, Tremblay M, Beauchamp D, Désormeaux A, Bergeron MG. Encapsulation of foscarnet in liposomes modifies drug intracellular accumulation, in vitro anti-HIV-1 activity, tissue distribution and pharmacokinetics. AIDS 1995;9:833–841.
102. Balzarini J, Perno C-F, Schols D, De Clercq E. Activity of acyclic nucleoside phosphonate analogues against human immunodeficiency virus in monocyte/macrophages and peripheral blood lymphocytes. Biochem Biophys Res Commun 1991;178:329–335.
103. De Clercq E. Antiviral therapy for human immunodeficiency virus infections. Clin Microbiol Rev 1995;8:200–239.
104. Kay J, Dunn BM. Viral proteinases: Weakness in strength. Biochim Biophys Acta 1990;1048:1–18.
105. Robins T, Plattner J. HIV protease inhibitors: Their anti-HIV activity and potential role in treatment. J Acquir Immune Defic Syndr 1993;6:162–170.
106. Kaplan AH, Manchester M, Swanstrom R. The activity of the protease of human immunodeficiency virus type 1 is initiated at the membrane of infected cells before the release of viral proteins and is required for release to occur with maximum efficiency. J Virol 1994;68:6782–6786.
107. Meek TD, Lambert DM, Dreyer GB, Carr TJ, Tomaszek TA Jr, Moore ML, Strickler JE, Debouck C, Hyland LJ, Matthews TJ, Metcalf BW, Petteway SR. Inhibition of HIV-1 protease in infected T-lymphocytes by synthetic peptide analogues. Nature 1990;343:90–92.
108. Ashorn P, McQuade TJ, Thaisrivongs S, Tomasselli AG, Tarpley WG, Moss B. An inhibitor of the protease blocks maturation of human and simian immunodeficiency viruses and spread of infection. Proc Natl Acad Sci USA 1990;87:7472–7476.
109. Erickson J, Neidhart DJ, VanDrie J, Kempf DJ, Wang XC, Norbeck DW, Plattner JJ, Rittenhouse JW, Turon M, Wideburg N, Kohlbrenner WE, Simmer R, Helfrich R, Paul DA, Knigge M. Design, activity, and 2.8 Å crystal structure of a C_2 symmetric inhibitor complexed to HIV-1 protease. Science 1990;249:527–533.
110. Roberts NA, Martin JA, Kinchington D, Broadhurst AV, Craig JC, Duncan IB, Galpin SA, Handa BK, Kay J, Kröhn A, Lambert RW, Merrett JH, Mills JS, Parkes KEB, Redshaw S, Ritchie AJ, Taylor DL, Thomas GJ, Machin PJ. Rational design of peptide-based HIV proteinase inhibitors. Science 1990;248:358–361.
111. Huff JR. HIV protease: A novel chemotherapeutic target for AIDS. J Med Chem 1991;34:2305–2314.
112. Kempf DJ, Marsh KC, Denissen JF, McDonald E, Vasavanonda S, Flentge CA, Green BE, Fino L, Park CH, Kong X-P, Wideburg NE, Saldivar A, Ruiz L, Kati WM, Sham HL, Robins T, Stewart KD, Hsu A, Plattner JJ, Leonard JM, Norbeck DW. ABT-538 is a potent inhibitor of human immunodeficiency virus protease and has high oral bioavailability in humans. Proc Natl Acad Sci USA 1995;92:2484–2488.
113. Gartner S, Markovits P, Markovitz DM, Kaplan MH, Gallo RC, Popovic M. The role of mononuclear phagocytes in HTLV-III/LAV infection. Science 1986;233:215–219.
114. Ho DD, Rota TR, Hirsch MS. Infection of monocyte/macrophages by human T lymphotropic virus type III. J Clin Invest 1986;77:1712–1715.
115. Thompson WJ, Fitzgerald PMD, Holloway MK, Emini EA, Darke PL, McKeever BM, Schleif WA, Quintero JC, Zugay JA, Tucker TJ, Schwering JE, Homnick CF, Nunberg J, Springer JP, Huff JR. Synthesis and antiviral activity of a series of HIV-1 protease inhibitors with functionality tethered to the P1 or P1′ phenyl substituents: X-ray crystal structure assisted design. J Med Chem 1992;35:1685–1701.
116. Pretzer E, Flasher D, Düzgüneş N. Inhibition of HIV-1 infection of macrophages and H9 cells by free or liposome-encapsulated L-689,502, an inhibitor of the viral protease. Antiviral Res 1995;26:A358 (Abstract).
117. Pretzer E, Flasher D, Düzgüneş N. Inhibition of human immunodeficiency virus type-1 replication in macrophages and H9 cells by free or liposome-encapsulated L-689,502, an inhibitor of the viral protease. Antiviral Res 1997;34:1–15.
118. Kageyama S, Anderson BD, Hoesterey BL, Hayashi H, Kiso Y, Flora KP, Mitsuya H. Protein binding of human immunodeficiency virus protease inhibitor KNI-272 and alteration of its in vitro antiretroviral activity in the presence of high concentrations of proteins. Antimicrob Agents Chemother 1994;38:1107–1111.
119. Bender AR, Von Briesen H, Kreuter J, Duncan IB, Rübsamen-Waigmann H. Efficiency of

nanoparticles as a carrier system for antiviral agents in human immunodeficiency virus-infected human monocytes/macrophages in vitro. Antimicrob Agents Chemother 1996;40:1467–1471.

120. Wagner RW, Flanagan WM. Antisense technology and prospects for therapy of viral infections and cancer. Mol Med Today 1997;3:31–38.
121. Matsukura M, Zon G, Shinozuka K, Robert-Guroff M, Shimada T, Stein CA, Mitsuya H, Wong-Staal F, Cohen JS, Broder S. Regulation of viral expression of human immunodeficiency virus in vitro by an antisense phosphorothioate oligodeoxynucleotide against *rev* (*art/trs*) in chronically infected cells. Proc Natl Acad Sci USA 1989;86:4244–4248.
122. Stein CA, Cheng YC. Antisense oligonucleotides as therapeutic agents—Is the bullet really magical. Science 1993;260:1004–1012.
123. Matsukura M. Antiviral activities of antisense oligonucleotides against human immunodeficiency virus (HIV). In: Crooke ST, Lebleu B, eds. Antisense Research and Applications. Boca Raton: CRC Press, 1993;505–533;
124. Lisziewicz J, Sun D, Klotman M, Agrawal S, Zamecnik P, Gallo R. Specific inhibition of human immunodeficiency virus type 1 replication by antisense oligonucleotides: An in vitro model for treatment. Proc Natl Acad Sci USA 1992;89:11209–11213.
125. Lisziewicz J, Sun D, Weichold FF, Thierry AR, Lusso P, Tang J, Gallo RC, Agrawal S. Antisense oligodeoxynucleotide phosphorothioate complementary to Gag mRNA blocks replication of human immunodeficiency virus type 1 in human peripheral blood cells. Proc Natl Acad Sci USA 1994;91:7942–7946.
126. Weichold FF, Lisziewicz J, Zeman RA, Nerurkar LS, Agrawal S, Reitz MS Jr, Gallo RC. Antisense phosphorothioate oligodeoxynucleotides alter HIV type 1 replication in cultured human macrophages and peripheral blood mononuclear cells. AIDS Res Hum Retroviruses 1995;11:863–868.
127. Matteucci MD, Wagner RW. In pursuit of antisense. Nature 1996;384(supp):20–22.
128. Cantin EM, Woolf TM. Antisense oligonucleotides as antiviral agents: prospects and problems. Trends Microbiol 1993;1:270–276.
129. Leonetti J-P, Leserman LD. Targeted delivery of oligonucleotides. In: Crooke ST, Lebleu B, eds. Antisense Research and Applications. Boca Raton: CRC Press, 1993;493–504.
130. Thierry AR, Dritschilo A. Liposomal delivery of antisense oligodeoxynucleotides. Application to the inhibition of the multidrug resistance in cancer cells. Ann NY Acad Sci 1992;660:300–302.
131. Zelphati O, Imbach J-L, Signoret N, Zon G, Rayner B, Leserman L. Antisense oligonucleotides in solution or encapsulated in immunoliposomes inhibit replication of HIV-1 by several different mechanisms. Nucleic Acids Res 1994;22:4307–4314.
132. Straubinger RM, Düzgüneş N, Papahadjopoulos D. pH-sensitive liposomes mediate cytoplasmic delivery of encapsulated macromolecules. FEBS Lett 1985;179:148–154.
133. Connor J, Huang L. Efficient cytoplasmic delivery of a fluorescent dye by pH-sensitive liposomes. J Cell Biol 1985;101:582–589.
134. Chu C-J, Dijkstra J, Lai M-Z, Hong K, Szoka FC. Efficiency of cytoplasmic delivery by pH-sensitive liposomes to cells in culture. Pharm Res 1990;7:824–834.
135. Konopka K, Pretzer E, Plowman B, Düzgüneş N. Long-term noncytopathic productive infection of the human monocytic leukemia cell line THP-1 by human immunodeficiency virus type 1 (HIV-1_{IIIB}). Virology 1993;193:877–887.
136. Ropert C, Lavignon M, Dubernet C, Couvreur P, Malvy C. Oligonucleotides encapsulated in pH sensitive liposomes are efficient toward Friend retrovirus. Biochem Biophys Res Commun 1992;183:879–885.
137. Ropert C, Mishal Z, Rodrigues JM, Malvy C, Couvreur P. Retrovirus budding may constitute a port of entry for drug carriers. Biochim Biophys Acta 1996;1310:53–59.
138. Konopka K, Pretzer E, Felgner PL, Düzgüneş N. Human immunodeficiency virus type-1 (HIV-1) infection increases the sensitivity of macrophages and THP-1 cells to cytotoxicity by cationic liposomes. Biochim Biophys Acta 1996;1312:186–196.
139. Ellens H, Bentz J, Szoka FC. pH-Induced destabilization of phosphatidylethanolamine-containing liposomes: Role of bilayer contact. Biochemistry 1984;23:1532–1538.
140. Slepushkin VA, Pretzer EL, Simões S, Antao VP, Collins ML, Düzgüneş N. Antisense oligonucleotides in sterically stabilized pH-sensitive liposomes inhibit human immunodeficiency virus type 1 replication in macrophages. Antimicrob Agents Chemother (submitted), 1998.
141. Slepushkin VA, Simões S, Dazin P, Newman MS, Guo LS, De Lima MCP, Düzgüneş N. Sterically stabilized pH-sensitive liposomes: Intracellular delivery of aqueous contents and prolonged circulation in vivo. J Biol Chem 1997;272:2382–2388.
142. Leonetti J -P, Machy P, Degols G, Lebleu B, Leserman L. Antibody-targeted liposomes contain-

ing oligodeoxyribonucleotides complementary to viral RNA selectively inhibit viral replication. Proc Natl Acad Sci USA 1990;87:2448–2451.
143. Renneisen K, Leserman L, Matthes E, Schröder HC, Müller WEG. Inhibition of expression of human immunodeficiency virus-1 in vitro by antibody-targeted liposomes containing antisense RNA to the *env* region. J Biol Chem 1990;265:16337–16342.
144. Felgner PL, Gadek TR, Holm M, Roman R, Chan HW, Wenz M, Northrop JP, Ringold GM, Danielsen M. Lipofection: a highly efficient, lipid-mediated DNA-transfection procedure. Proc Natl Acad Sci USA 1987;84:7413–7417.
145. Felgner PL, Tsai YJ, Sukhu L, Wheeler CJ, Manthorpe M, Marshall J, Cheng SH. Improved cationic lipid formulations for in vivo gene therapy. Ann NY Acad Sci 1995;772:126–139.
146. Düzgüneş N, Felgner PL. Intracellular delivery of nucleic acids and transcription factors by cationic liposomes. Methods Enzymol 1993;221:303–306.
147. Singhal A, Huang L. Gene transfer in mammalian cells using liposomes as carriers. In: Wolf JA, ed. Gene Therapeutics: Methods and Applications of Direct Gene Transfer. Birkhäuser, Boston, 1994;118–142.
148. Lasic DD, Templeton NS. Liposomes in gene therapy. Adv Drug Deliv Rev 1996;20:221–266.
149. Lee RJ, Huang L. Folate-targeted, anionic liposome-entrapped polylysine-condensed DNA for tumor cell-specific gene transfer. J Biol Chem 1996;271:8481–8487.
150. Remy J-S, Kichler A, Mordvinov V, Schuber F, Behr J-P. Targeted gene transfer into hepatoma cells with lipopoylamine-condensed DNA particles presenting galactose ligands: A stage toward artificial viruses. Proc Natl Acad Sci USA 1995;92:1744–1748.
151. Zabner J, Fasbender AJ, Moninger T, Poellinger KA, Welsh MJ. Cellular and molecular barriers to gene transfer by a cationic lipid. J Biol Chem 1995;270:18997–19007.
152. Konopka K, Harrison GS, Felgner PL, Düzgüneş N. Cationic liposome-mediated expression of HIV-regulated luciferase and diphtheria toxin A genes in HeLa cells infected with or expressing HIV. Biochim Biophys Acta 1997;1356:185–197.
153. Sarver N and Rossi J. Gene therapy: A bold direction for HIV-1 treatment. AIDS Res Hum Retroviruses 1993;9:483–487.
154. Bridges SH, Sarver N. Gene therapy and immune restoration for HIV disease. Lancet 1995;345:427–432.
155. Harrison GS, Maxwell F, Long CJ, Rosen CA, Glode LM, Maxwell IH. Activation of a diphtheria toxin A gene by expression of human immunodeficiency virus-1 tat and rev proteins in transfected cells. Hum Gene Ther 1991;2:53–60.
156. Lee S-W, Gallardo HF, Gilboa E, Smith C. Inhibition of human immunodeficiency virus type 1 in human cells by a potent Rev response element decoy comprised of the 13-nucleotide minimal Rev-binding domain. J Virol 1994;68:8254–8264.
157. Tuerk C, MacDougal-Waugh S. In vitro evolution of functional nucleic acids: high-affinity RNA ligands of HIV-1 proteins. Gene 1997;137:33–39.
158. Cheng PW. Receptor ligand-facilitated gene transfer: Enhancement of liposome-mediated gene transfer and expression by transferrin. Hum Gene Ther 1996;7:275–282.
159. Konopka K, Düzgüneş N, Rossi J, Lee NS. Receptor ligand-facilitated cationic liposome delivery of anti-HIV-1 Rev binding aptamer and ribozyme DNAs. J Drug Targeting 1998 (in press).
160. Haseltine WA. Molecular biology of the human immunodeficiency virus type 1. FASEB J 1991;5:2349–2360.
161. McDougal JS, Maddon PJ, Orloff G, Clapham PR, Dalgleish AG, Jamal S, Weiss RA, Axel RA. Role of CD4 in the penetration of cells by HIV. Adv Exp Med Biol 1991;300:145–158.
162. Smith DH, Byrn RA, Marsters SA, Gregory T, Groopman JE, Capon DJ. Blocking of HIV-1 infectivity by a soluble, secreted form of the CD4 antigen. Science 1987;238:1704–1707.
163. Cudd A, Noonan CA, Tosi PF, Melnick JL, Nicolau C. Specific interaction of CD4-bearing liposomes with HIV-infected cells. J Acquir Immune Defic Syndr 1990;3:109–114.
164. Flasher D, Konopka K, Chamow SM, Dazin P, Ashkenazi A, Pretzer E, Düzgüneş N. Liposome targeting to human immunodeficiency virus type 1-infected cells via recombinant soluble CD4 and CD4 immunoadhesin (CD4-IgG). Biochim Biophys Acta 1994;1194:185–196.
165. Papahadjopoulos D, Allen TM, Gabizon A, Mayhew E, Matthay K, Huang SK, Lee K-D, Woodle MC, Lasic DD, Redemann C, Martin FJ. Sterically stabilized liposomes: Improvements in pharmacokinetics and antitumor therapeutic efficacy. Proc Natl Acad Sci USA 1991;88:11460–11464.
166. Huang SK, Lee K-D, Hong K, Friend DS, Papahadjopoulos D. Microscopic localization of sterically stabilized liposomes in colon carcinoma-bearing mice. Cancer Res 1992;52:5135–5143.
167. Allen TM, Brandeis E, Hansen CB, Kao GY, Zalipsky S. A new strategy for attachment of

antibodies to sterically stabilized liposomes resulting in efficient targeting to cancer cells. Biochim Biophys Acta 1995;1237:99–108.

168. Zalipsky S. Chemistry of polyethylene glycol conjugates with biologically active molecules. Adv Drug Deliv Rev 1995;16:157–182.
169. Deen KC, McDougal JS, Inacker R, Folena-Wasserman G, Arthos J, Rosenberg J, Maddon PJ, Axel R, Sweet R. A soluble form of CD4 (T4) protein inhibits AIDS virus infection. Nature 1988;331:82–84.
170. Fisher RA, Bertonis JM, Meier W, Johnson VA, Costopoulos DS, Liu T, Tizard R, Walker BD, Hirsch MS, Schooley RT, Flavell RA. HIV infection is blocked in vitro by recombinant soluble CD4. Nature 1988;331:76–78.
171. Hussey RE, Richardson NE, Kowalski M, Brown NR, Chang HC, Siliciano RF, Dorfman T, Walker B, Sodroski J, Reinherz EL. A soluble CD4 protein selectively inhibits HIV replication and syncytium formation. Nature 1988;331:78–81.
172. Traunecker A, Luke W, Karjalainen K. Soluble CD4 molecules neutralize human immunodeficiency virus type 1. Nature 1988;331:84–86.
173. Daar ES, Li XL, Moudgil T, Ho DD. High concentrations of recombinant soluble CD4 are required to neutralize primary human immunodeficiency virus type 1 isolates. Proc Natl Acad Sci USA 1990;87:6574–6578.
174. Kennedy PE, Moss B, Berger EA. Primary HIV-1 isolates refractory to neutralization by soluble CD4 are potently inhibited by CD4-*Pseudomonas* exotoxin. Virology 1993;192:375–379.
175. Ashkenazi A, Smith DH, Marsters SA, Riddle L, Gregory TJ, Ho DD, Capon DJ. Resistance of primary isolates of human immunodeficiency virus type 1 to soluble CD4 is independent of CD4-rgp120 binding affinity. Proc Natl Acad Sci USA 1991;88:7056–7060.
176. Brighty DW, Rosenberg M, Chen ISY, Ivey-Hoyle M. Envelope proteins from clinical isolates of human immunodeficiency virus type 1 that are refractory to neutralization by soluble CD4 possess high affinity for the CD4 receptor. Proc Natl Acad Sci USA 1991;88:7802–7805.
177. Turner S, Tizard R, DeMarinis J, Pepinsky RB, Zullo J, Schooley R, Fisher R. Resistance of primary isolates of human immunodeficiency virus type 1 to neutralization by soluble CD4 is not due to lower affinity with the viral envelope glycoprotein gp120. Proc Natl Acad Sci USA 1992;89:1335–1339.

Lasic and Papahadjopoulos (eds.), Medical Applications of Liposomes

CHAPTER 4.1

Cancer therapy

DANILO D. LASIC[a] AND DEMETRIOS PAPAHADJOPOULOS[b]

[a]*Liposome Consultations, 7512 Birkdale Drive, Newark, CA 94560, USA*
[b]*Department of Cellular and Molecular Pharmacology, University of California, San Francisco, and California Pacific Medical Center Research Institute, San Francisco, CA 94115, USA*

Introduction and general strategies

Several different strategies have evolved to combat the growth of neoplastic tissues by using liposome delivered agents. Mostly, we can distinguish delivery of anti-cancer drugs, enhancement of body's own defense system by activation of macrophages (immunotherapy), as well as delivery of molecules with genetic or immunologic information, such as DNA, antisense oligonucleotides or ribozymes as well as antibodies and cytokines to cancer cells. In the case of delivery of drug molecules in chemotherapy, we can distinguish between small molecular weight drugs, polypeptides and proteins. In the case of chemotherapy with small molecular weight drugs, several approaches resulted in successful applications, culminating with two commercial products discussed in section 8 of this volume. Currently, the major scientific effort is in improved targeting of drug-carrying liposomes and in gene therapy of cancer. All these strategies are based on the destruction of neoplastic tissue and cells, either directly, or indirectly, such as blocking the cells drug resistance or their blood supply.

Activation of the host immune system, as is described in the chapter by Fidler and colleagues, is a strategy anticipated for low tumor and cancer cell burdens after surgery, radiation, chemotherapy or other treatments.[1] While macrophages can be activated by liposome-associated immunomodulators, such as muramyl peptides, the major problem remains the spatial separation of tumor cells and activated macrophages. This is also a problem in gene therapy because current delivery vehicles for nucleic acids are not selective with respect to the target site and even intra-tumor administration does not yield high level gene expression or effective delivery of shorter nucleic acids into cells. At present, the major effort is still in development of anticancer formulations of small molecules and improvement of their carriers.

The major problem in cancer chemotherapy is toxic side effects of the agents administered. Typically, less than 1% of the administered free drug reaches target cells, while the rest damages non-targeted cells. Most anti-cancer agents are usually small molecular weight hydrophilic molecules with significant solubility in hydrophobic environment, and can therefore permeate cell membranes readily. Examples are fluorouracil, which leaks from the bilayers very quickly, vincristine, cisplatin and anthracyclines. Only few agents which are hydrophobic, such as taxol, have been used.[2] Proper formulation of these requires lipid-based carriers. Because the molecules of taxol and its derivatives are rather bulky and have rather irregular molecular shape, lipid bilayers may not be the best solubilizing milieu and emulsions (or sterically stabilized emulsions) may be more appropriate.[3] Amphiphilic pro-drugs, such as (di)alkyl cisplatin, which can be easily incorporated in the membrane, and can revert to active drugs after hydrolysis, are also being tested.[4] Along those lines, however, retention of these molecules in the bilayer and the rate of their activation still have to be improved (see Chapter 4.5).

Early experiments with anthracyclines have generally shown reduced toxicity and, what was realized only later, reduced activity. This is due to the fact that liposomes were taken up by RES and therefore systemic toxicity was greatly reduced. Obviously, this sequestration of the drug also eliminated some of its activity because of unavailability of the drug. Drug leakage from liposomes while still in circulation and leakage from macrophages, produced a reduction of toxicity due to substantial reduction of high peak levels of free drug in the plasma, while it prolonged the presence of the agent in blood. These animal models studies with conventional liposomes led to some initial clinical trials, which verified the effect of lowering toxicity. However, they were not successful enough to warrant further investigation (see Chapter 8.1 by A. Gabizon).

It should be noted here that conventional small liposomes with solid neutral bilayers have been shown to have considerable stability and relatively long half-lives in blood (discussed by Gregoriadis in Chapter 1.2). This became the basis for further work by two different groups of researchers using anthracycines and also vincristin for anti-cancer treatment. Such use of conventional liposomes in cancer therapy is discussed by Mayer et al., in Chapter 4.2 of this section and by Schmidt et al., in Chapter 8.4. Although we do not agree with the conclusions of Mayer et al. (Chapter 4.2), we expect that the readers will arrive at their own conclusions concerning the relative advantages of conventional vs. sterically stabilized liposomes, and the future of targeting to tumors (for contrasting views, see Chapters 4.3, 4.6 and 4.7).

While the jury is still out concerning the question whether conventional liposomes can significantly improve chemotherapy to warrant commercialization, it has been established that small, sterically stabilized liposomes can increase the therapeutic efficacy of the encapsulated drugs due to their enhanced accumulation in tumor tissue. Because of its importance and the fact that this subject is not covered in any other chapter, we shall briefly describe the initial experiments which lead to clinical studies of doxorubicin encapsulated in sterically stabilized liposomes. These experiments are important, not only because they showed unprecedented results in murine tumor models but also because they resulted in the first commercial formulation relatively quickly: first experiments were performed in 1989 and a product was approved and marketed 6 years later (F. Martin, Chapter 8.2).

The first experiments leading to sterically stabilized liposomes involved the use of glycolipids such as GM1 ganglioside, cereboside sulfate, and phosphatidyl inositol.[5,6] Those studies showed that inclusion of such negatively charged lipids produced lengthening of the circulation time of liposomes in blood, contrary to earlier findings with other negatively charged lipids. More importantly, it was also established that such long circulating liposomes show a much increased uptake by tumors.[6] The next significant stage involved the use of polyethylene glycol conjugated on phospholipids as a more advantageous molecule for steric stabilization[7–10] and the demonstration that such liposomes were very effective against mouse carcinomas.[11] The superior anti-tumor efficacy of such liposomes was further demonstrated by a series of pharmacokinetic and therapeutic experiments in various animal tumor models including human xenographs. These are described in detail in Chapters 4.3 and 4.4 of this section and Chapter 7.3 of this volume. Following FDA approval, these liposomes were tested in clinical trials with cancer patients, and the results are described by F. Martin, in the the last section of this volume (Chapter 8.2).

The mechanism responsible for the increased tumor accumulation of sterically stabilized liposomes involves a variety of crucial factors. Tumor vasculature has been known to be relatively leaky.[12,13] Therefore, once liposomes were constructed with reasonably long circulation time in blood, their extravasation was conceivable, assuming a small enough particle size to pass through the gaps between endothelial cells (Figure 1). Lack of lymphatic drainage within the tumor mass, should help in their retention, while high interstitial pressure should be inhibiting. The first demonstration of the extravasation of liposomes into the tumor mass was shown by the presence of colloidal gold particles,[14] (Figure 2) and later by intra-vital fluorescence microscopy[15] (Figure 3) showing perivascular accumulation of liposome material in the post-capillary network of the tumor. It is reasonable to assume that doxorubicin, once released from such extravasated liposomes, will diffuse rapidly through the tumor mass, reaching distant cells over an extended period of time. This effect of a localized reservoir for doxorubicin diffusion within a tumor was demonstrated by Vaage et al., using confocal fluorescence microscopy (Chapter 4.4 in this section).

Long-circulating liposomes have also opened the possibility of selective targeting of accessible cells and sites. Conventional targeted liposomes, with the exception of perhaps intraperitoneal application were not too successful in vivo studies because of non-specific uptake by the RES.[16] Although some of the first experiments with ligand-bearing sterically stabilized liposomes look promising, further work will have to be done to reduce possible immunogenicity.[17,18] Further research envisages use of immunoglobulin fragments because the most immunogenic part (the Fc) can be deleted without affecting the binding properties. Targeting can also be achieved using small peptides and other small molecular weight ligands, such as folate.[19] On the other hand, we must also keep in mind that not many cellular sites are accessible in vivo, because of their location behind biological barriers such as the endothelium. It is therefore conceivable that non-targeted sterically stabilized liposomes may be optimal for localization in areas of increased endothelial permeability due to smaller size and lower reactivity. The subject of ligand-directed targeting of liposomes is discussed extensively by T. Allen in Chapter 4.6, and by D. Kirpotin et al. in Chapter 4.7 in this section.

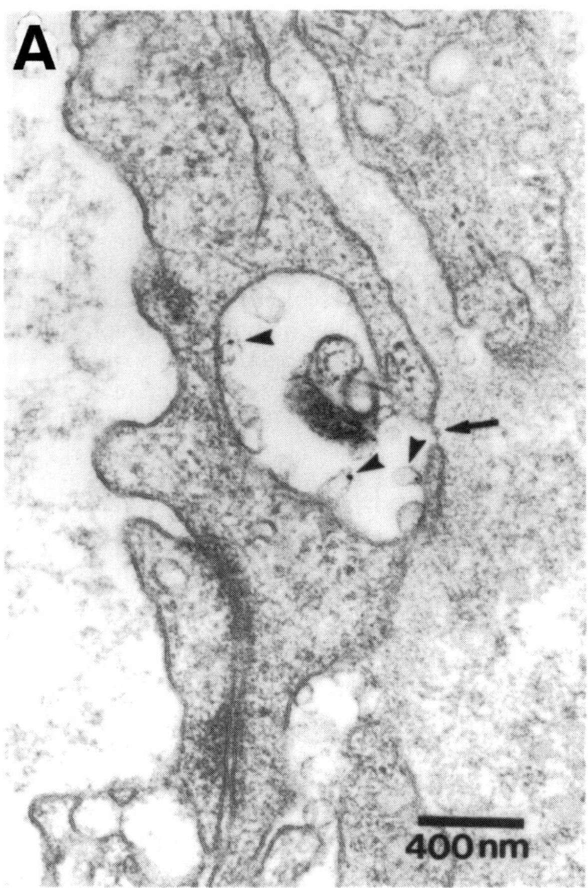

Fig. 1A.

Fig. 1. Thin section electron micrographs of liposomes containing colloidal gold 24 h after i.v. injection in mice: (A) within endothelial cells, in skin tissue; (B) in blood, near erythrocytes. With permission, from Huang et al., Amer J Pathol 1993;143:10–14.

Recent studies in the editor's own laboratory (D.P.) have produced very promising data on the possibility of targeting to solid tumors in vivo. Thus, sterically stabilized liposomes encapsulating doxorubicin and conjugated to an anti-HER2 immunoglobulin (Fab) fragment, have been shown to have higher anti-tumor efficacy against human breast tumor xenographs in nude mice, as compared to similar non-targeted liposomes. Moreover, microscopic analysis of the localization of such liposomes within the solid tumor, indicate that they are internalized within tumor cells in vivo. These new and exciting results are discussed in Chapter 4.7, this section.

Several laboratories are also developing liposomal formulations of various pro-

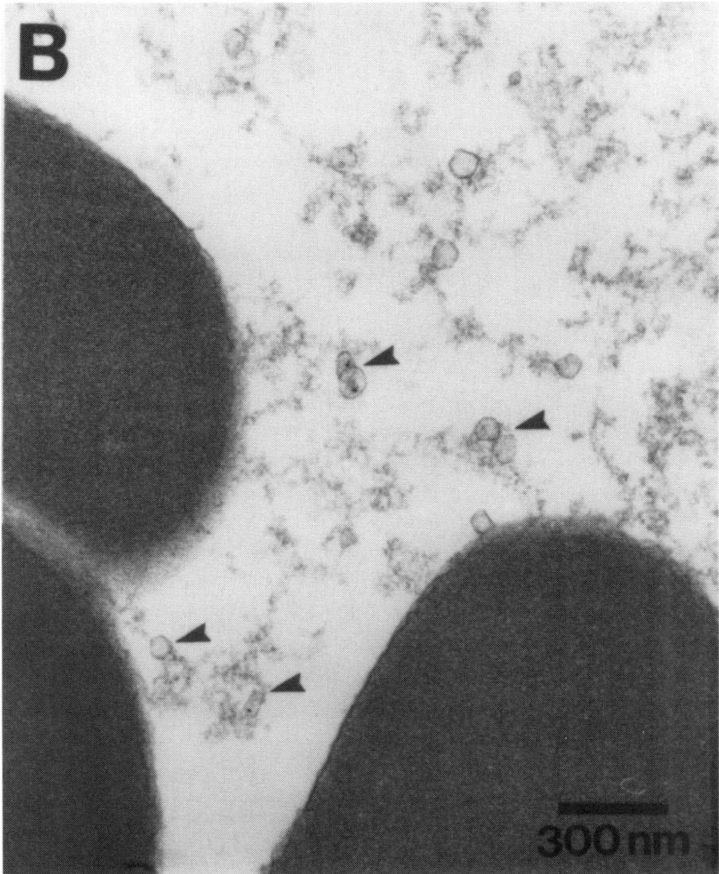

Fig. 1B.

teins, such as various interferons, interleukins, tumor necrosis factors, etc. These proteins, depending on their structure as well as origin can vary considerably. Efficient encapsulation of these large molecules with appropriate release kinetics is difficult, and at present we are not aware of clinical trials with such formulations. Certainly approaches such as long circulating microreservoirs with tailored release kinetics or passive targeting to the sites of increased vascular permeability are theoretically possible. Another possibility is combinational therapy with other (liposomal) drugs.

It should be noted here that sterically stabilized liposomes may also prove useful with respect to passive targeting to macrophage-like cells. Despite the fact that they are taken much less avidly by macrophages than conventional liposomes, it is still true that approximately half of the injected dose eventually ends up in some phagocytic cells, possibly also due to the gradual loss of PEG coating.

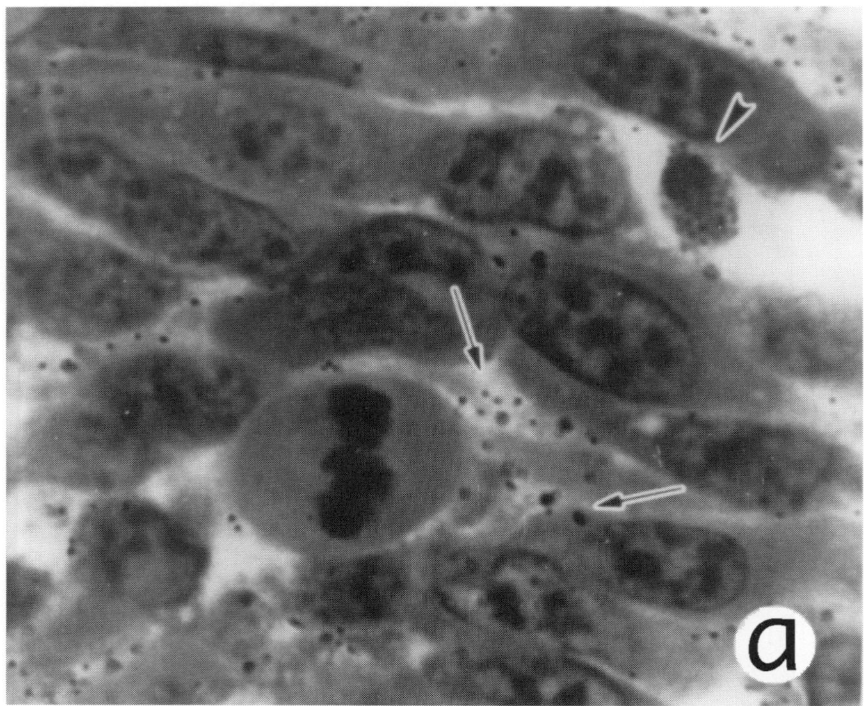

Fig. 2A.

Fig. 2. Localization of liposomes in implanted tumors in mice. Thick sections from C-26 carcinoma were processed with silver enhancement solution and then stained with Eosin-Hematoxilin. The dark particles represent colloidal gold particles, initially encapsulated in liposomes, enlarged by deposition of silver. (a) Dark particles are seen in the extracellular space next to cancer cells. (b) Dark particles are seen beyond the thin endothelial cells of a small vessel within the tumor tissue. With permission, from Huang et al., Cancer Res 1992;52:135–143.

However, a large fraction of these cells are not only macrophages in the liver and spleen, but also deep tissue macrophages, which are practically inaccessible by any other delivery system or route. Many infections, including HIV, are located in these cells and the potential of targeting these cells may be very significant. Another potentially promising route is targeting to lymph nodes by subcutaneous injection of small conventional or sterically stabilized liposomes. Lymph drainage from these sites brings such liposomes into lymph nodes where they can eradicate metastases.

Several approaches to treat cancer and metastases are based on gene therapy principles. Silencing of oncogenes, expressing tumor suppressor genes, transfecting cells with suicide genes, reversing multi drug resistance or making cells susceptible to various drugs (via expression of HSV tk gene, for instance) are the major approaches. The main problem is how to provide for efficient and durable gene expression. Additionally, in cancer every cell should be transfected in order to

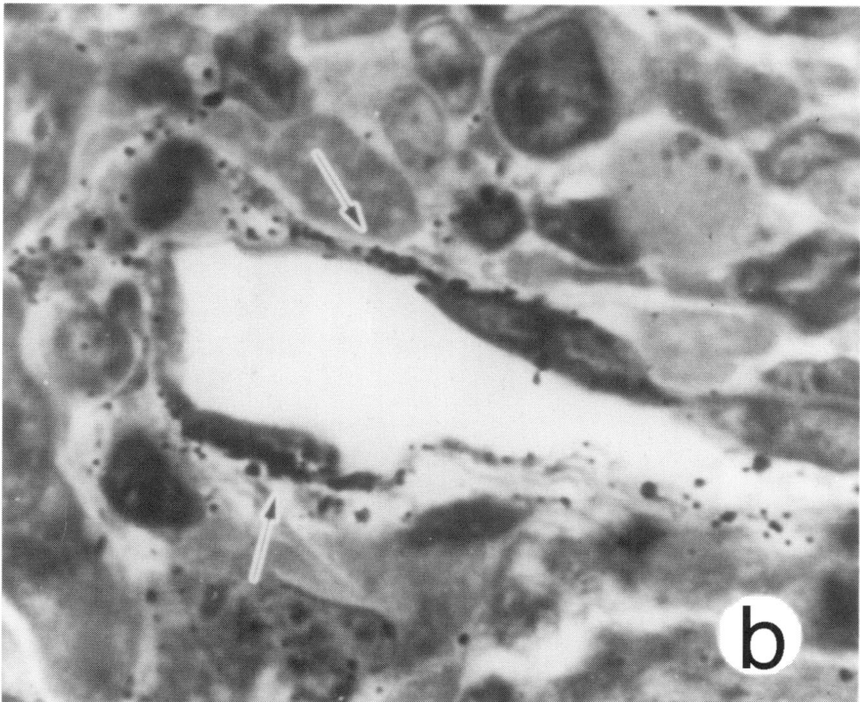

Fig. 2B.

eradicate the disease. Because this is virtually impossible, many treatments rely on the so-called by-stander effect, i.e., the transfected cells can secrete in the surrounding space cytotoxic or cytostatic substances. Genes can be switched off also by the so-called anti-sense technology, which is based on inactivation of mRNA and arresting the synthesis of vital proteins. Similarly, ribozymes can stop the translation process by cleaving RNA molecules. These concepts are sound and were proven in many in vitro experiments. Biological systems, however, are characterized by numerous compartments, permeability barriers and delivery of these large molecules into appropriate cells represents a serious problem which is still largely unsolved. Due to their characteristics of enhanced penetration, protection of associated agents and ability to transfer across the cell membrane, liposomes are being tested as delivery vehicles for genes and anti-genes. While all these approaches can use other delivery methods, liposomes represent a substantial part of all of these endeavors. While at present most of the technology is based on plain cationic liposomes which condense DNA and deliver it nonspecifically into various cells in vivo, we believe that the next generation of products will have to include some stabilization of these DNA-liposome complexes and eventually their targeting. This subject is discussed in Section 5 of this volume.

In anti-cancer therapy, we have witnessed the launch of a successful formulation

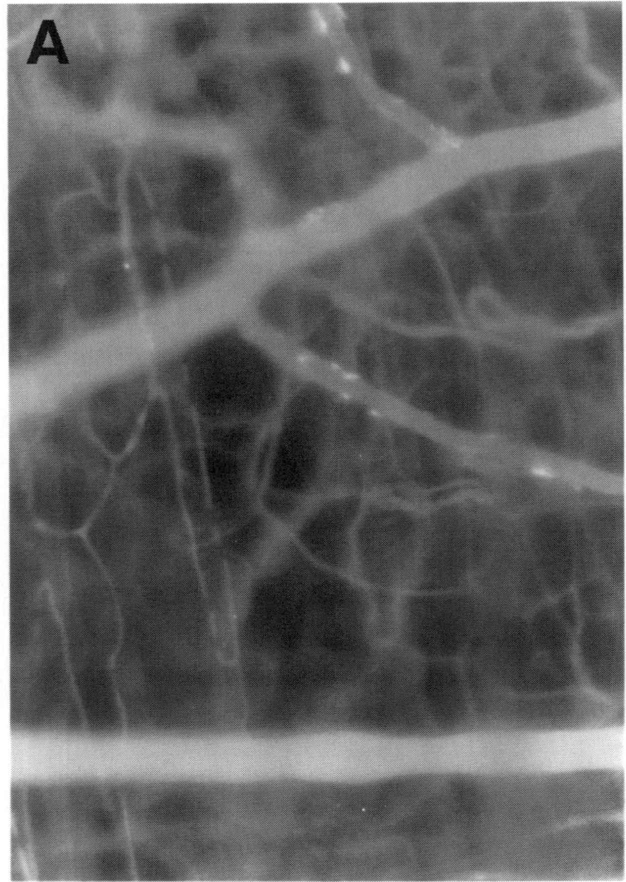

Fig. 3A.

Fig. 3. Extravasation of liposomes in normal and tumor tissue. Intravital fluorescence microscopy in a human tumor xenograft in SCID mice: (A) normal subcutaneous tissue adjacent to tumor; (B) human colon adenocarcinoma. With permission from Yuan et al., Cancer Res 1994;54:3352–3356.

Doxil, followed by DaunoXome. These formulations are likely to be followed by other drugs in similar liposomes. Sequus Pharmaceuticals, for instance, is working on cisplatin in sterically stabilized liposomes, NeXstar, Depotech and The Liposome Company on doxorubicin and other cytotoxic drugs. In addition to these formulations which are based on passive targeting, active targeting may become important, especially for blood borne cancers and metastases in conjunction with selective vasodilatation to increase extravasation. Enhanced extravasation of liposomes has been already observed with hyperthermia[20] and with the use of vasoactive compounds such as substance P.[21] Anti-angiogenesis is also a promising approach to stop tumor growth. One can use special ligands to access the tumor

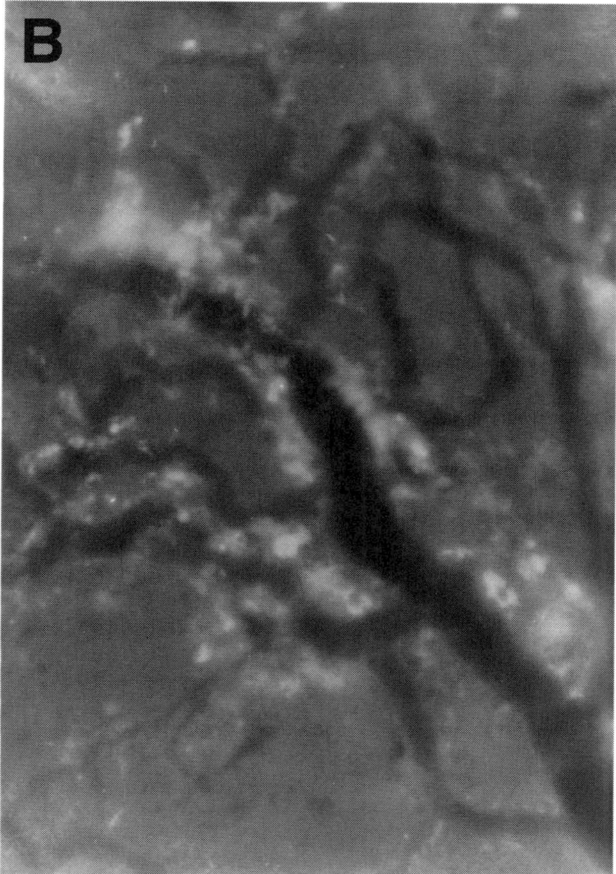

Fig. 3B.

endothelial cells, as well as deliver large amounts of toxic substances to these sites which may induce necrosis of these areas. Several such approaches already look promising[22,23] and could be modified using liposomes in order to take advantage of their special properties, such as high payload per ligand.

Another approach is the use of toxic amphiphiles to construct liposomes or micelles. For instance, alkyl phosphocholines exhibit anticancer activity and can form liposomes.[24] Such and other similar formulations have been shown to be effective in cancer chemotherapy. In addition, in the fight with this formidable opponent, we can hope that combination therapy, consisting for instance, of a liposomal anthracycline and perhaps a liposomal cytokine, may bring further advances in anticancer therapy, just as drug cocktails did in AIDS.

In conclusion, liposomes have the potential to play an important role in cancer therapy, both as carriers of small cytotoxic molecules, and as delivery vehicles of other macromolecules such as cytokines or genes.

References

1. Fidler IJ. Systemic macrophage activation with liposome-entrapped immunomodulators from therapy of cancer metastasis. Liposomes and Macrophage Functions. Res Immunol 1992;143:199–204.
2. Sharma A, Mayhew E, Straubinger RM. Antitumor Effect in Taxol-containing Liposomes in a Taxol-resistant Murine Tumor Model. Cancer Research 1993;53:5877–5881.
3. Wheeler JJ, Wong KF, Ansell SM, Masin D, Bally MB. Polyethylene Glycol Modified Phospholipids Stabilize Emulsion Prepared from Triacylglycerol. J Pharm Sci 1994;83:1558–1564.
4. Perez-Soler R, Han I, Al-Baker S, Khokhar AR. Lipophilic platinum complexes entrapped in liposomes: improved stability and preserved antitumor activity with complexes containing linear alkyl carboxylato leaving groups. Cancer Chemother and Pharmacol 1994;33:378–384.
5. Allen TM, Chonn A. Large unilamellar liposomes with low uptake into the reticuloendothelial system. FEB Letters 1987;223(1):42–46.
6. Gabizon A, Papahadjopoulos D. Liposome formulations with prolonged circulation time in blood and enhanced uptake by tumors. Proc Natl Acad Sci 1988;85:6949–6953.
7. Woodle MD, Lasic DD. Sterically stabilized liposomes. Biochim Biophys Acta 1992;1113:171–199.
8. Allen TM, Papahadjopoulos D. Sterically stabilized (stealth) liposomes: pharmacokinetic and therapeutic advantages. In: Gregoriadis G, ed. "Liposome Technology, 2nd edition, Volume III". Boca Raton, FL: CRC Press, Chapter 5, 1992;59–72.
9. Blume G, Cevc G. Molecular mechanism of the lipid vesicle longevity in vivo. Biochem Biophys Acta 1993;1146:157–168.
10. Huang L (ed). Forum on Stealth Liposome. J of Lip Res 1992;2(3):1–454.
11. Papahadjopoulos D, Allen T, Gabizon A, Mayhew E, Matthay K, Huang SK, Lee K-D, Woodle MC, Lasic DD, Redemann C, Martin FJ. Sterically stabilized liposomes: Improvements in pharmacokinetics, and anti-tumor therapeutic efficacy. Proc Natl Acad Sci (USA) 1991;88:11460–11464.
12. Jain RK. Barriers to Drug Delivery in Solid Tumors. Scientific American 1994;271(1):58–65.
13. Matsumura Y, Maeda H. A New Concept for Macromolecular Therapeutics in Cancer Chemotherapy: Mechanism of Tumoritropic Accumulation of Proteins and the Antitumor Agent Smancs. Cancer Research 1986;6387–6392.
14. Huang SK, Lee K-D, Hong K, Friend DS, Papahadjopoulos D. Microscopic localization of sterically stabilized liposomes in colon carcinoma-bearing mice. Cancer Res 1992;52:5135–5143.
15 Yuan F, Leunig M, Huang SK, Berk DA, Papahadjopoulos D, Jain RK. Microvascular permeability and interstitial penetration of sterically stabilized (Stealth) liposomes in a human tumor xenograft. Cancer Research 1994;54:3352–3356.
16. Debs RJ, Heath TD, Papahadjopoulos D. Targeting of anti-Thy 1.1. monoclonal antibody conjugated liposomes in Thy 1.1 mice after intravenous administration. Biochim Biophys Acta 1987;901:183–190.
17. Harding JA, Engbers CM, Newman MS, Goldstein NI, Zalipsky S. Immunogenicity and Pharmacokinetic Attributes of Poly(Ehtylene Glycol)-Grafted Immunoliposomes. Biochim Biophys Acta 1997;1327:181–192.
18. Phillips N, Emili A. Immunogencity of immunoliposomes. Immunol Letts 1991;30:291–296.
19. Lee RJ, Low PS. Folate-mediated tumor cell targeting of liposome-entrapped doxorubicin in vitro. Biochim Biophys Acta 1995;1233:134–144.
20. Huang SK, Stauffer PR, Hong K, Guo JWH, Phillips TL, Huang A, Papahadjopoulos D. Liposomes and Hyperthermia in Mice: Increased Tumor Uptake and Therapeutic Efficacy of Doxorubicin in Sterically Stabilized Liposomes. Cancer Research 1994;54:2186–2191.
21. Rosenecker J, Zhang W, Hong K, Lausier J, Geppetti P, Yoshihara S, Papahadjopoulos D, Nadel JA. Increased liposome extravasation in selected tissues: Effect of substance P. Proc Natl Acad Sci (USA) 1996;93:7236–7241.
22. Ferrara N. Vascular endothelial growth factor. Eur J Cancer 1996;32A:2413–2422.
23. Huang X, Molena G, King S, Watkins L, Edgington TS, Thorpe P. Tumor Infarction in mice by Antibody-directed Targeting of Tissue Factor to Tumor Vasculature. Science 1997;275:547–550.
24. Muschiol C, Berger MR, Schuler B, Scherf HR, Garzon FT, Zeller WJ, Unger C, Eibl HJ and Schmahl D. Alkyl phosphocholines: toxicity and anticancer properties. Lipids 1987;22:930–934.

Lasic and Papahadjopoulos (eds.), Medical Applications of Liposomes

CHAPTER 4.2

Designing therapeutically optimized liposomal anticancer delivery systems: Lessons from conventional liposomes

LAWRENCE D. MAYER,[a,b] PIETER R. CULLIS[c,d] AND MARCEL B. BALLY[a,e]

[a]*Department of Advanced Therapeutics, BC Cancer Agency, 600 W. 10th Ave., Vancouver, B.C. V5Z 4E3, Canada*; [b]*Pharmaceutical Sciences*; [c]*Biochemistry and* [e]*Pathology Departments, University of British Columbia, Vancouver, B.C.*; [d]*Inex Pharmaceuticals Corp., 100-8900 Glenlyon Parkway, Burnaby, BC V5J5J8*

Overview

I. Summary

Recent technological advances in the production, stability and biological (RES avoidance, targeting and intracellular delivery) properties of liposomes have greatly increased the degree of sophistication that can be designed into liposomes in order to improve their therapeutic/toxicity activity profile. Transmembrane ion gradients to control drug retention, surface stabilizing lipids to increase circulation longevity, targeting ligands to increase selectivity for disease sites/cells and fusogenic/internalizing components that increase intracellular delivery of liposome contents all have been employed for various therapeutic applications. However, the use of these technologies for particular disease states has often been based on rather intuitive projections of how such systems may act therapeutically in vivo.

For example, the increased delivery of liposome encapsulated anticancer agents to tumor sites relative to free drug is frequently used as a comparative characteristic to evaluate liposomal formulations. However, the improvements in efficacy obtained with liposomes are often small relative to the increase in disease site drug accumulation. Similarly, increases in liposome circulation lifetimes do not always translate into enhanced disease site localization and/or efficacy. Such examples point out that issues including drug bioavailability, disease site micro-environment and drug pharmacology (concentration vs. time dependence) must be addressed in order to efficiently optimize therapeutic liposomes. Further, the advantages provided by specific design components incorporated into conventional liposomes must be critically assessed in the context of what can be achieved by manipulating basic properties of the conventional liposomes themselves.

II. Liposomal anticancer drugs

Before considering the design attributes of liposomal carriers, it is useful to comment on some of the common rules that govern cancer chemotherapy and to reflect briefly on the rationale(s) for developing liposomal anticancer drugs. We would argue that there are two general reasons for developing a liposomal anticancer drug: (i) because the drug is hydrophobic and difficult or impossible to dissolve in aqueous solutions and thus requires a hydrophobic environment in order to stay in solution/suspension; and (ii) because the liposome can serve as a carrier that will improve drug specificity by increasing delivery to the site of disease and/or decrease delivery to a site where toxicities are manifested. The former is an important, perhaps underdeveloped, role for lipid-based carriers. However, the methods and characterization studies required for development of **liposomes formulated for optimal drug solubilization** should be clearly distinguished from those used in the development of **liposome drug carrier technology**. Differences in the two approaches can be defined primarily through in vivo studies that determine plasma elimination behavior of both drug and liposomal lipid. If the drug dissociates from the liposome immediately following administration then the lipid-based carrier is acting as an excipient for drug solubilization. When drug elimination parameters are dictated by the elimination behavior of the liposomes, then the systems are acting as a true delivery vehicle. This review will focus on the use of liposomes developed as drug carriers. In this regard, the primary consequence of anticancer drug encapsulation is liposome mediated changes in drug elimination and biodistribution. With this in mind, questions regarding why improvements in anticancer drug therapy should be achieved through the use of liposomal drug carrier technology can be addressed.

II.1. Optimal cancer chemotherapy requires the use of several drugs in combination

It is important to recognize that therapeutic responses obtained following administration of anticancer drugs, in free form or associated with a drug carrier, are dependent on tumor physiology and tumor cell heterogeneity. Ideally, an effective drug must access the target cell populations at levels sufficient to cause cytotoxic effects and should be effective in all microenvironments present within tumors. In humans, strategies designed to maximize the antitumor activity of chemotherapeutic agents must, therefore, contend with a heterogeneous population of proliferating cells that are: (i) in various stages of the cell cycle; (ii) proliferating at different rates; (iii) growing in different tissues and (iv) capable of adapting rapidly to the chemotherapeutic stresses exerted on them. In practical terms this means that chemotherapy typically involves the use of multiple drugs that exert antitumor activity via different mechanisms.[1,2] Vincristine, for example, is a cell cycle specific agent that acts by destabilizing microtubules and is almost always used in combination with two or three other anticancer drugs. The therapeutic action of vincristine is complemented by drugs such as doxorubicin, an anthracycline that is a DNA intercalator and topoisomerase II inhibitor, as well as cyclophosphamide, a nitrogen mustard pro-drug and strong alkylating agent. The mechanisms of therapeutic action of these drugs are complementary and, importantly, the toxicities of each drug are sufficiently different such that they can be used in combination without aggravation of any one specific target organ toxicity.

There are over 30 cytostatic and cytotoxic drugs commonly employed in the treatment of cancer and these drugs have been used in a variety of combinations that have been refined through years of clinical testing. Interestingly, drug carrier technology has been utilized for a limited number of drugs and there are only a few examples where pre-clinical studies evaluated a liposomal anticancer drug in combination with a second drug, either free or liposomal. Table 1 provides a list of drugs that have been formulated using liposomes. It is important to note that this list includes drugs that have been associated with liposomes in a manner that does not require chemical modification of the drug. We have not included, for

Table 1

Major antineoplastic agents evaluated in a liposomal drug carrier system

Class/Drug	# of different liposomal formulations	Pre-clinical evaluations	Clinical testing
Plant Alkaliods-			
Vincristine	<10	Extensive	Phase II
Vinblastine	<5	Very limited	–
Antibiotics-			
Doxorubicin	>10	Extensive	Approved
Daunorubicin	<5	Extensive	Approved
Antimetabolites-			
Methotrexate	<5	Limited	–
5-Fluorouracil	<5	Limited	–
Cytosine arabinoiside	<5	Limited	–
Alkylating Agents			
cis-diamminedichloroplatinum	<5	Limited	–
Other-			
Mitoxantrone	<5	Extensive	Phase II

example, the valuable work completed by Roman Perez-Soler and his associates on lipophilic derivatives of doxorubicin and dach-platinum or the studies from Dr. Schwendener's laboratory on lipophilic derivatives of cytosine arabinoside. We have also not included the extensive work completed on the lipophilic macrophage activator muramyl tripeptide phosphatidylethanolamine or the more recent studies evaluating biopharmaceuticals such as the immune modulator IL-2, plasmid DNA or antisense oligonuceotides. We have restricted discussions to conventional anticancer drugs that exhibit proven therapeutic activity against cancer in humans.

II.2. Maximizing dose intensity

In addition to the necessity of using multiple agents to achieve optimal therapy, another general principle of cancer chemotherapy concerns maximizing dose intensity.[3] Tumor cells must be exposed to the highest levels of drug for the longest time periods if maximum therapeutic effects are to be achieved.[4] The advantage of anticancer drug carrier technology is based on carrier characteristics that give rise to increased drug exposure in sites of tumor growth. An example of how liposome drug carrier technology can improve the pharmacodynamic behavior of an anticancer agent is evident when evaluating previous studies with doxorubicin. Efforts to maximize the dose intensity of this chemotherapeutic agent (in free form) have been limited due to non-specific toxic side effects. For example, doxorubicin is a potent myelosuppressive agent.[5–7] Therapeutic doses must, therefore, be limited to schedules and amounts that do not compromise regeneration of blood cells or cells of the immune system. In addition, doxorubicin exhibits a dose limiting cardiotoxicity[8,9] restricting the total dose to approximately 450 mg/m^2. Myelosuppression can be counteracted using the hematopoietic growth factor granulocyte-macrophage colony stimulating factor (GM-CSF).[10] Cardiotoxicity on the other hand can be reduced by administering the drug in a liposomally encapsulated form.[11,12] It has also been shown that the therapeutic activity of the liposomal drug is greater than or equal to free doxorubicin in a variety of preclinical and clinical studies.[13,14]

The pharmacodynamic alterations provided by liposomes appear to be well suited to basic principles of cancer chemotherapy. This review considers the relationship between circulation lifetime/plasma drug concentration and tumor drug accumulation as well as how liposome design is critical if optimal drug exposure at the disease site is to be achieved. The importance of achieving a balance between drug exposure and drug delivery to sites of tumor progression will become apparent.

II.3. Liposomal drug carriers versus drug infusion technology

For many applications, liposomal delivery systems are employed to improve the therapeutic index of encapsulated agents by selectively accumulating in extravascular disease sites. Further, there is an increase in evidence indicating that drug released from liposomes in the circulation does not contribute significantly to therapeutic activity of liposomal anticancer agents. There is no question that liposomes can provide sustained exposure of therapeutic agents in the blood compartment through controlled release kinetics of encapsulated drugs, however it is difficult to justify development of liposomal drugs using a rationale that involves sustained systemic exposure. This is largely due to significant advances made in the area of drug infusion technology. Compact and cost effective infusion pumps are now widely used and these can provide well controlled systemic drug exposure over several days. We maintain that the most significant advantage for the use of liposome drug carriers arises as a consequence of disease specific changes in vascular permeability that favor accumulation of the intact liposome and associated drug into the site of disease progression. We differentiate this property from the benefits of drug infusion technology, which are primarily concerned with maintenance of circulating blood levels of free drug.

III. The compartmental model of liposome fate in vivo

In vivo studies are usually initiated only after one has developed a formulation that exhibits the necessary chemical and physical stability properties to be considered pharmaceutically viable. Subsequent in vivo analysis must then consider the fact that the liposomal drugs will interact with a number of distinct physiological "compartments" and associated barriers between compartments. For the purpose of discussions here, we will focus on systemic administration and, in particular, on the fate of liposomes injected intravenously (i.v.). After injection, liposomes are exposed to a variety of circulating protein and cellular components that reside within the central blood compartment, many of which can destabilize the liposomes through interactions with the lipid bilayer or initiate biological processes that lead to increased liposome leakage and/or clearance via the RES. To gain access to a disease in an extravascular compartment liposomes must cross the vascular endothelium, the blood vessel lining which is composed primarily of endothelial cells and, in most cases, an underlying basement membrane and associated smooth muscle cells. This vascular barrier represents the greatest obstacle for liposomal drug delivery to extravascular disease sites, however, at the same time it offers properties that can be utilized to differentiate between normal and diseased tissue. Should liposomes traverse this barrier, a second compartment is encountered consisting of the interstitial space and associated fluids and cells. This compartment can vary significantly not only between normal and disease tissues but also among normal tissues in different organs of the body. Within this compartment, the barriers to liposome movement and distribution are varied and include factors such as interstitial volume, interstitial pressure, and the presence (or absence) of a lymphatic system. The final physiological compartment(s) are the cells into which liposomes and/or their encapsulated agents are taken up. This includes intracellular organelles that may be involved in processing of the administered agent or that contain the molecular target through which the drug exerts its therapeutic activity. The critical barrier that must be crossed in order to access this final compartment is the cell membrane. Similar to the vascular endothelium, crossing this barrier is a significant obstacle to the development of therapeutically optimized liposomal anticancer drugs.

In the following sections we will follow the fate of liposomes as they enter these physiological compartments and pass through the various barriers. We will focus on specific interactions between liposomes and the biological milieu in the various compartments that directly impact on the delivery of encapsulated agents to their therapeutic target. Further, we will highlight where strategies have been employed to augment conventional liposomes (defined as underivatized membrane bilayers composed of naturally-occurring or synthetic lipids) with components that alter these interactions. Importantly, we critically review the impact such manipulations have on meaningful therapeutic endpoints.

III.1. Liposomes in the central blood compartment

When liposomes are injected intravenously they are immediately exposed to a plethora of circulating cells, lipoproteins and soluble factors including proteins, carbohydrates and small ions (Na^+, Cl^-, Ca^{2+}, Mg^{2+}, etc.). Assuming that liposomes contain sufficient amounts of cholesterol to avoid the bilayer destabilization effects of lipoproteins,[15,16] the fate of liposomes in this compartment is dictated primarily by interactions between the liposome surface and serum protein components. Two deleterious responses can result when proteins adsorb to liposomes: (i) increased membrane permeability which compromises drug retention in the liposomes; and (ii) recognition and subsequent clearance of liposomes by the RES.

III.1.1. Liposome-protein interactions and membrane permeability

The ability of adsorbed blood proteins to increase liposome permeability properties has been demonstrated by several laboratories.[16–19] Such interactions can be simply modeled by determining the drug release kinetics for liposomes suspended in serum compared to protein-free buffer. An example of this is shown in Figure 1. The leakage of vincristine from DSPC/Chol liposomes is approximately 5-fold faster in the presence of serum. Interestingly, comparison of these results with the release kinetics of vincristine from DSPC/Chol liposomes after i.v. administration (as determined by monitoring changes in the circulating drug-to-lipid ratio) reveals that drug leakage is further increased in vivo (Figure 1). These differences are not simply due to the presence of a "tissue sink" into which the released vincristine is absorbed since increased dilutions or extended dialysis times in the presence of serum do not increase in vitro drug release rates (L. Mayer, unpublished observations). Consequently, we believe that in vivo drug retention properties as well as comparisons of drug release kinetics for different liposomes cannot always be predicted simply on the basis of in vitro data.

In addition to increasing the permeability of liposome bilayers in the blood, protein adsorption can also lead to increased susceptibility to transmembrane stresses caused by ion gradients or high levels of encapsulated drugs. The high concentrations of buffer components and/or drug entrapped in liposomes often result in significant osmotic gradients across the liposome membrane when exposed

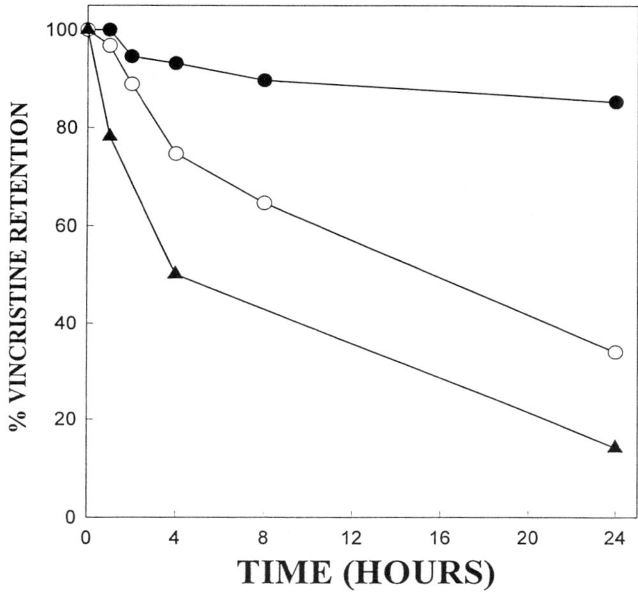

Fig. 1. Release of vincristine from 100 nm DSPC/cholesterol (55:45, mol:mol) liposomes encapsulated via pH gradient-dependent loading at a drug to lipid ratio of 0.05:1 (wt:wt). Liposomes were dialyzed in the presence of Hepes buffered saline, pH 7.5 (●) or bovine serum (○) at 37°C. Liposomal vincristine was injected i.v. into BDF1 mice at a drug dose of 2 mg/kg and the plasma drug to lipid ratio was determined and used to calculate the percent vincristine leakage from the liposomes in the circulation (▲).

to physiological fluids. While most liposomes can withstand a significant transmembrane osmotic gradient in the absence of extraneous proteins, exposure of liposomes exhibiting large osmotic gradients to plasma or purified lipoprotein fractions results in a burst of leakage from the liposomes while osmotic balance is re-established.[20] This effect is more pronounced with less ordered membranes where, for example, DSPC/Chol liposomes can withstand osmotic gradients of far greater magnitude than EPC/Chol liposomes in the presence of proteins.[20] This may, in part, explain the differences observed between DSPC/Chol and EPC/Chol liposomal doxorubicin formulations in vivo where the circulating drug-to-lipid ratio (used to assess drug leakage) observed for EPC/Chol liposomes drops approximately 50% within 1 h of injection and subsequently decreases to a release rate comparable to that observed for DSPC/Chol (Figure 2).

III.1.2. Liposome-protein interactions and circulation longevity

A significant amount of attention has focused on identifying the protein components in the circulation that, upon binding to liposomes, mark them for clearance from the circulation (for review see Ref. 21). Studies have demonstrated that increased protein binding to liposomes after i.v. administration is associated with

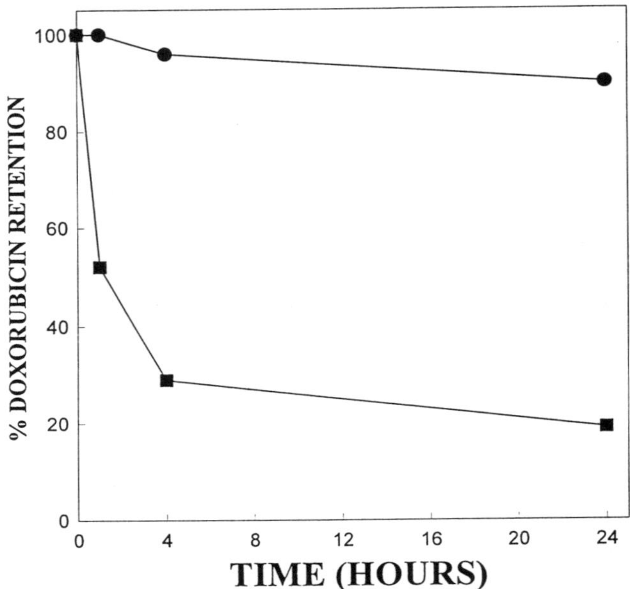

Fig. 2. Release of doxorubicin from DSPC/cholesterol (●) and EPC/cholesterol (■) liposomes in plasma after i.v. injection to mice. The 100 nm liposomes were prepared at a phospholipid to cholesterol molar ratio of 55:45. Doxorubicin encapsulation was completed using the pH gradient entrapment technique at a drug to lipid weight ratio of 0.2:1 (wt:wt). Lipid levels in plasma were determined using tritiated cholesterylhexadecyl ether and doxorubicin was quantified by fluorescence detection of extracted samples.

increased elimination from the blood.[19,21,22] Increased protein binding and clearance is observed for liposomes prepared with phosphatidylserine, cardiolipin and phosphatidic acid. Certain proteins, most notably complement proteins, serum albumin and beta 2 glycoprotein 1 have been associated with increased recognition or "opsonization" of these liposomes.[18,22–24] In contrast, liposomes prepared with other anionic lipids such as phosphatidylglycerol and phosphatidylinositol exhibit extended circulation lifetimes following i.v. administration despite having reasonable levels of adsorbed serum proteins. Variations in the fluidity (acyl chain composition) of neutral liposomes containing ⩾30% cholesterol do not result in substantial differences in the types of proteins adsorbed and correspondingly these liposomes are cleared from the circulation at similar rates.[18,22]

III.1.3. Inhibition of liposome-protein interactions

Clearly, rapid release of a liposomal encapsulated agent following i.v. administration negates the value of using liposomes as drug carriers. In addition, unless the target disease is localized in organs such as the liver and spleen,[25,26] rapid removal of liposomes from the central blood compartment seriously compromises their ability to provide pharmacological improvements as drug delivery systems. In this regard, one significant breakthrough in liposome technology over the past decade has been the identification of lipids, in particular polyethylene glycol (PEG)-derivatized PE, that can be incorporated into conventional liposomes which reduce protein interactions with liposomes.[19, 27–30] It is believed that these lipids act by providing a steric barrier that limits the exposure of the liposome surface to macromolecules in bulk solution.[31,32] Liposomes which are prepared using these lipids exhibit extended circulation times relative to conventional liposomes of similar bulk lipid composition. The application of this technology is discussed in greater detail in Chapter 4.3 by Goren and Gabizon. Of interest are recent observations where reduced protein binding and increased circulation longevity of neutral liposomes can be achieved without incorporating PEG by utilizing sphingomyelin rather than phosphatidylcholine as the main bilayer forming lipid.[33] Given that both lipids contain the same phosphorylcholine head group, we can suggest that the decreased protein binding and clearance result from altered lipid packing properties for sphingomyelin which may limit adsorption and insertion of protein domains into the hydrophobic region of the bilayer. It should be noted that work completed by Parr et al.[33] and Holland et al.[34] suggest that PEG-modified lipids exchange out of the liposomal membrane at a rate that is dependent on the acyl chain composition. PEG-modified lipids prepared using unsaturated or short (<14 carbons) acyl chains are lost rapidly following i.v. administration.

III.1.4. Is there a therapeutic advantage to increased circulation lifetimes?
Although it is generally believed that liposomes for systemic drug delivery should contain either PEG-lipids or other "stabilizing" lipids (e.g., sphingomyelin), such generalizations can be misleading and at times inappropriate depending on the disease site and drug being delivered. This is perhaps best exemplified by reviewing the biological properties of liposomal formulations developed for the anticancer drugs doxorubicin, mitoxantrone and vincristine.

III.1.4.1. Doxorubicin Doxorubicin is an anthracycline antineoplastic agent that is actively taken up by cells and is retained with high avidity by many tissues, most notably those associated with drug toxicity (heart and epithelial cells of the gut) and tumors. Liposome encapsulation can significantly reduce the toxicity of doxorubicin by decreasing drug accumulation in drug sensitive normal tissue, presumably by decreasing peak levels of free doxorubicin that are experienced after administration in conventional (unencapsulated) form.[11,12,35–37] The degree of toxicity buffering is directly related to the ability of the liposomes to retain their entrapped doxorubicin where increased phospholipid acyl chain saturation results in decreased toxicity.[37,38]

The antitumor activity of liposomal doxorubicin, however, is much less sensitive to drug leakage or circulation longevity. Liposomal formulations with widely varying doxorubicin retention properties have been shown in some preclinical models to exhibit comparable antitumor activities when compared on an equal dose basis.[38] In this case, increased efficacy for the less permeable liposomes is achieved by the ability to administer elevated drug doses due to their reduced toxicity. Further, while the inclusion of PEG-PE increases the circulation longevity of liposomal doxorubicin,[39,40] the magnitude of increased liposome levels in the blood (compared to conventional liposomes) is far less than that observed for empty (drug-free) liposomes.[40,41] This is related to an RES blockade effect that is observed for doxorubicin loaded conventional liposomes (see Section III.2.1 for further discussion). Controversy still exists as to the overall therapeutic benefit of incorporating steric stabilizing lipids like PEG-PE into conventional liposomes since examples of significant as well as negligible improvements in efficacy have been documented.[39–42] (For comparison, see Chapter 4.3 by Goren and Gabizon.)

III.1.4.2. Vincristine In contrast to the observations made with doxorubicin, altering the physical properties of liposomal vincristine formulations results in dramatic changes in antitumor activity while only minimally affecting drug toxicity characteristics. Increasing the retention of vincristine inside 100 nm liposomes by changing the phosphorylcholine-containing lipid component from EPC to DSPC to sphingomyelin (while maintaining cholesterol content at 45 mol%) leads to dramatic increases in antitumor activity, particularly when compared to the efficacy obtained with free vincristine.[43–45] This is consistent with the steep dependence of vincristine antitumor potency on the duration of drug exposure[45,46] as well as the fact that retention of vincristine in most tissues, including tumors, is rather poor.[47] In this case it appears that the ability to prolong the exposure of vincristine in vivo is more important than peak drug concentrations. Furthermore, although inclusion of PEG-PE in the liposomes increases the circulating liposomal lipid levels at extended time periods, this steric stabilizing lipid does not improve the vincristine pharmacokinetic or therapeutic properties over conventional DSPC/Chol or sphingomyelin/Chol systems.[45] This is due to the fact that PEG-PE increases the permeability of the lipid bilayer to vincristine, thus offsetting the potential benefits provided by increased longevity of the liposomal carrier. The reasons for this increased drug leakage are not well understood. It may be related to the fact that PEG-modified phosphatidylethanolamine is negatively charged and this may alter drug partitioning properties at the inner monolayer membrane surface. In addition, it is not yet clear whether this phenomenon is specific for vincristine encapsulated via pH gradient techniques employing citrate buffers, compared to ammonium sulfate entrapment systems.[48]

III.1.4.3. Mitoxantrone The final example, derived from recent reports describing liposomal formulations of mitoxantrone, illustrates how a balance between efficient liposome delivery to the disease site and controlled drug release can work synergistically to achieve optimum therapeutic results.[25,26] Mitoxantrone is less cardiotoxic than doxorubicin and is not capable of generating free radical mediated toxicities on non-dividing cell populations.[49] The liposome mediated increases in mitoxantrone MTD observed for formulations (phosphorylcholine and cholesterol based systems) described by Chang et al.[25] and Lim et al.[26] are comparable to those reported for liposomal mitoxantrone formulations prepared using an anionic lipid-drug complex.[50,51] In contrast to the results of Schwendener et al., liposomal mitoxantrone formulations prepared using DSPC or DMPC and cholesterol

(45 mol%) exhibit significantly better drug retention characteristics. This is reflected in higher blood levels and improved circulation lifetimes for mitoxantrone encapsulated in the PC/Chol based liposomal carriers. These differences may be due to protein binding and rapid clearance of anionic liposome formulations. Alternatively, differences in drug release characteristics may, as suggested above for vincristine, be a consequence of the use of anionic lipids, which have been shown to enhance release of the anthracycline doxorubicin even in the absence of serum.[52]

Studies evaluating the therapeutic activity of DSPC/Chol and DMPC/Chol liposomal mitoxantrone focused on treatment of an i.v. L1210 and/or P388 tumor model, where cells seeded primarily in the liver and spleen following i.v. administration.[26] These studies illustrated how controlled drug release effected significant improvements in therapeutic activity of the anticancer drug mitoxantrone. It is well established that the liver is a primary site of liposome accumulation, and that the rate of accumulation for DSPC/Chol liposomes in liver is comparable to DMPC/Chol liposomes. Based on this information, a relatively simple question was asked: Is a liposome (DSPC/Chol) which retains drug following i.v. administration therapeutically more active than a liposome (DMPC/Chol) that releases drug when tested against a tumor that progresses in the liver? Despite being less effective in terms of delivering drug to the site of tumor progression, the DMPC/Chol liposomes, which release drug steadily following administration, were strikingly more efficacious than the DSPC/Chol formulations. A natural extension of the previous question was: What effect would incorporation of PEG-modified lipids have on the therapeutic activity of either of these formulations when used to treat disease in the liver? For both formulations, addition of PEG-PE resulted in significant reductions in antitumor activity[25] (Lim et al., unpublished observation).

It can be concluded from such data that it is not necessarily sufficient to develop drug carriers that accumulate at the disease site to high levels; one must also engineer appropriate drug release rates. Controlled drug release must, however, be balanced with liposome mediated drug delivery to the site of tumor growth. Regardless, it is apparent that whether or not additional components (i.e., PEG-PE) should be incorporated into conventional liposomes must be re-evaluated for each therapeutic agent and one must consider the site of disease progression.

III.2. Barriers to extravasation

While in the circulation, liposomes are continually exposed to cells lining the vasculature. The inner lining, or intima, of blood vessels is composed primarily of endothelial cells which form a contiguous layer on the interior surface of all blood vessels. Underlying this layer is the basement membrane and in larger (non-capillary) vessels the vasculature is supported by smooth muscle cells.[53] The endothelial cells in most normal vasculature exhibit intact intercellular junctions and only small molecules are able to readily permeate across capillaries of this type. However, this structure is significantly altered in certain normal tissues, most notably the liver and spleen, as well as in disease sites such as infection and tumor growth. The latter are characterized by the presence of capillaries that exhibit fenestrae or larger intercellular openings and can be devoid of the basement membrane layer. The gaps in these endothelial layers can range in size from 30 nm for fenestrated capillaries to greater than 500 nm in liver, tumor and inflammation site vascular beds.[54–56] In the liver, these openings provide access to sinusoids wherein the phagocytic Kupffer cells lie. In disease sites, the fenestrated/discontinuous capillary beds and post-capillary venules allow direct exposure of the underlying epithelial cells to the circulation. It is the unique nature of vascular structures that exist in liver/spleen and disease tissues which significantly impacts the pharmacological behavior of liposomal drug delivery systems.

III.2.1. The reticuloendothelial system (RES) and liposome clearance

The RES has long been recognized as the major site of liposome accumulation after systemic administration. The primary organs associated with the RES are the liver, spleen and lung. The liver exhibits the largest capacity for liposome uptake while the spleen can accumulate liposomes such that the tissue concentration (liposomal lipid/gm tissue) is 10-fold higher than that which can be achieved in other organs. Assuming that liposomes are designed to minimize protein binding and cell interactions, the extent of liposome accumulation in the lung is typically below 1% of the injected dose. Early studies demonstrated that large as well as charged liposomes (particularly those containing negatively charged lipids like PS, PA or cardiolipin) were removed very rapidly by the liver and spleen with clearance half-lives of less than 1 hour.[57,58] The rate of clearance from the circulation could be reduced to some extent by increasing the administered lipid dose, however, only when small (approx. 100 nm), neutral liposomes containing $\geqslant$30% cholesterol were utilized at doses of at least 10 mg/kg or more could circulation lifetimes in the range of several hours be achieved.[59] The removal of liposomes from the blood is attributed to phagocytic cells that reside in the RES and appears to be mediated through direct interactions between the phagocytic cell and the liposomes. In vitro studies have shown that liposome uptake into macrophages can occur in the absence of serum proteins, however recognition mediated by protein elements that associate with liposome surfaces is likely playing a dominating effect on interactions with the RES (see Section III.1.2).

The identification of certain naturally occurring lipids (e.g., ganglioside GM_1 and PI) that increase the circulation lifetime of liposomes in which they are incorporated spawned what is often referred to as the "second generation" of liposome technology. Analogous to the development of polymer surfaces that exhibit reduced protein binding characteristics, it is believed that these carbohydrate containing lipids act by limiting the interaction of liposome surfaces with proteins and this, in turn, inhibits the rate of uptake by phagocytic cells.[29,30,60] As indicated in Section III.1.3, a variety of synthetic lipids have been developed to prevent protein binding. The most notable are based on hydrophilic polymers, such as PEG, which are attached to phospholipids such as PE. Perhaps the most widely utilized steric stabilizing lipid is one composed of 2,000 MW linear PEG attached to DSPE and it is incorporated at levels of 2 to 10 mol% in the bilayer of conventional liposomes. Inclusion of PEG-PE into conventional empty neutral (PC/cholesterol) liposomes can result in 3- to 20-fold increases in plasma liposome content 24 h after i.v. injection.[30,32,40] This is accompanied by significant decreases in liposome uptake by the liver and spleen at early times post-injection. It is important to note that the difference in cumulative uptake of liposomes by the RES organs between conventional and sterically stabilized liposomes become less significant over time, indicating that the effect of PEG-PE is to reduce the rate of liposome removal by cells of the RES. It has not been determined whether eventual removal of these liposomes by the RES is due to time dependent increases in protein association or the loss of PEG from the surface of the liposomes.[33]

Although liposome elimination rates differ greatly between conventional and sterically stabilized liposomes in the absence of encapsulated agents, this difference can be significantly reduced for liposomes containing entrapped drugs, particularly drugs that impair the ability of cells to accumulate or process liposomes.[61–63] This is perhaps best exemplified in the case of the anticancer drug doxorubicin. As shown in Figure 3, when empty 100 nm DSPC/Chol liposomes are injected i.v. at a lipid dose of 100 mg/kg into C3H mice, inclusion of PEG-DSPE results in circulating liposomal lipid levels 24 h post injection that are approximately 20-fold higher than that observed in the absence of the PEG-lipid. However, when the

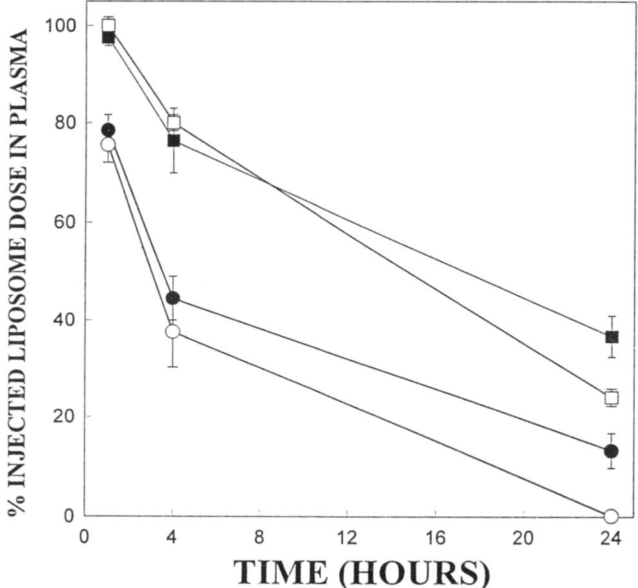

Fig. 3. Circulating levels of 100 nm empty (open symbols) or doxorubicin loaded (closed symbols) DSPC/cholesterol liposomes after i.v. injection to C3H mice. Liposomes were prepared in the absence (●, ○) or presence (■, □) of 5 mol% PEG_{2000}-DSPE. Doxorubicin was encapsulated using the pH gradient-dependent entrapment technique and quantification of plasma levels of liposomes and doxorubicin were accomplished as described in the legend to Figure 2.

DSPC/Chol liposomes are loaded with doxorubicin, the 24 h plasma liposome concentrations are significantly increased and are only 2.8-fold less than those observed for 5 mol% PEG-DSPC containing DSPC/Chol liposomal doxorubicin systems.[41]

Significant increases in circulating levels of empty liposomes can also be achieved by pre-dosing animals with a low dose (10 mg lipid/kg) of liposomal doxorubicin.[61,62] This effect, referred to as RES "blockade", has raised concerns over potential deleterious side effects resulting from altered RES phagocytic capacity. In vitro studies have demonstrated that liposomal doxorubicin uptake by cultured macrophages can result in cell death,[64] and exposure of macrophages in culture to concentrations of doxorubicin that are not cytotoxic significantly impairs the ability of these cells to accumulate particles (M. Bally, unpublished observation). Although a substantial amount of doxorubicin can accumulate in liver tissue,[65] indications of significant liver toxicity arising from this uptake have only been observed pre-clinically under conditions of extremely high doses (80 mg doxorubicin/kg) and in clinical situations where pre-existing liver impairment was a factor.[66]

Investigators have been able to demonstrate macrophage and Kupffer cell depletion following administration of high doses of large and/or negatively charged liposomes containing doxorubicin or other agents such as clodronate.[63] RES blockade induced by low doses (<10 mg/kg lipid and 2 mg/kg drug) of small, uncharged liposomal doxorubicin formulations, however, does not appear to result in reduced numbers of Kupffer cells.[65] This was determined by histological evaluations of thin sections of liver stained with hematoxylin and eosin as well as on the basis of carbon particle uptake in livers of mice that have been previously (4 days) treated with liposomal doxorubicin. This information suggests that our understanding of the mechanisms whereby liposomes (particularly small liposomes) are recognized, cleared from the blood and processed may be somewhat simplistic.

In light of the observations cited above, steric stabilizing lipids are likely to provide the greatest RES avoidance benefits at low liposome doses and for liposome formulations containing drugs that do not lead to reduced liposome clearance. Regarding the latter, it has been shown that encapsulation of vincristine, doxorubicin or cisplatin results in a reduction in liposome elimination. In contrast, liposomal mitoxantrone formulations exhibit circulation characteristics identical to liposomes without entrapped drug. It should also be stressed that the theoretical "benefits" arising from decreased liposome elimination by the RES is typically assumed to be related to the increased circulating concentrations of liposomes obtained. However, we suggest that it is not the plasma concentration of liposomes that dictates therapy, but rather the amount of liposomal drug that penetrates the vascular barrier and gains access to diseased tissue. In the following section we will focus on this extravasation event.

III.2.2. Liposome extravasation through vascular endothelium

If liposomes are designed in an appropriate manner, with respect to size, lipid composition, and/or use of PEG-modified lipids, they can remain in the blood compartment for a period of several days. The fact that under such circumstances the vast majority of liposomes administered can be accounted for in the blood, liver and spleen demonstrates that liposomes are inefficient at crossing the endothelial cell barrier present in most tissues. The property of long circulating liposomes that is exploited for therapeutic purposes relies on changes in the endothelial cell barrier, prevalent in many disease states, that allow liposomes to traverse out of the blood compartment and into the tissue.

Major diseases, such as bacterial infection, inflammation and tumors, have the common feature of altered vasculature permeability at the site of disease progression. The mediators that lead to increased permeability of the vascular barrier are quite distinct for different disease states. For example, chemotactic factors and adhesion molecules overexpressed at sites of inflammation attract infiltrating lymphocytes and granulocytes that subsequently release factors which can directly damage endothelial cells and/or cause defects in intercellular junctions.[67] In hypoxic environments, such as those that arise during rapid cell proliferation or through vascular injury, cells can release vascular endothelial growth factor (VEGF).[68,69] VEGF is an endothelial cell specific mitogen and its release can lead to the development of neovasculature. Interestingly, VEGF has proven to be identical to vascular permeability factor,[70,71] a protein first identified as a factor capable of inducing defects in the permeability barrier of blood vessels. Regardless of the mediator, the end result for all of these conditions is the presence of blood vessels that are permeable to large molecules. This may be a consequence of fenestrae or larger "gaps" occurring between adjacent endothelial cells through which macromolecules can pass[72] or, alternatively, may involve increases in endothelial cell mediated transcytosis.[73]

Increases in vascular permeability give rise to the accumulation of small liposomes in sites of infection, inflammation and tumor growth. However, this is not a selective process and there is also a general increase in extravascular fluids in these regions. The hydrostatic pressure within these sites is elevated relative to the vascular pressure, resulting in a pressure gradient that impedes movement of molecules from the blood into the tissue interstitium.[74,75] We must therefore assume that additional features lead to selective accumulation of macromolecules in the diseased extravascular space. Studies, for example, have demonstrated that the lack of a developed lymphatic system in conjunction with the large openings in the vascular endothelial cell lining may lead to an extravascular "trapping" phenomenon.[75] In the absence of lymphatic drainage, interstitial diffusion of molecules leads to egress from the disease site and this diffusion rate is dependent on molecule size, small molecules exiting more rapidly than large molecules.

Liposome extravasation and accumulation in solid tumors has been well studied and there is a great deal of phenomenological evidence demonstrating that liposomes can enter an extravascular site in regions of tumor growth following i.v. administration. Although evidence for endothelial cell uptake of liposomes and transcytosis across endothelial cells have been documented, videomicroscopy investigations in solid tumor models indicate that the majority of liposome extravasation occurs directly through the openings present in tumor neovasculature.[76,77]

Table 2

Tumor accumulation efficiency (T_e) for conventional and steric stabilized (PEG-containing) liposomal anticancer drug formulations

Tumor model	Preparation[a]	Plasma AUC[b]	Tumor AUC	T_e[c]
Lewis lung	DSPC/Chol[d]	2,118 μgh/ml	819 μgh/g	0.39
(murine solid tumor)	DSPC/PEG-PE/Chol[d]	7,910 μgh/ml	1,432 μgh/g	0.18
Fsa-N fibrosarcoma	DSPC/Chol[e]	10,560 μgh/ml	2,981 μgh/g	0.28
(murine solid tumor)	DSPC/PEG-PE/Chol[e]	18,500 μgh/m	2,892 μgh/g	0.16
P388	DSPC/Chol[e]	16,530 μgh/ml	1,720 μgh/peritoneum	0.10
(murine ascitic tumor)	DSPC/PEG-PE/Chol[e]	37,600 μgh/ml	2,037 μgh/peritoneum	0.05
	SM/Chol[f]	5,116 μgh/ml	206 μgh/peritoneum	0.041
	SM/PEG-PE/Chol[f]	6,762 μgh/ml	184 μgh/peritoneum	0.027

[a]Area under the curve (AUC) values were calculated as trapezoidal AUC over the time period 0–24 h.
[b]All liposomes were 100 nm in size and contained 45 mol% cholesterol. PEG-DSPE was incorporated at 5 mol% when utilized.
[c]Tumor accumulation efficiency was calculated as the 0–24 h liposome AUC in the tumor divided by the 0–24 h liposome AUC in plasma.
[d]Empty liposomes injected at a dose of 100 mg/kg.
[e]Liposomal doxorubicin preparations constituted by pH gradient encapsulation at a drug to lipid weight ratio of 0.2:1.
[f]Liposomal vincristine preparations constituted by pH gradient encapsulation at a drug to lipid ratio of 0.1:1.

This extravasation process appears to be quite heterogeneous within the tumor and does not appear to be associated with any specific histological characteristics in the tumor mass. The net result of this phenomenon is that peak drug concentrations achieved are greater and drug exposure as measured by concentration vs time AUCs is increased when the drug is administered in a liposome, compared to free form.

The design of liposomes that will exhibit maximal extravasation in disease sites associated with leaky vasculature has received considerable attention and is an area of some controversy. As summarized in Section III.1.3, the inclusion of PEG-modified lipids in conventional liposomes can significantly increase the circulating liposome levels over extended times by decreasing the rate of clearance by the RES. It has generally been assumed that increases in the concentration of liposomes in plasma over time will lead to increased accumulation of liposomes in extravascular disease sites and experimental evidence supporting this has been reported.[42] Videomicroscopy has also suggested that the permeability coefficient of tumor vasculature is greater for PEG-PE containing liposomes compared to conventional liposomes.[77] In contrast, studies conducted in our laboratories as well as others have demonstrated that although plasma levels of PEG containing liposomes are several fold higher than for comparable conventional liposomes, this often does not result in increased extravasation and accumulation in solid tumor tissue.[40,41,78]

As shown in Table 2, we have examined the tumor uptake properties for conventional and steric stabilized liposomal formulations of doxorubicin and vincristine in a variety of tumor models. Three important observations can be made on the basis of the comparative biological properties of conventional and sterically stabilized liposomes. First, sterically stabilized liposomes uniformly display increased circulation longevity compared to conventional liposomes, regardless of the presence of encapsulated drug. Second, the rate and extent of liposome accumulation in tumor tissue are often comparable for both conventional and sterically stabilized liposomes. Third, the tumor targeting efficiency or T_E (defined as the AUC in the tumor divided by the AUC in plasma) is higher for conventional liposomes compared to sterically stabilized systems. It is important to note that the relationship between tumor liposome uptake and plasma liposome AUC is linear for conventional and sterically stabilized liposomes, respectively (M. Bally, unpublished observation). This suggests that mass action does appear to drive the accumulation of specific types of non-targeted small liposomes into tumors. However, inclusion of lipids such as PEG-DSPE appears to decrease the efficiency of liposome extravasation from the blood into tumor tissue as indicated by the decreased T_E values observed for sterically stabilized liposomes in several solid tumor models (Table 2).

The basis for discrepancies in tumor extravasation comparisons between conventional and sterically stabilized liposomes may be related to one of several potential explanations. The tumor models utilized may exhibit different vascular structures[79] and it is reasonable to assume that increases or decreases in conventional liposome extravasation in comparision to sterically stabilized liposomes may be tumor specific. However, many different tumor types have been evaluated and it would appear that preferential accumulation efficiency of conventional liposomes is prevalent in most tumor types, regardless of differences in vascular structure. Another factor that may contribute to the discrepancies concerns the techniques utilized to monitor liposome extravasation. In the study by Dewhirst and coworkers, a fluorescent lipid label was employed to follow liposome distribution using fluorescent videomicroscopy in a breast carcinoma skin flap window chamber model.[76] Vascular permeability measurements were based on extravasation events that occurred over 90 minutes post injection. In such studies it is important to demonstrate that the fluorescent lipid label is equally representative as a marker for conventional and sterically stabilized liposomes, particularly when considering the potential for such lipids to exchange.[80]

Our comparisons are typically based on extended AUC measurements of total tumor liposome uptake (following a non-exchangeable, non-metabolizable lipid label and correcting for blood volume contributions) and we place great emphasis on measuring both liposomal lipid and drug over the specified time course. Simultaneous measurements of drug and liposomal lipid can be used to assess drug retention, which is a determining factor in terms of accumulation of entrapped contents in tumors. It should be noted that the lipid compositions utilized in our studies for conventional and sterically stabilized liposomes contain 45 mol% cholesterol whereas studies by others often utilize liposomes with 33 mol% cholesterol. Reduced cholesterol content will result in increases in drug permeability (see Section III.1). Several comparative studies demonstrating improved tumor accumulation for sterically stabilized liposomes have relied on the use of entrapped aqueous markers such as Ga^{67} which are rapidly cleared when released from the liposomes. Consequently, it is difficult to determine if differences in tumor accumulation are due to altered elimination and/or extravasation properties affected by lipid composition or a result of lipid composition effects on drug retention. It is important to resolve these issues and this will require a concerted effort to standardize the tumor models, liposome compositions and methods/parameters for evaluating liposome extravasation into tumors.

It should not be unexpected that conventional and sterically stabilized liposomes exhibit different efficiencies in extravasation. Videomicroscopy studies with steric stabilized liposomal doxorubicin systems have identified that some endothelial cells can take up liposomes.[76,77] Endothelial cell interactions may contribute to the extravasation process either directly via transcytosis or indirectly by facilitating an increase in the local liposome concentration at the endothelial cell surface, thereby increasing access to openings in the vasculature. Given the effects of PEG on inhibiting liposome-cell interactions, this polymer may reduce endothelial cell interactions and this, in turn, would reduce the rate of extravasation. In contrast, conventional liposome extravasation could be facilitated through increased interactions with the endothelial cell lining of the neovasculature in tumors. This is, of course, highly speculative but is consistent with the surface properties of conventional liposomes compared to steric stabilized liposomes. A logical extension of this argument, however, is that improved extravasation may be possible by designing liposomes which interact more extensively with vascular endothelium in tumors.

III.3. The behavior of liposomes in interstitial tissue compartments

Once liposomes have moved through the vascular endothelial barrier, their fate in the interstitial spaces is tissue specific. Generally, negligible levels of liposomes extravasate into tissues such as muscle and kidney.[45] Presumably the liposomes that have distributed into these sites migrate slowly through the intercellular matrix until they are removed via the lymphatics. Interestingly, liposomes administered i.v. do appear to accumulate to high levels in lymph nodes (on a per weight basis), where combined filtration and presence of phagocytic cells act to concentrate liposomes.[81] In liver and spleen, fixed macrophages actively take up liposomes and these cells process the carrier via the intracellular phagolysosomal system. However, for the purpose of this discussion we will focus on the behavior of liposomes that have extravasated into disease sites, and in particular, into solid tumors.

As cited in Section III.2, the distribution of liposomes which have extravasated into the tumor interstitium is heterogeneous. This is not unexpected given the irregular and often redundant organization of tumor vasculature. Tumor vascular structure often engenders highly variable blood flow properties and evaluation of histological sections from tumors reflect this heterogeneity. This would be more apparent for liposomes compared to unencapsulated small molecules due to the decreased diffusion through the interstitial space for large macromolecules. This slow diffusion after extravasation has been documented by fluorescence video microscopy where fluorescently labeled liposomes could be seen to accumulate in the perivascular spaces primarily associated with the roots of capillary sprouts.[77] Diffusion away from these sites was observed to be very slow and significant perivascular clustering was observed for several days. This is consistent with the data from several tumor models which demonstrate that tumor accumulation levels of liposomes reached a maximum approximately 24 h after injection and these levels are maintained for extended time periods. Importantly, evaluations of drug accumulation properties can suggest remarkably different behavior, where drug release from the liposomes in the extravascular site results in greater drug penetration into the tissue and more rapid loss of the drug from the site when compared with the loss of liposomal lipid.

The preferential extravasation and accumulation of liposome encapsulated anticancer drugs in solid tumors results in tumor drug levels that can be as much as 15-fold higher than achieved with free (non-liposomal) drug.[38,40,45] An example of this increased tumor drug delivery is shown in Figure 4. These fluorescent micrographs illustrate the dramatic increase in tumor doxorubicin levels obtained at 24 h after injection when the drug is encapsulated inside 100 nm DSPC/Chol liposomes compared to free doxorubicin. In addition, the prolonged residence of liposomes in tumors also significantly increases the duration of tumor drug exposure and AUC relative to free agents.[40] In some tumor models, such properties have been shown to correlate with increased antitumor activity for liposomal formulations of drugs such as doxorubicin and daunorubicin. It is not clear from these studies, however, what the relative increase in therapeutic potency is in the context of tumor drug delivery improvements. Specifically, studies have typically compared the efficacy and tumor drug accumulation following administration of equal doses of free and liposomal drug. A comparison of efficacy under conditions where tumor drug accumulation is comparable for free and liposomal drug has not been completed, but would likely demonstrate that the liposomal drug is less potent. Other studies have demonstrated comparable antitumor efficacy for free and liposomal doxorubicin under conditions where tumor drug levels were as much as 5-fold higher for liposomal systems.[40] Such observations have raised obvious questions about the bioavailability of anticancer drugs carried inside liposomes that have extravasated into solid tumors as well as the mechanisms that lead to drug release in the interstitial compartment.

The consensus emerging from studies in several laboratories on the mechanism of action of liposomal anticancer drug formulations is that liposomes exert their effect on therapeutic activity by providing an in situ drug infusion reservoir within the tumor. Once released, the anticancer drug can diffuse through the tumor and has direct access to tumor cells where it can act in a manner that presumably is similar to drug in the absence of a liposomal carrier. In vitro studies have demonstrated that macrophages can engulf doxorubicin loaded liposomes, process them and re-release doxorubicin extracellularly in free form.[82] In view of the high macrophage content residing in some tumors,[83] such phenomena led to the pro-

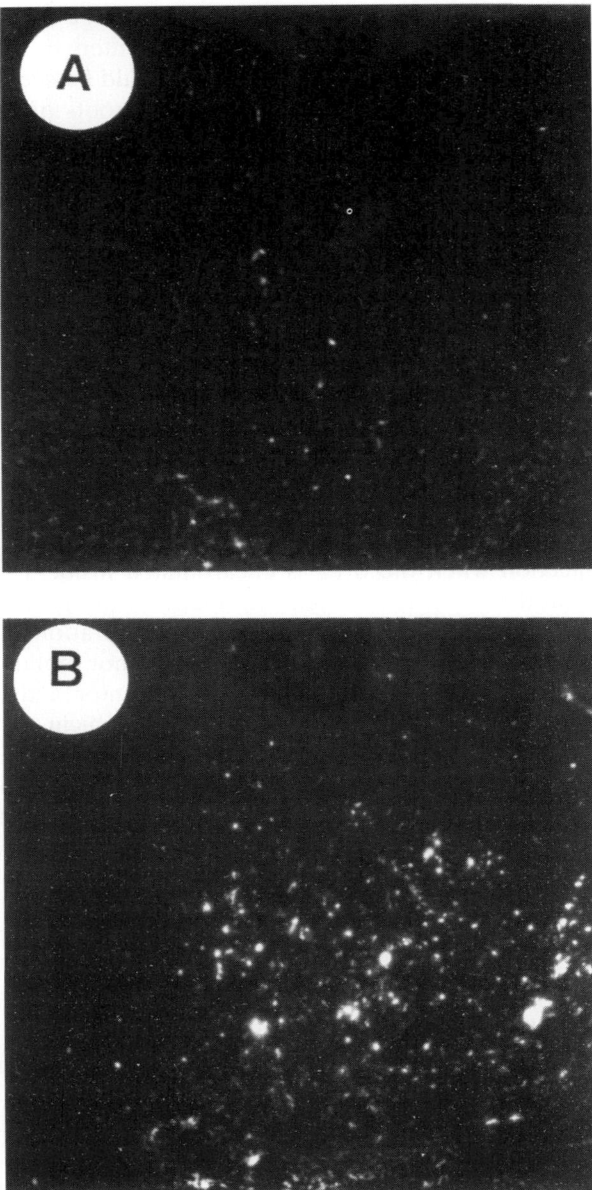

Fig. 4. Fluorescent micrographs of Fsa-N fibrosarcoma frozen thin sections 24 hours after i.v. injection of free (A) and 100 nm DSPC/cholesterol liposome encapsulated (B) doxorubicin at a drug dose of 20 mg/kg. Images were viewed employing a 10× objective lens.

posal that liposomal anticancer drug release may involve macrophage processing after extravasation. However, recent studies have shown that in solid tumors there are limited interactions between tumor associated macrophages and extravasated liposomes.[41] Although macrophage enriched tumors do accumulate higher levels of liposomal doxorubicin, this effect appears more related to increased vascular permeability rather than direct uptake and processing of the liposomes by the macrophages. This was further supported by the fact that both conventional and sterically stabilized liposomes displayed comparable distribution properties (as determined by fluorescence microscopy of tumor thin sections) after extravasation into the tumor.

III.4. Intracellular delivery and processing of liposomes and their contents

As mentioned in Section III.3, for most applications, investigators exploit the ability of liposomes to provide a disease site localized depot of drug, which is slowly released and taken up by target cells. However, current efforts to improve the therapeutic properties of liposomes are focusing on designing systems that will not only localize selectively in a disease site, but will also specifically deliver their encapsulated contents into a defined target cell population. Stratagies for targeting liposomal anticancer drugs are reviewed in the chapter by Theresa Allen and these will not be considered in any detail here. However, it is important to think about the general approaches that are being taken to accomplish cell specific delivery.

Targeting concerns the use of liposomes with surface associated targeting ligands that can bind molecules over-expressed on the surface of disease cells in an extravascular site. It is important that such targeting information does not inherently alter the extravasation events required for the liposomes to reach their cellular target. Consequently, the pharmacodistribution benefits provided by targeted liposomes should arise from a decreased rate of egress from the disease site (rather than increased influx) and cell specific binding. Given that liposomes migrate slowly through interstitial spaces in disease sites such as solid tumors, one must question whether liposomes within the interstitial compartment will be able to interact with a target cell. In addition, the avidity of liposome binding to target cells may actually inhibit liposome migration and subsequent drug exposure in areas more distant from blood vessels. This is anticipated on the basis of the binding barrier effects which have been described by Saga et al.[84] Targeting approaches may be most appropriate for small foci of disease where extensive interstitial diffusion is not required to expose all of the diseased cells to the therapeutic agent. This has been demonstrated with immunoliposomes targeted to lung cancer metastases growing in mice where small tumors could be treated much more effectively with targeted liposomes compared to conventional liposomes or free drug.[85] (See also Chapter 4.7 for contrasting results.)

When liposomes are being designed with targeting ligands there has been an emphasis on targeting cell surface molecules known to be internalized via the endocytic pathway.[86,87] When preparing these targeted carriers additional lipid components can be included to make the liposomal carrier pH sensitive[88] or fusogenic.[89] Release of entrapped contents or fusion with the organelle membrane occurs when the liposomes are exposed to the low pH of the late endosome/lysozome. This has been shown to dramatically increase the potency of liposomal anticancer drugs in vitro.[90] A second approach for intracellular delivery is based on the use of fusogenic lipids. In this case, introduction of liposome encapsulated agents into the cytoplasm of targeted cells is via membrane fusion of the liposome bilayer with the disease cell's plasma membrane. This first requires binding to the cell surface. Recent reports suggest that highly fusogenic lipid mixtures can be stabilized by incorporation of small amounts of exchangeable or cleavable PEG lipids.[91,92] Loss of the PEG moiety leads to destabilization of the liposome membrane which, in turn, will have the potential to fuse with nearby cell membranes. While these novel approaches for intracellular delivery of liposomal contents are providing exciting data in cell culture systems, their utility in vivo will depend on maintaining or increasing access and delivery of the liposomal carrier and encapsulated drug to the disease tissue.

IV. The dilemma faced when designing optimized liposomal anticancer drugs

Investigators designing liposomal anticancer drug carrier technology, rationalized on the basis of improved tumor drug delivery, must contend with a dilemma of opposing goals in the different biological compartments that the formulations experience. Since uptake of liposomes in tumors appears to be passive, extended circulation times (irrespective of conventional vs. steric stabilized comparisons) appear necessary to facilitate liposome accumulation. It follows that drug leakage from the liposomes must be minimized in order to avoid toxicities associated with free drug as well as to optimize drug delivery to the tumor. However, the characteristics that are well suited for these aims in the circulation (biologically inert, non-leaky liposomes) seriously limit the bioavailability of the encapsulated agent. This is due to the fact that tumor cells do not actively take up liposomes and, in the absence of any targeting or internalization information on the liposome surface, encapsulated drugs must be released in order to exert their antitumor activity. It should also be emphasized that the design of liposomes with optimal drug release kinetics, in either the blood compartment or disease site interstitial compartment, will also be highly dependent on the specific drug encapsulated. This could explain, in part, why the antitumor potency of vincristine, a drug whose activity is extremely sensitive to the duration of tumor cell exposure, is dramatically improved by reducing its leakage from liposomes whereas the antitumor potency of liposomal doxorubicin is much less dependent on drug release rates.

The inability to differentially control drug release rates in the plasma compartment and disease site is perhaps the most significant limitation of presently available liposomes. Ideally, one would be able to completely eliminate drug leakage in the circulation and then increase the release rate at the disease site to a level that would provide the optimal concentration vs. time profile for the specific drug being utilized. While this may seem to be a very onerous task, initial indications suggest that such approaches may be very fruitful. Early attempts to selectively increase drug leakage at tumor sites centered on the fact that liposomes can be constructed to become leaky in the acidic interstitial pH of some solid tumors,[93] which can drop to values of 6.5. More direct evidence of the importance of site-specific drug release has been obtained using localized hyperthermia.[94,95] Liposomal doxorubicin preparations, for example, can be prepared such that there is an increase in drug release at 42°C, compared to 37°C. These liposomes are administered i.v. to tumor bearing mice and the tumor site is then heated using a topical microwave heating device placed on the subcutaneous tumor. Application of a transient heating pulse after the liposomal doxorubicin had accumulated into the solid tumor resulted in a significant increase of therapeutic activity compared to free drug with hyperthermia and liposomal doxorubicin in the absence of heating. Although hyperthermia may not be applicable to many multifocal or deep seated tumors, this technique provides encouraging indications that liposomes exhibiting controlled or triggered release of their contents will significantly augment the pharmacological improvements provided by liposomes.

V. Closing comments

As our understanding of the processes that dictate the fate of liposomes after i.v. injection has increased, we have been better able to design formulations that will optimize the selectivity of action for encapsulated agents. Inclusion of additional components into conventional liposomes can now be done on the basis of extensive data describing the in vivo behavior of various liposome types. Although some questions still remain in areas such as the uptake and processing of liposomes in extravascular sites, we can now more reliably predict how such specific manipulations of liposomes should affect therapeutic activity. This increased understanding has also helped to identify new directions that may improve the therapeutic activity of liposomal drug formulations. Greater control of drug leakage rates within disease sites and the use of targeted and/or fusogenic liposomes for intracellular delivery offer opportunities to dramatically increase the efficiency and specificity of liposome encapsulated agents. The challenge for the future will be to develop systems that are actually therapeutically superior and not just technologically sophisticated.

References

1. DeVita VT, Young RC, Canellos GP. Combination versus single agent chemotherapy: Review of the basis of selection of drug treatement of cancer. Cancer 1975;35:98–108.
2. Goldie JH, Coldman AJ. A mathematical model for relating drug sensitivity of tumors to the spontaneous mutation rate. Cancer Treat Rep 1978;63:1727–1733.
3. Livingston RB. Dose intensity and high dose therapy. Cancer 1994;74:1177–1183.
4. Lin JH. Dose-dependent pharmacokinetics: experimental observations and theoretical considerations. Biopharmaceutics and Drug Disposition 1994;15:1–31.
5. Marsh JC. The effects of cancer chemotherapeutic agents on normal hematopoietic cells: a review. Cancer Res 1976;36:1863–1869.

6. Bally MB, Nayar R, Masin D, Cullis PR, Mayer LD. Studies on the myelosuppressive activity of doxorubicin entrapped in liposomes. Canc Chemother Pharm 1990;27:13–19.
7. Bonadonna G, Monfardini S, De Lena M. Phase I and preliminary phase II evaluation of Adriamycin (NSC 123127). Cancer Res 1970;30:2572–2582.
8. Rinehart JJ, Lewis RP, Balcerzak SP. Adriamycin cardiotoxicity in man. Ann Intern Med 1974;81:475–478.
9. Minow RA, Benjamin RS, Gottlieb JA. Adriamycin (NSC 123127)-cardiomyopathy: an overview with determination of risk factors. Cancer Chem Rep 1975;6:195–201.
10. Vose JM, Armitage JO. Clinical applications of hematopoietic growth factors. J Clin Oncol 1995;13:1023–1035.
11. Gabizon A, Dagan A, Goren D, Barenholz Y, Fuks Z. Liposomes as in vivo carriers of adriamycin: reduced cardiac uptake and preserved anti-tumor activity in mice. Cancer Res 1982;42:4734–4739.
12. Olson F, Mayhew E, Maslow D, Rustum Y, Szoka F. Characterization, toxicity and therapeutic efficacy of adriamycin encapsulated in liposomes. Eur J Cancer Clin Oncol 1982;18:167–175.
13. Conley BA, Egorin MJ, Whitacre MY, Carter DC, Zuhowski EG, Van Echo DA. Phase I and pharmacokinetic trial of liposome-encapsulated doxorubicin. Cancer Chem Pharm 1993;33:107–112.
14. Cowens JW, Creaven PJ, Greco WR, Brenner DE, Tung Y, Ostro M, Pilkiewicz F, Ginsberg R, Petrelli N. Initial clinical (phase I) trial of TLC D-99 (doxorubicin encapsulated in liposomes). Cancer Res 1993;53:2796–2802.
15. Kirby C, Clark J, Gregoriadis G. Cholesterol content of small unilamellar liposomes controls phospholipid loss to high density lipoproteins. FEBS Lett 1980;111:324–328.
16. Scherphof G, Roerdink F, Waite M, Parks J. Disintegration of phosphatidylcholine liposomes in plasma as a result of interaction with high density lipoproteins. Biochim Biophys Acta 1978;842:296–307.
17. Litzinger DC, Buiting AM, van Rooijen N, Huang L. Effect of liposome size on the circulation time and intraorgan distribution of amphipathic poly(ethylene glycol)-containing liposomes. Biochim Biophys Acta 1994;1190:99–107.
18. Semple SC, Chonn A, Cullis PR. Influence of cholesterol on the association of plasma proteins with liposomes. Biochemistry 1996;35:2521–2525.
19. Woodle MC, Newman MS, Cohen JA. Sterically stabilized liposomes: physical and biological properties. J Drug Targeting 1994;2:397–403.
20. Mui BL, Cullis PR, Pritchard PH, Madden TD. Influence of plasma on the osmotic sensitivity of large unilamellar vesicles prepared by extrusion. J Biol Chem 1994;269:7364–70.
21. Bonte F, Juliano RL. Interactions of liposomes with serum. Chem Phys Lipids 1986;40:359–372.
22. Chonn A, Semple SC, Cullis PR. Association of blood proteins with large unilamellar liposomes in vivo. Relation to circulation lifetimes. J Biol Chem 1992;267:18759–18765.
23. Oja CC, Semple SC, Chonn A, Cullis PR. Influence of dose on liposome clearance: critical role of blood proteins. Biochim Biophys Acta 1996;1281:31–37.
24. Chonn A, Semple SC, Cullis PR. Beta 2 glycoprotein I is a major protein associated with very rapidly cleared liposomes in vivo, suggesting a significant role in the immune clearance of "non-self" particles. J Biol Chem 1995;270:25845–25849.
25. Chang CW, Barber L, Ouyang C, Masin D, Bally MB, Madden TM. Comparison of the plasma clearance, biodistribution, and therapeutic properties of mitoxantrone encapsulated in conventional and sterically stabilized liposomes. British J of Cancer, 1997.
26. Lim HJ, Masin D, Madden TD, Bally MB. (1996) Influence of drug release characteristics on the therapeutic activity of liposomal mitoxantrone. J Pharmacology Exp Therapeutics 1997;281:566–573.
27. Lasic DD. Improved liposome stability and drug retention significantly increase the anticancer activity of encapsulated doxorubicin (Doxil), enhancing the effectiveness of chemotherapy and potentially reducing its toxicity. Nature 1996;380:561–562.
28. Allen TM. Long-circulating (sterically stabilized) liposomes for targeted drug delivery. Trends in Pharm Sci 1994;15:215–220.
29. Woodle MC, Lasic DD. Sterically stabilized liposomes. Biochim Biophys Acta 1992;1113:171–199.
30. Torchilin VP, Trubetskoy VS, Whiteman KR, Caliceti P, Ferruti P, Veronese FM. New synthetic amphiphilic polymers for steric protection of liposomes in vivo. J Pharm Sci 1995;84:1049–1053.
31. Torchilin VP, Omelyanenko VG, Papisov MI, Bogdanov AA Jr, Trubetskoy VS, Herron JN, Gentry CA. Poly(ethylene glycol) on the liposome surface: on the mechanism of polymer-coated liposome longevity. Biochim Biophys Acta 1994;1195:11–20.

32. Woodle MC, Matthay KK, Newman MS, Hidaya, JE, Collin LR, Redemann C, Martin FJ, Papahadjopoulos D. Versatility in lipid compositions showing prolonged circulation with sterically stabilized liposomes. Biochim Biophys Acta 1992;1105:193–200.
33. Parr MJ, Ansell SM, Choi LS, Cullis PR. Factors influencing the retention and chemical stability of poly(ethylene glycol)-lipid conjugates incorporated into large unilamellar vesicles. Biochim Biophys Acta 1994;1195:21–30.
34. Holland JW, Cullis PR, Madden TD. Poly(ethylene glycol)-lipid conjugates promote bilayer formation in mixtures of non-bilayer forming lipids. Biochemistry 1996;35:2610–2617.
35. Van Hossel QGCM, Steerenberg PA, Crommelin DJA, van Dijk A, van Oost W, Klein S, Douze JMC, de Wildt DJ, Hillen FC. Reduced Cardiotoxicity and nephrotoxicity with preservation of antitumor activity of doxorubicin entrapped in stable liposomes in the LOU/M Wsl Rat Cancer Res 1984;44:3698–3705.
36. Rahman A, White G, Moore N, Schein PS. Pharmacological, toxicological and therapeutic evaluation in mice of doxorubicin entrapped in liposomes. Cancer Res 1985;45:796–803.
37. Mayer LD, Tai LCL, Ko DSC, Masin D, Ginsberg RS, Cullis PR and Bally MB. Influence of vesicle size, lipid composition, and drug to lipid ratio on the biological activity of liposomal doxorubicin in mice. Cancer Res 1989;49:5922–5930.
38. Mayer LD, Cullis PR, Bally MB. The use of transmembrane pH gradient-drive drug encapsulation in the pharmacodynamic evaluation of liposomal doxorubicin. J Liposome Res 1994;4:529–553.
39. Huang SK, Mayhew E, Gilani S, Lasic DD, Martin FJ, Papahadjopoulos D. Pharmacokinetics and therapeutics of sterically stabilized liposomes in mice bearing C-26 colon carcinoma. Cancer Res 1992;52:6774–6781.
40. Parr MJ, Masin D, Cullis PR, Bally MB. Accumulation of liposomal lipid and encapsulated doxorubicin in murine Lewis Lung carcinoma: the lack of beneficial effects by coating liposomes with poly(ethylene glycol). J Pharmacol Exp Therapeut 1997;280:1319–1327.
41. Mayer LD, Dougherty G, Harasym TO, Bally MB. The role of tumor-associated macrophages in the delivery of liposomal doxorubicin to solid murine fibrosarcoma tumors. J Pharmacol Exp Therapeut 1997;280:1406–1414.
42. Williams SS, Alosco TR, Mayhew E, Lasic DD, Martin FJ, Bankert RB. Arrest of human lung tumor xenograt growth in severe combined immunodeficient mice using doxorubicin encapsulated in sterically stabilized liposomes. Cancer Res 1993;53:3964–3967.
43. Mayer LD, Nayar R, Thies RL, Boman NL, Cullis PR, Bally MB. Identification of vesicle properties that enhance the antitumor activity of liposomal vincristine against murine L1210 leukemia. Cancer Cemother Pharmacol 1993;33:17–24.
44. Mayer LD, Masin D, Nayat R, Boman NL, Bally MB. Pharmacology of liposomal vincristine in mice bearing L1210 ascitic and B16/BL6 solid tumours. Br J Cancer 1995;71:482–488.
45. Webb MS, Harasym TO, Masin D, Bally MB, Mayer LD. Sphingomyelin-cholesterol liposomes significantly enhance the pharmacokinetic and therapeutic properties of vincristine in murine and human tumor models. Br J Cancer 1995;72:896–904.
46. Jackson DV, Bender RA. Cytotoxic thresholds of vincristine in a murine and a human tumor leukemia cell line in vitro. Cancer Res 1979;39:4346–4349.
47. Houghton JA, Williams LG, Dodge RK, George SL, Hazelton BJ, Houghton PJ. Relationship between binding affinity, retention and sensitivity of human rhabdomyosarcoma xenografts to vinca alkaloids. Biochem Pharmacol 1987;36:81–88.
48. Allen TM, Newman MS, Woodle MC, Mayhew E, Uster PS. Pharmacokinetics and anti-tumor activity of vincristine encapsulated in sterically stabilized liposomes. Int J Cancer 1995;62:199–204.
49. Durr FE. Biological and biochemical effects of mitoxantrone. Semin Oncol 1984;11:3–10.
50. Schwendener RA, Fiebig HH, Berger MR, Berger DP. Evaluation of incorporation characteristics of mitoxantrone into unilamellar liposomes and analysis of their pharmacokinetics properties, acute toxicity and antitumor efficacy. Cancer Chemother Pharmacol 1991;27:429–439.
51. Schwendener RA, Horber DH, Rentsch K, Hanseler E, Pestalozzi B, Sauter C. Preclinical and clinical experience with liposome-encapsulated mitoxantrone. J Liposome Res 1994;4:605–639.
52. Mayer LD, Bally MB, Cullis PR. Strategies for optimizing liosomal doxorubicin. J Liposome Res 1990;1:463–480.
53. Jain RK. Transport of molecules across tumor vasculature. Cancer and Metastasis Rev 1987;6:559–593.
54. Dvorak HF, Nagy JA, Dvorak JT, Dvorak AM. Identification and characterization of the blood vessels of solid tumors that are leaky to circulating macromolecules. Amer J Pathol 1988;133:95–109.

55. Poste G, Kirsh R, Kuster T. The challenge of liposome targeting in vivo. In: Gregoriadis G, ed. Liposome Technology, Vol. III. Boca Raton, FL: CRC Press, 1987; pp. 1–28.
56. Yuan F, Dellian M, Fukumura D, Leunig M, Berk DA, Torchilin VP, Jain RK. Vascular permeability in a human tumor xenograft: molecular size dependence and cutoff size. Cancer Res 1995;55:3752–6.
57. Abra RM, Hunt CA. Liposome disposition in vivo: III Dose and vesicle size effect. Biochim Biophys Acta 1981;666:493–503.
58. Hwang KJ. Liposome pharmacokinetics. In: Ostro MJ, ed. Liposomes: from Biophysics to Therapeutics. New York: Marcel Dekker, 1987;109–156.
59. Senior J, Crawley JCW, Gregoriadis G. Tissue distribution of liposomes exhibiting long half lives in the circulation after intravenous injection. Biochim Biophys Acta 1985;839:1–8.
60. Gabizon AA. Selective tumor localization and improved therapeutic index of anthracyclines encapsulated in long-circulating liposomes. Cancer Res 1992;52:891–896.
61. Bally MB, Nayar R, Masin D, Hope MJ, Cullis PR, Mayer LD. Liposomes with entrapped doxorubicin exhibit extended blood residence times. Biochim Biophys Acta 1990;1023:133–139.
62. Parr MJ, Bally MB, Cullis PR. The presence of GM1 in liposomes with entrapped doxorubicin does not prevent RES blockade. Biochim Biophys Acta 1993;1168:249–252.
63. Buiting AMJ, Zhou F, Bakker JAJ, van Rooijen N, Huang L. Biodistribution of clodronate and liposomes used in the liposome mediated macrophage "suicide" approach. J Immunol Meth 1996;192:55–62.
64. Daemen T, Hofstede G, Ten Kate MT, Bakker-Woudenberg IA, Scherphof GL. Liposomal doxorubicin-induced toxicity: depletion and impairment of phagocytic activity of liver macrophages. Int J Cancer 1995;61:716–721.
65. Parr MJ. Circulation lifetimes and tumor accumulation of liposomal drug delivery systems. Ph.D. Thesis, University of British Columbia, 1995.
66. Hengge UR, Brockmeyer NH, Rasshofer R, Goos M. Fatal hepatic failure with liposomal doxorubicin, Lancet 1993;341:383–384.
67. Chen KR, Su WP, Pittelkow MR, Conn DL, George T, Leiferman KM. Eosinophilic vasculitis in connective tissue disease. J Amer Acad Derm 1996;35:173–182.
68. Hanahan D, Folkman J. Patterns and emerging mechanisms of the angiogenic switch during tumorigenesis. Cell 1996;86:353–364.
69. Rak JW, St Croix BD, Kerbel RS. Copnsequences of angiogenesis for tumor progression, metastasis and cancer therapy. Anticancer Drugs 1996;6:3–18.
70. Dellian M, Witwer BP, Salehi HA, Yuan F, Jain RK. Quantitation and physiological characterization of angiogenic vessels in mice: effect of basic fibroblast growth factor, vascular endothelial growth factor/vascular permeability factor, and host microenvironment. Amer J Path 1996;149:59–71.
71. Dvorak HF, Detmar M, Claffey KP, Nagy JA, van de Water L, Senger DR. Vascular permeability factor/vascular endothelial growth factor: an important mediator of angiogenesis in malignancy and inflammation. Int Arch Allergy Immunol 1995;107:233–235.
72. Kohn S, Nagy JA, Dvorak HF, Dvorak AM. Pathways of macromolecular tracer transport accross venules and small veins. Lab Invest 1992;67:596–607.
73. Huang SK, Martin FJ, Jay G, Vogel J, Paphadjopoulos D, Friend DS. Extavasation and transcytosis of liposomes in Kaposi' sarcoma-like dermal lesions of transgenic mice bearing the HIV tat gene. Am J Pathol 1993;143:10–14.
74. Jain RK. Physiological resistance to treatment of solid tumors. In: Teicher BA, ed. Drug Resistance in Oncology. New York: Marcel Dekker, 1993;87–105.
75. Buocher Y, Jain RK. Microvascular pressure is the principal driving force for interstitial hypertension in solid tumors: Implications for vascular collapse. Cancer Res 1992;52:5110–5114.
76. Wu NZ, Da D, Rudoll TL, Needham D, Whorton AR, Dewhirst MW. Increased microvascular permeability contributes to preferential accumulation of stealth liposomes in tumor tissue. Cancer Res 1993;53:3765–3770.
77. Yuan F, Leunig M, Huang SK, Berk DA, Papahadjopoulos D, Jain RK. Microvascular permeability and interstitial penetration of sterically stabilized (Stealth) liposomes in a human tumor xenograft. Cancer Res 1994;54:3352–3356.
78. Sadzuka Y, Nakai S, Miyagishima A, Nozawa Y, Hirota S. The effect of dose on the distribution of adriamycin encapsulated in polyethyleneglycol-coated liposomes. J Drug Targeting 1995;3:31–37.
79. Jain RK. Physiological barriers to delivery of monoclonal antibodies and other macromolecules in tumors. Cancer Res 1990;50:814s–819s.

80. Pagano RE, Martin OC, Schroit AJ, Struck DK. Formation of asymmetric phospholipid membranes via spontaneous transfer of fluorescent lipid analogues between vesicle populations. Biochemistry 1981;20:4920–4927.
81. Harvie P, Desormeaux A, Gagne N, Tremblay M, Poulin L, Beauchamp D, Bergeron MG. Lymphoid tissues targeting of liposome-encapsulated 2′,3′-dideoxyinosine. AIDS. 1995;9:701–707.
82. Storm G, Steerenberg PA, Emmen F, van Borssum Waalkes M, Crommelin DJ. Release of doxorubicin from peritoneal macrophages exposed in vivo to doxorubicin-containing liposomes. Biochim Biophys Acta 1988;965:136–145.
83. Leek RD, Lewis CE, Whitehouse R, Greenall M, Clarke J, Harris AL. Association of macrophage infiltration with angiogenesis and prognosis in invasive breast carcinoma. Cancer Res 1996;56:4625–4629.
84. Saga T, Neumann RD, Heya T, Sato J, Kinuya S, Le N, Paik CH, Weinstein JN. Targeting cancer micrometastases with monoclonal antibodies: A binding-site barrier. Proc Natl Acad Sci USA 1995;92:8999–9003.
85. Ahmad I, Longenecker M, Samuel J, Allen TM. Antibody-targeted delivery of doxorubicin entrapped in sterically stabilized liposomes can eradicate lung cancer in mice. Cancer Res 1993;53:1484–1488.
86. Lee RJ, Low PS. Delivery of liposomes into cultured KB cells via folate receptor-mediated endocytosis. J Biol Chem 1994;269:3198–3204.
87. Suzuki S, Uno S, Fukuda Y, Aoki Y, Masuko T, Hashimoto Y. Cytotoxicity of anti-c-erbB-2 immunoliposomes containing doxorubicin on human cancer cells. British J Cancer 1995;72:663–668.
88. Connor J, Huang L. pH-sensitive immunoliposomes as an efficient and target-specific carrier for antitumor drugs. Cancer Res 1986;46:3431–3435.
89. Mizuguchi H, Nakanishi M, Nakanishi T, Nakagawa T, Nakagawa S, Mayumi T. Application of fusogenic liposomes containing fragment A of diphtheria toxin to cancer therapy. British J Cancer 1996;73:472–476.
90. Archer A, Miller K, Reich E, Hautmann, R. Photodynamic therapy of human bladder carcinoma cells in vitro with pH sensitive liposomes, as carriers for 9-acetoxytetrapropylporphyrene. Urological Research 1994;22:25–32.
91. Holland JW, Hui C, Cullis PR, Madden TD. Poly(ethylene glycol)-lipid conjugates regulate the calcium-induced fusion of liposomes composed of phosphatidylethanolamin and phosphatidylserine. Biochemistry 1996;35:2618–2624.
92. Kirpotin D, Hong K, Mullah N, Papahadjopoulos D, Zalipsky S. Liposomes with detachable polymer coating: Destabilization and fusion of dioleoylphosphatidylethanolamine vesicles triggered by cleavage of surface-grafted poly(ethylene glycol). FEBS Letters 1996;388:115–118.
93. Hazemoto N, Harada M, Komatsubara N, Haga M, Kato Y. pH-sensitive liposomes composed of phosphatidylethanolamine and fatty acid. Chem Pharma Bulletin 1990;38:748–51.
94. Huang SK, Stauffer PR, Hong K, Guo JW, Phillips TL, Huang A, Papahadjopoulos D. Liposomes and hyperthermia in mice: Increased tumor uptake and therapeutic efficacy of doxorubicin in sterically stabilized liposomes. Cancer Res 1994;54:2186–2191.
95. Unezaki S, Maruyama K, Takahashi N, Koyama M, Yuda T, Suginaka A, Iwatsuru M. Enhanced delivery and antitumor activity of doxorubicin using long-circulating thermosensitive liposomes containing amphiphathic polyethylene glycol in combination with local hyperthermia. Pharmaceutical Res 1994;11:1180–1185.

Lasic and Papahadjopoulos (eds.), Medical Applications of Liposomes

CHAPTER 4.3

Pharmacologic advantages of anthracyclines encapsulated in poly-ethylene-glycol coated Stealth liposomes: Potential for tumor targeting

DORIT GOREN AND ALBERTO GABIZON

Department of Oncology, Hadassah Hebrew University Hospital, Jerusalem, Israel

Overview

I. Introduction

Chemotherapy is a powerful tool in cancer treatment, not only as the main treatment modality against metastatic cancer, but also as a useful adjuvant to surgery and radiotherapy in localized cancer. However, most of the cytotoxic drugs used in cancer chemotherapy have a narrow therapeutic window causing serious side effects, and impaired quality of life, which lead frequently to suboptimal dosing and reduced patient compliance to the therapy. Despite a huge effort in drug development, cancer chemotherapy remains largely non-specific, since most drugs are toxic for tumor cells as well as for normal cells. Therefore, drug carriers have been developed to modify the biodistribution of cytotoxic drugs, aiming at refined selectivity for tumors and reduced damage of normal tissues. The design of drug delivery systems for cancer therapy is faced with serious obstacles, since it is generally directed at a systemic disease, either in the form of occult micrometastases or clinically detectable macrometastases. To reach all tumor sites efficiently, the intravenous route is the most logical choice. Consequently, the interaction of the carrier system with plasma proteins, blood cells and the reticuloendothelial system (RES) must be considered. Tumor heterogeneity with regard to patterns of metastatic spread, microvascular architecture, macrophage infiltration, and mechanism of drug resistance adds an extra difficulty in predicting the therapeutic impact of a drug carrier approach.

As non-covalently bound, biocompatible and biodegradable carriers, liposomes have raised considerable interest as a drug delivery system in cancer chemotherapy.[1] Most applications of liposomes in cancer chemotherapy are directed at altering tissue distribution and various pharmacokinetic parameters of the drug in question in such a way that toxicity can be reduced and/or efficacy increased.[2] Reduced toxicity may be gained through site circumvention of drug sensitive tissues and by slow release of the cytotoxic agent from the carrier, avoiding peak plasma concentrations after bolus injection of free drug. Liposome- mediated decrease in toxicity could enable escalation of dose, which will result in increased tumor exposure to the drug.

One of the most encouraging areas in the liposome-anticancer drug field is the work with anthracyclines (doxorubicin-DOX; daunorubicin-DAU and epirubicin-EPI). DOX, a major anti-neoplastic anthracycline and one of the drugs most widely used in cancer chemotherapy, has a broad spectrum of anti-tumor activity against solid tumors and leukemias.[3] Anthracycline-induced cardiotoxicity,[4] a severe cumulative and in most cases irreversible effect, may be attenuated by a carrier system that decreases drug uptake by the heart muscle without lessening the anti-tumor activity, thereby improving the therapeutic index. Liposomes can adequately fulfill this task, given their relative inability to cross continuous capillaries and the lack of cells of the RES in the myocardial tissue. In addition, the slow-release effect of liposomal delivery of anthracyclines may reduce the peak plasma concentration of free drug, a factor which is directly correlated with cardiotoxicity.[5] Various liposome formulations have been tested as carriers of DOX in the early 80 s.[6–12] Examination of their pharmacologic properties, indicate that liposome entrapment of DOX decreases drug distribution to the heart, thus reducing the cardiotoxic effect of DOX. Furthermore, liposomal encapsulation of the drug enhances its therapeutic activity in a limited number of experimental tumor models.[13–16] The presence of negatively charged phospholipids such as cardiolipin, phosphatidylserine (PS), and phosphatidylglycerol (PG) increases and stabilizes considerably the association of DOX to the lipid bilayer, augmenting the drug load of liposomes.[15,17] The interaction of the drug with the bilayer is both electrostatic, (DOX is protonated at physiologic pH and below) and hydrophobic.[17]

Among the fluid negatively charged liposomes with high DOX loading capacity, the most extensively-tested liposome formulation in our laboratory, was of the following composition: egg Phosphatidylcholine (PC); egg-derived Phosphatidylglycerol (EPG); Cholesterol (Chol) at a molar ratio of 7:3:4 respectively.[15,17] Preclinical and clinical pharmacology studies with this formulation of oligolamellar vesicles (~200 nm size), indicated that they lack the stability needed for an efficient delivery of their cargo to the target organ. Their insufficient drug retention capacity, and rapid clearance from circulation through uptake by the reticuloendothelial

system (RES), impairs the performance required from these carriers. In an attempt to reduce RES uptake, and improve access to tumor tissues, we downsized our formulation using sonication. However the sonicated small unilamellar liposomal vesicles (size ~60 nm), showed low encapsulation capacity, and were structurally unstable resulting in rapid drug loss.[18]

In view of these drawbacks, the progress in liposome engineering accomplished in recent years has been remarkable resulting in the development of liposomal drug delivery systems with valuable pharmacologic properties: inhibition of the rapid clearance from circulation by the RES and reduction of the rate of drug leakage, leading to stable long-circulating liposome formulations. Particularly the coating of liposomes with polyethylene-glycol (PEG), a hydrophilic polymer, that generates a steric barrier preventing the hydrophobic interactions of plasma opsonins with the vesicle surface, has a major impact in conferring protection to the vesicles from RES-mediated clearance.[19] Moreover, bilayer rigidification using high Tm phospholipids reduces the rate of leakage of liposome content.[20] The term "Stealth liposomes" has been coined to designate these long-circulating, PEG-coated liposomes. (Also referred to as: sterically stabilized liposomes.)

In the design of stable liposome formulations for drug delivery the method of drug encapsulation is an essential determinant of the carrier properties. An efficient remote loading method of preformed liposomes with doxorubicin, compatible with Stealth formulations was developed.[21] By means of transmembrane pH and ammonium sulfate gradients that drive DOX and other cationic amphiphiles into the liposome water compartment,[21,22] extremely high drug load (150–200 μg DOX/μmol phospholipid) is obtained. Once inside, the drug becomes ionized, thus preventing its escape through the lipid bilayer. This is followed by the formation of a gel-like precipitate, as the drug in liposomes reaches a concentration exceeding its aqueous solubility.[23] For intact gradient maintenance, high phase-transition temperature phospholipids in combination with cholesterol are essential components. This type of formulations can remain for long periods in circulation with minimal leakage of drug.[24,25] A Stealth liposome formulation of DOX, known as Doxil*, has been approved for clinical use, underscoring the validity of this approach.

We will review here some of the recent findings of our laboratory with Stealth liposomal anthracyclines at the preclinical pharmacology level and discuss the rationale behind this approach and the potential for active targeting of Stealth liposomes. Examples of biodistribution and therapeutic studies will be presented for several murine and human tumor xenografts inoculated by various routes.

II. Correlation between liposome longevity in circulation and distribution to tumors

A key issue in cancer drug delivery is enhancing selectively drug accumulation in tumors. The relevance of the Stealth liposome formulation to this favorable

*Doxil is a registered trademark of Sequus Pharmaceuticals, Menlo Park, CA.

pharmacologic property was investigated in several murine and human tumor models.

*II.1. Mouse M*109 *carcinoma*

This model, derived from a mouse lung carcinoma, provides an example of an epithelial solid tumor model with metastatic ability, which represents the most lethal and common form of cancer in humans. BALB/c mice were inoculated subcutaneously with 10^6 cells of M-109 carcinoma. As tumors reached an approximate weight of 200 mg, mice were injected i.v. with free-DOX or PEG-hydrogenated PC(HPC)-Chol liposome-encapsulated-DOX. The drug levels in tumor and liver are presented in Figure 1. Peak concentrations of free-DOX in liver and tumor are observed 1 h post injection, whereas liposomal drug peaks in liver and tumor 24 h and 48 h after injection respectively, pointing at the fact that liposome accumulation in tumors is a slow process requiring a long circulating time. A 4-fold enhancement of peak DOX levels is obtained in subcutaneous tumor implants when PEG-coated liposomes are the drug carriers. It seems that the efficiency of drug delivery to the tumor is as high as to the liver when PEG-HPC-Chol liposomes are used, since liposomal DOX levels in tumor were roughly of the same magnitude as in liver. These observations are in accordance with those obtained with IM implanted J-6456 lymphoma,[26] where liposomal drug tumor concentration also peaked 48 h after injection.

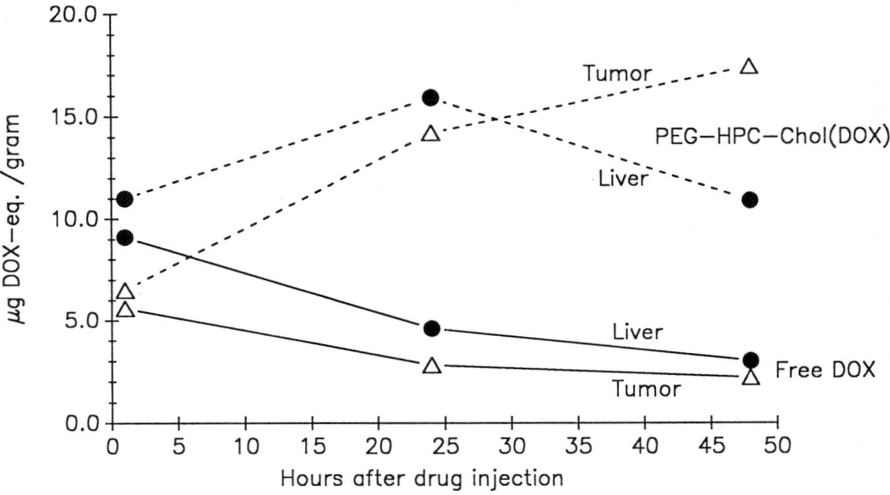

Fig. 1. Tissue distribution of liposomal DOX in tumor bearing mice. BALB/c female mice bearing subcutaneous implants of M-109 carcinoma received i.v. 10 mg/kg DOX in either free form or encapsulated in PEG-HPC-Chol liposomes. Solid lines, mice injected with free DOX, dashed lines, mice injected with liposomal DOX. ●, DOX concentration in liver; △, DOX concentration in tumor. (Adapted from J Liposome Res[25]).

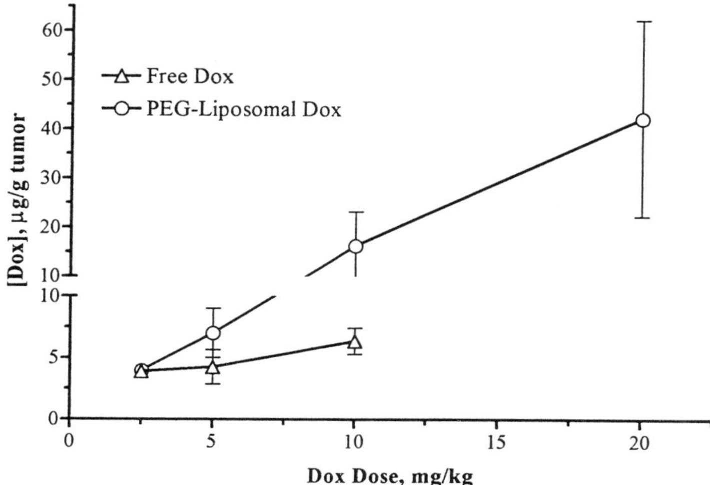

Fig. 2. The effect of dose on tumor levels of liposomal DOX. BALB/c mice inoculated subcutaneously with 10^6 M-109 cells. Two weeks later, free DOX or liposomal DOX (liposome composition: HPC/PEG-distearoylphosphatidylethanolamine/Chol, 92.5/7.5/70 molar ratio respectively) injected i.v. at the indicated doses. Mice were sacrificed 3 h after free DOX injection or 48 h after liposomal DOX injection. Each point is the mean of 3–4 mice. Average tumor weight: 53 to 153 mg for free DOX groups; and, 102 to 174 mg for liposomal DOX groups. The 20 mg/kg dose level of free DOX was not tested because of its lethal toxicity. (Adapted from Adv Drug Deliv Rev[26]).

Preliminary experiments with the same tumor model-subcutaneous implants of M109 carcinoma, point at a good correlation between the administered dose of liposomal DOX and the tumor drug levels in the dose range of 2.5 to 20 mg/kg (Figure 2). This observation is consistent with a passive process of liposome accumulation in tumors which follows non-saturable kinetics. In contrast, when free DOX was administered, dose increase resulted in a minimal advantage in tumor drug levels (Figure 2). This kinetics of liposomal drug accumulation in tumors, could be exploited for enhanced therapy through dose escalation, provided toxicity is not increased concomitantly to a prohibitive degree.

II.2. Mouse J-6456 lymphoma

Another interesting tumor model examined is the J-6456 murine lymphoma inoculated i.p. Biodistribution study of a variety of liposomal-DOX formulations intravenously administered, in mice with an ascitic tumor of ~1.0 ml (tumor load of 100×10^6 cells), revealed that PEG-HPC containing liposomes is the most advantageous formulation.[27] The highest plasma DOX levels as well as highest peak levels in the ascitic fluid are obtained with PEG-HPC-Chol formulation, a result which is apparently due to their long circulation time.[26,27] Interestingly, the drug levels in mice injected with free DOX are undetectable in plasma and ascitic fluid,[27,28] but only slightly lower than for liposome-encapsulated DOX in the case

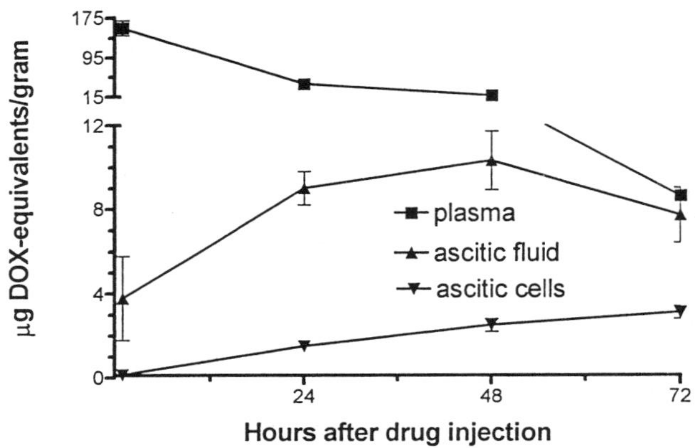

Fig. 3. Kinetics of DOX delivery to the J6456 ascitic tumor. Mice previously inoculated i.p. with J6456 tumor cells received i.v. 10 mg/kg of DOX encapsulated in PEG-HPC-Chol liposomes. Note that the drug peak in the ascitic fluid is obtained 48 hr after drug injection. Drug penetration in the cellular compartment is slower and equilibration with the extracellular compartment is still not reached 72 h after injection. Each time point is the mean of 3 to 4 mice. (Adapted from J drug Targ[27]).

of ascitic cells.[27] Despite this small difference in cell-associated DOX, it should be stressed that the reservoir of drug in plasma is huge in the case of formulations like PEG-HPC-Chol. Through a gradual process of extravasation of liposomal DOX from plasma into the ascitic fluid, followed by efflux of drug from liposomes, the ascitic cellular compartment is fed, resulting in a greater cell exposure to drug, i.e., AUC (area under the concentration-time curve). To illustrate this point, the kinetics of the process of accumulation in ascites using PEG-HPC-Chol(DOX) liposomes is shown in Figure 3. We found that the DOX levels in the ascitic fluid gradually increase reaching a peak at 48 h after injection, and equilibrating with plasma levels at 72 h after injection. This is indicative of a slow equilibration of liposomal drug between two compartments (intravascular and extracellular), probably mediated by the extravasation of circulating liposomes. Another important observation in Figure 3 is the late but steady increase in the levels of cell-associated drug, which have not yet reached their peak at 72 h after injection. Clearly, the pattern of drug accumulation in cells results from the slow efflux of DOX from liposomes present in the surrounding ascitic fluid. Since, even at 72 h after injection there is still a high concentration of liposomal drug in the ascitic fluid, it is likely that drug levels in cells will further increase. However, due to the advanced stage of the tumor and the deterioration of the animals, we did not examine the drug levels beyond 72 h. Interestingly, no significant accumulation of liposomal DOX was observed in peritoneal washes from tumor-free mice, indicating that extravasation depends on the increased microvascular permeability caused by the ascitic tumor.[28]

II.3. Other tumor models

In an experimental brain tumor model[26] (rat sarcoma implanted in Fischer rat brains), enhancement of drug delivery and of the therapeutic effect of DOX in PEG-HPC-Chol liposomes, was observed stressing the advantage of long-circulating liposomes in an intra-craneal tumor model[26] which otherwise shows reduced exposure to cytotoxic concentrations of most of the intravenously administered chemotherapeutic agents.

In the case of human N-87 gastric carcinoma and A-375 melanoma, implanted subcutaneously into nude mice, DOX in PEG-HPC-Chol liposomes, was superior to free drug as observed previously in other biodistribution studies. The tumor drug levels obtained after administration in the PEG-HPC-Chol formulation are about 4-fold greater than those obtained with free-DOX, while the increase in AUC is in the range of 8 fold.[26] The results are similar for the two tumor models.

III. Liposome composition and therapeutic efficacy

Evidently, the critical issue is to look into the relevance of differences in circulation longevity and drug delivery to tumors and their implications on anti-tumor activity. The effect of bilayer fluidity of PEG-coated liposomes on the therapeutic activity of encapsulated DOX (Figure 4) was examined, using the ascitic J6456 tumor model. Clearly, PEG- HPC-Chol(DOX) and PEG-dipalmitoylPC(DPPC)-Chol-(DOX) were more effective than PEG- eggPC(EPC)-Chol(DOX) in prolonging the survival of tumor-bearing mice, an observation that underscores the relevance of a high T_m bilayer for an efficient drug delivery to the tumor. This result is similar to a previous observation in the same tumor model in which a hydrogenated phosphatidylinositol(HPI)-HPC-Chol formulation of DOX was found therapeutically superior to a formulation of EPG-EPC-Chol liposomes.[28] Despite the longer circulation time of PEG-HPC-Chol-(DOX) over PEG-DPPC-Chol(DOX), there was no difference in anti-tumor activity between them. When these results are viewed together with previous ones comparing free-DOX efficacy with that of various HPC containing formulations (distearoylPG; HPI; PEG),[27] an improvement in anti-tumor activity of liposomal DOX against ascites J6456 lymphoma, is gained as manifested by an increase in median survival and even a few cases of cures.[27] Liposomal DOX increased anti-tumor activity is not related to a buffering of toxicity conferred to the drug through encapsulation, since it is obtained at sub-toxic dose levels, although the therapeutic effect is more pronounced at levels that are toxic for free DOX i.e., 15 mg/kg.[27] An important observation refers to therapeutic studies done with liposomal epirubicin,[27] in which a HPC-Chol formulation was found to be significantly inferior to HPI-HPC-Chol liposomes. Since HPC-Chol is a formulation with lower plasma and ascitic fluid levels than HPI-HPC-Chol, this suggests that circulation longevity does improve the anti-tumor effect. Additional experiments directed at a detailed examination of a possible difference in liposome formulations with close T_m but different circulation times: PEG-HPC-Chol(DOX) vs. HPC-Chol(DOX) were carried out, with i.v. inocu-

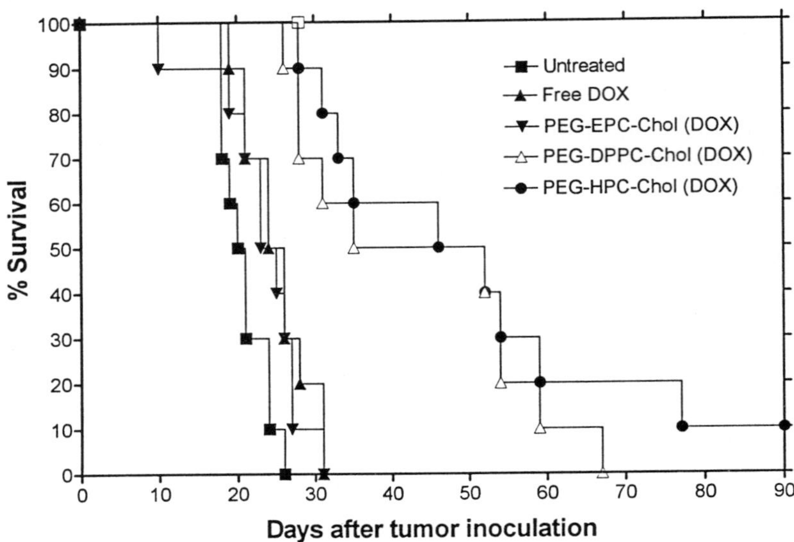

Fig. 4. Therapeutic efficacy of liposome-encapsulated DOX against the ascitic J-6456 lymphoma. BALB/c mice were inoculated i.p. with 10^6 J-6456 cells and treated 5 days later i.v. with 10 mg/kg of free or liposome-encapsulated DOX. There were 10 mice in each experimental group. The survival of PEG-HPC(DOX)-treated mice was significantly longer than that of Free DOX, and PEG-EPC(DOX)-treated mice (Log-rank test, $p \leqslant 0.0006$). There was no significant difference between the PEG-HPC(DOX) and PEG-DPPC(DOX) groups. (Adapted from J Drug Targ[27]).

lated M-109 carcinoma. This murine carcinoma line inoculated intravenously generates lung metastases, which appear to be well correlated with survival. As seen in Figure 5 the presence of PEG resulted in a significant lengthening of median survival and a significant number of cures, indicating that the increased longevity provided by the PEG coating is of biological relevance when liposomes of similar bilayer rigidity are compared.[27]

The tissue distribution pattern in nude mice indicating a large advantage for the liposomal formulation in terms of tumor exposure to the drug, provide a solid rationale for the claims on improved therapeutic activity of PEG-liposomal DOX in the numerous human xenogeneic models reported.[29–32] In most tumor models tested, PEG-liposomal DOX was found to be more effective than free Dox.[25,27,32–35] Therefore, it is conceivable that the enhanced drug delivery to tumors achieved with long-circulating liposomes underlies the therapeutic advantage observed. This advantage is apparently due to a preferred access of small (<100 nm) stable DOX-loaded long-circulating liposomes, to tumor tissue. Studies with a variety of liposome formulations in normal and tumor bearing mice point at direct correlation between prolonged circulation time and liposome localization in tumors,[36,37] stressing the key role of liposome longevity in circulation for enhanced therapeutic potency. In general, the permeability of tumor vascularisation is increased as compared with normal tissues.[38,39] The efficiency and kinetics

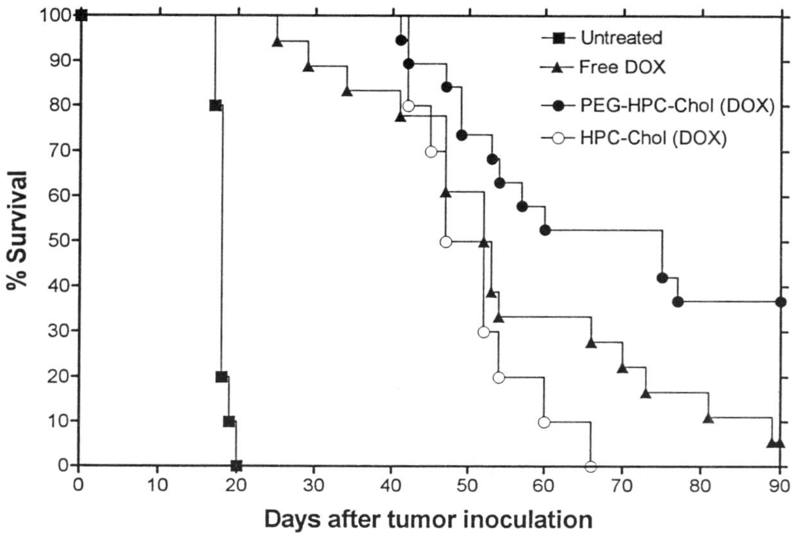

Fig. 5. Therapeutic efficacy of liposome-encapsulated DOX in the M-109 lung metastases model. BALB/c mice were inoculated i.v. with 10^6 M-109 carcinoma cells and treated 5 days later i.v. with 10 mg/kg of free or liposome-encapsulated DOX. The number of mice per experimental group was: Untreated, 10; Free DOX, 17; PEG-HPC(DOX), 12; HPC(DOX), 10. The survival of PEG-HPC(DOX)-treated mice was significantly longer than that of Free DOX ($p = 0.0228$), and HPC(DOX)-treated mice ($p = 0.0059$), as assessed by the Log-rank test. (Adapted from J Drug Targ[27]).

of this process are determined, among other factors, by the capillary and basement membrane permeability and the size or molecular weight of the particle in question. Liposomes could extravasate through a leaky endothelium by passive convective transport and the chances of particle extravasation will increase with a higher concentration in blood and a greater number of circulation passages through the tumor bed. In addition, factors such as the increased interstitial pressure and the lack of lymphatic drainage in tumors[40] will undoubtedly affect the influx/efflux of nano-particles such as liposomes into the tumor compartment. The EPR (Enhanced Permeability and Retention) model which has been proposed to explain the preferential accumulation of macromolecules in tumors, may also be applicable to nano-particles and liposomes.[26]

Another important aspect related to liposome composition and therapeutic efficacy is the balance between stability/longevity and drug release rate: Is longer circulation always advantageous for tumor drug delivery and therapeutic effect? Probably there is no simple answer, since the need for bioavailable drug which requires achieving a satisfactory rate of drug release in the tumor site is basically in conflict with the high stability requirements of a prolonged circulation time. Circulation longevity and drug release profile may need to be carefully balanced to achieve an optimal therapeutic effect. However, animal tumor models may be inadequate to resolve this issue since their growth kinetics are generally more rapid than in humans. As the comparative testing in humans of different liposome formulations is virtually an impossible task, the safest approach is to favor the formulation with the longest circulation residence time, i.e., PEG-HPC-Chol, which is likely to be the most useful in the commonplace human tumors with doubling times in the range of weeks.

IV. Active targeting of stealth liposomes for drug delivery

The recognition of PEG coated Stealth liposomes as long-circulating drug delivery systems with stable drug retention evoked the feasibility of ligand-mediated liposome targeting to tumor cell for selective enhancement of drug delivery. Active targeting using ligands on the surface of liposomes, would bring about a direct interaction of liposomes and their contents with tumor cells.[41–43] This is one step further beyond the passive accumulation of Stealth liposomes in the tumor interstitial fluid.[28,33,44] In our laboratory, we have attempted to investigate this approach using antibody-targeted-liposomes i.e., immunoliposomes, directed against erbB2/HER2 oncoprotein.[32] The erbB2 gene product is a membrane glycoprotein of 185kD ($p185^{HER2}$), with intrinsic kinase activity, amplified and over-expressed uniquely in malignancies. The protein, is observed in many epithelial malignancies, particularly in breast and ovarian carcinomas (15–20% of human carcinomas),[45,46] predicting poor prognosis,[47–49] whereas, in certain normal epithelial tissues, detected only at low levels.[50] Being a membraneous over-expressed antigen with ready accessibility and high level of tumor specificity, erbB2 offers an attractive target for cancer therapy. N12A5 (IgG1) monoclonal antibody having a high binding capacity to erbB2 positive cells,[51] and significant growth inhibition of erbB2 positive human tumors (N-87 implanted carcinoma[52]), was selected for our studies of targeted therapy. The N-87 cells, human erbB2 overexpressor gastric carcinoma line,[53] with proven growth capability in nude mice, was chosen as the tumor model.

Pharmacokinetic studies with antibody-targeted liposomes in mice, were done initially with α-erbB2 conjugated liposome preparations of relatively high protein/-phospholipid ratio, $\sim 100\ \mu g/\mu mol$. These high-protein immunoliposomes were cleared from plasma significantly faster than plain liposomes or low-protein immunoliposomes ($<60\ \mu g/\mu mol$) (data not shown). The differences in plasma DOX clearance rates between targeted and non-targeted liposomes are minimal when a liposome preparation with low protein to lipid ratio is used. We inferred that low levels of protein conjugated to liposomes are required to maintain Stealth qualities of immunoliposomes as reported by other investigators.[54] In vitro experiments demonstrated that these antibody-targeted liposomes bind avidly and selectively to Her/2 positive cells even at low protein to lipid ratio. Further experiments discussed here were done with these low-protein immunoliposomes.

Since the key point in designing immunoliposomes is enhancement of cancer drug delivery we will present the results of drug levels only in tumor tissue.[32] Contrary to our expectations liposome levels in N-87 subcutaneous implants were slightly higher for plain liposomes than for immunoliposomes when either DOX or ^{3}H-Chol ether are considered Figure 6. It should be noted that the levels of ^{3}H-Chol ether do not show any significant drop even as late as 4 days after injection. The reason is that ^{3}H-Chol ether is in a non-degradable form (ether-bond) and therefore, the ^{3}H-Chol ether values point actually at a cumulative liposome localization in tissues. It is also clear from Figure 6 that the tumor drug levels are by far higher when DOX is delivered by plain liposomes or immunoliposomes as compared to free DOX, pointing at a substantial advantage of liposome delivery with respect to tumor drug exposure.

Since pharmacokinetic studies pointed at a close pattern of in vivo distribution for the plain and immunoliposomes, we proceeded further examining whether the antibody targeting to tumor cells in itself would confer amplified therapeutic efficacy to immunoliposomes. Nude mice bearing subcutaneously implanted N-87 carcinoma were intravenously treated with 2×8 mg/kg (7 days interval) of free and liposomal (plain and immuno) DOX when tumor implants became palpable. As reported previously the relative changes of tumor volume during 60 days of follow up indicate that groups treated with plain (non-targeted) liposomes and immunoliposomes behave similarly with actually a slight advantage for plain liposomes.[32] The final therapeutic results shown in Figure 7 are the median tumor weights after sacrificing the mice two months after start of treatment. There is a significant and unequivocal greater tumor-inhibitory effect for liposome-delivered DOX than for Free DOX. However, there was no apparent difference in tumor weight when immunoliposome and plain liposome treated groups are compared. Addition of unconjugated, soluble, antibody to free DOX or plain liposomes, at doses equal to the amount of antibody given with immunoliposomes ($\sim$100 μg protein/mouse), had no impact on the therapeutic effect (data not shown). Thus, antibody targeting of liposomes did not endow any therapeutic advantage over plain liposomes, nor was there a significant loss of activity.

Despite optimal preparation of antibody-conjugated liposomes with high in vitro affinity to tumor cells and reduced RES uptake, no benefit in tumor targeting over plain liposome was noted. Therapeutic experiments correlated with biodistribution studies, i.e., no improvement in therapeutic efficacy was achieved with immunoliposomes. These studies suggest that the rate limiting factor of liposome accumulation in tumors is the liposome extravasation process, irrespective of liposome affinity or targeting to tumor cells. (But see Chapter 4.7 for contrasting results.)

V. Concluding remarks

Significant advances in liposome development as drug carriers have been accomplished over the last years. The broad versatility of liposome formulations has a strong impact on the pharmacokinetics and pharmacodynamics of liposome encapsulated drugs. The delivery of anthracyclines remains a leading project in the field. Though prevention of drug-induced cardiotoxicity may be profitable, the main issue is improved anti-tumor responses. To accomplish this goal liposome

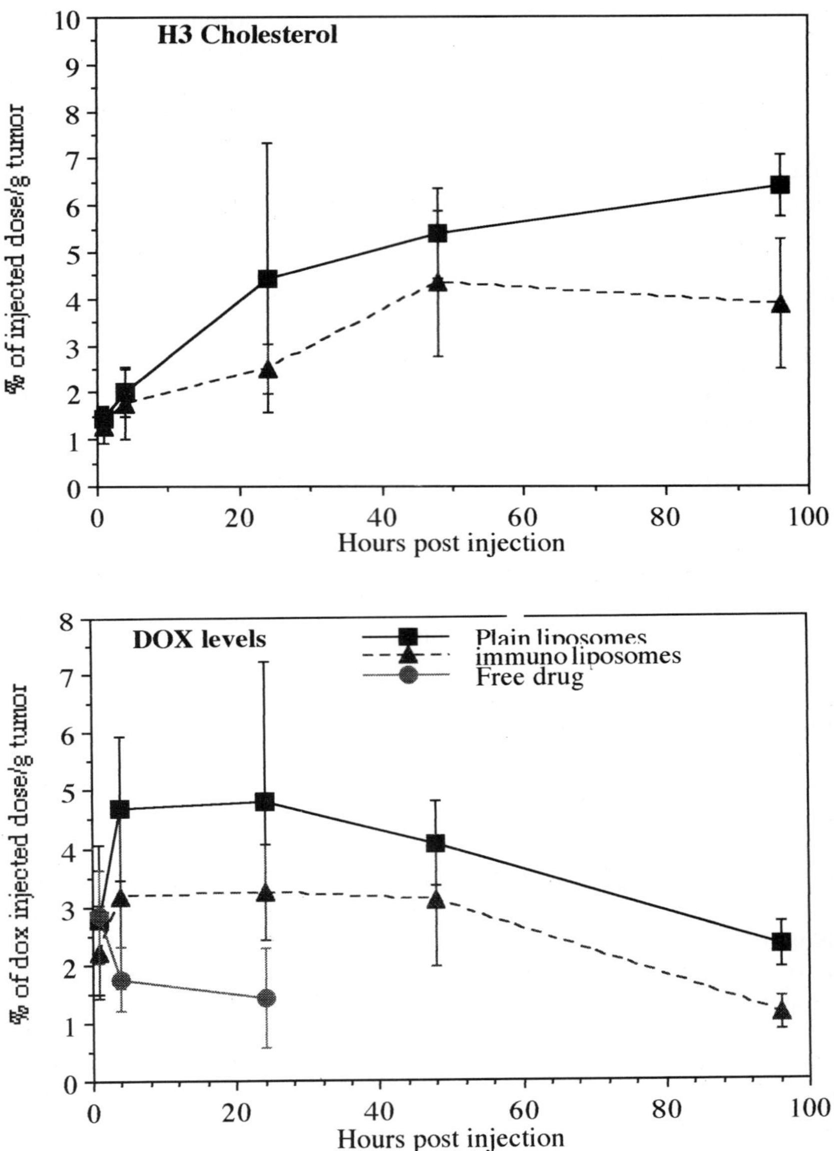

Fig. 6. Tumor accumulation of immunoliposomes and plain liposomes injected i.v. into tumor bearing mice. N-87 cells (6×10^6) were injected into both flanks of nude mice. On day 14 the mice were i.v. injected with 10 mg/kg DOX. ●, free DOX; ■, DOX in plain liposomes; ▲, DOX in immunoliposomes. (Modified from Br J Cancer[32]).

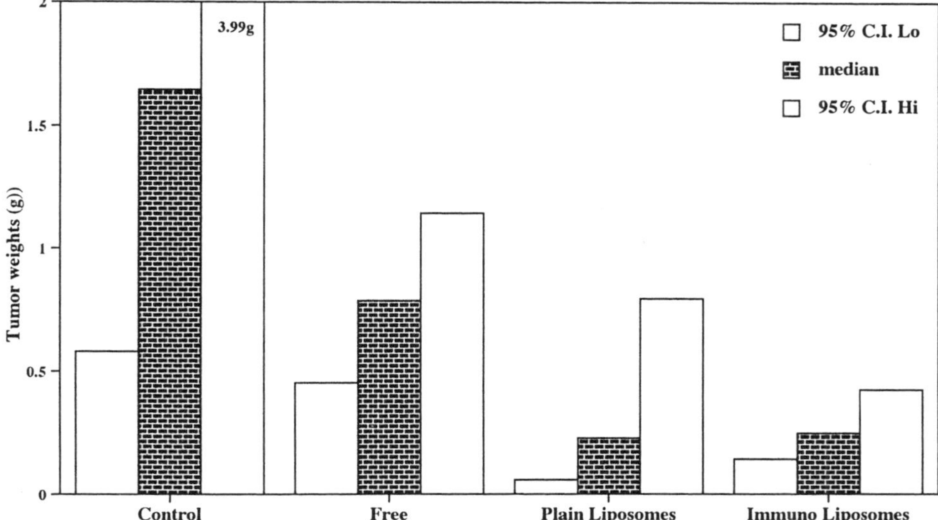

Fig. 7. Therapeutic efficacy of liposomal and free doxorubicin. S.C. N-87 implanted nude mice were sacrificed 75 days post tumor inoculation. i.v. treatments of 8 mg DOX/kg were given on days 15 and 22 after tumor implantation. (Modified from Br J Cancer[32]).

preparations should have increased capacity to localize in tumors. The encouraging clinical notes pointing at liposome accumulation in tumors,[55] suggest that we can expect an improved therapeutic effect in the clinic based on a tumor dose-response effect.

Stable encapsulation of DOX in long circulating liposomes coated with PEG, Stealth liposomes, results in a substantial change of the pharmacological profile of the drug, as compared to the free drug or encapsulated in conventional liposomes. This includes pharmacokinetic changes characterized by prolonged circulation half-life and reduced clearance, changes in tissue distribution including increased deposition in tumor, and slow drug release with delayed and prolonged bioavailability. PEG-HPC-Chol is the formulation with the longest circulation longevity. Stealth liposomes have delayed peak tumor concentrations (>48 h) resulting from prolonged circulation time. As long as plasma concentration of the drug remains high, liposomal drug level continues to build up in the tumor. The extravasation process of small stable vesicles (<100 nm) through tumor hyperpermeable endothelium is followed by drug release in the extracellular fluid with subsequent uptake of free drug by tumor cells. Data point at a process of extracellular drug release from liposomes, taken place mainly at the tumor interstitial fluid, where liposome vesicles may be trapped owing to lack of lymphatic drainage, whereas evidence for direct uptake of liposomes by tumor cells remains absent. To summarize:

- Long circulating PEG-coated liposomes as carriers of anthracyclines show superior therapeutic efficacy as compared to free DOX and conventional liposomes.

- Circulation longevity is an important factor in terms of anti-tumor activity but changes in pharmacokinetics have to be beyond a certain threshold to confer advantageous therapeutic efficacy.
- Anti-tumor activity and bilayer fluidity: High T_m of PC confers greater stability and therapeutic advantage for encapsulated anthracyclines. This is consistent with stability as prerequisite for circulation longevity and effective drug localization in tumors.
- Active targeting of liposomes mediated by a coupled ligand specific for tumor cells can in principle offer improved delivery of drug to target cells, aiming at enhanced cytotoxic efficacy. However, the rate limiting step of accumulation in tumors, remains the extravasation step.

References

1. Gregoriadis G (eds). Liposomes as drug carriers: Trends and progress. London: Wiley, 1988.
2. Mayhew E, Papahadjopoulos D. Therapeutic application of liposomes. In: Ostro MJ, ed. Liposomes. New York: Marcel Dekker, 1983;289–341.
3. Young RC, Ozols RF, Myers CE. The anthracycline anti-neoplastic drugs. N Engl J Med 1981;305:139–153.
4. Minow RA, Benjamin RS, Gottlieb JA. Adriamycin cardiomyopathy; an overview with determination of risk factors. Cancer Chemother Rep 1975;6:195–20.
5. Legha SS, Benjamin RS, Mackay B et al. Reduction of doxorubicin cardiotoxicity by prolonged continuous infusion. Ann Intern Med 1982;96:133–139.
6. Gabizon A, Dagan A, Goren D, Barenholz Y, Fuks Z. Liposomes as in vivo carriers of adriamycin: Reduced cardiac uptake and preserved antitumor activity in mice. Cancer Res 1982;42:4734–4739.
7. Forssen EA, Tokes ZA. Use of anionic liposomes for the reduction of chronic doxorubicin-induced cardiotoxicity. Proc Natl Acad Sci 1981;78:1873–1877.
8. Forssen EA, Tokes ZA. Improved therapeutic benefits of doxorubicin by entrapment in anionic liposomes. Cancer Res 1983;43:546–550.
9. Olson F, Mayhew E, Maslow D, Rustum Y, Szoka F. Characterization, toxicity and therapeutic efficacy of adriamycin encapsulated in liposomes. Eur J Cancer Clin Oncol 1982;18:167–176.
10. Van Hoesel QG, Steerenberg PA, Crommelin DJ, Van Dijk A, Van Oort W, Klein S, Douze JM, de Wildt DJ, Hillen FC. Reduced cardiotoxicity and nephrotoxicity with preservation of antitumor activity of doxorubicin entrapped in stable liposomes in the Lou/M Wsl Rat. Cancer Res 1984;44:3698–3705.
11. Rahman A, White G, More N, Schein PS. Pharmacological, toxocological and therapeutic evaluation in mice of doxorubicin entrapped in cardiolipin liposomes. Cancer Res 1985;45:769–803.
12. Gabizon A, Goren D, Ramu A, Barenholz Y. Design, characterization and anti-tumor acticity of adriamycin containing phospholipid vesicles, In: Gregoriadis G, Senior J, Poste G, eds. Targeting of drugs with synthetic systems. London: Plenum, 1986;229–238.
13. Gabizon A, Goren D, Fuks Z, Meshorer A, Barenholz Y. Superior therapeutic activity of liposome associated adriamycin in a murine metastatic tumor model. Br J Cancer 1985;51:681–689.
14. Gabizon A, Meshorer A, Barenholz Y. Comparative long term study of the toxicities of free and liposomes associated doxorubicin in mice after intravenous administration, J Natl Cancer Inst 1986;77:459–469.
15. Gabizon A, Goren D, Fuks Z, Barenholz Y, Dagan A, Meshorer A. Enhancement of adriamycin delivery to liver metastatic cells with increased tumoricidal effect using liposomes as drug carriers. Cancer Res 1983;43:4730–4735.
16. Rahman A, Fumagali A, Barbieri B, Schein PS, Casazza AM. Anti-tumor and toxicity evaluation of free doxorubicin and doxorubicin entrapped in cardiolipin liposomes. Cancer Chemother Pharmacol 1986;16:22–27.

17. Gabizon A, Barenholz Y. Adriamycin-containing liposomes in cancer chemotherapy. In: Gregoriadis G, ed. Liposomes as drug carriers: Trends and progress. London: Wiley, 1988;365–379.
18. Goren D, Gabizon A, Barenholz Y. The influence of physical characteristics of liposomes containing doxorubicin on their pharmacological behavior. Biochim Biophys Acta 1990;1029:285–294.
19. Torchilin VP. Long circulating drug delivery systems. Adv Drug Deliv Rev 1995;16:125–348.
20. Lasic D, Martin F (eds). Stealth liposomes (Pharmacology and Toxicology series). Boca Raton, FL: CRC Press, 1995.
21. Haran G, Cohen R, Bar LK, Barenholz Y. Transmembrane ammonium sulfate gradients in liposomes produce efficient and stable entrapment of amphipathic weak bases. Biochim Biophys Acta 1993;1151:201–215.
22. Mayer LD, Tai LC, Bally MB, Mitilens GN, Ginsberg, RS, Cullis PR. Characterization of liposomal systems containing doxorubicin entrapped in response to pH gardients. Biochim Biophys Acta 1990;1025:143–151.
23. Lasic DD, Fredrik PM, Stuart MC, Barenholz Y, McIntosh TJ. Gelation of liposome interior a novel method for drug encapsulation. FEBS Lett 1992;312:255–258.
24. Gabizon A, Barenholz Y, Bialer M. Prolongation of the circulation time of doxorubicin encapsulated in liposomes containing a polyethylene glycol-derivatized phospholipid: Pharmacokinetic studies in rodents and dogs. Pharm Res 1993;10:703–708.
25. Gabizon A, Pappo O, Goren D, Chemla M, Tzemach D, Horowitz AT. Preclinical studies with doxorubicin encapsulated in a polyethylene glycol-coated liposomes. J Liposome Res 1993;3:517–528.
26. Gabizon A, Goren D, Horowitz AT, Tzemach D, Lossos A, Siegal T. Long-circulating liposomes for drug delivery in cancer therapy: a review of biodistribution studies in tumor-bearing animals. Adv Drug Deliv Rev 1997;24:337–344.
27. Gabizon A, Chemla M, Tzemach D, Horowitz AT, Goren D. Liposome longevity and stability in circulation: Effects on the in vivo delivery to tumors and therapeutic efficacy of encapsulated anthracyclines. J Drug Targ 1996;3:391–398.
28. Gabizon A. Selective tumor localization and improved therapeutic index of anthracyclines encapsulated in long-circulating liposomes. Cancer Res 1992;52:891–896.
29. Vaage J, Mayhew E, Lasic D, Martin FJ. Therapy of primary and metastatic mouse mammary carcinomas with doxorubicin encapsulated in long-circulating liposomes. Int J Cancer 1992;51:942–948.
30. Williams SS, Alosco TR, Mayhew E, Lasic DD, Martin FJ, Bankert RB. Arrest of human lung tumor xenograft growth in severe combined immunodeficient mice using doxorubicin encapsulated in sterically stabilized liposomes. Cancer Res 1993;53:3964–3967.
31. Vaage J, Donovan D, Mayhew E, Abra R, Huang A. Therapy of human ovarian carcinoma xenografts using doxorubicin encapsulated in sterically stabilized liposomes. Cancer 1993;72:3671–3675.
32. Goren D, Horowitz AT, Zalipsky S, Woodle MC, Yarden Y, Gabizon A. Targeting of Stealth liposomes to erbB2 (Her/2) receptor: in vitro and in vivo studies. Br J Cancer 1996;74:1749–1756.
33. Papahadjopoulos D, Allen TM, Gabizon A, Mayhew E, Matthay K, Huang SK, Woodle MC, Lasic DD, Redemann C, Martin FJ. Sterically stabilized liposomes: Improvements in pharmacokinetics and anti tumor therapeutic efficacy. Proc Natl Acad Sci USA 1991;88:11460–11464.
34. Gabizon A. Stealth liposomes and cancer targeting: a realistic compromise in drug delivery. J Liposome Res 1995;5:705–710.
35. Siegal T, Horowitz AT, Gabizon A. Doxorubicin encapsulated in sterically stabilized liposomes for the treatment of a brain tumor model: biodistribution and therapeutic efficacy. J Neurosurg 1995;83:1029–1037.
36. Gabizon A, Papahadjopoulos D. Liposome formulations with prolonged circulation time in blood and enhanced uptake by tumors. Proc Natl Acad Sci USA 1988;85:6949–6953.
37. Gabizon A, Price DC, Huberty J, Bresalier RS, Papahadjopoulos D. Effect of liposome composition and other factors on the targeting of liposomes to experimental tumors: Biodistribution and imaging studies. Cancer Res 1990;50:6371–6378.
38. Hwang KJ, Luk KK, Beaumier PL. Hepatic uptake and degradation of unilamellar sphingomyelin/cholesterol liposomes: A kinetic study. Proc Natl Acad Sci USA 1980;77:4030–4034.
39. Peterson HI (ed). Tumor blood circulation, Angiogenesis, vascular morphology and blood flow of experimental and human tumors. CRC Press, Boca Raton, 1979.
40. Jain RK. Delivery of novel therapeutic agents in tumors:physiological barriers and strategies. J Natl Cancer Inst 1989;81:570–576.

41. Ahmad I.Allen TM. Antibody mediated specific binding and cytotoxicity of liposome entrapped doxorubicin to lung cancer cells in vitro. Cancer Res 1992;52:4817–4820.
42. Debs RJ, Heath TD, Papahadjopoulos D. Targeting of anti-thy 1.1 monoclonal antibody conjugated liposomes in Thy 1.1 mice after intravenous administration. Biochim Biophys Acta 1987;901:183–190.
43. Peeters PA, Brunink BG, Eling WM, Crommelin DJ. Therapeutic effect of chloroquine (CQ)-containing immunoliposomes in rats infected with *Plasmodium berghe*: parasitized mouse red blood cells: comparison with combination of antibodies and CQ or liposomal CQ. Biochim Biophys Acta 1989;981:269–276.
44. Huang SK, Hong K, Lee KD, Papahadjopoulos D, Friend, DS. Light microscopic localization of silver-enhanced liposome entrapped colloidal gold in mouse tissues. Biochim Biophys Acta 1991;1069:117–121.
45. Berchuck A, Kamel A, Whitaker R, Kerns B, Olt G, Kinney R, Soper JT, Dodge R, Clarke-Pearson DL, Marks P. Overexpression of HER-2/neu is associated with poor survival in advanced epithelial ovarian cancer. Cancer Res 1990;50:4087–4091.
46. Yonemura Y, Ninomiya I, YamaguchI A, Fushida S, Kimura H, Ohoyama S, Miyazaki I, Endou Y, Tanaka M, Sasaki T. Evaluation of immunoreactivity for erbB2 protein as a marker of poor short term prognosis in gastric cancer. Cancer Res 1991;51:1034–1038.
47. Slamon DJ, Clark GM, Wong SG, Levin WJ, Ullrich A, Mcguire WL. Human breast cancer: Correlation of relapse and survival with amplification of the HER-2 neu oncogene. Science 1987;235:177–182.
48. Slamon DJ, Godolphin W, Jones LA, Holt JA, Wong SG, Keith DE, Levin WJ, Stuart SG, Udove J, Ullrich A, Press MF. Studies of the HER-2/neu protooncogene in human breast and ovarian cancer. Science 1989;244:707–712.
49. Park JW, Stagg R, Lewis GD, Carter P, Maneval D, Slamon DJ, Jaffe H, Shepard HM. Advances in Cellular and Molecular Biology of Breast Cancer. In: Dickson RB, Lippman ME, eds. Genes, Oncogenes, Hormones. Boston: Kluwer Academic Publishing, 1992;193–211.
50. Press MF, Cordon-Cardo C, Slamon DJ. Expression of the HER-2/neu proto-oncogene in normal human adults and fetal tissues. Oncogene 1990;5:953–962.
51. Stancovski I, Hurwitz E, Leitner O, Ullrich A, Yarden Y, Sela M. Mechanistic aspects of the opposing effects of monoclonal antibodies to the erbB-2 receptor on tumor growth. Proc Natl Acad Sci USA 1991;88:8691–8695.
52. Hurwitz E, Stancovski I, Sela M, Yarden Y. Suppression and promotion of tumor growth by monoclonal antibodies to ErbB-2 differentially correlate with cellular uptake. Proc Natl Acad Sci USA 1995;92:3353–3357.
53. Park JG, Frucht H, Larocca RV, Bliss DPJ, Kurita Y, Chen TR, Henslee JG, Trepel JB, Jensen RT, Johnson BE, Bang YJ, Kim JP, Gazdar AF. Characteristics of cell lines established from human gastric carcinoma. Cancer Res 1990;50:2773–2780.
54. Allen TM, Agrawal AK, Ahmad I, Hansen CB, Zalipsky S. Antibody mediated targeting of long circulating (Stealth) liposomes. J. Liposome Res 1994;4:1–25.
55. Uziely B, Jeffers S, Isacson R, Kutsch K, Wei-Tsao D, Yehoshua Z, Muggia FM, Gabizon A. Liposomal doxorubicin: antitumor activity and unique toxicities during two complementary phase I studies. J Clin Oncol 1995;13:1777–1785.

Lasic and Papahadjopoulos (eds.), Medical Applications of Liposomes

CHAPTER 4.4

Cellular distribution of DOXIL® within selected tissues, assessed by confocal laser scanning microscopy

JAN VAAGE,[a] DOROTHY DONOVAN,[a] PETER WORKING[b] AND PAUL USTER[b]

[a]*Department of Molecular Immunology, Roswell Park Cancer Institute, Buffalo, NY 14263, USA;*
[b]*SEQUUS™ Pharmaceuticals, Inc., 960 Hamilton Court, Menlo Park, CA 94025–1430, USA*

Overview

Confocal laser scanning (CLS) microscopy is a research tool well suited for studying the spatial and temporal distribution of fluorescent compounds in cultured cells in vitro. In this chapter we highlight some of our recent CLS microscopic work probing the in vivo cellular distribution of pegylated liposomal doxorubicin (DOXIL®*). These studies suggest that CLS microscopic observation of drug uptake and elimination in tissues provides levels of information not available from a chemical analysis of homogenized tissue.

I. CLS microscopic observations of pegylated liposome uptake in liver, spleen, and kidney

The liver and spleen are the two major organs involved in the removal of sterically-stabilized STEALTH® liposomes from the blood after intravenous injection.[1–3] Because of the fenestrated structure of the endothelium of the hepatic portal system, the liver is the major site of the removal of liposomes and free drug from the blood.[2]

We have observed DOXIL uptake by parenchymal and Kupffer cells of the

*STEALTH liposomes and DOXIL are registered trademarks of SEQUUS Pharmaceuticals, Inc.

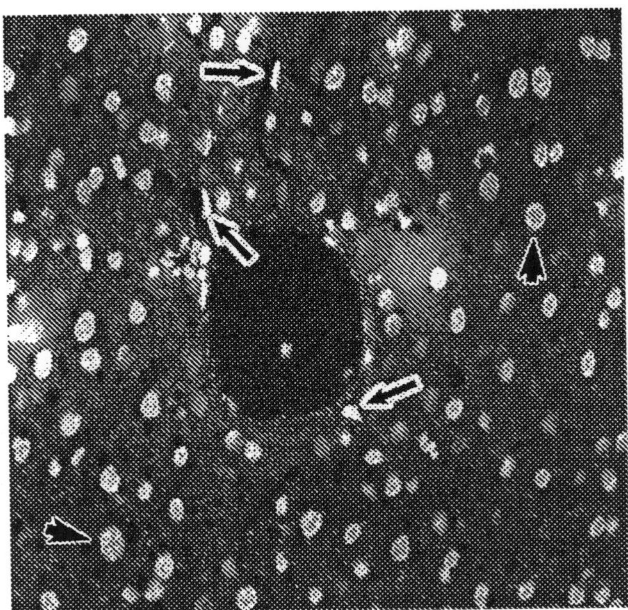

Fig. 1. Scanning laser microscope fluorescence image of doxorubicin uptake by hepatocytes (arrowheads point to typical nuclei) and Kupffer cells (arrows point to typical nuclei) 24 hr post injection of 3 mg/kg doxorubicin in STEALTH liposomes (DOXIL).

Liver (Figure 1). The Kupffer cells, being relatively fewer than the parenchymal in number, take up proportionately less of the dose of STEALTH liposomes and free drug. DOXIL localized primarily in the small fenestrated endothelium close to portal veins at one hour. At 24 hours, DOXIL had continued to accumulate, and was most concentrated in periportal areas. The uptake of free doxorubicin by the liver was at one hour almost entirely through the large fenestrated endothelium close to central veins. At 24 hours, most of the free doxorubicin was cleared from the liver.

Most of the liposomes and free drug found in the spleen localized in red pulp macrophages and dendritic reticular cells. In lymph nodes, dendritic reticular cells and macrophages in the cortex and medulla take up a significant quantity of liposomes and drug.

In the kidneys, DOXIL was at one hour more concentrated in the filtering renal tubules than was free doxorubicin, and excreted more slowly. We measured doxorubicin in the urine by the simple expedient of placing each individual mouse on absorbent paper after the tail vein injection. By noting the time of urination and examining the spot under u.v. light, the relative rates of drug elimination were estimated. We detected doxorubicin in the urine as early as 15 minutes after injection and was detectable for up to 48 hours. Doxorubicin released from DOXIL was first detected after 50 minutes and was detectable for up to five days.

II. CLS microscopic observations of pegylated liposome uptake in tumor

The therapeutic usefulness of pegylated STEALTH liposomes as an intravenous delivery system for anti-cancer drugs has been demonstrated in many implanted animal tumor models,[2–11] spontaneously-arising canine tumors[12] and in clinical trials.[13,14] However, the mechanisms accounting for the improved efficacy of DOXIL® are not yet well understood. Wu and coworkers found that the STEALTH formulation was taken up by the tumor in greater quantity and persisted in the tumor and in the circulation longer than rapidly cleared liposome formulations.[15] Huang et al. injected pegylated liposomes containing colloidal gold, and observed the particulate label in spaces between perivascular cells, but not intracellularly.[1] Recent CLS microscopy work from our laboratories injected placebo STEALTH liposomes labeled with a fluorescent phospholipid.[10] Within hours after injection, the marker distributed to stroma and tumor cell plasma membranes and cytoplasm, but not the nuclear membrane and nucleus.

The comparative accumulation of unentrapped ("free") doxorubicin and of DOXIL was studied with CLS microscopy by exploiting doxorubicin's fluorescent properties. Fluorescence and transmitted light micrographs were made of cryostat sections from xenografted human prostate carcinoma PC-3, normal liver, and normal kidney. Figures 2 through 5 show CLS micrographs of the distribution of free doxorubicin and DOXIL in human prostate carcinoma implants removed 1 and 24 hours after the i.v. injection of 3 mg/kg drug.[2] DOXIL and free doxorubicin were found around the capillaries and venules of tumors, but not around the arterioles. Intracellularly, the drug was primarily located in the nuclei of stroma and tumor cells. The amount of free doxorubicin in the tumor was greater at 1 hour, but drug levels in the tumor at 24 hours were greater with DOXIL than free doxorubicin. While DOXIL continued to increase in tumors over the first day or two, free doxorubicin was barely detectable in the tumors 24 hours post-injection.

Figure 6 shows a section of the human pancreatic carcinoma AsPC-1, removed two hours after the i.v. injection of 3 mg/kg DOXIL. The image contains a venule with several leukocytes, all of which have taken up the drug into the nuclei. Doxorubicin was found to persist in blood leukocytes for up to 48 hours after the i.v. injection of the free drug. After the injection of DOXIL, doxorubicin was visible in blood leukocytes for up to five days.

The relatively long persistence of doxorubicin in the stroma cells and tumor cells after administration in STEALTH liposomes raised the question of whether the loaded liposomes or the drug payload alone entered the cells. To study this question, placebo liposomes with a fluorescent red phospholipid were prepared.

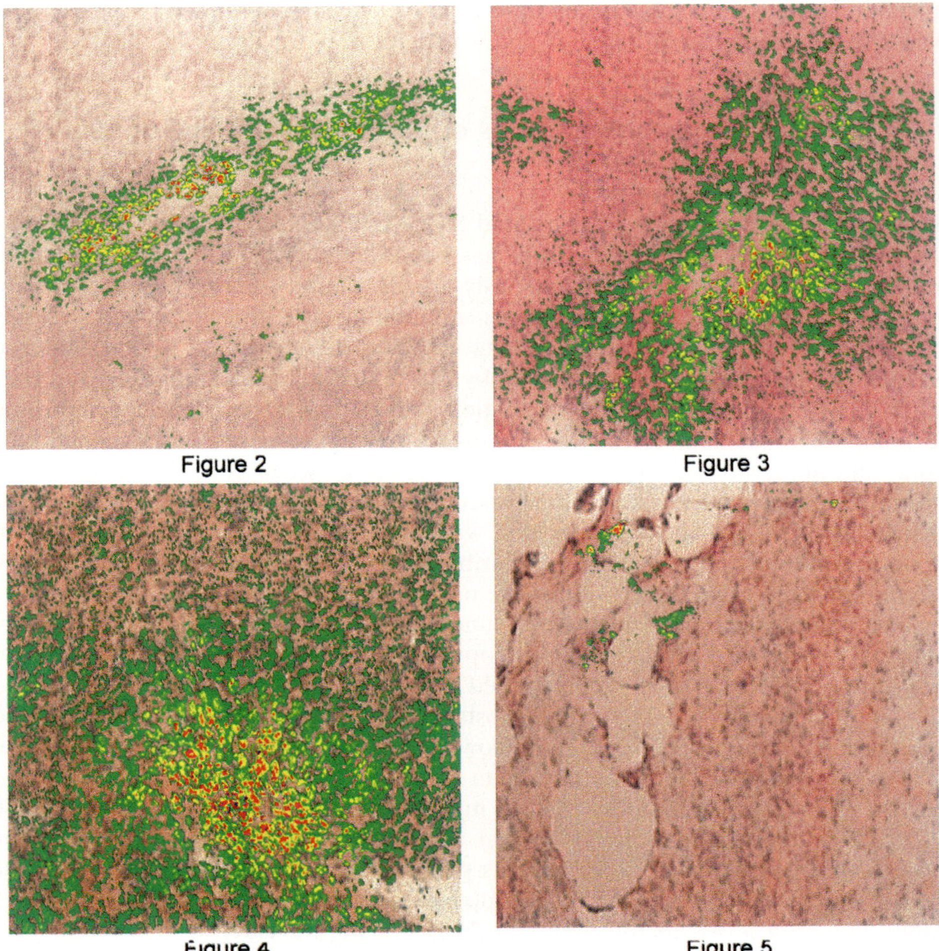

Figs. 2–5. Scanning laser microscope images of subcutaneous implants of PC-3 showing the uptake of doxorubicin in STEALTH liposomes (DOXIL) at 1 hour (Figure 2) and at 24 hours (Figure 3), and of free doxorubicin in saline at 1 hour (Figure 4) and at 24 hours (Figure 5). Laser scan images of doxorubicin have been overlaid on video images of the same tissue sections stained with H and E. The drug appears as green, yellow, and red, in a color scale of increasing concentrations.

Figure 7 shows that, twenty-four hours after the injection with the same lipid dose as in a 3 mg/kg dose of DOXIL, the lipid label fluorescence localized only in the plasma membrane and cytoplasm of stroma and tumor cells of the AsPC-1 xenograft. This location of liposome fluorescent lipid label was consistently observed in other tumors and in all normal tissues examined such as liver, lung, spleen, and lymph nodes.

The doxorubicin content per gram wet weight of tumor was determined by CLS

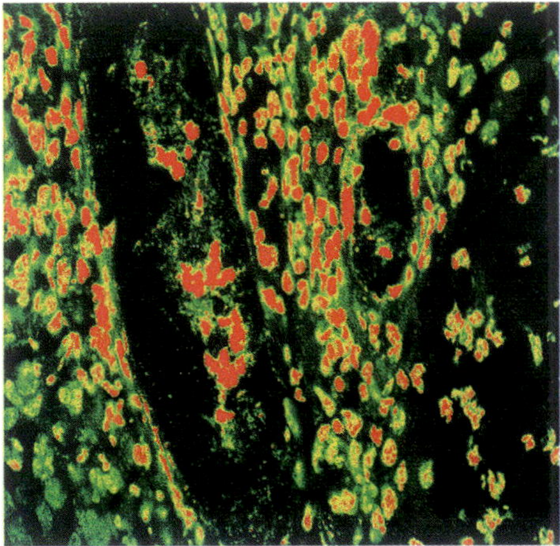

Fig. 6. Uptake of Doxorubicin by blood leukocytes and by tumor cells two hours after the intravenous injection of 3 mg/kg DOXIL.

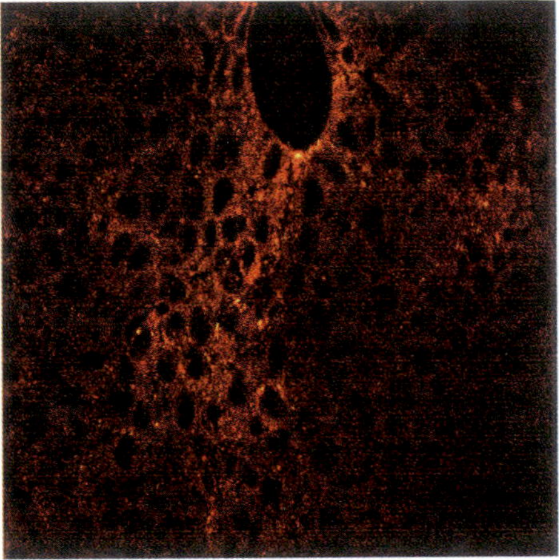

Fig. 7. Uptake of drug-free, Texas Red-labeled liposomes by an AsPC-1 implant removed 24 hours after the intravenous injection of the liposomes.

after the i.v. administration of DOXIL or free doxorubicin. The quantitation used standard curves of the fluorescence intensities of serial dilutions of DOXIL and free doxorubicin, both in agar gel. The auto-quenching factor for doxorubicin in STEALTH liposomes was 2.8 ± 0.15 for all fluorescence intensity levels. Drug fluorescence measurements were made on cryostat sections of human pancreatic carcinoma AsPC-1 xenografts removed at specified times after the i.v. injection

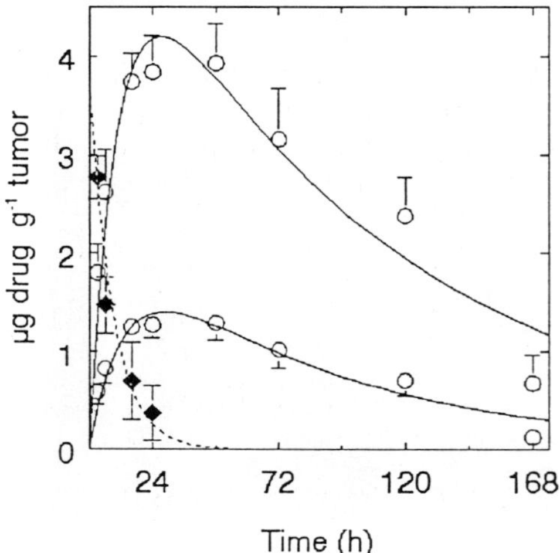

Fig. 8. Quantitation by micro-fluorimetry of free doxorubicin in saline (F-Dox, ◆) and doxorubicin encapsulated in polyethylene-glycol coated liposomes (DOXIL, ○) in subcutaneous implants of AsPC-1. The high-low spread in values for DOXIL reflects adjustments for the fluorescence auto-quenching of encapsulated doxorubicin assuming that all doxorubicin was released (low limit) or that all of the doxorubicin was encapsulated (high limit). The auto-quenching factor for encapsulated doxorubicin is 2.8. Each mouse received 3.0 mg/kg drug intravenously at 0 hour. (Data from Ref. 10).

of 3 mg/kg "free" doxorubicin or in STEALTH liposomes. Figure 8 shows the quantities of free doxorubicin and liposomal doxorubicin in the tumors. Free doxorubicin was detectable for only 24 hours, but doxorubicin in STEALTH liposomes was detectable for 168 hours. The value for the area under the curve (AUC) was 29 for free doxorubicin. The calculated AUC for liposomal doxorubicin gave a low limit of 165 (assuming that all of the drug had been released from the liposomes, with no adjustment for auto-quenching used in the calculations). The calculated high limit was 462 (assuming that all of the drug was encapsulated, and using the auto-quenching factor 2.8 in the calculations). The actual proportion of encapsulated doxorubicin was probably highest while the drug was accumulating in the tumor, and very low 168 hours after injection. This means that the STEALTH formulation had produced a 6-fold or greater increase in the AUC.[10]

The traditional view in vascular physiology is that exchanges across endothelial surfaces occur through openings at intercellular junctions, openings that vary with physiological and pathological circumstances. More recently, the mechanism of plasmalemmal vesicle transport through the endothelial cell cytoplasm (transcytosis) has had experimental support.[16–19] The plasmalemmal vesicles form at the luminal endothelial cell surface and discharge their content at the abluminal surface. The fusion of vesicles to form transitory trans-endothelial channels for the passage of macromolecules has also been suggested.[19] Accumulation of STEALTH liposomes is likely due to some combination of these mechanisms.

It is likely that the prolonged circulation time of STEALTH liposomes[20] makes it possible for more drug to enter a tumor. This, and the slow drug release from liposomes inside the tumor, are two reasons for the increased therapeutic efficacy of DOXIL. From the intra-tumor location, the slow release of doxorubicin from STEALTH liposomes would maintain effective intracellular and extra-cellular cytotoxic levels[21] for prolonged periods.

References

1. Huang SK, Lee K-D, Hong K, Friend DS, Papahadjopoulos D. Microscopic localization of sterically stabilized liposomes in colon carcinoma-bearing mice. Cancer Res 1992;52:5135–5143.
2. Vaage J, Barbera-Guillem E, Abra R, Huang A, Working P. Tissue distribution and therapeutic effect of intravenous free or encapsulated liposomal doxorubicin on human prostate carcinoma xenografts. Cancer 1994;73:1478–1484.
3. Senior JH. Fate and behavior of liposomes in vivo: a review of controlling factors. CRC Crit Rev Ther Drug Carrier Systems 1988;3:123–193.
4. Papahadjopoulos D, Allen TM, Gabizon A, Mayhew E, Matthay K, Huang SK, Lee K-D, Woodle MC, Lasic DD, Redemann C, Martin FJ. Sterically stabilized liposomes: Improvements in pharmacokinetics and antitumor therapeutic efficacy. Proc Natl Acad Sci, USA 1991;88:11460–11464.
5. Vaage J, Donovan D, Mayhew M, Uster P, Woodle M. Therapy of mouse mammary carcinomas with vincristine and doxorubicin encapsulated in sterically stabilized liposomes. Int J Cancer 1993;54:959–964.
6. Allen TM, Newman MS, Woodle MC, Mayhew E, Uster PS. Pharmacokinetics and anti-tumor activity of vincristine encapsulated in sterically stabilizaed liposomes. Int J Cancer 1995;62:199–204.
7. Siegal T, Horowitz A, Gabizon A. Doxorubicin encapsulated in sterically stabilized liposomes for the treatment of a brain tumor model: biodistribution and therapeutic efficacy. J Neurosurg 1995;83:1029–1037.
8. Vaage J, Donovan D, Loftus T, Uster P, Working P. Prophylaxis and therapy of mouse mammary carcinomas with doxorubicin and vincristine encapsulated in sterically stabilized liposomes. Eur J Cancer 1995;3:367–372.
9. Zhu G, Oto E, Vaage J, Quinn Y, Newman M, Engbers C, Uster P. The effect of vincristine-polyanion complexes in STEALTH liposomes on pharmacokinetics, toxicity and anti tumor activity. Cancer Chemother Pharmacol 1996;39:138–142.
10. Vaage J, Donovan D, Uster P, Working P. Tumor uptake of doxorubicin in polyethylene-glycol coated liposomes and therapeutic effect against a xenografted human pancreatic carcinoma. Br J Cancer 1997;75(4):482–48.
11. Working PK. Nonclinical Studies of Lipid-Complexed and Liposomal Drugs: AMPHOTEC®, DOXIL® and SPI-77. In: Lasic DD and Papahadjopoulos D, eds. Medical Applications of Liposomes. Amsterdam, New York: Elsevier, 1998;605–624.
12. Vail DM, Kravis LD, Cooley AJ, Chun R, MacEwan EG. Preclinical trial of doxorubicin entrapped in sterically stabilized liposomes in dogs with spontaneously arising malignant tumors. Cancer Chemother Pharmacol 1997;39:410–416.
13. Harrison M, Tomlinson D, Stewart S. Liposomal-entrapped doxorubicin: An active agent in AIDS-related Kaposi's sarcoma. J Clin Oncol 1995;13:914–920.
14. Uziely B, Jeffers S, Isacson R, Kutch K, Wei-Tsao D, Jehosua Z, Libson E, Muggia F, Gabizon A. Liposomal doxorubicin: Antitumor activity and unique toxicities during two complementary phase I studies. J Clin Oncol 1995;13:1777–1785.
15. Wu NZ, Da D, Rudoll TL, Needham D, Whorton R, Dewhirst MW. Increased microvascular permeability contributes to preferential accumulation of STEALTH liposomes in tumor tissue. Cancer Res 1993;53:3765–3770.
16. Audus KL, Raub TJ. Lysosomes of brain and other vascular endothelia. In: Pardridge WM, ed. The Blood-Brain-Barrier. New York: Raven Press, 1993;201–227.
17. Predescu D, Palade GE. Plasmalemmal vesicles represent the large pore system of continuous microvascular endothelium. Amer J Physiol 1993;265:H725–733.
18. Broadwell RD, Banks WA. Cell biological perspective for the transcytosis of peptides and proteins

through the mammalian blood-brain fluid barriers. In: Pardridge WM, ed. The Blood-Brain-Barrier. New York: Raven Press, 1993;165–199.

19. Simionescu M, Ghitescu L, Fixman A. How plasma macromolecules cross the endothelium. News Physiol Sci 1987;2:97–100.
20. Allen TM, Hansen C, Martin FJ, Redemann C, Yau-Young A. Liposomes containing a synthetic lipid derivative of polyethylene glycol show prolonged circulation half-lives in vivo. Biochim Biophys Acta 1991;1066:29–36.
21. Vichi P, Tritton TR. Adriamycin: Protection from cell death by removal of extracellular drug. Cancer Res 1992;52:4135–4138.

Lasic and Papahadjopoulos (eds.), Medical Applications of Liposomes

CHAPTER 4.5

Liposomes as carriers of lipophilic antitumor agents

Roman Perez-Soler and Yiyu Zou

Section of Experimental Therapy, Department of Thoracic/Head and Neck Medical Oncology, M.D. Anderson Cancer Center, Houston, TX, USA

Overview

I. Introduction

There is a compelling rationale for using liposomes as carriers of lipophilic antitumor agents. First, many of the most active antitumor agents within the different families of compounds with established antitumor activity have a low water solubility. Therefore, an appropriate dosage form for intravenous administration is needed. Some of these compounds may be incorporated within the lipid bilayers of liposomes thus providing an appropriate dosage form for intravenous administration that in addition alters their pharmacokinetics and organ distribution to suit better certain therapeutic indications. Multilamellar liposomes (0.5–5 μm) are a better choice than unilamellar vesicles (0.02–0.5 μm) as carriers of these drugs since they consist of solid concentric phospholipid bilayers. There are two rational and potential, although not validated, applications of multilamellar liposomes as carriers of lipophilic antitumor agents. One is the treatment of microscopic liver tumors that get their blood supply from the fenestrated liver capillaries and the other intracavitary therapy, because of the prolonged retention of the vesicles within the cavity. Second, many lipophilic compounds tend to bind to serum lipoproteins and other proteins, which may alter their metabolism and affect their antitumor activity and toxicity. When incorporated in liposomes, such compounds may be partially protected from enzymatic degradation by plasma proteins and reach the tumor tissue in their active form. In some cases, this can be exploited to enhance the therapeutic index of the agents. Finally, many chemotherapeutic agents are more effective when delivered in a continuous infusion rather than bolus because they are cytotoxic only when the cells are at a particular phase of the cell cycle. Multilamellar liposomes may provide a continuous and prolonged transfer of the drugs incorporated within their phospholipid bilayers and consequently enhance the cytotoxicity of S-phase specific agents.

II. Platinum compounds

Cisplatin is one of the most effective antitumor agents in the treatment of testicular, ovarian, head and neck, and lung cancer.[1] However, its use is limited by significant side effects, such as nausea and vomiting, nephrotoxicity, and chronic peripheral neuropathy. Our laboratory has pioneered for more than a decade the use of multilamellar liposomes for the delivery of non-cross resistant lipophilic platinum compounds. Our interest in these compounds arose from the realization that cisplatin can not be effectively incorporated or encapsulated in liposomes and the need to develop agents with activity against cisplatin-resistant tumors.

NDDP, a diaminocyclohexane platinum analog, formulated in multilamellar vesicles composed of DMPC and DMPG at a 7:3 molar ratio (L-NDDP), was developed and introduced in clinical trials in 1989 (see Figure 1 for chemical structure). It is currently being evaluated in a Phase II study by intrapleural administration in patients with malignant pleural mesothelioma and studies by intraperitoneal administration in patients with ovarian carcinoma are being planned.

NDDP was selected because of its lack of nephrotoxicity, lack of cross-resistance with cisplatin, and easiness of incorporation in multilamellar vesicles.[2–4] We have now demonstrated that NDDP is an inactive prodrug that is transformed into an active platinum species within the liposomes, but only when the acidic phospholipid DMPG is a component of the lipid bilayers.[5] DACH-Pt-Cl_2 has been recently identified as the active platinum species (Figure 1).[6] Interestingly, attempts to incorporate DACH-Pt-Cl_2 in liposomes had been unsuccessful in several laboratories because of its lack of solubility in commonly used organic solvents. DACH-Pt-Cl_2 is the first example to our knowledge of intraliposomal synthesis of an active antitumor agent from an inactive prodrug incorporated within the lipid bilayers. In all other examples of liposomal-prodrugs of antitumor agents, activation of the prodrug occurs after liposome destruction. The identification of DACH-Pt-Cl_2 as the active species of NDDP has opened the door for definitely resolving the formulation and characterization problems that have delayed the clinical development of L-NDDP.

After intravenous administration, NDDP incorporated in multilamellar liposomes distributes preferentially into the liver and spleen. In studies in New Zealand white rabbits, liver drug levels were 5-fold higher with L-NDDP than with cisplatin

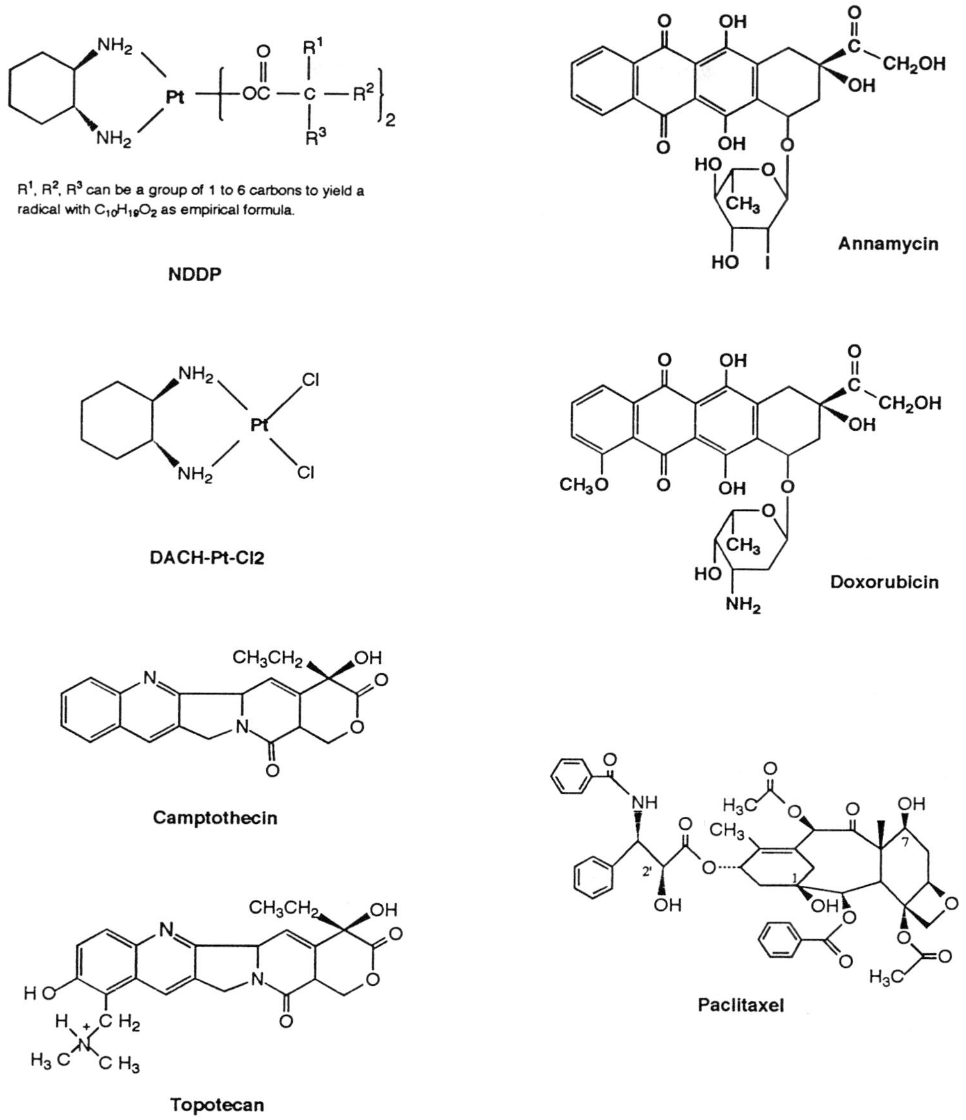

R^1, R^2, R^3 can be a group of 1 to 6 carbons to yield a radical with $C_{10}H_{19}O_2$ as empirical formula.

NDDP

Annamycin

DACH-Pt-Cl2

Doxorubicin

Camptothecin

Paclitaxel

Topotecan

Fig. 1.

when both drugs were administered intravenously.[7] Interestingly, administration into the proper hepatic artery did not increase the liver NDDP levels while it increased those of cisplatin but only by two-fold. Intraarterial administration may not, therefore, offer any additional advantage for the treatment of microscopic liver tumors in the case of L-NDDP. In accordance with these results, L-NDDP was more effective than cisplatin in the treatment of liver micrometastases in mice.[2] In a Phase I clinical study by intravenous administration conducted in 1989, the maximum tolerated dose of L-NDDP was 300 mg/m^2 and the dose limiting toxicity myelosuppression.[8] No nephrotoxicity was observed and nausea and vomiting were reduced compared with that caused by cisplatin.

After intraperitoneral administration, L-NDDP is cleared much more slowly than cisplatin from the peritoneal cavity.[9] This observation led to a clinical trial of L-NDDP by intrapleural administration in patients with malignant pleural effusions. The maximum tolerated dose was 450 mg/m^2 and the dose limiting toxicity was local chemical pleuritis.[10] In contrast with the Phase I study by intravenous administration, where myelosuppression was dose-limiting at 300 mg/m^2, no myelosuppression was observed at a dose of 550 mg/m^2 by the intrapleural route. These findings demonstrate the potentially beneficial depot effect conferred by incorporating the drug into multilamellar liposomes. Disappearance of the malignant pleural effusion for two years was observed in one of five patients with malignant pleural mesothelioma, a disease that grows confined to the pleural cavity and causes death by local growth and for which no standard therapy is available. A Phase II study of L-NDDP by intrapleural administration in patients with malignant pleural mesothelioma is in progress.

The use of intraperitoneal cisplatin has been recently shown to increase by about 10% the cure rate of patients with ovarian carcinoma and minimal residual disease after debulking surgery.[11] This finding is particularly impressive when one takes into consideration that cisplatin is not an ideal agent for intraperitoneal therapy. The peritoneum acts as a semipermeable membrane allowing the passage of small molecules from and into the systemic circulation. For this reason, cisplatin is absorbed very quickly into the systemic circulation from the peritoneal cavity after intraperitoneal administration. Ideal agents for intraperitoneal administration should remain for a long time inside the cavity and display an enhanced ability for tumor penetration. L-NDDP certainly meets the first criteria since its incorporation into the liposomes prevents its free passage into the systemic circulation. In addition, a variety of studies have also shown that it can enter the cells through vesicle-cell membrane fusion, and cross in and out cell membranes much faster than cisplatin because of its increased lipophilicity,[12] thus suggesting an enhanced ability to reach deep cellular layers. A Phase I study of L-NDDP by intraperitoneal administration in patients with ovarian carcinoma refractory to cisplatin is being planned.

III. Anthracyclines

The anthracycline antibiotic doxorubicin is one of the most effective antitumor agents against a variety of malignancies.[1] However, it causes chronic cardiotoxicity which limits its use in patients whose tumors are sensitive to the drug. Numerous investigators have been interested during the last decade in developing liposomal doxorubicin formulations to reduce its cardiotoxicity and enhance drug tumor targeting. For many years, the effort in this area was focused on solving formulation problems centered around the low encapsulation efficiency. This problem was finally solved by developing the so-called active drug loading method based on creating an internal acidic milieu that keeps the drug in its protonized form, thus preventing its efflux across the liposome membranes.[13] Several of these formulations are now in clinical trials and one, in long-circulating (Stealth) liposomes, has been approved for the treatment of Kaposi's sarcoma.[14]

Unfortunately, most tumors that respond to doxorubicin are rarely cured and eventually become resistant to it. The best known mechanism of resistance to doxorubicin and other structurally unrelated antitumor agents is through the overexpression in the cell membrane of proteins that work as drug efflux pumps and decrease the intracellular accumulation of the drug. The best characterized of these proteins are P-glycoprotein (PGP)[15] and the multidrug-resistance-associated protein (MRP).[16] There is now convincing evidence that both overexpression of PGP and MRP are clinically-relevant mechanisms of resistance in acute leukemia, breast carcinoma, osteosarcoma, and neuroblastoma. Inhibitors of PGP including verapamil and cyclosporin A have been shown to sensitize multidrug resistant cells to doxorubicin. Unfortunately, when used in vivo, they block the physiological excretory function of PGP in the liver and kidney, thus altering the drug's pharmacokinetics, increasing their toxicity, and forcing dose reduction.

An alternative way to overcome PGP and MRP mediated multidrug resistance is to design analogs of doxorubicin that are not substrates for PGP and MRP. Extensive structure-activity studies conducted for more than a decade have led to the identification of key modifications in the anthracycline molecule that result in compounds that circumvent PGP and MRP function. More specifically, deamination at position 3′ of the sugar portion was found to result in compounds that were as active against parental and MDR cells.[17] Unfortunately, deaminated compounds are poorly soluble in water. About a decade ago, we started exploring the use of liposome carriers to deliver non-cross resistant anthracycline antibiotics. The purpose was to select a compound incorporating a combination of structural changes that prevent its binding to PGP and that enhance its association with phospholipid bilayers. From screening a group of fifty compounds, we selected Annamycin as the anthracycline fulfilling these criteria (Figure 1).[18]

Liposomal-Annamycin (L-Ann) is now being tested by the intravenous route in a Phase I clinical trial at M.D. Anderson Cancer Center. Phase II studies in patients with refractory breast cancer have begun in mid-1997. The formulation consists of 150 nm oligolamellar vesicles composed of DMPC and DMPG obtained upon hydration of a lyophilized preliposomal powder. In preclinical studies, L-Ann showed remarkable lack of cross-resistance with doxorubicin in a panel of 6 pairs of parental and MDR positive cell lines as well as 2 pairs of parental and MRP positive cell lines.[19–21] In in vivo studies, L-Ann was markedly superior to doxorubicin against liver metastases of M5076 reticulosarcoma and lung tumors of Lewis lung carcinoma. More importantly, L-Ann had significant activity against

KB/MDR human xenografts in nude mice while doxorubicin was ineffective. Toxicity studies in mice treated weekly with equitoxic doses of doxorubicin or L-Ann indicated also that L-Ann did not cause cardiotoxicity while doxorubicin caused cardiac lesions in most animals.[22]

In the about to be completed Phase I study, the MTD of L-Ann appears to be around 200 mg/m^2 and the dose limiting toxicity is myelosuppression. No alopecia, nausea and vomiting nor mucositis have been observed. A Phase II study in patients with refractory breast cancer is about to be initiated and a Phase I-II study in patients with refractory acute leukemia is being planned.

IV. Camptothecins

Camptothecin is a natural compound isolated from the plant *camptotheca accuminata*. The sodium salt of camptothecin was tested in a Phase I clinical trial two decades ago and its development interrupted because of unacceptable bladder toxicity.[23] In the late 1980s, the discovery that camptothecin is a topoisomerase I poison triggered a renewed wave of interest in this family of compounds.[24] Because of the lack of water solubility of camptothecin neutral base, efforts to synthesize and develop hydrosoluble analogs were carried out by several laboratories. The tangible results of these efforts have been the approval of two hydrosoluble analogs, hycamptin (Topotecan) for the treatment of cisplatin resistant ovarian cancer and CPT-11 (Irinotecan) for the treatment of 5-FU refractory colon carcinoma (Figure 1). Camptothecins are phase specific agents since only cells in S phase, that actively synthesize DNA, are susceptible to their cytotoxic effect. It is also well-known that a closed lactone ring is essential for cytotoxicity and that at a physiological pH the lactone ring tends to open quickly, thus resulting in drug inactivation.

These three characteristics, i.e., lack of water solubility, S-phase specific cytotoxicity, and fast inactivation at physiological pH make the camptothecins potential good candidates for a delivery system that would provide for a vehicle for their intravenous administration, an in vivo slow drug release system, and protection from interaction with plasma components and inactivation. Burke et al.[25] studied the affinity for lipid bilayers of different camptothecin analogs and showed that complexation of camptothecins to lipid vesicles composed of DMPC and DMPG stabilizes the lactone moiety of the drugs, thus preventing drug inactivation in the presence of plasma. However, their experiments were performed using very high lipid:drug ratios because of the relative low affinity of these compounds for phospholipid bilayers. In a different study, the same investigators demonstrated that encapsulation of topotecan within the acidic aqueous space (pH 5) of liposomes composed of DSPC prolonged significantly the half-life of the drug in human plasma (pH 7.6).[26]

Several laboratories have been actively exploring the use of lipid-based drug carriers to develop pharmaceutically acceptable liposomal-camptothecin formulations. Success has been limited due to the low affinity of this group of compounds for phospholipid bilayers. Daoud et al.[27] reported the formulation and in vivo antitumor activity of liposomal-camptothecin formulations after intramuscular administration. The lipid composition used was DPPC:Sph:Chol:PI (2.4:6.6:1.0:0.05 M ratio). The highest incorporation efficiency was around 40% using a lipid:drug ratio of 40:1. Freeze-fracture studies demonstrated that the presence of camptothecin increased the interlamellar space of the vesicles as a result of intercalation between lipid bilayers but did not cause disruption of the bilayer structure. The formulation displayed significant in vivo antitumor activity against two human xenografts by the intramuscular route, probably as a result of a depot effect. Sugarman et al.[28] explored a wide variety of different lipid compositions using lipid:drugs ratios as high as 50:1. In no case, a complete incorporation of the drug into the bilayers was observed. Most preparations consisted of minimally fluorescent liposomes (indicating a low association of drug to lipid) and lipid:camptothecin complexes which probably correspond to aggregated micelles. No free drug crystals were observed by electron microscopy, which allowed the authors to conclude that the lipid complexation efficiency was close to 100%. The lipid:camptothecin complexes had, however, significant antitumor activity, as good if not better than free camptothecin in suspension, and displayed a two-fold increased antitumor potency compared with free camptothecin.

The successful development of a liposomal-camptothecin formulation that fulfills the potential advantages outlined above will probably require to use analogues specifically designed or suited for liposome incorporation as with the other two main families of anticancer compounds, the platinum drugs and the anthracyclines. This will require detailed studies of the effect of different structural changes in the activity of the molecule as well as its affinity for phospholipid bilayers. One possible approach is to attach a long fatty acid chain to one of the positions that are not crucial for antitumor activity. In the case of anthracyclines[29] and antimetabolites (see below) this approach resulted in compounds with an increased affinity for phospholipid bilayers. Such approach might be an easy way to solve the problems encountered with the formulation of the camptothecin drugs in liposomes.

V. Taxanes

The taxanes paclitaxel (Taxol) and docetaxel (Taxotere) are natural products isolated from the pacific and european yew tree, respectively (Figure 1).[30] The taxanes have become one of the most important families of antitumor agents in the last few years as a result of showing striking antitumor activity against a wide variety of solid tumors including breast, ovarian, lung, and head and neck cancer. They are however ineffective in the treatment of gastrointestinal malignancies. One of the problems in the use of taxanes is their poor water solubility. The formulations currently used for the treatment of cancer patients contain cremophor EL or Tween 80, which cause allergic reactions, sometimes severe, in a significant number of cases.

Extensive efforts to improve the formulation of these agents have been undertaken and are currently pursued by numerous laboratories. The strategies explored have included the potential use of liposomes as carriers of these agents. In contrast with the camptothecins, the main objective was to develop formulations devoid of detergent. The possibility to preferentially target certain organs or tumors (liver in the case of multilamellar vesicles, extrahepatic solid tumors in the case of Stealth liposomes) was not the major objective. Most efforts have only been partially successful due to the inability to obtain vesicles that can incorporate the drugs without causing significant bilayer distortion and loss of liposome structure, probably because of their very large size. Sharma et al.[31–33] have reported the in vivo antitumor activity of a liposomal-Taxol formulation using SUV's composed of DMPC and DMPG at a 1:9 molar ratio and a total lipid:drug molar ratio of 33:1. Significant in vivo antitumor activity was observed with the liposomal-Taxol formulation against colon-26, a murine tumor that is naturally resistant to free Taxol. The authors claimed an incorporation efficiency of 100% and a total absence of aggregates or crystals of free drug. Our personal experience with Taxol using similar lipid compositions, although preliminary, has been far less promising.[34] In all cases we observed the presence of free drug in the form of crystals outside the liposomes, thus leading to the conclusion that because of its large size Taxol can not be accommodated between the phospholipid molecules that form the bilayer. Potentially more successful have been the efforts to prepare micellar suspensions using a variety of phospholipids.[35] However, no lipid-based taxane formulation appears to have reached clinical evaluation.

VI. Antimetabolites

The use of prodrugs of the antimetabolites cytosine arabinoside (Ara C), 5-fluoruracil (5-FU) and methotrexate (MTX) delivered in liposome carriers has been explored during the last decade by different investigators. The general approach was to chemically couple the agents to a phospholipid molecule and to use the complex as a liposome component, or to attach through an ester bond a fatty acid chain that can act as a chemical anchor to the liposome membrane. Once inside the cell or in circulation, the liposomes are destroyed and unspecific esterases can cleave the ester bond and the active drug is slowly released. This approach works best for antimetabolites since they are S phase specific and, therefore, more effective when cells are continuously exposed to them. We have used this approach for the anthracyclines[29] and the platinum compounds, which are non S-phase specific agents. In spite of the promising results, none of the formulations described below has been developed for clinical evaluation.

Rubas et al.[36] have reported the liposome formulation and antitumor activity of N4 and 5′ oleyl and palmitoyl derivatized Ara-C. The incorporation efficiency was very high (85–97%) even at low lipid:drug ratios (4:1). The liposome prodrug formulations were 5–10 fold more potent than free Ara-C against L1210 leukemia. At the optimal doses, the prodrugs were also more effective than free Ara C in this tumor model but about as effective against B16 melanoma. The authors speculated that the incorporation of the Ara-C prodrugs into liposomes provides protection against fast degradation and systemic clearance which may explain the enhanced potency and improved antitumor activity. It is however very likely that the advantages observed may be just the result of slow bond cleavage and prolonged exposure of the tumor cells to the antitumor agent.

Kinsky et al.[37] reported the formulation and in vitro antitumor activity of methotrexate-γ-DMPE in cells sensitive and resistant to MTX because of a transport deficit or amplification of dihydrofolate reductase. The MTX prodrug was equally cytotoxic to parental cells and cells resistant due to a transport defect and was able to partially overcome resistance in the cells resistant due to enzyme amplification. No in vivo studies with such compounds were reported.

More recently, Borssum Waalkes et al.[38] have reported formulation and in vitro cytotoxicity studies with liposomal formulations of diacylated derivatives of 5-fluoro-2′-deoxyuridine (FUdR). Lipophilic products were prepared by esterifying the free hydroxyl groups in the sugar moiety with fatty acids of different chain length. FUdR-diplamitate and dioctanoate were synthesized and incorporated in liposomes. FUdR-dipalmitate was very efficiently incorporated in different types of liposomes and no exchange of the prodrug with plasma components or hydrolsysis was observed when the liposomes were incubated with plasma. The opposite was observed with FUdR dioctanoate. The in vitro cytotoxicity of the prodrug liposomal formulations was assessed against C26 colon adenocarcinoma cells which are highly sensitive to FUdR. FUdR dipalmitate was several fold less potent than FUdR dioctanoate and FUdR. The authors concluded that the differences in antitumor activity between the different prodrugs and formulations is probably due to the differences in the rate of hydrolysis of the prodrugs to FUdR by esterase activity in the serum or tumor cells.

Finally, Jorge et al.[39] reported the liposome formulation, pharmacokinetics, in vivo toxicity, and in vivo antitumor activity of a prodrug of L-asparaginase, palmitoyl-L-asparaginase. The liposomal formulation of the prodrug was compared with the free prodrug. The prodrug incorporated in liposomes displayed a remarkably prolonged blood half-life, was non-immunogenic, and had similar in vivo antitumor activity.

VII. Other agents

A wide variety of other lipophilic agents, including some commercially available in oral forms have been explored as potential candidates for liposome incorporation.

The first liposomal formulation of a lipophilic antitumor agent ever developed and tested in clinical trials was (2-[3′-(methoxycarbonylamino)-phenyl]-3-phenyl-6-methoxycarbonylamino-4-(3H)-quinazolone (NSC-251635) at the Institute Jules Bordet in Brussels.[40,41] In preclinical studies, the compound was found to be inactive in free form in Klucel or saline and significantly active in the liposomal formulation. In a reported Phase I study, 14 patients were treated with doses as high as 456 mg/m^2. Unfortunately, the maximum tolerated dose could not be reached because of limitations in the amount of liposomes that could be prepared at one time in the laboratory and the project was apparently abandoned. Another water-insoluble agent developed in a liposomal formulation (sonicated vesicles composed of egg PC, cholesterol, and stearylamine) at the same Center was 6-aminochrysene. A phase II study in patients with heavily pretreated breast cancer patients with liver metastases was closed after 14 patients were enrolled and no responses observed. No myelosuppression was observed. Biodistribution studies with radiolabeled liposomes failed to show any increased uptake at the tumoral sites.

Another class of compounds that has received some attention as potential candidates for liposome delivery are the nitrosureas.[42–44] Several papers from different investigators have in general reported a decreased toxicity and/or an enhanced antitumor activity in in vivo tumor models. None of these formulations was ever developed for clinical trials.

Finally, it is worth mentioning our exploratory screening studies with several other lipophilic antitumor agents including hexamethylmelamine, penclomedine, mitindomide, fazarabine, diaziquone, batracyclin, and trimelamol.[34] Among these, excellent formulations could be obtained with the first four compounds. No in vivo studies were performed. No common structural features could be identified within the compounds with good liposome incorporation.

VIII. Future challenges

Although in principle a sound and rational idea, the use of liposomes as carriers of lipophilic antitumor agents is faced with significant challenges. First, many water· insoluble drugs can not be efficiently incorporated in liposomes because they have a low affinity for phospholipid bilayers or can not physically be accomodated within phospholipid bilayers without disrupting their structure. If the problem is low affinity, a possible solution may be to increase the amount of carrier, i.e., the lipid : drug ratio, to ensure that all drug is incorporated within the liposomes upon hydration of the lipid film or lyophilized preliposomal powder. The highest possible lipid : drug ratio depends on the potency of the compound and the toxicity of the lipids used. In our clinical studies with different drugs incorporated in liposomes composed of DMPC and DMPG at a 7 : 3 molar ratio, we have administered up to close to 12 g/m^2 of lipid over a period of 2 hours without significant toxicity. Based on this data, a lipid : drug ratio as high as 1000 : 1 is theoretically possible for a compound whose optimal dose would be 12 mg/m^2. Improper accomodation of the drug within the lipid bilayers thus causing structural alterations or disruption of the bilayers is a more difficult problem to solve. In these cases, the agents tend to precipitate as crystals of free drug immediately or over a period of time after liposome formation or cause significant distortion of the bilayer structure that enhances vesicle disintegration after administration. Increasing the lipid : drug ratio or using phospholipids with different charge, transition temperature, or length of fatty acid chain, may improve the physical imcompatibility but may not solve it completely. The use of a chemically modified analog may be the right solution in these cases.

The identification of an agent and a lipid composition with a high degree of compatibility is a major step towards the rapid development and characterization of a pharmaceutically acceptable formulation. However, the next challenge is to secure that the drug stays incorporated within the liposomes while in circulation and that the vesicles are not quickly destroyed in the blood stream. This is essential to preserve the tumor targeting ability of the vesicles. Unfortunately, while a high affinity for phospholipid bilayers is essential for incorporation of the agent in the liposomes, it may also enhance its free transfer to other circulating lipid particles, i.e., the lipoproteins. If that occurs, lipoproteins may act as a second carrier and that by itself may not be therapeutically detrimental since it is well known that lipoproteins have a long circulation time and many tumors overexpress receptors for lipoproteins. Both NDDP and Annamycin fall into this category of compounds. They have an exquisite affinity for phospholipid bilayers (one of the major reasons they were selected) and tend to disassociate quickly from the liposomes while in circulation. In the case of Annamycin, we know that transfer to lipoproteins occurs quite rapidly. In spite of that, all preclinical studies suggest that the current formulations may still provide an advantage for the treatment of microscopic liver metastases since most liver uptake occurs after the first passage through the liver, which occurs within minutes of administration. This may not be the case for the treatment of established extrahepatic solid tumors. Obviously, the other main potential advantage of these compounds is not related to their delivery in a liposome carrier but to their ability to overcome cisplatin and doxorubicin resistance at the cellular level.

We have explored he use of GM1 containing liposomes and PEG-coated liposomes containing high transition temperature phospholipids as a potential way to minimize the transfer of NDDP and Annamycin from the liposomes in to other plasma components. Shi et al.[45] and Mori et al.[46] reported partial success by using NDDP entrapped in liposomes composed of DSPC, Chol, and PE-PEG3000. A 3 fold higher drug tumor uptake was observed in correlation with an enhanced antitumor activity. In contrast, we have been completely unsuccessful so far in prolonging the plasma circulation time of L-Ann by using liposomes composed of combinations of GM1, DSPC, and PEG-PE.[47] This is surprising in view of the prolonged plasma circulation time and enhanced tumor localization observed with doxorubicin and other compounds encapsulated in this type of liposomes. We do not have an explanation for this phenomenon. Possibilities include that the presence of PEG-PE and Annamycin in the bilayer causes structural changes that enhance vesicle disruption (this does not occur with drugs that are encapsulated in the central aqueous space), or that the protection provided by the presence of PEG-PE on the liposome surface is just insufficient to prevent the transfer of Annamycin from the liposome bilayer to the lipoprotein bilayer given its high affinity for phospholipids. In summary, it appears that methods that have been successfully applied to prevent liposome interaction with the RES and prolong their plasma circulation time may not be effective in enhancing drug tumor targeting for lipophilic drugs incorporated in the lipid bilayers and that new approaches should be explored for that purpose.

Anticancer drug discovery and development has been so far more based on empiricism or serendipity than rational design because of the overwhelming complexity of the physiology and biology of human cancer. If selective tumor targeting of anticancer agents is not an impossible dream, it must be accomplished through rational application of knowledge developed in a wide range of disciplines. The work performed during the last few years on the use of liposomes as carriers of lipophilic antitumor agents is just telling us that it is not impossible and actually quite feasible to incorporate highly selected or specifically designed antitumor agents within lipid bilayers and to use such lipid bilayers to control drug release, organ distribution, and metabolization. Selective targeting of these vesicles to the tumor remains the major challenge and may not be accomplished until technology to prepare vesicles coated with tumor specific ligands that may be able to recognize and/or cross tumor capillary endothelial cells becomes available.

Acknowledgments

This work was supported by NIH CA45423, CA50270, CA58342, CA60496, Texas Higher Education Commission, and Aronex Pharmaceuticals, Inc. Dr. Perez-Soler is a shareholder and a consultant for Aronex, which is developing L-NDDP and L-Ann for commercialization.

References

1. DeVita VT, Hellman S, Rosenberg S (eds). Principles and practices of oncology. 4th Ed. Philadelphia: J.B. Lippincott Co., 1993.
2. Perez-Soler R, Khokhar AR, Lopez-Berestein G. Treatment and prophylaxis of experimental liver metastases of M5076 reticulosarcoma with cis-bis-neodecanoato-trans-R,R-1,2-diaminocyclohexane platinum (II) encapsulated in multilamellar vesicles. Cancer Res 1987;47:6462–6466.
3. Perez-Soler R, Yang LY, Drewinko B, Lautersztain J, Khokhar AR. Increased cytotoxicity and reversal of resistance to cisplatin with entrapment of cis-bis-neodecanoato-trans-R,R-1,2-diaminocyclohexane platinum (II) in multilamellar lipid vesicles. Cancer Res 1988;48:4509–4512.
4. Perez-Soler R, Lautersztain J, Stephens LC, Wright K, Khokhar AR. Pharmacology and toxicity of liposome entrapped cis-bis-neodecanoato-trans-R,R-1,2-diaminocyclohexane platinum (II) in mice and dogs. Cancer Chemother Pharmacol 1989;24:1–8.
5. Perez-Soler R, Khokhar AR. Lipophilic cisplatin analogues entrapped in liposomes: Role of intraliposomal drug activation in biological activity. Cancer Res 1992;52:6341–6347.
6. Han I, Khokhar AR, Perez-Soler R. Intraliposomal conversion of lipophilic diamino cyclohexane (DACH)-dicarboxylato platinum (II) complexes into DACH-Pt-$C1_2$. Cancer Chemother Pharmacol 1996;39:17–24.
7. Khokhar AR, Wright K, Siddik ZH, Perez-Soler R. Organ distribution and tumor uptake of liposome entrapped cis-bis-neodecanoato-trans-R,R-1,2-diaminocyclohexane platinum (II) administered intravenously and into the proper hepatic artery. Cancer Chemother Pharmacol 1988;22:223–227.
8. Perez-Soler R, Lopez-Berestein G, Lautersztain J, Al-Baker S, Francis K, Macias-Kiger D, Raber Martin N, Khokhar AR. Phase I clinical and pharmacology study of liposome-entrapped cis-bis-neodecanoato-trans-R,R-1,2-diaminocyclohexane platinum (II). Cancer Res 1990;50:4254–4259.
9. Vadiei K, Siddik ZH, Khokhar AR, Al-Baker S, Sampedro F, Perez-Soler R. Pharmacokinetics of liposome entrapped cis-bis-neodecanoato-trans-R,R-1,2- diaminocyclohexane platinum (II) and cisplatin administered IV and IP in the rat. Cancer Chemother Pharmacol 1992;30:365–369.
10. Perez-Soler R, Shin DM, Siddik ZH, Murphy WK, Huber M, Lee JS, Khokhar AR, Hong WK. Phase I clinical and pharmacological study of liposome-entrapped NDDP (L-NDDP) administered intrapleurally in patients with malignant pleural effusions. Clin Cancer Res. (In press).
11. Alberts DS, Liu PY, Hannigan EV, O'Toole R, Williams SD, Young JA, Franklin EW, Clarke-Pearson DL, Malviya VK, DuBeshter B, Adelson MD, Hoskins WJ. NEJM 1996;335(26):1950–1955.

12. Han I, Khokhar AR, Perez-Soler R. Cellular accumulation and DNA damage induced by liposomal cis-bis-neodecanoato-trans-R,R-1,2-diaminocyclohexane platinum (II) (L-NDDP) in LoVo and LoVo/PDD cells. Anticancer Drugs 1994;5:64–68.
13. Mayer LD, Bally MB, Hope MJ, Cullis PR. Uptake of antineoplastic agents into large unilamellar vesicles in response to membrane potential. Biochim Biophys Acta 1985;816:294–302.
14. Gabizon A, Catane R, Uziely B, Kaufman B, Safra T, Cohen R, Martin F, Huang A, Barenholz Y. Prolonged circulation time and enhanced accumulation in malignant exudates of doxorubicin encapsulated in polyethylene-glycol coated liposomes. Cancer Res 1994;54:987–992.
15. Juliano RL, Ling V. A surface glycoprotein modulating drug permeability in Chinese hamster ovary cell mutants. Biochim Biophys Acta 1976;455:152–162.
16. Cole SPC, Bhardwaj G, Gerlach JH et al. Overexpression of a transporter gene in a multdrug-resistant human lung cancer cell line. Science 1992;258:1650–1654.
17. Priebe W, Van NT, Burke TG, Perez-Soler R. Removal of the basic amino group at position 3 from doxorubicin overcomes multidrug resistance and decreases cardiotoxicity. Anticancer Drugs 1993;4:37–48.
18. Perez-Soler R, Priebe W. Anthracycline antibiotics with high liposome entrapment: structural features and biological activity. Cancer Res 1990;50:4260–4266.
19. Zou Y, Ling YH, Van NT, Priebe W, Perez-Soler R. Antitumor activity of the lipophilic and partially non-cross resistant anthracycline annamycin entrapped in liposomes. Cancer Res 1994;54:1479–1484.
20. Ling YH, Priebe W, Yang LY, Burke TG, Pommier Y, Perez-Soler R. In vitro cytotoxicity, cellular pharmacology, and DNA lesions induced by Annamycin, an anthracycline derivative with high affinity for lipid membranes. Cancer Res 1993;53:1583–1589.
21. Ling YH, Priebe W, Perez-Soler R. Apoptosis induced by anthracycline antibiotics in P388 parental and multi-drug resistant cells. Cancer Res 1993;53:1845–1852.
22. Zou Y, Priebe W, Perez-Soler R. Preclinical toxicity of liposome-incorporated Annamycin: Selective bone marrow toxicity with lack of cardiotoxicity. Clin Cancer Res 1995;1:1369–1374.
23. Muggia FM, Creaven PJ, Hansen HS et al. Phase I clinical trial of weekly and daily treatment with camptothecin (NSC-100880): Correlation with preclinical studies. Cancer Chemother Rep 1972;56:515–521.
24. Slichenmyer WJ, Rowinsky EK, Donehower RC et al. The current status of camptothecin analogues as antitumor agents. J Natl Cancer Inst 1993;85:271–291.
25. Burke TG, Mishra AK, Wani MC, Wall ME. Lipid bilayer partitioning and stability of camptothecin drugs. Biochemistry 1993;32(20):5352–64.
26. Burke TG, Gao X. Stabilization of topotecan in low pH liposomes composed of distearoylphosphatidylcholine. J Pharma Sci 1994;83(7):967–969.
27. Daoud SS, Fetouh MI, Giovanella BC. Antitumor effect of liposome-incorporated camptothecin in human malignant xenografts. Anti-Cancer Drugs 1995;6(1):83–93.
28. Sugarman SM, Zou Y, Wassan K, Poirot K, Kumi R, Reddy S, Perez-Soler R. Lipid-complexed camptothecin: formulation and initial biodistribution and antitumor activity studies. Cancer Chemother Pharmacol 1996;37:531–538.
29. Perez-Soler R, Priebe W. Liposomal formulation and antitumor activity of 14-0-palmitoyl hydroxyrubicin. Cancer Chemother Pharmacol 1992;30:267–271.
30. Rowinsky EK, Cazenave LA, Donehower RC. Taxol: a novel investigational antimicrotubule agent. J Natl Cancer Inst 1990;82:1247–1259.
31. Sharma A, Straubinger RM, Ojima I, Bernacki RJ. Antitumor efficacy of taxane liposomes on a human ovarian tumor xenograft in nude athymic mice. J Pharma Sci 1995;84(12)1400–1404.
32. Sharma A, Straubinger RM. Novel taxol formulations: preparation and characterization of taxol-containing liposomes. Pharmaceutical Res 1994;11(6):889–96.
33. Straubinger RM, Sharma A, Murray M, Mayhew E. Novel taxol formulations: taxol-containing liposomes. Monogr Natl Cancer Inst 1993;15:69–78.
34. Sampedro F, Partika J, Santalo P, Molins-Pujol AM, Bonal J, Perez-Soler R. Liposomes as carriers of new lipophilic antitumor agents. A preliminary report. J Microencapsulation 1994;11:197–206.
35. Wheeler JJ, Wong KF, Ansell SM, Masin D, Bally MB. Polyethylene glycol modified phospholipids stabilized emulsions prepared from triacylglycerol. J Pharma Sci 1994;83(11):1558–1564.
36. Rubas W, Supersaxo A, Weder HG, Hartmann HR, Hengartner H, Schott H, Schwendener R. Treatment of murine L1210 lymphoid leukemia and melanoma B16 with lipophilic cytosine arabinoside prodrugs incorporated into unilamellar liposomes. International J Cancer 1986;37(1):149–154.
37. Kinshy SC, Hashimoto K, Loader JE, Knight MS, Fernandes DJ. Effect of liposomes sensitized

with methotrexate-(-dimyristoylphosphatidylethanolamine on cells that are resistant to methotrexate. Biochimica et Biophysica Acta 1986;885:129–135.
38. van Borssum Waalkes M, van Galen M, Morselt H, Sternberg B, Scherphof GL. In vitro stability and cytostatic activity of liposomal formulations of 5-fluoro-2′-deoxyuridine and its diacylated derivatives. Biochimica et Biophysica Acta 1993;1148(1):161–172.
39. Jorge JCS, Perez-Soler R, Morais JG, Cruz MEM. Liposomal palmitoyl-L-Asparaginase: characterization and biological activity. Cancer Chemother Pharmacol 1994;34:230–234.
40. Brassinne C, Atassi G, Fruhling J, Penasse W, Coune A, Hildebrand J, Ruysschaert J-M, Laduron C. Antitumor activity of a water-insoluble compound entrapped in liposomes on L1210 leukemia in mice. JNCI 1983;70(6):1081–1085.
41. Sculier JP, Coune A, Brassinne C, Laduron C, Atassi B, Ruysschaert JM, Fruhling J. Intravenous infusion of high doses of liposomes containing NSC 251635, a water-insoluble cytostatic agent. A pilot study with pharmacokinetic data. J Clin Oncol 1987;4:789–797.
42. Inaba M, Yoshida N, Tsukagoshi S. Preferential action of liposome-entrapped 1-(2-chloroethyl)-3-(4-methylcyclohexyl)-1-nitrosourea on lung metastasis of Lewis lung carcinoma as compared with the free drug. Gann 1981;72(3):341–345.
43. Mathe G, Bothorel P. In vivo enhancement of the experimental oncostatic effect of RFCNU by its encapsulation in liposomes. Biomedicine 1981;35:201–202.
44. Maral R, Bourut C, Chenu E, Mathe G, Bernon R, Lussan C, Imbach JL, Schein P, Bothorel P. Comparison of the experimental antitumor activities of three nitrosourea derivatives chlorozotocin, RFCNU and CNCC encapsulated in liposomes with those in the free state. Oncol 1985;42:122–128.
45. Li S, Khokhar AR, Perez-Soler R, Huang L. Improved antitumor activity of cis-bis-neodecanoato-trans-R,R-1,2-diaminocyclohexane platinum (II) entrapped in long circulating liposomes. Oncology Research 1995;7:611–617.
46. Mori A, Wu SP, Han I, Khokhar AR, Perez-Soler R, Huang L. In vivo antitumor activity of cis-bis-neodecanoato-trans-R,R-1,2 diaminocyclohexane platinum (II) formulated in long-circulating liposomes. Cancer Chemother Pharmacol 1996;37:435–444.
47. Zou Y, Ling YH, Reddy S, Priebe W, Perez-Soler R. Effect of lipid composition and vesicle size on the in vivo tumor selectivity of the non-cross resistant anthracycline Annamycin incorporated in liposomes. Int J Cancer 1995;61:666–671.

Lasic and Papahadjopoulos (eds.), Medical Applications of Liposomes

CHAPTER 4.6

Targeted sterically stabilized liposomal drug delivery

THERESA M. ALLEN, CHRISTIAN B. HANSEN AND DARRIN D. STUART

Department of Pharmacology, University of Alberta, Edmonton, AB T6G 2H7, Canada

Overview

I. Introduction

Drug carriers, such as liposomes, are used in attempts to improve the therapeutic index of associated therapeutic molecules by increasing their localization to specific target tissues or cells and by decreasing their localization to sensitive normal tissues. As seen in Chapters 4.3 and 4.4, passively targeted, sterically stabilized liposomes, with their ability to localize in regions of increased capillary permeability, e.g., localization of liposomal anticancer drugs to solid tumours undergoing angiogenesis, can result in considerable clinical benefit. An important question then arises: is it possible to further improve on the results achieved with non-targeted or 'passively' targeted liposomes through the strategy of attaching targeting molecules such as antibodies or other ligands to the liposome surface? Further questions include: What types of ligands should we be using and how should they be attached to the liposomes? Which therapeutic applications are the most rational in the light of liposomal physical properties and host biology? What are the major problems which can be anticipated for this approach and what tactics might overcome these problems? A growing volume of work on targeted liposomal drug delivery is beginning to address some of the above questions and we discuss these developments in this chapter.

Possibly the greatest potential for a targeted drug delivery vehicle lies in the field of cancer chemotherapy. Most chemotherapeutic agents have severe, dose limiting toxicities to normal, non-diseased cells. Doses are generally pushed to the limit of toxicity in order to obtain maximal therapeutic benefit in treating the cancer. A method of drug delivery which would deliver drug preferentially to the diseased cells and away from non-diseased cells would obviously increase the therapeutic index of the drug. To date, improvements in the therapeutic index of anticancer drugs have come from passive accumulation, i.e., 'passive' targeting, of the liposomal drug in the solid tumours through the process of 'selective' extravasation, as is the case with Stealth® liposomal doxorubicin (Doxil).[1,2] Another possibility, which has been called 'active' targeting, is to couple a targeting ligand to the surface of the drug carrier. The ligand should specifically bind to a surface epitope on the target cell, leading to the accumulation of the liposomal drug package at/in the target tissue as a result of this ligand receptor interaction. Many studies have shown that ligand-bearing liposomes will selectively bind to target cells in vitro, however few studies have demonstrated this specificity in vivo.

A decade ago, the greatest limitation to targeted liposomal drug delivery in vivo was the rapid clearance of ligand-bearing liposomes from the circulation resulting from their uptake into the mononuclear phagocyte system (MPS).[3,4] This rapid clearance prevented the ligand liposomes from reaching, and binding to, their target cells. The discovery that polymers such as polyethylene glycol (PEG) and monosialylganglioside GM_1 inhibited the uptake of liposomes, including ligand-liposomes, by the MPS has revived interest in targeted liposomal drug carriers.[5–15] Several studies demonstrating active targeting of these liposomes in vivo have been published.[16,24] Overcoming the problem of rapid clearance, however, has led to new questions and has uncovered a new set of potential problems. Questions include: how do the contents of the targeted liposome enter into the target cells following the binding step? Is it by passive diffusion, fusion of the liposome with the cell membrane, receptor-mediated endocytosis or some other mechanism? While we may be able to target specific cell populations (organs or tissues), can we deliver the liposomal drug into specific intracellular compartments? How do we strike a balance between the stability required for liposomes to retain their contents until they reach their target cells and the instability required for the release of drug once the liposomes bind, fuse, or are internalized into the target cell? Finally, the ability to increase circulation times of ligand-liposomes allows us to ask which tissues are accessible to targeted liposomes and which diseases will be candidates for liposomal drug therapy?

II. Choice of targeting moieties

Selective drug delivery to cancer cells requires the presence of markers on the cancer cell surface which distinguishes them from non-tumor cells. These markers are often referred to as tumor associated antigens.[25–27] In some cases, the antigen is well characterized and some are known to be receptors which are mutated or overexpressed on the tumor cells, while in other cases the antigen is simply characterized as a cell surface glycoprotein with uncharacterized structure or function.[28] Characterization of the antigen may be very important in determining the intracellular processing of the liposome and/or its associated drug. Several different molecules have been attached to liposomes for the purpose of targeting to specific cells. Monoclonal antibodies (mAb) and antibody fragments have been used extensively in vitro as well as in vivo.[3,29–41] In addition, various other endogenously occurring ligands such as peptides[42] proteins and lipoproteins,[43–45] growth factors and vitamins,[46,47] carbohydrates[48–50] and glycoproteins[51] have all been attached to liposomes for the purpose of targeting specific cells, mostly in vitro. For the purpose of this review, we will focus on those ligands which have been used specifically to target cancer cells.

II.1. Monoclonal antibodies

The most widely used molecules for targeting liposomes have been monoclonal antibodies.[3,29–41] Monoclonal antibodies, selected for their ability to bind to cancer-associated epitopes on the surface of cancer cells, have been coupled to liposomes to produce targeted drug carriers, often referred to as immunoliposomes.[35] Coupling strategies are discussed in the following section.

The use of whole antibodies (Ab), or antibody fragments as targeting ligands is not without problems. The production of monoclonal antibodies is time consuming and expensive. Antibodies are large proteins, which may be denatured or otherwise inactivated by some of the procedures used in the formation and sizing of liposomes, or in the procedures used in the formation of immunoliposomes. Since most monoclonal antibodies are generated in mice, or other animal species, their use in humans will lead to an immune response, e.g., the generation of human anti-mouse antibodies (HAMA).[52,53] The antibodies may suffer from a lack of specificity, as it is difficult to identify surface epitopes which are specific solely to cancer cells and are not also shared by some normal cells (albeit usually at much lower concentrations). Even if a monoclonal antibody is highly selective for cancer-associated epitopes, the antibody may be of low affinity. Furthermore, cancer cells are notorious for their ability to down-regulate, shed or alter their surface epitopes. Also the antibody may only recognize a small portion of the total tumour cell burden. Use of antibody fragments, as opposed to whole antibodies, can lead to reduced avidity, although formulating fragments into liposomes should restore multivalent binding and avidity. Some of the strategies which have been used to overcome some of the above problems are discussed below.

Whole mAbs are bivalent, with two antigen binding domains joined by a constant region (Fc). The Fc region is involved in recognition by the Fc receptor on phagocytic and effector immune cells.[54–56] The Fc portion can be enzymatically cleaved resulting in $F(ab')_2$ (bivalent) and F(ab′) (monovalent) antibody fragments which maintain affinity for the antigen, although some loss of avidity may result. Cleavage of the Fc region would have potential advantages in reducing clearance of the liposomes and in decreasing their immunogenicity, since the Fc portion is responsible for much of the immunogenicity of a whole antibody following its administration into a foreign host (e.g., murine antibodies into humans).[52,54,55,57]

Recombinant DNA technology has provided a new generation of monoclonal antibodies which may be utilized as homing devices on immunoliposomes. For example, chimeric antibodies, which consist of a human antibody framework combined with variable light and heavy chain regions of a mouse monoclonal antibody, can now be produced.[16] This strategy further decreases the quantity of 'foreign' protein in the antibody fragment, thus decreasing the potential for immune response upon repeat administration. An even more promising strategy is the production of totally human antibodies or antibody fragments in *E. coli*. Combinatorial phage libraries, which contain cloned genes from the human variable light and heavy chains, have been produced and can be quickly screened for specificity to almost any antigen.[58,59]

II.2. Other ligands

Malignant cells often overexpress receptors for growth factors or other molecules which will help to maintain the aggressive growth pattern characteristic of cancer cells. Ligands for growth factor receptors,[46,60] the folate receptor,[47,61] the transferrin receptor,[43,62] and the low density lipoprotein (LDL) receptor[44,63] have all been coupled to liposomes for targeting cancer cells in vitro. All of these receptors have been shown to be overexpressed on certain malignant cells and the endogenous ligands for these receptors serve as useful targeting molecules for several reasons. Growth factors, folate, and some proteins have a much lower molecular weight than antibodies, will be less immunogenic and cheaper than antibodies currently used for targeting. In many cases, the receptors for these ligands are well characterized with respect to their internalizing capacity, thus providing a route for active intracellular drug delivery. On the other hand, receptors for many of these ligands are expressed widely in the body, and are not confined to tumour cells. Thus, the specificity of drug delivery will be compromised.

II.3. Internalizing receptors

The capacity of a cell to internalize the targeted liposome is an important consideration when choosing a targeting ligand. Liposome binding to the target cell brings the drug package in contact with the cell, thus increasing the local concentration of drug. If target binding does not result in receptor-mediated endocytosis, or fusion mediated by a fusogenic protein, the liposome will remain attached to the outer surface of the cell until it dissociates and redistributes, is destroyed by phospholipases, proteases, low pH, or is engulfed by macrophages.[64] In the above examples the efficacy of the drug is limited by its need to escape from the liposome and enter the target cell by diffusion rate limited processes (with the exception of fusogenic liposomes). On the other hand, if targeted liposome-receptor binding results in internalization, efficacy is no longer limited by diffusion because the whole, concentrated drug package is delivered directly into the cell. This does, however, bring about other potential problems related to the intracellular distribution of the drug.

Studies have demonstrated that the most effective cytotoxicity occurs when the targeted liposome, and its contents, are internalized into the target cell.[36,40,65,66] The specific intracellular processing of ligand or antibody-targeted liposomes is not well elucidated, however, evidence suggests that it may begin through fusion of the liposome with the cell membrane (normally a rare event) and/or through the coated-pit pathway,[16] which initiates the process of receptor-mediated endocytosis. Following receptor-mediated endocytosis of the liposome-drug-receptor complex, it enters into an endosomal compartment and eventually into an acidic lysosomal compartment.

It has been demonstrated that the lysosomal stability of the encapsulated drug must be considered when targeted to an internalizing receptor.[67] The ability of the drug to reach its cellular site of action depends on the ability of the drug to survive the lysosomal apparatus, and escape into the cytoplasm (unless the lysosome is the site of action). It may be assumed that any drug which is acid labile, or susceptible to degradation by lysosomal enzymes will be less effective when delivered via an immunoliposome targeted to an internalizing receptor. This has been demonstrated in at least one case, where 1-β-D-arabinosylfuranosylcytosine (ara-C) was unable to cause target cell specific cytotoxicity when delivered using an immunoliposome targeted to an internalizing receptor.[67] Under the same conditions, methotrexate, which is more stable, caused target cell cytotoxicity. It was suggested that ara-C is inactivated by the lysosomal system, while methotrexate was able to escape degradation.[67] When using ara-C, non-internalizing immunoliposomes may be more desirable since there is an active cell membrane pump for nucleosides. As long as there is significant leakage from the liposome to increase the local concentration of ara-C, a non-internalizing targeting strategy may result in enhanced therapeutic efficacy. However, for drugs which are able to escape lysosomal degradation, targeting via an internalizing receptor may be the most efficient strategy.

Targeting via one of the receptors listed above offers the advantage that receptors which internalize upon ligand binding can be specifically chosen. The transferrin receptor,[68] LDL receptor,[44] folate receptor,[47] all demonstrate internalization upon ligand binding. The heterogeneity of tumor associated antigens, which are the targets for immunoliposomes, does not ensure internalization, although mAbs can be screened for this ability.

In some applications, particularly those involving two step targeting, it may be necessary to choose a non-internalizing ligand.[21,69] In this strategy, a bispecific antibody which recognizes both the target and a non-specific molecule such as biotin is administered as a first step. After allowing sufficient time for the antibody to find its target, a chase step follows, using a drug-containing liposome designed to bind to the biotin moiety on the target localized antibody, e.g., liposomes with coupled streptavidin. Removal of the ligand from the cell surface by internalization would prevent the liposome-drug package from recognizing and binding to the target. The down-side of this approach is the necessity for the drug to be released in order for there to be biological activity. Little is known about the rate of diffusion of the drug from the bound liposome into the target cell, versus the rate of diffusion and redistribution of the drug away from the target cell.

There is a pressing need for more research in the whole field of liposome-drug-cell interactions. Little or no experimental evidence exists which directly addresses the questions of how much, by what route and at what rate liposome-associated drugs enter cells, particularly following antibody- or ligand-mediated targeting.

III. Procedures for the formation of targeted long-circulating liposomes

Targeted liposomes must survive in the systemic circulation long enough to reach and bind to their target, and a critical step to achieving this was the development of liposomes which remained long-circulating following coupling of ligands at the liposome surface. Strategies to increase the circulation time of liposomes, like a reduction in liposome size[70] or the inclusion of cholesterol and/or high phase-transition lipids[71–73] provided a modest decrease in the clearance rate of liposomes. More success came from the tactic of including a hydrophilic molecule at the liposome surface, e.g., GM_1, and phosphatidylinositol[5,7–10] or lipid derivatives of polymers like PEG, [6,11–15] poly(acrylamide),[74] poly(vinyl pyrrolidone)[74] or poly(methyl or ethyl oxazoline).[75,76] While not all of the above polymers have been investigated for their ability to prolong the circulation time of antibody- or ligand-containing liposomes, it has been shown that both GM_1 and PEG have been successful in this regard.[42,77–81] As outlined below, there are many different methods which can be used to couple targeting molecules to long-circulating liposomes.

III.1. Coupling strategies

The principle strategies for the formation of ligand targeted liposomes are summarized in Figure 1. A ligand is coupled, often through a spacer molecule, to a hydrophobic anchor via cross linking molecules. The hydrophobic anchor is required for stable insertion of the conjugate into the lipid bilayer of a liposome. The hydrophobic anchor must be sufficiently strong to bind a ligand (e.g., an antibody) and a spacer molecule (e.g., a large hydrophilic polymer like PEG) securely to the liposome surface. There are a variety of anchors, including hydro-

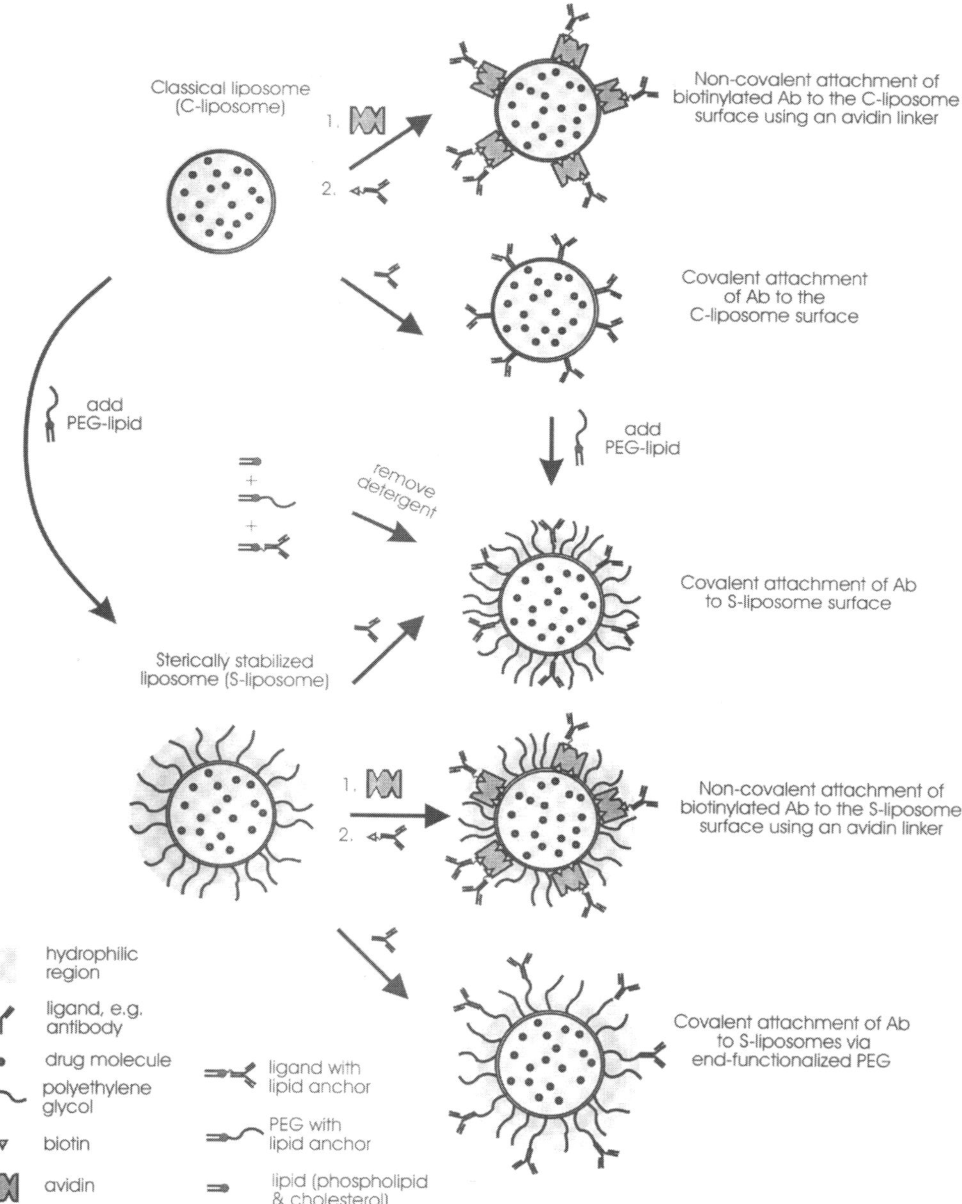

Fig. 1. Binding of targeted liposomes to cells. (1) Ligands conjugated on the surface of immunoliposomes containing PEG-lipids are sterically inhibited from binding to target epitopes on the cell surface. (2) Ligand can be conjugated to liposomes via a spacer molecule (e.g., avidin) and cell binding proceeds unhindered by PEG. However, initial ligand coupling efficiency may be reduced by the presence of PEG on the liposomes. (3) Two step non-covalent method of cell binding of ligand-liposomes. Initial binding of ligand-linker to cell associated antigens followed by liposome binding in situ. (4) Ligand can be conjugated at the distal end of a spacer molecule (e.g., functionalised PEG-lipid derivatives) and cell binding proceeds unhindered by PEG.

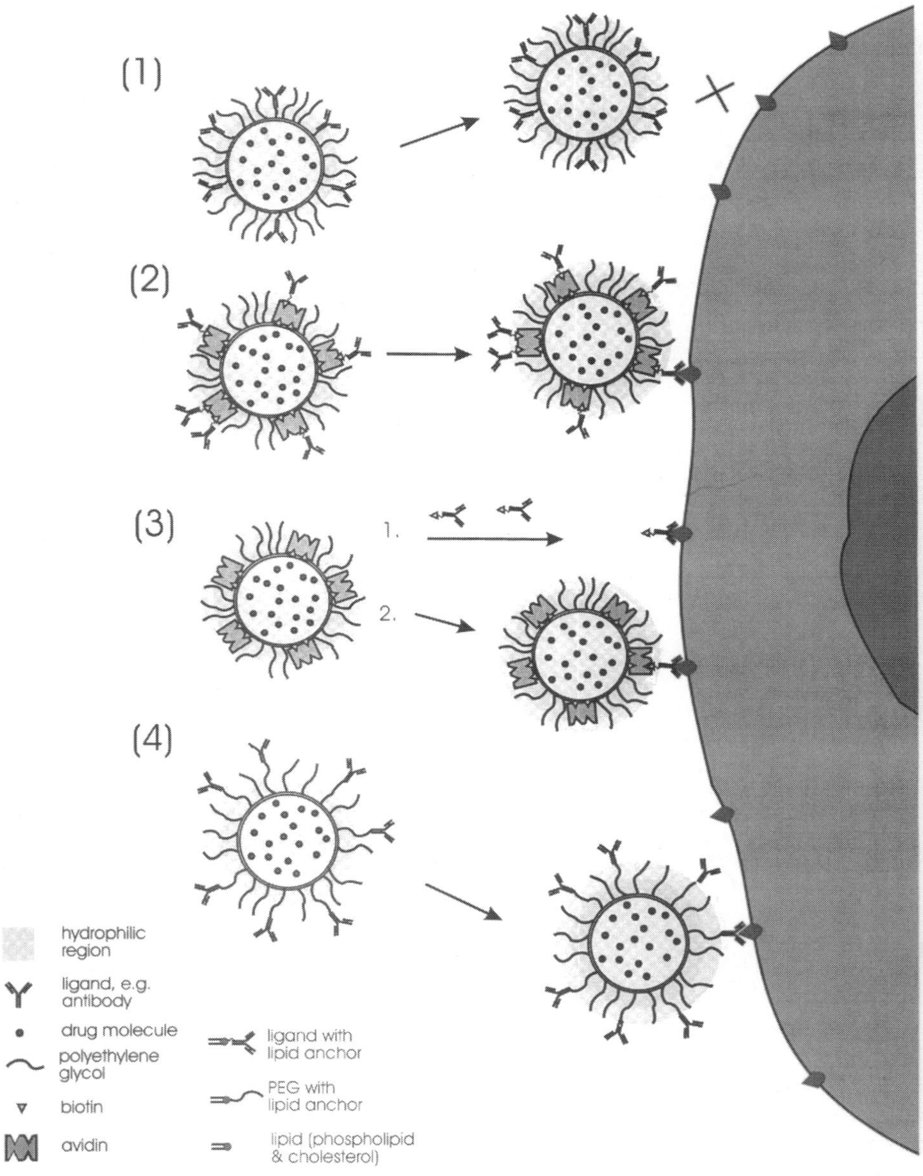

Fig. 2. General strategies for the formation of targeted liposomes. Ligands can be linked covalently to the surface of classical or sterically stabilized liposomes, or non-covalently through the use of avidin and biotin. For sterically stabilized liposomes, ligands can also be linked to the terminus of the polymer. In addition, classical ligand-liposomes can be converted to sterically stabilized ligand-liposomes by the addition of polymers in the form of micelles which will spontaneously incorporate into the liposomes. Immunoliposomes can also be formed in a one step procedure by through detergent dialysis of a mixture of lipids, PEG-lipid and ligand-lipid.

carbon acyl chains, hydrophobic peptides, and glyco- and phospholipids (reviewed in[82,83]). The choice of anchor depends on the general coupling strategies available, i.e., which reactive groups are presented on the anchor and the ligand, the availability of heterofunctional cross linking molecules, and the type of chemical bond desired (e.g., stable versus unstable, covalent versus non-covalent). Generally the anchor of choice has been phosphatidylanolamine (PE) because of the reactive amine in its head group and the availability of various acyl chain lengths of different degrees of unsaturation, e.g., dimyristoyl PE (DMPE), dipalmitoyl PE (DPPE), distearoyl PE (DSPE), dioleoyl PE (DOPE) or 1-palmitoyl-2-oleoyl PE (POPE).

To confer long circulation times, PEG-lipids are normally incorporated into liposomes at 4–10% of total lipid.[11–13,17,84–87] Also, the optimal weight range of the grafted PEG is between 2000–5000 Da (PEG-2000 to PEG-5000, 45–115 repeat units).[12,17,86–88] This amount of polymer, with long hydrophilic chains, when grafted onto the liposomes, hinders the interaction of proteins, including antibodies, with liposome surfaces.[14,89] This will reduce both the coupling efficiency of ligands and antibodies to the liposomes and their ability to bind to their target.[17,19,90,91] To overcome these problems, new strategies have been adopted which involve moving the coupling reaction from the liposome surface to the terminal end of a functionalised PEG lipid conjugate[19,42,47,61,80,81,90,92–105] (Figures 1, 2).

III.2. Coupling ligands to the liposome surface

III.2.1. Attaching ligands to the surface of pre-formed, PEG-containing liposomes

Several types of coupling chemistries exist. Some of the chemistry for coupling ligands to long-circulating liposomes were adapted from procedures for conjugating antibodies and/or antibody fragments (Fab′ and F(ab′)$_2$) to the liposome surface.[82,83,90,106–108] The first major class of linkage chemistry involves sulphydryl reactions. Two thiol-reactive PE derivatives, which can be incorporated into liposomes are N-pyridyldithiopropionyl-PE (PDP PE)[109–111] and maleimidophenyl butyrate-PE (MPB-PE).[110] PDP-PE is easily reduced to form free thiol groups at the liposome surface. The free thiol groups can then be coupled through thiol-ether bonds to maleimide groups on proteins (introduced onto exposed amino groups) through the use of a heterobifunctional cross-linking reagent like SMPB (H-succinimidyl-4-)ρ-maleimidophenyl)butyrate). In the reverse of this reaction, exposed thiol groups on proteins, introduced through reaction with SPDP (N-succinimidyl-3-(2-pyridyldithio)proprionate), can be cross-linked, via a thiol-ether bond, with maleimide-containing groups, e.g., MPB PE, at the liposome surface. Similar chemistry is used to couple free thiol groups on Fab′ antibody fragments to MPB-PE at the liposome surface.[110] The formation of a thiol-ether bond, by either conjugation of maleimide-lipids to thiolated antibodies or antibody fragments, or thiolated-lipids to maleimide-containing proteins, is a very efficient reaction and the bond is very stable.[110] Reactions between thiolated lipids and thiolated proteins will result in the production of disulfide bonds, which, despite high coupling efficiencies, result in bonds which are unstable in serum and thus difficult to use in vivo.[111]

Attempts to couple antibodies to the surface of PEG-2000-grafted liposomes with PDP-DOPE[90] or a maleimido-benzoyl-derivative of DPPE (MB-PE)[112] incorporated in the bilayer resulted in low antibody densities and low coupling efficiencies. PEG-lipids, at concentrations of approximately 5 mol% or higher in the bilayer, probably sterically interfere with the accessibility of the antibody to the liposomal surface.[90,112] Surprisingly, grafting PEG-2000 on liposomes did not reduce the efficiency of thiolated antibody binding to MPB-PE liposomes.[90] We have speculated that the addition of a hydrophobic phenyl group at the liposome surface leads to spreading of the hydrophilic PEG polymers, allowing access of the antibody to the liposome surface. When the size of the grafted polymer was increased to PEG-5000, conjugation of antibodies via MPB-PE was reduced about 5-fold, compared to PEG-2000.[79] Although high coupling efficiencies and good antibody densities could be achieved on PEG-grafted liposomes through the use of MPB-PE as a coupling lipid, this coupling method compromised the efficiency of remote-loading of doxorubicin (DXR) into the liposomes and increased its rate of efflux.[90] In a study of ascitic ovarian cancer in mice, these effects of MPB-PE on DXR loading and efflux have been suggested to contribute to the failure to observe increased therapeutic efficacy when targeted liposomes were compared to non-targeted liposomes.[113]

III.2.2. Addition of PEG after ligand coupling

To overcome the interference of PEG in the conjugation of antibodies onto the liposome surface, PEG could be incorporated into liposomes after ligand coupling occurs. PEG-2000-DSPE was effectively transferred from micelles and inserted into preformed liposomes, however, high temperatures were needed for efficient transfer, which may destroy protein ligands like antibodies.[114] In another method, PEG was covalently coupled onto the liposome surface after antibody conjugation.[112] Liposomes containing maleimido-benzoyl-DPPE (MB-PE) were conjugated first with thiolated antibodies, then PEG was grafted to the surface of the immunoliposomes through the use of PEG-succinylcystine (PEG-SC) of various polymer chain lengths.[112] This post-coating method resulted in both efficient antibody conjugation and efficient grafting of PEG-750-SC, PEG-2000-SC and PEG-5000-SC onto the liposome surface. However, only PEG-2000-SC-immunoliposomes retained extended circulation times and in vitro target binding, compared to control, non-PEG liposomes.[112] The advantage of this post-coating method of immunoliposome preparation is that both the ligand and the grafted polymer occur on the outside leaflet only, leaving the maximal interior space for drug loading.

III.2.3. Formation of ligand-anchor conjugates prior to liposome formation

Another tactic which has been used to bind ligands to the surface of liposomes is to create a ligand-anchor conjugate before liposome preparation. The liposomes can then be formed by either co-solubilizing the conjugates in detergent followed by dialysis,[17,19,23,78,115,116] or by hydrating a dry lipid film containing the lipid-ligand conjugate plus other lipids.[47,61,99,102,105] The first method uses carbodiimide activation of the carboxyl groups of N-glutaryl-PE (NGPE) followed by coupling to free amine groups on antibodies in an octyl glucoside solution. Additional lipids are added to the detergent solution and liposomes are formed following detergent removal through dialysis.[17,23,78,115,116] A portion of the incorporated antibody will be orientated to the interior aqueous space of the liposome, making it unavailable for target binding. Also, the internal antibody will also occupy internal space which will reduce the available volume for drug loading. Immunoliposomes formed by this method, in liposomes containing PEG-5000-PE, showed poor target binding, due to steric hinderance of antibody-antigen interactions.[17,19] However, the amount of antibody-lipid conjugate incorporated into PEG-grafted liposomes was reported to be independent of the polymer size and surface density.[17,115]

III.2.4. Non-covalent coupling methods

The non-covalent, but high affinity interaction of avidin or streptavidin with biotin has been adapted for coupling ligands to the liposome surface.[24,37,41,117–119] The cross linking of a ligand and liposomes can proceed via an avidin bridge either before[24,37,41,117–119] or after[21,69] target binding. In one variation, ligands are non-covalently bound to preformed liposomes by first binding avidin or streptavidin to liposomes containing a biotinylated lipid (usually a derivative of PE) and then incubating with a biotinylated ligand. In another variation, biotinylated-ligands (or an avidin-ligand conjugate) are first bound to the target epitope (Figure 2).[21,69] A chase step follows, using either streptavidin- or avidin-liposomes (or biotinylated liposomes). The advantage of this two step protocol is that many different ligand conjugates can be synthesized and bound to their intended targets, independent of the drug carrier.[21,69]

The ability to bind avidin or streptavidin and, subsequently, antibodies to liposomes is inversely related to the size and surface density of PEG grafted onto liposomes.[79,90,117,119] Optimal amounts of avidin or streptavidin could be coupled either covalently (thiolated with SPDP) to MPB-PE or non-covalently to a biotinylated-PE, at 2.0–2.5 mol% of PEG-2000 in the liposomes or 1 mol% of PEG-5000.[19,120,121] These PEG concentrations also represent the point at which avidin-induced liposome aggregation was prevented[120] and calcium-induced fusion is inhibited.[122] However, these grafting densities are below the optimal 5–10 mol% needed for maximal circulation times.[11–15,78,84,91] Incorporation of 5 mol% PEG-2000-DSPE into liposomes dramatically reduced the amount of avidin and therefore, the amount of biotinylated-IgG bound to the liposomes.[79,117,119–121] Increasing the size of PEG-DSPE to 5000 Da further reduced the amount of biotinylated-IgG bound to the bilayer.[19,79,120,121] However, the avidin bridge was able to extend the bound IgG beyond the PEG layer, such that some target binding was retained even with PEG-5000-DSPE.[79] Even with the inefficient avidin-mediated coupling chemistry and low antibody densities on PEG-2000-liposomes, a therapeutic advantage was shown for targeted therapy compared to non-targeted therapy with DXR-loaded liposomes in the treatment of mice with KLN-205 squamous cell carcinoma of the lung.[24,41] We can hypothesize at least two reasons for this success: at low antibody coupling densities the circulation times of immunoliposomes are long, and these preparations had good target binding.

III.3. Coupling ligands to the end of a spacer molecule

As stated above, binding ligands to sterically stabilized liposomes poses an interesting dilemma. How can we efficiently bind ligands to liposomes when steric barriers exist to the approach of these molecules to the liposome surface? Moving the coupling chemistry away from the liposome surface to a region beyond the influence of the steric barrier through use of spacer molecule (e.g., a short hydrocarbon spacer between biotin or maleimide and the PE anchor,[119,123] or through use of a bridging molecule (e.g., avidin)) accomplishes this to some degree. However, the most versatile solution is to bind ligands to the functionalised terminal ends of the actual polymer used to stabilize the bilayer, e.g., PEG.[19,42,47,61,79–81,90,92–105] The ligand-PEG-lipid conjugate may be constructed either before[47,61,99,102,105] or after[42,79–81,90,92,93,95–98,100,101] liposome formation and proceeds unencumbered by the steric barrier of the polymer, and without an increase in liposome aggregation. Both methods use heterobifunctional derivatives of PEG[124,125] as a bridge from the liposome surface to the PEG perimeter.

For convenience, the resulting conjugates can be loosely divided into four functionally-related groups; thiol-reactive,[80,90,93,99,100,102,103] carboxyl,[81,92,94] hydrazide[42,79,90,97,98,101,126] and amino-succinimidyl.[47,96,105,127]

Several thiol-reactive functionalised PEG lipids conjugates have been prepared that have been used to couple antibodies or their fragments to sterically stabilized immunoliposomes (SIL). These functionalised PEG-lipids include: N-pyridyldithiopropionyl-PEG-2000-PE (PDP-PEG-PE),[80,90,99] N-pyridyldithiopropionyl-PEG-800-DMPE (PDP-PEG-800-DMPE,[103] bromoacetyl-PEG-2000-PE (BA-PEG-PE),[102] N-β-maleimidopropionyl-PEG-PE (BMP-PEG-1000-PE and BMP-PEG-5000-PE),[100] 4-(N-maleimidomethyl)cyclohexane-1-carboxyl-PEG-2000-DSPE (MMC-PEG-DSPE),[128] β-(N-maleimido)propionyl-PEG-2000-DSPE (MP-PEG-DSPE)[128] and N-(6-maleiimide caproyloxy polyethyleneglycol succinyl)-DPPE (mal-PEG-DPPE).[93] Fab′ and $F(ab')_2$ antibody fragments were bound to maleimide-containing BMP-PEG liposomes,[100] MP-PEG-liposomes,[128] MMC-PEG-liposomes[128] or mal-PEG-liposomes[93] respectively. This method has the advantage of being very simple, with very few manipulations of either the fragments or liposomes. Stable coupling of antibody fragments proceeds efficiently, with the antigen binding domains remaining unmodified and oriented away from the liposome. Fc receptor-mediated uptake of sterically stabilised immunoliposomes by the cells of the MPS may be avoided when using antibody fragments where the Fc region is cleaved before coupling.[54–56,100] Whole antibodies were attached to PDP-PEG-liposomes either by a disulfide bond with a mAb activated by SATP (N-succinimidyl-3-(S-acetylthio)propionate),[103] or by a thiol-ether bond after introducing a maleimide group onto the Ab by coupling with SMPB.[80,90] With the latter method a very high Ab density with very high coupling efficiencies could be achieved at a PEG-lipid content of 5 mol% (1% PDP-PEG-2000-DSPE and 4% mPEG-2000-DSPE), with the conservation of target binding in vitro.[80,90] Increased cytotoxicity in vitro was also observed for targeted, DXR-loaded liposomes compared with their non-targeted counterparts.[80,90] Kirpotin et al. were able to efficiently conjugate Fab′ fragments to PEG-containing liposomes.[128] Binding and internalization of these immunoliposomes, which had Fab′ fragments conjugated to the terminal end of PEG-lipids, were unaffected by the surface density of PEG. However, when sterically stabilized immunoliposomes were formulated with the Fab′ fragments conjugated to the bilayer surface, the cell binding affinity and the subsequent internalization was reduced as the PEG densities increased.[128]

PDP-PEG-2000-DSPE was also used to construct ligand-PEG-lipid derivates of both a pentapeptide and an oligosaccharide.[99,102] A bromoacetyl derivative of the oligosaccharide sialyl-Lewis x, an E-selectin-mediated cellular adhesion inhibitor, was coupled to the thiolated lipid PDP-PEG-DSPE. When the sialyl Lewis x conjugate was incorporated into PEG-liposomes, it greatly enhanced the inhibition of cell adhesion, relative to the free ligand.[99,102] This is postulated to be due to the multivalent nature of a ligand attached to liposomes.

N-(3-carboxypropionoyl-PEG-3500-DSPE (DSPE-PEG-3500-COOH) (5.4 mol%) and N-(3-carboxypropionoyl-PEG-2000-DSPE (DSPE-PEG-2000-COOH) (6 mol%) were incorporated into liposomes then coupled to carbodiimide-activated proteins, glu-plasminogen[81] or antibodies.[92,94] These studies demonstrated an increase in target binding in vitro when the ligand was coupled to the distal end of PEG compared to PEG-liposomes with surface bound ligands.[81,92,94]

The Fc portion of an antibody can be oxidized with periodate to form a reactive aldehyde which can be coupled to liposomes containing hydrazide-PEG-2000-DSPE (HZ-PEG-DSPE) by formation of a hydrazone bond. This simple procedure, is one which produces antibody-liposome conjugates with unencumbered antigen binding regions, unlike many of the procedures described above.[42,79,97,98,101,126]

Another very simple conjugation procedure involves a single step for linking liposomes containing *p*-nitrophenylcarbonyl-PEG-3350-PE (NP-PEG-PE) to any ligand with a primary amine, via a stable urethane bond.[127] Quenching unreacted groups on NP-PEG-PE may be very important because these liposomes could bind to non-specific proteins in circulation. Amino-PEG-2000-DSPE has been developed for the dual purposes of coupling targeting ligands and conferring a net positive charge to the liposomes.[96] It has been suggested that cationic liposomes containing the amino-PEG-DSPE could be used to stabilize DNA/lipid complexes for systemic targeted gene delivery.[96]

Sterically stabilized, folate-targeted liposomes have been prepared in two different ways, using PEG-*bis*-amine (M.W. approx. 3350) as a precursor for amine-PEG-folate or amine-PEG-SH. First, amine-PEG-folate was synthesized and mixed with N-succinyl-PE.[47,61] Alternatively, liposomes were prepared with maleiimidecaproyl-PE (MC-PE) and egg-PC (4:1) and reacted first with amino-PEG-SH then with a N-hydroxysuccinimide ester of folate.[104,105] The resulting folate-PEG-PE conjugate (0.1 mol%) was mixed with 4 mol% PEG-2000-DSPE and other lipids to form liposomes. Folate targeted liposomes were shown to be internalized into cultured cells by receptor-mediated endocytosis after effective ligand mediated delivery.[105] Also, an increased in vitro cytotoxicity was shown for DXR-loaded folate targeted liposomes over non-targeted liposomal DXR and free DXR.[47]

Folate targeted liposomes were also used to make a lipid/DNA complex for gene transfer.[61] Poly-L-lysine (PLL) condensed DNA was complexed with pH-sensitive, folate targeted, anionic liposomes. Tumour cell uptake and transfection by negatively, but not positively charged complexes of lipid/DNA were shown to be dependent on both folate targeting and pH sensitivity. However, this study used only 0.1 mol% PEG-lipids in the lipid/DNA complex formulation.[61] Effective targeted gene therapy in vivo may require extended circulation times. An increase in circulation times of lipidic DNA vectors could be achieved by using PEG-lipids in the liposomes. However, PEG-lipids were shown to stabilize PE-containing bilayers, reducing bilayer destabilizing events which may be important for effective gene transfer.[122] Cleavable[129] and transferable[122] PEG-lipids have been developed. The possibility exists that a targeted lipid/DNA complex could be delivered to an in vivo target site where uptake and transfection proceeds after the PEG coat is cleaved or transferred from the vector.

It might be useful if all of the mPEG-PE could be substituted with functionalised PEG-lipids without incurring a loss of liposome integrity or impairing long circulation times. No significant loss of circulation time or alteration in biodistribution occurred when amino-PEG-PE,[96] HZ-PEG-PE[98] or PE-PEG-COOH[81] were incorporated into liposomes as the only PEG-lipid component. Furthermore, vesicle aggregation was not observed during ligand coupling to the liposomes using any of the aforementioned functionalised PEG-lipids.

Another consideration in the choice of linker lipids is the potential for increased immunogenicity of the immobilized cross-linking reactive groups present on the terminal ends of the flexible polymers. Maleimide residues bound to DPPE and incorporated into liposomes were found to be the most immunogenic of six thiol reactive reagents tested with bromoacetyl groups eliciting the lowest immune response.[130] The immunogenicity of the immunoliposomes is discussed in Section IV.6.

IV. Potential problems with targeted liposomes

Several problems can be identified which may require alterations in our strategies for the use of targeted liposomes, particularly in vivo. These problems, together with possible solutions are given theoretical consideration below.

IV.1. Rapid clearance of immunoliposomes

The use of targeted liposomes for in vivo applications was stalled for a number of years because "classical" liposomes, i.e., liposomes lacking surface sterically stabilizing molecules such as PEG, were rapidly removed from circulation when antibodies were attached at the liposome surface.[3] The clearance of immunoliposomes may occur primarily via recognition of the Fc portion of the antibodies by cells of the mononuclear phagocyte system.[56,57,93] It was realized that, if one was to achieve targeting in vivo (other than to the cells of the MPS which liposomes naturally target), it would be necessary to increase substantially the circulation half-lives of the immunoliposomes over that seen for "classical" liposomes.

Immunoliposomes formed by attaching antibodies to the surface of PEG-containing liposomes, or to the PEG terminus (see Section III), are cleared from blood with half-lives on the order of several hours.[17,22,23,79,80,93,98,112] This is an order of magnitude slower than the rate of clearance of "classical" liposomes and should be sufficiently long to allow target recognition of at least some non-MPS cells/tissues in vivo.

It has been demonstrated that the clearance rate of long-circulating immunoliposomes is proportional to the antibody density at the liposome surface, and above densities of approximately 5×10^{-4} μmol Ab (75 μg Ab)/μmol phospholipid (PL) the immunoliposomes are removed from circulation at a rate rapid enough that the potential exists to significantly interfere with most types of in vivo target recognition.[80] At low surface densities of bound antibody (10–25 Ab molecules per 100 nm vesicles) the immunoliposomes retained favourable circulation times.[20,79,80] Ligands like plasminogen and the YIGSR pentapeptide appear to be exceptions, as, at high densities (600 plasminogen molecules or 200 YIGSR peptides per vesicle), liposome circulation times were only slightly reduced from that observed in liposomes without bound ligands.[42,81] Several studies have shown no difference in circulation times between targeted and non-targeted liposomes when proteins or antibodies were bound to the surface of PEG-grafted liposomes, as opposed to the PEG terminus, but as discussed previously, in these liposomes target binding is compromised.[22,112]

It becomes a balancing act to achieve Ab densities which are not high enough to trigger rapid removal of the immunoliposomes, but which are high enough to lead to good target recognition and binding. Fortunately, it appears that target binding is not particularly sensitive to Ab density and good target recognition can be achieved in the Ab density range of approximately 20 to 40 μg Ab/μmol PL, an antibody density which is compatible with circulation half-lives of several hours for immunoliposomes.[80]

Other strategies which may decrease the clearance rate of targeted liposomes include the use of Ab fragments or non-antibody ligands, as discussed above. Little or no information is presently available on the effect of ligand or Ab fragment densities on target binding and pharmacokinetics of Stealth® liposomes.

IV.2. Target tissue heterogeneity

For some targeting applications, e.g., delivery of immune modulators, it may not be necessary to deliver therapeutic molecules to every diseased cell, while for other applications, e.g., delivery of antineoplastic drugs, this would be highly desirable. While some monoclonal targets exist, most in vivo target cells will exhibit a lesser or greater degree of heterogeneity in their expression of the target epitope. The question then becomes: how is it possible to deliver therapeutics to the target cells lacking the epitope being targeted by the immuno- or ligando-liposomes? The question of heterogeneity is particularly relevant for the targeted delivery of anticancer drugs, since cancer cells, with a few exceptions, are notorious for their heterogeneity.

One possible solution would be to have a stable of monoclonal (or polyclonal) antibodies or ligands, which are capable of recognizing all of the cells in a heterogeneous population of cells, available for attachment to liposomes. From a practical viewpoint, this would be difficult if not impossible to achieve.

Another approach is to take advantage of the 'bystander effect'. As seen in Chapters 4.2 and 4.3, passively targeted liposomes can localize in solid tumours and gradually leak their drug into the interstitial fluid where it will diffuse, reaching many tumour cells which are remote from the actual locale of the liposomes. Whether or not this so-called 'bystander effect' will be equally effective for ligand- or antibody-targeted liposomes needs to be considered further. In theory, there is no reason why there should not be a 'bystander effect', since targeted liposomes have drug leakage rates very similar to non-targeted liposomes. However, this naive view is complicated by considerations of whether or not the targeted epitope is internalizing or non-internalizing. A liposome targeted to an internalizing epitope may be ingested by a target cell before a significant amount of drug is released outside the cell, i.e., little or no bystander effect might occur in the case of an internalizing epitope. In the case of a non-internalizing epitope, binding of the targeted liposome to the cell surface will present the opportunity for leakage of the drug from the liposome and diffusion within the interstitial fluid of a solid tumour, increasing the chance of a bystander effect. However, still considering only solid tumours, the situation is further complicated by the presence of a 'binding site barrier' as discussed below.

IV.3. Binding site barrier

The 'binding site barrier' hypothesis, introduced by Weinstein and van Osdol,[131] suggests that antibodies (and other ligands) will bind to the first target cells they encounter, which in the case of solid tumours will tend to be to the cells at the periphery of the tumour. This will retard or prevent the penetration of the targeted therapy further into the tumour interior, although allowing more time for penetration, or increasing the concentration of targeting molecules, will overcome this to some extent.[131] According to this hypothesis, non-targeted (i.e., passively targeted) liposomes will have greater penetrability into solid tumours with leaky vasculatures compared to targeted liposomes. This greater penetrability, with its increased opportunity for a 'bystander effect', will lead to increased cytotoxicity relative to targeted formulations. As can be seen in Section V there is some evidence that this is indeed the case.

Well-developed solid tumours are not the only targets which we must consider. Passively targeted liposomes may work well on solid tumours which have grown to a point where angiogenic factors are produced. However, in micrometastases prior to angiogenesis, passively targeted liposomes cannot take advantage of this extravasation mechanism. Here antibody- or ligand-targeted liposomes may have an advantage since the 'binding site barrier' should not exist for micrometastases consisting of only a few cells. Similar arguments can be made for the treatment of other conditions where the target consists of either single cells or a small clumps of cells, particularly if they are readily accessible from the vasculature. In addition to early metastatic disease, examples include metastatic cells migrating in blood or lymph, haematological cancers or vascular targets like infarcted tissue or blood clots.

IV.4. Down-regulation or sloughing of surface epitopes

Obviously, targeting of liposomes to specific cells will be most effective in a monoclonal population of cells with constant expression of the target epitope. However, in reality, epitopes may be up- or down-regulated depending on the life cycle of the cell or the stage of growth of, for example, a tumour. It will be difficult to effectively target tissues when the targeted epitopes have been down-regulated, but a clear understanding of the factors leading to up- or down-regulation of the epitopes may allow an effective targeting strategy to be developed. For example, some mammary lung carcinomas are reported to express surface carbohydrate epitopes when they are newly seeded and growing rapidly, but as the tumour becomes established these epitopes are down-regulated.[132] In this case, the most effective time to treat the tumour would presumably be very early in the disease when the tumour cells can still be readily accessed by the immunoliposomes, and the target epitope is still strongly expressed.

Also complicating the use of targeted liposomal therapy is the observation that some cells slough considerable amounts of the target epitope into blood, e.g., the sloughing of the Muc-1 antigen from breast cancer cells.[133] The presence of these sloughed epitopes is the basis of a number of diagnostic assays. When the antigen is present free in blood or other fluids, immunoliposomes or ligandoliposomes will bind to the free antigen and be impeded from reaching the target cells. A possible solution to this problem would be to pre-inject free antibody or ligand to clear the antigen from blood prior to injecting the targeted liposomes.

IV.5. Diffusion and redistribution of released drug

As discussed in Section IV.2 above, release of free drug from immunoliposomes bound to the surface of target cells in solid tumours, and its diffusion throughout the tumour, may increase the 'bystander effect' against cells not expressing the target epitope. However, for some target cells, e.g., cancer cells circulating in blood or present in the peritoneal cavity, release of the free drug in these turbulent environments will lead to rapid diffusion of the drug away from the cell and to its redistribution throughout the body. Diffusion and redistribution would decrease the advantage of targeting. In this situation we hypothesize that liposomes conjugated to antibodies or ligands against internalizing epitopes will be more effective than those targeted to non-internalizing epitopes, but this hypothesis awaits definitive experimental proof.

IV.6. Immune reactions against immunoliposomes

The production of human anti-mouse antibodies (HAMA) has been shown to occur when mouse mAbs are administered to human patients,[53,134] and Phillips and Dahman have demonstrated the immunogenic nature of immunoliposomes in mice.[135] Aside from the obvious problems associated with the transfer of immunoliposomal technology to the clinic, there is the more immediate problem of whether or not repeated injections of immunoliposomes in mouse tumour models will result in dramatic decreases in circulating half-lives after the first injection as a result of immune reactions against the immunoliposomes. In our hands, two repeat injections, in mice, of mouse antibodies attached to Stealth liposomes did not lead to a significant alteration of the immunoliposomal pharmacokinetics. However, another paper has demonstrated the generation of isotype-specific antibodies in mice injected with three repeated doses of immunoliposomes.[135]

Almost all therapeutic mAbs being produced at present are murine. Clearly, use of mouse antibodies in other species will be problematic. If we can reduce the immunogenicity of these in humans using smaller Ab fragments, e.g., those lacking the Fc region, while retaining specific binding to target cells and therapeutic efficacy, this will be an important accomplishment. Although Ab avidity is lost as one uses smaller fragments, due to loss of multivalent binding, coupling the fragments to liposomes restores multivalency to the fragments. To date, no investigator has conducted a thorough set of experiments comparing whole Ab with Ab fragments attached to the PEG terminus of long-circulating liposomes for immunogenicity, pharmacokinetics, binding, internalization, etc.

One solution to the problem of immunogenicity of immunoliposomes might be the use of humanized or chimerized antibodies or antibody fragments for clinical applications. The technology presently exists to make such constructs, however, no experiment data is available for immunoliposomes to suggest whether or not this approach will be successful.

Another solution might be to avoid the use of antibodies altogether by targeting the liposomes by means of ligands against receptors which are either overexpressed or uniquely expressed on the target cells. Examples of such ligands include folate, transferrin, apolipoproteins, asialofetuin, peptides recognizing adhesion molecules and many others. A potential problem with this is the expression of many of these receptors on normal cells, albeit at lower levels, which may lead to an increase in the non-selective toxicity of the targeted liposomes relative to that seen with antibody-liposomes. This is another research area for targeted liposomes which requires more experimentation before one can reach a conclusion as to the potential for this approach.

V. Therapeutic applications of targeted liposomes

V.1. In vitro model systems for testing targeted liposomes

Several promising models involving targeted liposomes have been developed, although no therapeutic results have yet been described.

The folate receptor is often overexpressed on epithelial cancer cells and therefore it has been used as a target for folate-bearing liposomes containing doxorubicin and DNA.[47,61] Folate-bearing liposomes loaded with doxorubicin, were shown to be more cytotoxic to target cells than free doxorubicin. This model also demonstrated cell-specific cytotoxicity for HeLa cells which overexpress the folate receptor in a co-culture with human lung fibroblasts not expressing the folate receptor.[47] No in vivo results have been reported for this promising application.

Targeting liposomes to the low density lipoprotein receptor has also been achieved by coupling low density lipoproteins, or the receptor binding portion, apolipoprotein-B, to the liposomal surface.[44,63] Again, the rationale is that cancer cells have a higher LDL receptor activity than normal cells.[136] LDL itself, has been evaluated as a drug carrier in cancer chemotherapy, however, its use is restricted to lipophilic drugs. Liposomes offer the advantage that they can carry hydrophilic drugs in the interior aqueous space or lipophilic drugs in the bilayer. Covalent coupling of LDL to liposomes loaded with hygromycin B resulted in specific cytotoxicity to leukemic L2C lymphocytes in vitro.[63] The coupling of apolipoprotein B to the liposomal surface also resulted in active uptake and internalization of the liposomal contents into the target cells.[44] Again, it will be interesting to see the results of in vivo experiments.

Transferrin (MW 76–81 kD) is a non-heme iron-binding glycoprotein found in the sera of most vertebrates and is responsible for carrying iron to cells. The transferrin receptor is expressed at high densities on some malignant cells and therefore it has been suggested as a useful target for anti-transferrin receptor immunotoxin conjugates, and transferrin-toxin conjugates.[137] However, transferrin has also been coupled to liposomes and this may also represent an effective strategy for targeting malignant cells.[43] Transferrin labeled liposomes containing cytosine arabinoside (ara-C) were shown to be internalized by CV-1 cells to a greater extent than non-labeled liposomes, resulting in enhanced cytotoxicity.[62]

Receptors for different growth factors are often overexpressed on cancer cells and therefore represent an excellent target for specific drug delivery via liposomes bearing anti-receptor antibodies, or the actual growth factor receptor ligand. Liposomes bearing nerve growth factor were shown to be specifically taken up by human melanoma cells in vitro.[46] Epidermal growth factor (EGF) has also been coupled to liposomes[60] and may represent a useful ligand for targeting liposomes because many cancer cells have been shown to overexpress the EGF receptor.

V.2. In vivo targeting of ligand liposomes: No therapeutic endpoint

The first in vivo attempt at targeting long-circulating liposomes were the experiments of Maruyama et al.[138] They demonstrated that the addition of GM_1 into immunoliposomes, targeted to lungs with mAb 34A against the lung endothelial anticoagulant protein thrombomodulin, resulted in significant increases in the binding and retention of these liposomes in lung over immunoliposome lacking GM_1.[138] These experiments have been repeated for PEG liposomes, also demonstrating increased lung binding for liposomes containing PEG-2000, but not PEG-5000.[19,94,138,139] The potential exists for using this model to test the ability of targeted liposomal anticancer drugs to treat lung malignancies (see below).

Torchilin et al. have described the successful in vivo targeting of PEG immunoliposomes to experimentally infarcted rabbit myocardium. Liposomes were targeted to exposed myosin through use of antimyosin Fab′ fragments.[23] These liposomes could be useful in targeted applications to plug and seal cardiac lesions.[140]

V.3. Experiments with targeted liposomes: Therapeutic endpoint

One example of successful in vivo targeting and therapeutic efficacy of immunoliposomes in a cancer model came from our laboratory.[24] Mice were injected with a murine lung squamous carcinoma cell line which seeds in the lung three days following intravenous injection. Using immunoliposomes loaded with doxorubicin and tagged with a mAb which recognizes epitopes on the tumor cell surface, resulted in significant decreases in the number and size of tumours and increases in survival times. Significant numbers of long-term survivors were found. Treatment of more advanced tumours was, however, unsuccessful, a result of either receptor down-regulation or lack of tumour penetration.[24]

GM_1 immunoliposomes have been used to deliver the lipophilic prodrug dpFUdR (dipalmitoyl-fluoro-deoxyuridine) to EMT-6 mouse mammary tumours seeded into the lung of mice. A significant therapeutic effect (%T/C of 165%) was observed when the mAb 34A-immunoliposomes were injected into mice at days 1 and 3 after tumour inoculation.[141]

In another therapeutic experiment, an ascitic ovarian cancer was targeted with DXR-loaded liposomes conjugated to Fab′ fragments of the mAb OV-TL3, directed against the non-internalizing OA3 antigen on human ovarian cancers. In these experiments, i.p. treatment of the ascitic tumour-bearing mice resulted in no improvement in therapeutic outcome compared to non-targeted liposomal doxorubicin.[113] These negative results could be due to lack of internalization of the targeted drug-liposome package, or to rapid release rates of the drug from the liposomes related to the type of coupling chemistry used (see Section II.3).

We have described the treatment of subcutaneous Caov.3 human ovarian cancer xenografts in mice.[142] In these experiments, non-targeted liposomal DXR was more effective than mAb B43.13-liposomal DXR in reducing the rate of tumour growth. In this more advanced tumour model, differences in the rate of penetration of the targeted liposomes versus non-targeted liposomes into the tumour could account for the observations (see the discussion on 'binding site barrier' in Section IV.3).

More success has been achieved in the treatment of a haematological malignancy, where the target cell population is present in the blood stream, and easily accessible by injected immunoliposomes.[142] In this model, CD19+ human B cell lymphoma was treated in SCID mice following either i.p. or i.v. inoculation of the tumour. We observed up to a 77% increase in life span against the i.v. tumour when the mice were treated with DXR-loaded mAb anti-CD19-immunoliposomes compared to non-targeted liposomes or free drug. Binding of mAb anti-CD19 to its epitope results in receptor-mediated internalization. These results confirm the importance of having both an accessible target and an internalizing epitope in achieving successful in vitro therapy with targeted liposomes.

Another promising application has been described by Kirpotin et al.[128] Breast tumour xenografts, overexpressing the p185^{HER2} receptor were treated with Fab′ fragments of a humanized recombinant mAb against the extracellular domain of HER2/*neu*. The antibody fragments were coupled either to the liposome surface using a short spacer or to the terminus of maleimide-PEG. Cell binding, internalization, and antiproliferative activity of the immunoliposomes was readily demonstrated for Fab′ fragments coupled to the PEG terminus.[128] A recent abstract described the treatment of p185^{HER2}-overexpressing breast tumour xenografts in nude mice with DXR-loaded mAb anti-HER2-immunoliposomes. Increased antitumour activity of targeted compared to non-targeted liposomes was reported, but no further details are available.[143]

VI. Conclusions

Recent years have seen significant progress in our attempts to achieve specific targeting of drugs in vivo. These include: development of long-circulating immunoliposomes, adaptation of existing chemistries and the development of new strategies for coupling ligands to the surface of liposomes, and the development of model systems in which to test liposome targeting. We can appreciate both the promise inherent in this field and the need to continue research to solve the new problems, and answer the new questions which have arisen.

References

1. Gabizon A, Catane R, Uziely B, Kaufman B, Safra T, Cohen R, Martin F, Huang A, Barenholz Y. Prolonged circulation time and enhanced accumulation in malignant exudates of doxorubicin encapsulated in polyethylene-glycol coated liposomes. Cancer Res 1994;54:987–992.
2. Northfelt DW, Martin FJ, Working P, Volberding PA, Russell J, Newman M, Amantea MA, Kaplan LD. Doxorubicin encapsulated in liposomes containing surface-bound polyethylene glycol: pharmacokinetics, tumour localization, and safety in patients with AIDS-related Kaposi's sarcoma. J Clin Pharmacol 1996;36:55–63.
3. Debs RJ, Heath TD, Papahadjopoulos D. Targeting of anti-Thy 1.1 monoclonal antibody

conjugated liposomes in Thy 1.1 mice after intravenous administration. Biochim Biophys Acta 1987;901:183–190.

4. Papahadjopoulos D, Gabizon A. Targeting of liposomes to tumor cells in vivo. Ann NY Acad Sci 1987;507:64–74.
5. Allen TM. The use of glycolipids and hydrophilic polymers in avoiding uptake of liposomes by the mononuclear phagocyte system. Adv Drug Del Rev 1993;13:285–309.
6. Woodle M, Lasic D. Sterically stabilized liposomes. Biochim Biophys Acta 1992;1113:171–199.
7. Allen TM, Chonn A. Large unilamellar liposomes with low uptake into the reticuloendothelial system. FEBS Lett 1987;223:42–46.
8. Ghosh P, Bachhawat BK. Role of surface glycolipids—natural or synthetic origin in the biodistribution of liposomes. J Liposome Res 1992;2:369–382.
9. Gabizon A, Papahadjopoulos D. The role of surface charge and hydrophilic groups on liposome clearance in vivo. Biochim Biophys Acta 1992;1103:94–100.
10. Mumtaz S, Ghosh PC, Bachhawat BK. Design of liposomes for circumventing the reticuloendothelial cells. Glycobiol 1991;1:505–510.
11. Klibanov AL, Maruyama K, Torchilin VP, Huang L. Amphipathic polyethyleneglycols effectively prolong the circulation time of liposomes. FEBS Lett 1990;268:235–237.
12. Allen TM, Hansen C, Martin F, Redemann C, Yau-Young A. Liposomes containing synthetic lipid derivatives of poly(ethylene glycol) show prolonged circulation half lives in vivo. Biochim Biophys Acta 1991;1066:29–36.
13. Blume G, Cevc G. Liposomes for the sustained drug release in vivo. Biochim Biophys Acta 1990;1029:91–97.
14. Senior J, Delgado C, Fisher D, Tilcock C, Gregoriadis G. Influence of surface hydrophobicity of liposomes on their interaction with plasma protein and clearance from the circulation: studies with the poly(ethylene glycol)-coated vesicles. Biochim Biophys Acta 1991;1062:77–82.
15. Papahadjopoulos D, Allen TM, Gabizon A, Mayhew E, Matthay K, Huang SK, Lee KD, Woodle MC, Lasic DD, Redemann C, Martin FJ. Sterically stabilized liposomes: improvements in pharmacokinetics and antitumor therapeutic efficacy. Proc Natl Acad Sci USA 1991;88:11460–11464.
16. Park JW, Hong K, Carter P, Asgari H, Guo LY, Keller GA, Wirth C, Shalaby R, Kotts C, Wood WI, Papahadjopoulos D, Benz CC. Development of anti-p185^{HER2} immunoliposomes for cancer therapy. Proc Natl Acad Sci USA 1995;92:1327–1331.
17. Mori A, Klibanov AL, Torchilin VP, Huang L. Influence of the steric barrier of amphipathic poly(ethyleneglycol) and ganglioside GM_1 on the circulation time of liposomes and on the target binding of immunoliposomes in vivo. FEBS Lett 1991;284:263–266.
18. Mori A, Kennel SJ, Huang L. Immunotargeting of liposomes containing lipophilic antitumor prodrugs. Pharm Res 1993;10:507–514.
19. Klibanov AL, Maruyama K, Beckerleg AM, Torchilin VP, Huang L. Activity of amphipathic poly(ethyleneglycol) 5000 to prolong the circulation time of liposomes depends on the liposome size and is unfavorable for immunoliposome binding to target. Biochim Biophys Acta 1991;1062:142–148.
20. Emanuel N, Kedar E, Bolotin EM, Smorodinsky NI, Barenholz Y. Targeted delivery of doxorubicin via sterically stabilized immunoliposomes: pharmacokinetics and biodistribution in tumor bearing mice. Pharm Res 1996;13:861–868.
21. Longman SA, Cullis PR, Choi L, de Jong G, Bally MB. A two-step targeting approach for delivery of doxorubicin-loaded liposomes to tumour cells in vivo. Cancer Chemother Pharmacol 1995;36:91–101.
22. Torchilin VP, Narula J, Halpern E, Khaw BA. Poly(ethylene glycol)-coated anti-cardiac myosin immunoliposomes: factors influencing targeted accumulation in the infarcted myocardium. Biochim Biophys Acta 1996;1279:75–83.
23. Torchilin VP, Klibanov AL, Huang L, O'Donnell S, Nossiff ND, Khaw BA. Targeted accumulation of polyethylene glycol-coated immunoliosomes in infarcted rabbit myocardium. FASEB J 1992;6:2716–2719.
24. Ahmad I, Longenecker M, Samuel J, Allen TM. Antibody-targeted delivery of doxorubicin entrapped in sterically stabilized liposomes can eradicate lung cancer in mice. Cancer Res 1993;53:1484–1488.
25. Fleuren GJ, Gorter A, Kuppen PJ, Litvinov S, Warnaar SO. Tumor heterogeneity and immunotherapy of cancer. Immunol. Rev. 1995;145:91–122.
26. Linehan DC, Goedegebuure PS, Eberlein TJ. Vaccine therapy for cancer. Ann Surg Oncol 1996;3:219–228.

27. Urban JL, Schreiber H. Tumor antigens. Annu Rev Immunol 1992;10:617–644.
28. Urdal DL, Hakomori S. Tumor associated ganglio-N-triosylceramide: target for antibody-dependent, avidin-mediated drug killing of tumor cells. J Biol Chem 1980;255:10509–10516.
29. Weinstein J, Blumenthal R, Sharrow S, Henkart P. Antibody-mediated targeting of liposomes. Binding to lymphocytes does not ensure incorporation of vesicle contents into the cells. Biochim Biophys Acta 1978;509:272–288.
30. Leserman LD, Barbet J, Kourilsky F, Weinstein JN. Targeting to cells of fluorescent liposomes covalently coupled with monoclonal antibody or protein A. Nature 1980;288:602–604.
31. Heath TD, Fraley RT, Papahdjopoulos D: Antibody targeting of liposomes: cell specificity obtained by conjugation of $F(ab')_2$ to vesicle surface. Science 1980;210:539–541.
32. Huang A, Huang L, Kennel S. Monoclonal antibody covalently coupled with fatty acid. A reagent for in vitro liposome targeting. J Biol Chem 1980;255:8015–8018.
33. Barbet J, Machy P, Leserman LD. Monoclonal antibody covalently coupled to liposomes: specific targeting to cells. J Supramol Struct Cell Biochem 1981;16:243–258.
34. Martin F, Kung V. Binding characteristics of antibody-bearing liposomes. Ann NY Acad Sci 1985;446:443–456.
35. Connor J, Sullivan S, Huang L. Monoclonal antibody and liposomes. Pharmac Ther 1985;28:341–365.
36. Berinstein N, Matthay KK, Papahadjopoulos D, Levy R, Sikic BI. Antibody-directed targeting of liposomes to human cell lines: role of binding and internalization on growth inhibition. Cancer Res 1987;47:5954–5959.
37. Loughrey H, Bally MB, Cullis PR. A non-covalent method of attaching antibodies to liposomes. Biochim Biophys Acta 1987;901:157–160.
38. Matthay KK, Heath TD, Badger CC, Bernstein ID, Papahadjopoulos D. Antibody-directed liposomes: comparison of various ligands for association, endocytosis, and drug delivery. Cancer Res 1986;46:4904–4910.
39. Heath TD, Bragman KS, Matthay KK, Lopez-Straubinger NG, Papahadjopoulos D. Antibody-directed liposomes: the development of a cell-specific cytotoxic agent. Biochem Soc Trans 1984;12:340–342.
40. Suzuki H, Zelphati O, Hildebrand G, Leserman L. CD4 and CD7 molecules as targets for drug delivery from antibody bearing liposomes. Exp Cell Res 1991;193:112–119.
41. Ahmad I, Allen TM. Antibody-mediated specific binding and cytotoxicity of liposome-entrapped doxorubicin to lung cancer cells in vitro. Cancer Res 1992;52:4817–4820.
42. Zalipsky S, Puntambekar B, Bolikas P, Engbers CM, Woodle MC. Peptide attachment to extremities of liposomal surface grafted PEG chains: Preparation of long-circulating from of laminine pentapetide, YIGSR. Bioconjug Chem 1995;6:705–708.
43. Vidal M, Sainte-Marie J, Philippot JR, Bienvenue A. The influence of coupling transferrin to liposomes or minibeads on its uptake and fate in leukemic L2C cells. FEBS Lett 1987;216:159–163.
44. Lundberg B, Hong K, Papahadjopoulos D. Conjugation of apolipoprotein B with liposomes and targeting to cells in culture. Biochim Biophys Acta 1993;1149:305–312.
45. Schreier H, Moran P, Caras IW. Targeting of liposomes to cells expressing CD4 using glycosylphosphatidylinositol-anchored gp120. Influence of liposome composition on intracellular trafficking. J Biol Chem 1994;269:9090–9098.
46. Rosenberg MB, Breakefield XO, Hawrot E. Targeting of liposomes to cells bearing nerve growth factor receptors mediated by biotinylated nerve growth factor. J Neurochem 1987;48:865–875.
47. Lee RJ, Low PS. Folate-mediated tumor cell targeting of liposome-entrapped doxorubicin in vitro. Biochim Biophys Acta 1995;1233:134–144.
48. Spanjer HH, van Berkel TJ, Scherphof GL, Kempen HJ. The effect of a water-soluble tris-galactoside terminated cholesterol derivative on the in vivo fate of small unilamellar vesicles in rats. Biochim Biophys Acta 1985;816:396–402.
49. Medda S, Das N, Bachhawat BK, Mahato SB, Basu MK. Targeting of plant glycoside-bearing liposomes to specific cellular and subcellular sites. Biotechnol Appl Biochem 1990;12:537–543.
50. Murahashi N, Sasaki A, Higashi K, Morikawa A, Yamada H. Relationship between the anchor structure of the galactosyl ligand for liposome modification and accumulation in the liver. Biol Pharm Bull 1995;18:82–88.
51. Hara T, Aramaki Y, Takada S, Koike K, Tsuchiya S. Receptor mediated transfer of Psv2cat DNA to mouse liver cells using asialofetuin labeled liposomes. Gene Therapy 1995;2:784–788.
52. Miller RA, Oseroff AR, Stratte PT, Levy R. Monoclonal antibody therapeutic trials in seven patients with T-cell lymphoma. Blood 1983;62:988–995.

53. Courtenay-Luck NS, Epenetos AA, Moore R, Larche M, Pecatasides D, Ritter MA. Development of primary and secondary immune responses to mouse monoclonal antibodies used in the diagnosis and therapy of maligant neoplasms. Cancer Res 1986;46:6489–6493.
54. Ravetch JV, Kinet J-P. Fc receptors. Annu Rev Immunol 1991;9:457–492.
55. Morgan EL, W.O. W. Biological activities residing in the Fc region of immunoglobulin. Adv Immunol 1987;40:61–134.
56. Aragnol D, Leserman L. Immune clearance of liposomes inhibited by an anti-Fc receptor antibody in vivo. Proc Natl Acad Sci USA 1986;83:2699–2703.
57. Harding JA, Engbers CM, Newman MS, Goldstein NI, Zalipsky S. Immunogenicity and pharmacokinetic attributes of poly(ethylene glycol)-grafted immunoliposomes. Biochem Biophys Acta (in press), 1997.
58. Winter G, Griffiths AD, Hawkins RE, Hoogenboom HR. Making antibodies by phage display technology. Annu Rev Immunol 1994;12:433–455.
59. Dekruif J, Storm G, Vanbloois L, Logtenberg T. Biosynthetically lipid modified human scfv fragments from phage display libraries as targeting molecules for immunoliposomes. FEBS Lett 1996;399:232–236.
60. Ishii Y, Aramaki Y, Hara T, Tsuchiya S, Fuwa T. Preparation of EGF labeled liposomes and their uptake by hepatocytes. Biochem Biophys Res Commun 1989;160:732–736.
61. Lee RJ, Huang L. Folate-targeted, anionic liposome-entrapped polylysine-condensed DNA for tumor cell-specific gene transfer. J Biol Chem 1996;271:8481–8487.
62. Brown PM, Silvius JR. Mechanisms of delivery of liposome-encapsulated cytosine arabinoside to CV-1 cells in vitro. Fluorescence-microscopic and cytotoxicity studies. Biochim Biophys Acta 1990;1023:341–351.
63. Vidal M, Sainte-Marie J, Philippot JR, Bienvenue A. LDL-mediated targeting of liposomes to leukemic lymphocytes in vitro. EMBO J 1985;4:2461–2467.
64. Lasic DD. Liposomes: from physics to applications. Elsevier Science Publishers B.V., Amsterdam, 1993.
65. Machy P, Barbet J, Leserman L. Differential endocytosis of T and B lymphocyte surface molecules evaluated with antibody-bearing fluorescent liposomes containing methotrexate. Proc Natl Acad Sci USA 1982;79:4148–4152.
66. Matthay KK, Abai AM, Cobb S, Hong K, Papahadjopoulos D, Straubinger RM. Role of ligand in antibody-directed endocytosis of liposomes by human T-leukemia cells. Cancer Res 1989;49:4879–4886.
67. Huang A, Kennel SJ, Huang L. Interactions of immunoliposomes with target cells. J Biol Chem 1983;258:14034–14040.
68. Ciechanover A, Schwartz AL, Dautry-Varsat A, Lodish HF. Kinetics of internalization and recycling of transferrin and the transferrin receptor in a human hepatoma cell line. J Biol Chem 1983;258:9681–9689.
69. Trubetskoy VS, Berdichevsky VR, Efremov EE, Torchilin VP. On the possibility of the unification of drug targeting systems. Studies with liposome transport to the mixtures of target antigens. Biochem Pharmacol 1987;36:839–842.
70. Juliano RL, Stamp D. Effects of particle size and charge on the clearance of liposomes and liposome-encapsulated drugs. Biochim Biophys Acta 1975;1113:651–658.
71. Senior JH. Fate and behaviour of liposomes in vivo: a review of controlling factors. Crit Rev Therap Drug Carrier Sys 1987;3:123–193.
72. Hwang KJ. Liposome Pharmacokinetics. In: Ostro MJ, ed. Liposomes: from Biophysics to Therapeutics. New York: Marcel Dekker, 1987;109–156.
73. Harashima H, Kiwada H. Liposomal targeting and drug delivery: kinetic consideration. Adv Drug Del Rev 1996;19:425–444.
74. Torchilin VP, Shtilman MI, Trubetskoy VS, Whitman K, Milstein AM. Amphiphilic vinyl polymers effectively prolong liposome circulation time in vivo. Biochim Biophys Acta 1994;1195:181–184.
75. Woodle MC, Engbers CM, Zalipsky S. New amphipatic polymer-lipid conjugates forming long-circulating reticuloendothelial system-evading liposomes. Bioconjug Chem 1994;5:493–494.
76. Zalipsky S, Hansen CB, Oaks JM, Allen TM. Evaluation of blood clearance and biodistribution of poly(2-oxazoline)-grafted liposomes. J Pharm Sci 1996;85:133–137.
77. Emanuel N, Kedar E, Bolotin EM, Smorokinsky NI, Barenholz Y. Preparation and characterization of doxorubicin-loaded sterically stabilized immunoliposomes. Pharm. Res 1996;13:352–359.
78. Maruyama K, Kennel SJ, Huang L. Lipid composition is important for highly efficient target binding and retention of immunoliposomes. Proc Natl Acad Sci USA 1990;87:5744–5748.

79. Allen TM, Agrawal AK, Ahmad I, Hansen CB, Zalipsky S. Antibody-mediated targeting of long-circulating (Stealth®) liposomes. J Liposome Res 1994;4:1–25.
80. Allen TM, Brandeis E, Hansen CB, Kao GY, Zalipsky S. A new strategy for attachment of antibodies to sterically stabilized liposomes resulting in efficient targeting to cancer cells. Biochim Biophys Acta 1995;1237:99–108.
81. Blume G, Cevc G, Crommelin MD, Bakker-Woudenberg LA, Kluft C, Storm G. Specific targeting with poly(ethylene glycol)-modified liposomes: coupling of homing devices to the ends of the polymeric chains combines effective target binding with long circulation times. Biochim Biophys Acta 1993;1149:180–184.
82. Torchilin VP, Klibanov AL. Coupling of ligands with liposome membranes. Drug Target Del 1993;2:227–238.
83. Weiner AL. Chemistry and biology of immunotargeted liposomes. In: Tye P, Ram BP, eds. Targeted therapeutic systems. New York: Marcel Dekker, 1990;305–336.
84. Allen TM, Hansen C, Rutledge J. Liposomes with prolonged circulation times: factors affecting uptake by reticuloendothelial and other tissues. Biochim Biophys Acta 1989;981:27–35.
85. Maruyama K, Mori A, Bhahra S, Subbiah MTR, Huang L. Prolonged circulation time in vivo of large unilamellar liposomes composed of distearoylphosphatidylcholine and cholesterol containing amphipathic poly(ethylene glycol). Biochim Biophys Acta 1992;1128:44–49.
86. Kenworthy AK, Simon SA, McIntosh TJ. Structure and phase behavior of lipid suspensions containing phospholipids with covalently attached poly(ethylene glycol). Biophys J 1995;68:1903–1920.
87. Woodle MC, Matthay KK, Newman MS, Hidayat JE, Collins LR, Redemann C, Martin FJ, Papahadjopoulos D. Versatility in lipid compositions showing prolonged circulation with sterically stabilized liposomes. Biochim Biophys Acta 1992;1105:193–200.
88. Torchilin VP, Omelyanenko VG, Papisov MI, Bogdanov AA, Trubetskoy VS, Herron JN, Gentry CA. Poly(ethylene glycol) on the liposome surface: on the mechanism of polymer coated liposome longevity. Biochim Biophys Acta 1994;1195:11–20.
89. Chonn A, Cullis PR. Ganglioside GM_1 and hydrophilic polymers increase liposome circulation times by inhibiting the association of blood proteins. J Liposome Res 1992;2:397–410.
90. Hansen CB, Kao GY, Moase EH, Zalipsky S, Allen TM. Attachment of antibodies to sterically stabilized liposomes: evaluation, comparison and optimization of coupling procedures. Biochim Biophys Acta 1995;1239:133–144.
91. Maruyama K, Yuda T, Okamoto A, Ishikura C, Kojima S, Iwatsura M. Effect of molecular weight in amphipathic polyethyleneglycol on prolonging the circulation time of large unilamellar liposomes. Chem Pharm Bull 1991;39:1620–1622.
92. Takizawa T, Maruyama K, Iwatsuru M. Novel immunoliposomes modified with amphipathic polyethyleneglycols conjugated at their distal terminals to monoclonal antibodies. J Liposome Res 1996;6:261.
93. Maruyama K, Takizawa T, Takahashi N, Tagawa T, Nagaike K, Iwatsuru M. Factors influencing longevity and target binding of PEG-immunoliposomes conjugated antibodies at PEG's terminals. J Liposome Res 1996;6:206–207.
94. Maruyama K, Takizawa T, Yuda T, Kennel SJ, Huang L, Iwatsuru M. Targetability of novel immunoliposomes modified with amphipathic poly(ethylene glycol)s conjugated at their distal terminals to monoclonal antibodies. Biochim Biophys Acta 1995;1234:74–80.
95. Zalipsky S, Hansen BC, Lopes de Menezes DE, Allen TM. Long circulating, polyethylene glycol-grafted immunoliposomes. J Control Rel 1996;39:153–161.
96. Zalipsky S, Brandeis E, Newman M, Woodle MC. Long circulating, cationic liposomes containing amino-PEG-phosphatidylethanolamine. FEBS Lett 1994;353:71–74.
97. Zalipsky S. Synthesis of end-group functionalized polyethylene glycol-lipid conjugates for preparation of polymer-grafted liposomes. Bioconjug Chem 1993;4:296–299.
98. Zalipsky S, Newman M, Punatambekar B, Woodle MC. Model ligands linked to polymer chains on liposomal surfaces: application of a new functionalized polyethylene glycol-lipid conjugate. Polym Materials Sci Eng 1993;67:519–520.
99. Zalipsky S, Mullah N, Harding JA, Gittelman J, Guo L, DeFrees SA. Poly(ethylene glycol)-grafted liposomes with oligopeptide or oligosaccharide ligands appended to the termini of the polymer chains. Bioconjug Chem 1997;8:111–118.
100. Shahinian S, Silvius JR. A novel strategy affords high-yield coupling of antibody Fab′ fragments to liposomes. Biochim Biophys Acta 1995;1239:157–167.
101. Ansell SM, Tardi PG, Buchkowsky SS. 3-(2-pyridyldithio)propionic acid hydrazide as a cross-linker in the formation of liposome-antibody conjugates. Bioconjug Chem 1996;7:490–496.

102. DeFrees SA, Phillips L, Guo L, Zalipsky S. Sialyl Lewis x liposomes as a multivalent ligand and inhibitor of E-selectin mediated cellular adhesion. J Am Chem Soc 1996;118:6101–6104.
103. Haselgrübler T, Amerstorfer A, Schindler H, Gruber HJ. Synthesis and applications of a new poly(ethylene glycol) derivative for the crosslinking of amines with thiols. Bioconjug Chem 1995;6:242–248.
104. Vogel K, Wang S, Lee RJ, Chmielewski J, Low PS. Peptide mediated release of folate targeted liposome contents from endosomal compartments. J Am Chem Soc 1996;118:1581–1586.
105. Lee RJ, Low PS. Delivery of liposomes into cultured KB cells via folate receptor-mediated endocytosis. J Biol Chem 1994;269:3198–3204.
106. Heath TD, Martin FJ. The development and application of protein-liposome conjugation techniques. Chem Phys Lipids 1986;40:347–358.
107. Schuber F. Chemistry of ligand-coupling to liposomes. In: Philippot JR, Schuber F, eds. Liposomes as tools in basic research and industry. CRC Press, 1994;21–39.
108. Allen TM, Hansen CB, Zalipsky S. Antibody-targeted Stealth® liposomes. In: Lasic DD, Martin F, eds. Boca Raton, FL: CRC Press, Inc., 1995;193–202.
109. Martin FJ, Hubbell Wl, Papahadjopoulos D. Immunospecific targeting of liposomes to cells: a novel and efficient method for covalent attachment of Fab′ fragments via disulfide bonds. Biochemistry 1981;20:4229–4238.
110. Martin FJ, Papahadjopoulos D. Irreversible coupling of immunoglobulin fragments to preformed vesicles. An improved method for liposome targeting. J Biol Chem 1982;257:286–288.
111. Leserman LD, Barbet J, Kourilsky F, Weinstein JN. Targeting to cells of fluorescent liposomes covalently coupled with monoclonal antibody or protein A. Nature 1980;288:602–604.
112. Suzuki S, Watanabe S, Masuko T, Hashimoto Y. Preparation of long-circulating immunoliposomes containing adriamycin by a novel method to coat immunoliposomes with poly(ethylene glycol). Biochim Biophys Acta 1995;124:9–16.
113. Vingerhoeds MH, Steerenberg PA, Hendriks JJGW, Kekker LC, van Hoesel QGCM, Crommelin DJA, Storm G. Immunoliposome-mediated targeting of doxorubicin to human ovarian carcinoma in vitro and in vivo. Br J Cancer 1996;74:1023–1029.
114. Uster PS, Allen TM, Daniel BE, Mendez CJ, Newman MS, Zhu GZ. Insertion of poly(ethylene glycol) derivatized phospholipid into preformed liposomes results in prolonged in vivo circulation time. FEBS Lett 1996;386:243–246.
115. Maruyama K, Mori A, Bhadra S, Ravi Subbiah MT. Proteins and peptides bound to long-circulating liposomes. Biochim Biophys Acta 1991;1070:246–252.
116. Litzinger D, Huang L. Biodistribution and immunotargetability of ganglioside-stabilized dioleoylphosphatidylethanolamine liposomes. Biochim Biophys Acta 1992;1104:179–187.
117. Loughrey HC, Choi LS, Wong KF, Cullis PR, Bally MB. Preparation of streptavidin-liposomes for use in ligand specific targeting applications. In: Gregoriadis G, ed. Liposome Technology. 2nd ed., Vol. III. Boca Raton, FL: CRC Press, 1993;163–178.
118. Loughrey HC, Choi LS, Cullis PR, Bally MB. Optimized procedures for the coupling of proteins to liposomes. J Immunol Methods 1990;132:25–35.
119. Corley P, Loughrey HC. Bonding of biotinated-liposomes to streptavidin is influenced by liposome composition. Biochim Biophys Acta 1994;1195:149–156.
120. Harasym TO, Tardi P, Longman SA, Ansell SM, Bally MB, Cullis PR, Choi LS. Poly(ethylene glycol)-modified phospholipids prevent aggregation during covalent conjugation of proteins to liposomes. Bioconjug Chem 1995;6:187–194.
121. Litzinger DC, Buiting A, MJ, van Rooijen N, Huang L. Effect of liposome size on the circulation time and intraorgan distribution of amphipathic poly(ethylene glycol)-containing liposomes. Biochim Biophys Acta 1994;1190:99–107.
122. Holland JW, Hui C, Cullis PR, Madden TD. Poly(ethylene glycol)-lipid conjugates regulate the calcium induced fusion of liposomes composed of phosphatidylethanolamine and phosphotidylserine. Biochemistry 1996;35:2618–2624.
123. Frisch B, Boeckler C, Schuber F. Synthesis of short polyoxyethylene-bases heterobifunctional cross-linking reagents. Applications to the coupling of peptides to liposomes. Bioconjug Chem 1996;7:180–186.
124. Francis GE, Delgado C, Fisher D, Malik F, Agrawal AK. Polyethylene glycol modification: relevance of improved methodology to tumour targeting. J Drug Targeting 1996;3:321–340.
125. Zalipsky S. Functionalized poly(ethylene glycol) for preparation of biologically relevant conjugates. Bioconjug Chem 1995;6:150–165.
126. Goren D, Horowitz AT, Zalipsky S, Woodle MC, Yarden Y, Gabizon A. Targeting of stealth

liposomes to erB-2 (Her/2) receptor: in vitro and in vivo studies. Br J Cancer 1996;74:1749–1756.

127. Klibanov AL, Serbina N, Torchilin VP, Huang L. Attachment of ligands to liposomes via PEG spacer for prolonged liposome circulation and targeting. J Liposome Res 1996;6:195–196.
128. Kirpotin D, Park JW, Hong K, Zalipsky S, Li W-L, Carter P, Benz CC, Papahadjopoulos D. Sterically stabilized anti-HER2 immunoliposomes: design and targeting to human breast cancer cells in vitro. Biochemistry 1997;36:66–75.
129. Kirpotin D, Hong K, Mullah N, Papahadjopoulos D, Zalipsky S. Liposomes with detachable polymer coating: destabilization and fusion of dioleoylphosphatidylethanolamine vesicles triggered by cleavage of surface-grafted poly(ethylene glycol). FEBS Lett 1996;388:115–118.
130. Boeckler C, Frisch B, Muller S, Schuber F. Immunogenicity of new heterobifunctional cross-linking reagents used in the conjugation of synthetic peptides to liposomes. J Immunol Methods 1996;191:1–10.
131. Weinstein JN, van Osdol W. Early intervention in cancer using monoclonal antibodies and other biological ligands: micropharmacology and the "binding site barrier". Cancer Res 1992;52:2747–2751.
132. Samuel J, Noujaim AA, Willans DJ, Brzezinska GS, Haines DM, Longenecker BM. A novel marker for basal (stem) cells of mammalian stratified squamous epithelia and squamous cell carcinomas. Cancer Res 1989;49:2465–2470.
133. Regimbald LH, Pilarski LM, Longenecker BM, Reddish MA, Zimmermann G, Hugh JC. The breast mucin MUC1 as a novel adhesion ligand for endothelial intercellular adhesion molecule 1 in breast cancer. Cancer Res 1996;56:4244–4249.
134. Schroff RW, Foon KA, Beatty SM, Oldham RK, Morgan AC. Human anti-mouse immunoglobulin responses in patients receiving monoclonal antibody therapy. Cancer Res 1985;48:879–885.
135. Phillips NC, Dahman J. Immunogenicity of immunoliposomes: reactivity against species-specific IgG and liposomal phospholipids. Immunol Lett 1995;45:149–152.
136. Vitols S, Garthon G, Ost A, Peterson C. Elevated low density lipoprotein receptor activity in leukemic cells with monocytic differentiation. Blood 1984;63:1186–1193.
137. Yazdi PT, Murphy TM. Quantitative analysis of protein synthesis inhibition by transferrin-toxin conjugates. Cancer Res 1994;54:6387–6394.
138. Maruyama K, Kennel SJ, Huang L. Lipid composition is important for highly efficient target binding and retention of immunoliposomes. Proc Natl Acad Sci USA 1990;87:5744–5748.
139. Holmberg E, Maruyama K, Litzinger DC, Wright S, Davis M, Kabalka GW, Kennel SJ, Huang L. Highly efficient immunoliposomes prepared with a method which is compatible with various lipid compositions. Biochem Biophys Res Commun 1989;165:1272–1278.
140. Khaw B-A, Torchilin VP, Vural I, Narula J. Plug and seal: prevention of hypoxic cardiocyte death by sealing membrane lesions with antimyosin liposomes. Nature Med 1995;1:1195–1198.
141. Mori A, Kennel SI, Waalkes MVB, Scherphof GL, Huang L. Characterization of organ-specific immunoliposomes for delivery of 3′, 5′-O-dipalmitryl-5-fluoro-2′-deoxyuridine in a mouse lung-metastasis model. Cancer Chemother Pharmacol 1995;35:447–456.
142. Allen TM, Ahmad I, Lopes de Menezes DE, Moase EH. Immunoliposome-mediated targeting of anti-cancer drugs in vivo. Biochem Soc Trans 1995;23:1073–1079.
143. Kirpotin D, Park JW, Hong K, Keller G, Benz C, Paphadjopoulos D. Binding and endocytosis of sterically stabilized anti-HER2 immunoliposomes by human breast cancer cells. Proc Am Assoc Cancer Res 1996;37:3186.

Lasic and Papahadjopoulos (eds.), Medical Applications of Liposomes

CHAPTER 4.7

Targeting of sterically stabilized liposomes to cancers overexpressing HER2/neu proto-oncogene

Dmitri B. Kirpotin[a,c], John W. Park[b], Keelung Hong[a,c], Yi Shao[a,c], Gail Colbern[c], Wei-wen Zheng[a,c], Olivier Meyer[a,c], Christopher C. Benz[b], and Demetrios Papahadjopoulos[a,c]

[a]*Department of Cellular and Molecular Pharmacology*, [b]*Department of Medicine, Division of Hematology-Oncology, University of California San Francisco, San Francisco, CA 94143, USA, and* [c]*California Pacific Medical Center Research Institute, San Francisco, CA 94115, USA*

Overview

I. Introduction

The potential of liposomes as vehicles for targeted delivery of drugs to diseased tissues has been long recognized.[1] There is a copious and convincing experimental evidence for specific association of the cells exposing a characteristic molecular "tag" with liposomes bearing "tag" recognition molecules, such as antibodies or antigen-binding antibody fragments.[2–5] Specific association with the target should lead to better therapeutic efficacy and less systemic side effects of liposome-encapsulated pharmaceuticals[2,3] which is especially important in cytotoxic chemotherapy of cancer where therapeutic indices are narrow and systemic toxicities are high. Introduction of liposome designs allowing avoidance of early clearance by the cells of mononuclear phagocytic system (MPS), longer circulation times, and increased likelihood for a liposome drug to reach its intended destination in the body[6–9] brought about a new interest in the liposome targeting. Grafting of hydrophilic, flexible chains of poly(ethylene glycol) with M_r of 2,000–5,000 (PEG) on the liposome surface (sterically stabilized, of Stealth® liposomes) proved to be among the most successful methods for introduction of long-circulating properties.[10] Targeting of PEG-stabilized liposomes was at first achieved by attachment of antibodies, via a hydrophobic anchor, onto the surface of liposomes containing also phospholipids with poly-(ethylene glycol)-modified head groups (PEG-DSPE).[11] Such liposomes had long-circulating properties, but their interaction with the target was reduced presumably by the same "steric repulsion" phenomenon which contributed to the mechanism of MPS avoidance.[12,13] Later, target-specific molecules (antibodies, sugars, or peptides) were attached to the distal termini of liposome-grafted poly(ethylene glycol) chains, leading to targeted sterically stabilized liposomes with fully preserved target-binding capability.[14–16]

This fruitful approach will be illustrated here by the studies of sterically stabilized liposomes targeted to cancer cells overexpressing the HER2/neu protooncogene. In contrast to a number of previous liposomal targeting systems for anticancer agents, sterically stabilized anti-HER2 immunoliposomes (anti-HER2 SIL) were not only capable of target-specific binding and internalization by cancer cells in culture,[17,18] but were also able to cross the vascular barrier and to be internalized by the target cancer cells in solid tumor xenografts following intravenous injection.[19] Anti-HER2 SSL carried the anticancer drug doxorubicin to HER2 overexpressing cancer xenografts more efficiently than their non-targeted counterparts.[20,21] Inclusion of a cationic lipid in the composition of anti-HER2 SIL resulted in the specific delivery of nucleic acids into HER2-overexpressing cancer cells.[22]

II. HER2 oncoprotein as a target recognition molecule in cancer

Malignant phenotype is often associated with the expression of protooncogenes. The HER2 (c-erbB-2, neu) protooncogene encodes a 185 KDa (1255 amino acids) receptor tyrosine kinase (p185HER2, ErbB-2, or HER2) which belongs to the family of receptor tyrosine kinases including also the products of epidermal growth factor (EGFR), HER3 (erbB-3), and HER4 (erbB-4) genes.[23,24] Overexpression of HER2 was first observed in 20–30% of breast carcinomas and was associated with aggressive tumor growth, high recurrence rate, and poor prognosis for the patients.[25,26] Further studies showed ubiquitous overexpression of HER2 in a variety of malignancies, including cancers of the ovary,[27] endometrium,[28] lung (non-small cell),[29] stomach,[30,31] pancreas,[32] bladder,[33] and prostate.[34] Especially high incidence of HER2 overexpression (up to 50%) was found in the breast ductal carcinoma in situ (DCIS), particularly in the lesions having high risk of recurrence.[35,36] The accumulated evidence suggests that HER2 is an important mediator of tumor growth which directly contributes to tumor onset and progression, and confers especially aggressive malignant phenotype.[23]

HER2 offers a number of advantages as a recognition marker for targeted delivery of anticancer agents. HER2 is a readily accessible cell surface protein with substantial levels of overexpression (10^5–10^6 copies/cell) in various malignancies.[37] In normal adult tissues HER2 occurs only in certain epithelial types, and at very low levels.[38] HER2-overexpression is relatively homogenous within primary tumors, and is maintained at metastatic sites, suggesting continuous requirement for high levels of HER2 throughout the malignant process,[39] while many other tumor-associated antigens show variable expression patterns within the tumor tissue and/or in the course of tumor progression. Last, but not least, is the fact that HER2 activation is accompanied by its internalization into the cell, which may occur upon interaction with an agonistic antibody.[40–42] Therefore, ligands targeted to HER2 by means of such antibodies would have better chance to enter the cell rather than stay attached to the cell surface.

The role of HER2 in the malignant progression brought about considerable effort to achieve antitumor effect by blocking the function of this receptor protein. A variety of monoclonal antibodies[40–43] as well as phage-display library generated single-chain Fv's[44,45] reactive with the extracellular portion of HER2 have been reported, offering a vast palette of target-recognition molecules suitable for liposome attachment. The variants with the highest antiproliferative effect also demonstrated the highest rate of cell internalization.[41] One such antibody, muMAb4D5, is highly reactive toward HER2[43] and inhibitory for the growth of HER2 overexpressing tumor cells in vitro[37] and in animal models.[46] This antibody was engineered into a fully humanized version, rhuMAbHER2, to reduce the potential for immunogenicity,[47] and entered clinical trials showing objective, but infrequent (12%) antitumor responses in patients with metastatic HER2-overexpressing breast cancer.[48]

III. Design of sterically stabilized anti-HER2 immunoliposomes

Successful design of antibody-targeted pharmaceutical liposomes for the treatment of solid tumors should satisfy a number of requirements aimed at maximum targeting effect of immunoliposomes administered systemically in the bloodstream. Antigen binding sites of the liposome-conjugated antibody must be accessible for unperturbed interaction with antigens on the surface of target cells. To ensure that immunoliposomes will reach their target cells, the rate of MPS clearance or other "non-productive" elimination of blood-borne immunoliposomes must be minimized in comparison with the rate of extravasation in the tumor. Since the liposome-conjugated antibody is a foreign protein likely to elicit host immune response, their immunogenicity must be minimized. Immunoliposomes must allow efficient loading and retention of a selected anticancer drug. And finally, the drug and antibody incorporation must be stable enough to permit liposomal entry into the tumor tissue without the loss of either of these agents.

The elements of immunoliposome design were chosen to maximally satisfy these requirements. Lipid composition of anti-HER2 immunoliposomes was based on hydrogenated soy phosphatidylcholine ($T_m = 54°C$) so that at the body temperature the liposome bilayer maintained the "solid" (gel) state. Cholesterol (40 mol.%) was included to increase the bilayer stability in the presence of plasma.[49] To reduce MPS clearance rate, the liposomes contained up to 5.7 mol.% of DSPE modified with methoxypoly (ethylene glycol) with molecular mass 1900.[7] To assist extravasation, the liposomes were of uniform, small size (90–110 nm) achieved by extrusion of hydrated lipid suspension (multilamellar vesicles) through track-etched polycarbonate membranes (pore sizes 100 and 50 nm) after several cycles of freezing and thawing.[50,52]

To avoid immunogenic effects of Fc portion and increased MPS clearance through specific recognition of immunoliposomes by the phagocytic cells carrying Fc receptor, we used Fab′ fragments instead of the whole anti-HER2 MAb. Fab′ fragments also allowed better way of conjugation to the liposome through unique thiol groups in the hinge region[51,52] resulting in the definite, correct (outward) orientation of the antigen-binding sites. The binding affinity of liposome-conjugated Fab′ was not compromised in comparison with the binding of the whole antibody (see below) probably because the presence of multiple Fab′ fragments on the liposome restored the multicenter interaction characteristic for the whole antibody which carries two antigen-binding sites. The Fab′ fragments were a recombinant protein, a portion of the fully humanized anti-HER2 MAb (rhuMAbHER2) developed by Genentech, Inc. (South San Francisco, CA, USA) for the therapy of cancers overexpressing HER2.[47,48] This antibody was derived from a murine prototype muMAb4D5 which induces endocytosis of HER2 receptor upon binding to its extracellular domain.[43] The use of Fab′ antibody fragments with humanized sequence further reduced the risk of immune reaction in human patients; recombinant technology provided stable, reproducible and more economical source of this protein, and the ability to induce internalization upon binding to the target antigen was favorable for intracellular delivery of liposome-associated pharmaceuticals.

Two types of conjugation linkers were used for attachment of anti-HER2 Fab′ to the sterically stabilized liposomes. Both linkers contained a hydrophobic anchor (DSPE or DPPE), and a thiol-reactive maleimide function. The use of aromatic maleimides, such as maleimidophenylbutyric acid derivatives (MPB-PE)[51,53] was avoided because of their reported ability to cause leakage of encapsulated drugs from liposomes.[54] Linkers of the first type consisted of DPPE or DSPE with their amino groups acylated by hydroxysuccinimide esters of (N-maleimido)methylcyclohexylcarboxylic acid (MMC-PE) or β-(N-maleimido)propionic acid (MP-PE). These linkers positioned the conjugated Fab′ fragment "in parallel" with the surface-attached PEG chains. The second type of linkers additionally included PEG chain ($M_r = 2,000$) between DSPE and maleimido group (MMC or MP) (Figure 1).[18,52] Linkers of this type positioned Fab′ fragments "in series" with surface-attached PEG, i.e., outside of the PEG "cloud" surrounding the liposome. The second type of conjugation was essential for preservation of the target-binding and, to the less extent, internalization of the anti-HER2 liposomes containing more than 1.2 mol.% of PEG-DSPE, as illustrated below.

Liposome-Fab′ conjugations were performed by incubation of SSL containing 1.2 mol.% of appropriate linker with anti-HER2 Fab′ (approx. 30 μmol of liposomal phospholipid) for 2 hours to overnight at 4°C and pH 7.2–7.4 in neutal

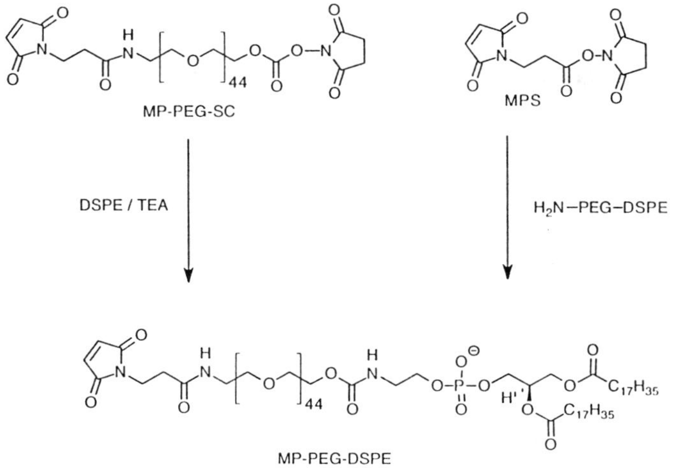

MP-PEG-SC MPS

DSPE / TEA H_2N–PEG–DSPE

MP-PEG-DSPE

Fig. 1. Synthesis of maleimido-terminated PEG-DSPE derivatives.

atmosphere. Under these conditions, major portion of the protein became covalently attached to the liposomes and co-eluted with the liposomes in the void volume fraction during gel-exclusion chromatography on Sepharose 4B. More than 90% of Fab′ bearing free thiol groups were coupled to the liposomes containing MP-PEG-DSPE as a conjugation linker, resulting in 50–60 Fab′/liposome, while the Fab′ coupling to MMCC-PE-containing liposomes somewhat decreased at PEG-DSPE content more than 2 mol.%, perhaps due to the increasing expulsion of the protein from the liposome surface by overlapping PEG chains. However, even at PEG densities characteristic for the emerging "brush" regime (at PEG(M_r = 2,000)-DSPE⟩5 mol.% of the liposome phospholipid, or, in our case, >3.3 mol.% or total lipid[55]), conjugation of Fab′ fragments (Mol. weight 46 kD) was efficient enough (40%), i.e., Fab′ molecules still were able to reach the liposome bilayer. Therefore, permeability of the PEG "cloud" for protein molecules may be higher than previously predicted.[12,56]

The anticancer drug of choice for this study was doxorubicin which is well characterized in a similar, non-targeted sterically stabilized liposome system (Doxil[57–59]). Practically quantitative drug loading (0.09–0.15 mg of doxorubicin/μmol of liposomal phospholipid) was performed prior to the conjugation of anti-HER2 Fab′ by ammonium sulfate gradient method[60] above the liposome T_m, using 250 mM ammonium sulfate in the inner space of the liposomes, and isoosmotic NaCl-MES-buffer, pH 5.5, as the outer buffer.

IV. Interaction of anti-HER2 SIL with cancer cells in vitro

Targeting properties of SSL with conjugated Fab′ fragments of rhuMAbHER2 were studied in the cultures of human breast carcinoma cells with high (SK-

BR-3, 10^6 HER2/cell) or low (MCF-7, 10^4 HER2/cell) expression of HER2.[37] Liposomes of HSPC and cholesterol, containing 2.6 mol.% of methoxyPEG-DSPE, 1.2 mol.% of MMC-PEG-DSPE, and 0.2 mol.% of a fluorescent lipid marker N-lissamine-rhodamine B-dihexadecanoyl phosphatidylethanolamine (Rh-PE) were prepared and conjugated to anti-HER2 Fab′ as described above. The liposomes were incubated with the cells at 37°C in the presence of fluorescein-labeled (FITC) transferrin which served as a marker for endosomes due to its ability to undergo endocytosis via transferrin receptors present at the surface of the cells. The liposomes were at first deposited on the surface of SKBR-3 cells, and further entered the cells and became co-localized with FITC-transferrin, as evidenced by yellow color produced by overlapping green (FITC) and red (Rh-PE) fluorescent colors (Figure 2A,B). Under the same experimental conditions, "target-negative" MCF-7 cells displayed only green punctate fluorescence of endocytosed FITC-transferrin, but not of the lipid label (Figure 2C); similar results were obtained with SKBR-3 cells incubated with Rho-PE labeled SSL lacking conjugated anti-HER2 Fab′.[18] These results confirmed the ability of anti-HER2 SIL to specifically bind to target cells and undergo endocytosis. Endocytosis of anti-HER2 liposomes without PEG-DSPE by SKBR-3 cells was also shown by electron microscopy using liposomes labeled with colloidal gold.[17]

Quantitative assessment of cell binding and endocytosis was performed with liposomes containing pH-sensitive probe 1-hydroxypyrene-3,6,8-trisulfonic acid (HPTS, pyranine). Fluorescence excitation spectrum of liposome-entrapped HPTS undergoes rapid changes in response to the liposome entry into acidic environment of endosomes and lysosomes, but it also has a pH-independent isosbestic point for accurate quantitation of the probe.[61] To ensure fast equilibration of proton concentrations across the liposome bilayer, in the studies using HPTS method we replaced HSPC with 1-palmitoyl-2-oleoylphosphatidylcholine (POPC) which forms bilayers maintaining more proton-permeable liquid crystalline structure rather than gel state at ambient conditions. Upon incubation with HER2-overexpressing SKBR-3 cells, HPTS-loaded anti-HER2 SSL became rapidly associated with the cells in neutral compartment (cell surface), followed by acidification of the liposome environment indicating endocytosis. The amount of cell-associated anti-HER2 SSL reached plateau after 3–4 hours of incubation, with more than 80% of liposomes endocytosed. At saturating concentrations, the uptake of anti-HER2 SIL by SKBR-3 cells, estimated from the fluorescence at HPTS isosbestic point, was in the range of 8,000–25,000 vesicles/cell, while the uptake of liposomes by "target-negative" MCF-7 cells was less than 100 vesicles/cell.[18]

Attachment of Fab′ to the liposomes via hinge thiol group and maleimide-activated hydrophobic linker[51] resulted in high conjugation yield without the loss of antigen-binding activity; however, the relative position of liposome-conjugated Fab′ and PEG was crucial for maintaining high uptake of anti-HER2 SIL by target cells. During incubation of SKBR-3 cells at the constant concentration of liposomes in the cell growth medium, the increasing PEG-DSPE content inhibited the uptake of anti-HER2 SSL prepared with MMC-DSPE linker, while no such inhibition occurred when anti-HER2 Fab′ were attached via MP-PEG-DSPE

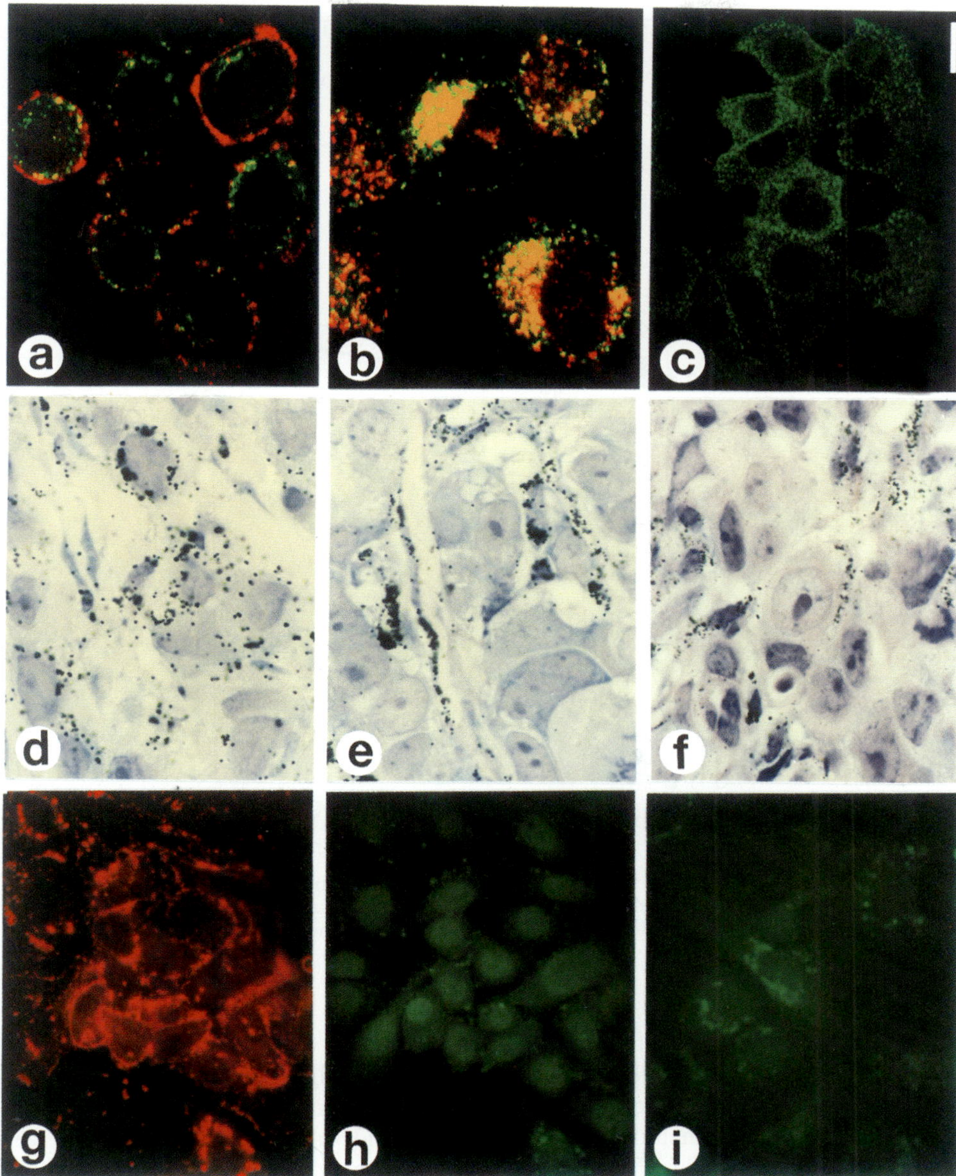

Fig. 2. *Panels A–C*: Confocal fluorescent microscopy of Rh-PE-labeled anti-HER2 SIL (red) and FITC-transferrin (green) co-incubated with breast cancer cells in cell culture at 37°C. SKBR-3 cells: 10 min. incubation (a); 30 min. incubation (b). MCF-7 cells: 30 min. incubation (c). *Panels d–f*: Localization of anti-HER2 SIL (d,f) and matched non-targeted SSL (e) in HER2-overexpressing (SK-BR-3, d,e) and low-expressing (MCF-7, f) breast cancer xenografts in nude mice 48 hours after intravenous administration. Liposomes are visualized as black grains by silver enhancement of liposome-entrapped colloidal gold. Staining with hematoxylin-eosin. *Panels g–i*: Uptake of FITC-ODN (green) and liposome lipid (Rh-PE-labeled, red) after incubation of SKBR-3 cells with PEG-coated anti-HER2 cationic liposome-ODN complex ([ODN]/[Lipid] = 0.007/22. (g) no free Fab′, lipid fluorescence; (h) no free Fab′, ODN fluorescence; (i) HER2 blocked by preincubation with excess free antiHER2 Fab′, ODN fluorescence.

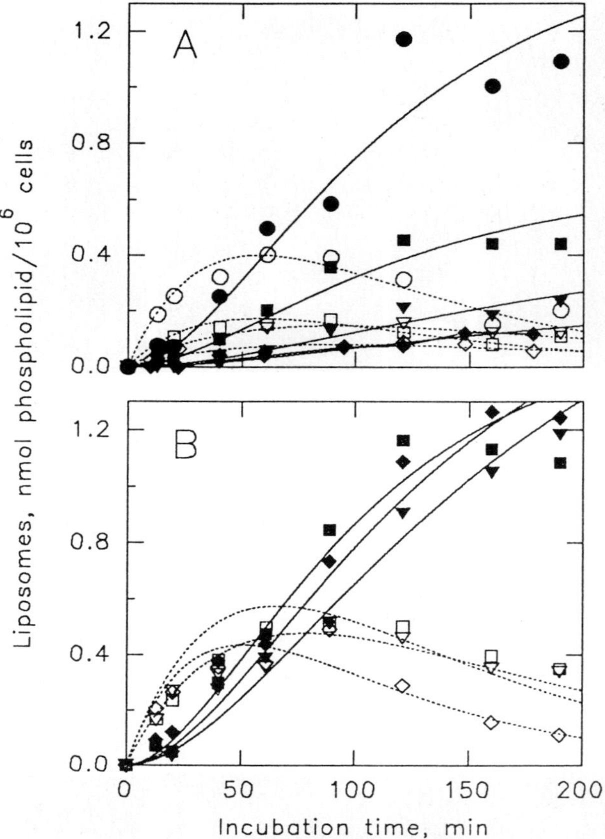

Fig. 3. Effect of PEG-DSPE content on the kinetics of anti-HER2 liposome association with SKBR-3 cells. (A) Anti-HER2 Fab′ attached via short linker MMC-DSPE (A) or to the distal end of PEG chain (linker MP-PEG-DSPE). Surface-bound liposomes: dashed line, hollow symbols. Endocytosed liposomes: solid line, filled symbols. PEG-DSPE (mol.%): none (○), 1.2 (□), 3.5 (▽), 5.7 (◇).

linker (Figure 3). Similar effect of the placement of a targeting antibody on the uptake of SIL by target cells/tissues was reported earlier[16,62] in the studies that used whole antibodies rather than Fab′ fragments. Since the cellular uptake of anti-HER2 liposomes includes steps of binding and subsequent endocytosis, we have studied the effect of PEG-DSPE and Fab′ placement on each of these steps separately. Liposome-cell binding was characterized by dissociation constants (K_d) estimated from the amounts of cell-associated anti-HER2 liposomes after incubation of SKBR-3 cells with various liposome concentrations at low temperature (4°C) that completely inhibits endocytosis. For anti-HER2 liposomes without PEG coating, K_d normalized to the amount of liposome-conjugated Fab′ was 12.0 ± 1.7 nM (mean ± SE), close to the reported values for the whole murine prototype anti-HER2 MAb 4D5 (6.0 nM) and its free Fab′ fragment (19.0 nM).[41]

When Fab′ fragments were conjugated through the "short" linker MMC-PE, 1.2 mol.% of PEG(M_r = 2,000)-DSPE decreased the binding affinity approximately three-fold, and at 3.5–5.7 mol.% of PEG-DSPE the binding affinity was 20 to 75 times lower (K_d = 320–900 nM). This is to be expected from the size of Fab′ fragment (6 nm in length[63]) and the thickness of PEG layer on the liposome surface (5–7 nm for PEG with M_r = 2,000 in "brush" regime[55,64]). On the contrary, conjugation of Fab′ to the termini of PEG chains did not affect liposome binding to the target cells at increasing PEG-DSPE content (K_d 13–15 nM for 1.2–5.7 mol.% PEG-DSPE).[18] First-order rate constants of liposome endocytosis (k_e) were determined from the kinetic curves of cell surface-bound and endocytosed liposomes obtained by HPTS method (Figure 3), as the ratio of liposome internalization rate to the steady-state surface concentration of liposomes.[65] Compared to K_d, k_e of anti-HER2 SIL prepared with the "short" linker MMC-PE was somewhat affected by increasing PEG-DSPE content (k_e decreased 2 times at 5.7 mol.% PEG-DSPE vs. 0%), and was not affected at all when MP-PEG-DSPE was used as a linker.[18] Thus, PEG interfered with the ability of liposome-conjugated Fab′ to bind to the cell surface antigen, and to the less extent, with its ability to induce endocytosis of the liposomes; however, this interference was completely abolished by conjugation of anti-HER2 Fab′ at the distal termini of liposome-grafted PEG chains.

Minimum requirements for the number of liposome-conjugated anti-HER2 Fab′ and for the cellular level of HER2 protein to achieve specificity and effectiveness of liposome uptake were established using anti-HER2 liposomes without PEG coating.[17,18] Binding of liposomes with the target cells increased in a linear manner as a function of Fab′ density, and reached saturation (plateau) at ~40 Fab′/vesicle; endocytosis of cell-bound liposomes occurred with 60% efficiency already at ~10 Fab′/vesicle.[18] Therefore, relatively few conjugated Fab′ were needed for the targeting. Similarly, the uptake of anti-HER2 liposomes by the cells with minimally elevated cellular levels of HER2 to allow classification as "HER2-positive" (MDA-MB-453: 44 ng HER2/mg cell protein, 6.52 ± 0.22 nmol of liposome phospholipid/mg cell protein)) was comparable to that by the cells with extremely high levels of HER2 expression (SKBR-3: 920 ng HER2/mg cell protein, 7.21 ± 0.45 nmol of liposome phospholipid/mg cell protein; BT-474: 550 ng HER2/mg cell protein, 4.47 ± 0.21 nmol of liposome phospholipid/mg cell protein). There was, however, pronounced difference in the uptake of anti-HER2 liposomes between the cells with elevated levels of HER2 and those with only basal HER2 expression (MCF-7: 7.3 ng HER2/mg cell protein, <0.01 nmol of liposome phospholipid/mg cell protein) (Figure 4).

Liposomes with Fab′ fragments of rhuMAbHER2 conjugated via MMC-PE linker were loaded with doxorubicin using ammonium sulfate gradient method. In contrast to the liposomes with aromatic maleimide derivative, MPB-PE,[16,54] the drug loading was practically quantitative even though the linker constituted 1.2 mol.% of total liposome lipid. Doxorubicin-loaded anti-HER2 liposomes showed efficient and specific in vitro cytotoxicity against HER2-overexpressing cancer cells. After 1 hour incubation, doxorubicin delivered by anti-HER2 lipo-

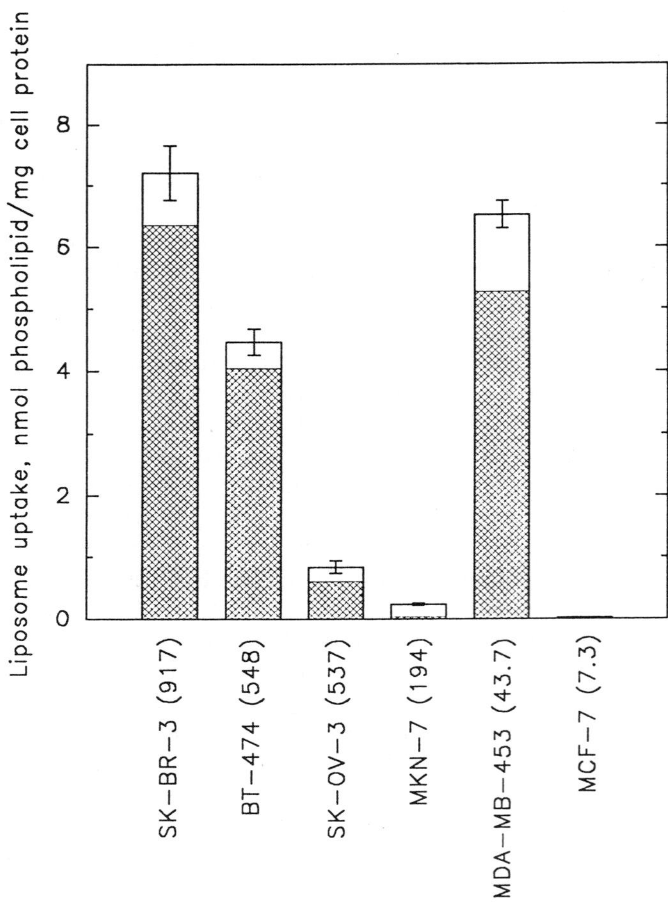

Fig. 4. Uptake of anti-HER2 liposomes by the cells with various levels of HER2 expression. Values in brackets indicate total cellular HER2 expression (ng/mg cell protein). Cross-hatched areas on the bars correspond to endocytosed liposomes; hollow areas, to surface-bound liposomes. Error bars, SEM ($n = 3$).

somes to SKBR-3 cells was as cytotoxic as free doxorubicin ($IC_{50} = 0.3$ μg/ml), while cytotoxicity of doxorubicin in the liposomes with conjugated irrelevant Fab′, or in anti-HER2 liposomes incubated with lung fibroblast cells (WI-38) expressing only minimal levels of HER2 was 20–30 times less than that of free drug.[17] Goren and co-workers[66] reported doxorubicin-loaded anti-HER2 SIL bearing whole antibodies attached to PEG terminal groups via hydrazone formation with periodate-oxidized carbohydrate moieties. These liposomes bound quite well to HER2-overexpressing gastric cancer cells (N-87), but their cytotoxicity was equal to that of non-targeted doxorubicin-loaded liposomes, and much less than that of the free drug, presumably because of the inability of these liposomes to be internalized by the target cells.[66] In contrast, doxorubicin-loaded immunoliposomes bearing conjugated anti-HER2 MAb SER4 were endocytosed by HER2 overexpressing cells SKBR-3 and MKN-7, and were 25 times more cytotoxic to these cells than the matching liposomes conjugated to an antibody against non-internalizable surface protein gp125.[67] Moreover, compared with anti-gp125 immunoliposomes, SER4-conjugated immunoliposomes required 4.3–4.5 times less doxorubicin association with HER2 overexpressing cells for equal cytotoxicity.[67] Evidently, endocytosis of doxorubicin-loaded anti-HER2 liposomes is important for increased cytotoxicity of the liposomal drug, presumably by creating acidic environment and transmembrane pH gradients which favor escape of the drug from the liposome and its further distribution throughout the cell.

V. Properties of anti-HER2 SIL in vivo

Plasma pharmacokinetics of doxorubicin loaded in anti-HER2 SIL was studied in healthy Lewis rats following intravenous injection at the dose of 5 μmol of liposomal phospholipid (0.8–1 mg of doxorubicin) per animal. The liposomes showed biphasic elimination profile with $t_{1/2}\alpha = 6.1$ min., $t_{1/2}\beta = 976$ min., AUC = 93,100 min%, and blood MRT = 1460 min.[21,67] This pharmacokinetic behavior was characteristic for long-circulating liposomes and similar to that of similarly designed non-targeted doxorubicin-loaded SSL, in contrast to free doxorubicin which had plasma half-life of about 5 min.[68] The use of Fab′ instead of the whole antibody was of importance, since the analogous constructs bearing conjugated whole antibodies show lower circulation half-lives than corresponding non-conjugated SSL.[16,66] To analyze possible drug leakage or dissociation of Fab′ from the liposomes in circulation, plasma pharmacokinetics of the anti-HER2 SIL-entrapped drug was compared to that of liposome-conjugated Fab′ fragments (Plasma concentration of anti-HER2 Fab′ was assayed by ELISA using microtiter plates coated with extracellular domain of HER2 for capture, and horseradish peroxidase-linked goat anti-human IgG for detection). These two markers showed identical pharmacokinetic profiles indicating excellent stability of the drug-loaded anti-HER2 SIL in circulation.[19,20]

Nude mice with established subcutaneous xenografts of HER2-overexpressing human breast carcinoma (BT-474) were used to study biodistribution and tumor localization of anti-HER2 SIL. For quantitation in tissues, the liposomes were prepared in the presence of a chelator (DTPA) and loaded with radiotracer ^{67}Ga using remote loading via oxine complex.[69] Biodistribution of anti-HER2 SIL in non-tumor tissues was characteristic for PEG-coated liposomes and was not significantly different from that of the similar non-targeted SSL (Table 1). There was also no statistically significant difference between the accumulation of HER2-targeted SIL or matching non-targeted SIL in HER2-overexpressing BT-474 tumors, or between the accumulation of anti-HER2 SIL in similarly established xenografts of MCF-7 tumors which express low levels of HER2 (Table 1). Nonetheless, treatment of animals with established (approximately 200 mm^3) xenografts of HER2-overexpressing human breast carcinomas(BT-474, MDA-MB-453) by

Table 1

Biodistribution of ^{67}Ga-labeled anti-HER2 SIL and matching non-targeted SSL in nude mice with breast cancer xenografts. Liposomes (1 μmol of phospholipid) were injected via tail vein 24 hours prior to sacrifice. Data: mean ± SE of 6 animals/group

Tissue	Anti-HER2 SSL	Non-targeted SSL % injected dose per g tissue
Blood	7.04 ± 0.82	7.98 ± 0.85
Skin	4.71 ± 0.86	3.97 ± 0.72
Muscle	0.37 ± 0.12	0.74 ± 0.20
Bone	2.06 ± 0.35	3.50 ± 0.82
Heart	0.58 ± 0.05	0.68 ± 0.16
Lungs	0.42 ± 0.10	0.61 ± 0.10
Liver	15.6 ± 3.8	14.2 ± 1.9
Spleen	41.6 ± 4.7	33.5 ± 5.5
Kidneys	3.21 ± 0.24	3.72 ± 0.24
Tumor: HER2-positive (BT-474)	8.34 ± 1.54	7.32 ± 1.05
Tumor: HER2-negative (MCF-7)	7.18 ± 0.60	8.59 ± 1.16

three weekly injections of free or SIL-encapsulated doxorubicin revealed superior activity of doxorubicin in anti-HER2 SIL.[20,21] The average ratio of the volume of BT-474 tumors at the end of experiment (48–60 days post tumor inoculation) to that at the beginning of experiment (12–14 days post inoculation) was 19.13 ± 1.14 in the group treated with free doxorubicin at a total maximum tolerated dose (MTD) of 7.5 mg/kg, 2.59 ± 0.28 in the group receiving non-targeted doxorubicin-loaded SSL (MTD, 15 mg/kg), and 0.63 ± 0.12 in the group injected with doxorubicin-loaded anti-HER2 SIL (15 mg/kg); in the case of MDA-MB-453 tumor (with lower expression of HER2), these values were 3.54 ± 0.53, 2.15 ± 0.29, and 1.17 ± 0.16, respectively.[19,21] The difference between growth rates of tumors in the groups receiving doxorubicin in HER2-targeted vs. non-targeted SIL was statistically significant at $p = 0.001$ (BT-474) and $p = 0.004$ (MDA-MB-453) according to a modified Norton-Simon model of tumor growth.[70] Administration of "empty" anti-HER2 SIL at equal dose/schedule did not produce antitumor effect in these models, ruling out the inhibitory effect of immunoliposome itself. Therefore the increased antitumor activity had to be attributed to the targeting.

The mechanisms by which the targeting of doxorubicin-loaded SIL to a surface antigen on cancer cells may increase therapeutic efficacy of the drug are at present not fully understood. The biodistribution data ruled out the increased uptake of targeted SIL over non-targeted ones in the tumor overexpressing target antigen; this observation is in accord with the view that the major bottleneck in the tumor accumulation of the circulating liposomes is crossing of the vascular wall,[71] a process on which this type of targeting evidently has no effect. One can not exclude that specific interaction with HER2-overexpressing tumor cells resulted in the extended residence time of anti-HER2 SIL in the tumor tissue. Another mechanism by which anti-HER2 SIL may be more efficient as carriers of cytotoxic drugs into HER2-positive tumors was revealed by the studies of liposome disposition in the tumor. To visualize the location and distribution of anti-HER2 SIL in the tumor tissue, the liposomes were labeled with entrapped colloidal gold as described.[72] Entrapment of gold has no effect on the liposome stability or anti-HER2 Fab′ conjugation. Two days after intravenous administration of colloidal gold-labeled anti-HER2 SIL (5 μmol of phospholipid/animal), tumors were excised, fixed, embedded in glycol metacrylate, and gold-labeled liposomes were visualized on tumor sections by silver enhancement method. In HER2-positive tumor (BT-474) anti-HER2 SIL were abundantly deposited in the intercellular spaces throughout the tumor tissue, while in MCF-7 tumors with low HER2 expression, or if non-targeted gold-labeled SIL were given, the label was concentrated mostly within tumor-resident macrophages and in perivascular areas, in agreement with previous observations on non-targeted SIL.[73] At higher magnification, anti-HER2 SIL were frequently revealed within the cytoplasm and in the perinuclear spaces of HER2-positive cancer cells within the tumor tissue (Figure 2d), while non-targeted SIL (Figure 2e), or anti-HER2 SIL in the HER2-negative tumor (MCF-7) (Figure 2f) showed no clear deposition of silver granules within the cancer cells, but prominent intercellular deposition and localization of the silver grains in macrophages was apparent. Thus, anti-HER2 SIL not only crossed the vascular barrier into the solid tumor, but, in the case of HER2-overexpression, frequently became endocytosed by the cancer cells as they would in vitro. Increased deposition in the intercellular spaces within the tumor tissue outside tumor-resident macrophages, as well as the intracellular delivery of the encapsulated drug in vivo may contribute to superior antitumor efficacy of doxorubicin-loaded anti-HER2 SIL.

VI. Targeted delivery of nucleic acids by cationic anti-HER2 SIL

Liposomes that incorporate cationic lipids and therefore bear overall positive charge (cationic liposomes) have been established as non-viral vectors for introducing functional DNA, RNA, and oligonucleotides into cells.[74–76] The use of cationic liposomes as delivery vehicles in gene therapy and antisense oligonucleotide therapy is attractive because of the high loading capacity driven by electrostatic interactions of DNA or RNA with the liposome surface and ability to deliver at least part of its DNA/RNA load into cytoplasmic/nuclear compartments necessary for gene expression or other appropriate function. This use is now limited, however, by the instability of cationic liposomes against aggregation and dissociation in the physiological media, and by unfavorable pharmacokinetic properties. Other limiting aspects of cationic liposomes as in vivo delivery carriers for nucleic acids are non-specific reactivity and lack of targeting. As mentioned above, stability and pharmacokinetics of "neutral" liposomes used to carry encapsulated drugs is improved by "steric stabilization" with liposome-grafted PEG.[10] The similar approach to cationic liposomes would face a principal difficulty because steric hindrance created by PEG-coating may interfere with the liposome ability to form complexes with nucleic acids and, most importantly, to interact with cell membrane and deliver them into appropriate intracellular compartments, either by fusion with plasma membrane,[74] endocytosis,[77,78] or formation of pores.[79] Combination of steric stabilization by amphiphilic PEG derivatives and HER2-targeting by an internalizable antibody construct may help to overcome these limitations and utilize the potential of steric stabilization for creation of cationic liposome-based systems for in vivo systemic delivery of therapeutic genes and/or oligonucleotides into HER2-overexpressing cancer cells.

The design of HER2-targeted SIL described in the preceding sections was used to demonstrate the feasibility of this approach. Cationic liposomes composed of equimolar amounts of dioctadecyldimethylammonium bromide (DDAB) and dioleoyl-phosphatidylethanolamine (DOPE) formed complexes with expression plasmid DNA carrying luciferase reporter gene (Lux) under the control of early CMV promoter. SKBR-3 cells incubated with such complexes for 4 hours at a dose of 1 μg DNA (16 nmol lipid)/10^6 cells showed high levels of Lux expression 48 hours later. Addition of 5 mol.% of methoxyPEG(M_r = 2,000)-DSPE substantially increased the stability of complexes against aggregation and destruction in the serum,[80] but the expression was down approximately 20-fold. When the same complex was prepared with cationic liposomes containing 4 mol.% of methoxy-PEG-DSPE and 1 mol.% of MP-PEG-DSPE, and conjugated to anti-HER2 Fab′ as described above, the level of Lux expression increased to the value observed without PEG-modification (Figure 5A) while the stability of such construct was the same as without conjugated Fab′ (K. Hong & W.-W. Zheng, unpublished data). In MCF-7 cells (low expression of HER2 receptor) modification of cationic liposomes with PEG led to an equal reduction of transfection by HER2-targeted or nontargeted complexes (Figure 5B). These results suggest that, first, PEG-

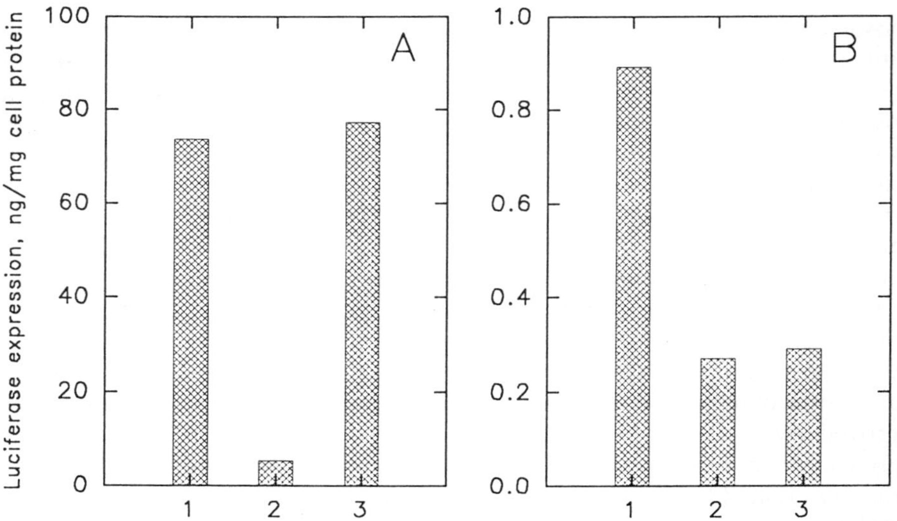

Fig. 5. Expression of luciferase reporter gene in SKBR-3 (A) and MCF-7 (B) cells transfected with DDAB/DOPE-plasmid DNA complexes: (1) no PEG-DSPE; (2) 5 mol.% of methoxyPEG (M_r = 1,900)-DSPE; (3) methoxyPEG-DSPE (5 mol.%) + anti-HER2 Fab′ conjugate of MP-PEG-DSPE.

modification at the above level did not preclude interaction between plasmid DNA and cationic liposomes, and, second, that internalization of PEG-modified anti-HER2 cationic liposome-DNA-complex through HER2-mediated endocytosis can substitute for internalization pathway of the complex between DNA and "conventional" cationic liposome (without PEG coating), leading to functional intracellular delivery of DNA.

Similar construct was developed for the delivery of oligonucleotides to HER2-overexpressing cancer cells, revealing more subtle targeting effects.[22] Cationic liposomes of 1,2-di(oleoyloxy)-3-trimethylammoniopropane (DOTAP), DOPE, and 6 mol.% of methoxyPEG-DSPE were capable of binding substantial amount of 18-mer antisense phosphorothiate oligodeoxynucleotide (ODN); unlike their prototypes without PEG-coating, these complexes were stable against aggregation and upon incubation in 50% human plasma. Interaction of such complexes with SKBR-3 cells resulted in the appearance of both lipid (followed by rhodamine-labeled lipid) and oligonucleotide (followed by FITC-labeled ODN) in the punctate cytoplasmic patterns and on the cell surface, but not in the cell nuclei. Conjugation of anti-HER2 Fab′ via MP-PEG-DSPE linker at such complexes resulted in the nuclear delivery of ODN into every cell (Figure 2h), while the lipid components remained associated with cytoplasmic, but not nuclear, compartments (Figure 2g). Nuclear delivery of ODN was specific to the uptake of the complexes mediated by specific interaction of the conjugated Fab′ with extracellular domains of HER2, since preincubation of the cells with excess of free anti-HER2 Fab′ completely abolished nuclear localization (Figure 2i). In "target-negative" MCF-7 cells neither HER2-targeted, nor non-targeted PEG-coated cationic liposome-ODN complexes could deliver ODN into the cell nuclei.[22] Nuclear localization of ODN is considered a prerequisite for its functional activity and can be normally achieved by incubation of the cells with ODN and cationic liposomes without PEG-coating.[76,81,82] Therefore, in the above two examples, the likely role of HER2-targeting in the delivery of DNA by sterically stabilized cationic liposomes was to substitute HER2-mediated endocytosis for the cell internalization otherwise mediated by the cationic lipid, while preserving the "correct" intracellular address for the delivered DNA.

VII. Conclusion

Over the last decade, introduction of "long-circulating" liposomes,[6–10] refinement of liposome preparation techniques[50,52] and "remote loading" methods for drug loading into liposomes[60,83] greatly advanced liposomal pharmacology. This advance is clearly evidenced by the recent appearance of anticancer liposomal drugs (see Section VIII) in the pharmaceutical market.[84] It also brought new enthusiasm to the area of liposome targeting.[1,2] Here we presented a case study that illustrates, in our view, a "rational design" approach to antibody-targeted pharmaceutical liposomes (Table 2). Each element of this design answers certain demand related to the ultimate medical use of the targeted liposomal drug carrier. This design may be readily applied to other targeted drug delivery systems utilizing different

Table 2

Components of the immunoliposome design

Component	Considerations for optimal design
Target Antigen	*Expression*: Highly and homogeneously overexpressed in target tissue. *Function*: Vital to tumor progression, so that down-modulation does not occur or is associated with therapeutic benefit. *Shedding of antigen*: Limited, to avoid binding to soluble antigen and accelerated clearance.
Antibody	*Affinity*: High enough to ensure binding at low liposome concentrations. *Immunogenicity*: Humanized MAb, to remove murine sequences. Use fragments without Fc portion (Fab′, scFv) *Internalization*: Efficiently endocytosed by target cells. *Biological activity*: Intrinsic antitumor activity may enhance antitumor effect. *Scale-up*: Easy and economical scale-up, e.g., by efficient bacterial expression system Stability during storage.
Linkage	*Stability*: Covalent attachment to hydrophobic anchor, stable in blood. *Attachment site*: Away from the binding site, to ensure correct orientation of antibody molecule. Well defined, to ensure reproducibility and uniformity of coupling. Avoids steric hindrance (e.g., from PEG) of MAb binding and internalization. *Chemical nature of the linker*: Non-toxic. Non-immunogenic. Avoids opsonization. Does not affect drug loading and membrane stability. Excess linker may be quenched to avoid non-specific coupling to biomolecules. Availability, economical manufacturing process.
Liposome	*Stability*: Stable as intact construct in vivo. *Pharmacokinetics*: Long circulating. *Tumor penetration*: Capable of extravasation in tumors. Small diameter improves tumor penetration.
Drug	*Encapsulation*: Efficient, high capacity (e.g., by remote loading). Encapsulated drug storage-stable and resists leakage. *Bystander Toxicity*: Drug affects tumor cells not directly targeted (bystander cells) *Interaction with target cells*: Effective against target cell population. Cytotoxicity enhanced by binding of MAb.

drugs and/or different target-specific molecules, such as, for example, single chain anti-HER2 antibody fragments produced by phage display libraries.[44,45] Doxorubicin-loaded HER2-targeted SIL constructed according to such design had superior antitumor activity compared to matched non-targeted liposomes in established solid tumor xenografts overexpressing HER2 oncoprotein. Unexpectedly, this phenomenon was not associated with an increased accumulation of targeted liposomes in HER2-overexpressing tumors, but rather resulted from a different pattern of liposome disposition (improved penetration and internalization into HER2-overexpressing cancer cells) within the tumor tissue. Finally, the same design showed promise in the development of HER2-targeted sterically stabilized cationic liposomes for the delivery of therapeutic genes and oligonucleotides. The "rational design" of cancer cell-targeted sterically stabilized liposomes leads to a re-evaluation of tumor targeting paradigms and opens new avenues for better treatment of cancer.

Acknowledgments

This work was supported by NIH grant P50CA58207, by the grants from the State of California Breast Cancer Research Program 2CB-0004 and 2CB-0250, Genta, Inc., and Bayer AG. The authors are grateful to Drs. Samuel Zalipsky (SEQUUS Pharmaceuticals, Inc.) and Martin Woodle (Genetic Therapy, Inc.) for valuable discussions.

References

1. Papahadjopoulos D. 30-Year progress for liposomes: from serendipity to molecular design. J Liposome Res 1995;5:ix–xiv.
2. Lasic DD, Papahadjopoulos D. Liposomes revisited. Science 1995;267:1275–1276.
3. Leserman L, Machy P. Ligand targeting of liposomes. In: Ostro MJ, ed. Liposomes. From Biophysics to Therapeutics. New York: Marcel Dekker, 1987;157–194.
4. Ranade VV. Drug delivery systems. 1. Site-specific drug delivery using liposomes as carriers. J Clin Pharmacol 1989;29:685–694.
5. Straubinger RM, Lopez NG, Debs RJ, Hong K, Papahadjopoulos D. Liposome-based therapy of human ovaruan cancer: parameters determining potency of negatively charged and antibody-targeted liposomes. Cancer Res 1988;48:5237–5245.
6. Gabizon A, Papahadjopoulos D. Liposome formulations with prolonged circulation time in blood and enhanced uptake by tumors. Proc Natl Acad Sci USA 1988;85:6949–6953.
7. Papahadjopoulos D, Allen T, Gabizon A, Mayhew E, Matthay, Huang, SK, Lee, K-D, Woodle MC, Lasic DD, Redemann C, Martin FJ. Sterically stabilized liposomes: Improvements in pharmacokinetics, and anti-tumor therapeutic efficacy. Proc Natl Acad Sci USA 1991;88:11460–11464.
8. Klibanov AL, Huang L. Long-circulating liposomes: development and perspectives. J. Liposome Res 1992;2:321–332.
9. Allen TM, Chohn A. Large unilamellar liposomes with low uptake into the reticuloendothelial system. FEBS Lett 1987;223:42.
10. Woodle MC, Lasic DD. Sterically stabilized liposomes. Biochim Biophys Acta 1992;1113:171–199.
11. Torchilin VP, Klibanov AL, Huang L, O'Donnell S, Nossiff ND, Khaw BA. Targeted accumulation of polyethylene glycol-coated immunoliposomes in infarcted rabbit myocardium. FASEB J 1992;6:2716.
12. Klibanov AL, Maruyama K, Beckerleg AM, Torchilin VP, Huang, L. Activity of amphipathic poly(ethylene glycol) 5000 to prolong the circulation lifetime of liposomes depends on the liposome

size and is unfavorable for immunoliposome binding to target. Biochim Biophys Acta 1991;1062:142–148.
13. Torchilin VP, Papisov MI, Bogdanov AA, Trubetskoy VS, Omelyanenko VG. Molecular mechanism of liposome and immunoliposome steric protection with poly(ethylene glycol). In: Lasic D, Martin F, eds. Stealth Liposomes. Boca Raton, FL: CRC Press, 1995;51–62.
14. Allen TM, Brandeis E, Hansen CB, Kao GY, Zalipsky S. A new strategy for attachment of antibodies to sterically stabilized liposomes resulting in efficient targeting to cancer cells. Biochim Biophys Acta 1995;1237:99–108.
15. Zalipsky S, Hansen CB, Lopes de Menezes DE, Allen TM. Long-circulating, polyethylene glycol-grafted immunoliposomes. J Controlled Release 1996;39:153–161.
16. Hansen CB, Kao GY, Moase EH, Zalipsky S, Allen TM. Attachment of antibodies to sterically stabilized liposomes: evaluation, comparison and optimization of coupling procedures. Biochim Biophys Acta 1995;1239:133–144.
17. Park JW, Hong K, Carter P, Asgari H, Guo LY, Wirth C, Shalaby R, Kotts C, Keller GA, Wood WI, Papahadjopoulos D, Benz CC. Development of anti-p185^{HER2} immunoliposomes for cancer therapy. Proc Natl Acad Sci USA 1995;92:1327–1331.
18. Kirpotin D, Park JW, Hong K, Zalipsky S, Li W-L, Carter P, Benz CC, Papahadjopoulos D. Sterically stabilized anti-HER2 immunoliposomes: desin and targeting to human breast cancer cells in vitro. Biochemistry 1997;36:66–75,
19. Park JW, Hong K, Kirpotin DB, Papahadjopoulos D, Benz C. Immunoliposomes for cancer treatment. Adv Pharmacol 1997;40:399–436.
20. Park JW, Hong K, Kirpotin DB, Meyer O, Papahadjopoulos D, Benz C. Anti-HER2 immunoliposomes for targeted therapy of human tumors. Cancer Lett, 1997;118:153–160.
21. Park JW, Colbern G, Hong K, Kirpotin D, Shao Y, Baselga J, Moore D, Papahadjopoulos D, Benz CC. Targeted intracellular drug delivery via anti-HER2 immunoliposomes yields superior antitumor efficacy. Proc ASCO 1997;16.
22. Meyer O, Kirpotin D, Hong K, Sternberg B, Park JW, Woodle MC, Papahadjopoulos D. Poly-(ethylene glycol)-modified cationic liposomes as carriers for cellular delivery of antisense oligonucleotides. J Biol Chem 1998; in press.
23. Hynes NE, Stern DF. The biology of erbB-2/neu/HER2 and its role in cancer. Biochim Biophys Acta 1994;1198:165–184.
24. Carraway KL, Cantley LC. A neu acquaintance for ErbB3 and ErbB4: a role for receptor heterodimerization in growth signaling. Cell 1994;78:5–8.
25. Slamon DJ, Clark GM, Wong SG, Levin WJ, Ullrich A, McGuire WL. Human breast cancer: correlation of relapse and survival with amplification of HER2/neu oncogene. Science 1987;235:177–182.
26. Slamon DJ, Godolphin W, Jones LA, Holt JA, Wong SG, Keith DE, Levin WJ, Stuart SG, Udove J, Ullrich A, Press M. Studies of the HER2/neu proto-oncogene in human breast and ovarian cancer. Science 1989;244:707–712.
27. Berchuck A, Kamel A, Whitaker R, Kerns B, Olt G, Kinney R, Soper JT, Dodge R, Clarke-Pearson DL, Marks P, McKenzie S, Yin S, Bast RC. Overexpression of HER2/neu is associated with poor susvival in advanced epithelial ovarian cancer. Cancer Res 1990;50:4087–4091.
28. Berchuck A, Rodriguez G, Kinney RB, Soper JT, Dodge RK, Clarke-Pearson DL, Bast RC. Overexpression of HER2/neu in endometrial cancer is associated with advanced stage disease. Am J Obstetrics Gynecol 1991;164:15–21.
29. Kern JA, Schwartz DA, Nordberg JE, Weiner DB, Greene MI, Torney L, Robinson RA. p185neu expression in human lung adenocarcinomas predict shortened survival. Cancer Res 1990;50:5184–5191.
30. Yokota J, Yamamoto T, Miyajima N, Toyoshima K, Nomura N, Sakamoto H, Yoshida T, Terada M, Sugimura T. Genetic alterations of the c-erbB-2 oncogene occur frequently in tubular adenocarcinoma of the stomach and are often accompanied by amplification of v-erbA homologue. Oncogene 1988;2:283–287.
31. Park JB, Rhim JS, Park SC, Kimm SW, Kraus MH. Amplification, overexpression, and rearrangement of the erbB-2 proto-oncogene in primary human stomach carcinomas. Cancer Res 1989;49:6605–6609.
32. Lei A, Appert HE, Nakata B, Domenico DR, Kim K, Howard JM. Overexpression of HER2/neu oncogene in pancreatic cancer correlates with shortened survival. Int J Pancreatology 1995;17:15–21.
33. Zhau HE, Zhang X, von Eschenbach AC, Scorsone K, Babaian RJ, Ro JY, Hung MC. Amplifi-

cation and expression of the c-erbB-2/neu proto-oncogene in human bladder cancer. Molec Carcinogen 1990;3:254–257.
34. Zhau HE, Wan DS, Zhou J, Miller GJ, von Eschenbach AC. Expression of c-erbB-2/neu proto-oncogene in human prostatic cancer tisues and cell lines. Molec Carcinogen 1990;3:254–257.
35. VandeVijver MJ, Peterse ML, Mooi WJ, Wisman O, Lomans J, Dalesio O, Nusse R. Neu-protein expression in breast cancer. New Engl J Med 1988;319:1239–1245.
36. Liu E, Thor A, He M, Barcos M, Ljung BM, Benz C. The HER2 (c-erbB-2) oncogene is frequently amplified in in situ carcinomas of the breast. Oncogene 1992;7:1027–1032.
37. Lewis GD, Figari I, Fendly BM, Wong W-L, Carter P, Gorman C, Shepard HM. Differential responses of human tumor cell lines to anti-p185HER2 monoclonal antibodies. Cancer Immunol Immunother 1993;37:255–263.
38. Press MF, Cordon-Cardo C, Slamon DJ. Expression of the HER-2/neu proto-oncogene in normal humad adult and fetal tissues. Oncogene 1990;5:953–962.
39. Niehans GA, Singleton TP, Dykoski D, Kiang DT. Stability of HER-2/neu expression over time and at multiple metastatic sites. J Natl Cancer Inst 1993;85:1230–1235.
40. Tagliabue E, Centis P, Campiglio M, Mastroiannini A, Martignone S, Pellegrini R, Casalani P, Lanzi C, Menard S, Colnaghi MI. Selecion of monoclonal antibodies which induce internalization and phosphorylation of p185HER2 and growth inhibition of cells with HER2/neu gene amplification. Int. J. Cancer 1991;47:933–937.
41. Sarup JC, Johnson RM, King KL, Fendly BM, Lipari MT, Napier MA, Ullrich A, Shepard HM. Characterization of an anti-p185HER2 monoclonal antibody that stimulated receptor function and inhibits tumor cell growth. Growth Regulation 1991;37:72–82.
42. Hurwitz E, Stancovski I, Sela M, Yarden Y. Suppression and promotion of tumor growth by monoclonal antibodies to ErbB-2 differentially correlate with cellular uptake. Proc Natl Acad Sci USA 1995;92:3553–3557.
43. Fendly BM, Winget M, Hudziak RM, Lipari MT, Napier MA, Ullrich A. Characterization of murine monoclonal antibodies reactive to either the human epidermal growth factor receptor or HER2/neu gene product. Cancer Res 1990;50:1550–1558.
44. Schier R, Marks JD, Wolf EJ, Apell G, Wong C, McCartney JE, Bookman MA, Huston JA, Houston LL, Weiner LM, Adams GP. In vitro and in vivo characterization of a human anti-c-erbB-2 single-chain Fv isolated from a filamentous phage antibody library. Immunotechnology 1995;1:73–81.
45. Schier R, McCall A, Adams GP, Marshall KW, Merritt H, Yim M, Craford RS, Weiner LM, Marks C, Marks JD. Isolation of picomolar affinity anti-c-erbB-2 single chain Fv by molecular evolution of the complementary determining regions in the center of the antibody binding site. J Mol Biol 1996;263:551–567.
46. Park JW, Stagg R, Lewis GD, Carter P, Maneval D, Slamon DJ, Jaffe H, Shepard HM. Anti-p185HER2 monoclonal antibodies: biological properties and potential for immunotherapy. In: Dickson RB, Lippman ME, eds. Genes, Oncogenes, and Hormones: Advances in Cellular and Molecular Biology of Breast Cancer. Boston, MA: Kluwer Academic Publishers, 1992;193–211.
47. Carter P, Presta L, Gorman CM, Ridgway JBB, Henner D, Wong WLT, Rowland AM, Kotts C, Carver ME, Shepard HM. Humanization of an anti-p185HER2 antibody for human cancer therapy. Proc Natl Acad Sci USA 1992;89:4285–4289.
48. Baselga J, Tripathy D, Mendelsohn J, Baughman S, Benz CC, Dantis L, Sklarin NT, Seidman AD, Hudis CA, Moore J, Rosen PP, Twaddell T, Henderson IC, Norton L. Phase II study of weekly intravenous recombinant humanized anti-p185HER2 monoclonal antibody in patients with HER2/neu-overexpressing metastatic breast cancer. J Clin Oncol 1996;14:737–744.
49. Papahadjopoulos D, Jacobson K, Nir S, Isac T. Phase transitions in phospholipid vesicles, fluorescence polarization and permeability measurements concerning the effect of temperature and cholesterol. Biochim Biophys Acta 1973;311:330–348.
50. Olson F, Hunt CA, Szoka FC, Vail WJ, Papahadjopoulos D. Preparation of liposomes of defined size distribution by extrusion through polycarbonate membranes. Biochim Biophys Acta 1979;557:9–23.
51. Martin F, Papahadjopoulos D. Irreversible coupling of immunoglobulin fragments to preformed vesicles. An improved method for liposome targeting. J Biol Chem 1982;257:286–288.
52. Shahinian S, Silvius JR. A novel strategy affords high-yield coupling of antibody Fab′ fragments to liposomes. Biochim Biophys Acta 1995;1239:157–167.
53. Loughrey HC, Choi LS, Cullis PR, Bally MB. Optimized procedures for the coupling of proteins to liposomes. J Immunol Methods 1990;132:25–35.

54. Bredehorst R, Ligler FS, Kusterbeck AW, Chang EL, Gaber BP, Vogel C-W. Effect of covalent attachment of immunoglobulin fragments on liposomal integrity. Biochemistry 1986;25:5693–5698.
55. Hristova K, Needham D. Physical properties of polymer-grafted bilayers. In: Lasic D, Martin F, eds. Stealth Liposomes. Boca Raton. FL: CRC Press, 1995;35–49.
56. Mori A, Klibanov AL, Torchilin VP, Huang L. Influence of the steric barrier activity of amphipathic poly(ethylene glycol) and ganglioside GM_1 on the circulation time of liposomes and on the target binding of immunoliposomes in vivo. FEBS Lett 1991;284:263–266.
57. Huang SK, Mayhew E, Gilani S, Lasic DD, Martin FJ, Papahadjopoulos D. Pharmacominetics and therapeutics of sterically stabilized liposomes in mice bearing C-26 colon carcinoma. Cancer Res 1992;52:6774–6781.
58. Working PW, Newman M, Huang SK, Vaage J, Mayhew E, Lasic DD. Pharmacokinetics, biodistribution and therapeutic efficacy of doxorubicin encapsulated in Stealth liposomes. J Liposome Res. 1994;4:667–687.
59. Bogner JR, Goebel F-D. Efficacy of Dox-SL (Stealth® liposomal doxorubicin) in the treatment of advanced AIDS-related Kaposi's sarcoma. In: Lasic D, Martin F, eds. Stealth Liposomes. Boca Raton. FL: CRC Press, 1995;267–278.
60. Haran G, Cohen R, Bar LK, Barenholz Y. Transmembrane ammonium sulfate gradients in liposomes produce efficient and stable entrapment of amphipathic weak bases. Biochim Biophys Acta 1993;1149:180–184.
61. Daleke DL, Hong K, Papahadjopoulos D. Endocytosis of liposomes by macrophages: binding, acidification, and leakage of liposomes monitored by a new fluorescence assay. Biochim Biophys Acta 1990;1024:352–366.
62. Maruyama K, Takizawa T, Yuda T, Kennel SJ, Huang L, Iwatsuru M. Targetability of novel immunoliposomes modified with amphipathic poly(ethylene glycol)s conjugated at their distal terminals to monoclonal antibodies. Biochim Biophys Acta 1995;1234:74–80.
63. Nezlin RC. Structure and Biosynthesis of Antibodies. New York: Consultants Bureau, 1977;174.
64. Kenworthy AK, Simon SA, McIntosh TJ. Structure and phase behavior of lipid suspensions containing phospholipids with covalently attached poly(ethylene glycol). Biophys J 1995;68:1903–1920.
65. Lee K-D, Nir S, Papahadjopoulos D. Quantitative analysis of liposome-cell interactions in vitro: rate constants of binding and endocytosis with suspension and adherent J774 cells and human monocytes. Biochemistry 1993;32:889–899.
66. Goren D, Horowitz AT, Zalipsky S, Woodle MC, Yarden Y, Gabizon A. Targeting of stealth liposomes to erbB-2 (Her/2) receptor: in vitro and in vivo studies. Br J Cancer 1996;74:1749–1756.
67. Suzuki S, Uno S, Fukuda Y, Aoki Y, Masuko T, Hashimoto Y. Cytotoxicity of anti-c-erbB-2 immunoliposomes containing doxorubicin on human cancer cells. Br J Cancer 1995;72:663–668.
68. Park JW, Colbern G, Baselga J, Hong K, Shao Y, Kirpotin D, Nuijens A, Wood W, Papahadjopoulos D, Benz C. Antitumor efficacy of anti-p185HER2 liposomes: enhanced therapeutic index due to targeted delivery. Proc ASCO 1996;15:501.
69. Gabizon A, Huberty J, Straubinger R, Price DC, Papahadjopoulos D. An improved method for in vivo tracing and imaging of liposomes using Gallium-67-deferoxamine complex. J Liposome Res 1988;1:123–135.
70. Heitjan DF. Generalized Norton-Simon models of tumour growth. Statistics in Medicine 1991;10:1075–1088.
71. Matzku S, Krempel H, Weckenmann H-P, Schirrmacher V, Sinn H, Stricker H. Tumour targeting with antibody-coupled liposomes: failure to achieve accumulation in xenografts and spontaneous liver metastases. Cancer Immunol Immunother 1990;31:285–291.
72. Hong K, Friend DS, Glabe CG, Papahadjopoulos D. Liposomes containing colloidal gold are a useful probe of liposome-cell interactions. Biochim Biophys Acta 1983;732:320–323.
73. Huang SK, Lee K-D, Hong K, Friend DS, Papahadjopoulos D. Microscopic localization of sterically stabilized liposomes in colon-carcinoma bearing mice. Cancer Res. 1992;52:5135–5143.
74. Felgner PL, Gadek TR, Holm M, Roman R, Chan HW, Wenz M, Northrop JP, Ringold GM, Danielsen M. Lipofection: a highly efficient, lipid-mediated DNA-transfection procedure. Proc Natl Acad Sci USA 1987;84:7413–7417.
75. Malone RW., Felgner PL, Verma IM. Cationic liposome-mediated RNA transfection. Proc Natl Acad Sci USA 1989;86:6077–6081.
76. Zelphati O, Szoka FC Jr. Cationic liposomes as an oligonucleotide carrier: mechanism of action. J Liposome Res 1997;7:31–49.

77. Friend DS, Papahadjopoulos D, Debs RJ. Endocytosis and intracellular processing accompanying transfection mediated by cationic liposomes. Biochim Biophys Acta 1996;1278:41–50.
78. Zelphati O, Szoka FC Jr. Intracellular distribution and mechanism of delivery of oligonucleotides mediated by cationic lipids. Pharm Res 1996;13:1367–1372.
79. Boulikas T. Cancer gene therapy and immunotherapy (review). Int J Oncology 1996;9:941–954.
80. Hong K, Zheng, W-W, Baker A, Papahadjopoulos D. Stabilization of cationic liposome-plasmid DNA complexes by polyamines and poly(ethylene glycol)-phospholipid conjugates for efficient in vivo gene delivery.FEBS Letters 1997;400:233–237.
81. Lewis JG, Lin K-Y, Kothavale A, Flanagan WM, Mateucci MD, DePrince RB, Mook Jr RA, Hendren RW, Wagner RW. A serum-resistant cytofectin for cellular delivery of antisense oligodeoxynucleotides and plasmid DNA. Proc Natl Acad Sci USA 1966;93:3176–3181.
82. Bennett CF, Chiang M-Y, Chan H, Shoemaker JEE, Mirabelli CK. Cationic lipids enhance cellular uptake and activity of phosphorothioate antisense oligonucleotides. Molec Pharmacol 1992;41:1023–1033.
83. Madden TD, Harrigan PR, Tai LC, Bally MB, Mayer LD, Redelmeier TE, Loughrey HC, Tilcock CP, Reinish LW, Cullis PR. The accumulation of drugs within large unilamellar vesicles exhibiting a proton gradient: a survey. Chem Phys Lipids 1990;53:37–46.
84. Janknegt R. Liposomal formulations of cytotoxic drugs. Supportive Care Cancer 1996;4:298–304.

Lasic and Papahadjopoulos (eds.), Medical Applications of Liposomes
Elsevier Science B.V.

CHAPTER 5.1

Gene therapy: Liposomes and gene delivery—a perspective

CLAUDE NICOLAU[a] AND DEMETRIOS PAPAHADJOPOULOS[b]

[a]*CBR Laboratories and Harvard Medical School, Boston, MA 02135, USA*; [b]*Department of Cellular and Molecular Pharmacology, University of California, San Francisco, CA and CPMCRI, San Francisco, CA 94115, USA*

In 1972, Gregoriadis and Ryman[1] showed, for the first time, that liposomes could be used to transport molecules into liver cells following an i.v. injection into rats. This observation was important in several respects. It was made at a time when a considerable body of physico-chemical work about liposomes had accumulated valuable information on permeability and stability of lipid bilayers, lipid-protein interactions, liposome fusion, as well as mobility of lipids in the bilayers.[2,3] The newly shown capacity of liposomes to deliver encapsulated material into cells, besides being a very exciting avenue of research, suggested a significant potential for medical use of liposomes (for a review, see Ref. 4).

The mechanism of liposome uptake by cells and the intracellular fate of the liposome-encapsulated material had to be understood, if liposomes should have any medical use. A study by Straubinger et al.[5] investigated both these aspects. Using gold particles, the authors followed the wandering of the liposomes from the coated pits on the cell plasma membrane to the lysosomes. Combined with the realization for the absence of fusion between phospholipid vesicles and eukaryotic cells,[6] this work suggested a mechanism of liposome uptake by cells.

Thus, a clear understanding emerged about the mechanism of endocytosis through which liposomes appeared to be taken up by cells. The similitude in specific steps of liposome uptake with the steps of enveloped virus-uptake by the cells were apparent. The first paper on the encapsulation of the polio virus in lipid vesicles from the laboratory of one of us (D.P.), showed the possibility of infection of virus-resistant cells by the encapsulated virus.[7] In other words, it was possible to overcome the barrier to entry into the cell using liposomes and to deliver viral RNA (and proteins) thus inducing infection in resistant cells. One year later, the same group reported the encapsulation of picorna viruses in liposomes summarizing and expanding the previous observations.[8] Two years later, two papers appeared at about the same time: one from Papahadjopoulos' laboratory and the other one from Nicolau's laboratory reporting the use of liposomes to transfer DNA into cultured cells.[9,10] It appeared that gene transfer and expression in vitro could be mediated by liposomes, but that the efficiency of transfection was quite low. The possibilities to enhance efficiency of DNA delivery by liposomes in vitro by changing the lipid composition of liposomes and the conditions of incubation were investigated.[11] Shortly afterwards, liposomes were used to transfer genes for transient expression in vivo in the cells of the liver.[12,13] Targeting to specific liver cells in vivo using glycolipids added to the liposome bilayer was demonstrated by detection of the intact insulin gene carried by the liposomes to specific liver cells.[13] Liposomes were also used for successful transfection of prokaryotic cells such as bacteria,[14] mycoplasma,[15] and of protoplasts.[16,17]

In spite of these observations, anionic or neutral liposomes were not very effective in transferring genes. In vitro, there were many other ways to transfer genes, but in vivo, there were no other carriers at that time, and therefore, work on this subject continued. A detailed study, using electron microscope autoradiography and subcellular fractionation[18] reported the intracellular fate of liposome-encapsulated DNA in liver cells after i.v. injection of such liposomes to rats and mice. A kinetic study of DNA-accumulation in different cellular organelles was made and DNA was detected in lysosomes and endosomes, in nuclei, and associated with mitochondria.[18] Further investigations showed that the biologically active DNA transported by liposomes into the liver cells after i.v. injection could be found in clathrin coated vesicles.[19] In vivo gene expression after i.p. injection of liposomes containing DNA was reported a little later also by Wang and Huang.[20] The gene was encapsulated in pH-sensitive immunoliposomes, targeted to lymphoma cells grown in the peritoneum of nude mice. The expression of the gene was, as in the previous cases low.[12]

A decisive advance in lipid/liposome-mediated gene transfer was made by Felgner and associates when they first reported the use of cationic lipids with high efficiency of DNA delivery into cells.[21,22] The first cationic lipid molecule of its kind, DOTMA shows parallel orientation of the aliphatic chains thus favoring bilayer formation rather than micelles. The polar head group bears a quarternary amine so that vesicles comprised of DOTMA are positively charged.[23] Furthermore, ether linkages afford greater chemical stability in aqueous solutions than the comparable ester derivatives. As expected, aqueous suspensions of DOTMA, alone or in combination with other phospholipids, results in the formation of multilamellar liposomes (MLV) which can be sonicated to form small unilamellar vesicles (SUV; 0.03 micrometer diameter by quasielastic laser light scattering). Multilayer structures in preparations of MLV are apparent by freeze/fracture electron microscopy as are vesicles in SUV preparations. These vesicles are capable of entrapping fluorescent dextran and have a typical liposome appearance as judged by freeze fracture electron microscopy.[23] And as predicted, positively charged liposomes of this type interact avidly with the negatively charged surface of tissue culture cells, and fluorescent lipid delivered to cells in this way rapidly enters cells.[23]

The problem of low efficiency for encapsulating large DNA molecules into liposomes has been an important technical obstacle to the utilization of liposomes for gene delivery. This low encapsulation efficiency problem can be overcome by using cationic liposomes, which interact spontaneously with the negatively charged nucleic acid polymers. Simply mixing positively charged liposomes with DNA, results in 100% of the polynucleotide to be found into a lipid-DNA complex. Furthermore, by carefully controlling the complexation conditions, relatively homogeneous and physically stable suspensions can be obtained.[24] This quantitative complexation eliminates the need for a separate step to remove unencapsulated material, and all of the polynucleotide, is utilized for each experiment.[25] Interestingly, simple complexes of this type, without any additional biological elements to improve delivery, can be sufficient to transfect cells both in vitro and in vivo.[25–31] Because of its convenience and efficacy, cationic lipid mediated gene delivery technology has become a promising system for in vivo gene therapy, as an alternative to viral-based vectors.

Studies on the structure of the cationic lipid DNA complexes[31–35] and on the release of oligonucleotides[36] and DNA from cationic liposome DNA complexes,[37] have added significant data towards the development of carriers for gene-transfer in humans. The clinical trials of the lipid-DNA-complexes conducted already[38] have mostly shown lack of adverse effects and moderate expression in a relatively low fraction of the cells, but no decisive clinical advantages.

Cell-specific targeting of liposomes carrying DNA remains critical, so that systemic administration could be followed by delivery to a specific cell type. Glycolipid targeting appeared feasible for hepatocytes and liver endothelial cells;[13] antibodies attached to liposomes can specifically attach and sometimes deliver encapsulated molecules to any type of cell expressing a surface antigen against which antibodies can be raised. The potential of immunoliposomes has been recognized many years ago.[39] If the cells surface antigen happens to be rapidly endocytosed upon ligand binding, as it happens in human hepatocellular carcinoma cells or in a large proportion of other human cancer cells,[40–44] then targeting of liposomes with a monoclonal antibody against this surface antigen enhances dramatically specific delivery of encapsulated molecules. The monoclonal antibody AF20, covalently attached to cationic liposomes significantly enhances the expression of β-glactosidase in hepatoma cells upon transfection with cationic immunoliposomes associated with the β-galactosidase gene.[45] Expression, quite specifically enhanced when transfection of cells occurs in the presence of serum, has significantly reduced level as compared to cells transfected with the same system in the absence of serum.[45] Immunoliposomes covalently conjugated to a monoclonal antibody raised against E-selectin, appear to be very effectively and specifically targeted to activated vascular endothelial cells.[46] Attachment to cells of immunoliposomes is enhanced several hundred fold over that of liposomes attached to an irrelevant antibody.[46] Their potential for drug, oligonucleotide or gene delivery is thus quite evident.[47,48] (See also Chapters 4.6 and 4.7.)

Liposomes, or lipid complexes appear as likely substitutes for virus in gene delivery. By suitable engineering, they can become specific for target-cells, the amount of drugs or DNA associated with them has been dramatically increased,[49] and so has been their lifetime in circulation.[50–52] Further engineering will be required in order to enhance the efficiency of DNA delivery. A more complete understanding of the intracellular fate of delivered DNA will be helpful in achieving this goal. Of course, many problems regarding the molecular biology aspect of gene transfer remain to be elucidated before successful gene therapies can be developed. Nonetheless, the results obtained thus far, justify a prudent optimism.

References

1. Gregoriadis G, Ryman BE. Fate of protein-containing liposomes injected into rats. Eur J Biochem 1972;24:485–491.
2. Papahadjopoulos D, Kimelberg HK. Phospholipid vesicles (liposomes) as models for biological membranes. In: Davidson SG, ed. Progress in Surface Science. Pergamon Press, Vol. 4, Part 2. 1973;141–232.
3. Bangham AD, Hill MW, Miller NGA. Preparation and use of liposomes as models of biological membranes. In: Korn ED, ed. Methods in Membrane Biology, Vol.1. Plenum Press, 1974;1–68.
4. Nicolau C, Alving C. Demetrios Papahadjopoulos and liposomes: From art to science. Liposome Research 1995;5:627–634.
5. Straubinger R, Hong K, Friend S, Papahadjopoulos D. Endocytosis of liposomes and intracellular fate of encapsulated molecules: Encounter with a low pH compartment after internalization in coated vesicles. Cell 1983;32:1069–1079.
6. Szoka F, Jacobson K, Derzko Z, Papahadjopoulos D. Fluorescence studies on the mechanism of liposome-cell interactions in vitro. Biochem Biophys Acta 1980;600:1–8.
7. Wilson T, Papahadjopoulos D, Taber R. Biological properties of poliovirus encapsulated in lipid vesicles: Antibocy resistance and infectivity in virus-resistant cells. Proc Natl Acad Sci USA 1977;74:3471–3475.
8. Taber R, Wilson T, Papahadjopoulos D. The encapsulation of picornaviruses by lipid vesicles: Physical and biological properties. Ann NY Acad Sci 1978;308:268–274.
9. Fraley R, Subramani S, Berg P, Papahadjopoulos D. Introduction of liposome-encapsulated SV40 DNA into cells. J Biol Chem 1980;255:10431–10435.
10. Wong TK, Nicolau C, Hofschneider P. Appearance of B-lactamase activity in animal cells upon liposome-mediated gene transfer. Gene 1980;10:87–94.
11. Fraley R, Straubinger RM, Rule G, Springer L, Papahadjopoulos D. Liposome-mediated delivery of DNA to cells: Enhanced efficiency of delivery related to lipid composition and incubation conditions. Biochemistry 1981;20:6978–6987.
12. Nicolau C, LePape A, Soriano P, Fargette F, Juhel M-F. In vivo expression of rat insulin after i.v. administration of the liposome-entrapped gene for rat insulin I. Proc Natl Acad Sci USA 1983;80:1968–1072.
13. Soriano P, Dijkstra J, Legrand A, Spanjer HH, Londos-Gagliardi D, Roerdink R, Scherphof G, Nicolau C. Targeted and non-targeted liposomes for in vivo transfer to rat liver cells of a plasmid containing the rat preproinsulin I gene. Proc Nal. Acad Sci USA 1983;80:7138–7131.
14. Fraley RT, Fornari CS, Kaplan S. Entrapment of a bacterial plasmid in phospholipid vesicles: potential for gene transfer. Proc Natl Acad Sci USA 1979;76:3348–3352.
15. Nicolau C, Rottem S. Expression of a βb-lactasme activity in Mycoplasma capricolum transfected with the liposomes-encapsulated E. coli pBR322 plasmid. Biochem Biophys Res Comm 1982;108:982–988.
16. Lurquin PF. Entrapment of plasmid DNA by liposomes and their interactions with plant protoplasts. Nucl Acids Res 1979;6:3773–3779.
17. Fraley RT, Dellaporta SL, Papahadjopoulos D. Liposome-mediated delivery of tobacco mosaic virus RNA into tobacco protoplasts: a sensitive assay for monitoring liposome-protoplast interactions. Proc Natl Acad Sci USA 1982;79:1859–1863.
18. Cudd A, Nicolau C. Intracellular fate of liposome-encapsulated DNA in mouse liver. Analysis using electron microscope autoradiography and subcellular fractionation. Biochim Biophys Acta 1985;845:477–491.
19. Nandi PK, Legrand A, Nicolau C. Biologically active, recombinant DNA in clathrin-coated vesicles isolated from rat livers after in vivo injection of liposome-encapsulated DNA. J Biol Chem 1986;261:16722–16728.

20. Wang C-Y, Huang L. pH-sensitive immunoliposome mediate cell-specific delivery and controlled expression of a foreign gene in mouse. Proc Natl Acad Sci USA 1987;84:7851–7855.
21. Felgner PL, Gadek TR, Holm M, Roman R, Chan HW, Wenz M, Northrop JP, Ringold, G-M, Danielse M. Lipofection: A highly efficient lipid-mediated transfection procedure. Proc Natl Acad Sci USA 1987;84:7413–7417.
22. Felgner PL, Ringold GM. Cationic liposome-mediated transfection. Nature 1989;331:461–462.
23. Felgner PL. The evolving role of liposomes in gene delivery. J Liposome Res 1995;5:725–734.
24. Hong K, Zheng W, Baker A, Papahadjopoulos D. Stabilization of cationic liposome DNA complexes by polyamines and poly(ehtylene glycol)-phospholipid conjugates for efficient in vivo gene delivery. FEBS Letters, 1997;400:233–237.
25. Düzgünes N, Felgner PL. Intracellular delivery of nucleic acids and transcription factors using cationic liposomes. Methods in Enzymology 1993;221:303–306.
26. Logan JJ, Bebok Z, Walter LC, Peng S, Felgner PL, Wheeler CJ, Siegal GP, Frizzell RA, Dong J, Howard M, Matalon S, Duvall M, Sorscher EJ. Cationic lipids for reporter gene and CFTR gene transfer to rat pulmonary epithelium. Gene Therapy 1995;2:39–49.
27. Hyde SC, Gell DR, Higgins CG, Trezise AE, Macvinish LJ, Cuthbert AW, Ratcliff R, Evans MJ, Colledge WH. Correction of the ion transport defect in cystic fibrosis transgenic mice by gene therapy. Nature 1993;362:209–211.
28. Zhu N, Liggit D, Lin Y, Debs R. Systemic gene expression after intravenous DNA delivery in adult mice. Science 1993;261:209–211.
29. Nabel EG, Yang Z, Muller D, Chang AE, Gao X, Huang L, Cho KJ, Nabel GJ. Safety and toxicity of catherer gene delivery to the pulmonary vasculature in a patient with metastatic melamona. Human Gene Therapy 1994;5:1089–1094.
30. Behr J-P. Gene transfer with synthetic cationic amphiphiles: prospects for gene therapy. Bioconjugate Chem 1994;5:382–289.
31. Legendre JY, Szoka FC. Delivery of plasmid DNA into mammalian cell lines using pH-sensitive liposomes. Pharm Res 1992;9,1235–1242.
32. Sternberg B, Sorgi FL, Huang L. New Structures in complex formation between DNA and cationic liposomes visualized by freeze-fracture E.M. FEBS Letters 1994;356:361–366.
33. Lasic DD, Strey H, Stuart MCA, Podgornik R, Frederik P. The structure of DNA-liposome complexes. J Am Chem Soc 1997;119:832–833.
34. Raedler JO, Koltover I, Sadditt T, Safinya C. Structure of DNA-cationic multilamellar membranes. Science 1997;275:810–814.
35. Gershon H, Ghirlando R, Guttman SB, Minsky A. Mode of formation and structural features of DNA-cationic liposome complexes used for transfection. Biochemistry 1993;32:7143–7151.
36. Zelphati O, Szoka FC. Mechanism of oligonucleotide release from cationic liposomes. Proc Natl Acad Sci USA 1996;93:11493–11498.
37. Xu Y, Szoka FC. Mechanism of DNA release from cationic liposome-DNA complexes used in cell transfection. Biochemistry 1996;35:5616–5623.
38. Nabel GJ, Chang A, Nabel EG, Plautz G, Fox BA, Huang L, Shu S. Immunotherapy of malignancy by in vivo gene transfer into tumors. Human Gene Therapy 1992;3:399–440.
39. Heath TD, Fraley R, Papahadjopoulos D. Antibocy targeting of liposomes: Cell specificity obtained by conjugation of $F(ab')_2$ to vesicles surface. Science 1980;210:602–604.
40. Wands JR, Blum HE. Primary hepatocellular carcinoma. New England J Med 1991;325:729–731.
41. Park JW, Hong K, Carter P, Asgari H, Guo LY, Keller GA, Wirth C, Shalaby R, Knotts C, Wood WJ, Papahadjopoulos D, Benz CC. Development of anti-$p185^{HER2}$ immunoliposomes for cancer therapy. Proc Natl Acad Sci USA 1995;92:1327–1331.
42. Kirpotin D, Park JW, Hong K, Zalipsky S, Li W-L, Carter P, Benz C, Papahadjopoulos D. Sterically stabilized anti-HER2 immunoliposomes: Design and targeting to human breast cancer cells in vitro. Biochemistry, 1997;36:66–75.
43. Ahmad I, Longenecker M, Allen T. Antibody targeted delivery of doxorubicin entrapped in sterically stabilized liposomes can eradicate lung cancer in mice. Cancer Res 1993;53:1484–1488.
44. Moradpour D, Compagnon B, Wilson BE, Nicolau C, Wands JR. Specific targeting of human hepatocellular carcinoma cells by immunoliposome in vitro. Hepatology 1995;22:1527–1537.
45. Compagnon B, Moradpour D, Alford DR, Larsen CE, Stevenson M, Mohr L, Wand J, Nicolau C. Enhanced gene delivery and expression in human hepatocellular carcinoma cells by cationic immunoliposomes. J Liposome Res 1997;7:127–141.
46. Sprague DD, Alford DR, Greferath R, Larsen CE, Lee K-D, Gurther GC, Cybulski MI, Tosi P-F, Nicolau C. Immunotargeting of liposomes to activated vascular endothelial cells: A strategy for

site selective delivery in the cardiovascular system. Proc Natl Acad Sci USA 1997;accepted for publication.

47. Park JW, Hong K, Kirpotin D, Papahadjopoulos D, Benz CC. Immunoliposomes for cancer treatment. Adv Pharmacol 1997;40:399–435.
48. Kirpotin D, Park JW, Hong K, Shao Y, Shalaby R, Colbern G, Benz CC, Papahadjopoulos D. Targeting of liposomes to solid tumors: The case of the sterically stabilized anti-HER2 immunoliposomes. J Liposome Res 1997;7(4):391–417.
49. Felgner PL. Improvements in cationic liposomes for in vivo gene transfer. Human Gene Therapy 1996;7:1791–1793.
50. Allen TM, Chonn A. Large unilamellar liposomes with low uptake by the reticuloendothelial system. FEBS Letters 1987;223:42–47.
51. Gabizon A, Papahadjopoulos D. Liposome formulations with prolonged circulation time in blood and enhanced uptake by tumors. Proc Natl Acad Sci USA 1988;85:6949–6953.
52. Papahadjopoulos D, Allen T, Gabizon S, Mayhew R, Matthay K, Huang SK, Lee K-D, Woodle MC, Lasic DD, Redemann C, Martin FJ. Sterically stabilized liposomes: Improvements in pharmacokinetics, and anti-tumor therapeutic efficacy. Proc Natl Acad Sci USA 1991;88:11460–11464.

Lasic and Papahadjopoulos (eds.), Medical Applications of Liposomes

CHAPTER 5.2

Cationic liposomes, DNA and gene delivery

Danilo D. Lasic[a] and David Ruff[b]

[a]*Liposome Consultations, 7512 Birkdale Dr, Newark, CA 94560, USA*
[b]*Perkin Elmer, Applied Biosystem Division, Foster City, California, USA*

Overview

I. Introduction

Gene therapy, the concept which was initially proposed some 30 years ago and which may come to fruition in the next decade, theoretically offers healing of human diseases at their cause rather than treating their symptoms. This approach corrects aberant bodily processes at the genetic level by providing the necessary genetic cues in the cells responsible for the disease process.

For a successful gene therapy effective and safe delivery of plasmids, antisense oligonucleotides, ribozymes or other nucleic acid sequences into appropriate cells, preferentially in vivo, is required. While vectors based on viral-based delivery systems are plagued with safety concerns, immune response and formulation issues, the problems of lipid based systems are mostly low efficiency of transfection and gene expression as well as its short duration.[1] Of course, some safety and immunogenicity issues of cationic lipids will have to be addressed as well.

Scientists have been using liposomes for nucleic acids and gene delivery since the late 1970s.[2] However, it was only after the introduction of cationic liposomes, which were shown to complex DNA,[3] which offered some promise for an easy and efficient liposomal gene delivery.[4] Upon complexation of DNA with various cationic liposomes, the colloidally soluble (suspended) DNA-liposome/lipid complexes (genosomes) can be formed and were shown to significantly improve transfection (delivery of plasmids into cell nuclei) and gene expression—the synthesis of the protein encoded in the DNA plasmid by the cell machinery. Nowadays, numerous liposome kits for in vitro gene transfection are commercially available. For commercially viable gene therapy, however, in vivo delivery is preferred. Commercial formulations have been found to be very inefficient for gene delivery in vivo. To increase transfection efficacy, the majority of researchers have concentrated on the synthesis of novel lipids, hoping that more efficient lipids will be discovered. Indeed, numerous novel cationic lipids which improved transfection and gene expression have been synthesized.[6–8] Typically they are mixed with neutral lipids to formulate liposomes. The role of neutral lipid was also carefully studied and numerous in vitro studies have shown that in general, but not always, equimolar mixture of cationic lipid with dioleoyl phosphatidylethanolamine (DOPE) gives rise to the optimal transfection.[7–9] This effect has been attributed to its ability to induce fusion with endosomal membrane and mediate DNA release from primary endosomes before its chemical degradation. In vitro studies have also demonstrated cytotoxic effects of cationic liposomes and genosomes. An important observation has been that toxicity does not parallel the transfection activity of cationic lipids. No relationships between the size of the complex and transfection activity have been established. Presently it is believed that for each cell line the transfection conditions, including liposome composition and DNA/lipid ratio, have to be optimized. While the presence of plasma typically significantly reduces the expression, it seems that an increase in positive charge of the complexes reduces the neutralizing effect of plasma.[1]

The results of in vivo studies are even less understood. In contrast to numerous transfection studies the functional roles of lipid composition and DNA/lipid ratio and the few in vivo safety evaluations (toxicity and immunogenicity of cationic lipids, liposomes and complexes), practically no physico-chemical characterizations and biological studies of the complexes have been performed. While it is a common knowledge that DOPE containing liposomes cause various specific effects upon intravenous administration, similarly to phosphatidylserines,[10] DOPE has been used practically as a sole neutral lipid for years before it was realized that other neutral lipids, most notably cholesterol, can form complexes which yield higher efficiencies of in vivo transfection.[11–13]

In addition to the poor understanding of physico-chemical properties of complexes, their biological characteristics, including stability in plasma, pharmacokinetics and biodistribution upon systemic or localized application, have not been studied adequately. Mechanistic studies typically concentrated on narrowly focused specific processes which were analyzed without global perspective. Gene expression upon administration of colloidal DNA, however, is a multiple step process in which high losses at each step occur and are highly relevant to development of successful gene therapies. Foreign gene expression in somatic cells is a reverse cascade process due to the ability of organisms to defend themselves against invading pathogens with a sequence of barriers and neutralizing agents. We must say, however, that the system seems to be extremely complicated and that it is very likely that multiple mechanisms operate in parallel and few experiments attempting to shed some light on particular steps in this multi-step process may have simply measured one of few, or several, parallel alleles.

Now, when it is becoming clear that approaches based on permutations, variations and combinations of various reagents, types of colloidal particles, administration routes and similar have not generated an efficient and safe delivery system, fundamental questions such as the mechanism of transfection[1] and the structure of the complexes, are gaining more attention in studies. In addition to several electron microscopy studies of genosomes,[14,15] some studies dealing with the cellular uptake and release of complexed DNA,[16] and interaction of genosomes with the plasma based complement system, at present there are only few rigorous studies of the structure of some genosomes.[17–20] Thorough studies to address genosome stability, interaction characteristics and their pharmacokinetics and biodistribution have yet to be performed.

II. Nucleic acids and cationic liposomes

Cationic lipids and DNA both exhibit rich polymorphic behavior at higher concentrations or upon interaction with specific agents. In concentrated solutions DNA forms a variety of liquid crystalline phases,[1,21] similar to lipids.[10] While lipid phases are mostly lamellar and inverse hexagonal, DNA is packed into hexatic phases whereby the DNA helices are arranged into normal hexagonal phase.[22] At concentrations below dense liquid crystalline phases, the DNA double helix is normally in the B configuration, in which ten base pairs form a pitch of the helix which measures 3.4 nm. The diameter of the helix is around 2 nm to which a shell of water of hydration of approximately 0.2–0.3 nm has to be added. At the backbone of the helix there are two negative charges per 0.34 nm or, when expressed in molecular weight units, per 660 Da.[1] An important structural change that DNA can undergo is condensation.[23] During this phase transition, random coil collapses into ordered or disorderd aggregate with up to a million fold reduced hydrodynamic volume (by this process approx. 1 meter long human genome fits in a nucleus with diameter of 5–8 μm). Interactions with polyvalent cationic species reduce electrostatic repulsion on the DNA chain and its structure commences folding into itself. Typically multivalent cations, polyamines or cationic polyelectrolytes are used and resulting condensates are of toroidal or cylindrical form of colloidal dimensions.[1,23] It is believed that one of the major effects of positively charged systems is their ability to effectively reduce DNA size which facilitates its internalization. On the other side, however, this interaction must not be too irreversible, because for transfection and gene expression DNA has to be decondensed. Energetics of DNA is directly related to electrostatic interactions with flexibility, bending elasticity, torsional stress as well as coiling and knotting factors contributing. Persistence length is defined as the length of DNA where a 180 degree turn requires 1 kT energy and under physiological conditions equals around 50 nm. Strong interactions and shielding of negative charge can reduce this 2–3 fold.

While the liquid crystalline behavior of DNA (longer than 1 kb) commences above approximately 15 mg/ml, at concentrations used in genosome preparations (<2–3 mg/ml) molecules of DNA form isotropic solutions of semi-rigid worm-like random coils with different degrees of supercoiling, linearization and nicking, depending on its history (preparation procedure and treatment). The electrostatic behavior of DNA can be approximated by Poisson Boltzmann (PB) equation. We must be aware, however, that at these high surface charge densities at low ionic strengths Debye Huckel approximation and theory of counterion condensation are not valid and very specific approximations have to be used.[1,24] Similarly, the electrostatics of cationic liposomes at high surface charges cannot be described by Gouy-Chapman approximation. At low ionic strengths potential decays as logarithm of the separation while at higher ionic strengths the decay is exponential. For high surface charges special boundary conditions have to be used (constant potential vs. constant charge). Despite this, an advanced Poisson Boltzmann formalism is still the most reasonable approach to estimate these interactions. Typically one solves DNA electrostatics in cylindrical coordinates and this potential interacts either with a plane (large liposome) or sphere (small liposome) around which the potential is defined by the solution of the PB equation in appropriate coordinate system.[1] Boundary conditions, however, have to be carefully defined for each reactant and must be self-consistent. Another problem is the effective dissociation in colloidal systems. It is well known that at higher surface charge densities the degree of ionization (pK) can be drastically reduced. Alternative theoretical approaches are molecular dynamics and other computer studies. Especially, on a small scale various lattice models can be constructed. While with present computer capabilities it is unlikely that genosome structures can be modeled, they can, perhaps, provide a basis for the interaction between negative and positive charge and therefore lipid structure—transfection activity (or DNA complexation and de-complexation) relationships. Such approaches run into trouble with defining potentials, especially around ill-characterized phosphate ion, not to mention novel cations where molecular coordinates and potential maps have not been measured and calculated yet.

Similar to DNA, cationic lipids exhibit rich phase behavior as well. While no phase diagram studies have been performed yet, preliminary observations indicate that these lipids form mostly lamellar or inverse hexagonal phases while at lower concentrations liposomes or hexasomes, which are dispersed lamellar or hexagonal phase, may be a stable phase. Diacyl lipids containing multiple positive charges can form also micellar phases. Cationic lipids which have relatively small polar head and/or pK values around 7 may be very hydrophobic and can form inverse hexagonal phases. Pure cationic lipids also tend to gel during swelling. This is due to strong electrostatic interactions, high lipid concentrations and can be reduced by adding electrolytes or neutral lipid. Also, sonication can disperse gel and a suspension of small unilamellar vesicles exhibits normal viscosity. Mostly these lipids have rather high values of critical micelle concentration. This is especially true for multivalent diacyl lipids which in many cases form micellar structures, even in mixtures with DOPE or cholesterol.

III. DNA vectors and gene expression

Biologists have developed techniques to efficiently introduce exogenous genetic material into living prokaryotic and eukaryotic cells, both in vitro and in vivo situations. Since the first studies with bacterial transformation, the utilization of marker genes which encode easily detectable protein products or cause phenotypic changes which allow selection of transformants have been the cornerstone of measuring the level of success of these endeavors. Recombinant DNA technology can be used to design any protein coding sequence in expression cassettes with genetic control elements directing the production of the exogenous protein. Expression vectors are developed in the structural context of plasmid DNA molecules. Plasmid DNA is propagated and manipulated in *Escheritia coli* bacterial hosts. *E. coli* is an excellent host for genetic engineered plasmids. Plasmids from bacteria are in a circular supercoiled structure conformation. The bacteria can produce large quantities of DNA and the technologies to manipulate genetic material based on *E. coli* systems have been well defined.[25] Sufficient understanding of the biological processes involved in gene regulation have enabled researchers to use constitutive, tissue-specific, developmental or environmental cues to selectively control transcription of these expression cassettes.

III.1. Expression vector plasmid DNA design

Plasmid DNA vectors for gene therapy uses contain several common features. First, an origin of replication (ori) that effectively interacts with *E. coli* DNA replication factors must be present. Second, it must contain a bacterial selection marker such as the antibiotic ampicillin resistance gene. Culturing *E. coli* harboring an ampicillin resistance marker on the plasmid allows for positive selection of the host cells and amplification of the plasmid vector. The best ori under ideal selection and growth conditions can produce up to a range of near a thousand plasmid molecules per cell. Third, a eukaryotic expression cassette(s) must be present. The basis of a single cassette is a promoter at the start (at the 5′ end), protein coding sequence with or without intervening non-coding sequences (introns) and a downstream termination sequence (poly A site, or the 3′ end) as shown in Figure 1. Introns usually have a dramatic positive effect on transcription in most systems. Multiple cistronic vectors code for two or more exogenous proteins and can be used in vivo. However, to obtain efficient expression the cistrons must be carefully designed. Internal ribosomal entry sites (IRES) derived from the encephalomyocardiovirus or picarnoviruses can be used to derive multiple protein sequences from one promoter.[26] Also, two complete cistrons each containing their own promoters, coding sequence and polyA sites can be spliced into one plasmid vector. However interference between the juxtaposed transcription units can create an inefficient expression system.

A key feature of the eukaryotic expression cassette is the selection of the

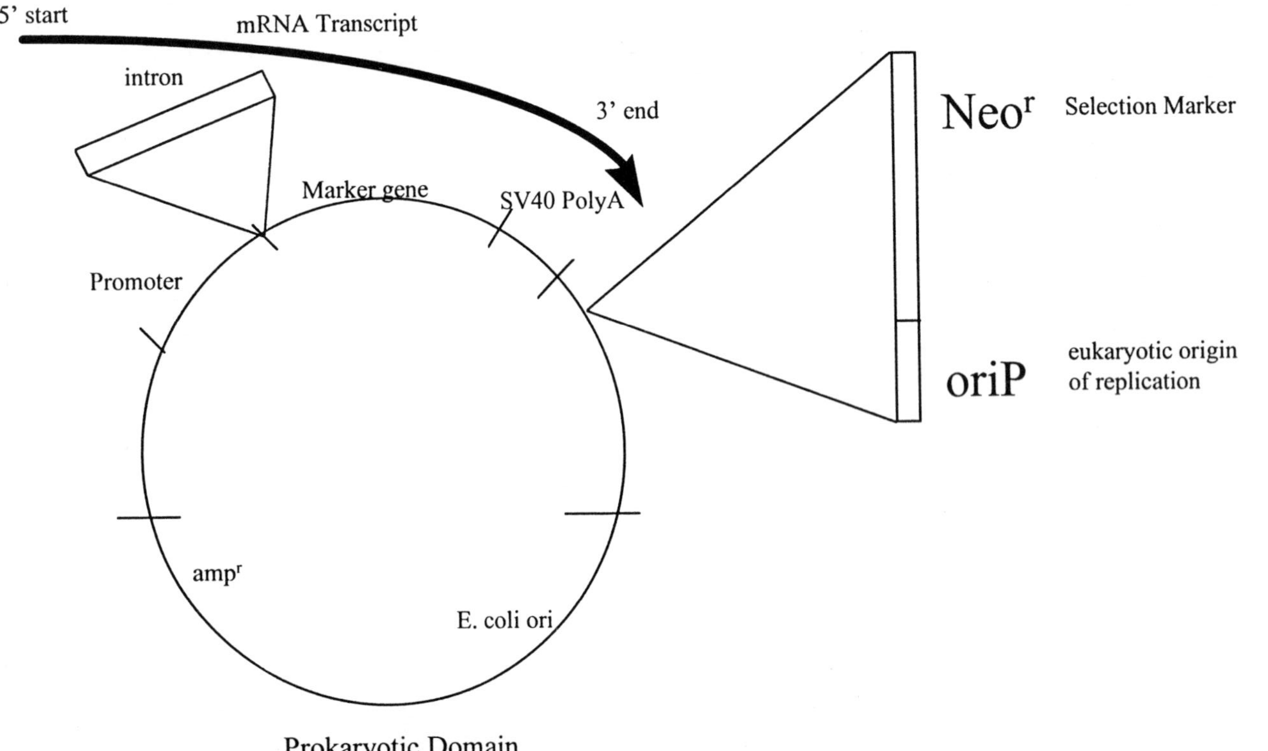

Fig. 1A. Diagrammatic view of typical expression vector. A typical plasmid construct is depicted. The plasmid consists of two major domains: (1) A prokaryotic cassette which ensures propagation in *E. coli* host cells. This cassette codes for a selection factor for ampicillin resistance and an origin of replication which functions to replicate the plasmid in bacterial cultures; (2) A eukaryotic cassette which directs expression of protein when introduced into mammalian cells by genosomes. This cassette typically contains an intron to increase mRNA levels. A selection marker (such as neomycin or dihydrofolate reductase) with its own promoter can be added to help prolong the longitivity of the construct and its expression levels in mammalian cells. A eukaryotic origin of replication can be added and when appropriate factors are present to interact with it, vector stability is extremely high.

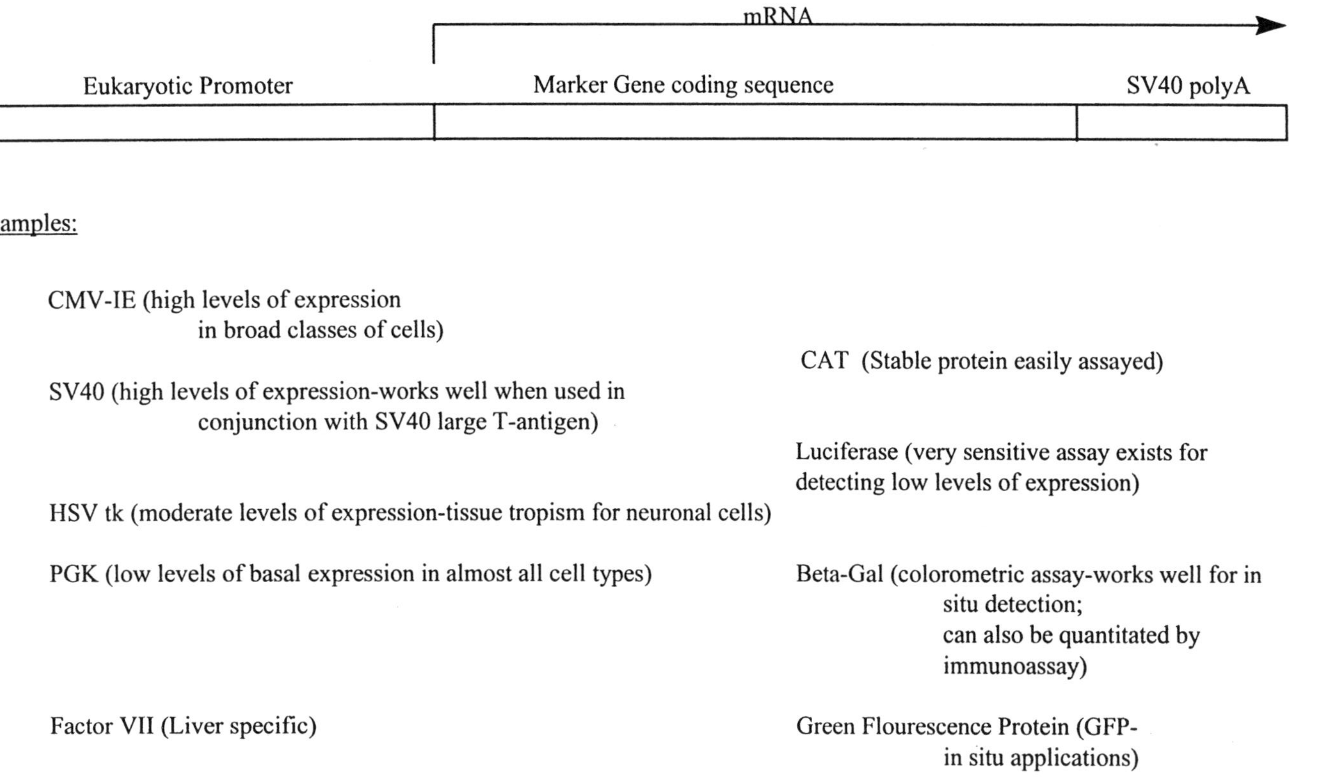

Fig. 1B. Eukaryotic domain cassette components and representative promoter and marker gene elements widely used. A promoter of choice in many systems is CMV-IE. However with more demands being placed on tissue specific approaches to gene expression, many genosomes are being designed to target expression to specific cells (such as Factor VII–Greenberg D et al., 1995. Liver-specific expression of the human factor VII gene. Proc Natl Acad Sci USA 92:12347–12351.). Four commonly used reporters are indicated. However, many researchers are now using quantitative RT-PCR techniques to measure gene expression and thereby eliminating the need to use marker genes.

promoter. The promoter directs the expression of mRNA. Promoters used in gene therapy are derived from well understood biological systems. Most vectors in current use utilize promoter elements from viruses. Viruses have evolved their genetic material under stringent selection conditions and many viruses contain powerful genetic elements capable of producing mRNA at levels far above that of endogenous host cell promoters. Well characterized viral promoter systems include the cytomegalovirus intermediate-early promoter-enhancer element (CMV-IE), simian virus 40 (SV40), retroviral elements and the herpes simplex thymidine kinase promoter (HSV-tk).[27] Other promoter elements derived from mammalian chromosomal sequences are capable of honing the expression to specific cellular developmental phases, environmental cues or tissue types.[28] House-keeping gene promoters that express stable levels of mRNA (such as the phosphoglycerate kinase (PGK)) have been used in eukaryotic expression vectors. Several inducible systems have been characterized. Among the widely used inducible systems based on environmental cues are steroid and metal inducible elements. And an inducible expression system which functions in vivo based on the antibiotic tetracycline has received wide interest in gene therapy applications. Despite the wide repertoire of promoters available, development of novel promoters continues to be a priority in many laboratories as the demands for gene therapy efficacy increase.

Most preliminary studies of new vector designs use reporter genes for characterizing the expression dynamics. Reporters are ideal for this purpose because rapid and reliable assays have been developed for in vitro and in vivo applications. Commonly used markers include the enzymes β-galactosidase (β-gal), chloramphenicol acetyltransferase (CAT), luciferase and the green flourescence protein (GFP). Assays for these markers can be based on enzyme activity (measuring light output-luciferase and GFP; chromogenic dye formation-β-gal; immunoassay quantitation-CAT and β-gal). Since these commonly used reporters are not found in untransfected mammalian cells, any marker activity detected in vitro or in vivo must have originated from gene expression activity from the introduced vector.

With the array of promoters available, a key concern becomes vector persistence in the host cells. Important parameters that effect persistence of expression are stability of the construct once it enters the cell, nuclear localization, resistance to nuclease degradation and retention. Each cell division results in a dilution of vector. Therefore, most expression systems have a transitory existence in the cell. Using selection factors (such as neomycin) in the plasmid construct backbone, one can significantly increase the in vitro half-life of expression because the integration in the chromosome can be achieved. With selective pressure and months of culture, the isolation of stable transfectants can be accomplished in vitro. Placing a eukaryotic origin of replication (ori) into the backbone of the plasmid vector can dramatically improve the expression. Mammalian chromosomal-derived ori have been difficult to characterize and are too large for convenient cloning and transfection. Two ori commonly used are from viral sources (SV40 and oriP). Co-expressing viral proteins which associate with viral ori, episomal expression can be increased to long-term expression. Unfortunately, co-expressing or co-delivering these proteins can be cytotoxic both in vitro and in vivo. Other viral elements in an expression vector can be used for enhancing the stability or interacting with viral packaging systems (such as adeno-associated virus, retroviruses, adenovirus and the cytoplasmic-based sindbis α-virus expression system).[29,30]

A robust purification of plasmid DNA is required to obtain expression in cationic lipid-based gene delivery systems. Large quantities of highly-purified plasmid DNA are usually obtained from growing the *E. coli* cultures in fermentors to an optimized cell density, lysing the cells by adding an alkaline solution to the cells, precipitating cellular debris with detergents and salt and then selectively isolating the plasmid DNA. Ion-exchange chromatography is typically used in the final purification step(s). Contaminants in plasmid DNA must be removed or the genosome formation process as well as expression become unreproducible and the contaminants can be highly toxic to cells. Contaminants commonly found in plasmid preparations are endotoxins, short oligonucleotides (both DNA and RNA), chromosomal DNA and bacterial proteins. Gene expression levels obtained in vivo are extremely sensitive to the purity of plasmid DNA used to make the lipid complexes.

IV. Genosome preparation and interactions

Genosomes are typically prepared by rapid mixing of the DNA and liposome suspensions. The initial interaction between negatively charged nucleic acids and positively charged liposomes is primarily electrostatic. When electrostatic interaction brings particles sufficiently close, other attractive interactions, such as van der Waals, hydrophobic and electrodynamic (ion correlation) attraction as well as the formation of hydrogen bonds, can strengthen molecular/colloidal assemblies and provide energy for DNA conformational changes and bilayer restructuration. Electrostatic interaction depends on the ionic strength. At Bjerrum length, which is around 0.7 nm at physiological ionic strength, the strength of the interaction between the opposite charges equals 1 kT (thermal energy at temperature T, k— Boltzmann constant). In low ionic strength (I) solutions which are normally used for genosome preparation the interaction range is much longer, as it decays proportionally to $I^{-1/2}$.

This thought analysis shows that when DNA plasmids interact with liposomes hundreds or thousands kT of energy are released per complex which consecutively may contribute to DNA and lipid structural changes. Disintegration of typical liposome requires 20–50 kT and forcing DNA into turns from 1 to 10 kT, depending on the degree of charge neutralization and curvature. Energy associated with liposome restructuration originates in bending rigidity of the bilayers as well as their stretching elasticity (lysis tension) and possible creation of hydrophobic defects and opening of the bilayers forced by interactions with the DNA.[10] A very simple experiment can show the lipid restructuring and solubilizing power of DNA: if DNA is added into a turbid suspension of cationic multi (oligo)lamellar vesicles at approximately five fold excess of charge, the suspension becomes transparent, indicating complete dissolution of large liposomes.

Complexation process was shown to be thermodynamically as well as kinetically controlled.[1,18] While temperature does not seem to significantly contribute to the size of the complexes formed, lipid and DNA concentrations and ionic strength do. Regularly, cationic complexes for cationic lipid concentration >1 mM precipitate, regardless of cationic or neutral lipid for ionic strengths above 10 mM. And within this range, it is only quick mixing of the reagents which assures stable colloidal solution and not precipitation or flocculation. In the solutions of nonelectrolytes or in distilled water up to around 1 mg DNA/ml can be colloidally suspended in the case of cationic complexes. Several labs also noticed that mixing the minor component into the major, i.e., for the preparation of anionic complexes pouring liposomes into DNA and vice versa for cationic genosomes, reduces precipitation. This can be easily explained[1,18] by the fact that such mixing avoids crossing the solubility gap which always accompanies electrically neutral complexes at sufficiently high concentrations (>0.1–0.2 mM cationic lipid).

The importance of quick mixing of reactants to avoid precipitation was qualitatively explained by the fact that a reaction far from equilibrium conditions typically generates an explosion of nucleation embrii resulting in a simultaneous growth of numerous complexes as opposed to closer to equilibrium growth of complexes which grow much smaller number (but of much larger and precipitation prone) complexes. Also, the translational and rotational diffusion of DNA and liposomes as well as lipids within bilayers is in millisecond range for the distances important in this reaction and therefore quick mixing can aid significantly to increase locally the homogeneity of reactants and assure that local concentrations are closer to the bulk. For this reason small unilamellar vesicles (which are the highest dispersal state of these lipids in aqueous phase) are preferred and so is equivolumetric mixing, concentration permitting.

V. Structure of genosomes

The first electron micrographs of genosomes were shown in late 1993. Cationic complexes were studied most intensely because they are predominantly used in transfection protocols. Metal staining technique showed that DNA condensed and became encapsulated in the DOTMA:DOPE bilayer.[31] DOTAP-DNA complexes were shown to be onion like multilamellar liposomes with some detached DNA fibers.[15] Somehow similar complexes were observed with DC-Chol-DOPE liposomes where spherical aggregates composed from smaller particles surrounded by many fibrils were observed.[14] Due to its appearance, this model is often referred to as "spaghetti and meatball" model. Fibrils were shown to be DNA (or double DNA) coated by a tubular lipid bilayer.[14] In another freeze fracture study it was shown that fibrilar structures are mostly associated with the presence of DOPE in the formulation.[32]

Cryo electron microscopy coupled with X-ray scattering showed that cationic genosomes exhibited short range order with characteristic lamellar symmetry. Locally, parallel DNA helices sandwiched between cationic bilayers formed intercalated lamellar array with short range order. Lamellar arrays of smectic DNA sandwiched between lipid bilayers may propagate over a substantial part of the complex. In these planes DNA seems to be 2 D (dimensional) ordered and superhelical twists are unwound. Topologically, DNA has to form some less ordered organization at the edges of these structures to preserve its closeness and supercoiling. Since this observation,[17,18] periodic structure of genosomes was observed in other systems as well.[19,33]

Cationic genosomes, for cationic lipid concentrations between 1.5 and 5 mM and negative/positive charge ratio, $\rho = 0.5–0.1$, are typically in the size range from 100–300 nm when prepared from SUV and have zetapotentials from +40 to +60 mV (zetapotentials of cationic liposomes in nonelectrolyte solutions are from 65 to 80 mV). Often, however, the measured zetapotential values are difficult to comprehend because they can decrease with increasing charge density or decreasing salinity. These effects, which were observed also in the force = f(distance) profiles, however, are due to different levels of counterion association with lipid polar heads at different surface charge densities within the electrical double layer.[34] Excess of cationic charge results in unreacted liposomes in the formulation. They may have a role of inactivating plasma components which neutralize genosomes in vitro and in vivo, because it is known that neutralizing effects of plasma components can be overcome by higher cationic charges.

Anionic complexes, which are sometimes used for intramuscular, subcutaneous, intratracheal or pulmonary DNA delivery, are much less characterized. Cryoelectron microscopy studies of these complexes either yielded a fraction of lamellar structures or snake-like particles. Zetapotentials vary from expected values (around −20 to −40 mV) to values identical to naked DNA (−60 mV) in other cases. Multilamellar vesicle dissolution experiment described above indicates, that at higher values of ρ the structure must consist of (partially) coated polymers. In turbid solutions at lower values of ρ qualitative relations between size and turbidity are not obeyed, possibly indicating nonspherical structures. Even transparent solutions when sized in quasielastic light scattering particle apparatuses give rise to diameters around 200–300 nm, indicating an artifact of the methodology. This phenomenon was also observed with long, flexible rodlike micelles.

Despite the lack of rigorous physico-chemical experiments, several models of DNA-lipid complexes have been presented. They are schematically shown in Scheme I. The original stoichiometric model, which showed a simple aggregate of liposomes and DNA[35] was amended into a model of condensed DNA surrounded by a cationic bilayer.[9] Similarly, it was proposed that lipid induces DNA condensation and in turn, DNA induces lipid restructuration, giving rise to elongated complexes in which condensed DNA is encapsulated in a lipid bilayer.[31,36] Following freeze fracture electron microscopy spherical aggregates surrounded by a halo of fibers were observed.[14,15,32,37,38] It was shown that fibers are DNA strands encapsulated by a tubular lipid bilayer.[14] Because DNA itself condenses into a hexagonal array, it was also proposed that DNA-lipid complexes form an inverse hexagonal phase.[39] However, we have investigated numerous DNA lipid complexes (but none with large excess of DOPE) by X-ray scattering and never saw any reflections which would indicate hexagonal symmetry. Rather either amorphous

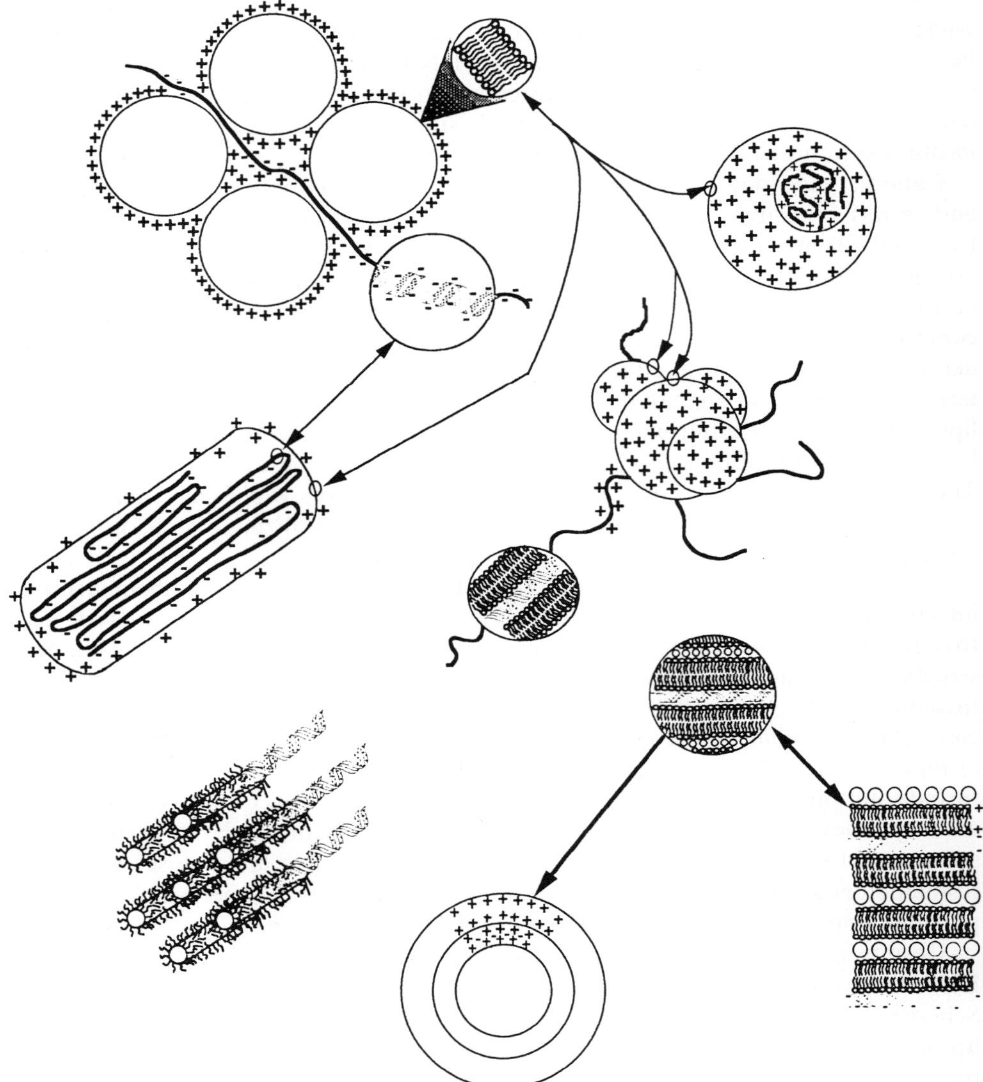

Scheme 1. Schematic presentation of various models of genosome structure. Top left: stoichiometric model, top right: condensed DNA coated by a lipid coat, middle left: condensed DNA and restructured liposomes forming an elongated structure with encapsulated DNA, middle right: spherical aggregates with a halo of fibrils (the so-called "spaghetti and meat-ball" model), low left: inverse hexagonal phase structure in which DNA helices form the aqueous channels in the liquid crystalline phase; bottom right: intercalated lamellar phase model: 2 dimensionally condensed DNA sandwiched between cationic bilayers forming an intercalated lamellar phase. It can form flat or spherical structures. For details see text. (Courtesy Stan Hansen.)

complexes or structures characterized by a lamellar symmetry were observed. Mostly, complexes showed local short range lamellar symmetry. Lamellar arrangement was predicted by a theoretical calculation as well.[40]

Currently, it is believed that complexes are heterogeneous with respect to size, shape, density and structure.[1,18] In several cases, structures with short range local order were observed. It is very likely that there is much more ordering in these complexes which simply cannot be observed by methods used in their characterization and that each model can perhaps approximately describe a portion of the complex. In addition to lipid phase behavior, a crucial parameter which causes ordered structures is the stiffness of DNA molecule. Complexation with less rigid polyelectrolytes, such as polyglutamic acid results in disordered complexes and amorphous precipitates, as confirmed by SAXS.[1] As will be shown below, effective size reduction may have important consequences on the activity of these complexes.

VI. Gene expression and structure activity relationships

In cell culture transfection studies only minute amounts of DNA are needed and also precipitated complexes can be used. Therefore the problem of sample precipitation does not abolish transfection. Often, commercially available liposomes are mixed with DNA in electrolyte solutions, resulting in large aggregated complexes. While these systems work satisfactorily in vitro, such genosomes yield very low level transfection in vivo. One reason is the size and morphology of such complexes and the other is the low dose of applied DNA because transfection, if experiment is performed properly, is dose dependent.

Structure activity studies were performed mostly in vitro. With the exception of the molecular structure—in vitro activity and cytotoxicity of DC-Chol analogues,[41] not much structure activity relationships (SAR) are known. When studying cationic cholesterol derivatives it was noted that tertiary amine gave the best transfection at lowest toxicity. For multivalent cholesterol derivatives it was shown that the site and the angle of the attachment of the polyamine was important, with molecules having perpendicular arrangement between either long axis of sterol and direction of polyamine being more active than the parallel ones.[8] In the case of diacyl lipids it was discovered that dioleoyl and dimyristoyl chains give rise to the highest expression.[7,8] This is hardly surprising, because for both, interaction of DNA with liposome as well as of genosomes with cells, fluid membranes are necessary. With respect to polar heads and number of charges no clear conclusions have been reported. Studies of DOTMA like molecules have shown that decorating polar head with hydroxyethyl group and with beta amines increased transfection effciencies.[7,42]

A conclusive observation of in vitro transfection is that increased amount of cationic lipids increases transfection as well as cytotoxicity. The balance between activity and toxicity therefore determines transfection activity. Unfortunately, not much is known with respect to colloidal structure—transfection activity relationships. Important information needs to be gained about activity as a function of genosome size at specified charge ratios.

Such a simple mechanism, however, cannot be applicable for in vivo cases, and obviously not for the systemic delivery. Therefore the first studies of in vivo transfection were not very successful. Researchers used commercially available liposome kits and prepared complexes according to the preparation procedures.[43] Although some transfection was observed, the expression levels were low and some results were often difficult to reproduce. Next, novel lipids were tried and in some models much higher transfection efficiencies were found.[44]

For in vivo applications, however, various routes of administration exhibit profound differences. While localized (subcutaneous, intramuscular or intratumor), intraperitoneal and topical (skin, oral and mostly intratracheal instillation and aerosol inhalation) administration can be eventually compared to in vitro systems, systemic delivery involves completely different interactions prior to cell transfection and has to be approached differently. At present, it is still often stated that in general there is little or no correlation between activity in vitro and in vivo. We believe, that activity in vitro is a necessary but not sufficient condition for activity in vivo. The latter one, obviously, depends on the route of administration. While topical delivery (including intratracheal instillation and inhalation of genosome aerosol) and intraperitoneal administration may resemble in vitro conditions, intravenous administration presents completely different conditions and gene expression depends on pharmacokinetics, biodistribution and stability of genosomes in biological environment to a much greater extent than in other administration routes. Intramuscular and subcutaneous injection may lay in between with respect to biological severity of the environment. Therefore one has to determine these correlations in order to improve efficiency of gene expression. It is likely that each administration route has different optimization characteristics. For instance, for pulmonary delivery it is not known if smaller or larger genosomes are preferred. This transfection may be similar to in vitro conditions and therefore DOPE may be a superior neutral lipid than cholesterol. The situation was partially explained for intravenous administration where it was shown that complexes have to be small, tightly packed (condensed DNA is sandwiched between fluid lipid bilayers) in order to express. This ensures protection of DNA in the bloodstream and allows some limited volume of biodistribution. In contrast, large, noncompact genosomes (which most often are prepared by using commercial liposome kits) are very likely broken in blood in milliseconds and the exposed DNA is degraded in seconds.[1] For systemic administration cholesterol was shown to be much more effective neutral lipid than DOPE.[1,11–13,20] Because physico-chemical characteristics of genosomes containing either neutral lipid are rather similar, this may imply that it is the stability of the structures in plasma which causes the difference and also eliminates the early endosomal release induced by DOPE as the mechanism of transfection.

This analysis implies that the correlation between in vitro and in vivo experiments has to account for genosome stability in plasma, pharmacokinetics and biodistribution. Cells in culture are also known to exhibit rather large differences in phagocytic activity upon subtle changes in the environment which further complicates comparisons.

For intratracheal (it) instillation of genosomes, lipids were found which increased transfection up to 1000-fold.[8,44] Spermine attached via a carbamate linker on a non-terminal amine to cholesterol (giving rise to a "T" shaped conformation) yielded the optimal lipid for gene expression upon it administration.[8] Gene expression in the range from 15 to 125 ng CAT protein/mg of protein was observed and dose-dependent pulmonary inflammation, which did not leave fibrotic lessions was reported.[45] For in vitro transfection, however, diacyl hydrophobic anchors were better. Genzyme scientists also reported that as a neutral lipid dilinoleyl phosphatidyl ethanolamine (at ratio 1:2) can augment gene expression up to 10 fold relative to that obtained by DOPE.

Levels obtained upon systemic administration are in general much lower, around 1–2 ng of CAT protein/mg protein in the lung.[11] In general, however, short duration levels of about ng of expressed protein per mg of protein do not promise, with exception in vaccination or cytokine biosynthesis, many medical applications and indicate that further improvements in gene delivery are probably necessary for commercialization of this technology.

Only recently,[20] very efficient gene expression upon systemic administration was reported. Expression of ca 0.2 μg CAT protein/mg protein in mouse lung was observed upon administration of DOTAP-Chol/DNA ($\rho = 0.5$) complexes. Expression in 12 other tissues, although at 100 to 1000 fold lower levels (in the decreasing order heart, muscle, lymph nodes, skin, thymus, colon, tail, spleen, liver, kidney, brain) could be also observed. Parallel cryo-electron microscopy and small angle x-ray scattering studies have revealed the unique structure of these complexes—condensed DNA is encapsulated in the middle of spherical liposomes. It is hypothesized that this particular structure protects DNA and allows better biodistribution than other complexes which are characterized by either stacks of lamellae with adsorbed DNA plasmids or lipid aggregates surrounded by a halo of fibrilar DNA coated by lipid tubules. The unusual spontaneous "self-encapsulation" of DNA was attributed to the use of specially prepared, invaginated liposomes which resemble spherical vase-like structures and have a large excess of free surface area. Upon DNA adsorption such liposomes can undergo inversion, resulting in complete DNA encapsulation. Because such an interaction neutralizes charges only on one side of DNA, often a second liposome adsorbs on the adsorbed DNA. In other words, briefly, the cohesive, but fluid, lipid bilayer with a large excess of free surface area allow much better DNA organization and better condensation than regular liposomes. Condensation, packing and structural events have more degrees of freedom and time to self-assemble and self-organize than in the case of small unilamellar liposomes which can only break or large multilamellar vesicles which inevitably give rise to large complexes. As a consequence of effective (2 dimensional) DNA condensation,[46] these liposomes can also colloidally suspend higher DNA concentrations than other systems. This is very important, because gene expression is dose dependent. An important observation was also that optimal size distribution of the complexes was between

200 and 450 nm. No toxicity was observed and that duration of expression was longer than in other similar experiments reported in the literature. For instance, gene expression dropped twofold in a week and 5-fold in three weeks. Furthermore, by attaching a targeting ligand asialofetuin, expression in the liver increased 7-fold. Safety of these formulations, which typically contain several fold less cationic lipid per amount of DNA than complexes described in the literature, was carefully studied too. Detailed pathologies of tissues did not show any adverse effects and complexes were toxic only at very high DNA concentrations (but never immediately upon administration). Surprisingly, the toxicity depended on the type of the plasmid indicating the potential toxicity of the synthesized bacterial proteins.

Because DOTAP was not considered as a cationic lipid with a high transfection efficiency, we believe, that this study shows the importance of the colloidal structure on the gene expression.

VII. Antisense oligonucleotides and ribozymes

While DNA plasmid delivery tends to turn certain cell function on by delivering wild type genes to appropriate cells, antisense oligonucleotide technology aims at stopping the synthesis of unwanted proteins by binding to and inactivating messenger RNA. Similarly, ribozymes stop the translation by cutting mRNA. While these technologies have recorded fascinating development in the chemistry of these agents, the delivery issue was largely neglected. Convincing test tube experiments and effective in vitro studies were not matched in vivo. Liposomes seem to be one of the more promising delivery vehicles. Not many papers were published and the contribution by Woodle and Leserman and reference 1 review some therapeutic results. Recently, however, it seems that these problems were realized and several academic groups and companies have started thorough studies of delivering antisense oligonucleotides and ribozymes via liposomes and anecdotally impressive results have been mentioned.

From the physico-chemical point of view the electrostatics is similar. However, ordered structures are not formed because these short segments cannot condense and do not have any stiffness (antisense oligonucleotides are single strand DNA fragments from 15 to 30 base pairs long while synthetic ribozymes are at most twice longer). In the anionic regime nucleic acids induce liposome fusion and large and giant unilamellar vesicles are observed. In the cationic regime adsorption of oligonucleotides induces shape changes and small oval liposomes are observed in cryo electron microscope.

VIII. Conclusion

While the majority of researchers are trying to improve transfection by synthesizing novel lipids, we believe, that colloidal properties of DNA-lipid complexes are at least as important. This claim can be strengthened by the fact that despite a decade of work no clear molecular structure—transfection activity correlations have been found.

As we have discussed above, it is the polymorphism of lipids as well as of DNA which gives rise to the novel structures which permit efficient transfection. We therefore believe that by having an improved understanding of thermodynamics, kinetics, and stability of these complexes better delivery vehicles for transfection will be constructed as happened with liposomes.[47]

Since it seems that further improvements in promoters, enhancers, introns, and terminating sequences of current plasmids will not dramatically improve gene expression, we may speculate that by co-delivering DNA binding proteins possessing nuclear localization sequences, nuclear transport would be facilitated. In such a system, DNA plasmid would be complexed with DNA binding proteins containing nuclear localization sequences and thereby increasing the delivery capability of plasmids into the cell nucleus.

References

1. Lasic DD. Liposomes in gene delivery. Boca Raton, FL: CRC Press, 1997.
2. Fraley RP, Papahadjopoulos D. Liposomes: the development of a new carrier system for introducing nucleic acids into plant and animal cells. Curr Top Microbiol Immunol 1982;96:171–187.
3. Behr JP. DNA strongly binds to micelles and vesicles containing lipopolyamines or lipointercalants. Tetrahedron Lett 1986;27:5861–5864.
4. Felgner PL, Gadek TR, Holm M, Roman R, Chan HS, Wenz M, Northrop JP, Ringold M, Danielsen H. Lipofection: a highly efficient lipid-mediated DNA transfection procedure. Proc Natl Acad Sci USA 1987;84:7413–7417.
5. Gao X, Huang L. A novel cationic liposome reagent for efficient transfection of mammalian cells. Biochem Biophys Res Commun 1991;179:280–285.
6. Leventis R, Silvius JR. Interactions of mammalian cells with lipid dispersions containing novel metabolizable cationic amphiphiles. Biochim Biophys Acta 1990;1023:124–132.
7. Felgner JH, Kumar R, Sridhar R, Wheeler C, Tsai YJ, Border R, Ramsay P, Martin M, Felgner PL. Enhanced gene delivery and mechanism studies with novel series of cationic lipid formulations. J Biol Chem 1994;269:2550–2561.
8. Lee ER, Marshall JM, Siegel CS, Jiang C, Yew NS, Nichols, MR, Nietupski JB, Ziegler JR, Lane MB, Wang KX, Wan NC, Scheule RK, Harris DJ, Smith AE, Cheng SH. Detailed analysis of structures and formulations of catiuonic lipids for efficient gene transfer to the lung. Human Gene Ther 1996;7:1701–1717.
9. Behr JP. Synthetic gene transfer vectors, Acc Chem Res 1993;26:274–278.
10. Lasic DD. Liposomes: From physics to applications. Amsterdam: Elsevier, 1993.
11. Zhu N, Liggitt D, Liu Y, Debs R. Systemic gene expression after intravenous DNA delivery in adult mice. Science 1993;261:209–211.
12. Lasic DD, Barenholz Y, eds. Liposomes: From gene therapy, diagnostics to ecology. Boca Raton, FL: CRC Press, 1996.
13. Hong K, Zheng W, Baker A, Papahadjopoulos D. Stabilization of cationic liposome-plasmid DNA complexes by polyamines and PEG-lipid conjugates for efficient in vivo gene delivery. FEBS Lett 1997;400:233–237.
14. Sternberg B, Sorgi F, Huang L. New structures in complex formation between DNA and cationic liposomes visualized by freeze-fracture electron microscopy. FEBS Lett 1994;356:361–366.
15. Gustaffson J, Almgrem, M, Karlsson G, Arvidson G. Complexes between cationic liposomes and DNA visualized by cryoTEM. Biochim Biophys Acta 1995;1235:305–317.
16. Xu Y, Szoka FC. Mechanism of DNA release from cationic liposome/DNA complexes used in cell transfection. Biochemistry 1996;35:5616–5623.
17. Lasic DD, Strey H, Podgornik R, Frederik PM. Recent developments in medical applications of liposomes: sterically stabilized and cationic liposomes. 5th European Symp Control Drug Del, Book of Abstracts. Nordwijk aan Zee, 1996;61–65.
18. Lasic DD, Strey H, Podgornik R, Frederik PM. The structure of DNA-liposome complexes. J Am Chem Soc 1997;119:832–833.
19. Raedler JO, Koltover I, Sadditt T, Safinya C, Structure of DNA-cationic liposome complexes:

DNA intercalation in multilamellar membranes in distinct interhelical packing regimes. Science 1997;275:810–8914.
20. Templeton NS, Lasic DD, Frederik, PM, Stery HH, Roberts DD, Pavlakis G. Improved DNA: liposome complexes for increased systemic delivery and gene expression, in press.
21. Podgornik R, Strey HH, Rau DC, Parsegian VA. Watching molecules crowd: DNA double helix under osmotic stress. Biophys Chem 1995;57:111–121.
22. Podgornik R, Strey HH, Gawrisch K, Rau DC, Rupprecht A, Parsegian VA. Proc Natl Aacd Sci USA 1996;93:4261–4266.
23. Bloomfield VA. DNA condensation. Curr Op Struct Biol 1996;6:334–341.
24. Frank-Kamenitskii MD, Anshelevich VV, Lukashin P. Polyelectrolyte model of DNA. Sov Phys Usp 1987;30:317–330.
25. Current Protocols in Molecular Biology, F.M. Ausubel et al. John Wiley and Sons, 1997
26. Veelken H. Systemic evaluation of chimeric marker genes on dicistronic transcription units for regulated expression of transgenes in vitro and in vivo. Human Gene Ther 1996;7:1827–1836.
27. Kreigler M. Gene Transfer and Expression: A Laboratory Manual. Stockton Press, 1990.
28. Miller N, Wheelan J. Progress in transcriptionally targeted and regulatable vectors for gene therapy. Human Gene Ther 1997;8:803–815.
29. Lee RJ, Huang L. Lipidic vectors for gene transfer. Crit Rev Ther Drug Carr Syst 1007;14:173–206.
30. Frolov I. Alphavirus based expression vectors. Proc Natl Acad Sci USA 1997;93:11371–11377.
31. Gerhson H, Ghirlando R, Guttman SB, Minsky A. Mode of formation and structural features of DNA-cationic liposome complexes used for transfection. Biochemistry 1993;32:7143–7151.
32. Xu Y, Hui SK, Szoka FC. Effect of lipid composition and lipid-DNA charge ratios on physical properties and transfection activity of cationic lipid-DNA complexes. Biophys J 1995;A432.
33. Podgornik R, Strey HH, Frederik PM, Lasic DD. Unpublished.
34. Campbell, S, Lasic, DD, Israelachvili JN. Unpublished.
35. Felgner PL, Ringold RG. Cationic liposome mediated transfection. Nature 1989;337:387–388.
36. Minsky A, Ghirlando R, Gerhson H. Features of DNA cationic liposome complexes and their implication for transfection. In: Lasic DD, Barenholz Y, eds. Liposomes: from Gene Therapy to Diagnostics and Ecology. Boca Raton, FL: CRC Press, 1996;7–30.
37. Farhood H, Huang L. Delivery of DNA, RNA and proteins by cationic liposoems.In: Lasic DD, Barenholz Y, eds. Liposomes: from Gene Therapy to Diagnostics and Ecology. Boca Raton, FL: CRC Press, 1996;31–42.
38. Sterneberg, B. Liposomes as models for membrane structures and structural transformations. In: Lasic DD, Barenholz Y, eds. Liposomes: from Gene Therapy to Diagnostics and Ecology. Boca Raton, FL: CRC Press, 1996;271–298.
39. Felgner PL, Tsai YJ, Felgner JH. Advances in the design and application of cytofectin formulations. In: Lasic DD, Barenholz Y, eds. Liposomes: from Gene Therapy to Diagnostics and Ecology. Boca Raton, FL: CRC Press, 1996;43–56.
40. Dan N. Formation of ordered domains in membrane bound DNA. Biophys J 1996;71:11267–1272.
41. Farhood H, Bottega R, Epand RM, Huang L. Effect of cholesterol derivatives on gene transfer and protein kinase C activity. Biochim Biophys Acta 1992;1111:239–246.
42. Wheeler CJ, Sukhu L, Yang G, Tsai Y, Bustamante C, Felgner P, Norman J, Manthorpe M. Converting an alcohol to an amine in a cationic lipid dramatically alters the co-lipid requirement, cellular transfection activity and the ultrastructure of DNA-cytofectin complexes. Biochim Biophys Acta 1996;1280:1–11.
43. Brigham KL, Meyrich B, Christman B, Magnusson M, Berry L. In vivo transfection of murine lungs with functioning prokaryotic gene using a liposome vehicle. Am J Med Sci 1989;298:278–281.
44. Felgner LP. Improvement in cationic liposome mediated transfection. Human Gene Ther 1996;7:1791–1793.
45. Scheule RK, St George JA, Bagley RG, Marshall J, Kaplan JM, Akita GJ, Wang KX, Lee ER, Harris DJ, Jiang C, Yew NS, Smith AE, Cheng SH. Basis of pulmonary toxicity associated with cationic lipid-mediated gene transfer to the mammalian lung. Human Gene Ther 1997;8:689–707.
46. Fang Y, Jie Y. Two dimensional condensation of DNA molecules in cationic lipid membrane. J Phys Chem B 1997;101:441–449.
47. Lasic DD, Papahadjopoulos D. Liposomes revisited. Science 1995;267:1275–1276.

Lasic and Papahadjopoulos (eds.), Medical Applications of Liposomes

CHAPTER 5.3

Cationic liposome-DNA complexes in gene therapy

SOUMENDU BHATTACHARYA† AND LEAF HUANG*

The Laboratory of Drug Targeting, Department of Pharmacology, University of Pittsburgh, School of Medicine, Pittsburgh, PA 15261, USA

Overview

I. Introduction

In the past decade, gene therapy is definitely one of the fastest developing fields in biomedical research. The ability to transfer genes to mammalian cells provides an opportunity for studying the biology of altered genotype. On the other hand, it offers a conceptually novel therapeutic strategy for the treatment and cure of acquired diseases like cancer[1] and inherited diseases such as cystic fibrosis.[2] Gene transfer is also being developed as a preventive measure, e.g., vaccines.[3,4] The advancements in the field of gene therapy is augmented by the rapid develop-

*Author to whom all correspondence should be addressed: Department of Pharmacology, University of Pittsburgh, School of Medicine, W1351 Biomedical Science Tower, Pittsburgh, PA 15261
†Present address: The Liposome Company, One Research Way, Princeton, NJ 08540, USA.
(Tel) 412.648.9667; (Fax) 412.648–1945
Email: Leaf@Prophet.pharm.pitt.edu

ment in molecular biology, which makes it possible to identify and construct therapeutic genes in sufficient quantities.

Paradoxically, the "bottle-neck" of the gene therapy concept, is the vector which efficiently packages the DNA, carries it through the membranes and deliver the gene to the nucleus for expression. An ideal vehicle or vector should be highly efficient in delivering the gene in a target-specific manner, stable in vitro as well as in vivo, protect the gene from nuclease degradation, non-toxic, non-immunogenic and easily prepared in large quantities.

Although a wide array of physical (e.g., electroporation, microinjection, particle bombardment), chemical (e.g., DEAE-dextran, polybrene-dimethyl sulfoxide, calcium phosphate precipitation, liposomes, polylysine conjugates) and biological (e.g., viruses) methods[5] are available for transferring genes into cells, none of the gene-delivery vectors known to date can be referred to as an "ideal" vector. Nevertheless, some of them show promising efficiency. The most widely used types of vehicles for gene delivery are: viral (e.g., adenovirus, retrovirus and adeno-associated virus) and non-viral (e.g., liposomes, polymers, peptides). Viral vectors, by far, are more efficient than their non-viral counterparts but they have the disadvantage of being immunogenic, potentially mutagenic, with low viral titers and limited loading capacity in terms of the size of the DNA. Among the non-viral vectors, cationic liposomes are the most widely used vectors. Although less efficient in delivering the genes than the virus, they have many important qualities such as being much less or nonimmunogenic and nontoxic, have no known limitation in the size of the DNA, can be custom-synthesized for targeting and easily scalable for large scale production. Moreover, the liposomes can deliver different kinds (supercoiled or linear) of DNA (or RNA) with or without proteins, even to non-dividing cells and are usually composed of bio-degradable lipids. Also covalent attachment of target specific ligands on the liposome can facilitate targeted delivery of genes. These advantages have prompted researchers to explore the applications of cationic liposomes in gene therapy clinical trials.

In a pioneering study by Felgner et al. in 1987,[6] a cationic lipid, *N*-[1-(2,3-dioleoyloxy)propyl]-*N*,*N*,*N*-trimethylammonium chloride (DOTMA), was reported to have transferred DNA into mammalian cells. The efficiency was shown to be much improved as compared to the naked DNA as well as to the DNA bound to DEAE-dextran or calcium phosphate. This report prompted other research groups to develop different cationic lipids which shows transfection activities.[7–13] Cationic lipids were used to deliver DNA,[2] RNA,[14] oligonucleotides,[15,16] antisense[17] and protein[18] to mammalian cells.

The transfection protocol using the cationic liposome is very simple. The lipid and the DNA are mixed to form a complex called "lipoplex" (according to recent system of nomenclature, see Ref. 19), by condensation of the DNA through electrostatic charge-charge interactions, usually in a ratio with a little excess of cationic lipid. This ensures an overall positive charge on the lipoplex and significantly improves the docking of the complex on the primarily negatively charged (sialic acid residues) on the plasma membrane of the cell.

The various steps of the transfection process, starting from the complexation

Table 1

Major steps of cationic lipid mediated gene transfer

1.	Complex formation by DNA condensation
2.	Binding with molecules in biological fluid such as serum
3.	Transport from the site of injection to target cell surface
4.	Complex binding to the cell surface
5.	Uptake into the cell by endocytosis
6.	Release of the complex from endosome
7.	Uncoating of DNA
8.	Uptake into the nucleus
9.	Expression of the gene

of the vector to DNA to the final step of transgene expression, is a very complicated process. The possible individual steps are summarized in Table 1. The mechanism of these individual steps is a poorly understood process (Figure 1).[20]

Often, a neutral, helper co-lipid, e.g., dioleoylphosphatidylethanolamine (DOPE) or cholesterol, is used along with the cationic lipid in the liposomal formulations. Both the lipids share a common structural feature that both of them has a smaller head group as compared to the hydrophobic part. So it is speculated that they can destabilize the bilayer by forming a hexagonal H_{II} phase.[21–23] It is generally believed that presence of such fusogenic lipids causes the disruption of endosome and releases the trapped DNA (bound or free) into the cytosol of the cell. However, it should be noticed that some cationic lipids (e.g., DOGS) do not require helper lipids for activity. In these cases, the cationic lipids themselves must cause endosome disruption with an unknown mechanism.

II. Formation and characterization of the cationic lipoplex

The formation of the lipoplex is a spontaneous process; the positive charge of the polar head group of the cationic lipid binds through charge-charge interactions with the negative charge of the DNA strand and thus condenses the DNA. This process occurs within a time scale of seconds to minutes.[24] The kinetics and thermodynamics of this complexation, which depend on the relative concentrations of the components, the rate and order of mixing, temperature, salt concentrations, etc., is a poorly understood process. Due to the spontaneous nature of binding, the heterogeneity of the complexes in terms of shape and size is significant. Also, the size of the complex formed was found to be apparently independent of the size of the cationic liposome[25] as well as the size of the gene.[26] This fact indicates that the process of complexation may go through a structural reorganization with the disruption of the liposomal morphology as an intermediate step.

Several attempts were made to elucidate the fine structures of the lipoplex using various techniques of electron microscopy, dynamic light scattering, etc. Gershon et al.[27] in a metal shadowing EM, showed photographs of the lipid coated DNA complexes which were roughly spherical at low lipid to DNA ratio but gradually

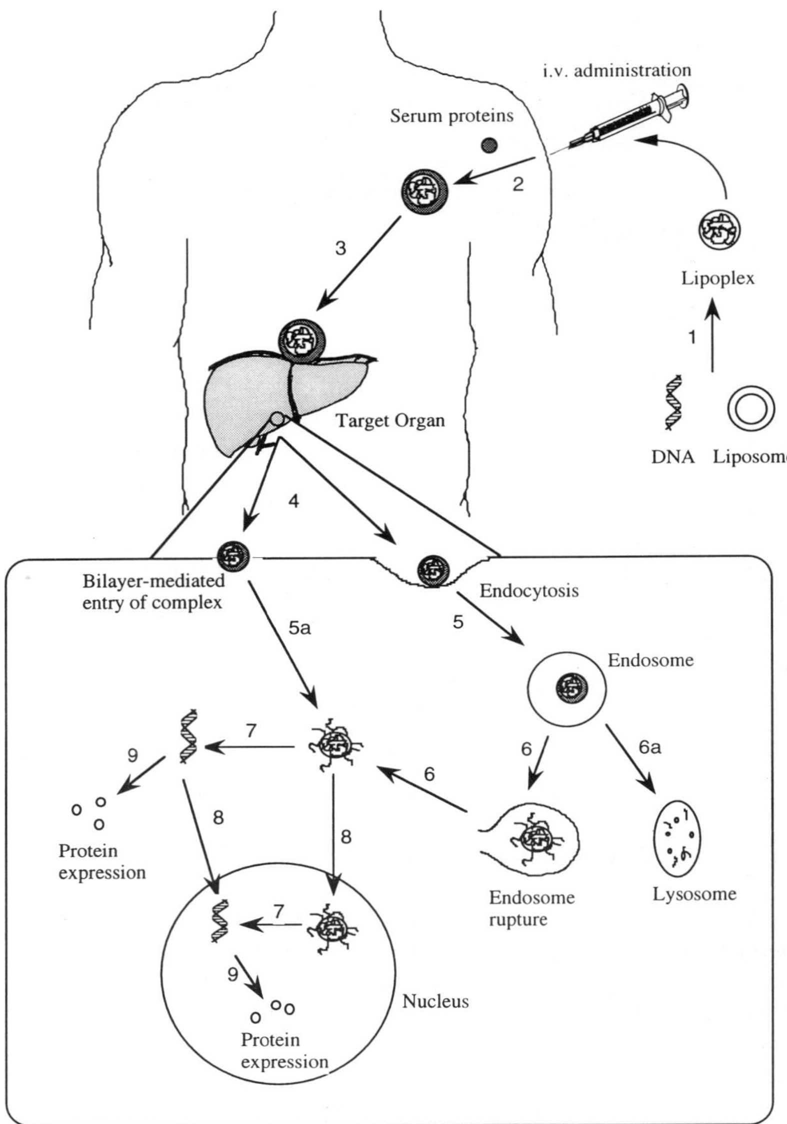

Fig. 1. Schematic diagram of major steps/barriers of cationic lipid mediated gene transfer. (1) Complex formation by DNA condensation; (2) binding with molecules in biological fluid such as serum; (3) transport from the site of injection to target cell surface; (4) complex binding to the cell surface; (5) uptake into the cell by endocytosis; (5a) escape of DNA into the cytoplasm; (6) release of the complex from endosome; (6a) degradation of DNA in lysosome; (7) uncoating of DNA; (8) uptake into the nucleus; (9) expression of the gene.

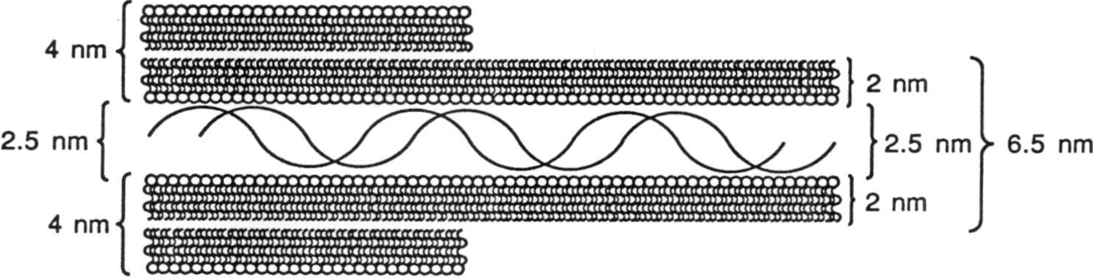

Fig. 2. Lipid coated DNA showing fracture plane in freeze-fracture electron microscopy. The theoretical thicknesses of the bilayer and DNA are shown on the left. The thicknesses on the right depict a theoretical thickness of 6.5 nm and is in agreement with the measured value of 7 nm. (Taken from Ref. 80.)

changed to rod-like structures at high lipid to DNA ratio. It also indicated fusion of liposomes. In a later study using cryo-transmission EM, spherical structures of the complex were seen at about 1:1 lipid to DNA ratio where the DNA was speculated to be trapped between the layers of lipidic oligolamellar structures.[28] But these methods of microscopy had some limitations regarding the resolution and often accompanied by artifacts due to sample preparation. Sternberg et al.[29] used freeze-fracture electron microscopy, probably the best microscopic method to study bilayer membranes, as a much more reliable method to probe the morphology of the complexes. Two types of structure were evident; first, spherical DNA-lipid complex ("meatball") and second, tubular structures of DNA coated with lipid ("spaghetti") (Figure 2). They also found the size of the complex to be increasing with higher lipid to DNA ratio. These globular meatball-like structures of lipid-DNA complexes were also confirmed by using fluorescence microscopy.[30]

Recently, a combination of studies with electron microscopy and X-ray diffraction was reported.[26,31] It was observed that when cationic liposomes are mixed with DNA, there is a rapid topological transition from the liposomal structure to a liquid-crystalline, condensed, globular structure. The lipoplex is of about 1 μm in size and was shown to consist of a higher ordered multilamellar structure with DNA sandwiched between the cationic bilayers. By using DNA molecules from various sources with very different sizes, it was also demonstrated that the size of lipoplex apparently does not depend on the length of DNA molecule. These observations are supported by a theoretical simulation model by Dan.[32] According to this model, ordering of DNA on a cationic lipid membrane is delicately balanced by the electrostatic repulsion between the phosphates of the DNA molecules and an attractive interaction due to undulations of the bilayer induced by the adsorbed DNA. The balance of these two forces results in the formation of ordered domains characterized by a finite spacing between the two consecutive lipid layers and interaxial distance between the two consecutive DNA helices.

In a separate study, Fang et al. have directly imaged the DNA on lipid membrane by atomic force microscopy (AFM) which shows distinct ordered domains.[33] They also observed that there was no ordered phase formed if the lipid bilayer is below the gel transition temperature. This means that the fluidity of the lipid bilayer is an absolute essential for DNA condensation, which supports the theory of DNA-induced membrane distortion by Dan's model.[32] Further studies are being carried out to show a 3-D ordering in DNA-lipid system due to undulation-induced coupling of adjacent bilayers and a periodic distortion of the bilayer along the long-axis of DNA.

III. Synthetic cationic lipids and their formulations

Following the first report of lipofection by Felgner et al. in 1987 using DOTMA, a cationic lipid, a wide array of cationic lipids were synthesized by various groups (Figure 3 and Table 2). In an attempt to classify the lipids in a broad sense, a lipid molecule can be divided into three major regions: (1) the hydrophobic moiety, (2) the polar head group and (3) the linker group between the hydrophobic chains and the head group.

III.1. The hydrophobic moiety

This part of the lipid molecule, as the name suggests, imparts all the hydrophobicity in the molecule. Although there are some scattered reports of using cationic lipids with single hydrophobic chain for gene transfer,[34,35] the efficiency is usually low to poor as compared to the double chain analogs. The class of double chain hydrocarbons can be divided into two major subclasses: unsaturated and saturated chains. Oleoyl chain (C18 : 1) is the most common of all the unsaturated class, e.g., DOTMA, DOSPA, DOTAP, Tfx-50, DOSPER, etc. Among the saturated class, C18 (e.g., DOGS, DDAB), C16 (e.g., TM-TPS), C14 (e.g., DMRIE), C12 (e.g., DLRIE) are known. All of them can form liposome on their own, but still DOPE is often used as a helper lipid with them.

Other than the class of double chain hydrocarbons, there are lipids synthesized where the backbone is made up of a cholesterol moiety, e.g., DC-Chol, lipid 67, BGSC, BGTC. First of these kind of lipids, ChoTB and ChoSC, were synthesized by Leventis and Silvius[12] but the transfection activity was not high. Later, Gao and Huang synthesized DC-Chol[10] which shows improved activity. The cholesterol derivatives usually are unable to form stable bilayers unless used in conjunction with DOPE or other neutral lipids as a helper lipid.

Another series of cholesterol derived lipids were synthesized by conjugating natural polyamines like spermine or spermidine molecules to the cholesterol moiety.[36] An optimized formulation of the Lipid 67 (with DOPE), a lipid with a cholesterol anchored to a spermine head group in a T-shape configuration, was shown to be very effective in expressing transgene. It can also deliver CFTR gene into lung cells in vivo with relatively high efficiency and rectify biochemical defects of chloride transport in cystic fibrosis.

Recently, Vigneron et al.[37] synthesized two other cholesterol-based lipids, BGSC and BGTC, by conjugating guanidinium groups with different spacer arms to cholesterol moiety. It was postulated that due to the high pK_a value of the

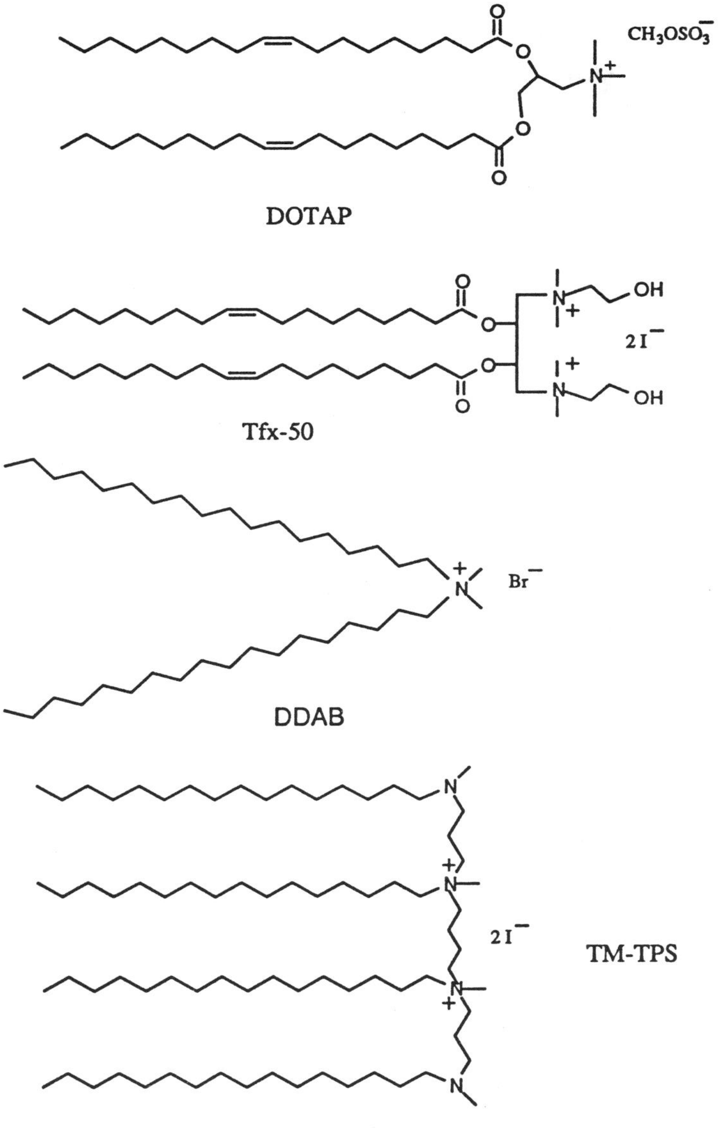

DOTAP

Tfx-50

DDAB

TM-TPS

Fig. 3A.

Fig. 3. Structures of some commonly used cationic lipids in gene therapy.

guanidinium group, the DNA will be tightly bound throughout the fluctuations of pH during the cell trafficking. Furthermore, the tertiary amine in BGTC, between the two guanidinium groups may have lower pK_a which will provide a buffer to the acidic environment and protect the DNA during the late endosomal stage.

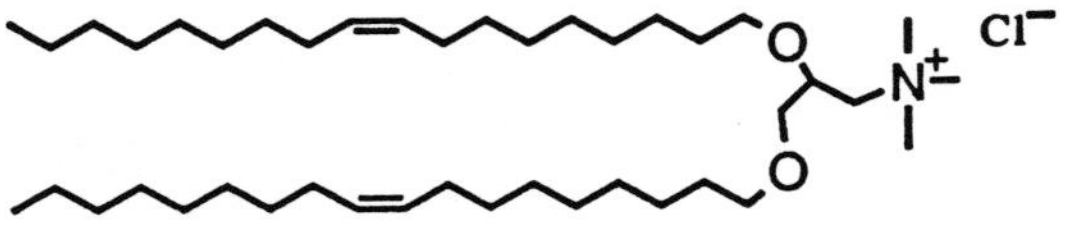

Cl^-

DOTMA

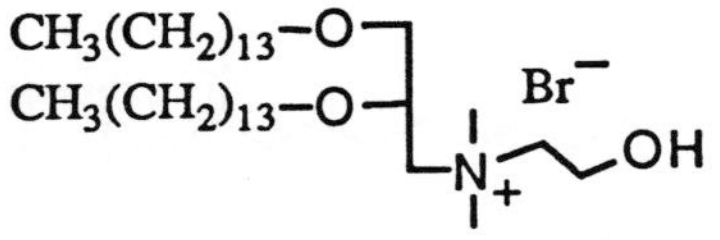

$CH_3(CH_2)_{13}$–O
$CH_3(CH_2)_{13}$–O
Br^-
OH

DMRIE

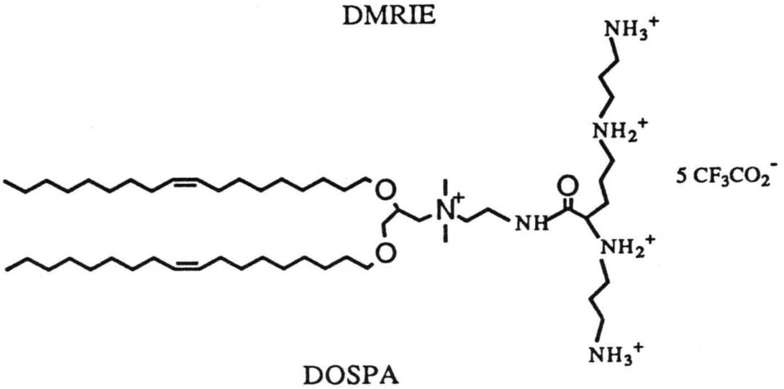

NH_3^+
NH_2^+
5 $CF_3CO_2^-$
O
NH
NH_2^+
NH_3^+

DOSPA

NH_3^+
NH_2^+
4 $CF_3CO_2^-$
O
N
NH
O
NH_2^+
NH_3^+

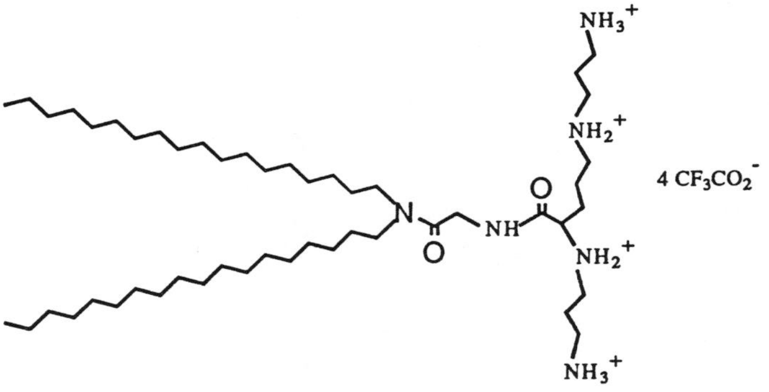

DOGS

Fig. 3B.

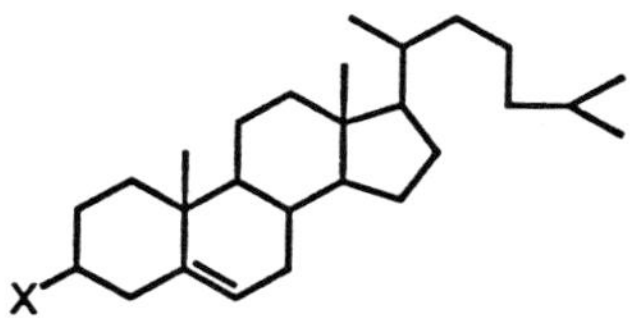

X

$-OOC(CH_2)_3N^+(CH_3)_3$	ChoTB
$-OOC(CH_2)_2COO(CH_2)_2N^+(CH_3)_3$	ChoSC
$-OCONH(CH_2)_2N(CH_3)_2$	DC-Chol
$-OCONH(CH_2)_2N^+(CH_3)_3\ Cl^-$	TC-Chol
	Lipid 67
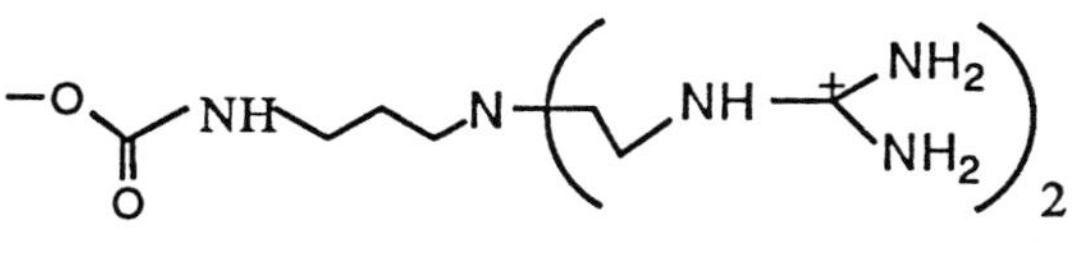	BGTC
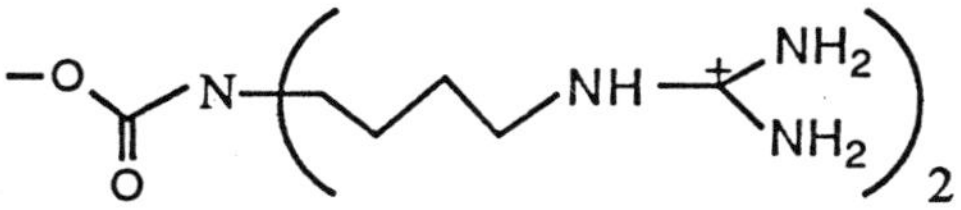	BGSC

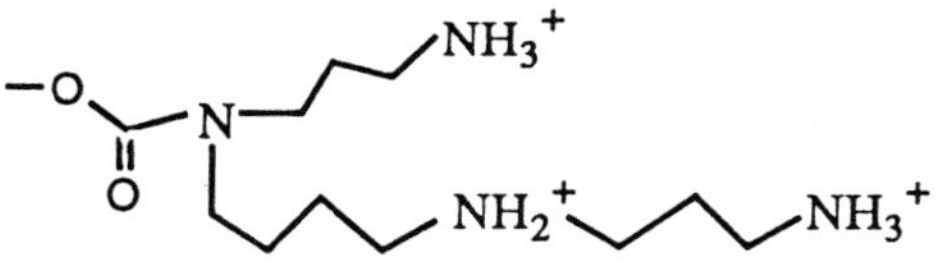

Fig. 3C.

Both lipids, BGSC and BGTC, forms a liposomal formulation with DOPE as a helper lipid (BGSC or BGTC:DOPE = 3:2 molar ratio) which shows high transfection capability in a variety of cell lines. The transfection activities of BGSC and BGTC liposomal formulations were found to be higher than that of Lipofectin (DOTMA/DOPE). Further, liposomes composed of BGTC and dioleoylphos-

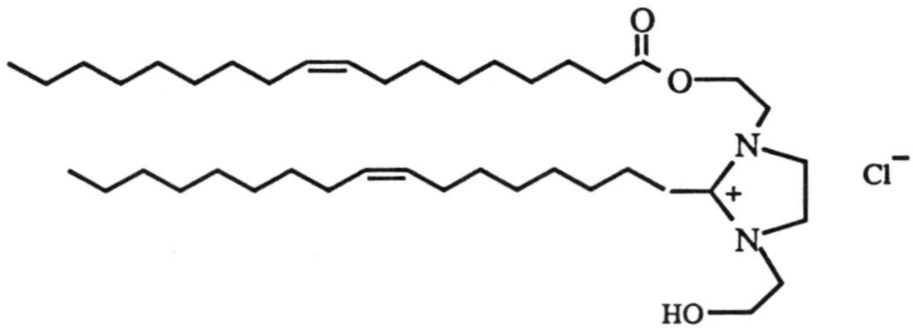

DOTIM

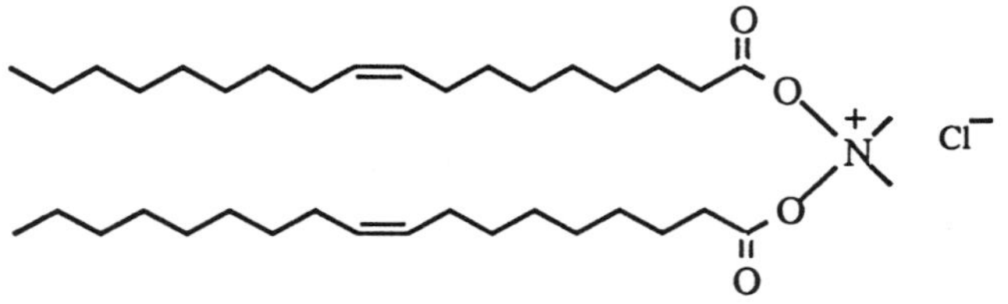

DODAC

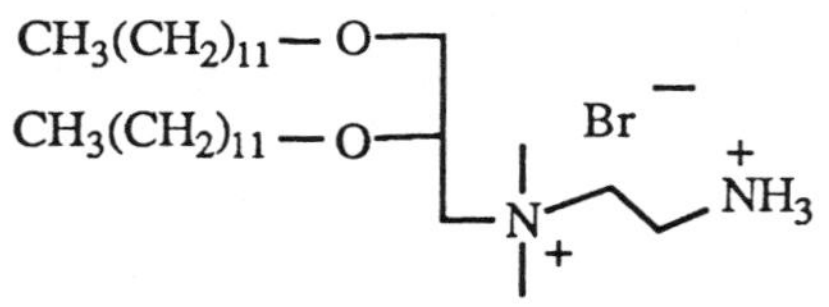

$CH_3(CH_2)_{11}-O$
$CH_3(CH_2)_{11}-O$
Br^-
$\overset{+}{N}H_3$

GAP-DLRIE

Fig. 3D.

phatidylethanolamine (DOPE) are efficient for gene delivery to the mouse airway epithelium in vivo. Transfected cells were detected both in the surface epithelium and in submucosal glands.[38]

Moreover, BGTC was shown[37] to exist as a true micellar solution in the concentration range where transfection experiments is usually carried out with a critical micellar concentration of 9×10^{-5} M. The BGTC lipid, when used as a micellar solution was efficient in transferring gene into a variety of mammalian cell lines. The transfection activity of micellar BGTC was relatively higher than that of Transfectam, which is a micellar solution of DOGS.[37]

Table 2

List of available lipids formulations

	Commercial name	Lipids	Molar ratio	Available from
1.	DMRIE-C	DMRIE:Cholesterol	1:1	GibcoBRL
2.	Lipofectin	DOTMA:DOPE	1:0.9	GibcoBRL
3.	Lipofectamine	DOSPA:DOPE	1:0.65	GibcoBRL
4.	DC-Chol	DC-Chol:DOPE	1:0.67	Sigma (only DC-Chol lipid)
5.	LipofectASE	DDAB:DOPE	1:2.1	GibcoBRL
6.	TransfectASE	DDAB:DOPE	1:3	GibcoBRL
7.	Transfectam	DOGS	–	Promega
8.	DOTAP	DOTAP	–	Avanti (only as lipid)
9.	Tfx-50	Tfx-50:DOPE	1:1	Promega
10.	Cellfectin	TM-TPS:DOPE	1:1.5	GibcoBRL
11.	GL 67	Lipid 67:DOPE	1:2	Genzyme (not sold)

III.2. The polar head group

The polar head group is the part which bears the positive charge(s) on the lipid molecule. The common feature among all the head groups studied in this review, is that in every case the positive charge(s) is due to amine nitrogen(s). The amine nitrogen may either have a labile positive charge due to the exchangeable proton, as in primary, secondary (e.g., DOGS, DPPES) or tertiary amines (e.g., DC-Chol) or may possess a permanent positive charge as in quaternary amine (e.g., DOTMA, DMRIE, DOTAP, Tfx-50, TC-Chol, DDAB). Some lipids are multivalent in terms of the number of positive charges and may contain more than one kind of nitrogen. For example, DOSPA head group has two primary, two secondary and one quaternary amine nitrogens. In Tfx-50, both the charges are due to quaternary nitrogens; whereas in TM-TPS there are two tertiary and two quaternary nitrogens. Among the multivalent functional groups on the cationic lipids, spermine moiety is very common (e.g., DOSPA, DOGS, DOSPER).

The chemical transformations of the head group may also trigger the transfection activity of some lipids. Converting the alcohol of the head group of DLRIE into a primary amine enhanced the efficiency of the lipid (GAP-DLRIE) for CAT expression in mouse lung[39,40] and catheter-mediated gene transfer in porcine arteries.[41] The authors postulate that GAP-DLRIE is a more hydrophilic molecule with a higher critical micellar concentration as compared to DOTAP, DOTMA, DOSPA, DMRIE mainly because of the two positive charges on the head group and shorter hydrophobic chains. Thus, GAP-DLRIE can bind to DNA more efficiently because of higher spontaneous rate of monomer lipid transfer through the aqueous phase.[39]

The transfection activity of a lipid may also critically depend on the orientation of the polyamine head group with respect to the lipid anchor. When the spermine head group was coupled with the cholesterol moiety by a carbamoyl linkage through the secondary amine nitrogen of the spermine to produce a T-shaped motif (Lipid 67), the lipid vector showed high activity in delivering genes to the lungs.[36] But, by the attachment of spermine through a terminal primary amine producing a linear structure, the lipid showed 100 times lower transfection activity as compared to the T-shaped Lipid 67. Recently, the safety and efficacy of the optimized formulation of Lipid 67/DOPE (1:2), named as GL-67, in connection with the gene transfer to the mammalian lung has been discussed in detail.[42]

III.3. The linker group

A linker group is the part of the lipid which links the hydrophobic chains with the polar head group. The length of this linker group is very critical for the activity of the lipid in binding DNA and generally is of intermediate polarity between the nonpolar chains and the polar head group. Often this linker is a glycerol unit as in DOTAP, DMRIE, DOSPA, DOTMA, etc., or a glycine unit as in DOGS. A few lipids have no linker group at all, e.g., DDAB, TM-TPS. In case of lipids with cholesterol backbone, a short linker of 3 atoms seems to be the most efficient (e.g., DC-Chol).

Another important aspect of the linker is how the linker group is connected to the hydrophobic residue. In some lipids like DOTMA, DMRIE or DOSPA, the two oleoyl chains are connected to the linker group through ether bonds which are stable, non-biodegradable and may be the cause of long-term toxicity to the cells. Whereas in cases of DOTAP or Tfx-50, it is relatively less stable ester bonds or in case of DOGS, it is an amide bond which are eventually biodegradable.

DC-Chol, 3β[*N*-(*N'*,*N'*-dimethylaminoethane)-carbamoyl] cholesterol, is unique in a sense that it was deliberately designed with a relatively labile carbamoyl linker[10] which is not hydrolyzed easily like the ester bond but once inside the cell, it is eventually biodegradable probably by the cellular esterases. This is apparently one of the reasons for the good pharmaceutical characteristics of DC-Chol.[43] Later, lipid 67,[36] BGSC and BGTC[37] were synthesized which share the same feature as the DC-Chol. The gene transfer activity of lipid 67 is apparently greater than that of DC-Chol.[36]

There are other useful features which puts DC-Chol high in the list of cationic lipids for gene delivery. First, it forms a stable liposomal formulation with DOPE (DC-Chol:DOPE = 3:2, mol/mol) which can be stored for months at 4°C without any change in size or lipid degradation.[43] Second, it is synthesized in a one-step coupling reaction and the product can be easily isolated and purified in an inexpensive manner. Third, DC-Chol has been approved by US FDA and the regulatory authorities of other countries for use in clinical trials. Fourth, DC-Chol/DOPE liposomes show a better transfection activity than other cationic lipid formulations in vivo.[44,45] Lastly, DC-Chol is now commercially available from Sigma. Since its first synthesis, DC-Chol liposome formulations have been used in seven different clinical trials involving human gene therapy for various diseases such as cystic fibrosis,[2] cancer[46] and Canavan's Leukodystrophy.[47]

IV. Lipopolyplex: Liposome/polycation/DNA complex

For most of the formulations, the lipoplex formed by complexation of cationic liposomes with DNA has some disadvantages, especially in vivo. First, they have a tendency to aggregate with DNA to form large and heterogeneous particles at high concentration. Second, cationic liposomes in general lack the ability of targeted delivery because of the non-specific charge interactions with the cells. Third, the overall excess of cationic lipid in the lipoplex renders it sensitive to serum as it tends to bind with the negatively charged serum proteins. The negatively charged serum proteins might also dissociate the lipoplex causing the premature release and enzymatic degradation of the DNA. Lastly, these lipoplexes are readily cleared from the blood circulation by the RES system.

The main reason for the aggregation of the complex of DNA and DC-Chol, lipofectin and probably other monovalent lipids into "spaghetti and meatball" structures could be due to poor condensation of DNA by the DC-Chol lipid. But polycations like poly-L-Lysine condense DNA far more effectively than the lipids mainly because of its high charge density. Gao and Huang used poly-L-Lysine along with DC-Chol/DOPE liposomes to condense DNA and form a self-assembled vector system named LPDI (a "lipopolyplex").[48] The LPDI was purified by sucrose density gradient ultracentrifugation to remove the excess of free liposomes to avoid cytotoxicity. Under optimal conditions, the transfection efficiency was shown to increase by 2–28 fold over the control DC-Chol/DOPE and DNA complex. Upon examination by negative stain electron microscopy, the purified LPDI appeared to be spherical particles (50–75 nm) with a dense core which probably represents polylysine condensed DNA.

The advantages of LPDI as compared to DC-Chol/DOPE and DNA complex is that LPDI is much smaller in size, DNA is more condensed and exhibits higher gene transfer activity. The formulation is quite stable at 4°C for months and can be stored as single-vial formulation. Moreover, the DNA in LPDI is better protected from the enzymatic degradation as compared to the partial protection of DNA by DC-Chol lipoplex. LPDI is currently used in a clinical trial for gene therapy of Canavan's Leukodystrophy.[47]

Another effective formulation was described recently by Sorgi et al.[49] This formulation is based on the idea that protamines are known in sperms to condense DNA effectively. It was found to be superior to polylysine (Figure 4) and with an established safety profile for human use. The protamines are small cationic peptides (MW $\sim$ 5000), with approximately 66% of residues being arginine. Moreover, protamine is thought to possess nuclear localization signals (NLS) which might facilitate the entry of the gene into the nucleus from the cytoplasm. For these reasons, it was hypothesized that protamine might be an improved replacement for polylysine. The complexation was carried out by premixing protamine sulfate, USP with DNA, followed by addition of DC-Chol/DOPE (3/2, molar ratio) liposomes in Hank's Balanced Salt Solution (HBSS). Unlike LPDI containing polylysine, there was no need of gradient purification in this case.

The potentiation of the luciferase reporter-gene expression varied considerably

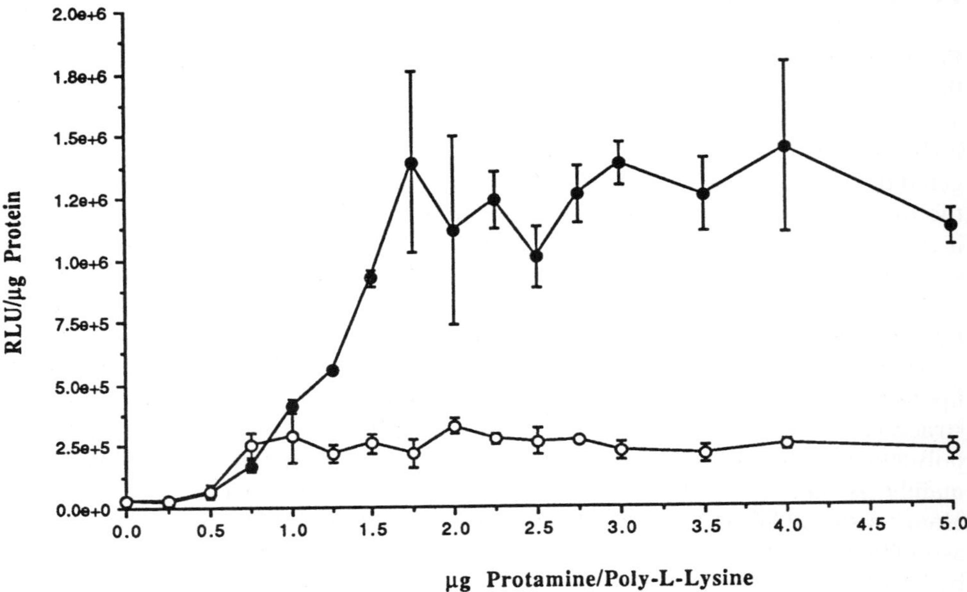

Fig. 4. Comparison of protamine sulfate, USP and poly-L-lysine on the ability to increase transfection activity in CHO cells. Varying amounts of protamine sulfate USP (●) or poly-L-lysine (○) were added to 1 µg pUK21-CMV-LUC DNA prior to complexing with 7.5 nmol of DC-Chol liposomes per well. Each data point represents the mean (with standard deviation) of triplicate samples and are normalized to protein content. (Taken from Ref. 49.)

with different types of protamines which differ from each other in the extent of lysine substitution in place of arginine.[49] Apparently, the activity of the protamine correlates inversely with the lysine content. Protamine phosphate (with 8.84% lysine) and free base (with 8.14% lysine) showed almost no improvement over that seen in the absence of any polycation. Protamine sulfate (with 0.23% lysine) and protamine chloride (with 1.49% lysine) showed the highest and moderate activity, respectively. It was hypothesized that the exchange of arginine residue with lysine may interfere with the binding of the DNA with protamine, resulting in a decrease in efficiency of DNA condensation. This hypothesis was supported by the amino acid analysis and fluorescence binding assays and transfection.[49]

Recently, Li and Huang[50] designed a new formulation of lipopolyplex which consists of DOTAP liposomes, Protamine Sulfate and DNA. The lipopolyplex was prepared by premixing the DNA with Protamine Sulfate followed by introduction of DOTAP liposomes into the complex. The size of the final ternary complex was found to be between 200–300 nm. The gene expression of this new lipopolyplex was found to be consistently higher than that of DOTAP/DNA lipoplex, proving once again the positive contribution from protamine. The luciferase gene expression was found in all organs with the highest expression in the lung; approximately 20 ng of luciferase protein per mg of extracted protein was found in lungs at an optimal dose of 50 μg of DNA per mouse. The gene expression in the lung was noticed within an hour of injection and peaked at 6 hours.

V. Stable cationic lipid/DNA formulations for intravenous administration

The major problem with the liposomal DNA complexes are basically the large size of the complex, inefficient DNA condensation especially by the monovalent cationic lipids and the complex rapidly aggregates into large structures and loses transfection activity when stored. These hurdles prompted a few groups to search for solutions so that a single-vial formulation can be developed. In clinical studies, a stable complex containing high concentrations of lipids and DNA is highly desirable. A single-vial formulation would allow the lipid-DNA complex to be stored and results obtained within the same or different batches of the lipid-DNA complex to be compared directly.

Hofland et al.[51] suggested that the cationic lipid in micellar form is a better choice than liposome, because in micellar structure the lipid head group is more accessible to all the binding sites on the DNA, mainly because the lipid has less motional constraint in a micelle as compared to the same in a rigid bilayer structure. Also, micelles are much smaller in size (usually less than 10 nm) than liposomes, which may allow the lipid to interact with DNA without much steric hindrance. The lipid is first dissolved in a detergent solution to form a micellar structure. The DNA is then added to the lipid solution to form lipid/DNA complex. As the detergent was subsequently removed by dialysis, the excess unbound lipid forms a further coating around the complex. The latter process is a consequence of hydrophobic interactions. They used Lipofectamine (DOSPA/DOPE, 1.53/1 molar ratio) as the cationic liposome and octylglucoside was used as a detergent. There is no observed loss in transfection activity even after storing the resulting complex, either frozen or at 4°C, for at least 90 days, whereas for the non-stabilized liposome/DNA complex the activity was completely lost 24 hours after mixing the two components. The transfection activity of the new complex is partially serum sensitive, whereas the unstabilized complex is much more sensitive to serum. The toxicity of the complex can be greatly reduced when the complex is further purified from the unbound excess lipids by centrifugation. Lastly, the complex can be concentrated with no loss of activity.

In another formulation,[52,53] 1,2-dioleoyl-*N*,*N*-dimethylammonium chloride (DODAC) was selected as the cationic lipid and DOPE or egg sphingomyelin was used as a co-lipid for the formulation. Octylglucoside was again used as a detergent. It was observed that at low detergent concentration (20 mM), the complexes formed spontaneously with a size distribution ranging from 55 to 70 nm. But at high detergent concentration (100 mM), large complexes were formed (>2 μm) after removal of the detergent. The transfection activity of the stable complex was greater when sphingomyelin was used instead of DOPE. It is hypothesized that at the initial stage, the cationic lipid/DNA complex formation is a consequence of electrostatic interactions. After the optimal amount of lipid has been bound to the DNA, the formation of the final complex is accompanied by further binding of lipids to the complex, governed by the classical hydrophobic interactions.

Recently, another formulation of a stable DNA/cationic lipid complex was reported.[54] As compared to the previously published results,[51] the serum resistivity in this study was reported to have been improved. The cationic lipid used was DDAB. Inclusion of cholesterol was shown to be more favorable than DOPE as a helper lipid and increased the stability of the complexes in presence of 50% serum. Two additional ingredients appeared to be important for the improved stability of the complex. First, poly (ethylene glycol)-PE conjugate provides steric stabilization to liposomes, increasing its circulation time in blood.[55,56] A small amount of PEG-PE (1% of cationic lipid) was added to the lipid/DNA complex, within a few minutes of their preparation. Due to the steric protection property of PEG-PE, this stabilized complex showed reduced transfection activity at first, but reclaimed its original activity after storage for a month at 4°C due to some unknown structural reorganization. The second ingredient is spermidine. The polyamines are known to condense DNA by electrostatic interactions.[57] It was hypothesized that pre-condensing the DNA with a polyamine (0.5 nmol of spermidine per μg of DNA), prior to mixing with liposomes, would reduce the amount of lipid required to form a stable complex. A low lipid-to-DNA ratio is always desirable, especially in vivo, for reduced toxicity. Again, the transfection activity was much higher when the complex was stored at 4°C for a month as compared to that of the freshly prepared ones, due to some unknown reason. The sizes of the complexes were measured by dynamic light scattering to be around 400 nm. The expression of the marker gene (luciferase) was the highest in lungs (up to 3 ng of luciferase per mg of tissue protein), which was approximately 3 orders of magnitude higher when compared with a recent study by Thierry et al.[58] using another stable lipid/DNA complex prepared in the presence of ethanol.

PEG-derivatized lipid has also been used by Eastman et al. to stabilize a highly concentrated lipoplex prepared for aerosol administration.[59] Only a small fraction (1.64 mol% of total lipid) of DMPE-PEG_{5000} facilitated formation of a stable lipoplex with DNA concentration exceeding 20 mM, at approximately 10-fold higher concentrations than previously reported.[60] Most of the DNA in these formulations was bound to the lipid component and thereby protected from nebulizer-induced shearing; the DNA also maintained full biological activity both in vitro and in vivo without precipitation.

In a recent report by Liu et al.,[61] DOPE is replaced by a non-ionic surfactant, namely mono-oleate polyoxyethylene (Tween 80).[62,63] To avoid aggregation and serum sensitivity, the DOTMA : Tween 80 : DNA ratio was found to be very important and was carefully controlled. The results showed that higher transfection efficiency was evident at higher DOTMA to DNA and DOTMA to Tween 80 ratios. Also, all the internal organs including lung, liver, spleen, heart, kidney expressed the transgene with highest expression (at least by 100-fold) in the lung. Furthermore, the biodistribution studies with ^{125}I-labeled DNA suggests that the highest expression of the transgene in lung is probably due to highest uptake of

DNA by the lung tissue and longer retention of the transgene in this organ. It was interesting to note that although lung and liver had similar levels of DNA accumulation, the gene expression was significantly lower in liver than in lungs. This may suggest that in addition to the factors of delivery and retention of the DNA, transfection efficiency of each formulation may also vary with different cells in different organs. In this work, stability of lipoplex is probably related to the presence of Tween 80 and the excess of cationic lipid in the complex.

In another report,[64] it was observed that a higher level of in vivo transfection was obtained with multilamelar vesicles (MLVs) instead of small unilamelar vesicles (SUVs). In this case, 1-[2-(9(Z)-Octadecenoyloxy)-ethyl]-2-(8(Z)-heptadecenyl)-3-(2-hydroxyethyl)-imidazolinium chloride (DOTIM),[13] an imidazole-based cationic lipid was used together with cholesterol. The results showed that the cholesterol containing MLV is more active in transgene expression in vitro and in vivo (i.v. administration) than using DOPE containing SUV. According to the authors, the reason MLVs are effective in transfection may be because the larger complex contains more DNA copies and thus delivers a larger amount of DNA to the cells than the small SUV. The factors determined for the i.v. administration are the retention of complex in the circulation, subsequent uptake followed by its retention in tissues and lastly, the host determinants of different tissue.

In general cationic lipid-DNA complexes, with or without polycations, can transfer genes to many different cell types in vivo in different routes of administration. The tissue-specific expression can partly be achieved by placing tissue-specific promoter elements in the transgene. To increase the specificity further, an approach would be to design cationic lipids with different structural motifs or new formulations to impart different physico-chemical properties to the vector that will be recognized by a particular type of tissue or cell. The other approach may be to functionalize the lipids by chemical modifications with targeting molecules, e.g., antibodies, oligosaccharides. This strategy has been used to target lipopolyamine-DNA complexes to the hepatocytes in vitro[65] and polylysine-DNA complexes to the liver in vivo.[66]

VI. Emulsions for gene transfer

VI.1. Cationic emulsions

In an attempt to solve the problems of aggregation and serum sensitivity of unstable cationic liposome-DNA complex, Liu et al. reported gene transfer using oil-in-water emulsions as an alternative to liposomes.[62–63] Different types of non-ionic surfactants including Tween, Span, Brij and pluronic copolymers were tested as co-emulsifiers for the preparation of the emulsions composed of Castor oil, DOPE and DC-Chol.[62] Tween 80, containing branched polyoxyethylene chains as the hydrophilic head group was the most effective for transfection. Moreover, in contrast to DC-chol liposome-DNA complex, Tween 80 containing emulsions, were resistant to serum, stable for at least ten days without any noticeable aggregation. Overall, it seems that the possible reason for the prevention of aggregation of the DNA/emulsion complexes and resistance to serum is because the branched, hydrophilic polyoxyethylene chain of Tween covers a large area on the surface of emulsions like an umbrella. This structural motif provides a steric hindrance on the surface of emulsions, which allows only one DNA molecule to bind to one emulsion particle and may also prevent serum proteins interacting with emulsions. In a further study to develop appropriate dosage forms of DNA using Tween 80-containing emulsions, these new formulations transfected different cell lines with an equivalent or higher transfection activity as compared to the cationic liposomes.[63] Moreover, the absence of DOPE in emulsion formulations had practically no effect on transfection activity; even the micellar formulations (DC-Chol/Tween 80) were shown to have high transfection activity. Overall, the cationic emulsions with Tween 80 appeared as a stable, efficient and serum resistant gene transfer reagent.

VI.2. Reconstituted chylomicron remnants

In another novel approach, reconstituted chylomicron remnants were designed by Hara and Huang to deliver DNA to the liver.[67] Dietary lipids absorbed by the intestine are packaged into triglyceride-rich lipoproteins, naturally occurring biological emulsions, termed chylomicrons.[68] In the blood circulation, chylomicrons are transformed into chylomicron remnants by hydrolysis of the core triglycerides by lipoprotein lipase and absorption of apolipoproteins.[69] The circulating chylomicron remnants are taken up by liver parenchymal cells via apolipoprotein-specific receptors.[70] Recently, using commercially available lipids, reconstituted chylomicron remnants (RCR) are made and reported to be taken up by the liver hepatocytes.[71] Therefore, the strategy adopted by Hara and Huang was to make a DNA-cationic lipid complex hydrophobic enough so that the complex can be solubilized inside the oil-core of the chylomicron emulsions. The advantages of designing chylomicrons as a gene delivery vector include their ability to evade recognition by the reticuloendothelial system (RES), physically stable due to their hydrophobic core, protection of the encapsulated DNA from the environment during circulation and ability to bind to specific receptors in the liver.

To achieve a more efficient complexation with DNA, the tertiary amino group was methylated to a quaternary ammonium group with a permanent positive charge. The new lipid, named TC-Chol, forms a hydrophobic complex with DNA at 1:1 molar ratio and can be extracted by chloroform. The DNA/TC-Chol complex can be encapsulated in RCR by emulsifying the complex with appropriate amounts of triglyceride (olive oil), L-α-phosphatidylcholine (egg PC), L-α-lyso-phosphatidylcholine (lyso PC), cholesteryl oleate and cholesterol in a 70:22.7:2.3:3.0:2.0 weight ratio. By determining the amount of DNA floatation after centrifugation, it was concluded that more than 65% of DNA added as the hydrophobic complex was incorporated into the RCR.

Hara and Huang also reported[67] that DNA/TC-Chol-RCR delivered intraportally in mice expressed a high amount of luciferase protein (about 5 ng of transgene product per mg of liver protein per 100 μg of injected DNA) in the liver (Figure

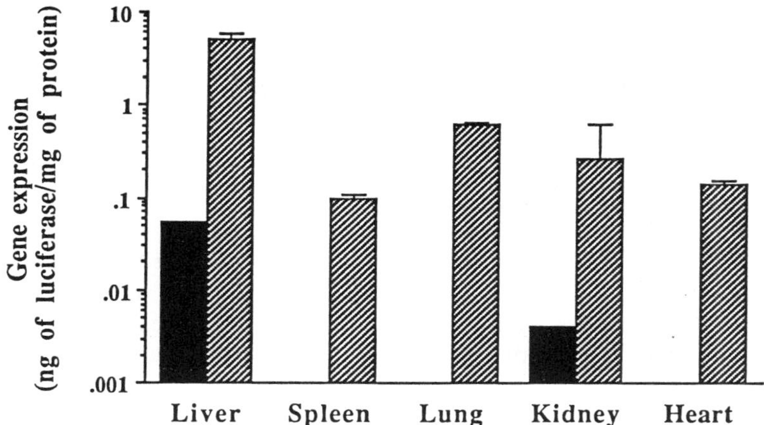

Fig. 5. In vivo gene expression following portal vein injection. CD1 mice were injected with 100 μg of naked DNA (■) or DNA in the form of reconstituted chylomicron (▨). Two days after injection, mice were sacrificed and luciferase activity and protein concentration of tissue extracts were assayed. (Taken from Ref. 67).

5). This level of expression was about 100-fold higher than that of naked DNA dissolved in an isotonic solution, injected into mice. Even as compared to a recent report of expression of naked DNA in a hypertonic solution (15% mannitol, 0.9% NaCl) in hepatocytes,[72] the expression level with RCR was still 10-fold higher. Transgene expression was also seen in spleen, heart and lung but the levels were 25- to 800-fold less than that of the liver. Histochemical examination by X-Gal staining revealed that approximately 10% of total liver cells expressed the LacZ gene.

The necessity of incorporation of DNA inside the RCR was critical as injection of a mixture of empty RCR (TC-Chol-RCR) with externally added DNA resulted in immediate aggregation and no gene expression in any organ. On the other hand, when a mixture of DNA and empty RCR without TC-Chol was injected, the expression level in all the organs was the same as in the case of naked DNA in isotonic solution. At a saturating dose of 50 μg of DNA, the expression level remained high for two days, followed by a gradual decrease to almost nil in seven days. However, the gene expression was completely regained by a second injection on day 7. Thus, it is theoretically possible to maintain a high level of gene expression by repetitive injections of the formulation by using a catheter for multiple portal vein infusion.[73] When the human α-1 antitrypsin (hATT) gene is formulated in the RCR, the injected mice show expression of hATT in the blood for up to 60 days.[74]

After a single tail vein injection of DNA/TC-Chol-RCR, there was some level of gene expression in liver but was significantly lower than the intraportal injection. The efficiency of delivery of genes may be further improved by the addition of apolipoprotein E due to receptor-mediated uptake. Also there are other kinds of lipoproteins whose surface can be modified by a ligand such as galactose, folate, transferrin, antibody, etc. Thus, a gene delivery system with such a high activity level can be potentially targeted not only to liver but to other organs as well.

VII. Clinical trials

After a series of steps of designing a vector, creating a new formulation of vector-gene complex, testing its efficiency in transgene expression in vivo, critically scrutinizing its safety profile, the formulation is then advanced to the next step of clinical trials (reviewed in Ref. 75). Cationic lipids have been used in clinical trials for gene therapy of cancer, cystic fibrosis and Canavan's disease.[1,2,46,47]

A human trial for therapy of cystic fibrosis,[2] was carried out after the successful results regarding the efficacy of DC-Chol liposome-DNA complex in a mouse model.[76,77] The DC-Chol liposome complexed with the therapeutic DNA containing the cDNA for the CFTR gene, was administered to the nasal epithelium of the CF individuals. In 6 out of the 7 patients, the CFTR mRNA was detected in the nasal epithelium cells with no apparent toxicity. The cAMP mediated chloride channel defect (typical of cystic fibrosis) was partially rectified in some patients. Two other recent similar studies using DOTAP and DC-Chol liposomes, respectively, show similar clinical results.[78,79]

In the treatment of melanoma, DC-Chol was used to complex with a plasmid which contained the cDNA for HLA-B7.[46] A few days after the intratumoral injection of the DC-Chol/DNA complex, cytotoxic T-lymphocytes specific for HLA-B7 were generated. In the case of one out of the five patients, complete regression of the primary tumor occurred.

In another ongoing clinical trial for the gene therapy of Canavan's disease, an autosomal recessive leukodystrophy, preliminary reports are successful.[47] After the direct intracranial injection of the DC-Chol liposomes complexed with poly-L-lysine and plasmid DNA which contains the aspartoacylase (ASPA) gene, there is no apparent sign of accumulation of *N*-acetylaspartate, a neurotoxic metabolite at or near the site of injection. No apparent toxicity related to the injection is observed.

VIII. Conclusion

The cationic lipid-mediated human gene therapy, since its first report in 1987, has come a long way. The major hurdles for this treatment are now to reduce the toxicity of the lipids and increase the level of transgene expression. To overcome these hurdles, will need a more in-depth understanding of the interplay between the structure-function relationships of the lipid, and its complex with DNA. Also a better understanding of the molecular mechanism of the action is needed, through the various steps of the transfection process, starting from the cellular association of the vector-gene complex to the final transgene expression (Table 1). These understandings will help a rational design of the gene therapy vectors of the future.

Acknowledgment

The original work of this laboratory was supported by NIH grants CA 64654, CA 71731 and DK 44935, and a grant from Targeted Genetics Corporation.

References

1. Nabel EG, Yang Z, Muller D, Chang AE, Gao X, Huang L, Cho KJ, Nabel GJ. Safety and toxicity of catheter gene delivery to the pulmonary vasculature in a patient with metastatic melanoma. Human Gene Therapy 1994;5:1089–1094.
2. Caplen NJ, Alton EWFW, Middleton PG, Dorin JR, Stevenson BJ, Gao X, Durham SR, Jeffery PK, Hodson ME, Coutelle C, Huang L, Porteous DJ, Williamson R, Geddes DM. Liposome-mediated CFTR gene transfer to the nasal epithelium of patients with cystic fibrosis. Nature Medicine 1995;1:39–46.
3. Gregoriadis G, Saffie R, de Souza JB. Liposome-mediated DNA vaccination. FEBS Lett 1997;402:107–110.
4. Wizel B, Rogers WO, Houghten RA, Lanar DE, Tine JA, Hoffman SL. Induction of murine cytotoxic T lymphocytes against *Plasmodium falciparum* sporozoite surface protein 2. Eur J Immunol 1994;24:1487–1495.
5. Deshmukh HM, Huang L. Liposome and polylysine mediated gene transfer. New J of Chem 1997;21:113–124.
6. Felgner PL, Gadek TR, Holm M, Roman R, Chan HW, Wenz M, Northrop JP, Ringold GM, Danielsen M. Lipofection: A highly efficient, lipid-mediated DNA transfection procedure. Proc Natl Acad Sci USA 1987;84:7413–7417.
7. Zhou X, Huang L. DNA transfection mediated by cationic liposomes containing lipopolylysine: characterization and mechanism of action. Biochim Biophys Acta 1994;1189:195–203.
8. Remy J-S, Sirlin C, Vierling P, Behr J-P. Gene transfer with series of lipophilic DNA-binding molecules. Bioconjugate Chem 1994;5:647–654.
9. Zhou X, Klibanov AL, Huang L. Liposophilic polylysines mediate efficient DNA transfectin in mammalian cells. Biochim Biophys Acta 1991;1065:8–14.
10. Gao X, Huang L. A novel cationic liposome reagent for efficient transfection of mammalian cells. Biochem Biophys Res Commun 1991;179:280–285.
11. Rose JK, Buonocore L, Whitt MA. A new cationic liposome reagent mediating nearly quantitative transfection of animal cells. Biotechniques 1991;10:520–525.
12. Leventis R, Silvius JR. Interactions of mammalian cells with lipid dispersions containing novel metabolizable cationic amphiphiles. Biochim Biophys Acta 1989;1023:124–132.
13. Solodin I, Brown CS, Bruno MS, Chow C-Y, Jang E-H, Debs RJ, Heath TD. A novel series of amphiphilic imidazolinium compounds for in vitro and in vivo gene delivery. Biochemistry 1995;34:13537–13544.
14. Malone RW, Felgner PL, Verma IM. Cationic liposome-mediated RNA transfection. Proc Natl Acad Sci USA 1989;86:6077–6081.
15. Jääkkeläinen I, Mönkkönen J, Urtti A. Oligonucleotide-cationic liposome interactions: A physiochemical study. Biochim Biophys Acta 1994;1195:115–123.
16. Litzinger DC, Brown JM, Wala I, Kaufman SA, Van GY, Farrell CL, Collins D. Fate of cationic liposomes and their complex with oligonucleotide in vivo. Biochim Biophys Acta 1996;1281:139–149.
17. Bennett CF, Chiang MY, Chan H, Shoemaker JE, Mirabelli CK. Cationic lipid enhance cellular uptake and activity of phosphorothioate antisense oligonucleotides. Mol Pharmacol 1992;41:1023–1033.
18. Sells MA, Li J, Chernoff J. Delivery of protein into cells using polycationic liposomes. Biotechniques 1995;19:72–78.
19. Felgner PL, Barenholz Y, Behr J-P, Cheng SH, Cullis P, Huang L, Jessee JA, Seymour L, Szoka F, Thierry AR, Wagner E, Wu G. Nomenclature for synthetic gene delivery systems. Human Gene Therapy 1997;8:511–512.
20. Li S, Huang L. Lipidic supramolecular assemblies for gene transfer. J Liposome Res 1996;6:589–608.

21. Litzinger DC, Huang L. Phosphatidylethanolamine liposomes: drug delivery, gene transfer and immunodiagnostic applications. Biochim Biophys Acta 1992;1113:201–227.
22. Cullis PR, De Kruijff B. Polymorphic phase behaviour of lipid mixtures as detected by ^{31}P-NMR. Evidence that cholesterol may destabilize bilayer structure in membrane systems containing phosphatidylethanolamine. Biochim Biophys Acta 1978;507:207–218.
23. Gruner SM, Cullis PR, Hope MJ, Tilcock CP. Lipid polymorphism: the molecular basis of nonbilayer phases. Ann Rev Biophysics Biophysical Chem 1985;14:211–38.
24. Eastman SJ, Siegel C, Tousignant J, Smith AE, Cheng SH, Scheule RK. Biophysical characterization of cationic lipid-dna complexes. Biochim Biophys Acta: Biomembranes 1997;1325:41–62.
25. Felgner JH, Kumar R, Sridhar CN, Wheeler CJ, Tsai YJ, Border R, Ramsey P, Martin M, Felgner PL. Enhanced gene delivery and mechanism studies with a novel series of cationic lipid formulations. J Biol Chem 1994;269:2550–2561.
26. Rädler JO, Koltover I, Salditt T, Safinya CR. Structure of DNA-cationic liposome complexes: DNA intercalation in multilamellar membranes in distinct interhelical packing regimes. Science 1997;275:810–814.
27. Gershon H, Ghirlando R, Guttman SB, Minsky A. Mode of formation and structural features of DNA-cationic liposome complexes used for transfection. Biochemistry 1993;32:7143–7151.
28. Gustafsson J, Arvidson G, Karlsson G, Almgren M. Complexes between cationic liposomes and DNA visualized by cryo-TEM. Biochim Biophys Acta 1995;1235:305–312.
29. Sternberg B, Sorgi FL, Huang L. New structures in complex formation between DNA and cationic liposomes visualized by freeze-fracture electron microscopy. FEBS Lett 1994;356:361–366.
30. Mel'nikov SM, Yoshikawa K. First-order phase transition in large single duplex DNA induced by a nonionic surfactant. Biochem Biophys Res Commun 1997;230:514–517.
31. Lasic DD, Strey H, Stuart MCA, Podgornik R, Frederik, PM. The structure of DNA-liposome complexes. J Amer Chem Soc 1997;119:832–833.
32. Dan N. Formation of ordered domains in membrane-bound DNA. Biophys J 1996;71:1267–1272.
33. Fang Y, Yang J. Two-dimensional condensation of DNA molecules on cationic lipid membranes. J Phys Chem B 1997;101:441–449.
34. Pinnaduwage P, Schmitt KL, Huang L. Use of quaternary ammonium detergent in liposome mediated DNA transfection of mouse L-cells. Biochim Biophys Acta 1989;985:33–37.
35. Ballas N, Zakai N, Sela I, Loyter A. Liposomes bearing a quaternery ammonium detergent as an efficient vehicle for functional transfer of TMV-RNA into plant protoplasts. Biochim Biophys Acta 1988;939:8–18.
36. Lee ER, Marshall J, Siegel CS, Jiang C, Yew NS, Nichols MR, Nietupski JB, Ziegler RJ, Lane MB, Wang KX, Wan NC, Scheule RK, Harris DJ, Smith AE, Cheng SH. Detailed analysis of structures and formulations of cationic lipids for efficient gene transfer to the lung. Human Gene Therapy 1996;7:1701–1717.
37. Vigneron JP, Oudrhiri N, Fauquet M, Vergely L, Bradley JC, Basseville M, Lehn P, Lehn JM. Guanidinium-cholesterol cationic lipids: Effecient vectors for the transfection of eukaryotic cells. Proc Natl Acad Sci USA 1996;93:9682–9686.
38. Oudrhiri N, Vigneron JP, Peuchmaur M, Leclerc T, Lehn JM, Lehn P. Gene transfer by guanidinium-cholesterol cationic lipids into airway epithelial cells in vitro and in vivo. Proc Natl Acad Sci USA 1997;94:1651–1656.
39. Wheeler CJ, Felgner PL, Tsai YL, Marshall J, Sukhu L, Doh SG, Hartikka J, Nietupski J, Manthorpe M, Nichols M, Plewe M, Liang X, Norman J, Smith A, Cheng SH. A novel cationic lipid greatly enhances plasmid DNA delivery and expression in mouse lung. Proc Natl Acad Sci USA 1996;93:11454–11459.
40. Wheeler CJ, Sukhu L, Yang GL, Tsai YL, Bustamente C, Felgner P, Norman J, Manthorpe M. Converting an alcohol to an amine in a cationic lipid dramatically alters the co-lipid requirement, cellular transfection activity and the ultrastructure of DNA cytofectin complexes. Biochim Biophys Acta: Biomembranes 1996;1280:1–11.
41. Stephan DJ, Yang ZY, San H, Simari RD, Wheeler CJ, Felgner PL, Gordon D, Nabel GJ, Nabel EG. A new cationic liposome DNA complex enhances the efficiency of arterial gene transfer in vivo: Human Gene Therapy 1996;7:1803–1812.
42. Scheule RK, Stgeorge JA, Bagley RG, Marshall J, Kaplan JM, Akita GY, Wang KX, Lee ER, Harris DJ, Jiang CW, Yew NS, Smith AE, Cheng SH. Basis of pulmonary toxicity associated with cationic lipid-mediated gene transfer to the mammalian lung. Human Gene Ther 1997;8:689–707.
43. Sorgi FL, Huang L. Large scale production of DC-chol liposomes by microfluidization. Int J Pharm 1996;144:131–139.

44. Egilmez NK, Iwanuma Y, Bankert RB. Evaluation and optimization of different cationic liposome formulations for in vivo gene transfer. Biochem Biophys Res Commun 1996;221:169–173.
45. Schreier H, Sawyer SM. Liposomal DNA vectors for cystic fibrosis gene therapy: Current applications, limitations, and future directions. Adv Drug Delivery Rev 1996;19:73–87.
46. Nabel EG, Yang Z, Plautz G, Forough R, Zhan X, Haudenschild CC, Maciag T, Nabel GJ. Recombinant fibroblast growth factor-1 promotes intimal hyperplasia and angiogenesis in arteries in vivo. Nature 1993;362:844–846.
47. During MJ, Zeng L, Sorgi FL, Mee E, Huang L, Kaul R, Wang J, Mayes LC, Novotny EJ, Seashore MR, Leone P. Liposome-polymer-DNA complex-mediated gene transfer in canavan disease. Unpublished study, 1997.
48. Gao X, Huang L. Potentiation of cationic liposome-mediated gene delivery by polycations. Biochemistry 1996;35:1027–1036.
49. Sorgi FL, Bhattacharya S, Huang L. Protamine sulfate enhances lipid mediated gene transfer. Gene Therapy, 1997;4:961–968.
50. Li S, Huang L. In vivo gene transfer via intravenous administration of cationic lipid/protamine/DNA (LPD) complexes. Gene Therapy, 1997;4:891–900.
51. Hofland HEJ, Shephard L, Sullivan SM. Formation of stable cationic lipid/DNA complexes for gene transfer. Proc Natl Acad Sci USA 1996;93:7305–7309.
52. Zhang YP, Reimer DL, Zhang GY, Lee PH, Bally MB. Self-assembling DNA-lipid particles for gene transfer. Pharm Res 1997;14:190–196.
53. Bally MB, Zhang YP, Wong FMP, Kong S, Wasan E, Reimer DL. Lipid-DNA complexes as an intermediate in the preparation of particles for gene transfer—an alternative to cationic liposome/DNA aggregates. Adv Drug Delivery Rev 1997;24:275–290.
54. Hong K, Zheng W, Baker A, Papahadjopoulos D. Stabilization of cationic liposome-plasmid dna complexes by polyamines and poly(ethylene glycol)-phospholipid conjugates for efficient in vivo gene delivery. FEBS Lett 1997;400:233–237.
55. Allen TM, Hansen C, Martin F, Redemann C, Yau-Young A. Liposomes containing synthetic lipid derivatives of poly(ethylene glycol) show prolonged circulation half-lives in vivo. Biochim Biophys Acta 1991;1066:29–36.
56. Papahadjopoulos D, Allen TM, Gabizon A, Mayhew E, Matthay K, Huang SK, Lee KD, Woodle MC, Lasic DD, Redemann C. Sterically stabilized liposomes: improvements in pharmacokinetics and antitumor therapeutic efficacy. Proc Natl Acad Sci USA 1991;88:11460–11464.
57. Plum GE, Arscott PG, Bloomfield VA. Condensation of DNA by trivalent cations. 2. Effects of cation structure. Biopolymers 1990;30:631–643.
58. Thierry AR, Lunardi-Iskandar Y, Bryant JL, Rabinovich P, Gallo RC, Mahan LC. Systemic gene therapy: biodistribution and long-term expression of a transgene in mice. Proc Natl Acad Sci USA 1995;92:9742–9746.
59. Eastman SJ, Lukason MJ, Tousignant JD, Murray H, Lane MD, Stgeorge JA, Akita GY, Cherry M, Cheng SH, Scheule RK. A concentrated and stable aerosol formulation of cationic lipid-DNA complexes giving high-level gene expression in mouse lung. Human Gene Therapy 1997;8:765–773.
60. Eastman SJ, Tousignant JD, Lukason MJ, Murray H, Cheng SH, Scheule, RK. Optimization of formulations and conditions for the aerosol delivery of functional cationic lipid-DNA complexes. Human Gene Therapy 1997;8:313–322.
61. Liu F, Qi H, Huang L, Liu D. Factors controlling the efficiency of cationic lipid-mediated transfection in vivo via intravenous administration. Gene Therapy 1997;4:517–523.
62. Liu F, Yang JP, Huang L, Liu D. Effect of non-ionic surfactants on the formation of DNA/emulsion complexes emulsion-mediated gene transfer. Pharm Res 1996;13:1642–1646.
63. Liu F, Yang JP, Huang L, Liu D. New cationic lipid formulations for gene transfer. Pharm Res 1996;13:1856–1860.
64. Liu Y, Mounkes LC, Liggitt HD, Brown CS, Solodin I, Heath TD, Debs RJ. Factors influencing the efficiency of cationic liposome-mediated intravenous gene delivery. Nature Biotech 1997;15:167–173.
65. Remy J-S, Kichler A, Mordvinov V, Schuber F, Behr J-P. Targeted gene transfer into hepatoma cells with lipopolyamine-condensed DNA particles presenting galactose ligands: a stage toward artificial viruses. Proc Natl Acad Sci USA 1995;92:1744–1748.
66. Perales JC, Ferkol T, Beegen H, Ratnoff OD, Hanson RW. Gene transfer in vivo: sustained expression and regulation of genes introduced into the liver by receptor-targeted uptake. Proc Natl Acad Sci USA 1994;91:4086–4090.

67. Hara T, Liu F, Liu D, Huang L. Emulsion formulations as a vector for gene delivery in vitro and in vivo. Adv Drug Delivery Rev 1997;24:265–271.
68. Windler E, Chao Y, Havel RJ. Regulation of the hepatic uptake of triglyceride-rich lipoproteins in the rat. Opposing effects of homologous apolipoprotein E and individual C apoproteins. J Biol Chem 1980;255:8303–8307.
69. Huettinger M, Retzek H, Eder M, Goldenberg H. Characteristics of chylomicron remnant uptake into rat liver. Clin Biochem 1988;21:87–92.
70. Hussain MM, Maxfield FR, Mas-Oliva J, Tabas I, Ji ZS, Innerarity TL, Mahley RW. Clearance of chylomicron remnants by the low density lipoprotein receptor-related protein/alpha 2-macroglobulin receptor. J Biol Chem 1991;266:13936–13940.
71. Rensen PC, van Dijk MC, Havenaar EC, Bijsterbosch MK, Kruijt JK, van Berkel TJ. Selective liver targeting of antivirals by recombinant chylomicrons—a new therapeutic approach to hepatitis B. Nature Medicine 1995;1:221–225.
72. Budker V, Zhang G, Knechtle S, Wolff JA. Naked DNA delivered intraportally expresses efficiently in hepatocytes. Gene Therapy 1996;3:593–598.
73. Vrancken Peeters MJ, Perkins AL, Kay MA. Method for multiple portal vein infusions in mice: quantitation of adenovirus-mediated hepatic gene transfer. Biotechniques 1996;20:278–285.
74. Hara T, Huang L. In vivo gene delivery to the liver using reconstituted chylomicron remnants as a novel non-viral vector. Proc Natl Acad Sci, USA 1997;94:14547–14552.
75. Ledley FD. Nonviral gene therapy: the promise of genes as pharmaceutical products. Human Gene Therapy 1995;6:1129–1144.
76. Alton EWFW, Middleton PG, Caplen NJ, Smith SN, Steel DM, Munkonge FM, Jeffery PK, Geddes DM, Hart SL, Williamson R, Fasold KI, Miller AD, Dickinson P, Stevenson BJ, McLachlan G, Dorin JR, Porteous DJ. Non-invasive liposome-mediated gene delivery can correct the ion transport defect in cystic fibrosis mutant mice [published erratum appears in Nat Genet 5:312.1993]. Nature Genetics 1993;5:135–142.
77. Hyde SC, Gell DR, Higgins CF, Trezise AE, MacVinish LJ, Cuthbert AW, Ratcliff R, Evans MJ, Colledge WH. Correction of the ion transport defect in cystic fibrosis transgenic mice by gene therapy. Nature 1993;362:250–255.
78. Porteous DJ, Dorin JR, Mclachlan G, Davidsonsmith H, Davidson H, Stevenson BJ, Carothers AD, Wallace WAH, Moralee S, Hoenes C, Kallmeyer G, Michaelis U, Naujoks K, Ho LP, Samways JM, Imrie M, Greening AP, Innes JA. Evidence for safety and efficacy of DOTAP cationic liposome mediated CFTR gene transfer to the nasal epithelium of patients with cystic fibrosis. Gene Therapy 1997;4:210–218.
79. Gill DR, Southern KW, Mofford KA, Seddon T, Huang L, Sorgi F, Thomson A, Macvinish LJ, Ratcliff R, Bilton D, Lane DJ, Littlewood JM, Webb AK, Middleton PG, Colledge WH, Cuthbert AW, Evans MJ, Higgins CF, Hyde SC. A placebo-controlled study of liposome-mediated gene transfer to the nasal epithelium of patients with cystic fibrosis. Gene Therapy 1997;4:199–209.
80. Sorgi FL, Huang L. Drug delivery applications of liposomes containing non-bilayer forming phospholipids. In: Lipid Polymorphs and Membrane Properties. In: Current Topics in Membranes, Vol. 44, ed. Epand R, Academic Press Inc., Chapter 12, 1997; pp. 449–475.

Lasic and Papahadjopoulos (eds.), Medical Applications of Liposomes

CHAPTER 5.4

Ultrastructural morphology of cationic liposome-DNA complexes for gene therapy

BRIGITTE STERNBERG*

University of the Pacific and Associate Scientist in Residence, California Pacific Medical Center, Research Institute, 2340 Clay Street, San Francisco, CA 94115; Tel: (415) 202-1576, Fax: (415) 831-2813, e-mail: brigitt@itsa.ucsf.edu

Overview

I. Introduction

Almost two decades have passed since the first efforts to develop liposomes for the transfer and expression of extracellular RNA[1] and DNA into mammalian cells.[2–8] Efficient delivery of functional DNA into eucaryotic cells is highly desir-

*Although in my private life my name has changed now to Papahadjopoulos-Sternberg I will continue using my former name for scientific publications.

able for adding missing or replacing defective genes in gene therapy.[9,10] When compared to viral-based carriers, liposomal gene delivery systems offer several advantages, including the absence of viral components, the protection of the DNA/RNA from inactivation or degradation, and the possibility for cell-specific targeting.

Furthermore, positively charged liposomes, or cationic liposomes[11–16] (reviewed in reference 13) as transfection agents show at least three additional advantages: Since nucleic acids are highly negatively charged molecules, they can interact spontaneously with the preformed cationic liposomes and 100% complexation followed by encapsulation is reached simply by appropriate mixing of both components. Since virtually all biological surfaces have a net negative charge, cationic liposome/DNA complexes (CLDC) can bind to cells by charge interaction with a 10-fold or greater improvement in cellular uptake.[17] Moreover, cationic liposomes as transfer vectors exhibit relatively low toxicity, non-immunogenicity, and are easy to produce. Earlier studies described formulations which showed rather good transfection efficacy in various cell culture systems but have been practically ineffective for *in vivo* applications. More recently, however, CLDC have been used successfully to express heterologous genes *in vivo*, by direct intra-tumoral injection,[18] by repeated intravenous injection,[19] by aerosol inhalation[20] or administration to the nasal epithelium.[21]

II. Historical perspective on ultrastructural investigations

Despite numerous studies and commercially available transfection kits based on cationic liposomes, the mechanism of DNA interaction with these liposomes and the morphology of the resulting complexes are still not well understood. It was initially assumed that there is no true encapsulation of the DNA by the cationic vesicles, but only binding at their surface while the size and shape of the vesicle are maintained.[12] This hypothetical model is inconsistent with observations presented in three more recent studies on the basis of electron micrographs, prepared by the Kleinschmidt-technique,[22] by freeze-fracture,[23] or by cryo-electron microscopy.[24] These studies represented evidence for DNA-induced fusion of the cationic lipid vesicles. The result of this fusion process, however, is visualized differently, such as "bead-on-string" structures,[22] oligolamellar structures,[24] and fibril-like images depicting DNA coated by a lipid bilayer.[23] Hexagonally packed DNA coated by lipid was also proposed.[25] In two very recent publications based on cryo-electron microscopy[26] and *in situ* optical microscopy,[27] both in combination with X-ray scattering data, the addition of DNA to cationic liposomes results in a transition from liposomes into heterogenous particles in the shape of flat, concentric, bent or amorphous stacks of bilayers in the size range 0.2–0.5 μm[26] or into birefringent liquid-crystalline condensed globules with sizes in the order of 1 μm.[27] In both publications the structure of the resulting complexes is described as particles/globules with a short-range lamellar order in which 2D layers of oriented DNA are sandwiched between lipid bilayers being apart from each other with an interlayer spacing of 6.5 nm.[26,27] In all publications it is claimed that a lipid coating is able to protect the DNA from being cut by restriction enzymes or other degradation processes[22–27] irrespective as to weather this coat is made of a lipid bilayer sheet,[22,24] tubule,[23] or stacks,[26,27] or even built by nonbilayer lipid arrangements.[25] It is also proposed that the lipid coat is able to enhance the uptake of CLDC by recipient cells possibly via endocytosis and/or fusion, and possibly also to deliver material into the nucleus.[28]

In the present review several electron microscopic techniques are chosen to record the morphology of CLDC and to demonstrate some of the crucial factors which determine the morphology of the formed complexes, such as lipid to DNA ratio, valency of the cationic component, ratio and type of the "helper lipid", type of the nucleotide component, the effect of sterically stabilized liposomes, as well as the effect of pre-condensed plasmid DNA. Our attention is especially focused at the morphology of these complexes in the suspension medium such as buffer and the effects of cell media as well as serum. Recently acquired results are shown about the relation between morphology and transfection activity of cationic liposome-DNA complexes as reveled from in vitro studies with SK-BR-3 cells and from in vivo studies, after i.v. injection in mice. Moreover, freeze-fracture electron microscopic snap shots are presented from the interaction of cationic liposome-DNA complexes with skin culture cells indicating a possible mechanism for DNA transfer across the plasma membrane and into the nucleus of the cells. It is hoped that this work will help to answer the question about a correlation between chemical composition, morphology, and transfection activity of cationic liposome-nucleotide complexes by a characterization of the active structure(s) in terms of transfection.

III. Morphology of cationic liposome-DNA complexes studied by several electron microscopic techniques

III.1 Formation of bilayer-coated DNA fibrils during interaction of DC-Chol/DOPE-liposomes with supercoiled plasmid DNA

Freeze-fracture electron microscopy (FFEM) has been chosen mainly to investigate the complex formation between plasmid DNA and preformed cationic liposomes in relation to incubation time and lipid to DNA ratio. Freeze-fracture electron micrographs of all preformed liposomes (here DC-Chol/DOPE-liposomes in a molar ratio of 3:2 as an example; *liposome control*), show mainly small (<200 nm) vesicles (Figure 1A) well separated from each other in the suspension buffer (HEPES buffer, 20 mM, pH 7.5). Due to their positive surface charge, contributed by the positively charged DC-Chol component, there is a repulsive force between the liposomes preventing aggregation.[23] Naked pRSV-LUC DNA in the same buffer (*DNA-control*) is hardly observable by freeze-fracture electron microscopy (Figure 1B, some very weak features are marked by an arrow head). With a width of 2.37 nm of the DNA strands and of about 5 nm in the supercoiled version,

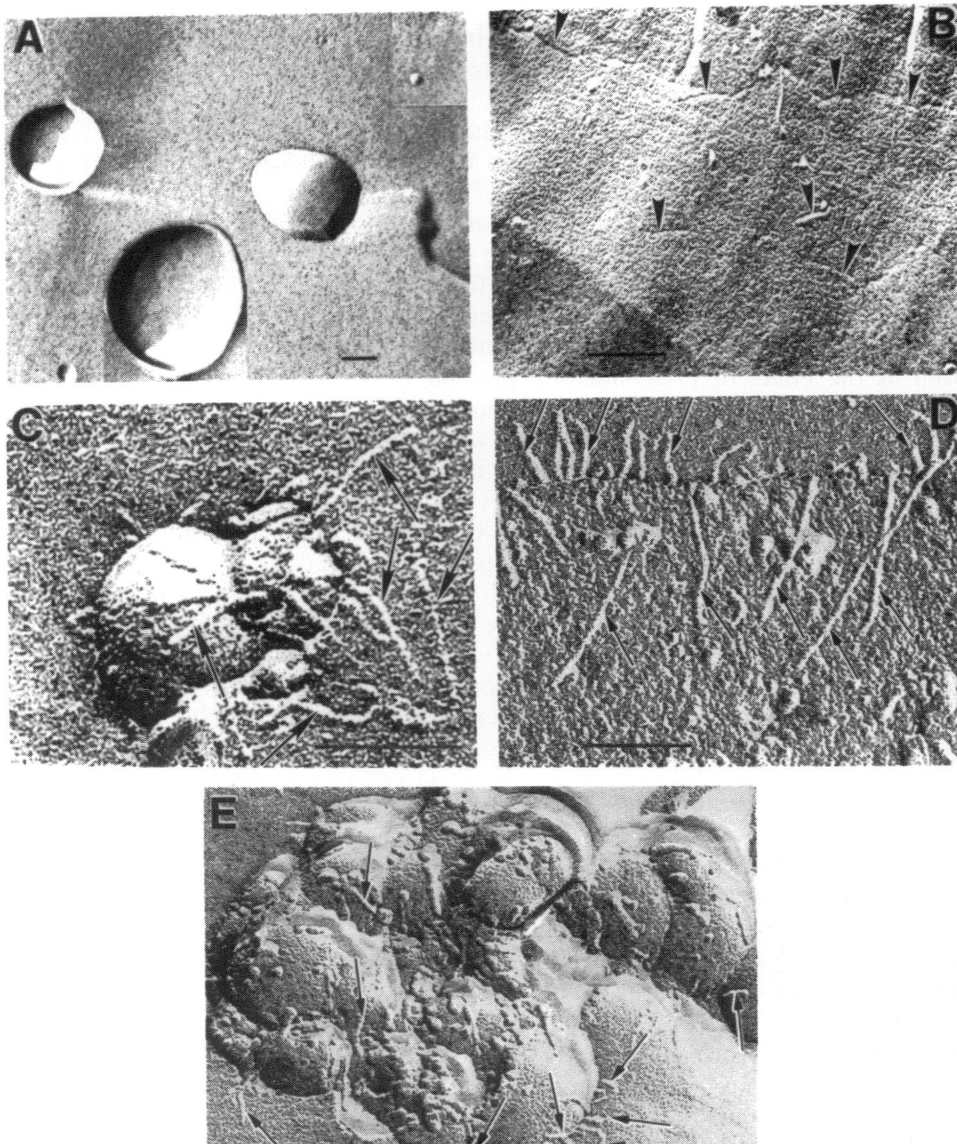

Fig. 1. Freeze-fracture electron micrographs of (A) liposomes, made of DC-Chol/DOPE (3:2 by mol; *liposome control*); (B) naked pRSV-Leu DNA (*DNA-control*, some of the very weak features are marked by an arrow head); and (C) and (E) DC-Chol-DNA-complexes, at various DNA-lipid ratios and incubation times: (C) *spaghetti-meatball*-complex at 24 hours incubation time and 2 μg DNA/20 nmol DC-Chol and (E) *spaghetti-meatball*-assembly at 30 min incubation time but 4 μg DNA/20 nmol DC-Chol (some *spaghetti*-like structures, connected with the complexes, are marked by an arrow in (C) as well as in (E)); (D) "free" *spaghetti*-like structures, attached to the metal foil of the sandwich at the same incubation time and concentration as in (C) (some of the "free" *spaghetti* are marked by an arrow).[23] The bar on all electron micrographs represents 100 nm and the shadow direction is running from bottom to top.

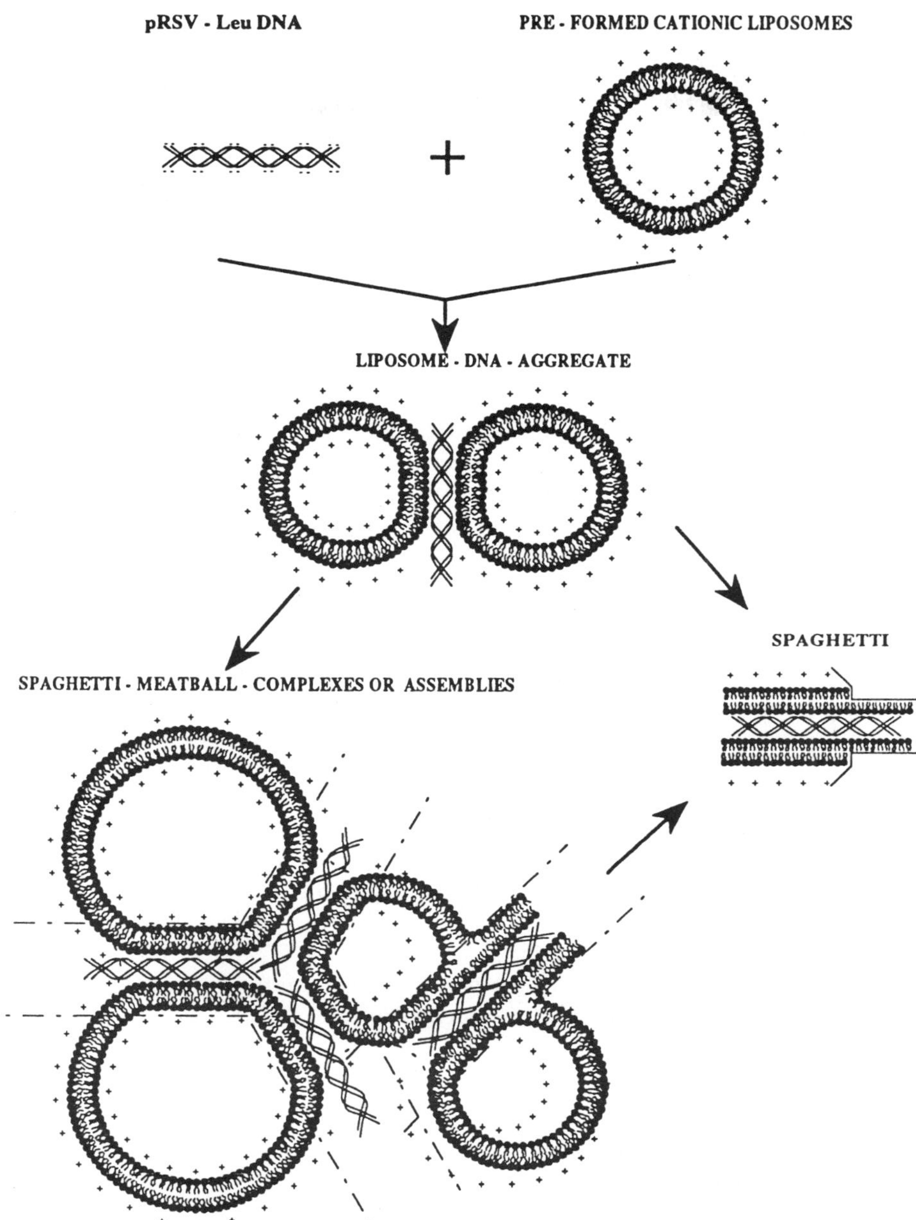

Fig. 2. Cartoon showing our personal suggestion, based on freeze-fracture electron micrographs, about the interaction of negatively charged and super-coiled plasmid DNA with the cationic DC-Chol-containing liposomes and the formation of Liposome-DNA-Aggregates without *spaghetti*-like structures, occurring at short incubation time and low DNA to lipid ratio, and *Spaghetti-Meatball*-Complexes, *Spaghetti-Meatball*-Assembles and *Spaghetti*-like structures, connected with the semi-fused liposomes but also "free" occurring in suspension at longer incubation time and higher DNA to lipid ratios (modified from Ref. 23).

these structures are at the resolution limit of freeze-fracture technique (about 2 nm for periodical structures).[29]

The DNA-liposome-interaction seems to follow a dynamic equilibrium (Cartoon in Figure 2): Negatively charged nucleic acid (not visible), adsorbed onto the outside of positively charged liposomes, may act as a fusogenic agent, drawing together the liposomes and forming complexes made of aggregated and fused liposomes. Some of them are presumably in an intermediate state (Figures 1C and 1E). At short incubation times (30 min) and low DNA concentrations (<3 μg DNA/20 nmol DC-Chol) the number of liposomes, involved in these complexes, is low (mainly 2–3 liposomes; Figure 1C), and their size is roughly the same compared to the control liposomes as shown in Figure 1A.

With higher DNA concentrations and longer incubation times (beginning with 4 μg DNA/20 nm DC-Chol and at 20 min) a proportion of the DNA-liposome complexes are observed as larger structures containing several partly or totally fused liposomes (Figure 1E). Most of the liposomes, involved in these assemblies, are grown in size compared to the controls (Figure 1A), presumably by fusion (Figure 1E). During the formation and growth of the complexes the membranes of the fused liposomes become at least transiently disrupted, allowing the self-encapsulation of the DNA.[23,30]

Additionally, during the process of interaction, some proportion of the plasmid DNA becomes clearly visible, as shown in Figures 1C–1E, and Figure 3A by freeze-fracture, in Figure 3B by negative staining,[31] and in Figure 3C by cryo-electron microscopy.[32] Obviously, its structure is enhanced, presumably by lipid-coating. *Spaghetti*-like structures[23] are found to be still connected to the liposomes (*spaghetti-meatball*-complex, Figure 1C; *spaghetti-meatball*-assembly, Figure 1E), but are also found separated from the liposomes, "free" in suspension (Figure 1D; cartoon of Figure 2). Possibly due to the residual positive charge at their surfaces, these fibrillar structures are often found in contact with the metal foil of the sandwich which was used for the rapid freezing of the samples (Figure 1D).[29]

III.2. Study of the fine-structure of the fibrils and complexes by freeze-fracture, negative staining and cryo-electron microscopy

The diameter of these fibrillar *spaghetti*-like structures (some are marked by arrows in Figures 1C–E) is approximately 7 nm, measured on freeze-fracture electron micrographs (Figures 1C–E and 3A)[23,30] or 13 nm, measured on electron micrographs prepared by negative staining technique (Figure 3B)[31] or cryo-electron microscopy (Figure 3C,D).[32] Both convex DNA-strands (shadow behind) as well as concave DNA-furrows (shadow in front; two of them are marked by an arrow in Figure 3A) are found. These findings support the possibility of a bilayer tubule covering the DNA strands. During freeze-fracturing, the fracture plane usually follows the hydrophobic interior of a bilayer, whether it is surrounding a cell or a liposome,[29] or in this case, most likely while surrounding a DNA-strand.[23,30] This is a reasonable explanation for finding a smaller diameter for the fibrils on

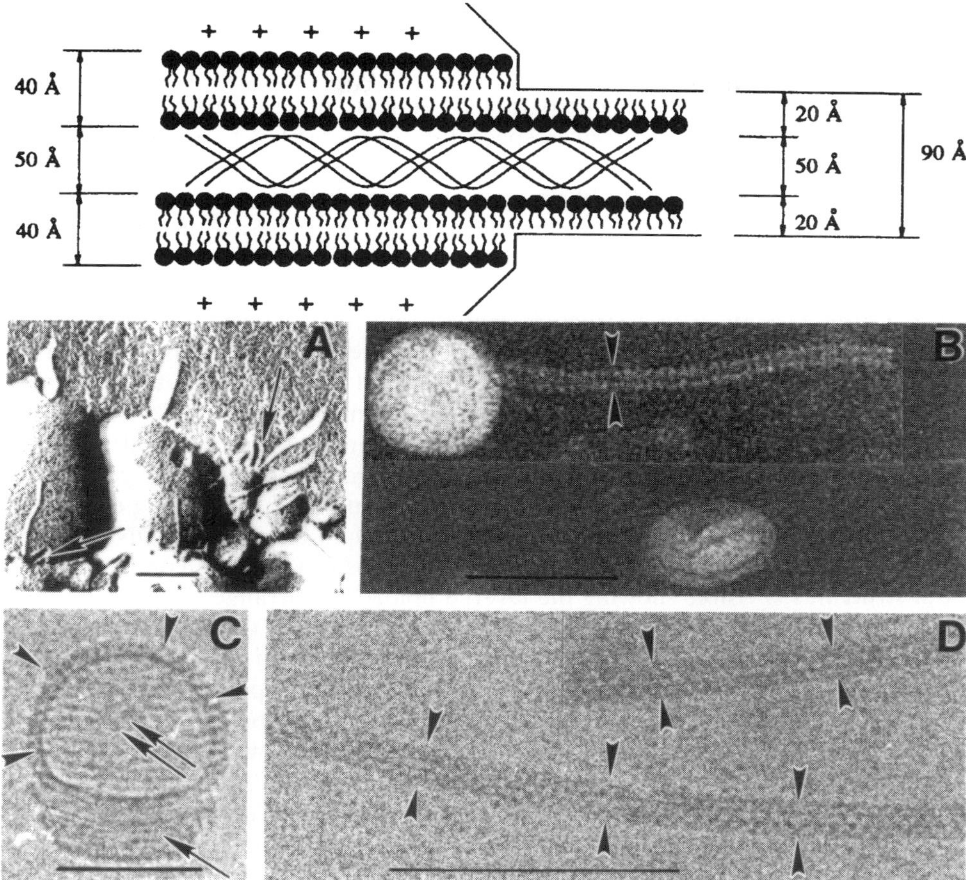

Fig. 3. Schematic drawing of a cross fraction of the *spaghetti*-like structure showing a bilayer tubule coated super-coiled plasmid DNA. The diameter of the whole arrangement represents 13 nm. Since the fracture plane is running along the hydrophobic area of a bilayer the diameter of the *spaghetti*-like structure is smaller on a freeze-fracture electron micrograph (A; about 9 nm) than on negative staining pictures (B; about 13 nm) or cryo-electron micrographs (C and D; about 13 nm) (modified from Refs. 23, 31).

(A) Cut-out of the freeze-fracture electron micrograph of a *Spaghetti-Meatball*-Assemble showing lipid-enhanced convex DNA-strands and two concave DNA-furrows, marked by an arrow. The shadow direction is running from bottom to top.

(B) Electron micrograph in negative staining technique of a *spaghetti*-like structure (presented between two arrow-heads) attached to a liposome-DNA complex. The tubular fibril between the arrow-heads displays a fine structure similar to a train-track.[31,32].

(C) + (D) Cryo-electron micrographs of a liposome/DNA complex (C) and of tubular fibrils (D). Within the liposome/DNA complex in (C) an area of regular periodicity is marked by two arrows displaying a smaller (~4 nm) and a wider (~7 nm) lamellar spacing. An area of less regular periodicity is marked by an arrow. These layered structures are coexisting with a rim of the fibrillar structure, marked by arrow heads, displaying a train-track-like fine structure. On the "free" tubular fibrils in (D) the fine structure like a train-track is also clearly visible.[32]. The bars on all electron micrographs represent 100 nm.

freeze-fracture than on negative staining electron micrographs (see also schematic drawing of the *spaghetti* in Figure 3).

Measuring a diameter of 13 nm for the bilayer-coated DNA on negative staining[31] as well as cryo-electron micrographs[32] and assuming a thickness of a fluid bilayer of 4 nm on each side gives space for a supercoiled plasmid DNA with roughly 2×2.5 nm within the middle of the bilayer tubule. Measured and calculated diameters of the tubular fibrils, presented in all three electron microscopy techniques, are in good agreement and support the hypothesis of a single bilayer tube coating the supercoiled plasmid DNA.[23,30–32]

Because of the higher resolution obtained by negative staining as well as cryo-electron microscopy, a fine structure like a train-track is visible on the fibrils: Using image analysis it became apparent that the bilayer-coat is arranged in toroidal segments around the supercoiled plasmid DNA leading to this train-track-like fine structure (marked by arrow heads in Figures 3B, 3C, and 3D).[32] Additionally, with both electron microscopic techniques there are also folded bilayer arrangements detectable (especially at bigger liposome/DNA complexes starting at a diameter of approx. 200 nm) showing an interlayer spacing consistent with a novel multilamellar structure of alternating lipid bilayers and DNA monolayers.[26,27] In some of our cryo-electron micrographs of CLDC, a combination of both structures is visible showing the layered lipid/DNA arrangement in the inner part (marked by two arrows in Figure 3C) with a rim of the fibrillar structure displaying the train-track-like fine structure (marked by arrow-heads in Figure 3C). The layered lipid/DNA arrangement within the complex (marked by two arrows in Figure 3C) shows clearly a smaller lamellar periodicity of about 4 nm and a wider one of about 7 nm what is in good agreement with the recent found multilamellar structure of alternating lipid bilayers and DNA monolayers.[26,27] The areas with lamellar periodicity are sometimes tighter packed (marked by one arrow in Figure 3C) and not always as regular arranged as seen in the interior of the particle (marked by the two arrows in Figure 3C).

The charge-charge interaction between the positively charged DC-Chol and the negatively charged DNA may stabilize the high curvature of the bilayer fibrils enclosing the supercoiled DNA strand, and the content of DOPE may also assist in the stabilization of the *spaghetti*-like structure. Based on its wedge-shaped molecular structure, DOPE can obviously adopt to highly curved structures, such as bilayer fibrils coating an approximately 5 nm thick, supercoiled DNA.[31]

From an examination of the optimal DNA to lipid ratio for transfection[33] and the dimension of the fibrils, it seems likely that the *spaghetti*-like structures may be the active DNA-lipid complex for the DC-Chol system. These fibrillar structures occur at DNA to lipid ratios which are typically used during transfection (2 μg of DNA to 20 nmol DC-Chol liposomes) and their diameter, comes close to the diameter of the nuclear pores.[23] Even so, the diameter of the DNA-lipid fibrils, which is about 13 nm (Figure 3) is too thick to pass freely through the nuclear pores, showing a diameter of about 7 nm. A process of un-coating may be necessary for passage of the DNA into the nucleus. On the other hand, the *spaghetti*-like structures, similar to microvilli, are extremely curved structures with relatively small radii (especially at their end-tips) and therefore are more likely to adhere and fuse more easily to flat cell membranes.[34]

IV. Factors determining the morphology of cationic liposome-DNA complexes

We have investigated the complex formation between supercoiled plasmid DNA or single strand 15-mer S-oligonucleotide and preformed cationic liposomes, made of monovalent amphiphiles such as DC-Chol, DOTMA, DIMRIE, DOTAP, and DDAB as well as the polyvalent amphiphile DOSPA by electron microscopy using mainly freeze-fracture technique but also cryo-electron microscopy. We studied the structural modifications of the cationic liposomes in relation to the DNA to lipid ratio and incubation time, type and charge of the cationic component, type and proportion of the helper lipid, and effect of the aqueous suspension medium, in order to identify the parameters and conditions favoring the occurrence of certain structures characteristic for the complexes formed.

IV.1. Lipid to DNA ratio and incubation time

The size of the complexes formed during interaction of cationic liposomes and the negatively charged nucleic acids is dependent upon charge neutralization and is smallest at a slightly positive net charge (± 0.8).[31] This is true for all cationic amphiphiles investigated such as DC-Chol[23] or DOTAP[33] as well as for quite different nucleic acids such as plasmid DNA[31] or 15-mer phosphorothioate oligonucleotides.[35]

The complex formation is a relatively fast process: After an incubation time of less than five minutes small liposome/DNA complexes are detectable. However, the formation of fibrils made of bilayer-surrounded plasmid DNA either free or connected to the complexes, needs approximately 20 minutes.[23]

IV.2. Cationic component

We investigated the morphology of complexes made of plasmid DNA and different *mono*valent and *poly*valent cationic amphiphiles.

IV.2.1. Monovalent cationic components

Formation of fibrillar *spaghetti-meatball*-like complexes, described above, is observed when supercoiled plasmid DNA interacts with cationic liposomes containing DOPE and amphiphiles bearing one positively charged group per molecule, such as DC-Chol (Figures 1C and E; 3A, B, C and D; 4A),[31] DOTMA (Figure 4B) or DMRIE (Figure 4C; in all electron micrographs some of the spaghetti-like structures are marked by arrow-heads). However, the formation of fibrils was less frequent when DOTMA was chosen instead of DC-Chol or DMRIE. It is interesting to notice that by using DMRIE as cationic amphiphile most of the liposomes are transformed into fibrils and only very few small liposomes are detectable

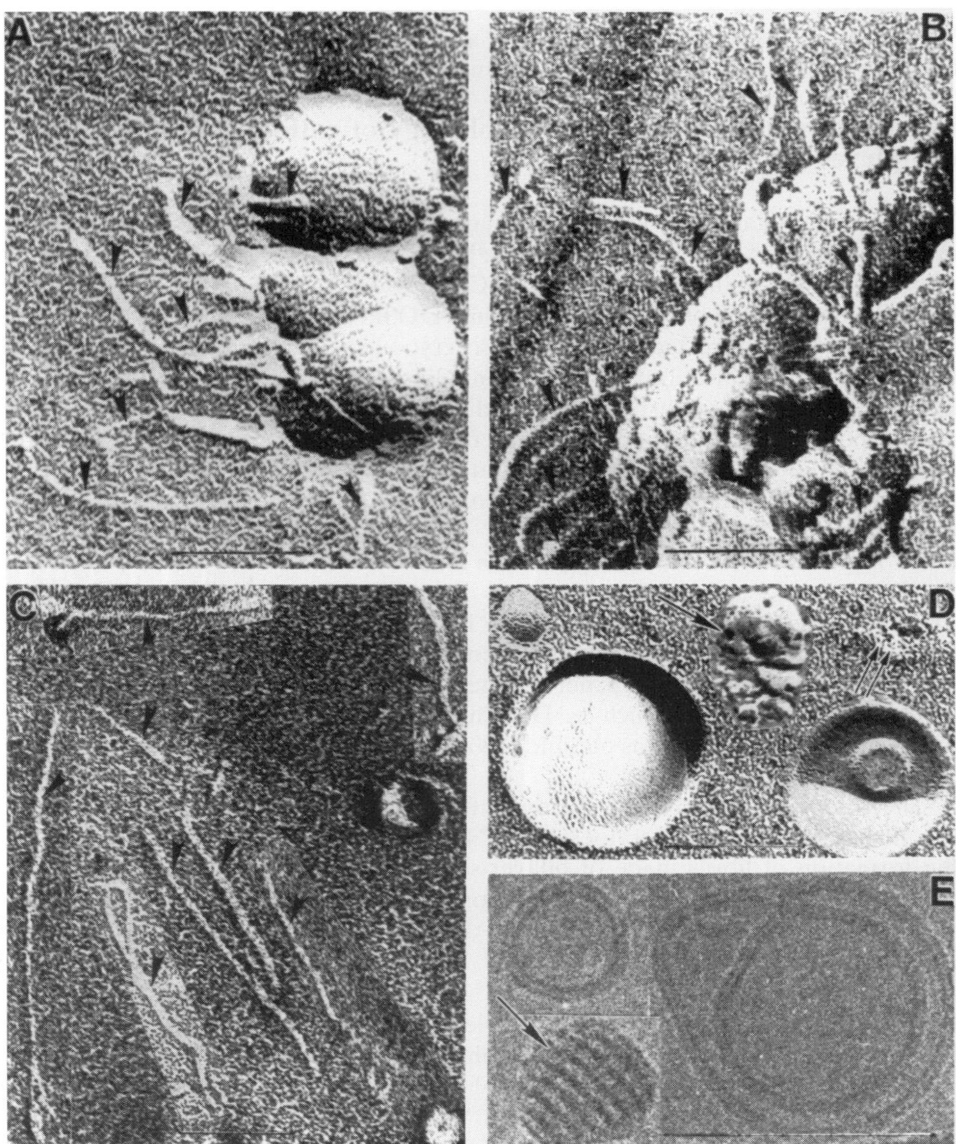

Fig. 4. Freeze-fracture electron micrographs of complexes made of plasmid DNA interacting with *mono*-valent (A)–(C) and *poly*-valent cationic liposomes (D). Whereas complexes made of *mono*-valent cationic amphiphiles such as DC-Chol (A), DOTMA (B) or DIMRIE (C) show formation of tubular fibrils (some of this spaghetti-like structures are marked by arrow-heads in (A–C)) the complexes made of *poly*-valent cationic amphiphiles such as DOSPA are very similar, especially on freeze-fracture electron micrographs (D), to the control liposomes and only some few disintegrated liposomes (one is marked by an arrow in (D)) or small structures (marked by double arrows in (D)) are visible.[33] On cryo-electron micrographs, however, some of the small complexes made of the *poly*-valent cationic amphiphile DOSPA show an interesting texture of wider and smaller periodicity marked by an arrow in (E).[36] Bars represent always 100 nm and the shadow direction is running from bottom to top of the electron micrographs.

on the freeze-fracture electron micrographs (Figure 4C). Furthermore DIMRIE belongs to the most active cytofectins as shown by *in vitro* transfection of COS. 7 cells.[25] Contrary, for amphiphiles with more than one positive charge per mole, such as DOSPA in no case was formation of fibrillar structures detected (Figure 4D and E).

IV.2.2. Polyvalent cationic component

The liposomes containing *poly*valent cationic amphiphiles such as DOSPA seem nearly unchanged in their morphology on freeze-fracture electron micrographs following interaction with negatively charged plasmid-DNA.[36] They appear free and single in the suspension, and are not attached or fused to each other (Figure 4D and E), in contrast to the *mono*valent cationic amphiphils DC-Chol (Figures 1C, D and E; 3A–D, and 4A–D), DOTMA (Figure 4B), and DIMRIE (Figure 4C).[23,31,33,36] Neither at a lower DNA-to-lipid ratio (1.2 μg/20 nmol), nor at a higher ratio (4 μg/20 nmol) where charge neutralization is reached, was it possible to observe *spaghetti*-like structures. Rather some very few condensed smaller structures were observed, marked by double arrows in Figure 4D, or what look like degraded liposomes, marked by one arrow in Figure 4D.[31,33,36] On cryo-electron micrographs, however, some of the single and separated complexes show an interesting inner texture of folded bilayers, marked by an arrow in Figure 4E, displaying a wider periodicity of about 7 nm and a smaller one of about 4 nm similar to that shown elsewhere.[26] Additionally, some few particles probably of condensed DNA are detectable (not shown here).[36] It is still an open question which of these structures may be responsible for the *in vitro* activity of the cytofectin DOSPA.[37]

IV.3. Ratio and type of the helper lipid

Unsaturated phosphatidylethanolamines such as DOPE are most commonly used as helper lipids in cationic liposome-mediated gene transfer.[38,39] Due to H-bond between the phosphate and the amino-group the head group of DOPE is much less hydrated and therefore much smaller compared to the headgroup of DOPC.[40] Because of its small head group compared to the larger area occupied by the hydrocarbon chains, DOPE molecules adapt an overall cone shape (dynamic shape concept[41]). This leads to a better packing of the molecules in the H_{II} phase. On the other hand, DOPE forms stable bilayers/liposomes when mixed with molecules occupying the opposite overall cone shape (big head group and small hydrophobic area) like detergents and/or cationic amphiphiles such as CTAB, DC-Chol or DDAB. DC-Chol alone for instance does not form stable liposomes. However, together with DOPE as a helper lipid well developed liposomes are observed (Figure 1A).[23,31] The presents of DOPE also reduces the cytotoxicity of several detergent-like cationic amphiphiles such as CTAB.[42]

In the majority of reported *in vitro* studies, the transfection potency of CLDC is increased when the cationic amphiphiles are mixed with a helper lipid.[43] It is generally believed that DOPE as a lipid forming nonbilayer structures facilitates the fusion of cationic liposome/DNA complexes with cell membranes, thereby promoting the release of DNA into the cytoplasm.[44] In most studies it is described that DOPE enhances the transfection efficiency, whereas DOPC reduces it.[38,39] In the case of chinese hamster ovary (CHO) cells, the addition at an equal molar ratio of DOPE improved the transfection efficiency of using DOTAP alone by a factor of 5–10.[43] However, there are also findings indicating that complexes containing the non-fusogenic DOPC are taken up more avidly by CHO cells and leading to higher transfection efficiency than those containing the fusogenic DOPE.[43] Especially in high ionic strength cell media and/or in the presence of serum the DOPE containing complexes tend to aggregate and form much bigger granules than DOPC containing complexes. These readily formed aggregated DOPE-containing granules may be too large to be internalized by the cells. For cells whose uptake rate is limited by endocytosis, such as CHO cells, the fusogenic properties of the helper lipid DOPE seems to play only a secondary role in determining the transfection efficiency.[43]

More recently, and especially for *in vivo* studies, Chol was used as a helper lipid instead of DOPE.[45–47] It was shown that intravenous injection of cholesterol-containing SUV increased the CLDC-mediated expression of the luciferase gene up to 100-fold, when compared to DOPE-containing SUV.[46]. Moreover, in cationic liposome-plasmid DNA complexes, stabilized by polyamines and poly(ethylene glycol)-phospholipid-conjugates, cholesterol and not DOPE was the helper lipid effective for sustaining high transfection activity in vivo over a longer period of time.[47]

We have studied the morphology of CLDC in relation to the molar ratio of DOPE to DC-Chol, ranging from 1:4 up to 4:1, and to the type of helper lipid including DOPE, DOPC, and Chol.

IV.3.1. DOPE/DC-Chol ratio

Aggregated and fused liposomes combined with fibrillar structures (some of them are marked by arrow heads in Figures 5A and 5B) are also formed in complexes of plasmid DNA interacting with liposomes made of DOPE : DC-Chol at different molar ratios, ranging from 1:4 (Figure 5A), to 2:3 (Figure 5B), and 3:2 (Figures 1C–E, and Figures 3A–D). This is especially true when the complexes were formed in buffers of low ionic strength.

However, at a higher DOPE-content of 4:1 molar ratio, the liposomal bilayers are transformed to non-bilayer hexagonal-tubular (H_{II}) structures as shown in Figures 5C and 5D by freeze-fracture electron microscopy. Based on its wedge-shaped molecular structure, DOPE can also adopt another, even higher curved structure, such as hexagonally packed tubules (some of the areas displaying H_{II} tubules are marked by two arrows in Figures 5C and 5D). They are usually observed at high lipid concentrations and/or high temperatures,[48] where DOPE has the ability to assume also non-bilayer structures.[49] FFEM is one of the best methods to distinguish between bilayer and non-bilayer structures[29] and to visualize hexagonal tubules.[29,49] Indeed, due to the high content of DOPE H_{II}-tubules are formed already in the control mixture, not interacting with the DNA,

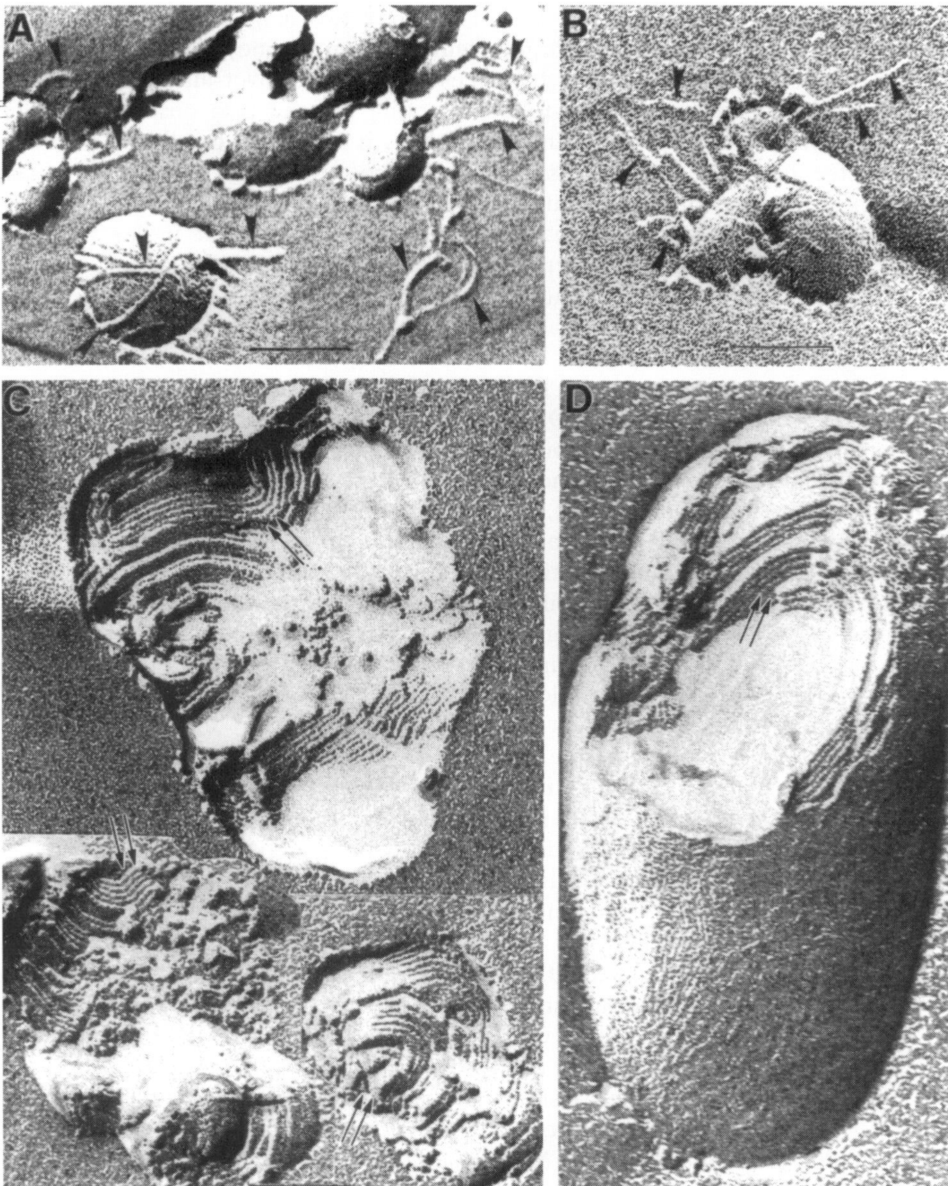

Fig. 5. Freeze-fracture electron micrographs of DOPE/DC-Chol mixtures with growing DOPE-content and interacting *with* (A–C) or *without* plasmid DNA (D). Molar ratios of DOPE:DC-Chol are 1:4 (A), 3:2 (B), and 4:1 (C and D). At low DOPE:DC-Chol ratios spaghetti-like structures are visible (some of them are marked by arrow-heads in (A) and (B)). At higher DOPE-content the lipid mixture is adopting the inverted hexagonal H_{II}-phase not depending upon the interaction *with* (C) or *without* plasmid DNA (control mixture, (D)). Some of the areas showing inverted hexagonal H_{II}-tubules are marked by double arrows in (C) and (D).[31] Bar represents always 100 nm and the shadow direction is running from bottom to top of the electron micrographs.

and containing DOPE:DC-Chol at 4:1 molar ratios (Figure 5D). These lipid areas, showing H_{II}-tubules, are large (Figure 5C) up to one micron (Figure 5D) and are obviously separated from the excess water by an over-all bilayer coat.[31]

Since this arrangement of H_{II}-tubules is quite big and also observed even in the control mixture, not interacting with DNA, it is a weak candidate for being the transfection-active structure as has been proposed by other authors.[24,50] Even more importantly, expression of luciferase gene, by pRSVL plasmid DNA in cells such as CHO, BHK as well as mouse lung cells, is best at molar ratios of DOPE:DC-Chol where the occurrence of the fibrillar structure is highest (around 2:3, molar ratios; personal communication with F.L. Sorgi & Leaf Huang).

IV.3.2. DOPE versus DOPC

We studied the morphology of CLDC containing plasmid DNA, DC-Chol, and the non-fusogenic helper lipid DOPC at a molar ratio of 3:2 by FFEM: In buffer of low ionic strength (HEPES-buffer, 20 mM, pH 7.5) small complexes were observed (Figure 6A) similar to those containing the fusogenic helper lipid DOPE (Figures 1C; 3A; 4A; 5A and 5B). Additionally, fibrillar structures were detected but by far less frequently than in DOPE containing complexes. The few fibrils seem to be also shorter and sometimes thicker when DOPE is replaced by DOPC (some of the shorter and thicker fibrils are marked by arrow heads in Figure 6A). Obviously, DOPC is not stabilizing the spaghetti-like structures as much as DOPE, which can adopt more easily to highly curved structures, such as a bilayer tubule around an approximately 5 nm thick, supercoiled plasmid DNA[23].

IV.3.3. DOPE versus cholesterol

Since DC-Chol/Chol mixtures do not form liposomes at any molar ratios we have studied the effect of cholesterol with the cationic amphiphile DDAB. We looked at the morphology of CLDC containing plasmid DNA complexed with DDAB, and Chol at a molar lipid ratio of 1:1. For comparison we studied the morphology of the related complexes containing the helper lipid DOPE instead of Chol in same molar ratio.

In buffer of low ionic strength (1 mM MES; pH 5.5) small complexes of DDAB/Chol were observed showing some fibrillar structures at the fracture planes of some aggregated liposomes as well as reaching into the water phase (some of them are marked by arrow heads in Figure 6B).[51] However, these fibrils are not as frequently found, and they are short, and not as well developed as in complexes made of DC-Chol/DOPE (Figures 1C and 1E; 3A; 4A; 5A and 5B), of DC-Chol/DOPC (Figure 6A), DOTMA/DOPE (Figure 4B), or DMRIE/DOPE (Figure 4C).

In contrast to the Chol containing complexes no fibrillar structures are depicted in DOPE containing DDAB complexes (Figure 6C).[51] They show fused liposomes clearly (at the fusion areas some of the lipidic particles are marked by arrows in Figure 6C) and form some bigger complexes than in Figure 6B. Obviously the relatively high DOPE to DDBA ratio (1:1 molar ratio) promotes fusion to bigger DNA-containing complexes.[52]

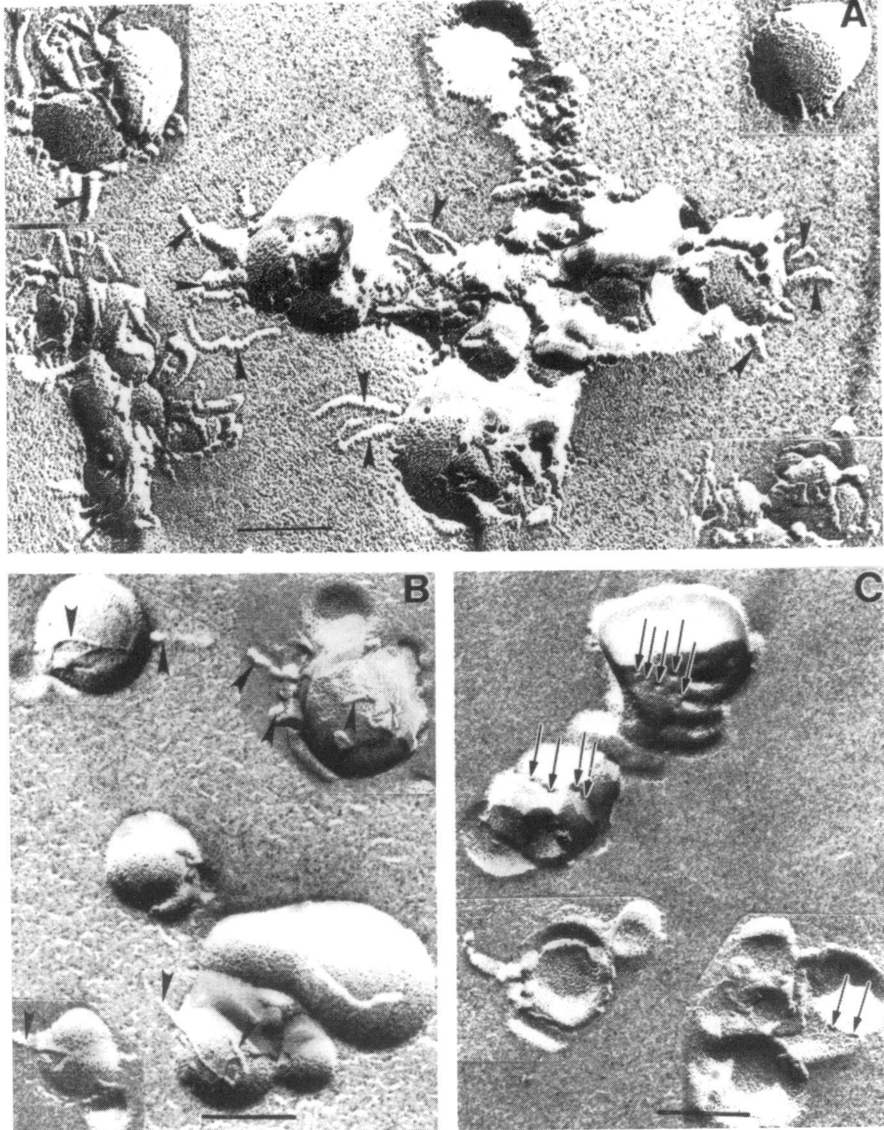

Fig. 6. Freeze-fracture electron micrographs of complexes made of plasmid DNA interacting with *mono*-valent cationic amphiphiles such as DC-Chol (A) and DDAB (B and C) but using different helper lipids such as DOPC (A), Chol (B), and DOPE (C). Some of the tubular fibrils formed with DC-Chol:DOPC (at 3:2 molar ratios) and with DDAB:Chol (at 1:1 molar ratios) are marked by arrow heads in (A) and (B). Bigger complexes of fused liposomes but without any fibrils are observed from DNA interacting with DDAB:DOPE liposomes (at 1:1 molar ratios; (C)). Lipidic particles at the fusion areas of the liposomes are marked by arrows in (C). Bar represents always 100 nm and the shadow direction is running from bottom to top of the electron micrographs.

IV.4. Nucleotide component

Oligonucleotides are potential therapeutics for blocking protein translation from mRNA (antisense effect) or transcription of RNA from a gene (antigen effect). The aim of this therapy is to deliver a short, single strand oligonucleotide with a complementary sequence to the part of the unwanted messenger RNA or gene, to switch genes off and to stop processes, such as cancer.[53] Cationic liposomes are efficient and simple nonviral systems to deliver oligonucleotides into cells.[54]

We looked at the morphology of complexes made of 18-mer oligonucleotides (ODN) interacting with liposomes, composed of DOTAP/DOPE/PEG-PE (1:1:0.12, by molar ratios), at a high ODN to lipid ratio of 0.16 ($\pm$ 0.17): Such negatively charged, single-strand, oligonucleotides also act as fusogenic agent similarly to plasmid DNA, drawing together the positively charged liposomes thereby forming small complexes (Figure 7A). Here, dotted boundaries of the complexes are visible (some of them are marked by arrows in Figure 7A) indicating that the lipid bilayers are somehow disrupted probably by the self-encapsulated oligonucleotides. However, no fibrillar structures are visible during interaction with the oligonucleotides as were seen with plasmid DNA. The absence of *spaghetti*-like structures at oligonucleotide/liposome complexes is very likely due to the fact that the diameter of a single-stand oligonucleotide of about 2 nm is too small and they are too short to support the formation of a lipid bilayer tubule.[55]

In support of this statement we studied for comparison, at the interaction of liposomes, made of DOTAP/DOPE/PEG-PE of the same molar ratio, with condensed plasmid DNA (Figure 7B): Here, as we usually observed for plasmid-DNA/liposome complexes, additionally to the aggregated and fused complexes some small fibrillar structures (some of them are marked by arrow heads in Figure 7B) are detectable even with lipids which were added to provide steric stabilization (PEG-PE)[56] and also when the plasmid-DNA was pre-condensed by spermidine before complex formation (see also paragraph 4.6).

IV.5. Composition of the aqueous medium

Up to this point we have investigated CLDC formed in buffer and at low ionic strength. In order to test their *in vitro* activity the complexes have to be applied to cell lines growing in culture media usually containing high concentrations of salts. This led us to a series of investigations where we compared the morphology of complexes exposed to Dulbecco's modified eagle medium (DMEM) as well as to HEPES-buffer (20 mM, pH 7.4). Here the complexes were composed of 15-mer phosphorothioate oligonucleotides (S-oligos) interacting with DOTAP or DOTAP/DOPE liposomes at molar ratio of cationic amphiphile to helper lipid of 1:1 and 1:2 and at a $\pm$ charge ratio of 0.3, 1.2, as well as 2.8.

Similarly to observations described in the previous paragraph, negatively charged, single-strand, short oligonucleotides such as (15-mer) S-oligos, act as a fusogenic agent, drawing together the positively charged liposomes made of DOTAP by itself (not shown) or DOTAP/DOPE 1:1 (control liposomes in

Fig. 7. Freeze-fracture electron micrograph of complexes made of 18-mer oligonucleotides (ODN) interacting with liposomes composed of DOTAP/DOPE/PEG-PE (1:1:0.12, by molar ratios) at a ODN to lipid ratio of 0.16 (A). Small complexes but no fibrillar structures are visible. The boundaries of some of the complexes are dotted and some of these areas are marked by arrows in (A).[55] Complexes made of the same lipid composition but interacting with pre-condensed plasmid DNA (B). Here small complexes and some small fibrillar structures, marked by arrow heads in (B), are visible even with PEG-PE added to provide steric stabilization and when the plasmid DNA was pre-condensed by spermidine before complex formation. Bar represents always 100 nm and the shadow direction is running from bottom to top of the electron micrographs.[51]

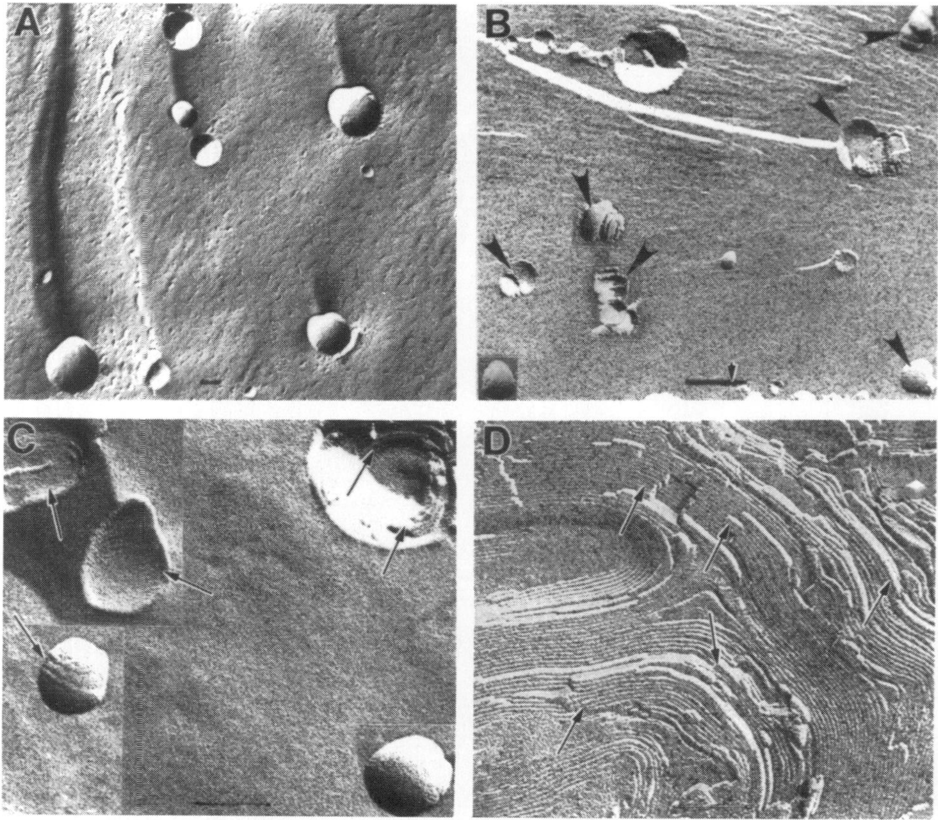

Fig. 8. Freeze-fracture electron micrographs of liposomes made of DOTAP:DOPE (1:**1**, molar ratio; (A) and (C)) interacting with a 15-mer S-oligonucleotide at ± of 1.2 (B) and (D) in HEPES-buffer (pH 7.4; (A) and (B)) and in DMEM (at high ionic strength; (C) and (D)). Some of the small complexes formed in HEPES-buffer are marked by arrow heads in (B). In DMEM even the control lipid mixture, not interacting with the S-oligo, shows some hints for H_{II} phase formation; some of them are marked by arrows in (C).[35] During interaction with the S-oligo extensive H_{II} phase formation is taking place. Some of the areas showing inverted hexagonal H_{II}-tubules are marked by arrows in (D).[35] Bar represents always 100 nm and the shadow direction is running from bottom to top of the electron micrographs.

HEPES-buffer; Figure 8A) thereby forming small complexes (some of the complexes in buffer are marked by arrow heads in Figure 8B). This observation is true for samples diluted in buffer at low ionic strength but the morphology of the complexes is different in DMEM, containing high concentrations of salt and glucose. Here, at high ionic strength, even the control lipid mixture show some hints for hexagonal lipid tubules (some of them are marked by arrows in Figure 8C) even when not interacting with the S-oligo. When interacting with the S-oligo, massive fusion of the lipid particles/liposomes, detected in the control mixture,

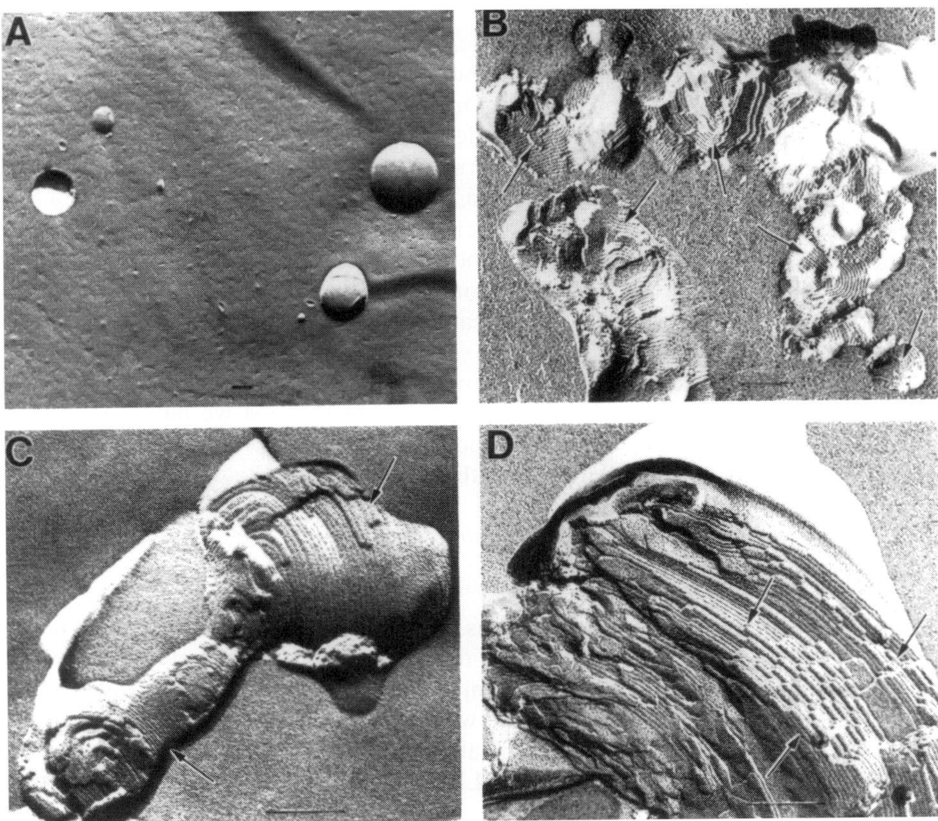

Fig. 9. Freeze-fracture electron micrographs of DOTAP:DOPE (1:**2**, molar ratio; (A) and (C)) interacting with a 15-mer S-oligonucleotide at ± of 1.2 (B) and (D) in HEPES-buffer (pH 7.4; (A) and (B)) and in DMEM (at high ionic strength; (C) and (D). Here, at high DOPE content, fusion of the liposomes, presented in (A) leads to the formation of H_{II}-tubules even in HEPES-buffer at low ionic strength (B). In DMEM even the control lipid mixture, not interacting with the S-oligo (C), as well as interacting with the S-oligo (D) shows extensive H_{II} phase formation. Some of the H_{II}-areas are marked by arrows in (C) and (D).[35] Bar represents always 100 nm and the shadow direction is running from bottom to top of the electron micrographs.

takes place and large extended non-bilayer lipid areas are formed, showing hexagonally packed (H_{II}) tubules (some of them are marked by arrows in Figure 8D).[35]

At a higher DOPE content (DOTAP/DOPE 1:2 by mole), extensive fusion of the liposomes, shown as a control in HEPES-buffer in Figure 9A, and formation of H_{II} lipid tubules are observed as well (some of the H_{II} areas are marked by arrows Figure 9B). In DMEM, a medium of high ionic strength, the lipid mixture itself shows the formation of well developed hexagonal lipid tubules even without interaction with 15-S-oligos (some of the areas displaying H_{II} tubules are marked by arrows in Figure 9C). In this case, the H_{II} formation is even more developed for the lipid mixture while interacting with the 15-S-oligos (some areas of H_{II} tubules are marked by arrows in Figure 9D).[35] This is in agreement with our observations of the formation of H_{II} lipid tubules in DC-Chol/DOPE mixtures with high DOPE content (1:4 by mole) in the presence (Figure 5C) and absence of added plasmid DNA (Figure 5D) as described in paragraph 4.3.1.[31]

The formation of hexagonal lipid structures in DMEM was observed at all DOTAP/DOPE mixtures investigated in the presence of S-oligos at all ± ratios studied ranging from 2.8 (not shown here), 1.2 (Figures 8D and 9D), 0.3 (not shown here) down to 0 (Figures 8C and 9C). This is in good agreement with the observation that fusion of DOPE-containing liposomes as well as H_{II}-phase formation can be triggered by certain cations or conditions inducing membrane contact.[52]

From these results obtained by comparing the morphology of cationic liposomes/nucleotide complexes formed in buffer or in DMEM we have to conclude that it is very important to undertake such studies under physiologically relevant conditions for *in vitro* studies including the cell medium.

IV.6. Effect of lipids providing steric stabilization and of pre-condensation of DNA

All parameters described in the pervious paragraphs, influencing the morphology of CLDC, show clearly that it is difficult to obtain homogeneous preparations with size distribution suitable for systemic injection. In most published studies metastable preparations of CLDC were used within a short period of time ranging from 30 min to few hours.[45,57,46] For clinical trials a solution for this situation has been to mix all components of the CLDC at the bed-side and to apply them immediately.[57]

Since structural instability is connected with the loss of transfection activity, for *in vivo* applications it is highly desired to develop well-defined formulations stable in buffer and in serum over a longer period of time. In a very recent publication, stabilization of cationic-plasmid DNA complexes was achieved by poly(ethylene glycol)-phospholipid conjugates and also by polyamines.[47] In order to study the influence of lipid components providing steric stabilization and the effect of pre-condensation of plasmid DNA on the morphology of CLDC, we investigated DDAB/Chol complexes either containing 1% PEG-PE and interacting with supercoiled plasmid DNA or DDAB/Chol complexes without PEG-PE but interacting with spermidine-condensed plasmid DNA. Furthermore we compared the morphology of these complexes with DC-Chol/DOPE complexes containing PEG-PE following their interaction with spermidine-condensed plasmid DNA (300:75:3:30; by mole) by freeze-fracture electron microscopy.

Small complexes of aggregated and partly fused liposomes as well as fibrillar structures were visible on electron micrographs of complexes made of DC-Chol/DOPE (4:1 by molar ratio) even when 1% PEG-PE was added shortly after interacting the liposomes with the pre-condensed DNA (Figure 10A). These images are similar to those obtained with DC-Chol/DOPE without PEG-PE or

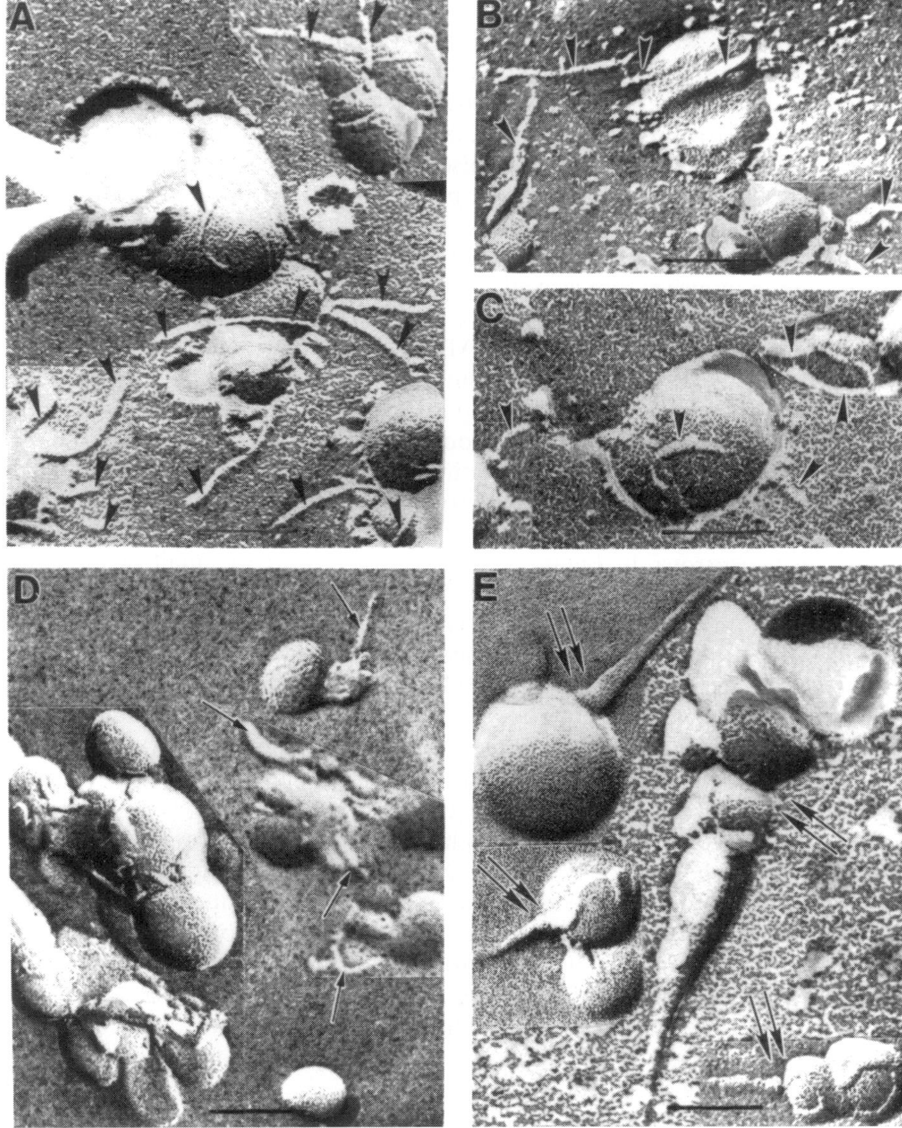

Fig. 10. Cationic liposome/DNA complexes made of DC-Chol : DOPE liposomes (4 : 1 by molar ratio) by interacting with spermidine-condensed plasmid DNA and 1% PEG-PE added, shortly after the interaction with the pre-condensed DNA, for steric stabilization. Small complexes and well developed but less flexible fibrils are observed in buffer (Mes-buffer pH 5.5; some of the fibrils are marked by arrow heads in (A)). When incubated in serum, the complexes are small and fibrils are slightly disintegrated (B). Some of the residual fibrils, found in mouse serum, are marked by arrow heads in (B). When incubated in cell medium the complexes are small but the fibrils are nearly destructed (C). Some of the residual fibrils, found in RPMI-1640 with 10% FCS, are marked with arrow heads in (C).[32]

[Caption continues overleaf.]

condensation with polyamines (Figure 5A). In buffer (MES-buffer pH 5.5) the fibrillar structures were well developed and quite frequent to observe. When spermidine was used for pre-condensation, however, the fibrils appeared less flexible than at complexes without pre-condensed DNA (some of the stiffer fibrils are marked by arrow heads in Figure 10A). The complexes, appear smaller when incubated in serum (Figure 10B) and also in cell medium (RPMI-1640, with 10% FCS, Figure 10C) compared to when suspended in buffer (Figure 10A). The fibrils, well-developed but stiff in buffer, are slightly disintegrated in serum (Figure 10B) and nearly destructed in cell medium (Figure 10C). Some of the residual fibrils still detected in mouse serum as well as in cell medium are marked by arrow heads in Figures 10B and 10D).

Experiments with complexes of DDAB/Chol (1:1, by mole) show the following morphology: The PEG-containing complexes are very similar to the control (Figure 6B) displaying small spherical particles (average diameter in the range of 0.1–0.25 μm) with some few protrusions only (marked by an arrow in Figure 10D). DDAB/Chol complexes interacting with spermidine-condensed DNA are also small, but showing additionally some few but well developed, stiff, and tapering protrusions. The "map-pin" structure of the complexes, as seen and marked by two arrows in Figure 10E, is characterized by small heads showing diameters of one or two liposomes (0.1–0.2 μm) and mainly short (about 0.2 μm) but sometimes up to 0.6 μm long and tapering "pins".[51] Different from the *spaghetti*-like structures these "pins" do not show normal fracture behavior when freeze-fracture technique is applied and they have a much thicker diameter at their base (near to the head) of about 30 nm. Interestingly, condensation methods such as applying spermidine have been found to condense DNA molecules into toroids or rods whereby the rods show rather similar diameters of 30 nm and lengths of 200–300 nm.[17,58] Therefore the tapering pins represent probably exposed but partly condensed DNA rods interacting at one end with the CLDC.[51]

In general, stabilization of the CLDC by adding 1% of PEG-PE does not change the morphology of the complexes (Figure 10D), but pre-condensation of the DNA has quite a remarkable effect and leads to the appearance of longer and stiffer productions (Figures 10A and 10E). Thus, DDAB complexes containing Chol as helper lipid and interacting with pre-condensed DNA tend to form a new structure, the "map pin" structure, as discussed above (Figure 10E).[51] The effect of medium and serum on these structures will be discussed below.

Fig. 10 (Continued). Cationic liposome/DNA complexes in Mes-buffer, pH 5.5, made of DDAB/Chol liposomes (1:1 by molar ratio) interacting with plasmid DNA and 1% PEG-PE added (D) or interacting with spermidine-condensed plasmid DNA but not sterically stabilized by PEG-PE (E). Some fibrillar structures are marked by arrows in (D). Some well developed "map-pin" structures, found at complexes where Chol was used as helper lipid and the plasmid DNA was pre-condensed by spermidine, are marked by two arrows in (E).[51]

V. Relation between morphology and transfection activity of cationic liposome-DNA complexes at studies *in vitro* and *in vivo*

During the last 10 years it was demonstrated convincingly by *in vitro* studies that CLDC can mediate gene delivery by showing detectable expression of a reporter gene in cultured cells.[11,38] More recently, CLDC have also been used for *in vivo* transfection in animals[16,19,45–47,59,60] as well as in humans.[21,61] Compared to recombinant viral vectors CLDC are well tolerated after systemic delivery into animals,[45,62,63] can transfect a wide variety of tissues and cell types,[45,59] do not induce host-immune responses and can therefore transfect also immunocompetent animals,[19] and can deliver very large DNA pieces into cells.[64] Moreover with about 5.5 mg expression per gram of packed cells in culture[62] and 1 μg expression per gram lung tissue[47] trans-gene expression levels have recently approached achievable with adenovirus. There is a growing awareness that CLDC which exhibit high levels *in vitro* activity are not necessarily active also *in vivo*.[62] Based on this observation *in vivo* assays were developed in the past few years to screen for new formulations.[16,19,45–47,59,60]

In a recent collaboration involving both morphology and functional studies,[51] we investigated the transfection activity both *in vivo* and *in vitro* of CLDC composed of DDAB and Chol or DOPE as helper lipids, and studied in parallel their morphology in serum as well as in cell medium. The *in vivo* studies were carried out in mice following i.v. injection and therefore the morphology of the CLDC was investigated in mouse serum. The *in vitro* transfection activity of the CLDC was measured on SK-BR-3 cells and therefore we studied in parallel their morphology in the same medium where these cells were kept. Additionally, the influence of stabilization of the complexes by PEG-PE and the effect of pre-condensation of the plasmid-DNA by spermidine on the transfection activity as well as morphological properties of the CLDC, were investigated under *in vivo* as well as *in vitro* conditions.[51]

When examined in mouse serum, CLDC composed of DDAB and Chol at a 1:1 molar ratio appear as loosely packed aggregates of liposomes where the number of attached liposomes is small, approx. 4–6 (not shown here). Such complexes show high transfection activity in mouse lungs after i.v. injection[57] and we take this level of activity as 100% for further comparisons. Residual fibrillar protrusions are not observed on DDAB/Chol complexes, stabilized by PEG-PE which otherwise look similar (Figure 11A). Interestingly, these stabilized complexes show about half the transfection activity observed with DDAB/Chol complexes where the DNA was pre-condensed by spermidine (Figure 11B). In the case of such liposomes interacting with pre-condensed DNA, the resulting complexes reveal map-pin structures, marked by two arrows in Figure 11B, and also show higher transfection activity *in vivo* (100%). In contrast to the above, the morphology in serum of DDAB complexes containing DOPE as a helper lipid (1:1 molar ratio) is quite different: In this case the liposomes are frequently tightly packed, fused to bigger units, and show a strong tendency to form hexagonal lipid

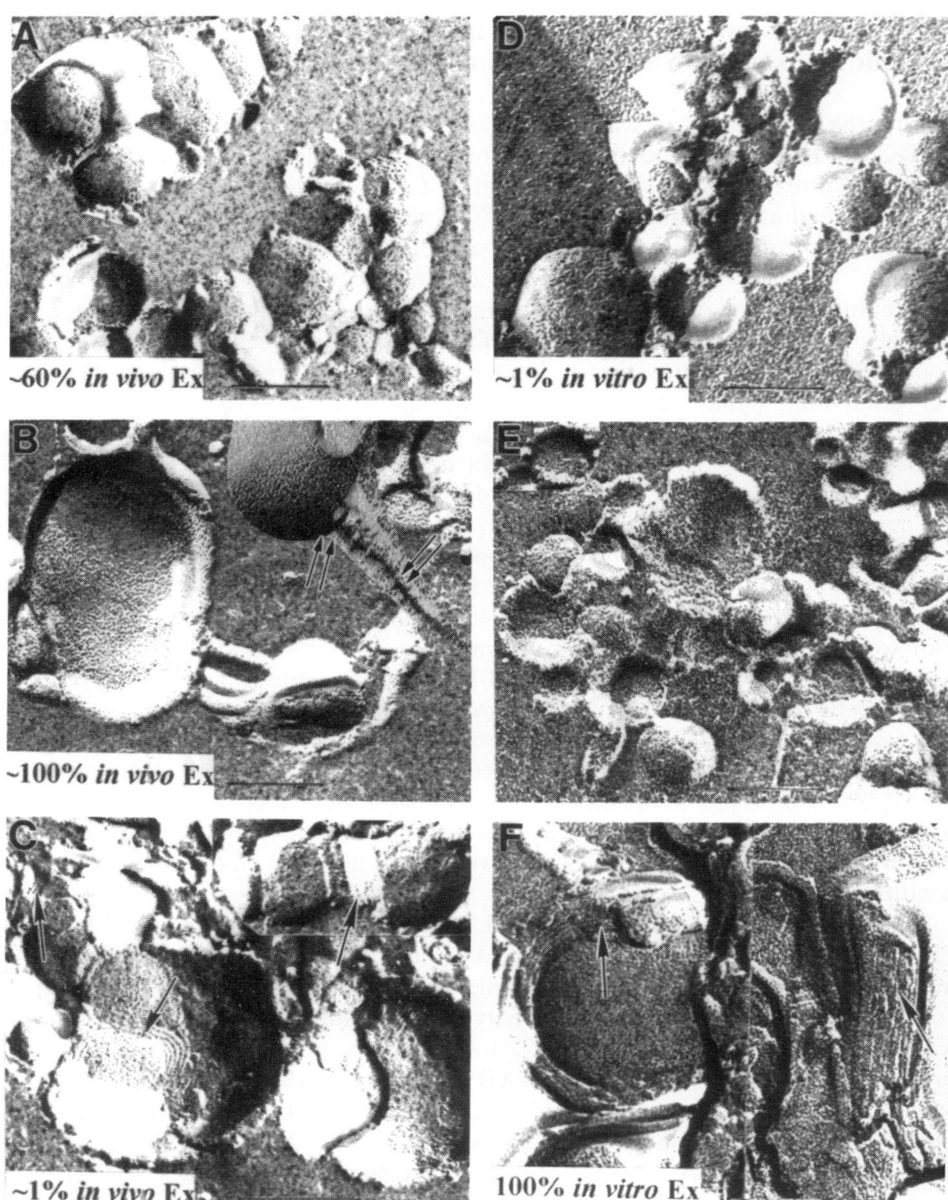

Fig. 11. Cationic liposome/DNA complexes made of DDAB/Chol liposomes (1:1 by molar ratio) interacting with plasmid DNA and 1% PEG-PE added (A) and (D) or interacting with spermidine-condensed plasmid DNA but not sterically stabilized by PEG-PE (B) and (E) incubated in mouse serum (A) and (B) or incubated in cell medium (RPMI-1640 with 10% FCS; (D) and (E). One of the "map-pin" structures, found at complexes where Chol was used as helper lipid and the plasmid DNA was pre-condensed by spermidine and well preserved in mouse serum, are marked by two arrows in (B).[51]

structures (marked with arrows in Figure 11C). No fibrillar structures of any type are observed. Interestingly, these complexes show only 1% of the *in vivo* transfection activity revealed by the DDAB complexes containing Chol as the helper lipid.[51]

When investigated in cell medium (RPMI-1640, with 10% FCS), CLDC containing DDAB and Chol (1:1 molar ratio) are larger than in mouse serum, more tightly packed and show hardly any fibrillar structures (not shown here). This is true for the DDAB/Chol complexes stabilized by PEG-PE (Figure 11D) as well as liposomes interacting with spermidine-condensed DNA (Figure 11E). In contrast, the DDAB/DOPE complexes at 1:1 molar ratio are very large, appear as a lipid precipitate, and show clearly the formation of well developed hexagonally packed lipid H_{II} tubules (marked by arrows in Figure 11F). Contrary to the *in vivo* studies described above the transfection activity in vitro of the DOPE containing complexes is very high while the complexes containing Chol show very little in vitro transfection activity, approx. 1% of the complexes containing DOPE as a helper lipid.[51]

These comparisons reveal that there is a fundamental difference between *in vitro* and *in vivo* activity of CLDC: For *in vitro* activity hexagonal lipid precipitates seem to be associated with high transfection rates, whereas *in vivo* activity seems to be associated with small, serum-stable complexes, connected with fibrillar structures.[51]

VI. Interaction of cationic liposome-DNA complexes with skin culture cells

Cationic liposome-nucleic acid complexes are currently used to deliver oligonucleotides, RNA and DNA into cells[65,66] although the mechanism of interaction and intercalation is still not well understood. There are three possible ways for CLDC to enter a cell: (A) Direct fusion with the plasma membrane,[11] (B) endocytosis.[67,68] or (C) transient lipid-mediated poration.[17,69] Early studies implicated liposome-plasma membrane fusion, but evidences obtained later suggested that liposomes do not fuse with the plasma membrane without perturbations such as polyethylene glycol treatment or the inclusion of viral proteins.[70] Recent results demonstrated clearly that binding to the cell surface is insufficient for CLDC-cell fusion and that uptake into the endocytic pathway is required.[67,68] In the case of endocytosis quick release from the endosomes is essential to protect the DNA from lysosomal degradation.[68]

To study the interaction of CLDC with cells in more detail, we used cultured human keratinocytes (HaCaT cells[71]) as a test model[72] and incubated them with

Fig. 11 (Continued). Cationic liposome/DNA complexes made of DDAB/DOPE liposomes (1:1 by molar ratio) interacting with plasmid DNA incubated in mouse serum (C) or cell medium (F) of the same composition as at (D) and (E). In (C) hints for the formation of H_{II} lipid phase and in (F) areas with well developed H_{II} tubules are marked by arrows.[51]

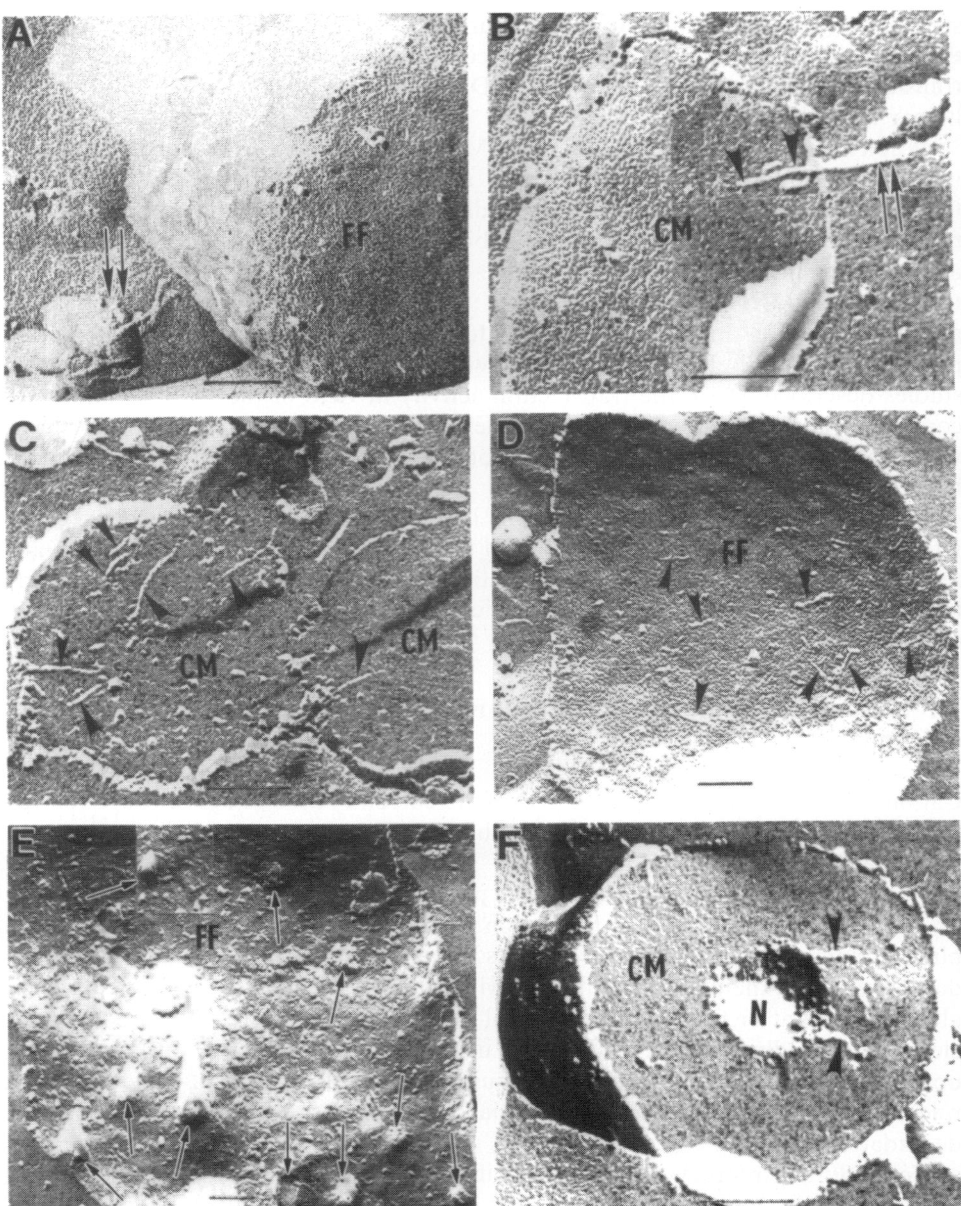

Fig. 12. Freeze-fracture electron micrographs of 1-day-old HaCaT cells, treated with cationic liposome/DNA complexes made of DC-Chol/DOPE liposomes (3:2 molar ratios) interacting with plasmid DNA. The *spaghetti/meatball*-complexes, formed under these conditions, are marked by double arrows in (A) and (B). The fracture faces of the HaCaT-cells are labeled with FF in (A), (D), and (E) and the cytoplasm, exposed at cross fractions of the cells, is labeled by CM in (B), (C), and (F).

After short-time treatment (for 30 min; (C) and (D)) free spaghetti-like structures, marked by arrow heads, are visible inside a cross-fractured HaCaT-cell (C) as well as at their fracture faces complexes, composed of plasmid DNA and DC-Chol/DOPE (3:2 molar ratio). As shown before[23,30–33,36] these CLDC display long fibrillar structures representing a single bilayer tube coating the supercoiled plasmid DNA. Taking this *spaghetti*-like structure as a morphological marker it is relatively easy to distinguish between *spaghetti-meatball*-complexes (two of them are marked by two arrows in Figures 12A and 12B) and cells (the immortalized human keratinocytes; the fracture faces of the cell membranes are marked by FF in Figures 12A, 12D and 12E).[30] In a preliminary study we looked for structural transformations involved in the interaction process of the CLDC with the HaCaT-cells in a time range between 10 min and up to 4 h by freeze-fracture electron microscopy

After short incubation times of 10 or 30 minutes we could frequently observe free *spaghetti*-like structures, not attached to liposomes, intact and inside of the cells (the cytoplasm inside of some cells is marked by CM in Figures 12B, 12C and 12F),[30,32] although these are very small structures with a diameter of about 7 nm and a length of approx. 100 nm as measured by freeze-fracture technique.[23,31] Here, some of the "free" spaghetti are marked by arrow-heads in Figure 12C, inside a cross-fractured HaCaT cell. They are also attached intact to the fracture faces of the cell bilayer as shown in Figure 12D.[30,32]

After longer incubation time (2 hours) endocytosis of the CLDC seems to take place (some endocytosis-events are marked by arrows in Figure 12E). *Spaghetti*-like structures as well as *spaghetti-meatball*-complexes, which were observed during our investigation, may still bear residual positive charges on their surfaces. This may also lead to an interaction with the cell membranes, thereby promoting the transfer of the DNA into the cytoplasm. Because of the high radius of curvature especially at the tips of the thin fibrillar structures, attachment and local fusion might take place[34] allowing the entrance and passage through the cell membrane of the whole intact fibre. A part of a *spaghetti*-like structure reached intact the cytoplasm of a HaCaT-cell is marked by arrow heads in Figure 12B. Eventually these fibrillar structures, containing presumably the DNA protected inside a bilayer tubule, end up at the nucleus (N) of the cell (marked by arrow heads in Figure 12F).[32] Indeed, in several different CLDC, with fibrillar protrusions of high radii curvature were associated with higher *in vitro*[69] as well as *in vivo* tansfection activity.[51]

Fig. 12 (Continued). (C).[30,32] After a longer incubation time (for 2 h; (A), (B), (E), and (F)), *spaghetti/meatball*-complexes, bigger and slower moving than the free *spaghetti*, are approaching the cells (A), interacting with them via their attached *spaghetti* (B, part of a spaghetti, still intact after crossing the cell membrane, is marked by arrow heads) or are taken up via endocytosis (E).[32] Endocytosed complexes, marked by arrows in (E) look similar to endocytosed liposomes.[72] Obviously, *spaghetti*, free or attached to the complexes, are able to cross the cell membrane intact (B) and (D) and stay intact in the cytoplasm inside the cells (B), (C) and (F). Eventually these fibrillar structures, (marked by arrow heads in (F)), end up at the nucleus (N) of the cell.[32]

VII. Concluding remarks

Complexes formed during interaction of cationic liposomes with plasmid DNA or oligonucleotides display a variety of polymorphic and metastable structures. These include smaller or larger, looser or tighter packed aggregates and semi-fused or fused products of the two oppositely charged partners; fibrillar structures, free or attached to the complexes, among them *spaghetti*-like tubules, displaying a fine structure like a train-track, or small protrusions and "map-pin" structures; and, last but not least, hexagonal (H_{II})-lipid arrangements. Conditions were found favoring the formation of an individual type from the large variety of the polymorphic structures such as: charge neutralization, incubation time, valency of the cationic component, type and ratio of the helper lipid, type of the nucleotide component and degree of condensation, as well as ionic strength of the aqueous medium. Moreover, the stability of the structures formed is strongly dependent on the charge conditions of the complexes in a medium of certain ionic strength and serum concentration, on storage time, and on the proportion of components providing steric stabilization. Electron microscopic techniques such as freeze-fracture and cryo-electron microscopy as well as negative staining provide an excellent tool to study the conditions favoring formation of individual structures, and those supporting their stability or leading to structural transitions.

Small complexes (<300 nm), consisting of few aggregated or semi-fused liposomes mainly of the original size are frequently observed in media of low ionic strength and in freshly prepared samples, especially at a slightly net positive charge (excess lipid to DNA charge) and short incubation times. They contain presumably condensed, self-encapsulated plasmid DNA or short, single-strand antisense oligonucleotides.

Additionally, fibrillar *spaghetti*-like structures, free or connected with these complexes are observed under certain conditions such as using supercoiled plasmid DNA, and *mono*-valent cationic amphiphiles mixed with DOPE at a molar ratio of <1.5. The evidence supports the hypothesis of a single bilayer tube coating the supercoiled plasmid DNA. Thus, the diameter is 13 nm as measured on electron micrographs prepared by negative staining technique or cryo-electron microscopy, and approximately 7 nm as measured on freeze-fracture electron micrographs. Further support is given by the finding of convex and concave fracture planes from these fibrillar *spaghetti*-like structures. Using negative staining and cryo-electron microscopy as well as image analysis it became apparent that the bilayer-coat is arranged in toroidal segments around the supercoiled plasmid DNA leading to a train-track-like fine structure. Additionally, folded bilayer arrangements are detectable with both electron microscopic techniques especially in larger CLDC (>200 nm) showing an interlayer spacing consistent with a recently described multilamellar structure of alternating lipid bilayers and DNA monolayers.[26,27] We observed co-existence of these layered lipid/DNA arrangements with the fibrillar *spaghetti*-like structures.

Fibrillar structures are also seen in some cases as short protrusions on the liposome surface, especially when plasmid DNA interacts with the cationic lipids

DDAB or DTAP mixed with Cholesterol as a helper lipid. However, no fibrillar structures are visible during interaction of cationic liposomes with negatively charged oligonucleotides. This is possibly due to the fact that the diameter of a single-strand oligonucleotide of about 2 nm is too small and they are also too short to promote the formation of a lipid bilayer tubule.

An important parameter that affects the morphology of the CLDC is the medium in which they are suspended. Most of the remarks above are based on observations obtained in low ionic strength buffer. Our studies indicate large changes obtained in the presence of culture media and, to a lesser extend, serum: For example, fibrillar structures are not visible at CLDC containing high proportions of the helper lipid DOPE (DOPE:cationic lipid >1.5 molar ratio) and especially in aqueous media at high ionic strength, such as cell media. In this case, massive fusion of the liposomes takes place and large extended non-bilayer lipid areas are formed, showing hexagonally packed (H_{II}) tubules.

Preparations of CLDC are metastable with time and their structural instability is connected with the loss of *in vivo* transfection activity. In order to keep the size distribution suitable for systemic injection, stabilization of CLDC was achieved by using Cholesterol instead of DOPE as helper lipid, by adding poly(ethylene glycol)-phospholipid conjugates, and also by pre-condensation of plasmid DNA using polyamines. Furthermore "map-pin" structures showing tapered rods of a length of 200–300 nm were observed in complexes containing some cationic lipids mixed with Cholesterol and interacting with pre-condensed DNA. Comparison of *in vitro* transfection activity of CLDC measured on SK-BR-3 cells with their *in vivo* transfection activity expressed in mouse lung following i.v. injection revealed a fundamental difference: Hexagonal lipid precipitates and, in some cases, fibrillar *spaghetti*-like structures seem to be associated with high transfection rates *in vitro*, whereas *in vivo* activity seems to be associated with small, serum-stable complexes, connected with short fibrillar structures, appearing like protrusions, or "map-pin" structures.

Studies on the interaction between CLDC and a variety of cultured mammalian cells *in vitro*, including skin culture cells, showed frequently endocytosis events after incubation times of 2–4 hours. However, after short incubation times of 10–30 minutes, fibrillar *spaghetti*-like structures were frequently observed intact and inside the cells. It is plausible that attachment and local fusion might be taking place through the residual positive charges and the high radii of curvature especially at the tips of the thin fibrillar structures. This may allow the entrance and passage through the cell membrane of the whole intact fibre. Such fibrillar structures, presumably containing the DNA protected inside a bilayer tubule, were also observed attached to the nucleus of the cell.

Abbreviations

CLDC: cationic liposome/DNA complexes;
DC-Chol: 3ß-[N-(N′,N′-dimethylaminoethane)-carbamoyl] cholesterol;
DOPE: 1,2-dioleoyl-*sn*-glycero-3-phosphoethanolamine;

DOTMA: N-[1-(2,3-dioleoyloxy)propyl]-N,N,N-trimethylammonium chloride;
DMRIE: N,N-dimethyl-1,2-dimyristoyloxy-3-aminopropane;
DOTAP: (N[1-(2,3-dioleoyloxy)propyl]-N,N,N, trimethylammoniummethylsulfate;
DDAB: dimethyl-dioctadecylammonium bromide;
DOSPA: 2,3-dioleoyloxy-N-[2(sperminecarboxamido)ethyl]-N,N-dimethyl-l-propanammonium trifluoroacetate;
CTAB: cetyl-trimethylammonium bromide;
Chol: cholesterol; H_{II}: inverse hexagonal lipid phase;
HEPES: N-(2-hydroxyethyl)-piperazine-N′-(2-ethane-sulphonic acid);
MES: 2-N-(morpholino)ethane-sulfonic acid:
PBS: phosphate-buffered saline: DMEM: Dulbecco's Modified Eagle Medium;
PEG-PE: N-[ω-methoxypoly(oxyethylene)-α-oxycarbonyl-DSPE;
FCS: fetal calf serum.

Acknowledgements

I wish to thank all collaborators in this work; especially Professor Leaf Huang and Dr. Frank L. Sorgi at University of Pittsburgh for providing the liposomes made of DC-Chol in all compositions with DOPE, Lipofectin and LipofectAMINE, and the plasmid pRSV-LUC; Dr. Philip Felgner at Vical Inc. for providing the cationic amphiphiles DOTMA, DMRIE, and DOSPA for liposome and complex formation; Dr. Ilpo Jääskeläinen and Dr. Jukka Mönkkönen at Kuopio University for providing the DOTAP liposomes in all compositions with DOPE and the 15-mer phosphorothioate oligonucleotides; Drs Keelung Hong and Weiwen Zheng at UCSF for providing the DDAB liposomes in all compositions with Chol or DOPE as helper lipids (some of them stabilized with PEG-PE interacting with pre-condensed plasmid DNA), Dr. Olivier Meyer at UCSF for providing DOTAP/DOPE/PEG-PE liposomes interacting with an 18-mer oligonucleotide; and Dr. C. Böttcher at Free University Berlin for looking at some of the complexes by negative staining and cryo-electron microscopy. I also thank Mrs. I.-M. Herrmann and Mrs. R. Kaiser for technical assistance in freeze-fracture, Mrs. G. Engelhardt and Mrs. G. Vöckler for their phototechnical work, and PhD-students U. Strohbach and M. Müller (all working at the Friedrich-Schiller-University Jena) for designing and modifying Figures 2 and 3 on the computer. I am grateful to Professor D. Papahadjopoulos at UCSF for many helpful discussions.

References

1. Wilson T, Papahadjopoulos D, Taber R. Biological properties of poliovirus encapsulated in lipid vesicles: Antibody resistance and infectivity in virus-resistant cells. Proc Natl Acad Sci USA 1977;74:3471–3475.
2. Dimitriadis GT. Translation of rabbit globin mRNA introduced by liposomes into mouse lymphocytes. Nature 1978;274:923–924.
3. Dimitriadis GT. Entrapment of plasmid DNA in liposomes. Nucleic Acids Res 1979;6:2697–2705.

4. Hoffman RM, Margolis LB, Bergelson LD. Binding and entrapment of high molecular weight DNA by lecithin liposomes. FEBS Lett 1978;93:365–368.
5. Lurquin PF. Entrapment of plasmid DNA by liposomes and their interactions with plant protoplasts. Nucleic Acids Res 1979;6:3773–3784.
6. Mannino RJ, Allenbach ES, Strohl WA. Encapsulation of high molecular weight DNA in large unilamellar phospholipid vesicles. Dependence on the size of the DNA. FEBS Lett 1979;101:229–232.
7. Fraley R, Subramani S, Berg P, Papahadjopoulos D. Introduction of liposome-encapsulated SV40 DNA into cells. J Biol Chem 1980;255:10431–10435.
8. Wong T, Nicolau C, Hofschneider PH. Appearance of -lactamase activity in animal cells upon liposome-mediated gene transfer. Gene 1980;10:87–94.
9. Crystal RG. Transfer of genes to humans: Early lessons and obstacles to success. Science 1995;270:404–410.
10. Mulligan RC. The basic science of gene therapy. Science 1993;260:926–932.
11. Felgner PL, Gadek TR, Holm M, Roman R, Chan HW, Wenz M, Northrop JP, Ringold GM, Danielsen M. Lipofection: A highly efficient, lipid-mediated DNA-transfection procedure. Proc Natl Acad Sci USA 1987;84:7413–7417.
12. Felgner PL, Ringold GM. Cationic liposome mediated transfection. Nature 1989;331:461–462.
13. Felgner PL. Particulate systems and polymers for *in vitro* and *in vivo* delivery of polynucleotides. Adv Drug Del Rev 1990;5:163–187.
14. Felgner PL, Rhodes G. Gene therapeutics. Nature 1991;349:351–352.
15. Gao XA, Huang L. A novel cationic liposome reagent for efficient transfection of mammalian cells. Biochem Biophys Res Commun 1991;179:280–285.
16. Solodin I, Brown CS, Bruno MS, Chow C-Y, Jang E-H, Debs RJ, Heath TD. High efficiency *in vivo* gene delivery with a novel series of amphilic imidazolinium compounds. Biochemistry 1995;34:13537–13544.
17. Lasic DD, Templeton NS. Liposomes in gene therapy. Advanced drug delivery reviews 1996;20:221–266.
18. Plautz GE, Yang ZY, Wu B, Gao X, Huang L, Nabel GJ. Immunotherapy of malignancy by *in vivo* gene transfer into tumors. Proc Natl Acad Sci USA 19??;90:4645–4649.
19. Liu Y, Liggitt D, Zhong W, Tu G, Gaensler K, Debs RJ. Cationic liposome-mediated intravenous gene delivery. J Biol Chem 1995;270:24864–24870.
20. Stribling R, Brunette EB, Liggitt DL, Gaensler K, Debs RJ. Aerosol gene delivery *in vivo*. Proc Natl Acad Sci USA 1992;89:11277–11281.
21. Caplen NJ, Alton EWFW, Middleton PG, Dorin JR, Stevenson BJ, Gao X, Durham SR, Jeffery PK, Hodson ME, Coutelle C, Huang L, Porteous DJ, Willjamson R, Geddes DM. Liposome-mediated CFTR gene transfer to the nasal epithelium of patients with cystic fibrosis. Nature Medicine 1995;1:39–46.
22. Gershon H, Ghirlando R, Guttman SB, Minsky A. Mode of formation and structural features of DNA-cationic liposome complexes used for transfection. Biochem 1993;32:7143–7151.
23. Sternberg B, Sorgi FL, Huang L. New structures in complex formation between DNA and cationic liposomes visualized by freeze-fracture electron microscopy. FEBS Lett 1994;356:361–366.
24. Gustafsson J, Arvidson G, Karlsson G, Almgreen M. Complexes between cationic liposomes and DNA visualized by cryo-TEM. Biochm Biophys Acta 1995;1235:305–312.
25. Felgner PL, Tsai YJ, Felgner JH. Advances in the design and application of cytofectin formulations. In: Lasic DD, Berenholz Y, eds. Handbook of Nonmedical Applications of Liposomes. IV: From Gene Delivery and Diagnostics to Ecology. Boca Raton, FL: CRC Press, 1996;43–56.
26. Lasic DD, Strey H, Stuart MCA, Podgornik R, Frederik PM. The structure of DNA-liposome complexes. J Amer Chem Soc 1997;119:832–833.
27. Rädler JO, Koltover I, Salditt T, Safinya CR. Structure of DNA-cationic liposome complexes: DNA intercalation in multilamellar membranes in distinct interhelical packing regimes. Science 1997;275:810–814.
28. Szoka Jr FC, Xu Y, Zelphati O. How are nucleic acids released in cells from cationic lipid-nucleic acid complexes? J Liposome Res 1996;6:567–589.
29. Sternberg B. Freeze-fracture electron microscopy of liposomes. In: Gregoriadis G, ed. Liposome Technology. 2nd Ed, Boca Raton, FL: CRC Press, 1992;21:363–383.
30. Sternberg B. Liposome as a model for membrane structures and structural transformations: A liposome album. In: Lasic DD, Barenholz Y, eds. Handbook of Nonmedical Applications of Liposomes. IV: From Gene Delivery and Diagnostics to Ecology. Boca Raton, FL: CRC Press, 1996;20:271–297.

31. Sternberg B. Morphology of cationic liposome/DNA complexes in relation to their chemical composition. J Liposome Res 1996;6:515–533.
32. Sternberg B. Böttcher C, Stark H. Fine-structure of cationic liposome/DNA complexes and their interaction with cells; in preparation.
33. Sorgi FL, Sternberg B, Huang L. Interaction of DNA with Liposomes Containing Different Types of Cationic Amphiphiles, in preparation.
34. Bangham AD. Surrogate cells or trojan horses. The discovery of liposomes. BioEssays 1995;17:1081–1088.
35. Jääskeläinen I, Sternberg B, Mönkkönen J, Urtti A. Physicochemical and morphological properties of complexes made of cationic liposomes and oligonucleotides. Accepted by Intern J Pharmac.
36. Sternberg B. Böttcher C. Electron microscopic examinations of monovalent and polyvalent cationic liposome-DNA complexes. In preparation.
37. Behr J. Gene transfer with synthetic cationic amphiphiles; Prospect for gene therapy. Bioconjugate Chem 1994;5:382.
38. Felgner JH, Kumar R, Sridhar CN, Wheeler CJ, Tsai YJ, Border R, Ramsey P, Martin M, Felgner PL. Enhanced gene delivery and mechanism studied with a novel series of cationic lipid formulations. J Biol Chem 1994;269:2550–2561.
39. Farhood H, Serbina N, Huang L. The role of dioleoyl phosphatidylethanolamine in cationic liposome mediated gene transfer. Biochim Biophys Acta 1995;1235:289–295.
40. Boggs JM. Intermolecular hydrogen bonding influence on structural organization and membrane function. Biochim Biophys Acta 1987;906:353–404.
41. Israelachvili NJ, Marcelja S, Horn RG. Physical principle of membrane organization. Quart Rev Biophys 1980;13:121–200.
42. Pinnaduwage P, Schmitt L, Huang L. Use of a quaternary ammonium detergent in liposome mediated DNA transfer of mouse L-cells. Biochim Biophys Acta 1989;985:33–37.
43. Hui SW, Langner M, Zhao Y-L, Ross P, Hurley E, Chan K. The role of helper lipids in cationic liposome-mediated gene transfer. Biophys J 1996;71:590–599.
44. Litzinger DC, Huang L. Phosphatidylethanolamine liposomes: drug delivery, gene transfer and immunodiagnostic applications. Biochim Biophys Acta 1992;1113:201–227.
45. Zhu N, Liggitt HD, Liu Y, Debs R. Systemic gene expression after intravenous DNA delivery into adult mice. Science 1993;281:209–211.
46. Liu Y, Mounkes LC, Liggitt HD, Brown CS, Solodin I, Heath TD, Debs RJ. Factors influencing the efficiency of cationic liposome-mediated intravenous gene delivery. Nature Biotechnology 1997;15:167–173.
47. Hong K, Zheng W, Baker A, Papahadjopoulos D. Stabilization of cationic liposome-plasmid DNA complexes by polyamines and poly(ethylene glycol)-phospholipid conjugates for efficient *in vivo* gene delivery. FEBS Lett 1997;400:233–237.
48. Cullis PR, De Kruijff B. Lipid polymorphism and functional roles of lipids in biological membranes. Biochim Biophys Acta 1979;559:399–420.
49. Seddon JM. Structure of the inverted hexagonal $H_{(II)}$ phase, and non-lamellar phase transitions of lipids. Biochim Biophys Acta 1990;1031:1–69.
50. Felgner PL. Structural and functional aspects of cytofectin mediated gene delivery. Vancouver Liposome Research Days Conference 1994;19.
51. Sternberg B, Hong K, Zheng W, Papahadjopoulos D. Relation between morphology and transfection activity of cationic liposome-DNA complexes; submitted.
52. Allen TM, Hong K, Papahadjopoulos D. Membrane contact, fusion, and hexagonal (H_{II}) transitions in phosphatidylethanolamine liposomes. Biochemistry 1990;29:2976–2985.
53. Stein CA and Cheng Y-C. Antisense oligonucleotides as therapeutic agents—is the bullet really magical? Science 1993;261:1004–1011.
54. Malone RW, Felgner PL, Verma IM. Cationic liposome-mediated RNA transfection. Proc Natl Acad Sci USA 1989;86:6077–6081.
55. Meyer O, Kirpotin D, Hong K, Sternberg B, Park JW, Woodle MC, Papahadjopoulos D. Cationic liposome coated with poly(ethylene glycol) as carriers for oligonucleotides. submitted
56. Sternberg B, Hong K, Zheng W, Papahadjopoulos D. Steric stabilization of cationic liposome-DNA complexes: Influence of morphology and transfection activity. In: Gregoriadis G ed. Targeting of Drugs 6: Strategies for Stealth Therapeutic Systems, Plenum Press, 1998 in press.
57. Gao X, Huang L. Cationic liposome-mediated gene transfer. Gene Ther 1995;2:710–722.
58. Lasic DD, Papahadjopoulos D, Podgornik R. Polymorphism of lipids, nucleic acids and their interactions. In: Kabanow AV, Seymour LW, Felgner PL, ed. Self-Assembling Complexes for

Gene Delivery: From Chemistry to Clinical Trial. New York & Chichester: J Wiley, 1997, in preparation.

59. Thierry AR, Lunardi-Iskandar Y, Bryant JL, Robinovich P, Gallo RC, Mahan LC. Systemic gene therapy: biodistribution and long-term expression of a transgene in mice. Proc Natl Acad Sci USA 1995;92:9742–9746.
60. Stephan DJ, Yang Z-Y, Simari RD, San H, Wheeler CJ, Felgner PL, Gordon D, Nabel GJ, Nabel EG. A novel cationic liposome DNA complex enhances the efficiency of arterial gene transfer *in vivo*. Human Gene Ther 1996;7:1803–1813.
61. Nabel GJ, Nabel EG, Yang Z-Y, Fox BA, Plautz GE, Gao X, Huang L, Shu S, Gordon D, Chang AE. Direct gene transfer with DNA-liposome complexes in melanoma: expression, biological activity, and lack of toxicity in humans. Proc Natl Acad Sci USA 1993;90:1130–1138.
62. Felgner PL. Improvements in cationic liposomes for *in vivo* gene transfer. Human Gene Ther 1996;7:1791–1793.
63. Canonico AE, Plitman JD, Conary JT, Meyrick BO, Brigham KL. No lung toxicity after repeated aerosol or intravenous delivery of plasmid-cationic liposome complexes. J Appl Physiol 1996;77:415–419.
64. Strauss WM, Dawsman J, Beard C, Johnson C, Lawrence JB, Jaenisch R. Germ line transmission of a yeast artificial chromosome spanning the murine alpha 1 (I) collagen locus. Science 1993;259:1904–1906.
65. Lasic DD, Papahadjopoulos D. Liposomes revisited. Science 1995;267:1275–1276.
66. Lasic DD. Liposomes in gene therapy. In: Lasic DD, Barenholz Y, eds. Handbook of Nonmedical Applications of Liposomes. IV: From Gene Delivery and Diagnostics to Ecology. Boca Raton, FL: CRC Press, 1996;20:1–5.
67. Wrobel I, Collins D. Fusion of cationic liposomes with mammalian cells occurs after endocytosis. Biochim Biophys Acta 1995;1235:296–304.
68. Friend DS, Papahadjopoulos D, Debs RJ. Endocytosis and intracellular processing accompanying transfection mediated by cationic liposomes. Biochim Biophys Acta 1996;1278:41–50.
69. Xu Y, Hui SK, Szoka FC. Effect of lipid composition and lipid-DNA charge ratios on physical properties and transfection activity of cationic lipid-DNA complexes. Biophys J 1995;68:A432.
70. Szoka F, Magnusson K-E, Wojcieszyn J, Hou Y, Derzko Z, Jacobson K. Use of lectins and polyethylene glycol for fusion of glycolipid-containing liposomes with eukaryotic cells, Proc Natl Acad Sci USA 1981;78:1685–1689.
71. Boukamp P, Petrussevska RT, Breitkreutz D, Hornung J, Markham A, Fusenig NE. Normal keratinization in a spontaneously immortalized aneuploid human keratinocyte cell line. J Cell Biol 1988;106:761–771.
72. Prüfer K, Merz K, Barth A, Wollina U, Sternberg B. Interaction of liposomal incorporated vitamin D3-analogues and human keratinocytes. J Drug Target 1994;2:419–429.

Lasic and Papahadjopoulos (eds.), Medical Applications of Liposomes

CHAPTER 5.5

Liposomal antisense oligonucleotide therapeutics

M.C. Woodle[a] and L. Leserman[b]

[a]*Genetic Therapy Inc., Gaithersburg, MD 20878 USA*; [b]*Centre d'Immunologie de Marseille-Luminy, Case 906, 13009 Marseille, France*

Overview

Abstract

Antisense oligonucleotides are a potentially important new class of therapeutic agents. These interact with specific mRNA sequences by Watson-Crick base pairing, resulting in reduced synthesis of the proteins those RNAs encode. Parenteral administration of native (phosphodiester) oligonucleotides is not feasible because of rapid degradation by nucleases. Phosphorothioate analogues of oligodeoxynucleotides (PS) have been shown to exhibit therapeutic activity in animal models of disease but require relatively high doses and/or undesirable means of administration, such as slow infusion. Structural modifications of nucleotides also increase undesired association with proteins and with nucleotide sequences that are imperfectly complementary, potentially increasing toxicity. Conventional liposomes might be useful for transport of oligonucleotides by reducing the administered dose. Recent work with sterically stabilized liposomes demonstrates that pharmacokinetic properties of oligonucleotides can be determined by the liposomes with which they are associated. Advances in oligonucleotide analogue chemistry and lipid formulations providing intracellular delivery should further expand the therapeutic application of antisense agents, especially if this permits minimally modified oligonucleotides to be used therapeutically by reducing their exposure to nucleases.

I. Introduction: Antisense oligonucleotide therapeutic agents

Oligonucleotides are being developed as therapeutic agents to selectively alter genetic functions through sequence specific interactions with intracellular RNA or DNA. This may occur by a variety of mechanisms including what are called antisense, ribozyme, triplex and decoy, of which only the former will be discussed in this review. When an oligonucleotide sequence is complementary by Watson-Crick base pairing to RNA, the oligonucleotide sequence is referred to as "antisense". Antisense oligonucleotides generally reduce protein production by inhibitory interactions with mRNA or its precursors. Oligonucleotide sequences of about 15–20 bases in length are considered to be specific for a single gene target, at least on a statistical basis. The ability of oligonucleotides to discriminate RNA sequences which differ by a single base has been demonstrated both in vitro and in vivo.[1] Antisense oligonucleotide binding to RNA can interfere with cellular processing required for protein synthesis by several mechanisms. The greatest inhibition may be through the use of DNA or analogues that can catalyze degradation of the RNA through activation of RNase H, an enzyme normally present in most cells which degrades RNA in DNA/RNA duplexes. This and other antisense mechanisms are described in many recent reviews.[2–8] Additional approaches include expression of antisense sequences from plasmids or viral vectors.[9,10] In these examples part of the gene sequence is inverted and an mRNA in antisense orientation is transcribed from the introduced gene.

Many studies have now demonstrated in vitro that selective inhibition of proteins can be achieved by antisense oligonucleotides complementary to sites on the target mRNA. However, with only a few exceptions, these in vitro results indicate a strong dependence on other agents to facilitate intracellular delivery of antisense oligonucleotides.[11,12] Some studies indicate that significant differences exist between cell culture (in vitro systems requiring exogenous agents facilitating intracellular delivery) and in animals (in vivo results which are independent of these agents).[12–16] Despite rapid progress in the elucidation of the mechanism(s) of action of antisense oligonucleotides in vitro, adequate measures to identify the exact mechanism of action in vivo are difficult to perform[17] and many have been inadequately interpreted.[16] Thus a continuing question is the extent to which successful formulations for antisense therapeutics will be needed to enhance cellular internalization by the target cells.[17–21]

Phosphorothioates (PS) are one of the oligonucleotide chemical analogues now being studied most actively for use as therapeutic agents. In the PS modification, sulfur is substituted for one of the two unesterified oxygens in the phospho-ribose backbone, shown in Figure 1. This modification confers stability to degradation by nucleases with retention of the ability to activate RNase H cleavage of complementary RNA. PS oligonucleotides are readily water soluble and chemically stable, important properties for therapeutic agents which permit initial evaluation simply as aqueous solutions or in more complex formulations such as liposomes. Importantly, methods for PS production exist at relatively large scale and low cost. Studies showing evidence of efficacy in animal models of disease have fostered PS as one of the most promising antisense analogues.

Other chemical analogues, in particular a form referred to as methylphosphonates (MP), also shown in Figure 1, have received considerable attention. However, this form has significant limitations: methylphosphonates are only sparingly soluble in water due to the neutral phospho-ribose "backbone"; they are not able to

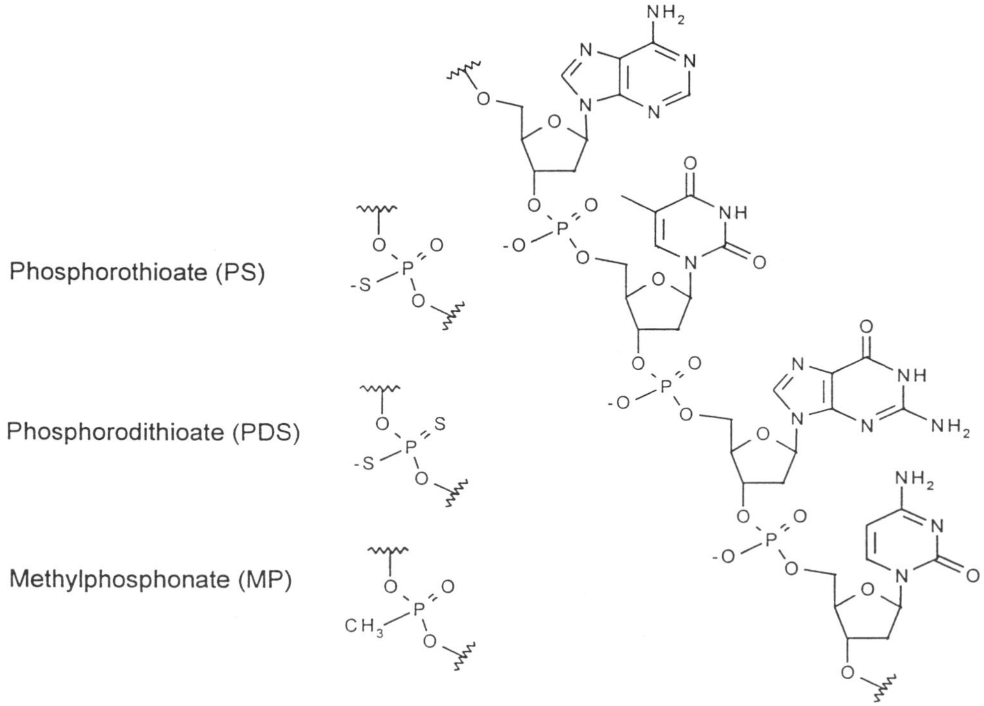

Phosphorothioate (PS)

Phosphorodithioate (PDS)

Methylphosphonate (MP)

Fig. 1. The chemical structure of phosphorothioate and other common chemical analogues of oligodeoxynucleotides with each of the four common bases found in DNA.

activate RNase H to catalytically degrade complementary RNA; and they require the same chiral form of the methyl group at each linkage for hybridization to target RNA under physiological conditions. Nonetheless, chiral forms prepared by use of an alternating chemistry, shown in Figure 2, and use of blocks of different chemical analogues, chimera or hybrid analogues, promise to provide significant improvements.[22] However, these and numerous other oligonucleotide chemical analogues have yet to be actively studied in formulations beyond saline solutions. Biological effects of oligonucleotides made with more recently developed nucleotide analogues are just beginning to appear in the literature.[23] Therefore, this chapter emphasizes primarily PS analogues.

PS oligonucleotides are more stable than DNA in vivo but they are rapidly cleared from plasma, distribute widely to most tissues, and are metabolized over periods of a few hours to a few days. The initial rapid plasma clearance, with a half life of a few minutes, is followed by a relatively long elimination half life of 10 to 40 hr depending on the study. These data are concordant for many species, including humans. This elimination phase appears to be indicative of intact oligonucleotide as well as metabolic product efflux from sites of distribution back into the plasma. Metabolic products rather than intact oligonucleotide are excreted in

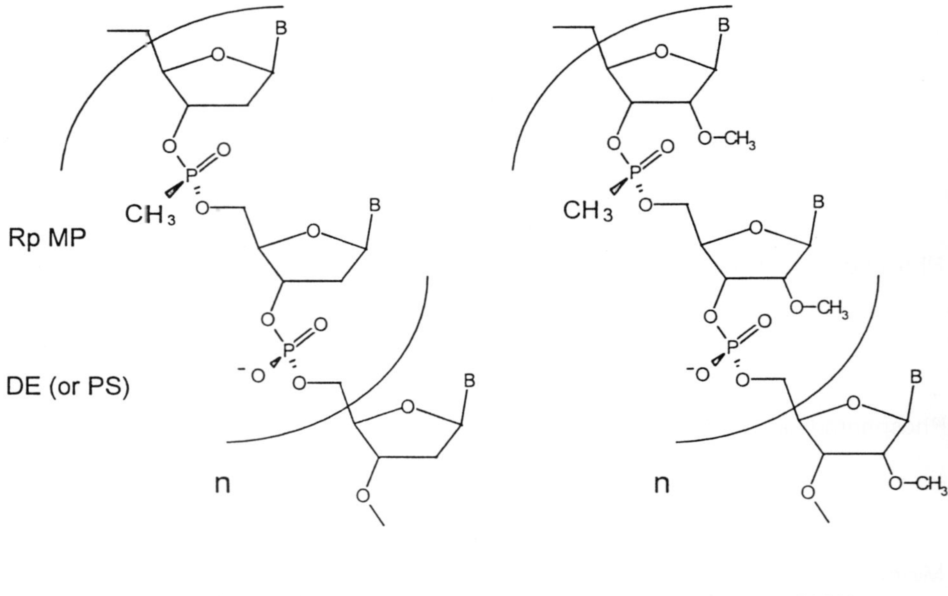

Fig. 2. The chemical structure of oligonucleotide backbone structure prepared as methylphosphonate linkages alternating with phosphodiester or phosphorothioate linkages and either as a DNA analogue or as 2′ modified ribose, an analogue of RNA. The base is represented as a B.

the urine even though glomerular filtration might be expected to result in excretion of the parent compound given their small size, typically a few thousand daltons. Plasma proteins appear to bind PS which somewhat increases plasma circulation beyond that expected for glomerular filtration. Such protein binding may contribute to the mechanism of immune stimulation in rodents by certain oligonucleotide sequences by a non-antisense mechanism.[24] Other manifestations of this non-specific pharmacological activity (PS class toxicities) include complement activation and hemodynamic effects.[25] Consequently, PS oligonucleotides represent a promising class of therapeutic agents but are presumably amenable to significant improvements by use of drug delivery systems such as liposomes.

The pharmacokinetics (PK) and tissue distribution of PS following a variety of administration routes and schedules have been studied most extensively in rodents, as described in several reviews.[26,27] Results in mice are shown in Figure 3. Studies in both rodents and primates, including humans, indicate that they can be administered parenterally as simple saline formulations achieving substantial tissue levels. The exact route, schedule, dose, and other parameters can be adjusted to affect changes in plasma levels but with only moderate effects on tissue distribution. Direct comparison of s.c. and i.p. with i.v. injections showed around 60% bioavailability for both routes with only slight differences in amount or timing of peak

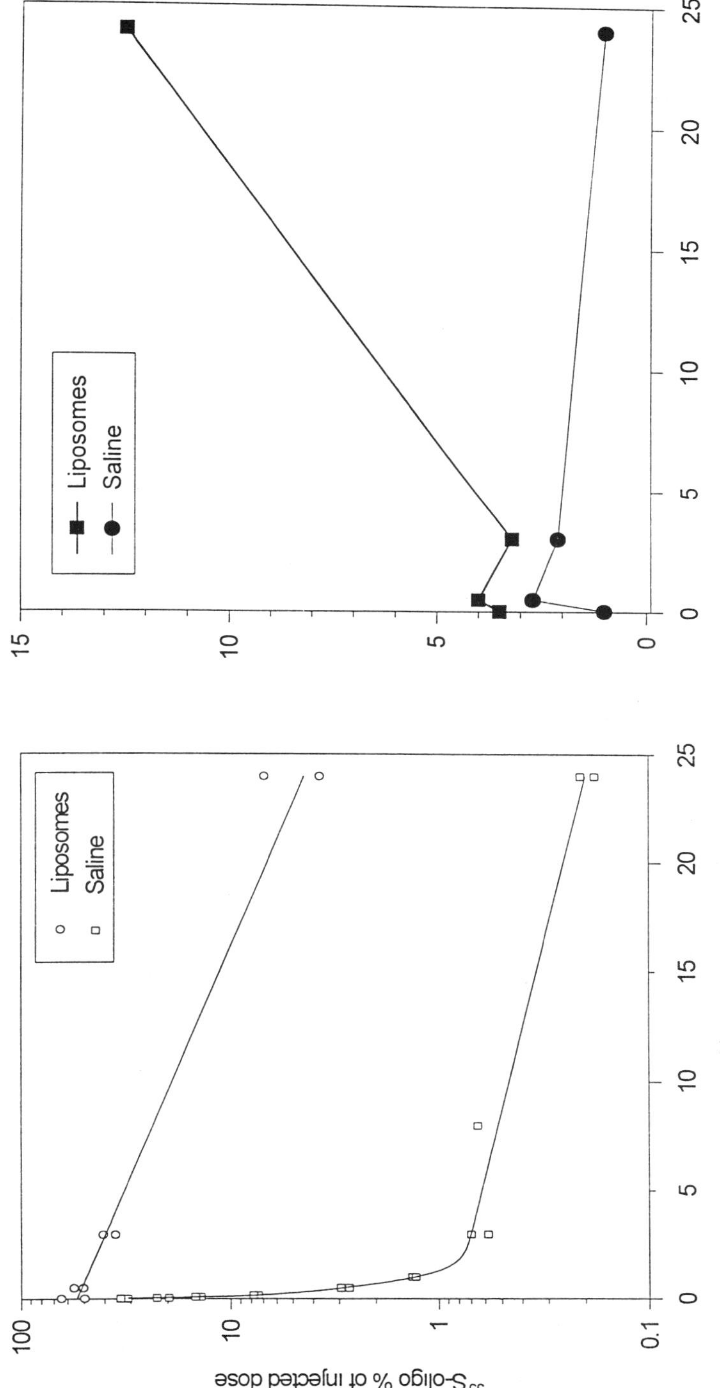

Fig. 3. Plasma clearance and localization into B16 melanoma tumors in the flank of mice. The results are measurement of total radioactivity using ^{35}S-labeled PS oligonucleotide either in saline solution or entrapped in sterically stabilized liposomes.[166]

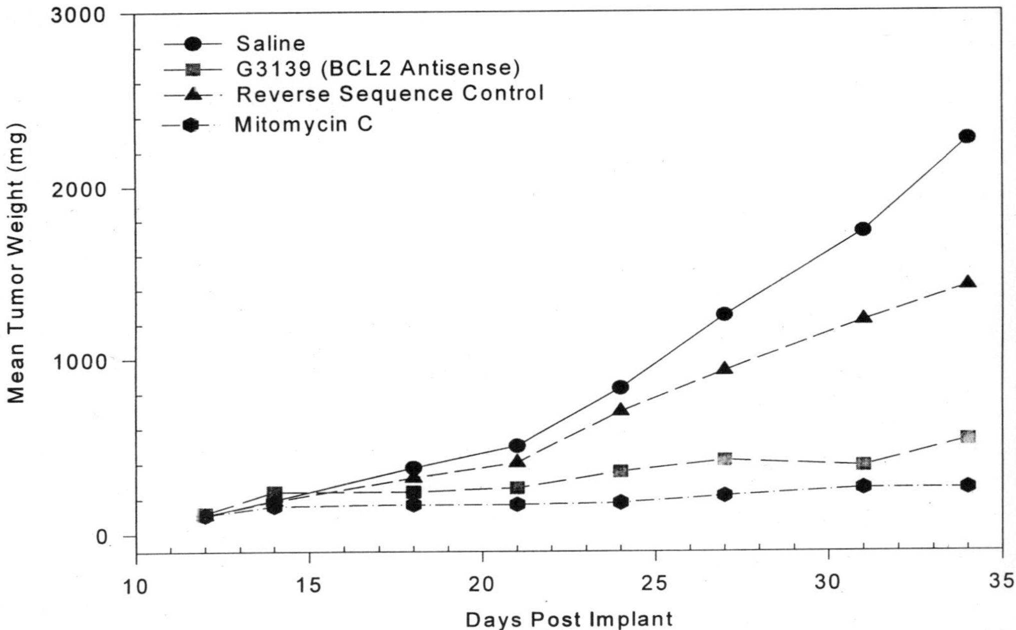

Fig. 4. Tumor growth inhibition of human tumor line implanted in the flank of nude mice by treatments administered for 14 days using subcutaneously implanted infusion pumps.[166]

plasma levels and most importantly little difference in tissue distribution.[28] It appears that PS exhibit fairly rapid movement through the lymphatics to the plasma compartment. Importantly, peak plasma levels are associated with toxicity in primates that has led to extensive use of slow i.v. infusion and more recently s.c. infusion. Efficacy studies with infusion of a PS oligonucleotide designed to reduce BCL2 expression and thereby induce tumor cell apoptosis shown in Figure 4 suggest the therapeutic potential of these agents. Levels of PS in tumors and other pathological sites are not enhanced over most healthy tissues by use of these simple aqueous formulations[28,29] but they do provide an opportunity to evaluate PS oligonucleotides for efficacy and toxicity as a proof of principle. Concentration of antisense oligonucleotides in pathologic sites would not be necessary if the target RNA were uniquely expressed there and the oligonucleotides generated minimal non-specific effects. Nevertheless, the high cost of oligonucleotides and especially the need to reduce potential toxicity via non-antisense mechanisms has stimulated an interest in advanced delivery formulations to enhance their commercial viability. Other motivations for carriers exist, especially those permitting oral administration or access to the brain, since oligonucleotides do not pass the blood brain barrier and current analogues are not orally available.[30] At the present time liposomes seem no more likely than any other existing carrier to satisfy these criteria.

Several PS antisense agents have reached clinical studies and initial results are being reported.[31–36] The clinical studies largely have been limited to plasma or tissue pharmacokinetics, safety, and to a lesser extent activity. The general success of these studies increases the impetus to identify better formulations.

Numerous approaches have been considered to favorably alter or control the biological fate of therapeutic agents with the intent to increase desired activity while diminishing toxicity, the therapeutic index. Of the wide ranging approaches considered many have utilized particulate and colloidal systems, such as liposomes, nanoparticles, emulsions, and micelles, which may also be associated with ligands designed to associate some target specificity to the carrier. Other approaches have utilized direct chemical modifications of oligonucleotides including reversible and irreversible covalent derivatives. This concept is to alter the pharmacological properties of oligonucleotides by coupling them to cholesterol, polymers, or receptor ligands. These covalent conjugates are proposed to alter cell interactions or association or permit enhanced incorporation into delivery systems based on changes to their physical properties such as hydrophobicity.[13,20,37–53] In addition, systems to provide a depot producing a controlled release into a chosen implant site are of interest. It is clear that these and other approaches need to be evaluated for their ability to improve the pharmacology of oligonucleotides.[54] This review considers liposome and cationic lipid complex formulations of PS antisense oligonucleotides.

II. Liposome formulations

Liposomes, reviewed extensively in other chapters, consist of lipid bilayer shells surrounding entrapped aqueous compartments. An important facet of liposomes is that they form spontaneously when the lipid mixture is dispersed in an aqueous solution and can be prepared controllably in several different forms varying in size, number of bilayers, entrapped aqueous phase, and other parameters. Also, they can be stable both to aggregation and especially to dilution, unlike most micelles which respond to such changes because they are at equilibrium. This type of response can have a tremendous consequence in parenteral drug delivery due to the rapid dilution into the blood following i.v. administration. For example, a micellular formulation of amphotericin B, Fungizone®, changes rapidly upon injection. As a consequence, liposomes have been investigated extensively and have proven to be a versatile and interesting system for therapeutic applications.

Liposomes were originally envisioned as being biocompatible, with low toxicity and immunogenicity, as well as being inert with respect to bodily fluids. However, it is now clear that liposomes are not inert and rapid interactions with plasma proteins including lipoproteins lead to significant changes in the liposomes, resulting in phagocytic uptake and thus rapid clearance of encapsulated agents from the blood by cells of the mononuclear phagocytic system (MPS).[55] Two approaches have been developed to overcome this barrier to liposome therapeutic formulations: small, neutral, and "rigid" liposomes composed of saturated lipids in combination with cholesterol administered at relatively high doses, which bind plasma proteins poorly[55] and surface modification by grafted PEG or other hydrophilic polymers.[56,57] The latter approach is referred to here as sterically stabilized liposomes (SSL) due to the proposed mechanism of the polymer coating sterically inhibiting the interactions with biological components including proteins and cell surfaces. The extensive development efforts with liposome formulations have culminated in several approved parenterally administered liposome (or lipid) based commercial therapeutic products.

The use of long circulating liposome formulations can provide a number of beneficial effects on the therapeutic properties of encapsulated drugs. Most studies reported have focused on liposomal formulations of already approved therapeutic agents. Many of these have demonstrated that these liposomes can greatly improve therapeutic agent localization into pathological sites, in some cases so dramatically that new therapeutic efficacy is possible.[58,59] The potential for such change of biodistribution of encapsulated agents means that a renewed evaluation of their efficacy and toxicity should be performed without bias from prior knowledge of the agent in other formulations. The success achieved through formulation of many existing therapeutic agents with liposomes indicates they have the potential to address some limitations in pharmacology of some drugs not only by enhancing desired or beneficial activities but also expanding their range. Note that the potential for new beneficial activities comes at a price; potential new toxicities must be ruled out. Liposomes containing conventional drugs appear to be particularly useful when the pathologic process locally augments vascular permeability, as in inflammation caused by infection, or in cancer. These are the major disease processes in which oligonucleotides are being evaluated clinically. Thus, the standard or classical liposomes are one of the drug delivery systems being evaluated for enhancing the therapeutic properties of oligonucleotide agents.

In addition to conventional and long circulating liposomes, a considerable effort has begun exploring forms with added capabilities. Foremost have been studies with added targeting ligands. A now standard approach to targeting is through the use of antibodies making "immuno-liposomes". Recently, this approach has been extended to other potential selective ligands, both large proteins and small molecules of various origins.[60–65] Application of most of these methods in vivo has yet to be demonstrated.

Many efforts for drug delivery are oriented to achieve intracellular delivery of nucleic acid agents. These have been focused upon liposomes and cationic lipid complexes.[66–69] With liposomes, the original optimistic expectation that they could fuse with cells to deliver all kinds of drugs directly to the cytoplasm has not usually been fulfilled. As a result, approaches are being developed to induce liposomes to fuse with the membranes of cells for the delivery of small oligonucleotides and large plasmids.

Both of these agents would seem to require intracellular delivery, with only a few exceptions.[11,12] However, some in vivo studies have identified therapeutic activities of oligonucleotides or conjugated forms independent of assisted intracellular delivery. This could be due either to direct oligonucleotide uptake, uptake following association with serum proteins, or activity via non-antisense mechanisms as the consequence of binding to some serum or cell-surface determinant. It seems reasonable to expect that even a few intracellular oligonucleotide molecules will be able to exert antisense effects, since the number of target mRNA molecules is small relative to the number of protein molecules they encode and because of the possibility that a single oligonucleotide could inactivate multiple mRNA molecules via the catalytic activity of RNase H. Therefore, traditional liposomal transport of oligonucleotides has received interest,[18,20,28,70–76] since these increase time in the circulation and probably also permit release of some of their contents intracellularly, albeit inefficiently. Ultimately, though, it seems likely the activity of antisense oligonucleotide therapeutic agents will be more predictable and enhanced by delivery systems with capabilities beyond localization at sites of pathology to include provisions for intracellular delivery.

Much of the work on cell fusion of liposomes has been motivated as basic studies designed to understand viral processes, as viruses have to penetrate cells in order to survive. In return, these studies have suggested application of viral fusion mechanisms to liposome drug delivery systems. Incorporation of viral membrane protein mixtures has been studied, especially through the use of Sendai virus.[54,77–81] Others have used components sensitive to the drop in pH following endocytosis, in order to transform the liposome in a manner that leads to fusion with the endosome membrane.[82–92] An important step exploited by many viruses is use of receptor mediated uptake to induce intracellular delivery. This also has been studied through the addition of ligands such as folate to the surface of PEG-grafted liposomes.[61,65] It may be possible to prepare long circulating liposome formulations to bind specific receptors with subsequent fusion triggered much like a virus. Essentially, these strategies represent efforts to develop synthetic viral envelope particles.

The ongoing work on these systems reveals that fusogenic liposome systems are limited by the same barriers as conventional drug delivery, in particular the need to reduce non-specific recognition while achieving blood persistence, extravasation through endothelial barriers to target tissues, and pathological site specific localization. Intracellular delivery, dependent upon protein and cellular interactions, represents a significant challenge. The addition of fusogenic materials to the surface of liposomes may simply destabilize the liposomes. Therefore, incorporation of tissue or cell binding sensitive ligands in liposomes may be a useful strategy, such as reported using target sensitive and pH sensitive immunoliposomes.[82,91,93–97] Liposome based synthetic virus drug delivery systems will depend on separation and control of mutually exclusive requirements: an inert particle possibly targeted to specific cells on one hand and cellular reactions resulting in delivery of encapsulated agents to the cytoplasm and nucleus on the other hand. Ultimately, these requirements must be accommodated. Approaches include coating the liposome with sterically stabilizing polymers and appropriate ligands in a manner that the coating is shed, exposing fusogenic components. As with ligand targeting, an adverse consequence of steric stabilization is inhibition of liposome surface interactions and cellular uptake, although evidence already exists that the barriers can be overcome.[98] Thus, even if the PEG presumably precludes intracellular delivery, approaches are being evaluated to selectively remove the PEG over time or under specific chemical conditions such as pH.[90] Another important issue will be to emulate the high efficiency of viral assembly by development of processes whereby the therapeutic agents are efficiently encapsulated.

To date none of the synthetic fusogenic liposome formulations have been sufficiently refined to demonstrate success in animal models of disease but some results are encouraging. The greatest systemic efficacy will probably require both prolonged circulation permitting time for tissue selectivity and intracellular delivery. Continuing work on liposome and other lipidic colloidal systems such as cationic lipid complexes (see below) promises to provide many exciting advances in the near future.

III. Lipid complexes

Electrostatic complexes of nucleic acids with cationic lipids are being evaluated as potential intracellular drug delivery systems for gene therapy. Since early findings that combination of lipid mixtures having an overall cationic charge with DNA can provide intracellular delivery of DNA, or transfection, at least in vitro,[99] numerous studies have been reported (reviewed recently in Ref. 100). These materials have also proven effective at intracellular delivery of antisense oligonucleotides in vitro.[13,86,94,101–104] Indeed, much of what we know about the action of antisense oligonucleotides derives from studies in which uptake by cells was aided by the use of cationic lipid. Nevertheless, the delivery is not very effective; only a tiny fraction of the DNA deposited onto the cells actually makes it to the nucleus to be expressed.[105] It must be kept in mind that use of cationic lipid, which probably disrupts cell membranes, may have consequences that either augment or reduce antisense effects. Aqueous dispersion of the lipid mixtures often results in liposome structures but the subsequent combination with anionic polynucleotides leads to what are currently called lipoplexes. Thus some reports have referred to these systems as liposomes while others have used more vague but less misleading terms such as lipid complexes. These colloidal dispersions may contain bilayer phase lipid structures but overall are quite dissimilar from liposomes[106–108] and thus they are referred to here as lipid complexes.

Regardless of nomenclature, cationic lipid complexes of DNA and its analogues such as small antisense oligonucleotides spontaneously form particulate dispersions with considerable heterogeneity when prepared by traditional procedures (mixing aqueous solutions of the lipids and DNA)[105,106,109]. Partly due to this inherent heterogeneity, the materials formed have not been well characterized and represent a significant challenge for use in commercial pharmaceutical products. The state of understanding and level of ongoing investigation is reminiscent of early studies with liposomes but enthusiasm was generated by initial reports that these complexes can function in vivo.[110] Hopefully, advances made in understanding liposomes will reduce the time required for understanding and developing DNA lipid complexes. For example, improvements in preparation and characterization using liposome methods have already been reported.[111–114] Studies of the fate and stability of cationic complexes have revealed some of their limitations.[115–118] Approaches to address these limitations include coating DNA-lipid complexes with sterically stabilizing polymers, again in a manner that the coating can be shed to allow intracellular delivery.[69]

IV. Loading and release

Both liposome and lipid complex formulations of antisense oligonucleotide are colloidal dispersions. These formulations are characterized by extent of the therapeutic agent associated with the particle versus that in solution or another state (% loading); the amount of agent captured in the preparation procedure (% loading efficiency or encapsulation); and the amount of agent released from the particles under various conditions (leakage or release). In general for parenteral routes of administration, the ideal system has high percentage loading and a low leakage rate during distribution followed by a high release rate once the intended site has been reached. Clearly, systems sensitive to their environment providing such performance will require considerable development. In the mean time, adjustment of the formulation components is used to achieve a compromise between these features. For small molecule agents such as doxorubicin and vincristine, the lipids used in forming the liposome bilayer can be selected to provide an acceptable compromise in certain cases.[119]

In the case of antisense oligonucleotides, early studies with liposomes revealed that these relatively large and highly charged molecules, typically at least 5 kD and ten or more ionized phosphodiesters, are not able to pass through intact lipid bilayers of liposomes in less than a few days.[120–122] In fact, analogues lacking the charge still are retained by liposomes, as found with peptide nucleic acids.[123] As a consequence, long circulating liposomes with rigid bilayers might release insufficient antisense to be active. In contrast, the sterically stabilized liposomes permit a greater range in lipid properties which might be amenable to achieving adequate release rates of oligonucleotides. Unfortunately, leakage and release of oligonucleotides from liposomes in vivo has not been adequately addressed. However, recent results would suggest that by controlling lipid composition release both in vitro and in vivo may be possible.[19,28,72,85,124] Cationic lipid complexes of nucleic acids appear to be at the other extreme in stability. While being similarly stable to leakage when dispersed in simple aqueous solutions, they typically are inhibited by the presence of serum and even more troubling is growing evidence that they rapidly dissociate into separate components in vivo.[115,117,125]

Perhaps equally important are percentage loading and encapsulation properties. An approximate rule for parenteral formulations might be that the therapeutic agent should compromise at least 1% of all components by weight and up to 10% would be preferable. A loading of 1% means that a potency requiring 1 to 10 mg per kg body weight would require doses of 100 mg to 1 gram of formulation components per kg body weight or 7 to 70 grams for an average human. This upper value is quite high for intravenous injection of lipids. Due to the highly charged and water soluble nature of most oligonucleotides, their encapsulation in liposomes so far has been restricted to a so called passive encapsulation process giving a maximum of about 10% loading efficiency. This means that only marginal levels of percentage loading and efficiency can be attained with published techniques. Fortunately, this level should be sufficient for feasibility tests. In contrast, cationic lipid complexes form by interaction of the lipid and nucleic acid components generally giving a very high "loading" efficiency. Alternative improved methods of loading are possible as shown by use of hydrophobic oligonucleotide conjugates to provide for interaction and binding with the bilayer.[126,127] As pointed out above, numerous other parameters have yet to be characterized including methods to reduce heterogeneity, while concerns remain about the biocompatibility of the lipid components. Again, liposomes and lipid complexes are at opposite extremes and the ideal may be somewhere in between but in this case much closer to that of the cationic lipid complex.

V. Cellular interaction and uptake

Ultimately the intent of antisense formulation is to enable intracellular delivery to target cells. This has been addressed primarily through in vitro studies which have demonstrated that poor cellular uptake is a key problem in antisense therapeutics, with the caveat concerning transposition of in vitro to in vivo results to be kept in mind. Many variations in oligonucleotide chemical structure have been examined to understand their cellular interactions as well as identify potential methods to control those interactions, including backbone chemistry, ligands, hydrophobic derivatives (conjugates), base sequence, and effects of cationic ions.[11,120,122,128–142] Perhaps the most successful approaches to date have been using cationic lipid "transfection" methods first adapted from studies with DNA.[101] Unquestionable intracellular uptake, cytoplasmic and nuclear localization, and the desired antisense activities, can be attained in most cell types by cationic lipid delivery and the mechanism appears to involve endocytosis.[13,86,94,143–145] Antisense activity of oligonucleotides has been demonstrated in vitro using transfection but is questionable without it. Clearly exceptions to this exist and may be largely cell line dependent.[146] One cell type potentially not requiring intracellular delivery is the keratinocyte[12] but this may only be true at late stages of development when antisense effects might be minimal or not useful.[147] Also, the hematopoietic cell types are particularly resistant to intracellular delivery by lipid transfection but some success is beginning to be reported.[50] At least in vitro, antisense oligonucleotide activity is greatly enhanced by, if not limited to, intracellular delivery.

For standard liposome formulations, encapsulation of oligonucleotides has been reported to have some beneficial affects on their stability and activity.[71,85,148–150] Note that these results are obtained despite the general understanding that such standard liposomes generally lack intracellular delivery beyond endocytic or phagocytic uptake in cells with those capabilities which often leads simply to digestion of their contents. Consequently, many studies of oligonucleotides with liposomes have been oriented toward ligand, usually antibody, targeting,[61,64,151–

[155] and to a lesser extent complex forms that might disrupt endosomes to provide intracellular delivery, such as pH dependent or sensitive.[124] The results of in vitro studies have suggested that intracellular delivery may be obtained by proper selection of the ligand,[61,156] but the extent to which these systems will prove selective in vivo remains unclear. On the other hand, some studies have been interpreted that the effects observed may be due to multiple mechanisms acting in parallel (and more than one mechanism possibly influenced by the immunoliposomes).[72]

Unlike liposomes, cationic lipid-DNA complexes overall have proven to be one of the best methods to achieve intracellular levels of antisense oligonucleotides which has resulted in successful demonstration of antisense activity, at least in vitro. Numerous lipid complexes have been reported with clear differences in performance, variations are reported for specific agents in each study but taken together it appears that variations can also be expected across different cell types. Nonetheless, these materials have become well accepted for intracellular delivery of antisense oligonucleotides.[50,101,103,157,158] The fate of these lipid-DNA complexes is thought to involve cell membrane interactions leading to endocytic uptake followed by a disruption of the normal endosome processing pathway eventually giving rise to cytoplasmic levels of biologically active oligonucleotide which are transported to the nucleus by energy independent cellular processes. Measurement of nuclear fluorescence of fluorescent labeled oligonucleotides has become accepted as indicating effective intracellular delivery. As pointed out above, cationic lipid complexes with antisense oligonucleotides provides good intracellular delivery which has permitted a demonstration that antisense activity can be attained but their use in vivo remains uncertain.

Intracellular delivery in vivo is more difficult both to study and to provide. Most of the studies rely upon measurements only of efficacy in disease with respect to some variety of oligonucleotide controls but this type of experiment is not adequate to prove the mechanism of action. A few studies have begun to address intracellular delivery in vivo with perhaps the most reasonable start being local administration, for example.[137] Despite a lack of proof for an antisense mechanism in vivo, many studies demonstrate efficacy in many disease models using oligonucleotides and in a few studies a spectrum of structure-activity relationships has been observed closely correlated with in vitro results via bona fide antisense mechanisms. Thus regardless of mechanism, these agents represent potential therapeutics which may be facilitated by traditional delivery systems without requiring advances providing intracellular delivery.

VI. Tissue and pathology localization: Disease targeting

As discussed above, antisense phosphorothioate oligonucleotides distribute widely to most tissues when administrated as saline solutions by parenteral systemic routes, with one clear exception being the brain. This general lack of apparent tissue specificity has a potential advantage to provide moderate exposure of many disease target tissues but only about equal to that of healthy tissues. Thus if the healthy tissue is not responsive to the antisense agent, and the antisense is self-sufficient to act at the disease site, then a moderate levels of activity can be elicited in vivo. This appears to be the case in numerous reports of efficacy by antisense agents in animal models of disease and even in clinical studies. Recent updates of clinical efficacy have been presented, including at the 12th International Roundtable on Nucleosides and Nucleotides.[31–34] Note that studies in which efficacy was lacking would probably not be reported and thus the consistency of antisense efficacy using saline solutions is not known at this time. However, even without a clear correlation in most studies between antisense design in vitro and efficacy toward in vivo models, there are numerous pharmacological properties of oligonucleotides, especially of phosphorothioates, which can be improved by drug delivery systems in general and liposomes in particular.

One of the principal improvements oligonucleotides require is to shift their distribution away from healthy tissues and increase exposure of disease tissues. For example, reduced kidney exposure to phosphorothioates appears to be desirable largely independent of their sequence as renal toxicity seems to be one of the major limitations.[25] But even more importantly, increased exposure of the disease tissue should increase efficacy at the same dose or reduce the dose required which would make the use of these still relatively expensive agents more feasible.

Essentially two approaches have been used to enhance disease exposure of antisense by liposomes: local administration at the site of disease and disease targeting from systemic administration. Antisense oligonucleotides have been used with various forms of local administration including many reports using intracerebroventricular administration;[159–161] administration with a catheter associated with angioplasty;[78,162,163] direct injection into tumors;[164] and administration into the portal vein for liver targeting.[80] Combination of local administration with liposome formulations can provide some advantages, in particular increasing stability, but ultimately may not justify the considerable added effort and cost. On the other hand, local administration of antisense in lipid complexes may prove a very useful step to achieving formulations capable of intracellular delivery in vivo. Again, though, whether this step is more demonstration of feasibility or of a useful platform for treating human disease remains unclear.

The greatest potential value of liposome formulations is their potential to enhance localization in sites of disease from systemic administration as now demonstrated for conventional small molecule therapeutic agents. The long circulating liposomes clearly can provide tumor localization to oligonucleotides[150] but clear value to enhanced efficacy has not yet been demonstrated. Also, given the apparent lack of stability of the current simple cationic complex, including both lipids and polymers, in the in vivo environment it is not surprising that they have yet to provide tumor targeting from systemic administration[165] despite activity generally with plasmids.[110] Consequently, the disease localization capabilities using systemic administration appears dependent upon more stable systems such as the sterically stabilized liposomes or even simple systems resembling lipoproteins.[40] Thus the issue of tumor targeting followed by specific intracellular delivery to target cells represents an as yet unrealized need and current goal in drug delivery system development. This is expected to greatly enhance oligonucleotide therapeutics as well as that of many other classes of compounds. Systems that accomplish such complex drug delivery function may utilize features of liposome and lipid complex methods.

References

1. Saison-Behmoaras T, Tocque B, Rey I, Chassignol M, Thuong NT, Helene C. Short modified antisense oligonucleotides directed against Ha-ras point mutation induce selective cleavage of the mRNA and inhibit T24 cells proliferation. Embo J 1991;10:1111–1118.
2. Marshall W, Caruthers, M. Phosphorothioate DNA as a potential therapeutic drug. Science 1993;259:1564–1570.
3. Milligan JF, Jones RJ, Froehler BC, Matteucci MD. Development of antisense therapeutics. Implications for cancer gene therapy. Ann NY Acad Sci 1994;716:228–241.
4. Robinson-Benion C, Holt JT. Antisense techniques. Methods Enzymology 1995;254:363–375.
5. Agrawal S, Akhtar S. Advances in antisense efficacy and delivery. Trends Biotechnology 1995;13:197–199.
6. Gold L. Oligonucleotides as research, diagnostic, and therapeutic agents. J Biol Chem 1995;270:13581–13584.
7. Askari FK, McDonnell WM. Antisense-oligonucleotide therapy. New Engl J Med 1996;335:316–318.
8. Tonkinson JL, Stein CA. Antisense oligonucleotides as clinical therapeutic agents. Cancer Invest 1996;14:54–65.
9. Aoki K, Yoshida T, Sugimura T, Terada M. Liposome-mediated in vivo gene transfer of antisense K-ras construct inhibits pancreatic tumor dissemination in the murine peritoneal cavity. Can Res 1995;55:3810–3816.
10. Phillips MI. Antisense inhibition and adeno-associated viral vector delivery for reducing hypertension. Hypertension 1997;29:177–187.
11. Vlassov VV, Balakireva LA, Yakubov LA. Transport of oligonucleotides across natural and model membranes. BBA 1994;1197:95–108.
12. Nestle FO, Mitra RS, Bennett CF, Chan H, Nickoloff BJ. Cationic lipid is not required for uptake and selective inhibitory activity of ICAM-1 phosphorothioate antisense oligonucleotides in keratinocytes. J Invest Dermatol 1994;103:569–575.
13. Bennett CF, Chiang MY, Chan H, Shoemaker JE, Mirabelli CK. Cationic lipids enhance cellular uptake and activity of phosphorothioate antisense oligonucleotides. Mol Pharmacol 1992;41:1023–33.
14. Crooke S, Lebleu B, ed. Antisense research and applications. Boca Raton, FL: CRC Press, 1993.
15. Dean NM, McKay R. Inhibition of protein kinace C-alpha expression in mice after systemic administration of phosphorothioate antisense oligodeoxynucleotides. Proc Natl Acad Sci USA 1994;91:11762–11766.
16. Crooke ST. The future of antisense technology. Pharmaceutical News 1995;2:8–11.
17. Stein CA. Does antisense exist? Nat Med 1995;1 11.
18. Leserman L. Liposomes as transporters of oligonucleotides. Philippot JR, Schuber F, ed. Boca Raton, FL: CRC Press, 1994.
19. Ropert C, Okruszek A, Couvreur P, Malvy C. Retroviral inhibition by antisense oligonucleotides determined by intracellular stability. Antisense Res Dev 1994;4:207–210.
20. Alahari SK, Dean NM, Fisher MH, Delong R, Manoharan M, Tivel KL, Juliano RL. Inhibition of expression of the mutidrug resistance-associated P-glycoprotein by phosphorothioate and 5′ cholesterol-conjugated phosphorothiorate antisense oligonucleotides. Mol Pharmacol 1996; 50:808–819.
21. Ma L, Calvo F. Recent status of the antisense oligonucleotide approaches in oncology. Fundam Clin Pharmacol 1996;10:97–115.
22. Reynolds MA, Beck TA, Say PB, Schwartz DA, Dwyer BP, Daily WJ, Vaghefi MM, Metzler MD, Klem RE, Arnold LJ. Antisense oligonucleotide containing an internal, non-nucleotide-based linker promote site-specific cleavage of RNA. Nucleic Acids Res 1996;24:760–765.
23. Baker BF, Lot SS, Condon TP, Cheng-Flournoy S, Lesnik EA, Sasmor HM, Bennett CF. 2′-O-(2-methoxy)ethyl-modified anti-intercellular adhesion molecule 1 (ICAM-1) oligonucleotides

selectively increase the ICAM-1 mRNA level and inhibit formation of the ICAM-1 translation initiation complex in human umbilical vein endothelial cells. J. Biol Chem 1997;272:11994–12000.

24. Krieg AM, Yi A-K, Matson S, Waldschmidt TJ, Hornbeck P, Teasdale R, Koretzky GA, Klinman D. CpG motifs in bacterial DNA direct B-cell activation. Nature 1995;374:546–549.
25. Black L. Recent perspectives on oligonucleotide drug development: analysis of nonclinical safety data. Antisense Therapeutics Program Feb. 1996, Coronado, CA,.1996.
26. Agrawal S, Temsamani J, Galbraith W, Tang J. Pharmacokinetics of antisense oligonucleotides. Clin Pharmacokinet 1995;28:7–16.
27. Srinivasan SK, Iversen P. Review of in vivo pharmacokinetics and toxicology of phosphorothioate oligonucleotides. J Clin Lab Anal 1995;9:129–137.
28. Woodle MC, Raynaud FI, Dizik M, Meyer O, Huang SK, Jaeger JA, Brown BD, Orr R, Judson IR, Papahadjopoulos D. Oligonucleotide pharmacology and formulation: G3139 anti-BCL2 phosphorothioate in Stealth® liposomes and gel implants. Nucleosides Nucleotides, 1997;16:1731–1734.
29. Plenat F, Klein-Monhoven N, Marie B, Vignaud JM, Duprez A. Cell and tissue distribution of synthetic oligonucleotides in healthy and tumor-bearing nude mice: and autoradiographic, immunohistological, and direct fluorescence microscopy study. Am J Pathol 1995;147:124–135.
30. Crooke S. Delivery of oligonucleotides and polynucleotides. J Drug Target 1995;3:185–190.
31. Webb A, Cotter F, Cunningham D. BCL-2 antisense therapy in lymphoma. Annals Oncology 1996;7:
32. Arnold LJ. The preclinical and clinical experience of antisense oligonucleotides that target BCL2 expression. XII Int Roundtable Nucleoside Nucleotide Program Sept. 1996, San Diego, CA, 1996;63.
33. Schechter PJ, Martin RR, Grindel JM. Clinical results of drug candidates. XII Int Roundtable Nucleoside Nucleotide Program Sept. 1996, San Diego, CA, 1996;65.
34. Kisner D. Development of phosphorothioate oligonucleotides. XII Int Roundtable Nucleoside Nucleotide Program Sept. 1996, San Diego, CA, 1996;61.
35. Cossum PA. AR177, an anti-HIV G-tetrad oligonucleotide. XII Int. Roundtable Nucleoside Nucleotide Program Sept. 1996, San Diego, CA, 1996.
36. Shanahan JWR. Clinical status of ISIS 2302, an ICAM-1 antisense phosphorothioate oligonucleotide. XII Int Roundtable Nucleoside Nucleotide Program Sept. 1996, San Diego, CA, 1996;64.
37. Lemaitre M, Bayard B, Lebleu B. Specific antiviral activity of a poly(L-lysine)-conjugated oligodeoxynucleotides sequence complementary to vesicular stomatitis virus N protein mRNA initiation site. Proc Natl Acad Sci USA 1987;84:648–652.
38. Kabanov AV, Vinogradov SV, Ovcharenko AV, Krivonos AV, Melik NN, Kiselev VI, Severin ES. A new class of antivirals: antisense oligonucleotides combined with a hydrophobic substituent effectively inhibit influenza virus reproduction and synthesis of virus-specific proteins in MDCK cells. FEBS Lett 1990;259:327–30.
39. Shea RG, Marsters JC, Bischofberger N. Synthesis, hybridization properties and antiviral activity of lipid-oligodeoxynucleotide conjugates. Nucleic Acids Res 1990;18:3777–83.
40. van Berkel TJC. Association of antisense oligonucleotides with lipoproteins prolongs the plasma half-life and modifies the tissue distribution. Nuc Acid Res 1991;19:4695–4700.
41. Akhtar S, Shoji Y, Juliano RL. Pharmaceutical aspects of the biological stability and membrane transport characteristics of antisense oligonucleotides. IzantJG, RPEa, ed. New York: Raven Press, 1992.
42. Boado RJ, Pardridge WM. Complete protection of antisense oligonucleotides against serum nuclease degradation by an avidin biotin system. Bioconjugate Chemistry 1992;3:519–523.
43. Bonfils E, Depierreux C, Midoux P, Thuong NT, Monsigny M, Roche AC. Drug Targeting–Synthesis and endocytosis of oligonucleotide-neoglycoprotein conjugates. Nucleic Acids Research 1992;20:4621–4629.
44. Svinarchuk FP, Konevetz DA, Pliasunova OA, Pokrovsky AG, Vlassov VV. Inhibition of HIV proliferation in MT-4 cells by antisense oligonucleotide conjugated to lipophilic groups. Biochimie 1993;75:49–54.
45. Chavany C, Saison-Behmoaras T, Le Doan T, Puisieux F, Couvreur P, Helene C. Adsorption of oligonucleotides onto polyisohexylcyanocrylate nanoparticles protects them against nucleases and increases their cellular uptake. Pharm Res 1994;11:1370–1378.
46. Schwab G, Chavany C, Duroux I, Goubin G, Lebeau J, Helene C, Saison-Behmoaras T. Antisense oligonucleotides adsorbed to polyalkylcyanocrylate nanoparticles specifcally inhibit mutated Ha-ras-mediated cell proliferation and tumorigenicity in nude mice. Proc Natl Acad Sci USA 1994;91:10460–10464.

47. Zhou JH, Pai BS, Reed MW et al. Discovery of short, 3′-cholesterol-modified DNA duplexes with unique antitumor cell activity. Can Res 1994;54:5783–????.
48. Guy-Caffey JK, Bodepudi V, Bishop JS, Jayaraman K, Chaudhary N. Novel polyaminolipids enhance the cellular uptake of oligonucleotides. J Biol Chem 1995;270:31391–31396.
49. Mishra RK, Moreau C, Ramazeilles C, Moreau S, Bonnet J, Toulme JJ. Improved leishmanicidal effect of phosphorothioate antisense oligonucleotides by LDL-mediated delivery. Biochim Biophys Acta 1995;1264:229–237.
50. Albrecht T, Schwab R, Peschel C, Engels HJ, Fischer T, Huber C, Aulitzky WE. Cationic lipid mediated transfer of c-abl and bcr antisense oligonucleotides to immature normal myeloid cells: uptake, biological effects and modulation of gene expression. Ann Hematol 1996;72:73–79.
51. Kim SG, Nakashima H, Shoji Y, Inagawa T, Yamamoto N, Kinzuka Y, Takai K, Takaku H. 5′-linked lipid-oligodeoxyribonucleotide derivatives as inhibitors of human immunodeficiency virus replication. Bioorg Med Chem 1996;4:603–608.
52. Williams SA, Chang L, Buzby JS, Suen Y, Cairo MS. Cationic lipids reduce time and dose of c-myc antisense oligodeoxynucleotides required to specifically inhibit Burkitt's lymphoma cell growth. Leukemia 1996;12:1980–1989.
53. Berton M, Sixou S, Kravtzoff R, Dartigues C, Imbertie L, Allal C, Favre G. Improved oligonucleotide uptake and stability by a new drug carrier, the supramolecular bio vector (SMBV). Biochim Biophys Acta 1997;1355:7–19.
54. Bertling WM, Gareis M, Paspaleeva, V, Zimmer A, Kreuter J, Nurnberg E, Harrer P. Use of liposomes, viral capsids and nanoparticles as DNA carriers. Biotech and Applied Biochem 1991;13:390–405.
55. Senior JH. Fate and behavior of liposomes in vivo: a review of controlling factors. Critical Reviews in Therapeutics and Drug Carrier Systems 1987;3:123–193.
56. Lasic D, Martin F, ed. Stealth Liposomes. Boca Raton, FL: CRC Press, 1995.
57. Woodle MC. Sterically stabilized liposome therapeutics. Adv Drug Del Rev 1995;16:249–265.
58. Vaage J, Donovan D, Mayhew E, Uster P, Woodle MC. Therapy of mouse mammary carcinomas with vincristine and doxorubicin encapsulated in sterically stabilized liposomes. Int J Cancer 1993;54:959–???.
59. Allen TM, Newman MS, Woodle MC, Mayhew E, Uster PS. Pharmacokinetics and antitumor activity of vincristine encapsulated in sterically stabilized liposomes. Int J Cancer 1995;62:199–204.
60. Toth G, Schlammadinger J, Szabo G. DNA uptake stimulating protein from *Neurospora crassa* enhances DNA and oligonucleotide uptake also in mammalian cells. Biochimica et Biophysica Acta 1994;1219:314–320.
61. Wang S, Lee RJ, Cauchon G, Gorenstein DG, Low PS. Delivery of antisense oligodeoxyribonucleotides against the human epidermal frowth factor receptor into cultured KB cells with liposomes conjugated to folate via polyethylene glycol. Proc Natl Acad Sci USA 1995;92:3318–3322.
62. Murohara T, Margiotta J, Phillips LM, Paulson JC, DeFrees SA, Zalipsky S, Guo LSS, Lefer AM. Cardioprotection by liposome-conjugated sialyl Lewis(x)-oligosaccharide in myocardial ischaemia and reperfusion injury. Cardiovasc. Res. 1995;30:965–974.
63. DeFrees SA, Phillips L, Guo L, Zalipsky S. Sialyl lewis x liposomes as a multivalent ligand and inhibitor of e-selectin mediated cellular adhesion. J Am Chem Soc 1996;118:6101–6104.
64. Henderson GB, Stein CA. 5Boca Raton,-cholesteryl-phosphorothioate oligodeoxynucleotides: potent inhibition of methotrexate transport and antagonism of methotrexate toxicity in cells containing the reduced-folate receptor. Nucleic Acids Research 1995;23:3726–3731.
65. Lee RJ, Low PS. Delivery of liposomes into cultured KB cells via folate receptor-mediated endocytosis. J Biol Chem 1994;269:3198–3204.
66. Remy J-S, Sirlin C, Behr J-P. Gene transfer with cationic amphiphiles. Philippot JR, Schuber F, ed. Boca Raton, FL: CRC Press, 1994.
67. Schreier H. The new frontier: gene and oligonucleotide therapy. Pharm Acta Helv 1994;68:145–159.
68. Sobol RE, Scanlon KJ, ed. The internet book of gene therapy. Stamford, Connecticut: Appleton & Lange, 1995.
69. Lasic DD. Liposomes in gene delivery. Boca Raton, FL: CRC Press, 1997;1–320.
70. Juliano RL, Akhtar S. Liposomes as a drug delivery system for antisense oligonucleotides. Antisense Res Dev 1992;2:165–176.
71. Thierry AR, Rahman A, Dritschilo A. Overcoming multidrug resistance in human tumor cells using free and liposomally encapsulated antisense oligonucleotides. Biochem Biophys Res Commun 1993;190:952–???.

72. Zelphati O, Imbach JL, Signoret N, Zon G, Rayner B, Leserman L. Antisense oligonucleotides in solution or encapsulated in immunoliposomes inhibit replication of HIV-1 by several different mechanisms. Nucleic Acids Res 1994;22:4307–4314.
73. Hatta T, Nakagawa Y, Takai K, Nakada S, Yokota T, Takaky H. Inhibition of influenza virus RNA polymerase and nucleoprotein genes expression by unmodified, phosphothioated, and liposomally encapsulated oligonucleotides. Biochem Biophys Res Commun 1996;223:341–346.
74. Hughes JA, Aronsohn AI, Avrutskaya AV, Juliano RL. Evaluation of adjuvants that enhance the effectiveness of antisense oligodeoxynucleotides. Pharm Res 1996;13:404–410.
75. Tari AM, Andreeff M, Kleine HD, Lopez-Berestein G. Cellular uptake and localization of liposomal-methylphosphonate oligodeoxynucleotides. J Mol Med 1996;74:623–628.
76. Wielbo D, Simon A, Phillips MI, Toffolo S. Inhibition of hypertension by peripheral administration of antisense oligodeoxynucleotides. Hypertension 1996;28:147–151.
77. Kaneda Y, Iwai K, Uchida T. Introduction and expression of the human insulin gene in adult rat liver. J Biol Chem 1989;264:12126–12129.
78. Morishita R, Gibbons G, Kaneda Y, Ogihara T, Dzau VJ. Pharmacokinetics of antisense oligodeoxyribonucleotides (cyclin B1 and CDC 2 kinase) in the vessel wall in vivo: enhanced therapeutic utility for restenosis by HVJ-liposome delivery. Gene 1994;149:13–19.
79. Dzau V, Mann M, Morishita R, Kaneda Y. Fusigenic viral liposome for gene therapy in cardiovascular diseases [comment]. Proc Natl Acad Sci USA 1996;93:11421–11425.
80. Tomita N, Morishita R, Higaki J, Aoki M, Nakamura Y, Mikami H, Fukamizu A, Murakami K, Kaneda Y, Ogihara T. Transient decreae in high blood pressure by in vivo transfer of antisense oligodeoxynucleotides against rat angiotensiogen. Hypertension 1995;26:131–136.
81. Ellison KE, Bishopric NH, Webster KA, Morishita R, Gibbons GH, Kaneda Y, Sato B, Dzau VJ. Fusigenic liposome-mediated DNA transfer into cardiac myocytes. J Mol Cell Cardiol 1996;28:1385–1399.
82. Wang CY, Huang L. Highly efficient DNA delivery mediated by pH-sensitive immunoliposomes. Biochem 1989;28:9508–9514.
83. Wilschut J, Hoekstra D, ed. Membrane fusion. New York: Marcel Dekker, 1991.
84. Schoen P, Bron R, Wilschut J. Delivery of foreign substances to cells mediated by fusion-active reconstituted influenza virus envelopes (virosomes). J Liposome Res 1993;3:767–792.
85. Ropert C, Malvy C, Couvreur P. Inhibition of the friend retrovirus by antisense oligonucleotides encapsulated in liposomes: mechanism of action. Pharm Res 1993;10:1427–1433.
86. Legendre JY, Szoka FC. Delivery of plasmid DNA into mammalian cells lines using pH-sensitive liposomes: comparison with cationic liposomes. Pharm Res 1992;9:1235–1242.
87. Lasic DD. Liposomes: from physics to applications. Amsterdam: Elsevier, 1993;1–575.
88. de Lima MCP, Hoekstra D. Liposomes, viruses, and membrane fusion. Philippot JR, Schuber F, ed. Boca Raton, FL: CRC Press, 1994.
89. Düzgüneş N, Nir S. Liposomes as tools for elucidating the mechanism of membrane fusion. Philippot JR, Schuber F, ed. Boca Raton, FL: CRC Press, 1994.
90. Kirpotin D, Hong K, Mullah N, Papahadjopoulos D, Zalipsky S. Liposomes with detachable polymer coating: destabilization and fusion of dioleoylphophatidylethanolamine vesicles triggered by cleavage of surface-grafted poly(ethylene glycol). FEBS Lett 1996;388:115–118.
91. Ma DD, Wei AQ. Enhanced delivery of synthetic oligonucleotides to human leukaemic cells by liposomes and immunoliposomes. Leuk Res 1996;20:925–930.
92. Slepushkin VA, Simoes S, Dazin P, Newman MS, Guo LS, deLima MCP, Düzgüneş N. Sterically stabilized pH-sensitive liposomes--intracellular delivery of aqueous contents and prolonged circulation in vivo. J Biol Chem 1997;272:2382–2388.
93. Ho RJY, Rouse BT, Huang L. Target-sensitive immunoliposomes as an efficient drug carrier for antiviral activity. J Biol Chem 1987;262:13973.
94. Zhou F, Huang L. Liposome-mediated cytoplasmic delivery of proteins: an effective means of accessing the MHC class I-restricted antigen presentation pathway. Immunomethods 1994;4:229–235.
95. Tari AM, Fuller N, Boni LT, Collins D, Rand P, Huang L. Interactions of liposome bilayers composed of 1,2-diacyl-3-succinylglycerol with protons and divalent cations. Biochim Biophys Acta 1994;1192:253–262.
96. Babbitt B, Burtis L, Dentinger P, Constantinides P, Hillis L, McGirl B, Huang L. Contact-dependent, immunecomplex-mediated lysis of hapten-sensitized liposomes. Bioconj Chem 1993;4:199–205.
97. Collins D. pH-sensitive liposomes as tools for cytoplasmic delivery. Philippot JR, Schuber F, ed. Boca Raton, FL: CRC Press, 1994.

98. Kirpotin D, Park JW, Hong K, Zalipsky S, Li W-L, Carter P, Benz CC, Mullah N, Papahadjopoulos D. Sterically stabilized anti-HER2 immunoliposomes: design and targeting to human breast cancer cells in vitro. Biochem 1997;36:66–75.
99. Felgner PL, Gadek TR, Holm M, Roman R, Chan HW, Wenz M, Northrop JP, Ringold GM, Danielsen M. Lipofection: a highly efficient lipid-mediated DNA-transfection procedure. Proc Natl Acad Sci USA 1987;84:7413–7417.
100. Schreier H, Sawyer SM. Liposomal DNA vectors for cystic fibrosis gene therapy. Current applications, limitations, and future directions. Adv Drug Del Rev 1996;19:73–87.
101. Chiang M-Y, Chan H, Zounes MA, Freier SM, Lima WF, Bennett CF. J Biol Chem 1991;266:18162–18171.
102. Lappalainen K, Urtti A, Soderling E, Jaaskelainen I Syrjanen K, Syrjanen S. Cationic liposomes improve stability and intracellular delivery of antisense oligonucleotides into CaSki cells. Biochim Biophys Acta 1994;1196:201–208.
103. Lappalainen K, Urtti A, Jaaskelainen I, Syrjanen K, Syrjanen S. Cationic liposomes mediated delivery of antisense oligonucleotides to HPV 16 E7 mRNA in CaSki cells. Antiviral Res 1994;23:119–130.
104. Le Bolch G, Le Bris N, Yaouanc J-J, Clement J-C, des Abbayes H. Cationic phosphonolipids as non viral vectors for DNA transfection. Tetrahedron Letters 1995;36:6681–6684.
105. Zabner J, Fasbender A, Moninger T, Poellinger K, Welsh M. Cellular and molecular barriers to gene transfer by a cationic lipid. Proc Natl Acad Sci USA 1995;270:18997–19007.
106. Gershon H, Ghirlando R, Guttman SB, Minsky A. Mode of formation and structural features of DNA-cationic liposome complexes used for transfection. Biochem 1993;32:7143–7151.
107. Lasic DD, Strey H, Stuart M, Podgornik R, Frederik PM. The structure of DNA-lipid complexes. J Am Chem Soc 1997;119:832–833.
108. Radler JO, Koltover I, Salditt T, Safinya CR. Structure of DNA-cationic liposome complexes: DNA intercalation in multilamellar membranes in distinct interhelical packing regimes. Science 1997;275:810–814.
109. Sternberg B, Sorgi FL, Huang L. New structures in complex formation between DNA and cationic liposomes visualized by freeze-fracture electron microscopy. FEBS Lett 1994;356:361–366.
110. Zhu N, Liggitt D, Liu Y, Debs R. Systemic gene expression after intravenous DNA delivery into adult mice. Science 1993;261:209–211.
111. Campbell MJ. Lipofection reagents prepared by a simple ethanol injection technique. Biotechniques 1995;18:1027–1032.
112. Reimer DL, Zhang Y, Kong S, Wheeler JJ, Graham RW, Bally M. Formation of novel hydrophobic complexes between cationic lipids and plasmid DNA. Biochem 1995;34:12877–12883.
113. Hofland HEJ, Shephard L, Sullivan SM. Formation of stable cationic lipid/DNA complexes for gene transfer. Proc Natl Acad Sci USA 1996;93:7305–7309.
114. Brown BD. Stable cationic lipid complexes at high concentration. Personal Communication, 1995.
115. Litzinger DC, Brown JM, Wala I, Kaufman SA, Van GY, Farrell CL, Collins D. Fate of cationic liposomes and their complex with oligonucleotide in vivo. Biochim Biophys Acta 1996;1281:139–149.
116. Zelphati O, Szoka JFC. Intracellular distribution and mechanism of delivery of oligonucleotides mediated by cationic liposomes. Pharm Res 1996;13:1367–1372.
117. Zelphati O, Szoka JFC. Mechanism of oligonucleotide release from cationic liposomes. Proc Natl Acad Sci USA 1996;93:11493–11498.
118. Lewis JG, Lin KY, Kothavale A, Flanagan WM, Matteucci MD, DePrince RB, Mook RA, Hendren RW, Wagner RW. A serum-resistant cytofectin for cellular delivery of antisense oligodeoxynucleotides and plasmid DNA. Proc Natl Acad Sci USA 1996;93:3176–3181.
119. Webb MS, Harasym TO, Masin D, Bally MB, Mayer LD. Sphingomyelin-cholesterol liposomes significantly enhance the pharmacokinetic and therapeutic properties of vincristine in murine and human tumour models. Br J Cancer 1995;72:896–904.
120. Akhtar S, Basu S, Wickstrom E, Juliano RL. Interactions of antisense DNA oligonucleotides analogues with phospholipid membranes (liposomes). Nucleic Acids Res 1991;19:5551–5559.
121. Akhtar S, Juliano RL. Liposome delivery of antisense oligonucleotides: adsorption and efflux characteristics of phophorothioate oligodeoxynucleotides. J Control Rel 1992;22:47–56.
122. Hughes JA, Bennett CF, Cook PD, Guinosso CJ, Mirabelli CK, Juliano RL. Lipid membrane permeability of 2′-modified derivatives of phosphorothioate oligonucleotides. J Pharm Sci 1994;83:597–600.

123. Wittung P, Kajanus J, Edward K, Haaima G, Nielsen PE, Norden B, Malmstrom BG. Phospholipid membrane permeability of peptide nucleic acid. FEBS Lett 1995;375:27–29.
124. Ropert C, Lavignon M, Dubernet C, Couvreur P, Malvy C. Oligonucleotides encapsulated in pH sensitive liposome are efficient toward Friend retrovirus. Biochem Biophys Res Commun 1992;183:879–885.
125. Dizik M, Mather E, Nero P, Rao V, Metzler M, Woodle M. Pharmacokinetics of phosphorothioate antisense oligonucleotide-lipid complexes in mice. Seventh Int Symp Recent Adv Drug Del Sys Abstracts Feb. 1995: Salt Lake City, UT, 1995;187–188.
126. Oberhauser B, Wagner E. Effective incorporation of 2′-O-methyl-oligoribonucleotides into liposomes and enhanced cell association through modification with thiocholesterol. Nucleic Acids Research 1992;20:533–538.
127. Zelphati O, Wagner E, Leserman L. Synthesis and anti-HIV activity of thiocholesteryl-coupled phosphodiester antisense oligonucleotides incorporated into immunoliposomes. Antiviral Res 1994;25:13–25.
128. Loke SL, Stein CA, Zhang XH, Mori, K, Nakanishi M, Subasinghe C, Cohen JS, Neckers LM. Characterization of oligonucleotide transport into living cells. Proc Natl Acad Sci USA 1989;86:3474–8.
129. Chin DJ, Green GA, Zon G, Szoka FC. Rapid nuclear accumulation of injected oligodeoxynucleotides. New Biologist 1991;2:1091–1100.
130. Thierry AR, Dritschilo A. Intracellular availability of unmodified, phosphorothioated and liposomally encapsulated oligodeoxynucleotides for antisense activity. Nucleic Acids Res 1992; 20:5691–5698.
131. Boutorine AS, Kostina EV. Reversible covalent attachment of cholesterol to oligodeoxyribonucleotides for studies of the mechanisms of their penetration into eukaryotic cells. Biochimie 1993;75:35–41.
132. Krieg AM, Tonkinson J, Matson S, Zhao Q, Saxon M, Zhang L-M, Bhanja U, Yakubov L, Stein CA. Modification of antisense phosphodiester oligodeoxynucleotides by a 5′ cholesteryl moiety increases cellular association and improves efficacy. Proc Natl Acad Sci USA 1993;90:1048–1052.
133. Krieg AM. Uptake and efficacy of phosphodiester and modified antisense oligonucleotides in primary cell cultures. Clinical Chemistry 1993;39:710–712.
134. Bongartz JP, Aubertin AM, Milhaud PG, Lebleu B. Improved biological activity of antisense oligonucleotides conjugated to a fusogenic peptide. Nucleic Acids Res 1994;22:4681–4688.
135. Hughes JA, Avrutskaya AV, Juliano RL. Influence of base composition on membrane binding and cellular uptake of 10-mer phosphorothioate oligonucleotides in Chinese hamster ovary (CHRC5) cells. Antisense Res Dev 1994;4:211–215.
136. Vinogradov SV, Suzdaltseva Y, Alakhov VY, Kabanov Av. Inhibition of herpes simplex virus 1 reporduction with hydrophobized antisense oligonucleotides. Biochem Biophys Res Commun 1994;203:959–966.
137. Yee F, Ericson H, Reis DJ, Wahlestedt C. Cellular uptake of intracerebroventricularly administered biotin- or digoxigenin-labeled antisense oligodeoxynucleotides in the rat. Cell Mol Neurobiol 1994;14:475–486.
138. Soukchareun S, Tregear GW, Haralambidis J. Preparation and characterization of antisense oligonucleotide-peptide hybrids containing viral fusion peptides. Bioconjug Chem 1995;6:43–53.
139. Clark RE. Poor cellular uptake of antisense oligodeoxynucleotides: an obstacle to their use in chronic myeloid leukaemia. Leuk Lymphoma 1995;19:189–195.
140. Demirhan I, Hasselmayer O, Hofmann D, Chandra A, Svinarchuk FP Vlassov VV, Engels J, Chandra P. Gene-targeted inhibition of transactivation of human immunodeficiency virus type-1 (HIV-1)-LTR by antisense oligonucleotides. Virus Gene 1995;9:113–119.
141. Walker I, Irwin WJ, Akhtar S. Improved cellular delivery of antisense oligonucleotides using transferrin receptor antibody-oligonucleotide conjugates. Pharm Res 1995;12:1548–1553.
142. Wu-Pong S. The role of multivalent cations in oligonucleotide cellular uptake. Biochem Mol Biol Intl 1996;39:511–519.
143. Quattrone A, Papucci L, Schiavone N, Mini E, Capaccioli S. Intracellular enhancement of intact antisense oligonucleotide steady-state levels by cationic lipids. Anticancer Drug Des 1994;9:549–553.
144. Wrobel L, Collins D. Fusion of cationic liposomes with mammalian cells occurs after endocytosis. Biochim Biophys Acta 1995;1235:296–304.
145. Friend DS, Papahadjopoulos D, Debs RJ. Endocytosis and intracellular processing accompanying transfection mediated by cationic liposomes. Biochim Biophys Acta 1996;1278:41–50.

146. Wu-Pong S, Weiss TL, Hunt CA. Antisense c-myc oligonucleotide cellular uptake and activity. Antisense Res Dev 1994;4:155–163.
147. Giachetti C, Chin DJ. Increased oligonucleotide permeability in kerotinocytes of artificial skin correlates with differentiation and altered membrane function. J Invest Dermatol 1996;106:1–7.
148. Thierry AR, Rahman A, Dritschilo A. Liposomal delivery as a new approach to transport antisense oligonucleotides. Izant JG, R.P.E.a. ed. New York: Raven Press, 1992.
149. Clarenc JP, Degols G, Leonetti JP, Milhaud P, Lebleu B. Delivery of antisense oligonucleotides by poly(L-lysine) conjugation and liposome encapsulation. Anticancer Drug Des 1993;8:81–94.
150. Huang SK, Quinn Y, Brown BD, Engbers C, Newman MS, Dizik M, Vaghefi M, Woodle MC. Oligonucleotide-encapsulated in stealth liposomes as antisense carriers against solid tumors. Proc Am Ass Caner Res 1996;37:302.
151. Machy P, Lewis F, McMillan L, Jonak ZL. Gene transfer from targeted liposomes to specific lymphoid cells by electroporation. Proc Natl Acad Sci USA 1988;85:8027–8031.
152. Milhaud PG, Machy P, Lebleu B, Leserman L. Antibody-targeted liposomes containing poly (rc) exert a specific antiviral and toxic effect on cells primed with interferons a/b or g. Biochim Biophys Acta 1989;987:15–20.
153. Leonetti JP, Machy P, Degots G, Lebleu B, Leserman L. Antibody-targeted liposomes containing oligodeoxynucleotides complementary to viral DNA selectively inhibit viral replication. Proc Natl Acad Sci USA 1990;87:2448–2451.
154. Leserman L, Degols G, Machy P, Leonetti J-P, Mechti N, Lebleu B. Targeting and intracellular delivery of antisense oligonucleotides interfering with oncogene expression. Wickstrom E, ed. New York: Wiley-Liss, 1991.
155. Leserman L, Machy P, Zelphati O. Immunoliposome-mediated delivery of nucleoic acids: a review of our laboratory's experience. J Liposome Res 1994;4:107–119.
156. Michael SI, Curiel DT. Strategies to achieve targeted gene delivery via the receptor-mediated endocytosis pathway. Gene Therapy 1994;1:223–232.
157. Jaaskelainen I, Monkkonen J, Urtti A. Oligonucleotide-cationic liposome interactions. A physicochemical study. Biochim Biophys Acta 1994;1195:115–123.
158. Gao X, Huang L. Cationic liposome-mediated gene transfer. Gene Therapy 1995;2:710–722.
159. Skutella T, Stohr T, Probst JC, Ramalho-Ortigao FJ, Holsboer F, Jirikowski GF. Antisense oligodeoxynucleotides for in vivo targeting of corticotropin-releasing hormone mRNA: comparison of phosphorothioate and 3′-inverted probe performance. Horm Metab Res 1994;26:460–464.
160. Ogo H, Hirai Y, Miki S, Nishio H, Akiyama M, Nakata Y. Modulation of substance P/neurokinin-1 receptor in human astrocytoma cells by antisnse oligodeoxynucleotides. Gen Pharmacol 1994;25:1131–1135.
161. Szklarczyk A, Kaczmarek L. Antisense oligodeoxynucleotides: stability and distribution after intracerebral injection into rat brain. J Neurosic Methods 1995;60:181–187.
162. Morishita R, Gibbons G, Ellison KE, Nakajima M, Zhang L, Kaneda Y, Ogihara T, Dzau VJ. Single intraluminal delivery of antisense CDC 2 kinase and proliferating-cell nuclear antigen oligonucleotides) results in chronic inhibition of neointimal hyperplasia. Proc Natl Acad Sci USA 1993;90:8474–8478.
163. Villa AE, Guzman LA, Poptic EJ, Labhasetwar V, D'Souza S, Farrell CL, Plow EF, Levy RJ, DiCorleto PE, Topol EJ. Effects of antisense c-myb oligonucleotides on vascular smooth muscle cell proliferation and response to vessel wall injury. Circ Res 1995;76:505–513.
164. Cammilleri S, Sangrajrang S, Perdereau B, Brixy F, Calvo F, Bazin H, Magdelenat H. Biodistribution of iodine-125 tryamine transforming growth factor alpha antisense oligonucleotide in athymic mice with a human mammary tumour xenograft following intratumoral injection. Eur J Nucl Med 1996;23:448–452.
165. Pimm MV, Hudecz F. Biodistribution in tumour-bearing mice of polycationic, amphoteric and polyanionic branched polypeptides with a poly(L-lysine) backbone labelled with 125I and 111In: tumour accumlation less than that of labelled serum proteins. J Cancer Res Clin Oncol 1996;122:45–54.
166. Woodle MC. Physicochemical and biological requirements to achieve drug targeting in gene therapy. Walsh B, ed. Southborough, MA: IBC, 1996.

Lasic and Papahadjopoulos (eds.), Medical Applications of Liposomes

CHAPTER 6.1

Other applications

Danilo D. Lasic[a] and Demetrios Papahadjopoulos[b]

[a]*Liposome Consultations, 7512 Birkdale Drive, Newark, CA 94560, USA*; [b]*Department of Cellular and Molecular Pharmacology, University of California, San Francisco, CA and California Pacific Medical Center Research Institute, San Francisco, CA 94115, USA*

Introduction

A vast majority of liposomal medical applications uses parenteral injection (systemic intravenous, or localized, subcutaneous, intramuscular, or intra-site). There has also been considerable work on other routes of administration, such as topical and pulmonary, intra-articular, intra-arterial, dermal, and oral. For the latter applications, liposomes can be formulated as a colloidal suspension, nebulized aerosol or dry powder for pulmonary applications, encapsulated into a gel matrix for sustained release, while dermal administration makes use of various semi-liquid or semi-solid lotions, creams, or gels. A system can be gelled either on the basis of lipid formulation (concentration, unscreened repulsive forces) or simply by adding hydrophilic thickener, such as carbopol, cellulose polymers or polyhyaluronic acid at low weight fraction ($<$1–0.5%).

The major modes of action of liposomes include (i) solubilizer (or suspending agent) for difficult-to-dissolve drugs, (ii) sustained release system, (iii) passive accumulation (RES, or tumors and sites of inflamation for small long circulating liposomes) or (iv) active targeting carrier. In addition to these, liposomes or their constituents can serve as skin penetration enhancers, systems to reduce surface tension in the lung (see Chapter 6.2 by A.D. Bangham), systems to adsorb cholesterol, buffering system to reduce toxicity of surfactant drugs, lubricant of joints, and also for carrying oxygen in synthetic blood (see Chapter 6.4 by Beissinger and collaborators). For the majority of these applications, it is still too early to predict the real potential and eventual development of pharmaceutically acceptable products.

In parallel to therapeutic applications, liposomes have also become useful in medical diagnostics (see Chapter 6.6 by V. Torchilin) as well as high tech cosmetics and nutrition. While we will not go into details, we can briefly mention clinical diagnostic studies of imaging as well as diagnostic kits for blood coagulation. In the latter case, which was for years based on various biological reagents, such as aqueous suspensions of rabbit brain lipid extract, well defined liposomal dispersions have yielded significant improvements in detection and accuracy.

For instance, a rapid screening test for coagulation system, Innovin[R] (Dade International, Miami) is a lyophilized product, which consists of liposomes containing recombinant human tissue factor, and is used for monitoring deficiencies in blood clotting. The company has two more liposome-based products as does Johnson and Johnson. In addition to such products which are based on simple agglutination, the property of liposomes that can contain hundreds of marker molecules incorporated and a single specific ligand can be used for ELISA type of assays. At the academic level, antigen coupled liposomes are used in immunodiagnostics[1] such as for homogeneous immunoassays of polyclonal antibodies.[2] Steric stabilization of such systems may reduce non-specific interactions[3] while fragments of DNA coupled to liposomes were used in PCR systems.[4] We are aware of several others and it is likely that more products which can be used in medical diagnostics will soon be commercially available. One of the possible next developments of liposomal (bio)sensors would be phase immobilization of liposomes into a transparent gel matrix.

We believe that pulmonary applications are more promising than currently anticipated (see Chapter 6.3 by H. Schreier). The same is probably true for the case of transdermal delivery. Definitively different formulations show different skin penetration fluxes[5] (see Chapter 6.5 by N. Weiner et al.). At present, the mechanisms for skin penetration are still not understood but it is possible that many liposome components, especially the ones with high values of hydrophilic-lipophilic balance can act as penetration enhancers. In addition, some of these preparations also contain classical penetration enhacers such as ethanol, fatty acids, bile salts, various laurates etc. One of them may be transfersomes, introduced by Cevc and colleagues,[5] which according to some compelling data, can achieve large fluxes of associated agents through the mouse skin. More sophisticated cosmetic products based on liposomes also contain various antibiotics, antifungals, antiinflammatory agents, and even recombinant proteins.[6] Liposomal formulations were also recently introduced in nutrition. Sublingual liposomal sprays are used to deliver vitamins, minerals, and other active agents (BioZone Labs, Pittsburg, California). With improved understanding of liposome stability, some oral applications may become feasible.

A significant barrier to the development of liposomes for oral or mucosal delivery of macromolecules has been the susceptibility of liposomes formed from commonly available lipids to enzymatic degradation, low pH in the stomach and bile-salt dissolution in the intestine. Hence common liposomes fail to protect macromolecules against enzymatic degradation in the gastrointestinal tract. Mechanically and sterically stabilized liposomes can survive these conditions, but are consequently too stable in the intestine to deliver the encapsulated drug via the normal absorption process. A possible solution to this problem would be to design liposomes with time-dependent stability or using other vehicles to deliver them past the stomach. Direct uptake by M cells in Peyer's patches, which sample the surroundings for antigens, seem to be too low for substantial absorption of drugs, but may be effective for vaccination where low concentration of antigens may be adequate to stimulate an immune response. Recently, the potential of utilizing polymerized liposomes as oral drug and vaccine vehicles has been explored.[7,8] Such liposomes survive intestinal transit intact and can be rapidly taken up by M cells and subsequently by macrophages. Providing M-cell targeting molecules by the covalent attachment of lectins and other ligands produces further increases in intestinal liposome uptake.[8] Very stable liposomes and polymerized liposome vehicles may provide a route for the commercial development of liposome technology for oral vaccination and, eventually, for a wide variety of macromolecular drugs. With respect to the development of polymerized liposomes, however, several safety issues, such as excretion of polymerized lipids and toxicity of unpolymerized polymerizable lipids, will have to be critically addressed.

The chapters included in this section exemplifiy the most successful areas for the use of liposomes in a variety of applications ranging from pulmonary to in vivo diagnostics and topical applications.

References

1. Singh A, Carbonell RG. Liposomes as immunodiagnostics. In: Lasic DD, Barenholz Y, eds. Handbook of Nonmedical Applications of Liposomes, Vol IV: From Gene Delivery and Diagnostics to Ecology. Boca Raton, FL: CRC Press, 1996;209–228.
2. Katoh S, Kishimura M, Fukuda H. Use of antigen coupled liposomes for homogeneous immunoassays of polyclonal antibody. In: Lasic DD, Barenholz Y, eds. Handbook of Nonmedical Application of Liposomes, Vol IV: Gene delivery and diagnostics to ecology. Boca Raton, FL: CRC Press, 1996;199–208.
3. Emanuel N, Kedar E, Toker O, Bolotin E, Barenholz Y. Steric stabilization of liposomes improves their use in diagnostics, In: Lasic DD, Barenholz Y, eds. Handbook of Nonmedical Applications of Liposomes. Boca Raton, FL: CRC Press, 1996;229–244.
4. Rule GS, Montagna RA, Durst RA. Rapid method for visual identification of specific DNA sequences based on DNA-tagged liposomes. Drug Carr Syt 1996;13:257–388.
5. Cevc G. Lipid suspensions on the skin. Permeation enhancement, vesicle penetration and transdermal drug delivery. Crit Rev Ther Drug Carr Sys 1996;13:257–388.
6. Yarosh D, Bucana D, Alas L, Kibitel J, Kripke M. Localization of liposomes containing a DNA repaid enzyme in murine skin. J Invest Dermat 1994;103:461–468.
7. Okada J, Cohen S, Langer R. Pharm Res 1995;12:576–582.
8. Chen H, Torchilin V, Langer R. Pharm Res 1996;13:1378–1383.

Lasic and Papahadjopoulos (eds.), Medical Applications of Liposomes

CHAPTER 6.2

Artificial lung expanding compound (ALEC™)

A.D. BANGHAM

The Old Garage, 17 High Green, Great Shelford, Cambridge CB2 5EG, UK

Abstract

The mechanics of the lung depend upon two, unrelated, properties of phospholipids. At birth, they reduce the burden of expanding an air/water interface from 1 cm^2 to 2 m^2 by lowering surface tension. Subsequently, plaques of condensing dipalmitoylphosphatidylcholine help ratchet the alveolar spaces open and keep them dry for the rest of life.

ALEC™ is a dry protein-free powder consisting of a 7:3 mole/mole mixture of dipalmitoylphosphatidylcholine (DPPC) and phosphatidylglycerol (PG) and seems to function as a reasonably good substitute for natural lung surfactant in very premature babies. It has the following necessary properties: 1. It spreads rapidly and spontaneously at an air/water interface at 37°C reducing the surface tension of water by about 2/3. 2. The unsaturated PG moiety can be squeezed out of the mixed monolayer by rapid overcompression (equivalent to exhalation) and is irreversibly lost to the surface system by reassembly into liposomes. 3. The residual dipalmitoylphosphatidylcholine (DPPC) becomes progressively enriched to the point that, at 37°C, it condenses out as a solid phase so rigid that it prevents the alveolae from collapsing; a structure resembling and sharing the properties of a geodesic dome. 4. The preparation is protein-free. 5. New rapid acting delivery techniques are described. 6. ALEC™ is now the most frequently prescribed prophylactic formulation in the UK. Clinical experience and commentaries within the public domain are listed.

The collective wisdom which maintains that zero or near-zero surface tension accounts for the anti-oedema effect of lung surfactant is challenged and considered to be misleading.

The Old Garage, 17 High Green
Great Shelford, Cambridge CB2 5EG

Saturday, August 1996

Dear Demetri,

Thank you so much for the opportunity of contributing to this book. I am doing so, as you have by now realised, by writing you a letter, perhaps even a second to Danilo, a literary genre which lends itself to an informality of presentation you seem to invite and in which I have already indulged[1] and encouraged.[2] Its virtue, provided no-one is offended, is that it permits the writer to indulge in asides which no editor of a peer-reviewed journal would ever sanction. Its weakness is that a scientific manuscript which has not been honed by criticism is the lighter authority, or is it?

The identifying acronym derived from the title of this article was somewhat abruptly proposed by my colleague Colin Morley at a meeting of neonatologists held on Mackinac Island in 1982. ALEC is no more or no less than a sophisticated freeze-dried formulation of liposomes and thus, surely, qualifies for space in this book! The adjective *artificial* was chosen to differentiate our formulation, which consisted of just two synthetic, but nevertheless, naturally occurring phospholipids from those of so-called *natural* lung surfactants typically containing a miscellany of phospholipids and proteins and derived from human or animal (calf, beef, porcine) lungs.

Preoccupied as we were, with the expanding properties of liposomes[3–5] I nevertheless found it difficult to resist the challenge of explaining how Clement's phospholipid lung surfactant (LS) squared with the infinitesimal critical micelle concentration (CMC) of lecithins. Our hands-on experience with the lecithins was that the purer the source, the less tendency there was for froth to form when dispersions were shaken in water; no monomers, no froth. What trick had Nature devised that enabled the phospholipids, so securely locked into the lamellar bodies of Type II cells, to become available and facilitate the life-depending and rapidly extending air/water interface of the newborn?

Jeoff Watkins invited himself back from Australia (recollect, you published a liposome paper with him[6]) and I asked him to confirm the major surface properties of DPPC.[7] At that time we were naive enough to believe that the pull on a Wilhelmy plate could be converted to a surface pressure as justified by Clements and Tierney[8] from the expression $\gamma_{\text{film}} = \gamma_{\text{water}} - \pi_{\text{film}}$, where γ_{film} is the surface tension of an interface covered by a film, γ_{water} is the surface tension of air/water and π_{film} is the surface pressure of the insoluble phospholipid.

With hindsight, it is noteworthy that we preferred to report our measurements as surface pressure rather than surface tension, failing to realize then that the relationship of tension and pressure does not hold beyond equilibrium for insoluble surfactant substances.[7] Later still, we were to distinguish between the orthodox use of the term surface tension and diminishing "contractile force". Today I would suggest that the pull registered by a strain gauge attached to a dipping plate be considered simply as an "operational" value. Watkin's paper fully confirmed Clement's claim that DPPC was the active agent in natural lung surfactants and not the protein suggested by Pattle.[9]

Our experience, with phospholipids, surface balances and such-like devices invited visitors to our laboratory, more often than not, to assay their particular lung surfactant preparations and the ALEC saga started! To quote from my enduring colleague, Colin Morley, who described[10] his first visit to Babraham in 1976 as follows:

> I turned up in his laboratory unannounced and received a friendly but rather cool reception, particularly when I told him that I was asking for help with studying the surface tension of lung surfactant. I learnt in retrospect that this was a topic he had been faced with before and had been frustrated by the curious behaviour of surfactant monolayers. However, he graciously offered to help and the apparatus was

duly set up. I remember the experiment very clearly. Bangham took some of the freeze dried surfactant, rather casually shook it up with some saline, poured it into a circular surface tension trough about 5 cm in diameter and measured the surface tension. His whole manner changed immediately and he became quite excited, confirming that the substance I had produced was very surface active. We sat down at coffee to discuss more details about what we had produced and then returned to repeat the experiment. This time the surface tension fell only slightly. We then spent the rest of the day trying to repeat the initial result. Bangham then made the observation that in the original experiment, the surfactant had not been very well mixed and that there had been particles floating on the surface but when we had repeated the second experiment all the surfactant had become thoroughly wetted. . . . Bangham's conclusion was that the dried lung surfactant had a molecular configuration which allowed molecules to spread rapidly on the surface of water and thereby lower the surface tension whereas once the surfactant had become thoroughly mixed with water it would be very largely in the form of a smectic mesophase and predominately as liposomes.

About a year later Morley presented our work to participants at a meeting of neonatologists and concluded[11] with these words:

We believe that the surfactant monolayer solidifies during full expiration, and, acting like an archway of bricks, splints the alveoli open. The mechanism should be understood at the molecular level; when there is an excess of phospholipid molecules, like DPL (DPPC), on a wet surface they will spread until they are tightly packed in a monolayer having a high equilibrium surface pressure. If this close packed film is compressed beyond the equilibrium spreading pressure two things may happen. Firstly, the film area is reduced with little increase in surface pressure as molecules are squeezed out of the surface leaving pure DPL (DPPC) in a sort of survival of the fittest, or more most stable molecules. The contractile forces within the alveolus force the molecules together and due to their long symmetrical shape the stearic hindrance allows them to set solid and withstand high pressures without buckling. It was calculated by Gaines[20] that a pressure of 1 dyne/cm acting on a monolayer 20 A high corresponds to a three dimensional pressure of 5 atmospheres (5 atms × 70 = 350 atms). The thermodynamic equilibrium of surface pressure (π_{film}) and the reduction of surface tension $\gamma_{\text{film}} = \gamma_{\text{water}} - \pi_{\text{film}}$ breaks down because the molecules have now changed in state from liquid to solid. Shah and Schulman [1967] showed by the talc test that the surface monolayer of DPL (DPPC) solidifies on compression. Thus it is incorrect to refer to the 'surface tension' of a monolayer which is condensed or hypercompressed.

We stand by such concepts paraphrased 20 years ago and illustrated in Figures 2 and 5, ALEC was formulated accordingly.

The properties of 'dry' surfactant (Figure 1) were more formally reported in 1978 remarking that there had been many attempts to treat respiratory distress syndrome (RDS) with surfactant substances as a mist nebulised with water and wondered whether our serendipitous 'dry' surfactant would work better.[12] Our suggestion was dismissed as being quite unrealistic despite a lone report that the phospholipids of lamellar bodies gave NMR signals compatible with their being 'dry'.[13]

Encouraged by our experience with the aggressive spreading properties of dry ALEC, Morley courageously gave it to babies as a dust "bagged" through an endotracheal tube.[14] Alas, the method was abandoned because the dust tended to block the moist end of the tube and all formulations are now routinely delivered, very inefficiently, as voluminous dispersions in saline or nebulized. All methods, not surprisingly, encourage the formation of liposomes and reduce the population of monomers available to adsorb to the air/water interface. The protocol for the delivery of ALEC™ attempted to mitigate this problem by dispersing the freeze-dried material (some liposomes, some coagel) in cold saline (<25°C) and relying upon the release of monomers upon warming above $T_c = 33°C$ to body temperature. This idea has recently been revived by Perkins, Dause. Parente, Minchey,

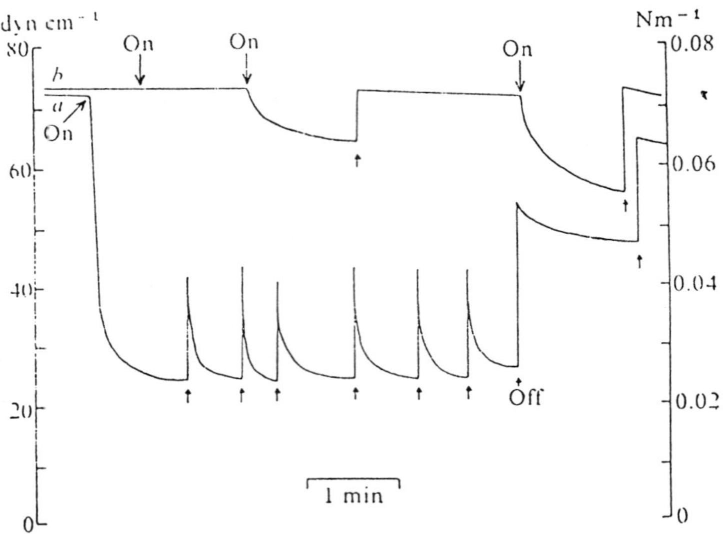

Fig.1. The surface tension lowering properties of *a*, dry surfactant and *b*, wet surfactant placed on saline. In both instances surfactant was placed on the surface at 'On'. The dry particle was removed from the surface at 'Off'. Upwards-pointing arrows indicate when the surface monolayer was partly removed by aspiration. From Reference 12.

Neuman, Gruner, Taraschi and Janoff[15] who alloyed DPPC with dioleylphosphatidylethanolamine (DOPE) and cholesterol to permit a lamellar to hexagonal$_{II}$ phase change to take place with release of monomers when the temperature is raised. It is not, however, clear that the presence of cholesterol enables the DPPC to fulfill its most important function namely to prevent the alveolae from collapsing by forming solid plaques. In other words, can a film of the DPPC/DOPE/cholesterol sustain a surface pressure of >70 mN/m?

You would, I'm sure, be interested to know that we are again looking at methods that bypass the hydration-repulsion forces that limit the availability of monomers when LS is delivered in the form of liposomes. For children and adults, for example, the contents of a vial of ALEC™ can be easily blown or inhaled as an exquisite dust which can be experienced as being no more irritating to mucus membranes or the eye than melting snowflakes! It is likely that the unimpeded flight of a mgm of dust is more effective than 100 times it's weight delivered as an aqueous dispersion. Another method[16] is to evaporate a solution of lung surfactant onto the tip of an endotrachael tube thus presenting surfactant in the form of a coagel, a completely dry source of ALEC available within seconds. Clinical trials with this device are taking place, as I write.

But 20 years ago it was generally accepted that the melting temperature of DPPC in the presence of water was 41°C and would thus fail to spread at body temperature. In Babraham, however, we knew that the hydrocarbon region of bilayers behaved as a bulk phase, albeit hydrocarbon with regard to the presence

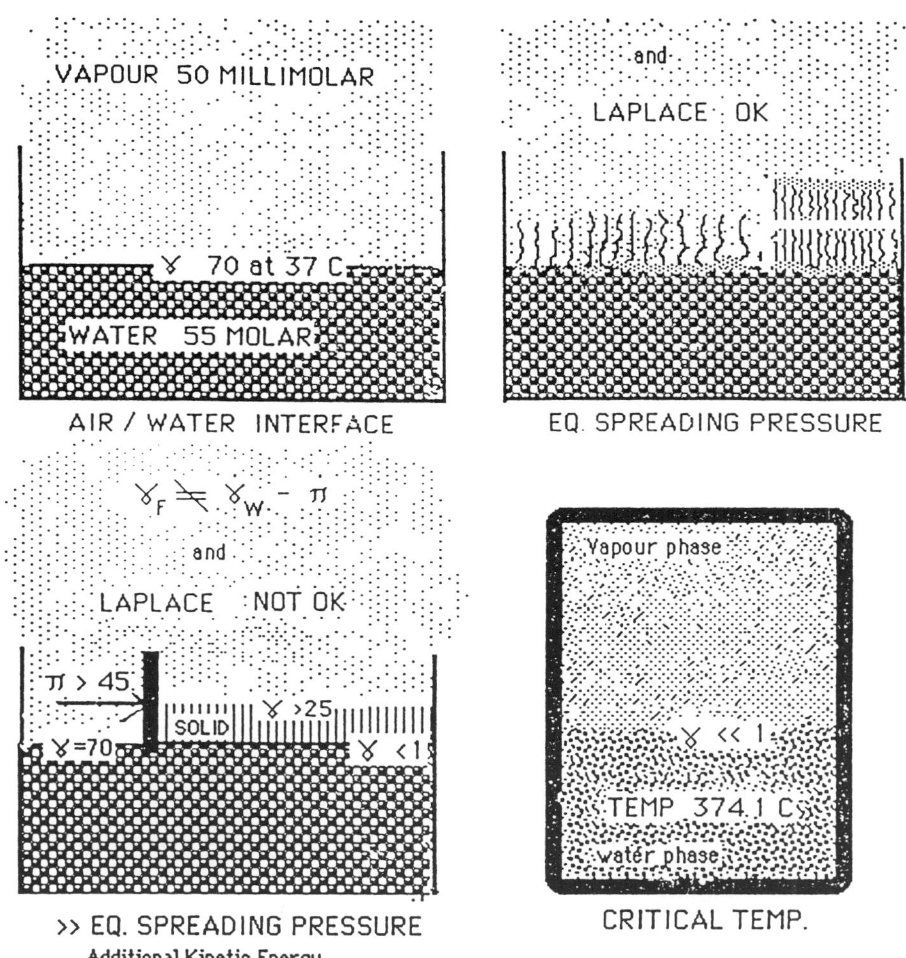

Fig. 2. Top left: Air/water interface illustrating concentration difference between phases, uniform surface energy (70 mJ/m^2). Top right: Water surface covered by spreading phospholipid in equilibrium with dry crystal, uniform surface energy (25 mJ/m^2); values of γ_{film} valid; so can be used in Laplace equation. Bottom left: Film of phopholipids is being compressed on a substrate below the melting temperature. The surface tension cannot be derived from values of surface pressure π_{film} and are not valued for use in the Laplace equation. Bottom right: The only situation which the surface free energy (surface tension) of a vapour/water interface approaches zero. Reproduced from Reference 22.

of solutes, e.g., chloroform, butanol.[17] It became obvious to suggest that there were molecules in natural lung surfactant that, acting as hydrophobic solutes, could also lower the melting temperature of the DPPC. Morley and I tried many combinations of DPPC with various compounds before we achieved the desired colligative effect whilst retaining the unique property of the monolayer to withstand high surface pressures at 37°C. In the end we settled for a 7:3 mixture of

DPPC and egg PG (PG), the two most abundant phospholipids found in natural lung surfactants. It should be noted that EXOSURF is formulated with two alien ingredients, hexadecanol and tyloxapol both of which might, if accumulated, give rise to complications.

Parenthetically, I recollect the Oxford Professor of Child Health warning us that he would not, under any circumstances, use our proposed formulation if it contained any animal protein. A wise warning in the light of recent human growth hormone and bovine spongiform encephalopathy epidemics leading to versions of Creutzfeld Jacob disease. To quote Haywood[18]

> any pharmaceutical product that is isolated from animals (especially from primates) has the potential to be carrying a persistent virus.

In 1979 we were confident enough to publish a paper entitled The Physical Properties of an Effective Lung Surfactant.[19] Our reasoning was based upon two important properties which a 7:3 mixture of DPPC/PG possessed namely, the ability to spread rapidly (dry) and to become solid (under compression) at 37°C.

It was as a result of observing just how solid a film of DPPC/PG became under pressure that we started to worry seriously about the true meaning of 'surface tension' and of the universal acceptance of claims that lung surfactants, natural or artificial, could lower surface tensions at air/water interfaces to near-zero values. We invited Michael Phillips, a bone-fide surface chemist, to endorse our suspicions and in our paper we chose to refer to the pull registered by our Wilhelmy dipping plate as 'surface contractile force' thus avoiding an ambiguity which we felt might confuse physical chemists. Our reasons for rejecting the claims by other authors to have reduced the surface tension of an air/water interface to less than 20 mN/m were supported by a paragraph in Gaines's classic book[20] and by the realisation that a zero or near-zero claim was unimaginable at a phase boundary concentration step greater than 1000:1 (water is 55 molar, vapour 50 millimolar). A full discussion of these fundamental matters and of their implications regarding their role in replacement therapy can be found in my letter to Danilo Lasic which follows and in the references.[21–24]

But how were we to make a product with the two phospholipids which could be delivered in dry form to premature infants down an endotrachael tube no more than 2.5 mm diameter? We decided that we wanted a dry, light, crushed meringue-like substance that would easily blow into a dust. Nigel Miller achieved it brilliantly by freeze-drying liposome suspensions.

The very first 200 mg dose of the freeze-dried formulation, was given by me (5th April 1978) to a premature foal born by Caesarian section from an injured mare (Newmarket is a mere 12 miles from Cambridge). The dry powder was blown directly down a long catheter inserted in an endotrachael tube which reached the division of the bronchi. Anecdotal it may have been but the benefit was dramatic, and the control of RDS now shown for the first time in Figure 3. It was a very important result and encouraged Morley to treat his first distressed baby in September 1979. The results of his pilot trial were published in the Lancet[14] and were no less encouraging than those published a year earlier by Fujiwara,

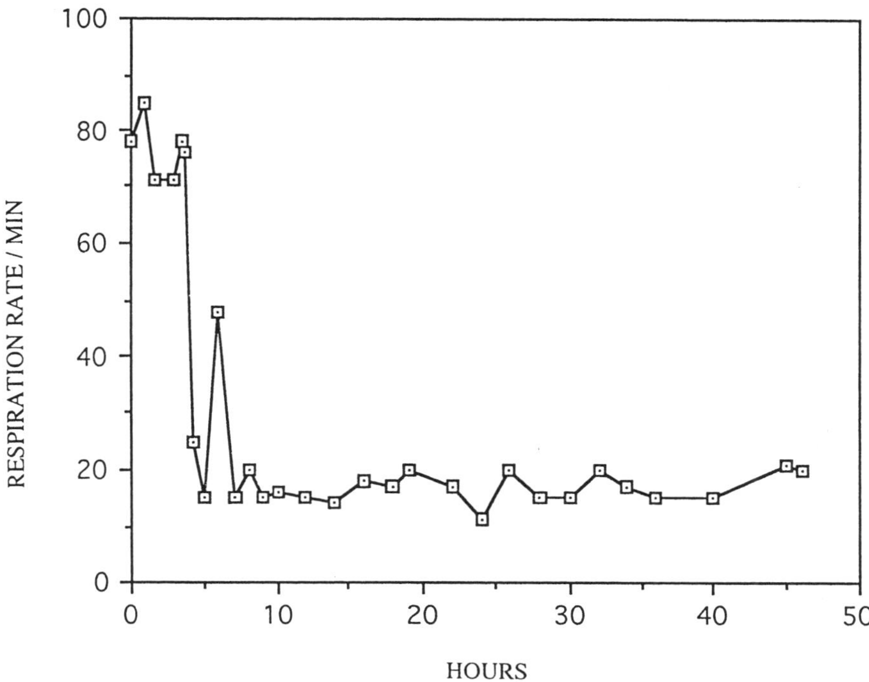

Fig. 3. On the 5th April 1978, foal Gitane: became the first case of respiratory distress syndrome (RDS) treated with dry artificial lung expanding compound (ALEC). 200 mgs were blown through a polythene catheter introduced down the length of an endotrachael tube. It was estimated that the catheter reached the bifurcation of the bronchi. The weight of the foal was approximately × 10 that of a premature human infant but nevertheless seemed to respond to what is now considered a normal infant dose. Courtesy Peter Rossdale, Beaufort Cottage Stables, Newmarket, Suffolk.

Maeta, Chida, Morita, Watabe and Abe[25] using an enriched bovine lung wash out formulation, now marketed world-wide as SURVANTA.

In all, Miller prepared something like 500 doses over a period of 10 years enabling Morley and colleagues to undertake both pilot and multicentred trials.[14,26,27] The results, better than a 40% reduction in mortality, encouraged us to seek commercial manufacture and distribution. In 1987 Britannia Pharmaceuticals acquired the rights to make and market ALEC™ from a small limited liability company Morley and I had formed. Clinical experience and comment in the public domain can be examined in the following papers by my colleague, Colin Morley, his assistants and his wife.[28–32]

From 1987 to 1994 we endured, with Britannia Pharmaceuticals Ltd, the agony of acquiring Master files for the ingredients, good manufacturing practice clearances and, of course, a Product Licence. About halfway through this nightmare, EXOSURF (Wellcome) was granted a UK license and proceeded to establish itself as the major formulation within the UK as well as in many other parts of the world. Britannia marketed ALEC at half the price and coupled with an instruction to use half the volume of saline! In two years ALEC achieved market dominance in the UK.

But there were physicians who, reading the literature, believed that protein containing formulations did what they believe was required faster and preferred to use them. A "pink" baby became a benchmark but until someone, somewhere carries out a double-blind trial to establish this claim, I remain doubtful. The effectiveness of lung surfactant depends primarily upon an efficient method of delivering DPPC. I believe this can be achieved with pure phospholipids and without animal or human proteins, alien detergents and their inherent dangers. I think the future lies in the simple expedient of presenting the phopholipids in the form of a coagel or as dust particles. We are currently investigating both these suggestions and clinical trials are in progress.

Yours sincerely,

PS: Postscripts are irresistable reading and chronicle, more often than not, the most important points that the correspondent wishes to make.

To begin with I cannot resist recalling that soon after Britannia Pharmaceuticals Ltd took over the ALEC project they requested about 50 g of egg PG for quality testing etc. This was early 1989 and I was now well into retirement and Nigel Miller was committed to run the Babraham Institute cell sorter, full time. I rashly offered to undertake what Miller had spent a lifetime doing and found myself standing at the back of the Stores queue, humping heavy Winchesters of chloroform back to the tiny, temporary laboratory I had occupied 31 years previously, patiently collecting fractions from off a large silicic acid column (which often went dry) and desperately hoping that the quantity and quality of the PG justified all the lovely fresh eggs and cauliflower's, the litres of chloroform, acetone and methanol, not to mention the time and physical effort that has gone into the earlier stages of the preparation. My efforts were pathetic and if Chris Evans (Enzymatics), starting up his now multimillion pound biotechnology company (Chiroscience) had not been renting the laboratory next door I doubt whether ALEC™ would ever have been produced. Amersham International perfected the freeze-drying methodology and Britannia Pharmaceuticals sold it and turned it into the best seller in the UK. I have learned a great deal.

The Old Garage, 17 High Green
Great Shelford, Cambridge CB2 5EG

Tuesday 17th December 1996

Dear Danilo

You should know that my interest in lung surfactant was first alerted by John Clements (and his in liposomes) when we both attended a Harden Conference in 1963 on the role of phospholipids in biological systems. A year later I spent a memorable week in his department highlighted by an experiment in which a monolayer of very pure dipalmitoylphosphatidylcholine (DPPC) was squeezed to an extent that a strain gauge (of which I was very envious) connected to a Langmuir barrier registered 70 mN/m or was the gauge measuring a near-zero pull on a Wilhelmy dipping plate? Significantly for Clement's understanding of lung mechanics, the measurement was interpreted as demonstrating a near-zero surface tension at an air water interface. I was impressed but my critical senses, non-existent!

As a physical chemist, Danilo, I forgive you for not knowing that LS is an acronym of Lung Surfactant, a deficiency of which gives rise to a condition known as RDS or Respiratory Distress Syndrome, prevalent in infants, and ARDS, the adult version of RDS. In my opinion, LS should be considered to participate in two, quite different, yet rarely distinguished physico-chemical processes. The first occurs at birth and involves the rapid extension of an air/water interface from a mere 2 cm^2 (surface area of the baby's larynx) to some 2 m^2 of an expanded baby's lung. The work required for this two-dimensional extension of a fluid/air interface is not inconsiderable and the work required naturally increases as the lungs expand. Comroe[33] gives a fine account of von Neergaard's seminal observations[34] relating to this aspect of respiratory physiology and I commend this publication for many other reasons. Thus, any material that lowers the surface tension of an air-water interface would facilitate this process but, in practical terms, the recruitment of such molecules must be at least as rapid as occurs with soapy bathwater. It is this requirement (necessity) of rapid recruitment that has posed the fundamental puzzle surrounding the initial process of lung expansion (the first few breaths after birth) because there is little evidence that the tracheal fluid of even mature new-borns, release material that will rapidly lower an air/water surface tension. Quite frankly, I was astounded at the number of samples we analysed only to find that the surface tension (γ_{water}) of full-term amniotic fluid (the continuous fluid phase outside and inside the foetus before birth) was close to that of water despite the presence of massive quantities of phospholipids in the form of liposomes, DPPC being the most abundant.

So, dealing with the first property of LS, how important is it to lower the surface tension of the extending air/water interface? Certainly, if the tension can be reduced for that of a typical soap (20–25 mN/cm) the amount of work is reduced by about 2/3. On the other hand the approximate air pressure available to support the air/fluid interface at inspiration is about 20 cm of water which, from the Laplace equation, limits the minimum curvature of any liquid meniscus in the alveolus to a radius of 70 microns if the near-instantaneous surface tension of the tracheal fluid is 72 mN/cm. This minimum curvature reduces to about 25 microns if the surface tension is reduced to 25 mN/cm. Thus, there appear to be no theoretical reasons why absence of LS in the amniotic fluid should hinder the extension of the air/water interface down the newborn's trachea, bronchi and smaller branches, to the limit of an advancing air "bubble" of radius 70 microns.

The second, and in my view the more important biophysical property of lung surfactants (natural and artificial) is to ensure that the alveolar spaces are progressively ratcheted open to whatever size is compatable with their anatomical site and to keep them open and dry for a lifetime. This is known as the anti-oedema effect. Nature has anticipated these problems by providing lung surfactants, recruited from crystalline sources, viz. the lamellar bodies found in Type II cells or by mixtures of pure phospholipids as formulated in ALEC and other lung surfactant preparations.

But if these alveolar spaces are indeed, lined by liquid (water with salts, proteins etc.), are of small dimension (<10 microns) and the air pressure close to zero (as might be at full expiration) why do they not fill with water and thereby drown the individual? Contemporary wisdom suggests that lung surfactants solve this problem by reducing the surface tension at the alveolar air/water interface to whatever value (γ_{film}) is necessary to satisfy the Laplace ratio $\Delta P \cdot r/2$ where ΔP is the difference in pressure between the inside and outside of the alveolar space, r the radius. For many years the target value was considered to be <10 >5 mN/m but subsequent formulations vied with each other to lower values and claiming improved activity. The ultimate formulation[35] registered 0 mN/m! This cannot be so, or can it?

As I have pointed out in a number of articles[21–24] surface tension or surface stress (i.e. a solid surface with a shape not at equilibrium) must exceed 20 mN/cm wherever two adjacent phases, one of which is hydrocarbon, differ greatly in density as with a air/liquid or air/solid. Moreover the situation in the alveolus is likely to be complex with a solid (duplex), liquid expanded mono and oligolayers of phospholipids and clean water interfaces to air. No measuring device known to man is capable of offering honest values for surface tension/pressure, for both compression and expansion cycles, of such a complex film. And yet we have all naively interpreted the "pull" experienced by a high energy plate claiming that it is measuring "surface tension". We did so until, one day, we saw a spinning dipping plate stop when the "pull" became, zero (Figure 4).

As a Senior Citizen one seems to have rather more time to contemplate, laterally and in depth, the problems one has been investigating over the years, and that is not a bad thing to do. 'Nullius in verba' is the motto of the Royal Society of London for Improving Natural Knowledge, and so, long before ALEC was conceived as a commercial formulation, I found myself questioning the current widely held notion that the surface tension of a water/air interface within an alveolus can be reduced to values below 20 mN/m. For example, Comroe[33] refers to a "unique substance formed and secreted by pulmonary Type II alveolar cell, that decreases surface tension to very low levels at low alveolar volume (<10 mN/m) and so serve as an anti-collapse factor".

My objections to the prevailing theory are threefold and consider them neither pedantic nor semantic since a failure to recognise the original misconception, perpetuated consistently during the past 30 years, has obfuscated a useful and more realistic understanding of lung mechanics. It has also resulted in a number of commercial formulations that are, in my opinion, unnecessarily complex being based (naively) upon the speed at which their dispersions reduced the pull on a Wilhelmy plate (proteins, alone, could be expected to lower the surface energy of a plate). If for no other reason, formulations containing protein should have heeded the warning given by Haywood[18] that animal proteins can also be dangerous.

So as far as I can ascertain the original sin was committed by Pattle who whilst undoubtedly making the most significant observation upon the oddness of lung bubbles, failed to distinguish between the terms surface tension, surface stress and interfacial tension. In para 2 line 6 of his original letter to Nature[35] he wrote "The *surface tension* of the lung bubbles is therefore zero;" and later in the last 3 lines of the same para "It is thus evident that the alveoli are lined with an insoluble protein which can abolish the *tension of the alveolar surface*". Thus the word "tension" carried two distinct meanings, the first as a manifestation of surface free energy (mJ/m^2) and defined above and secondly something entirely different namely, an absence of "tension" between the liquid phase of the alveolus and the air space due to the presence of solid material. Pattle imagined it to be denatured protein, Clements showed it to be phospholipid but, in my view, chose Pattle's wrong explanation as to why the bubbles failed to behave like soap bubbles. To quote from a later paper[9]

> It is evident from these facts that bubbles obtained from the lung in various ways are lined with a film which can exert a surface pressure sufficient to reduce the net surface tension nearly to zero

Now, unless I am much mistaken, Pattle and Clements must have read the same two chapters of Harkin's book[37] for it is Clements and Tierney who quote that a

> film is said to exert a pressure equal and opposite to the fall in surface tension, i.e. film pressure equals surface tension without film minus surface tension with film

or $\pi_{\text{film}} = \gamma_{\text{water}} - \gamma_{\text{film}}$ where π_{film} is the pressure, γ_{water} and γ_{film} the surface tension of water and of the film respectively and the relationship illustrated in Fig. 2, p. 1567, of their chapter.

But seeing a heavy, spinning, platinum plate immobilised by a monolayer of phospholipids, ostensibly, subjected to a pressure of 70 mN/m but at the same time registering zero "surface tension", challenged credulity (Figure 4). It was when I came across a paragaph in George Gaines monograph[20] spelling out a warning that not all surface pressures could be converted to surface tension that I began to understand a misunderstanding! In a compelling paragraph which read:

> It should be noted that, at pressures above the monolayer stability limit (approx 45 mN/m for DPPC

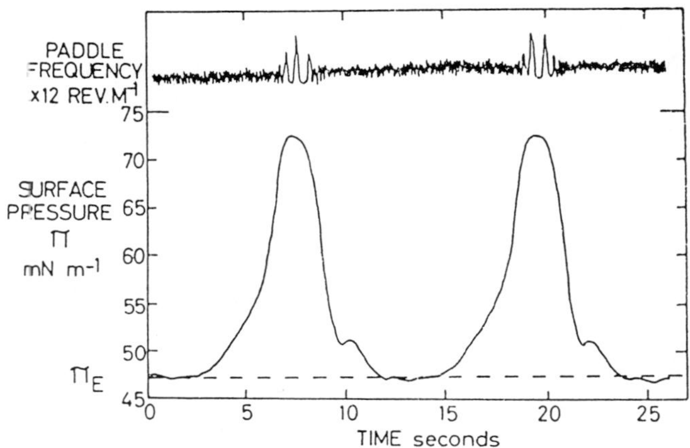

Fig. 4. Simultaneous tracings of surface pressure (π) and the rotational freedom of a platinized platinum paddle during expansion and compression of a DPPC/PG monolayer. The paddle 2.5 cm wide together with a light beam chopper (12 revs/s) weighed 0.75 g. It was spun at a steady 30 rev./min yielding 6 cycles/s on clean water. Enough dry DPPC/PG was sprinkled on the surface of the aqueous phase to achieve a $\pi_{equilibrium}$ of 48 mN/m and a reciprocating surface piston was started; $\omega = 0.08$ Hz and at full compression the initial surface area was reduced by a factor of 2. It can be seen that the paddle rotated freely at constant, steady-state, frequency until $\pi > \sim 70$ mN/m at which point, and for a period of approx 2 seconds its freedom to rotate was restricted. Reproduced from Reference 19.

above 41°C) the thermodynamic equality of the surface pressure, P and the reduction of surface tension, γ_{water} breaks down. We have seen that higher surface pressures can in many cases be produced and measured. Surface tensions of clean and film (insoluble monolayers) covered surfaces are, *by definition* equilibrium properties of liquids, and the difference between them cannot exceed the highest equilibrium surface pressure value, which we have defined as the monolayer stability limit. The surface pressure (the tangible, physical force described by Adam[37] can be raised, at least temporarily, above this point, and when it is, the observed pressure is actually the sum of two contributions. One of these is the thermodynamic term, $\gamma_{water} - \pi_{film}$ while there is an additional pressure which is a true two-dimensional compression forcing the molecules together. The compressive forces which some monolayers can withstand are very high, since a pressure of 1 dyne/cm acting on a layer 20 A high corresponds to a three dimensional pressure of 5×10^6 dynes cm^2 or 4.9 atm.

With hindsight, it would seem apparent that the relationship was never intended to apply to surface pressures in excess of about 45 mN/m and so I was motivated to write to John Clements in September 1976 suggesting that we had all been writing nonsense about "surface tensions" below 20–24 mN/m. I suggested that by the time a condensed monolayer had formed on the surface of alveolar liquid, the functional interface with the air phase was effectively a solid one of phospholipid hydrocarbon; a Wilhelmy dipping plate was being gripped as in a vice. We presented these observations to the members of the Pre-Term Labour Group of the Royal College of Obstetricians and Gynaecologists in September[11] and as an experiment at a Conversazione held at the Royal Society (Figure 5). Subsequent to these meetings, Morley and I invited Mike Phillips to co-author a paper[19] which described our convictions in formal terminology (see later).

A second reason for doubting claims that surface tension at a liquid/air interface can be reduced to near-zero by LS or replacement material is that the two phases differ in density by a factor of >1000 (water is 55 molar whereas water vapour at NTP is 22 millimolar). It is only close to the critical temperature (374.1°C at 228.5 atms) that the interface between liquid water and its vapour nears extinction and the surface tension might be described as being near-zero (Figure 2). To emphasise this point and to remind you of a definition of surface tension, I would like to quote from my English guru, NK Adam,[38]

> In the interior [of a phase e.g., liquid, solid, vapour] each molecule is surrounded by others on every side; it is therefore subject to attraction in all directions. On the average, over periods of time long compared with molecular vibrations, the attraction on any molecule is uniform in all directions. *At the surface, however, conditions are entirely different.* The molecules at the surface are attracted inwards, and to each side, by their neighbours; but there is no outward attraction to balance the inward pull, *because there are few molecules outside.* Hence every surface molecule is subject to a strong inward attraction, perpendicular to the surface. This inward attraction causes the surface to diminish in area, because the surface molecules are continually moving inwards more rapidly than others move outwards to take their places; the number of molecules in the surface is therefore continually diminishing, and the contraction of the surface continues until the maximum possible number of molecules are in the interior, i.e., until the surface is the smallest possible for a given volume, subject to the external conditions or forces acting on the drop.

and then on the next page,

> The fact that a liquid surface contracts spontaneously shows that there is free energy associated with it, *that work must be done to extend the surface.* The origin of this work, in terms of the molecules, is that when the surface is extended molecules must be brought from the interior to the surface against the inward attractive forces; work must be done against these inward attractive forces for each molecule which is brought to the surface. Since the molecules have definite size there will always be a definite number of them in the surface (we need not at the moment discuss whether the surface is one or more molecules in depth); *provided the surface is of the same nature and structure everywhere*, the work done in extending it will be definite. This free energy in the surface is of fundamental importance; a vast number of problems relating to the equilibrium of surfaces can be solved without knowing more than the magnitude of this free energy. On the solution of such problems a *mathematical device* is almost invariably employed to simplify the calculations; *it is to substitute for the free energy a hypothetical tension*, acting in all directions parallel to the surface, equal to the free surface energy. This is what is generally known as the *surface tension.*

Since, by no stretch of imagination, can the surface material of an alveolus be considered to be of the same "nature and structure", applying classical terms with their implicit meaning, is misleading. Or have the goal posts been moved for this particular field of interest?

My last reason for rejecting claims that films of lung surfactant whether natural or artificial exhibit a 'surface tension' of less than about 20 mN/m is that I doubt whether the devices used are reporting what the measurer thinks it is measuring! I concede, for example, that a barrier balance would give valid measurements of surface pressure on compression and has been shown to do so by Fontange, Bonte, Taupin and Ober.[39] Should the film condense at high pressure, as do both natural and formulated surfactants, it is likely that a barrier balance will only report on the pressure of very adjacent liquid zone and this might become as free of surface molecules as the control side and consequently accounting for the observed hysteresis. Wilhelmy dipping plates merely weigh the meniscus drawn up by presence of a high(er) energy surface. Contamination of such a surface is inevitable with heterogeneous surface films and for that reason alone I fail to see how the device distinguishes between the pull arising from pure water gripping short lengths of clean (high energy) plate from the pull arising from a low tension film acting on the full length of the plate. In effect, we have been using it as though it were a diviner's rod to find what we think is there, irrespective of a scientific understanding as to how it works! Unless, and it occurs to me, the diminishing pull on the plate is due to a transfer from measuring the air/water interface (70 > 23 mN/m) to the headgroup/water interface which could well experience values below 1 mN/m. After all, dipping plates are used to measure oil/water interface tensions. All these problems are to do with contact angle and can be difficult if not impossible to interpret away from equilibrium and with trailing meniscii. If I was still in my laboratory I would like to see whether the pressure/area profile perceived by a barrier balance (contact angles being less important) follow the potential/area profile reported by Watkins; technically a difficult measurement to make because of leaks occurring at the hinge region of the barrier as a consequent of the very high pressures being imposed upon the monolayer (>200 atms). Alternatively, if I could find my old polonium surface electrode I could take it over to France where they claim to have a leak-proof barrier balance.[38]

This concept of a heterogeneous surface material giving rise to artefactual surface pressure measurements was discussed at length by Colacicco and Scarpelli.[40] Their comments regarding the observed hysteresis observed with films of lung surfactant, as with many others, are entirely warranted despite their adherence at that time to the belief that the 'surface tension' at the water/air interface within the alveolus becomes 'zero'. The extraordinary pictures published by Weis and McConnell[41] and more recently by Lipp, Lee, Zasadzinsky and Waring[42] must have thrilled Colacicco and Scarpelli if and when he saw them as they did me!

So the work of extension of the alveolar surface, i.e., one covered by condensed phases, liquid expanded films and/or clean water will depend upon the extent (area) and nature of the material available at any one time of the respiratory cycle. Recruitment of monomers (sublimation) from a condensed film of DPPC is unlikely at 37°C, on the other hand, recruitment from a surface store of 'dry' phospholipids would be spontaneous, arising from the free energy of solvation of the head groups up to the equilibrium spreading pressure. In the absence of either, water molecules would have to be recruited into the surface from the subphase at an energy rate of 72 mJ/m^2! Thus unless one is aware of the actual extent and composition of the domain being extended, the work cannot be expressed as a meaningful surface energy nor, by the same argument, as a surface tension.

How then does one envisage DPPC ($T_c = 41$°C), materialising as a condensed duplex surface film from a mixture of phospholipids and proteins spreading on water at 37°C? To quote from our 1979 paper;[19]

. . . the dry mixture (commercial ALEC™ for example) spreads spontaneously above 35°C (the rate increases with T) to form a liquid-expanded monolayer with $\pi_{\text{equi}} = 48\ \text{mN m}^{-1}$.

The film is liquid in that it has a low surface shear viscosity (η_s) as monitored with either talc or a rotating paddle (see Figure 3). The rheological property of prime importance in the lung is the

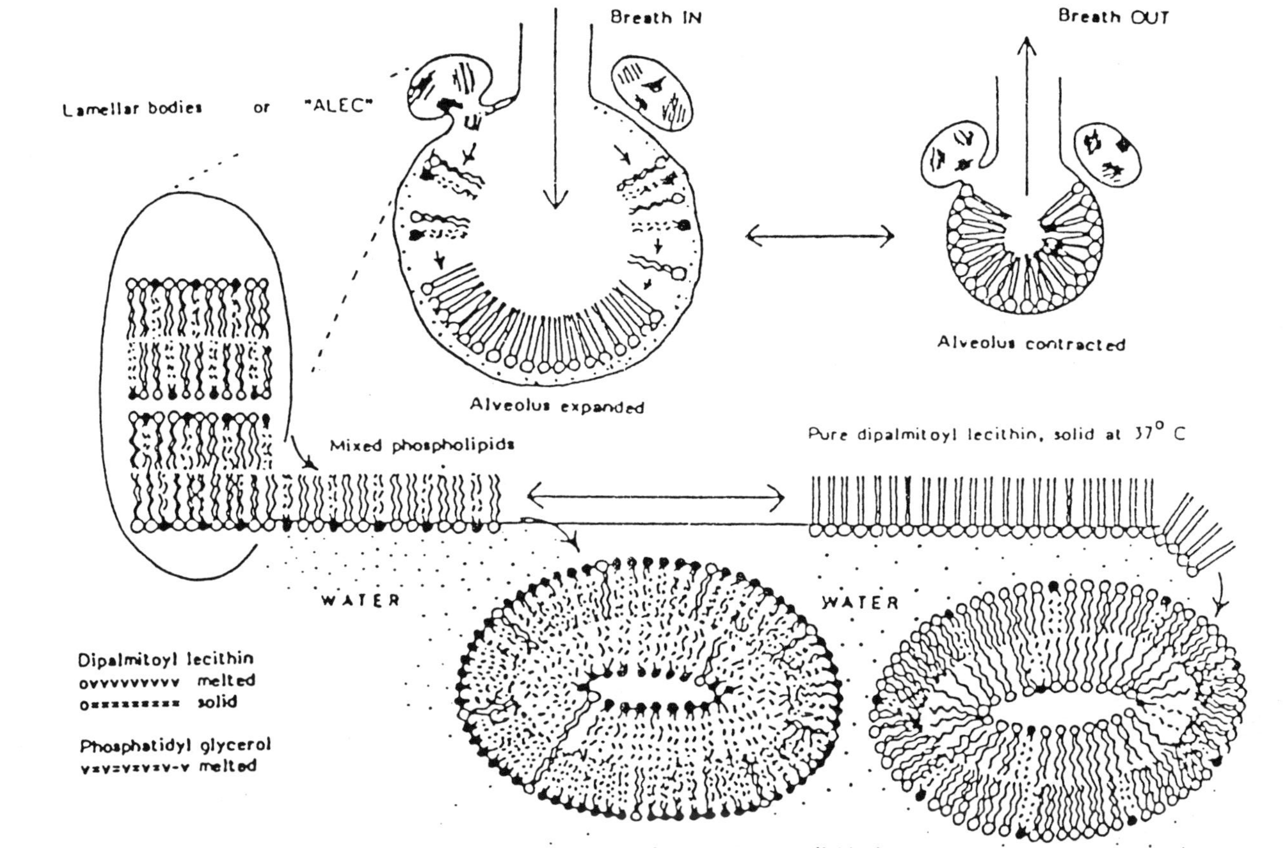

Fig. 5. Flow diagram shows how a semihydrous lung surfactant (exocytosed) from a lamellar body or a fragment of ALEC might spread, refine, condense and ablate on a trough and of the consequences of such manifestations in an alveolus. Original cartoon shown at a Royal Society Converzatione (June 1979).

surface dilatational modulus (ϵ) which is the response of a surface of area (A) to expansion and contraction at a frequency of ω . ϵ is defined by Equation (1).

$$\epsilon = A\frac{d\gamma}{dA} = \epsilon_d + i\omega\eta_d \tag{1}$$

where ϵ_d = surface dilatational elasticity, $i = (-1)^{1/2}$ and η_d = surface dilatational viscosity. Differentiation of Equation (1) shows that η_d is a measure of the frequency-dependence of ϵ and, as expected for a liquid-expanded phospholipid monolayer of low η_s, ϵ for egg phosphatidylcholine monolayers is independent of ω indicating that $\eta_d \sim 0$. This indicates that the liquid-expanded monolayer behaves as an elastic, insoluble film for which the relaxation processes are fast compared to ω. When such a liquid-expanded monolayer of dipalmitoyl phosphatidylcholine undergoes the transition to the condensed state on compression η_s increases substantially. Figure 1, shows that as π increases to $\sim 70\,\text{mN/m}^{-1}$, η_s increases rapidly, indicating that the film solidification is due to condensation of phospholipid molecules. Since the PG derived from egg phosphatidylcholine is unsaturated and has T_c below room temperature, this solidification (crystallization) can be due only to dipalmitoyl phosphatidylcholine molecules. η_d (= $d\epsilon/\delta\omega$) increases in parallel with η_s and gives a measure of the relaxation phenomena which occur in such metastable ($\pi_{\text{equi}} = 0\,\text{mN m}^{-1}$ for dipalmitoyl phosphatidylcholine when $T < 41°\text{C}$) films. Movement of phospholipid molecules into and out of the monolayer in the time scale of the cycling frequency leads to a positive η_d and hysteresis effects during the expansion/compression cycles. These kinetic effects probably involve movement of phospholipid molecules with respect to the collapse phases.

see Figure 5.

Over the years I have embellished the above scenario and from time to time I imagine myself to be in the centre of an alveolus which is rhythmically contracting and expanding. At the point of full expiration I would look around and see a mosaic of DPPC plates or plaques fitting more or less together, a geodesic dome looking, perhaps, like the remarkable electron micrographs of DPPC and LS by Tchoreloff, Gulik, Denizot, Proust and Puisieux.[44] As the alveolar volume increased I would see the plates moving apart creating surfaces of clean water (γ_w) which may or may not be covered by phospholipids but whose radius of curvature might be negligible (planes between geodesic lines). From the Laplace equation, minimal pressure for any γ_{water} or γ_{film} is needed to deform and thus increase the volume of the space? Or am I dreaming?

I concur with Brian Hill[44] when he exclaims that the emphasis on (extreme) reduction of surface tension "may be something of a red herring".

Yours sincerely

A further postscript relates to the thought that perhaps it is just as well that DPPC is present in the alveolae in condensed phase ensuring that it would not be hydrolysed by itinerant phospholipases. Many, many years ago Dawson and I showed that tight packed and very pure lecithins were "out of reach" so to speak of their hydrolytic enzymes[45]

References

1. Bangham AD. Liposomes: An Historical Perspective. Liposomes. Ed., Marcel Dekker New York, 1983.

2. Gregoriadis G. How liposomes Influenced my Life and got Away with it, pp 405–407 and 51 others Liposome Letters Ed., AD Bangham, Academic Press, London, 1983.
3. Bangham AD, Standish MM, Watkins JC. Diffusion of univalent ions across the lamellae of swollen phospholipids. J Mol Biol 1965;13:238–252.
4. Bangham AD, Papahadjopoulos D. Biophysical properties of phospholipids. I. Interaction of phosphatidyl-serine monolayers with metal ions. Biochim Biophys Acta 1955;126:181–184.
5. Papahadjopoulos D, Bangham AD. Biophysical properties of phospholipids. II. Permeability of phosphatidylserine liquid crystals to univalent ions. Biochim Biophys Acta 1955;126:185–188.
6. Papahadjopoulos D, Watkins J. Phospholipid model membranes. II. Biochim Biophys Acta 1967;135:639–645.
7. Watkins, JC. The surface properties of pure phospholipids in relation to those of lung surfactants. Biochim Biophys Acta 1968;152:293–306.
8. Clements JA and Tierney DF. Alveolar instability associated with altered surface tension. Handbook of Physiology, Respiration Chapter 69 1964;2:1565–1583.
9. Pattle RE: Properties, function and origin of the alveolar lining layer. Proc Roy Soc B 1958;148:217–240.
10. Morley CJ. Artificial surfactant: from bench to baby. Applied Cardiopulmonary Pathophysiology 1995;1:111–118.
11. Morley CJ, Bangham AD, Thorburn GD, Johnson P, Jenkin G. Parry D. The biochemistry and physiology of fetal pulmonary surfactant. In: anderson A, Beard R, Brudenell JM, Dunn PM, eds. Pre-Term Labour. 1977;261–272. Proc. 5th Study Group of the Royal College of Obstetricians & Gynaecologists.
12. Morley CJ; Bangham AD; Johnson P; Thorburn GD & Jenkin G. Physical and physiological properties of dry lung surfactant. Nature 1978;271:162–163.
13. Gratwohl C, Newman GE, Phizackerly PJR, Town MH. Phizackerley: Structural studies on lammelated osmiophilic bodies isolated from pig lung. Biochim Biophys Acta 1979;552:509–518.
14. Morley CJ, Bangham AD, Miller N, Davis JA. Dry artificial lung surfactant and its effect on very premature babies. Lancet 1081;i:64–68.
15. Perkins WR, Dause RB, Parente RA, Minchey SR, Neuman KC, Gruner SM, Taraschi TF, Janoff AS. Role of lipid polymorphism in pulmonary surfactant. Science 1996;273:330–332.
16. Bangham AD, Morley CJ. UK Patent application No. 9623669.0, 1996.
17. Hill MW. The effect of anaesthetic-like molecules on the phase transition of smectic mesophases of dipalmitoyl lecithin. Biochim Biophys Acta 1974;356:117–124.
18. Haywood AM. Patterns of persistent viral infection. New Eng J Med 1986;315:939–948.
19. Bangham AD, Morley CJ, Philips MC. The physical properties of an effective lung surfactant. Biochim Biophys Acta 1979;573:552–556.
20. Gaines G. Insoluble Monolayers at Liquid-Air Interfaces. Interscience Publishers, John Wiley. New York, 1966.
21. Bangham AD. Lung surfactant: how it does and does not work. Lung 1987;165:17–25.
22. Bangham AD. Pattle's bubbles and Von Neergaard's lung. Med Sci Res 1991;19:795–799.
23. Bangham AD. "Surface tension" in the lungs. Nature 1992;359:110.
24. Bangham AD. "Surface tensions" in the lung. Biophysical Journal 1995;68:1630–1631.
25. Fujiwara T, Chida S, Watabe Y, Maeta H, Morita T, Abe T. Artificial surfactant therapy in hyaline membrane disease. Lancet 1980;i:55–59.
26. Bangham AD, Miller NG, Davies RJ, Greenough A, Morley CJ. Introductory remarks about artificial lung expanding compound (ALEC). Colloids and Surfaces 1984;10:337–341.
27. Ten Centre Study Group: Ten centre trial of artificial surfactant (artificial lung expanding compound) in very premature babies. Brit Med J 1987;294:991–996.
28. Morley CJ, Greenough A, Miller NG, Bangham AD, Pool J, Wood S, South M, Davis JA, Vyas H. Randomized trial of artificial surfactant (ALEC) given at birth to babies from 23–34 weeks gestation. Early Human Development 1987;17:41–54.
29. Morley CJ. Surfactant Current Obstetrics and Gynaecology 1996;6:46–51.
30. Morley CJ. Surfactant treatment for premature babies—a review of clinical trials. Arch Dis Child 1991;445–450.
31. Morley JC, Morley R. Follow up of premature babies treated with artificial surfactant (ALEC). Arch Dis Child 1990;65:667–669.
32. Morley RJ Surfactant replacement in RDS artificial surfactant. European Respiratory J 1989;2:81–86.
33. Comroe JH Jr. Premature science and immature lungs Am Rev Resp Dis 1977;116:127–135 and 311–317.

34. Neergaard K von. Neue auffassungen uber einen grundbegriff der atemmechanik. Die retraktionskraft der lunge, abhangig von der oberflachenspannung in der alveolen. Z ges Exp Med 1929;66:373–394.
35. Cochrane CG, Revak SD. Pulmonary surfactant protein B (SP-B): Structure function relationship. Science 1991;254:566–568.
36. Pattle RE Properties, function and origin of the alveolar lining layer. Nature 1955;175:1125–1126.
37. Harkins WD. Physical Chemistry of Surface Films. Chapters 1 and 2, 1952;1–98. Reinhold, New York.
38. Adam NK. The Physics and Chemistry of Surfaces. Oxford University Press, 1930.
39. Fontanges A de, Taupin C, Bonte F, Ober R. Pressure-area curves of phospholipid monolayers in relation to pulmonary surfactant. Colloids and Surfaces 1985;14:309–316.
40. Colacicco G, Scarpelli EM. Pulmonary surfactants: Molecular structure and biological activity. In: Prince LM, Sears DF, eds. Biological Horizons in Surface Science. Academic Press New York, 1973.
41. Weis RM, McConnell HM Two dimensional chiral crystals of phospholipid. Nature 1984;310:47–49.
42. Lipp MM, Lee KYC, Zasadzinski JA, Waring AJ. Phase and morphologychanges in lipid monolayers induced by SP-B protein and its amino terminal peptide. Science 1996;273:1196–1199.
43. Tchoreloff P, Gulik A, Denizot B, Proust JE, Puisieux F. A structural study of interfacial phospholipid and lung surfactant layers by transmission electron microscopy after Blodgett sampling: influence of surface pressure and temperature. Chemistry and Physics of Lipids 1991;151–165.
44. Hills BA. What is the true role of surfactant in the lung? Thorax 1981;36:1–4.
45. Bangham AD, Dawson RMC. Electrokinetic requirements for the reaction between *Cl. perfringens* α toxin (phospholipase C) and phospholipid substrates. 1962;59:103–115.

Lasic and Papahadjopoulos (eds.), Medical Applications of Liposomes

CHAPTER 6.3

Pulmonary applications of liposomes

HANS SCHREIER
Advanced Therapies, Inc., Novato, CA 94949, USA

Overview

I. Introduction

Pulmonary application of liposomes has generated promising results with respect to both, prolonged and targeted delivery to the lungs and reduced systemic toxicities, invariably resulting in enhanced therapeutic efficacies. However, only very recently has the field 'matured' to the clinical level although preformulation and formulation, and later preclinical studies have been conducted and reported in the literature on a regular basis since the mid-1980's. The major activities in the field have always been in three categories, i.e., infectious diseases (antibiotics), asthma (corticosteroids), and lung injury (antioxidants); in addition to these 'classic' areas, a fourth has emerged only recently, i.e., genetic diseases of the lung (cystic fibrosis, α_1-antitrypsin deficiency) and their treatment by gene therapy using nonviral, i.e., liposomal vectors.

Progress in the field has been documented rather extensively in a series of review articles beginning with Mihalko et al.,[1] continuing with Kellaway and Farr,[2] Schreier et al.,[3] Gonzalez-Rothi and Schreier,[4] and most recently Schreier and Sawyer[5] who reviewed the status of liposome vectors for cystic fibrosis gene therapy.

As to the technological advances in the field, milestones have been the series of papers on the physical characterization of liposome aerosols by Niven and Schreier[6–8] and Taylor et al.,[9] the development of dry liposome powder aerosols by Schreier et al.[10] including more recently dry liposome : DNA powder aerosols,[11] and the development and characterization of nebulized cationic liposome : DNA complexes by Schwarz et al.,[12] Eastman et al.,[13] and Gagné and Schreier.[14]

Preclinical studies established the innocuousness of inhaled liposomes by a number of measures including macrophage function and lung histology in mice[15,16] and lung function studies in sheep.[17] Normal lung function and no toxic side effects were documented in several studies employing small numbers of human volunteers.[18,19] Pharmacokinetic and deposition studies in sheep[17] and dogs[20] as well as in human volunteers[21–24] showed that liposomes were retained in the lungs for prolonged periods of time and could potentially serve as sustained-release carriers.

In the following, recent developments in the main areas of activity mentioned above will be discussed and analyzed in more detail.

II. Current applications

II.1. Corticosteroid therapy

Gonzalez-Rothi and Schreier[4] have critically reviewed the small body of literature on the formulation and use of corticosteroid-containing liposomes for the treatment of asthma.

Waldrep et al.[25] have for several years focused on the development of a beclomethasone dipropionate liposome aerosol. They surveyed a total of 18 continuous-flow jet nebulizers ranging in droplet size generation from 0.9 μm (Respirgard) to 7.2 μm (Heart), with the majority in the 1.3–2.7 μm size range, and provided estimates for pulmonary distribution characteristics and drug output; they suggested that most devices tested, with particle size in the 1–3 μm MMAD range, would deliver significant doses of liposomes to peripheral lung sites. In a follow-up study they characterized in vivo in mice and in normal volunteers the distribution and pharmacokinetics of the same formulation labelled with ^{99m}technetium.[22] Not unexpectedly, lung clearance of ^{99m}technetium was greatly prolonged in experimental animals and in human volunteers. The authors measured both corticosteroid and lipid concentration by HPLC and found an excellent correlation between distribution of the three components, technetium label, lipid and drug, in the Andersen cascade impactor. Tolerance and safety of the beclomethasone liposome aerosol were tested in volunteers.[24] Doses of 0.56 and 1.29 mg beclomethasone, corresponding to lipid doses of 14 and 34.6 mg, respectively, were tolerated without any clinical side effects or abnormal laboratory values.

However, counter to what one would intuitively accept as a perfect combination of lipophilic drug and matrix carrier system, it has been shown that upon disturbance of the partition equilibrium between the liposome matrix and bulk solution (i.e., upon dilution), corticosteroids redistribute, i.e., 'leak' rapidly from the carrier into the bulk medium driven by their partition coefficient. This has been shown for hydrocortisone liposomes[26] and for triamcinolone acetonide liposomes.[27] Furthermore, comparative glucocorticoid receptor occupancy studies suggested that triamcinolone acetonide (without lipid carrier) delivered directly to the lungs occupies lung and liver receptors with essentially identical kinetics and magnitude, and, hence, that no lung targeting occurs.[28] Based on these observations, the authors cautioned that pulmonary targeting of lipophilic corticosteroids, be it in solution, suspension or formulated as liposomes, may indeed be unattainable.

As a consequence, Gonzalez-Rothi et al.[29] have formulated a liposome containing a water-soluble prodrug of triamcinolone acetonide, triamcinolone acetonide phosphate. In contrast to the lipophilic parent drug, the hydrophilic prodrug is effectively retained within the liposome carrier upon delivery to the lung. Again using the above described pharmacodynamic receptor occupancy in lung vs. liver as monitoring method, they showed unequivocally that the liposome membrane is the rate-limiting barrier for triamcinolone acetonide phosphate, resulting in clearly distinguishable magnitude and duration of receptor occupancy in lung vs. liver (Figure 1).

Hence, we suggest that corticosteroids for pulmonary targeting benefit from a liposomal formulation only if the liposome membrane is the rate-limiting barrier for release and thus receptor binding; this can only be accomplished with water-soluble corticosteroid prodrugs that are effectively encapsulated within the liposome payload compartment, rather than intercalated within the lipid membrane.

II.2. Anti-infectious therapy

The use of liposomes for pulmonary delivery has primarily focused on antifungal and antimycobacterial therapy. Locally administered liposome-encapsulated antibiotics may offer advantages over antibiotics in aqueous solution; these would include sustained therapeutic concentrations, minimal systemic absorption, greatly reduced toxicity concomitant with an increase in apparent efficacy.

Aminoglycosides were thought to be ideal candidates for pulmonary delivery, not only because of their broad spectrum and high potency against mycobacteria and *pseudomonas* species, but also because direct delivery to the lungs could potentially overcome their poor penetration ratio from serum to lung sputum, even more so in the case of obligatory intracellular microbes such as *Mycobacterium avium*, where active intracellular delivery to the infected alveolar macrophages is required. Furthermore, any aerosolized dose that would be swallowed by the patient would not be absorbed from the gastrointestinal tract, hence toxicity could be expected to be greatly reduced vis-a-vis systemic administration.

We have shown that the lungs provide a rate-limiting barrier for the absorption of amikacin when instilled in sheep (absorption/elimination rate flip-flop).[17] However, when the amikacin was encapsulated within both fluid-phase and rigid (cholesterol-rich) liposomes the absorption rate was further reduced and, concomitantly, the drug mean residence time was increased 5-fold compared to the instilled solution and 2-fold in the rigid-type liposomes relative to the fluid-phase liposomes.[17]

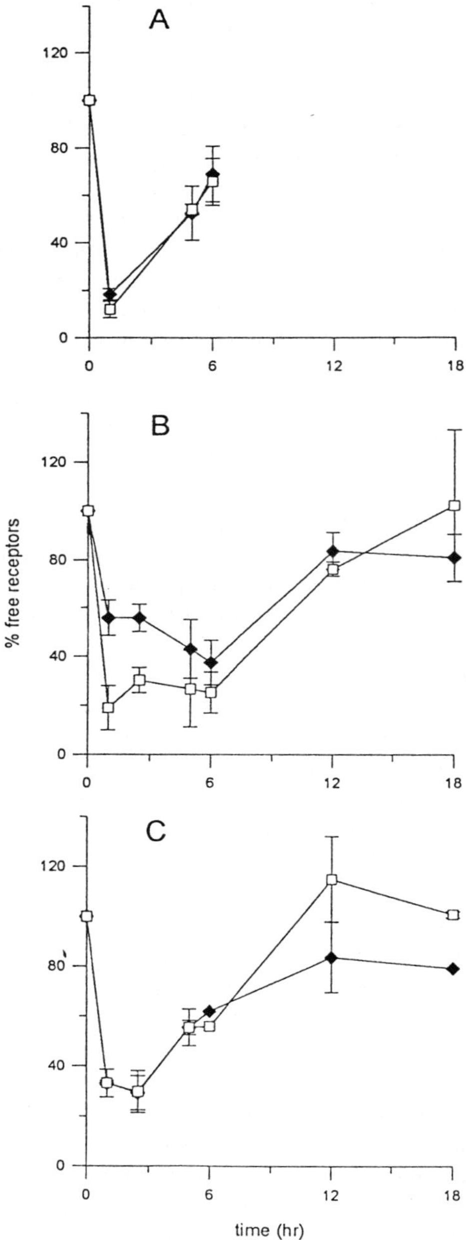

Fig. 1. Lung (□) and liver (◆) glucocorticoid receptor occupancy profiles after administration of 160 μg/kg of triamcinolone acetonide phosphate (TAP); (a) intratracheal instillation of TAP-sol; (b) intratracheal instillation of TAP-lip; (c) intravenously administered TAP-lip. Error bars represent mean ± S.D. (from Reference 29; with permission).

Similar to our results with amikacin, Omri et al.[30] found greatly prolonged retention of liposome-encapsulated tobramycin in rat lungs that were infected with *Pseudomonas aeruginosa*. Interestingly, a shorter mean retention time was found in infected animals compared to healthy animals, although it was not explored in this study if this had therapeutic consequences. In a follow-up study, the same group[31] confirmed that prolonged mean residence time in the lungs resulted in improved reduction of colony forming units (CFU) of *Pseudomonas aeruginosa*, however, only fluid-phase liposomes generated therapeutic concentrations in the lungs, whereas cholesterol-rich liposomes were retained longer but released subtherapeutic concentrations of tobramycin. Concurrent with prolonged high concentrations in the lung, kidney concentrations, as a measure for systemic spill-over were less than 1 μg/mg tissue, whereas equivalent concentrations of free tobramycin resulted in over 5 μg/mg tissue. The authors concluded that the local administration of fluid liposomes with encapsulated tobramycin could greatly improve the management of chronic pulmonary infection in cystic fibrosis patients.

Fielding et al. have developed a liposomal amikacin formulation (Mikasome®) whose intravenous form is currently in Phase II clinical trials. They have also tested this formulation for pulmonary delivery in rats and dogs.[32] In rats, doses of 10 mg/kg of free and liposomal amikacin were instilled and pulmonary mean residence time, plasma pharmacokinetics and kidney concentrations (as a 'predictor' for systemic toxicity) compared. As expected, free amikacin was extensively absorbed with a bioavailability of 70% compared to the i.v. dose. The liposomes employed in this study were very small, i.e., <70 nm average diameter. This may explain the surprising finding that Mikasome was also quite extensively absorbed, although with a much slower rate (half-life >10 hrs) (Table 1), reaching a maximum plasma concentration of 80 μg/ml after 20 hrs, with terminal pharmacokinetics similar to the i.v. dose (Figure 2). Retention of amikacin in the liposomal form was confirmed via Sephadex gel separation of plasma-derived samples. Kidney concentrations were significantly lower in animals receiving liposomal formulations than in the ones receiving a free amikacin dose.

A similar study in beagle dogs[32] generated equivalent data. A dose of 5 mg/kg Mikasome administered via bronchoscopic instillation resulted in prolonged retention in the lungs. At 48 hrs post-instillation, lung levels were >300 μg/g tissue

Table 1

Pharmacokinetics of free and liposomal amikacin after intratracheal instillation (10 mg/kg) in rats

Parameter	Amikacin solution (sucrose buffer)	Mikasome[1] (sucrose buffer)	Mikasome[1] (saline)
T_{max} (hrs)	0.5	20	6
C_{max} (μg/ml)	17	80	84
F^2	0.7	0.8	0.9
$T_{1/2\beta}$ (hrs)	0.7	43	33

[1]Mikasome = amikacin liposome (HSPC, cholesterol, DSPC 2:1:0.1 molar ratio).
[2]Bioavailability of i.t. amikacin calculated relative to i.v. amikacin; bioavailability of i.t. Mikasome calculated relative to i.v. Mikasome.

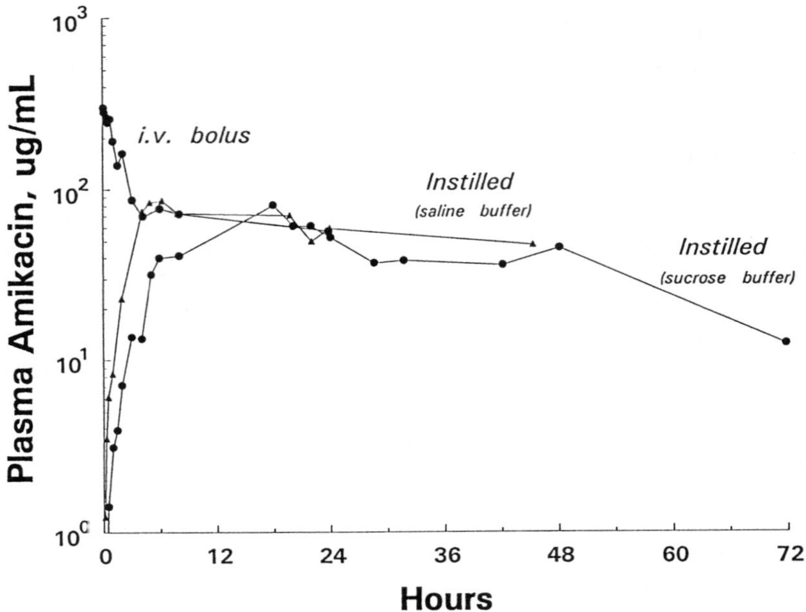

Fig. 2. Amikacin plasma elimination profile following i.v. injection and intratracheal instillation of liposomal amikacin (Mikasome^{R}) (10 mg/kg) in rats (4 animals per time point) (R. Fielding, NeXstar Pharmaceuticals; with permission).

with 50% of the dose recovered in the lungs, while kidney concentrations were <10 μg/g tissue and plasma levels were undetectable. The lung tissue levels can be considered bactericidal as the bactericidal dose of Mikasome for *M. avium* was determined to be 64 μg/ml. In contrast, only 0.3% of the dose was recovered from the lungs following an equivalent i.v. dose of Mikasome. The authors concluded from these two studies that intratracheal administration of liposomal amikacin provided the expected high and prolonged local concentrations in the lung and greatly reduced systemic exposure relative to an i.v. dose.

Gilbert et al.[33] treated mice that had been intranasally inoculated with *Cryptococcus neoformans* with aerosolized liposomal amphotericin B formulations and found dramatic enhancement of the apparent efficacy of the aerosolized vis-a-vis i.v. administration of the same formulation. Aerosols generated with a Collison apparatus that produced droplets with a 1.8 μm mass median aerodynamic diameter (MMAD) delivered 10.3 μg/L amphotericin B; for a 2-hr administration a dose of 0.3 mg/kg was estimated. A single 2-hr treatment within 24 hours of inoculation was reported effective in reducing pulmonary *Cryptococcus* infection and was more effective than an equivalent i.v. dose given for three continuous days. The same dose was effective when treatment commenced on days 7, 14, or 21 post-inoculation, including a reduction of cocci in the brain at day 21. Three doses delivered on days 7, 14, and 21 prolonged survival but provided only limited protection as mice died eventually from residual cocci. A more recent study in mice lethally infected with *Candida albicans* that followed an identical protocol also reported significant increase in both mean time of survival and percent survival.[34] A recent case report of a single patient with bronchopulmonary aspergillosis and deteriorating condition indicated prompt improvement following substitution of i.v. and intrapleural amphotericin B with nebulized liposomal amphotericin B.[35]

Hence, of all pulmonary applications of liposomes in the preclinical phase, liposomal formulations of antibiotics appear to have the greatest impact on anti-infectious therapy of pulmonary microbes. Theoretical benefits based on the physicochemical properties and biological fate of aminoglycosides have certainly been substantiated experimentally.

III.3. Antioxidant therapy

Shek and co-workers have been active in the design and testing of pulmonary antioxidant-carrying liposome formulations that could intervene with oxidation damage, i.e., fulminant and uncontrollable generation of radicals, in both a preventive and curative fashion.

The incorporation of α-tocopherol into a liposome matrix and direct pulmonary instillation of this compound prolonged the residence time of radioactively labelled glutathione (GSH); >30% of the instilled dose were found in the lungs after 24 hrs, vs. 18% with GSH encapsulated in liposomes without α-tocopherol and 2% with glutathione solution.[36] This was taken as an indication that liposomes would provide a reservoir for α-tocopherol in the lungs and this α-tocopherol would in turn be available to inactivate oxidative radicals in lung tissue. It was, however, not assessed in this study if oxidative protection of GSH in the liposome would 'translate' into prolonged and therapeutic free concentrations of GSH in the lungs.

Paraquat poisoning which inflicts extensive damage to lung tissue was found to be attenuated if rats were pretreated with α-tocopherol liposomes with respect to enzyme activities, including angiotensin converting enzyme, alkaline phosphatase and myeloperoxidase; however, increase in lung weight, i.e., edema formation was not affected by α-tocopherol treatment.[37] More importantly, it was found that α-tocopherol liposomes at a dose of 8 mg/kg reversed paraquat-related damage to the lungs, suggesting that α-tocopherol liposomes may not only be of preventive but of curative value in this case.[38] In a parallel study with phorbol myristate acetate similar protective effects of α-tocopherol formulated as liposomes and directly instilled into the lungs were found.[39]

As a logical extension of this study, the authors formulated bifunctional liposomes containing both α-tocopherol and GSH which was instilled into the lungs of rats 30 min prior to an i.p. challenge with paraquat dichloride. Again it was found that the liposomal α-tocopherol/GSH formulation protected the lungs significantly from paraquat damage, however, similar to the results with α-tocopherol liposomes alone, that development of edemas was not affected by the therapeutic intervention.[40]

A variation of this theme was introduced by Walther et al.[41] who formulated surfactant liposomes that carried CuZn-superoxide dismutase and catalase. These liposomes were shown to increase type II cell antioxidant activity and to protect the lungs of premature rabbits when exposed to hyperoxia.

Shek et al.[42] have reviewed their work in pulmonary antioxidant therapy and the use of liposomes therein and concluded that liposome-associated antioxidants can protect the lung against an oxidant challenge, and the extent of protection appears to be related to the characteristics of the antioxidant formulation.

II.4. Gene therapy

Cystic fibrosis, a genetic disease resulting from a mutation of the cystic fibrosis transmembrane conductance regulator (CFTR) gene, has become the paradigm for gene therapy and concomitantly for pulmonary delivery of DNA.[43] Direct delivery to the lungs would be highly desirable for two reasons: (i) systemic distribution and thus transfection of other tissues could be avoided, and (ii) the quantities deployed by aerosol could likely not be achieved by systemic administration. Although aerosol delivery of DNA formulations was the explicit goal from the beginning, most studies to date have employed intranasal or intratracheal instillation in order to be able to control the delivered dose and eliminate potential degradation of the DNA during aerosolization (see below).

Seminal studies by Rosenfeld et al.[44] and Crystal et al.[45] employed genetically modified adenovirus as vector to deliver CFTR complementary DNA (cDNA) to the pulmonary epithelium. More recently, cationic liposome plasmid vectors[46,47] have been used rather than viral vectors in order to overcome therapy-limiting immune and inflammatory responses. The first report of successful CFTR gene expression in the nasal cavity of cystic fibrosis patients using DC-Chol/DOPE liposomes complexed with plasmid CFTR has been published recently.[48] The status of liposome-based gene vectors in cystic fibrosis has been critically reviewed by Schreier and Sawyer.[5] Other lung-related diseases that could potentially benefit from pulmonary gene therapy approaches include α_1-antitrypsin deficiency[49] and adult respiratory distress syndrome.[50]

There have been earlier successful attempts at transfecting the lungs of animals with aerosolized cationic liposome:DNA complexes.[49,51] Although fairly large quantities of DNA were used in those studies, the transfection efficiency was rather low which might be ascribed to DNA damage, poor delivery, or low intrinsic expression efficacy (or a combination thereof). However, neither the nebulization conditions nor the physical state, i.e., the integrity of the aerosolized DNA have been assessed in detail in these studies and they are, therefore, only of anecdotal value. Only very recently have we seen a series of publications that describe the performance characteristics of liposome:DNA aerosols and monitor the stability of liposome:DNA complexes during the nebulization process, and these shall be discussed in greater detail in the following.

Schwarz et al.[12] systematically investigated two types of jet aerosols, an Aerotech II and a Puritan-Bennett 1600, and two types of cationic liposomes, *N*-(2-hydroxyethyl)-*N*,*N*-dimethyl-2,3-bis (tetradecyloxy)-1-propanaminium bromide (DMRIE) and 2,3-dioleyloxy-*N*-[2(sperminecaroxamido)ethyl]-*N*,*N*-dimethyl-1-propanaminium trifluoroacetate (DOSPA), both mixed with dioleoyl phosphatidylethanolamine (DOPE). They selected fixed DNA : lipid weight ratios of 1 : 4 (DMRIE) and 1 : 3 (DOSPA) as at these ratios the complex appeared to be charge neutral. It was found that material taken from the reservoir of the Aerotech II during aerosolization lost transfection efficiency rapidly within the first three minutes of operation, while the decrease in transfection activity was much more gradual with the Puritan-Bennett 1600 nebulizer. This observation was attributed to the more efficient through-put of the Aerotech which results in more frequent cycling and thus a higher probability for damage of the DNA liposome product. The authors could demonstrate that both increasing the total reservoir volume from 5 ml to 10 ml or a decrease of the flow rate from 15 L/min to 8 L/min reduced denaturation and increased transfection efficacy over time, i.e., from less than 3 min towards the 10 min range. The output yield was estimated to be about 12–14 μg of DNA over 6 min. The original lipid DNA complexes were of a size range of 3–5 μm diameter, whereas the major fraction (58%) of aerosol droplet sizes was found in the 0.5–2 μm range. This would have afforded some 'resizing' in the nebulizer as the authors point out correctly, a process that may well have contributed to the observed denaturation. Interestingly, no degradation of DNA nor of the lipid components was found which led the investigators to speculate that perhaps lipid phase separation may be responsible for the observed effects.

Eastman et al.[13] extended these studies in a systematic fashion. Perhaps the most significant finding was the fact that maximal aerosol transfer efficiency of cationic lipid DNA complexes was accomplished in the presence of a minimum salt concentration of 25 mM NaCl. This was likely related to the apparent zeta potential that remained net negative in the range of −40 to −50 mV for lipid : DNA molar ratios of 1 and smaller, and even for a 1.25 lipid : DNA ratio when the salt concentration was kept at 150 mM NaCl; the actual zeta potential of the latter formulation was approximately +20 mV at low salt (1 mM) concentration. In addition, the lipid : DNA ratio needs to be optimized in order to avoid excess of non-complexed DNA in the formulation. As had been found before, naked DNA was rapidly degraded when aerosolized while coating with cationic lipid effectively protected DNA from shearing. The authors caution that optimal complexation, i.e., a high lipid : DNA ratio needs to be balanced against the increasing instability of the system resulting in aggregation and precipitation when the highest lipid : DNA molar ratio of 1.25 was employed. Degradation was again a function of recycling in the aerosol reservoir and was found more pronounced with a PAIR L Jet with a mass output of nominally 0.55 ml/min compared to a Puritan Bennett Raindrop device with a mass output of approximately 0.175 ml/min. Most importantly, a reasonable transfection efficiency of the optimized formulation was demonstrated both in vitro and in vivo in mice. An inherent problem in the aerosolization of small rodents is the poor delivery efficiency which was estimated to be

2 μl for every 10 ml aerosolized, although this may not be a limiting factor with a large animal or a compliant human patient. Furthermore, it was pointed out that formulations optimized for intranasal instillation differed grossly from formulations optimized for aerosol delivery in their respective transfection efficiencies. While an optimized aerosol formulation was found 20- to 40-fold less active by instillation (425 pg of chloramphenicol acetyl transferase expression/100 mg lung tissue vs. 10 ng/100 mg for the intranasal formulation), the intranasal form was found much less or completely inactive when aerosolized.

We have studied the protective effects of solute, condensing agents and liposome carrier on plasmid DNA of varying sizes, ranging from 6.2 kb to 10.2 kb, when aerosolized with a Pair L Jet.[14] Similar to the studies discussed above, we found that large plasmids of 10.2 kb size were severely sheared and destroyed when aerosolized without protection for 5 min; smaller plasmids (6.2–6.7 kb) were clearly damaged, but to a much lesser degree than large plasmids and both large and small plasmids were protected from shear damage in the presence of cationic liposomes (Figure 3A–D). When plasmids were condensed with protamine sulfate at ratios of 2 : 1 to 3 : 2, protection from degradation was further improved. In contrast, presence of 250 mM sucrose did not protect plasmids from shearing. The mass output, ranging from 0.5–0.6 g/min, was on average not affected by the presence of lipid : DNA complexes in this study, although the results were erratic, pointing to the fact that these liposome formulations are prone to aggregation and precipitation as has been also been reported by all other investigators.

An inherent problem in the aerosol delivery of these intrinsically metastable colloidal formulations is the duration required to deliver therapeutically useful doses of DNA to the lungs; the problem is compounded by the still exceedingly low transfection efficacy of nonviral systems. As an example, in the above discussed study by Eastman et al.,[13] the mice were exposed to uninterrupted aerosolization for 40 min in order to deliver a 10 ml dose (or 4×40 min to deliver 40 ml, respectively). In a related study, although not delivered by aerosol, nasal instillations of lipid : DNA complexes has to be delivered to human probands over time intervals ranging from 2.5 to 7.5 hours, depending on the dose.[48] Clearly, we are faced with a potentially unsurmountable compliance problem. It is generally believed that a dry powder formulation may overcome the mass : volume limitations of aqueous dispersions and provide a more acceptable means of delivering therapeutic doses of DNA to the lungs.

To this end, we have commenced work to formulate and characterize DNA-containing dry powder liposomes that could be applied using dry powder inhalers.[11] To date, we have investigated the pharmaceutical parameters of manufacturability and stability of the dry powder liposomes. It was found that the liposome DNA formulations retained their original size during manufacture, lyophilization, milling and storage over at least 6 months; concomitantly, no degradation of DNA was found when assayed by agarose gel electrophoresis, although 'tailing' of the bands following jet-milling to small particle powders indicated gradual destruction of the plasmids.

Although beyond the scope of this review, it should be noted that similar

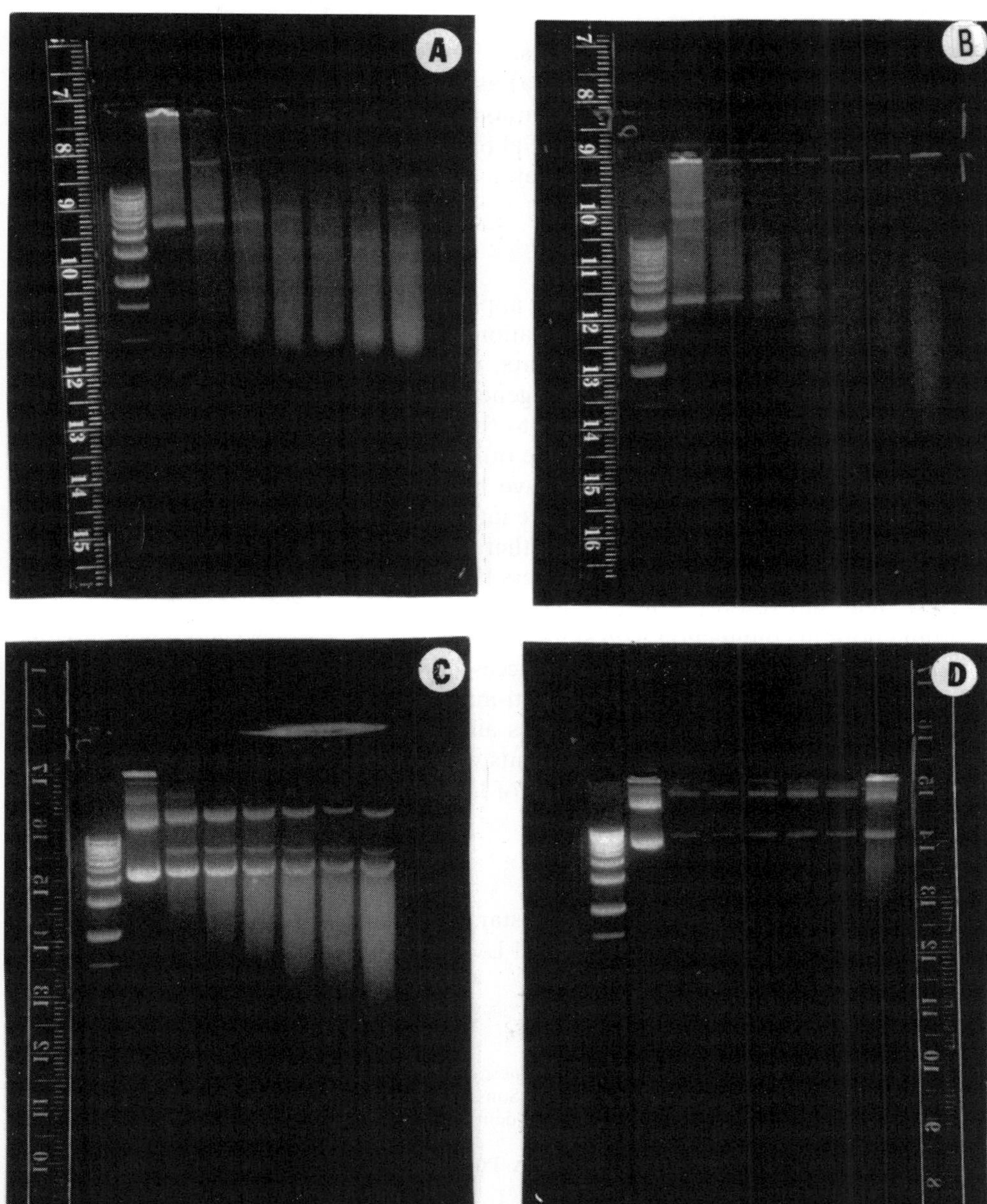

Fig. 3. Effect of nebulization with a PARI LL-Jet nebulizer on the integrity of plasmid DNA with and without cationic liposomes (DC-Chol/DOPE). (a) β-galactosidase plasmid (pRSV-β-GAL) (10.2 kb); (b) pRSV-β-GAL/DC-Chol/DOPE 3:1 w/w; (c) prostaglandin G/H synthase plasmid (pCMV$_4$-PGH) (6.7 kb); (d) pCMV$_4$-PGH/DC-Chol/DOPE 3:1 w/w. Lane 1: DNA standard; lane 2: plasmid DNA untreated; lanes 3–7: 1,2,3,4,5 min nebulization; lane 8: plasmid DNA remaining in nebulizer reservoir. Agarose gels (1%) were stained with 0.05% ethidium bromide.

studies have been published using adenoviral CFTR vectors in primates[52] and more recently in cystic fibrosis patients.[53] Interestingly, one such study employed exogenous bovine surfactant (Survanta) as adjuvant to facilitate spreading of the DNA vector and reported a more uniform lobar distribution and concomitant greater transfection efficiency of the -Gal reporter gene.[54] This has not been tested with nonviral DNA formulations to date.

III. Conclusions

The focus of this review of pulmonary applications of liposomes has been on four areas that are distinguished by both a rational basis for development and promising preclinical and clinical experimental data. These areas are anti-asthma, anti-infectious, antioxidant, and most recently gene therapy of pulmonary diseases. While the latter is by far the newest entry into the field, it has in some ways progressed more rapidly and aggressively than the other areas for which technological, pharmacological and toxicological data have been generated over roughly a decade since the late 1980's. The rapid advance into the clinic of liposomal DNA formulations has not resulted in memorable therapeutic successes to date. Fortunately, frustration generated by lack of progress has prompted numerous investigators to invest time and resources in the design and characterization of suitable aerosol formulations in conjunction with suitable aerosol devices which should now greatly and with much higher probability of success advance the field.

As to the 'classic' applications in anti-infectious, antiasthmatic and antioxidant therapy, more extensive clinical studies and product introductions appear imminent; preclinical data suggest that patients would indeed benefit greatly from these three forms of pulmonary applications of liposomes.

Acknowledgements

I would like to thank R. Fielding, NeXstar Pharmaceuticals, for providing unpublished data on amikacin liposomes, and L. Gagné for technical assistance.

References

1. Mihalko PJ, Schreier H, Abra, RM. Liposomes: a pulmonary perspective. In: Gregoriadis G, ed. Liposomes as Drug Carriers. John Wiley & Sons. New York, NY, 1988;679–694.
2. Kellaway IW, Farr SJ. Liposomes as drug delivery systems to the lung. Adv Drug Deliv Rev 1990;5:149–161.
3. Schreier H, Gonzalez-Rothi RJ, Stecenko AA. Pulmonary delivery of liposomes. J Control Release 1993;24:209–223.
4. Gonzalez-Rothi RJ, Schreier H. Pulmonary delivery of liposome encapsulated drugs in asthma therapy. Clin Immunother 1995;4:331–337.
5. Schreier H, Sawyer SM. Liposomal DNA vectors for cystic fibrosis gene therapy. Current applications, limitations, and future directions. Adv Drug Del Rev 1996;19:73–87.
6. Niven RW, Schreier H. Nebulization of liposomes. I. Effects of lipid composition. Pharm Res 1990;7:1127–1133.
7. Niven RW, Speer M, Schreier H. Nebulization of liposomes. II. The effects of size and modeling of solute release profiles. Pharm Res 1991;8:217–221.

8. Niven RW, Carvajal MT, Schreier H. Nebulization of liposomes. III. Effect of operating conditions. Pharm Res 1992;9:515–520.
9. Taylor KMG, Taylor G, Kellaway IW, Stevens J. The stability of liposomes to nebulization. Int J Pharmaceut 1990;58:57–61.
10. Schreier H, Mobley WC, Concessio N, Niven RW, Hickey A. Formulation and in vitro performance of liposome powder aerosols. S T P Pharma Sciences 1994;4:38–44.
11. Allon N, Erdos GW, Schreier H. Formulation and characterization of cDNA containing dry powder liposomes. Proceed Intern Symp Control Rel Bioact Mater 1997;24:35–36.
12. Schwarz LA, Johnson JL, Black M, Cheng SH, Hogan ME, Waldrep JC. Delivery of DNA-cationic liposome complexes by small-particle aerosol. Hum Gene Ther 1996;7:731–741.
13. Eastman SJ, Tousignant JD, Lukason MJ, Murray H, Siegel CS. Constantino P, Harris DJ, Cheng SH, Scheule RK. Optimization of formulations and conditions for the aerosol delivery of functional cationic lipid:DNA complexes. Hum Gene Ther 1997;8:313–322.
14. Gagné L, Schreier H. Aerosolization of plasmid DNA. Protective effects of solute, condensing agents, and liposome carriers. Proceed Intern Symp Control Rel Bioact Mater 1997;24:641–642.
15. Gonzalez-Rothi RJ, Cacace J, Straub L, Schreier H. Liposomes and pulmonary alveolar macrophages: functional and morphological interactions. Exp Lung Res 1991;17:687–705.
16. Myers MA, Thomas DA, Straub L, Soucy DW, Niven RW, Kaltenbach M, Hood CI, Schreier H, Gonzalez-Rothi RJ: Pulmonary effects of chronic exposure to liposome aerosols in mice. Exp Lung Res 1993;19:1–19.
17. Schreier H, McNicol KJ, Ausborn M, Soucy DW, Derendorf H, Stecenko AA, Gonzalez-Rothi RJ. Pulmonary delivery of amikacin liposomes and acute liposome toxicity in the sheep. Int J Pharmaceut 1992;87:183–193.
18. Gilbert BE, Six HR, Wilson SZ, Wyde PR, Knight V. Small particle aerosols of enviroxime-containing liposomes. Antiviral Res 1988;9:355–365.
19. Thomas DA, Myers MA, Wichert BM, Schreier H, Gonzalez-Rothi RJ. Acute effects of liposome aerosol inhalation on pulmonary function in healthy human volunteers. Chest 1991;99:1268–1270.
20. Bennett DB, Tyson E, Mah S, de Groot JS, Hedge SG, Terao S, Teitelbaum Z. Sustained delivery of detirelix after pulmonary administration of liposomal formulations. J Control Release 1994;32:27–35.
21. Farr SJ, Kellaway IW, Parry-Jones DR, Woolfrey SG. ^{99m}Technetium as a marker of liposomal deposition and clearance in the human lung. Int J Pharmaceut 1985;26:303–316.
22. Taylor KMG, Taylor G, Kellaway IW, Stevens J. The influence of liposomal encapsulation on sodium cromoglycate pharmacokinetics in man, Pharm Res 1989;6:633–636.
23. Vidgren M, Waldrep JC, Arppe J, Black M, Rodarte JA, Cole W, Knight V. A study of ^{99m}technetium-labelled beclomethasone diproprionate dilauroylphosphatidylcholine liposome aerosol in normal volunteers. Int J Pharmaceut 1995;115:209–216.
24. Waldrep JC, Gilbert BE, Knight CM, Black MB, Scherer P, Knight V, Eschenbacher W. Pulmonary delivery of beclomethasone liposome aerosol in volunteers. Chest 1997;111:316–323.
25. Waldrep JC, Keyhani K, Black M, Knight V. Operating characteristics of 18 different continuous-flow jet nebulizers with beclomethasone dipropionate liposome aerosol. Chest 1994;105:106–110.
26. Farr SJ, Kellaway IW, Carman-Meakin B. Comparison of solute partitioning and efflux of liposomes formed by a conventional and aerosolized method. Int J Pharmaceut 1989;51:39–46.
27. Schreier H, Lukyanov AN, Hochhaus G, Gonzalez-Rothi RJ. Thermodynamic and kinetic aspects of the interaction of triamcinolone acetonide with liposomes. Proceed Intern Symp Control Rel Bioact Mater 1994;21:228–229.
28. Hochhaus G, Gonzalez-Rothi RJ, Lukyanov A, Derendorf H, Schreier H, Dalla Costa T. Assessment of glucocorticoid lung targeting by ex vivo receptor binding studies in rats. Pharm Res 1995;12:134–137, 1995.
29. Gonzalez-Rothi RJ, Suarez S, Hochhaus G, Schreier H, Lukyanov A, Derendorf H, Dalla Costa T. Pulmonary targeting of liposomal triamcinolone acetonide. Pharm Res 1996;13:1699–1703.
30. Omri A, Beaulac C, Bouhajib M, Montplaisir S, Sharkawi M, Lagace J. Pulmonary retention of free and liposome-encapsulated tobramycin after intratracheal administration in uninfected rats and rats infected with *Pseudomonas aeruginosa*. Antimicrob Agents Chemother 1994;38:1090–1095.
31. Beaulac C, Clement-Major S, Hawari J, Lagace J. Eradication of mucoid *Pseudomonas aeruginosa* with fluid liposome-encapsulated tobramycin in an animal model of chronic pulmonary infection. Antimicrob Agents Chemother 1996;40:665–669.
32. Fielding RM, Feistner B, Moon-McDermott L, Gill SC, Snipes MB, Bendele R A. Instilled

liposomal amikacin (Mikasome[R]) prolongs antibiotic residence in lungs and airways of dogs. Pharm Res 1996;13:S-167.
33. Gilbert BE, Wyde PR, Wilson SA. Aerosolized liposomal amphtericin B for treatment of ulmonary and systemic *Cryptococcus neoformans* infections in mice. Antimicrob Agents Chemother 1992;36:1466–1471.
34. Gilbert BE, Wyde PR, Lopez-Berestein G, Wilson SA. Aerosolized amphtericin B-liposomes for treatment of systemic *Candida* infections in mice. Antimicrob Agents Chemother 1994;38:356–359.
35. Purcell IF, Corris PA. Use of nebulised liposomal amphotericin B in the treatment of *Aspergillus fumigatus* empyema. Thorax 1995;50:1321–1323.
36. Suntres ZE, Shek PN. Incorporation of alpha-tocopherol in liposomes promotes the retention of liposome-encapsulated glutathione in the rat lung. J Pharm Pharmacol 1994;46:23–28.
37. Suntres ZE, Shek PN. Intratracheally administered liposomal alpha-tocopherol protects the lung against long-term toxic effects of paraquat. Biomed Environ Sci 1995;8:289–300.
38. Suntres ZE, Shek PN: Prevention of phorbol myristate acetate-induced acute lung injury by alpha-tocopherol liposomes. J Drug Target 1995;3:201–208.
39. Suntres ZE, Shek PN. Liposomal alpha-tocopherol alleviates the progression of paraquat-induced lung damage. J Drug Target 1995;2:493–500.
40. Suntres ZE, Shek PN. Alleviation of paraquat-induced lung injury by pretreatment with bifunctional liposomes containing alpha-tocopherol and glutathione. Biochem Pharmacol 1996; 22:1515–1520.
41. Walther FJ, David-Cu R, Lopez SL. Antioxidant-surfactant liposomes mitigate hyperoxic lung injury in premature rabbits. Am J Physiol 1995;269:L613–L617.
42. Shek PN, Suntres ZE, Brooks JI. Liposomes in pulmonary applications: physicochemical considerations, pulmonary distribution and antioxidant delivery. J Drug Target 1994;2:431–442.
43. Alton EWFW, Geddes DM. Gene therapy for cystic fibrosis: a clinical perspective. Gene Ther 1995;2:88–95.
44. Rosenfeld MA, Yoshimura K, Trapnell BC, Yoneyama K, Rosenthal ER, Dalemans W, Fukayama M, Bargon J, Stier LE, Stratford-Perricaudet L, Perricaudet M, Guggino WB, Pavirani A, Lecocq JP, Crystal RG. In vivo transfer of the human cystic fibrosis transmembrane conductance regulator gene to the airway epithelium. Cell 1992;68:143–155.
45. Crystal RG, McElvaney NG, Rosenfeld MA, Chu CS, Mastrangeli A, Hay JG, Brody SL, Jaffe AH, Eissa NT, Danel C. Administration of an adenovirus containing the human CFTR cDNA to the respiratory tract of individuals with cystic fibrosis. Nat Genet 1994;8:42–51.
46. Hyde SC, Gill DR, Higgins CF, Trezise AEO, MacVinish LJ, Cuthbert AW, Ratcliff R, Evans MJ, Colledge WH. Correction of the ion transport defect in cystic fibrosis transgenic mice by gene therapy. Nature 1993;362:250–255.
47. Alton EWFW, Middleton PG, Caplen NJ, Smith SN, Steel DM, Munkonge FM, Jeffery PK, Geddes DM, Hart SL, Williamson R, Fasold KI, Miller AD, Dickinson P, Stevenson BJ, McLachlan G, Dorin JR, Porteous DJ. Non-invasive liposome-mediated gene delivery can correct the ion transport defect in cystic fibrosis mutant mice. Nature Genet 1993;5:135–142.
48. Caplen NJ, Alton EWF, Middleton PG, Dorin JR, Stevenson BJ, Gao X, Durham SR, Jeffery PK, Hodson ME, Coutelle Ch, Huang L, Porteous DJ, Williamson R, Geddes DM. Liposome-mediated CFTR gene transfer to the nasal epithelium of patients with cystic fibrosis. Nature Medicine 1995;1:39–46.
49. Canonico AE, Conary JT, Meyrick BO, Brigham KL. Aerosol and intravenous transfection of human α_1-antitrypsin gene to lungs of rabbits. Am J Respir Cell Mol Biol 1994;10:24–29.
50. Brigham KL, Canonico AE, Meyrick BO, Conary JT, Schreier H. Prospects for gene therapy for endotoxin-induced lung injury. In: Brigham KL, ed. Endotoxin and the Lungs. Lung Biology in Health and Disease, Vol 77. Marcel Dekker, New York, 1994;457–470.
51. Stribling R, Brunette E, Liggitt D, Gaensler K, Debs R. Aerosol gene delivery in vivo. Proc Natl Acad Sci USA 1992;89:11277–11281.
52. Sené C, Bout A, Imler JL, Schultz H, Willemot JM, Hennebel V, Zurcher C, Valerio D, Lamy D, Pavirani A. Aerosol-mediated delivery of recombinant adenovirus to the airways of nonhuman primates. Hum Gene Ther 1995;6:1595–1601.
53. Bellon G, Michel-Calemard L, Thouvenot D, Jagneaux V, Poitevin F, Malcus C, Accart N, Layani MP, Aymard M, Bernon H, Bienvenu J, Courtney M, Dring G, Gilly B, Gilly R, Lamy D, Levrey H, Morel Y, Paulin C, Perraud F, Rodillon L, Sené, C, So S, Rouraine-Moulin F, Schatz C, Pavirani A. Aerosol administration of a recombinant adenovirus expressing CFTR to cystic fibrosis patients: A phase I clinical trial. Hum Gene Ther 1997;8:15–25.
54. Katkin JP, Husser RC, Langston C, Welty SE. Exogenous surfactant enhances the delivery of recombinant adenoviral vectors to the lung. Hum Gene Ther 1997;8:171–185.

Lasic and Papahadjopoulos (eds.), Medical Applications of Liposomes
Elsevier Science B.V.

CHAPTER 6.4

Toxicity of liposome-encapsulated hemoglobin: Effect of liposomal membrane composition on host defense, platelet activation and hemostases during laminar shear flow

J. Jato,[a] R. Beissinger,[a] S. Zheng,[a] V. Shankey,[b] J. Fareed,[b] R. Sherwood,[c] D. McCormick,[c] D. Lasic and F. Martin[d]

[a]*Department of Chemical Engineering, Illinois Institute of Technology, Chicago, IL 60616, USA;*
[b]*Dept. of Pathology, Loyola Medical Center, Maywood, IL, USA;*
[c]*Life Sciences, IIT Research Institute, Chicago, IL 60616, USA;*
[d]*Liposome Technology, Inc., Menlo Park, CA 94025, USA*

Overview

Abstract

A safe and effective red blood cell (RBC) substitute would have broad implications in the practice of emergency medicine, trauma management, surgery, and several other areas of medicine. Hemoglobin-based RBC substitutes have been developed that can deliver oxygen to peripheral tissues.[1] However, although these RBC substitutes have desirable biophysical properties, their in vivo efficacy is limited by their toxicity. In view of the very large doses of blood substitute that are likely to be used clinically, important work on various safety issues have been started that include immunotoxicity and host defense,[2] platelet activation (p-selectin) and platelet aggregation, and hemostatic,[3] while maintaining efficacy, are critical considerations for the development and ultimate application of artificial RBCs. So far the results suggest that the sterically-stabilized liposome-encapsulated hemoglobin (LEH) is less immunotoxic than conventional LEH, less platelet activating (p-selectin), less platelet aggregating and less hemostatic (with respect to thrombin formation and thromboxane B_2 generation).

I. Introduction

Liposome technology provides a mechanism for encapsulation and in vivo delivery of drugs, specifically hemoglobin (Hb) in these applications, probably would otherwise be degraded, cleared rapidly, or toxic to the host. Techniques for Hb encapsulation within liposomes include film hydration followed by sonication,[6] membrane extrusion,[7,8] french press extrusion,[9] microfluidization[10–13] or homogenization,[14] detergent dialysis,[15] reverse-phase evaporation,[16] reverse-micelle;[17] and other mechanisms.[18] Recently, we have demonstrated that a multiple emulsion approach for encapsulation of Hb can be used to generate an efficient oxygen-carrying red cell substitute.[19,20] Most recently, our laboratories have developed a modified film hydration technique to make LEH.[1] Potential safety problems may exist in the use of LEH and these are being evaluated.[46] It is estimated that the commercial market for a safe and effective RBC substitute could be from $2 billion to $18 billion per year.[4]

Aspects of clinical concern with respect to cardiovascular infusion of LEH include tissue/organ toxicity and immunotoxicity,[21,22] effect on coagulation system and activation and aggregation of platelets,[23–30] and effect on reticuloendothelial (i.e., the mononuclear phagocyte) host defense system.[31–38] Several reports have suggested that administration of large quantities of lipid-containing materials results in overloading of the RES.[35,36] Host phagocytes (monocytes, macrophages) are crucial in inactivating and removing bacteria, fungi and viruses, and other foreign material from the body.[38] They also interact extensively with lymphocytes in modulating normal immune responses. Thus, impairment of host phagocytic cells could result in increased susceptibility to pathogenic or opportunistic infections or enhance improper immune modulation resulting in immunosuppression or allergy.

The biological effects of LEH may be due to the nature of the phospholipid bilayer. The phospholipid part of the platelet membrane, when it is chemically and *physically* (sheared) perturbed, results in the immediate release of arachidonic acid, which are "subsequently converted to biologically active compounds known as eicosanoids".[39] Thromboxane A_2, an eicosanoid, is considered a strong aggregator of platelets.[39] For the whole blood samples, there may be an enhancement of platelet reactivity and modulation of eicosanoid production by the intact erythrocytes.[40]

It was found that negatively-charged liposomes produced in vivo a transient reduction in platelet count,[30,41] which is recovered within 60 minutes post-infusion. This effect was most striking for multilamellar vesicles containing phosphatidylglycerol (PG). The thrombocytopenic effect diminished as vesicle size decreased to 0.22 μm. This liposomal-induced transient thrombocytopenia suggests that LEH with negative charge may activate platelets in-vivo. Larsen et al. reported that P-selectin mediates the binding of activated platelets to both neutrophils and monocytes.[42] Just a little later on another study by Hamburger and McEver, which confirmed that activated platelets (P-selectin) mediated adhesion of white blood cells, specifically neutrophils.[43] Also, P-selectin is required for PMN adhesion to a pathophysiologic surface of activated adherent platelets at physiologic shear rates.[44]

The approaches used in these studies, which are based on the development and characterization of LEH as an oxygen-carrying RBC substitute, incorporate the diverse characteristics required for a practical RBC replacement fluid. They address those aspects involving evaluation of biophysical and biochemical properties of the LEH, efficacy, toxicity, biocompatibility, effect on reticuloendothelial host defense system, stability and storability (i.e., in the wet and/or dry state), and feasibility of scaling-up production (using aseptic techniques to obtain pyrogen-free, sterile and virus-free production) in order to support the extensive animal testing required before clinical testing can be seriously considered.

Specifically, in these studies we are interested in knowing if P-selectin activation can be expected for platelets exposed to flow in a low stress, shear field. We plan to do this by examining the shear-induced activation of platelets (CD-62), along with markers for platelets (CD-61), white blood cells (CD-45) and their interactions. Shear-induced hemostasis, will also be investigated, through markers of thromboxane B_2 generation (RIA) and thrombin formation (F1 + 2, Elisa).

II. Results and discussion

Liposome-encapsulated hemoglobin (LEH) products are being investigated as potential blood substitutes. It was one objective in these studies to determine if changes in LEH composition can modify the immune response. Red blood cell substitutes based on conventional lipids containing phosphatidylinositol (PI) or phosphatidyglycerol (PG), i.e., LEH1, and sterically stabilized lipid vesicles containing polyethyleneglycol(1900)-phosphatidylethanolamine (PEG-PE), i.e., LEH2, were tested for effects on host resistance. On Day 0, groups of 18 to 20 female CD-1 mice were given an intravenous (i.v.) infectious challenge with a 20% lethal dose of *Listeria monocytogenes*. Mice received a single i.v. dose of LEH1, LEH2, or albumin vehicle on Day +1 or Day −3 relative to infectious challenge. Mice dosed with LEH1 and LEH2 on Day +1 died rapidly from *Listeria* infection; but mice dosed with LEH2 lived significantly longer than did mice receiving LEH1. By contrast, when administered on Day −3, LEH1 had no significant effect on host immunity, while LEH2 increased susceptibility to *Listeria* infection.[2]

In addition, LEH1 and LEH2 both caused significant reduction of phagocytic activity as measured by rat alveolar macrophage (AM) ingestion of latex microspheres. AM incubated 4 hr with either LEH1 or LEH2 prior to addition of microspheres ingested fewer beads in a dose-dependent manner. No difference in in vitro phagocytic activity was observed between LEH1 or LEH2.[2] The inability to differentiate LEH formulations based on in vitro phagocytic activity suggests that the in vivo *Listeria* infection model may be more relevant in discerning the immunotoxicity of the LEH formulations tested.

It was a second objective of these studies to determine the effects of lipid composition of biological membranes (liposomes) and flow, and how these can modify platelet activation and platelet particle fragmentation, thromboxane generation and thrombin formation. Whole blood, anticoagulated with recombinant (*r*)-hirudin,[45] was investigated for its interactions with human hemoglobin encapsulated in lipid vesicles (LEH).[3] Polyethyleneglycol-phosphatidylethanolamine

(PEG-PE) liposomes versus PG liposomes were compared for their hemostatic effects in static and rotational cone-and-plate reactors, operating at 37°C. Activation was measured using flow cytometric analysis of CD62 monoclonal antibody (for P-selectin) expression on CD61 monoclonal antibody (GP IIIa) positive platelets. Thrombin formation and thromboxane B_2 generation were also evaluated, especially under pathological (shear rates $>1100\ s^{-1}$) flow conditions.[6]

Electron Microscopy: Electron micrographs were prepared, e.g., a 0.45 μm filtered PG LEH, using negative staining-thin section methodology at a magnification of 25,000.[1] Liposomes seen at this magnification seem to have slightly irregular surfaces with a wide particle size distribution; a few large irregular particles, which could be aggregates, are also seen.

Platelet Fragment Generation: The laminar flows were that of simple shear flow generated in a cone and plate viscometer at 37°C. Shear stresses along with liposomal membrane compositions seem to play an important role in formation of three distinct major populations of platelet-antibody labeled with CD61-FITC. After gating out the negative events for non-specific binding, CD61 FITC antibody to GpIIIa fragments of platelet population appeared to be generated as function of shear rate.[3] The generation of this population was more significant for PG LEH than PEG-PE LEH. Platelet fragments appeared to occupy the first log of the FSC and the second log of FL1 (CD61 FITC) of FSC VS FL1 sheared samples coordinates.[3] As the exception, it had been noted in a previous study that platelet fragments were generated over the entire shear rate range used in this study (0–5400 s^{-1}).[5]

Platelet Activation During Sheared Flow: Activation was assessed by measuring the expression of P-selectin on the outer plasma surface of the platelet. Platelet activation of LEH-whole blood and whole blood samples (without LEH) were evaluated during laminar shear flow as function of time and shear rate.[3] Note that platelets get activated only at shear rates above 1100 s^{-1}. Again it was noted that the PG LEH-blood sample that is significantly poorer, especially for shear rates above 1100 s^{-1}.

Interaction of Activated Platelets with Leukocytes: The increase in intensity of expression of P-selectin on the CD45 positive leukocytes, was seen at high shear rates (2200–5400 s^{-1}) where about 20% of the leukocytes were CD62 positive after exposure of the blood samples to the laminar shear flow for 45 s^3. We were not able to measure the intensity of p-selectin expression on leukocytes during the 2 min shear study due to clot formation in the LEH-whole blood samples.[3] These results suggest that shear stress activates platelets and forms activated platelet-leukocyte aggregates.

Thromboxane B_2 Formation During Shear Flow: LEH modulated platelet eicosanoid production such as thromboxane B_2 in a unique manner with a shear flow.[3] The TXB_2 plasma levels of whole blood sample were in the range of 4–8 pg/μl. PG LEH caused a significant rise in the plasma levels of PG LEH-whole blood sample only at higher shear rates (1100–5400 s^{-1}) and the enhancement in TXB2 plasma levels of PG LEH-whole blood was 7–10 fold higher than PEG-PE LEH-whole blood sample.[3] TXB2 responses was time dependent only at high shear rates and at the physiological low shear rates, i.e., below 1100 s^{-1}, LEH evoked no response.

Thrombin (F1 + 2) Formation During Shear Flow: The thrombin generation obtained by incubation of various blood samples in batch flow system (cone-and-plate viscometer) at 37°C and as a function of shear rates and time.[3] It is seen that under these conditions thrombin (F1 + 2) was significantly measurable at high shear rates (1100–5400 s^{-1}) during the 2 minutes shear flow study.[3] Both liposomal membrane compositions seemed not to be thrombogenic at the low shear rates below 1100 s^{-1}. However, at higher shear rates of the PG LEH-whole blood sample appeared to enhance the thrombin activity by two fold in comparison to the PEG-PE LEH-whole blood sample.

III. Summary

The results obtained suggest that neither PEG-PE LEH nor PG LEH activated platelets for nonpathological flow conditions (shear rates of less than 1100 s^{-1}) and also, there was neither thrombin generation nor thromboxane formation. However, platelets were activated and platelet particles were generated at pathological shear rates (above 1100 s^{-1}) reactions at 37°C and residence times up to 2 minutes. Also, PG LEH was more thrombogenic than PEG-PE LEH due to generation of significant amounts of thrombin during pathological shear flow conditions (above shear rates greater than 1100 s^{-1}). The above observations were similar for the prostaglandin synthesis in platelets as evaluated by the thromboxane B_2 measurement, except that PG LEH showed a more pronounced effect.

The approaches used in these studies, which are based on the development and characterization of LEH as an oxygen-carrying RBC substitute, incorporate the diverse characteristics required for a practical RBC replacement fluid. It appears that we are still a long way off from an acceptable artificial RBC substitue for use in emergency situations, but we are on the right track.

References

1. Zheng S, Zheng Y, Beissinger RL, Fresco R. Microencapsulation of hemoglobin in liposomes using a double emulsion, film dehydration/rehydration approach. Biochimica et Biophysica Acta 1994;1196:123–130.
2. Sherwood RL, McCormick DL, Zheng S, Beissinger, RL. Influence of steric stabilization of liposome-encapsulated hemoglobin on *Listeria monocytogenes* host defense. Artificial Cells, Blood Substitutes, and Immobilization Biotechnology 1995;23:665–679.
3. Jato J, Beissinge R, Shankey V, Fareed J. The effect of liposome-encapsulated hemoglobin on platelet activation and thrombin generation in sheared whole blood. Submitted 1996.
4. Glanz J. R & D Magazine. September, 1994;55–58.
5. Cooney D. Biomedical Engineering Principles, Marcel Dekker, New York, 1976.
6. Djordjevich D, Miller IF. Exp Hemat 1980;8:584–92.
7. Gaber BP, Yager P, Sheridan JP, Change. FEBS Letters 1983;153:295d, 285–288.
8. Farmer MC, Gaber BP. Methods in Enzymology 1987;149:184–200.
9. Brandl M, Becker D, Bauer KH. Drug Dev Ind Pharm 1989;15:655–69.
10. Beissinger RL, Farmer MC, Gossage JL. Trans Am Soc Art Int Organs 1986;32:58–63.
11. Farmer MC, Rudolph AS, Vandegriff KD, Hayre MD, Bayne SA, Johnson SA. Biomater, Artificial Cells, Artif Organs 1988;16:289–299.

12. Rudolph AS. Cryobiology 1988;25:277–284.
13. Gossage JL, Alkhamis T, Beissinger RL, Farmer MC. In: Chang TMS, Geyer RP, eds. Blood Substitutes. Marcel Dekker Inc., New York, 1989.
14. Vidal-Naquet A, Sullivan TP, Gossage JL, Hanes JW, Giruth BH, Beissinger RL, Sehgal LR, Rosen AL. Biomater, Artificial Cells, and Artif Organs 1989;17:531–552, 1989.
15. Jopski B, Pirkl V, Jaroni HW, Schubert R, Schmidt KH. Biochim Biophys Acta 1989;978:79–84.
16. Hunt CA, Burnett RR, MacGregor RD, Strubbe AE, Lau DT, Taylor N, Kawada H. Science 1985;230:1165–8.
17. Kato A, Kondo T. Advances in Biomedical Polymers. CG, 1987.
18. Hayward JA, Levine DM, Neufeld L, Simon SR, Johnston DS, Chapman D. FEBS Letters 1982;187:261.
19. Borwanker CM, Beissinger RL, Wasan DT, Sehgal LR, Rosen AL. Biotechnology Progress 1989;4,:210–217.
20. Zheng S, Beissinger RL, Wasan DT. Hemoglobin-in-oil-in-water multiple emulsion as a blood substitute. J. Colloid and Interface Science 1991;144:72–85.
21. Vercellotti GM, Hammerschmidt DE, Craddock PR, Jacob HS. Blood 1982;59:1299–1304.
22. Bucala R, Kawakami M, Cerami A. Science 1983;220:965–7.
23. Bidwell E, Biggs. J Physiol 1957;138:37–38.
24. Bangham AD. Nature 1961;192:1197–1198.
25. Daemen FJM, van Arkel C, Hart HCh, Van der Drift C, Van Deenen LLM. Thromb Diath Haemorrh 1965;13:194–217.
26. Zwaal RFA, Comfurius P, Van Deenen LLM. Nature 1977;268:358–360.
27. Anderson LO, Brown JE. Biochem J 1981;200:161–167.
28. Gitel SN, Owen WG, Esmon CT, Jackson CM. Proc Na Acad Sic USA 1973;70(5):1344–1348.
29. Rawala-Sheikh R, Ahmad SS, Monroe DM, Roberts HR, Walsh PN. Blood 1990;76(10):435a.
30. Reinish LW, Bally MB, Loghrey HC, Cullis PR. Thrombosis and Hasmostasis 1988;60:518–523.
31. Allen TM, Hansen C, Rutledge J. Biochimica et Biophysica Acta 1989;981:27–35.
32. Saba TM. Arch Intern Med 1970;126:1031–1052.
33. Altura BM. Adv Microcirc 1980;9:252–294.
34. Allen TM, Austin GA, Chonn A, Lin L, Lee KC. Biochimica et Biophysica Acta 1991;1061:56.
35. Allen TM, Murray L, Alving CR, Moe J. Can J Physiol Pharmacol 1987;65:185–190.
36. Merion RM. Transplantation 1985;40:86–90.
37. Nugent KM. Intralipid effects on reticuloendothelial function. J Leuk Biol 1984;36:123–132.
38. Eisen HN. The cellular basis for immune responses. In: Immunology. Second Ed. Harper and Row, Hagerstown, PA, 1980.
39. Marcus AJ. Multicellular eicosanoid and other metabolic interactions of platelets and other cells. In: Colman RW, Hirsh J, Marder VJ, Salzman EW, eds. Hemostasis and Thrombosis. 3rd edition, Chapter 28, J.B. Lippincott Company, Philadelphia, 1994.
40. Santos MT, Valles J, Marcus AJ. Enhancement of platelet reactivity and modulation of eicosanoid production by intact erythrocytes. A new approach to platelet activation and recruitment. J Clin Invest 1991;87:571.
41. Jalal Jato. PhD Thesis, Illinois Institute of Technology, Chicago, IL, 1997.
42. Larsen E, Celi A, Gilbert GE, Furie BC, Erban JK, Bonfanti R, Wagner DD, Furie B. PADGEM protein: A receptor that mediates the interaction of activated platelets with neutrophils and monocytes. Cell 1989;59:305.
43. Hamburger SA, McEver RP. GMP-140 mediates adhesion of stimulated platelets to neutrophils. Blood 1990:75:550.
44. Yeo EL, Sheppard JI, Feuerstein IA. Role of p-selectin and leukocyte activation in polymorphonuclear cell adhesion to surface adherent activated platelets under physiologic shear conditions. Blood 1994 83:2498–2507.
45. Jato J, Beissinge R, Shankey V, Fareed J. Anticoagulant effects on platelet activation and thrombin generation in sheared whole blood. Submitted 1996.
46. Rabinovici R, Rudolph A, Ligler FS, Smith III EF, Feuerstein G. Circulatory Shock 1992;37:124–132.

Lasic and Papahadjopoulos (eds.), Medical Applications of Liposomes

CHAPTER 6.5

Developing uses of topical liposomes: Delivery of biologically active macromolecules

NORMAN WEINER[a] AND LINDA LIEB[b]

[a]*University of Michigan, College of Pharmacy, 428 Church St., Ann Arbor, MI 48109-1065, USA*; [b]*University of Utah Health Sciences Center, 50 North Medical Drive, Salt Lake City, Utah 84132, USA*

Overview

I. Barrier function of the skin

For most substances, the main resistance to transport is encountered in the stratum corneum which is the outermost layer of the skin, the so-called "barrier layer". The stratum corneum is a dead tissue layer comprised of many (15–25) sheet-like layers of cells all held together by transcellular desmosomes and cementing intercellular substances. The layer is approximately 10 μm thick over most of the body, but may be as much as 100 times thicker or more (1 mm) at friction surfaces like the soles of the foot or the palms of the hand. Sometimes the stratum corneum is histologically displayed as stacks of cells in neat columns. More often and over most of the body, there is a considerable randomness apparent within the cellular configuration. The stratum corneum cell has a volume of about 300 μm^3. The physicochemical properties of this tissue are critical to understanding how liposomes, or for that matter, any topically applied vehicle, can affect delivery of a drug into the skin. The structure of the stratum corneum has been likened to bricks and mortar. The bricks are the cellular units which are packed full with a protein known as keratin. This protein is laid down in the form fibers which lie in the plane of the skin and crisscross the cell. For the most part, the keratin has a helical, crystalline configuration. There is no other human protein even remotely like it, and most importantly, even when hydrated at normal levels, the matrix formed from the keratin is highly dense and therefore difficult to diffuse through. Because of both restricted solubility and slow diffusion, few molecules actually permeate the keratin field. Thus, the cellular building blocks of the stratum corneum can actually be viewed as the principal source of the barrier resistance of the horny tissue (and not the intercellular lipid which, according to most recent thought, is in fact the phase of diffusive conduction). The mortar of the stratum corneum is its intercellular lipid. Until relatively recently, it was believed that most of the lipid was intracellular and intimately associated with keratin. It is now generally accepted that at least 90% of the lipid is intercellular. Though the lipid organizes into bilayer structures, it does not have a phospholipid content typical of such structure. Rather, another class of polar lipids, the ceramides, seems responsible for the gel-like organization. One also finds sterols, sterol esters, and free fatty acids blended within the lipid milieu.

Although the stratum corneum is widely acknowledged as the main barrier to percutaneous absorption, it is also regarded as the main pathway for penetration. However, recent reports have suggested that in addition to the transepidermal route, hair follicles and sebaceous glands may contribute significantly to topical or transdermal delivery. In the past, doubt has been cast upon the actual significance of the follicular pathway based on the fact that the orifices of hair follicles occupy only about 0.1% of the total skin surface area.[1] However, the hair follicle is an invagination of the epidermis extending deep into the dermis, providing a much greater actual area for potential absorption below the skin surface. Release of sebum by sebaceous glands associated with the hair follicle may also influence absorption by providing a lipoidal pathway.[2]

The mammalian hair follicle is a complex, dynamic structure in which unique biochemical and immunological reactions dictate cyclic phases of growth, regression and activity throughout life. Several epithelial cell types, specialized structures and immunocompetent cells co-exist within the structure. Hormones, aging, growth factors, ultraviolet radiation and some pharmacological agents are known to exert varied effects upon the hair follicle. Recent new approaches to molecular and cellular biology may be useful in elucidating molecular signals that control the onset and duration of hair follicle growth and development, which still are not fully understood.[2–8] Greater understanding of cellular interactions within the structure and the biochemical mechanisms that govern it may enable rational design of targeted delivery systems.

Heightened interest in the pilosebaceous unit as a potential drug delivery target lies in the fact that the etiologies of several dermatological abnormalities relate to the hair follicle. Acne, androgenetic alopecia, alopecia areata and some skin cancers are among these conditions.[2,7,9,10] Besides localized delivery, systemic delivery via the hair follicle may also be desirable.

II. Topical delivery issues (conventional vehicles vs. liposomes)

The skin is the largest organ of the body and one of the most anatomically heterogeneous (Figure 1). The target site for delivery of a drug is highly dependent on the pharmacological activity the drug is supposed to influence (Table 1). As can be seen from this table, the intended target objective varies from complete non-penetration to systemic absorption. An additional dilemma we face is that we are often unable to accurately determine drug levels at specific tissue strata within the skin. For conventional topical formulations (i.e., creams, lotions, gels), the only control the formulator has with respect to the extent of drug deposition into the skin is: (i) the concentration of drug in the vehicle; (ii) the volume of application; (iii) the number of applications per day; and (iv) optimization of the vehicle with respect to the drug partitioning from the vehicle into the stratum corneum. However, with such conventional vehicles, although we can control, to some

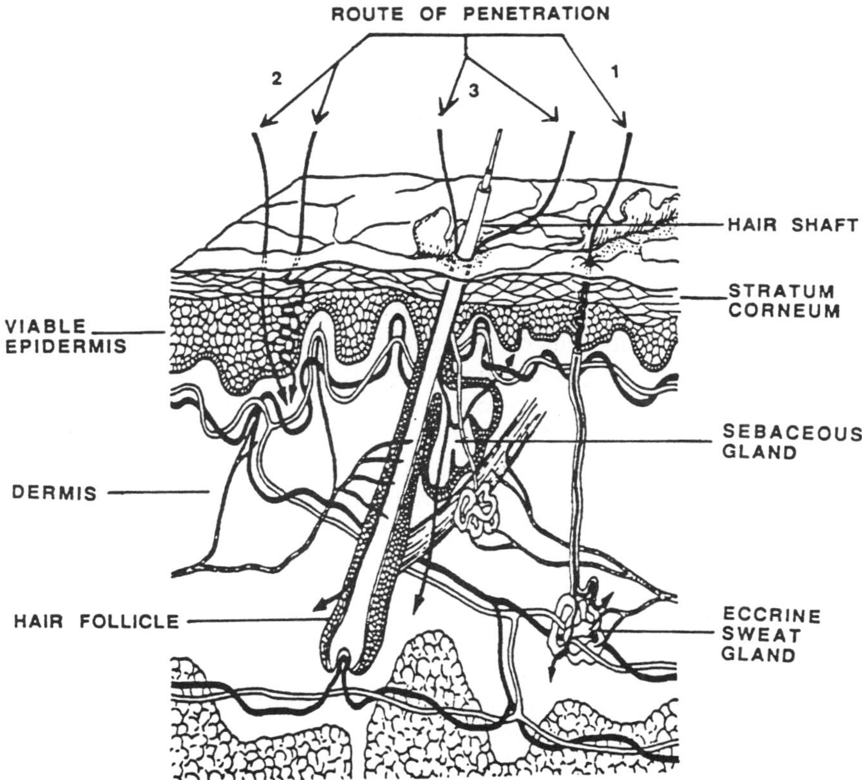

Fig. 1. Three potential routes of penetration of a drug into the skin: (1) via the sweat glands; (2) across the continuous stratum corneum; and (3) through the hair follicles with their associated sebaceous glands.

Table 1.

Skin strata target site for various pharmacological classes of topically applied drugs

Skin strata target	Class	Example
Surface	Cleansers	Soaps
	Protectants	Sunscreens
	Occlusants	Petrolatum
Stratum corneum	Moisturizers	Glycerin
	Emollients	Mineral oil
	Keratolytics	Salicylic acid
Living epidermis, dermis	Anti-inflammatories	Hydrocortisone
	Antibacterials	Clindamycin
	Anesthetics	Lidocaine
Subcutaneous	Analgesics	Methyl salicylate
Systemic (transdermals)	Vasodilators	Nitroglycerin
Sweat glands	Antiperspirants	Aluminum salts
Sebaceous glands, hair follicles, pilosebaceous units	Antiacne	Retinoic acid
	Antiandrogens (Alopecia)	Minoxidil

extent, the amount and rate of penetration of drug into the stratum corneum, further deposition of the drug molecule into deeper strata of the living skin, skin appendages and systemic circulation is strictly a function of the physicochemical properties of the drug molecule and not the vehicle (i.e., the drug does not remember where it came from). Herein lies the enormous potential of liposomes as a nonconventional topical delivery system. Although the use of liposomal drug formulations for topical application has been steadily increasing, few studies have been undertaken in order to explain the mechanism of liposomal action on drug transfer into the skin and ultimately, its improved therapeutic effect. Most in-vitro transport studies, which typically concern themselves with permeation of drug *through* the skin, do not focus on the extent of drug accumulation in the various skin strata. In order to evaluate formulation effects on the treatment of dermatological diseases by topical application, a knowledge of such tissue levels is crucial since it is expected that for a formulation to be most effective, it should facilitate increased drug levels in the appropriate skin strata.

How do liposomes actually promote drug deposition into the various skin strata? A substantial amount of evidence suggests there are a number of mechanisms in play, each exerting an influence dependent on liposomal composition. The most important mechanisms appear to be:[11–21]

1. Transfer of drug from liposomal bilayers into stratum corneum bilayers. This transfer mechanism occurs upon partial dehydration so the application site must be non-occluded. This is an extremely important observation since with conventional vehicles, occlusion of the site promotes penetration. Also, the liposomal bilayers must be in the liquid crystalline (not gel) state.
2. Permeation enhancement due to liposomal bilayer components altering phase transition properties of stratum corneum bilayers. Nonionic liposomes provide much greater penetration enhancement than phospholipid liposomes since a

number of the commonly used bilayer components of nonionic liposomes are powerful penetration enhancers.

3. Deposition of bilayers on surface of stratum corneum resulting in depot effect. Here, the gel phase liposomes are preferred and it appears that the liposomal bilayers upon dehydration following application to the skin forms a second skin providing a depot effect.
4. Deposition of bilayers and drug into hair follicle. Once again, deposition occurs upon partial dehydration (must be non-occluded) and the liposomes must be in liquid crystalline (not gel) state.

III. Historical perspective

The first vesicles tested on skin were formed from nonionic surfactants, referred to as niosomes.[22] While most of the studies using niosomes focused on cosmetic applications, a number of papers report application of nonionic liposomes encapsulating a variety of drugs for potential treatment of diseased skin. A recent example of the potential application of niosomes for transdermal delivery of estradiol was reported by Hofland et al.[23] Data illustrating skin deposition as a result of dermal delivery from phospholipid liposomes first appeared in the early 1980s in a series of papers by Mezei et al.[24,25] They reported that deposition of triamcinolone acetonide within rabbit skin was found to be localized to epidermal and dermal regions. It was hypothesized that phospholipid liposomes facilitated targeting of drugs to specific skin strata. Since that time, hundreds of papers and abstracts have been published on the topical application of liposomally entrapped drugs encompassing every imaginable pharmacological category with potential application to skin treatment. Analgesics,[26] antibiotics,[27] antifungals,[28] antipsoriatic agents,[11] antivirals,[14] non-steroidal antiinflamatory agents[29] and steroids[30] are the most studied of the topically applied liposomally encapsulated drugs since it is generally recognized that conventional topical dosage forms of these drugs are far from optimized. The rationale for the use of liposomal vehicles as opposed to conventional vehicles was to improve the extent of deposition into the living skin tissue while reducing systemic uptake. An additional goal was to alter the pharmacokinetics of drug deposition into the skin to provide a metered, prolonged therapeutic effect. Overall, the results of these studies indicate that with proper optimization, one may better control deposition of certain drugs into the skin from liposomes (as opposed to creams, lotions, etc.) since the liposomes act as a reservoir at the skin surface. However, one needs to be very cautious about making generalizations since the physicochemical properties of the drug, the lipid composition of the liposome, the degree of occlusion and a host of other variables come into play. One would in fact expect to see examples where encapsulation of certain drugs would result in a reduction of deposition to the extent where it is no longer therapeutic. To expect a liposomal delivery system to provide therapeutic systemic levels of drug upon topical application appears to be unreasonable unless one can design liposomes that can actually traverse the barrier layers of the skin.

IV. Delivery of biologically active macromolecules

There are a number of compelling reasons why biologically active macromolecules (e.g., polypeptides, proteins, antibodies, DNA) are the best drug candidates for topical liposomal delivery. Many difficult-to-treat skin diseases are caused by infectious agents (e.g., herpes, genital warts) or by immunological aberrations (e.g., psoriasis, alopecia areata, atopic dermatitis). Although present topical treatments for these and other skin diseases are palliative at best, there have been a number of clinical observations wherein either systemic or intralesional administration of specific polypeptides have resulted in a pronounced amelioration of clinical symptoms. However, when systemic routes of administration are used to deliver drugs to specific extravascular sites, far more drug than actually necessary to resolve the situation has to be administered to account for dilution of the drug. In situations where a drug's actions are non-selective, systemic regimens adequate to suppress skin symptomology invariably result in adverse effects. Moreover, administering a drug systemically still may not overcome the inaccessibility of the skin tissue to the drug. In these regards, drug delivery remains one of the most, if not the singularly most limiting factor to the effective treatment of a variety of skin diseases.[31,32]

Unfortunately, topical treatments with interferon-α for the treatment of herpes and genital warts; interferon-γ for the treatment of atopic dermatitis and cyclosporin-A for the treatment of psoriasis and alopecia areata have all been tried and have a disappointing history at the clinical level.[33–36] However, during the previous few years, it has been demonstrated, that therapeutic levels of the above-mentioned molecules can be reached in the dermis and pilosebaceous unit upon topical application of liposomal formulations.[11–21] As an exciting sequel to these findings, it has also been shown that antibodies grown against doxorubicin can be delivered into the hair follicles in doses sufficient to prevent doxorubicin-induced alopecia.[37]

Psoriasis is a common skin disease characterized by epidermal hyperplasia and inflammatory cell infiltrate in both the dermis and epidermis. The disease wreaks havoc with people's lives and currently used topical treatment regimens (corticosteroids, salicylic acid, tars) are so ineffective that indiscriminately acting drugs such as methotrexate (orally administered) still find their way into clinical use.[38] Participation of the immune system in the etiology and pathogenesis of this disease has been observed with many markers (e.g., proliferation of T-cells in lesions and immune-dependent expression of adhesion molecules on psoriatic keratinocytes) by numerous investigators.[39–40] Most interesting from a drug delivery perspective are the antipsoriatic effects of treatments known to profoundly influence the immune system, e.g., cyclosporin-A (CsA). While CsA is orally effective,[41–42] renal and hepatic toxicities limit its serious consideration of long term clinical use. Moschella and Hurley[38] state that "Owing to erratic absorption and serious potential toxicity, its use should be restricted to those patients with severe disease failing to respond to other forms of therapy".

Topical delivery of CsA for the treatment of psoriasis has been investigated by many groups, but so far the results have been disappointing[43–46] in that they all failed to show topical clinical responsiveness. When specifically looked at (punch biopsies), the levels of CsA reached in the dermis upon topical application of CsA to humans and animals were at least an order of magnitude lower than those obtained upon oral administration of CsA. Yet, in a series of in vivo animal studies, we have shown that, upon topical application of a nonionic liposomal formulation, CsA levels in the tissue comparable to those achieved on oral administration of doses used to prevent tissue rejection were achieved.[16]

Alopecia areata (AA) has long been recognized as an autoimmune disease.[38] While CsA has been shown to be effective for the treatment of AA upon its oral administration,[47] its adverse side effects again argue against its long term clinical use. Again, topical application of CsA from a variety of vehicles proved to be clinically ineffective for the treatment of AA.[48–50] Once again we see a drug delivery problem, in this case one which requires targeting of CsA to the hair follicle rather than the dermis. But, in an in vivo hamster ear model previously shown to be predictive for human pilosebaceous units, our most efficient nonionic liposomal formulation with respect to delivering CsA into living skin was also the most effective one in depositing the drug into the pilosebaceous unit.[17]

As another example, chemically-induced alopecia is a side effect produced by anti-cancer treatment with doxorubicin. Balsari et al. examined the effect of liposomally-entrapped monoclonal antibodies on alopecia induced rats[37] and demonstrated that topical treatment with liposome-incorporated monoclonal antibodies prevented alopecia. Our recent collaborative studies with this group suggest strongly the follicular route is the primary pathway for penetration. Of great importance, animal studies on the deposition of radiolabeled and fluorescent antibodies from a series of formulations indicated that the liposomal composition of the delivery system must be custom-tailored to the antibody for it to be delivered to the active site.

We view gene therapy as a promising area of research that has been hampered by the lack of studies focusing specifically on drug (DNA) delivery. Transient expression of interferons by cells of the pilosebaceous unit (an easily accessible delivery target) would provide local and regional delivery of these antiviral proteins. The theoretical advantages of such a system would include the appropriate processing (e.g., glycosylation) and secretion of the transgenic interferons, and prolonged levels of protein delivery to the regional microenvironment relative to those that might be achieved by topical application of the recombinant protein itself. The two approaches (use of recombinant proteins and gene therapy) are complementary and not mutually exclusive. Many of the problematic issues associated with the delivery of macromolecules to the pilosebaceous unit are shared by recombinant proteins and plasmid DNA.

Many of the early gene therapy studies have focused on disease states that will require regulated transgene expression for prolonged periods of time. Expression plasmid DNA and in vivo transfection are not generally effective for these applications. On the other hand, high level transient expression of therapeutic proteins is well suited for the potential treatment of dermatologic conditions in general. We therefore need to place greater emphasis on optimizing delivery of expression plasmid DNA to accessible target cells, most probably those of the pilosebaceous unit. The optimal DNA expression vector has yet to be developed, and regardless of whether this optimal expression vector takes the form of a plasmid, YAC or a recombinant viral genome, the liposomal formulations that will be developed will be applicable to the eventual formulation of a topical gene delivery system.

The ability to deliver expression plasmid DNA to cells of the pilosebaceous unit via topical application of a liposomal formulation would have a significant impact on the treatment of many dermatologic diseases. Several categories of dermatologic diseases would be amenable to treatment with topical gene therapy including infectious diseases, neoplastic diseases, autoimmune diseases and acquired conditions. For all of these diseases, the general strategy would be similar — transfect accessible cells (probably of the pilosebaceous unit) in order to mediate the expression of a transgenic protein that would have biological effects proximal to the site of application. In the case of transgenic secreted proteins, the biological effects would extend regionally beyond the immediate microenvironment of the transduced cells. Through manipulation of the physical characteristics of the liposomes, it is even possible to envision liposomal formulations that would result in systemic delivery of soluble transgenic proteins.

In addition to the established data showing the therapeutic utility of recombinant proteins in the treatment of infectious and neoplastic disease, we can envision numerous theoretical examples of effective therapeutic strategies for a topical gene delivery system to the pilosebaceous unit. This could include the modulation of cell adhesion molecules and pro-inflammatory cytokines via the expression of soluble receptors or receptor antagonist proteins for the treatment of autoimmune skin diseases (e.g., psoriasis, alopecia areata). The expression of soluble or trans-dominant negative androgen receptors by transfected follicular cells might have special applicability to the treatment of acne and male pattern baldness.

For the remainder of this chapter, work from this and other laboratories will be presented that best demonstrate the potential to topically deliver therapeutic doses of biologically active molecules to living skin strata.

V. Local and systemic delivery of proteins

Over the last few years, there has been a number of papers describing topical delivery of a variety of proteins from liposomal formulations. One potential application for topically applied liposomes would be for application of agents that alter UV-dependent effects in skin. Superoxide dismutase (SOD) is a known "superoxide anion scavenger" enzyme counteracting the photooxidative damaging effects of UV light.[51] SOD levels are normally decreased under the effects of UV light and this decrease has been shown to be reduced in the presence of topically applied liposomal SOD in mice.[52] It has been shown, for example, that topical formulations of SOD can suppress the formation of fibrosis resulting from gamma irradiation.[53]

A most successful application of liposomes for targeting and localizing biological viable proteins into skin has been with T4 endonuclease V enzyme.[54] This is another enzyme which inhibits associated UV damage by repairing UV-induced DNA damage. Recent studies have suggested that topically applied liposome encapsulated DNA excision repair enzyme, T4 endonuclease V (T4N5 liposomes) may minimize biological effects of UV irradiation contributed by DNA damage in mice.[54–57] Biological effects resulting from the use of T4N5 liposomes include an increase in the removal of cyclobutyl pyridimine dimers (CPD), and a reduction of the incidence of skin cancer in UV-irradiated mice.[56] Recent studies show inhibition of UV-induced systemic suppression of contact hypersensitivity (CHS) and delayed type hypersensitivity with topical application of T4N5 liposomes directly after UV exposure.[57] T4N5 liposomes show promise in phase II studies for the treatment of Xeroderma Pigmentosum, an autosomal recessive disease where cells are deficient in nucleotide excision repair of solar UV-induced photoproducts.

The most promising studies dealing with topical delivery of proteins for systemic (as opposed to local) delivery were reported by Cevc et al. They developed a unique liposomal delivery system termed Transfersomes™ that are claimed to "deform" so as to adapt to the size and shape of skin pores.[58] He demonstrated an effective lowering of blood glucose levels after topical application of insulin associated with ultradeformable vesicles (Transfersulin™). The biological action is reported to be the same even within interspecies differences (mice, minipigs and humans). Formulations containing mixed micelles and standard liposomes were not able to produce the same result as the Transfersome™ under these conditions. Cevc compares the effect of a number of topically applied molecules in a suspension of transfersomes to a subcutaneous injection of the same. Clearly, "conventional" liposomes, whether they be phospholipid-based or nonionic, fail by orders of magnitude to deliver therapeutic quantities of protein systemically.

VI. Topical delivery of interferon-γ into human skin

The purpose of these studies[18,59] was to assess the ability of liposome-encapsulated IFN-γ to penetrate the stratum corneum of normal human skin grafted onto nude mice, and to established whether IFN-γ in this formulation remains biologically active. For the IFN-γ dermal absorption study in human skin-grafted nude mice, the experimental protocol employed both morphological and immunological approaches to establish IFNs active presence in the tissue. Human skin sections, 2.25 cm^2, were grafted to athymic mice kept in special, aseptic housing. The graft take rate was approximately 90%. The surrounding mouse area excised at the time of harvest was used to differentiate the human-mouse-skin border during microscopic analysis. The epidermis in human split-thickness skin is relatively flat with a dermis approximately two to three times as thick as the mouse dermis. Antimouse IgG peroxidase reaction shows mouse remodeling of dermis, but not epidermis.

In vitro studies of the transport of IFN-γ into and through split-thickness human skin using a new low-level ELISA showed steady-state transport of the cytokine within the first 5 hours of exposure with approximately 10% transported demonstrating activity in a cell-based bioassay.[59] Skin exposed to IFN-γ also demonstrated regions of ICAM-1 induction by keratinocytes in the basal region of the skin, providing further evidence that at least some of the transported cytokine was biologically active.

For the in vivo studies,[18] interferon solutions and phospholipid-based liposomal interferon formulations were applied to the human skin-grafted nude mouse twice daily for 3 days. The animals were sacrificed and the tissue was placed in OCT embedding compound and snap frozen in liquid isopentane. The frozen specimens were sectioned, air-dried, fixed, rinsed and then incubated with either anti-human ICAM-1 antibody, an irrelevant antibody or no antibody. A Vectasstain ABC-AP Kit and Vector Red II staining was then used to visualize labeled ICAM-1.

Importantly, ICAM-1 expression was not induced by either blank liposomes or the solution of IFN-γ when applied to intact skin. ICAM-1 responses of grafted human skin to IFN-γ treatment either encapsulated in liposomes or mixed with pre-formed liposomes were strong and similar. Although ICAM-1 induction is not evenly distributed through the skin, the extent of ICAM-1 induction is in general greatest at the basal layer. ICAM-1 induction did not appear to be associated with hair follicles. Epidermal adnexa (hair follicles, sweat and sebaceous glands) and other dermal substructures, such as piloerector muscles develop during fetal gestation[60] and are not generated from split-thickness skin during wound healing.[61] Only poorly defined follicular remnants are apparent.

The most important finding of these studies is that there is now clear evidence that hydrophilic macromolecules can be transferred into and across deeper strata of human skin and maintain biological activity following topical application.

VII. Topical delivery of peptide drugs into pilosebaceous units

The purpose of this study was to test the hypothesis that nonionic liposomes facilitate the topical delivery of peptide drugs into pilosebaceous units.[16] The hamster ear was used as a model for human pilosebaceous units. The deposition of a hydrophilic protein, alpha-interferon (α-IFN), into pilosebaceous units and other strata of the hamster ear 12 hours after topical in vivo application of three nonionic liposomal formulations, one composed of glyceryl dilaurate/cholesterol/polyoxyethylene-10-stearyl ether (Non-1), the second composed of glyceryl distearate/cholesterol/polyoxyethylene-10-stearyl ether (Non-2) and the third composed of polyoxyethylene-10-stearyl ether/cholesterol (Non-3), a phospholipid-based liposomal formulation (PC) and an aqueous control solution (AQ) was determined. We also determined the deposition of a hydrophobic peptide, cyclosporin-A (CsA), into pilosebaceous units and other strata of the hamster ear after topical in vivo application of these liposomal formulations and a hydroalcoholic control solution (HA).

The liposomal formulations used are summarized in Table 2. The total lipid concentration in all preparations was 50 mg/ml. The total α-IFN concentration in all interferon formulations was 1×10^8 IU/ml and the formulations also contained 0.1% HSA. The liposomal suspensions were examined using a Nikon Diaphot

Table 2

Summary of liposomal formulations used in the studies

CsA formulations			
Liposomal formulation	Lipid composition	Mole or weight ratio	Saturation level of entrapped CsA (mg/ml)
Non-1	GDL:CH:POE	57:15:28 (wt)	2.2
Non-2	GDS:CH:POE	57:15:28 (wt)	1.4
Non-3	POE:CH	60:40 (wt)	1.4
PC	PC:CH:PS	1:0.5:0.1 (mole)	1.1
***α*-Interferon formulations**			
Liposomal formulation	Lipid composition	Mole or weight ratio	α-IFN concentration (IU/ml)
Non-1	GDL:CH:POE	57:15:28 (wt)	1×10^8
Non-2	GDS:CH:POE	57:15:28 (wt)	1×10^8
Non-3	POE:CH	60:40 (wt)	1×10^8
PC	PC:CH:PS	1:0.5:0.1 (mole)	1×10^8

light microscope to assure integrity and quality of the liposomal preparations. If lipid particulates were present or if the liposomes were not uniform and spherical the preparation was discarded and a fresh batch was prepared. The CsA liposomal systems were prepared so that the bilayers of each of the formulations were saturated with respect to CsA. This procedure was used so that comparisons of drug deposition could be made using formulations of equal thermodynamic activity and equal total lipid concentration (50 mg/ml). The entrapment percent of CsA in the liposomal systems was determined using size exclusion chromatography with Sephadex G-75 columns. Unseparated CsA liposomal formulations containing both entrapped and non-entrapped drug were used in all experiments. All formulations were stored at 4°C overnight before use in in vivo experiments.

Male Golden Syrian hamsters were anesthetized with sodium pentobarbital and 50 ml of the test formulation were applied to the ventral surface of each ear. All experiments were carried out under non-occluded conditions. At 12 hours, the hamsters were sacrificed and the ears removed by cutting across the base and processed to separate the ventral ear strata (dermis, pilosebaceous unit and cartilage) from the dorsal ear.

Table 3 shows the distribution of radiolabeled -IFN marker in the various compartments of golden Syrian hamster ear 12 hr after topical in vivo application of various liposomal formulations and an aqueous control solution. The recovery of total radioactivity was greater than 90% in all cases. The amount of α-IFN found in the pilosebaceous units was in the order: Non-1 $\gg$ PC $>$ Non-2 $>$ Non-3 = AQ. The amounts of α-IFN found in the cartilage and in the dorsal ear were negligibly low for all formulations except Non-1. Overall, the Non-1 liposomal formulation is far more efficient than the other four formulations tested in facilitating deposition of α-IFN into all of the strata of the hamster ear ($p < 0.01$, two-tailed t-test).

Table 4 shows the distribution of radiolabeled CsA in the various compartments

Table 3

Distribution of α-IFN in various strata of Syrian hamster ear (expressed as IU ± sd) 12 h after topical in vivo application of various formulations. (n = 4–7). Applied amount = 5 × 10^6 IU

Strata	Formulation				
	AQ	PC	Non-3	Non-2	Non-1
Pilosebaceous units	2400 ± 2200	11000 ± 3600	2300 ± 900	6100 ± 2500	49500 ± 13000
Dermis	400 ± 400	700 ± 400	200 ± 100	400 ± 100	2000 ± 1500
Cartilage	300 ± 200	1200 ± 1600	1200 ± 1300	2300 ± 1300	50500 ± 34000
Dorsal	300 ± 200	1100 ± 1500	900 ± 1000	1600 ± 500	37500 ± 32500

Table 4

Distribution of CsA in various strata of Syrian hamster ear (expressed as μg ± sd) 12 h after topical in vivo application of various formulations. (n = 3–6). Applied amount = 125 μg

Strata	Formulation				
	AQ	PC	Non-2	Non-3	Non-1
Pilosebaceous units	0.77 ± 0.12	0.51 ± 0.06	0.41 ± 0.11	0.38 ± 0.12	2.16 ± 0.52
Dermis	0.18 ± 0.11	0.19 ± 0.11	0.04 ± 0.01	0.04 ± 0.01	0.33 ± 0.09
Cartilage	0.16 ± 0.19	0.00 ± 0.00	0.04 ± 0.03	0.58 ± 0.64	6.09 ± 2.54
Dorsal	0.14 ± 0.18	0.03 ± 0.01	0.01 ± 0.01	1.03 ± 1.19	3.46 ± 1.74

of golden Syrian hamster ear 12 hr after topical in vivo application of various formulations. The recovery of total radioactivity was greater than 95% in all cases. The amount of CsA found in the pilosebaceous units was in the order: Non-1 ≫ HA > PC > Non-2 = Non-3. The amounts of CsA found in the cartilage and in the dorsal ear were negligibly low for all formulations except for the Non-1 and Non-3 liposomes. Overall, the Non-1 liposomal formulation is again more efficient than all the other formulations tested in delivering CsA into all of the strata of the hamster ear ($p < 0.01$).

The low levels of both α-IFN and CsA found in the ventral dermis following topical application to hamster ventral ear appears to be incongruent with the rather significant and large amounts of the drugs found in the cartilage and dorsal ear especially from the Non-1 liposomal formulation. It is well known that the pilosebaceous unit has a rich and elaborate plexus of capillaries that deliver blood to this highly metabolically active area.

An examination of the data in Tables 3 and 4 reveals that the amounts of drug label found in the cartilage and dorsal ear are generally proportional to the level of drug found in the sebaceous glands. It appears, therefore, that increased deposition into the cartilage and dorsal ear may have resulted from the clearance of the drug by the vast vasculature network from the vicinity of the glands. The presence of substantial amounts of the drug found in the glands themselves coupled with the curiously low amounts in the dermis further suggests a predominant and preferred follicular route of drug deposition from the Non-1 liposomal formulation.

It is interesting to note the parallel behavior of the liposomal formulations with respect to the amounts of α-IFN or CsA found in the pilosebaceous units. The excellent correlation ($r^2 = 0.996$) between the two, despite major differences in hydrophobicity/hydrophilicity, suggests that the relative ability of the liposomal formulations in facilitating deposition of a given drug is independent of the drug. The greater extent of CsA deposition compared to that for α-IFN (based upon percent of applied formulation) for a given formulation may indicate the greater ease of partitioning of the highly hydrophobic CsA into a sebum-rich environment.

Thus, Non-1 liposomal formulations facilitate the deposition of both hydrophilic and hydrophobic drugs into pilosebaceous units via the follicular route. This study also demonstrates the potential for the use of Non-1 liposomal formulations in targeted drug delivery into the follicles. Although a simple explanation for their action is proposed, the driving force for deposition into the follicles and beyond (cartilage and dorsal ear) is a complex phenomenon greatly dependent on formulation factors.

VIII. Topical delivery of monoclonal antibodies into the hair follicle

Chemically-induced alopecia is a side effect produced by anti-cancer treatment with doxorubicin. Balsari et al.[37] examined the effect of liposomally-entrapped monoclonal antibodies on alopecia induced rats and demonstrated that topical treatment with liposome-incorporated monoclonal antibodies prevented alopecia. Our recent collaborative studies with this group suggest strongly the follicular route is the primary pathway for penetration.

For these in vivo studies, liposome formulations containing MAD-11 were evaluated with the intent of optimizing their lipid composition and concentration, liposome particle size and charge and extent of drug entrapment using quantitative deposition of liposomal MAD-11 into hairless rat skin. This was determined by radiolabel assay of an ^{125}I-F(ab$'$)$_2$ IgG antibody. Formulations were applied to the dorsal skin surface for up to 12 hours. The rat was euthanized and the skin was excised, stripped and analyzed for radiolabel. The hamster ear was also used to assess deposition of MAD-11 directly into the sebaceous glands from liposomal formulations.

In another in vivo approach, the effect of formulation on deposition of antibody in fully developed follicles was studied in the hairy rat by assessing the localization of a fluorescent antibody, FITC-MAD-11 by confocal laser microscopy. Rats were euthanized and their excised skins were frozen in OCT solution with liquid nitrogen following application of the delivery systems. The frozen skin was then cryosectioned into 20 mm vertical sections and examined under the confocal microscope to access the depth of penetration of the antibody into the follicle.

Based on our promising studies involving the delivery of other substances into follicles, the Novasome I liposome was the first system tested. A preliminary clinical study carried out in Milano was disappointing in that nonionic liposomes containing MAD-11 offered no protection against doxorubicin-induced alopecia. A concurrent rat study in our laboratories indicated that these liposomes failed to deposit MAD-11 into the deeper skin strata. Based on the intriguing results that

Table 5

Distribution of MAD-11 (expressed as a percent of applied dose standard deviation) in various strata of rat skin 12 hours after in vivo topical application of various formulations containing MAD-11, 0.5 mg/ml ($n = 3$)

Formulation	Strips 4–9	Strips 10–25	Viable skin	% Recovery
Aqueous	7.31 ± 6.80	0.73 ± 0.42	0.03 ± 0.00	95.5 ± 2.42
PC/CH/PG (sonicated w/MAD-11)	16.7 ± 2.73	3.45 ± 1.37	0.17 ± 0.03	95.3 ± 1.69
PC/CH/PG (with entrapped MAD-11)	6.31 ± 3.75	1.00 ± 0.64	0.04 ± 0.03	101 ± 0.44
Neutral novasome GDL/CH/POE	18.0 ± 5.12	2.80 ± 1.23	0.04 ± 0.00	95.4 ± 4.15
Negative novasome GDL/CH/POE/PS	43.9 ± 4.09	9.48 ± 0.37	0.20 ± 0.08	96.0 ± 1.74

a crude phospholipid liposomal preparation similar to the one used in the rat studies had been somewhat active clinically, we began work on optimizing our *in-house* formulation as a function of lipid concentration, sonication effects and drug entrapment. Our basic phospholipid liposomal formulation contained phosphatidylcholine (PC), cholesterol (CH) and phosphatidylserine (PS) at a mole ratio of 1.0 : 0.5 : 0.1, respectively. Table 5 summarizes how formulation parameters affect antibody deposition into the skin. The greatest deposition of MAD-11 from phospholipid liposomes into the deeper skin strata was attained by using 75 mg/ml lipid and by sonicating the MAD-11 with the liposomes. Addition of free MAD-11 to empty phospholipid liposomes without sonication or sonication of liposomes before adding MAD-11 both resulted in significantly less deposition into the viable skin of hairless rats. An aqueous MAD-11 formulation, used as a control, was ineffective in transporting MAD-11 into the deeper skin strata of hairless rats. Introduction of a negative charge to the Novasome liposomes by the addition of phosphatidylglycerol (PG) (GDL/CH/POE : PG 52/15/28/5; weight ratio) resulted in increased deposition into the deeper skin strata of hairless rats, with a remarkably elevated amount in the last stratum corneum strips. Sonication of this formulation with MAD-11 did not significantly alter deposition. The results from these liposome studies collectively suggest that a charge-charge interaction may occur between the negatively charged liposome and the positively charged antibody, and this interaction may be needed to transport MAD-11 into the skin.

In our hamster ear studies, the results were similar in that only charged liposomes led to deposition of MAD-11 into the sebaceous glands (0.25% of applied dose from all negatively charged liposomes tested and no deposition from neutral liposomes or aqueous solution), thereby suggesting a follicular route of delivery. Confocal microscopy was also used to view hair follicles to which FITC-MAD-11 in liposomal formulations was applied. In all sections, fluorescent label was localized in the stratum corneum, hair follicle openings, and within the hair follicle. Most importantly, there appears to be a high level of antibody deep within the follicle at the level of the matrix cells.

These studies provide evidence that the composition of liposomes must be custom-tailored to a drug for to be transported effectively to the active site. Whereas the nonionic liposomes described in previous studies have little difficulty in penetrating deep into the hair follicle, they failed to facilitate deposition of the high molecular weight, positively charged antibody. The addition of a negatively charged lipid in the bilayer greatly facilitates the deposition of the antibody to the target sites deep within the hair follicle. These findings suggest that the observed prevention of doxorubicin-induced alopecia may have been mediated through direct penetration of the hair follicle.

IX. Topical application of a novel liposome-plasmid DNA formulations in vivo

Topical delivery of gene vectors to cells within the skin is an attractive strategy for gene therapy of many human diseases, including a number of dermatological conditions thought to be mediated by abnormal regulation of soluble cytokines. While it is highly unlikely that macromolecules can permeate the stratum corneum, the presence of follicles and associated structures may not only allow localized delivery to viable skin cells, but may also promote diffusion of transgenic soluble proteins into the surrounding tissue and/or the systemic circulation. The development of pharmaceutical reagents that can mediate transfection of epidermal cells would have far reaching experimental and therapeutic applications. For topical gene therapy to be successful, it will be necessary to optimize delivery of recombinant DNA to accessible target cells within living skin strata using vehicles that can overcome the formidable permeability barriers of the skin and its appendages.

We hypothesized that expression plasmid DNA could be substituted as the charged macromolecule in nonionic liposomal formulations. The goal of this substitution was the development of a topical formulation with two essential physicochemical properties required for transfection of perifollicular skin cells in vivo; (1) transdermal delivery of large amounts of plasmid DNA proximal to perifollicular cells, and (2) intracellular delivery of the DNA into the target cells. Because successful gene delivery in vivo is best assessed by the use of theoretically relevant and biologically active transgenes (as opposed to marker transgenes), the cDNA for human interleukin-1 receptor antagonist protein was used as a transgene in our studies. The purpose of this study[21] was to show that an expression plasmid encoding the cDNA for human IL-1ra protein formulated with nonionic and cationic lipid components can be used as a topical pharmaceutical reagent for the transient transfection of skin cells in vivo.

Expression plasmid DNA for the human interleukin-1 receptor antagonist (IL-1ra) protein was formulated with nonionic : cationic (NC) liposomes or phosphatidylcholine : cationic (PC) liposomes and applied to the auricular skin of hamsters in single and multiple dose protocols. Confocal microscopy identified delivery of plasmid DNA proximal to perifollicular cells, and successful transfection of perifollicular cells was identified by immunohistochemistry and ELISA. Skin treated for three days with the NC liposomes had statistically significant levels of transgenic IL-1ra present for 5 days post-treatment. Expression of transgenic IL-1ra was specific to areas of skin treated with NC liposomes but not PC liposomes. The results indicate that the NC liposomes can deliver expression plasmid DNA to perifollicular cells and mediate transient transfection in vivo.

The nonionic/cationic (NC) liposomal formulations used in the experiments contained glyceryl dilaurate (GDL), cholesterol (CH), polyoxyethylene-10-stearyl ether (POE-10), and 1,2-dioleoyloxy-3-(trimethylammonio) propane (DOTAP) at a weight percent ratio of 50 : 15 : 23 : 12. The lipid mixture also contained α-tocopherol (1% by weight of total lipids). Appropriate amounts of the lipids were mixed and melted at 70°C in a sterile polystyrene centrifuge tube. The lipid melt was then filtered through a 0.22 mm filter (Nucleopore®) and the filtrate was reheated in a water-bath at 70°C prior to being drawn into a sterile syringe. A second syringe containing sterile, autoclaved, double-distilled water was preheated to 65°C and connected via a 3-way sterile stopcock to the lipid phase syringe. The aqueous phase was then slowly injected into the lipid phase syringe. The mixture was rapidly passed back and forth between the two syringes while being cooled under cold tap water until the mixture was at room temperature and stored at 4°C until use. The total lipid concentration in the suspension was 100 mg/ml.

The ventral side of the male hamster ears were carefully shaved one day prior to the experiments. The hamsters were anesthetized and 50 ml of the test formulation containing the pSG5IL-1ra plasmid DNA were applied to the ventral surface of one ear, twice daily for three days. The contralateral ear was treated with an equivalent amount of liposomes without plasmid DNA (control). Additionally, a set of control animals was treated as described above with NC liposomes containing pSG5lacZ plasmid DNA. The total amount of lipid applied per ear of NC based liposomes was 15 mg (2.5 mg/dose), and the total lipid applied per ear with the PC-based liposomes was 11.25 mg (1.875 mg/dose). For both NC and PC based formulations the total amount of DNA applied was 1.05 mg (0.175 mg/dose). One day later (fifteen hours after the last application of the test formulations), the hamsters were sacrificed and the ears excised by dissection across the base. Kinetics of transgene expression following topical application of NC liposomal pSG5IL-1ra plasmid DNA and blank NC liposomes was studied by sacrificing treated animals at 1, 3, 5 and 8 days after the last application. Ears of untreated animals were also used as negative controls. All experiments were carried out under non-occluded conditions. At the time of sacrifice the ears were isolated by sharp dissection, weighed and measured along each border (in order to calculate the surface area exposed to treatment), then processed for either confocal laser scanning microscopy using fluorescently labeled plasmid DNA, Southern analysis, detection IL-1ra by immunohistochemistry or assay of soluble expressed protein from various strata of the hamster ear.[16]

The fluorescent studies showed that the delivery of the labeled DNA into the hair follicles and perifollicular glands appeared to be complete by 24 hours post administration. Control animals treated with an aqueous formulation containing an equivalent dose of the fluorescently labeled plasmid failed to show evidence of DNA beyond the superficial epidermis 24 hours after topical application. This indicates that perifollicular delivery is a physicochemical property specific to NC liposomal formulations.

Southern analysis data revealed that most of the plasmid DNA present within the skin was in the form of closed circular or linearized plasmid. Analysis of skin samples obtained at various times after the topical application of a single dose showed similar amounts of expression plasmid DNA present within the skin from 12–24 hours after treatment. These results indicate that for the first 24 hours post-administration, plasmid DNA was not subjected to progressive degradation and suggests that some of the plasmid was delivered intracellularly and may have been protected from digestion by extracellular nucleases.

We next tested the ability of the NC liposomal formulation to mediate transfection of the perifollicular cells proximal to the in vivo location of the delivered expression plasmid DNA. The in vivo expression of transgenic human IL-1ra was initially detected by in situ immunohistochemical staining using a monoclonal antibody specific for the human IL-1ra protein. The NC liposomal formulation also functioned as a transfecting reagent. Transfected human Il-1ra expressing cells were identified within the follicles in the proximal third of the hair shaft and occasionally at the base of the hair shaft. Negative controls treated with aqueous formulations of expression plasmid DNA, or with liposomes alone, failed to show evidence for IL-1ra expressing perifollicular cells.

We then examined the kinetics of hIL-1ra expression within treated skin over an 8 day period following a multiple dose (twice daily for three days) topical application protocol. Transgenic expression of human IL-1ra in the skin of the ventral ear was detected at its highest levels on day 1 after application of the final topical dose. The levels of transgene expression remained significantly above control values ($p < 0.005$) on days 1–5, and had returned progressively to baseline levels by day 8 (Figure 2). Ear cartilage and dorsal skin were also assayed for transgenic human IL-1ra, however all of these values were at or below the detection limits of the ELISA (29 pg/ml) and no significant differences were observed between animals treated with NC liposomal DNA and those treated with NC liposomes alone. It was also found that samples of ventral ear, glands, cartilage and dorsal skin obtained from control animals treated with NC liposomes + pSG5lacZ plasmid DNA exhibited transgenic human IL-1ra levels that were below the detection limits of the assay. In addition, no transgenic human IL-1ra was detected in the serum of the treated or control animals. These results suggest that expression of transgenic protein is confined to tissues locally targeted by the NC liposomal pSG5IL-1ra plasmid DNA formulation, and that the diffusion of transgenic IL-1ra protein is largely confined to the microenvironment proximal to the point of topical application. These results corroborate immunohistochemical analysis of treated skin showing expression of transgenic hIL-1ra in vivo.

X. Summary

In summary, the successful treatment of cutaneous diseases with a variety of macromolecules relies on an ability to effectively deliver them to appropriate sites

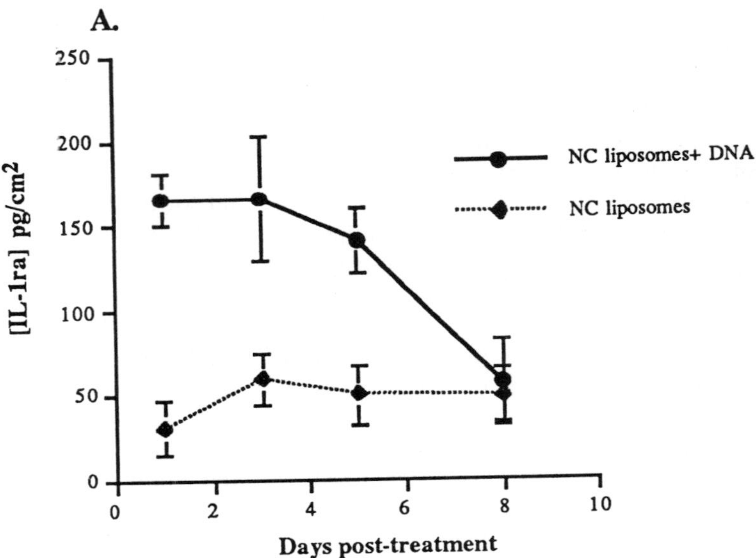

Fig. 2. Expression of human IL-1ra in the ventral skin of the hamster ear following topical in vivo application of nonionic/cationic (NC) liposomes with and without plasmid DNA.

within the skin. So far it has proven almost impossible to control many of these skin disorders using conventional dermatological formulations. Work done to date, performed in these and other laboratories, suggests topical delivery of therapeutically adequate amounts of such molecules by way of liposomes is feasible. Since the nonionic liposomes developed for these studies are stable, inexpensive and easily scaleable to quantities of mass production, they also appear to offer a pharmaceutically practical system for formulating active macromolecules. Consequently, continuous efforts with such delivery systems offer the hope that a generally effective means of topically controlling a number of skin diseases is a reachable goal through future systematic research on the liposomal delivery of therapeutic macromolecules.

References

1. Schaefer H, Watts F, Brod J, Ille Bl. Follicular penetration. In: Scott RC, Guy RH, Hadgraft J, eds. Prediction of Percutaneous Penetration: Methods, Measurements, and Modelling. London: IBC Technical Services, 1990;163–173.
2. Ebling FJG, Hale PA, Randall VA. Hormones and hair growth. In: Goldsmith LA, ed. Physiology, Biochemistry, and Molecular Biology of the Skin. Oxford:Oxford Press, 1991;660–696.
3. Weinberg WC, Goodman LV, George C, Morgan DL, Ledbetter S, Yuspa SH, Lichti V. Reconstitution of hair follicle development in vivo: determination of follicle formation, hair growth, and hair quality by dermal cells. J Inv Dermatol 1993;100:229–236.
4. Sawaya ME. Steroid chemistry and hormone controls during the hair follicle cycle. Ann NY Acad Sci 1991;642:376–384.
5. Randall VA, ThorntonM J, Hamada K, Redfern CPF, Nutbrown M, Ebling FJG, Messenger AG.

Androgens and the hair follicle; cultured human dermal papilla cells as a model system. Ann NY Acad Sci 1991;642:355–375.
6. Gibson WT, Westgate GE, Craggs RI. Immunology of the hair follicle. Ann NY Acad Sci 1991;642:291–300.
7. Cotsarelis G, Sun T, Lavker RM. Label-retaining cells reside in the bulge area of pilosebaceous unit: implications for follicular stem cells, hair cycle, and skin carcinogenesis. Cell 1990;61:1329–1337.
8. Bertolino AP. Hair growth regulation: a molecular biologic approach. J Invest Dermatol 1991;96:82S–83S.
9. Price VH. Alopecia areata: clinical aspects. J Invest Dermatol 1991;101:68S.
10. Plewig G. Models to study follicular diseases. In: Plewig G, ed. Skin Models to Study Function of and Disease of Skin. Berlin: Springer-Verlag, 1986;13–23.
11. Hu Z, Niemiec SM, Ramachandran C, Wallach DFH, Weiner N. Topical delivery of ciclosporin-A from nonionic liposomal systems: An in vivo/in vitro correlation study using hairless mouse skin. STP Pharma Sci 1994;4:466–469.
12. Weiner N, Lieb L, Niemiec S, Ramachandran C, Hu Z, Egbaria K. Liposomes: a novel topical delivery system for pharmaceutical and cosmetic applications. J Drug Targeting 1994;2:405–410.
13. Lauer A, Lieb L, Ramachandaran C, Flynn G, Weiner N. Transfolicular drug delivery. Pharm Res 1995;12:179–186.
14. Hu Z, Wu H, Weithoff C, Ramachandranm C, Weiner N. Topical delivery of alpha-interferon from liposomal systems: an in vivo study with hairless mouse. Drug Delivery 1995;2:94–97.
15. Fleisher D, Niemiec SM, Oh CK, Hu Z, Ramachandran C, Weiner N. Topical delivery of growth hormone releasing peptide using liposomal systems: An in vitro study using hairless mouse skin. Life Sci 1995;57:1293–1298.
16. Niemiec SM, Ramachandran C, Weiner N. Influence of nonionic liposomal composition on topical delivery of peptide drugs into pilosebaceous units: an in vivo study using the hamster ear model. Pharm Res 1995;12:1184–1188.
17. Lauer AC, Ramachandran C, Lieb L, Niemiec S, Weiner N. Targeted delivery to the pilosebaceous unit via liposomes. Adv Drug Deliv Revs 1986;19:311–325.
18. Short SM, Paasch BD, Turner JH, Weiner N, Daugherty A, Mrsny RJ. Percutaneous absorption of biologically-active interferon-γ in a human skin graft-nude mouse mode. Pharm Res 1996;13:1020–1027.
19. Jayaraman SC, Ramachandran C, Weiner N. Topical delivery of erythromycin from various formulations: an in vivo hairless mouse study. J Pharm Sci 1996;85:1082–1085.
20. Waranuch N, Ramachandran C, Weiner N. Effect of lipid composition on topical delivery of cyclosporin-A from nonionic liposomal formulations: an in vitro study with hairless mouse skin. J Liposome Res, in press.
21. Niemiec S, Ramachandran C, Weiner N, Roessler B: Perifollicular transgenic expression of human interleukin-1 receptor antagonist protein following topical application of novel liposome-plasmid DNA formulations in vivo. J Pharm Sci, in press.
22. Handjani-Vila RM, Guesnet JH. Liposomes: a promising future in dermatology. Ann Dermatol Venerol 1989;116:423–430.
23. Hofland HEJ, VanderGeest R, Bouwstra JA. Estradiol permeation from non-ionic surfactant vesicles through human stratum corneum in vitro, Pharm Res 1994;11:659–666.
24. Mezei M, Gulasekharam V. Liposomes: a selective drug delivery system for topical route of administration-gel dosage form. J Pharm Pharmacol 1981;34:473–474.
25. Mezei M, Gulasekharam V. Liposomes: a selective drug delivery system for topical route of administration. I-lotion dosage form. Life Sci 1980;26:1473–1477.
26. Planas ME, Gonzales P, Rodriguez L, Sanches S, Cevc G. Noninvasive percutaneous induction of topical analgesia by a new type of drug carrier, and prolongattion of local pain insensitivity by anesthetic liposomes. Anesth Analges 1992;75:615–621.
27. Skalko N, Cajkovac M, Jelsenjak I. Liposomes with clindamycin hydrochloride in the therapy of acne vulgaris. Int J Pharm 1992;85:97–101.
28. Hanel H, Braun B, Jovic N Comparative Activity of a Liposomal and a Conventional Econazole Preparation for Topical Use According to a Guinea Pig Tinea Model. Liposome Derm, Griesbach Conference, 1992;251–255.
29. Natsuki T, Tomomichi S, Matsuo R, Takabatake E, Nakanishi M. Absorption and excretion of indomethacin gel ointment containing egg lecithin. J Pharmacobio-Dyn 1986;9:s-12.
30. Vermorken, AJ, Hukkelhoven MW, Vermeesch-Markslag AM, Goos CM, Wirtz P, Ziegenmeyer

J. The use of liposomes in the topical application of steroids. J Pharm Pharmacol 1984;36:334–340.

31. TKuhls TL, Sachar J, Pineda E, Santomauro D, Wiesmeier E. Suppression of recurrent genital herpes simplex virus infection with recombinant alpha 2 interferon. J Infect Dis 1986;154:437–442.
32. Pazin GJ, Harger JJ, Armstrong JA, Breinig MK. Leukocyte interferon for treating first episodes of genital herpes in women. J Infect Dis 156, 891–900, 1987.
33. Eron LJ, Toy C, Salsitz B, Scheer RR, Wood DL, Nadler PI. Therapy of genital herpes with topically applied interferon. Antimicrob Agents Chemother 1987;31:1137–1142.
34. Freeman DJ, McKeough MB, Spruance SL. Recombinant human interferon-αA treatment of an experimental cutaneous herpes simplex virus infection of guinea pigs. J Interferon Res 1987;7:213–221, 1987.
35. Schulze HJ, Mahrle G, Steigleder GK. Topical cyclosporin A in psoriasis. Br J Dermatol 1990;122:113–120.
36. Gilhar A, Pillar T, Etzioni A. Topical cyclosporin A in alopecia areata. Acta Derm Venereol 1989;69:252–259.
37. Balsari AL, Morelli D, Menard S, Veronesi U, Colnaghi MI. Protection against doxorubicin-induced alopecia in rats by liposome-entrapped monoclonal antibodies. Res Commun 1994;8:226–230.
38. Moschella S, Hurley H. In: Moschella S, ed. Dermatology (Third edn). Philadelphia, PA: W.B. Saunders, Co. , 1992.
39. Lafferty KJ, Paris LL. Cyclosporine A and the regulation of autoimmune disease. In: Bach JF, ed. Immunointervention in Autoimmune Diseases. New York: Academic Press, 1989;23–47.
40. Fry L. An atlas of psoriasis. In: The Encyclopedia of Visual Medicine Series, Parthenon Publishing Group. New Jersey, 1992;21–24.
41. Ellis CN, Gorsulowsky DC, Hamilton TA, Billings JK, Brown MD, Voorhees JJ. Cyclosporine improves psoriasis in a double-blind study. JAMA 1996;256:3110–3117.
42. Ellis CN. Long-term management of patients taking cyclosporin A for psoriasis. In: Shuster S, ed. A Practical Guide to Cyclosporin A in the Treatment of Psoriasis. New York: Royal Society of Medicine Services Limited, 1993;35–47.
43. Schauder CS, Gorsulowsky DC. Topical cyclosporine A in the treatment of psoriasis [Abstract]. Clin Res 1986;34:1007A.
44. Griffiths CE, Powles AV, Baker BS. Topical cyclosporin and psoriasis. Lancet 1987;i:806–814.
45. Gilhar A, Winterstein G, Golan DT. Topical cyclosporine in psoriasis [Letter]. J Am Acad Dermatol 1988;18:378.
46. Bousema MT, Tank B, Heule F, Naafs B, Stolz E, van Joost T. Placebo-controlled study of psoriasis patients treated topically with a 10% cyclosporine gel. J Am Acad Dermatol 1990;22:126–131, 1990.
47. Gupta AK, Ellis CN, Cooper KD, Nickoloff BJ, Ho VC. Oral cyclosporine for the treatment of alopecia areata: A clinical and immunohistochemical analysis. J Am Acad Dermatol 1990;22:242–249.
48. DeProst Y, Teillac D, Paquez F, Carrugi L, Bachelez H. Placebo-controlled trial of topical cyclosporone in severe alopecia areata [Letter]. Lancet 1986;2:803.
49. Gilhar A, Pillar T, Etzioni A. Topical cyclosporin A in alopecia areata. Acta Derm Venerol 1989;69:252–258.
50. Coulson IH, Holden CA. Topical cylosporine A in alopecia totalis: Failure of therapeutic effect due to lack of penetration. Br J Dermatol 1989;121:53–60.
51. Miyachi, Y. In: Hayaisi O, Imamura S, Miyachi Y, eds. Reactive oxygen species in photodermatology. Tokyo: University of Tokyo Press, 1987;37–41.
52. Miyachi Y, Imamura S, Niwa Y. Decreased skin superoxide dismutase activity by a single exposure of ultraviolet radiation is reduced by liposomal superoxide dismutase pretreatment. J Invest Dermatol 1987;89:111–112.
53. Lafaix JL, Delanian S, Leplat, JJ. Radiation induced cutaneous, muscular fibrosis. II. Major therapeutic efficacy of liposomal Cu/Zn superoxide dismutase. Bull Cancer 1993;80:799–806.
54. Yarosh D, Bucana C, Cox P. Localization of liposomes containing a DNA repair enzyme in murine skin. J Invest Dermatol 1994;103(4):461–468.
55. Yarosh D, Yee V. SKH-1 hairless mice repair UV-induces pyrimidine dimers in epidermal DNA. J Photochem Photobiol 1990;B7:173–179.
56. Yarosh D, Alas G, Yee V. Pyrimidine dimer removal enhanced by DNA repair liposomes reduces the incidence of UV skin cancer in mice. Cancer Res 1992;52(15):4227–4231.

57. Kripke M, Pa C, Lg A. Pyrimidine dimers in DNA initiate systemic immunosuppression in UV-irradiated mice. Proc Natl Acad Sci USA 1992;89:7516–7520.
58. Cevc G. Transfersomes, liposomes and other lipid suspensions on the skin: Permeation enhancement, vesicle penetration, and transdermal drug delivery. Critical Reviews in Therapeutic Drug Carrier Systems 1996;13:257–388.
59. Short SM, Rubas W, Paasch BD, Mrsny R. Transport of Biologically active interferon-gamma across human skin in vitro. Pharm Res 1995;12:1140–1145.
60. Moore KL, ed. The Developing Human. 3rd ed. Philadelphia: WB Saunders Co., 1982;432–436.
61. Boyce ST, Foreman TJ, English KB. Skin wound closure in athymic mice with cultured human cells, biopolymers, and growth factors. Surgery 1991;110:866–871.

Lasic and Papahadjopoulos (eds.), Medical Applications of Liposomes

CHAPTER 6.6

Liposomes as carriers of contrast agents for in vivo diagnostics

VLADIMIR P. TORCHILIN

Center for Imaging and Pharmaceutical Research, Massachusetts General Hospital and Harvard Medical School, 149 13th Street, Charlestown, MA 02129, USA

Overview

Abstract

The current status of application of liposomes as carriers for diagnostic imaging agents in experimental and clinical medicine is considered. Liposomes loaded with the appropriate contrast agents have been shown to be suitable for all used imaging modalities, including γ-, magnetic resonance (MR), computed tomography (CT) and ultrasound imaging. The methods are briefly described to prepare liposomes loaded with various contrast agents, as well as some basic data on their in vitro and in vivo properties and biodistribution. The application of contrast-loaded liposomes in different modalities for the experimental and clinical imaging of reticulo-endothelial system (RES) organs (liver and spleen); components of lymphatic system; tumors; cardio-vascular system including the blood pool; and infection and inflammation sites is briefly reviewed together with some data available on the use of contrast liposomes for more exotic miscellaneous imaging. New trends in the use of contrast-loaded liposomes are also considered, such as the application of targeted immunoliposomes and long-circulating polymer-modified liposomes for imaging purposes.

I. Introduction

I.1. Imaging modalities

Diagnostic imaging, widely used in contemporary medicine, requires that an appropriate intensity of signal from an area of interest is achieved in order to differentiate certain structures from surrounding tissues, regardless of the modality used. As noted by G. Wolf,[1] imaging involves the relationship between the three spatial dimensions of the region of interest and a fourth dimension, time, which relates to both the pharmacokinetics of the agent and the period necessary to acquire the image. The physical properties that can be used to create an image include emission or absorption of radiation, nuclear magnetic moments and relaxation, and transmission or reflection of ultrasound. Imaging modalities can be further subdivided depending on the type of probes used, equipment, and detection methods. According to the physical principles applied, currently used imaging modalities include γ-scintigraphy (involving the application of γ-emitting radioactive materials); magnetic resonance (MR, phenomenon based on the transition between different energy levels of atomic nuclei under the action of radiofrequency signal); computed tomography (CT, the modality which utilizes ionizing radiation with the aid of computers to acquire cross-images of the body and three-dimensional images of areas of interest); and ultra-sonography (US, the modality using irradiation with ultrasound and based on the different rate at which ultrasound passes through various tissues). All four imaging modalities differ in their physical principles, sensitivity, resolution (both spatial and temporal), ability to provide images without contrast agent-mediated enhancement, and some other parameters, such as cost and safety. Usually, the imaging of different organs and tissues for early detection and localization of numerous pathologies cannot be successfully achieved without appropriate contrast agents (see further) in different imaging procedures. However, non-enhanced local CT and MR imaging (MRI) are occasionally used for certain practical purposes. Unfortunately, non-enhanced imaging techniques are useful only when relatively large tissue areas are involved in the pathological process. For example, using CT, only those metastases that are larger than 15 mm in any cross-sectional diameter are detectable, whereas in patients with colon cancer 88% of metastatic lymph nodes are smaller than 1 cm in diameter. For such small lesions invasive radiological techniques utilizing contrast media are therefore recommended.

Attenuations (i.e., the ability of a tissue to absorb a certain signal, such as X-ray, sound waves, radiation, or radiofrequencies) of different tissues differ, however, as was already mentioned, in the majority of cases this difference is not sufficient for clear discrimination between various tissues (for example, between normal and pathological ones). To solve a problem and to achieve a sufficient attenuation, contrast agents are used. These are the substances which are able to absorb certain types of signal (irradiation) much stronger than surrounding tissues. The contrast agents are specific for each imaging modality (see Table 1), and as a result of their accumulation in certain sites of interest, those sites may be easily

Table 1

Imaging modalities and required concentration of diagnostic moieties

Imaging modality	Diagnostic moiety	Required concentration
1. Gamma-scintigraphy	Diagnostic radionuclides, such as ^{111}In, ^{99m}Tc, ^{67}Ga	10^{-10} M
2. Magnetic resonance (MR) imaging	Paramagnetic ions, such as Gd and Mn, and iron oxide	10^{-4} M
3. Computed tomography (CT) imaging	Iodine, bromine, barium	10^{-2} M
4. Ultrasonography	Gas (air, argon, nitrogen)	

visualized when the appropriate imaging modality is applied. As one can easily understand, different chemical nature of reporter moieties used in different modalities and different signal intensity (sensitivity and resolution) require various amounts of a diagnostic label to be delivered into the area of interest (Table 1).

In many cases, contrast agent-mediated imaging is based on the ability of some tissues (i.e., macrophage-rich tissues) to absorb the particulate substances. This process is particle size-dependent and relies on a fine balance between particles small enough to enter the blood or lymphatic capillaries, yet large enough to be retained within the tissue. In any of imaging modalities, two main routes of administration of contrast agent are used: systemic and via local circulation. Each has its own advantages and disadvantages. By varying the physico-chemical properties of a contrast, or contrast carrier, the rate of its disappearance from the injection site upon local administration can be modulated. A disadvantage of systemic administration is that it increases the exposure of non-target organs to potentially toxic contrast agent. As the tissue concentration that must be achieved for successful imaging varies between diagnostic moieties, for this reason it was a natural progression to use microparticulate carriers for an efficient delivery of contrast agents selectively into the required areas.

1.2. Liposomes

To facilitate the accumulation of contrast in the required zone, various microparticulates have been suggested as carriers for contrast agents. Among those carriers, liposomes, microscopic artificial phospholipid vesicles, draw special attention because of their easily controlled properties and good pharmacological characteristics.

Many individual lipids and their mixtures, when suspended in an aqueous phase, spontaneously form bilayered structures (liposomes) in which the hydrophobic parts of their molecules face inwards and the hydrophilic parts are exposed to the aqueous phase surrounding them. Several different types of liposomes exist; each type has specific characteristics and can be prepared by specific methods.[2] Usual classification of liposomes is based on their size and number of concentric bilayers forming a single vesicle (such as MLVs, SUVs, LUVs). The methods for producing

LUVs can be easily scaled up and used for industrial production of large batches of liposomes with a predictable size and a narrow size distribution.

For almost two decades liposomes have been recognized as promising carriers for drugs and diagnostic agents[2,3] for the following reasons: (1) Liposomes are completely biocompatible; (2) they can entrap practically any drug or diagnostic agent into either their internal water compartment or into the membrane itself depending on the physico-chemical properties of the drug; (3) liposome-incorporated pharmaceuticals are protected from the inactivating effect of external conditions, yet at the same time do not cause undesirable side-reactions; (4) liposomes also provide a unique opportunity to deliver pharmaceuticals into cells or even inside individual cellular compartments.[4] Pursuing different in vivo delivery purposes, the size, charge and surface properties of liposomes can be easily changed simply by adding new ingredients to the lipid mixture before liposome preparation and/for by variation of preparation methods.

Unfortunately, phospholipid liposomes, if introduced into the circulation, are very rapidly (usual half-clearance time is within 30 min) sequestered by the cells of the reticuloendothelial system (RES). Liver cells are primarily responsible,[5] and the sequestration is relatively dependent on their size, charge, and composition of the liposomes. Circulating peripheral blood monocytes can also endocytose liposomes and later infiltrate tissues and deliver endocytosed liposomes to certain pathological areas in the body.[6] The elimination of the conventional liposomes from the blood is the dose-dependent process, large doses being removed at a slower rate than smaller doses. Generally speaking, one can increase the RES uptake of liposomes by decreasing liposome doses, increasing liposome size, or modifying liposome surface with lectins, sugar moieties, and negative charge. The suppression of RES uptake can be achieved by decreasing liposome size, increasing liposome dose, pre-saturating RES with "empty" liposomes or other particles, or modifying liposome surface with certain "protective" polymers.

To increase liposome accumulation in the 'required' areas, the use of targeted liposomes has been suggested. Liposomes with a specific affinity for an affected organ or tissue might increase the efficacy of liposomal pharmaceutical agents, and also decrease the loss of liposomes, and their contents, resulting from either liposome destruction by blood components or their capture by cells. To obtain targeted liposomes, different methods have been developed to bind specific ligands to the liposome surface. Immunoglobulins, primarily of the IgG class, are the most promising and widely used targeting moieties for various drugs and drug carriers including liposomes. Targeted liposomes with immunoglobulins as the targeting moieties are called immunoliposomes. Numerous methods for antibody coupling to liposomes are reviewed in Ref. 7. In general, immunoliposomes have to meet the following, most important, requirements: (1) antibody specificity and affinity should not change upon binding to the liposome; (2) a sufficient quantity of antibody molecules should be firmly bound to the liposome surface; (3) the liposomal integrity has to be preserved during the binding procedure; (4) the binding procedure should be simple and with a high yield of antibody binding to the liposome. At present, as much as 50 to 1000 antibody molecules can be bound per single 200–250 nm liposome.

Despite evident success in the development of antibody-to-liposome coupling technique and improvements in the targeting efficacy, the majority of immunoliposomes still ends in the liver, which is usually a consequence of insufficient time for the interaction between the target and targeted liposome. This is especially true in cases when a target of choice has diminished blood supply (ischemic or necrotic areas). Even high liposome affinity towards the target could not provide high liposome accumulation because of small quantity of liposomes passing through the target with the blood during the time period when liposomes are present in the circulation. The same lack of targeting occurs if the concentration of the target antigen is very low, and even sufficient blood flow (and consequently, liposome passage) through the target still does not result in good accumulation effect due to the small number of productive collisions between target antigens and immunoliposomes. It is quite evident that in both cases much better accumulation can be achieved if liposomes can stay in the circulation long enough. This will increase the total quantity of immunoliposomes passing through the target in the first case, and the number of productive collisions between immunoliposomes and target antigen in the second.

Different methods have been suggested to achieve this, including coating the liposome surface with inert, biocompatible polymers, such as polyethylene glycol (PEG), which form a protective layer over the liposome surface and slows the liposome recognition and clearance by opsonins.[8–10]

I.3. Liposomes and imaging

The use of liposomes for the delivery of imaging agents has quite a long history.[12–13] The ability of liposomes to entrap different substances into both the aqueous phase and the liposome membrane compartment made them suitable for carrying the diagnostic moieties used with all imaging modalities: γ-scintigraphy, MR imaging, CT imaging and even sonography. The different chemical nature of reporter moieties used in different modalities requires different protocols to load liposomes with the given contrast agent. Besides, all the imaging modalities listed not only differ in their sensitivity and resolution, but also require different amounts of a diagnostic label to be delivered into the area of interest. These general considerations, taken together, led to the development of the whole family of liposomal contrast agents for various purposes.

To design an appropriate diagnostic agent, one has to understand how the liposome and liposome-associated reporter moiety interact with the local environment. If the contrast liposome is rapidly taken up by the RES, the investigator will be able to observe the liver and other macrophage-rich tissues. If the liposome contains a pH- or temperature-sensitive agent it might become possible to measure these parameters within the zone which is in close proximity to the location of the contrast-loaded liposome.

This Chapter will discuss how the advantages of liposomes have been used so far in the fast growing field of diagnostic medical imaging.

II. Loading of liposomes with contrast agents

Independent or the supposed imaging modality, there exist several general approaches to liposome loading with a contrast agent (reporter group, label). The contrast agent might be: (1) added and incorporated into the aqueous interior of liposome or into the liposome membrane during the manufacturing process to liposomes; (2) adsorbed onto the surface of preformed liposomes; (3) incorporated into the lipid bilayer of preformed liposomes; (4) loaded into preformed liposomes using membrane-incorporated transporters or ion channels.

An important moment should be taken into account in the case of contrast liposomes for γ-imaging, which results from the fact that most clinically relevant radioisotopes have rather short half-life (not exceeding 3 days). As a result, the last step of the preparation of contrast liposomes for gamma-imaging normally takes place directly before the application moment. With this in mind, liposomes have to be prepared which can be sufficiently loaded with a contrast label by applying simple and fast labeling protocol. An excellent review of problems arising in the area of γ-imaging with liposomes has been published recently by W.T. Phillips and B. Goins.[14]

The relative efficacy of entrapment of radiopaque materials into different liposomes as well as advantages and disadvantages of liposome types were analyzed by C. Tilcock.[15] The maximum entrapment into the inner liposome interior might be achieved for the vesicles prepared by reverse phase evaporation, dehydration/rehydration, and interdigitation/fusion methods. In the latter case, the 4:1 ratio of iodine to lipid (CT contrast) was achieved. However, all these methods are difficult to control and scale-up. The simplest to prepare and scale-up MLVs provide the lowest entrapment (efficacy is less than 1%). So, from the practical point of view, the optimum method for liposome loading with contrast material by entrapment is still to be developed.

CT contrast agents (primarily, heavily iodinated organic compounds) were included in the inner water compartment of liposomes or incorporated into the liposome membrane during the liposome preparation.[12,16] Thus, Iopromide-containing liposomes were prepared by ethanol evaporation method with an encapsulation efficiency up to 40%.[17,18] Obtained liposomes were ca. 500 nm in diameter and demonstrated relatively low toxicity in mice and rats. The main route of iodine elimination from the body was via the kidneys. The efficacy of liposome loading with iodinated compounds depends to a great extent on the method of liposome preparation.[15] Reverse phase evaporation, dehydration/rehydration, and interdigitation/fusion methods provide the highest load. Still, liposome instability and iodine leakage together with difficulties in the scaling up procedures might cause certain problems.

Contrast liposomes for the ultrasound diagnostics[19–21] were prepared by incorporating gas bubbles (which are efficient reflectors of sound) into the liposome, or by forming the bubble directly inside the liposome as a result of a chemical reaction, such as bicarbonate hydrolysis yielding carbon dioxide. Gas bubbles stabilized inside the phospholipid membrane demonstrate good performance and low toxicity of these contrast agents in rabbit and porcine models. The authors claim that clinical studies of these agents may begin shortly. In more broad terms, some other lipid-based sonographic agents are known based on gas-containing lipid micelles and emulsions.[22–24]

Gamma-scintigraphy and MR imaging both require the sufficient quantity of radionuclide or paramagnetic metal to be associated with liposome (liposomes with radioiodine did not attract any noticeable interest as diagnostic tools). Though, attempts have been made to load liposomes with metals by encapsulation of certain metals or their adsorption onto the surface of liposomes,[25–28] two very general approaches turned out to be the most efficient to prepare liposomes for γ- and MR-imaging. According to the first approach, metal was chelated into the appropriate chelate (such as, for example, diethylene triamine pentaacetic acid or DTPA) and than included into the water interior of a liposome.[29–30] Alternatively, DTPA or a similar chelating compound may be chemically derivatized by the incorporation of a hydrophobic group, which can anchor the chelating moiety on the liposome surface during or after liposome preparation.[31,32] Different chelators and different hydrophobic anchors were tried for the preparation of ^{111}In, ^{99m}Tc, Mn-, and Gd-liposomes.[31–35] Sometimes, the use of chelates was combined with more exotic methods, such as ionophore-mediated active entrapment of a metal (Gd) by the intraliposomal DTPA.[36] The same approach can be applied to liposome loading with such radioactive metals as ^{111}In and ^{67}Ga.[37,38] However, ionophoretic loading can not be used with ^{99m}Tc because pertechnetate anion non-specifically adsorbs onto the membrane and then desorbs from it in the blood.

Because of short half-life and ideal radiation energy, isotope ^{99m}Tc is most clinically attractive for γ-scintigraphy. However, the stability of its association with liposomes was usually a problem. Recently, a new method for labeling preformed liposomes with ^{99m}Tc was developed[39] which is extremely effective (labeling efficiency is >90%) and results in very stable product. The method is based on the use of hexamethylpropyleneamine oxime, which is reconstituted with $^{99m}TcO_{4-}$ and then incubated with preformed liposomes containing glutathione. Other isotopes of choice are: ^{111}In, which is usually coupled to liposome-associated chelates by transchelation mechanism from its citrate or oxime complex,[40] and ^{67}Ga.[41,42] Numerous methods of liposome loading with those isotopes, including so called "afterloading" approaches involving the use of active metal transporters (ion channels, ionophores) incorporated into the liposomal membrane are reviewed in Ref. 14.

Magnetic resonance imaging with contrast liposomes required elaborate theoretical background. Normally, liposomal contrast agents act by shortening relaxation times (T_1 for spin-lattice and T_2 for spin-spin relaxation) of surrounding water protons resulting in the increase (T_1 agents) or decrease (T_2 agents) of the intensity of a tissue signal. The same agent may serve as both T_1 agent (at low concentration) and T_2 agent (at high concentration). For a better signal, all reporter metal atoms should be freely exposed for interaction with water. This requirement makes metal encapsulation into the liposome less attractive than metal coupling with polymeric chelators exposed into the outer water space. The detailed analysis of the behavior of liposomal MR contrasts can be found in Refs. 43–45. Mn or Gd are most frequently used to prepare liposomal contrasts for MR imaging, and properties of Mn[35] and Gd[29,30] containing liposomes were thoroughly investigated.

Liposomes loaded with chelated paramagnetic ions (Gd, Dy, Mn, Fe) have been demonstrated to be useful as MRI contrast agents mostly for the visualization of the macrophage-rich tissues such as organs of the reticuloendothelial system.[46,47] It is important to note here that all in vivo studies of paramagnetic ion-containing liposomes were limited mainly to systemic applications. Among MRI contrast agents incorporated into liposomes, $MnCl_2$ was used because of its easy incorporation into liposomes and good relaxivity enhancement.[48] Besides, the liver, which would receive a large dose of the contrast agent, was the organ responsible for maintaining Mn^{2+} homeostasis and was capable of excreting Mn^{2+} in a non-toxic form into the bile.[49]

Still, because of toxicity and poor solubility of many free paramagnetic heavy metal cations at physiologic pH, chelated complexes are used in most MRI T_1 contrast agent designs. Among different paramagnetic ion-containing compounds low molecular weight Gd-diethylenetriaminepentaacetic acid (DTPA) complexes were the first to be incorporated inside the liposomal aqueous compartment.[29] Proteins and NH_2-group-containing polymers can be easily modified with chelating DTPA moiety by treatment with the acid cyclic anhydride. Low molecular weight compounds Gd-DTPA (Magnevist, Schering AG), Gd-DTPA-BMA (Omniscan, Sanofi-Winthrop), along with relatively new macrocyclic chelator ProHance (Gd-HP-DO3A, Bristol-Myers Squibb) remain the only approved MR contrast media for clinical use. The majority of T_1 macromolecular agents developed so far are based on DTPA-chelated gadolinium.

Low-molecular-weight water-soluble paramagnetic probes may leak from liposomes upon the contact with body fluids, which destabilizes most liposomal membranes. Moreover, it has been shown that when too high concentrations of Gd-DTPA are encapsulated inside liposomes for the better enhancement, the relaxivity of the compound might be even lower than for non-encapsulated Gd-DTPA complex, probably because of decreased residence lifetime of water molecules inside vesicles.[50] The next step in the development of the usable Gd-based liposomal contrast media was the creating of membranotropic chelating agents such as DTPA-stearylamine,[51] DTPA-phosphatidyl ethanolamine,[52] and amphiphilic acylated paramagnetic complexes of Mn and Gd.[53] These amphipathic agents consist of the polar head containing chelated paramagnetic atom, and the lipid moiety which anchors the metal-chelate complex in the liposome membrane. This approach has been shown to be far superior in terms of the relaxivity of the final preparation when compared with liposome-encapsulated paramagnetic ions[50] due to the decrease in the rotational correlation times of the paramagnetic moiety rigidly connected to a relatively large particle. Liposomes with membrane-bound paramagnetic ions also demonstrate reduced risk of leakage upon contact with body fluids.

As in terms of MR contrast properties the membranotropic chelator-metal complexes are superior to the entrapped ones due to the enhanced relaxivity,[50,53,54] we suggested the method to increase the number of chelated Gd atoms attached to a single lipid anchor, capable of incorporating into the liposomal membrane. As a result, one might sharply increase the number of membrane-bound Gd atoms per vesicle and decrease the dosage of a lipid administered without compromising the image signal intensity. It might be also important for achieving high quality MR imaging, when high local concentration of the paramagnetic metal is required to obtain a good image. Figure 1 shows the pathway for the synthesis of amphiphilic polychelator *N*,α-(DTPA-polylysyl)glutaryl phosphatidyl ethanolamine (DTPA-PL-NGPE) and the schematic structure of a liposome containing such a component. The use of polylysine *N*-terminus modification chemistry originates from our previous work where similar technique was employed for the design of the chelating polymer-antibody conjugates.[55,56]

Amphiphilic polymeric modifiers, where a hydrophilic polymer is tail-to-head bound to a lipid anchor, are widely-used in liposome research (e.g., poly(ethylene glycol)-PE[8–10,57] or neoglycolipids[58]). Upon the incorporation into the bilayer, the NGPE anchor grafted with a chelating polymer forms a "coat" of chelated metal atoms around the liposomal membrane. These metal atoms are directly exposed to both interior and exterior water. Outer paramagnetic ion-containing polymeric chains, protruding from the liposome, form a more developed surface compared to conventional spherical vesicles. Paramagnetic ions located on these chains have better access to the adjacent tissue water protons. This may lead to the enhancement of the relaxivity of the paramagnetic ions and the corresponding enhancement of the vesicle contrast properties.

In case of starting poly-ϵ-CBZ-L-lysine with polymerization degree 11, used by us, the elemental analysis has revealed that after the saturation with Gd^{3+} ions, Gd-DTPA-PL-NGPE contains ca. 40% (w/w) Gd, which corresponds to 8–10 metal atoms per single lipid-modified polymer molecule. This is superior to one metal atom per one lipid molecule for previously used amphiphilic chelator Gd-DTPA-PE[52] and Gd-DTPA-SA[51] probes. The higher Gd content should lead to better relaxivity parameters and, consequently, to greater MR signal intensity (if the Gd tissue concentration does not exceed millimolar range[50]). This was proved by the proton relaxivity measurements for the different liposomal preparations each containing 3 mol% of the individual amphiphilic Gd-containing probe (Figure 2). The results demonstrated that polychelator-containing liposomes have higher relaxation influence on water protons compared with conventional liposomal preparations at the same phospholipid content. Clinically this would mean that one can considerably reduce the total lipid dose of the contrast material required for the diagnostic procedure without decreasing MR signal intensity.

To investigate the dependence of probe membrane density on the preparation

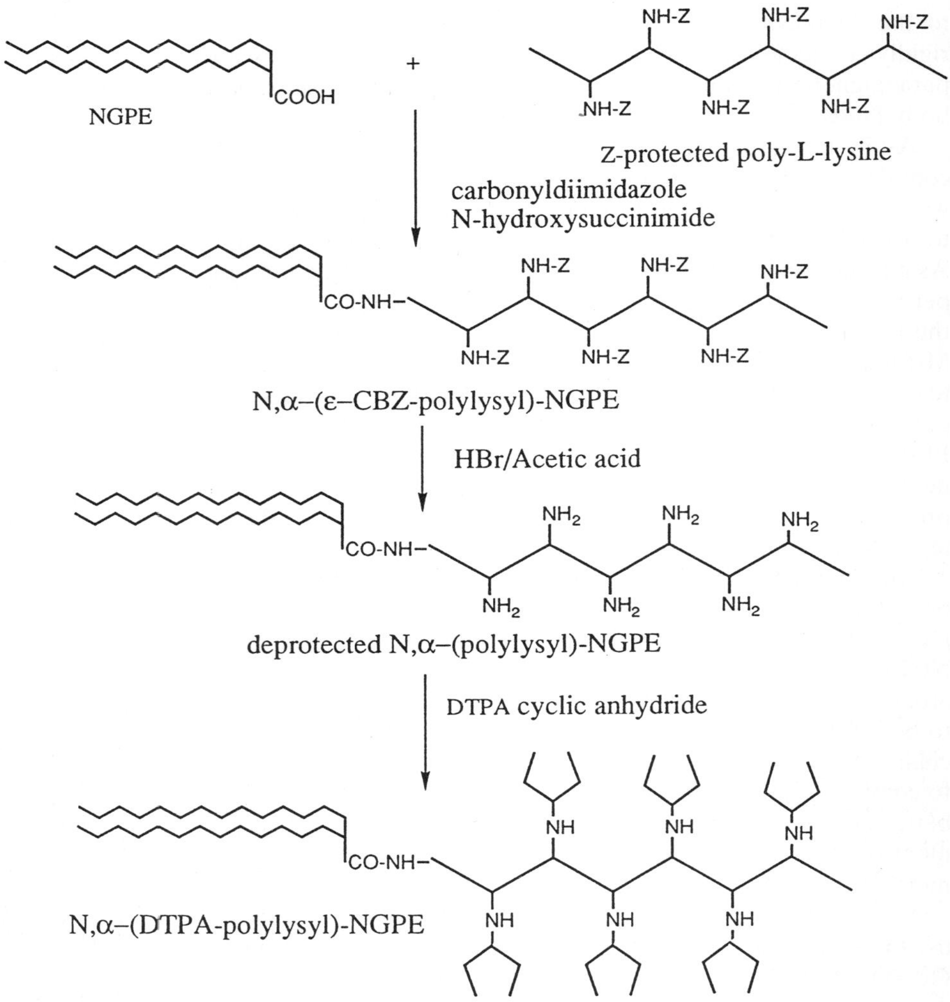

NGPE + Z-protected poly-L-lysine

carbonyldiimidazole
N-hydroxysuccinimide

N,α–(ε–CBZ-polylysyl)-NGPE

HBr/Acetic acid

deprotected N,α–(polylysyl)-NGPE

DTPA cyclic anhydride

N,α–(DTPA-polylysyl)-NGPE

Fig. 1A. Synthesis of amphiphilic polychelator, *N,a*-(DTPA-polylysyl)-NGPE.

relaxivity, we have studied the inverse T_1 response on amphiphilic chelator content. Gd-DTPA-PE was found to have an optimum relaxivity at approximately 15 mol% for egg lecithin/cholesterol liposomes. This finding is consistent with the results of Grant et al.,[52] who found that liposomes with 12.5 mol% of Gd-DTPA-PE demonstrate the maximal relaxivity. These authors have explained the phenomenon observed by the closeness of Gd atoms to one another at elevated Gd-DTPA-PE concentrations. However, Gd-poly-NGPE liposomes do not possess a relaxivity maximum at least within the concentration range studied (0–20 mol%), suggesting an increase in inter-metal atom distances on the liposome membrane.

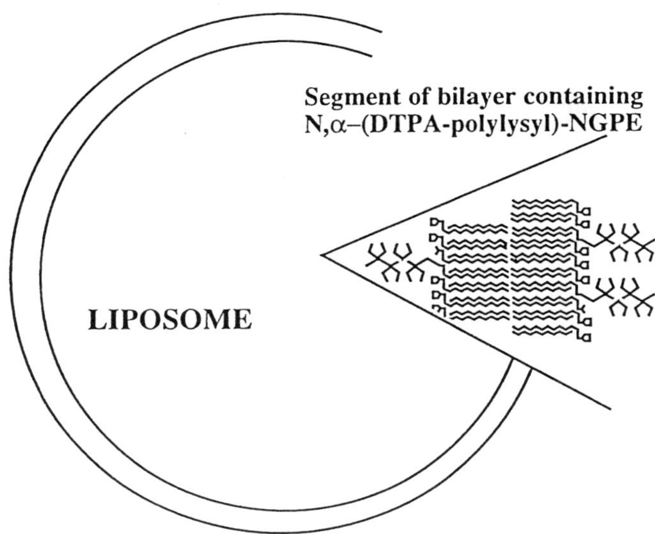

Fig. 1B. Schematic representation of liposome with incorporated amphiphilic polychelator.

It must also be mentioned that liposome surface modification with different polymers has been used to modify both in vivo and in vitro properties of the vesicles. Modification with polyethylene glycol (PEG) is known to prolong the circulation times of the vesicles[8] by inhibition of their interaction with macrophage-recognizable but not yet identified serum proteins.[59] With this in mind, many liposomal contrast agents have been modified with PEG to extend their blood circulation and to create a blood pool imaging agent, see for example.[42] In our experiments[60] we found also that the presence of a grafted polymer on the liposome surface can favorably influence spectral properties of the liposomal contrast. Thus, the relaxivity ($1/T_1$) measurements of preparation of "plain" and PEG-modified Gd-liposomes demonstrates that $1/T_1$ values of PEG-Gd-liposomes are ca. 2 times higher than the corresponding parameters for plain Gd-liposomes (see Figure 3). This fact might be explained by the presence of an increased amount of PEG-associated water protons in close proximity to chelated Gd ions located on the liposomal membrane. This observation opens interesting possibilities in regulating contrast liposome physical properties and biological behavior.

In conclusion, let us summarize the general requirements which have to be met by clinically acceptable diagnostic liposomes. Those are: (1) The labeling procedure should be simple and efficient; (2) The reporter group should be affordable, stable and safe/easy to handle; (3) The liposomes should be stable in vivo with no release of free label; (4) the liposomes need to be stable on storage—within acceptable limits.

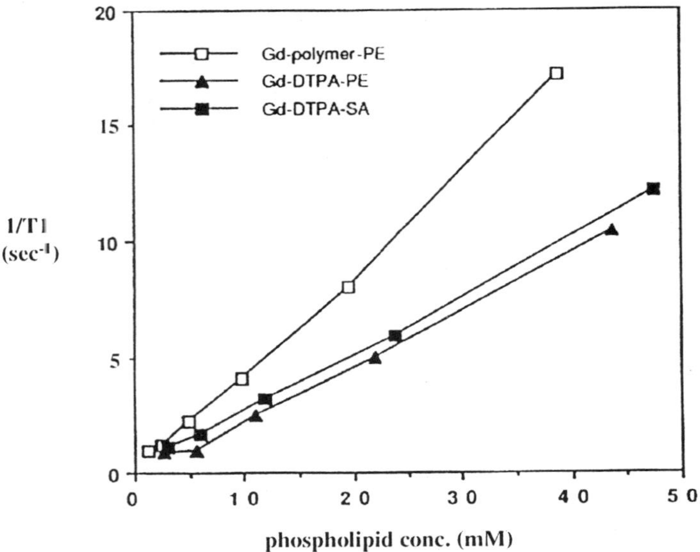

Fig. 2. Molecular relaxivities (T_1) of liposomes with different Gd-containing membranotropic chelators (Gd-polymer-PE, Gd-DTPA-PE, and Gd-DTPA-SA). Liposomes (egg lecithin:cholesterol:chelator = 72:25:3 molar ratio) were prepared by consecutive extrusion of lipid suspensions in HEPES buffered saline, pH 7.4, through the set of polycarbonate filters with pore size of 0.6, 0.4 and 0.2 μm. Final liposome size was ca. 200 nm. The relaxation parameters of all preparations were measured at 20°C using a 5 MHz RADX NMR proton spin analyzer. Notice higher relaxivity of polychelate-containing liposomes compared to liposomes with low-molecular-weight DTPA-PE or DTPA-SA at a similar molar content of chelators (as lipid moieties) because of higher content of Gd in Gd-polymer-PE. Adapted from Ref. 80.

III. Liver and spleen imaging with contrast liposomes

The imaging of the most macrophage-rich organs of RES, liver and spleen, was the earliest one performed with contrast-loaded liposomes, as RES organs are the natural targets for liposomes and accumulate them well upon intravenous administration. To ensure even more rapid uptake of the liposomes by RES macrophages (liver Kupffer cells), liposomes were enriched with phosphatidylserine (PS).[61] This approach is based on the observation that negatively charged lipids increase liver uptake of liposomes.[5] If it is necessary to increase the rate of liposome degradation in the liver and intraliposomal marker release, an additional unsaturated phospholipid, such as egg phosphatidylcholine (PC) might be included in the liposome formulation, whereas the stability of liposomes in the blood is improved by the addition of cholesterol (up to 40 mol%). The diagnostic imaging of liver and spleen is usually aimed at discovering tumors and metastases in those organs, as well as certain blood flow irregularities and inflammatory processes. The use of at least three different imaging modalities for this purpose was described, namely, γ-, MR-, and CT-imaging.

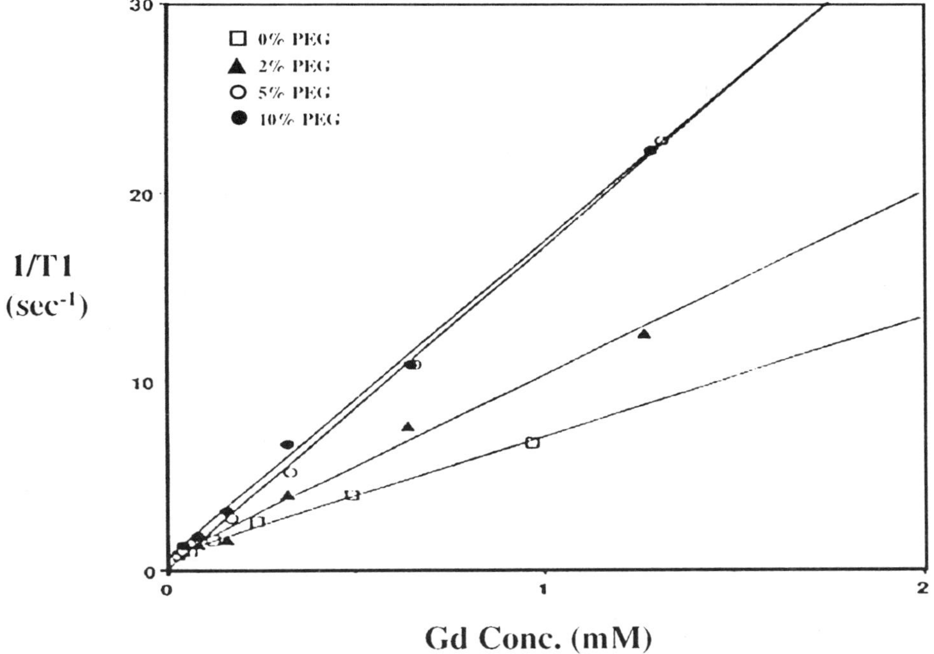

Fig. 3. Molar relaxivities of plain and PEG-modified Gd-liposome at similar concentrations of Gd. Liposome composition is the same as in Figure 2 with the addition of a specified quantity of PEG-PE. Liposome preparation and relaxivity measurements—see legend to Figure 2. Gd determinations were performed on a commercial basis by Galbraith Laboratories, Inc. (Knoxville, TN). Notice relaxivity enhancement in the presence of up to 5 mol% of PEG.

Gamma-scintigraphy of the liver and spleen with liposomal contrast is used rather rarely. However, small unilamellar vesicles loaded with ^{99m}Tc-glucoheptonate were found to accumulate well in the liver of mice and rabbits,[34,62] whereas free radiopharmaceutical demonstrated strong and undesirable accumulation in the kidneys. Negatively charged liposomes were found to be the most effective followed by neutral and positively charged ones. Good scintigraphic images were obtained 90 min upon intravenous administration of the liposomal contrast.

The optimum composition of liposomes for the liver delivery of Mn-based MR imaging agent is described in Ref. 63, where the authors discovered that the contrast agent delivered to the liver is not effective when still encapsulated within liposomes. These data confirmed the importance of free water access to the MR contrast agent in order to get a sufficient relaxation enhancement.[64] Combining the NMR and ESR techniques, the authors were able to prove the mechanism of liver signal enhancement with Mn-liposomes which is connected with the metal release from liposomes and redistribution between Kupffer's cells and hepatocytes.[65,66]

Currently, MR and CT imaging of liver and spleen with diagnostic liposomes are well developed. Liposomes containing a soluble Gd-DTPA complex were demonstrated to be effective contrast agents for MR imaging of liver, spleen, and hepatic metastases.[46,67,68] In Ref. 68, small unilamellar liposomes with membrane-incorporated Gd-DTPA-stearylamide accumulated well in the liver of rats and dogs (60% of injected dose in 4 hours). T_1-weighted MR images of the liposomal Gd-DTPA in rats and dogs gave a strong signal enhancement of the abdominal organs, liver and spleen. Another reason for choosing the liposomal form of Gd-DTPA is that although soluble Gd-DTPA is clinically effective for improving detection of many tumors, it is not well suited for detecting hepatic metastases.[46,69] However, recent studies by Unger and coworkers have confirmed that Gd-DTPA encapsulated in liposomes composed of egg phosphatidylcholine (PC) and cholesterol resulted in improved detection of liver metastases. They demonstrated that Gd-DTPA in 70 nm vesicles resulted in greater enhancement of liver and major blood vessels in the rat than the equivalent dose of free Gd-DTPA.[29,70] The work continues to show promise and may allow the improved imaging of liver pathology in the clinic in the future.

Liposomes containing X-ray contrast agents (such as verografin, ioxaglate, iohexol, iopromide, different nonionic contrast media, etc.) were used both for planar X-ray imaging of the liver and spleen,[16] and for CT imaging.[18,71,72] Iopromide-containing liposomes were even obtained in the lyophilized state, and then used for effective detection of liver tumors in mice[17] and demonstrated storage longevity benefits of liposomes. Iodine-containing interdigitation/fusion vesicles[15,73,74] with high iodine content and a size of more than 1 μm were shown to provide sufficient parenchyma enhancement at a dose of 0.1 g of iodine/kg.

IV. Liposomes for lymphatic imaging

The imaging of lymph nodes plays a major role in the early detection of neoplastic involvement in cancer patients and cannot be successfully achieved without appropriate contrast agents in different imaging procedures, since intranodal tissue has nearly the same attenuation in all imaging modalities. Nonenhanced local CT and MR imaging techniques are useful only when relatively large lymph nodes are involved. For small tissue pieces, invasive radiological techniques utilizing contrast media are used based on the ability of the lymph nodes (as any other macrophage-rich tissues) to absorb the particulate substances. Lymph nodes with malignancies contain less normal macrophage-rich tissue that actively absorbs particulates, and so, on the image (no matter which imaging modality is used), an abnormal lymph node should have a certain "filling defect" (see Figure 4).

Historically, the first radiological procedure to be used for lymph node visualization was indirect lymphography using Lipiodol, an iodine-containing oil, as an X-ray contrast agent.[75] Nuclear medicine methods have also been used for the detection of malignant involvement in lymph nodes. Lymphoscintigraphy is based on the subcutaneous injection of mainly inorganic radioactive colloids (like ^{198}Au or $^{99m}TcSb_2S_3$) following examination of the lymph nodes using γ-camera. Due to the excellent radionuclide detection limits the method is relatively fast:

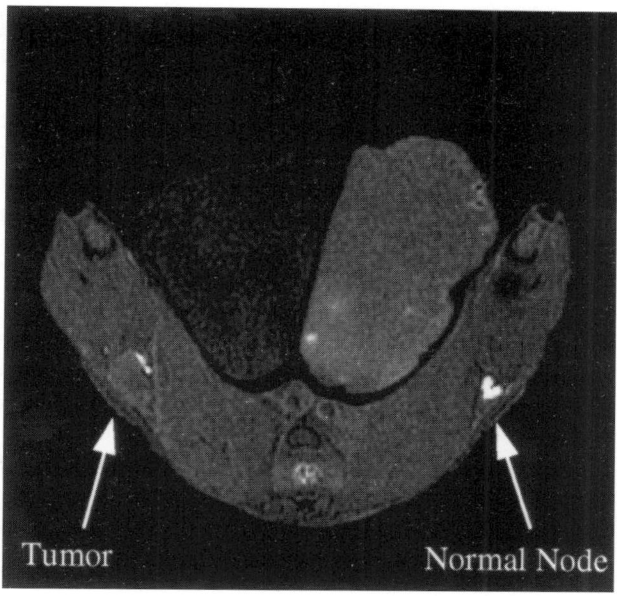

Fig. 4. MR visualization of VX_2 human sarcoma tumor in rabbit popliteal lymph node using Gd-polymer-PE-liposomes. Ten minutes postinjection time. See experimental conditions in the legend to Figure 5. The abnormal lymph node (left) can easily be distinguished from the normal node (right) based on the deficit in liposomal uptake ("filling defect"). Modified from.Ref. 13.

the accumulation in the nodes is generally completed within 2 h post injection time in rabbits and within 3 h in humans.[76] However, this method suffers from insufficient image spatial resolution, which is crucial for successful malignancy detection.

Contrast agent-mediated MRI is suitable for lymphography as any other invasive imaging procedure based on the accumulation of particulates in compartmentalized phagocytic cell-containing tissue. This imaging modality has been used for experimental lymph node visualization since the recent emergence of iron oxide based super-paramagnetic particles AMI-227[77] and USPIO.[78] USPIO particles, a potent T_2 agent, have been shown to accumulate in the lymph nodes even after intravenous administration.

Liposomes were suggested as carriers for contrast agents in lymph node visualization in a hope that, as particulates, they would be actively adsorbed by nodal macrophages. It was shown that conventional liposomes accumulate in macrophages of regional lymph nodes after subcutaneous injection. It takes 6–24 h in rat to get the nodes visualized with γ- or paramagnetic-labeled liposomes.[79]

There are two possible routes to improve the efficacy of paramagnetic liposomes as MR contrast media for the visualization of lymph nodes: to increase the quantity of liposome-associated paramagnetic metal (usually, Gd), and to enhance the signal intensity. We have tried to solve the first task by using membrane-anchored chelating polymers and the second—by modifying the liposome surface with certain polymers.[80–82] It was found that liposome modification with Gd-DTPA-polylysine-based chelating polymer (shown in Figure 1) can increase the metal load per vesicle by several fold (the approach can also result in a decrease of lipid load in patients), while surface modification with polyethylene glycol (PEG) might lead to the increased relaxivity of paramagnetic vesicles. The use of PEG-coating on liposomes in this case may change Gd water surroundings due to the presence of water molecules tightly associated with PEG molecule, and to increase thus the possible signal.

In order to prove the efficiency of Gd-DTPA-PL-NGPE liposomes as MR contrast agents[60,80] we have performed lymph node visualization in rabbit using subcutaneous administration of 20 mg of egg PC/cholesterol (75:25) liposomes containing 5 mol% of amphipathic polychelator. The rabbit transverse MRI scan (Figure 5) demonstrates that axillary lymph nodes can be seen on the scan taken only 5 min after injection, demonstrating thus the suitability of this preparation for fast MR lymph node visualization.

In another attempt to increase the overall performance of paramagnetic ion-containing liposomes for lymph node MR visualization by using liposomes grafted with polymers, we have compared relative MR signals from lymph nodes after subcutaneous administration of plain and PEG(5,000)-coated Gd-containing liposomes. PEG was incorporated into the liposome membrane as a hydrophobized derivative prepared using the modification of phosphatidylethanolamine with PEG succinimidosuccinate.[8] The study of the kinetics of the relative MR signal intensity of axillary and subscapular lymph nodes in rabbit after the subcutaneous administration of Gd-containing polymer-modified liposomes showed that plain liposomes only slightly enhance both lymph nodes; node-to-muscle intensity ratio being around 1.5 even after 80 min of observation. PEG-coated Gd-containing vesicles produce the MR signal increase in both lymph nodes fast and effectively (node-to-muscle ratio reaches 2.5 within 5 to 10 min), and can substantially increase MR signals from target tissue. It is especially remarkable that with PEG-modified Gd-containing liposomes, because of better relaxivity, the lymph nodes can be made visible within minutes of administration. To illustrate this, Figure 6 shows a typical rabbit transverse MR slice image where both axillar and subscapular lymph nodes substantially enhanced with PEG-Gd-liposomes. This is in sharp contrast with other imaging modalities, where it takes substantially longer times to get a good lymph node image. For example, it takes 24–48 h, when using X-ray contrast agent Ethiodol.[75]

The measurements of the actual delivery of liposomes using the surface-bound ^{111}In radiolabel, demonstrated the decreased accumulation of PEG-Gd-liposomes in the lymph nodes under the study because of macrophage-evading properties of PEG-modified liposomes[8] and the fact that lymphoid tissue contains considerable amounts of macrophages which serves as scavengers for filtered particulates from the local lymph.[83] Thus, in this particular case a smaller amount of the contrast material provides greater signal enhancement. The observed phenomenon can be explained by increased $1/T_1$ values of PEG-Gd-liposomes because of the presence

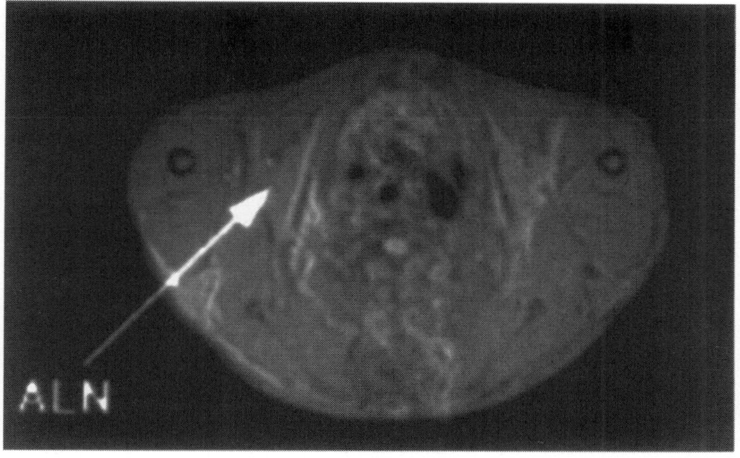

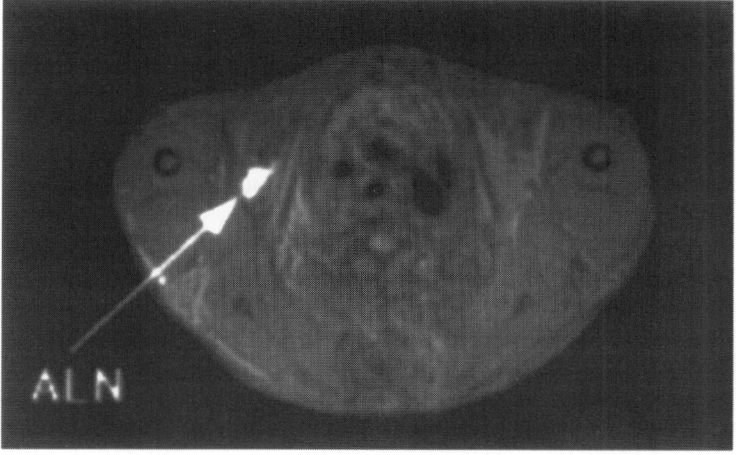

Fig. 5. Transverse scan of axillary lymph node (ALN) area in rabbit (I - preinjection, II - 5 min postinjection of Gd-containing liposomes). Liposomes (egg lecithin : cholesterol : Gd-polychelator = 70 : 25 : 5 molar ratio, 20 mg total lipid) were injected subcutaneously into the forepaw of anesthesized rabbit in 0.5 ml of HEPES buffered saline. Images were acquired using 1.5 Tesla GE Signa MRI scanner operated at fat suppression mode and T_1-weighted pulse sequence. High content of Gd in polychelate-modified liposomes permits to obtain image much faster than in case of liposomes loaded with Gd via monomeric chelators (within minutes compared to several hours). From Ref. 80.

of increased amounts of PEG-associated water protons in close proximity to chelated Gd ions located on the liposomal membrane. In addition to enhanced relaxivity, coating of the liposome surface with PEG polymer can help in avoiding the contrast agent uptake in the site of injection by resident phagocytic cells.

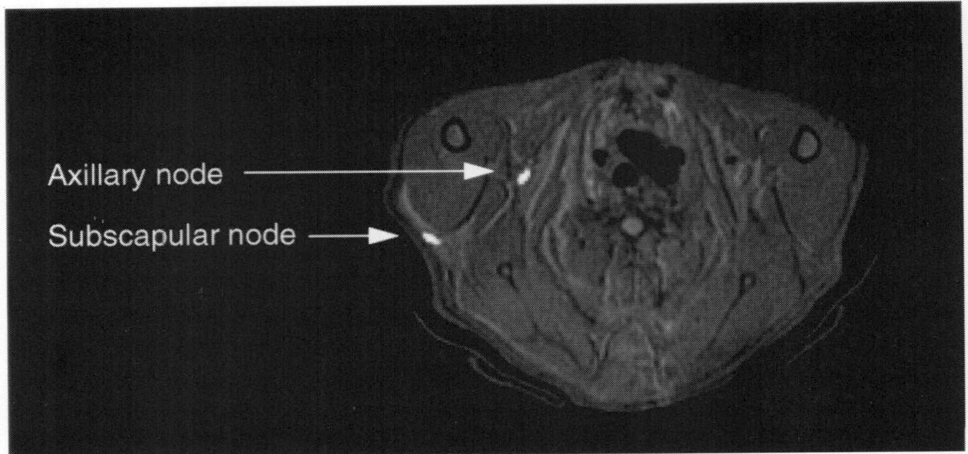

Fig. 6. Transverse scan of axillary/subscapular lymph node area in rabbit 22 min after subcutaneous injection of PEG-containing Gd liposomes. Liposomes (egg lecithin : cholesterol : Gd-DTPA-PE : PEG-PE = 70 : 25 : 5 : 5 molar ratio) were injected, and images acquired as described in the legend to Figure 5. Enhancing effect of PEG on Gd relaxivity provides faster visualization compared to liposomes without PEG. Adapted from Ref. 80.

This circumstance might increase penetration of the vesicles into initial lymphatic capillaries and further down the chain of lymph nodes.

V. Tumor imaging with contrast liposomes

The area of important potential application of contrast-loaded liposomes is tumor imaging. The main mechanism of liposome accumulation in tumors is via extravasation through leaky tumor capillaries into the interstitial space.[84] As in many other cases, the efficacy of such accumulation can be sharply increased by using long-circulating PEG-coated liposomes.[84a] Liposome-based imaging agents have already been successfully used for γ-, MR-, CT-, and sonographic imaging of tumors. ^{111}In-labeled liposomes for tumor imaging (VesCan®, Vestar, Inc.) are already in Phase II–III clinical trials. The results obtained so far with the preparation based on small unilamellar liposomes are quite encouraging; sensitivity of this preparation is up to 82%, and specificity 98% in the detection of various carcinomas as well as for melanoma, sarcoma and lymphoma.[85] ^{111}In-liposomes were also used for γ-visualization of human tumor xenografts in mice.[86] Clinical data with ^{111}In-labeled liposomes are already available on the visualization of primary lung cancer,[87] Kaposi sarcoma and lymphoma in AIDS patients,[33] and some other tumors.[88] In human phase II/phase III clinical studies, tumors were detected in 22 of 24 patients with various types of cancer using γ-scintigraphy and ^{111}In-labeled small neutral liposomes containing the A23187 ionophore (ionophore facilitates liposome loading with the metal label) (see references in Ref. 14).

In some early studies involving the rat model, negatively charged ^{99m}Tc-labeled liposomes were found to accumulate in experimental tumors better than neutral or positively charged liposomes, which was explained by better stability of the negatively charged vesicles in the blood.[28,89] The data, however, were not confirmed in clinical studies. The use of ^{99m}Tc-DTPA-liposomes for experimental tumor imaging is also described,[90,91] as well as the use of liposomes labeled with ^{67}Ga.[41] Liposomes loaded with lipophilic ^{99m}Tc-hexamethylpropyleneamine oxime were successfully used for γ-scintigraphy of tumors in mice.[92] In the latter case, neutral liposomes demonstrated the best liver-to-tumor and spleen-to-tumor ratios.

As soon as low efficacy of contrast liposomes in tumor imaging was often explained by fast clearance of such liposomes from the blood and their inability to accumulate in tumors, one of the first studies on the use of long-circulating liposomes for γ-imaging also involved tumor visualization. Radiolabeled ganglioside GM1- and phosphatidylinositol-liposomes were used for imaging B16 mouse melanoma; liposome accumulation in the footpad implanted tumor was up to 10% dose/g at 24 h.[93] Palmityl-glucuronate also prolongs circulation time of ^{99m}Tc-radiolabeled liposomes and causes them to accumulate in tumors.[94]

Interestingly, in certain cases contrast liposomes accumulate in tumors indirectly, via loading resident macrophages.[88,95–97] This was already proved in animal experiments (with mice bearing melanoma, mammary adenocarcinoma, and lung carcinoma), as well as in clinical studies with various types of cancer.[88,96,97] The sensitivity of the method reaches 85% far exceeding that for traditional contrast agents.

The use of lipid-coated stable gas-containing microbubbles for sonographic imaging of experimental brain glyomas in rats is also described.[98]

VI. Blood pool imaging with long-circulating liposomes

Blood pool imaging is of special interest for the evaluation of the current state of blood flow and for the discovery of its irregularities caused by atherosclerotic lesions, thrombi or tumors. Blood pool imaging (experimental so far) requires prolonged circulation of diagnostic liposomes and is usually based on the utilization of sterically-protected polymer-modified liposomes. Some authors[14] consider blood pool imaging as an "antitargeting" application, though it might also be considered as liposome targeting to the blood compartment. Based on the longevity of liposomes in blood, blood perfusion and various cardiac parameters (ventricular ejection fraction, cardiac output, and wall motion being among them) can be evaluated.[19,42,68,79,94,99] Different imaging modalities can be used for blood pool imaging, such as γ-scintigraphy, CT and MR imaging, and sonography. Long-circulating liposomes loaded with ^{111}In, ^{67}Ga, Gd, heavily iodinated organic compounds, and gas bubbles are already described as promising contrast agents for this purpose. It seems that contrast-loaded liposomes may serve as a good alternative for currently used ^{99m}Tc-labeled red blood cells, which do not allow acquisition of desirable quality images because of the fast accumulation of the latter in the liver. Besides, the label in contrast liposomes, such as ^{99m}Tc-liposomes loaded by the glutathione-mediated method, might be more stable than the label associated with red blood cells, which permits those liposomes to be used for the visualization of gastrointestinal bleeding or deep vein thrombosis.[100,101]

VII. Liposomes for imaging cardio-vascular pathologies

Plain, non-modified liposomes long ago were shown to accumulate to some extent in infarcted areas (experimental myocardial infarction) via the impaired filtration mechanism (extravazation through ischemically damaged blood vessels), in particular, liposomes loaded with ^{99m}Tc-DTPA.[102,103] However, the degree of this accumulation is insufficient to create high enough local concentrations of imaging agents and to obtain good, diagnostically significant images. The main reason for this is the relatively small size of the area of interest and diminished blood flow through this area. To solve the problem, immunoliposomes were used for targeted delivery of drugs and imaging agents.[7]

The first visualization of myocardial infarction with liposomes was described Torchilin et al.,[104] who successfully used $^{111}InCl_3$-loaded immunoliposomes modified with infarct-specific, anti-myosin antibody to perform γ-imaging of an experimental myocardial infarction in dogs. Despite the evident success in increasing the targeting efficacy, the majority of mAb-modified liposomes (and corresponding imaging agent) still ends up in the liver and does not have sufficient time for interaction with the target. The problems created by this fact for the delivery of contrast liposomes into areas with low blood flow or insufficient antigen concentration, were discussed already in the Introduction Section. So, further attempts to use liposomes as imaging agents for myocardial infarction involved long-circulating immunoliposomes (including those modified with antimyosin antibody). ^{111}In-labeled PEG-modified antimyosin liposomes have been shown to accumulate successfully in the areas of experimental myocardial infarction in rabbits and dogs and provide good γ-immunoscintigraphic images of the infarct within several hours upon intravenous administration.[40,105,106]

To prove that long-circulating PEG-coated liposomes can be targeted by co-incorporation of PEG and an antibody onto the liposome surface, we have studied in vivo PEG-liposomes with anti-myosin antibody[105] for targeting of experimental myocardial infarction in rabbits. The antibody used (mAb R11D10) effectively binds myosin inside ischemic and/or necrotic cardiomyocytes with affected or destroyed cellular membranes, but does not interact with normal cells being unable to penetrate the intact plasmic membrane.[107,108] This forms the basis for the targeted delivery of radiolabeled PEG-coated long-circulating liposomes in the region of ischemically compromised myocardium. Infarcts in rabbits were generated (as described in Ref. 105) and different radiolabeled liposome preparations were injected intravenously after 30 to 60 min of reperfusion. Liposomes sized around 150 nm were prepared by detergent dialysis method from a mixture of phosphatidyl choline and cholesterol in 3:2 molar ratio. Additionally liposomes contained 1 mol% of ^{111}In-DTPA-SA, and, when necessary, 4 to 10 mol% of PEG(5,000)-PE.[8,105] For incorporation into the liposomal membrane anti-myosin

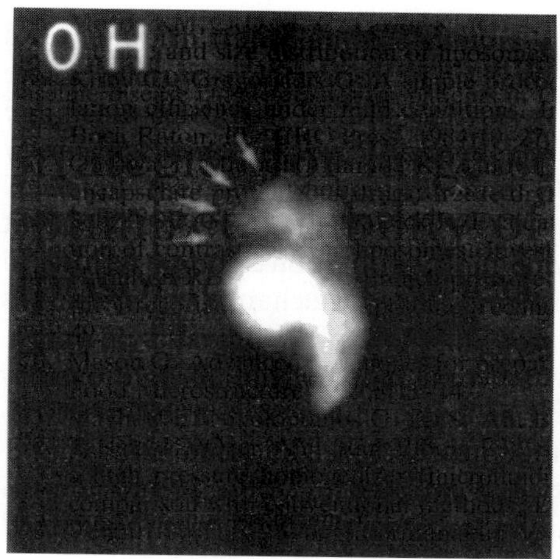

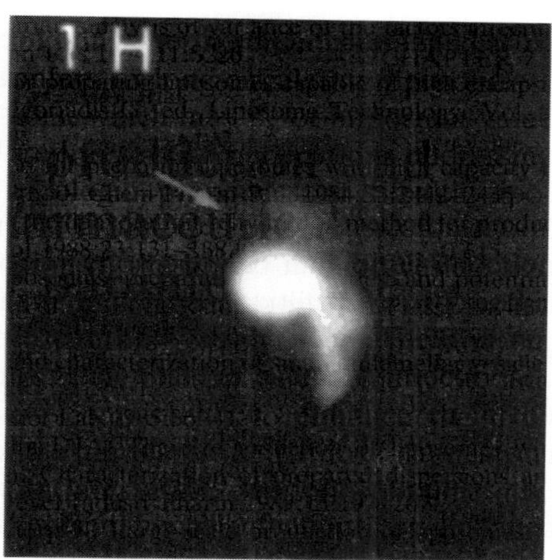

Fig. 7. Radioimmunoscintigraphy of experimental myocardial infarction in dog with ^{111}In-PEG-anti-myosin-liposomes. Infarct localization can already be seen as early as in 1 h postinjection (single arrow), while when using routinely labeled antimyosin antibodies it requires up to 24 hours to obtain good images. Multiple arrows show the area of the initial blood flow pattern in the heart area immediately upon liposomal contrast administration. For experimental details, see Ref. 105.

antibody was first modified with hydrophobic "anchor", NGPE, as described by us in Ref. [109]. Five to six hours after liposome injection, animals were killed by an overdose of pentobarbital, and samples of normal and infarcted myocardium were weighed and counted in a γ-counter. The data on the liposome accumulation in the heart were expressed as infarct-to-normal myocardium radioactivity ratio.

It was found that the half-life of PEG-free immunoliposomes in rabbits was 40 min, and increased to 200 min for liposomes with 4 mol% PEG and to ca. 1000 min for liposomes with 10 mol% PEG. The highest uptake ratio was achieved for PEG-immunoliposomes with 4 mol% of PEG and reached 20:1. Very high PEG concentration (10 mol%) on immunoliposomes were found to diminish the uptake ratio probably because of sterically "shielding" antibodies. All acute myocardial infarctions have been confirmed by histochemical staining with triphenyl tetrazolium chloride (TTC), which is specific for delineation of normal tissues following dehydrogenase activity, whereas infarcted myocardium remains unstained.[107] It appeared that lack of TTC staining coincided with increased ^{111}In radioactivity accumulation, and the uptake of PEG-immunoliposomes really corresponded to the areas of infarction. Thus, long-circulating immunoliposomes with the properly chosen PEG-to-mAb ratio provide the most effective accumulation in the regions where not all cells are dead, and the target antigen concentration might be relatively low. Using similar protocol[110] we have performed experimental visualization of a myocardial infarction in a dog using PEG-antimyosin-^{111}In-labeled liposomes (see Figure 7). The results prove that gamma-labeled PEG-immunoliposomes can effectively accumulate in the target zone (the necrotic areas of the reperfused infarcted myocardium) and permit its quite efficient and fast visualization. The relative importance of such parameters as liposome size and antibody and/or PEG presence on the liposome surface for contrast liposome biodistribution and targeting was described by us in Ref. 111. We quite expectedly found that the combination of Fab and PEG on the surface of relatively small (ca. 150 nm) liposomes gives the best accumulation of the liposomal label in the target due to the fact that both specific (antibody-mediated) and non-specific (mediated by the impaired filtration of long-circulating liposomes) mechanism work in this particular case resulting in additive effect.

Myocardial infarct was also visualized by sonography, when gas-filled liposomes were used for infarct imaging in rabbits.[20] Still, infarct imaging with liposomes remains at the level of animal experimental imaging.

Multilamellar liposomes with encapsulated ^{99m}Tc-DTPA have also been shown to accumulate within 2 hours in necrotic intestinal areas in rats after experimental mesenteric occlusion.[112] As noted by Phillips,[14] if this study were repeated with long-circulating smaller liposomes, and better labeling techniques, the observed effect is likely to be even greater.

VIII. Visualization of inflammation and infection sites

The use of microparticulate imaging agents for the visualization of infection and inflammation sites is based on the ability of microparticulates to extravasate from the circulation and accumulate in those sites similar to what we already described for tumors and infarcted tissues. With this in mind, ^{99m}Tc-labeled liposomes were used in several studies. In early studies, surface-labeled liposomes were shown to accumulate within 30 min in experimental staphylococcal abscess in rats.[113] Same approach was successfully used in preliminary clinical studies, where deep-seated infections were visualized with labeled liposomes in several patients.[114] Later on, more advanced ^{99m}Tc-liposomes labeled using lipophilic chelator and reduced glutathione were shown to accumulate in rats in sites injected with *Staphylococcus aureus*.[115] Abscesses were visualized within 2 hours after the injection of ^{99m}Tc-liposomes with maximum contrast accumulation over a period of 24 hours. The authors also mentioned the possible role of local phagocytosis by neutrophils and macrophages, which are present in higher concentration in inflammed tissues, in enhanced liposome accumulation. In all cases, positively charged liposomes demonstrated better accumulation in target areas than neutral or negatively charged liposomes. The visualization of abscesses in vivo was also achieved by γ-scintigraphy with multilamellar ^{99m}Tc-liposomes[116] and by MR imaging with lipid-coated iron oxide particles.[117]

Despite the fact that tumors also accumulate liposomes via an impaired filtration mechanism, there exists a possibility to discriminate between infection and tumor using positively-charged ^{67}Ga-liposomes[41] which do accumulate in tumors and do not accumulate in infection sites.

Inflammation sites have also been shown to accumulate contrast liposome. Thus, γ-imaging demonstrated specific accumulation of ^{99m}Tc-loaded negatively charged liposomes in the inflamed paws of rats with adjuvant induced arthritis.[118] Liposomes labeled with ^{67}Ga well enough detected experimentally induced inflammations in rat, the inflammation-to-muscle accumulation ratio reaching maximum (32 : 1) in case of neutral liposomes.[41]

IX. Miscellaneous imaging with liposomes

Some other organs and tissues of interest were imaged with contrast-loaded liposomes. Thus, MR imaging of rabbit brain was achieved after intracarotid administration of large multivesicular liposomes loaded with paramagnetic metals chelated by DTPA.[119] Contrast-loaded liposomes were also used for the visualization of bone marrow.[120] ^{99m}Tc-labeled liposomes have also been used for intra-articular administration.[121] It was already mentioned in connection with the inflammation imaging that certain data are available on the accumulation of liposomes in sites of experimentally induced arthritis in a rat model.[118] The pilot studies utilizing ^{99m}Tc-labeled negatively charged liposomes demonstrated increased uptake of the liposomes by all clinically involved joints in patients with rheumatoid arthritis (see references in Ref. 14). Phagocytes within the inflammatory lesion may also play a role in the enhanced liposome accumulation.

A very interesting approach to investigate vascular network in an eye using carboxyfluorescein-loaded liposomes was suggested in Ref. [63]. The authors obtained impressive data on both venous and arterial retina blood flow in experimental monkeys following fluorescent liposome distribution as well as distribution of the dye released from liposomes (including temperature-sensitive ones) into different eye compartments. This technique, shows great promise for the measurement of blood flow in any blood vessels where liposomes can be visualized, but especially in the eye. The authors[63] list some of the key advantages of their temperature-dependent release over the method of laser-induced dye release from liposomes containing a fluorescent dye in self-quenching concentrations: (1) the laser pulse is not needed to release the liposome-entrapped dye; (2) the lower amount of light needed to image the vesicles is now in the range that is safe for humans; (3) a lower concentration of dye is entrapped in the vesicles since the goal is to have the fluorescence of the entrapped dye visible; (4) a much lower dosage of liposome is required for good visibility in the retina; and (5) the potential of being able to measure the blood flow in very small vesicles. The authors are absolutely right that this new technology has the potential to revolutionize the imaging and quantitation of blood flow in the retina. However, they point out that the existing paucity of animal models to test the capabilities of this system may inhibit the development of this technology.

X. Conclusion: New trends and approaches; future directions

If liposome based diagnostics are to develop and become important in clinical medicine, they must demonstrate clear advantages over currently employed methods. Perhaps the key benefit to be gained by employing a liposome-based diagnostic is the ability to "image" (i.e., measure) the functional characteristics of a tissue or other region of interest. As long as an understanding of the characteristics of the liposome carrier and how it interacts with the biological milieu exists, it should be possible to design liposome-based diagnostics that make these functional measurements possible. Everything discussed above is just a small part of the information currently available on the use of liposomes as vehicles for contrast agents. The data obtained so far leads to a definite conclusion that contrast-loaded liposomes may be and will be successfully used for visualization of numerous organs, tissues, and disease sites, and liposome-based imaging agents are quite close to becoming the contrast agents of choice in clinical conditions in all imaging modalities.

However, the progression from experimental animals to human patients may proceed differently for different liposomal preparations. While liposomes for CT imaging and sonography still are mainly the subjects of laboratory research, the advantages of γ and MR imaging with liposomes have been already demonstrated in many clinical studies. Gamma-imaging with radiometal-loaded liposomes seems to be the most attractive because of relatively small quantities of both phospholipid carrier and reporter metal needed for successful imaging, which greatly minimizes the problems of side-effects, possible metal toxicity and the cost of preparation. Such problems as liposome uptake by the RES and the lack of targeting are already at least partially solved by the use of immunoliposomes, long-circulating liposomes and even long-circulating immunoliposomes.

PEG- and PEG-like polymer-coated liposomes and immunoliposomes are already considered as promising diagnostic (imaging) agents. They can be easily loaded with sufficient quantities of radioactive (for γ-imaging) or MR-active (for MRI) agents. PEG can perform two different functions on liposomes. It can serve as a capture-avoiding agent permitting effective accumulation of the diagnostic label in the target and high target-to-normal ratio, as we have seen with ^{111}In-PEG-immunoliposomes used for myocardial infarction imaging. On the other hand, surface modification of Gd-liposomes might improve their properties as MR-imaging agents, as we have seen with Gd-PEG-liposomes used for lymph node imaging.

The main obstacles that remain for the clinical use of liposome-based contrast agents for MR-imaging are: (1) the possibility of toxicity of metal-loaded liposomal preparations owing to the intracellular liberation of metal, such as Gd, from the liposome-incorporated chelating moieties, and (2) the increased cost of the product owing to the amount of liposomes (lipids) required for a single diagnostic procedure. However, these problems apply to systemic applications only, such as liver and spleen imaging after intravenous injection of contrast agent. The local administration of small doses of reporter metal-containing liposomes may be free from the above mentioned shortcomings. In general, there is no doubt that within the next two to three years there will be extensive data on the clinical use of liposome-based imaging agents.

Still, before this happens, we have to answer several important questions, such as is it possible to increase liposome loading with reporter moieties still further in order to minimize the overloading patient with a phospholipid and to decrease a cost of the preparation? What is the fate of liposome-associated reporter metal in vivo? How to optimize liposome properties to make them clinically useful ultrasound imaging agents? How can the large scale production of homogenous preparations of diagnostic liposomes be achieved so that they may be used in routine clinical practice cost effectively? By answering these questions properly, we might be able to introduce into practice a new generation of high performance imaging agents.

References

1. Wolf G. Targeted delivery of imaging agents: an overview. In: Torchilin VP, ed. Handbook of Targeted Delivery of Imaging Agents. Boca Raton, FL: CRC Press, 1995;3–22.
2. Lasic DD. Liposomes from physics to applications. Amsterdam: Elsevier Science Publishers, 1993.
3. Gregoriadis G. Liposomes as drug carriers. London: John Wiley, 1988.
4. Torchilin VP, Zhou F, Huang L. pH-Sensitive liposomes. J Liposome Res 1993;3:201–255.
5. Senior JH. Fate and behavior of liposomes in vivo: A review of controlling factors. CRC Crit Rev Ther Drug Carriers Syst 1987;3:123–193.
6. Fidler I. Incorporation of immunomodulators on liposomes for systemic activation of macrophages and therapy of cancer metastasis. In: Gregoriadis G, ed. Liposome Technology, 2nd ed., Vol 2. Boca Raton, FL: CRC Press, 1993;45–63.
7. Torchilin VP. Liposomes as targetable drug carriers. CRC Crit Rev Ther Drug Carriers Syst 1985;2:65–115.
8. Klibanov A, Maruyama K, Torchilin V, Huang L. Amphipathic polyethyleneglycols effectively prolong the circulation time of liposomes. FEBS Lett 1990;268:235–237.
9. Lasic D, Martin F, eds. Stealth liposomes. Boca Raton, FL: CRC Press, 1995.
10. Torchilin VP, Trubetskoy VS. Which polymers can make nanoparticulate drug carriers long-circulating? Adv Drug Deliv Rev 1995;16:141–155.
11. Seltzer SE. The role of liposomes in diagnostic imaging. Radiology 1989;171:19–21.
12. Torchilin VP, Trubetskoy VS. In vivo visualizing of organs and tissues with liposomes. J Liposome Res 1995;5:795–812.
13. Torchilin VP. Liposomes as delivery agents for medical imaging. Mol Med Today 1966;2:242–249.
14. Phillips WT, Goins B. Targeted delivery of imaging agents by liposomes. In: Torchilin VP, ed. Handbook of Targeted Delivery of Imaging Agents. Boca Raton, FL:CRC Press, 1995;149–173.
15. Tilcock C. Imaging tools: liposomal agents for nuclear medicine, computed tomography, magnetic resonance, and ultrasound. In: Philippot JR, Schuber F, eds. Liposomes as Tools in Basic Research and Industry. Boca Raton, FL: CRC Press, 1995;225–240.
16. Rozenberg OA, Hanson KP. Radiopaque liposomes for imaging of the spleen and liver. Radiology (Russ) 1983;149:877–881.
17. Sachse A, Leike JU, Robling GL, Wagner SE, Krause W. Preparation and evaluation of lyophilized iopromide-carrying liposomes for liver tumor detection. Invest Radiol 1993;28:838–844.
18. Krause W, Leike J, Sachse A, Schuhmann-Giamprieri G. Characterization of iopromide liposomes. Invest Radiol 1993;28:1028–1032.
19. Unger E, Fritz T, Shen DK, Lund P, Sahn D, Ramaswami R, Matsunaga T, Yellowhair D, Kulik B. Gas filled lipid bilayers as imaging contrast agents. J Liposome Res 1994;4:861–874.
20. Unger E, Shen DK, Fritz T, Lund P, Wu GL, Kulik B, DeYoung D, Standen J, Ovott T, Matsunaga T. Gas-filled liposomes as echocardiographic contrast agents in rabbits with myocardial infarction. Invest Radiol 1993;28:1155–1159.
21. Unger EC. Liposomes as contrast agent for ultrasound imaging and methods for preparing the same. US patent no 5,088,499, 1992.
22. Swanson DP. Enhancement fundamentals for ultrasound. In: Swanson DP, Chilton HM, Thrall JH, eds. Pharmaceuticals in Medical Imaging. Macmillan, New York, 1990;682–687.
23. Lund P, Fuller L, Fritz T, Kulik B, Herres B, Tilcock C. "Aerosomes" liposomes as ultrasound contrast agents. J Ultrasound Met 1991;10:S44.

24. Matheson IBC, King AD. Solubility of gases in micellar solutions. J Coll Interf Sci 1978;66:464–469.
25. Barratt GM, Tuzel NS, Ryman BE. The labeling of liposomal membranes with radioactive technetium. In: Gregoriadis G, ed. Liposome Technology, 1st ed, Vol 2. Boca Raton, FL: CRC Press, 1984;93–105.
26. Patel HM, Boodle KM, Vaughan-Jones R. Assessment of the potential uses of liposomes for lymphoscintigraphy and lymphatic drug delivery. Biochim Biophys Acta 1984;801:76–86.
27. Hnatowich D, Friedman B, Clancy C, Novak M. Labeling of preformed liposomes with Ga-67 and Tc-99m by chelation. J Nucl Med 1981;22:810–814.
28. Richardson VJ, Jeyasingh K, Jewkes RF, Ryman BE, Tattersall MHN. Properties of 99m-Tc-labeled liposomes in normal and tumor-bearing rats. Biochem Soc Trans 1977;5:290–293.
29. Tilcock C, Unger E, Cullis P, MacDougall P. Liposomal Gd-DTPA: preparation and characterization of relaxivity. Radiology 1989;171:77–80.
30. Unger E, Cardenas D, Zerella A, Fajardo LL, Tilcock C. Biodistribution and clearance of liposomal gadolinium-DTPA. Invest Radiol 1990;25:638–644.
31. Kabalka GW, Davis MA, Holmberg E, Maruyama K, Huang L. Gadolinium-labeled liposomes containing amphiphilic Gd-DTPA derivatives of varying chain length: targeted MRI contrast enhancement agents for the liver. Magn Res Imaging 1991;9:373–377.
32. Schwendener RA, Wuethrich R, Duewell S, Westera G, Von-Schulthess GK. Small unilamellar liposomes as magnetic resonance contrast agents loaded with paramagnetic manganese-, gadolinium-, and iron-DTPA-stearate complexes. Int J Pharm 1989;49:249–259.
33. Presant CA, Blayney D, Proffitt RT, Turner AF, Williams, LE, Nadel HI, Kennedy P, Wiseman C, Gala K, Cossley RJ. Preliminary report: imaging of Kaposi sarcoma and lymphoma in AIDS with indium-111-labeled liposomes. Lancet 1990;335,1307–1309.
34. Jaggi M, Khar R, Chauhan U, Gangal S. Liposomes as carriers of technetium-99m glucoheptonate for liver imaging. Int J Pharm 1991;69:77–79.
35. Unger E, Fritz T, Shen DK Wu G. Manganese-based liposomes: comparative approaches. Invest Radiol 1993;28:933–938.
36. Tilcock CP, Ahkong QF, Parr M. An improved method for the preparation of liposomal gadolinium-DTPA. Ionophore-mediated active entrapment of gadolinium. Invest Radiol 1991;26:242–247.
37. Mauk MR, Gamble RC. Preparation of lipid vesicles containing high levels of entrapped radioactive cations. Anal Biochem 1979;94:302–307.
38. Boaumier PL, Hwang KJ. An efficient method for loading indium-111 into liposomes using acetylacetone. J Nucl Med 1982;23:810–815.
39. Phillips WT, Rudolph AS, Goins B, Timmons JH, Klipper RT, Blumhardt R. A simple method for producing a technetium-99m-labeled liposome which is stable in vivo. Nucl Med Biol 1992;19:539–547.
40. Torchilin VP, Trubetskoy VS, Narula J, Khaw BA. PEG-modified liposomes for gamma- and magnetic resonance imaging. In: Lasic D, Martin F, eds. Stealth Liposomes. Boca Raton, FL: CRC Press, 1995;219–231.
41. Ogihara I, Kojima S, Jay M. Differential uptake of gallium-67-labeled liposomes between tumors and inflammation lesions in rats. J Nucl Med 1986;27:1300–1307.
42. Woodle M. ^{67}Gallium-labeled liposomes with prolonged circulation: preparation and potential as nuclear imaging agents. Nucl Med Biol 1993;20:149–155.
43. Matwiyoff NA. Magnetic resonance workbook. New York: Raven Press, 1990.
44. Barsky D, Putz B, Schulten K, Magin RL. Theory of paramagnetic contrast agents in liposome systems. Magn Reson Med 1992;21:1–13.
45. Putz B, Barsky D, Schulten K. Mechanisms of liposomal contrast agents in magnetic resonance imaging. J Liposome Res 1994;4:771–808.
46. Unger E, Winokur T, MacDougall P, Rosenblum J, Clair M, Gatenby R, Tilcock C. Hepatic metastases: liposomal Gd-DTPA-enhanced MR imaging. Radiology 1989;171:81–85.
47. Schwendener R, Wurthrich R, Duewell S, Wehrli E, von Schulthess G.A pharmacocinetic and MRI study of unilamellar gadolinium-, manganese- and iron-DTPA-stearate liposomes as organ-specific contrast agents. Invest Radiol 1990;23:922–932.
48. Koenig SH, Goldstein E, Burnett K, Wolf G. Magnetic field dependence of proton relaxation rates in tissue with added Mn^{2+}: Rabbit liver and kidney. Magn Reson Med 1985;2:159–167.
49. Papavasiliou PS, Miller ST, Corzias GC. Role of liver in regulatory distribution and excretion of manganese. Amer J Physiol 1966;211:211–216.

50. Tilcock C. Liposomal paramagnetic MR contrast agents. In: Gregoriadis G, ed. Liposome Technology, 2nd ed. Vol 2. New York: CRC Press, 1993;65–87.
51. Kabalka G, Buonocore E, Hubner K, Davis M, Huang L. Gadolinium-labeled liposomes containing paramagnetic amphipatic agents: targeted MRI contrast agent for the liver. Magn Res Med 1988;8:89–95.
52. Grant C, Karlik S, Florio E. A liposomal MRI contrast agent: phosphatidyl ethanolamine-DTPA. Magn Res Med 1989;11:236–243.
53. Unger E, Shen D, Fritz T. Liposomes as MR contrast agents: pros and cons. Magn Res Med 1991;22:304–308.
54. Tilcock C, Ahkong Q, Koenig S, Brown 3rd R, Davis M, Kabalka G. The design of liposomal paramagnetic MR agents: effect of vesicle size upon the relaxivity of surface-incorporated lipophilic chelates. Magn Res Med 1992;27:44–51.
55. Slinkin M, Klibanov A, Torchilin V. Terminal-modified polylysine-based chelating polymers: highly efficient coupling to antibody with minimal loss in immunoreactivity. Bioconjugate Chem 1991;2:342–348.
56. Trubetskoy VS, Torchilin VP, Kennel SJ, Huang L. Use of N-terminal modified poly (L-lysine)-antibody conjugate as a carrier for targeted gene delivery in mouse lung endothelial cells. Bioconjugate Chem 1992;3:323–327.
57. Torchilin VP, Shtilman MI, Trubetskoy VS, Whiteman K, Milstein AM. Amphiphilic vinyl polymers effectively prolong liposome circulation time in vivo. Biochim Biophys Acta 1994;1195:181–184.
58. Stoll M, Mizuochi T, Childs R, Feizi T. Improved procedure for the construction of neoglycolipids having antigenic and lectin-binding activities, from reducing oligosaccharides. Biochem J 1988;256:661–664.
59. Patel H, Moghimi S. Tissue specific opsonins and phagocytosis of liposomes. In: Gregoriadis G, Allison A, Poste G, eds. Targeting of Drugs—Optimization Strategies. New York: Plenum Press, 1990;87.
60. Trubetskoy VS, Cannillo JA, Milshtein A, Wolf GL, Torchilin VP. Controlled delivery of Gd-containing liposomes to lymph nodes: surface modification may enhance MRI contrast properties. Magn Res Imaging 1995;13:31–37.
61. Spanjer H, VanGalen M, Roerdink FH, Regts J, Scherphof GL. Intrahepatic distribution of small unilamellar liposomes as a function of liposomal lipid composition. Biochim Biophys Acta 1986;863:224–230.
62. Khar RK, Jaggi M. Liposomes for liver scintigraphy. J Liposome Res 1994;4:939–957.
63. Niesman MR, Khoobehi B, Magin RL, Webb AG. Liposomes and diagnostic imaging: the potential to visualize both structure abd function. J Liposome Res 1994;4:741–768.
64. Koenig SH, Brown RD Jr. The importance of the motion of water for magnetic resonance imaging. Invest Radiol 1985;20:297.
65. Basic G, Niesman M, Magin R, Swartz H. NMR and ESR study of liposome delivery of manganese to murine liver. Magn Reson Med 1990;13:44–61.
66. Niesman MR, Bacic GG, Wright SM, Swartz HM, Magin RL. Liposome encapsulated $MnCl_2$ as a liver specific contrast agent for magnetic resonance imaging. Invest Radiol 1990;25:545–551.
67. Kabalka G, Davis M, Moss T, Buonocore E, Hubner K, Holmberg E, Maruyama K, Huang L. Gadolinium-labeled liposomes containing various amphiphilic gadolinium-DTPA derivatives: tsrgeted MRI contrast enhancement agents for the liver. Magn Res Med 1991;19:406–415.
68. Schwendener RA. Liposomes as carriers for paramagnetic gadolinium chelates as organ specific contrast agents for magnetic resonance imaging (MRI). J Liposome Res 1994;4:837–855.
69. Carr DH, Graif M, Nienedorf HP, Brown J, Steiner RE, Blumgart LH, Young IR. Gadolinium-DTPA in the assessment of liver tumors by magnetic resonance imaging. Clin Radiol 1986;37:347–353.
70. Unger EC, Shen DK, Fritz TA. Status of liposomes as MR contrast agents. J Magn Reson Imaging 1993;3:195–198.
71. Passariello R, Pavoner P, Patrizio G, Di Renzi P, Mastantuono M, Giuliani S. Liposomes loaded with nonionic contrast media: hepatosplenic computed tomographic enhancement. Invest Radiol 1990;25:S92.
72. Revel D, Corot C, Carrilon Y, Dandis G, Eloy R, Amiel M. Ioxaglate-carring liposomes: computed tomographic study as hepatosplenic contrast agent in rabbits. Invest Radiol 1990;25:S95.
73. Janoff AS, Minchey SR, Perkins WR, Boni LT, Seltzer SE, Adams DF, Blau M. Interdigitation fusion vesicles: a new approach to selective opacification of the RES. Invest Radiol 1991;26:S167.

74. Seltzer SE, Janoff AS, Blau M, Adams, DF, Minchey SR, Boni LT. Biodistribution and image characteristics of iotrolan-carrying interdigitation-fusion vesicles. Invest Radiol 1991;26:S169.
75. Swanson D, Shetty P. Lymphographic contrast media. In: Swanson D, Chilton H, Thrall J, eds. Pharmaceuticals in Medical Imaging. New York: Macmillan, 1990;236–242.
76. Strand S-E, Persson BRR. Quantitative lymphoscintigraphy I: Basic concepts for optimal uptake of radiocolloids in the parasternal lymph nodes of rabbits. J. Nucl. Med. 1979;20:1038–1046.
77. Tanoura T, Bernas M, Darkazanli A, Elam E, Unger E, Witte M, Green A. MR lymphography with iron oxide compound AMI-227. Am J Radiol 1992;159:875.
78. Weissleder R, Elizondo G, Josephson L, Compton CC, Fretz CJ, Stark DD, Ferrucci JF. Experimental lymph node metastases: enhanced detection with MR lymphography. Radiology 1989;171:835–839.
79. Osborne MP, Richardson VJ, Jeyasingh K, Ryman BE. Radionuclide-labeled liposomes - a new lymph node imaging agent. Nucl Med Biol 1979;6:75–83.
80. Trubetskoy VS, Torchilin VP. New approaches in the chemical design of Gd-containing liposomes for use in magnetic resonance imaging of lymph nodes. J Liposome Res 1994;4:961–980.
81. Torchilin VP. Immunoliposomes and PEGylated immunoliposomes: possible use for targeted delivery of imaging agents. Immunomethods 1994;4:244–258.
82. Torchilin VP, Trubetskoy VS, Milshteyn AM, Cannillo J, Wolf GL, Papisov MI, Bogdanov AA, Narula J, Khaw BA, Omelyanenko VG. Targeted delivery of diagnostic agents by surface-modified liposomes. J Contr Release 1994;26:45–58.
83. O'Driscoll C. Anatomy and physiology of the lymphatics. In: Charman WN, Stella VJ, eds. Lymphatic Transport of Drugs. Boca Raton, FL: CRC Press, 1992;1–35.
84. Yuan F, Leunig M, Huang SK, Berk DA, Papahadjopoulos D, Jain RK. Microvascular permeability and interstitial penetration of sterically stabilized (Stealth) liposomes in a human tumor xenograft. Cancer Res 1994;54:3352–3356.
84a. Gabizon A. Liposome circulation time and tumor targeting: implications for cancer chemotherapy. Adv. Drug Delivery Rev. 1995;16:285–294.
85. Presant CA, Turner AF, Proffitt RT. Potential for improvement in clinical decision-making: tumor imaging with In-111 labeled liposomes. Results of a Phase II–III study. J Liposome Res 1994;4:985–1008.
86. Nakamura K, Kuba A, Katayama M, Hashimoto S, Takatoku K, Wakui Y, Bando, K. Indium-111-labeled liposomes: quality and biodistribution in mice bearing human tumor xenografts. Drug Delivery 1993;1:69–74.
87. Tonthat T, Taillefer R, Duranceau A, Lafontaine E, Deschamps C, Lambert R, Labonte C, Norman L. Indium-111-labeled phospholipid vesicles liposomes imaging in detection of primary lung cancer: preliminary results. J Nucl Med 1991;32(Suppl 5):1085.
88. Briele B, Graefen M, Bockisch A, Hartlapp JP, Roedel W, Hotze A, Biersack HJ. Indium (111) labeled liposomes as a tumor imaging agent: first clinical results. Eur J Nucl Med 1990;16:411–416.
89. Richardson VJ, Jeyasingh K, Jewkes RF, Ryman BE, Tattersall MHN. Possible tumor localization of Tc-99m-labeled liposomes: effects of lipid composition, charge, and liposome size. J Nucl Med 1978;19:1049–1054.
90. Goto R, Kubo H, Okada S. Effect of reticuloendothelial blockade on tissue distribution of technetium-99-labeled synthetic liposomes in Ehrlich solid tumor-bearing mice. Chem Pharm Bull 1991;39:230–232.
91. Goto R. Tissue distribution of 99m-Tc-labeled liposomes prepared from synthetic amphiphiles containing amino acid residues. J Liposome Res 1994;4:877–905.
92. Goins B, Klipper R, Rudolph AS, Phillips WT. Use of technetium-99m-liposomes in tumor imaging. J Nucl Med 1994;35:1491–1498.
93. Gabizon A, Price DC, Huberty J, Bresalier RS, Papahadjopoulos D. Effect of liposome composition and other factors on the targeting of liposomes to experimental tumors: biodistribution and imaging studies. Cancer Res 1990;50:6371–6378.
94. Oku N, Namba Y, Takeda A, Okada S. Tumor imaging with technetium-99m-DTPA incapsulated in RES-avoiding liposomes. Nucl Med Biol 1993;20:407–412.
95. Patel KR, Tin GN, Williams, LE, Baldeschweiler JD. Biodistribution of phospholipid vesicles in mice bearing Lewis lung carcinoma and granuloma. J Nucl Med 1985;26:1048–1055.
96. Turner AE, Presant CA, Proffitt RT, Williams LE, Winsor DW, Werner JL. In-111 labeled liposomes: dosimetry and tumor depiction. Radiology 1988;166:761–765.
97. Presant CA, Blayney D, Proffitt RT, Turner AF, Williams LE, Nadel HI, Kennedy P, Wiseman

C, Gala K, Cossley RJ. Preliminary report: imaging of Kaposi sarcoma and lymphoma in AIDS patients with indium-111-labeled liposomes. Lancet 1990;335:1307–1309.
98. Simon RH, Ho S-Y, Perkins C, D'Arrigo JS. Quantitative assessment of tumor enhancement by ultrastable lipid-coated microbubbles as a sonographic contrast agent. Invest Radiol 1992;27:29–34.
99. Tilcock C, Ahkong QF, Fisher D. Polymer-derivatized technetium 99m-labeled liposomal blood pool agents for nuclear medicine applications. Biochim Biophys Acta 1993;1148:77–84.
100. Winzelberg GG, McKusik KA, Forelich JW, Callahan RJ, Strauss HW. Detection of gastrointestinal bleeding with ^{99m}Tc-labeled red blood cells. Semin Nucl Med 1982;12:139–146.
101. Lisbona R, Derbekyan V, Novales-Diaz JA, Rush CL. Tc-99m red blood cell venography in deep venous thrombosis of the lower limb. Clin Nucl Med 1987;10:208.
102. Caride VJ, Zaret BL. Liposome accumulation in regions of experimental myocardial infarction. Science 1977;198:735–738.
103. Palmer T, Caride V, Caldecourt M, Twicjler J, Abdullah V. The mechanism liposome accumulation in infarction. Biochim Biophys Acta 1984;797:363–368.
104. Torchilin VP, Khaw BA, Smirnov VN, Haber E. Preservation of antimyosin antibody activity after covalent coupling to liposomes. Biochem Biophys Res Commun 1989;89:1114–1119.
105. Torchilin VP, Klibanov AL, Huang L, O'Donnell S, Nossiff ND, Khaw BA. Targeted accumulation of polyethylene glycol-coated immunoliposomes in infarcted rabbit myocardium. FASEB J 1992;6:2716–2719.
106. Torchilin VP. Immunoliposomes as targeted carriers of pharmaceuticals in the cardiovascular system. In: Khaw BA, Narula J, Strauss HW, eds. Monoclonal Antibodies in Cardiovascular Diseases. Malvern: Lea and Febiger, 1994;257–267.
107. Khaw BA, Mattis JA, Melnicoff G, Strauss HW, Gold HK, Haber E. Monoclonal antibody to cardiac myosin: imaging of experimental myocardial infarction. Hybridoma 1984;3:11–23.
108. Khaw BA, Yasuda T, Gold HK, Strauss HW, Haber E. Acute myocardial infarct imaging with indium-111-labeled monoclonal antimyosin Fab. J Nucl Med 1987;28:1671–1678.
109. Weissig V, Lasch J, Klibanov AL, Torchilin VP. A new hydrophobic anchor for the attachment of proteins to liposomal membranes FEBS Lett 1986;202:86–90.
110. Khaw BA, Klibanov A, O'Donnell SM, Saito T, Nossif N, Slinkin MA, Newell JB, Strauss HW, Torchilin VP. Gamma imaging with negatively charge-modified monoclonal antibody: modification with synthetic polymers. J Nucl Med 1991;32:1742–1751.
111. Torchilin VP, Narula J, Halpern E, Khaw BA. Poly(ethylene glycol)-coated anti-cardiac myosin immunoliposomes: factors influencing targeted accumulation in the infarcted myocardium. Biochim Biophys Acta 1996;1279:75–83.
112. Palmer TN, Caride VJ, Fernandez LA, Twickler J. Liposome accumulation in ischaemic intestine following experimental mesenteric occlusion. Biosci Rep 1981;1:337–344.
113. Morgan JR, Williams KE, Davies RL, Leach K, Thomason M, Williams MHN. Localization of experimental staphylococcal abscesses by ^{99m}Tc-technetium-labeled liposomes. J Med Microbiol 1981;14:213.
114. Morgan JR, Williams LA, Howard CB. Technitium-labeled liposomes for deep-seated infection. Br J Radiol 1985;58:35–39.
115. Goins B, Klipper R, Rudolph AS, Cliff RO, Blumhardt R, Phillips WT. Biodistribution and imaging studies of technetium-99m-labeled liposomes in rats with focal infection. J Nucl Med 1993;34:2160–2168.
116. Kasi LP, Lamki LM, Podoloff DA. The potential of imaging abscesses with technetium-99m multilamellar liposomes. J Nucl Med 1988;29(Suppl 5):888.
117. Chan TW, Eley C, Liberty P, So A, Kressel HY. Magnetic resonance imaging of abscesses using lipid-coated iron oxide particles. Invest Radiol 1992;27:443–449.
118. Love WG, Amos N, Kellaway IW, Williams BD. Specific accumulation of technetium-99m radiolabelled, negative liposomes in the inflamed paws of rats with adjuvant induced arthritis: effect of liposome size. Ann Rheum Dis 1989;48:143–148.
119. Turski P, Kalinke T, Strother L, Perman W, Scott G, Kornguth S. Magnetic resonance imaging of rabbit brain after intracarotid of large multivesicular liposomes containing paramagnetic metals and DTPA. Magn Res Med 1988;7:184–196.
120. Shermatova SZ. Using labeled liposomes for the visualization of bone marrow. Med Radiol 1990;35:36.
121. Shetty HGM, Williams BD, Facey PE, Hayward MH, Evans W, Williams K, Morgan J. Fate of intra-articular Technetium-99m-labelled liposomes. Br J Rheumatol 1988;27(Suppl 2):118.

Lasic and Papahadjopoulos (eds.), Medical Applications of Liposomes

CHAPTER 7.1

Design of liposome-based drug carriers: From basic research to application as approved drugs

YECHEZKEL BARENHOLZ

Department of Biochemistry, Hebrew University – Hadassah Medical School, Jerusalem, Israel

Overview

I. Introduction

Today five liposome or liposome-like dosage forms for intravenous administration have been approved and are being marketed for clinical use in the EC countries: Abelcet, AmBisome, and Amphotec — in which amphotericin B is the active ingredient — and DaunoXome and Doxil (CAELYX) — which include the anthracyclines daunorubicin and doxorubicin, respectively, as the active ingredient. Four of the above (Abelcet, Amphotec, DaunoXome, and Doxil) are marketed as approved drugs in the US. These formulations were developed by the three American companies which are fully dedicated to development of liposomal products: Ablcet by The Liposome Company, Princeton, NJ; AmBisome and DaunoXome by NeXstar, San Dimas, CA; and Amphotec and Doxil (CAELYX) by SEQUUS, Menlo Park, CA. AmBisome, Abelcet, and Amphotec are all used to treat systemic fungal infection.[1,2] DaunoXome and Doxil (CAELYX) are designed to treat cancer and have been approved so far for treating Kaposi's sarcoma.[3–7] It is not incidental that major pharmaceutical companies were reluctant to have major investments in liposomal dosage forms and left it for much smaller, dedicated start-up companies. The only liposomal project that was developed by a major pharmaceutical company (Ciba-Geigy) was using the muramyl tripeptide coupled with phosphatidylethanolamine (MTP-PE) to treat cancer.[8,9] This project was discontinued, although MTP-PE may still be used as an effective vaccine adjuvant.

The reason for the low level of involvement of the major pharmaceutical companies is that they recognized the many technological difficulties and unknowns of the field. Liposomes are a "baby" of the academia which has studied them as such (mainly as a membrane model) since 1965.[10–13] Academic interest and the knowledge obtained by academic researchers were not sufficient for the minimal technological needs which are prerequisites of the pharmaceutical industry. This discrepancy between the interests of academia and pharmaceutical industry needs can best be exemplified by: (a) academia's lack of interest in long-term physical and chemical stability, (b) the use of raw materials (lipids) of high price and low availability which often are poorly defined, such as egg phosphatidylcholine (PC), (c) lack of reproducible production procedures which can be scaled up and applied under good manufacturing practice (GMP) conditions, (d) Minimal efforts devoted to sterility and pyrogenicity, (e) lack of means to obtain stable and therapeutic levels of drug loading in liposomes and especially in small liposomes, (f) lack of suitable quality control assays.

It was clear that the major pharmaceutical companies were waiting until the above (and other) issues would be resolved in order to become seriously involved in development of liposomal drugs. We can say it in a different way—that in the early eighties, the time that the dedicated liposome companies were born as start-up—the scientific know-how was much ahead of the technological development. Also some major scientific issues, such as means to escape the reticuloendothelial system (RES) and to improve stability of drug-loaded liposomes in vivo were not understood. The scientific know-how at that time included: understanding the basic physicochemical properties of liposomes; various methods of liposome preparation at the laboratory scale, basic knowledge on the biofate, and good tolerability of "conventional" liposomes (see Refs. 14–16); also basic (but what turned out to be incomplete) knowledge of the "solubilizing" and microencapsulation potential of liposomes, and proof of the concept as drug carriers for treatment of infectious diseases (such as leishmaniasis[15]) and tumors (for review see Refs. 10, 15, 17). This proof of concept is what caused a big "hype" in the early eighties which led to the birth of a few liposome-dedicated companies, among them NeXstar (founded as Vestar), The Liposome Company, and SEQUUS (founded as Liposome Technology). It took 15 more years for the first liposomal product (AmBisome) and another 3 years for the other four products (see above) to be approved and marketed. The five approved liposomal drugs are based on two drugs only—amphotericin B and anthracycline (doxorubicin and daunorubicin). From following the work done in development of liposomal drug carriers either from the scientific literature[10,15,16,18] or from the various publications and prospectuses of the companies dedicated to liposomal products it is clear that most formulations developed by these companies failed, although over half a billion dollars was invested. Many of them did not make it to clinical trials and most others that underwent clinical trials failed or were discontinued for other reasons.[16,18] Therefore, the five successful formulations of liposomal products for intravenous administration that made it to the marketplace should serve as an important take-home lesson which, together with understanding the reasons for the many failures, should lead to rational design of liposomes as drug carriers in order to improve the future success rate of the development of liposome products.

II. What is needed for rational design of a liposomal drug carrier?

The design of liposomes as a drug carrier is much more complicated and "full of mines" than it seemed at the birth of the field ~30 years ago. Major scientific input in the fields of physical chemistry, chemistry, life and medical sciences, including pharmacy, should be used in a cross-talk with technology. This field is an excellent example of the need for science and technology to move together hand-in-hand in order to achieve a product. The design requires laying out a concept map which consists of a hierarchy of steps, many of them needing specific answers, and points of go/no-go decisions. The use of such an approach may save much effort, investment, and frustration. The design of the liposome as a drug (agent) carrier should be dependent on the function it should perform. Therefore, to simplify the complex picture we will introduce a functional classification of liposomes (Table 1). The user has to be aware that, for the reasoning described below, this classification is not absolute and there is overlapping; also some liposomes that are sterically stabilized in some mammalian species, show characteristics of conventional liposomes in others.[20]

III. Sequence of steps in the design of a liposome-based dosage form

III.1. Theoretical aspects—no experimental work required

III.1.1. Evaluation of whether the benefits justify the development

The first step in the design is an evaluation of whether the expected benefits justify the development. In many cases when one is dealing with improving the performance of well established drugs, the above evaluation is crucial. Only large added value will justify the "go" answer. The criteria for selection of a candidate system for design should include answering the following specific questions:

- Disease characteristics?
- Tissue specificity?
- Drug toxicity profile?
- Does the desired drug slow down (inhibit) or reverse the disease state?
- Drug dose needed to achieve therapeutic efficacy?
- Is it feasible to deliver the therapeutic level of the desired drug via liposomes?
- What are the alternative therapies available and how does their performance compare with the liposomal dosage form?

The evaluation should take into consideration that the expected performance

Table 1

Functional classification of liposomes

Liposome type	Main featuresa[a]	Clearance for large liposomes	Clearance for small liposomes
Conventional liposomes	High accessibility to interact with factors from extraliposomal medium (i.e., opsonins, enzymes, low molecular mass agents), high RES uptake. For µm size range, high lung uptake	Fast ($t_{1/2} < 30$ min) Saturation dependent	Medium ($t_{1/2} > 1$ h) Saturation dependent Extravasation is feasible
Sterically stabilized liposomes	Low accessibility to extraliposomal factors due to grafted steric barrier, low RES uptake	Slow ($t_{1/2} > 5$ h) Saturation independent	Slow ($t_{1/2} > 8$ h) Saturation independent Extravasation into tumors and sites of inflammation
Activsomes	Vesicular to nonvesicular transition upon contact with agent (cationic liposomes—nucleic acid) or medium components. High lung uptake for cationic lipid–DNA complexes	Very fast ($t_{1/2} \sim$ min range) Saturation dependent	Fast ($t_{1/2} \sim$ min range Saturation dependent

[a]Based on Barenholz and Lasic.[19]

of the liposomal drug should be much better than that of the drug in a simpler dosage form in order to justify investment in design and development.

The economic considerations such as the expected price of the product may be important. However, the marketed AmBisome, Abelcet, and Amphotec, which are all much more expensive than Fungizone (the conventional dosage form of amphotericin B), penetrate very nicely the market of systemic antifungal drugs, indicating that this consideration is overcome for improved performance in diseases which are life threatening. Assuming the answers to all the above questions justify initiating the development program, then the next go/no-go point design scheme should include two parallel arms: the model system in which therapeutic efficacy, toxicity, and pharmacokinetics will be evaluated, and the engineering of the formulation.

III.1.2. Availability of suitable animal models

This is a very important aspect of the formulation design, although it is beyond the scope of this review. The "designer" has to keep in mind that the animal model has to resemble the actual disease with respect to anatomy and microanatomy (i.e., location and vascularization) as well as pathology, and that expected pharmacokinetics, biodistribution, and therapeutic efficacy in the animal model system are relevant to the actual disease in humans. Namely, differences between various mammalian species have to be taken into consideration,[20] including species-specific toxicity.[21] (For more details of relevancy of animal model systems see Refs. 5, 20, 22–24.)

III.1.3. Pharmaceutical acceptability

(a) *Drug relevant considerations.* This issue is much more specific than the parts related to the liposomes (see below). In this review we will touch on only the drug aspects which are relevant to the drug–lipid interaction. For simplicity and convenience we classified all drugs into 3 categories based on their oil/buffer and octanol/buffer partition coefficients (K_p). Molecules of very low values in both partition coefficients are by definition water soluble; molecules having low oil/buffer partition coefficient and medium to high octanol/buffer K_p are amphipathic, while molecules having high oil/buffer K_p are lipophilic.[5,25]

The first stage in evaluating drug suitability requires collecting all the available information on the chemical and physicochemical properties of the drug, including parameters such as: (i) oil/buffer and octanol/buffer partition coefficients (K_p).[5,25,26] For example, drugs having high oil/buffer K_p are inappropriate for a liposomal carrier if a high level of drug loading is required and will be better off in emulsion carrier.[5] As many of the drugs include ionizable moieties, it is recommended to have the partition coefficient values at three different pHs (below, above, and close to the pK_a). Knowing the electrostatic properties of the agent may help one to select the liposome lipid composition (improving drug loading and/or preventing liposome aggregation). Solubility and solubility product in the medium of liposome preparation and the intraliposomal aqueous phase (if possible at the three pHs above)[5,27] are all important parameters for assessing drug suitability.

(b) *Data available on chemical stability of the lipids and the drug used in the formulation.*[26,28,29] From these data, estimates can be made of the sensitivity of phospholipid ester bonds to hydrolysis and of phospholipids and sterols to oxidative damage during preparation and various storage conditions.

(c) *Data on physical stability of liposomes, which include:*

(i) Thermotropic behavior, especially effects of phase transitions in monocomponents, liposomes, and phase transition and/or phase separations of multicomponent systems.[30–32]
(ii) Permeability coefficients of a broad spectrum of molecules in liposome membranes of various compositions, including the effect of saturation and level of cholesterol[5,33,34] and the effect of liposome curvature on the permeability.[35] The permeability coefficient has a major contribution to the rate at which molecules which are present in the intraliposome aqueous phase are released from the liposomes (K_{off}).[5]
(iii) Intraliposomal forces such as osmotic gradients, or development of defects due to differences in curvature of inner and outer monolayers ("frustration").[36]
(iv) Interliposomal forces related to interaction between particles,[36,37] as well as between particles and their medium,[38] especially the colloidal nature of the liposomes and the involvement of forces affecting aggregation and fusion, such as those explained by the theories of Derjaguin–Landau–Verwey–Overbeek (DLVO) and hydration forces.[37,39,40]

Combining all available data on chemical and physical stability of the carrier and the drug will help in selection of optimal lipid composition and method of liposome preparation, as was exemplified for Doxil[25,41,42] and for liposome–cytokine formulations.[43–45]

III.1.4. Tolerability of liposomes

The information available on tolerability of liposomes is much less than that on physicochemical and chemical properties of liposomes. Still, the data available to the public are sufficient to predict the window of opportunity; namely, in the triangle of lipid composition, size, and dose for various routes of administration, what is the safe zone.[46] For example, it was found that in dogs, for egg PC SUV up to at least 250 mg/kg, tolerability is good and no side effects could be detected.[47] At the dose of 675 mg/kg in the same dogs, the same egg PC SUV liposomes induced transient but large plasma increases in certain liver enzymes and ~25% reduction in cholesterol/phospholipid mole ratio in red blood cells.[47] Even at a higher dose of 6 administrations of 1,000 mg/kg each, tolerability was good except for the transient plasma increases in some liver enzymes (Zomber and Barenholz, in preparation). A dose-dependent transient hypercholesterolemia due to the vesicles is expected for i.v. administration of cholesterol-free or cholesterol-poor liposomes;[47] this is related to the ability of the circulating small unilamellar vesicles (SUV) to act as a cholesterol sink (see Ref. 47 and references therein) and to be involved in reverse cholesterol transport. The opposite movement of cholesterol from liposomes to red and white blood cells and lipoproteins is therefore expected when a large amount of lipid vesicles (SUV) (compared with 4–7.5 mM phospholipids present in plasma lipoproteins[24]) is administered. The implication of this effect requires special attention. SUV composed of disaturated phospholipids, such as distearoyl phosphatidylcholine (DSPC), or hydrogenated soy PC (HSPC) large unilamellar vesicles (LUV) when injected i.v. induce significant undesired hemodynamic and cardiovascular side effects in rats[48] at the dose of 300 mg/kg. No side effects for the same dose in rats or dogs were observed for egg PC SUV (size <100 nm) which are rich in unsaturated acyl chains.[47]

Other parameters to watch include: level of phosphatidylserine (PS) injected as part of liposomes which may induce coagulation in plasma,[49,50] the risk for complement activation (which is species dependent), platelet activation, RES saturation,[46,51] and disseminated coagulation (Barenholz and Cohen, unpublished).

III.1.5. Biofate and biodistribution of liposomes

Data related to clearance rates (k_c), $t_{1/2}$ volume of distribution, and other pharmacokinetic parameters, as well as on organ biodistribution in animals of many liposome preparations differing in their composition, size, and dose are available.[14,20,52,53] Most of these data are available for small rodents (mice and rats) (see above references). However, sufficient data are also available for dogs and humans.[47,54] It became clear that for some 100-nm liposome compositions, such as those composed of PC : cholesterol and GM_1, and PC : cholesterol : PS, clearance is highly species dependent (being much faster in rats than in mice).[20] For other 100-nm liposome compositions (such as PC : cholesterol or PC : cholesterol : ^{5000}PEG-PE) there are much smaller differences.[20] To a large extent the clearance is determined by the balance between degree and type of opsonins, on the one hand, and the dysopsonins, on the other.[55]

For *conventional liposomes* (Table 1):

(i) Clearance for larger liposomes is faster than for smaller liposomes (of identical composition). The cut-off between large and small is a diameter of ~100 nm; for liposomes <100 nm there are packing differences between the inner and outer monolayers of the external bilayer.[40]

(ii) Clearance of liposomes from blood circulation when the gel to liquid crystalline phase transition temperature (T_m) of the liposome matrix lipid is higher than the blood temperature may be faster or slower than liposomes having T_m < blood temperature. This will depend on the level of membrane defects, which will determine level of plasma protein adsorption[56] (see below). Usually the larger the circulation time the larger are the differences in clearance rate (k_c), as is exemplified for small liposomes, which usually circulate longer than large ones.

(iii) Due to the involvement of the RES in the clearance, there is a clear saturation effect which is dependent on the balance between rate of RES uptake and RES ability to process the liposomes. Liposomes having $T_m >$ body temperature are processed slower than those having $T_m <$ body temperature. Sphingolipids are processed slower than ester phospholipids. Liposomal components which inhibit and slow down the enzymatic hydrolysis of the lipids by lysosomal enzymes (i.e., grafted PEG) will slow down processing, thereby decreasing clearance rate (k_c).

For *sterically stabilized liposomes* (Table 1):[3,52,53,57,59]

It seems that some steric stabilizers such as ^{2000}PEG-DSPE are universal and act as such in all mammalian species tested[3,53,57] almost irrespective of liposome size or of other liposomal lipids,[53] except PS which overrides the steric stabilization imposed on liposomes by the ^{2000}PEG-DSPE.[60] Other steric stabilizers such as GM_1 and hydrogenated phosphatidylinositol (HPI) are not universal and their effect is species dependent.[20]

In general, for both conventional liposomes[47] (and Barenholz and Shmeeda, in preparation) and sterically stabilized liposomes,[3,4,58] when they are small their k_c in various mammalian species is in the order mice > dogs > human.

Activsomes (Table 1): The information available on the biofate of activsomes such as cationic liposomes or cationic lipid–DNA complexes is much less than on conventional and sterically stabilized liposomes. For example, the biofate of complexes of cationic lipids and nucleic acids (such as DNA or oligonucleotides) are just now starting to be determined. It is clear that cationic liposomes including those containing neutral phospholipids (DOPC or DOPE) even when they are small (≤100 nm) will acquire large amounts of proteins (500 g protein/mole lipid) compared with a much smaller protein adsorption of 1/3 or less for anionic liposomes. This may explain their extremely fast k_c.[56] Cholesterol, but not DOPE, prolongs the circulation time and tissue retention of the cationic lipid–DNA complexes (CLDC) which results in improved in vivo transfection by 3 orders of magnitude.[57] The electrical charge of CLDC is dependent on the (DNA^-/cationic $lipid^+$) charge ratio, which varies depending on the complex used.[61] It seems that in addition to charge, the exact lipid composition of the complex has a very large impact on CLDC circulation time, biodistribution, and transfection efficacy. (See Ref. 62 for a recent review on the involvement of cationic lipids in gene delivery.)

With respect to the many factors contributing to k_c there are accumulating indications that the role of the adsorbed proteins is important, and many (but not all) of the effects described above for liposome composition and steric stabilization can be correlated with protein adsorption to the liposome surface. This will also affect the integrity of (and therefore leakage from) the circulating liposomes. The range of adsorption (described in grams protein per mole lipid) is broad, from 500 for cationic liposomes to 13 for DSPC:cholesterol:^{2000}PEG-DSPE. In spite of the large difference in protein adsorbed, the composition of proteins adsorbed is similar irrespective of the level adsorbed. It was proposed that liposomes adsorbing $\geq$50 g proteins/mole lipid will be cleared rapidly ($t_{1/2} \leq 0.5$ h), while those adsorbing $\leq$20 g proteins/mole lipids will be cleared much slower ($t_{1/2} > 2$ h).[56] However, it is important to note that other factors, such as liposome membrane defects, dose (in relation to RES saturation), and rate of processing in the RES may have major input into the overall clearance rate.[14]

III.2. Further steps in the design of liposome-based dosage forms — experimental work

As was demonstrated above, a major part of the work leading to the development of a liposome-based drug carrier as a pharmaceutical product can be performed without "wetting hands". At the end of the "dry" stage the "designer" should have sufficient information to select a few potential liposome formulations for experimental evaluation. The first stage in the experimental evaluation is the preparation of the selected liposome formulations by a method and under conditions which should be relevant to the proposed pharmaceutical production (Tables 2 and 3). For this, various quality control assays have to be implemented.

III.2.1. Quality control of liposomal formulations

Many of these assays (see Table 4) are well established, having been developed through 30 years of intensive research on liposomes. Major contributions to the development of Q.C. assays have been made in the last 5 years, when the 5 liposomal products reached the last stages of their development. Following the assays described in Table 4 with storage time is required for stability characterization. Among the assays mentioned in Table 4 only phospholipid concentration, lipid degradation, concentration of trapped drug (agent), and size distribution are important for the first stage of assessment.

III.2.2. In vitro assessment to predict in vivo performance

Having the above Q.C. data and theoretical evaluation, the next stage is in vitro tests so designed that they serve as a good indicator for the in vivo performance of the liposomal formulation. The main predictors are two parameters obtained from assays which determine the drug liposome/medium partition coefficient, $K_{p,c}$, and the k_{off}, which is the rate of release of the drug (agent) from the liposome upon dilution similar to the dilution which the liposome dosage form faces in vivo (Table 5). The significance of $K_{p,c}$ and k_{off} to the go/no-go decision can be exemplified by our own previous failure in the development of liposomal doxorubicin in which the doxorubicin was membrane associated.[17,54,64] Pharmacokinetics[65] in mice which suggested no fast drug release in plasma was misleading, as was demonstrated from the pharmacokinetics in humans in which doxorubicin release from liposomes was very fast.[54] It is now clear to us that comparing mice and humans is misleading. The explanation for the differences between mice and humans was rather trivial: A dilution-dependent release assay demonstrated that the k_{off} is very large (fast release) and $K_{p,c}$ although relatively high (concentration dependent 4.5×10^3 to 60×10^3), is not high enough to prevent drug release in

Table 2

Hydration methods in liposome production and their upscaling feasibility

Method	Product	Upscaling feasibility
Thin lipid hydration by mechanical shaking after evaporation or lyophilization (from cyclohexane or *tert*-butanol)	MLV	Very good
Dehydration–rehydration of SUV	MLV (*having exceptionally large trapped volume*)	Very good but expensive due to lyophilization step
Organic solvent(s) replacement		
(a) Solvents miscible with water, such as ethanol, methanol, and isopropanol	MLV, OLV, UV (LUV, SUV)	Very good
(b) Solvents immiscible with water, such as ether and alkyl halides	MLV, OLV, UV (LUV, SUV)	Feasible but industrially problematic
Formation of lipid–detergent mixed micelles, followed by detergent removal	OLV, LUV, SUV	Good for OLV Reasonable for SUV

For more details on methods see Ref. 19 and references listed therein.

Table 3

Sizing methods in liposome production and their upscaling feasibility

Method	Upscaling feasibility	Disadvantage
Ultrasonic irradiation; gives mainly small unilamellar vesicles (including those of minimal size, ca. 20 nm diameter)	Poor	Further fractionation is required in order to achieve homogeneous vesicle populations
High-pressure extrusion (above 35 MPa, ca. 5000 psi) gives mainly small (but not minimal size), mostly unilamellar vesicles. ***Examples:*** **French pressure cell, micro-fluidizer, high-pressure homogenizers**	For high pressure homogenizers feasibility is very good	Limited to vesicles obtaining ≤100 nm
Low- or medium-pressure extrusion (up to 14 MPa, ca. 2000 psi) through pores of defined sizes. ***Examples:*** **Polycarbonate membranes, ceramic filters, stainless steel filters.** (Vesicle size distribution and number of lamellae are determined by the combination of pore size and pressure)	Very good	Tricky

For more details on methods see Ref. 19 and references listed therein.

Table 4

Summary of quality control assays of liposomal formulations

	Methodology
Basic characterization assays	
pH	pH meter
Conductivity	Conductivity meter
Osmolarity	Osmometer
Phospholipid concentration	Lipid phosphorus content (Bartlett method)
Phospholipid composition	TLC (combined with the Bartlett method), HPLC
Cholesterol concentration	Cholesterol oxidase assay, HPLC (normal and/or reverse phase
Phospholipid acyl chain autoxidation	Conjugated dienes, lipid peroxides, TBARS, and fatty acid composition (GC)
Phospholipid hydrolysis	TLC, HPLC, total PL
Nonesterified fatty acid concentration	HPLC or enzymatic assay
Cholesterol autoxidation	TLC, HPLC
Antioxidant degradation	TLC, HPLC
Agent degradation	TLC, HPLC, spectrophotometry, spectrofluorometry
Trapped volume	Measure of osmotic intraliposomal aqueous phase
Agent concentration	Spectrophotometry, HPLC, GC, spectrofluorometry
Residual organic solvents and heavy metals	NMR, GC Standard protocols
Agent/phospholipid ratio	Determination of drug and phospholipid concentrations
$[H]^+$ or ion gradient for remote loading	Fluorescent indicators, ESR indicators, ^{31}P-NMR, ^{19}F-NMR

Physical characterization assays	
Vesicle size distribution	
Submicron range	Photon correlation spectroscopy (PCS), gel exclusion chromatography and specific turbidity, electron microscopy (various methods)
Micron range	Coulter counter, light microscopy, light diffraction, and specific turbidity
Electrical surface potential and surface pH	Use of membrane-bound electrical field probes and pH-sensitive probes
Zeta potential	Electrophoretic mobility
Thermotropic behavior, phase transition on phase separation	DSC, NMR, fluorescence methods, FTIR, Raman spectroscopy, ESR
Percentage of free drug	Gel exclusion chromatography, ion exchange chromatography, precipitation by polyelectrolyte
Liposome/aqueous phase partition coefficient ($K_{p.c}$) by dilution-dependent drug release assay	Dilution effect (up to 10,000-fold) on liposomal Agent/PL ratio
k_{off} (rate of desorption from liposomes)	Effect of time on Agent/Liposome ratio
$[H]^+$ or ion gradient	Fluorescent pH indicators, ^{31}P-NMR, ^{19}F-NMR, intraliposomal ion concentration
Biological assays	
Sterility	Aerobic and anaerobic bottle cultures
Pyrogenicity	Rabbit and/or limulus amebocyte lysate (LAL) tests
Toxicity	Related to agent and use of liposomal product
Medium-induced leakage	Gel exclusion chromatography, ion exchange chromatography, precipitation by polyelectrolyte

NOTE: For more details on the methodologies, see Refs. 41, 63, and references listed therein.
Abbreviations: PL, phospholipids; TLC, thin layer chromatography; HPLC, high performance liquid chromatography; NEFA, nonesterified fatty acid; GC, gas liquid chromatography; NMR, nuclear magnetic resonance; PCS, photon correlation spectroscopy; DSC, differential scanning calorimetry; FTIR, Fourier transformed infra-red spectroscopy; ESR, electron spin resonance; TBARS, thiobarbituric acid reactive substances.
K_{off}, constant which describes the rate of agent release of the liposomes.[5,24,26]

Table 5

Relationship between liposome features and performance upon intravenous injection

Characteristics			Performance		
k_{off}	$K_{p,c}$	k_c[a]	Agent clearance[b]	Agent biodistribution	Dependent on
Slow	High	Slow	Slow	Plasma[c]	k_{off}/k_c
Slow	Low	Slow	Slow	Plasma[c]	k_{off}/k_c
Fast	High	Slow	Fast (k_{off} dependent)	$K_{p,c}$ dependent	k_{off}
Fast	Low	Slow	k_{off} dependent	k_{off} dependent	$K_{p,c}$
Slow	High	Fast	Fast	RES	k_c
Slow	Low	Fast	Fast	RES	k_c

$K_{p,c}$ = Carrier-to-medium partition coefficient.
k_{off} = Agent release rate.
k_c = Carrier clearance rate.
[a] k_c is determined mainly by the properties of the carrier.
[b] Assuming free agent clearance is very fast.
[c] For colloidal systems having slow clearance, small size may permit extravasation in tumors and inflamed tissues.

humans.[24] This $K_{p,c}$ was sufficient to prevent massive drug release in mice where the dilution was rather smaller (<10-fold). However, in humans the very large dilution (3,500-fold for the first ml infused) when combined with the large k_{off} induced massive fast release during the reperfusion. The pharmacokinetic data were in very good agreement with the dilution release assay.[5]

The performance of the dilution release assay requires determining the drug/liposome ratio as a function of dilution, and for each dilution as a function of time. Such experiments permit determining two important parameters—k_{off}, the desorption rate of the drug/agent from the liposome (a complex umbrella, under which rate constants of various processes exist, such as desorption of the lipid bilayer) and dissolution of drug precipitated in the intraliposomal aqueous phase.[5,24] The drug (agent) liposome to medium partition coefficient ($K_{p,c}$) can be determined from the kinetic curves for a broad range of dilutions using the value of [drug/liposome] after release was completed and equilibrium was achieved, as we exemplified for doxorubicin.[24] For water-soluble drugs (agents) encapsulated in the intraliposomal aqueous phase, the rate of permeability through the liposome membrane is the actual rate-limiting factor which will determine k_{off}. $K_{p,c}$, being very low, will have no effect on drug/agent biodistribution.

For drugs which precipitate and/or form a gel, or drugs which are noncovalently attached to a matrix in the intraliposomal aqueous phase, it is of interest to determine the contribution of rate of dissolution (or release from the matrix) to k_{off}.[5,27] The importance of the dissolution rate of the drug and of its solubility product is well demonstrated by a comparison of doxorubicin and ciprofloxacin encapsulated in sterically stabilized liposomes of identical size (~100 nm) and composition: (hydrogenated PC, phospholipids of high T_m, cholesterol, and ^{2000}PEG-DSPE) by means of an ammonium sulfate gradient.[27] The solubility product of the ciprofloxacin is higher and its release rate (k_{off}) is faster so that the value added of the liposome encapsulation is questionable unless the formulation will be improved.

Another important parameter which was not fully investigated and deserves special attention is the energy of activation of the drug release of liposomes which is obtained from the slope of the Arrhenius plot describing release rate as a function of the reciprocal of the absolute temperature, $1/T(K)$.[25,27] The energy of activation involves various processes, some of them related to the barrier properties of the membrane,[66] others to dissolution of aggregated drugs.[67] However, we found that when the energy of activation of the release kinetics is high, the release is slower than for systems in which it is low.[27,67]

It is always recommended to compare the dilution release assay in buffer and in physiological fluid such as plasma or blood. Such comparison will reveal the role of a "sink" in the release of liposomes.[5,24] For liposomal doxorubicin we found a minimal effect of serum components on $K_{p,c}$ and k_{off}.

However, for amphotericin B, which is very poorly water soluble, there is a strong effect of the serum as it interacts with lipoproteins.[68,69] A similar phenomenon was found for the unglycosylated interleukin 2 (Chiron, Emeryville, CA) delivered via sterically stabilized LUV,[44] which in vivo is desorbed from the liposomes into other compartment(s) whose nature is not yet known.

These and other examples[5,24] raise a very important issue that tends to be ignored: it is a misconception to assume that so-called hydrophobic or amphipathic agents (drugs) are superior to water-soluble ones in their suitability for being delivered via liposomes using the intravenous route. Every system has to be evaluated separately.

III.2.3. Loading of liposomes with drugs/agents

In many cases there is a problem of reconciling the optimal use of liposomes of <150 nm with the need for sufficient loading of drugs. This can be overcome for weak amphipathic bases by remote loading procedures such as using a proton gradient,[70,71] ammonium sulfate gradient,[25,27,42,66] or calcium acetate gradient for loading amphipathic weak acids.[34] The remote loading approach enables reaching sufficient therapeutic concentrations of drugs in the intraliposomal aqueous phase. However, in order to optimize the performance of the liposomes the following criteria have to be met:[25,27,42]

- The counterion of the charged drug (anion for loading amphipathic weak bases, and cation for loading amphipathic weak acids) should practically not efflux. During loading the drug should influx at a reasonable rate (≤1 hour).
- It is preferable that the loaded drug will "aggregate" or "gel" in the intraliposomal aqueous phase.
- The aggregation (or gelation) should not be a "dead end", and the drug should have the potential of being released. An in vitro release assay should therefore be available. For Doxil we are using the collapse of the ammonium sulfate gradient as a release assay. Using the ionophore nigericin[25,72] we demonstrated that the doxorubicin which was precipitated in the intraliposo-

mal aqueous phase became fully available to cells in culture.[72] The doxorubicin release and bioavailability was also confirmed by the formation of drug metabolites in the urine and in the tumors in mice and humans.[3,4,73]

The design of the loading procedure should take into consideration three important aspects:

(i) Medium-related properties, which include the gradient to be used and permeability coefficients of the species involved in the gradient;
(ii) Vesicle-related properties, which include lipid composition, surface charge, thermotropic behavior, trapped volume;
(iii) SUV or LUV loaded molecules which include: pK_a, oil/buffer and octanol/buffer K_p which describe lipophilicity and amphipathy; $K_{p,c}$, the permeability coefficient through the liposome membrane. The energy of activation calculated from the Arrhenius plots describing the k_{off} as a function of the reciprocal of the absolute temperature. (See Refs. 25,34 for more details.)

In general, we found that high energy of activation means better stability of the loaded liposome on the shelf and, more important, while circulating in the plasma.[25,27,67] No solutions equivalent to remote loading are available for loading of agents which are neither amphipathic weak bases nor amphipathic weak acids. If these molecules are lipophilic or amphipathic, high loading could be obtained by association of the drug with the lipid bilayer. However, only agents with very high ($>10^7$) $K_{p,c}$ and very low k_{off} have a chance of staying liposome-associated during circulation after i.v. injection (or infusion) to humans (Ref. 5 and references therein). The problem is worse for liposomes of long circulation time, as even liposomes of relatively low k_{off} may lose their loaded drug. A good example of this aspect are the various amphiphile-based dosage forms used to deliver amphotericin B. Those dosage forms which have a lipid "trap" for the amphotericin B (DMPG for Abelcet, DSPG and cholesterol for AmBisome, and cholesteryl sulfate for Amphotec) are all superior to the formulations lacking such a trap (Fungizone and amphotericin B emulsions).[1,5,74]

It is of interest that, although not planned, there is an adjustment between k_{off} and k_c of Abelcet, AmBisome, and Amphotec in the following order: AmBisome > Amphotec > Abelcet.[74] Another example of a trap is the inclusion of negatively-charged lipid to enhance and stabilize the association of anthracyclines with liposomes.[75] However, although the inclusion of phosphatidylglycerol (or other negatively-charged lipids) improved the doxorubicin encapsulation, the improvement is not sufficient to significantly prevent or slow down the dilution-induced doxorubicin release in vitro[5,24] or in vivo in humans.[54]

Other methods besides gradients for active loading (defined as $[\mathrm{Drug}]_{\mathrm{liposome}} > [\mathrm{Drug}]_{\mathrm{medium}}$) of drugs which are not liposome amphipathic or not charged include various little-studied trapping mechanisms. Examples of such approaches are inclusion of polyelectrolytes having the opposite charge to the drug (agent) charge, or providing an environment which will induce precipitation inside the liposome aqueous phase, such as loading by hydration in the presence of drug at high temperature under conditions that the drug is soluble, followed by quenching at lower temperature, which will lead to drug precipitation inside the liposome.[52] For all water-soluble drugs which cannot be actively loaded, the main feasible loading procedure is either classical passive loading by hydrating the lipids by a solution containing the drug[26] or by co-lyophilizaton of the drug/agent using either the dehydration/rehydration approach[45,76] or, for molecules that are stable in the presence of *tert*-butanol, by co-lyophilization of lipids and drug (agent) from a *tert*-butanol/buffer mixture. The main advantages of *tert*-butanol is its absolute miscibility with aqueous solutions, its convenient freezing temperature, and its high vapor pressure.[77–79]

III.3. Animal studies

Animal studies are justified only for liposomal formulations that:

- Pass successfully the in vitro tests which can predict their in vivo pharmaceutical performance, pharmacokinetics, release profile, biodistribution.[5,24]
- Prove to have the potential (after further pharmaceutical development) to become pharmaceutically acceptable dosage forms.[26,29,41,80]

Animal studies require a suitable animal model, such as mice, in which pharmacokinetics, efficacy, and toxicity can be performed. However, the "designer" should include at least pharmacokinetic studies in larger animals, such as dogs, in which the blood volume and hemodynamics are closer to humans than in mice.[47,58] The use of the dog is also a good test that the in vitro prediction assays indeed operate well.

The details on animal studies and toxicity studies are not in the scope of this review. For more details the reader is referred to the vast literature on this subject present in this book and elsewhere.

Acknowledgments

The work described in this paper was supported in part by SEQUUS Pharmaceuticals, Menlo Park, CA.
The help of Mr. Sigmund Geller in editing the manuscript and of Mrs. Beryl Levene in typing it is acknowledged with pleasure.

References

1. de Marie S, Janknegt R, Bakker-Woudenberg IAJM. Clinical use of liposomal and lipid-complexed amphotericin B. J Antimicrob Chemotherap 1994;33:907–916.
2. Szoka FC, ed. Forum: Liposomal amphotericin B. J Liposome Res 1994;3:363–492.
3. Gabizon A, Catane R, Uziely B, Kaufman B, Safra T, Cohen R, Martin F, Huang A, Barenholz Y. Prolonged circulation time and enhanced accumulation in malignant exudates of doxorubicin encapsulated in polyethylene-glycol coated liposomes. Cancer Res 1994a;54:987–992.
4. Gabizon A, Isacson R, Libson E, Kaufman B, Uziely B, Catane R, Gera Ben Dor C, Rabello E,

Lass Y, Peretz T, Sulkes A, Chisin R, Barenholz Y. Clinical studies of liposomal encapsulated doxorubicin. Acta Oncol 1994b;33:779–786.

5. Barenholz Y. Cohen R. Rational design of amphiphile-based drug carriers and sterically stabilized carriers. J Liposome Res 1995;5:905–932.
6. Gill PS, Esping BM, Muggio F, Carbriales S, Talpoule A, Esplin JA, Liebman HA, Forssen E, Ross ME, Levine AM. Phase I/II clinical and pharmacokientic evaluation of pharmacokinetic daunorubicin. J Clin Oncol 1995;13:996–1003.
7. Coukell A, Spencer CM. Polyethylene glycol-liposomal doxorubicin: A review of its pharmacodynamic and pharmacokinetic properties and therapeutic efficacy in management of AIDS-related Kaposi's sarcoma. Drugs 1997;53:520–530.
8. van Hoogevest P, Fankhauser P. An industrial liposomal dosage form for muramyl-tripeptide-phosphatidyl ethanolamine (MTP-PE). In: Fidler IJ, Lopez-Berestein G, eds. Liposomes in the Therapy of Infectious Diseases and Cancer. Alan R. Liss, New York, 1989;453.
9. Nii A, Utsugi T, Fan D, Denkins Y, Pak C, Brown D, Van Hoogevest P, Fidler IJ. Optimization of the liposomes encapsulating a new lipopeptide CGP 31362 for efficient activation of tumoricidal properties in monocytes and macrophages. J Immunother 1991;10:236.
10. Barenholz Y. Liposome production: Historic aspects. In: Braun-Falco O, Korting HC, Maibach HJ, eds. Liposome Dermatics. Springer Verlag, Berlin, 1992;69–81.
11. Bangham AD. Liposomes, the Babraham connection. Chem Phys Lipids 1993;64:275–285.
12. Barenholz Y, Lasic DD, eds. Handbook of Nonmedical Applications of Liposomes. Vols I–IV. CRC Press, Boca Raton, FL, 1996.
13. Lasic DD, Barenholz Y. Liposomes: Past, present and future. In: Barenholz Y, Lasic DD, eds. Handbook of Nonmedical Applications of Liposomes. Vol IV. CRC Press, Boca Raton, FL, 1996;299–315.
14. Senior J H: Fate and behaviour of liposomes in vivo. Crit Rev Ther Drug Carrier Syst 1987;3:123, 1987.
15. Gregoriadis G, ed. Liposomes as Drug Carriers. Recent Trends and Progress. Wiley, New York, 1988.
16. Fidler IJ, Lopez-Berestein G, eds. Liposomes in the Therapy of Infectious Diseases and Cancer. Alan R Liss, New York, 1989.
17. Gabizon AA, Barenholz Y. Adriamycin-containing liposomes in cancer chemotherapy. In: Gregoriadis G, ed. Liposomes as Drug Carriers. Wiley, New York, 1988;365–379.
18. Szoka FC. Liposomal drug delivery. In: Wilschut J, Hoekstra D, eds. Current Status and Future Prospects in Membrane Fusion. Marcel Dekker, New York, 1991;845–890.
19. Barenholz Y, Lasic DD. An overview of liposome scaled-up production and quality control. In: Barenholz Y, Lasic DD, eds. Handbook of Nonmedical Applications of Liposomes. Vol III. CRC Press, Boca Raton, FL, 1996;23–30.
20. Liu D. Animal species dependent liposome clearance. J Liposome Res 1996;6:77–97.
21. Cornelius CE. In: Kaneko JJ, ed. Clinical Biochemistry of Domestic Animals. 4th Edition. Academic Press, San Diego, 1989.
22. Snell K, Mullock B, eds. Biochemical Toxicology: a Practical Approach. IRL Press, Oxford, 1987.
23. Sacher RA. Widmann's Clinical Interpretation of Laboratory Test. 10th Edition. FA Davis, Philadelphia, 1991.
24. Amselem S, Cohen R, Barenholz Y. In vitro tests to predict in vivo performance of liposomal dosage forms. Chem Phys Lipids 1993;64:219–237.
25. Haran G, Cohen R, Bar LK, Barenholz Y. Transmembrane ammonium sulfate gradients in liposomes produce efficient and stable entrapment of amphipathic weak bases. Biochim Biophys Acta 1993;1151:201–215.
26. Barenholz Y, Crommelin DJA. Liposomes as pharmaceutical dosage forms. In: Swarbrick J, Boylan JC, eds. Encyclopedia of Pharmaceutical Technology. Vol 9. Marcel Dekker, New York, 1994;1–39.
27. Lasic DD, Ceh B, Stuart MCA, Guo L, Frederik PM, Barenholz Y. Transmembrane gradient driven phase transitions within vesicles: Lessons for drug delivery. Biochim Biophys Acta 1995;1239:145–156.
28. Barenholz Y, Amselem S, Goren D, Cohen R, Gelvan D, Samuni A, Golden EB, Gabizon A. Stability of liposomal-doxorubicin formulation: Problems and prospects. Med Res Rev 1995;13:449–461.
29. van Winden ECA, Zuidam NJ, Crommelin DJA: Strategies for large scale production and optimised stability of pharmaceutical liposomes developed for parenteral use. In: Lasic DD, Papahad-

jopoulos D, eds. Medical Applications of Liposomes. Elsevier Science Publishers, New York, 1998; Chapter 7.2.
30. Silvius JR. Thermotropic phase transitions of pure lipids in model membranes and their modification by membrane proteins. In: Jost PC, Griffith OH, eds. Lipid Protein Interactions. Vol 2. Wiley, New York, 1982;239–281.
31. Marsh D. Handbook of Lipid Bilayers. CRC Press, Boca Raton, FL, 1990.
32. Caffrey M, Koynova R, Hogan J, Moynihan D. LIDIDAT: A database of lipid phase transition temperatures and enthalpy changes. In: Barenholz Y, Lasic DD, eds. Handbook of Nonmedical Applications of Liposomes. Vol II. CRC Press, Boca Raton, FL, 1996;85–104.
33. de Gier J. Osmotic behaviour and permeability properties of liposomes. Chem Phys Lipids 1995;64:187–196.
34. Clerc S, Barenholz Y. Loading of amphipathic weak acids into liposomes in response to transmembrane calcium acetate gradients. Biochim Biophys Acta 1995;1240:257–265.
35. Epand RM, Polozov I. Liposome and membrane stability. In: Barenholz Y, Lasic DD, eds. Handbook of Nonmedical Applications of Liposomes. Vol II. CRC Press, Boca Raton, FL, 1996;105–111.
36. Leung D, Lasic DD. Stability and stabilization of liposomes. In: Barenholz Y, Lasic DD, eds. Handbook of Nonmedical Applications of Liposomes. Vol III. CRC Press, Boca Raton, FL, 1996;31–41.
37. Israelachvili JN. Intermolecular and Surface Forces. Academic Press, London, 1992.
38. Tenchov B, Koynova R. Effect of solutes on the membrane lipid phase behavior. In: Lasic DD, Barenholz Y, eds. Handbook of Nonmedical Applications of Liposomes. Vol I. CRC Press, Boca Raton, FL, 1996;237–245.
39. Wilschut J, Hoekstra D. Membrane fusion: lipid vesicles as a model system. Chem Phys Lipids 1986;40:145–166.
40. Lichtenberg D, Barenholz Y. Liposomes: Preparation, characterization and preservation. In: Glick D, ed. Methods of Biochemical Analysis. Vol 33. Wiley, New York, 1988;337–462.
41. Barenholz Y, Amselem S. Quality control assays in the development and clinical use of liposome-based formulations. In: Gregoriadis G, ed. Liposome Technology. Liposome Preparation and Related Techniques. Vol I. 2nd Edition. CRC Press, Boca Raton, FL, 1993;527–616.
42. Bolotin EM, Cohen R, Bar LK, Emanuel N, Ninio S, Lasic DD, Barenholz Y. Ammonium sulfate gradients for efficient and stable remote loading of amphipathic weak bases into liposomes and ligandoliposomes. J Liposome Res 1994;4:455–479.
43. Barenholz Y, Palgi O, Golod G, Emanuel N, Rutkoski Y, Braun E, Kedar E. Liposomal delivery of cytokines: A means to improve their therapeutic performance. In: Gregoriadis G, Allison AC, Poste G, eds. New Generation Vaccines: The Role of Basic Immunology. Plenum Press, New York, 1993;201–210.
44. Kedar E, Rutkowski Y, Braun E, Emanuel N, Barenholz Y. Delivery of cytokines by liposomes. I. Preparation and characterization of interleukin-2 encapsulated in long-circulating sterically stabilized liposomes. J Immunother 1994;16:47–59.
45. Kedar E, Palgi O, Golod G, Babai I, Barenholz Y. Delivery of cytokines by liposomes. III. Liposome-encapsulated GM-CSF and TNFa show improved pharmacokinetics and biological activity and reduced toxicity in mice. J Immunother 1997;20:180–193.
46. Storm G, Oussoren C, Peeters PAM, Barenholz Y. Tolerability of liposomes *in vivo*. In: Gregoriadis G, ed. Liposome Technology. Vol 3. 2nd Edition. CRC Press, Boca Raton, FL, 1993;345–383.
47. Zomber G, Bogin E, Barenholz Y Effect of i.v. injection of small unilamellar liposomes of egg phosphatidylcholine on cholesterol in plasma and erythrocytes, serum enzymes and liver function in dogs. J Liposome Res 1996;6:455–477.
48. Muschick P, Sachse A, Leike J, Wehrmann D, Krause W. Lipid dependent cardio-haemodyanmic tolerability of liposomes in rats. J Liposome Res 1995;5:933–993.
49. Borenstain V, Barenholz Y, Hy-Am E, Rachmilewitz EA, Eldor A. Phosphatidylserine in the outer leaflet of red blood cells from β-thalassemia patients may explain the chronic hypercoagulable state and thrombotic episodes. Am J Hematol 1993;44:63–65.
50. Lentz BR. Liposomes as a tool to study blood coagulation. In: Barenholz Y, Lasic DD, eds. Handbook of Nonmedical Applications of Liposomes. Vol I. CRC Press, Boca Raton, FL, 1996;123–136.
51. Parnham MJ, Wetzig H. Toxicity screening of liposomes. Chem Phys Lipids 1993;64:263–274.
52. Lasic DD. Liposomes: from Physics to Applications. Elsevier, Amsterdam, 1993.

53. Woodle MC. Surface-modified liposomes: assessment and characterization for increased stability and prolonged blood circulation. Chem Phys Lipids 1993;64:249–262.
54. Gabizon A, Chisin R, Amselem S, Druckmann S, Cohen R, Goren D, Fromer I, Peretz T, Sulkes A, Barenholz Y. Pharmacokinetic and imaging studies in patients receiving a formulation of liposome-associated adriamycin. Br J Cancer 1991;64:1125–1132.
55. Liu D, ed. Forum: Mechanism of Liposome Clearance. J Liposome Res 1996;6:1–140.
56. Semple SC, Chonn A. Liposome-blood protein interactions in relation to liposome clearance. J Liposome Res 1996;6:33–60.
57. Liu D, Mounkes LC, Liggit HD, Brown CS, Solodin I, Heath TD, Debs RJ. Factors influencing the efficiency of cationic liposome-mediated intravenous gene delivery. Nature Biotechnology 1996;15:167–173.
58. Gabizon AA, Barenholz Y, Bialer M. Prolongation of the circulation time of doxorubicin encapsulated in liposomes containing a polyethylene glycol-derivatized phospholipid: Pharmacokinetic studies in rodents and dogs. Pharm Res 1993;10:703–708.
59. Lasic D, Martin F, eds. Stealth Liposomes. CRC Press, Boca Raton, FL, 1995.
60. Scherphof GL, Crommelin DJA. Cells involved in removing liposomes from the blood circulation: why are they so special. J Liposome Res 1996;6:19–32.
61. Zuidam NJ, Barenholz Y. Electrostatic parameters of cationic liposomes commonly used for gene delivery as determined by 4-heptadecyl-7-hydroxy coumarin. Biochim Biophys Acta 1997; 1329:211–222.
62. Lasic DD. Liposomes in Gene Delivery. CRC Press. Boca Raton, FL, 1997.
63. Shmeeda H, Even-Chen S, Nissim R, Cohen R, Weintraub C, Barenholz Y. Enzymatic assays for quality control and pharmacokinetics of liposomal formulations. In: New RRC, ed. Liposomes: A Practical Approach. 2nd Edition. IRL Press, Oxford, 1998, in press.
64. Gabizon A, Peretz T, Sulkes A, Amselem S, Ben Yosef R, Ben Baruch N, Catane R, Biran S, Barenholz Y. Systemic administration of doxorubicin containing liposomes in cancer patients: A phase I study. Eur J Cancer Clin Oncol 1989;25:1795–1803.
65. Goren D, Gabizon A, Barenholz Y. The influence of physical characteristics of liposomes containing doxorubicin on their pharmacological behavior. Biochim Biophys Acta 1990;1029:285–294.
66. Stein WD. Transport and Diffusion across Cell Membranes. Academic Press, Orlando, FL, 1986;69–112.
67. Clerc S, Barenholz Y. A simple model for the loading of amphipathic weak base into preformed liposomes in response to ammonium gradients. In: Gregoriadis G, Florence AT, eds. Liposome and Vaccines, Progress in Drug and Vaccine Delivery. London, 16–20th December, 1996;63.
68. Wasan KM, Lopez-Berestein G. Targeted liposomes in fungi: Modifying the therapeutic index of amphotericin B by its incorporation into negatively charged liposomes. J Liposome Res 1995;5:883–903.
69. van Etten EWM. Liposomal Amphotericin B for Invasive Fungal Infections, Ph.D. thesis, Erasmus University, Rotterdam, Chap 4, 1995.
70. Madden TD, Narrigan PR, Tai I, Bally MB, Mayer LD, Redelmeir TE, Longhrey HC, Tilock CPS, Renish IW, Cullis PR. The accumulation of drugs within large unilamellar vesicles exhibiting protein gradient: A survey. Chem Phys Lipids 1990;53:37–46.
71. Parr MJ, Cullis PR. Transbilayer transport induced by transmembrane pH gradients in liposomes: Implications for biological systems. In Barenholz Y, Lasic DD, eds. Handbook of Nonmedical Applications of Liposomes. Vol II. CRC Press, Boca Raton, FL, 1996;291–301.
72. Horowitz AT, Barenholz Y, Gabizon AA: In vitro cytotoxicity of liposome-encapsulated doxorubicin: Dependence on liposome composition and drug release. Biochim Biophys Acta 1992;393–395.
73. Emanuel N, Kedar E, Bolotin EM, Smorodinsky NI, Barenholz Y. Targeted delivery of doxorubicin via sterically stabilized immunoliposomes: Pharmacokinetics and biodistribution in tumor-bearing mice. Pharm Res 1996;13:861–868.
74. Cohen R, Polacheck I, Benita S, Levi M, Barenholz Y. Comparative evaluation of five types of amphiphile-based assemblies containing amphothericin B. In: Gregoriadis G, Florence AT, eds. Liposome and Vaccines, Progress in Drug and Vaccine Delivery. London, 16–20th December, 1996;71.
75. Amselem S, Gabizon A, Barenholz Y. Optimization and upscaling of doxorubicin containing liposomes for clinical use. J Pharm Sci 1990;79:1045–1052.
76. Kirby C, Gregoriadis G. Dehydration rehydration vesicles: A simple method for high yield drug entrapment. Biotechnology 1984;979–984.
77. Adler A, Schachter J, Barenholz Y, Bar LK, Klein T, Korytnaya R, Sulkes A, Cohen Y, Kedar

E. Allogeneic human liposomal melanoma vaccine with or without IL-2 in metastatic melanoma patients: Clinical and immunobiological effects. Cancer Biother 1995;10:293–306.
78. Diminsky D, Reimann ZJ, Schirmbeck R, Barenholz Y. Structural and functional characterization of liposomal recombinant hepatitis B vaccine. J Liposome Res 1996;6:289–304.
79. Eckstein M, Barenholz Y, Bar LK, Segal E. Liposomes containing *Candida albicans* ribosomes as a prophylactic vaccine against disseminated Candidiasis in mice. Vaccine 1997;15:220–224.
80. Barenholz Y. Quality control of liposomes. Special issue of Chem Phys Lipids 1993;64.

Lasic and Papahadjopoulos (eds.), Medical Applications of Liposomes

CHAPTER 7.2

Strategies for large scale production and optimized stability of pharmaceutical liposomes developed for parenteral use

E.C.A. van Winden,[1] N.J. Zuidam and D.J.A. Crommelin

Department of Pharmaceutics, Utrecht Institute for Pharmaceutical Sciences (UIPS), Utrecht University, Utrecht, The Netherlands*
[1]*Present address: OctoPlus B.V., Leiden, The Netherlands*

Overview

I. Introduction

A large number of studies on the pharmaceutical application of liposomes has appeared in recent years. From these studies one can derive that liposomes are versatile drug carrier systems that can be tailor-made to accommodate a large variety of drugs for a wide range of therapies. Both lipophilic and hydrophilic drugs can be incorporated in these vesicles, in the phospholipid bilayer and in

*UIPS participates in the Groningen Utrecht Institute for Drug Exploration (GUIDE).

the aqueous core, respectively. (Pro)drugs can be covalently bound to bilayer components as well. Moreover, the behaviour of liposomes in vivo and in vitro can be controlled within certain limits by selecting the proper liposome characteristics such as vesicle size and number of bilayers, bilayer fluidity, charge and hydrophilicity of the external surface, and the attachment of targeting molecules to the bilayer surface.

A variety of preparation techniques for varying types of liposomes on a laboratory scale have been described.[1,2] However, for the use in clinical practice the large scale production of vesicle dispersions is required. The product must be sterile, well characterised and stable. When the stability of the aqueous liposome dispersion is insufficient, freeze-drying may provide a solution. Unfortunately, both the suitability of certain production methods and the stability of the liposomes depend on characteristics such as vesicle size and bilayer composition. This limits the choice of liposome types from which one can select when optimising liposome based drug therapy. This chapter describes the possibilities for large scale production of different types of liposomes as a pharmaceutical formulation and efforts currently made to further develop this technology. In addition, it reviews the stability of different liposome formulations and recently gained insights into factors influencing this stability. Special attention will be paid to liposomes for parenteral administration, e.g., those loaded with cytostatics.

II. Selection of an optimal liposome type

II.1. Liposome characteristics

The liposome characteristics which are relevant when discussing large scale production of clinical grade liposomes, will be described briefly. For more detailed information one is referred to handbooks or reviews.[1,3,4]

Liposome characteristics of major importance are (1) vesicle size (2) number of bilayers and morphology, (3) bilayer fluidity, and (4) surface characteristics such as charge and hydrophilicity. Vesicle size can range between 0.03 and 10 μm. The larger vesicles may have more than 10 bilayers, but this number can be manipulated by the chosen preparation method.[5,6] Apart from multilamellar liposomes with a number of bilayers concentrically surrounding an aqueous core, so called multivesicular liposomes have been described, where one can find vesicles inside the liposomes, not organised in a concentric pattern. Vesicles smaller than 0.1 μm are mainly unilamellar, independent of the preparation technique. Dependent on temperature, a phospholipid bilayer may exist in several physical phases which differ in molecular packing of the phospholipids.[7] Two phases most often discerned in studies on the biological effect of liposomes are the rigid, gel phase, with ordered acyl chains, and the more disordered and fluid, liquid-crystalline phase. At the transition temperature (T_m) the bilayer passes from one phase into the other. The T_m depends mainly on the (phospho)lipid composition, but can also be affected by the presence of solutes (e.g., Ref. 8), or dehydration.[9–11] In general, fluid bilayers are more permeable to solutes than rigid bilayers, but the permeability shows a local maximum around T_m, where phase separation occurs. Compounds that interact with the bilayer surface or are located between the phospholipids (e.g., cholesterol) can affect or abolish phase transitions. In addition, they may change the molecular packing of these phospholipids and the bilayer fluidity.[4] The charge density of the bilayers can be tailored by adding negatively or positively charged (phospho)lipids. The surface characteristics of liposomes may also be altered by attachment of hydrophilic moieties to phospholipids or cholesterol, such as monosialogangliosides (GM1), phosphoinositol, polyoxethyleneglycol (e.g., Refs. 12, 13), or sugars (e.g., Refs. 14–16).

II.2. Relation between liposomes characteristics and therapeutic effect

For liposome formulations the disposition of the drug after parenteral administration is determined by: (1) the behaviour of the vesicles in the body, (2) the release kinetics of the drug associated with these carriers, and, of course, (3) the in vivo behaviour of the drug itself. An abundance of evidence is available to support these points. Below, a brief overview is presented of factors that affect the performance of liposomes and liposome associated drugs in vivo.

Manipulation of the blood circulation time after intravenous injection is a powerful tool in controlling the fate of the encapsulated drug. A long circulation time allows a sustained release of the drug into the blood for an extended period of time. Liposomes normally do not pass through the endothelial barrier of the vascular system.[17] They can only escape from the circulation to a substantial extent in sinusoids of the liver (e.g., see Ref. 18). However, under pathological conditions increased endothelial permeability has been observed and prolonged circulation promotes accumulation of small liposomes in the tissues with an increased blood vessel permeability by passive extravascularisation to the interstitial space. Such malfunctioning endothelial barriers may be found in tumours or inflammation sites and enables targeting to these tissues.[12,19–21]

However, unless special measures are taken, liposomes are rapidly taken up by cells of the mononuclear phagocyte system (MPS). This uptake can lead to depletion of the macrophage population when the liposomes are loaded with a macrophage eliminating drug.[22,23] Especially in cancer therapy avoidance of the MPS is important, since accumulation of cytostatics in the cells of the MPS may result in a decreased functional activity jeopardising its metastatic surveillance potential. Recently, it has been shown that liposomes loaded with doxorubicin can deplete the macrophage population and impair the phagocytic activity of liver macrophages.[24] The rate and extent of liposome uptake by the MPS, and therefore the circulation time, depend on size, bilayer rigidity, hydrophilicity, and charge of the liposomes. Large vesicles are rapidly eliminated from the bloodstream, whereas small liposomes circulate much longer, especially those with neutral or positively charged, rigid bilayers.[25,26] An example of these types of liposomes currently in development for the treatment of cancer is DaunoXome®, liposomes loaded with daunorubicin (Table 1). A significant reduction of uptake by the MPS is also obtained by attachment of hydrophilic chains to the phospholipid headgroups. The

Table 1

Examples of liposomal formulations for parenteral application currently in late stages of clinical development or already on the market. CHOL: cholesterol, DSPG: distearoylphosphatidylglycerol, DSPC: distearoylphosphatidylcholine, HSPC: hydrogenated soy phosphatidylcholine, PEG-DSPE: distearoylphosphatidylethanolamine derivatised with polyethyleneglycol (MW 1900), TC: α-tocopherol. For a more extensive overview see Ref. 31.

Product name (company)	Therapeutic application	Incorporated drug/ lipid composition (molar ratio)	Vesicle size (mm)	Remarks	Reference
Dox-SL® (DOXIL®) (Liposome Technology Inc.) SEQUUS	Cancer	Doxorubicin/ PEG-DSPE : HSPC : CHOL : TC = 5.5 : 56.1 : 38.2 : 0.2	0.08–0.12	Stable for 2 years at 4–8°C (aqueous dispersion)	(32) and F. Martin, personal communication (1995) (No stability data published yet)
DaunoXome® (NeXstar)	Cancer	Daunorubicin / DSPC : CHOL = 2 : 1	0.05–0.08	Stable for 20 weeks at 2–8°C (aqueous dispersion)	(33, 34)
Ambisome™ (NeXstar)	Fungal infections	Amphotericin B/HSPC : DSPG = 2 : 1	<0.08	Freeze-dried, stable for 2.5 year at 2–8°C, marketed in Europe	(35, 36)
TLC-D99 (The Liposome Company)	Cancer	Doxorubicin/EPC : CHOL = 55 : 45	0.10		(37)
ABLC™ (The Liposome Company)	Fungal infections	Amphotericin B/ DMPC : DMPG = 7 : 3	0.12		(38)

mechanism of avoidance of MPS uptake is under discussion,[13,26–30] but an important factor probably is the reduced adsorption of plasma components compared to more "hydrophobic" liposomes, which minimizes recognition by cells of the MPS. This principle is applied in Dox-SL®, liposomes loaded with doxorubicin (Table 1), which are being tested in clinical trials for the treatment of Kaposi's sarcoma, breast and ovarian carcinoma.

In contrast, the use of large liposomes may be advantageous when targeting to the macrophage is desired, e.g., when parasites reside in macrophages as in the case of Leishmaniasis,[39] or when the macrophage can function as a slow release depot after phagocytosis of the liposomes.[40,41]

Vesicle fate in muscular and subcutaneous tissue after i.m. or s.c. administration were shown to be dependent on vesicle size: large vesicles predominantly stay at the injection site and slowly release their contents, whereas small vesicles are mainly taken up by the draining lymph system. Uptake of liposomes by blood capillaries is expected to be limited by the continuous basement membrane and the tight junctions between the endothelial cells. In contrast to blood capillaries, lymphatic capillaries have a poorly organised basement membrane and junctions between the endothelial cells may be open.[42,43] This anatomical difference is being employed in targeting to the lymph nodes. Parameters controlling the accumulation of liposomes in the lymph nodes are currently under investigation.[44–46] Lymph node targeting will be advantageous in the therapy of lymph node metastases and of diseases where infected cells reside in the lymph nodes, as in the case of HIV.

The release of the entrapped drug from the liposome is dependent on the number of bilayers, and the bilayer permeability. As described above the bilayer permeability depends on its fluidity, and is greatly enhanced by phase separation, as occurs at the phase transition temperature. This feature is being exploited in thermosensitive liposomes,[47,48] which show a slow release of their contents at normal body temperatures, but a fast release at slightly elevated temperatures. This opens the way to trigger the release of the drug from the liposomes by warming up, preferably only relevant parts, of the patients body after allowing the carriers to accumulate at their target site.

Until now most liposomes that were developed as drug carrier systems were either neutral or negatively charged. By proper selection of the negative charge inducing molecule, the necessary long term physical stability against aggregation and fusion can be ensured. Positively charged liposomes were hardly used in the past because of safety concerns.[49] However, positively charged liposomes are now avidly studied as non-viral vectors for plasmids to be used in gene therapy.[50,51]

II.3. Choice of phospholipids

The above indicates that careful selection of the type of phospholipids is essential to achieve the desired therapeutic goals. A further distinction between phospholipids with similar physical characteristics can be made on the basis of their origin.[2] Several classes can be discerned: (1) phospholipids purified from natural sources, or (2) modified natural phospholipids, (3) semi synthetic phospholipids and (4) fully synthetic phospholipids.

Examples of natural phospholipids are phosphatidylcholine (PC), phosphatidylethanolamine (PE), phosphatidylinositol (PI) and sphingomyelin (SPM), which are obtained from egg yolk or soy beans. These are mixed acyl ester phospholipids with unsaturated acyl chains, containing double bonds in the cis isomeric form. Source dependent differences in acyl chain type have been found (e.g., between egg and soybean PC).[6]

The acyl chains of natural phospholipids can be modified by partial or complete hydrogenation. The resulting partially saturated phospholipids are less sensitive to oxidation, and have an increased main transition temperature.[52] Partially hydrogenated phospholipids may contain double bonds both in the cis and the trans form, due to isomerisation during the hydrogenation process. It is not clear what implications this isomerisation has on the bilayer properties of the vesicles and how it may influence the biological effect. Another possible modification of natural phospholipids is the conversion of the choline headgroup into a glycerol, serine or ethanolamine headgroup.[53] The composition of natural phospholipids must be strictly controlled, since apart from source-dependent differences, considerable interbatch variation has been observed, e.g., for egg PC.[6] Besides, care must be taken that the level of pyrogens in the phospholipids does not exceed pre-set limits.

Fully synthetic phospholipids have the advantage of a defined fatty acid composition and can be tailored to specific needs. For example, the acyl chain length and the degree of saturation can be selected in such a way to provide the required bilayer characteristics (T_m, rigidity) in combination with a low sensitivity to oxidation.[54] Quality control for these lipids is focused on the presence of by-products and residues of organic solvent used in the synthesis procedures. A disadvantage of synthetic phospholipids compared to natural phospholipids is their high costs. In an alternative approach, semi-synthetic lipids are prepared from natural phospholipids by removal of the acyl chains and replacement by (well defined) synthetic acyl chains.

III. Large scale preparation of liposomes

For the preparation of liposomes on a laboratory scale numerous techniques have been developed and they have been the subject of many publications and reviews.[1,2,4,55] In general, it is recognised that the large scale production of liposomes may present specific problems and relatively little attention has been paid to this issue up until now (see also Refs. 56–60).

Several steps can be discerned in the preparation process of liposomes: (1) Hydration of the lipids resulting in vesicle formation, (2) drug encapsulation, and (3) sizing of the vesicles. In some cases, these steps are linked together in the production process. Below, different aspects related to the large scale production of liposomes to be used for parenteral administration are discussed.

III.1. Liposome formation

III.1.1. Hydration of the lipids using mechanical forces

Factors that influence the hydration of phospholipids and the vesicle formation process are the particle size of the raw lipid material (relation with exposed surface area), the hydration temperature and the bilayer phase transition temperature, shear forces, composition of the hydration medium and time. Vesicle formation is greatly accelerated when the hydrated phospholipids are in the liquid crystalline state above their main transition temperature (T_m). The vesicles formed are predominantly multilamellar vesicles (MLVs). When a drug is dispersed with the dry phospholipids or in the hydration buffer, encapsulation will occur during the vesicle formation process (see Section III.3.1). Although the hydration of phospholipids by agitation with an aqueous medium is a simple and popular preparation method for MLVs, the mechanistic details of the vesicle formation process are complex and still not fully understood.[61] Examples of findings which may have practical implications are effects of lipid composition[62] and ionic strength on the vesicle size,[63] and exclusion of solutes from the interbilayer space.[64] In addition, passing the T_m several times was reported to improve the reproducibility of the vesicle formation.[65] When MLVs are the preferred liposome type and the encapsulation efficiency is satisfactory, no further preparation steps are necessary after hydration.[54,66–68] This provides the opportunity for liposome formation from freeze-dried lipids at the bed site of the patient, and it diminishes stability problems, as will be discussed below.

The use of shear forces speeds up the vesicle formation process and can result in a reduced vesicle size. High shear homogenisation techniques are already applied on a large scale in e.g., the food and cosmetic industry, and now several homogenisers have been described for sizing of liposomes: the French Press® (Aminco, USA, Illinois) was the first example,[69] but the working volumes are relatively small (50 ml). In the Microfluidizer® (Microfluidics, Newton, Massachusetts, USA) technique cavitational processing is used, which results in pronounced size reductions.[70] The liposome dispersion is recirculated after passage of the interaction chamber via a reservoir, subjecting the dispersion to a number of cycles through the interaction chamber.[71–73] Larger volumes can be processed than with the French Press. Other instruments which can be mentioned in the context of high shear homogenisers are the Gaulin Micron®LAB 40,[74,75] the Nanojet® (Nanojet Engineering, Dortmunt, Germany),[76] the Emulsiflex® B3 Homogenizer (Avestin, Inc. Ottawa, Canada)[77] and the continuous flow ultrasonic cavitation technique (Heat systems Ultrasonics, Farmingdale NY). The latter two have recently been compared with the Microfluidizer® technique for the preparation of oil in water emulsions.[78]

The obtained average vesicle size depends on the apparatus and the instrument settings, and factors such as lipid concentration, temperature, and homogenisation time. It should be noted that only dispersions with a small average vesicle size have a relatively narrow, well defined size distribution, whereas populations of large liposomes are generally heterogeneous. Potential problems mentioned in literature that may arise when using high shear homogenisation techniques are the poor temperature control in interaction chambers and the erosion of metal components in these instruments. An example is given by Brandle et al.,[75,79] who observed damage to metal and ceramic parts of the interaction chamber of the Gaulin homogeniser after prolonged usage of the instrument. However, it was claimed that contamination of the dispersion remained within acceptable limits, and was low compared to other homogenisers. If necessary, the use of appropriate filters may help to overcome contamination problems. High shear homogenisation may be less suitable for the preparation of liposomes which contain high molecular weight proteins in this step of the preparation process, since those proteins may be damaged by the applied shear forces.

A more gentle method by which liposome dispersions with different average sizes could be produced is described by Talsma et al.[80] Liposomes with an average diameter of around 0.3 μm could be formed by bubbling nitrogen through a dispersion of lipid material. Scaling up of this process may be feasible, but considerable time was needed to obtain small vesicles and elevated temperatures had to be used. However, the process time was shortened when the raw material was first homogenised into MLVs by mechanical forces. The advantage of this technique is the remote chance of contamination by eroded particles as the preparation can be performed under "all glass" conditions. A disadvantage is that the size reduction process stopped around 0.3 μm.

In principle, all high shear homogenisation techniques can be performed without using organic solvent, although in most of the described studies organic solvents were still used for initial mixing of the bilayer components. In the following section we will discuss techniques in which organic solvents are essential.

III.1.2. Hydration of the lipids using organic solvents

Another preparation approach is the injection of a lipid containing solution which is miscible with water (e.g., ethanol and DMSO), into an aqueous phase. The resulting dilution of the organic solvent induces vesicle formation. The obtained vesicle size depends on factors such as stirring rate, vesicle charge, ionic strength, etc. A general description of such an approach in scaled up form has been given by Martin,[56] and more specifically for the production of ecanozol containing liposomes developed to treat topical fungal infections by Kriftner.[57] Isele et al. used the same technique for the production of liposomes containing monomeric zinc phthalocyanine, which are aimed for the treatment of cancers of epithelial origin.[81] Inevitably, ethanol residues will remain in the liposomal bilayer, but low residual levels will not influence the bilayer structure, and ethanol has been approved as an solvent for parenteral drugs. Main disadvantages are the limited choice of (phospho)lipids as bilayer components, since sufficient solubility in ethanol is required, and the need for strong drug bilayer interaction or an active loading procedure to obtain an acceptable loading efficiency.

III.1.3. Liposome formation by dilution or removal of detergent

Liposome formation can also be achieved by dispersing mixed micelles of the phospholipids and the detergent, and if preferable, the drug to be encapsulated in an aqueous medium, followed by removal or dilution of the detergent. The detergent can be removed in several ways: ultrafiltration, dialysis, gel filtration, or the addition of polymeric adsorbants. Allen discussed the pros and cons of these methods.[82] Weder described the use of a (dead end) ultrafiltration unit with scaling-up potential.[83] Peschka studied the preparation of liposomes using a tangential (cross flow) filtration unit.[84] The influence of process parameters such as filter cut-off value, lipid concentration, selected detergent, filtration area and flow rate was evaluated. It was concluded that tangential filtration can be used for the preparation of homogeneous dispersions of varying sizes, and can be easily scaled up for commercial use. Factors which determine the obtained vesicle size after dilution of the mixed micellar dispersion were studied by Jiskoot et al.[85] Small vesicles (average size around 0.1 μm) were obtained when applying high dilution rates, or using a low cholesterol content, and charged phospholipids. Concern is expressed about the safety of the residual detergent in liposomes prepared by this technique. As this aspect strongly depends on the type of detergent used, no general rule can be given. Experience with detergents in other clinically applied parenteral formulations is available. For instance, for a number of years, glycocholate/soybean PC mixed micelles have been used as solubilisers for lipophilic, poorly water soluble drugs to be administered parenterally, e.g., Refs. 1, 86.

III.2. Sizing techniques

Particle size has proven to be a critical factor in the disposition of liposomes in the body. Therefore, the size of liposomes used as a pharmaceutical preparation should be well defined and reproducible upon production of different batches.

'High pressure' homogenisation, an attractive technique to obtain a reduced vesicle size of large batches, was discussed above. A recent example of this application was described by Sorgi and Huang,[87] who obtained liposomes smaller than 0.2 μm by the microfluidising hydrated cationic lipids (batch size: 0,5 l). Another regularly used sizing technique is "low" pressure extrusion of the liposome dispersion with pressures up to about 1 MPa through polycarbonate membranes with well defined pore diameters.[88,89] Polycarbonate filters with pore sizes ranging from 8 down to 0.03 μm are available. Alternatively, inert ceramic or stainless steel filters can be used as well. The resulting vesicle size and lamellarity are related with the pore size, although effects of the medium and phospholipid composition are generally observed as well. Extrusion rates depend on the bilayer phase:[88] the highest rates are obtained with phospholipids heated above the T_m, even when cholesterol is present, which abolishes the transition from the gel to the liquid crystalline state. The possible effects of the phospholipid composition on the obtained vesicle size can be minimised by repeated extrusion cycles. Until now, clogging of the filters when using large volumes was considered as a major limitation. However, a recent example of successful application of the extrusion technique on a large scale was described by Schneider et al.[90] Using a new continuous flow through, high pressure extruder, pressures up to 10.5 MPa could be applied, resulting in flow rates up to 500 ml/min. Several lipid compositions were tested and no clogging of the polycarbonate filters was observed. The obtained average vesicle sizes differed slightly between the studied lipid compositions: the smallest vesicles with average sizes around 60 nm were obtained with soybean PC in combination with 10 mol% of the charged soy bean PG or soybean phosphatidic acid. Addition of cholesterol to soybean PC liposomes up to 50 mol% resulted in an increase in the vesicle size from 75 to 94 nm, but it prevented fusion of liposomes made of the hydrogenated phospholipids, resulting in an average of 93 nm with cholesterol, compared to 0.20 μm without cholesterol. High lipid concentrations (up to 400 mg SPC/ml) could be processed even through 30 nm filters, and resulted in only a small increase of the obtained vesicle size as compared to dispersions of 50 mg/ml. The authors concluded that extrusion of liposomes with this extrusion device is easily done on a large scale.

Various medium or low pressure extrusion devices can be made by simple modification of existing equipment. For example, for extrusion of pressures below 0.7 MPa the UP110 stirrer of Schleicher and Schuell (Dassel, Germany) is suitable. Another apparatus for extrusion over a broad range of pressures (up to 10 MPa) is the Nucleopore GH 76–400 cell. The latter two instruments have both efficient stirring capabilities and can be connected to reservoirs for processing large volumes. For more details on large-scale extrusion, the reader is referred to reviews by Martin[56] and Amselem et al.[60,91,92] and references listed therein.

Alternative sizing methods are the bubbling technique[80] (see Section III.1.1.) or size alterations achieved by pH jumps. Small vesicles spontaneously form when aqueous phosphatidic acid or phosphatidylglycerol (PG) dispersions are exposed to transient alkaline pH conditions.[69,93–95] The average size depends strongly on the phospholipid structure (phosphatidic acid/PG), phospholipid mixture (PC/PG), cholesterol content, ionic strength and timing of the pH adjustment procedure. Despite attractive features of this technique (e.g., absence of organic solvents and detergents) its application is seriously limited by the stringent requirements on the lipid composition, and the size heterogeneity of the obtained vesicles.

Numerous other techniques for liposome preparation and sizing have been described and the reader is referred to the overviews written by e.g., Barenholz and Crommelin[2] and Crommelin and Schreier.[1] In some of those techniques disruption and reformation of vesicles with a reduced average size is accomplished by applying jumps in ionic strength, or dehydration and rehydration steps (using freezing/thawing cycles). The large scale clinical use of these methods may be hampered by technological barriers (freezing/thawing of large volumes). In other cases, the necessary high concentrations of solutes and the required isotonicity of parenteral formulations appear to be mutually exclusive.

III.3. Encapsulation of the drug into the liposomes

Methods for encapsulation of a drug into liposomes should preferably result in (1) a high encapsulation efficiency (= ((amount of drug in the liposomes)/(total amount of drug added to the dispersion)) × 100%), minimising the amount of extraliposomal drug that has to be removed from the dispersion, and (2) a high drug/phospholipid ratio (encapsulation capacity) in order to limit the dose of phospholipids administered to the patient.

Roughly, two different ways for encapsulation of the drug in the liposomes can be distinguished: (1) drug entrapment during the vesicle formation process (passive encapsulation) and (2) loading the drug into intact vesicles (active loading).

III.3.1. Passive encapsulation

For water soluble compounds which do not interact with the bilayer, the encapsulation efficiency after passive encapsulation is proportional to the aqueous volume enclosed by the vesicles, which itself depends on the phospholipid concentration of the dispersion and the lamellarity and morphology of the vesicles. For liposomes with a certain average number of bilayers, large vesicles will have a higher encapsulated volume than small vesicles. For liposomes of a certain average size, the encapsulated volume per mole lipid decreases with an increasing number of bilayers. The encapsulation capacity, but not the efficiency can be improved by using a high drug concentration during the vesicle formation process. Interaction of water soluble drugs with the bilayer enhances both the encapsulation efficiency and capacity. As with drugs which are less water soluble, and also interact with the bilayer, both encapsulation parameters will depend more on the phospholipid concentration and selection than on morphological parameters.

Several strategies have been developed to improve the encapsulation efficiency or capacity for drugs. The lipophilicity of the drug can be increased by attaching lipophilic chains to the drug molecule, e.g., MTP-PE (muramyltripeptide-phosphatidylethanolamine),[54] or by complex formation of the drug with a lipophilic chelating agent.[96] In addition, bilayer properties such as the surface charge may be altered in order to enhance the interaction between the drug and the bilayer, e.g., Refs. 97–99. Obviously, the effect of such adaptations in formulations for clinical usage should not interfere with the therapeutic effect aimed for. This approach is applied in the commercially available "freeze-dried empty liposomes for non-clinical application" (NOF Corporation, Japan), which show a high encapsulation efficiency for several drugs after rehydration with a solution containing the drug. For example, DPPC:DPPG:CHOL = 3:3:4 liposomes (0.13–0.15 μm, [phospholipid] = 60 mM) are claimed to contain 100% of the added doxorubicin (1 mg/ml) after rehydration,[100] which was ascribed to the interactive electrostatic forces between the drug and the bilayer components.[101]

An alternative method to achieve a high drug/phospholipid ratio was described by Chapman et al.,[102,103] who subjected a liposome dispersion to freezing/thawing cycles in the presence of tetraethylammoniumperchlorate (TEAP). Although the process by which the high loading is achieved is complex and not fully understood, the following mechanism was proposed. During freezing, crystals of pure ice are formed, between which a fluid remains with an increasing solute concentration. The liposomes are disrupted by osmotic or mechanical forces during this process, which allows the solutes to enter the liposomes. Upon thawing, the extraliposomal medium is diluted by the melting ice, resulting in a concentration gradient of the solutes and an osmolaric pressure in the vesicles. Diffusion of TEAP through the bilayer out of the vesicles (partly) prevents the rupture of the liposomes, leaving a high concentration of the non-bilayer permeating drug inside the vesicles. Another method using freezing/thawing steps which resulted in increased drug entrapment in liposomes of e.g., doxorubicin, is based on pH changes during these freezing/thawing steps.[104] The scaling up of the required freezing/thawing steps has not been described yet.

Another approach to obtain a high encapsulation efficiency and capacity is freeze-drying the lipids followed by rehydration with a small volume of aqueous medium, resulting in the so called dehydration/rehydration vesicles. The drug to be encapsulated can be present either in the freeze-dried cake or in the rehydration medium.[66–68,105] Because of the high lipid concentration, a high percentage of encapsulated volume and thus a high encapsulation efficiency of the drug is obtained.

III.3.2. Active loading

Partition coefficients of drugs can depend on the pH, and to a lesser extent on the ionic strength of the aqueous medium. For example, basic compounds carrying aminogroups are relatively lipophilic at high pH and hydrophilic at low pH. In a two chamber aqueous system separated by a membrane, accumulation will occur at the low pH side under dynamic equilibrium conditions. This offers opportunities for the loading of these drugs into intact liposomes, as first described by Deamer and coworkers.[106] A transmembrane pH-gradient causes accumulation of doxorubicin (pKa ≈8.1) at the low pH side, as shown in Figure 1. In the unprotonated

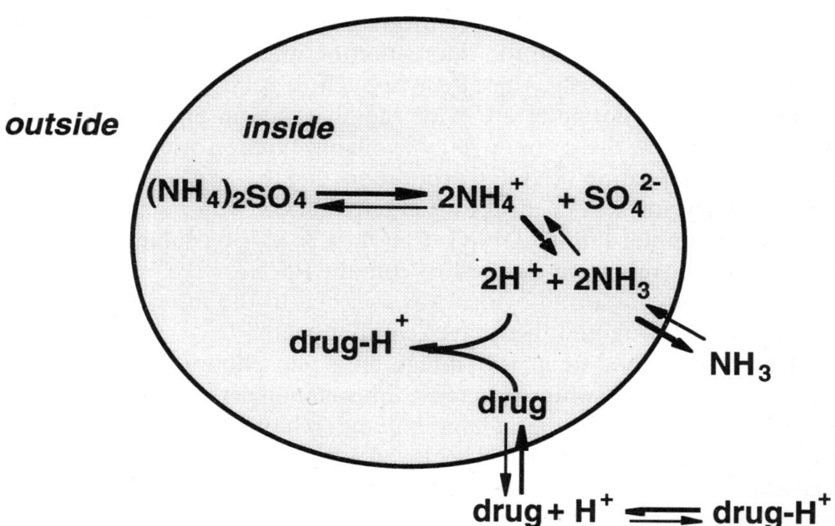

Fig. 1. Active loading of a drug into liposomes by a pH gradient which is obtained by a $(NH_4)_2SO_4$-gradient. Solid arrows indicate the shifts in equilibria resulting in an increased entrapment of the drug.

form, doxorubicin can diffuse through the bilayer. At the low pH-side the molecules are predominantly protonated, which lowers the concentration of this drug in the unprotonated form, and thus promotes the diffusion of doxorubicin to the low-pH side of the bilayer. This concept was extensively used by Cullis and co-workers, as reviewed earlier.[107] The pH-gradient was created by preparing liposomes with a low pH in- and outside the vesicles, followed by the addition of base to the extraliposomal medium. Another method to create such a pH-gradient is the preparation of liposomes with an ammonium sulfate gradient, as also illustrated in Figure 1. Diffusion of NH_3 out of the liposomes, e.g., upon dilution, creates a pH-gradient, by which agents can be entrapped.[108–110] An additional role is ascribed to the sulfate anions, which (1) stabilise the ammonium sulphate gradient by their low bilayer permeability, and (2) enhance anthracycline accumulation and improve its retention due to formation of gel-like anthracycline sulfate aggregates.[111] Both active loading methods have the advantage of (1) a high encapsulation efficiency and capacity, and (2) a reduced leakage of the encapsulated compound as compared to passive entrapment.[111,112] A third advantage is that loading of the drug can be performed "at the bed side", limiting loss of retention of the drug by diffusion, or chemical degradation during storage. This is especially important when drug and phospholipids require different conditions for optimal stability. For example, the optimal stability conditions for doxorubicin in aqueous solution are found around pH 4,[113] whereas phospholipids are most stable at pH 6.5.[114–116] Finally, (4) no biologically active compounds are present in the dispersion during the preparation steps, which can be advantageous for safety reasons. In this way safety hazards can be reduced e.g., in the case of aerosol production when high pressure is needed (extrusion/homogenisation), or when cleaning the production equipment.

III.4. Removal of the nonencapsulated drug

When the drug is for less than 100% associated with the liposomes upon liposome formation, removal of the free drug may be desirable. This is especially the case for toxic agents like antineoplastics (doxorubicin) and antibiotics, or biological response modifiers such as IL-2 or TNF. Another reason for a drug removal step could be that the free drug interferes with liposome stability. For example, free doxorubicin induces aggregation of negatively charged vesicles followed by fusion.[117] A problem may arise when the drug is mainly, but not completely associated with the bilayer, and is able to partition into the aqueous phase on a relatively short time scale (like some proteins). Whenever the concentration of the free drug in the outside aqueous phase is lowered, release from the liposome will occur. In such a case enhancement of the interactive forces between the drug and the bilayer may help to avoid loss of retention, or in some cases a pH-gradient may contribute to the association of the drug with the liposomes, as was demonstrated for encapsulated doxorubicin (see above).

Many approaches are available to separate the free drug from liposome-associated and encapsulated drug, but only a few can be scaled up easily. The laboratory scale approaches include gel permeation chromatography, which is labour intensive and leads to dilution of the liposomes, and ultracentrifugation, which is a convenient technique for large liposomes (>100 nm). However, ultracentifugation may be a "traumatic" process.[118] Vesicles are subjected to high forces which may modify them physically. In addition, shear forces may be needed to redisperse the vesicles, and some vesicle loss may occur as spinning down may be incomplete. Finally, the application of both techniques under aseptic conditions requires skilled personnel and strict protocols.

There are a few methods that can be easily scaled up and do not cause considerable perturbation of the vesicle structure. Separation of the free drug by ion exchange can be used if the encapsulated or associated molecule and the liposome have an opposite electrical charge. This method was used extensively to separate positively charged drugs such as anthracyclines and aminoglycosides from negatively charged or neutral liposomes.[119,120] It requires pretreatment of the ion exchange resin to obtain the appropriate ionic form, followed by the establishment of its exchange capacity. The latter is essential, as it is highly drug dependent.[92,119,120]

Dialysis is a rather general approach and does not have the inherent limitations of the ion exchange concept. Conventional dialysis membranes can be used with molecular weight cut-off characteristics depending on the molecular weight of the free compound to be removed from the liposome dispersion. A variety of dialysis and ultrafiltration equipment is available from lab scale to industrial production scale (hollow fiber, spiral wound). By selecting proper conditions, separation is fast and, if desired, concentration of the liposome dispersion can be achieved simultaneously. During ultrafiltration the dispersion is stirred or circulated by a pump. This convection process should be validated for not-inducing leakage of the encapsulated material.

III.5. Pharmaceutical requirements for liposomes

To ensure a predictable therapeutic effect, a strict control of factors such as vesicle size distribution, the drug/phospholipid ratio, and the percentage encapsulation, is required. No FDA guidelines exist for acceptable variations of liposome specific characteristics. For each product, the acceptance limits of these characteristics have to be defined. It must be demonstrated that the variations in efficacy and toxicity of the formulations are acceptable when the liposome characteristics are varied within these ranges. The same holds for the presence of unwanted by-products, such as residues of organic solvents, or degradation products. The acceptability of organic solvent residues will probably depend on the toxicity of the organic solvent involved, relative to the therapeutic advantage of the drug-liposome formulation as a whole, and the therapy that is aimed for. For example, more side effects may be acceptable when the liposomes are clearly therapeutically beneficial in anti-cancer therapy than in the case of marginal advantages in anti-inflammatory therapy.

For degradation products, which are non-toxic a common acceptance limit in the pharmaceutical industry is a 10% maximum for degradation. The value of 10% indicates that the percentage degradation should not exceed 5% after the final production step, so that a margin for further degradation upon storage is still available.

Liposomes containing pharmaceuticals administered via the parenteral route, on damaged skin or in the eyes must be apyrogenic and sterile.[1] Special attention must be paid to sterilisation methods applied to liposomes. Careful evaluation of all liposome characteristics before and after sterilisation should demonstrate the suitability of a certain sterilisation technique in manufacturing protocols.

The same holds for the effect of storage on the liposome formulation. The stability of liposomes strongly depends on their characteristics. Therefore, storage stability must be taken into consideration in the early stages of product development. Only a combination of optimal therapeutic efficacy and stability of the liposomes will result in a successful drug formulation. Aspects of liposome stability upon sterilisation and storage are discussed below.

III.6. Apyrogenic and sterile production of liposomes

Pyrogens can cause fever and shock. Common sources for pyrogens are micro-organisms, in particular those producing endotoxins (lipopolysacharides). It is extremely difficult to remove all pyrogens from formulated liposome dispersions.[121] Depyrogenation of fluids (including organic solvents holding the lipids) is possible by ultrafiltration through filters with cut-offs of 10 kDa.[121] Therefore, the manufacturer should check the quality of the raw materials, and design the liposome formulation process in such a way that the generation of pyrogens by micro-organism growth or contact with contaminated equipment during the production process is avoided. The reader is referred to the book of Pearson[122] and pharmacopeia[123,124] for further detailed information about pyrogens and pyrogenicity tests.

A product is considered sterile if the chance to find a unit that is contaminated with living microorganisms is less than 1 in 10^6 sterilized units of that product.[123,124] Recently, our group has published an overview of the different sterilisation techniques which may be considered to sterilise liposome dispersions.[125] It was stressed in this article that sterility can not be guaranteed by testing the final product, but should be assured by validated, well-defined preparation procedures.

Sterility will be achieved if a low degree of contamination (100 or less colony forming units (cfu)/ml) is combined with an effective sterilisation step immediately after finishing the preparation procedure. An effective sterilisation process gives at least a 10^{12} fold reduction of the test organism known to be highly resistant to that particular sterilisation method (worst-case assumption). The following approaches to achieve sterile liposomes have been considered: (1) autoclaving[125–128] (2) high pressure sterilisation,[129] (3) use of ethylene oxide[130], (4) γ-irradiation[131–133] and (5) filtration.[121,134]

Autoclaving (121°C, 15 minutes) is a preferred sterilisation method for several reasons. First of all, it is relatively simple and has been extensively validated. In addition, autoclaving can be applied to the end product. Under neutral, buffered pH conditions, liposomes without encapsulated agents or with heat-stable, bilayer interacting (lipophilic) agents can be sterilized.[127,135] Oxidation of egg phospholipids is not a problem when using EPC with a low peroxide value.[127] However, in other cases the chemical and physical stability of the liposomes and the drug during the heat treatment can be insufficient. Autoclaving may not be acceptable for liposomes (1) in a basic or acid medium, or (2) loaded with a water soluble, non-bilayer interacting drug which can leak out of the liposomes.[127,135] However, autoclaving of liposomes may still be an option if the free drug does not interfere with the desired therapeutic effect, or can be loaded after the autoclaving process (e.g., by active loading techniques).

High pressure sterilisation[129] (e.g., 5 hours at 60°C and 2.5×10^8 Pa or 21 hours at 40°C and 2×108 Pa) could be an attractive option for sterilising temperature sensitive and high pressure resistant drugs such as proteins. However, application of this technique is hindered by its limited efficacy against the spores of *Bacillus stearothermophilus*.

For the sterilisation of heat labile drugs, several other options exist. Treatment with ethylene oxide has been applied to freeze-dried cakes.[130] However, its sterilising capacity for freeze-dried liposome dispersions has not been proven yet. Moreover, the possibility that toxic residues remain in the cakes has to be excluded.

The use of γ-irradiation[131–133] as a sterilisation technique is still under debate.[124] After γ-irradiation of aqueous liposome dispersions with the sterilisation dose of 25 kGy, which is generally used for this purpose according to Pharmacopeias such as the U.S.P. and B.P., too much degradation of the selected liposomal phospholipids has been found.[136] More studies are necessary to evaluate this method in combination with a powerful antioxidant and/or freeze-drying or freezing in the presence of a safe and effective cryoprotectant.

Filtration through filters with a pore size of $\leqslant 0.22\ \mu m$ is not a sterilisation technique to be considered as a first option, because it is not the last step in the production process. This calls for carefully validated production protocols and well-trained personnel. According to the USP the probability of non product-related contamination may be about 10^{-3} during an aseptic operation,[123] much higher than the contamination level accepted after heat sterilization (10^{-6}). In spite of these disadvantages, filtration is still widely used to produce parenteral products that can not be sterilized with other techniques.[121,134] The major advantage of filtration is that it is not destructive (except possibly for inducing leakage) for small liposomes. To minimise initial contamination it is recommended to routinely filtrate media used for the preparation of liposomes and the liposomes "in statu nascendi" at different stages of the production process through filters with a pore size of $\leqslant 0.22\ \mu m$, even when autoclaving is used as a final sterilisation technique. In a study by Sorgi and Huang[87] it was found that the loss of lipid during filtration may depend on the choice of filter material and the selected lipids. A 18% loss of DOPE (1,2-dioleoyl-sn-glycero-3-phosphatidylethanolamine) was observed after filtration of small ($<0.2\ \mu m$) liposomes consisting of DOPE and DC-Chol (3β[*N*-(*N′N′*-dimethylaminoethane)-carbamoyl]cholesterol) through

0.2 μm filters composed of cellulose acetate. However, the use of nylon or cellulose nitrate filters reduced the loss of DOPE to circa 3%.

It would be interesting to study a combination of different approaches to reach acceptable reduction factors that are not achievable with one approach alone. The susceptibility of microorganisms to a certain sterilisation technique depends on the subpopulation. This susceptibility pattern is different for the different sterilization techniques. Therefore, treatment of liposome dispersions with more than one technique under relatively mild conditions, might result in an acceptable contamination level and little damage to the liposomes. For instance, a low dose of γ-irradiation (±5–10 kGy) in combination with filtration might be a suitable sterilisation method for those dispersions which can not be sterilised by autoclaving.

IV. Stability of liposome formulations

IV.1. Stability of liposomes in aqueous dispersions

Liposomes to be used as carriers for drugs must be stable over prolonged periods of time. For pharmaceuticals a shelf life of at least 2 years is normally requested. Liposome stability studies should address both chemical and physical stability issues. In the following section chemical and physical stability will be dealt with separately.

IV.1.1. Chemical stability

Two types of chemical degradation processes can occur upon storage: (1) hydrolysis of the liposomal phospholipids and (2) oxidation of the unsaturated acyl chains of the phospholipids (if present). Normally, oxidation is hardly a problem in practise, because it can be minimised by using an inert atmosphere (e.g., nitrogen), metal-complexing agents (e.g., EDTA) and antioxidants (e.g., α-tocopherol). More problematic is the effect of hydrolysis on the liposome dispersions.

In an aqueous liposome dispersion, the liposomal phospholipids can hydrolyse to free fatty acids and 2-acyl and 1-acyl lysophospholipids.[137–140] Further hydrolysis of both lysophospholipids results in glycero phospho compounds. The hydrolysis of liposomal phospholipids is catalysed by protons and hydroxyl ions and the hydrolysis rate reaches a minimum at pH 6.5 (see Figure 2).[114,115,137,139,141] The effect of temperature on the hydrolysis rate of phospholipids can be adequately described by an Arrhenius equation, if no phase transitions occur in the experimental temperature range.[114,115,137–142] However, if such a phase transition occurs, as with liposomes composed of hydrogenated soybean PC, dipalmitoylphosphatidylcholine (DPPC) or distearoylphosphatidylcholine (DSPC), biphasic Arrhenius plots were obtained.[114,115,140] Then, at low temperatures lower rate constants were found than expected when extrapolating from data obtained for liposomal phospholipids in the fluid state. Interestingly, the addition of cholesterol to DPPC bilayers (DPPC/CHOL 10/4, molar ratio) abolished the break in the Arrhenius curve for the hydrolysis of DPPC at pH 4.0 and this also caused an increase of

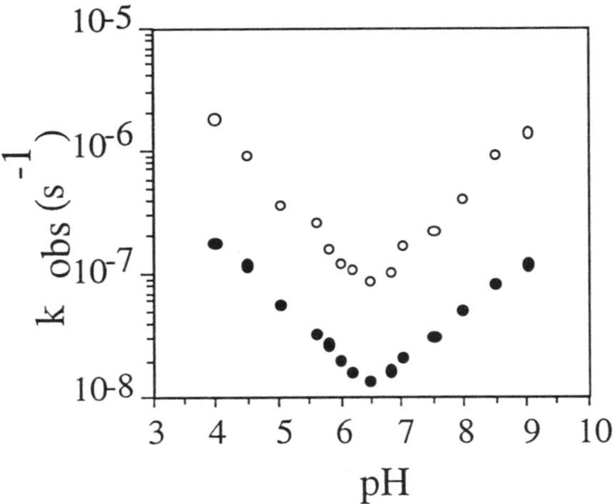

Fig. 2. Effect of pH on the hydrolysis of saturated soybean PC at 40°C (●) and 70°C (○) (from Ref. 115).

the rate constant at lower temperatures compared to hydrolysis rate constants of pure DPPC.[142] Table 2 shows the values for $t_{90\%}$ (time span to reach 10% hydrolysis level) for different phospholipids as obtained in long term storage experiments at pH 4.0 and at 4°C.

The presence of a number of studied charged buffer species resulted in a small increase in the hydrolysis rate of phospholipids.[114,115,141] However, incorporation of a charge inducing component into the liposomal bilayers had major effects on the hydrolysis kinetics of liposomal phospholipids.[116,142] Addition of charge to the liposomal bilayers causes a redistribution of cations and anions, including protons and hydroxyl ions, at the bilayer-water interface. Upon hydrolysing liposomes composed of partially hydrogenated EPC and 'natural' EPG, Grit and Crommelin showed that hydrolysis was a function of the surface pH and not the bulk pH.[116] Under acid conditions incorporation of a negative charge into the bilayers enhanced the hydrolysis of liposomal phospholipids. Under basic conditions incorporation of negative charge decreased the hydrolysis rate. When a positive charge is introduced into liposomal bilayers, the opposite pattern occurs.

Table 2

Time necessary for 10% degradation ($t_{90\%}$) of phosphatidylcholine with different saturated fatty acid chains at pH 4.0 and 4°C ('real time' measurements, $n = 3$), calculated from data in Ref. 142.

Liposome composition	$t_{90\%}$ in days (mean ± S.D.)
Dimyristoylphosphatidylcholine (DMPC)	122 ± 1
Dipalmitoylphosphatidylcholine (DPPC)	277 ± 5
Distearoylphosphatidylcholine (DSPC)	442 ± 13
Ddipalmitoylphosphatidylcholine/cholesterol (DPPC/CHOL) 10/4	132 ± 1
Egg phosphatidylcholine (EPC)	143 ± 5

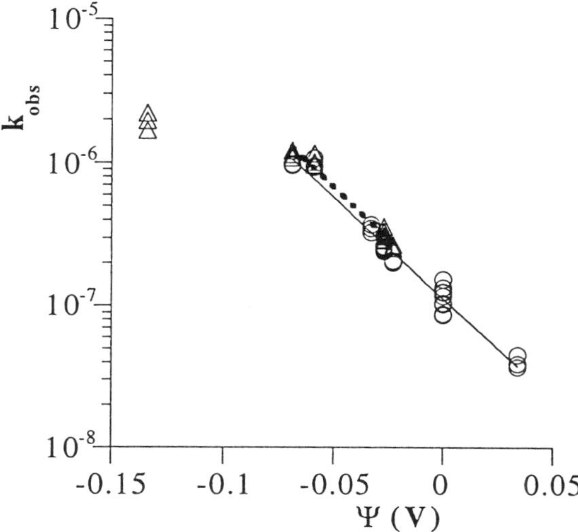

Fig. 3. The values for the observed first-order rate constant (k_{obs}) of liposomal DPPC (○) and DPPG (△) plotted against the surface potential (Y) of liposome dispersions stored in 50 mM acetate buffer (pH 4.0) and 0.12 M NaCl at 30°C (from Ref. 142).

A linear relationship between the logarithmic values of the hydrolysis rate constant of liposomal phospholipids and the surface potential at pH 4.0 and 30°C was established by Zuidam and Crommelin (see Figure 3).[142] When corrected for this charge effect, the differences between the degradation rate constant for DPPC, dipalmitoylphosphatidylglycerol (DPPG) and dipalmitoylphosphatidylethanolamine (DPPE) were small.

Incorporation of neutral additives in the liposomal bilayers, such as cholesterol and α-tocopherol (unless the aggregation state is affected, see above and Table 2), or addition of sodium chloride or cryoprotectants (glucose, trehalose, sucrose and propylene glycol) to the aqueous medium did not affect the hydrolysis kinetics of liposomal phospholipids.[114,142]

IV.1.2. Physical stability

The pharmacokinetic behaviour of liposome associated drugs depends on a number of physical parameters as briefly discussed in the section "relation between liposome characteristics and therapeutic effect". Size and the ratio of free drug vs. entrapped drug (retention) are parameters that have received considerable attention in studies on the physical shelf life of liposomes, and will be reviewed below. The effect of the chemical stability of bilayer components on the physical stability of liposomes has been subject of a number of studies as well.

Particle size

Changes in the average particle size and size distribution by aggregation and/or fusion of the liposomes are strongly dependent on (phospho)lipid composition, temperature and hydration medium (pH, presence of polyvalent ions, etc.).[6] Aggregation and fusion can be minimised by the addition of charge to liposomal bilayers. Moreover, polyethyleneglycol can be coupled to phosphatidylethanolamine (PEG-PE) to improve stability. Mixing of this lipid with an excess of PC not only results in formation of socalled long-circulating liposomes, but also in physically stable liposomes upon storage.[143]

Below 40% hydrolysis of liposomal phospholipids, Zuidam et al. hardly found any influence of the chemical degradation on the particle size of liposomes (non-sized or 0.2 μm), unless (1) the liposomes passed through a gel-to-liquid phase transition during a heating or cooling curve and (2) chemical hydrolysis exceeded a critical hydrolysis percentage (or critical % LPC/PC).[144] Then, the organisation of the lipid assembly can change from a bilayer into a micellar system (resulting into a pronounced drop in average size). The critical degree of hydrolysis (% of PC at $t = 0$) depended on the chain length and headgroup of the liposomal phospholipids. It was about 6% for DMPC-liposomes, 9% for DPPC-liposomes and circa 18% for DSPC-liposomes. This critical percentage did not depend on the phospholipid concentration (4–200 mM), pH (4.0, 7.4 or 10.5) nor size of the liposomes (non-sized or 0.2 μm) within the experimental range.

Loss of entrapped drug

Water soluble drugs can leak out of the liposomes. In general, leakage of drugs out of liposomes with bilayers in the gel state is slower than out of liposomes with bilayers in the fluid state. Storage around the T_m enhances leakage of encapsulated agents.[145,146] Therefore, the use of only DMPC for the preparation of liposomes is not recommended (T_m (23.5°C) around ambient temperature). Certain bilayer interacting compounds may be used to enhance the physical stability upon storage, such as cholesterol[147] and α-tocopherol.[148]

Oxidation of unsaturated phospholipids increases the permeability of bilayers composed of those phospholipids.[149] Because LPC is formed, an increase in permeability of liposomal bilayers was also expected upon chemical hydrolysis of phospholipids. It was found in many publications that exogenous LPC increased the permeability of the liposomal bilayers.[150] However, Grit and Crommelin showed that this is not necessarily true when LPC was present in the liposomal bilayers as a result of chemical hydrolysis. Lower permeability values were found than expected for liposomes with up to 15% hydrolysis of the selected phospholipids: partially hydrogenated EPC and EPG.[151] This was explained by the counteracting effect of the other hydrolysis product present in the bilayer, the free fatty acids. The liposomes used in this experiment did not contain a phase transition in the experimental temperature range. However, when liposomes do have such a phase transition and the above described conversion of bilayer assembly from a bilayer into a micellar system takes place, a substantial leakage of the encapsulated drug may occur.

As stated above, storage around the phase-transition temperature is, in general, not recommended. However, liposomes which have a gel-to-liquid phase transition are also used in hyperthermia treatment protocols. In vivo drug leakage is triggered by local heating of the tissue to a temperature in the phase transition temperature range of the liposomes. Thus, it is important that the phase transition remains stable upon storage. However, it was found by our group that the presence of even low percentages of the hydrolysis degradation products palmitic acid and LPC had a dramatic, pH-dependent effect on the melting characteristics of DPPC.[152] The pre-transition melting peak disappeared at low degrees of hydrolysis. Upon further hydrolysis, the main transition broadened and decreased in enthalpy and a second peak came up at its left hand side (at pH 10.5) or at its right hand side (at pH 4.0). This pH-dependent difference in position of the second peak is probably related to a difference in ionisation state of palmitic acid (at pH 4.0 non-charged and at pH 10.5 negatively charged).

IV.2. Freeze-drying of liposomes

A strategy to circumvent problems related to hydrolysis is storage of the liposome dispersion in the dry state. Freeze-drying of a liposome dispersion may result in a cake with a large surface area, which can easily be reconstituted at the bed-side of the patient. The process consists of a freezing step and subsequent sublimation of the water. However, liposomes can be damaged both by the freezing and the drying process. Numerous studies have provided insight into the mechanisms involved, but still critical questions remain unanswered. It should be made clear that most of the work done so far has been on a lab scale. Relatively little information is available on large scale freeze-drying of liposomes. The possibilities and difficulties of freeze-drying of liposomes are evaluated below. In addition, the stability of the freeze-dried product will be discussed.

IV.2.1. Freeze-drying of liposomes without lyoprotection

Freeze-drying of a liposome dispersion from an aqueous medium without additives for lyoprotection (= protection against damage by freeze-drying) results in the formation of large vesicular systems after rehydration, independent of the vesicle size before freeze-drying (e.g., Refs. 14, 153). In addition, encapsulated water soluble compounds will leak out of the vesicles. However, if MLVs are the liposome type aimed for, and if the drug associated with the liposomes strongly interacts with the bilayers, freeze-drying of just the lipids and the drug may be an attractive option.

An example of large scale preparation of MLVs loaded with a drug derivative with increased lipophilicity by linkage to a phospholipid was described by Van Hoogevest and Fankhauser.[54] Muramyltripeptide-posphatidylethanolamine (MTP-PE) was freeze-dried from a sterile-filtered solution with the mono-unsaturated phospholipids dioleylphosphatidylserine (DOPS) and 1-palmitoyl-2-oleylphosphatidylcholine (POPC) in *t*-butanol. The resulting cake was porous, and the T_m of the lipids in aqueous dispersion after reconstitution of the dry cake amounted to

−20°C. Both characteristics permit vesicle formation after adding sterilised water at room temperature. The number average vesicle size after hydration of the cake ranged between circa 2.4–3.1 μm, and around 5% of the vesicles was larger than 11 μm. No vesicles larger than 50 μm were observed. Another more recent example of freeze-dried liposomes without cryoprotectant was described by Sachse et al.[154] MLVs containing iopromide were prepared for liver tumour detection. Up to 42% of the material was liposome associated after rehydration. This percentage depended on lipid composition and rehydration volume. In this case the presence of free iopromide in the final preparation was considered non-critical, because of its fast renal elimination and excellent tolerance.

IV.2.2. Maintaining the liposome structure during freeze-drying

Cryoprotection (= protection against damage by freezing)
During freezing of a liposome dispersion ice crystals are formed and the solute concentration in the remaining fluid increases. Several processes may damage the liposomes during freezing.[155–161] Crystal formation may rupture the bilayer, or bilayer damage may be caused by the high concentration of the remaining solutes. The latter may result in disruptive osmotic forces on the bilayer, or damage (e.g., phase separation) caused by the high ionic strength or pH changes. Freezing was also shown to result in a partial release of 'bound water' by the bilayer. The reported amounts of water that remained uncrystallised in a frozen liposome dispersion range between 0.35 and 0.1 g/g phospholipid[158,162,163] and appeared to depend on the thermal history of the sample.[162] The damage to liposomes has been shown to depend on process parameters such as freezing rate and vesicle size.[157] The damage can be minimised by the addition of cryoprotectants such as carbohydrates which form an amorphous glass upon freezing. In this way a spacer is provided between the vesicles which prevents fusion, and damaging crystallisation processes of water or salts may be inhibited.[155] Other mechanisms may be involved as well, depending on the cryoprotectant involved (e.g., Ref. 164), but this is beyond the scope of this review. Under optimal cryoprotective conditions, no effect of the lipid composition on the freezing damage to liposomes was found.[165,166]

Storage of frozen liposomes may be considered as a way to inhibit degradation during storage, but is not preferred because of the high costs of storage and transport and its 'user unfriendly' character. In addition, recrystallisation processes in the frozen state may affect the liposome integrity.[157]

Lyoprotection (= protection against damage by dehydration)
In an aqueous liposome dispersion, water interacts with the polar headgroups of the phospholipids by hydrogen bond formation. This, in combination with the hydrophobicity of the acyl chains, is an essential contribution for organisation of these lipids in a bilayer. In addition, water may be considered as a spacer between

the vesicles. Therefore, it is not surprising that removal of water affects the liposome integrity, resulting in changes in vesicle size and leakage of the encapsulated compound. Since both parameters are important for the therapeutic activity of a liposome formulation, many studies have been performed to reveal the mechanism of damage by drying and to provide adequate solutions.

Damaging processes that may occur during drying from the frozen or unfrozen state until subsequent rehydration include: (1) fusion of the vesicles, (2) loss of bilayer structure, (3) passage of the phase transition temperature range with concomitant (temporary) phase separation and increased bilayer permeability, and (4) bilayer damage due to osmotic forces during rehydration. Some cryoprotectants also help to prevent damage during the drying and rehydration process, although freezing and dehydration were shown to be different stress vectors.[163] Currently, the most successful lyoprotectants are the disaccharides, which also provide protection for the liposomes against freezing and thawing stress. Lyoprotectants may exert their protective action by one or more of the following mechanisms. (1) Glass formation: this provides an amorphous matrix between the vesicles which prevents fusion or bilayer damage by crystal formation. (2) Interaction with the phospholipid headgroups: the hydroxyl groups of the sugars form hydrogen bonds with the phosphate group of the phospholipids in the dry state, and thus replace water (water substitution theory). This latter theory was investigated extensively by Crowe and co-workers, see e.g., Ref. 167. During drying of bilayers, the water molecules between the phospholipid headgroups are (partly) removed, which decreases the space between the phospholipid headgroups. As a result, the interaction between the acyl chains is stronger and the energy needed to heat these chains from the ordered gel phase to the disordered liquid crystalline phase increases. This is reflected by a dramatically increased T_m in the dry state as compared to the hydrated state. This phenomenon can be inhibited by the presence of cryoprotectants in sufficiently high carbohydrate/phospholipid ratios during drying (e.g., Refs. 168, 169). As pointed out by Crowe and co-workers, suppression of T_m may be crucial to avoid the liposome dispersion from passing the T_m during rehydration, since this may be accompanied by leakage. This is illustrated in Figure 4. Based on this theory, it can be predicted that glass formation itself is sufficient for protection during freeze-drying and rehydration when no bilayer transition is expected. The occurrence of such a transition is dependent on the T_m of the bilayer in the dry and hydrated state and the rehydration temperature. For DPPC vesicles, no transition is expected (see Figure 4), neither for cholesterol containing liposomes, which have no clear bilayer transition. In some cases, this prediction holds, as demonstrated recently.[170] Dextran, did not suppress the rise in T_m after freeze-drying DPPC liposomes. However, at high carbohydrate/phospholipid ratios, it prevented leakage of the non-bilayer interacting compound 6-carboxyfluorescein (CF) during a freeze-drying and rehydration cycle, which was ascribed to glass formation upon freezing. The presence of trehalose, but not of dextran, prevented leakage of CF from EPC vesicles after freeze-drying and rehydration, and resulted in a suppression of the phase transition in the dry state.

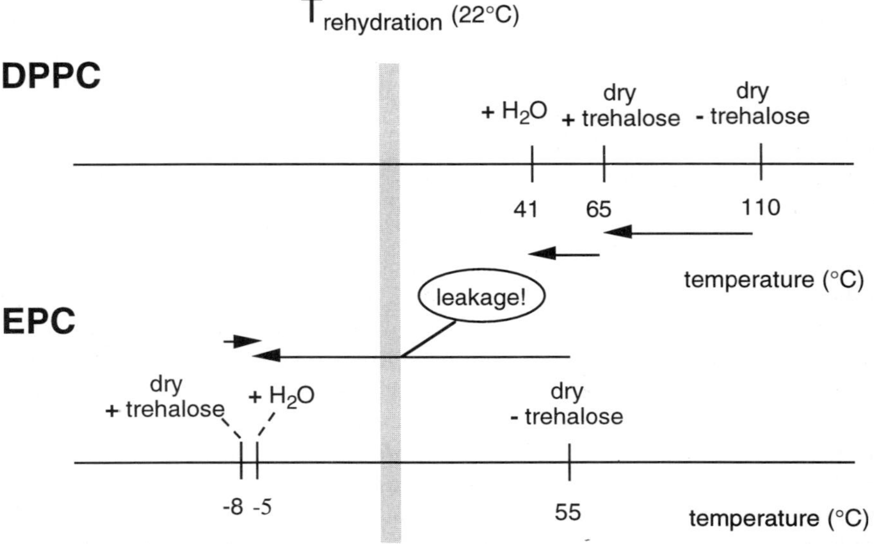

Fig. 4. Main transition temperatures (T_m) of the bilayer of EPC and DPPC liposomes during a drying and rehydration cycle. The arrows indicate the shift in T_m during the rehydration process. Leakage during rehydration at 22°C is predicted for EPC liposomes with insufficient cryoprotection because of the occurrence of a phase transition (data Refs. 169, 170).

However, not only glass formation and reduction of the phase transition temperature in the dry state are critical parameters. For successful stabilisation of liposomes containing non-bilayer interacting, hydrophilic compounds such as CF, during the freeze-drying process more parameters play a role.

One such parameter is the vesicle size. For instance, high retention of CF is only achieved with dispersions containing vesicles with an average size of around 0.1 μm; vesicles around 0.2 μm show significantly more leakage of CF after a freeze-drying and rehydration cycle.[166,171] The occurrence of this size effect may depend on the cryoprotectant/phospholipid ratio as suggested by Crowe.[171] We demonstrated that the difference in behavior between 0.2 μm vesicles and 0.1 μm vesicles in this respect was independent of the phospholipid composition.[172] Until now, only fusion and/or aggregation, but not leakage of CF can be prevented when freeze-drying MLVs, despite the presence of lyoprotectant in- and outside the vesicles. Several explanations for this size dependency were discussed in the literature: vesicles below 0.2 μm may be less vulnerable to mechanical stress by ice crystals or osmotic pressure.[157] Secondly, the amount of cryoprotectant between the adjacent bilayers of MLVs might be insufficient because of the small aqueous volume between these bilayers, or an unequal distribution of the cryoprotectant in the vesicles because of solute exclusion effects.[64,167]

However, even when sufficient carbohydrate phospholipid ratios are applied and liposomes of optimal size are used, not all liposome types will show a high retention. This clearly depends on their lipid composition. A CF retention of about 70% or more after freeze-drying and rehydration has been reported for several lipid compositions: DPPC (88%),[170] DOPC (82%),[166] POPC(69%),[166] EPC (73–80%)[166,170] EPC:CHOL (several ratio's: 75%),[166] POPC:PS = 9:1 (100%),[173] EPC:EPS = 10:1 (90%),[171] and DPPC:DPPG:CHOL = 10:1:4 or 10:1:8 (75% and 79% respectively).[165]

Harrigan et al.[166] found that addition of EPG to EPC liposomes (average size 0.1 μm) decreased the retention of CF after freeze-drying and rehydration. More leakage of CF after addition of cholesterol was found by Crommelin and Van Bommel for MLVs[174] and by Tanaka for SUV's.[175] Later we found that the lipid composition showed unexpected effects on the CF retention after a freeze-drying/rehydration cycle under optimal cryoprotective conditions: abundance of sucrose, quick freezing in boiling liquid nitrogen and small vesicle size (average size 0.1 μm). No bilayer transitions were expected during rehydration in any of the dispersions. Addition of CHOL to DPPC:DPPG = 10:1 liposomes increased the retention, and EPC:EPG = 10:1 liposomes containing CHOL showed more leakage than DPPC:DPPG = 10:1 liposomes with the same CHOL content. This lipid composition dependent damage occurs during the dehydration or rehydration step in the freeze-drying process, since no differences were found between these dispersions of different lipid compositions after freezing/thawing cycles.[165] Addition of phosphatidic acid to DPPC liposomes also decreased the CF retention after freeze-drying.[176] Another less successful lipid composition was hydrogenated soybean phosphatidylcholine (HSPC) with dicetylphosphate (DCP) with a molar ratio of 10:1 (sucrose as a lyoprotectant).[177] Only 25% and 5% CF was retained in vesicles of circa 0.13 μm and 0.28 μm, respectively. In a recent study we found that the freezing protocol of the freeze-drying process can strongly affect the retention of an encapsulated marker after rehydration, depending on the lipid composition.[178] Especially for the rigid DPPC liposomes (size: 0.1 μm), slow freezing (0.5°C/min) resulted in a much higher retention (90%) after freeze-drying and rehydration as compared to quick freezing in boiling nitrogen (50%).

In conclusion, when using the proper conditions, at the present freeze-drying of a wide range of liposomes loaded with hydrophobic, bilayer interacting drugs seems to be feasible. However for successful freeze-drying of liposomes loaded with hydrophilic, non-bilayer interacting drugs, constraints were defined in terms of phospholipid composition and vesicle size, even in the presence of lyoprotectants such as disaccharides. The current insights in the mechanism of damage by the freeze-drying process can not fully explain all findings and still do not allow to rationally design successfull freeze-drying protocols. Considering the limited stability of liposomes stored as aqueous dispersions, more studies are needed to make freeze-drying a flexible technique to be routinely applied to many different types of liposomes.

IV.3. Stability in the solid state

A lot of attention has been paid to the stability of liposomes during the freeze-drying process, but, so far, in only a few studies data were given on the solid state stability of freeze-dried liposomes. Relevant information may be obtained from studies on the stability of freeze-dried proteins, e.g., Refs. 179–181, or from literature on amorphous carbohydrate water systems in general, e.g., Refs. 182–184. Some information from these sources which may be relevant for liposome stability is briefly reviewed below.

IV.3.1. Lipids without lyoprotectants

In those cases where no lyoprotectant is used for preserving the bilayer organisation of the phospholipids in the dry state (see above), the product stability concerns the stability of the mixture of the lipids and the (lipophilic) drug. The most frequently observed chemical degradation of phospholipids in the solid sate is oxidation, especially when the acyl chains are unsaturated. In one study,[54] phospholipids with mono-unsaturated acyl chains (POPC and DOPS) were selected as the bilayer components of liposomes with incorporated MTP-PE as the active agent. The presence of only one double bond in the acyl chain resulted in a low bilayer transition temperature of −9.3°C, enabling rehydration at room temperature, and a minimised susceptibility for oxidation. A good storage stability was reported: no changes in vesicle size, phospholipid or MTP-PE content after rehydration could be observed after storage of the freeze-dried cake for 2 years at 4°C, or 1 year at 23°C.

IV.3.2. Liposomes with lyoprotectant

The amorphous state

Carbohydrates, at the present the most effective and most often used cryoprotectants for liposomes, form a major part of the cake, since at least circa 2 g/g PL is needed to prevent leakage of the encapsulated compound. The exact limits depend on the carbohydrate involved.[173] The physical and chemical stability of the total liposome product will largely depend on the stability of this excipient. Important in this respect is the physical state of the lyoprotectant after freeze-drying. During the freezing step, the carbohydrates, other solutes and the liposomes are concentrated between the ice crystals. The viscosity of this concentrate increases with concentration and lowering of the temperature, and leads to the formation of an amorphous glassy phase. After sublimation of the ice crystals, the glass is dried to a certain residual water content, depending on a variety of factors such as sample volume, vial properties, and the time, temperature and pressure program of the drying process (e.g., Refs. 185–187). Confirmation of the amorphous state of, e.g., sucrose and lactose after freeze-drying from an aqueous solution was obtained by the X-ray diffraction technique.[180,183,188] An exception is mannitol, which is often allowed to partly crystallise during the freezing step, in order to accelerate the drying process and improve the stability of the freeze-dried cake. However, crystal formation may damage the liposomes, and it is not clear yet whether mannitol is an appropriate lyoprotectant for these carriers.

The role of the glass transition temperature

A glass can be considered as a metastable state which is kinetically arrested due to its high viscosity. The physical and chemical stability of a compound (embedded) in a glassy phase, is mainly determined by the relevant glass transition temperature (T_g).[179,182,189] Below T_g, the viscosity can be circa 10^{11} Pa s and higher, corresponding to average diffusion rates as low as 50 nm per century. Therefore, it is generally assumed that little chemical or physical degradation will take place in this phase. However, discussion on this subject is still ongoing and recent findings show that chemical degradation can take place during storage below T_g.

When heated through its glass transition, the material shows two important changes: the heat capacity increases, and the viscosity decreases dramatically. The T_g can be defined as the temperature at which half the increase in heat capacity has occurred. Another parameter to consider when studying the long term stability, is the onset glass transition temperature ($T_{g,\text{onset}}$), at which the transition starts. In the rubbery phase above T_g, physical and chemical degradation processes are accelerated because of the increased molecular mobility.

Factors that affect the T_g of the freeze-dried product are: (1) choice of cryoprotectant; the T_g of a specific carbohydrate is determined both by its molecular weight and its molecular structure[190] and (2) the residual water content. Water is known to act as a plasticizer for carbohydrates and lowers the T_g.[191] In addition, (3) the $T_{g'}$ (subzero glass transition temperature in the presence of ice) after freezing of a multicomponent product depends on its excipient composition. A non-linear relationship was found between the solute fraction of sucrose and citrate, and the collapse temperature during the sublimation step of the freeze-drying process.[181]

In conclusion, it is very important to create conditions during the freeze-drying process and subsequent storage that ensure the cake to maintain its glassy state.

Physical degradation above T_g

Physical changes that may occur in the vicinity of or above T_g are stickiness, shrinkage and collapse of the cake, crystallisation of carbohydrates and buffer salts, fusion of liposomes, or annealing processes. For temperatures in the range between T_g and T_g + 100°C diffusion controlled processes can be described by the William Ferry-Landle kinetics as shown for example for crystallisation of sugars:[183,184]

$$\log (k) = [C_1 \times (T - T_g)]/[C_2 + (T - T_g)]$$

with k = rate constant, T = temperature (K), T_g = glass transition temperature (K), C_1 and C_2 = universal constants. Below T_g and above T_g + 100, the Arrhenius kinetics are assumed to apply.[189]

Apart from crystallisation of the lyoprotectant, other excipients such as buffer salts may crystallise as well.[188] Crystal formation could impose mechanical stress on the liposomes and result into leakage of the encapsulated compound, similar to the effect of recrystallisation of ice in frozen liposomes.[156]

Collapse or shrinkage of the cake is the result of a decreased viscosity and lowers the surface area of the cake. Consequently, the time needed for complete rehydration is raised tremendously. This effect, together with the changed cake appearance, also makes the product unacceptable for medical use. In addition, a slow rehydration process may affect the physical integrity of the liposomes upon reconstitution.

Special attention has been paid to physical changes of liposomes which occurred upon heating freeze-dried liposomes in the presence of lyoprotectants (see Refs. 169, 192–195). The heat treatment resulted in a further reduction of the membrane transition temperature as studied by DSC. The time scale and the temperatures used in these studies did not induce visible collapse or shrinkage of the cake. Based on this observation it is assumed that the interaction between the phospholipids and the carbohydrates is enhanced upon heating. This process is referred to as 'annealing', suggesting a relaxation of the interacting molecules to a thermodynamically more stable state. Recently, triggering of the annealing process in liposomes freeze-dried with disaccharides has been shown to occur at T_m, even when the sugar matrix is below T_g (see Refs. 193, 196). The physical changes that take place during annealing of the dry cake result in a decreased retention of water soluble marker encapsulated in the liposomes after rehydration. Thus, annealing can be considered as a physical degradation process, which makes it necessary to store the freeze-dried liposomes below T_g.

Chemical degradation above T_g

In the vicinity or above T_g, the decreased viscosity of the material enhances the chemical degradation. A study demonstrating chemical degradation during storage of freeze-dried liposomes above the T_g of the lyoprotectant was described by Friede et al.[197] Immunogenic liposomes of unsaturated phospholipids with a conjugated peptide freeze-dried in the presence of sorbitol were stored at −20°C and at room temperature. Anhydrous sorbitol has a T_g of 2°C[198] and storage at room temperature resulted in the development of a yellow colour of the liposomes, suggesting oxidation was taking place. However, when stored at −20°C, the vesicle size (0.11 μm) and immunogenicity were only slightly affected after a 2 months storage period. For the selection of a cryoprotectant, differences in chemical reactivity may be important. Two classes of carbohydrates are discerned in this respect: the reducing sugars such as lactose and maltose, which contain a hemiacetal carbon, and the non reducing sugars such as sucrose and trehalose, with no hemiacetal carbon. Several papers describe non-enzymatic browning of food and proteins, which is ascribed to the Maillard reaction between proteins and reducing sugars.[199] It should be noted that a non reducing sugar as sucrose can undergo degradation during storage, depending on the pH of the solution before freeze-drying.[200] It was suggested that under such conditions (low pH), fructose and glucose could be formed, the latter being able to bind to a protein via a Maillard reaction.[181]

Stability below T_g
A few studies have appeared in literature on the storage of freeze-dried, lyoprotected liposomes. Isele et al.[81] described the production and freeze-drying of liposomes loaded with the hydrophobic compound monomeric zinc phtalocyanine for photodynamic therapy of tumours. The liposomes (POPC:DOPS = 9:1) were freeze-dried in the presence of lactose to a residual water content of 1.6%, and no physical or chemical changes were reported after storage for 6 months at 8, 25 and 40°C. However, Vermuri and Rhodes[201] reported that the retention of the water soluble model drug orciprenaline sulphate in EPC:EPG:CHOL = 5:1:4 liposomes after rehydration decreased and the vesicle size increased when storing the freeze-dried cakes at 2–8°C. The lyoprotectant used in this study was lactose, but no data on the residual water content were given. Therefore, it can not be excluded that the dry lactose cakes were stored above or in the vicinity of T_g, resulting in physical degradation of the liposomes during storage. One should be aware that the residual water content of hygroscopic freeze-dried cakes may even increase by release of water from rubber stoppers,[202] which would result in a lowering of T_g. In two other studies, storage at low temperatures resulted in a sufficient stability. Freeze-dried immunogenic liposomes were stored at −20°C for 2 months (lyoprotectant: sorbitol),[197] or at 4°C for 12 months (lyoprotectant: trehalose).[203] In both cases the vesicles retained the immunogenicity aimed for, but showed a slight increase in average size after rehydration. In a recent study we observed both physical and chemical degradation of freeze-dried liposomes upon storage at temperatures well below T_g.[195,196] These findings were ascribed to the heterogeneous nature of the liposome-sugar cakes. Indications were found that the solid phase inside liposomes which contain large amounts of doxorubicin, has a different molecular mobility than the extraliposomal sugar matrix. The stability of these samples was improved by minimising the residual water content. The most stable formulations in this study were freeze-dried doxorubicin liposomes containing lactose, maltose or trehalose as a lyoprotectant with a residual water content below 0.7%. In these samples, no degradation was observed upon storage at temperatures up to 30°C for at least 6 months.

In conclusion, in some studies acceptable stability of lyoprotected, freeze-dried liposomes has been demonstrated for cakes stored below T_g. However, recent findings make clear that the T_g of the sugar matrix is not necessarily an appropiate indicator for the maximal temperature at which long term stability is preserved, especially when high concentrations of encapsulated compounds destabilise the solid phase inside the liposomes.

V. Concluding remarks and prospects

The title of this chapter is 'Strategies for large scale production and optimised stability of pharmaceutical liposomes developed for parenteral use'. This chapter is written to bring the reader up to date with the 'state of the art'. Large scale production of liposomes is possible. Liposomal products for parenteral administration containing doxorubicin or amphotericin B are already on the market, or will be introduced on the market soon. They have been approved by the regulatory authorities. That means that convincing data could be produced on chemical, physical and biological characterization of the product and by-products, reproducibility of the production process, sterility and apyrogenicity. Moreover, they have acceptable shelf lives. This success is the result of a joint effort of a number of industrial and academic groups, where widely different problems such as the quality of raw materials, environmental acceptability of preparation protocols, efficient extrusion technologies and long term stability were addressed and often successfully solved.

There has been enormous progress compared with the situation in the early eighties, when there was a great naivity concerning the typical pharmaceutical problems encountered in the development of liposomal preparations. Frank Fildes[204] could quite rightly write in 1981 as a warning: '. . . The lack of suitable manufacturing techniques could conceivably become the dominant factor in the industrial feasibility of a particular liposome application . . .'

Although nowadays liposomes have made it to the clinic, this chapter provides ample evidence that many, but not all technological problems have been solved. For instance, in terms of stability a number of issues to be dealt with are left, such as the freeze-drying possibilities of liposomes loaded with low molecular weight, hydrophilic and non-bilayer interacting drugs.

References

1. Crommelin DJA, Schreier H. Liposomes. In: Kreuter J, eds. Colloidal drug delivery systems. New York: Marcel Dekker, 1994;73–189.
2. Barenholz Y, Crommelin DJA. Liposomes as pharmaceutical dosage forms. In: Swarbrick J, Boylan JC, eds. Liposomes as pharmaceutical dosage forms to microencapsulation. New York: Marcel Dekker, 1994;1–39.
3. Cevc G, ed. Phospholipids Handbook, New York: Marcel Dekker, 1993.
4. New RRC: In: New RRC, eds. Liposomes: a practical approach. Oxford: Oirl Press, 1990;1–32.
5. Hope MJ, Bally MB, Mayer LD, Janoff AS, Cullis PR. Generation of multilamellar and unilamellar phospholipid vesicles. Chem Phys Lipids 1986;40:89–107.
6. Lichtenberg D, Barenholz Y. Liposomes: Preparation, characterization and preservation. In: Glick D, eds. Methods of Biological Analysis, Vol. 33. New York: John Wiley, 1988;337–461.
7. Marsh D. General features of phospholipid phase transitions. Biochim Biophys Acta 1991;57:109–120.
8. Cevc G. How membrane chain melting properties are regulated by the polar surface of the lipid bilayer. 1987;26:6305–6310.
9. Chapman D, Williams RM, Ladbrooke BD. Physical studies of phospholipids. IV Thermotropic and lyotropic mesomorphism of some 1,2-diacyl-phosphatidylcholines (lecithins). Chem Phys Lipids 1967;1:445–475.
10. Strauss G, Schurtenberger P, Hauser H. The interaction of saccharides with lipid bilayer vesicles: stabilization during freeze-thawing and freeze-drying. Biochim Biophys Acta, 1986;858:169–180.
11. Kodama M, Kuwubara M, Seki S. Successive phase-transition phenomena and phase diagram of the phosphatidylcholine-water system as revealed by differential scanning calorimetry. Biochim Biophys Acta 1982;689:567–570.
12. Gabizon AA. Selective tumor localization and improved therapeutic index of anthracyclines encapsulated in long circulating liposomes. Cancer Res 1992;52:891–896.
13. Huang L. Covalently attached polymers and glycans to alter the biodistribution of liposomes. J Liposome Res 1992;2:289–454.
14. Ausborn M, Schreier H, Brezesinsky G, Fabian H, Meyer HW, Nuhn P. The protective effect

of free and membrane-bound cryoprotectants during freezing and freeze-drying of liposomes. J Control Rel 1994;30:105–116.
15. Goodrich RP, Handel TM, Baldeschwieler JD. Modification of lipid phase behavior with membrane-bound cryoprotectants. Biochim Biophys Acta 1988;938:143–154.
16. Muramatsu K, Maitani Y, Machida Y, Nagai T. Effect of soybean-derived sterol and its glucoside mixtures on the stability of dipalmitoylphosphatidylcholine and dipalmitoylphosphatidylcholine/cholesterol liposomes. Int J Pharm 1994;107:1–8.
17. Poste G. In: Gregoriades G et al., eds. Receptor-Mediated Targeting of Drugs, New York: Plenum Press, 1985;427–474.
18. Spanjer HH, Scherphof G: Targeting of lactosylceramide-containing liposomes to hepatocytes in vivo. Biochim Biophys Acta 1983;734:40–47.
19. Bakker-Woudenberg IAJM, Lokerse AF, Ten Kate MT, Storm G. Enhanced localization of liposomes with prolonged blood circulation time in infected lung tissue. Biochim Biophys Acta 1992;1138:318–326.
20. Peeters PAM, Storm G, Crommelin DJA. Immunoliposomes in vivo: state of the art. Adv Drug Del Rev 1987;1:249–266.
21. Boerman OC, Storm G, Oyen WJG, Van Bloois L, Van der Meer JWM, Claessen RAMJ, Crommelin DJA, Corstens FHM. Sterically stabilised liposome labeled with ^{111}In for imaging local in infection in rats. submitted, 1995.
22. Delemarre FGA, Kors N, Kraal G, Van Rooijen N. Repopulation of macrophages in popliteal lymph nodes of mice after liposome mediated depletion. J Leukocyte Biol 1990;47:251–257.
23. Juliano RL, Hsu MJ, Regen SL. Interaction of polymerized phospholipid vesicles with cells. Uptake processing and toxicity. Biochim Biophys Acta 1985;812:42–48.
24. Daemen T, Hofstede G, Ten Kate MT, Bakker-Woudenberg IAJM, Scherphof GL. Liposomal doxorubicin induced toxicity: depletion and impairment of phagocytic activity of liver macrophages. Int J Cancer (in press), 1995;16.
25. Hwang KJ. Liposome pharmacokinetics. In: Ostro MJ, eds. Liposomes: From Biophysics to Therapeutics. New York: Marcel Dekker, 1987;109–156.
26. Senior J. CRC Crit Rev Therap Drug Carrier systems 1987;3:123–193.
27. Woodle MC, Lasic DD. Sterically stabilized liposomes. Biochim Biophys Acta 1992;1113:171–199.
28. Woodle MC, Lolling LR, Sponsler E, Kossovsky N, Papahadjopolous D, Martin FJ. Sterically stabilized liposomes. Reduction in electrophoretic mobility but not electrostatic potential. Biophys J 1992;61:902–910.
29. Patel HM. Serum opsonins and liposomes: their interaction and opsonophagocytosis. Crit Rev Therap Drug Carrier Systems 1992;9:39–90.
30. Lasic DD, Martin FJ, Gabizon A, Huang SK, Papahadjopolous D. Sterically stabilized liposomes: a hypothesis on the molecular origin of the extended circulation times. Biochim Biophys Acta 1991;1070:187–192.
31. Gregoriadis G. Engineering liposomes for drug delivery: progress and problems. TIBTECH 1995;13.
32. Northfelt DW, Martin FJ, Kaplan LD, Russel J, Anderson M, Lang J, Volberding PA. Safety pharmacokinetics and tumor localization of Doxil™ (Stealth liposomal doxorubicin) in AIDS patients with Kaposi's sarcoma. Abstract presented at 8th Int Conf on AIDS, Amsterdam, July, 1992.
33. Product information provided by NexStar (former Vestar, San Dimas, CA, USA), 1995.
34. Forssen EA, Coulter DM, Proffitt RT. Selective in vivo localization of Daunorubicin small unilamellar vesicles in solid tumors. Cancer Res 1992;52:3255–3261.
35. Product information on Ambisome™ (Vestar Inc., San Dimas, CA, USA), 1992.
36. Enclosure to Ambisome™, Vestar Inc., 1993.
37. Cowens JW, Creaven PJ, Greco et al. WR. Initial clinical (Phase 1) trial of TLC D-99 (doxorubicin encapsulated liposomes). Cancer Res 1993;53:2796–2802.
38. Janoff AS, Perkins WR, Saleton SL, Swenson CE. Amphotericin B Lipid complex (ABLC™): A molecular rationale for the attenuation of Amphotericin B related toxicities. J Liposome Res 1993;3.
39. Alving CA, Steck EA, Chapman J W.L., Waits VB, Hendriks LD, Swartz JGM, Hanson WL. Therapy of Leishmaniasis: superior efficacies of liposome encapsulated drugs. Proc Natl Acad Sci USA 1987;75:2959–2963.
40. Storm G, Regts J, Beijnen JF, Roerdink FH. Processing of doxorubicin-containing liposomes by liver macrophages in vitro. J Liposome Res 1989;1:195–210.

41. Storm G, Nässander UK, Steerenberg PA, Roerdink FH, De Jong WH, Crommelin DJA. Studies on the mode of action of doxorubicin liposomes. In: Fidler IJ, Lopez Berenstein G, eds. Liposomes in the therapy of infectious diseases and cancer. New York: Alan R. Liss, 1989;105–116.
42. Kadir F, Zuidema J, Crommelin DJA. Liposomes as drug delivery systems for intramuscular and subcutaneous injections. In: Rolland A, eds. Pharmaceutical particulate carriers in medical applications. New York: Marcel Dekker, 1993.
43. Takakura Y, Hashida M, Sezaki H. Lymphatic transport after parenteral drug administration. In: Charman WN, Stella VS, eds. Lymphatic transport of drugs. Boca Raton, FL: CRC press, 1992;255–277.
44. Yaguchi T, Yamauchi M, Takagi H, Ohishi N, Yagi K. Effect of sulfatide-inserted liposomes containing entrapped adriamycin on metastasized cells in lymph nodes. J Clin Biochem Nutr 1990;9:79–85.
45. Oussoren C, Zuidema J, Crommelin DJA, Storm G. Factors influencing lymphatic uptake of liposomes after subcutaneous administration. Parmacy World Sci 1994;16:F9.
46. Akamo Y, et al. Delivery of lymph node-targeted adriamycin by gastric submucosal liposomal injection in rabbits. Jpn J Cancer Res 1993;84:208–213.
47. Merlin J-L. Encapsulation of doxorubicin in thermosensitive small unilamellar vesicle liposomes. Eur J Cancer 1991;27:1026–1030.
48. Merlin J-L, Marchal S, Ramacci C, Notter D, Vigneron C. Antiproliferative activity of thermosensitive liposome-encapsulated doxorubicin combined with 43°C hyperthermia in sensitive and multidrug-resistent MCF-7 cells. Eur J Cancer 1993;29A:2264–2268.
49. Storm G, Oussoren C, Peeters PAM, Barenholz Y: Tolerability of liposomes in vivo. In: Gregoriadis G, eds. Liposome Technology, Vol. 3. Boca Raton, FL: CRC Press, 1993.
50. Felgner PL. Particulate systems and polymers for in vitro and in vivo delivery of polynucleotides. Adv Drug Del Rev 1990;5:163–187.
51. Felgner PL. Forum on "Cationic liposomes". J Liposome Res 1993;3:3–306.
52. Lang JK, Vigo-Pelfrey C, Martin F. Liposomes composed of partially hydrogenated egg phosphatidylcholine. Chem Phys Lipids 1990;53:91–101.
53. Eible H. Phospholipid synthesis. In: Knight CG, ed. Liposomes: from physical structure to therapeutic applications. Amsterdam: Elsevier/North Holland, 1981;19–50.
54. Van Hoogevest P, Fankhauser P. An industrial liposome dosage form for muramyl-tripeptide-phosphatidylethanolamine (MTP-PE). In: Liposomes in the Therapy of Infectious Diseases and Cancer. Alan R. Liss, Inc., 1989;453–466.
55. Nässander UK, Steerenberg PA, Peeters PAM, Crommelin DJA. Liposomes. In: Chasin M, Langer R, eds. Biodegradable polymers as drug delivery systems. New York: Marcel Dekker, 1990;261–338.
56. Martin FJ. Pharmaceutical manufacturing of liposomes. In: Tyle P, ed. Specialized drug delivery systems: manufacturing and production technology. New York: Marcel Dekker, 1990;267–316.
57. Kriftner RW. Liposome production. The ethanol injection technique and the development of the first approved liposome dermatic. In: Braun-Falco O, Korting HC, Maibach HI, eds. Liposome Dermatics. Berlin: Springer Verlag, 1992;91–100.
58. Weder HG, Zumbühl O. The preperation of various sized homogeneous liposomes for laboratory, clinical and industrial use by controlled detergent dialysis. In: Gregoriadis G, eds. Liposome Technology, Vol. 1. Boca Raton, FL: CRC Press, 1984;79–107.
59. Schwenderer RA. The preparation of large volumes of homogeneous, sterile liposomes containing various lipophilic cytostatic drugs by the use of a capillary dialyzer. Cancer Drug Deliv 1986;3:123–129.
60. Amselem S, Gabizon A, Barenholz Y. Optimization and up-scaling of doxorubicin containing liposomes for clinical use. Pharm Res 1990;79:1045–1052.
61. Lasic DD. The mechanism of vesicle formation. Biochem J 1988;256:1–11.
62. Talsma H, Gooris G, Van Steenbergen MJ, Salomons MA, Bouwstra J, Crommelin DJA. The influence of the molar ratio of cholesteryl hemisuccinate/dipalmitoylphosphatidylcholine on 'liposome' formation after lipid film hydration. Chem Phys Lipids 1992;62:105–112.
63. Racey TJ, Singer MA, Finegold L, Rochon P. The influence of fatty acyl chain length and head group on the size of multilamellar vesicles. Chem Phys Lipids 1989;49:271–288.
64. Gruner SM, Lenk SM, Janoff AS, Ostro MJ. Novel multilayered lipid vesicles: comparison of physical characteristics of multilamellar liposomes and stable plurilamellar vesicles. Biochemistry 1985;24:2833–2842.

65. Morris RM, D.Q.M. C, Taylor KMG. A multivariate analysis of variance of the factors affecting the size and size distribution of liposomes. Pharm Res 1994;11:S226.
66. Kirby CJ, Gregoriadis G. A simple procedure for preparing liposomes capable of high encapsulation efficiency under mild conditions. In: Gregoriadis G, ed. Liposome Technology, Vol. 1. Boca Raton, FL: CRC Press, 1984;19–27.
67. Ohsawa T, Miura H, Harada K. A novel method for preparing liposomes with high capacity to encapsulate proteineous drugs: freeze-drying method. Chem Pharm Bull 1984;32:2442–2445.
68. Seltzer SE, Gregoriadis G, Dick R. Evaluation of the dehydration-rehydration method for production of contrast carrying liposomes. Invest Radiol 1988;23:131–138.
69. Hamilton RL, Guo LSS. French pressure cell liposomes: preparation, properties, and potential. In: Gregoriades G, ed. Liposome Technology, Vol. 1. Boca Raton, FL: CRC Press, 1984;37–49.
70. Mason G. Advanced techniques for preparation and characterization of small unilamellar vesicles. Food Microstructure 1989;8:11–14.
71. Mayhew E, Nikolopoulos G, A. S. Am Biotechnol Lab 1985;36–41.
72. Talsma H, Özer AY, Van Bloois L, Crommelin DJA. The size reduction of liposomes with a high pressure homogenizer (microfluidizer™). Characterization of prepared dispersions and comparison with conventional methods. Drug Devel Industr Pharm 1989;15:197–207.
73. Vemuri S, Yu C, Wangsatorntanakun V, Roosdorp N. Large-scale production of liposomes by a Microfluidizer. Drug Devel Industr Pharm 1990;16:2243–2256.
74. Braun-Falco O, Korting HC, ed. Liposome Dermatics. Berlin: Springer Verlag, 1992.
75. Brandle M, Bachman D, Drechster M, Bauer KH. Liposome preparation by a new high pressure homogenizer Gaulin Micron LAB 40. Drug Develop Industr Pharm 1990;16:2167–2191.
76. Purmann T, Mentrup E, Kreuter J. Preparation of SUV-liposomes by high-pressure homogenization. Eur J Pharm Biopharm 1993;39:45–52.
77. Brandle M. Hochdruckhomogenisierte Liposome. Arch Pharmazie 1993;326:753.
78. Zimmermann JA, Gelotte KM. Evaluation of processing equipment for the preparation of submicron emulsions on a small scale. Pharm Res 1994;11:S-137.
79. Brandle MM, Bachmann D, Drechsler M, Bauer KH. Liposome preparation using high-pressure homogenizers. In: Gregoriades G, ed. Liposome Technology, Vol. 1. Boca Raton, FL: CRC Press, 1993;49–65.
80. Talsma H, Van Steenbergen MJ, Borchert JHC, Crommelin DJA. A novel technique for the one-step preparation of liposomes and nonionic surfactant vesicles without the use of organic solvents. Liposome formation in a continuous gas stream: the 'bubble' method. J Pharm Sci 1993;83:276–280.
81. Isele U, Van Hoogevest P, Hilfiker R, Capraro H, Schieweck K, Leuenberger H. Large-scale production of liposomes containing monomeric zinc phthalocyanineby controlled dilution of organic solvents. J Pharm Sci 1994;83:1608–1616.
82. Allen TM. Removal of detergent and solvent traces from liposomes. In: Gregoriadis G, ed. Liposome Technology, Vol. 1. Boca Raton, FL: CRC Press, 1984;109–122.
83. Weder HG. Liposome production: The sizing up technology starting from mixed micelles and the scaling up procedure for the glucocorticoid bethamethson dipropionate and betamethasone. In: Braun-Falco O, Korting HC, Maibach HI, eds. Liposome Dermatics. Berlin: Springer Verlag, 1992;101–109.
84. Peschka R, Production and development of liposomal preparations with hydrophilic drugs for topical application. (in german), Eberhardt-Karls-Universität, Tübingen, Germany,1994.
85. Jiskoot W, Teerlink T, Beuvery EC, Crommelin DJA. Preparation of liposomes via detergent removal from mixed micelles by dilution. The effect of bilayer composition and process parameters on liposome characteristics. Pharm Weekbl Sci Ed 1986;8:259–265.
86. Teelmann K, Sclaeppi B, Schuepbach M, Kistler A. Preclinically safety evaluation of intravenously administered mixed micelles. Arzneim Forsch 1984;34:1517–1523.
87. Sorgi FL, Huang L. Large scale production of DC-Chol cationic liposomes by microfluidization. Int J Pharm 1996;144:131–139.
88. Olson F, Hunt CA, Szoka F, Vail WJ, Papahadjopolous D. Preparation of liposomes of defined size distribution by extrusion through polycarbonate membranes. Biochim Biophys Acta 1979;557:9–23.
89. Szoka F, Olson F, Heath T, Vail W, Mayhew E, Papahadjopolous D. Preparation of unilamellar liposomes of intermediate size (0.1–0.2 μm) by a combination of reversed phase evaporation and extrusion through polycarbonate membranes. Biochim Biophys Acta 1990;601:559–571.
90. Schneider T, Sachse A, Rössling G, Brandle M. Large scale production of liposomes of defined

size by a new continuous high pressure extrusion device. Drug Develop Industr Pharm 1994;20:2787–2807.

91. Amselem S, Gabizon A, Barenholz Y. Evaluation of a new extrusion device for the production of stable oligolamellar liposomes in a liter scale. J Liposome Res 1989;1:287–301.
92. Amselem S, Gabizon A, Barenholz Y. A large-scale method for the preparation of sterile and nonpyrogenic liposomal formulations of defined size distributions for clinical use. In: Gregoriadis G, ed. Liposome Technology, Vol. 1. Boca Raton, FL: CRC Press, 1993;501–525.
93. Hauser H, Gaines N. Spontaneous vesiculation of phospholipids: a simple and quick method of forming unilamellar vesicles. Proc Natl Acad Sci USA 1982;79:1683–1687.
94. Hauser H. Mechanism of spontaneous vesiculation. Proc Natl Acad Sci USA 1989;86:5351–5355.
95. Li W, Haines TH. Uniform preparations of large unilamellar vesicles containing anionic lipids. Biochemistry 1986;25:7477–7483.
96. Rahman A. Encapsulation of a physiologically active compound—using a liposome—forming lipid and lipophilic chelating agent. EP 198765-A, 1986;
97. Yagi K, Kojima N, Takano M. Adriamycin-entrapping liposome preparation. Patent application E0226370B1, 1991.
98. Kurono M, Noda H, Ogasawara T, Yamakawa H, Lida T. Antineoplastic agent entrapping liposomes. Patent application E0278465B1, 1992.
99. Kikuchi H, Yachi K, Hirota S. Liposome products. Patent application E0467275A1, 1992.
100. Product description of Coatsome EL series, NOF Corporation, Japan, 1995.
101. Kikuchi H, Yachi K, Morita H, Hirota S. Method of producing liposomal products from freeze or spray-dried preparations of liposomes. US Patent 5,376,380, 1994.
102. Chapman CJ, Erdahl WE, Taylor RW, Pfeiffer DR. Factors affecting solute entrapment in phospholipid vesicles by the freeze-thaw extrusion method: a possible general method for improving the efficiency of entrapment. Chem Phys Lipids 1990;55:73–83.
103. Chapman CJ, Erdahl WE, Taylor RW, Pfeiffer DR. Effects of solute concentration on the entrapment of solutes in phospholipid vesicles prepared by freeze-thew extrusion. Chem Phys Lipids 1991;60:201–208.
104. Production of liposome using buffer solution—comprises freezing multilamellar liposomes with phosphate buffer to entrap e.g., adriamycin. Patent application J03081216-A, 1989.
105. Gregoriadis G, Florence AT. Efficient entrapment of solutes in microfluidized small dehydration-rehydration liposomes. In: Gregoriadis G, ed. Liposome Technology, Vol. 1. Boca Raton, FL: CRC Press, 1993;37–48.
106. Deamer DW, Prince RC, Crofts AR. Response of fluorescent amines to pH gradients across liposome membranes. Biochim Biophys Acta 1972;274:323–335.
107. Madden TD, et al. The accumulation of drugs within large unilamellar vesicles exhibiting a proton gradient: a survey. Chem Phys Lipids 1990;53:37–46.
108. Bally MB, Ginsberg RS, Mitilenes GN, Cullis PR, Mayer LD. Liposomal formulations with a high antineoplastic agent/lipid ratio. Patent application E0290296A2, 1988.
109. Barenholz Y. Loading and controlled release of amphiphatic molecules to and from liposomes. Patent application E0361894A2, 1989.
110. Haran G, Cohen R, Bar LK, Barenholz Y. Transmembrane ammonium sulfate gradients in liposomes produce efficient and stable entrapment of amphipathic weak bases. Biochim Biophys Acta 1993;1151:201–215.
111. Bolotin EM, Cohen R, Bar LK, Emmanuel N, Ninio S, Lasic DD, Barenholz Y. Ammonium sulfate gradients for efficient and stable remote loading of amphipatic weak bases into liposomes and ligando liposomes. J Liposome Res 1994;4:455–479.
112. Mayer LD, Bally MB, Hope MJ, Cullis PR. Uptake of antineoplastic agents into large unilamellar vesicles in response to a membrane potential. Biochim Biophys Acta 1985;816:294–302.
113. Beijnen JH, Van der Houwen OAGJ, Underberg WJM. Aspects of the degradation kinetics of doxorubicin in aqueous solution. Int J Pharm 1986;32:123–131.
114. Grit M, Zuidam NJ, Underberg WJM, Crommelin DJA. Hydrolysis of partially saturated egg phosphatidylcholine in aqueous liposome dispersions and the effect of cholesterol incorporation on hydrolysis kinetics. J Pharm Pharmacol 1993;45:490–495.
115. Grit M, Underberg WJM, Crommelin DJA. Hydrolysis of saturated soybean phosphatidylcholine in aqueous liposome dispersions. J Pharm Sci 1993;84:362–366.
116. Grit M, Crommelin DJA. The effect of surface charge on the hydrolysis kinetics of partially hydrogenated egg phosphatidylcholine and egg phosphatidylglycerol in aqueous dispersions. Biochim Biophys Acta 1993;1167:49–55.
117. Nikolai K, Van der Neut R, Fok JJ, DeKruyff B. Effects of adriamycin on lipid polymorphism

in cardiolipin-containing model and mitochondrial membranes. Biochim Biophys Acta 1985;819:55–65.
118. Heeremans JLM, Gerritsen HR, Meusen SP, Mijnheer FW, Gangaram Panday RS, Prevost R, Crommelin DJA. The preparation of tissue-type Plasminogen Activatore (t-PA) containing liposomes: entrapment efficiency and ultracentrifugation damage. (in preparation), 1995.
119. Moro L, Neri G, Rigamonti A. Stable liposome containing e.g., doxorubicin hydrochloride—purified by mixing with ion-exchange resin to remove unbound drug, filtration and lyopholization. US patent 4746516, 1986.
120. Storm G, Van Bloois L, Brouwer M, Crommelin DJA. The interaction of cytostatic drugs with adsorbents in aqueous media. The potential implications for liposome preparation. Biochim Biophys Acta 1985;818:343–351.
121. Barenholz Y, Amselem S. Quality control assays in the development and clinical use of lipsome-based formulations. In: Gregoriadis G, ed. Liposome Technology, Vol. I, 2nd ed. Boca Raton, FL: CRC Press, 1993.
122. Pearson FC. Pyrogens. Adv. Parent. Sci. New York: Marcel Dekker, 1985.
123. European Pharmacopeia, second edition, part II. European Treaty Series No 50.
124. The United States Pharmacopeia (USP) XXII, 1989.
125. Zuidam NJ, Talsma H, Crommelin DJA. Sterilisation of liposomes. In: Barenholz Y, Lasic DD, eds. Handbook of non-medical applications of liposomes, Vol.3: From design to microreactors. Boca Raton, FL: CRC Press, 1995.
126. Lukyanov AN, Torchilin VP. Autoclaving of liposomes. J Microencapsulation 1994;11:669–672.
127. Kikuchi H, Carlsson A, Yachi K, Hirota S. Possibility of heat sterilization of liposomes. Chem Charm Bull 1991;39:1018–1022.
128. Cherian M, Lenk RP, Jedrusiak JA. Heat treating liposomes. PCT Int. Appl. WO 90/03808, 1990.
129. Mentrup E, Butz P, Stricker H, Ludwig H. Hochdrucksterilisation von Liposomen. Pharm Res 1988;50:363–366.
130. Ratz H, Freise J, Magerstedt P, Schaper A, Preugschat W, Keyser D. Sterilization of contrast media (Isovist) containing liposomes by ethylene oxide, J. Microencapsulation. J Microencapsulation 1989;6:485–492.
131. Gregoriadis G. Medical applications of liposome-entrapped enzymes. Meth Enzym 1976;44:698–709.
132. Brassinne C, Atassi G, Frühling J, Penasse W, Coune A, Hildebrand J, Ruysschaert JM, Laduron C. Antitumor activity of a water-insoluble compound entrapped in liposomes on L1210 Leukemia in mice. JNCI 1983;70:1081–1086.
133. Coune A, et al. I.v. administration of a water-insoluble antimitotic coupound entrapped in liposomes. Preliminary report on infusion of large volumes of liposomes to man. Cancer Treat Rep 1983;67:1031–1033.
134. Freise J. The preparation of sterile drug-containing liposomes. In: Gregoriadis G, ed. Vol. I. Boca Raton, FL: CRC Press, 1984;131–137;
135. Zuidam NJ, Lee SSL, Crommelin DJA. Sterilization of liposomes by heat treatment. Pharm Res 1993;10:1591–1596.
136. Zuidam NJ, Lee SSL, Crommelin DJA. Gamma-irradiation of non-frozen, frozen and freeze-dried liposomes. Pharm Res 1995;12:1761–1768.
137. Grit M, Zuidam NJ, Crommelin DJA. Analysis and hydrolysis kinetics of phospholipids in aqueous liposome dispersions. In: Gregoriadis G, (Eds.), Liposome Technology, Vol. I. Boca Raton, FL: CRC Press, 1993;455–487.
138. Grit M, Crommelin DJA. Chemical stability of liposomes: implications for their stability. Chem Phys Lipids 1993;64:3–18.
139. Frøkjaer S, Hjorth EL, Wørts O. Stability and storage of liposomes. In: Bundgaard H, A. BH, Kofod H, eds. Optimization of Drug Delivery. Copenhagen: Munksgaard, 1982;384–404.
140. Kensil CR, Dennis EA. Alkaline hydrolysis of phospholipids in model membranes and the dependence on their state of aggregation. Biochemistry 1981;20:6079–6085.
141. Grit M, De Smidt J, Struijke HA, Crommelin DJA. Hydrolysis of natural soybean phosphatidylcholine in aqueous liposome dispersions. Int J Pharm 1989;50:1–6.
142. Zuidam NJ, Crommelin DJA. Chemical hydrolysis of phospholipids. J Pharm Sci 1995;84:1113–1119.
143. Blume G, Cevc G. Liposomes for the sustained drug release in vivo. Biochim Biophys Acta 1990;1029:91–97.
144. Zuidam NJ, Gouw HKME, Barenholz Y, Crommelin DJA. Physical (in)stability of liposomes

upon chemical hydrolysis; the role of lysophospholipids and fatty acids. Biochim Biophys Acta 1995;1240:101–110.
145. Fabrie CHJP, de Kruijff B, De Gier J. Protection by sugars against phase transition-induced leak in hydrated dimyristoylphosphatidylcholine liposomes. Biochim Biophys Acta 1990;1024:380–384.
146. Magin RL, Weinstein J. The design and characterization of temperature-sensitive liposomes. In: Gregoriadis G, ed. Liposome Technology, Vol. 3. Boca Raton, FL: CRC Press, 1984;137–155.
147. Crommelin DJA, Van Bommel EMG. Stability of liposomes on storage: freeze-dried, frozen or as an aqueous dispersion. Pharm Res 1984;1:159–164.
148. Hernández-Caselles T, Villalaín J, Gómez-Fernández JC. Stability of liposomes on long term storage. J Pharm Pharmacol 1990;42:397–400.
149. Asayama K, Aramaki Y, Yoshida T, Tsuchiya S. Permeability changes by peroxidation of unsaturated liposomes with ascorbid acid/Fe^{2+}. J Liposome Res 1992;2:275–287.
150. Weltzien HU. Cytolytic and membrane perturbing properties of lysophosphatidylcholine. Biochim Biophys Acta 1979;559:259–287.
151. Grit M, Crommelin DJA. The effect of aging on the physical stability of liposome dispersions. Chem Phys Lipids 1992;62:113–122.
152. Zuidam NJ, Crommelin DJA. Differential scanning calorimetric analysis of dipalmitoylphosphatidylcholine-liposomes upon hydrolysis. Submitted for publication.
153. Crowe LM, Spargo BJ, Crowe JH. Preservation of dry liposomes does nor require retention of residual water. Proc Natl Acad Sci USA 1987;84:1537–1540.
154. Sachse A, Leike JU, Rößling GL, Wagner SE, Krause W. Preparation and evaluation of lyophilized iopromide-carrying liposomes for liver tumor detection. Invest Radiol 1993;28:838–844.
155. Kristiansen J. Leakage of a trapped fluorescent marker from liposomes: effects of eutectic crystallization of NaCl and internal freezing. Cryobiology 1992;29:575–584.
156. Talsma H, Van Steenbergen MJ, Crommelin DJA. The cryopreservation of liposomes. 2. Effect of particle size on crystallization behavior and marker retention. Cryobiology 1992;29:80–86.
157. Talsma H, Van Steenbergen MJ, Crommelin DJA. The cryopreservation of liposomes: 3. Almost complete retention of a water-soluble marker in small liposomes in a cryoprotectant containing dispersion after a freezing/thawing cycle. Int J Pharm Sci 1991;77:119–1126.
158. Grünert M, Börngen L, Nimtz G. Structural phase transition due to a release of bound water in phospholipid bilayers at temperatures below 0°C. Ber Bunsenges Phys Chem 1986;88:608–612.
159. Quinn PJ. A lipid-phase separation model of low temperature damage to biological membranes. Cryobiologogy 1985;22:128–146.
160. Quinn PJ: Principles of membrane stability and phase behavior under extreme conditions. J Bioenerg Biomembr 1989;21:3–19.
161. MacDonald RC, Jones FD, Qiu R. Fragmentation into small vesicles of dioleylphosphatidylcholine bilayers during freezing and thawing. Biochim Biophys Acta 1994;1191:362–370.
162. Bronshteyn VL, Steponkus PL. Calorimetric studies of freeze-induced dehydration of phospholipids. Biophys J 1993;65:1853–1865.
163. Crowe JH, Carpenter JF, Crowe LM, Anchordoguy TJ. Are freezing and dehydration similar stress factors? A comparison of modes of interaction of stabilizing solutes with biomolecules. Cryobiology 1990;27:219–231.
164. Anchordoguy T, Carpenter JF, S.H. L, Crowe JH. Mechanism of interaction of amino acids with phospholipid bilayers during freezing. Biochim Biophys Acta 1988;946:299–306.
165. Van Winden ECA, Deketh BJH, Crommelin DJA. Physical stability of liposomes during freeze-drying and rehydration as a function of bilayer composition and vesicle size. Submitted.
166. Harrigan PR, Madden TD, Cullis PR. Protection of liposomes during dehydration or freezing. Chem Phys Lipids 1990;52:139–149.
167. Crowe JH, Crowe LM. Preservation of liposomes by freeze-drying. In: Gregoriadis G, ed. Liposome Technology, Vol. I. Boca Raton, FL: CRC Press, 1993.
168. Crowe JH, Crowe LM, Carpenter JF, Rudolph AS, Wistrom CA, Spargo BJ, Anchordoguy TJ. Interactions of sugars with membranes. Biochim Biophys Acta 1988;947:367–384.
169. Crowe LM, Crowe JH. Trehalose and dry dipalmitoylphosphatidylcholine revisited. Biochim Biophys Acta 1988;946:193–201.
170. Crowe JH, Leslie SB, Crowe LM. Is vitrification sufficient to preserve liposomes during freeze-drying? Cryobiology 1994;31:355–366.
171. Crowe JH, Crowe LM. Factors affecting the stability of dry liposomes. Biochim Biophys Acta 1988;939:327–334.

172. Van Winden ECA, Talsma H, Crommelin DJA. Physical stability of liposomes during the freeze-drying process. Pharm Res 1993;10:S-367.
173. Crowe LM, Wormersley C, Crowe JH, Reid D, Appel L, Rudolp A. Prevention of fusion and leakage in freeze-dried liposomes by carbohydrates. Biochim Biophys Acta 1986;861:131–140.
174. Crommelin DJA, Van Bommel EMG. Stability of doxorubicin-liposomes on storage: as an aqueous dispersion, frozen or freeze-dried. Int J Pharm 1984;22:299–310.
175. Tanaka K, Takeda T, Fuji K, Miyama K. Cryoprotective mechanism of saccharides on freeze-drying of liposomes. Chem Pharm Bull 1992;40:1–5.
176. Crowe JH, McKersie BD, Crowe LM. Effects of free fatty acids and transition temperature on the stability of dry liposomes. Biochim Biophys Acta 1989;979:7–10.
177. Crommelin DJA, Talsma H, Grit M, Zuidam NJ. Physical stability on long-term storage. In: Cevc G, ed. Phospholipids handbook. New York: Marcel Dekker, 1993;335–348.
178. Van Winden ECA, Zhang W, Crommelin DJA. Effect of freezing rate on the stability of liposomes during freeze-drying and rehydration. Submitted.
179. Izutsu K, Yoshioka S, Takeda Y. The effects of additives on the stability of freeze-dried β-galactosidase stored at elevated temperature. Int J Pharm 1991;71:137–146.
180. Townsend MW, DeLuca PP. Use of cryoprotectants in freeze-drying of a model protein, ribonuclease A. J Parent Sci Technol 1988;42:190–199.
181. Skrabanja ATP, De Meere ALJ, De Ruiter RA, Van Den Oetelaar PJM. Lyophilization of biotechnology products. Technology/Applications 1994;48:311–317.
182. Roos Y, Karel M. Applying state diagrams to food processing and development. Food Technol 1991;12:66–107.
183. Roos Y, Karel M. Crystallization of amorphous lactose. J Food Sci 1992;57:775–777.
184. Roos Y, Karel M. Amorphous state and delayed ice formation in sucrose solutions. Int J Food Sci Technol 1991;26:553–566.
185. Patel SD, Gupta B, Yalkowsky SH. Acceleration of heat transfer in vial freeze-drying of pharmaceuticals. I: corrugated aluminum quilt. J Parent Sci Technol 1989;43:8–14.
186. Pikal MJ, Shah S, Roy ML, Putman R. The secondary drying stage of freeze-drying: drying kinetics as a function of temperature and chamber pressure. Int J Pharm Sci 1990;60:203–217.
187. Schoen MP, Braxton BK, Gatlin LA, Jefferis III RP. A simulation model for the primary drying phase of the freeze-drying cycle. Int J Pharm Sci 1995;114:159–170.
188. Franks F, Van Den Berg C. A rational approach to the drying of labile biologicals: the role of salts and stabilizers. In: Topics in Pharmaceutical sciences. Stuttgart: Medpharm Scientific Publishers, 1992;233–240.
189. Levine H, Slade L. Principles of "cryostabilization" technology from structure/property relationships of carbohydrate/water systems. A review. Cryo-Letters 1988;9:21–63.
190. Levine H, Slade L. Thermochemical properties of small-carbohydrate-water glasses and 'rubbers'. J Chem Soc, Faraday Trans 1988;1:84:2619–2633.
191. Roos Y, Karel M. Plasticizing effect of water on thermal behavior and crystallization of amorphous food models. J Food Sci 1991;56:38–43.
192. Mobley WC, Schreier H. Phase transition temperature reduction and glass formation in dehydroprotected lyophilized liposomes. J Control Rel 1994;31:73–87.
193. Crowe JH, Hoekstra FA, Nguyen KHN, Crowe LM. Is vitrification involved in depression of the phase transition temperature in dry phospholipids? Biochim Biophys Acta 1996;1280:187–196.
194. Sun WQ, Leopold AC, Crowe LM, Crowe JH. Stability of dry liposomes in sugar glasses. Biophys J 1996;70:1769–1776.
195. Van Winden ECA, Crommelin DJA. Long term stability of freeze-dried, lyoprotected liposomes. Eur J Pharm Biopharm 1997;43:295–307.
196. Van Winden ECA, Crommelin DJA. Short term stability of freeze-dried, lyoprotected liposomes. Submitted.
197. Friede M, Van Regenmortel MHV, Schuber F. Lyophilized liposomes as shelf items for the preparation of immunogenic liposome-peptide conjugates. Am Biotechnol Lab 1993;211:117–122.
198. Van den Berg C. Glass transitions in carbohydrates. Carbohydrates in the Netherlands 1992;8:23–25.
199. Hageman MJ. Water sorption and solid state stability of proteins. In: Ahern T, Manning MC, eds. Stability of protein pharmaceuticals, Part A: Chemical and physical pathways of protein degradation. New York: Plenum Press, 1992;273–309.
200. Te Booy MPWM, De Ruiter RA, De Meere ALJ. Evaluation of the physical stability of freeze-

dried sucrose-containing formulations by differential scanning calorimetry. Pharm Res 1992;9:109–114.
201. Vemuri S, Rhodes CT. Development and characterization of a liposome preparation by a pH-gradient method. J Pharm Pharmacol 1994;46:778–783.
202. Vromans H, Van Laarhoven JAH. A study on the water permeation through rubber closures of injection vials. Int J Pharm 1992;79:301–308.
203. Chong-Kook K, Ju Jeong E. Development of dried liposome as effective immuno-adjuvant for hepatitis B surface antigen. Int J Pharm 1995;115:193–199.
204. Fildes FJT. Liposomes: the industrial viewpoint. In: Knight CG, ed. Liposomes: from physical structure to therapeutic applications. Elsevier/North-Holland, 1981;465–485.

Lasic and Papahadjopoulos (eds.), Medical Applications of Liposomes

CHAPTER 7.3

Pre-clinical studies of lipid-complexed and liposomal drugs: AMPHOTEC®, DOXIL® and SPI-77

PETER K. WORKING

SEQUUS Pharmaceuticals, Inc., 960 Hamilton Court, Menlo Park, CA 94025, USA

Overview

I. Introduction

Years have passed since the idea of encapsulating drugs in liposomes to improve their toxicity and efficacy profiles was first proposed. Two of the most commonly encapsulated drugs formulated have been amphotericin B and doxorubicin or related anthracyclines, and, in fact, the all three major liposome-based companies have a product in each of these categories (The Liposome Company: Abelcet®, amphotericin B-based and D-99, doxorubicin-based; NeXstar: AmBisome®, amphotericin B-based and DaunoXome®, daunorubicin-based; and SEQUUS: AMPHOTEC®, amphotericin B-based and DOXIL®, doxorubicin-based).

Though both amphotericin B and doxorubicin have been formulated as liposomes or, in two cases, lipid-complexes for the stated reason of improving their therapeutic ratio (by lowering toxicity and, hopefully, improving efficacy), an underlying reason for the unusual concordance of products among the three companies is that amphotericin B and doxorubicin were, in theory, "easy" to make into liposomes. Theory, as is often the case, departed from practice, and it has taken many years for even the first of the drugs to be approved for use in the U.S. (DOXIL and Abelcet, both near the end of 1995). Approvals came somewhat earlier and more piecemeal in Europe, with AmBisome first in the U.K. in

1991, followed by AMPHOTEC in 1993, and with DOXIL, DaunoXome and Abelcet approvals following in 1995 and 1996. Other liposomal or lipid-complex drugs are also in development, including MiKasome, a liposomal amikacin from NeXstar in Phase I trials; Ventus®, a lipid-complexed prostaglandin E from The Liposome Company in Phase III trials; a liposomal nystatin from Aronex in Phase II/III trials; a liposomal vincristine from Inex in Phase II trials; and SPI-77, a liposomal cisplatin from SEQUUS in Phase I/II trials.

In the pages that follow, the formulations, pharmacokinetics, therapeutic efficacy and toxicity of three of these products, AMPHOTEC, DOXIL and SPI-77, will be described. AMPHOTEC, of course, is a colloidal particle, not a liposome, whereas DOXIL and SPI-77 are true liposomes, being, respectively, doxorubicin and cisplatin encapsulated in long-circulating, pegylated ("Stealth") liposomes. All are produced by SEQUUS. DOXIL and AMPHOTEC are both approved products in the U.S. and European Community, and SPI-77 is currently in clinical trials in the U.S. and Europe.

II. The formulations

AMPHOTEC. AMPHOTEC (amphotericin B cholesteryl sulfate complex for injection) is not a liposome. Rather, it is a formulation of amphotericin B and cholesteryl sulfate in a near 1:1 molar ratio that forms uniform disc-shaped particles of approximately 115 nm in diameter and 4 nm in thickness. The formulation is a stable colloid, meaning that the particles do not separate from the water phase in suspension. The formulation is lyophilized and can be stored at room temperature for at least 2 years.

DOXIL. DOXIL (doxorubicin HCl liposome injection), known as CAELYX® in Europe, is a liposomal drug. DOXIL is a formulation of doxorubicin HCl encapsulated in long-circulating liposomes containing methoxypoly(ethylene glycol) (MPEG). The DOXIL liposomes are formulated with three lipids, fully hydrogenated soy phosphatidylcholine (HSPC), N-(carbamoyl-methoxypolyethylene glycol 2000)-1,2-distearoyl-*sn*-glycero-3-phosphoethanolamine sodium salt (MPEG-DSPE), and cholesterol, in a target molar ratio of 56:5:39. Total lipid content of DOXIL is approximately 16 mg/mL, and doxorubicin concentration is 2 mg/mL. Other ingredients of the formulation are sucrose, ammonium sulfate, and histidine.

Two of the three liposome components (HSPC and cholesterol) are ubiquitous dietary lipids and constituents of the mammalian plasma membrane. DSPE is also a dietary lipid, but in DOXIL it has been modified by covalent bonding to a synthetic MPEG moiety of approximately 2000 molecular weight. DOXIL liposomes have a diameter of approximately 100 nm. The formulation is provided as a liquid, which can be stored refrigerated for up to 20 months.

DOXIL liposomes, and indeed all Stealth liposomes, are long-circulating in the bloodstream by virtue of evading uptake by macrophages of the mononuclear phagocytic system (MPS). These liposomes are sometimes termed sterically stabilized, in analogy to stabilization of inorganic colloid particles, in which steric modifications of the particle surface result in reductions in particle-to-particle interactions that could lead to aggregation or fusion. In colloids, steric stabilization is accomplished by addition of surface charge or via coating of the surface with various molecules, including starch and PEG.[1] In the case of liposomes in biological environments, steric stabilization not only reduces particle-to-particle interactions, but also decreases adsorption of various macromolecules onto the liposome surface, the loss of liposomal components to other particles, and interactions of the liposomes with cells, all of which provide greater liposome stability.[2,3]

SPI-77. SPI-77 (Stealth® liposomal cisplatin) is also a true liposome. Its liposome is very similar to that of DOXIL, containing HSPC, MPEG-DSPE, and cholesterol in a 51 : 5:44 molar ratio. The total lipid content of the SPI-77 formulation is approximately 71 mg/mL, approximately 4-fold higher than DOXIL, and the cisplatin concentration is 1 mg/mL. SPI-77 liposomes average 110 nm in diameter. Other components of the formulation are sucrose and histidine, but no ammonium sulfate. As with DOXIL, the sterically stabilized liposomes of SPI-77 are long-circulating. The formulation is a refrigerated liquid. Preliminary findings indicate a shelf-life in excess of 18 months.

III. Therapeutic efficacy

AMPHOTEC. In vitro studies have shown that amphotericin B is bioavailable when formulated as AMPHOTEC and that it has broadly similar activity as conventionally formulated amphotericin B deoxycholate (Fungizone).[4,5] AMPHOTEC activity shows some minor species-dependent differences from Fungizone, but generally appears to be equipotent in vitro. The efficacy of AMPHOTEC and Fungizone has been compared in vivo in murine models of *Aspergillus fumigatus*,[6] *Candida albicans*,[6] *Coccidioides immitis*[7] and *Cryptococcus neoformans*[8] infections, a rabbit model of *A. fumigatus* infection[9,10] and a hamster model of *Leishmania donovani* infection.[11] Since systemic fungal infections often occur in immunocompromised patients, some studies were performed with animals that were genetically immunodeficient or pharmacologically immunosuppressed. A recent study compared the therapeutic efficacy of Fungizone to that of all three lipid-based amphotericin B formulations, AMPHOTEC, AmBisome and Abelcet.[12]

The therapeutic activity of AMPHOTEC and Fungizone was compared in immunocompromised rabbits with invasive pulmonary aspergillosis.[9,12] Granulocytopenia (granulocyte counts $< 500/\mu L$) was induced by using concurrent treatment with cytosine arabinoside. Rabbits were infected with a dose of *Aspergillus* conidia administered directly into the trachea beyond the vocal cords, and, beginning 1 day later, were treated with 1, 5, or 10 mg/kg AMPHOTEC or 1 mg/kg Fungizone daily for 10 consecutive days. Survival of rabbits that received 1 or 5 mg/kg AMPHOTEC (56 and 64%) was significantly greater than untreated controls (0%) and was also greater than in rabbits treated with 1 mg/kg Fungizone (33%), although not statistically significant.

Antifungal activity, as measured by the concentration of Aspergillus organisms in the lungs, frequency of culture-positive lobes, or reduction in lung weight, was

Table 1

Comparative activity of AMPHOTEC and Fungizone in treatment of pulmonary aspergillosis in immunosuppressed rabbits

Treatment	Survival (%)	% of culture-positive lobes	Log_{10} CFU/g of lung tissue	Lung weight (g)
AMPHOTEC (10 mg/kg)	27	4.5 ± 3.2	0.11 ± 0.56	12.7 ± 1.7
AMPHOTEC (5 mg/kg)	64	9.5 ± 7.1	0.25 ± 0.81	14.4 ± 1.4
AMPHOTEC (1 mg/kg)	56	66.7 ± 11.1	1.60 ± 1.20	22 ± 3.6
Fungizone (1 mg/kg)	33	13.3 ± 5.4	0.28 ± 0.73	15.3 ± 1.5
Untreated controls	0	57.6 ± 5.2	1.41 ± 1.45	41.3 ± 3.1

[b]ND=Not determined.
From Ref. 9.

greatest in rabbits given 5 or 10 mg/kg AMPHOTEC, or 1 mg/kg Fungizone (Table 1). In rabbits treated with 1 mg/kg AMPHOTEC, the tissue burden was not different from untreated controls. The frequency of hemorrhagic pulmonary lesions was related to dose after AMPHOTEC treatment, and was lowest in animals that received 5 or 10 mg/kg AMPHOTEC. However, survival in the 10 mg/kg AMPHOTEC group (27%) decreased markedly relative to groups that received 1 or 5 mg/kg AMPHOTEC, apparently as a result of nephrotoxicity, but was comparable to survival in the Fungizone treatment group. Antifungal activity of 5 mg/kg AMPHOTEC was comparable to that of 1 mg/kg Fungizone. Fungizone, however, was more nephrotoxic than AMPHOTEC at 1 or 5 mg/kg and was associated with reduced survival relative to AMPHOTEC treatment. The differences in survival between rabbits treated with 5 mg/kg AMPHOTEC and 1 mg/kg Fungizone were attributable to the combined effects of enhanced rate of tissue clearance of fungi, decreased pulmonary injury, and reduced nephrotoxicity in the AMPHOTEC treatment group.

These observations are supported by the results of a concurrent study that incorporated diagnostic imaging of rabbits for therapeutic monitoring of the invasive aspergillosis and demonstrated a more rapid clearance of pulmonary lesions in rabbits treated with 5 mg/kg AMPHOTEC than those treated with 1 mg/kg Fungizone.[10] Ultrafast computerized tomography (UFCT) scans were performed periodically to monitor clearance of pulmonary aspergillosis, and a mean pulmonary lesion score was established by evaluating the infiltrate in each lung lobe. Monitoring of pulmonary lesions by UFCT demonstrated a significant dose-response relationship; lesions continued to progress in untreated controls, but initially increased and then decreased in response to antifungal therapy with AMPHOTEC or Fungizone. The same trend in resolution of lesions was also evident by postmortem examination and by microbiological clearance of *A. fumi-*

gatus evaluations also conducted in these animals. The rapid rate of lesion resolution in rabbits treated with 5 or 10 mg/kg/day of AMPHOTEC was compared to slower and less effective lesion resolution in animals that received 1 mg/kg/day of Fungizone. In addition, lesions stabilized in animals treated with 1 mg/kg/day AMPHOTEC and showed a significant reduction compared to untreated controls by the end of therapy, but at this dose AMPHOTEC appeared to be slightly less effective than Fungizone in lesion clearance, despite improved survival (Table 1). Based on these results, AMPHOTEC is more effective than 1 mg/kg/day Fungizone at a dose of 5 or 10 mg/kg/day and is still effective at 1 mg/kg/day, but slightly less so than the conventional formulation, in treating invasive pulmonary aspergillosis in neutropenic rabbits.

There is only one published study that directly compares the activity of the three lipid-based drugs.[12] A model of experimental systemic murine cryptococcosis was established in female CD-1 mice by intravenous injection of 6.25×10^5 viable yeasts of *C. neoformans* strain 9759. Therapy began four days later in groups of mice ($n = 10$ per group) that received no treatment, 1 mg/kg Fungizone or 1, 5 or 10 mg/kg AMPHOTEC, AmBisome or Abelcet 3 times weekly for 2 consecutive weeks (6 treatments). Survival was followed for 49 days, and the number of viable CFU of *C. neoformans* remaining in the brain, spleen, liver, kidneys and lungs of surviving mice was determined.

Ninety percent of control mice died between days 15 and 34 (Figure 1). All

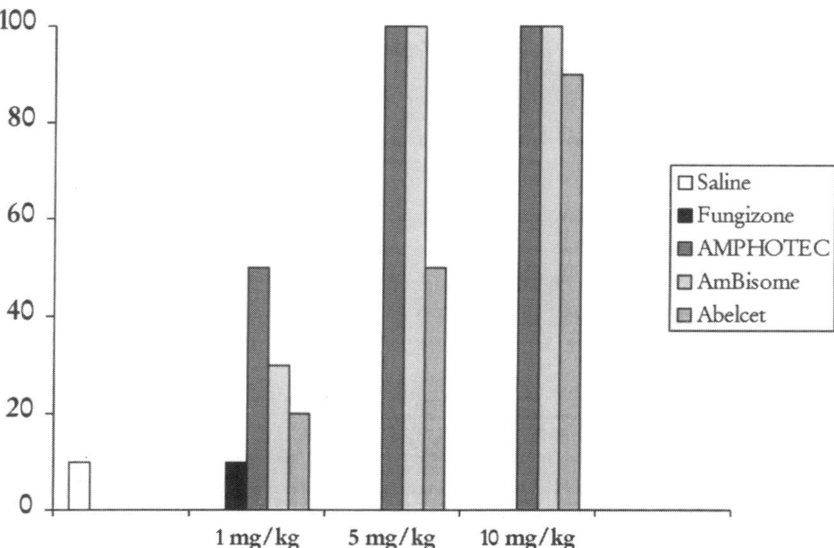

Fig. 1. Percent survival of cryptococcus-infected mice treated with 1, 5 or 10 mg/kg Fungizone, AMPHOTEC, AmBisome or Abelcet at day 39 post-infection. Data are from Ref. 12.

treatments except 1 mg/kg Abelcet significantly prolonged survival as compared to no treatment; survival in the low dose Abelcet group was no better than in the untreated group. AMPHOTEC and AmBisome were equivalent to Fungizone at 1 mg/kg. Fungizone was least effective in prolonging survival, with 5 or 10 mg/kg AMPHOTEC, AmBisome or Abelcet significantly better. Among the lipid formulations, AMPHOTEC and AmBisome were equivalent and both better than Abelcet at equal doses; 1 mg/kg AMPHOTEC or AmBisome were equivalent, in fact, to 5 mg/kg Abelcet.

Comparison of residual infection showed that all three drugs had some dose-responsive efficacy in three or more organs with increasing dose. AMPHOTEC and AmBisome showed therapeutic efficacy equivalent to Fungizone on a milligram-per-kilogram basis, but Abelcet was less effective. In the brain, 5 or 10 mg/kg AMPHOTEC or AmBisome or 10 mg/kg Abelcet were better than 1 mg/kg Fungizone, and 1 mg/kg AMPHOTEC or AmBisome were equivalent to 10 mg/kg Abelcet, with similar results in kidney and lung. In the spleen, 10 mg/kg AMPHOTEC cured all mice of infection and was superior to all other treatments; AmBisome showed a similar activity in liver at this dose, curing all mice of infection, but its activity was not significantly better than that of AMPHOTEC, which cured 8/9 animals (Figure 2). In the liver, 10 mg/kg doses of each treatment were equivalent, but 5 mg/kg AmBisome was superior to equal doses of AMPHOTEC or Abelcet.

Based on the survival and organ clearance data, the authors assigned an overall

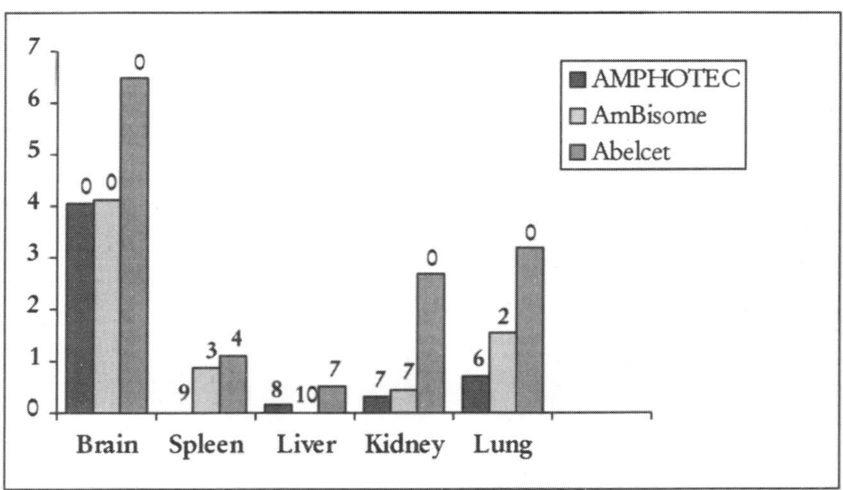

Fig. 2. Clearance of infection from brain, spleen, liver, kidney and lung in cryptococcus infected mice by treatment with 10 mg/kg AMPHOTEC, AmBisome or Abelcet at day 39 post-infection. Numbers on the top of each bar represent the number of animals completely cleared of infection. Data are the geometric mean of the $\log_{10}$ CFU/organ. Groups had 9 (AMPHOTEC) or 10 (AmBisome and Abelcet) animals each. Data are from Ref. 12.

rank order of efficacy of AMPHOTEC ≅ AmBisome > Abelcet ≫ Fungizone. AMPHOTEC and AmBisome had equal efficacy to Fungizone at 1 mg/kg and were about 10-fold better than Abelcet, particularly in the brain.

DOXIL. The therapeutic effectiveness of DOXIL has been examined in a series of efficacy studies in murine tumor models, xenografts of human tumors implanted in immune deficient mice and in a rat brain tumor model. In each study, the antitumor efficacy of DOXIL was compared to that of nonliposomal doxorubicin HCl (Adriamycin RDF®).

In a human xenograft study, mice were implanted subcutaneously or injected intraperitoneally with the human ovarian carcinoma HEY.[13] DOXIL given at 6 or 9 mg/kg was significantly better than the same doses of Adriamycin at inhibiting tumor growth and effecting cures in mice bearing subcutaneous implants. Overall, 70% of mice treated with 9 mg/kg DOXIL survived until study termination with no evidence of tumor, compared to just 18% of the mice that received the same dose of Adriamycin. After intraperitoneal injection of tumor cells, 80% of control mice developed palpable tumors between 18 and 25 days after injection. All ten DOXIL-treated (9 mg/kg), compared to four of ten Adriamycin-treated mice, were tumor-free at necropsy on Day 70. The other six Adriamycin mice died due to drug toxicity between Days 7 and 52. In this ovarian carcinoma model, DOXIL was more effective and less toxic than conventionally formulated doxorubicin HCl.

The biodistribution and preclinical efficacy of DOXIL were also compared to Adriamycin in a secondary brain tumor model.[14] Fischer rats were stereotactically implanted with a suspension of tumor cells (rat methycholanthrene-induced malignant fibrous histiocytoma) into the right cerebral hemisphere. After 8 days, each animal received an intravenous dose of 6 mg/kg of either Adriamycin or DOXIL for evaluation of drug biodistribution. For therapeutic experiments a single dose of 8 mg/kg was given 6 or 11 days after tumor implantation, or alternatively, weekly doses (5 mg/kg) were given on Days 6, 13, and 20. DOXIL was slowly cleared from plasma with a $t_{1/2}$ of 35 hours (see below for a more detailed description of DOXIL pharmacokinetics). Following Adriamycin treatment, maximum tumor drug levels reached a mean value of 0.8 (μgram/gram tissue and were identical in the brain tissue adjacent to the tumor and the contralateral hemisphere. DOXIL tumor levels were 14-fold higher at peak levels (48 hours), declining to nine-fold increased levels at 120 hours. A gradual increase in drug levels in the brain adjacent to tumor was noted between 72 and 120 hours (up to 4 μg/g). High-performance liquid chromatography analysis identified a small amount of aglycone metabolites within the tumor mass from 96 hours and beyond after DOXIL injection. Cerebrospinal fluid levels were barely detectable in tumor-bearing rats treated with Adriamycin up to 120 hours after drug injection (≤0.05 μg/ml), whereas the levels found after DOXIL were 10- to 30-fold higher. Adriamycin given as single-dose treatment 6 days after tumor inoculation increased the rats' life span (ILS) by 135% over controls ($p < 0.05$) but was not effective if given on Day 11. In contrast DOXIL treatment resulted in an ILS of 168% ($p < 0.0003$) with similar increases when given after 6 or 11 days. Treatment with three weekly doses of DOXIL produced an ILS of 189% compared to 126% for

Adriamycin ($p < 0.0002$). These results indicate that DOXIL administration to rats bearing a brain tumor results in enhanced drug exposure and improved therapeutic activity, with equal effectiveness against early small- and large-sized brain tumors.

DOXIL has also been shown to have superior efficacy to Adriamycin in syngeneic models of colon[15] and breast[16] cancers and in xenograft models of small cell lung[17] and pancreatic[18] tumors. In addition, DOXIL is therapeutically effective in a prophylactic model of spontaneous mammary tumors,[19] in preventing metastases of mammary tumors[20] and in treating spontaneous tumors in dogs, both in dogs that have failed all previous treatments[21] and as first line therapy for canine non-Hodgkin's lymphoma.[22] In every tumor model studied, DOXIL was more effective than the same dose of Adriamycin in inhibiting or halting tumor growth, in effecting cures and/or in prolonging the survival of tumor-bearing animals. Most often, all three endpoints were improved by DOXIL, and in no case was DOXIL less effective than Adriamycin. In many cases, Adriamycin was ineffective in the models tested and the relatively lower toxicity of DOXIL permitted treatment at higher doses, further increasing its therapeutic advantage.

SPI-77. SPI-77 has demonstrated antitumor activity in both murine and human xenograft tumor models, generally equivalent or greater than that exhibited by cisplatin (Platinol®) and much greater than that of carboplatin (Paraplat®) solutions.

In one study, male Balb/c mice were inoculated subcutaneously with 10^6 murine C26 colon carcinoma cells. Treatment began 6 days later with saline, 6 mg/kg SPI-77, 6 mg/kg cisplatin or 100 mg/kg carboplatin.[23] Tumor size was measured twice weekly, and standard measures of therapeutic activity were determined.

SPI-77 inhibited tumor growth more effectively than either cisplatin or carboplatin (Figure 3). Time to 3 doublings in saline-treated control mice was 3.4 days. Treatment with SPI-77 resulted in a tumor growth delay of 30.7 days, compared to growth delays of just 3.7 and 5.5 days in mice treated with solutions of cisplatin and carboplatin, respectively. T/C%, a measure of relative tumor growth that is considered an indicator of significant antitumor activity when less than 42, was 11.3 in SPI-77-treated mice, 48.4 in cisplatin-treated mice and 63.9 in carboplatin-treated mice. Thus, solutions of cisplatin and carboplatin were relatively ineffective in this syngeneic colon tumor model, whereas SPI-77 had a significant and prolonged anti-tumor effect.

In a second study, the activity of SPI-77 was compared to that of cisplatin in a Lewis lung carcinoma, another syngeneic model.[23] Male B6C3F1 mice were inoculated subcutaneously with approximately 10^6 Lewis lung tumor cells on day 1. On days 5, 12 and 19, mice received intravenous or intraperitoneal injections of 12 mg/kg SPI-77 or 6 mg/kg cisplatin solution. Survival and tumor size were assessed for 30 days post-inoculation. Treatment with SPI-77 resulted in tumor growth delays of 6.6 days administered intravenously and 3.6 days given intraperitoneally. Cisplatin treatment caused tumor growth delays of 1.9 and 0.8 days by the intravenous and intraperitoneal routes, respectively. T/C% values were 6.3 and 12.2 in the SPI-77 treatment groups, compared to 53.6 and 54.5 in the cisplatin groups.

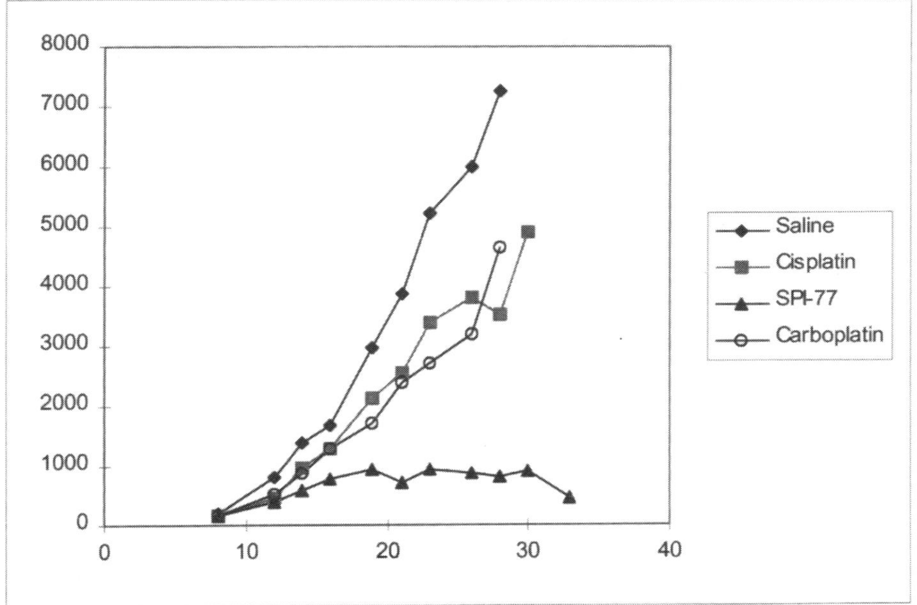

Fig. 3. Median tumor diameters (mm^3) in mice bearing the murine C26 colon carcinoma treated with 6 mg/kg SPI-77, 6 mg/kg cisplatin or 100 mg/kg carboplatin. Data are from Ref. 23.

Thus, SPI-77 showed meaningful anti-tumor activity in two different tumor models, the murine C26 colon carcinoma and Lewis lung tumors. Its therapeutic activity was reflected in tumor growth delay compared to saline-treated controls and in the different therapeutic indices calculated for various dosing regimens. SPI-77 induced a persistent inhibition of tumor growth during and after treatment. In many animals, tumors grew slowly to intermediate size and then were apparently arrested, with little additional growth evident. Cisplatin (and carboplatin) were ineffective in the C26 colon carcinoma model, in which weekly treatment with SPI-77 resulted in marked inhibition of tumor growth. SPI-77 was also effective in the Lewis lung model, and cisplatin again had only modest activity. Though early in the development of this drug, SPI-77, Stealth liposomal cisplatin, appears to have the same increased therapeutic activity compared to nonliposomal cisplatin that DOXIL exhibits relative to Adriamycin.

IV. Pharmacokinetics

AMPHOTEC. Following single intravenous injections in rats, plasma concentrations of amphotericin B were lower in animals that received AMPHOTEC than in those that received a comparable dose of Fungizone for up to 48 hours after treatment.[24] At later time points, the plasma concentration of amphotericin B was higher in AMPHOTEC-treated animals because of its longer terminal elimination

Table 2

Amphotericin B tissue distribution in rats and dogs: ratio of tissue concentration of AMPHOTEC to Fungizone

Tissue	Rat (A/F)[a]	Dog (A/F)[a]
Liver	5.2	1.69
Kidney	0.3	0.15
Spleen	0.4	1.09
Lung	0.3	0.02
Heart	0.3	0.01
Bone marrow	ND	2.78
Brain	0.4	0.07
Skeletal muscle	0.6	ND

[a]A/F = ratio of amphotericin B concentration in tissues of AMPHOTEC-treated animals to Fungizone-treated animals; μg/g tissue after 14 daily doses.
[b]ND = not determined.
From Refs. 24,27.

half-life when dosed as AMPHOTEC (27.1 hours compared to 10.4 hours for Fungizone). The concentration of amphotericin B in the kidneys of AMPHOTEC-treated animals was nearly seven-fold lower than in animals that received Fungizone (Table 2), correlating with the reduced renal toxicity of AMPHOTEC. Hepatic concentrations of amphotericin B were two- to eight-fold higher in AMPHOTEC animals, but were not accompanied by increased hepatotoxicity. The rate of release (washout) of amphotericin B from the liver was nearly identical to its plasma terminal elimination half-life in both AMPHOTEC- and Fungizone-treated animals (liver washout half-lives of 27.7 and 10.6 hours, respectively), suggesting that the plasma pharmacokinetic profile of amphotericin B may related to its washout from the liver and/or other tissues, which serve as a depot for the tissue.

In repeat dose study, rats received daily dose of AMPHOTEC or Fungizone for 14 consecutive days.[25] Peak plasma levels of amphotericin B were again significantly lower in animals that received AMPHOTEC. The plasma terminal elimination half-life of amphotericin B was longer and dose-dependent in AMPHOTEC-treated rats and the volume of distribution was larger (Table 3), consistent with the observed five-fold greater accumulation of amphotericin B in the liver (Table 2). Recent studies have shown that over 80% of the amphotericin B found in the liver is located in the Kupffer cells and that the amount of amphotericin B in the hepatocytes is nearly identical after treatment with AMPHOTEC or Fungizone.[26] Biliary and renal clearance of amphotericin B were proportional to plasma concentrations and similar in the AMPHOTEC and Fungizone groups.

The pharmacokinetics of AMPHOTEC and Fungizone were also evaluated in dogs. In a 14-day repeat dose study, plasma levels of amphotericin B were consistently lower in AMPHOTEC-treated dogs than in dogs that received comparable

Table 3

Plasma pharmacokinetic parameters of AMPHOTEC and Fungizone in rats treated for 14 days

	AMPHOTEC 1 mg/kg		Fungizone 1 mg/kg	
Parameter	Day 1	Day 14	Day 1	Day 14
C_{max} (ng/mL)	105	81	290	283
$t_{1/2\beta}$ (hr)	33.6	47.4	14.1	19.7
AUC_{∞} (ng-hr/mL)	–	–	4450	5315
CL (L/hr)	–	–	0.69	0.29

From Ref. 25.

doses of Fungizone.[27] AMPHOTEC-treated dogs exhibited a secondary rise in plasma concentration of amphotericin B that may have been related to feeding and concomitant increased hepatic blood flow and biliary excretion. There was no evidence of systemic accumulation of amphotericin B in the AMPHOTEC groups. In contrast, plasma concentrations of amphotericin B increased throughout treatment in the animals that received Fungizone, suggesting that amphotericin B accumulation had occurred. Following AMPHOTEC treatment, decreased concentrations of amphotericin B were measured in the kidneys and were reflected in decreased renal toxicity. The increased concentration of amphotericin B detected in the livers of AMPHOTEC-treated dogs were not associated with increased hepatotoxicity.

Tissue distribution was also different between amphotericin B formulations. Amphotericin B concentrations in most organs, including the kidney, were lower in animals that received AMPHOTEC, but were higher in the organs of the mononuclear phagocytic system, namely the liver, spleen and bone marrow. Liver concentrations tended to increase and plateau with repeated AMPHOTEC dose, while other tissue concentrations continued to increase. The high recovery of amphotericin B from the liver after dosing with AMPHOTEC suggests that, although the liver uptake is rapid, the amphotericin B is not subjected to extensive metabolism. This is consistent with the finding that most of the AMPHOTEC colloidal particles are phagocytosed by the Kupffer cells,[26] which subsequently release unmetabolized amphotericin B over time. Recent studies have confirmed that the AMPHOTEC complex is rapidly removed from the blood after administration, and that the circulating form of amphotericin B during the long terminal phase of the plasma concentration-time curve is the same as found after treatment with Fungizone. Less than 5% of the circulating amphotericin B is present as the AMPHOTEC complex within 1 hour after a single intravenous dose in rats.[28] This latter finding suggests that the efficacy of the AMPHOTEC and amphotericin B will be equivalent if similar blood levels can be attained in the terminal phase.

In both the rat and dog, the terminal half-life of amphotericin B in the plasma was controlled by the washout of the drug from the liver and other tissues. In most organs, clearance of amphotericin B from the tissues was slower after AMPHOTEC treatment. These observations suggest that the tissues may serve as a reservoir for AMPHOTEC, from which amphotericin B is slowly released, giving rise to the observed plasma pharmacokinetics. The dose proportionality of the plasma terminal elimination half-life of amphotericin B in AMPHOTEC-treated animals is most likely because of the dose-proportionate accumulation of amphotericin B in the tissues. Thus, the plasma pharmacokinetics reflect both the amount of amphotericin B accumulated, and its rate of washout from sites of accumulation. Furthermore, the rate of elimination of amphotericin B from the plasma must be faster than the rate of release from the tissues, since no plasma accumulation is seen during the washout phase of the studies.

In summary, these studies suggest that AMPHOTEC achieves reduced nephrotoxicity but maintains efficacy compared to Fungizone because the AMPHOTEC complex is stable in plasma and is rapidly removed from circulation. Studies have shown that following the infusion of AMPHOTEC, the amphotericin B-cholesteryl complex is rapidly cleared from the blood, such that most of the complex (≥95%) has been removed from the blood by 1 hour after treatment. The AMPHOTEC complex is stable in plasma and, compared to Fungizone, significantly amphotericin B less binds to the plasma lipoproteins;[29] other investigators have shown that binding of amphotericin B to low density lipoproteins significantly increases its nephrotoxicity.[30] Preclinical studies have demonstrated that the rapid removal of the complex is related to nearly complete uptake by the liver (specifically the Kupffer cells), spleen, and bone marrow without evidence of increased organ toxicity. Free amphotericin B is then gradually re-released from these cells, achieving therapeutic levels in the blood. The circulating form of amphotericin B is identical after treatment with either AMPHOTEC or Fungizone. These data explain how AMPHOTEC maintains its therapeutic activity, since plasma levels equivalent to Fungizone are achieved during the terminal plasma elimination phase. Infusion of doses of AMPHOTEC therapeutically equivalent to infusions of Fungizone appear to cause less nephrotoxicity because the stability of the complex prevents the high peak levels of amphotericin B which occur following the infusion of Fungizone. As a result, there is both reduced binding of amphotericin B to plasma lipoproteins and decreased amphotericin B concentration in the kidney.

DOXIL. The plasma pharmacokinetics of DOXIL are markedly different from those of non-liposomal doxorubicin because of its encapsulation in sterically stabilized liposomes.[31,32] Single and multiple dose pharmacokinetic studies were performed in rats, rabbits and dogs to characterize the plasma pharmacokinetics of DOXIL.[33] Tissue levels of doxorubicin were determined in tumor-bearing mice and in single dose and dogs multiple dose studies in normal rats.

The plasma pharmacokinetics of DOXIL and Adriamycin were significantly different in all species evaluated. The plasma concentration of doxorubicin was up to 2000-fold higher in DOXIL-treated animals after intravenous injection of equivalent doses of DOXIL and Adriamycin.[34] Plasma concentration by time data were best fit with a biexponential curve, with a relatively short first phase (half-life = 1–3 hours), and a more prolonged second phase, which represented the majority of the AUC. Half-life ranged from 20–30 hours. DOXIL has a significantly higher AUC, a lower rate of clearance and a smaller volume of distribution compared to the same dose of Adriamycin (Table 4).

Table 4

Pharmacokinetic parameters of DOXIL and Adriamycin in rat, rabbit and dog after a single dose

	$t_{1/2}$ (hr)	AUC_∞ (μg-hr/mL)	CL (ml/hr)	V_β (mL)
Rat				
DOXIL, 1 mg/kg	λ_1: 1.8 λ_2: 23.6	683	0.4	13
Adriamycin, 0.9 mg/kg	λ_1: 0.16 λ_2: 29.1	11.1	24.3	1014
Rabbit				
DOXIL, 1 mg/kg	λ_1: 0.5 λ_2: 21.3	368	6.0	176
Adriamycin, 1 mg/kg	λ_1: 0.03 λ_2: 4.07	1	2536	13651
Dog				
DOXIL, 1.5 mg/kg	λ_1: 0.20 λ_2: 25.9	656	15.5[a]	596
Adriamycin	ND[b]	ND	ND	ND

[a]Vss = Volume of distribution at steady state.
[b]ND = Not determined.

More than 93% and perhaps more than 99% of the doxorubicin measured in plasma during the two phases is liposome-encapsulated. Plasma doxorubicin concentration and AUC were dose-dependent after treatment with DOXIL, but plasma half-life, mean residence time, volume of distribution and clearance were not.

After a single intravenous dose of DOXIL, peak tissue concentrations of doxorubicin were lower and occurred later than after a single Adriamycin treatment. Doxorubicin persisted in the tissues in DOXIL-treated animals compared to animals treated with Adriamycin. Tumor levels and AUCs were higher in DOXIL-treated animals after equivalent doses of DOXIL and Adriamycin.[35] Doxorubicin levels rose in tissues of rats with repeated dosing of DOXIL, but tissue levels remained dose-proportionate even at the end of the dosing period, suggesting that saturation of tissues had not occurred. No accumulation of drug was seen in the heart, and, despite the apparent accumulation of doxorubicin in other tissues after multiple DOXIL treatments, no evidence of increased toxicity of DOXIL was observed, with the exception of cutaneous lesions. Doxorubicin concentrations were higher than in normal skin in rats and dogs, but decreased rapidly as the lesions healed, approaching levels seen in normal skin by five weeks after the last dose. In the multiple dose studies in rats, rabbits and dogs, DOXIL was less cardiotoxic and marginally less myelotoxic than was an equivalent dose of Adriamycin. The decreased relative toxicity of DOXIL is believed to be due to the decreased peak concentration of free, non-liposomal doxorubicin in the plasma and tissues of treated animals.

SPI-77. The plasma pharmacokinetics of SPI-77 are significantly different than those of cisplatin. They are similar to the pharmacokinetics of DOXIL, with substantially increased plasma concentrations and AUC, as well as reduced

Table 5

Pharmacokinetics of SPI-77 and cisplatin in rats after a single dose

Parameter	SPI-77 5 mg/kg	Cisplatin 5 mg/kg
C_{max} (μg/mL)	79.7 ± 6.7	20.5 ± 6.7
Half-life (hr)	40.6 ± 4.3	α 0.03 ± 0.01 β 93.3 ± 9.8
$AUC_{(0-\infty)}$ (μg-hr/mL)	5358 ± 540	51.1 ± 2.5
Clt (mL/hr)	0.52 ± 0.1	28.1 ± 0.4
Vss (mL)	18.5 ± 1.2	1909 ± 190.1

clearance and volume of distribution compared to non-liposomal drug (Table 5). In monkeys, for example, the pharmacokinetics of total plasma platinum following SPI-77 were best described by a one-compartment model with elimination by both a nonlinear and linear process.[36] The pharmacokinetics of cisplatin, in contrast, were best characterized as a linear two-stage model. The maximum plasma concentration of cisplatin measured at the end of the infusion was 50.1 ± 5.3 μg/mL in monkeys given 2.5 mg/kg SPI-77, compared to 8.0 ± 0.8 μg/mL in monkeys treated with 2.5 mg/kg cisplatin. In the SPI-77-treated animals, AUC was increased 14-fold in the SPI-77 group (3946 vs. 278 μg/mL h), volume of distribution (Vss) was decreased 17-fold (186 vs. 3182 mL) and clearance was decreased 70-fold (0.53 vs. 37 mL/h). Thus, SPI-77 pharmacokinetics are consistent with the pharmacokinetics of sterically stabilized liposomes, with substantially increased plasma concentrations and AUC, as well as reduced clearance and volume of distribution relative to non-liposomal drug.[37] Plasma clearance was monoexponential in all three species evaluated, with apparent half-lives ranging from 41 hours in rats (5 mg/kg dose) to 98 hours after three doses in monkeys (30 mg/kg dose).

Cisplatin does not leak from SPI-77 liposomes in plasma in vitro, suggesting that the prolonged clearance represents the clearance of liposome-encapsulated drug, not free or protein-bound cisplatin or platinum. Supporting this notion, the volume of distribution approximated blood volume in all species evaluated and did not change with increasing dose levels, suggesting that the disposition of cisplatin in SPI-77 liposomes is controlled by its liposome carrier, i.e., the drug is retained in plasma, within the liposomes, for a prolonged period, as is characteristic of drugs encapsulated in Stealth liposomes. Studies on the tissue disposition of SPI-77 have yet to be completed, but it is anticipated that it will be similar to that of DOXIL and other compounds encapsulated in Stealth liposomes.

V. Comparative toxicity

AMPHOTEC. Studies in three species (mice, rats and dogs) that employed daily dosing for up to 13 weeks have demonstrated that AMPHOTEC induces a similar spectrum of adverse effects to Fungizone, but only at dose levels four- to five-fold higher.[24,25,27] No toxicities unique to AMPHOTEC were observed, despite the markedly different disposition of amphotericin B after administration of each product.

The most common finding in AMPHOTEC-treated animals was dose-related renal tubular nephrosis similar to the renal damage produced by Fungizone except that it occurred at higher dose levels. The renal toxicity, as assessed by clinical chemistry and histopathological changes, was observed after as few as 14 daily doses and, in general, was partially or fully reversible, particularly at the dose level of AMPHOTEC below 5 mg/kg/day in rats and 1 mg/kg/day in dogs. The kidney is a known target organ of amphotericin B, so the occurrence of nephrotoxicity after treatment with AMPHOTEC is not unexpected. The histology of the renal changes were very similar in AMPHOTEC- and Fungizone-treated animals, suggesting that both were caused by the action of amphotericin B. Renal toxicity of comparable severity and kidney concentrations of amphotericin B of comparable magnitude were only seen after doses of AMPHOTEC four- to five-times higher than those of Fungizone. Partial or complete functional and histological reversal was seen in most AMPHOTEC-treated animals, particularly at the lower dose levels. Hepatotoxicity, also common after Fungizone treatment, was observed at high doses of AMPHOTEC. However, despite significant accumulation of amphotericin B in the liver (up to 20-fold), hepatotoxicity was often less notable after AMPHOTEC treatment.[24,25,27] There was no evidence of lipid storage lesions in the RES.

The second component of AMPHOTEC is sodium cholesteryl sulfate (SCS). No adverse effects unique to the AMPHOTEC formulation have been observed, suggesting that the addition of SCS does not impart any additional or novel toxicity to AMPHOTEC. Separate safety studies have not been conducted with SCS alone, since it cannot be formulated separately from amphotericin B in an equivalent aqueous formulation that could be used in relevant toxicity studies. SCS occurs naturally in mammalian tissues, body fluids and erythrocytes.[38,39] It is believed to act as a membrane stabilizer, have a role in capacitation in human sperm and as a marker of differentiation in a variety of cultured cell lines. There are many active enzyme systems for the metabolism of SCS, including the sterol sulfatases, that are difficult to saturate because of their redundancy.[38] SCS appears to be a relatively non-toxic substance that can readily be tolerated at the dose levels that are administered during AMPHOTEC treatment.

The decreased toxicity of AMPHOTEC appears to be due to its greater stability in plasma and its rapid and nearly complete uptake by the liver compared to the conventional formulation of amphotericin B.[29] Nearly the entire dose of AMPHOTEC is found in the liver within 30 minutes of treatment in rats, with only a minimal level of drug in other tissues.[24] This results in a markedly decreased peak plasma concentration of amphotericin B and significantly reduced distribution of the drug to the kidney and other known target organs of amphotericin B. Despite the storage of amphotericin B in the liver in a releasable form after AMPHOTEC treatment, there has been no evidence of a hepatotoxic action except after very high doses of AMPHOTEC

In summary, single and repeat dose toxicity studies show that intravenous administration of AMPHOTEC induces minimal to mild renal toxicity at most dose levels evaluated. The renal toxicity, as assessed by clinical chemistry and histopathological changes, is partially or fully reversible, particularly at lower doses of AMPHOTEC. Despite the increased levels of amphotericin B in the liver, spleen, and bone marrow of AMPHOTEC-treated animals, there was no evidence of increased toxicity in these tissues. Overall, the toxicity of AMPHOTEC was comparatively less than that of Fungizone at the same dose levels. In general, dose levels of AMPHOTEC 4- to 5-fold higher than Fungizone induced adverse effects of similar incidence and severity. In no study were any toxicities unique to AMPHOTEC observed; those seen were common to both the AMPHOTEC- and Fungizone-treated groups.

DOXIL. Formal safety studies supported the observation in the tumor model studies that DOXIL was significantly less toxic than Adriamycin. Toxicity observed following iv administration of single doses of DOXIL to mice, rats, rabbits and dogs was qualitatively similar in the nature of the response in each species.[33,40] Dogs were the most sensitive species, exhibiting severe gastrointestinal toxicity in response to high doses or rapid infusions of DOXIL; slowing the infusion rate almost completely alleviated the gastrointestinal toxicity. Treatment-related toxicity included hematologic changes, myelotoxicity, marginal clinical symptoms, gastrointestinal toxicity, reversible cutaneous lesions and alopecia, which, with the exception of the dermal lesions, were generally less severe in DOXIL-treated animals.

The toxicity profile of DOXIL following repeated administration was similar in rats and dogs and was an extension of the findings seen in the acute studies. Treatment-related effects included skin ulcers, particular on the legs and feet, body weight and food consumption changes, mild alopecia, myelotoxicity (bone marrow cellularity changes), testicular atrophy, hematologic effects (leukopenia and anemia) and, in rats and rabbits, limited cardiomyopathy. Dogs, again the more sensitive species, also exhibited gastrointestinal toxicity, but no pathologic signs of cardiotoxicity. With the exception of cardiotoxicity, testicular atrophy and persistent alopecia, adverse effects were fully resolved during the recovery periods that followed each repeat dose study.

DOXIL was less nephrotoxic than Adriamycin all species in which the drugs were compared. There was no evidence of nephrotoxicity in DOXIL-treated animals, although nephrotoxicity is a known adverse effect of Adriamycin in animals, particularly rodents and dogs. Myelotoxicity was observed in rats and dogs following DOXIL treatment, but at reduced level of severity compared to animals that received the same cumulative dose of Adriamycin.

Doxorubicin-related cardiotoxicity seen in DOXIL-treated animals was both less severe and less frequent, and was not simply due to a longer latency period prior to its development. As much as 50% more DOXIL could be administered than Adriamycin without incurring an equivalent risk of cardiomyopathy. Doxorubicin-associated nephrotoxicity and cardiotoxicity are well correlated with high peak plasma levels. The decreased relative toxicity of DOXIL is believed to be related to the decreased peak concentration of doxorubicin that has been measured in the plasma and tissues of treated animals compared to that measured after administration of the same dosage of Adriamycin.[40] Because anthracycline-related cardiotoxicity is the most significant toxicity associated with doxorubicin-based drugs, a study was conducted in rabbits to evaluate the comparative cardiotoxicity of DOXIL and Adriamycin.[41] The rabbit is frequently used to evaluate the effects of anthracycline-related cardiotoxicity and is considered a clinically relevant model because of the similarity of histologic changes in rabbits treated with anthracyclines to those that occur in humans after prolonged doxorubicin HCl administration.[42,43] Overall, cardiotoxicity occurred in 67% of Adriamycin-treated rabbits (cumulative dose = 12–14 mg/kg), compared to 16% of DOXIL-treated animals (cumulative dose = 14–21 mg/kg). Among animals that died prior to scheduled necropsy, five Adriamycin-treated rabbits died of congestive heart failure or with histologic evidence of moderate to severe cardiotoxicity. No rabbits treated with DOXIL died of congestive heart failure, though two animals that died prior to scheduled necropsy showed histopathologic evidence of mild cardiotoxicity. Cardiomyopathy increased both in incidence and severity with time after Adriamycin treatment, but not after treatment with DOXIL.

In general, the toxicity observed in animals following DOXIL administration was similar to but less severe than the known toxicity of Adriamycin, with several significant exceptions. At the same cumulative dose of Adriamycin and DOXIL, cardiotoxicity was present at substantially decreased incidence and severity or entirely absent in rats, rabbits and dogs administered DOXIL. There was no evidence of nephrotoxicity in DOXIL-treated animals, although nephrotoxicity is a known adverse effect of Adriamycin. Myelotoxicity was observed in rats and dogs following DOXIL administration, but at reduced level of severity compared to animals that received the same cumulative dose of Adriamycin. The decreased relative toxicity of DOXIL is likely related to the decreased peak concentration of free, non-liposomal doxorubicin in the plasma and tissues of treated animals.

SPI-77. An extensive series of GLP safety studies were conducted in rodents, rabbits, dogs and monkeys with SPI-77 prior to its entry into the clinic. These studies have shown that SPI-77 has a markedly different safety profile than cisplatin. The primary adverse effects induced by single and multiple-dose treatments of SPI-77 are hepatic and biliary tract toxicity and, somewhat more variably, effects on the erthyroid cell series. Reversible lipoidal glomerulopathy was seen in rats only. All SPI-77-related changes showed a clear dose-response relationship within species, and, accounting for rather marked differences in species sensitivity, were similar among all five species tested. There was no evidence of the renal toxicity associated with the administration of non-liposomal cisplatin in any species tested. SPI-77 did not cause local irritation when injected intravenously, nor was it hemolytic in human blood.

SPI-77 is significantly less acutely toxic than cisplatin in mice, rats, rabbits, dogs and monkeys. The LD_{50} of SPI-77 could not be determined due to its relative lack of single dose toxicity. Rats and mice readily tolerated single doses of SPI-77 3- to 4-fold higher than lethal doses of cisplatin (Table 6). Dogs and monkeys

Table 6

LD_{50} (mg/kg) of SPI-77 in mice and rats

Species	SPI-77 LD_{50}	Cisplatin LD_{50}
Mouse	>50	14.7 ± 1.3
Rat	>30	8.1 ± 1.0

Survival for 14 days post-dose; values calculated by probit analysis.

also tolerated single doses of SPI-77 (up to 15 and 25 mg/kg, respectively) that were many fold higher than lethal and/or toxic doses of cisplatin.

SPI-77 was similarly much less toxic than cisplatin with repeated doses. In dogs and monkeys, for example, multiple doses of SPI-77 do not induce emesis at dose levels up to 10-fold cisplatin dose levels that are associated with significant rates of emesis, are not nephrotoxic at cumulative doses of up to 1500 mg/m^2 (body surface area equivalent dose), and are up to 10-fold less neurotoxic than cisplatin.[44] The primary adverse effect associated with repeated treatments with SPI-77 is minimal decreases in red blood cell counts and minimal increases in alanine aminotransferase and indirect bilirubin levels in the plasma. These changes rarely went outside normal ranges, and all indices returned to baseline prior to each subsequent dose.

Though early in its development, SPI-77 shows potential as a much less toxic alternative to cisplatin, with similar or greater antitumor activity in animal models.

VI. Conclusions

Studies conducted with one lipid-complexed drug and two true liposome-encapsulated drugs demonstrate the potential advantages of these types of formulations. Typically, toxicity is reduced and, in some cases at least, efficacy is actually increased, particularly in the case of the "second generation" Stealth liposomes described here. In either case, the therapeutic ratio of the re-formulated drug is improved. Though shown in this chapter only for three products from SEQUUS, other formulations of other drugs from other pharmaceutical companies show similar promise, both in animals and in human patients. Though long delayed, it seems that the liposome may finally be realizing its therapeutic potential.

References

1. Napper DH. Polymeric Stabilization of Colloidal Dispersions. Academic Press, New York, NY, 1983.
2. Lasic DD, Martin FJ, Gabizon A, Huang SK, Papahadjopoulos D. Sterically stabilized liposomes: a hypothesis on the molecular origin of the extended circulation times. Biochim Biophys Acta 1991;1070:187–192.
3. Woodle MC; Lasic DD. Sterically stabilized liposomes. Biochim Biophys Acta 199?;199,1113:171–199.
4. Hanson HL, Stevens DA. Comparison of antifungal activity of amphotericin B deoxycholate

suspension with that of amphotericin B cholesteryl sulfate colloidal dispersion. Antimicrob Agents Chemother 1992;36:486–488.
5. Manuscript in preparation.
6. Unpublished data, SEQUUS Pharmaceuticals.
7. Hostetler JS, Clemons KV, Hanson LH, Stevens DA. Efficacy and safety of amphotericin B colloidal dispersion compared with those of amphotericin B deoxycholate suspension for treatment of disseminated murine cryptococcosis. Antimicrob Agents Chemother 1992;36:2656–2660.
8. Clemons KV, Stevens DA. Comparative efficacy of amphotericin B colloidal dispersion compared with those of amphotericin B deoxycholate suspension in treatment of murine coccidiodomycosis. Antimicrob Agents Chemother 1991;35:1829–1833.
9. Allende MC, Lee JW, Francis P, Garrett K, Bacher J, Berenguer J, Lyman CA, Pizzo PA, Walsh, TJ. Dose-dependent antifungal activity and nephrotoxicity of Amphotericin B Colloidal Dispersion (ABCD) in experimental pulmonary aspergillosis. Antimicrob Agents Chemother 1994;38:518–522.
10. Walsh TJ, Garrett K, Feuerstein E, Girton M, Allende M, Bacher J, Francesconi A, Schaufele R, Pizzo PA. Therapeutic monitoring of experimental pulmonary aspergillosis by ultrafast computerized tomography, a novel noninvasive method for measuring responses to antifungal therapy. J Antimicrob Agents Chemother 1995;39:1065–1069.
11. Berman JD, Ksionski G, Chapman WL, Waits VB, Hanson WL. Activity of Amphotericin B Colloidal Dispersion (ABCD) in experimental visceral leishmaniasis. Antimicrob Agents Chemother 1992;36:1978.
12. Clemons KV, Stevens DA. Comparison of Fungizone, AMPHOTEC, AmBisome and Abelcet against systemic marine cryptococcosis. Antimicrob Agents Chemother 1998; in press.
13. Vaage J, Donovan D, Mayhew E et al. Therapy of human ovarian carcinoma xenografts using doxorubicin encapsulated in sterically stabilized liposomes. Cancer 1993;72:3671–3675.
14. Siegal T, Horowitz A, Gabizon A Doxorubicin encapsulated in sterically stabilized liposomes for the treatment of a brain tumor model: biodistribution and therapeutic activity. J Neurosurg 1995;83:1029–1035.
15. Huang SK, Mayhew E, Gilani S, Lasic DD, Martin FJ, Papahadjopoulos D. Pharmacokinetics and therapeutics of sterically stabilized liposomes in mice bearing C-26 colon carcinoma. Cancer Res 1992;52:6774–6781.
16. Vaage J, Mayhew E, Lasic DD, Martin F. Therapy of primary and metastatic mouse mammary carcinomas with doxorubicin encapsulated in long-circulating liposomes. Int J Cancer 1992;51:942–948.
17. Williams SS, Alosco TR, Mayhew E, Lasic DD, Martin FJ, Bankert RB. Arrest of human lung tumor xenograft growth in severe combined immunodeficient mice using doxorubicin encapsulated in sterically stablized liposomes. Cancer Res 1993;53:3964–3967.
18. Vaage J, Donovan D, Uster P, Working P. Tumour uptake of doxorubicin in polyethylene glycol-coated liposomes and therapeutic effect against a xenografted human pancreatic carcinoma. Brit J Cancer 1997;75:482–486.
19. Vaage J, Donovan D, Loftus T, Uster P, Working P. Prophylaxis and therapy of mouse mammary carcinomas with doxorubicin and vincristine encapsulated in sterically stabilised liposomes. Eur J Cancer 1995;31A:367–372.
20. Vaage J, Donovan D, Loftus T, Working, P. Prevention of metastasis from spontaneous mouse carcinomas using long-circulating pegylated liposomes carrying doxorubicin. Brit J Cancer 1995;72:1074–1075.
21. Vail DM, Kravis LD, Cooley AJ, Chun R, McEwen EG. Preclinical trial of doxorubicin entrapped in sterically stablized liposomes in dogs with spontaneously arising malignant tumors. Cancer Chemother Pharmacol 1997;39:410–416.
22. Vail DM, Chun R, Thamm DH, Garrett LD, Cooley AJ, Obradovich JE. Efficacy of pyridoxine to ameliorate the cutaneous toxicity associated with doxorubicin-containing pegylated (Stealth) liposomes: A randomized double-blind clinical trial using a canine model. Clin Cancer Res, submitted.
23. Engbers CM, Daniel BE, Newman MS, Working PK. Efficacy of SPI-77 (Stealth liposomal cisplatin) in murine colon carcinoma C26 tumor bearing mice. Amer Assoc. Pharm Sci, Western Regional Meeting, 1997.
24. Fielding RM, Smith PC, Wang LH, Porter J, Guo LSS. Comparative pharmacokinetics of amphotericin B after administration of a novel colloidal delivery system, ABCD, and a conventional formulation in the rat. Antimicrob Agents Chemother 1991;35:1208–1211.
25. Wang LH, Fielding RM, Smith PC, Guo LSS. Comparative tissue distribution and elimination of

amphotericin B colloidal dispersion (Amphocil) and Fungizone after repeated dosing in rats. Pharmaceu Res 1995;12:275–283.

26. Daniel B, Guo L, Chou E, Najafi A, Gittelman J, Working PK. Cellular localization of ABCD in the liver. Amer Assoc Pharm Sci, Western Regional Meeting, 1995.
27. Fielding RM, Singer AW, Wang LH, Babbar S, Guo LSS. Relationship of pharmacokinetics and drug distribution in tissue to increased safety of amphotericin B colloidal dispersion in dogs. Antimicrob Agents Chemother 1992;36:299–307.
28. Newman MS, Daniel B, Engbers C, Brest N, Guo LSS, Working PK. Determination of circulating form of ABCD in rat plasma. Amer Assoc Pharm Sci, Western Regional Meeting, 1995.
29. Guo LSS, Fielding RM, Gantz DL, Steiner J, Small DM. Novel antifungal drug delivery: stable amphotericin B-cholesteryl sulfate discs. Int J Pharmaceu 1991;75:45–54.
30. Koldin MH, Kobayashi GS, Brajtburg J, Medoff G: Effects of elevation of serum cholesterol and administration of amphotericin B complexed to lipoproteins on amphotericin B-induced toxicity in rabbits. Antimicrob Agents Chemother 1985;28:144–145.
31. Papahadjopoulos D, Allen TM, Gabizon A, Mayhew E, Matthay K, Huang SK, Lee KD, Woodle MC, Lasic DD, Redmann C, Martin FJ. Sterically stabilized liposomes: Improvements in pharmacokinetics and antitumor therapeutic efficacy. Proc Natl Acad Sci USA 1991;88:11460–11464.
32. Gabizon A, Barenholz Y, Bialer M. Prolongation of the circulation time of doxorubicin encapsulated in liposomes containing a polyethylene glycol-derivatized phospholipid: Pharmacokinetic studies in rodents and dogs. Pharm Res 1993;10:703–708.
33. Working PK, Dayan AD. Pharmacological-toxicological Expert Report: CAELYX (Stealth® liposomal doxorubicin HCl). Human Exp Toxicol 1996;15:752–785.
34. Working PK, Newman MS, Carter JL, Kerl RE, Kiopes AL. Comparative target organ toxicity of free and liposome-encapsulated doxorubicin in rats. The Toxicologist 1994;14:217.
35. Engbers C, Newman MS, Working PK. Blood and tissue distribution of doxorubicin in C26 tumor-bearing mice after treatment with DOX-SL or Adriamycin. American Association of Pharmaceutical Scientists, Western Regional Meeting, 1995.
36. Daniel BE, Newman MS, Working PK, Amantea MA. Pharmacokinetics of SPI-77 (Stealth liposomal cisplatin) after a single intravenous dose in monkeys. Amer Assoc Pharm Sci, Western Regional Meeting, 1997.
37. Woodle MC, Newman MS, Working PK. Biological properties of sterically stabilized liposomes. In: Lasic D, Martin F, eds. Stealth Liposomes. CRC Press, Boca Raton, FL, 1995;103–117.
38. Iwamori M, Moser HW, Kishimoto Y. Cholesterol sulfate in rat tissues. Tissue distribution, developmental changes and brain subcellular localization. Biochim Biophys Acta 1976;441:268–279.
39. Lalumiere G, Longpre J, Trudel J, Chapdelain A, Roberts KD. Cholesterol sulfate. II. Studies on its metabolism and possible function in canine blood. Biochim Biophys Acta 1975;394:120–128.
40. Working PK, Newman MS, Huang SK, Mayhew E, Vaage J, Lasic DD. Pharmacokinetics, biodistribution and therapeutic efficacy of doxorubicin encapsulated in Stealth® liposomes (Doxil®). J Liposome Res 1994;4:667–687.
41. Newman MS, Sullivan T, Yarrington J, Kiorpes A, Carter J, Working PK. Reduced cardiotoxicity of pegylated liposomal doxorubicin (DOXIL) compared to non-liposomal doxorubicin hydrochloride (Adriamycin®) in animals. Human Exp. Toxicol. submitted.
42. Jaenke RS. An anthracycline antibiotic-induced cardiomyopathy in rabbits. Lab Investig 1974;30:292–304.
43. Iatropoulos MJ. Anthracycline cardiomyopathy: Predictive value of animal models. Cancer Treat Symp 1984;3:3–17.
44. Working PK, Newman MS, Sullivan T, Turner N. Reduced toxicity in cynomolgus monkeys of multiple doses of cisplatin encapsulated in pegylated, long-circulating liposomes. Toxicol Sci, submitted.

Lasic and Papahadjopoulos (eds.), Medical Applications of Liposomes

CHAPTER 8.1

Clinical trials of liposomes as carriers of chemotherapeutic agents: Synopsis and perspective

ALBERTO A. GABIZON*

Hadassah-Hebrew University Medical Center, Sharet Institute of Oncology, Jerusalem, Israel

The development of a plethora of new cancer chemotherapeutic agents in the post-war era caused pharmacologists to face many active but toxic drugs without target specificity. Finding drugs with unequivocal target specificity in situations like cancer, in which there are only subtle biologic differences from normal cells, is a major and long-term challenge with, sometimes, frustrating results. The short-term alternative is of course the use of a delivery system that will confer to nonspecific and highly toxic drugs, a "tamed" behavior by redirecting their biodistribution and controlling their rate of release. Thus, as far as cancer therapy is concerned, liposomes can fill in an important vacuum, especially given their intrinsic advantages as drug delivery systems:[1] versatility, biocompatibility, and lack of immunogenicity.

Despite the typical shortcomings of early studies, liposomes soon became attractive devices for drug delivery for large segments of the scientific community. This somewhat magical appeal was largely based on a nebulous scientific rationale and a handful of naivety. Why should liposomes go to sites of disease? Why should liposomes spare some specific tissues? Erlich's magic bullet proposal requires the target to interact with a specific receptor which is unknown for liposomes. The works of the pioneering groups in the 70's were still plagued with the childhood diseases of any novel field. Significant limitations in the ability of liposomes to control drug biodistribution and bioavailability became evident. In addition, a focus on efficient techniques for stable and reproducible entrapment of drugs was still lacking. To deal with these issues, a rational approach linking liposome formulation with drug delivery and its pharmacologic consequences was needed.[2] Depending on the drug and on the target disease, these factors may vary and demand a different strategy. This process is an essential pre-requisite for the development of any drug delivery system. Fortunately enough, the liposome field

*Corresponding Address: Hadassah Medical Ctr., Oncology Dept., P.O.B. 12,000, Jerusalem il-91120, Israel. (Fax: 972-2-643-0622)

did face this challenge and a collective effort from academic and industrial scientists together with significant technological advances has brought about a number of approved liposomal products for systemic administration in humans for the treatment of cancer and infectious diseases. The culmination of this process has been the approval by the U.S. F.D.A. of several liposomal anthracycline and amphotericin-B formulations for clinical use (Doxil, DaunoXome, Abelcet, Amphotec, AmBisome). In addition, there is growing interest in the use of liposomes as carriers of vaccines because of their safety and potent adjuvanticity, and a commercial, liposome-based vaccine is already available in Europe.[2] This is a major accomplishment for a field born only about 30 years ago.

In this introductory article to the clinical developments with liposome-encapsulated drugs, I will concentrate on liposomal anthracyclines. This is probably the area where the largest body of work on liposome-encapsulated drugs has been done. In fact, it is likely that the success or failure of liposomal anthracyclines in taking a sizable niche of the chemotherapy field will be an important test for the future of liposomes in pharmaceutics.

The last twenty years of liposome research have helped to frame a sound rationale for liposomes in cancer drug delivery. This is based on the three following pharmacological principles:

1. Slow release
2. Site avoidance for sensitive tissues
3. Accumulation in tumors

In most cases, liposome encapsulation of a drug will delay bioavailability and confer some degree of slow release, although the magnitude of this effect may vary depending on the stability of liposome-drug association and rate of liposome disruption in blood and tissues. Slow release may decrease the distressing acute side-effects of many cytotoxic drugs, and even attenuate the cardiotoxicity of anthracyclines by reduction of peak levels of free drug.[3]

The relative bulkiness of lipid vesicles will hinder extravasation in tissues with a tight endothelium and continuous basement membrane such as muscle and nervous tissue.[4] As a result, liposomes may decrease the amount of drug available to some tissues. This phenomenon is probably an important factor in the basis for the reduced cardiotoxicity of anthracycline preparations and may potentially reduce the neurotoxicity of agents such as liposomal vincristine or liposomal cisplatin.

Liposome accumulation in tumors is probably the most critical element of the strategy behind the use of liposome-mediated delivery of chemotherapeutic agents. A positive correlation between circulation longevity of liposomes and liposomal drug levels in tumors has been found.[5] The mechanism by which liposomes accumulate in transplantable tumors is apparently related to the increased microvascular permeability of tumors. Liposomes extravasate through the highly permeable microvessels of tumors and remain locked in the interstitial fluid compartment due to a lack of functional lymphatic drainage. This results in a preferential accumulation of liposomes in tumors as opposed to normal tissues. This process, also known as the enhanced permeability and retention effect, has also been described for other macromolecular carriers.[6] Enhanced drug delivery to tumors should result in improved anti-tumor responses in a similar way to high-dose chemotherapy, but without the side-effects of an indiscriminate exposure of body tissues to toxic doses of chemotherapy. The long-circulation profile of polyethylene-glycol coated (also known as Stealth) liposomes[7] endow them with this tumor-localizing property and makes them a promising vehicle for drug delivery in cancer.

The most prolific area of clinical research with liposome-entrapped cytotoxic agents has been with anthracyclines. This family of drugs is well known for its broad spectrum of anti-tumor activity and wide use in cancer therapy.[8] Doxorubicin (Dox), the most widely used anthracycline, is a critical component of chemotherapy regimes in the treatment of breast cancer, sarcomas, lymphomas, and has significant anti-tumor activity against many other neoplasms. Dox, as most other anthracyclines, causes cumulative cardiac toxicity which limits the total dose that can be administered with a relatively low risk to less than 550 mg/m^2.

The potential of liposomal anthracyclines to buffer a number of undesirable side effects of doxorubicin is a clinical fact.[9] Toxicity buffering stems mainly from the slow release of drug from liposomes and from the tissue distribution changes. There are however significant differences among the various liposome formulations which may have a major clinical weight. The unique toxicity of a PEGylated (polyethylene-glycol coated), long-circulating liposome formulation, known as Doxil, to the skin[10] underscores the contention that drastic pharmacokinetic changes may result in a compound with a qualitatively different toxicity profile. Ultimately, the use of liposomal drugs will depend on an assessment of their therapeutic index, i.e., the benefits/risks odds equation. For AIDS-related Kaposi's Sarcoma, Doxil[11] and DaunoXome[12] offer an improved therapeutic index, judged to be sufficient by regulatory authorities for their approval. In the case of common forms of cancer, promising antitumor activity has been observed with Doxil in Phase I-II trials,[10,13] but we still have to wait for the results of definitive Phase III clinical trials that will determine the "benefits/risks" ratio, in reference to the standard forms of therapy currently available.

A historical review of the development of liposomal anthracyclines will provide us with a perspective of the field and with an evaluation of the clinical potential of the liposome approach in cancer therapy. A tabular summary of the major clinical trials with liposomal anthracyclines as published in original articles during the last decade is presented in Table 1. In addition, there is a large number of Proceedings abstracts which will probably mature into publications in the coming years.

The first formulations to be tested in humans were based on liposomes with negatively-charged lipids (phosphatidylglycerol-PG, cardiolipin-CL) with a size range in the order of several hundred nanometers. Dox was entrapped in the liposome bilayer by a combination of electrostatic and hydrophobic interactions. The rationale guiding us during this initial phase of clinical testing was twofold:

1. Buffering of toxicity and, especially, the possibility of reducing cardiotoxi-

Table 1

A summary table of clinical studies with liposomal anthracyclines[a]

Ref. #, year	Liposome comp/size	Drug/dose mg/m^2	No. pts.	Diagnosis group	Study type	Toxicity/efficacy and PK results
Greene et al.[35] 1983[b]	None	Free Dox 75	10	Breast, Sarcoma	PK	$t_{1/2}$ 27 hr; Cl 48 L/hr; V_{ss} 1764 L; V_{β} 1868 L
Delgado et al.[36] 1989[c]	CL:PC:Ch:SA 90 nm	Dox 13–63 q3wk	15	Ovary	IP Phase I/II PK	MTD not reached; 3 PR; Periton./plasma AUC = ~50
Gabizon et al.[17] 1989	PG:PC:Ch 300–500 nm	Dox 20–120 q3wk	32	Colon, hepatoma other	Phase I	MTD 100 mg/m^2 leukopenia/stomatitis; 1 PR
Rahman et al.[28] 1990[c]	CL:PC:Ch:SA 90 nm	Dox 30–90 q3wk	14	NSC lung, unknown 1^{ry} other	Phase I PK	MTD 60–90 mg/m^2 neutropenia; $t_{1/2}$ 7 hr; Cl 8.2 L/hr; V_{ss} 81 L
Treat et al.[37] 1990[c]	CL:PC:Ch:SA 90 nm	Dox 60–75 q3wk	20	Breast	Phase II	Main toxicity: leukopenia 9 PR + CR;
Gabizon et al.[18] 1991	PG:PC:Ch 300–500 nm	Dox 50–120	16	Colon, hepatoma	PK, imaging	$t_{1/2}$ 28 hr; Cl 44 L/hr; V_{ss} 1400 L.
Pestalozzi et al.[38] 1992	PA:PC:Ch 100 nm	Mitox 3–24 q3wk	19	Breast	Phase I/II	MTD 16 mg/m^2 neutropenia; 3 PR
Owen et al.[39] 1992	PG:PC:Ch 100–250 nm	Dox 60–80q3wk /15–25q1wk	22	Stomach, colon, hepatoma	Phase I	MTD 70 mg/m^2 q3wk & 22.5 mg/m^2 q1wk leukopenia; 3 PR
Simpson et al.[40] 1993[d]	PEG/DSPE: HPC:Ch 80–100nm (Doxil®)	Dox 20 q2–3wk	16	AIDS-related Kaposi's sarcoma	Phase II	Main toxicity: neutropenia 11 PR (incl. 5 PR in pretreated pts.);
Cowens et al.[29] 1993	PC:Ch 180 nm (TLC D-99®)	Dox 20–90 q3wk	38	Colon, NSC lung, other	Phase I PK	MTD 75–90 mg/m^2 leukopenia; $t_{1/2}$ 17 hr; Cl 19 L/hr; V_{ss} 160 L.
Conley et al.[30] 1993[e]	TLC D-99®	Dox 20–37.5 q1wkx3	21	Colon, head & neck other	Phase I PK	MTD 20–30 mg/m^2/wk leukopenia; $t_{1/2}$ 12 hr; Cl 10 L/hr; V_{β} 234 L.

Embree et al.[31] 1993[e,f]	TLC D-99®	Dox 60–75 q3wk	12	NSC lung	PK	$t_{1/2}$ 22 hr; Cl 7.4 L/hr; V_γ 208 L.
Presant et al.[41] 1993	DSPC:Ch 50–80 nm (DaunoXome®)	Dauno 40 q2wk	25	AIDS-related Kaposi's sarcoma	Phase II	Main toxicity: neutropenia 62.5% (15/24) PR + CR
Gabizon et al.[33] 1994	Doxil®	Dox 25–50 q3wk	15	Breast, NSC lung, other	Exploratory, PK	Main toxicity: stomatitis; 2 PR; $t_{1/2}$ 46 hr; Cl 0.09 L/hr; V_{ss} 5 L.
Gill et al.[12] 1995	DaunoXome®	Dauno 10–60 q2wk	40	AIDS-related Kaposi's sarcoma	Phase I/II PK	MTD 40–60 mg/m^2 neutropenia; 55% (12/22) PR + CR; $t_{1/2}$ 4–8 hr; Cl 0.4–0.7 L/hr; V_{ss} 3–4 L.
Harrison et al.[11] 1995	Doxil®	Dox 20 q3wk	34	AIDS-related Kaposi's sarcoma	Phase II	Main toxicity: neutropenia; 74% (25/34) PR + CR
Uziely et al.[10] 1995	Doxil®	Dox 20–80 q3–4wk	56	Ovary, breast, colo-rectal, head & neck, other	Phase I	MTD 50–60 mg/m^2 stomatitis, hand-foot tox.; 33% (15/44) PR + Improv.
Gill et al.[42] 1996	DaunoXome®	Dauno 40 q2wk	116	AIDS-related Kaposi's sarcoma	Phase III randomized with standard	Main toxicity: neutropenia 25% PR + CR
Northfelt et al.[43] 1996	Doxil®	Dox 10–40 q2–3wk	9	AIDS-related Kaposi's sarcoma	Exploratory, PK	Main toxicity: neutropenia $t_{1/2}$ 39 hr; Cl 0.1 L/hr; V_c 3.5 L.
Northfelt et al.[44] 1997	Doxil®	Dox 20 q3wk	53	AIDS-related Kaposi's sarcoma	Phase II as 2nd line chemotherapy	Main toxicity: leukopenia 38% PR + CR
Muggia et al.[13] 1997	Doxil®	Dox 50 q4wk	35	Ovarian CA	Phase II as ≥3rd line chemotherapy	Main toxicity: stomatitis and hand-foot 26% (9/35) PR + CR

[a]Comp = composition, not including minor components such as α-tocopherol; pts. = patients; PK = pharmacokinetics including the following parameters, terminal $t_{1/2}$, Cl (plasma clearance), and V_c, V_{ss}, V_β or V_γ (volume of distribution of central compartment, at steady state, β or γ phases, respectively), presented as mean/median values of all individual or group data available; MTD defined as highest dose level causing dose-limiting toxicity. PR/CR = partial/complete tumor regression (i.e., treatment efficacy); qXwk = every X week(s); IP = intraperitoneal; SA = stearylamine. If needed for calculations, body wt. and surface estimated at 70 kg and 1.6 m^2 respectively. For other abbreviations, see text. DaunoXome, Doxil, and TLC D-99 are registered trademarks of NeXstar Inc., Sequus Pharmaceuticals Inc., and The Liposome Co. This table has been updated and modified based on Ref. 45.

[b]Presented as comparator for PK data.

[c]Information on liposome size obtained from Ref. 46.

[d]Information on the formulation obtained from Ref. 32.

[e]Information on liposome size obtained from Ref. 29.

[f]The patients of this PK study were treated and clinically evaluated under a phase I/II study reported by Lohri et al. (Abstract in Proc ASCO 1991;10:106.)

city, which was based on the decreased uptake of liposomal Dox by the heart and on the actual or reduced systemic and cardiac toxicities in preclinical studies.[14]

2. Enhancement of therapeutic efficacy in liver tumors. We perceived the large increase of drug delivered to the liver by liposomes as a tool of potential therapeutic value in the treatment of liver tumors. This was strengthened by the observations in some mouse liver tumor models of a superior efficacy of liposomal Dox over free Dox.[15,16]

In line with this rationale, we performed a Phase I study in the mid 1980's with PG-PC-Ch (phosphatidylglycerol-phosphatidylcholine-cholesterol) liposomes in the size range of ~300 to 500 nm in a group of patients with primary and secondary liver tumors.[17] The initial toxic dose and maximal tolerated dose were found to be 120 and 100 mg/msq respectively, which were considered as an indication of decreased toxicity of liposomal Dox, since the MTD of free Dox is only 75 mg/msq. The dose-limiting toxicities were similar to those of free doxorubicin, i.e. myelosuppression and stomatitis. Disappointingly, there was minimal anti-tumor activity in this population of patients with hepatomas and metastatic liver tumors. Imaging studies with In-111 labeled liposomes in a similar group of patients showed rapid accumulation in the RES of liver and spleen with no evidence of significant tumor uptake in most instances, including intra-hepatic metastases.[18] An implication of these observations with regard to delivery of liposome-encapsulated drugs is that the drug biodistribution may be shifted toward the RES, at the expense of tumor and probably other tissues. Thus, while the avid RES uptake of liposomes may help to buffer toxicity, it will also interfere with tumor exposure to drug and therefore with anti-tumor activity. In retrospect, it was obvious that the encouraging observations on therapeutic efficacy with the PG-based formulations were due to the choice of specific tumor models, mouse leukemia or lymphoma with diffuse spread in immediate proximity to the RES of liver, spleen, and bone marrow, which are anatomic sites readily accessible to circulating liposomes. In contrast, the common tumors involving the liver in humans are of epithelial type and form solid nodules devoid of Kupffer cells. An additional hurdle is the leakage of the drug from circulating liposomes as indicated by human studies using Dox encapsulated in PG-PC-Ch liposomes.[18] Clearly, improved formulation properties were needed to enable liposomes to stay longer in circulation without leaking their drug contents.

Recent studies have indicated that specific types of liposomes, referred to as Stealth or sterically stabilized liposomes can circulate in blood for prolonged periods of time without being trapped in the RES. The best example of Stealth liposome is given by formulations of small-sized vesicles (<100 nm diameter) containing a small fraction of a phospholipid derivatized with a hydrophilic polymer, polyethyleneglycol (PEG), which have been shown to alter dramatically the pharmacokinetic properties of liposome-encapsulated Dox in rodents and dogs leading to long distribution half-lives without detectable leakage of drug in circulation. One key feature of these long-circulating liposomes is their ability to accumulate in transplantable tumors, by exploiting the increased permeability of tumor microvascularization. This feature confers to long-circulating liposomes a significant selectivity in drug delivery to tumors, which may be of important value in cancer chemotherapy. The reader is referred to several publications addressing in detail the physicochemical and pharmacologic properties of Stealth liposomes.[7, 19–21]

No less important than the lipid composition of the liposomes and vesicle size characteristics, is the method of drug loading.[22] Given the high affinity of Dox for negatively-charged phospholipids, most investigators exploited this property to achieve high loading of Dox into the vesicles. PG, phosphatidylserine (PS), and CL have been successfully used as liposome components to enhance the drug loading efficiency. It has also been shown that mitoxantrone (Mitox) is encapsulated very efficiently in liposomes through its interaction with phosphatidic acid (PA), another negatively charged phospholipid.[23] However, when liposomes are loaded with drug intercalated or associated to the vesicle bilayer, the in vivo stability becomes very difficult to control due to the ability of the drug to rapidly exchange with plasma proteins and equilibrate with the external milieu upon large volume dilution.[24] A critical development in the design of Dox liposomes was the use of remote loading methods enabling the encapsulation of drug in the water phase of preformed liposomes. These are exemplified by transmembrane pH and ammonium sulfate gradients[25–26] that drive Dox into the liposome water phase. Once inside, the drug is ionized thus preventing its escape through the lipid bilayer. This is followed by formation of a gel-like precipitate when the drug reaches a concentration in liposomes exceeding its aqueous solubility.[27]

Regarding the pharmacokinetics of liposomal Dox, several important pieces of information can be obtained from the human studies reported (Table 1), especially when attention is paid to the clearance (Cl) and volume of distribution (V) values. Pharmacokinetic data are available on five different liposome formulations. The Georgetown University formulation[28] shows relatively low Cl and V values, but with a short $t_{1/2}$. This is consistent with a low leakage rate but also with a fast removal of liposomes from the circulation. The results that we obtained with a PG-based formulation[18] point at a large V_{ss}, much like that of free Dox, suggesting a significant leakage of drug in free form, obviously an unwanted effect. The results obtained with D-99[29–31] suggest a modest reduction of Cl and V, but when a side-by-side comparison was made with free Dox,[30] the differences do not appear to be major and probably reflect a mixed pattern of liposome clearance and drug leakage. Gill et al.[12] and Forssen and Ross[32] reported on the pharmacokinetics of daunorubicin (Dauno), a Dox analog, encapsulated in small vesicles made of distearoyl-PC (DSPC) and Ch. The Cl and V values are quite small. However, $t_{1/2}$ is relatively short (4–8 hr) suggesting that these liposomes do not have a prolonged circulation time, although they appear to be very stable in terms of preventing drug leakage. In a study in which the pharmacokinetics of free Dox and liposomal Dox were compared side-by-side, we reported striking changes in the pharmacokinetic parameters using a Stealth liposomal formulation (Doxil) containing a PEG-derivatized phospholipid.[33] Thus, Cl is reduced dramatically

(<0.1 L/hr) and V_{ss} becomes roughly equivalent to the blood volume. These values indicate that there is no significant drug leakage from Doxil; and furthermore, in view of the long $t_{1/2}$ (46 hr), one may conclude that Doxil circulates in intact form for prolonged time in agreement with preclinical data. A simple explanation for the difference between Doxil and DaunoXome is that the latter is cleared by the liver (RES) faster than the former, due to the lack of the PEG coating. This is supported by the results of a recent animal study addressing the role of PEG coating on liposome biodistribution.[34]

The significance of pharmacokinetic changes on the therapeutic index of Dox is uncertain although significant differences have been observed in toxicity profile with Doxil[13] (increased mucocutaneous toxicity, less alopecia, less myelosuppression), and the M.T.D. appears to increase slightly in the case of DaunoXome[32] (100 to 120 mg/msq vs. 90 mg/msq for free Dauno). It is obvious that the opportunities to achieve a change in the drug pharmacodynamics will improve as the pharmacokinetic differences between free and liposomal drug become consistently greater and sharper. Therefore, it is not surprising that these two very stable formulations are clinically the most different ones from free Dox.

The progress in the liposome industry field leading to large-scale, pharmaceutically-acceptable liposome preparations will enable during the next few years to test the therapeutic activity of liposomal anthracyclines in wide phase III studies against the common types of solid tumors. The most logical choice for these studies would be to focus on tumors moderately sensitive to Dox to detect whether improved response rate and survival are attainable. Another important area to explore is the combination of liposomal anthracyclines with other chemotherapeutic agents or other therapeutic modalities (radiotherapy hyperthermia, cytokines). The key issue for success of a liposome-delivered chemotherapeutic agent is increased efficacy. Prevention of the anthracycline-induced cardiotoxicity may have definitely some utility, but improved antitumor activity in solid tumors remains the major aim for such a sophisticated technology.

References

1. Szoka FC. Liposomal drug elivery: Current status and future prospects. In: Wilschut J, Hoekstra D, eds. Membrane Fusion. Marcel Dekker Inc, New York, 1991;845–890.
2. Gregoriadis G. Enginering liposomes for drug delivery: progress and problems. Trends in Biotechnology (Elsevier) 1995;13:527–536.
3. Legha SS, Benjamin RS, Mackay B, Ewer M, Wallace S, Valdivieso M, Rasmussen SL, Blumenschein GR, Freireich EJ. Reduction of doxorubicin cardiotoxicity by prolonged continuous infusion. Ann Intern Med 1982;96:133–139.
4. Gerlowski LE, Jain RK. Extravascular transport in normal and tumor tissues. Crit Rev Oncol Hematol 1986;5:115–170.
5. Gabizon A, Papahadjopoulos D. Liposome formulations with prolonged circulation time in blood and enhanced uptake by tumors. Proc Natl Acad Sci-USA 1988;85:6949–6953.
6. Maeda H. SMANCS and polymer-conjugated macromolecular drugs: advantages in cancer chemotherapy. Adv Drug Deliv Rev 1991;6:181–202.
7. Lasic D, Martin F. *Stealth Liposomes*. CRC Press, Boca Raton, 1995.
8. Doroshow JH. Anthracyclines and anthracenediones. In: Chabner BA, Longo DL, eds. Cancer Chemotherapy and Biotherapy. Lippincott-Raven, Philadelphia, 1996;409–434.
9. Gabizon A. Liposomal anthracyclines. Hematol Oncol Clin North Am 1994;8:431–450.

10. Uziely B, Jeffers S, Isacson R, Kutsch K, Wei-Tsao D, Yehoshua Z, Muggia FM, Gabizon A. Liposomal doxorubicin: Antitumor activity and unique toxicities during two complementary phase I studies. J Clin Oncol 1995;13:1777–1785.
11. Harrison M, Tomlinson D, Stewart S. Liposome-entrapped doxorubicin: An active agent in AIDS-related Kaposi's Sarcoma. J Clin Oncol 1995;13:914–920.
12. Gill PS, Espina BM, Muggia F, Cabriales S, Tulpule A, Esplin JA, Liebman HA, Forssen E, Ross ME, Levine AM. Phase I/II clinical and pharmacokinetic evaluation of liposomal daunorubicin. J Clin Oncol 1995;13:996–1003.
13. Muggia FM, Hainsworth JD, Jeffers S, 12 co-authors. Phase II study of liposomal doxorubicin in refractory ovarian cancer: Antitumor activity and toxicity modification by liposomal encapsulation. J Clin Oncol 1997;15:987–993.
14. Gabizon A, Meshorer A, Barenholz Y. Comparative study of the toxicities of free and liposome-associated doxorubicin after intravenous injection in mice. J Natl Cancer Inst 1986;77:459–469.
15. Gabizon A, Goren D, Fuks Z, Meshorer A, Barenholz Y. Superior therapeutic efficacy of liposome-associated adriamycin in a murine metastatic tumor model. Br J Cancer 1985;51:681–689.
16. Mayhew E, Rustum Y, Vail WJ. Inhibition of liver metastases of M5076 tumor by liposome-entrapped Adriamycin. Cancer Drug Deliv 1983;1:43–58.
17. Gabizon A, Peretz T, Sulkes A, Amselem S, Druckmann S, Ben-Yosef R, Ben-Baruch N, Catane R, Biran S, Barenholz Y. Systemic administration of doxorubicin-containing liposomes in cancer patients: A phase I study. Eur J Cancer Clin Oncol 1989;25:1795–1803.
18. Gabizon A. Chisin R, Amselem S, Druckmann S, Cohen R, Goren D, Fromer I, Peretz T, Sulkes A, Barenholz Y. Pharmacokinetic and imaging studies in patients receiving a formulation of liposome-associated adriamycin. Br J Cancer 1991;64:1125–1132.
19. Papahadjopoulos D, Allen TM, Gabizon A, Mayhew E, Matthay K, Huang SK, Lee KD, Woodle MC, Lasci DD, Redemann C, Martin FJ. Sterically stabilized liposomes: improvements in pharmacokinetics and antitumor therapeutic efficacy. Proc Natl Acad Sci USA 1991;88:11460–11464.
20. Lasic DD, Martin FJ, Gabizon A, Huang SK, Papahadjopoulos D. Sterically stabilized liposomes: A hypothesis on the molecular origin of the extended circulation times. Biochim Biophys Acta 1991;1070:187–192.
21. Woodle MC, Lasic DD. Sterically Stabilized Liposomes. Biochim Biophys Acta 1992;1113:171–199.
22. Barenholz Y, Amselem S, Goren D, Cohen R, Gelvan D, Samuni A, Golden EB, Gabizon A. Stability of liposomal doxorubicin formulations: Problems and prospects. Medicinal Res Rev 1993;13:449–491.
23. Schwendener RA, Fiebig HH, Berger MR, Berger DP. Evaluation of incorporation characteristics of mitoxantrone into unilamellar liposomes and analysis of their pharmacokinetic properties, acute toxicity, and antitumor efficacy. Cancer Chemother Pharmacol 1991;27:429–439.
24. Amselem S, Cohen R, Barenholz Y. In vitro tests to predict in vivo performance of liposomal dosage forms. Chem Phys Lipids 1993;64:219–237.
25. Mayer LD, Tai LC, Bally MB, Mitilens GN, Ginsberg RS, Cullis PR. Characterization of liposomal systems containing doxorubicin entrapped in response to pH gradients. Biochim Biophys Acta 1990;1025:143–151.
26. Haran G, Cohen R, Bar LK, Barenholz Y. Transmembrane ammonium sulfate gradients in liposomes produce efficient and stable entrapment of amphipathic weak basis. Biochim Biophys Acta 1993;1151:201–215.
27. Lasic DD, Frederik PM, Stuart MC, Barenholz Y, McIntosh TJ. Gelation of liposome interior-A novel method for drug encapsulation. FEBS Lett 1992;312:255–258.
28. Rahman A, Treat J, Roh JK, Potkul LA, Alvord WG, Forst D, Woolley PV. A phase I clinical trial and pharmacokinetic evaluation of liposome-encapsulated doxorubicin. J Clin Oncol 1990;8:1093–1100.
29. Conley BA, Egorin MJ, Whitacre MY, Carter DC, Zuhowski EG, Van Echo DA. Phase I and pharmacokinetic trial of liposome-encapsulated doxorubicin. Cancer Chem Pharm 1993;33:107–112.
30. Cowens JW, Creaven PJ, Greco WR, Brenner DE, Tung Y, Ostro M, Pilkiewicz F, Ginsberg R, Petrelli N. Initial clinical (phase I) trial of TLC-D99 (doxorubicin encapsulated liposomes). Cancer Res 1993;53:2796–2802.
31. Embree L, Gelmon KA, Lohr A, Mayer LD, Coldman AJ, Cullis PR, Palaitis W, Pilkiewicz F, Hudson NJ, Heggie JR, Goldie JH. Chromatographic analysis and pharmacokinetics of liposome-encapsulated doxorubicin in non-small lung cancer patients. J Pharm Sci 1993;82:627–634.

32. Forssen EA, and Ross ME. DaunoXome® treatment of solid tumors: Preclinical and clinical investigations. J Liposome Res 1994;4:481–512.
33. Gabizon A, Catane R, Uziely R, Kaufman B, Safra T, Cohen R, Martin F, Huang A, Barenholz Y. Prolonged circulation time and enhanced accumulation in malignant exudates of doxorubicin encapsulated in polyethylene-glycol coated liposomes. Cancer Res 1994;54:987–992.
34. Unezaki S, Maruyama K, Ishida O, Suginaka A, Hosoda J, Iwatsuru M. Enhanced tumor targeting and improved anti-tumor activity of doxorubicin by long-circulating liposomes containing amphipathic poly(ethylene glycol). Int J Pharm 1995;126:41–48.
35. Greene RF, Collins JM, Jenkins JF, Speyer JL, Myers CE. Plasma pharmacokinetics of adriamycin and adriamycinol: Implications for the design of in vitro experiments and treatment protocols. Cancer Res 1983;43:3417–3421.
36. Delgado G, Potkul RK, Treat JA, Lewandowski GS, Barter JF, Forst D, Rahman A. A Phase I/II study of intraperitoneally administered doxorubicin entrapped in cardiolipin liposomes in patients with ovarian cancer. Am J Obstet Gynecol 1989;160:812–819.
37. Treat J, Greenspan A, Forst D, Sanchez JA, Ferrans VJ, Potkul LA, Woolley PV, Rahman A. Antitumor activity of liposome-encapsulated doxorubicin in advanced breast cancer: phase II study. J Natl Cancer Inst 1990;82:1706–1710.
38. Pestalozzi R, Schwendener R, Sauter C. Phase I/II study of liposome-complexed mitoxantrone in patients with advanced breast cancer. Ann Oncol 1992;3:445–449.
39. Owen RR, Sell RA, Gilmore IT, New RR, Stringer RE. A Phase I clinical evaluation of liposome-entrapped doxorubicin (Lip-dox) in patients with primary and metastatic hepatic malignancy. Anticancer Drugs 1992;3:101–107, 1992.
40. Simpson JK, Miller RF, Spittle MF. Liposomal doxorubicin for treatment of AIDS-related Kaposi's sarcoma. Clin Oncol 1993;5:372–373.
41. Presant C, Scolaro M, Kennedy P, Blayney DW, Flanagan B, Lisak J, Presant J. Liposomal daunorubicin treatment of HIV-associated Kaposi's sarcoma. Lancet 1993;341:1242–1243.
42. Gill PS, Wernz J, Scadden DT, 13 co-authors. Randomized phase III trial of liposomal daunorubicin versus doxorubicin, bleomycin, and vincristine in AIDS-related Kaposi's sarcoma. J Clin Oncol 1996;14:2353–2364.
43. Northfelt DW, Martin FJ, Working P, Volberding PA, Russell J, Newman M, Amantea MA, Kaplan L. Doxorubicin encapsulated in liposomes containing surface-bound polyethylene glycol: tumor localization, and safety in patients with AIDS-related Kaposi's sarcoma. J Clin Pharmacol 1996;36:55–63.
44. Northfelt DW, Dezube BJ, Thommes JA, Levine R, Von Roenn JH, Dosik GM, Rios A, Krown SE, DuMond C, Mamelok RD. Efficacy of Pegylated-liposomal doxorubicin in the treatment of AIDS-related Kaposi's sarcoma after failure of standard chemotherapy. J Clin Oncol 1997;15:653–659.
45. Gabizon A. Liposomal delivery of doxorubicin: A review of recent preclinical and clinical studies. In: Puisieux F, Couvreur P, Delattre J, Devissaguet JP, eds. Liposomes: New Systems and New Trends in their Applications. Editions de Sante', Paris, 1995;507–522.
46. Rahman A, Carmichael D, Harris M, Roh JK. Comparative pharmacokinetics of free doxorubicin and doxorubicin entrapped in cardiolipin liposomes. Cancer Res 1986;46:2295–2299.

Lasic and Papahadjopoulos (eds.), Medical Applications of Liposomes

CHAPTER 8.2

Clinical pharmacology and antitumor efficacy of DOXIL (pegylated liposomal doxorubicin)

Francis J. Martin

SEQUUS Pharmaceuticals, Inc., 960 Hamilton Court, Menlo Park, CA 94025, USA

Overview

I. Introduction

I.1. Scientific rationale for liposomal encapsulation of doxorubicin

DOXIL (also known as CAELYX in Europe) is a long-circulating ("Stealth®") liposome formulation. The active drug substance encapsulated in DOXIL is doxorubicin hydrochloride, a cytotoxic anthracycline antibiotic isolated from cultures of *Streptomyces peucetius* var. *caesius*. Doxorubicin interferes with DNA synthesis by interacting strongly with the phosphate backbone of nucleic acids, presumably by specific intercalation of the planar anthracycline nucleus with the DNA double helix.[1,2]

Doxorubicin (which is often referred to by its trade name Adriamycin®) is an approved antineoplastic agent and has been in clinical use for over 20 years. Human tumors shown to be responsive to doxorubicin include acute leukemia, resistant Hodgkin's and non-Hodgkin's lymphomas, sarcoma, neuroblastoma, ovarian and endometrial carcinoma, breast carcinoma, bronchogenic carcinoma, lung cancer and thyroid and bladder carcinoma.[3] AIDS-related Kaposi's sarcoma (KS) is somewhat responsive to doxorubicin as a single agent and in combination regimens.[4,5] Dose dependent toxicities, including stomatitis/mucositis, nausea/vomiting, bone marrow suppression and cardiomyopathy, limit the amount of doxorubicin patients are able to tolerate.

Conventional liposomal formulations of doxorubicin have been proposed as a means to reduce doxorubicin-related toxicities and thereby improve the drug's therapeutic index. The scientific rationale for the use of liposomal formulations of doxorubicin is discussed below, first for conventional liposomes and then for Stealth liposomes.

I.1.1. Conventional liposomes

Conventional liposomes used for drug delivery purposes are generally small in size (<300 nm) and composed of naturally occurring or synthetic phospholipids, with or without cholesterol. The exposed outer surfaces of such liposomes are susceptible to opsonization and destabilization by components present in biological fluids. For example, following intravenous injection, a liposome of this type is rapidly recognized as a foreign body and cleared from circulation in a dose-dependent fashion by elements of the immune system: primarily by specialized phagocytic cells residing in the liver and spleen, the mononuclear phagocyte system (MPS). It is believed that binding of plasma proteins (lipoproteins, immunoglobulins, complement) to the liposome surface triggers such macrophage uptake.

Internalization of liposome-encapsulated antitumor agents by MPS cells has the potential to diminish exposure of other body tissues to the toxic effects of such drugs. Liposomal encapsulation of doxorubicin has been proposed as a means of reducing the side effects of this highly active antitumor agent. By taking advantage of MPS clearance of encapsulated drug, exposure of other healthy tissues to high plasma concentrations of doxorubicin is reduced. Doxorubicin-related nausea/vomiting and cardiomyopathy are believed to be related to the drug's peak levels in plasma. By using liposome encapsulation to sequester the majority of an injected dose in the MPS, in theory, plasma levels of free drug are attenuated and safety improved. The drug is eventually released from MPS organs and distributes to peripheral tissues in free form. In this case, the pharmacokinetic pattern is intended to mimic that seen following administration of doxorubicin as a divided-dose or prolonged infusion, regimens known to reduce drug-related side effects.[6–11] Indeed, it has been shown that administration of liposome-encapsulated doxorubicin reduces the drug's acute and chronic toxicities in preclinical animal models.[12] Moreover, results from animal models indicate that doxorubicin delivered in this fashion retains its activity against systemic tumors.[13] The pharmacokinetics and safety of various clinical formulations of conventional liposomal doxorubicin have been reported in the scientific literature.[12,14–32] Clinical pharmacokinetic measurements confirm that conventional liposome formulations are cleared rapidly from plasma. These data also suggest that a considerable amount of encapsulated doxorubicin is released into plasma *prior* to MPS uptake.[18,29]

I.1.2. Long-circulating "Stealth" liposomes

Recognizing that rapid liposome clearance, coupled with release of *encapsulated* drug, severely limits the potential of liposomes to transport encapsulated drug to systemic tumors, strategies have been sought to stabilize liposomes in plasma and prolong their circulation following administration. Similarly, efforts have been made to optimize liposome size and thus their potential to extravasate at sites of disease.[33–36]

DOXIL is a long-circulating "Stealth" liposomal formulation of doxorubicin. This type of liposome contains surface-grafted segments of the hydrophilic polymer methoxypolyethylene glycol (MPEG). These linear MPEG groups extend from the liposome surface creating a protective coating that reduces interactions between the lipid bilayer membrane and plasma components. A schematic representative of a Stealth liposome, not drawn to scale, is presented in Figure 1.

The critical design features of the Stealth liposome include:

- Polyethylene glycol ("Stealth" polymer) coating: reduces MPS uptake and provides long plasma residence times.
- Average diameter of approximately 85 nm: balances drug carrying capacity and circulation time, and allows extravasation through endothelial defects/gaps in microvasculature of tumors.[36]
- Low permeability lipid matrix and internal aqueous buffer system: provide high drug loading and stable encapsulation, i.e., drug retention during residence in plasma.

The "steric stabilization" effect provided by MPEG is believed to be responsible for the remarkable stability of DOXIL in plasma. The MPEG coating also inhibits the interaction (close approach) of liposomes with macrophage cells, thus reducing hepatic uptake and prolonging liposome residence time in the circulation.[37] Comparative pharmacokinetic measurements in rodents and dogs indicate that doxorubicin has a prolonged plasma residence time when administered as DOXIL relative to doxorubicin (15 to 30 hours, compared to a distribution half-life of 10 minutes for doxorubicin).[38] The long residence time of DOXIL was confirmed in studies conducted in cancer patients and AIDS patients with KS.[39–40] In these studies, which are described in detail below, DOXIL remained in circulation with a distri-

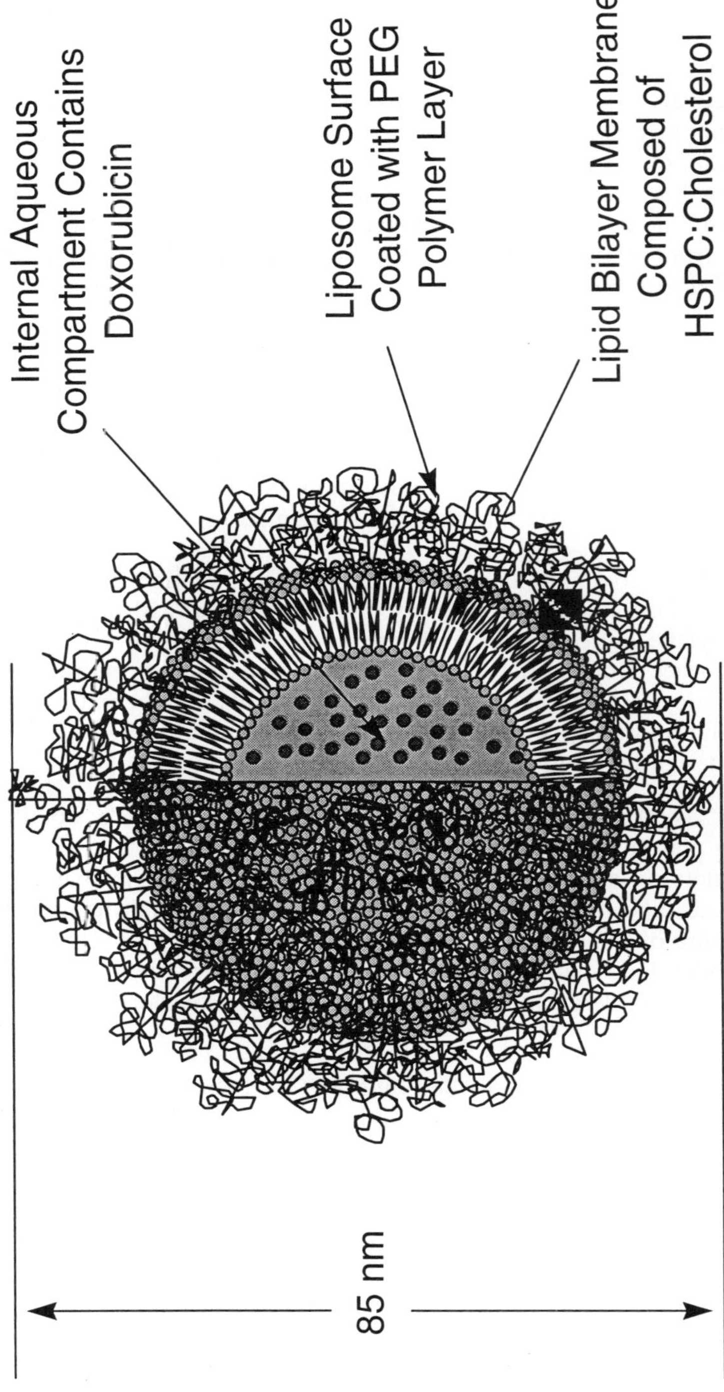

Fig. 1. Illustration of a DOXIL liposome. A single lipid bilayer membrane composed of hydrogenated soy phosphatidyl choline (HSPC), and cholesterol separates an internal aqueous compartment from the external medium. Doxorubicin is encapsulated in the internal compartment. Polymer groups (linear 2000 dalton segments of polyethylene glycol) are grafted to the liposome surface (although not shown, the polymer also extends from the inner monolayer of the membrane). The mean diameter of particle, including the PEG layer, is 85 nm.

bution half-life of 40 to 60 hours while the distribution half-life of doxorubicin is reported to be less than 10 minutes.

Solid tumors, including the cutaneous and visceral lesions characteristic of KS, depend on blood vessels for exchange of gases, nutrients, and metabolic waste products. Neo-vascularization is necessary to support tumors larger than a few millimeters in diameter. The permeability of vessels in tumors is significantly higher than those residing in normal tissues.[41] Vessels supplying KS lesions are particularly permeable as evidenced by edema and extensive extravasation of formed blood elements (perivascular streams of extravasated red blood cells are typically seen in KS lesions).[42] This increased vascular permeability has been attributed to several factors: the existence of fenestrated and discontinuous capillaries, the existence of blood channels without an endothelial lining,[43] increased occurrence of trans-endothelial channels and higher trans-endothelial pinocytotic transport.[44]

Light and electron microscopic examination of C-26 colon carcinoma and KS-like lesions show high concentrations of liposomes in interstitial areas surrounding capillaries in mice treated with Stealth liposomes containing colloidal gold particles as a liposome marker. These findings suggest that such Stealth liposomes circulate for a sufficient period of time and are small enough to extravasate through the capillaries supplying tumors.[36,45,46]

Following treatment of tumor-bearing mice with DOXIL, doxorubicin concentrations achieved in tumors are higher and anti-tumor activity is greater compared to animals receiving comparable doses of unencapsulated drug.[47–53] These findings suggest that DOXIL, by virtue of its plasma stability and slow clearance, might have a higher therapeutic ratio than earlier liposome formulations of doxorubicin. Relative to free doxorubicin, DOXIL treatment resulted in a 4- to 16-fold enhancement of drug levels in malignant effusion (fluid accumulated in tissues or body cavities as result of malignant growth) obtained from cancer patients.[39] Similar tumor localization results were also obtained in AIDS patients with KS.[40]

In addition to increasing doxorubicin localization in tumor tissues, the encapsulation of doxorubicin in Stealth liposomes could also result in the reduction of some of the adverse reactions associated with doxorubicin administration. For example, as detailed below, the cardiotoxicity caused by high cumulative doses of doxorubicin is believed to be related to the high peak plasma concentration of doxorubicin after its administration using the standard 3-week schedule. It is well established in the literature that the incidence of cardiomyopathy is significantly reduced when the drug is administered using a 1-week or a prolonged infusion schedule.[6–11] The encapsulation of doxorubicin will effectively reduce the peak drug concentration in plasma, therefore mimicking the prolonged infusion regimen. Indeed, a significant reduction in cardiotoxicity was observed in rats, dogs and rabbits after a multiple dose treatment with DOXIL when compared to an equal dose of doxorubicin. Reduced cardiotoxicity has also been reported in AIDS-KS patients receiving high cumulative doses of DOXIL relative to a control group of cancer patients given comparable cumulative doses of doxorubicin (see "Cardiac Safety of DOXIL" below).

I.2. Antitumor activity in animal models

The efficacy of DOXIL has been evaluated in a variety of different tumor models, including several human xenograft models.[47–53] These results are reviewed in the Chapter authored by P. Working in this volume. In every model examined DOXIL was more effective than the same doses of doxorubicin at inhibiting or halting tumor growth, at effecting cures and/or at prolonging survival times of tumor-bearing animals. Most often, all three endpoints were improved by DOXIL, and in no case was DOXIL less effective than doxorubicin. DOXIL was more active in both solid and dispersed tumors, and was more effective than doxorubicin in preventing spontaneous metastases from intramammary implants of two different mammary tumors in mice. These findings are also supported by studies done with DOXIL in several murine tumor models and a human xenograft model.

In general, the efficacy of doxorubicin in these models was limited by its toxicity at high doses. Typically, DOXIL could be used at a higher dose, offering an increased therapeutic advantage. Pharmacokinetic and tissue distribution studies suggest that the greater persistence, particularly in tumor tissue, achieved with DOXIL compared to conventional doxorubicin also contributes a therapeutic advantage. The efficacy of DOXIL compared to that of conventional liposomal (non-Stealth) doxorubicin indicated that DOXIL was significantly more effective than conventional liposomal doxorubicin, demonstrating the impact of the long-circulating Stealth liposome. Based on the results of these nonclinical studies, DOXIL appears to be an effective agent for the treatment of both solid and dispersed tumors.

II. Clinical pharmacology of DOXIL

II.1. Pharmacokinetics of DOXIL

The pharmacokinetics of DOXIL were investigated in four studies (Table 1).

Table 1

Pharmacokinetic studies of DOXIL formulations

Reference	Design	Patients	Dose range
39	Single dose crossover, DOXIL to adriamycin	Solid tumor	25 and 50 mg/m^2
40	Randomized, single dose, DOXIL or adriamycin	Kaposi's sarcoma	10, 20 and 40 mg/m^2
54	Randomized, crossover of single doses of DOXIL separated by 3-week washout period	Kaposi's sarcoma	10 and 20 mg/m^2
55	Population PK analysis of combined patient population	KS and solid tumor	10–60 mg/m^2

II.1.1. Pilot phase 1 study in solid tumor patients

Gabizon et al.,[39] reported results of a pilot phase 1 study of an early formulation of DOXIL in advanced cancer patients. The purpose of this study was to characterize the pharmacokinetics of DOXIL in patients in comparison with free doxorubicin. The pharmacokinetics of doxorubicin and/or doxorubicin encapsulated in pegylated containing 150 mM ammonium sulfate (rather than the 240 mM ammonium sulfate used in the commercial formulation of DOXIL) were analyzed in seven patients after injections of equivalent doses of free doxorubicin and DOXIL and in an additional group of nine patients after injection of DOXIL only. Two dose levels were examined, 25 and 50 mg/m^2. The plasma elimination of DOXIL followed a biexponential curve with half-lives of 2 and 45 h (median values), with >95% of the dose being cleared from plasma by a rate described by the longer half-life (Figure 2). Nearly 100% of the drug detected in plasma after DOXIL injection was in liposome-encapsulated form. A slow plasma clearance (0.056 L/h/m^2 for DOXIL vs. 27 L/h/m^2 for free doxorubicin) and a small volume of distribution (2.35 L/m^2 for DOXIL vs. 149 L/m^2 for free doxorubicin) were characteristic of DOXIL. Doxorubicin metabolites were detected in the urine of DOXIL-treated patients with a pattern similar to that reported for free doxorubicin, although the overall urinary excretion of drug and metabolites was significantly reduced. The results of this study were consistent with preclinical findings

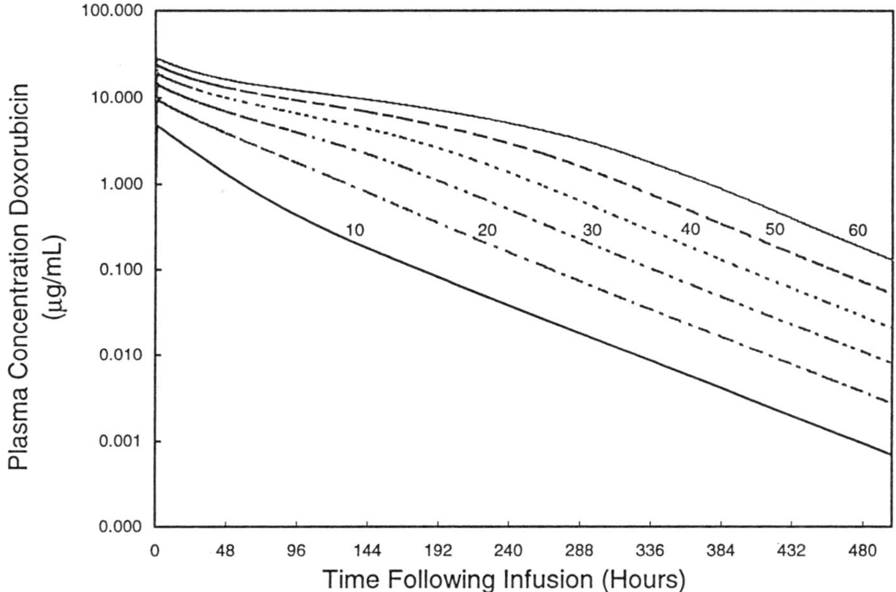

Fig. 2. Simulated plasma clearance kinetics of doxorubicin after a single 30 minute infusion of DOXIL at doses ranging from 10 to 60 mg/m^2.

indicating that the pharmacokinetics of doxorubicin are drastically altered using DOXIL and follow a pattern dictated by the liposome carrier.

II.1.2. Pharmacokinetics in KS patients

Northfelt et al. examined the pharmacokinetics of this early DOXIL formulation in AIDS KS patients.[40] A total of 18 patients were randomly assigned to receive a single dose of either DOXIL or doxorubicin at three dose levels, 10, 20 and 40 mg/m^2 (3 patients per dose group). In excellent agreement with the Gabizon study of the same formulation in cancer patients, the median values for DOXIL clearance and volume of distribution were 0.067 L/h/m^2 and 2.3 L/m^2, respectively. The clearance was biexponential with half-lives of 3.8 and 41 h, respectively. Once again, the vast majority of the dose was cleared under the curve described by the second half-life. Area under the plasma concentration vs. time curve and C_{max} for DOXIL were proportional dose in the dose range examined.

The pharmacokinetics of the commercial formulation of DOXIL was measured by Northfelt et al. in a two-period, randomized, crossover study of single doses of DOXIL 10 or 20 mg/m^2 administered intravenously (i.v.) over 30 minutes.[54] The doses were separated by a 3-week washout period. Twenty-six male patients with AIDS-related KS enrolled into the study; plasma was sampled over 10 days following dosing and KS lesion and normal skin tissue samples were collected 96 hours after the first dose.

DOXIL displayed linear pharmacokinetics best described by a two-compartment model. There were no carryover effects, although a few patients had low baseline doxorubicin plasma levels at Period 2.

Peak plasma concentrations (C_{max}) and AUC were dose proportional (Table 2). The mean C_{max} following 10 mg/m^2 was 4.1 μg/mL, and it was 8.9 μg/mL following 20 mg/m^2. Thirteen patients manifested a second peak plasma concentration after the end of infusion. The cause of this second peak is not known. Mean $AUC_{0\rightarrow240}$ was 232 μg/mL h for the 10 mg/m^2 dose group and 532 μg/mL h for the 20 mg/m^2 group. Mean $AUC_{0\rightarrow\infty}$ was 252 μg/mL h for the 10 mg/m^2 group and 577 μg/mL h for the 20 mg/m^2 group.

DOXIL dispositional pharmacokinetics were independent of dose. Disposition of drug occurred in two phases after DOXIL administration, with a relatively

Table 2

DOXIL pharmacokinetic parameter estimates[1]

Parameter	10 mg/m^2	20 mg/m^2
V_{ss} (L/m^2)	2.84 (0.124)	2.77 (0.129)
CL_t (L/h/m^2)	0.0572 (0.0103)	0.0419 (0.00401)
$t_{1/2}\lambda_1$	5.79 (1.07)	5.61 (1.21)
$t_{1/2}\lambda_2$	50.1 (5.71)	56.6 (5.74)
AUC_{0-240} (μg/mL h)	232 (22.7)	532 (45.6)
$AUC_{0-\infty}$(μg/mL h)	252 (28.5)	577 (57.2)

[1]Mean values (S.E.) for 23 patients.

short first phase and a prolonged second phase. Mean $t_{1/2}\lambda_1$ was 5.8 and 5.6 hours in the 10 and 20 mg/m^2 dose groups, respectively. Mean $t_{1/2}\lambda_2$ was 50.1 and 56.6 hours in the 10 and 20 mg/m^2 dose groups, respectively. Mean CL_t was low (0.06 L/h/m^2 for the 10 mg/m^2 dose group and 0.04 for the 20 mg/m^2 dose group) and mean V_{ss} was small (2.8 L/m^2 for both dose groups).

Very low levels of doxorubicinol, the primary metabolite of doxorubicin, were detected in the plasma. Levels ranged from 0.8 to 11.9 ng/mL in patients who received DOXIL 10 or 20 mg/m^2 after both the first and second dose groups. This represented approximately 0.3% of the measured doxorubicin levels in plasma.

II.1.3. Combined experience in KS and solid tumor patients

Since the initial results of Northfelt et al.[54] were presented (which included 42 HIV-infected patients), the plasma pharmacokinetics of DOXIL have been evaluated in an additional 41 patients being treated under various solid tumor protocols, for a total database of 83 patients (17 females, 66 males).[55] See Tables 3 and 4 for the listing of the protocols, the respective patient demographics and the summary statistics for all patients, including patients from the Northfelt study. The dose of DOXIL for the population ranged from 10–60 mg/m^2 and was administered over 0.5 to 1.0 hours.

Over this wider dosage range, DOXIL displayed nonlinear pharmacokinetics as evidenced by a disproportionate increase in the area-under-the-plasma concentration vs. time curve (AUC) with increasing dose amounts. The earlier pharmacokinetic results of Northfelt et al.[44,54] did not reveal the nonlinearity since only doses of 10, 20 and 40 mg/m^2 were evaluated. However, with escalating doses upwards of 60 mg/m^2, the disappearance rate of total doxorubicin from the plasma decreased. In general, drugs that display nonlinear pharmacokinetics have a potential to accumulate to toxic levels in the plasma if not monitored regularly (e.g., phenytoin). In the case of DOXIL, this is not a concern since the drug is administered a minimum of every three weeks, after which time no drug is detectable in the plasma of patients. The other pharmacokinetic parameter values from this recent analysis (e.g., volume of distribution, distributional clearance) were unchanged from the earlier results of Northfelt et al.[54]; see Table 5 for the statistics of selected pharmacokinetic parameters for all 83 patients. There was no evidence of accumulation at dose intervals of $\geqslant$3 weeks.

Utilizing the fitted pharmacokinetic parameter results from the recent analysis, simulated plasma concentration vs. time profiles of DOXIL were generated at doses of 10–60 mg/m^2 (Figure 3). The nonlinearity of DOXIL pharmacokinetics at higher doses is most evident at doses greater than 40 mg/m^2.

With this larger patient population, the effect of the following variables upon the pharmacokinetics of DOXIL were evaluated: age, weight, body surface area, tumor type, sex, and renal (as determined by serum creatinine) and hepatic function (as determined by total bilirubin levels). No correlations were observed between the studied variables and the pharmacokinetics of DOXIL. The impact of hepatic function on DOXIL pharmacokinetics is discussed in more detail below.

Table 3

Patient demographics sorted by study number

Study[1]	n	Dose (mg/m^2)	Age (years)	Sex	Weight (kilograms)	BSA (m^2)	Total Bilirubin (mg/dL)
30–06	2	40	46.5 (44–49)	1f, 1m	75.9 (69.5–82.2)	1.83 (1.80–1.86)	0.9 (0.6–1.2)
30–07	1	60	70	1m	76.2	1.96	0.3
30–14	42	10/20	40 (28–50)	42m	71.5 (52.3–102.5)	1.89 (1.46–2.30)	0.5 (0.22–1.9)
30–16	20	20–50	60 (37–75)	3f, 17m	74.1 (53.8–113)	1.85 (1.42–2.42)	0.8 (0.4–4.0)
30–18	5	60	60 (47–67)	3f, 2m	62.7 (50.4–73.6)	1.65 (1.54–1.89)	0.5 (0.4–0.5)
30–22	9	50	70 (46–75)	9 f	58 (43–107)	1.66 (1.35–2.15)	0.8 (0.6–0.9)
30–29	4	40–60	36 (21–66)	1f, 3m	68.9 (65.9–90)	1.77 (1.66–2.15)	0.4 (0.3–0.4)

[1]Study 30–06—"Maximal Tolerated and Pharmacokinetics of DOXIL in Patients with Cancer: A Phase I Clinical Study"; Study 30–07—"DOXIL for the Treatment of Advanced Non-Small Cell Lung Cancer in Patients Who Have Failed Prior Chemotherapy"; Study 30–14—"Randomized, Crossover, Pharmacokinetics and Tumor Localization Study of DOXIL in Patients with AIDS-Related Kaposi's Sarcoma"; Study 30–16—"Maximum Tolerated Dose, Pharmacokinetics, and Efficacy of DOXIL in Patients with Hepatocellular Neoplasm and in Patients with Hepatic Dysfunction"; Study 30–18—"DOXIL for the Treatment of Advanced Non-Small Lung Cancer in Patients Who Have Not Received Prior Chemotherapy"; Study 30–22—"Pharmacokinetics and Response to DOXIL in Patients with Recurrent or Persistent Epithelial Ovarian Cancer After Initial Therapy with Platinum and Taxol-based Regimens"; Study 30–29—"Treatment of Soft Tissue Sarcomas in Chemotherapy-Naïve Patients with DOXIL.

Table 4

Demographics of the pharmacokinetic patient population

$n = 83$	Age (years)	Weight (kilograms)	BSA (m^2)	Total Bilirubin (mg/dL)
Mean	48.4	71.2	1.84	0.732
CV%	29.4	18.5	11.1	76.5
Median	44.0	70.7	1.84	0.600
Minimum	21.0	43.0	1.35	0.224
Maximum	75.0	113	2.42	4.00

Table 5

DOXIL pharmacokinetic parameter estimates $n = 83$

Statistic	V_{ss} (L/m^2)	CL_i (L/h/m^2)	K_m (mg/L)	AUC_{50} (mg/L h)
Mean	3.40	0.108	2.01	3260
CV%	18.2	54.2	66.3	54.8
Median	3.42	0.0950	1.85	3018
Minimum	2.20	0.0269	0.428	535
Maximum	5.67	0.393	8.84	9520

V_{ss} – volume of distribution at steady-state, CL_i – intrinsic clearance, K_m – Michaelis Menton Constant, AUC_{50} – area-under-the-curve normalized for a 50 mg/m^2 dose of DOXIL.

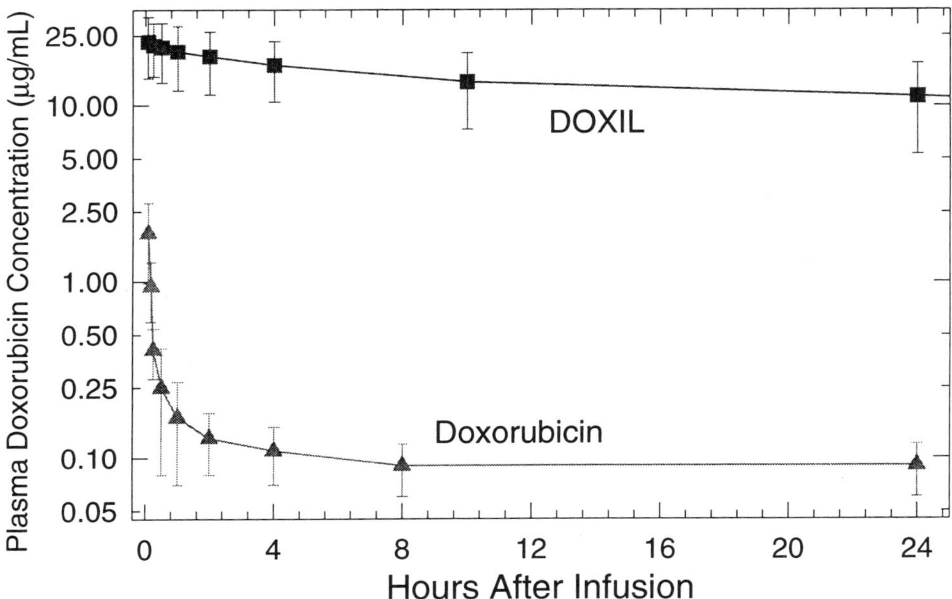

Fig. 3. Plasma clearance of doxorubicin after a single 50 mg/m^2 dose of an early formulation of DOXIL (containing 150 nm ammonium sulfate) in cancer patients. Data points represent mean values ± standard deviation for 14 patients in the DOXIL group and 4 patients in the doxorubicin group (adapted from Ref. 39).

Table 6

Comparison of DOXIL pharmacokinetic estimates by total bilirubin values

Pharmacokinetic parameter	Total bilirubin <1.2 mg/dL $n = 77$	Total bilirubin ⩾1.2 mg/dL $n = 6$	p value
V_{ss} (L/m^2)	3.41 (0.629)	3.36 (0.532)	0.965
CL_i (L/h/m^2)	0.109 (0.0607)	0.0999 (0.0144)	0.647
K_m (mg/L)	2.04 (1.36)	1.48 (0.721)	0.176
AUC_{50} (mg/L h)	3215 (1802)	3792 (1595)	0.276

II.1.4. Pharmacokinetics in patients with hepatic impairment or hepatocellular carcinoma

Six of the 83 patients included in the pharmacokinetic analysis reported baseline total bilirubin levels ⩾1.2 mg/dL (range: 1.2 to 4.0), a level at which conventional doxorubicin requires dose reduction. Patients with impaired liver function typically exhibit changes in pharmacokinetics of non-liposomal doxorubicin, characterized by increased half-life and AUC and decreased clearance, resulting in the recommendation for dose reduction to avoid high plasma concentrations of doxorubicin and potential drug accumulation. However, despite evidence of impaired liver function in these 6 DOXIL patients, as a group their plasma pharmacokinetics did not differ from patients with normal total bilirubin levels (Table 6).

In addition, there was no difference in the pharmacokinetics of the 20 patients with hepatocellular carcinoma (Study 30–16), who might be expected to have altered liver function, when compared to all other studied patients. The hepatic laboratory indices [median (range)] for the patients from Study 30–16 were the following: total bilirubin-0.850 mg/dL (0.4 to 4.0), AST-64 IU/L (21 to 323), alkaline phosphatase-215 IU/L (81 to 1090), LDH-202 IU/L (122 to 748). Given the unchanged pharmacokinetics, these findings, though preliminary, suggest that dose adjustment may not be necessary for DOXIL in patients with hepatic impairment

II.2. Amount of non-liposomal doxorubicin in plasma

Several lines of evidence support the conclusion that the majority of the doxorubicin (between 93% and 99%) in plasma is encapsulated within the liposome after i.v. administration of DOXIL. In the Gabizon study,[39] the fraction of the liposome-encapsulated and free, non-liposomal drug in circulation after DOXIL administration was quantitated directly using a Dowex column separation method that is able to accurately and reproducibly quantitate ⩾7% free drug in the plasma Figure 4).[56] Using this method, essentially all the doxorubicin measured in plasma was liposome-associated (Figure 5). These findings suggest that at least 90 to 95% of the doxorubicin measured in plasma, and possibly more, is liposome-encapsulated.

Doxorubicinol levels have been reported to range from 40 to 60% of the doxorubicin levels after administration of doxorubicin due to its rapid and extensive metabolism.[57,58] In the study of DOXIL pharmacokinetics, plasma doxorubici-

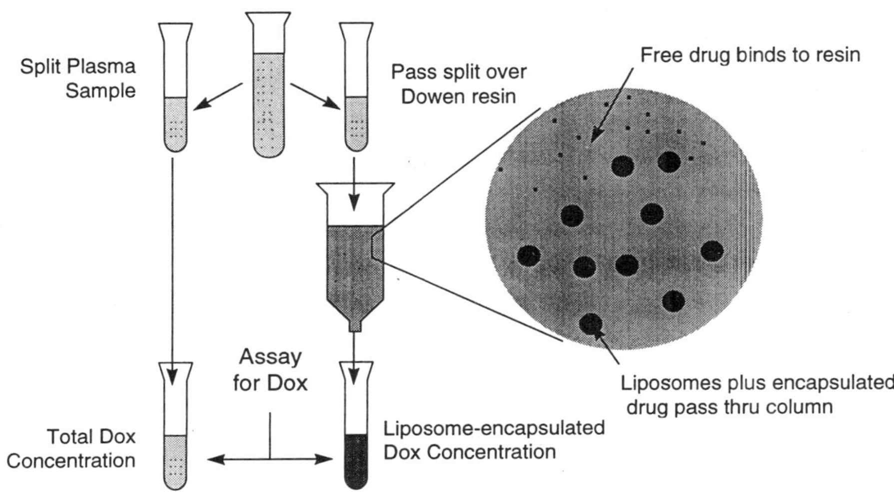

Fig. 4. Diagram illustrating the schema used to separate liposome-encapsulated doxorubicin from free doxorubicin in plasma samples.[56] A plasma sample is split into two fractions. Total doxorubicin concentration is measured in one split by HPLC analysis. The second is passed over a mini-column packed with Dowex resin. Free drug (i.e., the drug that has been released from liposomes) binds to the column material while the liposomes elute. Doxorubicin is measured after disrupting the eluted liposomes with solvents. Encapsulated drug is estimated by subtracting the amount measured after passing the column (and adjusting for dilution) from the total.

nol concentrations ranged from 0.1 to 0.5% of the doxorubicin concentrations in plasma. Based on the conservative doxorubicinol:doxorubicin concentration ratio reported in the literature, free doxorubicin levels would be expected to be approximately 0.25 to 1.25% of the total measured drug concentration after DOXIL treatment (Figure 6). Since only non-liposomal doxorubicin can be metabolized, these findings suggest that 99% of the drug remains liposome-encapsulated after DOXIL treatment.

The amount of doxorubicin that remains liposome-associated while circulating in plasma is an important point that deserves further emphasis from a safety perspective. Acute adverse reactions associated with doxorubicin administration including nausea and vomiting and chronic cardiotoxicity are believed to be directly related to peak concentrations of the drug in plasma. As pointed out above, while in the circulation, DOXIL liposomes remain intact, retaining virtually all of the doxorubicin in encapsulated form.[43] Although total plasma levels of doxorubicin may be relatively high for several days after DOXIL administration, the majority of the dose is sequestered within the liposome during this period and thus is not bioavailable to distribute (as free drug molecules) to tissues, including the GI tract

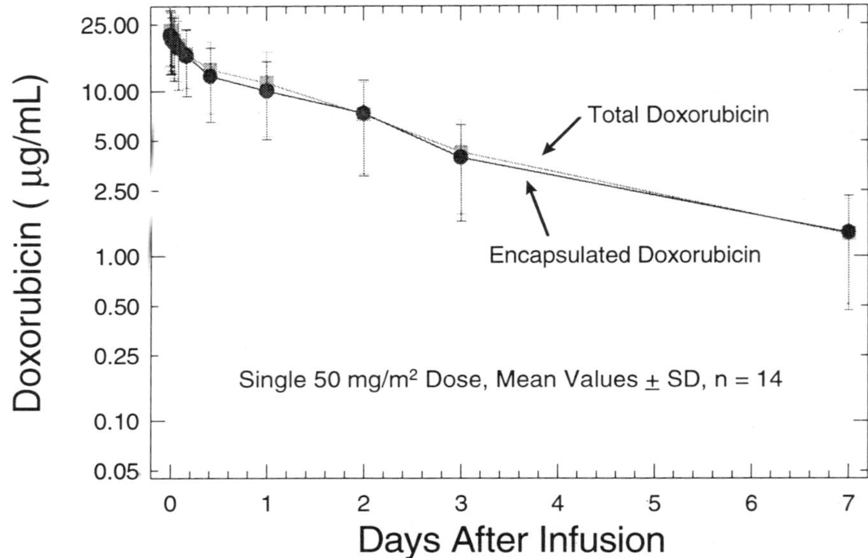

Fig. 5. Clearance over a one week period of total vs. encapsulated doxorubicin after a single 50 mg/m^2 dose of an early formulation of DOXIL (containing 150 nm ammonium sulfate) in cancer patients. Data points represent mean values ± standard deviation for 14 patients in the DOXIL group and 4 patients in the doxorubicin group (adapted from Ref. 39). The method described in the legend to Figure 4 was used to separate the encapsulated from released drug fractions.

and myocardium. With respect to level of available drug in plasma, DOXIL resembles more that of a 96 hour continuous infusion of doxorubicin than the usual 30 minute infusion. Prolonged infusion of doxorubicin is known to reduce cardiotoxicity and GI irritation.

II.3. Comparison of pharmacokinetic parameters: DOXIL vs. doxorubicin

According to literature reports, an i.v. bolus injection of doxorubicin in humans produces high plasma concentrations of doxorubicin that decline quickly due to rapid and extensive distribution into tissues.[59] Apparent volumes of distribution range from 1400 to 3000 L, reflective of the drug's extensive tissue distribution. The doxorubicin plasma concentration-time curve in humans is biphasic, with a distribution half-life of 5 to 10 minutes and terminal phase elimination half-life of 30 hours.[60–62] A triphasic curve has also been described with a terminal plasma half-life of approximately 30 hours.[63] Clearance of doxorubicin after doxorubicin administration ranges from 24 to 73 L/hour.[59] No accumulation in plasma occurs after repeated injections.[60–62]

The pharmacokinetics of DOXIL are significantly different from those reported for doxorubicin. Administration of DOXIL results in a significantly higher doxorubicin AUC, lower rate of clearance (approximately 0.1 L/hour) and smaller volume

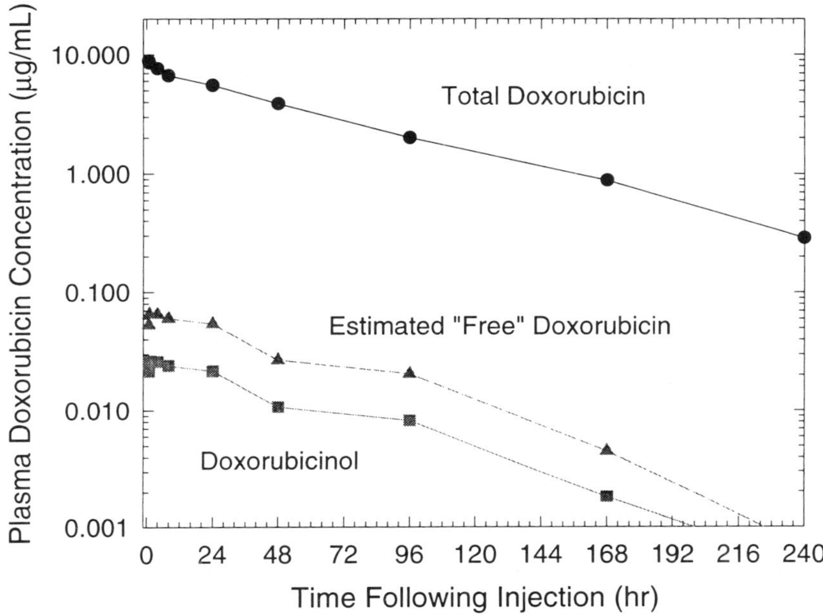

Fig. 6. Estimation of amount of "free" doxorubicin after DOXIL administration in cancer patient based on measured doxorubicinol concentration in plasma. Total doxorubicin (upper curve) and (lower curve) concentrations were measured by HPLC analysis after a single 25 mg/m^2 injection of DOXIL. The "estimated" free doxorubicin curve was generated using the assumption (based on literature values) that the amount of doxorubicinol in plasma represents 1% of the total doxorubicin available for metabolism (i.e., the amount of drug released from the liposomes).

of distribution (5 to 7 L) relative to administration of doxorubicin. The first phase of the biexponential plasma concentration-time curve after DOXIL administration is relatively short (approximately 5 hours), and the second phase, which represents the majority of the AUC, is prolonged (half-life 50 to 55 hours).

Doxorubicin C_{max} after DOXIL administration is 15- to 40-fold higher than after the same dose of doxorubicin, and the ratio quickly increases as doxorubicin is rapidly cleared from circulation. Importantly, the vast majority of the total plasma doxorubicin remains liposome-encapsulated after DOXIL treatment. Because of the high percentage of liposome encapsulation in DOXIL, the amount of free (i.e., "bioavailable") drug in the plasma appears to be significantly lower than that measured after administration of an equal dose of doxorubicin.

This conclusion is supported by the same type of calculations presented above, which derive the apparent concentration of free doxorubicin based on the reported relationship between doxorubicinol and doxorubicin concentrations in plasma. For example, five minutes after the end of the infusion, the mean doxorubicinol level following a 20 mg/m^2 dose of DOXIL was approximately 22 ng/mL. Using the doxorubicinol:doxorubicin concentration ratio reported in the literature, as described above, predicted free doxorubicin concentration at this time point would be 54 ng/mL in DOXIL-treated patients (the total plasma concentration measured at this time point was 8863 ng/mL). Comparatively, patients in the Northfelt et al. study[40] who received a dose of doxorubicin 20 mg/m^2, had initial plasma concentrations of doxorubicin of approximately 500 ng/mL.

II.4. Doxorubicin levels in KS lesions

Biopsies of KS lesion tissue and adjacent normal skin were obtained in 22 patients (Table 7).[54] Doxorubicin levels in KS lesions were higher than the levels in normal skin in 20 of the 22 patients; in 14 patients normal skin levels were below the lower limit of quantitation (0.4 μg/g tissue), whereas all KS lesion levels were quantifiable. Forty-eight hours after DOXIL administration, median doxorubicin levels in biopsies of KS lesions ranged from 3-fold to 16-fold higher than in normal skin from the same patients. The median doxorubicin concentration in KS lesions was 1.3 μg/g tissue in 7 patients receiving 10 mg/m^2 and 15.2 μg/g tissue in 7 patients receiving 20 mg/m^2; normal skin concentrations were 0.4 and 0.9 μg/g tissue in the 10 and 20 mg/m^2 dose groups, respectively. Biopsy data 48 hours after DOXIL injection in the 7 patients receiving 20 mg/m^2 are shown in Figure 7. Ninety-six hours after drug treatment, KS lesion doxorubicin levels were 3-fold and 5-fold greater than in normal skin from the same patients in the 10 and 20 mg/m^2 groups, respectively. Median doxorubicin concentration in KS lesions was 4.3 and 3.3 μg/g tissue in 4 patients receiving 10 mg/m^2 and 4 patients receiving 20 mg/m^2 dose, respectively; median concentration in the normal skin was 1.4 μg/g tissue for the 10 mg/m^2 dose group, and 0.7 μg/g tissue in the 20 mg/m^2 group.

Although too few time points were studied to allow determination of an AUC for doxorubicin in KS lesions or skin, these data suggest that doxorubicin accumulates in KS lesions after DOXIL treatment.

The increased doxorubicin in KS lesions compared to normal skin is not simply due to increased vascularization and/or blood content of the lesions. All samples were blotted after collection to remove excess blood, which would minimize differences due to blood content. However, even the most conservative comparison of doxorubicin levels in KS lesions and normal skin suggests that doxorubicin

Table 7

Concentration of doxorubicin in KS lesions and normal skin after DOXIL administration

Time after infusion	No. of patients	Dose (mg/m^2)	Doxorubicin concentration (μg/g tissue) Median (range)		KS/normal skin
			KS lesion	Normal skin	
48 hr	7	10	1.32 (0.17–22.43)	0.40 (0.26–1.55)	2.43
	7	20	15.21 (2.98–25.56)	0.92 (0.38–1.74)	20.89
96 hr	4	10	4.26 (1.91–36.44)	1.42 (0.70–2.78)	3.20
	4	20	3.28 (1.03–4.17)	0.73 (0.55–1.14)	4.94

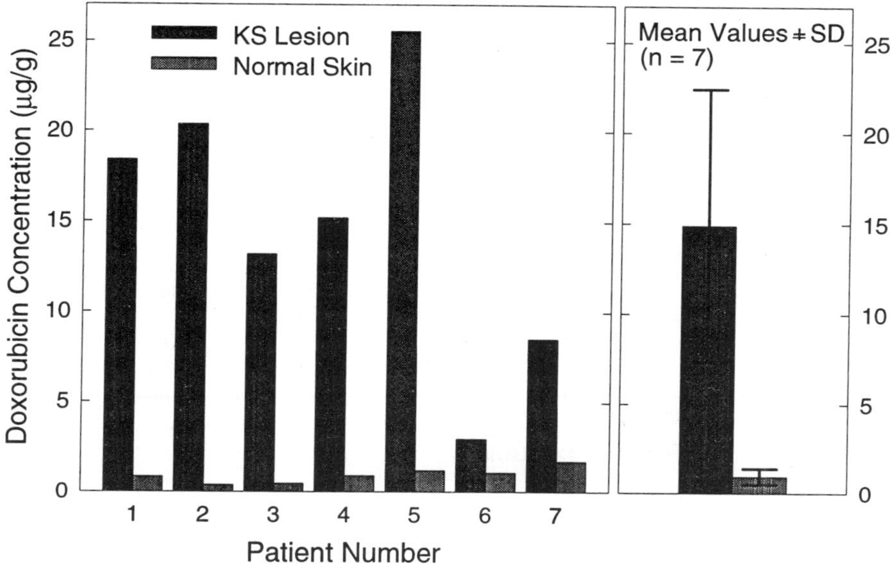

Fig. 7. Doxorubicin concentration in KS lesion tissue and adjacent normal skin tissue. Seven KS patients were given a 20 mg/m^2 dose of DOXIL and, 96 hours later, biopsies were taken of a representative cutaneous KS lesion and normal skin near the lesion. The tissue was homogenized, extracted with solvents and total doxorubicin measured by HPLC.

selectively accumulates in KS lesions. A comparison based only on blood content is made when the entire weight of the lesion or skin sample is assumed to be due to blood (i.e., there is no weight contribution from tissue). Because plasma constitutes 60% of blood, lesion doxorubicin levels due to blood content will be a maximum of 0.6-fold the plasma level of doxorubicin at the time of collection. At 48 hours after DOXIL treatment, median ratios of the doxorubicin concentration in KS lesions to predicted levels if all drug was due to blood were 2.8 and 5.5 in the 10 and 20 mg/m^2 treatment groups, respectively, compared to 0.6 and 3.4 in normal skin (Table 8). At 96 hours after treatment, the median ratio of KS lesions to blood was 5.5 and 1.9 in the 10 and 20 mg/m^2 treatment groups, respectively, compared to 0.4 and 1.6 in the normal skin biopsies.

Northfelt also reported higher doxorubicin concentrations in KS lesions after DOXIL.[40] Doxorubicin levels in KS lesions 72 hours after treatment with DOXIL were found to range from five- to 11-fold higher than after the same dose of doxorubicin (Figure 8). These levels may not represent the peak tumor levels achieved. Considering the shorter half-life of the major disposition phase of doxorubicin after delivery as doxorubicin, it is possible that peak tumor levels occurred earlier than 72 hours in the doxorubicin-treated patients. However, animal studies have shown that tumor levels of doxorubicin are higher after treatment with DOXIL even at 1 hour post-treatment.[49] These observations

Table 8

Predicted accumulation of doxorubicin in KS lesions and skin samples (median values)

Time point (hours)	DOXIL dose (mg/m^2)	KS lesion Ratio[1]	Skin Ratio[1]	KS Ratio / Skin Ratio
48	10	2.8	0.6	4.7
	20	5.5	3.4	1.6
96	10	5.5	0.4	13.8
	20	1.9	1.6	1.2

[1]Ratio of measured doxorubicin concentration to calculated concentration due to blood content. Lesions and skin sample weight assumed to be due to blood content only.

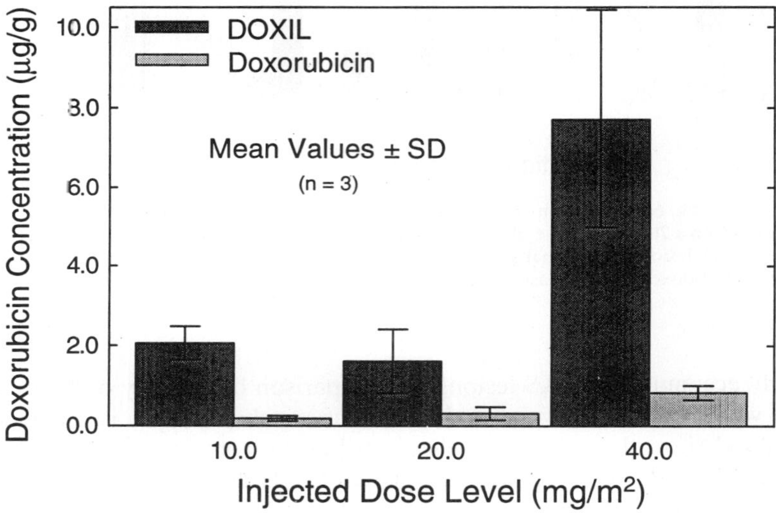

Fig. 8. Doxorubicin concentration measured in cutaneous KS lesion tissue 48 hours after a single injection of DOXIL and doxorubicin at three dose levels, 10, 20 and 40 mg/m^2. Mean values ± standard deviation for three patients per group.

suggest that doxorubicin levels in the KS lesions of doxorubicin-treated patients probably never attained the levels measured after DOXIL treatment.

III. Mechanism of enhanced DOXIL accumulation in tumors

An understanding of the mechanisms by which liposome-encapsulated doxorubicin accumulates within solid tumors after DOXIL administration, and how this deposition pattern and subsequent slow release of drug improve the antitumor activity of DOXIL relative to treatment with the free drug, is now emerging (refer to Figure 8).

III.1. Plasma stability and long plasma residence times are critical requirements

DOXIL liposomes are intend to carry their payload of doxorubicin directly to tumors. So, any premature release of the drug, while the liposomes are still in route (i.e., in the circulation), would detract from the total amount of encapsulated doxorubicin able to reach the desired target. This requirement highlights the importance of engineering plasma stability into DOXIL liposomes. As mentioned earlier, conventional liposome formulations of doxorubicin have been shown to release a significant proportion of their payload into the bloodstream soon after injection.[18,29] Drug release appears to follow protein adsorption/intercalation into the liposome which disrupts the barrier properties of the membrane. Moreover, the liposomes, together with any remaining drug, are removed by cells of the MPS within several minutes to a few hours after injection. As a consequence of this rapid clearance, doxorubicin delivered in conventional liposomes has little opportunity to reach tumors in *encapsulated* form.

By virtue of the PEG groups grafted to their surface, DOXIL liposomes are stable in plasma and release very little drug while in the circulation (see discussion above). Moreover, the PEG coating provides slow clearance; after a single injection, DOXIL can be detected in the circulation for 2–3 weeks. Slow clearance kinetics provide an opportunity for these liposomes to reach sites of disease such a tumors. Measurements made in tumor-bearing animals and cancer patients indicate that uptake of pegylated liposomes by tumors is also slow process. In preclinical tumor models, the peak uptake of DOXIL is reached 24–48 hours after injection.[49,64] In cancer patients given ^{111}Indium encapsulated in pegylated liposomes of the same composition and size as DOXIL, peak uptake in tumors is seen 48–72 hours after injection (Figure 9).[65] Slow uptake in tumors highlights the importance of long circulations times; if liposomes are to have an opportunity to reach and enter tumors in significant numbers, they must circulate for periods of days after injection.

III.2. Liposomes extravasate through gaps in the endothelium of tumor vessels

Stealth liposomes of the same size and lipid composition as DOXIL, but containing entrapped colloidal gold designed to serve as a marker to follow liposome distribution by microscopic techniques, have been shown to enter solid colon tumors implanted in mice[45] and KS-like lesions in HIV-transgenic mice.[46] In these mouse models, movement of liposomes from the vascular lumen into the tumor interstitium was visualized by light and electron microscopy. Transcytosis of liposomes from the lumen of blood vessels, through endothelial cells, and into the extravascular compartment of KS lesions was seen, as was intracellular uptake of liposomes by some spindle cells within lesions. However, these processes appear to be restricted to a minority of the particles entering the tumor.[46] The vast majority of the liposomes were seen to enter through gaps in the endothelial cell wall. This finding is consistent with results reported by Yuan et al., who used pegylated liposomes ranging in size from 100–600 nm to probe the cut off size of the gaps

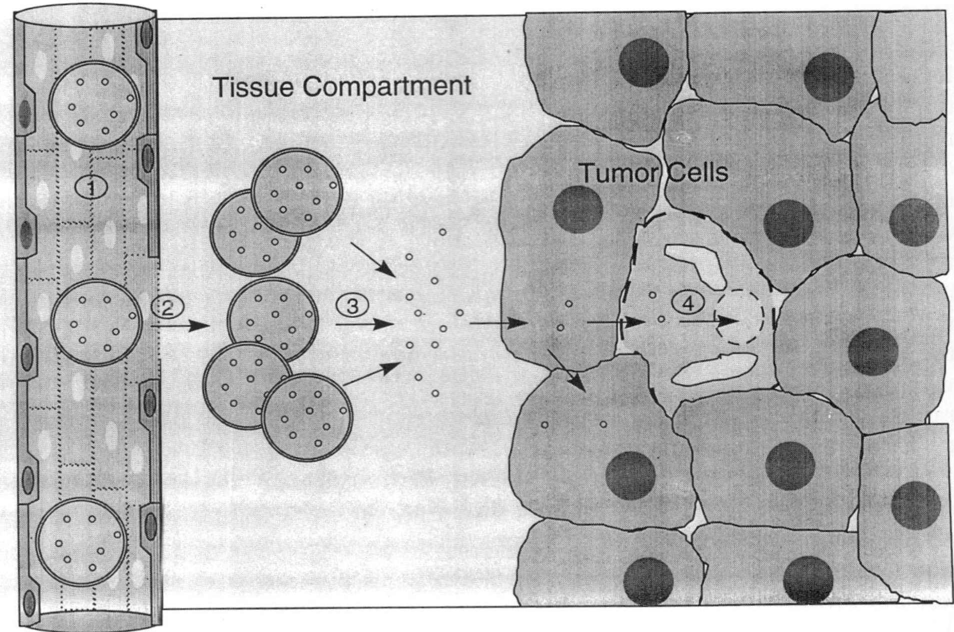

Fig. 9. Proposed mechanism for DOXIL accumulation in tumors. (1) Liposomes containing doxorubicin circulate for 2–3 weeks after injection. During this period virtually all of the drug remains encapsulated. The liposomes pass many times through the blood vessels feeding growing tumors. (2) Intact liposomes extravasate through defects/gaps present in newly sprouting vessels and enter the tissue compartment; lodging in the tumor interstitium near the vessel. (3) Drug molecules are released from the extravasated liposomes. Liposome leakage is believed to be the consequence of conditions present in the interstitial fluid surrounding tumors which lead to physical/chemical breakdown of the liposome membrane (low pH, oxidizing agents, enzymes, uptake by macrophages). (4) Free drug molecules penetrate deeply into the tumor, enter tumor cells, bind to nucleic acids and kill tumor cells. Note that such a mechanism does not require a close physical encounter between a liposome and target cell, since free drug molecules are able to diffuse through barriers that may intercept liposomes.

present in a human adenocarcinoma xenograft implanted in nude mice.[36] This tumor was permeable to liposomes up to 400 nm in diameter, suggesting the cut off size in this tumor is between 400–600 nm. Given their small size (85 nm) and long circulation times, DOXIL liposomes would be expected to extravasate in tumors that exhibit gaps of such dimensions. Gaps/defects are known to be present in solid tumors[43,44] and KS lesions.[42,46,66] Indeed, fluorescent pegylated liposomes of <100 nm in diameter have been visualized by video microscopy extravasating in real time into the interstitium of implanted tumors using window chamber models.[67–69]

III.2.1. Release of drug following extravasation

Encapsulated doxorubicin is released from the DOXIL liposomes after extravasation in tumors.[69] Several possible factors may contribute to liposome breakdown and drug release in tumors: (1) conditions present in the interstitial fluid surrounding tumors may cause breakdown of the liposomes, such as low pH,[70] and lipases released from dead or dying tumor cells;[71] (2) inflammatory cells (which are often found in tumors[72]) may release factors that lead to liposome destabilization such as enzymes or superoxide and other oxidizing agents,[73] or (3) phagocytic cells residing in tumors[74] which are known to engulf liposomes,[68] may digest the lipid matrix intracellularly and release doxorubicin (or its active metabolites) back into the interstitial fluid.[18] A combination of these possibilities may well be responsible for the observed release of doxorubicin after extravasation of DOXIL liposomes in tumors.

The rate of release of doxorubicin within a tumor has yet to be measured directly. In order to do so, it would be necessary to separate encapsulated drug (i.e., drug molecules that have not been released from intact liposomes) from free drug in a solid tissue. Although such a separation is possible in biological fluids (such as plasma)[56] it is technically difficult to conduct in solid tissues such as tumors; the conditions needed for quantitative extraction of doxorubicin lead to liposome disruption. Despite the difficultly of directly measuring release kinetics, indirect methods suggest that the release of doxorubicin from DOXIL liposomes occurs over a period of days to perhaps weeks following administration. In a recent study using a human pancreatic xenograft model in nude mice, Vaage et al showed that tumor levels of doxorubicin peak at 24–48 hours after DOXIL, and fall slowly over a period of a week.[75] These results suggest that the liposomes entering the tumor release their drug locally at quite a slow rate.

The improved antitumor activity of DOXIL relative to a comparable dose of free doxorubicin can be partially attributed to these slow in situ release kinetics. Consider the distribution kinetics after a dose of free doxorubicin. Drug molecules enter the tumor (and other tissues) quickly, reaching maximal exposure (i.e., peak concentrations) within minutes.[64] During the subsequent 24 hours, tumor doxorubicin concentration drops precipitously to undetectable levels. During this brief "pulse" of doxorubicin, those cells not exposed to a cytotoxic concentration for a sufficient amount of time, or which are not at a sensitive point in the cell cycle, can escape therapy and continue to proliferate. A typical course of doxorubicin is given on a three week cycle. This length of time between injections is needed to allow for recovery from the hematologic toxicity associated with doxorubicin therapy. Following such a schedule, it is quite likely that tumor cells are exposed to cytotoxic levels of drug for only a few hours during the 3 week interval between injections. In the case of DOXIL which is also given in a 2–4 week cycle, not only does more drug reach the tumor, but, by virtue of the slow in situ release kinetics provided by the liposomes, tumor cells are exposed to drug over a period of several days to perhaps a week or more after a single dose. Such a release pattern may contribute to DOXIL's antitumor response.

III.2.2. Tumor cell penetration and cytotoxicity

Given its amphipathic nature, a doxorubicin molecule that is released from a liposome can quickly diffuse through surrounding fluids and connective tissue, enter tumor cells, bind to nucleic acids and inhibit DNA synthesis. Indeed, it is quite likely that drug molecules released from DOXIL can penetrate many cell layers into the tumor, well beyond the point that the liposome itself has reached. Early findings suggest that penetration of "free" drug in this fashion may be essential for DOXIL's antitumor activity.

As mentioned above, microscopic observations indicate that liposomes extravasate in tumors at particular sites; primarily through vessels forming at the advancing edge of angiogenesis.[67] The deposition of extravasated liposomes in these areas is perivascular and focal, occurring primarily at the roots of capillary sprouts where weak spots (possibly defects or gaps) in the endothelium are believed to occur. Given the geometry of the system, liposomes that enter through such gaps may not be able to penetrate deeply into the tumor interstitium. Liposome penetration may be limited by a range of physical obstacles including tight cell-cell junctions (often found in highly differentiated epithelial cell tumors), dense connective tissue stroma, small extracellular volume and high interstitial fluid viscosity (that may be caused by fibrin cross-linking).[76] Ideally all tumor cells, regardless of their proximity to blood vessels or the liposome depots that may from near them, would be exposed to a cytotoxic dose of drug. So, the observation that drug molecules released from focal, perivascular deposits of liposomes are able to penetrate deeply into the tumor mass may be a critical requirement for expression of DOXIL's antitumor activity.

IV. Clinical antitumor activity

DOXIL has emerged as the treatment of choice for advanced AIDS-related Kaposi's sarcoma (KS) based upon the results of a series of clinical trials in over 800 KS patients.[77–83] The results from the key studies, including three comparative trials, are summarized below. Moreover, as will be discussed, DOXIL has shown impressive activity in epithelial-cell tumors including ovarian and breast carcinoma. In breast cancer, DOXIL appears to be as active as doxorubicin at a lower dose intensity, suggesting that targeting of drug to tumors is provided by encapsulation in pegylated liposomes.

The safety profile of DOXIL is quite different than doxorubicin. On a dose equivalent basis, less alopecia, nausea, vomiting, neutropenia and cardiotoxicity are seen with DOXIL relative to doxorubicin, but a greater incidence of skin toxicity is observed on standard 3 weekly dosing schedules. Skin toxicity is manageable by lengthening the interval between DOXIL infusions from 3 to 4 or more weeks. This extra time limits accumulation of drug in skin; it is believed that >3 weeks is required for drug that is delivered to skin by the DOXIL liposomes to clear from the interstitial spaces of the tissue via lymphatics. Lengthening the interval between DOXIL administrations to ≥ 4 weeks dose not appear to detract from its antitumor activity.

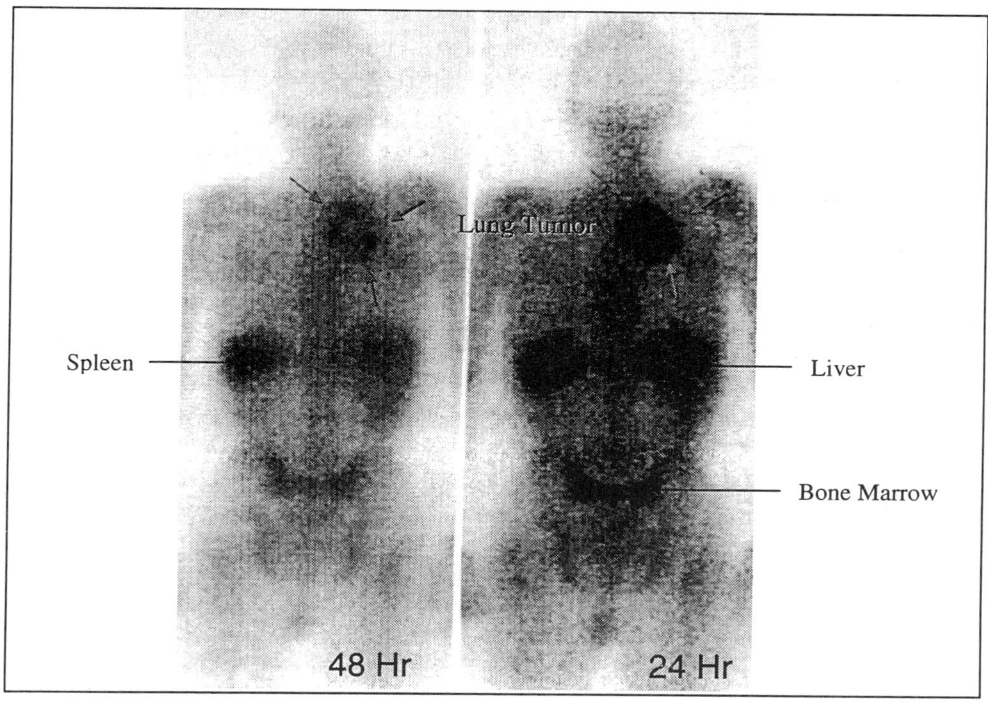

Fig. 10. Gamma scintigraphic image of a cancer patient 48 and 96 hours after administration of DOXIL liposomes containing ^{111}Indium. Note that both images are posterior views. Uptake of the radioactive liposomes is seen in certain normal tissues including spleen, liver, bone marrow. The activity visible in the central chest (substernal) and upper abdomen represent liposomes that are still circulating in the heart and major vessels at these time points. The liposomes are taken up by a large tumor in the left upper lung. The density of radioactivity is as high or higher in the tumor than in any normal organ.

IV.1. Kaposi' sarcoma: The disease setting

Kaposi's sarcoma is the most frequently reported opportunistic neoplasm in patients with acquired immunodeficiency syndrome (AIDS). Among all AIDS patients in the United States, 15–20% will develop KS at some point during the course of their HIV disease.[84] The prevalence of KS is estimated to be 20,000 cases in the US. Epidemiologic data have suggested that an infectious agent could spread the disease through sexual contact, which may explain the high incidence of KS among the male homosexual population. A number of viruses, including cytomegalovirus, hepatitis B virus and human papillomavirus have been found in patients with KS. Recent findings indicate that >90% of AIDS-associated KS patients were positive for herpesvirus-like DNA sequences[85] and a new member of the gamma herpesvirus family, referred to as Kaposi's sarcoma associated herpesvirus (KSHV) has been proposed as a putative etiologic agent for the disease.[86] With the demographics of the AIDS epidemic in the US shifting away from male homosexuals and toward intravenous drug uses and sex workers, the overall incidence of KS appears to be declining. At the same time, however, the actual mortality ascribed to KS has increased, probably due to improved prophylaxis and therapy of opportunistic infections and new antiviral agents such as protease inhibitors which slow the progression of HIV disease, without concurrent improvements in the outcome of therapy for KS itself.[88] KS can be disfiguring, debilitating and painful and no curative therapy exists. The goal of therapy is palliation of the signs and symptoms of the disease without exacerbating the underlying HIV disease or adversely impacting patients' quality of life. Favorable outcomes of therapy include improvements in the appearance of cosmetically troublesome lesions by reducing their size, number, color or nodularity, reduction in edema and pain and resolution of visceral disease symptoms.

Therapeutic options for localized KS include excision, X-ray irradiation, intralesional chemotherapy or cryotherapy. For patients with progressing disseminated disease, systemic chemotherapy is indicated. Single agents known to be active against KS include taxol, etoposide and Vinca alkaloid such as vincristine or vinblastine. For advanced KS, the combination regimens of vincristine with bleomycin (BV) or BV plus doxorubicin (ABV) are superior to single agents and are considered to be standard "first-line" chemotherapy.

Although these agents are generally effective in controlling KS, the disease will often progress despite therapy. Moreover, many patients cannot continue therapy due to side effects including bone marrow suppression, peripheral neuropathies, nausea and vomiting. Until the introduction of DOXIL, no clearly proven treatment options existed for this population of patients.

IV.2. DOXIL in KS patients after failure of standard chemotherapy

Treatment failure patients were identified from a cohort of 383 patients enrolled in a open-label study of DOXIL which was conducted at 17 US centers in 1993–1994.[89] Following review of individual patients' medical records, an expert panel of three AIDS KS clinical specialists agreed that 53 these had failed standard multiagent cytotoxic chemotherapy due either to disease progression or unacceptable toxicity.

A typical patient in this treatment-failure category was a white (85%) 38-year-old male (100%) homosexual (96%). Most (68%) had greater than 25 cutaneous KS lesions and 39% greater than 50. Almost half (44%) had oral lesions. Symptoms of pulmonary and or intestinal KS were reported by 29% and 15% of the patients, respectively. According to staging criteria developed by the AIDS Clinical Trials Group (ACTG),[77] the majority were in a poor risk category for tumor burden (85%), systemic illness (61%) and immune system status (95%). The median CD^{4+} lymphocyte count was 10 cells/mm^3. Despite the poor risk nature of this population, a mean Karnofsky performance status of 76% indicated that these patients could care for themselves, carry on normal activity and do active work. All patients had received bleomycin, 52 (98%) received a Vinca alkaloid (either vincristine or vinblastine), 29 (55%) received doxorubicin and 11 (21%)

received etoposide. Additionally, 14 (26%) had received interferon and 4 (8%) liposomal daunorubicin. Progression was documented among 28 patients (53%) of the patients who had received prior doxorubicin. The 53 patients in this treatment-failure group received median cumulative dose of 140.0 mg/m^2 of DOXIL.

Clinical response was assessed according to ACTG recommended criteria.[90] Complete response (CR) was defined as the absence of detectable residual disease. If residual pigmentation persisted in one or more cutaneous lesions, negative biopsy of one such lesion was required to qualify for CR. For patients with visceral symptoms, appropriate endoscopic or radiographic procedures were required to document the absence of KS. Clinical CR (CCR) was defined as in the case of CR except that biopsy and/or restaging by endoscopy or radiologic procedures was not required if these were contraindicated for some reason.

Partial response (PR) was defined as the absence of new lesions or new visceral disease or worsening edema or effusions. In addition, at least one of the following must apply:

- A 50% decrease in the number of all previously existing lesions.
- Complete flattening of >50% of all previously raised lesions.
- A 50% decrease in the sum of products of the largest perpendicular diameters of five prospectively selected marker lesions.
- Patients met the criteria required for CR, except residual tumor-associated edema or effusion was present.

These criteria were required to be met for at least 6 weeks and could not be preceded by progressive disease.

Stable disease (SD) was defined an any response not meeting the criteria for CR CCR or PR or Progressive disease.

Progressive disease (PR) was defined as any new site of visceral involvement or evidence of progression, evidence of new edema, a ⩾25% increase in number or size of lesions or ⩾25% of all previously flat lesions becoming raised.

Clinical response for these patients is presented in Table 9. Of the 53 patients, 19 (36%) achieved PR and one (2%) CCR for an overall response rate of 38%. Importantly, of the 28 patients whose disease had progressed on a combination

Table 9

Response to DOXIL in KS patients who failed first-line therapy

	All patients	Doxorubicin failure
Number evaluable	53	28
Partial response (PR)	36%	32%
Stable disease	36%	50%
Progressive disease	26%	18%
Median duration of PR in days	128	127
Median time to PR in days	109	109

Table 10

Clinical benefits of DOXIL therapy

Response	Benefit			
	Complete flattening $N = 48$	Color improvement $N = 48$	Pain reduction $N = 22$	Edema reduction $N = 23$
Partial response	68%	82%	70%	100%
Stable disease	38%	39%	25%	78%
Progressive disease	31%	46%	25%	60%
All patients	48%	56%	45%	83%

regimen containing doxorubicin, 9 (32%) experienced PR. The median time to treatment failure, which was defined as the time from initiation of therapy to progression, was 134 days for the entire group and 148 days for the cohort whose disease had progressed on prior doxorubicin.

Evidence of patient benefit was documented in this study by assessing the characteristics of indicator lesions during therapy. Flattening, color improvement, reduction of pain and reduction of edema were followed and favorable changes were regarded as benefiting the patient.

Table 10 presents the clinical benefits accruing to patients in this study. These benefits correlated with the tumor response. Of those achieving CR or PR, 68% of the 48 patients with raised lesions at baseline experienced complete flattening of their indicator lesions, 82% of the 48 with unfavorable color had color improvement, 70% of the 22 patients with pain at baseline experience pain reduction and all of the 22 patients with edematous indicator lesions at study entry showed improvement. Fewer improvements in indicator lesion characteristics were seen among patients who achieved only stable disease as their best response or who progressed on DOXIL.

DOXIL was generally well tolerated. Among the adverse events thought to be possibly or probably related to DOXIL, bone marrow suppression, which primarily manifested as neutropenia, was the most frequently reported, with 76% of patients experiencing at least one episode of leukopenia. One patient's ANC dropped below 500 cells mm^3 and 5 septic episode were reported. Nausea/vomiting and alopecia were relatively infrequent occurring in 15% and 9% of patients, respectively. As expected in a population with a median CD^{4+} lymphocyte count of 10 cells/mm^3, opportunistic infections were common, occurring in 23 patients (43%). One patient experienced a severe skin rash which resolved when therapy was withheld. An acute reaction was reported by one patients during his first infusion characterized by sudden onset of flushing associated with abdominal pain. The reaction was self-limited and this patient was subsequently able to tolerate 12 cycles of DOXIL without further incident. No clinical evidence of cardiotoxicity was seen.

Single agent DOXIL is active in patients who fail standard first line combination chemotherapy, including those who progressed on regimens containing doxorubicin. At the dose and schedule used (20 mg/m^2 q 2–3 weeks) DOXIL is well tolerated and provides meaningful benefit to patients.

IV.3. DOXIL as first-line therapy of KS: Comparative trial results

IV.3.1. DOXIL vs. ABV

A comparison of DOXIL with a triple-drug combination considered by many US AIDS specialists to be the standard of care for advanced KS was conducted by a consortium of 25 academic and community-based AIDS oncology practices in 1993–1994.[91] Patients enrolled in this study had progressive, biopsy-proven AIDS-KS with ≥25 cutaneous lesions or documented visceral involvement. Patients with active opportunistic infection, a history of heart disease and those who had received chemotherapy within one month were excluded.

After having met the entry criteria, 258 patients were prospectively randomized to receive either DOXIL (20 mg/m^2) or the combination of doxorubicin (20 mg/m^2), bleomycin (10 mg/m^2), and vincristine (1 mg) ("ABV") every 14 days for six cycles. The statistical hypotheses of the study were to show: (1) that DOXIL provides equivalent efficacy to ABV (i.e., an overall response rate no less than 15% below that of ABV, using a 95% one-tailed confidence interval) and (2) that DOXIL is better tolerated (i.e., 20–25% reduction in the incidence of pre-selected key adverse events within 80% power and a two tailed significance level of 0.05). Investigators prospectively identified 2–5 cutaneous indicator lesions that were assessed at baseline for thickness, nodularity, color, size, pain and edema and every two weeks throughout the study. Quality of Life (QoL) questionnaires were administered at study entry and every two weeks thereafter. Two QoL instruments were used; (1) a previously validated questionnaire by Wu at al.,[92] measuring AIDS-related quality of life, and (2) an AIDS-KS assessment tool created specifically for the study. The same standard ACTG criteria detailed in Section IV.2 above were used to assess tumor response.[90]

Patient characteristics and baseline disease status were well balanced between the two treatment groups (Tables 11 and 12). Among the patients receiving DOXIL, one had a clinical complete response (CCR) and 60 achieved a partial response (CCR + PR = 61/133, 45.9%; 95% confidence interval 37%–54%). Among ABV patients, 31 had a PR (PR = 31/125, 24.8%; 95% confidence interval 17%–32%). This difference was statistically significant, $p < 0.001$ (Table 13). Although DOXIL patients achieved their response earlier, the duration of response and time to treatment failure were similar between the two groups (Table 13).

With respect to indicator lesion characteristics, the size of lesions (i.e., the sum of the products of two perpendicular diameters) of all patients decreased 24% in the DOXIL group and 15% for patients receiving ABV ($p < 0.034$). Among the DOXIL patients, 96% had raised lesions at baseline, whereas at the end of treatment 51% had all their lesions flattened. In the ABV group, 97% had raised

Table 11

Characteristics of patients enrolled in DOXIL vs. ABV trial

	DOXIL, $n = 133$	ABV, $n = 125$	p-value
Sex			
• Male	133 (100%)	122 (97.6%)	0.112
• Female	0	3 (2.4%)	
Prior cytotoxic chemotherapy			
• Yes	104 (78.2%)	96 (76.8%)	0.882
• No	29 (21.8%)	29 (23.2%)	
Median age	36.0	38.0	0.661
Median weight	70.3	68.6	0.464
Race			
• White	107 (80.5%)	89 (71.2%)	0.425
• Black	11 (8.3%)	14 (11.2%)	
• Hispanic	12 (9.05)	17 (13.6%)	
• Asian	0	1 (0.8%)	
AIDS risk factor			
• Homosexuality	118 (88.7%)	107 (88.5%)	0.453
• Other/unknown	15 (11.3%)	18 (14.4%)	
Number of lesions			
• 1–9	12 (9.0%)	14 (11.2%)	0.939
• 10–24	20 (15.0%)	18 (14.4%)	
• 25–50	52 (39.1%)	44 (35.2%)	
• >50	25 (18.8%)	25 (20.0%)	
• >100	9 (6.8%)	11 (8.8%)	
• >200	6 (4.5%)	4 (3.2%)	

lesion at entry, but only 28% had all lesions flatten at the end of therapy. This difference in likelihood of flattening of previously raised lesions during therapy with the respective treatments was also statistically significant in favor of DOXIL ($p < 0.001$). At baseline, greater than 95% of patients in both groups had lesions that were erythematous or violaceous and therefore considered more disfiguring than brown lesion color. By the end of therapy 50% of the DOXIL patients and 30% of ABV patients had all indicator lesions turn brown, again a difference favoring DOXIL ($p < 0.002$). Both regimens were associated with a reduction in pain. Among DOXIL and ABV patients at baseline, 33% and 31%, respectively, reported moderate to severe pain. By the end of treatment 83% and 78%, respectively, were without lesion-associated pain. In both cases, pain relief was achieved without increases or new additions of either analgesics or antidepressants (Table 14).

Signs and/or symptoms of intestinal KS were present in 17% of DOXIL patients and 18% of ABV patients at study entry. At end of treatment these symptoms were seen in 8% of DOXIL patients and 16% of patients given ABV ($p \geqslant 0.05$). The percentage of patients in both groups with signs of pulmonary KS did not change significantly during the course of the study.

Table 12

Baseline disease characteristics for patients in DOXIL vs. ABV trial

	DOXIL $n = 133$	ABV $n = 125$	p-value
KS stage—Tumor burden			
• Good risk	59 (44.4%)	51 (40.8%)	0.615
• Poor risk	74 (55.6%)	74 (59.2%)	
KS stage—Immune system status			
• Good risk	19 (14.3%)	17 (13.6%)	1.000
• Poor risk	113 (85.0%)	107 (85.6%)	
KS stage—Systemic illness			
• Good risk	66 (49.6%)	62 (49.6%)	1.000
• Poor risk	67 (50.4%)	63 (50.4%)	
Median CD4 count (cells/mm^3)	12.5	13.0	0.940
KS lesion sites			
• Skin/mucocutaneous	132 (99.2%)	124 (99.2%)	
• Mouth	58 (43.6%)	57 (45.6%)	
• Lung	29 (21.8%)	27 (21.6%)	
• Stomach/intestine	18 (13.5%)	23 (18.4%)	
• Lymph nodes	14 (10.5%)	13 (10.4%)	
• Other	8 (6.0%)	5 (4.0%)	

Table 13

Tumor response among DOXIL and ABV patients

	DOXIL ($N = 132$)[1]	ABV ($N = 121$)[1]	p-value
Complete/partial response	61 (46.2%)	31 (25.6%)	<0.001
Time to PR/CR—median (days)	39	50	0.014
Duration—median (days)	90	92	0.234
Median time to treatment failure (Days)	124	128	0.259
Stable disease	70 (52.6%)	84 (67.2%)	NS
Progressive disease	2 (1.5%)	10 (8.0%)	NS

[1]One DOXIL patient and 4 ABV patients had no response data and therefore are not included.

In general, the improvements in QoL as assessed by the two instruments favored DOXIL (Tables 15 and 16).

Patients remained on therapy longer in the DOXIL group (a mean of 5.2 cycles or 90 days) relative to ABV (3.8 cycles or 47 days). In the ABV group, 37% discontinued treatment due to adverse events compared with 11% for DOXIL ($p < 0.001$). A Kaplan-Meier analysis of death rate differences between the groups was not statistically different. Seven ABV and no DOXIL patients discontinued

Table 14

Comparison of lesion characteristics at end of treatment on DOXIL vs. ABV

Indicator Lesion characteristic[1]	DOXIL			ABV			
	N^2	EOT value # Pts (%)	p-value[3]	N^2	EOT value # Pts (%)	p-value[3]	p-value[4]
Thickness flat	109	57 (51.4%)	>0.001	108	31 (28.2%)	>0.001	>0.001
Nodularity absent	109	78 (70.3%)	>0.001	110	65 (58.6%)	>0.001	0.035
Edema absent	109	96 (86.5%)	>0.001	110	96 (86.5%)	>0.001	0.561
Color brown	108	55 (50.0%)	>0.001	110	33 (29.7%)	>0.001	0.002
Pain none	118	100 (83.3%)	>0.001	116	91 (78.4%)	>0.001	0.417
Size (mm^2) Mean	114	863.2	0.765	113	1285.2	0.732	0.034

[1]The indicator lesion characteristic of interest (flat, absent, brown, none) is the best value a patient could attain.
[2]Number of patients having both baseline and end of treatment value on which comparisons are based.
[3]Comparing the change from baseline within treatments.
[4]Comparing the change from baseline between treatments.

Table 15

Change in quality of life parameters during DOXIL vs. ABV therapy

Domain	DOXIL			ABV			
	Baseline value (mean)	Mean change[1]	p-value[2]	Baseline value (mean)	Mean change[1]	p-value[2]	ABV or DOXIL p-value[3]
General health	2.57	+0.2	0.055	2.61	−0.2	0.048	0.020
Pain	3.46	+0.5	0.015	3.75	−0.0	0.777	0.020
Social functioning	3.28	+0.6	0.004	3.41	0.0	0.975	0.026
Mental health	4.14	+0.2	0.032	4.17	−0.0	0.613	0.076
Energy/fatigue	3.27	+0.2	0.114	3.46	−0.3	0.004	0.002
Health distress	3.87	+0.5	<0.001	3.92	+0.1	0.371	0.040
Cognitive functioning	4.86	+0.3	0.008	4.63	0.0	0.878	0.107
Overall quality of life	3.16	+0.2	0.057	3.28	−0.1	0.512	0.085
Health transition	2.97	+0.2	0.158	2.92	+0.1	0.405	0.832

[1]Positive changes represent improvement in health.
[2]Comparing the change from baseline within treatments.
[3]Comparing the change from baseline between treatments.

Table 16

Change in KS symptoms during DOXIL vs. ABV therapy

Domain	DOXIL			ABV			
	Baseline value (mean)	Mean change[1]	*p*-value[2]	Baseline value (mean)	Mean change[1]	*p*-value[2]	ABV or DOXIL *p*-value[3]
Pulmonary dysfunction & pain	3.35	+0.3	<0.001	3.35	0.0	0.413	0.038
GI pain & sensitivity	3.53	+0.1	0.346	3.50	−0.0	0.889	0.660
Difficulty walking	3.18	+0.3	0.003	3.21	+0.2	0.036	0.240
Difficulty wearing clothes	2.58	+0.7	0.003	2.68	+0.6	0.005	0.435
Eye dysfunction	3.30	+0.2	0.109	3.29	+0.3	0.098	0.422
Head & limb movement	2.50	+0.8	0.017	2.86	+0.1	0.973	0.045
Exercise limitations	2.21	+0.4	<0.001	2.49	0.0	0.866	0.034
Sleep disturbance	2.90	+0.8	<0.001	3.07	+0.3	0.053	0.001
Social well being	2.88	+0.5	<0.001	2.98	+0.3	<0.001	0.081

[1]Positive changes represent improvement in health.
[2]Comparing the change from baseline within treatments.
[3]Comparing the change from baseline between treatments.

due to disease progression. The most common adverse event in both groups was leukopenia occurring in 42% of patients receiving DOXIL and 44% of patients given ABV. Growth factor support was provided to 44% of the DOXIL patients and 53% of ABV patients. Severe neutropenia (absolute neutrophil count <500 cells/mm^3) was seen in 6% of DOXIL patients and 14% of the ABV group and the time to onset of severe neutropenia was significantly longer in the DOXIL group (Figure 11). Given that DOXIL patients were on study longer, the rates of neutropenic fever, documented septic events and opportunistic infections were not significantly different among the DOXIL and ABV patients (Table 17).

Six patients experienced an acute reaction during their first infusion of DOXIL. The reaction has been reported for other liposome/colloidal preparations and it is characterized as pseudoallergic with a range of symptoms including flushing, chest/back pain, difficulty swallowing and occasionally dyspnea and hypotension. The reaction was self-limiting, resolved when the infusion was stopped or slowed, and all six patients received additional courses of DOXIL without incident. Skin rashes were more frequent among DOXIL patients and three cases of palmar-plantar erythrodysesthesia was reported in the DOXIL group and one in the ABV group.

Table 18 lists selected non-hematological adverse events. Fewer DOXIL patients reported these adverse events compared with ABV patients, 54% vs. 79%, respectively, despite the fact that DOXIL patients were on study longer. Significantly less peripheral neuropathy and alopecia and a trend toward less nausea/vomiting were seen among DOXIL patients whereas the incidence of mucositis was greater. No difference was seen in clinical cardiotoxicity.

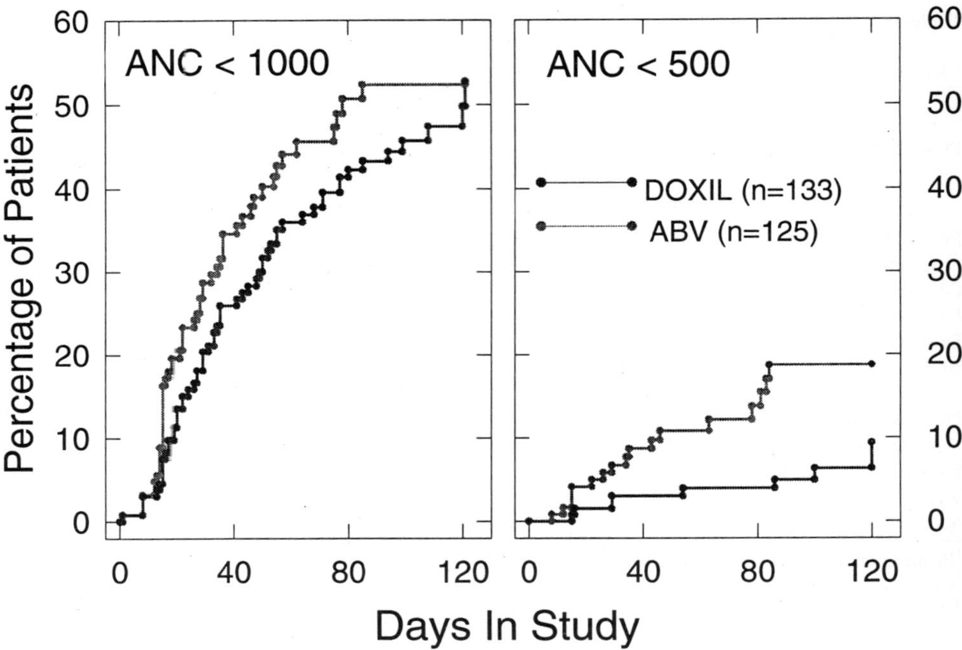

Fig. 11. Time to development of severe (grade 3 or 4) neutropenia in KS patient randomized to receive DOXIL or doxorubicin, bleomycin, vincristine (ABV). Grade 3 neutropenia is defined as absolute neutrophile count (ANC) < 1000 cells/mm^3 and grade 4 as ANC < 500 cells/mm^3.

Table 17

Incidence of severe adverse events related to neutropenia

	DOXIL ($N = 133$)	ABV ($N = 125$)
ANC (10^3/mm^3)		
Grade 3 ANC (0.5–<1.0)	51 (38.3%)	36 (28.8%)
Grade 4 ANC (<0.5)	8 (6.0%)	17 (13.6%)
Neutropenia, fever	0	5 (4.0%)
Sepsis, documented	8 (6.0%)	3 (2.4%)
Opportunistic infection	49 (36.8%)	38 (30.4%)
CSF usage	58 (43.6%)	66 (52.8%)

The findings of this randomized comparative trial demonstrate that single-agent DOXIL is effective in the treatment of advanced KS and is superior to the combination of ABV with a response rate of 46% for DOXIL compared to 25% for ABV. Moreover, because DOXIL is able to produce results superior to a proven combination regimen containing doxorubicin, these findings demonstrate that delivery of doxorubicin in pegylated liposomes enhances the therapeutic effect

Table 18

Selected adverse events during DOXIL vs. ABV therapy

Number reporting selected adverse event	DOXIL (n = 133)	ABV (n = 125)	p-value
	72 (54.1 %)	99 (79.2%)	
Nausea/vomiting	45 (34%)	72 (58%)	0.184
Hair loss	15 (11%)	53 (42%)	<0.001
Peripheral neuropathy	16 (12%)	35 (28%)	<0.090
Mucositis/stomatitis	24 (18%)	10 (8%)	<0.001
Palmar/plantar syndrome	1 (<1%)	1 (<1%)	1.000
Cardiotoxicity	4 (3.0%)	5 (4.0%)	0.057

of doxorubicin. Overall, DOXIL was better tolerated than ABV. The only adverse events that were reported with greater frequency on DOXIL were a self-limiting infusion reaction the occurred in a few patients, skin rashes and mucositis. Severe neutropenia, nausea/vomiting, peripheral sensory neuropathies, alopecia, fever, and anorexia were encountered less frequently with DOXIL relative to ABV. Quality of life, as recorded by the patients themselves, was more favorably impacted by DOXIL than ABV.

IV.3.2. DOXIL vs. BV

Among European AIDS specialists, particularly those practicing in Germany, the United Kingdom and the Netherlands, a combination of bleomycin and vincristine/vinblastine (BV) has often been preferred to that of ABV for the treatment of AIDS-KS. Also, European investigators generally consider using systemic chemotherapy earlier in the course of the KS compared to their U.S. counterparts. To assess the activity of DOXIL compared to BV, a second randomized trial was conducted in parallel to the DOXIL vs. ABV study described above in less severely ill KS patients.[93] A total of 22 investigational sites (17 European and 5 U.S.) participated from 1993–1995. A total of 241 patients with moderate to severe KS were prospectively randomized: 121 patients received single agent DOXIL (20 mg/m^2) and 120 received bleomycin (15 mg/m^2) plus vincristine (1.4 mg/m^2) every 3 weeks. Patient demographics and baseline disease characteristics were similar between the two groups. All but two patients were male, the mean age was 38 years, and 89% of the patients were Caucasian. Homosexuality was an AIDS risk factor in more than 93% of patients.

The response to DOXIL treatment, as measured by the ACTG criteria,[77] was significantly better than the response to BV, with complete or partial response achieved in 59% of the DOXIL patients and 23% of the BV group (Table 19). The proportion of patients responding in the DOXIL arm was significantly greater than those in the BV group at each evaluation (Figure 12).

Two QofL questionnaires completed by the patients confirmed better response to DOXIL compared to BV. In the Wu instrument,[79] DOXIL patients improved

Table 19

KS tumor response among DOXIL vs. BV patients

	DOXIL ($N = 121$)	BV ($N = 120$)	p-value
Complete/partial response	71 (58.7%)	28 (23.3%)	<0.001
Time to PR/CR—median (days)	44	64	0.025
Duration—median (days)	142	123	0.572
Stable disease	46 (38.0%)	81 (67.5%)	NS
Progressive disease	0	5 (4.2%)	NS

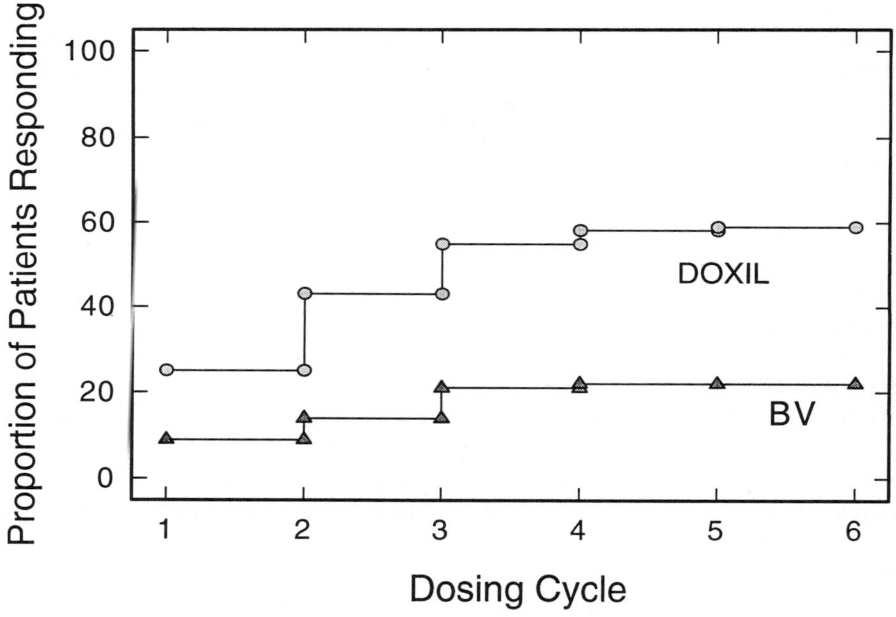

Fig. 12. Proportion of KS patients achieving complete, clinically complete or partial responses at each treatment cycle among patients randomized to receive DOXIL or bleomycin and vincristine (BV).

significantly in cognitive functioning, overall quality of like, and health transition. Patients treated with BV, on the other hand, showed significant improvement only in one domain, heath distress. For the KS Questionnaire, there was significantly improvement in ease of walking, relief from sleep disturbance, and social well being in the DOXIL group. BV patients showed significant improvement only in social well being (Table 20).

Table 20

Trends in quality of life measurements during DOXIL vs. BV therapy

Domain	DOXIL			BV		
	Improved	Worsened	Trend	Improved	Worsened	Trend
Pulmonary dysfunction & pain	4/30 (13.3%)	3/30 (10.0%)	↑	3/30 (10.0%)	6/30 (20.0%)	↓
GI pain & sensitivity	2/31 (6.5%)	3/31 (9.7%)	↓	3/29 (10.3%)	6/29 (20.7%)	↓
Difficulty walking	4/31 (12.9%)	0/31 (0%)	↑	2/29 (6.9%)	4/29 (13.8%)	↓
Difficulty wearing clothes	5/31 (16.1%)	2/31 (6.5%)	↑	0/29 (0%)	4/29 (13.8%)	↓
Eye dysfunction	3/31 (9.7%)	0/31 (0%)	↑	0/29 (0%)	2/29 (6.9%)	↓
Head & limb movement	5/31 (16.1%)	1/31 (3.2%)	↑	3/29 (10.3%)	7/29 (24.1%)	↓
Exercise limitations	6/31 (19.4%)	6/31 (19.4%)	–	4/29 (13.8%)	9/29 (31.0%)	↓
Sleep disturbance	6/31 (19.4%)	1/31 (3.2%)	↑	5/29 (17.2%)	4/29 (13.8%)	↑
Social well being	12/28 (42.9%)	1/28 (3.6%)	↑	7/29 (24.1%)	1/29 (3.4%)	↑

Table 21

Selected adverse events during DOXIL vs. BV therapy

	DOXIL (n = 121)	BV (n = 120)	p-value
Minimum ANC (10^3 cells/mm^3)			
<0.5	14 (12%)	4 (3%)	<0.001
0.5–<1.5	78 (64%)	59 (49%)	
>1.5	27 (22%)	56 (47%)	
Alopecia	4 (3%)	10 (8%)	0.107
Nausea/vomiting	19 (16%)	30 (25%)	0.080
Constipation	2 (2%)	13 (11%)	0.003
Peripheral neuropathy	10 (8%)	32 (27%)	<0.001
Mucositis/stomatitis	9 (7%)	6 (5%)	0.595
Rash	14 (12%)	11 (9%)	0.673
Fever	19(16%)	30 (25%)	0.080

A total of 55% of DOXIL patients and 31% of BV patients were able to complete the six cycles of therapy specified in the protocol. More BV discontinued the study prematurely to adverse events than DOXIL patients (27% vs. 11%). Adverse events were frequently reported in both groups; 97% of DOXIL and 96% of BV patients reported at least one adverse event. More AEs were thought to be probably or possibly related to drug in the BV group than in the DOXIL group (21% vs. 10%, respectively). In both treatment groups, the majority of adverse events were considered to be unrelated to the study drug (68% DOXIL, 53% BV).

Both drugs were associated with nausea and vomiting, but less frequently on DOXIL (16% with DOXIL vs. 25% with BV); most events were either mild or moderate in severity Table 21). Alopecia and mucositis/stomatitis were infrequent in both treatment arms. Treatment with BV caused a significantly higher incidence of peripheral sensory neuropathy, occurring in more than three times as many BV Patients as DOXIL patients (27% vs. 8%, $p < 0.001$).

DOXIL was more myelosuppressive than BV (Table 21), although this did not appear to impact patients' ability to complete therapy; significantly more patients discontinued treatment early in the BV arm (69%) relative to the DOXIL group (45%) ($p < 0.001$).

A greater number of patients developed opportunistic infections (OI) in the DOXIL group relative to those given BV (50% for DOXIL vs. 30% for BV). But when adjusted for time on study, the rate of development of OI and the time to development of OI were similar in both treatment arms (Table 22).

The finding of this study were consistent with those of the DOXIL vs. ABV trial detailed above. DOXIL as a single agent produced a superior response rate

Table 22

Development of opportunistic infections during DOXIL vs. BV therapy

	DOXIL $n = 121$	BV $n = 120$	p-value
Patients with any OI	60 (49.6%)	36 (30.0%)	0.002
Mean time to development of OI (SE)	111.8 (6.05)	121.1(5.99)	0.071
Median time on study	106	62	
OI per 1000 patient days	5.24	4.53	

to BV and showed an equivalent safety profile, but without the common dose-limiting toxicities associated with BV treatment.

IV.3.3. DOXIL vs. DOXIL + BV

The AIDS Clinical Trails Group (ACTG) conducted a Phase III comparison of DOXIL (20 mg/m^2) with and without bleomycin (10 mg/m^2) and vincristine (1 mg) (DBV) in patients with advanced KS.[94] Patients who had received prior chemotherapy were excluded. Patients must have had ⩾25 mucocutaneous lesions and/or visceral KS or ⩾10 lesions and CD^{4+} cell lymphocyte count of ⩽100 cells/mm^3. Toxicity assessment was conducted very two weeks and KS was evaluated using revised ACTG criteria every four weeks. A total of 129 patients were randomized from May, 1995 to July 1996 and 126 were assessable at the time of a protocol mandated interim analysis which was done in December, 1996 (62 on DOXIL and 64 on DBV).

The two groups were well balanced with respect to baseline disease characteristics and demographics. The median time to first grade ⩾3 toxicity was 10 weeks for DOXIL and 7 weeks for DBV ($p = 0.0095$), and more patients went off their treatment due to toxicity with DBV (16 patients) vs. DOXIL alone (4 patients). The overall tumor response was similar between the two treatment arms. In the DOXIL, group 5 patients achieved complete response (CR) and 43 a partial response (CR + PR = 79%) compared to 5 CR and 45 PR (CR + PR = 80%) on the DBV arm. The median time to tumor progression was also similar between the two groups (29 weeks on DOXIL and 32 weeks on DBV). At the time of the protocol mandated interim analysis, a trend toward better survival was observed for DOXIL vs. DBV (11 vs. 18 deaths, $p = 0.079$) and as a result of this finding, all patients were switched to continued treatment on DOXIL alone.

The investigators concluded that single agent DOXIL is an effective initial treatment for advanced-stage KS and that the addition of BV to DOXIL may cause more toxicity without substantially improving clinical outcomes.

IV.4. DOXIL in ovarian carcinoma

IV.4.1. Rationale for salvage therapy of recurrent disease

The management of recurrent ovarian cancer has improved with the introduction of paclitaxel-cisplatin combinations as first-line treatment, however, the majority of patients relapse and require additional therapy. Options for salvage therapy include re-treatment with platinum compounds, but the impact of such therapy on survival is modest, particularly in patients whose disease relapses within 6 months from their last cycle of chemotherapy.[95] Paclitaxel provided significant responses in platinum refractory patients and, in combination with cisplatin, has moved to front line therapy in many treatment centers. Other drugs such as ifosfamide, etoposide, epirubicin, topotecan, and gemcitabine have shown activity in the second or third line settings. Salvage treatments, unfortunately, are often poorly tolerated and remissions of meaningful duration are infrequent.

During the conduct of a phase I study of DOXIL, several heavily pretreated patients with advanced ovarian cancer achieved objective responses.[96] Preclinical studies showed that DOXIL provided a superior tumor response relative to free doxorubicin against a human ovarian cancer xenograft in nude mice. Therefore, a phase 2 study of DOXIL in patients with ovarian cancer refractory to the platinum agents and paclitaxel was conducted.[97]

The objective was to arrive at a dose and schedule of DOXIL that was tolerated in this patient group and to determine objective tumor response rate and duration of response.

IV.4.2. Patient characteristics and methods

A cohort of 35 patients with platinum- and paclitaxel refractory ovarian cancer with measurable or evaluable disease were entered at two institutions from October 1994 to July 1995. Patients had a Karnofsky performance status $\geqslant$50%, were required to have adequate bone marrow reserve (platelets $\geqslant$100,000/rnm^3, granulocytes $\geqslant$1500/mm^3, hemoglobin $\geqslant$8.0 g/dL), renal creatinine $\leqslant$2.0 mg/dL, liver bilirubin $<$1.5 mg/dL and adequate cardiac function (LVEF $>$50%).

In order to minimize the possibility of occasional pseudoallergic reactions, patients were routinely premedicated (hydrocortisone, diphenhydramine and cimetidine). The dose on the first cycle was 50 mg/m^2 every 3 weeks, with dose reductions for grade 3 and 4 toxicities to 40 mg/m^2, and lengthening of the interval to 4 weeks in the event of persistent toxicities (even if grade 1 or 2), not resolving by week 3 or 4. Treatment was continued until progression or unacceptable toxicity.

Standard oncologic response criteria were used in this study. A complete response (CR) was defined as the disappearance of all known disease, and a partial response (PR) a 50% or greater decrease in the sum of the product of cross-sectional diameters of the tumor(s), and no reappearance of new lesions. Both definitions include the need for confirmation at least 4 weeks from the initial assessment.

All patients had persistent or clinically recurrent epithelial ovarian cancer hav-

Table 23

Objective responses to DOXIL among refractory ovarian cancer patients

Number of patients	35
Complete response	1 (2.9%)
Partial response	8 (22.8%)
Overall response	9 (25.7%)
Stable disease	13 (37.1%)
Progression	9 (25.7%)

ing received platinum- and paclitaxel-based regimens without achieving a pathologic complete response. The median duration of drug-tree interval prior to DOXIL was 62 days for all patients enrolled in the study, and 70 days for patients who subsequently responded to DOXIL.

IV.4.3. Tumor response

Nine of 35 patients (25.7%) were documented to have objective tumor responses (Table 23).

The site of responses included the liver alone in 3, liver and pelvis in 3, and pelvis and/or retroperitoneal lymph nodes in 3. The median time to response was 5.8 months, and the median duration of response was 11 months (range 1.5 to 24+ months). The median progression-free survival was 6.7 months, whereas the median overall survival had not been reached (range 1.5 to 16.2+ months).

IV.4.4. Safety profile

During the first few minutes of two infusions, acute flushing reactions were seen, in one accompanied by back pain. Re-treatment was without incident. Four patients were hospitalized for fever and Grade 3 neutropenia accompanied by varying degrees of stomatitis (Table 24). One other patient experienced grade 3 stomatitis on week one after her first treatment followed by grade 2 neutropenia on week 2. Stomatitis led to dose reduction in 1 (accompanied by grade 2 neutropenia) and to dose delays in 4 others.

Grade 3 hand-foot syndrome was seen in 10 patients receiving DOXIL on a three-week schedule. All patients still receiving 50 mg/m^2 eventually required dose reductions and one week or greater dose delays because of skin toxicities after a median of 3 cycles (range 1–9 cycles).

Nausea and vomiting were infrequent, and no patient developed alopecia. Interestingly, patients who entered the study with paclitaxel-related hair loss experienced regrowth while undergoing DOXIL therapy. Among the 9 responding patients, cumulative doses of DOXIL ranged from 380–730 mg/m^2 (9–17 cycles). No significant declines in cardiac ejection fraction or signs of clinical cardiotoxicity were seen, including 7 patients who received $>$450 mg/m^2. There was no evidence of phlebitis or local problems even in several patients who received the drug via

Table 24

Toxicities among ovarian cancer patients receiving DOXIL

Number of Patients in whom doses were modified	27/35	
Grade 3 or 4 toxicities		
Neutropenia	7	all during cycles 1 and 2
Stomatitis	5	all during cycles 1 and 2+ one on cycle 3
Palmar-plantar erythrodysesthesia	10	
cycle 1	2	
cycle 2	3	
cycle 3	4	
cycle 4	1	
Grade 2 toxicities		
Neutropenia	1	
Stomatitis	2	persisting to week 4 of cycle
Palmar-plantar erythrodysesthesia	4	

peripheral veins exclusively. There was no evidence of drug-related liver dysfunction, and all patients with preexisting neuropathy experienced no worsening of symptoms during DOXIL administration.

IV.4.5. Conclusions

DOXIL has substantial activity in advanced ovarian cancer (26% among all 35 patients entered). Moreover, the duration of responses was meaningful (5.7 months) in a group of patients who had failed both platinum- and taxane-based therapy.

These response results have been duplicated in a follow-up Phase 2 trial of DOXIL in refractory ovarian patients.[97a] Moreover, the responses to DOXIL compare favorably with those reported for single-agent paclitaxel in advanced ovarian cancer. Response rates for paclitaxel ranging from 19–40% have been reported in early Phase 1–2 trials (Table 25), with durations of response on the

Table 25

Response rates in ovarian cancer reported in Paclitaxel registration trials

Trial	Ref.	*N*	Number responding			Mean duration (months)	Median survival (months)
			CR	PR	ORR[1]		
John's Hopkins (013)	98	46	1	9	22%	7.2	8.9
GOG (014)	99	46	5	9	31%	7.5	15.9
Albert Einstein (012)	100	32	2	4	19%	8.7	6.5
NCI (011)	101	15	2	4	40%	6.0	19.6
NCI (026)	101	50	1	17	36%	6.8	NR[2]
Total		189	11	43	29%	NR	11.1

[1]Percent of total evaluable patients achieving either CR or PR.
[2]Not reported.

Table 26

Response rates in platinum-refractory ovarian cancer[1]

Agent	Ref.	N	RR	Duration (months)
Carboplatin	95	90	7%	<7
Paclitaxel	102	652	22%	5

[1]Platinum resistance defined as recurrence within 6 months of platinum-based therapy.

order of 7 months. In a large series of platinum-resistant ovarian cancer patients, an overall response rate of 22% has been reported for paclitaxel, with a median duration of less than 5 months (Table 26).[102]

Ovarian cancer patients who relapse within 6 months of their last course of platinum-based therapy are less responsive to subsequent chemotherapy. It is interesting to note that 7/9 patients responding to DOXIL had relapsed with 2 month of their prior chemotherapy (4/9 within one month, 2/9 within 2 months and 1/9 within 3 months). Moreover, 4/9 patients who responded to DOXIL achieved disease stabilization as their best response on prior platinum and/or paclitaxel therapy.

The toxicity pattern indicates that mucositis and hand-foot syndrome are the dose-limiting toxicities for DOXIL in this disease setting. Dose reductions (to 40 mg/m^2) and an increase in the dosing interval from 3 weeks to 4–5 weeks allowed for continued dosing and was sufficient to manage these toxicities in most patients without sacrificing activity. DOXIL appears to be a very useful agent in second-line treatment and the safety profile suggests it may be able to be combined with other important drugs in the treatment of ovarian cancer—such as carboplatin or cisplatin and paclitaxel.

In conclusion, DOXIL has substantial antitumor activity resulting in meaningful response durations for patients who represent one of the most challenging groups to treat, those with epithelial ovarian cancer whose disease has progressed after treatment with platinum compounds and paclitaxel. Toxicity is minimal at a dose 40 mg/m^2 q4 weeks. Ongoing studies of DOXIL in combination regimens with paclitaxel will determine whether DOXIL may represent a valuable addition to front-line therapy of ovarian cancer.

IV.5. DOXIL in metastatic breast carcinoma

IV.5.1. Clinical rationale for single agent therapy

Anthracyclines such as doxorubicin are among the most active single agents used in the treatment of advanced breast cancer. However, the duration of treatment with these agents is limited by acute toxicities including nausea and vomiting, and subacute toxicities such as alopecia, mucositis and bone marrow suppression. Treatment-related neutropenia is frequently encountered with doxorubicin at the dose intensities needed to achieve a meaningful response rate. Severe neutropenia can lead to febrile episodes and occasionally to septic infections that can be life-threatening. Moreover, these agents have the potential to cause irreversible cardiac damage as cumulative doses reach and exceed about 500 mg/m^2 (see below).

Treatment strategies for advanced breast cancer vary geographically and among treatment centers in the same country. The goals of systemic chemotherapy in this setting can range from palliation to an intent to cure. Many medical oncologists believe that improved survival is not a realistic objective for systemic chemotherapy, particularly in patients who present with a high disease burden and/or multiple visceral metastatic sites. Although clinical data support the anti-tumor activity of agents such as doxorubicin, some argue that the benefit of marginally improved survival among a small number of patients does not outweigh the risk of toxicities to all patients who receive such therapy. In some treatment centers, therefore, patients with advanced breast cancer are not treated with chemotherapy, but rather given supportive care to palliate the signs and symptoms of the disease. This debate has raised awareness of the importance of patients' quality of life and has driven a search for treatment options which provide responses in a significant number of patients and/or delay disease progression for a meaningful period of time with minimal toxicity. In this context, a single-agent regimen would be preferred to combinations of drugs—provided that it is well tolerated and provides a respectable tumor response rate.

Preclinical studies of DOXIL and experience in the treatment of KS patients detailed above suggested that pegylated liposomes deliver a greater proportion of an injected dose of doxorubicin to tumor sites relative to unencapsulated doxorubicin. If this were also the case in breast cancer, one might reasonably expect DOXIL to have similar anti-tumor activity to that of doxorubicin, but at a lower dose intensity, and thus would produce less severe toxicity. This expectation provides a rationale for developing DOXIL as single-agent therapy for advanced breast cancer. The goal would be a tumor response rate comparable to doxorubicin but with a dose and schedule of DOXIL that minimizes the frequency and severity of nausea, vomiting, neutropenia, mucositis alopecia and cardiotoxicity.

IV.5.2. Patients and methods

Ranson et al.[103] have reported results of a multicenter, non-randomized phase 2 of DOXIL in patients with confirmed metastatic or locally advanced breast cancer. Patients were required to have measurable disease and were allowed to have received one cycle of prior chemotherapy (non-anthracycline). Standard objective oncologic response criteria were applied. The objectives of the study were to establish a dosing schedule of DOXIL that provided antitumor activity and was well tolerated by these patients. Three dose intensities were investigated in cohorts of patients: 60 mg/m^2 q3 weeks, 45 mg/m^2 q3 weeks and 45 mg/m^2 q4 week. A maximum of six cycles of DOXIL was administered. Dose reductions were required of patients receiving 60 mg/m^2 every 3 weeks for any garde 3 toxicity (except nausea, vomiting or alopecia), with treatment being delayed until recovery to grade 1 or better. In both of the cohorts given 45 mg/m^2 DOXIL, treatment was delayed if skin toxicity of grade $\geqslant 2$ developed, but therapy was restarted when this resolved to grade 1 or less. Dose reductions of 25% were required for any grade 3 toxicities (except nausea, vomiting or alopecia).

IV.5.3. Anti-tumor activity

A total of 71 patients were entered into the trial. All had stage IV disease with a mean Karnofsky status of 80% (range 60–100%). Just under 40% of patients had received prior chemotherapy (usually CMF) and 80% had been given prior hormonal therapy. This group of patients had a relatively high tumor burden: 73% had multiple visceral metastatic sites, predominantly in liver and lung and 20% had bone involvement.

Eight patients received only one cycle of DOXIL and therefore were not considered assessable for response. Of the remaining 64 assessable patients, 4 achieved complete response (6%) and 16 a partial response (25%) for an overall response rate of 31%. Another 20 patients had stable disease while on DOXIL therapy (31%). An objective response rate of 32% was seen among the 22 assessable patients who had received prior chemotherapy, a figure that was no different than the response rate for the entire group. The median overall survival was 7 months and the median time to disease progression among responders was 9 months.

IV.5.4. Safety profile

Skin toxicity, manifest principally as palmar-plantar erythrodysesthesia (hand-foot syndrome), was the most frequent toxicity encountered during this study. Importantly, the incidence and severity of this syndrome was found to be critically dependent upon the interval between DOXIL cycles. Among the 39 patients assigned to the two cohorts receiving DOXIL on a 3 week interval, grade 3 or 4 hand foot syndrome developed in 19 (27%). In contrast, of the 32 patients receiving DOXIL on a 4 week cycle, only 5 (15%) developed grade 3 skin toxicity and none developed grade 4 skin toxicity. Expressed in terms of the number of cycles administered that were associated with hand foot syndrome, grade 3 was seen in 12% and grade 4 in 8% of the 129 cycles given on the three week schedule, while only 5% of the 126 cycles given at the 4 weeks interval lead to grade 3 skin toxicity with no instances of grade 4 toxicity. Figure 13 presents these skin toxicity data graphically. It is clear from these data that a 3 week interval between DOXIL treatments leads to severe skin toxicity in some patients. However, lengthening the interval to 4 weeks reduced the incidence and severity of skin toxicity to an acceptable level.

Neutropenia was also more frequent on the 3 week dosing cycle, occurring at a level of grade 3 or 4 in 12% of the 129 cycles of DOXIL administered at 60 mg/m^2 q3 weeks or 45 mg/m^2 every 3 weeks. Of the 126 cycles given at 45 mg/m^2 every 4 weeks, however, grade 4 neutropenia was seen in only 2 (<2%) and grade 3 neutropenia was encountered in 7 cycles (<6%).

Alopecia, nausea and vomiting and other hematological toxicities such as thrombocytopenia and anemia were infrequent and no liver or kidney abnormalities were seen. Although mucositis of grade 3 or 4 was reported in 15/129 (12%) of

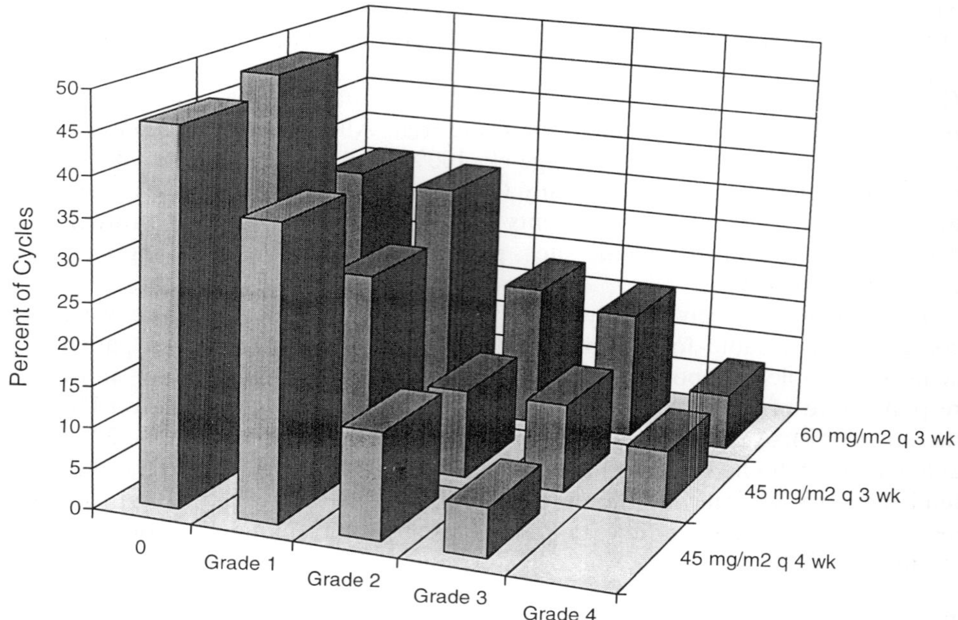

Fig. 13. Relationship between incidence and severity of palmar plantar erythrodysesthesia (hand-foot syndrome) and dosing schedule of DOXIL.

cycles given on the 3 week interval, the incidence of grade 4 mucositis was reduced to zero and grade 3 to 5% among patients in the 45 mg/m^2 q3 week dose group. No evidence of clinical cardiotoxicity associated with DOXIL was observed in any patient.

IV.5.5. Conclusions

Acute side effects normally associated with doxorubicin therapy that negatively impact the quality of life in most patients receiving the drug including alopecia, nausea and vomiting were infrequent on DOXIL and mild if they occurred at all.

Epithelial toxicity (hand-foot syndrome/mucositis) represent the most serious side effects associated with DOXIL therapy and their incidence and severity were directly related to the dose schedule. At 45 mg/m^2 q4 weeks, DOXIL was very well tolerated with a low incidence of skin toxicity (<10% grade 3, no grade 4). This unusual toxicity pattern is likely the result of redistribution of the drug to skin by encapsulation in pegylated liposomes.

At a dose that was well tolerated (45 mg/m^2 q4 weeks), single agent DOXIL was active in metastatic breast cancer with an overall response rate and duration of response comparable to doxorubicin at a higher dose intensity (60–75 mg/m^2 q3 weeks). This suggests that encapsulation of doxorubicin in pegylated liposomes also delivers more drug to tumors, in agreement with preclincial findings.

The results of this study provide evidence that DOXIL is a reasonable option in those breast cancer cases where "gentler" single agent therapy is indicated (i.e., settings where maintenance of quality of life is important such as elderly patients and those in whom cure or prolonged survival is not likely) but where a meaningful response rate and response duration is the objective.

Given the relatively mild myelosuppression seen with DOXIL and an understanding of how to manage skin toxicity gained here, combinations of DOXIL with other agents active against breast cancer such as taxanes and cyclophosphamide should be considered. It is likely that dose and dosing schedules of such combinations can be optimized to boost response rates with manageable toxicity.

V. Cardiac safety of DOXIL

V.1. Assessment by cardiac biopsy in KS patients

The length of therapy with doxorubicin is limited by cumulative cardiotoxicity. Studies of single-agent doxorubicin have demonstrated that CHF is very uncommon at cumulative doses of less than 500 mg/m^2, but rises to about 25% by 550 mg/m^2.[104–105] Dosing schedules of doxorubicin designed to lower peak plasma concentration are known to reduce cardiotoxicity, suggesting that the peak dose (C_{max}) in plasma after a typical doxorubicin infusion contributes to development of this toxicity.[6–11,106] As detailed above, peak levels of *bioavailable* doxorubicin are substantially reduced following DOXIL administration and preclinical results indicate that DOXIL causes less morphological changes in heart tissues relative to free doxorubicin in three species; rat, dog and rabbit.

These encouraging preclinical findings are supported by results of a clinical study done by Berry et al.[107] These authors have reported the results of endomyocardial biopsies performed on AIDS-KS patients who received cumulative doses of DOXIL in excess of 400 mg/m^2.

Standard noninvasive methods of myocardial function such as left ventricular ejection fraction (LVEF) and clinical evaluation, while useful for identifying patients who have developed heart failure, are not sufficiently sensitive to detect mild to moderate degrees of change resulting from anthracycline therapy. The most reliable technique to predict doxorubicin-induced cardiac toxicity in this setting is percutaneous endomyocardial biopsy. Biopsy changes assessed by electron microscopy are characteristic of anthracycline damage and not confounded by other chemotherapies, age, sex, primary disease site, underlying heart disease, or Karnofsky performance status.[108] The histologic pattern of HIV-related myocardial lesions is similar to that found in noninfected patients and thus can be differentiated on an ultrastructural level from anthracycline induced cardiac damage.[109–111] Because of the sensitivity and specificity of changes assessed by biopsy, significant differences in the cardiotoxicity associated with different drugs or drug regimens can be detected with a relatively small patient sample size.

Ten AIDS KS patients who had received a cumulative dose of >400 mg/m^2 of DOXIL and had no prior anthracycline exposure were recruited into this study.

All patients received DOXIL at a dose of 20 mg/m^2/2–3 weeks as an intravenous infusion over 30 minutes. At entry, patients had Karnofsky performance status ⩾60%, platelet count >100,000 cells/mm^3, Hgb >10 gm/dL, and PT and PTT within normal limits. Severely ill patients or those with an active opportunistic infection, mental disorder or a history of anticoagulation disorder were excluded.

Control patients were selected from a historical database of 131 patients having biopsies between 1975 and 1983 at the Stanford University Medical Center. All patients in the control group treated with cardiac irradiation were eliminated from this study. The control group thus included 100 patients, 76 treated at a dose rate of 60 mg/m^2 every 3 weeks and 24 at 20 mg/m^2 weekly.

Two control groups were assembled. First, each DOXIL treated patient was matched to the single patient among the 100 potential matches that most closely approximated the patient's cumulative doxorubicin dose (control group 1). When more than one doxorubicin patient was identified with a cumulative dose within 10 mg/m^2 of that of the DOXIL patient, an attempt was made to match the patients with respect to peak dose (i.e., 60 mg/m^2 or 20 mg/m^2). Second, all 10 patients treated with DOXIL were compared to the 24 patients treated with doxorubicin at 20 mg/m^2/wk (control group 2).

Endomyocardial biopsies, cardiac catheterization and specimen preparation were performed as described earlier.[111–113]

To avoid bias, micrographs from DOXIL patients and control group 1 were arranged as sets, coded, randomized and read blinded by two pathologists. A consensus morphologic grade was derived for each set of micrographs based on a 7-point scale previously described by Billingham et al.[114] Scores for control group 2 were used as found in the historical database.

V.2. Adjustment of cardiac biopsy scores for administered dose

To address the possible impact of the difference of peak dose between the DOXIL patients and control group 1, the biopsy scores of the five matched patients in group 1 who were treated with 60 mg/m^2 were adjusted downwards by 0.8 biopsy units. This correction factor was calculated by comparing biopsy scores retrieved from the historical data base for matched pairs of patients who received doxorubicin either at a dose of 20 mg/m^2/week or 60 mg/m^2/3 weeks.[106]

V.3. Results

The biopsy scores of each of the matched DOXIL and control group 1 patients, including unadjusted and adjusted scores for the five doxorubicin patients treated with 60 mg/m^2, are shown in Table 27. The mean and median cumulative dose of DOXIL was higher than that of control group 1 and 2. However, the mean and median biopsy scores of the DOXIL patients were significantly lower than either of these two groups.

The groups were similar in other regards. All DOXIL patients were male with a median age of 38 years (range 32–51), reflecting characteristics expected of the

Table 27

Blinded biopsy scores for pairs of patients receiving either DOXIL or doxorubicin matched by cumulative dose

	Cumulative dose (mg/m^2)			Biopsy score						
				DOXIL	Doxorubicin					
					Unadjusted			Adjusted for dose		
	DOXIL	Doxorubicin	Difference*		Score	Difference[†]	p_1-value[‡]	Score	Difference[†]	p_2-value[§]
Pair 1	469	472	3	1.0	3.0	2.0	$p < 0.001$	2.2	1.2	$p < 0.001$
Pair 2	482	487	5	0	3.0	3.0		2.2	2.2	
Pair 3	500	497	−3	1.0	1.5	0.5		0.7	−0.3	
Pair 4	541	542	1	0	1.5	1.5		1.5	1.5	
Pair 5	801¶	565	−256	0.5	2.5	2.5		2.5	2.5	
Pair 6	578	583	5	0	3.0	2.5		3.0	2.5	
Pair 7	610	585	−25	0.5	3.0	2.5		2.2	1.7	
Pair 8	660	600	−60	1.5	3.0	1.5		2.2	0.7	
Pair 9	780	645	−135	0	3.0	3.0		3.0	3.0	
Pair 10	860	671	−189	0	1.5	1.5		1.5	1.5	
Mean ± SD	628 ± 142	565 ± 66		0.5 ± 0.6	2.5 ± 0.7			2.1 ± 0.7		
Median	559.5	553.5		0.3	3.0			2.2		

*DOXIL patient cumulative dose subtracted from matched doxorubicin patient cumulative dose.
[†]DOXIL patient score subtracted from matched doxorubicin patient score.
[‡]Analysis of variance.
[§]Analysis of variance adjusted for peak dose.
¶After the study was completed, it was observed that DOXIL patient #5 had previous exposure to DOXIL that was not taken into account when the match to the doxorubicin patient was made. When previous exposure was taken into account, the actual cumulative dose waas 801 mg/m^2 rather than 571 mg/m^2 as originally thought.

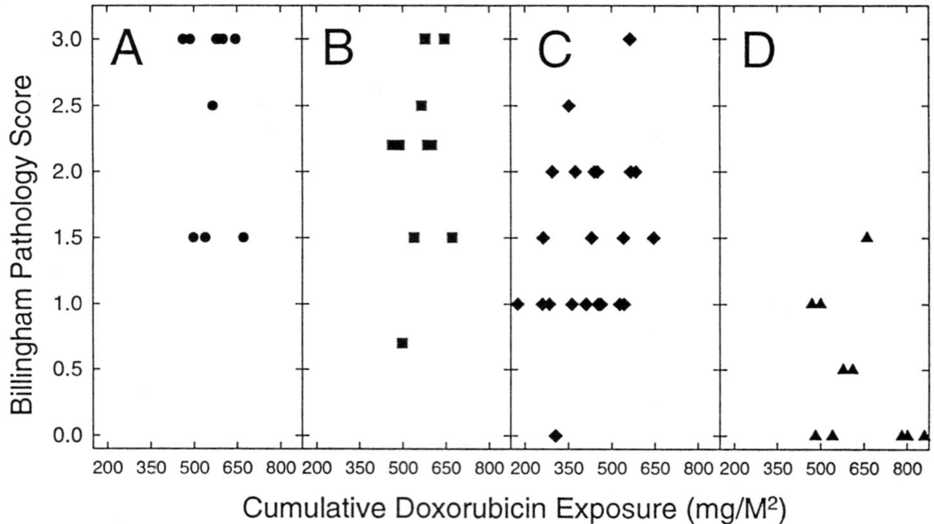

Fig. 14. Biopsy Scores vs. Cumulative Doxorubicin Dose. The relationship between biopsy scores and cumulative doxorubicin exposure in doxorubicin control groups assembled on the basis of matching cumulative dose (A) or peak dose (C) and DOXIL patients (D). Panel B, presents scores for patients in the doxorubicin cumulative dose control group (panel A) adjusted to compare with DOXIL patients with respect to peak dose (i.e., 20 mg/m^2, see text).

Kaposi's sarcoma population. None of the DOXIL patients had a history of heart disease. The median age of the control group 1 patients was higher at 56 years (range 31–68) with 4 females and 6 males. Histologies in control group 1 included breast (3), sarcoma (2), lung (2), urogenital (1), gynecological (1), and lymphoma (1). Half the patients in this group received concurrent cyclophosphamide. None of the selected control patients had a history of heart disease or congestive heart failure (CHF); 3 had a history of hypertension. None of the control patients had cardiac ejection fraction determinations.

The biopsy data are plotted in Figure 14. Comparison of the groups by an analysis of variance model based on the biopsy score indicated a significant difference between the DOXIL and doxorubicin patients, whether the comparison is to matched doxorubicin patients using unadjusted scores, matched patients using adjusted scores, or only those patients given doxorubicin on a lower dose schedule. The mean biopsy score for the DOXIL group was 0.5 (±0.6). For the three controls groups the mean biopsy scores were 2.5 (±0.7), 2.1 (±0.7) and 1.4 (±0.65), respectively. The comparisons between the mean DOXIL scores and the control groups were highly significant, $p < 0.001$ in each case.

V.4. Conclusions

Cumulative dose of doxorubicin is the single most important factor determining cardiac toxicity, and for this reason the primary analysis used in this study was based on matched cumulative dose. The highest cumulative dose given to any of the patients in control group 1 was 671 mg/m^2 but 3 DOXIL patients received higher doses, the highest level reached being 860 mg/m^2. Although five of the DOXIL doxorubicin matches were nearly identical and an additional two were very close, the average cumulative dose of DOXIL was significantly higher. Any bias resulting from mismatch should have favored the doxorubicin treated patients. Also the authors adjusted downward the biopsy scores of the control group 1 patients treated on a 60 mg/m^2 dose schedule by 0.8 biopsy units to better approximate both cumulative dose and dose schedule. This degree of correction is conservative when compared to the approximately 0.5 biopsy score reduction seen in a group of patients receiving 20 mg/m^2 relative to those given 60 mg/m^2 reported by Torti et al.[106]

Despite this conservative approach, the DOXIL patients had significantly lower biopsy scores than any of the control group patients, regardless of which comparison was made, suggesting that DOXIL causes severe cardiac toxicity than doxorubicin.

The mechanism by which DOXIL protects the heart from doxorubicin toxicity may reside in the altered clearance and tissue distribution pattern of the drug provided by encapsulation in pegylated liposomes. While in the circulation, DOXIL liposomes remain intact, retaining virtually all of the doxorubicin in encapsulated form.[39] Although total plasma levels of doxorubicin may be relatively high for several days after DOXIL administration, the majority of the dose is sequestered within the liposome during this period and thus is not bioavailable to distribute (as free drug molecules) to tissues, including the myocardium. With respect to level of available drug in plasma, DOXIL resembles more that of a 96 hour continuous infusion of doxorubicin than the usual 30 minute infusion. Prolonged infusion of doxorubicin is known to reduce cardiotoxicity.[7–11]

Tissue distribution studies in rodents and rabbits suggest that after DOXIL administration doxorubicin slowly enters the heart muscle in the form of intact DOXIL liposomes, reaching a peak at 24–48 hours.[64] Relative to comparable dose of doxorubicin, the peak concentration in heart muscle after DOXIL is lower but the AUC is greater, reflecting a slow rate of clearance of the liposome encapsulated drug once it has entered the heart tissue. Clearance of DOXIL from heart muscle may occur via several pathways. It is quite likely that two species of doxorubicin exist in heart muscle (and other normal tissues) after DOXIL administration, free (bioavailable) and encapsulated (bio-unavailable). A proportion of the total drug entering the tissue is likely to be released from the liposomes (as a consequence of wear and tear and/or loss of the PEG coating) and thus is free to enter cells or reenter the circulation. This fraction of drug, just as conventional doxorubicin, would have the potential to cause cardiotoxicity. The remaining fraction of drug may remain encapsulated and be cleared via lymphatics as intact DOXIL liposomes. This fraction would theoretically not pose a threat to the muscle cells as it passed through the tissue in encapsulated form. Unfortunately, it is not yet possible to separate free from encapsulated drug in solid tissues, so the relative rate of appearance and disappearance of these two species in heart muscle cannot be accurately measured.

References

1. Di Marco A, Arcamone F. DNA complexing antibiotics: daunomycin, adriamycin and their derivatives. Arzneimittel-Forschung 1975;25:368–375.
2. Meriwether W, Bachur N. Inhibition of DNA and RNA metabolism by daunorubicin and adriamycin in L1210 mouse leukemia. Cancer Res 1972;32:1137–1142.
3. Dorr R, Fritz W. Doxorubicin Cancer Chemotherapy Handbook, 1980;388–401.
4. Fischl MA., Krown SE, O'Boyle KP et al. Weekly doxorubicin in the treatment of patients with AIDS-related Kaposi's sarcoma. J AIDS 1993;6:259–264.
5. Gill PS, Rarick M, McCutchan JA et al. Systemic treatment of AIDS-related Kaposi's sarcoma: results of a randomized trial. Am J Med 1991;90:427–433.
6. Bielack SS, Erttmann R, Winkler K, Landbeck G. Doxorubicin: effect of different schedules on toxicity and anti-tumor efficacy. Eur J Cancer Clin Oncol 1989;25:873–882.
7. Legha SS, Benjamin RS, MacKay B, Ewer M, Wallace S, Valdivieso M, Rasmussen SL, Blumenschein GR, Freireich EJ. Reduction of doxorubicin cardiotoxicity by prolonged continuous intravenous infusion. Ann Int Med 1982;96:133–139.
8. Hortobagyi GN, Frye D, Buzdar AU, Ewer MS, Fraschini G, Hug V, Ames F, Montague E, Carrasco CH, MacKay B, Benjamin RS. Decreased cardiac toxicity of doxorubicin administered by continuous intravenous infusion in combination chemotherapy for metastatic breast cancer. Cancer 1989;63:37–45.
9. Shapira J, Gotfried M, Lishner M, Ravid M. Reduced cardiotoxicity of doxorubicin by a 6-hour infusion regimen: a prospective randomized evaluation. Cancer 1990;65:870–873.
10. Lokich J, Bothe A, Zipoli T, Green R, Sonneborn H, Paul S, Philips D. Constant infusion schedule for Adriamycin: a phase I-II clinical trial of a 30-day schedule by ambulatory pump delivery system. J Clin Oncol 1983;1:24–28.
11. Garnick MB, Weiss GR, Steele GD, Israel M, Schade D, Sack MJ, Frei E. Clinical evaluation of long-term, continuous-infusion doxorubicin. Cancer Treat Rep 1983;67:133–142.
12. Gabizon A, Amselem S, Goren D et al. Preclinical and clinical experience with a doxorubicin-liposome preparation. J Liposome Res 1990;1:491–502.
13. Olson F, Mayhew E, Maslow D, Rustum Y, Szoka F. Characterization, toxicity and therapeutic efficacy of Adriamycin encapsulated in liposomes. Eur J Cancer Clin Oncol 1982;18:167.
14. Sells RA, Gilmore IT, Owen RR et al. Reduction in doxorubicin toxicity following liposomal delivery. Cancer Treat Rev 1987;14:383–387.
15. Sells RA, Owen RR, New RRC et al. Reduction in toxicity of doxorubicin by liposomal entrapment. The Lancet 1987;11:624–625.
16. Gabizon A, Peretz T, Sulkes A et al. Systemic administration of doxorubicin-containing liposomes in cancer patients: a Phase I study. Eur J Cancer Clin Oncol 1992;25:1795–1803.
17. Owen RR, Sells RA, Gilmore IT et al. A Phase I clinical evaluation of liposome-entrapped doxorubicin (Lip-dox) in patients with primary and metastic hepatic malignancy. Anti-cancer Drugs 1992;3:101–107.
18. Gabizon A, Chisin R, Amselem S, Druckmann S, Cohen R, Goren D, Fromer I, Peretz T, Sulkes A, Barenholz Y. Pharmacokinetic and imaging studies in patients receiving a formulation of liposome-associated Adriamycin. Br J Cancer. 1991;64:1125–1132.
19. Delgado G, Potkul RK, Treat JA et al. A Phase I/II study of intraperitoneally administered doxorubicin entrapped in cardiolipin liposomes in patients with ovarian cancer. Am J Obstet Gynecol 1989;160:812–819.
20. Rahman A, Treat J, Roh JK, Potkul LA, Alvord WG, Forst D, Woolley PV. A Phase I clinical trial and pharmacokinetic evaluation of liposome-encapsulated doxorubicin. J Clin Oncol 1990;8:1093–1100.
21. Treat J, Greenspan A, Forst D, Sanchez JA, Ferrans VJ, Potkul LA, Woolley PV, Rahman A. Antitumor activity of liposome-encapsulated doxorubicin in advanced breast cancer: a Phase II study. J Natl Cancer Inst 1990;82:1706–1710.
22. Cowens JW, Kanter P, Brenner DE et al. Phase I study of doxorubicin encapsulated in liposomes (Abstract). Proc Am Soc Clin Oncol 1989;8:69.
23. Cowens JW, Creaven. PJ, Brenner DE et al. Phase I study of doxorubicin encapsulated in liposomes (Abstract). Proc Am Soc Clin Oncol 1990;9:7.
24. Cowens JW, Greco W, Ginsberg R. Pharmacokinetics of doxorubicin encapsulated in liposomes in patients with advanced cancer (Abstract). Proc Am Soc Clin Oncol 1990;8:87.

25. Cowens JW, Creaven PJ, Greco DE et al. Initial clinical (phase I) trial of TLC-D99 (doxorubicin encapsulated liposomes). Cancer Res 1993;53:2796–2802.
26. Creaven PJ, Cowens J.W, Ginsberg R, Ostro M, Browman G. Clinical studies with liposomal doxorubicin. J Liposome Res 1990;1:481–490.
27. Batist G, Ahlgren P, Panasci J et al. Phase II study of liposomal doxorubicin (TLC-D99) in metastatic breast cancer (Abstract). Proc Am Soc Clin Oncol 1992;11:82.
28. Mazanet R, Seidenberg S, Bartel S, Elias A, Saletan S, Faber D. A phase I study of liposome encapsulated doxorubicin (TLC-99) with G-CSF. (Abstract). Proc Amer Soc Clin Oncol 1993;12:154.
29. Conley B, Egorin M, Whitacre M, Carter D, Zuhowski E, Echo D. Phase I and pharmacokinetic trial of liposome-encapsulated doxorubicin. Cancer Chem Pharm 1993;33:107–112.
30. Embree L, Gelmon K, Lohr A, Mayer L, Coldman A, Cullis P, Palaitis W, Piliewicz F, Hudon N, Heggie J, Goldie J. Chromatographic analysis and pharmacokinetics of liposome-encapsulated doxorubicin in non-small lung cancer patients. J Pharm Sci 1993;82:627–634.
31. Kumai K, Takahashi T, Tsubouchi T et al. Selective hepatic arterial infusion of liposomes containing antitumor agents. Jpn J Cancer Chemother 1985;12:1946–1948.
32. Akamo Y, Mizuno I, Ichino T, Yamamoto T. et al. Endoscopic injection of liposomal Adriamycin targeting lymph node metastasis of gastric cancer. Jpn J Cancer Chemother 1991;18:1822–1824.
33. Juliano R, Stamp D. The effect of particle size and charge on the clearance rates of liposomes and liposome encapsulated drugs. Biochem Biophys Res Comm 1975;63:651.
34. Gregoriadis G, Senior J. The phospholipid component of small unilamellar liposomes controls the rate of clearance of entrapped solutes from the circulation. FEBS Lett 1980;119:43.
35. Presant C, Proffitt R, Turner A, Williams L, Winsor D, Werner J, Kennedy P, Wiseman C, Gala K, McKenna R, Smith J, Bouzaglou S, Callahan R, Baldeschweiler J, Crossley R. Successful imaging of human cancer with indium-111-labeled phospholipid vesicles. Cancer 1988;62:905.
36. Yuan F, Dellian M, Fukumura D et al. Vascular permeability in human xenograft: molecular size dependence and cut off size. Cancer Res 1995;55:3752–3756.
37. Allen T.M, Hansen C, Martin F, Redemann C, Yau-Young A. Liposomes containing synthetic lipid derivatives of poly(ethylene glycol) show prolonged circulation half-lives in vivo. Biochim Biophys Acta 1991;1066:29–36.
38. Gabizon A, Barenholz Y, Bialer M. Prolongation of the circulation time of doxorubicin encapsulated in liposomes containing a polyethyleneglycol-derivatized phospholipid: Pharmacokinetic studies in rodents and dogs. Pharm Res 1993;10:703–708.
39. Gabizon A, Catane R, Uziely B, Kaufman B, Safra T, Cohen R, Martin F, Huang A, Barenholz Y. Prolonged circulation time and enhanced accumulation in malignant exudates of doxorubicin encapsulated in polyethylene-glycol coated liposomes (DOXIL®). Cancer Res 1994;54:987–992.
40. Northfelt DM, Martin FJ, Working P et al. Doxorubicin encapsulated in liposomes containing surface-bound polyethylene glycol: pharmacokinetics, tumor localization and safety in patients with AIDS-related Kaposi's sarcoma. J Clin Pharm. 1996;36:55–63; Gerlowski LE, Jain RK. Microvascular permeability of normal and neoplastic tissues. Microvasc Res 1986;31:288–305.
41. Jain RK. Transport of molecules across tumor vasculature. Cancer Metastasis Reviews 1987;6:559–593.
42. Francis ND, Parkin JM, Weber J, Boylston AW. Kaposi's sarcoma in acquired immune deficiency syndrome (AIDS). J Clin Pathol 1986;39:469–474.
43. Seymour LW. Passive tumor targeting of soluble macromolecules and drug conjugates. Crit Rev Ther Drug Carrier Systems 1992;9:135–187.
44. Jain R. Delivery of novel therapeutic agents in tumors: physiological barriers and strategies. J Natl Cancer Inst. 1989;81:570–576.
45. Huang SK, Lee KD, Hong K, Friend DS, Papahadjopoulos D. Microscopic localization of sterically stabilized liposomes in colon carcinoma-bearing mice. Cancer Res 1992;52:5135–5143.
46. Huang SK, Martin FJ, Jay G et al. Extravasation and transcytosis of liposomes in Kaposi's sarcoma-like dermal lesions of transgenic mice bearing the HIV tat gene. Am J Pathol 1993;143:10–14.
47. Papahadjopoulos D, Allen TM, Gabizon A, Mayhew E, Matthay K, Huang SK, Lee KD, Woodle MC, Lasic DD, Redemann C, Martin FJ. Sterically stabilized liposomes: improvements in pharmacokinetics and anti-tumor therapeutic efficacy. Proc Natl Acad Sci (USA) 1991;88:11460–11464.
48. Huang SK, Mayhew E, Gilani S, Lasic DD, Martin FJ, Papahadjopoulos D. Pharmacokinetics and therapeutics of sterically stabilized liposomes in mice bearing C-26 colon carcinoma. Cancer Res 1992;52:6774–6781.

49. Vaage J, Barbera-Guillem E, Abra R, Huang A, Working P. Tissue distribution and therapeutic effect of Intravenous free or encapsulated liposomal doxorubicin on human prostate carcinoma xenografts. Cancer 1994;73:1478–1484.
50. Vaage J, Donovan D, Mayhew E, Abra R, Huang A. Therapy of human ovarian carcinoma xenografts using doxorubicin encapsulated in sterically stabilized liposomes. Cancer 1993;72:3671–3675.
51. Williams S, Alosco T, Mayhew E, Lasic D, Martin F, Bankert R. Arrest of human lung tumor xenograft growth in severe combined immunodeficient mice using doxorubicin encapsulated in sterically stabilized liposomes. Cancer Res 1993;53:3964–3967.
52. Siegal T, Horowitz A, Gabizon A. Doxorubicin encapsulated in sterically stabilized liposomes for the treatment of a brain tumor model: biodistribution and therapeutic efficacy. J Neurosurg 1995;83:1029–1037.
53. Vaage J. Mayhew E, Lasic D, Martin F. Therapy of primary and metastatic mouse mammary carcinoma with doxorubicin encapsulated in long circulating liposomes. Int J Cancer 1992;51:942–948.
54. Amantea MA, Forrest A, Northfelt DW, Mamelok R. Population pharmacokinetics and pharmacodynamics of pegylated liposomal doxorubicin in patients with AIDS-related Kaposi's sarcoma. Can Chemother Pharmacol, 1997, in press.
55. Amantea MA, Gabizon A. Population pharmacokinetics of DOXIL. Personal communuication, 1997.
56. Druckmann S, Gabizon A, Barenholz Y. Separation of liposome-associated doxorubicin from non-liposome-associated doxorubicin in human plasma. Biochim Biophys Acta 1989;980:381–384.
57. Benjamin RS, Riggs CE, Bachur NR. Plasma pharmacokinetics of Adriamycin and its metabolites in humans with normal hepatic and renal function. Cancer Res 1977;37:1416–1420.
58. Garnick MB, Weiss GR, Steele GD. et al. Clinical evaluation of long-term continuous infusion doxorubicin. Cancer Treat Rep 1983;67:133–142.
59. Speth PAJ, van Hoesel QGCM, Haanen C. Clinical pharmacokinetics of doxorubicin. Clin Pharmacokin 1988;15:15–31.
60. Greene RF, Collins JM, Jenkins JF et al. Plasma pharmacokinetics of Adriamycin and adriamycinol: implications for the design of in vitro experiments and treatment protocols. Cancer Res 1982;43:3417–3421.
61. Speth PAJ, Linssen PCM, Holdrinet RSG, Haanen C. Plasma and cellular adriamycin concentrations in patients with myeloma treated with ninety-six-hour continuous infusion. Clin Pharmacol Therap 1987;41:661–665
62. Speth PAJ, Linssen PCM, Boezeman JBM et al. Cellular and plasma Adriamycin concentrations in long-term infusion therapy of leukemia patients. Cancer Chemother Pharmacol 1987;20:305–310.
63. Benjamin RS, Wiernik PH, Bachur NR. Adriamycin chemotherapy—efficacy, safety, and pharmacologic basis of an intermittent singe high-dosage schedule. Cancer 1974;33:19–27.
64. Working PK, Newman MS, Huang SK et al Pharmacokinetics, biodistribution and therapeutic efficacy of doxorubicin encapsulated in Stealth liposomes (DOXIL). J Liposome Res 1994;4:667–687.
65. Stewart Simon. Personal communication.
66. Vogel J, Hinrichs SH, Reynolds RK, Luciw PA, Jay G. The HIV tat gene induces dermal lesions resembling Kaposi's sarcoma in transgenic mice. Nature 1988;335:606–611.
67. Yuan F, Leunig M, Huang SK et al. Microvascular permeability and interstitial penetration of sterically stabilized (Stealth) liposomes in human tumor xenografts. Cancer Res 1994;54: 3352–3356.
68. Hunag SK, Martin FJ, Friend DS, Papahadjopoulos D. Mechanism of Stealth liposome accumulation in some pathological tissues. In: Lasic D, Martin F, eds. Stealth Liposomes. Boca Raton: CRC Press, 1995;119–126.
69. Dewhirst MW, Needam D. Extravasation of Stealth liposomes into tumors: direct measurement of accumulation and vascular permeability using a window chamber. In: Lasic D, Martin F, eds. Stealth Liposomes. Boca Raton: CRC Press, 1995;127–137.
70. Stubbs M, Bhujwalla ZM, Tozer GM et al. An assessmant of ^{31}P MRC as a method of measuring pH in rat tumors. NMR Biomed 1992;5:351–359.
71. Sakayama K, Masuno H, Miyazaki T et al. Exostance of lipoprotein lipase in human sarcomas and carcinomas. Jpn J Cancer Res 1994;85:515–521.
72. Dvorak HF, Dickerson GR, Dvorak AM et al. Human breast carcinoma: fibrin deposits, and

desmoplasia, inflammatory cell type and distribution, microvasculature and infarction. J Natl Cancer Inst 1981;67:335–345.
73. Cobbs CS, Brenman JE, Aldape KD et al. Expression of nitric oxide synthase in human central nervous system tumors. Cancer Res 1995;15:727–730.
74. Pupa SM, Bufalino. R Invernizzi, AM. Macrophage infiltrate and prognosis in c-erbB-2-overexpressing breast carcinomas. J Clin Oncol 1996;14:85–94.
75. Vaage J, Donovan D, Uster P, Working PK. Tumour uptake of doxorubicin in polyethylene glycol-coated liposomes and therapeutic effect against xenografted pancreatic carcinoma. Brit J Cancer 1997;75:482–486.
76. Nagy JA, Morgan ES, Herzberg KT et al. Pathogenesis of ascites tumor growth: angiogenesis, vascular remodeling, and stroma formation in the peritoneal cavity. Cancer Res 1995;55:376–385.
77. Goebel F-D, Goldstein D, Gross M et al. Efficacy and safety of Stealth liposomal doxorubicin in AIDS-related Kaposi's sarcoma. Br J Cancer 1996;73:989–994.
78. Hengge UR, Brockmeyer N, Baumann M, Reimann G, Goos M. Liposomal doxorubicin in AIDS-related Kaposi's sarcoma. Lancet 1993;342:497.
79. Goebel FD, Bogner JR, Spathling S, Held M, Sandor P, Trubenbach K, Rauch CH, Kronawitter U. Quantitative ultrasound volume measurement serves as non-invasive method to follow response of cutaneous Kaposi's sarcoma lesions to Doxil therapy. (Abstract). Proc Amer Soc Clin Oncol 1993;12:51.
80. Bogner JR, Zietz C, Held M, Spathling S et al. Ultrasound as a tool to evaluate remission of cutaneous Kaposi's sarcoma. AIDS 1993;7:1081–1085.
81. Bogner JR, Kronawitter U, Rolinski B et. al. Liposomal doxorubicin in the treatment of advanced AIDS-related Kaposi's sarcoma. J Acquir Immune Def Synd 1994;7:463–468
82. Wagner D, Kern WV, Kern P. Liposomal doxorubicin in AIDS-related Kaposi's sarcoma: long-term experiences. Clin Invest 1994;72:417–423.
83. Simpson JK, Miller RF, Spittle MF. Liposomal doxorubicin for treatment of AIDS-related Kaposi's sarcoma. Clin Oncol 1993;5:372–374.
84. Renolds P, Saunders LD, Layefsky ME et al. The spectrum of acquired immunodeficiency syndrome (AIDS)-associated malignancies in San Francisco, 1980–1987. Am J Epidemiol 1993;137:19.
85. Moore PS, Chang Y. Detection of herpes-like sequences in Kaposi's sarcoma patients with and without AIDS infection. N Engl J Med 1995;332:1181–1185.
86. Su IJ, Hsu TS, Chang YC et al. Herpes virus like DNA sequences in Kaposi's sarcoma from AIDS and non-AIDS patients in Taiwan. Lancet 1995;345:722–723.
87. Foreman KE, Friborg J, Kong W-P et al. Propagation of a human herpesvirus from AIDS-associated Kaposi's sarcoma. N Eng J Med 1997;336:163–171.
88. Peters BS, Beck EJ, Coleman DG et al. Changing disease patterns in patients with AIDS in a referral centre in the United Kingdom: the changing face if AIDS. Brit Med J 1991;302:203–206.
89. Northfelt D Dezube B, Thommes J et al. Efficacy of pegylated liposomal doxorubicin in the treatment of AIDS-related Kaposi's sarcoma after failure of standard chemotherapy. J Clin Oncol 1997;15:653–659.
90. Krown SE, Metroka C, Wernz JC et al. Kaposi's sarcoma in the acquired immunddeffieiency syndrom: a proposal for uniform evaluation, response and staging criteria. AIDS Clinical Trials Oncology Committee. J Clin Oncol 1989;7:1201–1207.
91. Northfelt D Dezube B, Miller B, Mamelok R, DuMond C, Henry D. Randomized comparative trial of DOXIL vs. adriamycin, bleomycin, and vincristine (ABV) in the treatment of severe AIDS-related Kaposi's sarcoma. Blood 1995;86(suppl 1):382.
92. Wu AW, Rubin HR, Mathews WC et al. A health stauts questionnaire using 30 items from the medical outcomes study. Preliminary validation in persons with early HIV infection. Med Care 1991;29:786–798.
93. Stewart S, Jablonowski H, Goebel F et al. Randomized comparative trial of pegylated liposomal doxorubicin vs bleomycin and vincristine in the treatment of AIDS-related KS. J Clin Oncol 1998;16:683–691.
94. Mitsuyasu R, von Roenn J, Krown S et al. Comparison study of liposomal doxorubicin alone or with bleomycin and vincristine for treatment of advanced AIDS-associated Kaposi's sarcoma: AIDS Clinical Trial Group (ACTG) Protocol 286. Proc Am Soc Clin Oncol, 1997;16:55a (abstract).
95. Christan ME, Trimble EL. Salvage chemotherapy for epithelial ovarian carcinoma. Gynecol Oncol 1994;55:S143–S150.

96. Uziely B, Jeffers S, Lsacson R et. al. Liposomal doxorubicin: antitumor activity and unique toxicities during two complementary phase 1 studies. J Clin Oncol 1995;13:1777–1785.
97. Muggia FM, Hainsworth JD, Jeffers S et. al. Phase 2 study of DOXIL in refractory ovarian cancer: antitumor activity and toxicity modification by liposomal encapsulation. J Clin Oncol, 1997;15:987–993.
97a. Gordon AN. Sammons PA, Hainsworth J et al. DOXIL (doxorubicin Hcl liposome injection) in the treatment of patients with refractory advanced epithelial ovarian carcinoma. Proc Am Soc Clin Oncol, 1998 (in press).
98. McGuire WP, Rowinsky EK, Rosenshein NB et al. Taxol: a unique agent with significant activity in advanced ovarian epithelial neoplasms. Ann Int Med 1989;111:273–279.
99. Thigpen JT. Blessing JA, Ball H et al. Phase II trial of paclitaxel in patients with progressive ovarian carcinoma after platinum-based chemotherapy: a Gynecologic Oncology Group study. J Clin Oncol 1994;12:1748–1753.
100. Einzig AI, Wiernik PH, Sasloff J et al. Phase II study and long-term folllow-up of patients treated with Taxol for advanced ovarian adenocarcinoma. J Clin Oncol 1992;11:1748–1753.
101. Bristol-Myers Squibb Company. 1992 submission to FDA. Data on file.
102. Trimble EL, Adams JD, Vena D et al. Paclitaxel for platinum-refractory ovarian cancer: results from the first 1000 patients registered to National Cancer Institute Treatment Referral Center 9103. J Clin Oncol 1993;11:2405–2410.
103. Ranson MR, Carmichael J, O'Byrne K et. al. Treatment of advanced breast cancer with Stealth liposomal doxorubicin (CAELYX): results of a multicenter phase 2 trial. J Clin Oncol, 1997;15:3185–3191.
104. Von Hoff DD, Layard MW, Basa P et al. Risk factors for doxorubicin-induced congestive heart failure. Ann Intern Med 1979;91:710–717.
105. Shan K, Lincoff A, Young J. Anthracycline-Induced Cardiotoxicity. Ann Intern Med 1996;125:47–58.
106. Torti FM, Bristow MR, Howes AE et al. Reduced cardiotoxicity of doxorubicin delivered on a weekly schedule: assessment by endomyocardial biopsy. Ann Intern Med 1983;99:745–749.
107. Berry G, Billingham M, Alderman E et al. The use of cardiac biopsy to demonstrate reduced cardiotoxicity in AIDS Kaposi's sarcoma patients treated with pegylated liposomal doxorubicin. Annals Oncol, in press.
108. Billingham M. Endomyocardial changes in anthracycline-treated patients with and without irradiation. Front Radiation Ther Oncol 1979;13:67–81.
109. Billingham M, Mason J, Bristow M, Daniels J. Anthracycline cardiomyopathy monitored by morphological changes. Cancer Treat Reports 1978;62:865.
110. Buja L, Ferrans V, Mayer R, Robert W, Henderson E. Cardiac ultrastructural changes induced by daunorubicin therapy. Cancer 1973;32:771–787.
111. Rowan R, Masek M, Billingham M. Ultrastructural morphometric analysis of endomyocardial biopsies. Am J of Cardiovasc Path 1988;2(2):137–144.
112. Tilkian A, Daily E. Endomyocardial Biopsy. In: Tilkian A, Daily E, eds. Cardiovascular Procedures: Diagnostic Techniques and Therapeutic procedures. Washington, D.C.: CV Mosby, 1986;180–202.
113. Mason J. Techniques for right and left ventricular endomyocardial biopsy. Am J Cardiol 1978;41:887–92.
114. Billingham M, Bristow M. Evaluation of anthracycline cardiotoxicity: predictive ability and functional correlation of endomyocardial biopsy. Cancer Treat Symp 1984;3:71–6.

Lasic and Papahadjopoulos (eds.), Medical Applications of Liposomes

CHAPTER 8.3

The Liposome Company: Lipid-based pharmaceuticals in clinical development

CHRISTINE E. SWENSON, JEFFREY FREITAG AND ANDREW S. JANOFF
The Liposome Company, Inc., One Research Way, Princeton, New Jersey 08540, USA

Overview

I. Introduction

The Liposome Company (TLC) has focused its efforts on the development of novel lipid-based pharmaceuticals since its founding in 1981. Despite its name, TLC has not confined itself to the production of "classical" liposomes wherein a drug is encapsulated within the internal aqueous spaces of the lipid vesicle. Indeed, each of the three products currently in advanced clinical development have somewhat unique elements that distinguish them from the more traditional concept of liposomes. The first product, ABELCET® (ABLC or Amphotericin B Lipid Complex Injection) currently marketed in the United States and Europe, is actually an interdigitated complex of lipid and the drug, amphotericin B. It is not a true liposome and has no captured volume. The second product, VENTUS™ (TLC™ C-53 or Liposomal Prostaglandin E1 Injection), consists of prostaglandin E1 and egg phosphatidylcholine vesicles. Althoughthe PGE1 is associated with the lipid membrane of the vesicles when prepared, the PGE1 rapidly dissociates from the membrane upon injection, enabling the dual occupation of both prostaglandin receptors and receptors recognized by liposomal opsonins. The third product, TLC™ D-99 (Doxorubicin HCl Liposome Injection), also in Phase III human clinical trials, is loaded via a pH gradient that drives the drug into the liposome where it achieves concentrations that exceed its aqueous solubility. This review will provide an overview of the chemistry and preclinical profile as well as a summary of the clinical experience with each preparation.

All doses of ABELCET®, VENTUS™ and TLC™ D-99 described in this review refer to the active ingredient and not the lipid component. For example, a dose of 5 mg/kg of ABELCET® refers to 5 mg/kg amphotericin B delivered as ABELCET®.

II. ABELCET®

As of March, 1997, fifteen countries have approved ABELCET® for the treatment of various severe systemic fungal infections. Six of the countries have approved ABELCET® as first-line therapy for candidiasis. In the United States, ABELCET® is indicated for the treatment of invasive fungal infections in patients who are refractory to or intolerant of conventional amphotericin B therapy.[1]

The safety and efficacy of ABELCET® have been extensively evaluated in laboratory and clinical studies. In general, these studies have shown that ABELCET® is consistently and markedly less toxic than the conventional micellar formulation, Fungizone® or Amphotericin B-deoxycholate (AmB-d), with at least comparable and sometimes enhanced antifungal activity.

ABELCET® is a suspension of amphotericin B complexed with the phospholipids L-α-dimyristoylphosphatidylcholine (DMPC) and L-α-dimyristoylphosphatidylglycerol (DMPG). The antifungal activity of ABELCET® is due to amphotericin B. This antibiotic has been in use for over 35 years and the spectrum of its activity has been well characterized. Complexation of amphotericin B with lipids as in ABELCET® is not believed to change the intrinsic activity of the drug nor its mechanism of action. Nonetheless, ABELCET® is both biophysically and biologically distinct from AmB-d. Complexation with lipids appears to stabilize amphotericin B in a self-associated state so that it is not available to interact with cellular membranes (the presumed major site of its anti-fungal activity and its mammalian toxicity). Active (monomeric) amphotericin B can be released from ABELCET® by phospholipases.[2] It is hypothesized that the enhanced therapeutic index of ABELCET® relative to AmB-d is due, in part, to the selective release of active amphotericin at sites of fungal infection through the action of phospholipases released by the fungus itself or from activated host cells such as phagocytic cells, vascular smooth muscle cells or capillary endothelial cells. In addition, the particulate nature of ABELCET® assures that it is rapidly taken up by tissues, especially those of the reticuloendothelial system or in areas where the normal capillary endothelium is disrupted (by fungal invasion or inflammation, for example). ABELCET® has also been shown to be endocytosed by macrophages and to retain its activity after entering the cell.[3] These cells may then transport the drug to sites of localized infection.[4] The altered tissue distribution and "macrophage-loading" may also play roles in the enhanced therapeutic index of ABELCET®.

II.1. Chemistry and structure

The size of the amphotericin B molecule is similar to that of phosphatidylcholine (approximately 2.5 nm in length). The molar ratio of amphotericin B to lipid in ABELCET® is approximately 1:1. Based on this and other biophysical, chemical and morphological data,[2,5] a model for the supramolecular structure of ABELCET® was proposed.[6] In the model, lipid and amphotericin are arranged in a 1:1 interdigitated complex in which amphotericin-lipid pairs are arranged in cylinders. The hydrophobic polyene region of amphotericin is aligned with the lipid hydrocarbon chains and the polar hydroxyl groups face towards the center of the pore. The cylinders align side by side and possess two polar ends. To accommodate this, lipid headgroups and the amino end of amphotericin orient in the same or opposite direction. Alternatively, amphotericin-lipid pairs within the complex could alternate in direction. It is believed that this stable complexation of amphotericin with lipid is responsible for the enhanced selective toxicity (that is, unaltered toxicity to fungal cells with diminished toxicity to mammalian cells) of ABELCET®.

II.2. Pharmacology and toxicology

The safety of ABELCET® has been studied in mice, rats, rabbits and dogs with dosage regimens consisting of a single bolus dose or repeated daily doses for up to 27 weeks.[7] Evidence of dose-dependent (apparently reversible) renal toxicity was present in all animals (rats and dogs) that received repeated daily doses (for 2 weeks to 6 months) of ABELCET®. However, the nephrotoxicity of ABELCET® (as well as other toxicities) must be viewed in comparison with the only other widely available therapeutic option for the treatment or prevention of serious, systemic fungal infection in humans—amphotericin B deoxycholate. In single dose, acute toxicity studies in rodents, the dose of ABELCET® causing lethality in 50% of animals (LD50) was 10- to 25-fold higher than the LD50 of AmB-d. There were no serious adverse effects or gross pathologic lesions in rats or mice given ABELCET® at single doses of 20 mg/kg, a dose that is 4 to 10 times greater than the dose of AmB-d causing 50% lethality in these species. In repeated-dose studies in dogs, the renal cortical tubular damage (assessed histologically) in animals given 0.5 to 0.625 mg/kg/day of AmB-d was comparable to that seen in those given 5.0 mg/kg/day of ABELCET®. Dogs given lower doses of ABELCET® showed less evidence of renal toxicity. Thus, in dogs, ABELCET® was 10-fold less nephrotoxic than AmB-d.

ABELCET® was found to be without mutagenic effects in four different assay systems (bacterial reverse mutation assay, mouse lymphoma forward mutation assay, chromosomal aberration assay in Chinese hamster ovary cells and the in vivo mouse micronucleus assay[1]). Studies in rats demonstrated no impact on fertility in males or females with repeated daily doses up to 10 mg/kg/day. There were no embryotoxic, fetotoxic or teratogenic effects in pregnant rats and rabbits treated during gestation with doses in the range of the maternal maximum tolerated dose of ABELCET®.[1]

ABELCET® has been tested for efficacy in over 40 experimental models of fungal infections including aspergillosis,[8,9] candidiasis,[8,10,11,12] blastomycosis,[13,14] cryptococcosis,[8,15] coccidioidomycosis[16,17] and histoplasmosis.[8] In no case was ABELCET® shown to be less effective than AmB-d. Although higher doses of ABELCET® (in terms of amphotericin B content) were sometimes required to achieve antifungal efficacy similar to AmB-d, the overall therapeutic benefit was at least comparable. In some models where efficacy could not be achieved with the maximum tolerated dose of AmB-d, it was possible to achieve efficacy with higher doses of ABELCET® which could be given safely. There was no model where intravenous AmB-d was effective but adequate doses of ABELCET® were not.

Most pharmacokinetic studies reported to date with ABELCET® have measured total amphotericin B levels in blood; that is, they have not distinguished amphotericin B that is complexed with the phospholipids of ABELCET® from amphotericin B that is un-complexed. Nonetheless, studies in both animals and humans indicate that the pharmacokinetics of total amphotericin B is markedly different when administered as ABELCET® compared to AmB-d.[18–20] Experiments in mice, rats and dogs suggest that ABELCET® behaves pharmacokinetically like other particulate drug carriers. After intravenous administration of ABELCET®, blood concentrations of amphotericin B are lower but tissue concentrations are higher than that seen after a similar dose of AmB-d. It has been suggested that the decreased nephrotoxicity of certain liposomal amphotericin B formulations is due to an alteration in amphotericin B binding to lipoproteins in the circulation,[21] but we have found no evidence for this with ABELCET® in our studies in rats.[22] In human studies, the pharmacokinetics of amphotericin B after administration of ABELCET® have been shown to be nonlinear. Volume of distribution and clearance from the blood increase with increasing dose of ABELCET®, resulting in less than proportional increases in blood concentrations of amphotericin B over a dose range of 0.6–5 mg/kg/day. The large volume of distribution and high clearance from blood probably reflect uptake by tissues. The long terminal elimination half-life probably reflects a slow redistribution from tissues. Although amphotericin B is excreted slowly, there is little accumulation in the blood after repeated dosing.[1]

II.3. Clinical experience

To date, ABELCET® has been administered to an estimated 6,000 to 7,000 patients. Several recent reviews describe the key trials and emergency use protocols in patients with invasive fungal infections that supported the approval of ABELCET®.[23–25] Most patients in the open-label, emergency use studies were suffering from hematologic malignancies or were immunocompromised and were judged by their physicians to be refractory to or intolerant of conventional amphotericin B or to have preexisting nephrotoxicity. The usual dose of ABELCET® was 5 mg/kg/day for several weeks with a total dose generally about 7 g. ABELCET® demonstrated efficacy in patients infected with *Aspergillus*, *Candida*, *Zygomycetes*, *Cryptococcus* and *Fusaria* species. Small numbers of patients infected with other organisms (*Histoplasma*, *Coccidioidomyces*, *Blastomyces*) were also successfully treated. ABELCET®, at doses up to 5 mg/kg/day was shown to be significantly less nephrotoxic than the usually administered 0.5–1.0 mg/kg/day doses of AMB-d.

An increasing number of individual case reports are now appearing noting the efficacy and safety of ABELCET® in patients with rare or severe refractory, invasive fungal infections. Two reports have described complete clinical and radiographic resolution or clinical improvement in patients with Aspergillus osteomyelitis.[26,27] A small study of eight patients with life-threatening fusariosis who had failed treatment with AmB-d because of lack of efficacy ($n = 5$) or toxicity ($n = 3$) reported successful outcomes in 7 patients after four weeks therapy with ABELCET® at 5 mg/kg/day.[28] ABELCET® has been used successfully in a neutropenic patient with disseminated zygomycosis,[29] and one with disseminated *Trichosporon asahii* infection;[30] in a renal transplant patient with *Apophysomyces elegans* infection,[31] and in patients with hepatic[32] and rhinocerebral[33] mucormycosis. A recent abstract describes effective therapy of an *Absidia* wound infection in a solid organ transplant recipient.[34]

The recommended daily dosage for children and adults is 5 mg/kg given as a single intravenous infusion at a rate of 2.5 mg/kg/hr.[1] Despite generally less nephrotoxicity of ABELCET® at a dose of 5 mg/kg/day compared with conventional AmB-d at doses between 0.6–1.0 mg/kg/day, renal toxicity may still be observed with ABELCET®, thus patients are usually monitored for renal function during therapy.

III. VENTUS™

VENTUS™ may be useful as a treatment for acute respiratory distress syndrome (ARDS) and other inflammatory or vaso-occlusive diseases that involve over-activation of neutrophils or endothelial cells. VENTUS™ is liposome-associated prostaglandin E1 (PGE1) in an acetate-buffered aqueous medium. It consists of egg phosphatidylcholine (EPC) liposomes, 0.05 to 0.5 μm in diameter with maltose monohydrate as a cryoprotectant and butylated hydroxytoluene as an antioxidant. The PGE1 (10 μg per 4 mg EPC or 1:200 molar ratio) is protonated at pH 4.5 and is associated with the lipid bilayer in the formulation. At physiologic pH, the PGE1 rapidly dissociates from the vesicles.[35] This combination of liposomal particles with PGE1 has been shown to toggle the activity of PGE1 from pro- to anti-inflammatory in animal models[36] and has also shown significant effects in human clinical trials.[37]

III.1. Mechanism of action

It is postulated that VENTUS™ exerts its anti-inflammatory effects by modulating neutrophil and/or endothelial cell function. This effect of VENTUS™ appears to be due to the enhancement of the anti-inflammatory effects of PGE1 by liposomes.[36]

PGE1 is a naturally occurring eicosanoid synthesized from dihomo-γ-linolenic acid in virtually all mammalian tissues. PGE1 is biologically unusual in that it has both anti- and pro-inflammatory activity. It can be anti-inflammatory by reason of its ability to downregulate neutrophil-mediated responses.[38] Its pro-inflammatory activity is likely a consequence of its vasodilatory activity.

VENTUS™, however, differs from free PGE1 in its ability to modulate neutrophil function. The effect of VENTUS™ appears to be due to the facilitation by liposomes of the anti-inflammatory neutrophil-modulating effects of PGE1 thus shifting the balance of activity of PGE1 from pro-inflammatory to anti-inflammatory. Liposomes are endogenously opsonized in vivo with C3bi and thus target to the CR3 (CD11b/CD18) receptors of stimulated neutrophils. If these receptors are occupied when PGE1 occupies its receptor (called the EP2 receptor) then, through an intracellular signalling cascade involving cAMP, the neutrophil is significantly inactivated and the anti-inflammatory activities of PGE1 predominate. Occupation of the EP2 receptor alone does not produce significant neutrophil inactivation and the inflammatory activity of PGE1, mediated by vasodilation and vascular leak, predominate. Liposomes alone produce no meaningful therapeutic effect.[36] Additionally, liposomes are opsonized with fibronectin in vivo resulting in their targeting to venous endothelium by engagement of the αyB1 integrin receptor. The dual occupation of this receptor and the endothelial EP2 receptor results in a down regulation of ICAM-1.[39] The ability of VENTUS™ to both inactivate neutrophils and make capillary surfaces less responsive to neutrophil extravasation is the hallmark and distinguishing characteristic of this drug.

III.2. Pharmacology and toxicology

The safety of VENTUS™ has been studied in rats, dogs and rabbits in single, intravenous, bolus doses of up to 200 μg/kg PGE1 and as repeated intravenous infusions (20 minute infusion, 4 infusions per day) with total daily doses as high as 600 μg/kg (approximately 40 times the proposed human clinical dose). Overall, the drug was well tolerated with reversible physiological effects (e.g., peripheral vasodilation, decreased blood pressure, lethargy) at the highest doses being the most notable effect in the majority of studies. There have been no gross pathological changes at necropsy that could be directly attributed to the drug.[40]

The antiinflammatory activity of VENTUS™ has been demonstrated in several animal models. In an experimental endotoxemia study in rats where 50% of non-treated animals die within 24 hours of lipopolysaccharide (LPS) injection, treatment with VENTUS™, at a dose of 40 μg/kg given at the time of LPS injection, resulted in 100% survival.[36] VENTUS™ has also been shown to markedly reduce acute inflammation induced with various mediators (monosodium urate crystals, interleukin 1-β and tumor necrosis factor α) in subcutaneous air pouches experimentally produced in rats.[41] A single dose of 12 μg/kg VENTUS™ given intravenously 2 hours after inflammation was induced, reduced the accumulation of cells and exudate by 31–49%. Free PGE1 at similar doses was far less effective.

Intravenous administration of liposome associated PGE1 (VENTUS™) but not free PGE1, was shown to abolish platelet-aggregation associated cyclic flow reductions in experimentally stenosed and endothelium injured canine coronary arteries.[42] It is known that certain inhibitors of platelet aggregation can abolish cyclic flow variations in this model[43] and it is possible that VENTUS™ was effective because of its effect on platelets. However, other mechanisms (e.g., inhibition of neutrophil or platelet adhesion to the endothelium) may also have been involved.

VENTUS™ has been shown to decrease the severity of cardiac injury in ischemic-reperfusion models. Successful reperfusion of occluded arteries results in improved survival in patients with acute myocardial infarction. However, complete infarct salvage often fails to occur, apparently due to "reperfusion injury". Regional myocardial blood flow progressively deteriorates after reperfusion with accumulation of neutrophils and capillary plugging in the microvasculature. VENTUS™ has been shown to decrease the severity of cardiac injury in canine ischemic-reperfusion models at doses that did not produce significant hemodynamic effects.[44,45] This was accompanied by a significant reduction in neutrophil infiltration (as assessed by myeloperoxidase activity) in myocardial tissue. Free PGE1 or empty liposomes alone did not have this effect. Administering VENTUS™ intravenously just prior to thrombolytic therapy resulted in a reduced time to thrombolysis, improvement in coronary blood flow and a reduction in infarct size in an experimental canine thrombosis model.[46]

Acute lung injury (adult respiratory distress syndrome or ARDS) is characterized by increased pulmonary vascular permeability and interstitial edema, accompanied by massive neutrophil infiltration into the lung. The syndrome occurs in patients with predisposing conditions including aspiration, hypertransfusion, diffuse pulmonary infection, near drowning, and toxic inhalation. Intratracheal instillation of IL1-α in rats causes a rapid increase in lung neutrophils and leakage of protein-rich fluid into the alveoli and thus mimics the pathology of ARDS. Intravenous administration of 6–12 μg/kg of VENTUS™ to rats 2.5 hours after IL-1α instillation significantly reduced the fluid leak and neutrophil accumulation.[47]

III.3. Clinical experience

The incidence of ARDS in the U.S. is estimated to be 150,000 cases per year with a mortality rate approximating at least 40%.[48] There currently is no available licensed pharmacologic therapy to specifically treat ARDS and its management is primarily supportive.

Phase I human studies demonstrated that VENTUS™ was safe and that administration of doses in which the PGE1 component was ≤4.8 μg/kg/hr was infrequently associated with hypotension.[40]

The efficacy of VENTUS™ was studied in a phase II, placebo-controlled, randomized, prospective, double-blind multicenter clincal trial in patients with ARDS. Patients were enrolled within 24 hrs of their diagnosis and were randomly assigned to receive either VENTUS™ or placebo. VENTUS™ was administered intravenously over 60 minutes every 6 hours for 7 days. The starting dose was 0.15 μg/kg and was escalated every 12 hrs up to a maximal dose of 3.6 μg/kg. There were 25 evaluable patients; 17 who received VENTUS™ and 8 who received placebo. Treatment with VENTUS™ resulted in improvement in oxygenation, increased lung compliance, and a significant decrease in ventilator dependency compared with placebo treatment.[37] Eight of the 17 patients receiving VENTUS™ were healthy enough by day 8 of the trial to be removed from their ventilators while none of the placebo patients had been removed at this time.

A phase III trial did not confirm these effects, although certain sub-groups of patients benefited with VENTUS treatment. TLC is reviewing all data to determine what role the drug can play in the therapy of inflammatory diseases.

IV. TLC™ D-99

TLC D-99 (EVACET™) is a liposome-encapsulated form of the widely used antineoplastic drug, doxorubicin. Numerous preclinical studies in a variety of laboratories have shown that encapsulation of doxorubicin inside liposomes decreases the cumulative dose-limiting cardiotoxicity associated with the free form of the drug while maintaining or, in some cases, increasing the antitumor potency (see 49 for review). TLC D-99 is being developed for the treatment of metastatic breast cancer as well as other doxorubicin-sensitive tumors.

The liposomal component of TLC D-99, egg phosphatidylcholine/cholesterol (55:45, mol:mol), is a sterile, non-pyrogenic suspension of EPC/chol liposomes in a pH 4 citrate buffer. The nominal size of the vesicles is approximately 150 nm. After reconstitution with the doxorubicin provided in the TLC D-99 kit, the drug to lipid ratio is 0.25:1 (wt:wt).

IV.1. Chemistry and structure

The basis for TLC D-99 preparation uses a novel drug-entrapment procedure. Empty liposomes produced in an acidic citrate buffer (pH 4.0) are alkalinized externally with a sodium carbonate solution, creating a transmembrane pH gradient. Addition of doxorubicin to the liposomal suspension (with mild heating) drives the drug across the membrane into the internal aqueous compartment of the liposome.[49] The citrate concentration inside the liposome induces doxorubicin aggregation such that the actual amount of doxorubicin that is loaded into the internal volume of the liposomes exceeds the aqueous solubility of doxorubicin.[50] The aggregation state of the internalized doxorubicin and the pH gradient across the liposomal membrane also reduce the rate of drug release from the liposome. Because this loading procedure is so efficient (>95% of the added doxorubicin can be entrapped), there is no need to remove unentrapped drug and the loading procedure can be performed by the pharmacist just prior to administration.

The current understanding of how TLC D-99 ameliorates the gastrointestinal and cardiotoxicity of doxorubicin and exerts its antitumor activity suggests that one or a combination of mechanisms may be operative. After intravenous injection, liposomes such as TLC D-99 cannot escape the vascular space in sites which have tight capillary junctions such as the heart muscle and gastrointestinal tract. They generally exit the circulation only in tissues with discontinuous endothelia lining their capillaries such as the liver, spleen and bone marrow or areas where capillaries are disrupted by inflammation or tumor growth. Therefore, TLC D-99 directs its encapsulated doxorubicin away from sites of potential toxicity (heart, intestines) and towards sites of metastasis and tumor growth. TLC D-99 may act directly on tumors, or the liposome may be degraded and doxorubicin released once it is present at the tumor site. This degradation may occur intra- or extracellularly. Liposomes in the circulation may also act as reservoirs, slowly releasing doxorubicin.

IV.2. Pharmacology and toxicology

Single dose lethality studies in mice showed that TLC D-99 was approximately twofold less toxic than conventional doxorubicin. The LD50 was 17.5 mg/kg ($\sim$52.5 mg/m^2) for conventional doxorubicin and 32 mg/kg ($\sim$96 mg/m^2) for TLC D-99.[51] Multiple doses in mice (daily injections for 5 consecutive days) produced a 50% lethal dose of 7.5–10 mg/kg for TLC D-99 and 3.7–5.0 mg/kg for conventional doxorubicin. Empty liposomes (containing all components of TLC D-99 except doxorubicin HCl) were acutely toxic at levels approximately 2.5 times the equivalent amount of liposomes and sodium carbonate in an LD50 single dose of TLC D-99. The toxicity of the empty liposomes, where death occurs within 2 minutes of bolus injection, was probably due to a decrease in serum calcium caused by the citrate in the buffer formulation.

The toxicity of single and multiple (5 consecutive days) doses of TLC D-99 and conventional doxorubicin was compared in two studies in dogs.[51] Toxicities included anorexia, adipsia, weight loss, diarrhea, vomiting, fever, neutropenia, lymphopenia, anemia, thrombocytopenia, elevated liver enzymes and cholesterol and electrolyte imbalance. For both preparations, myelosuppression and gastrointestinal effects were the major toxicities. In both studies, dogs that received TLC D-99 had a lower incidence and less severe symptoms of gastrointestinal toxicity than dogs that received conventional doxorubicin. Some of the dogs that received TLC D-99 showed increased body temperatures after infusion, usually peaking at approximately 15 hours post-dose. This pyrexia was not noted in animals that received conventional doxorubicin or empty liposomes.

In a study that compared TLC D-99 with conventional doxorubicin under conditions more closely related to the actual intended clinical use of the drug (intravenous dosing once every three weeks for a total of eight doses), a clear difference was demonstrated in the cardiac toxicities of the two preparations.[52] All of the dogs treated with conventional doxorubicin, but none of those that received TLC D-99, exhibited histologic evidence of myocardial toxicity. There also appeared to be less gastrointestinal toxicity and alopecia with TLC D-99 than with conventional doxorubicin.

Inadvertant extravasation from the site of infusion of potent cytotoxic drugs such as doxorubicin can lead to severe necrosis and ulceration. This localized skin reaction can be demonstrated by injecting free doxorubicin subcutaneously in mice. In comparison to free doxorubicin, subcutaneous injection of TLC D-99 caused only a localized inflammation (erythema, edema) at the site of injection but no necrosis or ulceration.[53]

The antitumor activity of TLC D-99 has been compared to that of conventional doxorubicin in both ascitic (L1210, P388) and solid (B16, M5076, SC115) murine tumor models. In all cases, TLC D-99 showed an enhanced therapeutic activity because the optimal dose (that dose giving the greatest therapeutic effect without evidence of drug induced lethality) was greater.[54] In the SC115 murine mammary tumor model, intravenous administration of TLC D-99 resulted in tumor doxorubicin levels that were 1.3 to 2.9 times greater than the levels obtained with the same dose of free doxorubicin. Treatment with TLC D-99 in this model resulted in enhanced tumor growth inhibition compared to an equivalent dose of free drug.[55] Thus, in some models, the activity of TLC D-99 is greater than that of free doxorubicin on a mg per mg basis.

IV.3. Clinical experience

The primary objective of the clinical development program with TLC D-99 is to demonstrate that TLC D-99 is less toxic (primarily less cardiotoxic) with at least similar efficacy when compared with conventional doxorubicin. Studies have also been designed to determine if doxorubicin dose-intensification using TLC D-99 can increase the anti-tumor response without increasing the toxicity normally expected with conventional doxorubicin.

The starting dose of TLC D-99 initially used in clinical studies was based on the typical doses of conventional doxorubicin (20–90 mg/m^2 given by infusion once every 3 weeks). Three Phase I/II studies, involving 88 patients, defined the dose-limiting toxicity under these conditions to be myelosuppression with a maximum tolerated dose of 75–90 mg/m^2.[56–58] Measurement of doxorubicin in these patients showed that TLC D-99 administration resulted in the maintenance for 24 hours of total (both free and encapsulated) doxorubicin levels up to 30 times higher than would be expected after the same dose of conventional doxorubicin.[59] Free doxorubicin (but not total doxorubicin) and doxorubicinol plasma levels were lower after TLC D-99 administration than after similar doses of conventional doxorubicin.

A Phase II study of TLC D-99 (60 mg/m^2 every 3 weeks) in combination with other antineoplastic agents (5-fluorouracil and cyclophosphamide) in patients with metastatic breast cancer has shown a response rate of 73% with few cardiac events in 41 evaluable patients.[63]

To assess the non-hematological toxicities of TLC D-99, two phase I studies of TLC D-99 with granulocyte colony stimulating factor (G-CSF) support in a total of 53 patients with advanced tumors were conducted.[60,61] The MTD under these conditions was 105–150 mg/m^2. Preliminary results from a phase II study in metastatic breast cancer using high dose TLC D-99 (135 mg/m^2) with G-CSF support have shown an overall response rate of 65%.[62]

On the basis of these studies, several Phase III trials were initiated and are ongoing in the US and Europe to evaluate TLC D-99 alone and in combination therapy in metastatic breast cancer. These are multi-center, comparative trials, each with a planned enrollment of about 300 patients.

V. Summary

In conclusion, the recent approval of ABELCET® and continued clinical success with TLC D-99, as well as the progress made with other liposomal and lipid-based human pharmaceuticals, proves that this technology can make a meaningful contribution to human medicinal therapy. The Liposome Company's technology is particularly broad and versatile, allowing it not only to target drugs to reduce toxicity and enhance efficacy, but also to utilize liposomes as active facilitators of new drugs. TLC is continuing research with the three products described here to further understand their pharmacology and mechanism(s) of action. In addition, TLC has ongoing research programs aimed towards gene delivery, pulmonary surfactants, derivatives of paclitaxel and the discovery and use of novel, bioactive lipids.

References

1. ABELCET® package insert 10/96
2. Perkins WR, Minchey SR, Boni LT, Swenson CE, Popescu MC, Pasternack RF, Janoff AS. Amphotericin B-phospholipid interactions responsible for reduced mammalian cell toxicity. Biochim et Biophys Acta 1992;1107:271–282.
3. War G, Clark J, Kessler R. Eukaryotic cell interactions (in vitro) with a fluorescent-labeled amphotericin B lipid complex (ABLC). Annual Meeting of the Amer Soc of Microbiol; Abstr 1991;30:23.
4. Olsen SJ, Swerdel MR, Blue B, Williams S, Clark JM. Localized *Candida* infections in mice treated with Amphotericin B lipid Complex (ABLC) and putative mode of ABLC transport. Annual Meeting of the Amer Soc Microbiol 1990;Abst A-111:19.
5. Janoff AS, Boni LT, Popescu MC, Minchey SR, Cullis PR, Madden TD, Taraschi T, Gruner SM, Shyamsunder E, Tate MW, Mendelsohn R, Bonner D. Unusual lipid structures selectively reduce the toxicity of amphotericin B. Proc Natl Acad Sci USA 1988;85:6122–6126.
6. Janoff AS, Perkins WR, Saletan SL, Swenson CE. Amphotericin B Lipid Complex (ABLC): A molecular rationale for the attenuation of amphotericin B related toxicities. J Liposome Res 1993;3:451–471.
7. Swenson CE, Janoff AS. Preclinical studies with Amphotericin B Lipid Complex. Drugs Today 1996;32:397–402.
8. Clark JM, Whitney RR, Olsen SJ, George RJ, Swerdel MR, Kunselman L, Bonner DP: Amphotericin B lipid complex therapy of experimental fungal infections in mice. Antimicrob Agents Chemother 1991;35:615–621.

9. Cicogna CE, White MH, Bernard EM, Ishimura T, Sun M, Tong WP, Armstrong D. Efficacy of prophylactic aerosol Amphotericin B Lipid Complex in a rat model of pulmonary aspergillosis. Antimicrob Agents Chemother 1997;41:259–261.
10. Lee J, Allende M, Amantea M, Francis P, Peter J, Thomas V, Francesconi A, Lyman C, Schaufele R, Bacher J, Pizzo P, Walsh T. Pharmacokinetics and efficacy of Amphotericin B Lipid Complex (ABLC) against chronic disseminated candidiasis in rabbits. 31st Intersci Conf Antimicrob Agents Chemother 1991;Abst 579.
11. Lee J, Allende M, Dollenberg H, Garret K, Berenguer J, Francesconi A, Sein T, Lyman C, Francis P, Pizzo P, Walsh T. Reticuloendothelial loading with Amphotericin B Lipid Complex (ABLC)—A novel pharmacodynamic approach to the treatment of experimental hepatosplenic candidiasis. 32nd Intersci Conf Antimicrob Agents Chemother 1992;Abst 172.
12. Perfect J, Wright KA. Amphotericin B lipid complex in the treatment of experimental cryptococcal meningitis and disseminated candidosis. J Antimicrob Chemother 1994;33:73–81.
13. Clemons KV, Stevens DA. Comparative efficacies of amphotericin B lipid complex and amphotericin B deoxycholate suspension against murine blastomycosis. Antimicrob Agents Chemother 1991;35:2144–2146.
14. Krawiec DR, McKiernan BC, Twardock AR, Swenson CE, Itkin RJ, Johnson LR, Kurowsky LK, Marks CA. Use of an amphotericin B lipid complex for treatment of blastomycosis in dogs. JAVMA 1996;209:2073–2075, 1996.
15. Whitney RR, Kunselman L, Clark JM, Bonner DP. Efficacy of amphotericin B lipid complex (ABLC) in cryptococcal meningitis in normal and immunocompromised mice. 29th Intersci Conf Antimicrob Agents Chemother 1989;Abst 166.
16. Clemons KV, Stevens DA. Efficacies of amphotericin B lipid complex (ABLC) and conventional amphotericin B against murine coccidioidomycosis. J Antimicrob Chemother 1992;30:353–363.
17. Allendoerfer R, Yates R, Sun SH, Graybill JR. Comparison of amphotericin B lipid complex with amphotericin B and SCH 39304 in the treatment of murine coccidioidal meningitis. J Med Vet Mycology 1992;30:377–384.
18. Olsen SJ, Swerdel MR, Blue B, Clark JM, Bonner DP. Tissue distribution of Amphotericin B Lipid Complex in laboratory animals. J Pharm Pharmacol 1991;43:831–835.
19. Bhamra R, Sa'ad A, Bolcsak L, Swenson C, Janoff A. Behavior of ABLC (Amphotericin B Lipid Complex) in blood and plasma, in vitro and in vivo with comparison to Fungizone in the rat. FASEB 1995;9:2309.
20. Grasela D, Echevarria J, Summerill R, Gonzalez D, Christafalo B, Chang J, Campos P, Llanoscuentas A. Pharmacokinetics of amphotericin B following administration as Amphotericin B Lipid Complex or Fungizone in adults with mucocutaneous leishmaniasis. 33rd Intersci Conf Antimicrob Agents Chemother 1993;Abst 1439.
21. Wasan K, Lopez-Berestein G. Modification of amphotericin B's therapeutic index by increasing its association with high-density lipoproteins. Ann NY Acad Sci 1994;730:93–106.
22. Bhamra R, Sa'ad A, Bolcsak LE, Janoff AS, Swenson CE. Behavior of Amphotericin B Lipid Complex (ABLC) in plasma in vitro and in the circulation of rats. Antimicrob Agents Chemother 1997;41:886–892.
23. Dix SP, Wingard JR. Amphotericin B Lipid Complex: Review of safety, pharmacokinetics and efficacy. Drugs Today 1996;32:411–416.
24. Lister J. Amphotericin B Lipid Complex (ABELCET) in the treatment of invasive mycoses: the North American experience. Eur J Haematol 1996;56:18–23.
25. Karyotakis NC, Anaissie EJ. Amphotericin B Lipid Complex: Recent progress. Drugs Today 1996;32:423–431.
26. Kline MW, Bocobo FC, Paul ME, Rosenblatt HM, Shearer WT. Successful medical therapy of *Aspergillus* osteomyelitis of the spine in an 11-year old boy with chronic granulomatous disease. Pediatrics 1994;93:830–835.
27. Jones RS, Barman A, Suh B, Heifets M, Bocobo FC. Successful treatment of *Aspergillus* vertebral osteomyelitis with amphotericin B lipid complex. Infect Dis Clin Pract 1995;4:237–239.
28. Anaissie EJ, Ramphal R. Efficacy and safety of amphotericin B lipid complex (ABLC) in the treatment of patients with life-threatening fusariosis. 34th Intersci Conf Antimicrob Agents Chemother 1994;Abst M87.
29. Gonzalez CE, Couriel DR, Walsh TJ. Disseminated zygomycosis in a neutropenic patient: Successful treatment with Amphotericin B Lipid Complex and Granulocyte Colony-Stimulating Factor. Clin Infect Dis 1997;24:192–6.
30. Letscher V, Herbrecht R, Six-Kieffer I, Eyer D, Koenig H, Waller J. Successful treatment with amphotericin B lipid complex (ABLC) followed by itraconazole of a disseminated *Trichosporon*

asahii infection failing conventional amphotericin B. Ninth International Symposium on Infections in the Immunocompromised Host, Assisi, Italy, June 23–26 1996;Abstr 178.

31. Naguib MT, Huycke MM, Pederson JA, Pennington LR, Burton ME, Greenfield RA. *Apophysomyces elegans* infection in a renal transplant recipient. Am J Kidney Dis 1995;26:381–4.
32. Oliver MR, Van Voorhis WCV, Boeckh M, Mattson D, Bowden RA. Hepatic mucormycosis in a bone marrow transplant recipient who ingested naturopathic medicine. Clin Infect Dis 1996;22:521–4.
33. Strasser MD, Kennedy RJ, Adam RD. Rhinocerebral mucormycosis: Therapy with Amphotericin B Lipid Complex. Arch Intern Med 1996;156:337–339.
34. Veldkamp PJ, Miller GG. Treatment of *Absidia* wound infection with liposomal amphotericin B in a heart transplant recipient. Abstracts of the 34th Annual Meeting of the Infect Dis Soc Amer 1996;132.
35. Davidson SMK, Cabral-Lilly D, Maurio FP, Franklin JC, Minchey SR, Ahl PL, Janoff AS Association and release of prostaglandin E1 from liposomes. Biochim Biophys Acta 1997; 1327:97–106.
36. Eierman DF, Yagami M, Erme SM, Minchey SR, Harmon PA, Pratt KJ, Janoff AS. Endogenously opsonized particles divert prostanoid action from lethal to protective in models of experimental endotoxemia. Proc Natl Acad Sci USA 1995;92:2815–2819.
37. Abraham E, Park YC, Covington P, Conrad SA, Schwartz M. Liposomal prostaglandin E1 in acute respiratory distress syndrome: A placebo-controlled, randomized, double-blind, multicenter clinical trial. Crit Care Med 1996;24:10–15.
38. Chopra J, Webster RO. PGE1 inhibits neutrophil adherence and neutrophil-mediated injury to cultured endothelial cells. Am Rev Respir Dis 1988;138:915–920.
39. Smalling R, Uthman M, Ramanna N, Yeh ET, Amirian JH, Felli PR, Feld S, Accad M, Collen A, Smith CW, Emmerman ML, Janoff A. Suppression of ICAM-1 and P-selectin adhesion molecule expression by bolus iv liposomal PGE1 (TLC C-53) immediately prior to reperfusion in a two-hour canine infarct/reperfusion model. J Am Coll Cardiol Feb, 248A. [Abstract], 1995.
40. Data on File, The Liposome Co., Inc.
41. Rossetti RG, Brathwaite K, Zurier RB. Suppression of acute inflammation with liposome associated prostaglandin E1. Prostaglandins 1994;48:187–195.
42. Willerson JT, Yao S-K, McNatt J, Cuj K, Anderson HV, Swenson C, Ostro M, Buja LM. Liposome-bound prostaglandin E1 often prevents cyclic flow variations in stenosed and endothelium-injured canine coronary arteries. Circulation 1994;89:1786–1791.
43. Grolino P, Buja LM, Ashton JH, Kulkarni P, Taylor A, Willerson JT. Effect of thromboxane and serotonin antagonists on intracoronary platelet deposition in dogs with experimentally stenosed coronary arteries. Circulation 1988;78:701–711.
44. Smalling RW, Feld S, Ramanna N, Amirian J, Felli P, Vaughn WK, Swenson C, Janoff A. Infarct salvage with liposomal prostaglandin E1 administered by intravenous bolus immediately before reperfusion in a canine infarction-reperfusion model. Circulation 1995;92:935–943.
45. Feld S, Li G, Wu A, Felli P, Amirian J, Vaughn WK, Gornet T, Swenson C, Smalling RW. Reduction of canine infarct size by bolus intravenous administration of liposomal prostaglandin E1: Comparison with control, placebo liposomes and continuous intravenous infusion of prostaglandin E1. Am Heart J 1996;132:747–7.
46. Feld S, Li G, Amiran J, Felli P, Vaughn WK, Accad M, Tolleson TR, Swenson C, Ostro M, Smalling RW. Enhanced thrombolysis, reduced coronary reocclusion and limitation of infarct size with liposomal prostaglandin E1 in a canine thrombolysis model. J Am Coll Cardiol 1994;24:1382–1390.
47. Leff JA, Baer JW, Kirkman JM, Bodman ME, Shanley PF, Cho OJ, Ostro MJ, Repine JE. Liposome-entrapped PGE1 posttreatment decreases IL-1-induced neutrophil accumulation and lung leak in rats. J Appl Physiol 1994;76:151–157.
48. Kolleff MH, Schuster DP. The acute respiratory distress syndrome. N Eng J Med 1995;332:27–37.
49. Tardi PG, Boman NL, Cullis PR. Liposomal Doxorubicin. J Drug Targetting 1996;4:129–140.
50. Li X, Mayhew E, Janoff AS, Perkins WR. Doxorubicin-citrate interactions: The effect of doxorubicin aggregation upon its release from liposomes. Biophys J 1997;72:A406.
51. Kanter PM, Bullard, Pilkiewicz FG, Mayer LD, Cullis PR, Pavelic ZP. Preclinical toxicology study of liposome encapsulated doxorubicin (TLC D-99): comparison with doxorubicin and empty liposomes in mice and dogs. In vivo 1993;7:85–96.
52. Kanter PM, Bullard GA, Ginsberg RA, Pilkiewicz FG, Mayer LD, Cullis PR, Pavelic ZP.

Comparison of the cardiotoxic effects of liposomal doxorubicin (TLC D-99) versus free doxorubicin in beagle dogs. In vivo 1993;7:17–26.
53. Balazovits JAE, Mayer LD, Bally MB, Cullis PR, McDonell M, Ginsberg RS, Falk RE. Analysis of the effect of liposome encapsulation on the vesicant properties, acute and cardiac toxicities and antitumor efficacy of doxorubicin. Cancer Chemother Pharmacol 1989;23:81–86.
54. Data on File. The Liposome Company, Inc.
55. Mayer LD, Bally MB, Cullis PR, Wilson SL, Emerman JT. Comparison of free and liposome encapsulated doxorubicin tumor drug uptake and antitumor efficacy in the SC115 murine mammary tumor. Cancer Lett 1990;53:183–190.
56. Cowens JW, Creaven PJ, Greco WR, Brenner DE, Tung Y, Ostro M, Pilkiewicz F, Ginsberg R, Petrelli N. Initial clinical (phase I) trial of TLC D-99 (Doxorubicin encapsulated in liposomes). Cancer Res 1993;53:2796–2802, 1993.
57. Lohri A, Gelmon KA, Embree L, Mayer L, Cullis P, Saletan S, Goldie J: Phase I/II study of liposome encapsulated doxorubicin (TLC D-99) in non-small cell lung cancer. Proc ASCO 1991;10:A292.
58. Viens P, Baume D, Touinssi M, Catalin J, Camerlo J, Sauniere JF, Durand A, Maraninchi D. Tolerance and pharmacokinetic study of liposomal doxorubicin (TLC D-99) in patients with refractory cancer. Proc ASCO 1994;13:A432.
59. Embree L, Gelmon KA, Lohr A, Mayer LD, Coldman AJ, Cullis PR, Palatis W, Pilkiewicz F, Hudon NJ, Heggie JR, Goldie JH. Chromatographic analysis and pharmacokinetics of liposome encapsulated doxorubicin in non-small cell lung cancer patients. J Pharm Sci 1993;82:627–634.
60. O'Day SJ, Mazanet R, Skarin AT, Salgia R, Gordon D, Elias AD. Dose escalation of liposome-encapsulated doxorubicin (D-99) with granulocyte colony stimulating factor support in patients with advanced malignancies. Proc ASCO 1994;13:A406.
61. Early E, Shorter S, Sugarman A, Schwartz GK, Woodruff J, Brennan MF, Casper ES. Phase I trial of dose-intense liposomal encapsulated doxorubicin with G-CSF in patients with advanced soft tissue sarcoma. Proc ASCO 1996;15:A524.
62. Shapiro CL, Ervin T, Azarnia N, Keating J, Suppers V, Ayash L, Hayes D. Phase II trial of high dose liposome-encapsulated doxorubicin (D-99) with G-CSF in metastatic breast cancer. Proc ASCO 1996;15:A112.
63. Fonseca GA, Valero V, Buzdar A, Walters R, Willey J, Theriault G, Fraschini G, Booser D, Herrada J, Gordon D, Hortobagyi G. Phase II study of TLC D-99 (Liposomal Doxorubicin) in patients with metastatic breast carcinoma. Proc ASCO 1995;14:A99.

Lasic and Papahadjopoulos (eds.), Medical Applications of Liposomes
Elsevier Science B.V.

CHAPTER 8.4

Unilamellar liposomes for anticancer and antifungal therapy

PAUL G. SCHMIDT, JILL P. ADLER-MOORE, ERIC A. FORSSEN AND RICHARD T. PROFFITT

NeXstar Pharmaceuticals, Inc., 2860 Wilderness Place, Boulder, CO 80301, USA

Overview

I. Introduction

NeXstar Pharmaceuticals, Inc. (formerly Vestar, Inc. and NeXagen, Inc.) has commercialized liposome preparations for systemic antifungal therapy (AmBisome®) and cancer treatment (DaunoXome®). Clinical trials are underway for a third product (MiKasome®) being tested in patients with gram negative bacterial infections and mycobacterial disease. These products share common formulation features in that they are relatively small unilamellar liposomes (40–50 nm diameter) whose bilayers contain long chain, saturated phospholipids plus cholesterol for in vivo stability. Formulation differences include the mode of drug entrapment/encapsulation, choice of phospholipids and mole ratios and in the surface charge.

Research correlating in vivo performance with liposome design conducted since the late 1970s has confirmed the advantage of cholesterol containing liposomes

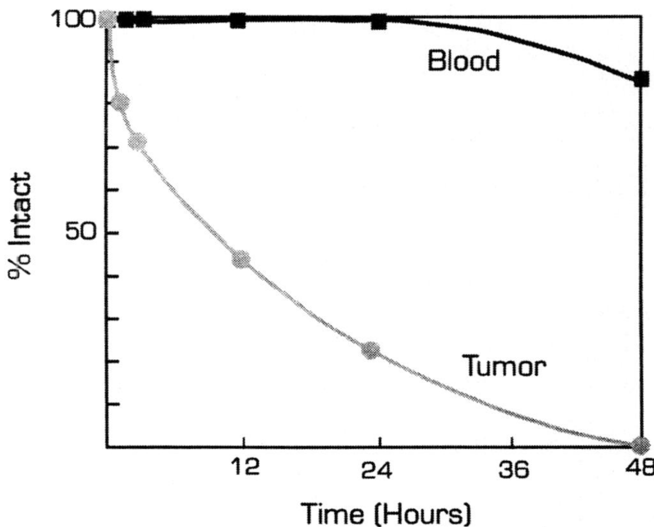

Fig. 1. The technique of perturbed angular correlation spectroscopy was used to detect the fractions of In-111 in intact liposomes (and released due to breakdown). The intact fraction is plotted for blood and tumor (EMT6) of mice after tail vein injection of DSPC:chol (2:1) unilamellar liposomes (about 60 nm mean diameter) containing In-111 bound to nitrilotriacetate in the aqueous interior. Blood and tumor tissues were harvested at various points from a cohort of mice receiving the liposome injections (from Reference 4).

for retention of contents and sustained blood circulation.[1,2] Measurements with the technique of γ-ray perturbed angular correlation spectroscopy showed that liposomes composed of distearoylphosphatidylcholine (DSPC) and cholesterol (chol) are stable to loss of contents while circulating in blood up to 24 hours, but break down relatively quickly when taken up in tissues such as tumor (Figure 1).[3,4] These features are desirable for liposome therapy, especially where cytotoxic drugs are encapsulated.

Of course, while these NeXstar products have some similar design features, they differ in important ways. DaunoXome is formulated with pure DSPC and cholesterol to give a neutral (Zwitterionic) bilayer charge. Such liposomes were shown to have remarkably high uptake in murine solid tumors (Figure 2).[5] The mechanism includes liposome transport by selective convection (flow) through the leaky vasculature of rapidly growing tumors.[3] Liposomes with circulation clearance rates that are slower than tumor uptake rates will collect preferentially in the tumors. In some cases the tumor concentration may be higher than in any other tissue, including liver and spleen. DaunoXome also has this tumor targeting property and shows the ability to deliver exceptionally high concentrations of its encapsulated drug, daunorubicin, to experimental solid tumors.[6]

Size is an important determinant of biodistribution and circulation lifetimes. Larger liposomes (>100–200 nm) are generally cleared more quickly by stationary

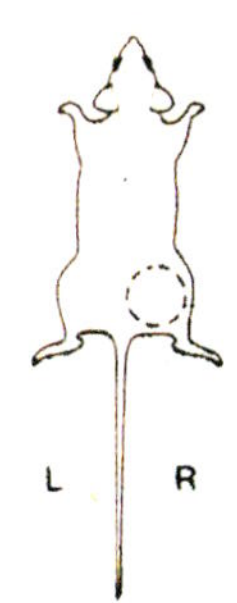

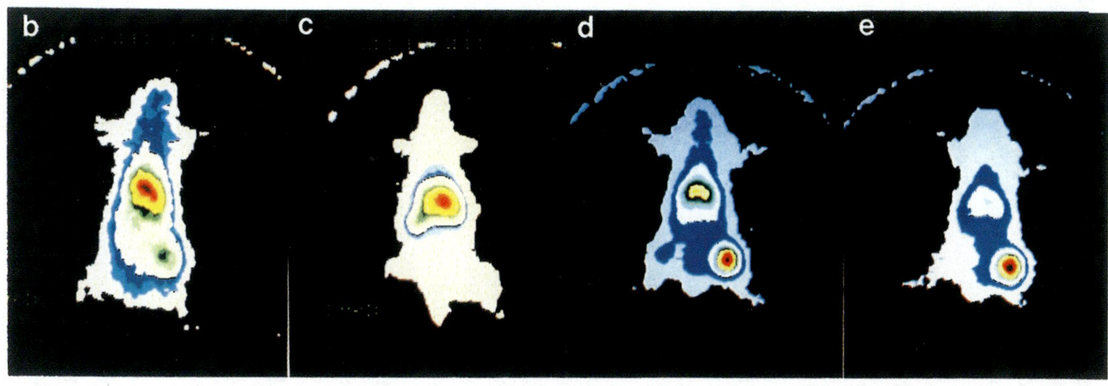

Fig. 2. Gamma camera scans of mice injected with liposomes containing the γ-emitting isotope ^{111}In. Each mouse had a tumor implanted in its right hind leg, in the position shown in drawing (a). When the mouse was injected directly with $^{111}InCl_3$, the isotope was distributed more or less uniformly throughout the animal 24 hours later, as seen in drawing (b). Red, orange and black, represent areas with the highest level of radioactivity, light blue and white, the lowest. Injected $^{111}In^{3+}$ encapsulated in liposomes with 6-aminomannose bound to the surface produced the image in drawing (c), where the liposomes accumulated preferentially in the liver and spleen. Encapsulation of the radioisotope in unmodified liposomes, shown in drawing (d), caused the liposomes to accumulate readily in the tumor, but also significantly in the liver and spleen. This unwanted effect can be reduced, as in drawing (e), by first injecting unlabeled liposomes modified with 6-aminomannose, followed by unmodified liposomes containing $^{111}In^{3+}$. (Some of these figures were published in Reference 5).

macrophage of the mononuclear phagocytic system (MPS) (e.g., Kupffer cells in the liver). Much smaller liposomes can pass through liver sinusoid fenestrations where they may be taken up by hepatocytes. An optimal size appears to be in the 40–80 nm range which minimizes both MPS and hepatocyte uptake. This size range allows access to tumors in areas of neovasculature where gaps in the rapidly proliferating endothelial cells can pass such particles.

Formulations for antifungal and antibacterial drugs have different tissue targets than anticancer liposomes. A negatively charged lipid, distearoylphosphatidylglycerol (DSPG) incorporated into the AmBisome and MiKasome bilayers makes these liposomes more attractive to stationary and circulating monocytes than if the bilayer is neutral. The mechanism of action for these liposomal drugs, in part, involves macrophage uptake both for therapy of intracellular disease and to sustain availability of drug near sites of extracellular infection. AmBisome, and especially MiKasome, also have relatively slow clearance rates (compared to free drug) and may be able to extravasate intact at infection sites where leaky vasculature is common.

Liposome encapsulation often decreases toxicities due to altered tissue biodistribution and retention of drug in circulating liposomes. If these toxicities are dose-limiting for the free drug, liposomes may allow higher dosing for improved therapy. This has clearly been the experience with AmBisome, which has been used safely at 15 times the maximum tolerated dose of conventional amphotericin B. DaunoXome and MiKasome also allow higher dosing than can be tolerated with the free drug. This chapter focuses mainly on AmBisome and DaunoXome because there are numerous publications on clinical results for these products. MiKasome has just entered systematic clinical trials, so published results are minimal.

II. AmBisome (Liposomal Amphotericin B)

II.1. Physicochemical Properties

AmBisome is a liposomal form of amphotericin B (AmB), the only true liposome among the new generation lipid-associated amphotericin B formulations being developed throughout the world. AmBisome is a suspension in buffered 9% sucrose of small unilamellar liposomes whose composition is HSPC (hydrogenated soy phosphatidylcholine)/chol/DSPG/AmB (2:1:0.8:0.4 mole ratio). Amphotericin B is anchored tightly in the AmBisome bilayer due to favorable interactions of the macrolide with the surrounding lipids. DSPG probably interacts directly with AmB; cholesterol may also play a role. The exact nature of these interactions is not known, but the data are consistent with a structure in which amphotericin B molecules form a barrel-like structure. Two barrels fit together tail-to-tail to span the lipid bilayer and form a pore that is permeable to ions and other solutes (Figure 3). The product is stored as a lyophilized powder that is reconstituted with the addition of water for injection followed by a few seconds of shaking to produce

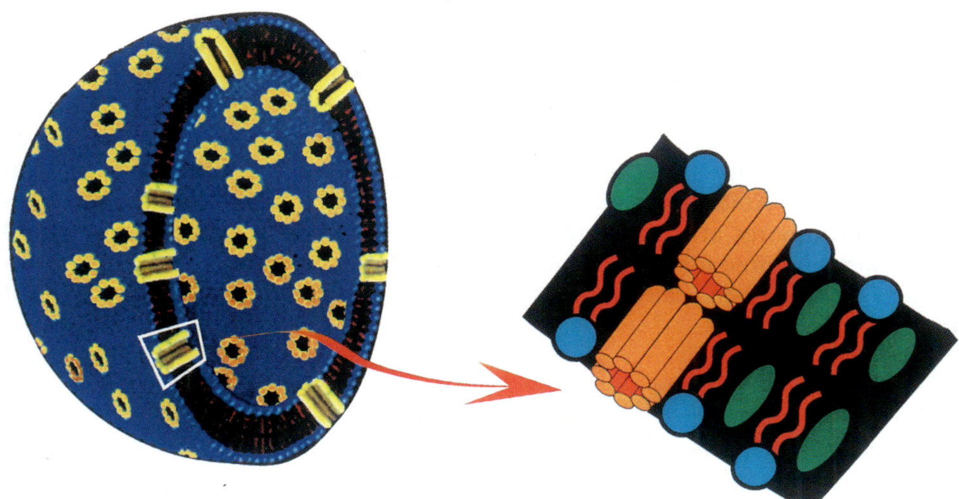

Fig. 3. Proposed arrangement of amphotericin B molecules (yellow) in the AmBisome bilayer. This structure accounts for the observation of rapid ion fluxes across the AmBisome bilayer in response to imposing a pH gradient from inside to outside. The individual amphotericin B molecules form a "barrel", two of which fit tail-to-tail to form a pore spanning the bilayer. This structure is believed to contribute to the exceptional stability of AmBisome to loss of drug in buffer or plasma (from Reference 12).

a slightly opalescent, yellow solution. In its lyophilized presentation, stored at 4°C, AmBisome has a shelf life in excess of 30 months.

II.2. In Vitro Studies

II.2.1. Efficacy of AmBisome compared to amphotericin B

Pathogenic yeast cultures were isolated from blood or CSF of 104 cancer patients with fungal disease including species of *Candida*, *Cryptococcus*, *Fusarium*, and *Aspergillis*.[7] Minimum inhibitory (MIC) and minimum fungicidal concentrations (MFC) for AmBisome and amphotericin B were determined using incubation times of 24 to 48 hours. AmBisome MICs and MFCs did not differ significantly from those of the free drug.[7] Other comparisons of AmBisome and AmB against fungal cultures have shown similar results for incubations of 24 and 48 hours. Data discussed below show that AmBisome remains largely intact for 24 hours under conditions similar to the culture medium and in the concentration range tested. The implication is that an interaction between AmBisome and fungal cells or a substance secreted by the cells liberates amphotericin B from the liposome at sufficiently long incubation times. On the other hand, with shorter incubation times (6 hours), AmBisome had lower activity than amphotericin B against extracellular *C. albicans*,[8] although the clinical relevance of this observation is not clear.

II.2.2. Studies with neutrophils

Neutrophils play a key role in clearance of fungal infections. Amphotericin B toxicity could compromise this activity. A comparison was made of neutrophil inhibition with Ambisome and AmB.[9] The free drug inhibited uptake of *C. albicans* blastospores at concentrations of 1 μg/ml and above, while 20 μg/ml and above were required for AmBisome to exhibit the inhibitory activity. Since AmBisome is often used clinically at doses 3 to 5 times higher than free drug, this result is reassuring.

II.3. Mechanism of action

The mechanism of action for fungal cytotoxicity due to AmB is generally ascribed to formation of pores in the cell membrane with subsequent release of ions, but additional activities such as oxidative degradation have been postulated.[10] AmB has well-known toxicities due to collateral damage to mammalian cells. The existence of any therapeutic benefit is likely due to a stronger interaction of the drug with fungal membrane ergosterol over the mammalian cell counterpart, cholesterol. Systemic toxicities include a host of debilitating infusion-related responses such as nausea; rigor; chills and fever; and cellular damage, particularly kidney toxicity, that limits dosages and can lead to suspension of therapy for many patients.

AmBisome has been tested in mammalian cell toxicity assays and has proved to be remarkably benign. Red cell lysis assays are a measure of free (or readily available) amphotericin B. Fungizone®, the standard free drug preparation (deoxycholate micelle solubilized, hereafter referred to as AmB-DOC) produced 92% lysis of red cells in 2 hours at 37° at a drug concentration of 1 μg/ml. At a concentration of 100 μg/ml (amphotericin B equivalent), AmBisome produced only 5% lysis under the same conditions and time of incubation.[11] These data suggest that AmBisome retains amphotericin B sufficiently tightly so that less than 1% of the drug is free (or loose enough to be transferred to mammalian cells) in buffer. Potentially the association of amphotericin B with AmBisome is dependent on the concentration of the liposomes, if there exists an equilibrium between free and liposome-bound drug. But, in buffer, even as low as 1 μg/ml, the drug remains exclusively with the liposome as evidenced by circular dichroism studies over a range of concentrations.[12]

In vitro studies in human and mouse serum show complete retention of amphotericin B by AmBisome for 6–24 hours.[11,13] For AmBisome in vivo, there is evidence that amphotericin B is largely retained by the liposome over several hours of circulation in mice.[14] Certainly the drug is not available in a free or toxic form since the LD50 of AmBisome is greater than 160 mg/kg in this species, as compared to 2.3 mg/kg for AmB-DOC.

There is evidence that AmBisome (and liposomes of the same composition without drug) can gain direct access to sites of fungal infection as intact structures probably because of leaky vasculature.[15] The hypothesis has been made that, with the prolonged circulation lifetime seen for AmBisome, uptake into infected tissue and direct action of the liposomal drug may contribute to therapy.[15] Indeed, AmBisome is highly active against cultured fungal species,[7] although the liposomal drug may be somewhat slower acting than AmB-DOC.[8] The liposomes, with or without, drug bind to fungal cells, and AmBisome (but not drug-free liposomes) disintegrates slowly. Gold labeled lipids incorporated in AmBisome-like liposomes can be located by electron microscopy (after silver enhancement).[16] Initially, intact liposomes are seen gathered around and bound to the cell wall of *Candida glabrata*. After 14 hours incubation, gold-labeled lipid is seen inside the cell membrane. The cell structure appears disrupted at this point, presumably due to action of amphotericin B that accompanies breakup of the liposome.[16]

While it appears feasible for AmBisome to act directly on systemic fungal infections, the quantitative contribution of intact liposomes to the success of systemic treatment with AmBisome needs further study.

Macrophages, including Kupffer cells of the liver and stationary macrophage in the spleen are a major cellular site for uptake of AmBisome and other lipid-associated amphotericin B preparations.[10] It is likely that macrophages, and possibly neutrophils, play key roles as depots for amphotericin B, although the details have not been elucidated.

Several authors have postulated that lipoprotein binding of amphotericin B is important for modulating drug toxicity.[17] In this model, LDL-bound drug is more toxic than HDL-bound amphotericin B. The experimental proof of this hypothesis is not complete for any formulation of amphotericin B, and its importance for the low acute and dose-limiting toxicities that AmBisome displays is not known.

II.4. Preclinical studies

II.4.1. Pharmacology

AmBisome pharmacokinetics (PK) and biodistribution were measured in mice and rats.[18] For a single dose of 5 mg/kg the peak plasma levels reached 118 and 87 mg/L for rats and mice, respectively. The plasma PK was described by a two exponential fit with half lives of 1.07 and 7.56 h for the rat, and values of 0.57 and 3.36 h for the mouse. The area under the curve was between 350 and 400 mg h/L for both species.[18]

Rabbit PK data were obtained at AmBisome doses of 0.5, 1, 2.5, 5, and 10 mg/kg and 0.5, 1, and 1.5 mg/kg for AmB-DOC.[19] The AUC of AmBisome at 5 mg/kg was 840 μg hr/ml and at 1 mg/kg it was 60 compared to 31 μg h/ml for AMB-DOC. The ratio AUC/dose for AmBisome increased with dose over the range studied (0.5–10 mg/kg), which was interpreted to indicate uptake in a saturable compartment, presumably the MPS.[19] This is consistent with the biodistribution data in rodents showing preferential uptake in liver, and reduction in kidney levels relative to free amphotericin B.[18] After 28 days of dosing AmBisome at 5 mg/kg/day and AmB-DOC at 1 mg/kg/day rabbit kidney levels were 14 times greater for the free drug than for AmBisome.[19] This is consistent with the reduced nephrotoxicity observed for AmBisome, even at a dose five times that of AmB-DOC.

II.4.2. Safety

A marked reduction in toxicity of AmBisome compared to AmB-DOC was observed in in vivo studies.[18] When the single dose toxicity of AmBisome in C57BL/6 mice was measured, the intravenous LD50 was found to be >160 mg/kg. When the drug was tested in Sprague-Dawley rats, the acute single dose LD50 was 50 mg/kg. For comparison, LD50 values for AmB-DOC were 2.3 mg/kg for mice and 1.6 mg/kg for rats.[18]

Multiple dose testing of AmBisome in mice, rats, and dogs also demonstrated AmBisome's reduced toxicity. The LD50 in mice given AmBisome for 14 consecutive days was >75 mg/kg and the LD50 in rats was 50 mg/kg after 30 days of daily dosing.[18]

Blood urea nitrogen (BUN) and creatinine levels were much higher in AmB-DOC treated dogs than for those treated with AmBisome. BUN levels were about four times higher for dogs given 0.6 mg/kg AmB-DOC (229 mg/dL) compared with those given 4 mg/kg AmBisome (58 mg/dL) and creatinine levels were more than two and one half times higher for 0.6 mg/kg AmB-DOC-treated animals (6.0 mg/dL) compared with those receiving 4 mg/kg AmBisome (2.3 mg/dL). These data emphasize the reduction of nephrotoxicity with AmBisome treatment.

II.4.3. Efficacy

In animal studies, AmBisome produced therapeutic responses that depended upon the infection model. Compared to free drug, AmBisome was significantly better than AmB-DOC for treating murine Leishmaniasis,[20,21] while it showed similar efficacy for candidiasis,[22] aspergillosis[19] and histoplasmosis.[16,23] In blastomycosis[24] and paracoccidioidomycosis[25] AmBisome was not as effective as comparable concentrations of amphotericin B. However, decreased toxicity in severely sick, infected animals allowed much higher doses of AmBisome to be administered safely, significantly reducing or eradicating the CFU/g in infected tissue and producing sustained drug levels in tissues above the MIC for the infecting organisms.[16]

II.5. Clinical pharmacology

II.5.1. Serum pharmacokinetics

AmBisome pharmacokinetics were measured in a group of 11 patients at doses of 3, 4, and 5 mg/kg.[26] Peak serum levels increased with increasing dose as did the AUC values. The ratio AUC/dose increased from 70 to 105 h/(L kg) suggesting that uptake occurs at least partially in a saturable compartment, in line with the preclinical observation on rabbits.[19]

In the human study, clearance rates were relatively slow (16–21 ml/min) over the dose range studied. AmBisome has higher AUC values and slower clearance rates than free AmB-DOC and other lipid-associated amphotericin B.[26] In comparing 1 mg/kg AmB-DOC with 3 mg/kg AmBisome (a typical human dose for therapy) AmBisome peak serum concentration was 29 μg/ml versus 3.6 μg/ml for AmB-DOC and steady state volume of distribution was 0.37 L for AmBisome versus 1.6 L for free drug.[27]

II.5.2. Biodistribution

Autopsy tissues have provided values for organ biodistribution of AmBisome. Liver and spleen are the chief destinations for liposomal amphotericin B; levels in heart and brain were low.[28]

II.6. Clinical efficacy and safety

Amphotericin B is an old drug with significant toxicities that often require discontinuing treatment, but its broad spectrum of activity against the main fungal pathogens has kept it in use despite the development of newer imidazole and triazole antifungal agents. Particularly with systemic mycoses of severely ill and immunocompromised patients, amphotericin B is the treatment of choice.

AmBisome is much better tolerated than conventional amphotericin B and is indicated in the treatment of severe systemic fungal infections where patients fail to respond to amphotericin B, are intolerant to its side effects or who have renal impairment precluding the use of conventional drug.

AmBisome was first used clinically in 1987 when a heart transplant patient developed pulmonary aspergillosis, which due to nephrotoxicity, could not be treated with conventional amphotericin B.[29] After 34 days of treatment with AmBisome at 1 mg/kg/day, the infection was eradicated, and no evidence of recurrence was reported during a 16 month follow-up period. Also, during the treatment period, kidney function improved and there were no acute side effects such as fever and chills. Since then AmBisome has been developed throughout the world and is presently licensed in more than 30 countries, including the US where it has received approval for empiric use (fever of unknown origin).

II.6.1. Safety

AmBisome therapy is associated with a low degree of acute and chronic side effects. The earliest safety data on AmBisome were reported by Meunier et al.[30] This multicenter study included 126 patients receiving 133 episodes of AmBisome treatment. The majority of these patients had failed previous conventional amphotericin B therapy due to toxicity. The mean duration of AmBisome administration was 21 days at an average daily dose of 2.1 mg/kg (range = 0.45–5 mg/kg). The most common side effect observed in 24 cases was hypokalemia. Nausea, vomiting, fever, chills and rigors were observed in a total of only five instances. Serum chemistries were monitored closely throughout the study for indications of organ toxicity. Although many patients entered this study with elevated creatinine levels, creatinine levels become elevated in only 11 cases during AmBisome treatment. In 17 episodes, creatinine was initially high, but returned to normal. Glutamyl oxaloacetate transaminase became elevated in 19 instances, and elevation in alkaline phosphatase was observed in 22 instances. However, there was no report of discontinuation of AmBisome therapy due to adverse side effects. Thus, AmBisome was well tolerated even in severely ill patients.[30]

More recently, AmBisome safety was assessed in a series of 187 transplant recipients.[31] AmBisome was administered daily at dose levels between 1 and 4 mg/kg for a median of 11 days (range of 1–112 days). Side effects attributed to AmBisome therapy including low serum potassium, low back pain during infusion, dyspnea, allergic reaction, and nausea and vomiting were observed in only 7% of the cases, and resulted in discontinuation of therapy in six cases. Serum creatinine increases were not statistically significant. Other side effects possibly related to AmBisome included elevated serum urea and alkaline phosphatase which normalized after AmBisome therapy was discontinued. In this context, with patients receiving a variety of potentially toxic drugs, the AmBisome side effect profile was mild and manageable in the vast majority of patients.[31]

Recent multicenter randomized trials compared AmB-DOC at 1 mg/kg/day to AmBisome at 1 and 3 mg/kg/day in adults[32] and children[33] with febril neutropenia unresponsive to broad-spectrum antibiotics. A group of 193 adult patients was prospectively randomized into the three treatment groups. Fifty-two patients had confirmed mycosis; seven were not classifiable and the rest were stratified as having fever of unknown origin (FUO).

Adverse events in the adult study were significantly lower for the AmBisome groups.[32] Incidence of nephrotoxicity in the AmB-DOC arm was 50% compared to 16% and 18% in the AmBisome 1 and 3 mg/kg/day groups ($p = 0.001$). Also, hypokalemia was significantly less in the AmBisome cohorts.[32]

The pediatric study produced a similar picture, but differed in detail.[33] Nephrotoxicity was lower in the AmBisome arms versus AmB-DOC, but the differences were not statistically significant. Significant advantages were seen for AmBisome in incidences of hypokalemia, treatment delay and resolution of fever.[33]

II.6.2. Treatment of confirmed and suspected mycoses

Amphotericin B is the treatment of choice for patients with severe systemic fungal infections. The drug is often used in immunocompromised patients with fever that does not respond to antibiotics (FUO) because of the strong suspicion of fungal disease even when cultures and microscopy are negative.

AmBisome has been used in this setting in a number of clinical trials.[32,34,35] An early multicenter trial reported on the use of AmBisome as salvage therapy in immunocompromised patients with suspected or proven fungal infections who had failed previous antifungal therapy, or had renal insufficiency or toxicity.[34] The median daily dose level was 2.2 mg/kg (range: 0.5–5.0 mg/kg). Out of a total of 126 patients, there were 64 cases with proven invasive infections. Of these, 37 (58%) were cured, 12 (19%) improved, and 15 (23%) failed to respond. From these results, it was concluded that AmBisome was an effective agent in the majority of patients with invasive or superficial fungal infections.[34]

There has been an analysis of AmBisome efficacy in confirmed mycoses covering data in nine published studies.[36] Most patients had either hematological malignancies or were transplant patients. Of 147 infections, 48 were candidosis, 65 aspergillosis, 21 cryptococcosis, and 14 other fungi. Mycological efficacy was 79% for candidosis, 67% for aspergillosis, 67% for cryptococcosis and 50% for other fungi. Cure rates for AmB-DOC in these fungal infections are normally 20–40% showing a clear efficacy benefit for AmBisome.[36]

II.6.3. AmBisome for treatment of leishmaniasis

Visceral leishmaniasis (VL) is endemic in many areas of the world, more prevalently nearer to the equator and including the Mediterranean. AmBisome is highly effective in VL[37] and has been approved as a first-line treatment for this disease in several countries.

AmBisome treatment is likely due in large part to facile uptake by macrophages of the liposomal drug. *Leishmania spp*. are harbored in macrophage and are killed by the accumulated amphotericin B. At the same time, the safety profile of AmBisome is highly favorable, making it the drug of choice for very young children and patients in poor general condition.[38]

Davidson et al. considered the optimum dose and schedule for AmBisome treatment of VL.[39] A group of 88 patients, mostly children, was treated with four different dose regimens. Eighty-four patients were completely cured of their disease by the initial treatment course lasting 10 days (4 or 5 days daily treatment at 3 or 4 mg/kg/day and one follow-up on day 10). Four relapsing children received an additional 10 day course of treatment at 3 mg/kg/day which cured them all.[39]

This study is remarkable not only short-course treatment and high-cure rate of VL patients, but also for the favorable safety profile. There were no significant adverse events. A small rise in serum creatinine levels was observed during therapy which stayed well within normal limits. Liver function was not impaired. Serum electrolytes did not change significantly during or after treatment in any patient group. One factor contributing to the therapeutic success of AmBisome in VL may be persistence of relatively high levels of drug in liver and spleen for extended time periods.[40]

II.6.4. AmBisome for invasive Aspergillosis

Aspergillosis is the most common fungal infection in neutropenic patients and is usually (90%) fatal even with conventional antifungal treatment.[41]

In one analysis of nine clinical trials involving 147 patients with prior fungal infections, there were 65 cases of Aspergillosis.[36] Mycological cure was achieved with AmBisome in 67% of these cases. A recent trial sponsored by EORTC compared AmBisome doses of 1 mg/kg/day and 4 mg/kg/day for treatment of pulmonary *Aspergillus* infections. Surprisingly, the cure rate was the same for both doses and was higher than historical results with AmB-DOC.[42]

II.6.5. Other fungal infections

a. *Candida spp*. In a report of 137 episodes of fungal infection treated with AmBisome, 108 were clinically evaluable. There were 52 proven cases of *Candida* species with an eradication rate of 83%.[34] Other reports document successful use of AmBisome to treat *Candida* endocarditis,[43] and hepatic candidosis.[44,45]

b. Cryptococcosis. A multicenter study examined the safety and efficacy of AmBisome for therapy of Cryptococcus in AIDS patients.[46] Twenty-three patients were evaluable for clinical efficacy and 21 for mycological efficacy. Treatment was given for at least 42 days and an average of 4.4 g cumulative dose was administered. Clinical cures were achieved in 14 of 23 (61%) patients and 17% were improved. Mycological cure was achieved in 66% of the 21 evaluable patients. The drug was well tolerated and a host of presenting clinical features seen at the start of the study were resolved by the end.[46]

III. DaunoXome

III.1. Physicochemical properties

III.1.1. Formulation

DaunoXome contains an aqueous solution of the citrate salt of daunorubicin, encapsulated within liposomes. The liposomes comprise a single bilayer membrane of the chemically pure phospholipid, DSPC and cholesterol in a 2:1 molar ratio, with a mean diameter of approximately 45 nm (range: 35 to 65 nm). The lipid-to-drug weight ratio is 12.7:1 (total lipid: daunorubicin base), equivalent to a 10:5:1 molar composition of DSPC: cholesterol: daunorubicin. DaunoXome is supplied as an aqueous suspension in a single-use vial containing daunorubicin at a concentration of 2 mg/mL.

III.1.2. Development of tumor targeted liposomes

For a number of years, investigators have focused on applications of liposomes in cancer treatment to enhance tumor delivery and to ameliorate drug toxicities. Mauk and Gamble were the first to develop active loading techniques for specific liposome formulations that were used in tumor imaging studies.[47] They prepared liposomes containing a weak chelator entrapped in the inner aqueous space and the cationic ionophore A23187 within the membrane bilayer. Liposomes could be loaded efficiently (approaching 100%) with the gamma emitting radioisotope In-111.

Radiolabeled liposomes were evaluated to determine formulation parameters that would preserve physical stability under physiological conditions in circulation, but which would nevertheless allow liposome breakdown and release of entrapped contents in tissues in vivo.[48] Monitoring In-111 radioactivity using perturbed angular correlation (PAC) techniques, the investigators identified critical parameters, including use of pure phospholipids with phase-transition temperatures above physiological (>37°C), a net neutral to slightly negative surface charge, and the inclusion of cholesterol in the lipid membrane (to minimize lipid exchange and the leakage of entrapped substances).

Measurements using the PAC technique to monitor In-111 release showed that liposomes composed of cholesterol and phosphatidylcholine molecules containing fully saturated long-chain fatty acids (e.g., DSPC), in a 1:2 molar ratio, were stable to leakage of the In-111 under physiological conditions.[48] This result was confirmed by Wallingford and Williams using PAC measurements to demonstrate the DSPC:cholesterol liposomes maintain their physical integrity with a half-life of approximately 130 hours at 37°C in freshly drawn human plasma.[49] In contrast, a liposome formulation prepared from egg lecithin rapidly leaked In-111, and had a half-life for physical integrity of only 7 hours under these conditions. Proffitt et al. demonstrated that small, stable liposomes were able to remain in circulation and deliver a greater percentage of the encapsulated In-111 to the tumor than did negatively or positively charged liposomes.[50] Also, tumors could be imaged clearly by gamma camera scintigraphy on mice that had been injected with stable In-111 labeled liposomes.[50] It was hypothesized that small stable liposomes could escape intact from the circulation due to increased capillary permeability at the tumor site. In a subsequent paper by these researchers, it was shown that MPS blockage (with unlabeled liposomes that were cleared rapidly from the circulation) improved In-111 tumor accumulation by over 50% after subsequent administration of liposomes that were small, stable, and neutral.[5] This result supports the hypothesis that intact circulating liposomes deliver their contents directly into the tumor.

Extensive screening studies have demonstrated that liposomes composed of DSPC and cholesterol in a 2:1 mole ratio and prepared with diameters between about 40 and 80 nm (Nicomp) are particularly effective for delivery of encapsulated agents to solid tumors in vivo.[3,4,50,51] Such DSPC:cholesterol vesicles preferentially deliver In-111 to in vivo murine tumors including: mammary adenocarcinomas (EMT-6 and MA16C), B16 melanoma, Lewis lung carcinoma, P1798 lymphosarcoma, sarcoma 180, and colon carcinoma 51.[5,50,52] Delivery of In-111 to these tumors was found to range from 10 to 40 percent of the injected In-111 dose per gram tissue.

These formulation efforts led to development of an In-111-based tumor imaging agent (VesCan®). VesCan was the first liposome product to enter into clinical studies where a liposome was designed to deliver its contents selectively to solid tumors in vivo.[51,53] This imaging agent was investigated in clinical trials in about 375 patients and found to be capable of imaging a wide range of solid neoplasms, Table 1.

During the course of clinical studies with VesCan, investigators observed that Kaposi's sarcoma (KS) demonstrated one of the highest levels of In-111 uptake of all tumors imaged.[54] This was believed to be a function of the high degree of neoangiogenesis and/or vascular smooth muscle cell proliferation associated with KS.[55,56] Although its derivation is under current investigation, KS is thought to arise either from endothelium (vascular or lymphatic) or from vascular smooth muscle cells. Not only are KS lesions highly vascular, they are also very permeable, allowing particulates such as small liposomes to extravasate readily.[57] This observation of significant accumulation of liposome-entrapped In-111 within KS lesions led directly to the clinical evaluation of DaunoXome against this disease. The DSPC:cholesterol vesicles of VesCan have been well tolerated without signi-

Table 1

Tumors imaged with In-111-labeled DSPC: cholesterol liposomes

Tumor	
Breast	Oropharyngeal
Cervix	Ovary
Kaposi's sarcoma	Prostate
Kidney	Small cell and non-small cell lung carcinomas
Larynx	Soft tissue sarcoma
Malignant lymphoma	Spindle cell sarcoma
Malignant melanoma	Testicular
Metastases to liver, lung, bone marrow, lymph nodes, and soft tissues	Thyroid

ficant adverse events. Clinical parameters including hematologic values, respiratory function, urinary output, and hepatic functions, have all remained normal.

III.1.3. Daunorubicin liposome development (DaunoXome)

Successful demonstrations of the use of In-111 SUVs for the imaging of a variety of clinical neoplasms led to the formulation of a cytotoxic drug into liposomes of similar composition.[6,58] Selecting an agent with potential for a significantly improved therapeutic index, however, was crucial to achieving a meaningful expression of enhanced in vivo efficacy. Liposome entrapment can result in variable and sometimes unpredictable effects upon drug activity. Increased toxicities have been noted for some drugs such as methotrexate[59] and cytosine arabinoside,[60,61] while decreased toxicities have been observed for others including the anthracyclines in particular.[50,62,63] Previous investigators have demonstrated that when anthracyclines are entrapped in liposomes, decreased toxicities can be achieved without loss of antineoplastic activity.[4,64,65]

Daunorubicin was the first of the anthracyclines in common use to enter clinical trials, followed closely by doxorubicin. While not as widely used as doxorubicin, daunorubicin has demonstrated efficacy in the treatment of several adult solid tumors.[66–68] In spite of activity against solid tumors during the early clinical investigations, daunorubicin was superseded by doxorubicin, which had displayed greater potency in some preclinical solid tumor models. Clinical investigations have continued to demonstrate significant activity for daunorubicin against various solid tumors.[68] Cytotoxicity of daunorubicin in tissue culture is comparable to, or somewhat greater than, that of doxorubicin.[69,70] Being less polar, the cellular uptake of daunorubicin is more rapid than it is for doxorubicin, potentially leading to better tumor penetration. An additional advantage of daunorubicin is that, on a cumulative basis, it is notably less cardiotoxic than doxorubicin.[71] A final consideration in selecting daunorubicin was its higher stability over doxorubicin in aqueous solutions.[72]

III.2. Mechanism of action

Mechanism of action studies suggest that a portion of DaunoXome liposomes enter tumor cells intact by endocytosis. In one study, the indigo dye precursors 5-bromo-4-chloro-3-indoly-β-D-galactopyranoside (X-gal) or 5-bromo-4-chloro-3-indolyl-β-D-phosphate (BCIP) was encapsulated in DSPC:cholesterol liposomes.[73] In their free state, these compounds are poorly taken up by cells either in vitro or in vivo. In the presence of the lysosomal enzymes *β-galactosidase* (X-gal) or *alkaline phosphatase* (BCIP), activated indoxyl intermediates are generated, which dimerize to form an insoluble blue-violet dye. When liposomes containing these preparations were tested in tissue culture, formation of a compartmentalized, intense blue-violet color was noted within the tumor cell cytoplasm.[13] Formation of similar pigmented regions in cells treated with the corresponding free substances was not observed. These observations are consistent with cellular uptake of DSPC:cholesterol liposomes by endocytosis followed by lysosomal fusion and release of liposome contents within lysosomes.

In vitro cytotoxicity testing of DaunoXome and free drug against P1798 tumor cells indicates that the relative potencies of these two forms of daunorubicin vary widely as a function of incubation times.[74] For short incubation times ($\leq$8 hours), DaunoXome is *less* cytotoxic than free daunorubicin. This is not surprising since the free drug is highly permeable to cell membranes and is fully available when added to the tissue culture medium. For longer incubation times, on the other hand, DaunoXome is *more* cytotoxic than the free drug. Confocal microscopy of P1798 cells incubated with free drug or DaunoXome has demonstrated differences in intracellular availability and trafficking.[74] While free daunorubicin is taken up rapidly by the cells, entering the nucleus in about one to two hours, liposome-delivered daunorubicin is released slowly from vesicles over a prolonged time, not becoming visible within the nucleus at levels comparable to free drug until more than twenty-four hours later.

More recent studies take advantage of daunorubicin's fluorescence and its self-quenching properties while encapsulated in liposomes.[74] Liposome entrapment quenches about 99% of daunorubicin's fluorescent signal; this signal increases to full intensity, however, as the drug is released from the vesicles when they break open. As with the in vitro investigations noted above, results from these fluorescence studies indicate that, following in vivo uptake, daunorubicin liposomes release their drug contents within the tumor cells over a prolonged period. This is illustrated for DaunoXome in vivo delivery of daunorubicin to P1798 tumor in which total extractable drug is compared to in situ fluorescence. The extractable fluorescence is the sum of all daunorubicin-associated fluorescent species (parent drug and fluorescent metabolites) whether free or still entrapped within the liposomes. On the other hand, in situ fluorescence[74] (measured non-invasively by monitoring daunorubicin-associated fluorescence at the skin surface (Figure 4), is proportional to the amount of daunorubicin *released* from the vesicles. Differences in the peak times for these two profiles indicate the delay between delivery (8 hours) and liposome release (30 hours). In parallel to the in vitro findings, liposo-

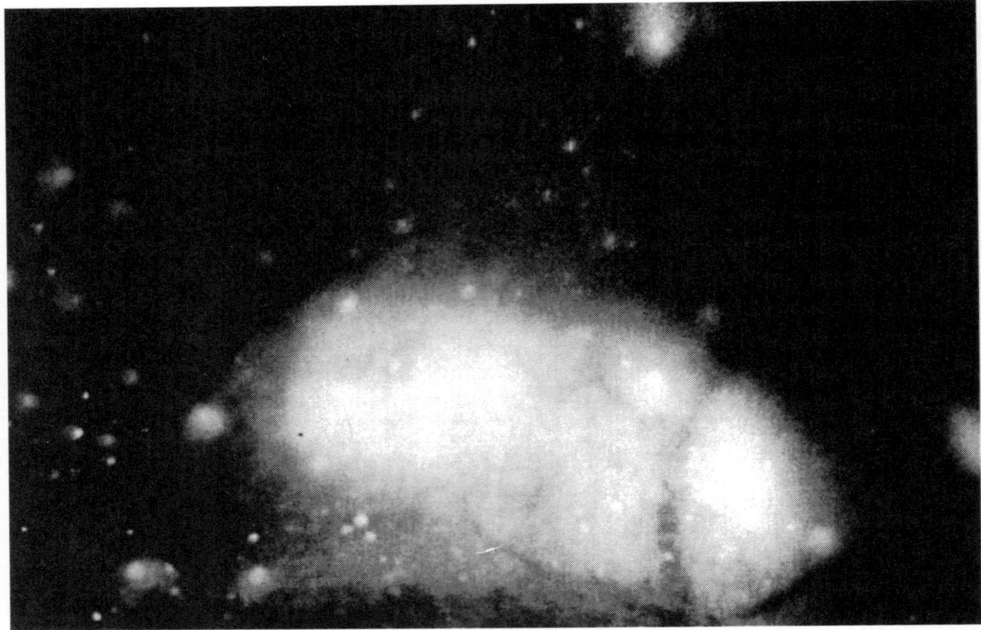

Fig. 4. In situ fluorescence image of an implanted P1798 tumor and surrounding skin from a $CD2F_1$ mouse. The image was taken 28 hours after i.v. injection of DaunoXome and shows fluorescence from daunorubicin that was released from liposomes. The tumor is highly fluorescent while the surrounding skin is not (except for porphyrin signals from razor nicks due to shaving) showing the tumor specificity of DaunoXome and the long persistence of high drug levels in the tumor (from Reference 74).

mal daunorubicin thus appears to be taken up by tumor cells intact and then breaks down slowly to release drug within the cells over a prolonged period, providing sustained high levels of active drug at the tumor site.

III.3. Preclinical studies

III.3.1. Pharmacology

Significant differences between the pharmacokinetics of free daunorubicin and DaunoXome are observed in plasma and in P1798 solid tumor.[6] DaunoXome produces one hour plasma levels of daunorubicin-eq (daunorubicin fluorescent equivalents; parent drug plus fluorescent metabolites) of 268 μg/ml, an increase of 185-fold over free drug levels of 1.4 μg/ml. DaunoXome increases plasma AUC values by a similar amount, 227-fold over free drug. Tumor levels for daunorubicin-eq at one hour post injection of free drug were 9.6 μg/g and demonstrated no significant increase thereafter.[6] In contrast, DaunoXome treatment produces an accumulation of daunorubcin-eq within the tumor that continues through eight hours, peaking at 100 μg/g. The kinetics for tumor accumulation of daunorubicin-eq followed an apparent first order absorption process. Comparison of tissue daunorubicin AUC ratios for DaunoXome to free drug indicate that DaunoXome produces the greatest increase (10-fold) in tumor tissue. Hepatic and spleenic tissues showed smaller AUC ratio increases (1.6-fold each). The increased brain AUC (2.1-fold) may not be significant since the absolute levels were low for both free and entrapped drug.[6] Additionally, previous reports have indicated that liposomes do not cross the blood-brain barrier. Minor AUC increases were noted for kidneys and small intestines. DaunoXome produced decreased AUC values for both heart and lungs (0.4 and 0.7-fold, respectively). In contrast, the selectivity of free drug for tumor tissue is comparatively low, with only brain tissue demonstrating AUC values below that for tumor. Although the pharmacokinetic data in this study did not include time points earlier than one hour, it is clear that peak tumor drug levels occur earlier for free drug (about 1–2 hours, post injection) than for DaunoXome (8–9 hours).[6] The slower daunorubicin tumor accumulation for DaunoXome illustrates the importance of maintaining high plasma levels of encapsulated drug for enhanced tumor delivery.

III.3.2. Preclinical efficacy

P1798 lymphosarcoma was used to compare the single-dose efficacies of DaunoXome and free daunorubicin.[6] This model is particularly useful as an initial formulation screen since it grows rapidly and the tumor take rate approaches 100%. For untreated controls the median survival time (MST) was 13 days. When compared at equivalent daunorubicin doses, DaunoXome produced significantly greater MST values than free drug at all dose levels tested. The maximum MST observed for free daunorubicin was 18.5 days at 30 mg/Kg. At 40 mg/Kg, free drug was toxic, reducing the MST to 12.5 days. No long-term survivors (LTS) were observed in any of the free drug treatment groups. In contrast, the maximum MST for DaunoXome was 21.5 days, at 30 mg/Kg. Of particular note for DaunoXome treatment was the observation of three LTS, >60 days.

A more slowly growing mammary adenocarcinoma, MA16C, was used as a second tumor model.[75] In this study, untreated control mice had an MST of 36 days. Two of twenty control mice survived tumor-free through the end of the study (114 days) while the remaining eighteen controls developed tumors that eventually proved fatal. In the free drug treatment groups, tumor-free LTS were noted at only 10 and 20 mg/Kg (one and four animals, respectively). All other free daunorubicin treated mice either developed detectable tumors or died of drug toxicity at higher doses. The incidence of tumor-free LTS for the DaunoXome treatment groups was dramatically higher. Of the 49 LTS treated with DaunoXome at doses ranging from 2 to 35 mg/Kg, nearly all were tumor free (three mice had small tumors: two at 2 mg/Kg and one at 20 mg/Kg). Comparison of equivalent daunorubicin doses indicated that DaunoXome produced significantly greater tumor growth inhibition than did free drug. DaunoXome was associated with apparent complete cures at 25 mg/Kg, with 10/10 of this group becoming tumor-free LTS. In contrast, only 40% of the mice treated with free drug at 20 mg/Kg were tumor-free LTS.[75]

The direct cytotoxic effect of DaunoXome on tumors in vivo is supported by the significant reduction of tumor volume, relative to free drug. Across all free drug dose levels, each established tumor eventually grew to considerable size (>500 mg). Of forty mice treated with DaunoXome at therapeutic doses (15–30 mg/Kg), 14 had confirmed tumors that regressed and disappeared completely. When compared to free drug at equivalent daunorubicin doses, DaunoXome significantly increased tumor growth suppression throughout the dose range of 2 to 25 mg/Kg ($p < 0.05$, at 2 and 5 mg/Kg for tumors measured on day 29 of the study; at 15, 20 and 25 mg/Kg on day 50). Tumor growth inhibition for DaunoXome at 2 mg/Kg was approximately equivalent to that observed for free drug at 15 mg/Kg. DaunoXome was thus more effective than free daunorubicin in terms of MST and tumor growth inhibition, throughout the 2–25 mg/Kg dose range.[6] Of particular interest was the observation that the MST and percent LTS observed for the lowest DaunoXome dose (2mg/Kg) were equivalent to those seen at the optimal free drug dose (20 mg/Kg).

III.4. Clinical pharmacology

III.4.1. Serum pharmacokinetics

The plasma pharmacokinetic profile of DaunoXome differs significantly from that of free drug.[75,76] Following IV administration of free drug at a dose of 80 mg/m^2 daunorubicin exhibits a low peak plasma concentration, a short initial distribution half-life and prolonged terminal elimination half-life, a large volume of distribution (reflecting extensive tissue distribution and binding), and rapid plasma clearance [76] (Table 2). After 12 hours, plasma daunorubicin concentrations are considerably lower, resulting in low AUC values; these observations are consistent with rapid removal of free daunorubicin from the circulation and extensive tissue uptake.

In contrast to free daunorubicin, DaunoXome plasma concentrations decline in a monoexponential fashion following a single dose of 40 mg/m^2, with a longer

Table 2

Pharmacokinetic parameters of free daunorubicin following a single dose of 80 mg/m^2 in patients with disseminated malignancies and of DaunoXome® following a single dose of 40 mg/m^2 in AIDS patients with Kaposi's sarcoma

Pharmacokinetic parameter, units	DaunoXome Mean ± SD (n = 30)	Free Daunorubicin Mean ± SD (n = 4)
Peak plasma concentration, mg/mL	18.26 ± 3.03	0.4
Plasma clearance, mL/min	17.3 ± 6.1	236 ± 181*
Volume of distribution (steady state), L	6.4 ± 1.5	1006 ± 622
Distribution half-life, h	4.41 ± 2.33	0.77 ± 0.3
Elimination half-life, h	–	55.4 ± 13.7

* Calculated.

Table 3

Pharmacokinetic parameters of DaunoXome®[75]

DaunoXome® single dose, mg/m^2	Peak plasma concentration, μg/mL	$t_{1/2}$, hr	AUC, μg hr/mL	Plasma clearance, mL/min	Volume of distribution, L
10 ($n = 3$)	5.9	2.8	16.9	15.7	3.75
20 ($n = 4$)	8.2	3.8	57.2	14.3	4.1
40 ($n = 4$)	18.15	4.0	120.1	10.5	3.7
60 ($n = 2$)	36.2	8.3	301.1	6.7	2.9
80 (n = 4)	43.6	5.2	375.3	6.6	2.9

apparent elimination half-life, a smaller steady-state volume of distribution (much lower than for free drug because the liposomes are confined to vascular fluid volume), and a much slower plasma clearance. The differences in volume of distribution and clearance result in a higher daunorubicin exposure (in terms of plasma AUC) from DaunoXome than with free daunorubicin. Following IV administration of DaunoXome at a dose of 80 mg/m^2, mean peak plasma concentrations are approximately 44 μg/mL, nearly 100-fold greater than for free daunorubicin.[75] The AUC for DaunoXome at this dose is approximately 36-fold greater than free daunorubicin, despite the shorter observed terminal half-life.

In Phase I studies, the pharmacokinetics of a range of IV DaunoXome single doses (10, 20, 40, 60, and 80 mg/m^2) were evaluated in patients with solid tumors, including some patients with Kaposi's sarcoma (Table 3).[75,77,78] In most patients, plasma concentrations of DaunoXome declined in a monoexponential fashion; however, in some patients a biexponential decline was observed. In a few patients dosed at $\geqslant 60$ mg/m^2, saturation kinetics were evident. Peak plasma concentrations were dose dependent and increased with dose in a nearly linear manner. AUC values also were dose dependent and increased linearly with dose. Clearance rates decreased with increasing dose, but dose dependency was not strong.

III.4.2. Distribution

DaunoXome has a small steady-state volume of distribution (approximately 6.4 L), suggesting that it is confined mainly to the vascular fluid volume. In vivo studies in animals indicate that DaunoXome selectively results in greater accumulation and sustained levels of daunorubicin in tumor versus normal cells.[6,74,75]

Only limited data on the distribution of DaunoXome in humans are available. In a study of normal skin samples and Kaposi's sarcoma lesions taken 24 hours after DaunoXome administration,[75] daunorubicin was not detected in normal skin samples of several evaluable patients. In contrast, following administration of DaunoXome, daunorubicin was detected in some samples from Kaposi's lesions at concentrations of 1.06 and 1.07 μg/g. These data are consistent with the results of the animal studies.

III.5. Clinical efficacy and safety

III.5.1. Efficacy in Kaposi's sarcoma (KS)

KS has become an aggressive disease in HIV patients suffering immunosuppression. The initial site of disease is usually cutaneous with nodular lesions potentially progressing rapidly to coalescence. This, in turn, may result in lymphatic obstruction and lymphedema. In addition, cutaneous disease may ulcerate, become infected, or produce local pain. Oral cavity and gastrointestinal tract involvement is common, and may involve as many as 50% of patients. Pulmonary involvement is also common, and can produce a range of symptoms including pulmonary insufficiency. In a few patients, visceral disease may precede cutaneous involvement and be the sole cause of symptomatology. KS can be the cause of death in many patients.

Chemotherapy has become the principal treatment alternative for disseminated KS in patients who have failed interferon or who have low $CD4^+$ T-cell counts. A wide-range of cytotoxic agents has activity against KS; these include bleomycin, etoposide, doxorubicin, vinblastine, and vincristine. In responding patients, chemotherapy can induce regression of cutaneous and visceral disease. Such tumor regression often is associated with improvement in tumor-associated symptomatology. Maintenance of responses requires continued therapy, however and relapse without maintenance therapy is the rule. Combination chemotherapy regimens can result in high response rates. For example, a response rate exceeding 80% has been reported for the ABV combination of doxorubicin, bleomycin, and vinca-alkaloid (vinblastine or vincristine).[79] Long-term therapy with aggressive combinations such as ABV is extremely difficult. The risks of cumulative cardiotoxicity with doxorubicin, pulmonary toxicity with bleomycin, and neutotoxicity with the vinca-alkaloids are superimposed on the possible increased risk of infectious complications and gastrointestinal toxicity observed with each treatment cycle. In addition, these agents are associated with significant myelosuppresion, worsening the immunosuppression already present from the HIV infection.

Due to the findings of high In-111 uptake by KS lesions in patients receiving the tumor imaging agent, patients with AIDS-related KS were included in the early Phase I/II dose escalation studies. In one study involving twenty-two patients, response rates (partial responders, PR and complete responders, CR) of 55 percent were observed at doses of 50 and 60 mg/m^2 every two weeks.[77]

Phase II Studies: Six Phase II studies of DaunoXome 40 mg/m^2 administered IV every two weeks were conducted across the United States and Europe in patients with advanced, HIV-associated Kaposi's sarcoma.[75,80,81] Most patients received concomitant antiretroviral therapy and necessary medications for the prevention and treatment of opportunistic infections. Concomitant therapy with other local or systemic medications for Kaposi's sarcoma was not allowed. Efficacy was evaluated by determining overall response, duration of response, and effects on quality of life. Unlike the Phase III trial described below in which responses were evaluated by an independent central reviewer according to ACTG criteria, responses in the Phase II studies were assessed by each individual study investigator.

Phase II studies demonstrated that DaunoXome had significant antitumor activity in advanced, HIV-associated Kaposi's sarcoma.[75] In a pooled analysis of 91 evaluable patients, an overall response rate of 64.8% (59/91) with 2.2% complete responses was observed. The mean duration of DaunoXome treatment was 20 weeks of 9.6 cycles. Quality of life was assessed by analyzing patients' Karnofsky Performance Status as a function of the cumulative DaunoXome dose. No change in mean Karnofsky Performance Status was observed for cumulative doses up to 1000 mg/m^2 (25 cycles).[75]

Money-Kyrle et al. investigated the efficacy and toxicity of DaunoXome at 40 mg/m^2 every two weeks in 25 patients with advanced AIDS-associated KS of poor prognosis, achieving a partial response rate of 40 percent (4/10). For patients with no previous chemotherapy, the observed response rate was 57 percent.[82] There has been minimal myelosuppression, no evidence of cardiac toxicity, and an overall decrease in the frequency and severity of side effects commonly related to chemotherapy.[75]

Phase III studies were designed as open-label, multicenter, randomized trials in 232 patients with advanced HIV-associated KS. The efficacy and safety of DaunoXome (40 mg/m^2) were compared with a combination of therapy regimen of doxorubicin (Adriamycin®, 10 mg/m^2), bleoymcin (15 U), and vincristine (1.0 mg) (ABV) given every two weeks.[83] The dose and schedule of ABV were chosen to provide an equivalent level of myelosuppression. Advanced disease was defined as the presence of ⩾25 mucocutaneous lesions, the development of ⩾10 mucocutaneous lesions in a one-month period, symptomatic visceral involvement, and/or tumor-related edema. The patients with advanced HIV-associated KS evaluated in this study are representative of those Kaposi's sarcoma patients who require palliative therapy in a normal clinical setting.

One hundred-sixteen patients were treated with DaunoXome 40 mg/m^2 and 111 patients with ABV. Patient demographic and baseline tumor characteristics were comparable in the two treatment arms. Response to therapy was assessed by an independent central reviewer according to the AIDS Clinical Trials Group (ACTG) criteria and required a minimum of the following for at least 28 days.

⩾50% decrease in number of total lesions
⩾50% decrease in the sums of the products of the largest perpendicular diameters of bidimensionally measurable marker lesions
Complete flattening of ⩾50% of all previously raised lesions

DaunoXome was comparable to ABV in the treatment of advanced, HIV-associated Kaposi's sarcoma with respect to tumor response, duration of response, and time to progression of the disease (Table 4). The overall response rate was 23% (27/116) with DaunoXome and 30% (33/111) with ABV. The lower response

Table 4.

Efficacy results of first-line therapy for HIV-associated Kaposi's sarcoma[18,21]

	DaunoXome (n = 116)	ABV (n = 111)
Response rate, %	23*	30
Median duration of response, days	110	113
Median time to progression, days	92	105
Survival, days	342	291

* The 95% CI for difference in response rates (ABV-DaunoXome) was (−5%, 18%).

rate of Phase III relative to Phase II is likely due to the different response assessment method for the trial designs. Twenty of 33 ABV responders and 11 of 27 DaunoXome responders had responses according to criteria other than flattening of lesions (i.e., shrinkage of lesions and/or reduction in the number of lesions). Photographic evidence of tumor response was comparable for DaunoXome and ABV across all anatomic sites (e.g., face, oral cavity, trunk, legs, and feet). Time to first evidence of response was rapid for both treatment groups, occurring at the first or second follow-up visit (4 or 8 weeks after initiation of treatment). A trend toward longer survival was observed with DaunoXome therapy (342 days) compared with ABV (291 days).

Patients treated with DaunoXome had significantly ($p < 0.0001$) less premedication use (antiemetic, antipyretic) than those treated with ABV. Dexamethasone (11.5% vs 40.9%) and acetaminophen (12.8% vs 35.2%) were required in a lower proportion of treatment cycles in DaunoXome patients than in ABV patients, respectively.

Patient quality-of-life evaluations also demonstrated comparable effects of treatment with DaunoXome and ABV. Quality-of-life assessments included evaluation of changes from baseline in Karnofsky Performance Status, quality-of-life scores based on patient-recorded diaries, and patient body weight over the study period. While patients who received ABV appeared to exhibit progressively greater decreases in weight and quality-of-life scores after five cycles, patients receiving DaunoXome had no changes in weight, quality-of-life scores, or Karnofsky Performance Status over 15 cycles of treatment.[83]

The results from this large multicenter, randomized, controlled Phase III trial demonstrating the efficacy of DaunoXome as first-line therapy for advanced, HIV-associated Kaposi's sarcoma are supported by the results of previous Phase II studies[80,84] in similar patient populations.

III.5.2. Efficacy in other solid tumors

Phase II trials have been conducted in solid tumors including colon,[85,86] breast[87,88] non-small cell lung cancer,[89] and brain.[90] Most of these tumor types have shown indications of activity and trials are continuing. The colon and lung cancer trials failed to show activity in 14 patients; further studies of these tumors have not been pursued.

III.5.3. DaunoXome for non-Hodgkin's lymphoma

Presant et al. reported that the tumor imaging liposome preparation VesCan was strongly taken up by tumor in a patient with non-Hodgkin's lymphoma (NHL).[54] Later a Phase II trial was conducted with DaunoXome in NHL patients showing activity.[91] Not enough direct comparison exists with VesCan imaging efficacy and tumor response to liposomal chemotherapeutics, but the possible correlation is intriguing and deserves a prospective study.

Fourteen patients were enrolled in a Phase II study designed to test DaunoXome in the treatment of relapsed/refractory low-intermediate grade non-Hodgkin's lymphoma.[91] The study design called for 100 mg/m^2 DaunoXome infusions every three weeks with a provision for dose escalation. The majority of patients had low grade small lymphocytic disease and 11/14 patients were Stage IV. All patients had received prior chemotherapy of various regimens (1–4 cycles). At the time of the cited report,[91] DaunoXome had been administered a median of four cycles (range 1–15).

Six of fourteen patients (43%) achieved a partial response, three (21%) had stable disease and five (36%) had progressive disease. These are promising, although early results; further studies are going forward in NHL to follow-up this observation.

DaunoXome is being tested as a substitute for doxorubicin in the cyclophosphamide, doxorubicin, vincristine, prednisone (CHOP) standard combination therapy for relapsed or refractory lymphoma.[92] In this study, nine patients were treated with DaunoXome 120 mg/m^2 every three weeks and nine with the modified CHOP regime where DaunoXome was given also at 120 mg/m^2. Neutropenia was experienced by all patients in the modified CHOP treatment arm; other toxicities were mild. The authors concluded that DaunoXome has clinically beneficial activity against refractory and/or relapsed lymphoma, particularly in combination therapy.

II.5.4. Safety and tolerability

a. Overview. Knowledge of the common side effects of free daunorubicin is important to any analysis of the safety and tolerability profile of DaunoXome. The dose-limiting side effect of free daunorubicin is acute myelosuppression, manifested primarily on the granulocytic series. In addition, free daunorubicin causes alopecia as well as nausea and vomiting in a significant number of patients. Chronic administration of daunorubicin therapy also has been associated with a cardiomyopathy manifested as congestive heart failure that appears to increase in incidence in adults after a cumulative dosage above 550 mg/m^2.[71] Extravasation of free daunorubicin can cause severe local tissue necrosis.

b. Phase III safety summary. Safety was evaluated in the open-label, randomized, controlled Phase III trial that compared the effects of DaunoXome and ABV as first-line therapy in 227 patients with advanced, HIV-associated Kaposi's sarcoma.[93] Patients with advanced, HIV-associated Kaposi's sarcoma are seriously ill and immunocompromised due to their underlying HIV infection. Because these patients are receiving several concomitant medications, including potentially toxic antiviral and antiretroviral agents, the contribution of the study drugs to the adverse experience profile is difficult to establish.

DaunoXome 40 mg/m^2 was generally well-tolerated in the Phase III study.[93] As with free daunorubicin, the most important acute toxicity of DaunoXome was myelosuppression, which was manifested primarily on the granulocytic series. The incidences of neutropenia were similar between the DaunoXome and ABV treatment groups.

The incidence of alopecia was significantly ($p < 0.001$) lower in DaunoXome-treated patients (8%) than in ABV-treated patients (36%). This is an important consideration for the AIDS patient, for whom alopecia can be particularly stigmatizing.

Similarly, the incidence of neuropathy was significantly ($p < 0.001$) lower in DaunoXome-treated patients (13%) than in ABV-treated patients (41%). This can be an important consideration for patients who may take other HIV/AIDS drugs associated with neuropathy.

Of particular importance is the lack of cardiotoxic effects with DaunoXome in the Phase III trial at cumulative doses that historically have been associated with significant cardiotoxicity with free daunorubicin. In contrast to free daunorubicin and other anthracyclines, reports of clinical cardiotoxicity have been rare in Phase II and III trials at cumulative DaunoXome doses >600 mg/m^2.[94]

IV. Conclusions

The efficacy and safety of DaunoXome in treatment of advanced AIDS-related Kaposi's sarcoma suggest that the liposomal drug will be useful for therapy of other cancers. Phase II clinical trials are underway to test DaunoXome as a single agent against lymphoma, leukemia, myeloma and solid tumors including breast, ovarian, and brain. There is already evidence of activity from these early studies, e.g., non-Hodgkin's lymphoma.[91] Finally on the basis of the Phase III efficacy and safety data, DaunoXome has been approved for first-line therapy of advanced AIDS-related Kaposi's sarcoma in the United States and 16 other countries.

References

1. Gregoriadis G, Davis C. Stability of liposomes in vivo and in vitro is promoted by their cholesterol content and the presence of blood cells. Biochem Biophys Res Commun 1979;89:1287–1293.
2. Papahadjopoulos D, Jacobson, K, Nir, S, Isac, T. Phase transitions in phospholipid vesicles: fluorescence polarization and permeability measurements concerning the effect of temperature and cholesterol. Biochim Biophys Acta 1973;311:330–348.
3. Williams LE, Duda, RB, Proffitt, RT, Beatty, BG, Beatty, JD, Wong, JY, Sively, JE, Paxton, RJ: Tumor uptake as a function of tumor mass: a mathematic model. J Nucl Med 1988;29:103–109.

4. Williams LE, Proffitt, RT, Lovisatti, L. Possible applications of phospholipid vesicles (liposomes) in diagnostic radiology. J Nucl Med Allied Sci 1984;28:35–45.
5. Proffitt RT, Williams, LE, Presant, CA, Tin, GW, Uliana, JA, Gamble, RC, Baldeschwieler, JD. Liposomal blockade of the reticuloendothelial system: improved tumor imaging with small unilamellar vesicles. Science 1983;220:502–505.
6. Forssen EA, Coulter, DM, Proffitt, RT. Selective In Vivo Localization of Daunorubicin Small Unilamellar Vesicles in Solid Tumors. Cancer Research 1992;52:3255–3261.
7. Anaissie E, Paetznick, V, Proffitt, R, Adler-Moore, JP, Bodey, GP. Comparison of the In Vitro Antifungal Activity of Free and Liposome-Encapsulated Amphotericin B. European Journal of Clinical Microbiology and Infectious Diseases 1991;10:665–668.
8. van Etten EWM, Changer, HR, Snijders, SV, Bakker-Woudenberg, AJM. Interactions of liposomal amphotericin B with extracellular and intracellular *Candida albicans*. J Antimicro Chemother 1995;36:961–974.
9. Pallister CJ, Johnson, EM, Warnock, DW, Elliot, PJ, Reeves, DF. In-vitro effects of liposome-encapsulated amphotericin B (AmBisome) and amphotericin B-deoxycholate (Fungizone) on the phagocytic and candidacidal function of human polymorphonuclear leucocytes. J Antimicro Chemother 1992;30:313–320.
10. Hartsel S, Bolard, J. Amphotericin B: new life for an old drug. Trends in Pharm. Sci. 1996;17:445–449.
11. Adler-Moore JP, Proffitt, RT. Development, Characterization, Efficacy and Mode of Action of AmBisome, A Unilamellar Liposomal Formulation of Amphotericin B. J Liposome Research 1993;3:429–450.
12. Fujii G. Liposomal Amphotericin B (AmBisome): Realization of the Drug Delivery Concept. Vesicles 1996;12:491–526.
13. NeXstar Pharmaceuticals, Inc. Data on file.
14. van Etten EWM, Otte-Lambillion, M, van Vianen, W, ten Kate, MT, Bakker-Woudenberg IAJ: Biodistribution of liposomal amphotericin B (AmBisome) and amphotericin B-desoxycholate (Fungizone) in uninfected immunocompetent mice and leucopenic mice infected with *Candida albicans*. Journal of Antimicrobial Chemotherapy 1995;35:509–519.
15. Adler-Moore JP, Fujii, G, Lee, MJA, Satorius, A, Bailey, A, Proffitt, RT. In vitro and in vivo interactions of AmBisome with pathogenic fungi. J Liposome Res 1993;3:151–156.
16. Adler-Moore JP. AmBisome targeting to fungal infections. Bone Marrow Transplantation 1994;14:53–57.
17. Wasan KM, Lopez-Berestein G. Modification of amphotericin B's therapeutic index by increasing its association with serum high-density lipoproteins. Ann NY Acad Sci 1994;730:93–106.
18. Proffitt RT, Satorius, A, Chiang, SM, Sullivan, L, Adler-Moore, JP. Pharmacology and toxicology of a liposomal formulation of amphotericin B (AmBisome) in rodents. J Antimicro Chemother 1991;28(Suppl B):49–61.
19. Francis PWLJ, Hoffman A, Peter J, Francesconi A, Bacher J, Shelhamer J, Pizzo PA, Walsh TJ. Efficacy of Unilamellar Liposomal Amphotericin B in Treatment of Pulmonary Aspergillosis in Persistently Granulocytopenic Rabbits: The Potential Role of Bronchoalveolar D-Mannitol and Serum Galactomannan as Markers of Infection. J Infectious Diseases 1993;169:356–368.
20. Croft SL, Davidson RN, Thornton EA. Liposomal amphotericin B in the treatment of visceral leishmaniasis. J Antimicrob Chemother 1991;28(Suppl B):111–118.
21. Gradoni L, Davidson RN, Orsini S, Betto P, Giambenedetti M. Activity of Liposomal Amphotericin B (AmBisome) Against Leishmania Infantum and Tissue Distribution in Mice. Journal of Drug Targeting 1993;1:311–316.
22. Adler-Moore JP, Chiang S, Satorius A, Guerra D, McAndrews B, McManus EJ, Proffitt RT. Treatment of murine candidosis and cryptococcosis with a unilamellar liposomal amphotericin B formulation (AmBisome). J Antimicrob Chemother 1991;28(Suppl B):63–71.
23. Graybill JR, Bocangera R. Liposomal amphotericin B therapy of murine histoplasmosis. Antimicro Agents and Chemother 1995;39:1885–1887.
24. Clemons KV, Stevens, DA. Therapeutic efficacy of a liposomal formulation of amphotericin B (AmBisome) against murine blastomycosis. Journal of Antimicrobial Chemotherapy 1993;32:465–472.
25. Clemons KV, Stevens DA. Comparison of a liposomal amphotericin B formulation (AmBisome) and deoxycholate amphotericin B (Fungizone) for the treatment of murine paracoccidioidomycosis. Medical and Veterinary Mycology 1993;31:387–394.
26. Janknegt R, de Marie S, Bakker-Woudenberg IAJM, Crommelin DJA. Liposomal and lipid formulations of amphotericin B. Clin Pharmacokinet 1992;23:279–291.

27. Heinemann V, Kahny B, Debus A, Wachholz K, Jehn U. Pharmacokinetics of liposomal amphotericin B (AmBisome) versus other lipid-based formulations. Bone Marrow Transplantation 1994;14:58–59.
28. Tollemar J, Ringden, O. Early Pharmacokinetic and Clinical results from a Noncomparative Multicentre Trial of Amphotericin B Encapsulated in a Small Unilammelar Liposome (AmBisome). Drug Invest 1992;4:232–238.
29. Katz NM, Pierce PF, Anzeck RA, Visner MS, Canter HG, Foegh ML, Pearle DL, Tracy C, Rahman A. Liposomal amphotericin B for treatment of Pulmonary Aspergillosis in a Heart Transplant Patient. The Journal of Heart Transplantation 1990;9:14–17.
30. Meunier F, Prentice HG, Ringden O. Liposomal amphotericin B (AmBisome): safety data from a phase II/III clinical trial. J Antimicrob Chemother 1991;28(Suppl B):83–91.
31. Ringden O, Andstrom E, Remberger M, Svahn BM, Tollemar J. Safety of liposomal amphotericin B (AmBisome) in 187 transplant recipients treated with cyclosporin. Bone Marrow Transplantation 1994;14(Suppl B):S10–S14.
32. Prentice HG, Hann IM, Herbrecht R, Aoun M, Kvaloy S, Catovsky D, Pinkerton CR, Schey SA, Jacobs F, Oakhill A, Stevens RF, Darbyshire PJ, Gibson BE. A randomized comparison of liposomal versus conventional amphotericin B for the treatment of pyrexia of unknown origin in neutropenic patients. Brit J Haematol 1997;98:711–718.
33. Hann IM, Stevens RF, Pinkerton CR: Safety and efficacy of two dose regimes of AmBisome versus amphotericin B as empiric antifungal treatment in neutropenic paediatric patients. 2nd International Symposium on Febrile Neutropenia, Brussels, 1995.
34. Ringden O, Meunier F, Tollemar J, Ricci P, Tura S, Kuse E, Vivani MA, Gorin NC, Klasterksy J, Fenaux P Prentice HG, Ksionski G. Efficacy of amphotericin B encapsulated in liposomes (AmBisome) in the treatment of invasive fungal infections in immunocompromised patients. J Antimicro Chemother 1991;28(Suppl B):73–82.
35. Mills W, Chopra R, Linch DC, Goldstone AH. Liposomal amphotericin B in the treatment of fungal infections in neutropenic patients: a single-centre experience of 133 episodes in 116 patients. British Journal of Haematology 1994;754–760.
36. Tollemar J, Ringden O. Formulations of Amphotericin B. Drug Safety 1995;13:207–218.
37. Davidson RN, Croft SL, Scott A, Maini M, Moody AH, Bryceson ADM: Liposomal amphotericin B in drug-resistant visceral leishmaniasis. The Lancet 1991;337:1991.
38. Gradoni L, Bryceson A, Desjeux P: Treatment of Mediterranean visceral leishmaniasis. Bulletin of the World Health Organisation 1995;73:191–197.
39. Davidson RN, di Martino L, Gradoni L, Giacchino R, Gaeta BG, Pempinello R, Scotti S, Cascio A, Castagnola E, Maisto A, Gramiccia M, di Caprio D, Wilkinson RJ, Bryceson AD: Short-course treatment of visceral leishmaniasis with liposomal amphotericin B (AmBisome). Clinical Infectious Diseases 1996;9:38–43.
40. Gradoni L, Davidson RN, Orsini S, Betto P, Giambenedetti M. Activity of Liposomal Amphotericin B (AmBisome) Against Leishmania infantum and Tissue Distribution in Mice. Drug Targeting 1993;1:311–316.
41. Wilson M, Denning DW: The commonest life threatening mould infection: invasive aspergillosis. Hosp Update 1993;4:225–233.
42. Ellis M, Spence D, Meunier F, De Pauw B, Bogaerts M, Van Der Cam C, Doyen C, Marinu A, Collette L, Sylvester R. Randomized Multicentre Trial of 1 mg/kg (LD) Versus 4 mg/kg (HD) Liposomal Amphotericin B (AmBisome) (LAB) in the Treatment of Invasive Aspergillosis (IA). Abstracts of the 36th ICAAC, New Orleans, LA, 1996.
43. Maggiolo F, Pellegata G, Marchetti G, Novali A, Bossetti F, Viviani M, Suter F. Liposomal amphotericin in a case of Candida Endocarditis, Abstract, Della Societa Italiana di Chemoterapia, 1991.
44. Sharland M, Hay RJ, Davies EG: Liposomal amphotericin B in hepatic candidosis. Archives of Disease in Childhood 1994;546–547.
45. Hudson J, Scott GL, Warneck DW: Treatment of hepatic candidosis with liposomal amphotericin B in patient with acute leukaemia. Letter, The Lancet 1991;339:374.
46. Coker RJ, Vivani M, Gazzard BG, Du Pont B, Pohle HD, Murphy SM, Atougia J, Champalimaud JL, Harris JRW: Treatment of cryptococcosis with liposomal amphotericin b (AmBisome) in 23 patients with AIDS. AIDS 1993;7:829–835.
47. Mauk MR Gamble RC. Preparation of lipid vesicles containing high levels of entrapped radioactive cations. Anal Biochem 1979;94:302–307.
48. Mauk MR, Gamble RC. Stability of lipid vesicles in tissues of the mouse: a gamma-ray perturbed angular correlation study. Proc Natl Acad Sci USA 1979;76:765–769.

49. Wallingford RH, Williams LE. Is Stability a key parameter in the accumulation of phospholipid vesicles in tumors? J. Nucl Med 1985;26:1180–1185.
50. Proffitt RT, Williams LE, Presant CA, Tin GW, Uliana JA, Gamble RC, Baldeschwieler JD: Tumor-imaging potential of liposomes loaded with In-111-NTA: biodistribution in mice. J Nucl Med 1983;24:45–51.
51. Turner AF, Presant, CA, Proffitt RT, Williams LE, Winsor DW, Werner JL. In-111-labeled liposomes: dosimetry and tumor depiction. Radiology 1988;166:761–765.
52. Patel KR, Tin GW, Williams LE, Baldeschwieler JD: Biodistribution of phospholipid vesicles in mice bearing Lewis lung carcinoma and granuloma. J Nucl Med 1985;26:1048–1055.
53. Presant CA, Proffitt RT, Turner AF, Williams LE, Winsor D, Werner JL, Kennedy P, Wiseman C, Gala K, McKenna RJ et al. Successful imaging of human cancer with indium-111-labeled phospholipid vesicles. Cancer 1988;62:905–911.
54. Presant CA, Blayney D, Proffitt RT, Turaner FA, Williams LE, Nadel HI, Kennedy P, Wiseman C, Gala K, Crossley RJ, Preiss SJ, Ksionski GE, Presant SL: Preliminary report: imaging of Kaposi's sarcoma and lymphoma in AIDS with indium-111-labelled liposomes. Lancet 1990;335:1307–1309.
55. Dictor M, Bendsoe, N, Runke, S, White, M. Major basement membrane components in Kaposi's sarcoma, angiosarcoma and benign vascular neogenesis. J Cutan Pathol (Denmark) 1995;22:435–441.
56. Albini A, Barillari G, Benelli R, Gallo R, Ensoli B. Angiogenic properties of human immunodeficiency virus type 1 Tat protein. Proc Natl Acad Sci, USA 1995;92:4838–4842.
57. Wu NA, Klitzman B, Rosner G, Needham D, De whirst MW. Measurement of material extravasation in microvascular networks using fluorescence video-microscopy. Microvasc Res 1995;46:231–253.
58. Forssen EA: Chemotherapy with anthracycline liposomes. In: G. Gregoriadis (Ed.) Liposomes as Drug Carriers: Recent Trends and Progress. J. Wiley and Sons, Chichester, 1988;355–364.
59. Kaye SB, Goden JA, Ryman BE. The effect of liposome (phospholipid vesicle) entrapment of actinomycin D and methotrexate on the in vivo treatment of sensitige and resistant solid murine tumours. Eur J Cancer 1981;17:279–289.
60. Ganapathi R, Krishan A, Woodinsky I, Zubrod CG, Lesko JJ. Effect of cholesterol content on anti-tumor activity and toxicity of liposome-encapsulated 1-beta-D-arabinofuranosylcytosine in vivo. Cancer Res 1980;40:630–633.
61. Rustum YM, Dave C, Mayhew E, Papahadjopoulos D. Role of liposome type and route of administration in the antitumor activity in liposome-entrapped 1-beta-D-arabinofuranosylcytosine against mouse L1210 leukemia. Cancer Res 1979;39:1390–1395.
62. Gabizon A, Meshorer A, Barenholz U. Comparative long-term study of the toxicities of free and liposome-associated doxorubicin in mice after intravenous administration. J Natl Cancer Inst 1986;77:459–469.
63. Hwang KJ, Mauk MR. Fate of lipid vesicles in vivo: a gamma-ray perturbed angular correlation study. Proc Natl Acad Sci USA 1977;74:4991–4995.
64. van Hoesel QG, Steerenberg PA, Crommelin DG, van Kijk A, van Oort W, Klein S, Douze JM, de Wildt DJ, Hillen FC. Reduced cardiotoxicity and nephrotoxicity with preservation of antitumor activity of doxorubicin entrapped in stable liposomes in the LOU/M Wsl rat. Cancer Res 1984;44:3698–3705.
65. Hwang KJ, Luk KFS, Braumier PL. Volume of distribution and transcapillary passage of small unilamellar vesicles. Life Sci 1982;31:949–955.
66. Weiss RB, Bruno S. Daunorubicin treatment of adult solid tumors. Cancer Treat Rep 1981;4:25–28.
67. Von Hoff DD, Rozencweig M, Slavik M, Muggia FM. Activity of daunomycin in solid tumors (letter). J. Am. Med. Assn. 1976;236:1693.
68. Von Hoff DD. Use of daunorubicin in patients with solid tumors. Semin Oncol 1984;11:23–27.
69. Nagasawa K, Natazuka T, Chihara K, Kitazawa F, Tsumura A, Takara K, Nomiyana M, Ohnishi N, Yokoyama T et al. Transport mechanism of anthracycline derivatives in human leukemia cell lines: uptake and efflux of pirarubicin in HL60 and pirarubicin-resistant HL60 cells. Cancer Chemother Pharmacol 1996;37:297–304.
70. Michieli M, Michelutti A, Damiani D, Pipan C, Raspadori D, Lauria F, Baccarani, M. A comparative analysis of the sensitivity of multidrug resistant (MDR) and non-MDR cells to different anthracycline derivatives. Leuk Lymphoma 1993;9:255–264.
71. Dorr RT, Von Hoff, DD. Cancer Chemotherapy Handbook, 2nd ed. Appleton and Lange: Norwalk, CT, 1994;1020.

72. Bosanquet AG. Stability of solutions of antineoplastic agents during preparation and storage for in vitro assays II. Assay methods, adriamycin and the other antitumour antibiotics. Cancer Chemother Pharmacol 1986;17:1–10.
73. Tsou KC, Lo KW, Ledis SL, Miller EE. Indigogenic phosphodiesters as potential chromogenic cancer chemotherapeutic agents. J Med Chem 1972;15:122–1225.
74. Forssen EA, Male-Brune R, Adler-Moore JP, Lee MJA, Schmidt PG, Krasieva TB, Shimizu S, Tromberg BJ. Fluorescence Imaging Studies for the Disposition of Daunorubicin Liposomes (DaunoXome®) within Tumour Tissue. Cancer Research 1996;56:2066–2075.
75. Forssen EA, Ross ME: DaunoXome® Treatment of Solid Tumours: Preclinical and Clinical Investigations. Journal of Liposome Research 1994;4(1):481–512.
76. Alberts DS, Bachur NR, Holtzman JL. The pharmacokinetics of daunomycin in man. Clin Pharmacol Ther 1971;12:96–104.
77. Gill PS, Espina BM, Muggia F, Cabriales F, Tulpule S, Esplin A, Liebman JA, Forssen E et al. Phase I/II clinical and pharmacokinetic evaluation of liposomal daunorubicin. J Clin Oncol 1995;13:996–1003.
78. Guaglianome P, Chan K, DelaFlor-Weiss E et al. Phase I and pharmacologic study of liposomal duanorubicin (DaunoXome). Invest New Drugsa 1994;12:103–110.
79. Gill PS, Naidu Y, Salahuddin SZ. Recent advances in AIDS-related Kaposi's sarcoma. Curr Opin Oncol 1990;2:1161–1166.
80. Presant CA, Scolaro M, Kennedy P, Blayney DS, Flanagan B, Lisak JJP. Liposomal Daunorubicin Treatment HIV-Associated Kaposi's Sarcoma. The Lancet 1993;341:1242–1243.
81. Money-Kyrle JF, Bates F, Ready J, Gazzard BG, Phillips RH, Boag FC. Liposomal daunorubicin in advanced Kaposi's sarcoma: a phase II study. Clin Oncol 1993;5:367–371.
82. Money-Kyrle JF, Bates F, Ready J, Gazzard BG, Phillips RH, Boag FC. Liposomal Daunorubicin in Advanced Kapsoi's Sarcoma: A Phase II Study. Clinical Oncology 1993;5(6):367–371.
83. Gill PS, Wernz J, Scadden DT, Cohen P, Mukwaya GM, Ross ME. A Randomized Phase III Trial of Liposomal Daunorubicin (DaunoXome®) Versus Doxorubicin, Bleomycin, Vincristine (ABV) in Advanced AIDS-Related Kaposi's Sarcoma. NCI-EORTC, Amsterdam, 1996.
84. Gill PS, Rarick M, McCutchan JA, Slater JA, Parker B, Muchmore E, Bernstein-Singer M, Akil B et al. Systemic treatment of AIDS-associated Kaposi's sarcoma: results of a randomized trial. Am J. Med 1991;90:427–433.
85. Thurmann AM, Eckardt JR, Burris HA, Rodriguez GI, Cobb P, Bowen K, Peacock NW, Campbell L, Ross ME, Weiss GR et al. A Phase II Trial of (DaunoXome®) (DX) in Patients with Advanced Adenocarcinoma of the Colon. Proceedings of the American Society of Clinical Oncology 1993;12.
86. Eckardt JR, Campbell E, Burris HA, Weiss GR, Rodriguez GI, Fields SM, Thurman AM, Peacock NW, Cobb P, Rotherberg ML et al. A Phase II Trial of (DaunoXome®), Liposome-Encapsulated Daunorubicin, in Patients with Metastatic Adenocarcinoma of the Colon. American Journal of Clinical Oncology 1994;17:498–501.
87. Erdkamp LG, Hupperts PSGJ, Ten Bokkel-Huinink WW, Neyts GD, Eestermans GH. Phase II Study of Liposomal Encapsulated Daunorubicin (DaunoXome®) in Advanced Breast Cancer. A Phase II Pilot Trial. 18th Annual San Antonio Breast Cancer Symposium, 1995.
88. Hupperets PSGJ et al. Phase II Study of Liposomal Encapsulated Daunorubicin (DaunoXome®) in Advanced Breast Cancer (Abstract). Proceedings of American Society of Clincial Oncology, 1995.
89. Gatzemeier U et al. Single Agent, High Dose (DaunoXome®) for the Treatment of Stage IIIB and IV Non Small Cell Lung Cancer (NSCLC). A Phase II Pilot Trial. Interscience Conferences on Antimicrobial Agents and Chemotherapy, 1995.
90. Lippens R. Liposomal Daunorubicin in Childhood Brain Tumours, Preliminary Results of a Phase II Study. The Canadian Journal of Infectious Diseases 1995;6.
91. Tulpule A, Rarick MU, Kolitz J, Bernstein J, Traynorm A, Myers A, Harvey-Buchanan L, Vergel de Dios-Salvosa M, Espina BM, Mukwaya G, Ross M, Levine AM. Liposomal Encapsulated Daunorubicin (DaunoXome®) has Activity in Relapsed/Refractory Low Grade and Intermediate Grade Non-Hodgkin's Lymphoma (NHL). 38th Annual Meeting and Exposition of the American Society of Hematology, Orlando, 1996.
92. McBride NC, Richardson DS, Johnson S, Schey S, Gray A, Newland AC, Kelsey SM. Liposomal Daunorubicin (DaunoXome) as a Single Agent and in Combination Therapy for Poor Prognosis Lymphoma. Abstract, British Society for Haematology Meeting, April, 1997.
93. Gill PS, Wernz J, Scadden DT, Cohen P, Mukwaya GM, von Roenn JH, Jacobs M, Kempin S, Silverberg I, Gonzales G, Rarick MU, Myers AM, Shepherd F, Sawka C, Pike MC, Ross ME.

Randomized Phase III Trial of Liposomal Daunorubicin versus Doxorubicinm, Bleomycin, and Vincristine in AIDS-Related Kaposi's Sarcoma. Journal of Clinical Oncology 1996;14:2353–2364.
94. Gill PS, Wernz J, Scadden DT, Cohen P, Mukwaya GM, Ross ME. Lack of Cardiac Toxicity of Liposomal Encapsulated Daunorubicin (DaunoXome®) After Long Term Use in AIDS-Related Kaposi's Sarcoma. NCI-EORTC, Amsterdam, 1996.

Lasic and Papahadjopoulos (eds.), Medical Applications of Liposomes

CHAPTER 8.5

Medical applications of multivesicular lipid-based particles: DepoFoam™ encapsulated drugs

JUDITH H. SENIOR

DepoTech Corporation, 10450 Science Center Drive, San Diego, CA 92121, USA

Overview

I. General introduction

This chapter describes recent advances in medical applications of multivesicular lipid-based particles commercially known as the "DepoFoam™ sustained-release drug delivery system". These lipid-based particles have also been referred to as multivesicular liposomes (MVL). Particles of the DepoFoam sustained-release drug delivery system have a highly characteristic physical structure, distinguishing them from other types of liposomes and other lipid-based drug delivery systems.

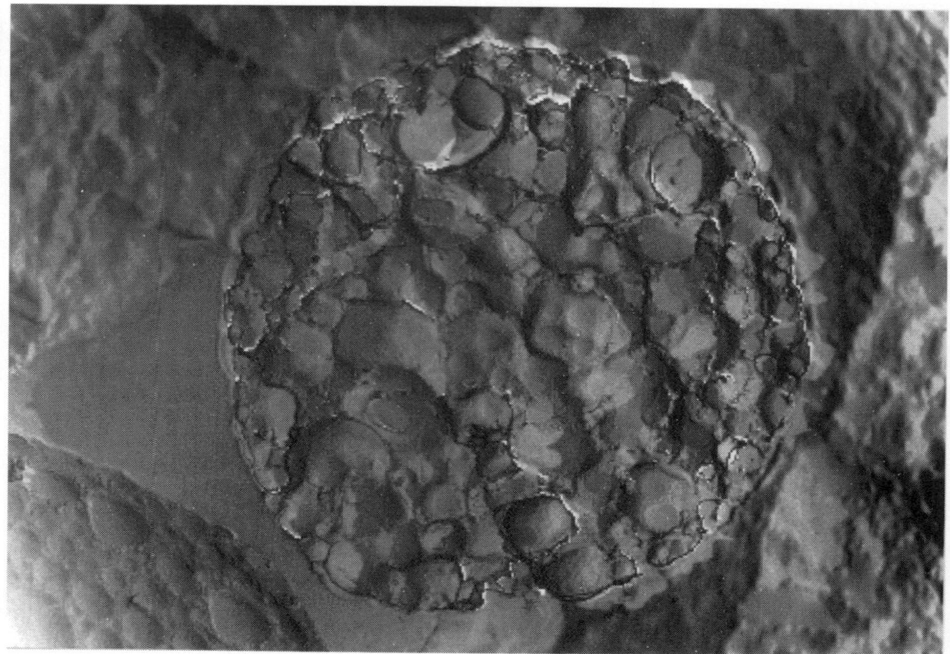

Fig. 1. Typical image of the fracture plane through an MVL particle which shows the bilayer walls of the multiple interior compartments, tightly packed into a roughly 10 μm diameter sphere (From Spector et al.,[3] with permission).

DepoFoam particles are microscopic and spherical, and each particle encloses multiple nonconcentric aqueous chambers bounded by a single bilayer lipid membrane with a "foam"-like appearance under the microscope (Figures 1 and 2). This drug delivery system has advantages over other lipid-based systems in aqueous suspension in that the composition and structure results in good stability during storage, control over drug release rate and highly efficient entrapment of hydrophilic molecules. Like other liposomes and lipid-based systems, the particles are made from lipids commonly found in biological membranes, and appear to be biodegradable by the usual lipid metabolic pathways.

DepoFoam formulations are manufactured by a process shown schematically in Figure 3, and depicted in Figure 4. Manufacture begins by emulsifying a mixture of an aqueous phase containing the drug to be encapsulated with the organic phase containing the lipids in chloroform[1,2] to form a water in oil emulsion. A lipid combination commonly used is:1,2-dioleoyl-sn-glycero-3-phosphocholine (DOPC), 1,2-dipalmitoyl-sn-glycero-3-phospho-rac-(1-glycerol) (DPPG), cholesterol, and triolein. The first water-in-oil emulsion is dispersed and emulsified in a solution of second aqueous phase (such as glucose/lysine) to make a water in oil in water emulsion (w/o/w).[3] The w/o/w is sparged with nitrogen to remove the

Fig. 2: Similar image to Figure 1, shown at a higher magnification (From Spector et al.,[3] with permission).

chloroform, at which time numerous sub-micron to micron-sized water compartments, separated by lipid bilayers, take on a close-packed polyhedral structure (Figures 1, 2).[3] The resulting MVL are then diafiltered by cross flow filtration to exchange buffers and remove unencapsulated drug. DepoFoam formulations are stored under refrigeration in a ready-to-use, injectable form, and are stable under the recommended storage conditions of 2–8°C for at least 18 months. The rate of drug release from DepoFoam particles in vitro and in vivo can be modified by changes in lipid composition of the particles, chemical properties of the drug to be encapsulated, and by changes in the manufacturing parameters used in production.

DepoFoam formulations are primarily being developed for local, depot, sustained release of drugs. Routes of administration being used in the clinic include intrathecal, epidural, and subcutaneous routes. Intramuscular, intraocular, intraarticular, intraperitoneal and direct injection at the disease site are also useful routes. DepoFoam formulations are not generally suitable for IV administration because of the relatively large particle size (over 5 μm). However the large particle size and unique vesicular architecture contribute to retention of the particles at the injection site in vivo. The type of water stable drugs that can be entrapped ranges from small molecules such as cytarabine, amikacin, and morphine sulfate, to

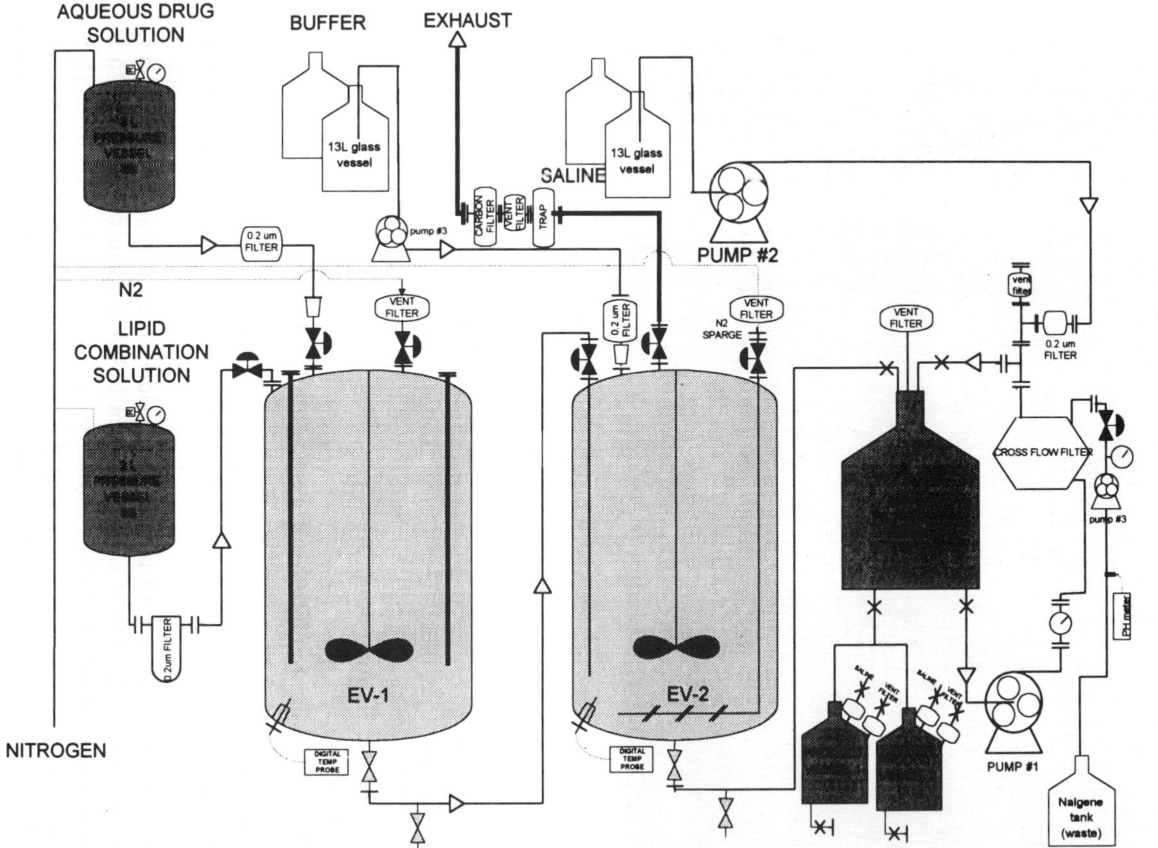

Fig. 3: A generalized scheme of the process used for the sterile, bulk manufacturing of DepoCyt, encapsulated cytarabine. The manufacturing process is designed to maintain an aseptic environment at every step of the process. Formulated components are introduced into the system by use of sterilizing filters. The process consists of two stainless steel emulsification vessels (EV-1 and EV-2), additional support vessels, a diafiltration cross flow system, support pumps and filters for sterile filtration of formulated components and vessel venting.

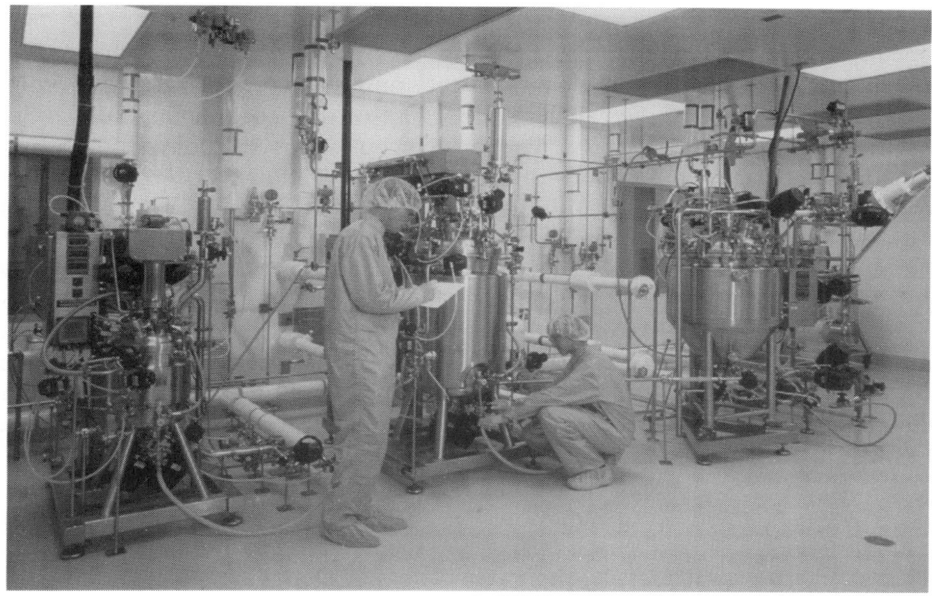

Fig. 4: Equipment for automated, commercial scale bulk sterile manufacture of DepoCyt, encapsulated cytarabine. Depicts manufacturing process equipment described in Figure 3.

larger molecules such as oligonucleotides and peptides, to macromolecules such as proteins and nucleic acids.[1,2,4–7] This article describes the development and clinical progress of DepoFoam encapsulated drugs in these unique lipid particles.

II. DepoFoam™ encapsulated sustained release cytarabine (DepoCyt™/DTC 101)

II.1. Introduction

Kinetic and pharmacological properties of the anti-cancer agent, cytarabine make it an excellent candidate to test the effectiveness of the DepoFoam sustained release drug delivery system in a clinical application. Neoplastic meningitis (carcinomatous meningitis, lymphomatous meningitis and leukemic meningitis), which results from the metastatic infiltration of the meninges by malignant cells, most commonly acute leukemias, lymphomas or carcinomas, is one example of how this technology can be applied to achieve therapeutic benefits. Standard treatment for this disease includes radiation therapy and single-agent or combination chemotherapy with compounds such as methotrexate, thio-TEPA, and cytarabine, delivered directly into the cerebrospinal fluid (CSF) by an intraventricular injection via an Ommaya reservoir or by lumbar puncture. Direct delivery of drug into the CSF (I-CSF administration) is considered to be more effective than intravenous

(IV) administation because chemotherapeutic agents administered IV have difficulty in crossing the blood-brain barrier.[8] Cytarabine (ara-C; cytosine arabinoside) is a cell-cycle specific agent that kills cells only when the cells are synthesizing DNA, in the S-phase of the cell cycle.[9] Because of the short elimination half-life (3.4 h) of cytarabine,[10] frequent injections or continuous infusions are necessary to maintain efficacious CSF levels of drug. An optimized delivery schedule would involve a continuous infusion of the drug into the subarachnoid space. Such infusion schedules are impractical because they are a discomfort to the patient, time consuming for the physician, and increase the risk of infection. A new approach to the treatment of neoplastic meningitis using I-CSF administration of a sustained release depot formulation of cytarabine presents a number of potential advantages. Such a formulation would require fewer injections which would in turn reduce infection risk, discomfort, and cost, and would be more effective at maintaining therapeutic concentrations of drug in the CSF over an extended period of time.

II.2. Overview of the clinical studies

The overall strategy for testing the effectiveness of DepoFoam-encapsulated sustained- release cytarabine (DepoCyt™) compared with standard therapy involved a series of clinical studies as described below. The expectation was that DepoCyt would be at least as effective, and certainly more convenient than standard therapy based on a less frequent dosing schedule. Objectives for the initial safety/dosing study (Phase I/II trial) were:

(a) To demonstrate the safety of DepoCyt over a range of doses, escalating from 12.5 to 125 mg;
(b) To establish a suitable clinical dose;
(c) To assess the pharmacokinetics (PK) of free and encapsulated cytarabine in ventricular and lumber CSF after DepoCyt administration by the intraventricular route of administration or by lumbar puncture.

The pivotal (Phase III) efficacy trial focused on drug safety and efficacy specifically for the clinical dose established as a result of the Phase I study. Details of the efficacy study findings will be published elsewhere, however an interim analysis of the Phase III clinical findings are given here for solid tumor metastases carcinomatous meningitis.

II.3. Methodology

Each vial of DepoCyt contains 50 mg of cytarabine at a concentration of 10 mg/mL encapsulated in the DepoFoam sustained release particles and suspended in 0.9% preservative-free saline.

In the Phase I study, 19 patients received multiple doses of 12.5, 25, 37.5, 50, 75, or 125 mg DepoFoam encapsulated cytarabine, by the intrathecal route. Doses were 2 or more weeks apart. PK data were available from 15 patients with ages ranging from 6–73 years (mean age was 44 ± 18 years). CSF samples were collected either by ventricular (14 patients) or lumbar (10 patients) routes. CSF and plasma samples were collected at various times up to 21 days post-dosing, were analyzed for free (unencapsulated) and encapsulated cytarabine by HPLC or LC/MS/MS. In addition, DepoFoam particle counts in the CSF were determined microscopically.

The Phase III trial was an open label study in patients with neoplastic meningitis (NM). The part of the study which will be described is for patients with carcinomatous meningitis (CM) confirmed by CSF cytology. Patients with solid tumors were randomized to DepoCyt or MTX standard therapy with dosing via the ventricular route (through a previously-implanted Ommaya reservoir) or by lumbar puncture. Patients in the DepoCyt treatment arm received DepoCyt at 50 mg (as a bolus injection administered over 5 min or less) every 14 days for 2 doses (induction); and 50 mg every 14 days for 3 doses, and 50 mg every 28 days for 1 dose (consolidation). Those patients with solid tumors randomized to standard MTX therapy received 10 mg of drug twice a week for 28 days (induction). Those responding at the end of induction were to receive 10 mg once every 7 days (5 doses), followed by 10 mg once every 14 days (3 doses). In addition the protocol called for all patients to be given dexamethasone (DM), 4 mg bid × 5 days, to prevent chemical arachnoiditis.

All CSF and plasma samples were collected in tubes containing tetrahydrouridine (THU) at a final concentration of 40 μmol/L to prevent in vitro catabolism of cytarabine to uracil arabinoside (ara-U) by cytidine deaminase.[10] For those patients in which PK data was to be collected (8 in all), two mL of CSF was collected at each time point, and analysed for free and encapsulated cytarabine, ara-U, protein, glucose and cell counts (RBC and WBC). Samples for pharmacokinetic evaluation were collected on both the first and second treatment cycles. Active follow-up consisted of evaluation of CSF cytology, physical examinations, neurological examinations, laboratory tests, and evaluation of adverse experiences.

Complete response rate was the primary endpoint of the Phase III study. Complete response was prospectively defined as conversion from a positive examination of CSF for malignant cells to two consecutive negative cytological examinations taken at least 3 days apart after 2 doses of DepoCyt without any evidence of clinical disease progression by neurological examination. Retrospectively, complete response was expanded to include patients with single post-treatment negative cytological examination or with a negative cytological examination obtained after ⩾3 doses of DepoCyt. An independent cytopathologist, blinded to study drug treatment and the chronology of CSF samples, reviewed all CSF cytology slides after patients completed the study; this evaluation was used for determination of complete response.

II.4. Results/discussion

II.4.1. Phase I/II safety/dose-finding study

In the Phase I dose-escalation study[2,11,12] patient dosing ranged from 12.5 to 125 mg of DepoCyt, although most of the pharmacokinetic data available are for the 75 mg dose (Table 1).[13,14] For this dose, intraventricular and lumbar injections of DepoCyt both resulted in peaks of free cytarabine within 5 hours in both the ventricular and lumbar spaces (Table 1). The peaks were followed by a roughly biphasic decline consisting of an initial sharp decline and subsequent slower decline with a terminal-phase half-life of 100 to 263 h. Encapsulated cytarabine concentrations and MVL particle counts followed a similar pattern.

The PK findings are significant in that free cytarabine concentrations equal to or higher than 0.1 μg/mL were maintained in both the ventricular and lumbar sites for 2 weeks (336 h) after ventricular injections of 25–125 mg DepoCyt or intralumbar injections of 75 mg DepoCyt. Typical minimum cytotoxic concentrations of cytarabine in CSF are reported to be between 0.02 and 0.1 μg/mL, depending on duration of exposure.[9,10,15] Irrespective of the route of CSF administration (ventricular or lumbar puncture), cytarabine concentrations were minimal in plasma, being undetectable or only detected sporadically at very low levels (0.3–5 ng/mL). Thus systemic exposure to cytarabine was minimal in patients administered DepoCyt by the intraventricular or lumbar routes in this study.

Dose limiting toxicity in the Phase I study was encephalopathy at the 125 mg dose level and the maximum tolerated dose was determined by investigators to be 75 mg when administered by the intraventricular or lumbar routes. The pattern of toxicity of DepoCyt was qualitatively similar to that found in previously published studies using unencapsulated cytarabine and included nausea, vomiting, headache, meningismus, fever, and encephalopathy. These toxicities were transitory and usually resolved in 1 to 7 days with a short course of coadministered dexamethasone.

For all patients, complete cytological responses in the Phase I/II study, were noted in 62.5% of patients (10 of 16 evaluable for response). There was a 62.5% complete response rate in the solid tumor group (5 of 8 patients), a 50.0% response rate in the lymphoma group (3 of 6 patients, 1 multiple myeloma and 2 AIDS-related lymphomas), and a 100% response rate in the leukemia group (2 of 2 patients). Although no patients showed neurological improvement, all 19 patients remained in relatively stable neurological status. The duration of responses ranged from 15 to 181 days with a median of 111 days.[2,11,12]

Based on the PK and toxicity data from the Phase I/II study, the dose chosen for Phase III study was 50 mg.

II.4.2. Phase III pivotal efficacy study

Final efficacy results are not yet available from this trial. However interim results analysed for 31 patients suggested that patients given DepoCyt (17 patients) had a higher response rate and longer survival than patients given MTX (18 patients). A PK study was also conducted as part of the pivotal efficacy study. The observed data were consistent with the half-lives (100–263 h) estimated in the Phase I study. The cytarabine levels in the second cycle of dosing were not visibly different from those in the first cycle. Free cytarabine CSF concentrations of approximately 0.02 μg/mL or higher were maintained for 2 weeks after a single intrathecal injection of 50 mg DepoCyt. Systemic cytarabine concentrations (<5 ng/mL) after

Table 1: Pharmacokinetic parameters for free and encapsulated cytarabine after IVT or lumbar administration of 75 mg of DepoCyt (from the Phase I study)

Dose route	Sample route	No. of patients	No. of samples	T_{max} (h)	C_{max} μg/mL	C_{min} (at 336 h) μg/mL	AUC (0–336 h) μg h/mL	AUC (0-∞) μg h/mL	$t_{1/2}$ (h)
Free									
IVT	Vent	9	114	0.25	257.75	0.49	838	999	218
IVT	Lumb	6	61	3	7.43	0.49	120	432	100
LP	Vent	7	94	3.25	6.10	0.38	288	333	263
LP	Lumb	8	105	0.25	225.42	0.74	2477	2729	238
Encap									
IVT	Vent	9	110	0.25	1460.25	0.38	4217	4542	122
IVT	Lumb	6	57	3.67	65.00	1.19	3753	3828	57
LP	Vent	7	91	3.25	77.00	0.54	2988	3043	59
LP	Lumb	8	107	0.25	2206.75	0.96	24344	24453	65

Cytarabine and ara-U analysis was carried out using LC-MS/MS on both CSF and plasma samples. The limit of detection (LOD) for cytarabine in CSF was 0.003 μg/mL for cytarabine, and 0.016 μg/mL for ara-U. In plasma, the LOD was 0.002 μg/mL for cytarabine, and 0.005 μg/mL for ara-U.

intrathecal doses of 50 mg DepoCyt in the Phase III trial are much lower than plasma cytarabine levels that are known to result in myelosuppression (50–100 ng/mL). Conventional systemic doses of cytarabine (100–200 mg/m^2/day) for several days can cause myelosuppression. As in the Phase I study, cytarabine concentrations in plasma were undetectable or were only detected sporadically (0.3–5 ng/mL). The expected corresponding plasma cytarabine concentrations after i.v. dosing are approx 0.05–0.1 μg/mL after a 50 mg dose of DepoCyt.

In terms of any possible toxicity arising from the DepoFoam carrier itself, a study was conducted in rats to determine the fate and metabolism of the predominant lipid component of the DepoFoam matrix (^{14}C-DOPC) and of the drug (^{3}H-cytarabine) administered as DepoFoam-encapsulated sustained release cytarabine.[16] In this study, the absorption, distribution, metabolism, and excretion (ADME study) of ^{14}C-DOPC and ^{3}H-cytarabine was determined after intrathecal injection of radiolabelled DepoFoam encapsulated cytarabine. The fate of the ^{3}H-cytarabine was as expected in that a high percentage of the original dose could be accounted for in the urine by 96 h with the excretion sustained out to 504 h. The fate of ^{14}C-DOPC was consistent with the hypothesis that, apart from a small percentage of the dose (<15%) which was incorporated into the CNS tissue early after injection, the majority of the ^{14}C-DOPC undergoes hydrolysis such that ultimately most of the radiolabel is expired by the animal as $^{14}CO_2$. Thus the data provide support for the contention that DepoFoam particles are biodegradable after intrathecal administration.

II.5. Conclusions

The administration of cytarabine in the form of DepoCyt, directly into the CSF is beneficial for targeting the meninges, with minimal systemic exposure of cytarabine. Because of the sustained release of cytarabine from the particles, drug exposure is prolonged over time, resulting in lower peak cytarabine levels compared with standard dosing with cytotoxic agents. In patients with carcinomatous meningitis, intrathecally administered DepoCyt provides a more convenient dosing schedule than MTX, with response rates that may be at least comparable with MTX. DepoCyt administration may also result in a longer time to progression of key signs and symptoms of the disease and perhaps longer survival than conventional intrathecal chemotherapy.

III. DepoFoam™ encapsulated sustained release amikacin (C0201)

III.1. Introduction

C0201 is multivesicular lipid-based particulate suspension of the aminoglycoside antibiotic amikacin, which is intended for the local treatment of soft tissue and closed body compartment infections caused by micro-organisms susceptible to amikacin. Localized bacterial infections occur in a wide variety of clinical settings. Modern clinical practice with its increased reliance on implanted devices, long-term intravenous catheters, and chemotherapies (with associated immunosuppression and direct toxicities), assures that such soft tissue infections will remain a significant problem despite the efficacy of available therapeutic and prophylactic modalities. The goal of anti-microbial therapy for such infections is to maintain effective free drug concentrations in the contaminated or infected tissue long enough to eliminate all pathogens. Since the length of time the drug concentration is above the minimum inhibitory concentration (MIC) has been shown to be the most significant parameter determining the efficacy of antibiotics,[17] effectiveness of antibiotics administered directly at the site of the infection is limited by their rapid clearance from the local site and into the systemic circulation.[18] In most situations, effective concentrations are present for only a few hours.[19]

Amikacin was selected for encapsulation because it is particularly effective against staphylococci and the gram-negative bacteria which commonly cause localized infections. However its utility is limited by significant risk of nephro- and oto-toxicity attributable to exposure to high levels of circulating drug. By providing effective local amikacin concentrations and low systemic drug levels, C0201 may make it possible to avoid these toxicities.

Some potential therapeutic targets for a sustained-release formulation of amikacin are:

Surgical wound infections: These are the third most frequent nosocomial infections in most hospitals. They often result in significant prolongation of hospital stay and may require additional surgical procedures. Amikacin can beneficially impact such infections.[20]

Prophylaxis of surgical wound infections or body cavities contaminated by surgery or trauma: A potential exists for the prophylactic treatment of incisions during surgery, wounds produced by contaminated foreign bodies, etc.

Infections related to implanted foreign bodies and soft tissue infections related to catheters: Infected arthroplasties have been successfully treated with amikacin delivered by an implanted pump,[21] an expensive and invasive method to achieve analogous kinetics.

Infections of body compartments not readily accessible to IV route of administration such as Osteomyelitis,[22] bacterial endophthelmitis[23] gram-negative meningitis.[24]

III.2. Overview of the clinical/preclinical studies

C0201 is designed to provide a sustained release of amikacin over a 7 to 14 day period after subcutaneous (sc) administration. These parameters were tested in preclinical toxicity testing/PK and efficacy studies and in a Phase I clinical study. In addition to a brief overview of preclinical toxicity and PK findings, two efficacy models are described: a mouse model of soft tissue bacterial infection[4,25] and a vascular graft infection model in rabbits.[26] The Phase I clinical trial was a double blind, randomized, placebo-controlled, dose-escalating study of the safety, tolerability, and pharmacokinetics of a single dose of DepoFoam encapsulated amikacin by the subcutaneous route in healthy volunteers. The objectives of the study were to: (i) Assess safety and tolerability of a single dose of C0201 administered by the sc route in doses up to 240 mg, or the maximum tolerated dose whichever came first; (ii) Assess the serum and urine levels of amikacin following sc administration of C0201.

III.3. Methodology

C0201 formulation is manufactured with amikacin at a final concentration of in the product of 15 mg/mL. The lipid composition and manufacturing process is similar to that described above for DepoCyt.

Efficacy of C0201 *in the "foreign body infection* (*FBI*) *model*": A piece of Teflon tubing 1 cm in length is implanted into the subcutaneous tissue in mice and 3 days later, the local site inoculated with 10^7–10^8 CFU of Staphylococcus aureus. Inoculation was followed by either no treatment or a local injection with C0201 (1 mg amikacin), free amikacin sulfate (1 mg), "blank" DepoFoam, or systemic free amikacin sulfate (1 mg; administered IV). Animals were sacrificed 10 days after implantation of the foreign body. Infection was measured by excising the foreign body from the infection site and culturing in liquid media for 7 days. CFU were also determined directly from plating homogenized excised tissue.[25]

Efficacy of C0201 *in the vascular graft infection model*: A rabbit model was developed in which a vascular graft prosthesis was placed as a 3 mm PTFE interposition graft in a 1 cm segment of distal descending aorta. This surgical field was infected with application of suspended Staphylococcus aureus (10^5 to 10^8 CFU) directly onto the graft. Nineteen rabbits underwent contaminated aortic graft placement with no subsequent local or systemic antibiotic therapy (infected controls). Twelve rabbits underwent contaminated aortic graft placement with 2.5 mL of C0201 applied directly onto the graft. All rabbits were observed for morbidity/ mortality over 2 weeks and survivors sacrificed to evaluate graft infection.[26]

Toxicological evaluations of C0201 were conducted in two species, mouse and dog. Animals were given doses of up to 250 mg/kg (mice) and 40 mg/kg (dog). Dosing was weekly for 28–29 days (5 doses). Standard toxicity testing was carried out with special emphasis on potential damage to kidneys or hearing because of the known toxicities of the drug.

Phase I clinical trial evaluated the tolerance of C0201 administered subcutaneously in 30 normal volunteers who each received a single administration of C0201 at doses of 15, 30, 60, 120, or 240 mg, placebo (saline control), or blank DepoFoam (without active ingredient). All doses were diluted with 0.9% normal saline in a total volume of 16 mL, including blank DepoFoam and saline control doses. Each dose was administered in a series of 8 subcutaneous injections at once, infused in a circular pattern into the lower abdomen. Active follow up consisted of physical examinations, laboratory tests, and evaluation of adverse events, including hearing tests.

III.4. Results of preclinical/phase I studies

In the *foreign body infection (FBI) model*, C0201 inhibited S. aureus colonization of murine soft tissue surrounding an implanted Teflon body and inhibited infection of the foreign body in a dose dependent manner.[25] C0201 exhibited activity in this model at a dose of 0.1 mg per mouse (or ~5 mg/kg). Although amikacin in not commonly used in the clinic for treatment of gram-positive bacterial infections,[27] it was found to be very effective in this mouse model when administered in the DepoFoam vehicle. This result coupled with previous findings that local or systemic administration of free amikacin was not effective in this model,[25] supports the contention that C0201 injected at a site of subcutaneous infection, can enhance the therapeutic effect of the antibiotic. In the *vascular graft model*,[26] a significantly lower rate of post operative infection and death was noted in rabbits treated locally with C0201 (25%) versus untreated animals (63%) with $p < 0.05$. Infection was evident by gross retroperitoeal abscess at sacrifice and/or positive bacterial cultures. Cultures were also used to verify the absence of organisms in those rabbits without gross infections. Thus C0201 applied locally to contaminated aortic PTFE grafts increased survival and decreased the incidence of postoperative graft infections. Adjunctive use of sustained release locally-applied C0201 in contaminated or infected vascular beds may allow safer clinical use of prosthetic vascular grafts.[26]

Toxicology studies in mice indicated no ototoxicity and only indications of a slight nephrotoxic response after repeated dosing at the highest dose levels (250 mg/kg). No histopathologic evidence of renal damage was observed even at the highest doses. Comparable results were obtained for dogs given weekly doses of 40 mg/kg C0201 for 28 days.

In the *Phase I study*, C0201 was well-tolerated at all dose levels. The most frequently reported drug-related adverse event was mild rash (erythema), most commonly observed at the site of the injection, in 7 of 20 (35%) of C0201-treated subjects. It was noted within 15 minutes of injection and resolved within 1 hour. Mild injection site reactions were noted in 13 of 20 of the C0201-treated subjects in 3 of 5 (60%) of placebo subjects, and in 2 of 5 (40%) of blank DepoFoam-treated subjects. These reactions were noted within 15 minutes of the injection and resolved within 2 hours. The adverse reactions were noted with similar frequency in all groups, including placebo controls, and were therefore more likely attributable to the procedure (subcutaneous injection) than to C0201 or blank DepoFoam. There were no clinically significant laboratory abnormalities or vital signs.

III.5. Conclusions

Clinical results show that C0201 is well tolerated at all dosage levels studied in human subjects. Efficacy findings in animal models show that C0201 provides potential therapeutic benefit from sustained release of amikacin at a site of local infection while reducing overall systemic exposure.

IV. DepoFoam™ encapsulated sustained release morphine (C0401)

IV.1. Introduction

It is well accepted by physicians that patients whose pain is well controlled recover more quickly. Poor management of pain not only requires longer periods of hospitalization, but also significant additional attention by nurses and physicians. Thus, more efficient management of pain can be a positive factor in reducing hospital care costs. The therapeutic goal for C0401 is to develop a single epidural injection that will provide a safe and effective alternative to repeated epidural injections, epidural infusions or systemic opioid therapy as a treatment for acute post operative pain. The potential advantages of C0401 over conventional morphine therapy include reduced need for physician intervention, reduction of side effects associated with epidural catheters, and improved duration of analgesia provided by sustained levels of morphine at the localized site of action.

Morphine itself is a frequently used and effective opiate analgesic for the management of postoperative pain. Current use of morphine in pain management modalities has a number of disadvantages. For example, the use of indwelling epidural catheters require significant attention from an anesthesiologist, and patients may not get all the pain relief they need because they have to wait for the anesthesiologist to adjust the medication. Also, the catheters can become dislodged and the tips may break off. A major advantage of C0401 would be the elimination of the need for an indwelling catheter which would reduce the incidence of these problems as well as the incidence of infection.

IV.2. Overview of the clinical/preclinical studies

DepoFoam encapsulated sustained release morphine has been tested in preclinical studies in both rat and dog models. The preclinical studies in the rat were designed to determine whether a single dose of epidurally administered C0401 could provide sustained analgesia. Preclinical studies in the dog were designed to determine the pharmacokinetics of epidural C0401 in order to define the minimum drug exposure intervals and the maximum tolerable dose, and to determine the safety of repeated delivery of the maximum tolerable dose of epidural C0401. These studies were used as the basis for design of a Phase I clinical study.

The Phase I study, was designed to assess safety and antinociceptive action of a single dose of C0401 administered epidurally up to a maximum tolerated dose, in normal, healthy volunteers. The principal objectives of Phase I investigations are to: determine the safety of a single dose of C0401 administered epidurally; determine the plasma, CSF, and urine PK profiles of morphine and its metabolites following epidural administration; make a preliminary characterization of the dose/antinociception activity relationship of C0401 using an experimental electrical pain model; and establish an appropriate dose range for evaluation in further studies.

IV.3. Methodology

IV.3.1. Preclinical studies in a rat model

Initial studies were conducted in a rat model to determine the effectiveness of a single epidural dose of DepoFoam encapsulated morphine sulfate and to look for any supraspinal toxic effects.[5] Rats were implanted with an epidural catheter and after overnight recovery, animals were dosed with approx 50 μl of appropriately-diluted C0401 containing morphine sulfate at approx 25 mg/mL. Antinociception studies were performed by subjecting the animals to standard hot plate (52–53°C) testing.[28] Response latencies (in seconds) to nociception were measured from the time the animal was placed on the hot plate to the time when the animal either licked the hind paw or jumped. A cut-off time of 60 seconds was used to prevent damage to the footpad. Toxicity monitoring included measurement of hemoglobin oxygen saturation by pulse oximetry and observations for catalepsy or absence of corneal reflex as signs of supraspinal toxicity.[5]

IV.3.2. Preclinical studies in a dog model

To determine epidural pharmacokinetics of C0401, dogs were prepared with chronic lumbar intrathecal and epidural catheters.[29] After a 3 day recovery period, dogs received an epidural injection of morphine (5 mg) followed 2 days later by an epidural injection of C0401 (30 mg). Venous blood and lumbar CSF were sampled at intervals after each drug injection.

Epidural safety of C0401 was evaluated as follows: After 7 days of conditioning during which baseline data was collected, dogs (9 male/9 female) were surgically prepared with lumbar epidural catheters. After a 2-day recovery, animals were assigned to receive one of three treatments a total of 4 times at 8 day intervals. Epidural treatments were: (a) C0401 (30 mg in 3 ml); (b) DepoFoam vehicle; or (c) saline. Behavioral and physiological effects were monitored during the study. On day 25 or day 32 (one dog from each group) animals were anesthetized and underwent whole body perfusion for histopathological examination and necropsy.[29]

IV.4. Results of preclinical studies

IV.4.1. Studies in a rat model

The data described by Kim et al.,[5] show that single epidural doses of DepoFoam encapsulated morphine result in equivalent time to peak analgesia compared with free morphine sulfate at doses ranging from 10 to 250 μg per animal. However the duration of analgesia was significantly prolonged, with peak analgesia occurring 60 minutes after a single epidural dose and then gradually decreasing throughout the next several days. The area under the analgesia effect vs time curve was increased 3 to 19-fold compared with morphine sulfate. Hemoglobin oxygen saturation was decreased minimally after administration of DepoFoam encapsulated morphine and the incidences of catalepsy and loss of corneal reflex were minimal even at large doses. In contrast, the larger doses of morphine sulfate significantly decreased hemoglobin oxygen saturation and caused catalepsy and loss of corneal reflex. For a 250 μg dose, the peak cisternal CSF and serum morphine concentrations after epidural administration of DepoFoam encapsulated morphine were 32% and 6% respectively of that following morphine sulfate epidural administration. The terminal half-life for C0401 was increased 32-fold in the CSF compared with morphine sulfate.

IV.4.2. Studies in a dog model

In the four week dog toxicity study, all dogs survived to scheduled date of necropsy. Observed behavioral changes included: alertness, coordination, and muscle tone decreased within 4 to 6 h of each injection lasting up to 72 h in C0401 group only; respiratory rate was decreased after each injection lasting up to 72 h in C0401 group only. No change was observed in the heart rate. Clinical pathology measurements indicated no drug-related changes in hematology, serum chemistry, urinalysis, heart rate, or CSF values. The histological findings showed that all animals had evidence of mild to moderate inflammation in the epidural space. Minimal changes were observed in the intrathecal space. The relatively modest reactive changes were not unexpected given the presence of chronically implanted catheters in the epidural and intrathecal spaces. The concentration of the drug in serum 24 h after each injection of C0401 was 3.7 to 10.1 ng/mL for males and 9.3 to 43.5 ng/mL for females (reflecting gender-related differences in body weight). Concentration of drug in cisternal CSF at 24 h after the last injection of C0401 ranged from 1656 to 584 ng/mL in all treated animals reflecting a greater than 10-fold concentration differential between lumbar CSF and placenta. Overall the study showed that multiple epidural injections (a total of 4 doses at 8 day intervals) of C0401 in dogs at the maximum injectable dose (30 mg/3mL) resulted in systemic effects typical of epidural morphine and minimal local effects (inflammation) with no evidence of spinal cord toxicity.[29]

IV.5. Conclusions

The studies described above indicate that the DepoFoam encapsulated form of morphine, C0401, is capable of sustained release of morphine sulfate over a 3–4 day period with minimal local effects. On the basis of these studies an IND was filed and a Phase I clinical trial was initiated in early 1997.

V. Other molecules

Although only cytarabine, amikacin, and morphine have entered the clinic as of writing, many other drugs are possible candidates for this technology. These include bupivacaine as a single-use sustained release depot formulation for local infiltration. The primary target patient population would be patients undergoing surgical procedures under local anesthesia. Other small molecules in the feasibility stage include clonidine, an $\alpha 2$ adrenergic agonist, for management of intractable chronic pain for a defined period of time following intrathecal injection; low-molecular weight heparin (LMW-Heparin); and several proprietary compounds being pursued with corporate partners.

A number of macromolecules including proteins have been successfully encapsulated in DepoFoam sustained release formulations. Proteins, peptides and nucleic acid based therapeutic products are particularly well suited to this technology since they are all water soluble compounds and very potent. However their shortcoming as therapeutic agents is their very limited in vivo half lives. Insulin-like Growth Factor-1 (IGF-1), interleukin-2 (IL-2), the antisense oligonucleotide, ISIS 2922, and various antigens for use in vaccine development have been formulated successfully and are being tested in model systems.

The protein therapeutic furthest along in the development process is DepoFoam encapsulated sustained release IGF-1, which is being developed together with Chiron Corporation. IGF-1 is being evaluated for its therapeutic benefit in several disease states including amyotrophic lateral sclerosis (ALS), diabetes, and acute renal failure. DepoFoam sustained-release formulations of IGF-1 have been successfully scaled up to allow cGMP production at a small scale, and the product is currently in GLP toxicology studies.

References

1. Kim S, Kim DJ, Geyer MA, Howell SB. Multivesicular liposomes containing 1-β-D-arabinofuranosylcytosine for slow-release intrathecal therapy. Cancer Research 1987;47:3935–3937.
2. Kim S. Liposomes as carriers of cancer chemotherapy. drugs 1993;46(4):618–638.
3. Spector MS, Zasadzinski, JA, Sankaram MB. Topology of multivesicular liposomes, a model biliquid foam. Langmuir 1996;12:4704–4708.
4. Grayson LS, Hansbrough JF, Zapata-Sirvent, Ramon; Roehrborn AJ, Kim T, Kim S. Soft tissue infection prophylaxis with gentamicin encapsulated in multivesicular liposomes. Critical Care Medicine 1995;23:84–91.
5. Kim T, Murdande S, Gruber A, Kim S. Sustained-release morphine for epidural analgesia in rats. Anesthesiology 1996;85:331–8.
6. Grayson LS, Hansbrough JF, Zapata-Sirvent RL, Kim T, Kim S. Pharmacokinetics of DepoFoam gentamicin delivery system and effect on soft tissue infection. J Surg Res 1993;54:1–6.
7. Kim S. DepoFoam-mediated drug delivery into cerebrospinal fluid. Methods in Neurosciences 1994;21:118–131.
8. Shapiro WR, Young DF, Mehta M. Methotrexate: distribution in cerebrospinal fluid after intravenous, ventricular and lumbar injections. N Engl J Med 1975;293:161–166.
9. Graham FL, Whitmore GF. The effect of 1-β-D-arabinofuranosyl-cytosine on growth, viability, and DNA synthesis of mouse L-cells. Cancer Res 1970;30:2627–2635.
10. Zimm S, Collins JM, Miser J et al. Cytosine arabinoside cerebrospinal fluid kinetics. Clin Pharmacol Ther 1984;35:826–830.
11. Chamberlain MC, Khatibi S, Kim JC et al. Treatment of leptomeningeal metastasis with intraventricular administration of depot cytarabine (DTC 101). A phase I study. Arch Neurol 1993;50:261–64.
12. Kim S, Chatelut E, Kim JC, Howell SB, Cates C, Kormanik PA, Chamberlain MC. Extended CSF cytarabine exposure following intrathecal administration of DTC 101. J Clin Oncol 1993;11:2186–2193.
13. Chamberlain MC, Kormanik P, Howell S et al. Pharmacokinetics of intralumbar DTC-101 for the treatment of leptomeningeal Metastases. Arch Neurol 1995;52:912–917.
14. Kohn F et al: In preparation.

15. Paull KD. Activity of cytababine against tumor lines. Personal Communication, 1995.
16. Kohn F, Malkmus S, Brownson E, Rossis, Yaksh T. Fate of predominant component of DepoFoam^{TM} drug delivery matrix after intrathecal administration of sustained-release encapsulated cytarabine in rats. Drug Delivery 1998. In Press.
17. Vogelman B et al. Correlation of antimicrobial pharmacokinetic parameters with therapeutic efficacy in an animal model. J Infect Dis 1988;158:831–847.
18. Matushek KJ, Rosin E. Pharmacokinetics of cefazolin applied topically to the surgical wound. Arch Surg 1991;126(7):890–893.
19. Rosin E, Ebert S, Uphoff T, Evans MH, Schultz-Darken NJ. Penetration of antibiotics into the surgical wound in a canine model. Antimicrob Agents Chemother 1989;33:700–704.
20. McEvoy GK. Antibiotics-aminoglycosides. AHF's drug information. American Society of health-system pharmacists 1995;54–61.
21. Perry CR, Hulsey RE, Mann FA, Miller GA, Pearson RL. Treatment of acutely infected arthroplastises with incision, drainage and local antibiotics delivered via an implantable pump. Clin Orthopaedics Related Res 1992;281:216–223.
22. Perry CR, Davenport K, Vossen MK. Local delivery of antibiotics via an implantable pump in the treatment of osteomyelitis. Clin Orthopaedics Related Res 1988;226:222–230.
23. Talamo JH, D'amicom DJ, Kenyon KR. Intraviteral amikacin in the treatment of bacterial endophthalmitis. Arch Ophthamol 1986;104:1483–1485.
24. Wright PF, Kaiser AB, Bowman CM, McKee Jr, Trujillo H, McGee ZA. The pharmacokinetics and efficacy of an aminoglycoside administered into the cerebral ventricles in neonates. J Inft Dis 1981;143:141–147
25. Roerhborn A, Hansbrough JF. Gualdoni B, Kim S. Lipid-based slow-release formulation of amikacin sulfate reduces foreign body-associated infections in mice. Antimicrob Agents Chemother 1995;39:1752–1755
26. Huh J, Chen JC, Kafie F, Furman GM, Milliken JC, Wilson SE. A multivesicular liposomal amikacin formulation in the treatment of prosthetic vascular graft infection: In preparation
27. Physicians Desk Reference, 49th Edition 1995;518–520, 973–975.
28. Wallace MS, Yanes AM, Ho RJY, Shen DD, Yaksh TL. Antinoceiception and side effects of liposome-encapsulated alfentanil after spinal delivery in rats. Anesth Analg 1994;79:778–86
29. Yaksh T, Provencher J, Rathbun M, Myers R, Richter P, Kohn F. Evaluation of safety of epidural sustained release encapsulated morphine in dogs. Submitted.

Lasic and Papahadjopoulos (eds.), Medical Applications of Liposomes

CHAPTER 9.1

Future of liposome applications: Serendipity vs. design

Demetrios Papahadjopoulos[a] and Danilo D. Lasic[b]

[a]*Department of Cellular and Molecular Pharmacology, University of California, San Francisco, and California Pacific Medical Center Research Institute, San Francisco, CA 94115, USA*
[b]*Liposome Consultations, 7512 Birkdale Drive, Newark, CA 94560, USA*

The initial interest in liposomes as a model membrane system and also as a drug carrier was based on their innate properties. This is the part in liposome development that can be characterized as serendipity, and includes a long list of useful properties: self assembly to a closed, relatively permeable membrane system; recognition by the RES macrophage system in blood, which can result in antigen presentation, macrophage activation, macrophage killing, and elimination of intracellular parasitic infections; lowering of the surface tension in the lung alveoli; penetration into the skin through hair folicles. Most of these properties have been described in detail in this volume.

Liposomes may also exhibit a variety of other properties which are not innate, but can be introduced by design. Such properties can control the physicochemical characteristics of lipid bilayers and their interaction with the biological environment. We feel that this is the most interesting direction for the future of liposome applications. The list of such "properties by design" is impressive and we will mention a few as example of past accomplishments pointing the way to the future.

1. Control of liposome size, initially by sonication[1,2] more recently by extrusion[3,4] has provided control of circulation time and enhanced extravasation.
2. Control of liposome permeability by defined lipid composition (solid-fluid transitions,[5] cholesterol stabilization[6]) has advanced liposome stability in biological fluids[7] and has allowed for temperature[8,9] and proton[10–12] induced sensitivity.
3. Control of liposome aggregation and fusion has produced insights into membrane fusion phenomena[13] and resulted in cochleated structures capable of carrying antigens and DNA as vaccine formulations.[14]
4. Steric stabilization by surface grafted hydrophilic molecules has changed entirely and beneficially the pharmacokinetics and tissue disposition of liposome and their contents.[15]
5. Remote loading methods have allowed stable encapsulation at high drug to lipid ratios with important practical applications.[16,17]

6. Formulation of deformable liposomes by inclusion of certain amounts of detergent molecules has allowed skin penetration at reasonably high rates for certain pharmaceuticals.[18]
7. Ligand directed targeting of liposomes, combined with steric stabilization have increased the potential for reaching efficiently specific cells and tissues in vivo, with many foreseeable applications (chapters in section 4).
8. Cationic lipid-DNA complexes, are potentially an excellent candidate as a non-viral vector for gene therapy. Especially when coupled with steric stabilization and ligand-directed targeting, such particles may act like artificial virus and have a high potential in a variety of medical applications (chapters in section 5).

If commercial activity is a measure of success for a particular discipline, we can safely say that medical applications of liposomes (and related lipid-based drug delivery systems) have finally matured into a respected and profitable enterprise. Doxorubicin encapsulated in sterically stabilized liposomes and daunorubicin encapsulated into small conventional liposomes are commercially available in US, and physicians in Europe can select among three lipid-based Amphotericin B formulations, two of which are also available in USA. In addition, several smaller scale anticancer preparations and liposomal vaccines are available in some European countries.

Currently, there are 6 larger liposome enterprises (50–300 people),[19] several smaller ones, as well as half a dozen companies which are exploiting lipid based carriers for gene therapy.[20] In addition to the liposomal products already commercially available, several are in advanced phases of clinical trials. These include encapsulated drugs doxorubicin and prostaglandin E1 by the Liposome Company, Inc.; cisplatin by Sequus Pharmaceuticals, Inc.; amikacin by NeXstar; vincristine by Inex and nystatin, tretinoin, and annamycin by Aronex. Several cationic liposomes have already been used in Phase I trials, and one in Phase II, for gene therapy.

This impressive development is mostly due to rational design of liposomal formulations based on biophysical and biochemical studies of lipid bilayers. While the first medical applications were undoubtedly attempted too far ahead of their time to yield valuable therapies, the safety of liposome administration was established. Subsequent work on stabilization of liposomes and control of their interaction characteristics contributed critically for the development of the above mentioned commercial applications. Technology already developed for these formulations is being transferred to other drugs with similar physico-chemical or biological characteristics. An example is transfer of the development of epirubicin encapsulated into sterically stabilized liposomes to doxorubicin (Doxil), and this technology to cisplatin, respectively.

Longer term improvements will be based on molecular design of lipid molecules leading to liposomes with specifically pre-determined properties. However, such programmable liposomes will not be based only on protective coating and targeting ligands. In drug delivery, two important problems are spatial and temporal control of drug release. Temperature- and proton-sensitive liposomes have already been used as examples of triggered release. Because these two principles are not always operational and/or effective, novel approaches for triggered bilayer destabilization are being sought. Lipid instabilities may depend on lamellar-micellar or lamellar-hexagonal II phase transitions, either due to hydrolysis of lipids or detachment of lipid bound polymers, or simply dissociation of molecules with higher value of critical micelle concentration. Chemical cleavage of a specific bond, either fatty acid to the lipid backbone, or polymer to the lipid, can be based on reaction kinetics, or can be induced by local conditions (pH), or triggered externally by heat or radiation. Upon such reaction, bilayer-forming lipid transforms into micelle-forming lipid (either into normal or inverse) and the original bilayer becomes much more leaky, or disintegrates or fuses with an adjacent membrane. Other liposome instabilities can be based on osmotic pressure or membrane asymmetry, which can be sustained for some time by a pH gradient across the closed membrane.

Loss of surface-attached polymers can induce lamellar-hexagonal II phase transition or stealth-non-stealth transition (steric destabilization), which can expose various reactive groups.[21] As we have pointed out, this transition can occur either by chemical cleavage of the linker,[21] physical dissociation of PEG-lipid molecule from the bilayer[22] or due to random coil-helix/condensation phase transition of the polymer chain.[23] In the case of surface attached polyelectrolyte, pH changes can induce membrane leakage and/or fusogenic activity.

Other chemical designs include (partial) lipid polymerization, inclusion of liposomes into programmable biogels[24] or formation of supra liposomal aggregates.[25] In the latter case, chemically linked individual liposomes could be prepared, which carry various functions, such as cytotoxic load, targeting, steric protection and others.[25] Improved mechanical and colloidal stability, as well as incorporation into specific capsules, can also render liposomes suitable for oral delivery. Some reports are already emerging showing activity of liposome-encapsulated agents upon absorption through gastrointestinal tract.[26] As in anticancer therapy, where liposome forming lipids which are themselves tumoricidal have been synthesized and used in chemotherapy,[27] it is possible that some lipids can be found which exhibit germicidal activity.

It is clear that, as with liposome applications which have already achieved success, future applications will depend on the proper biochemical and biophysical manipulation of the properties of liposomal lipids and their interactions with the biological milieu. In this respect, we are likely to depend heavily on the fundamental research on membrane systems, which has been the driving force behind many of the applications discussed in this volume. While it took almost thirty years from Bangham's initial discovery to the first liposome products useful in medicine, we believe, as this volume can testify, that liposomes have finally come of age.

References

1. Papahadjopoulos D, Miller N. Phospholipid model membranes. I. Structural characteristics of hydrated liquid crystals. Biochim Biophys Acta 1967;135:624–638.
2. Huang CH. Studies of Phosphatidylcholine vesicles. Formulations and Physical Characteristics. Biochemistry 1969;8:344–354.
3. Olson F, Hunt CA, Szoka FC, Vail WJ, Papahadjopoulos D. Preparation of liposomes of defined

size distribution by extrusion through polycarbonate membranes. Biochim Biophys Acta 1979;557:9–23.

4. Hope MJ, Nayar R, Mayer LD, Cullis PR. Reduction of Liposome Size and Preparation of Unilamellar Vesicles by Extrusion Techniques. In: Liposome Technology, 2nd Edition. G. Gregoriadis, Ed. CRI Press, Boca Raton, FL, Volume 1, 1993;124–139.
5. Papahadjopoulos D, Jacobson K, Nir S, Isac T: Phase transitions in phospholipid vesicles: Fluorescence polarization and permeability properties concerning the effect of temperature and cholesterol. Biochim Biophys Acta 1973;311:330–348.
6. Papahadjopoulos D, Cowden M, Kimelberg HK: Role of cholesterol in membranes: Effects on phospholipid-protein interactions, membrane permeability and enzymatic activity. Biochim Biophys Acta 1973;330:8–26.
7. Mayhew E, Rustum Y, Szoka F, Papahadjopoulos D: Role of cholesterol in enhancing the antitumor activity of cytosine arabinoside entrapped in liposomes. Cancer Treat Rep 1979;63(11–12):1923–1928.
8. Weinstein JN, Magin RL, Yatvin MB, Zaharko DS: Liposomes and Local Hyperthermia: Selective Delivery of Methotrexate to Heated Tumors. Science 1979;204:188–191.
9. Huang SK, Stauffer PR, Hong K, Guo JWH, Phillips TL, Huang A, Papahadjopoulos D. Liposomes and Hyperthermia in Mice: Increased Tumor Uptake and Therapeutic Efficacy of Doxorubicin in Sterically Stabilized Liposomes. Cancer Research 1994;54:2186–2191.
10. Connor J, Yatvin MB, Huang L. pH-sensitive liposomes: Acid-induced liposome fusion. Proc Natl Acad Sci USA 1984;81:1715–1718.
11. Ellens H, Bentz J, Szoka FC: H^+ and Ca^{2+}-Induced Fusion and Destabilization of Liposomes. Biochemistry 1985;24:3099–3106.
12. Düzgünes N, Straubinger RM, Baldwin PA, Friend DS, Papahadjopoulos D. Proton-induced fusion of oleic acid/phosphatidylethanolamine liposomes. Biochemistry 1985;24:3091–3098.
13. Papahadjopoulos D, Nir S, Düzgünes N. Molecular mechanisms of calcium-induced membrane fusion. J Bioenergetics and Membranes 1990;22(2):157–179.
14. Mannino RJ, Gould-Fogerite S. Liposome Mediated Gene Transfer. Bio Techniques 1988;6(7):682–690.
15. Papahadjopoulos D, Allen T, Gabizon A, Mayhew E, Matthay K, Huang SK, Lee K-D, Woodle MC, Lasic DD, Redemann C, Martin FJ. Sterically stabilized liposomes: Improvements in pharmacokinetics, and anti-tumor therapeutic efficacy. Proc Natl Acad Sci USA 1991;88:11460–11464.
16. Mayer LD, Bally MB, Hope MJ, Cullis PR: Techniques for encapsulating bioactive agents into liposomes. Chem Phys Lip 1986;40:333–345.
17. Haran G, Cohen R, Bar LK, Barenholz Y. Transmembrane ammonium sulfate gradients in liposomes produce efficient and stable entrapment of amphiphilic weak bases. Biochi Biophys Acta 1993;1151:201–215.
18. Cevc G. Lipid suspensions on the skin. Permeation enhancement, vesicles penetration and transdermal drug delivery. Crit Rev Ther Drug Carr Syt 1996;13:257–388.
19. Medium-size liposome enterprises. Aronex, Woodlands, TX; DepoTech, San Diego, CA; Inex, Vancouver, BC; NexStar, Boulder, CO; Sequus, Menlo Park, CA; The Liposome Company, Princeton, NJ.
20. Among many companies which are investigating viral and non-viral gene delivery the following concentrate mostly on lipid-based non-viral systems. Genzyme, Farmington, MA, GeneMedicine, Woodlands, TX; Vical, San Diego, CA; Advanced Therapies, Novato, CA; Megabios, Burlingame, CA; Genetic Therapy, Inc., Gaithersburg, MD.
21. Kirpotin D, Hong K, Mullah N, Papahadjopoulos D, Zalipsky S: Liposomes with detachable polymer coating: destabilization and fusion of dioleoylphosphatidylethanolamine vesicles triggered by cleavage of surface-grafted poly(ethylene glycol). FEBS Lett 1996;388:115–118.
22. Holland JW, Cullis PR, Madden TD. PEG-lipid conjugates promote bilayer formation in mixtures of non-bilayer forming lipids. Biochemistry 1997;35:2610–2617.
23. Thomas LJ, Tirrell DA. Polyelectrolyte Sensitized Liposomes. Acc Chem Res 1996;25:336–342.
24. Cohen S, Alonso MJ, Langer R. Novel approaches to controlled-release antigen delivery. Int J Technol Assess and Health Care 1994;10:121–130.
25. Walker SA, Kennedy MT, Zaszadinski JA. Encapsulation of bilayer vesicles by self-assembly. Nature 1997;387:61–64.
26. Aramaki Y, Tomizawa H, Hara T, Yachi K, Kikuchi H, Tsuchiya S. Stability of liposomes in vitro and their uptake by rat Peyer's patches following oral administration. Pharm Res 1993;10:1228–12231.
27. Muschiol C, Berger MR, Schuler B, Scherf HR, Garzon FT, Zeller WJ, Unger C, Eibl HJ, Schmahl D. Alkyl phosphocholine: toxicity and anticancer properties. Lipids 1987;22:930–934.

Lasic and Papahadjopoulos (eds.), Medical Applications of Liposomes

Appendix 1. List of abbreviations

ABCD	amphotericin B colloidal dispersion
ABV	adriamycin bleomycin vincristine/vinblastine
AChE	acetylcholinesterase
AFM	atomic force microscopy
AIDS	auto immune deficiency syndrome
ALEC	artificial lung expanding compound
AM	alveolar macrophages
AmB	amphotericin B
APC	antigen presenting cell
AraC	cytosine arabinose
ARDS	adult respiratory distress syndrome
ASPA	aspartoacylase
AUC	area under curve
AZT	azidodeoxythymidine
BD	biodistribution
BM	bone marrow derived macrophages
BMP	maleimidopropionyl
BUN	blood urea nitrogen
BV	bleomycin vincristine/vinblastine
C	complement
CAT	chloramphenicol acetyltransferase
CF	carboxy fluorecscein
CF	cystic fibrosis
CFA	complete Freund's adjuvant
CFTR	cystic fibrosis transmembrane condustance regulator
CFU	colony forming units
CHEMS	cholesterol hemisuccinate
CHO	chinese hamster ovary
CHOL	cholesterol
CHOP	cyclophosphamide,doxorubicin,vincristine,prednisone
CHS	contact hypersensitivity
CL	cardiolipin
CLDC	cationic liposome/DNA complex
CLS	confocal laser microscopy
CMC	critical micelle concentration
CPD	cyclobutyl pyridimine dimer
CR	complete responder
CS	circumsporozoite
CsA	cyclosporin A
CSF	cerebrospinal fluid
CT	cholera toxin
CT	computer tomography
CTAB	cetyl trimethylammonium bromide
CTB	cholera toxin subunit B
CTL	cytotoxic T lymphocyte
D	dimensional
DAU	daunorubicin hydrochloride
DC	Chol - dimethylaminoethane carbamoyl cholesterol
DC	dendritic cells
DCP	dicetyl phosphate
DDAB	dioctadecylammonium bromide
DDAC	dioctadecylammonium chloride
ddC	dideoxycitosine
ddCTP	dideoxycytidine triphosphate
ddI	dideoxyinosine
ddUMP	dideoxyuridine monophosphate
ddUTP	dideoxyuridine triphosphate

DLVO	Derjaguin-Landau-Verwey-Overbeek
DMEM	Dulbeccos modified Eagle medium
DMPC	dimyristoyl phosphatidylcholine
DMPE	dimyristoyl phosphatidylethanolamine
DMRIE	dimethyl-tetradecyloxy-propanaminium bromide
DMSO	dimethyl sulphoxide
DNA	deoxyribonucleic acid
DOC	deoxycholate
DODAC	dioleoyldimethylammonium chloride
DOGS	dioctadecyl amido glycyl spermine
DOPC	dioleoyl phosphatidylcholine
DOPE	dioleoyl phosphatidylethanolamine
DOPS	dioleoyl phosphatidylserine
DOSG	dioleoyl succinyl glycerol
DOSPA	dioleyloxy-sperminecarboxamido ethyl-dimethyl-propanaminium trifluoroacetate
DOTAP	dioleoyl trimethylammoniopropane
DOTIM	octadecenoyloxy-ethyl-heptadecenyl hydroxyethyl imidazolinium chloride
DOTMA	dioleoyloxy propyl trimethylammonium chloride
DOX	doxorubicin hydrochloride
DPPG	phosphatidylcholine
DPPG	dipalmitoyl phosphatidyl glycerol
DPSG	dipalmitoyl succinyl glycerol
DRV	dehydration rehydration vesicles
DSC	differential scanning calorimetry
DSPC	distearoyl phosphatidylcholine
DSPG	distearoyl phosphatidyl glycerol
DSPS	distearoyl phosphatidylserine
DTPA	diethylene triamine pentaacetic acid
DTPA-PL-NGPE	dtpa polylysyl glutaryl phosphatidylethanolamine
dU	deoxyuridine
DXR	doxorubicin hydrochloride
EDTA	ethylene diamino acetic acid
EGF	epidermal growth factor
EIA	enzyme immunoassay
ELISA	enzyme linked immunoassay
EM	electron microscopy
EPC	egg phosphatidylcholine
ER	endoplasmic reticulum
ESR	electron spin resonance
EYPC	egg yolk phosphatidylcholine
FACS	fluorescence-activated cell sorting
FBI	foreign body infection
FCS	fetal calf serum
FDA	food and drug administration
FF	freeze fracture
FITC	fluorescein isothiocyanate
FU	fluoruracil
FUO	fever of unknown origin
G-CSF	granulocyte colony stimulating factor
gal	galactose
GDL	glyceryl dilaurate
GFP	green fluorescent protein
GI	gastro intestinal
GM-CSF	granulocyte-macrophage colony stimulating factor
GM1	ganglioside (GM1 type)
GMT	geometric mean titre
GOV	giant oligolamellar vesicle
GSH	glutathione
H	hexagonal
HA	hemagglutinin

HAMA	human mouse antibodies
HAV	hepatitis A virus
Hb	hemoglobin
HBBS	Hanks balanced buffer solution
HBsAg	hepatitis B surface antigen
HDL	high density lipoprotein
HIV	human immunodeficiency virus
HLA	human leukocyte antigen
HPC	hydrogenated phosphatidylcholine
HPI	hydrogenated phosphatidylinositol
HPLC	high pressure liquid chromatography
HSPC	hydrogenated soy phosphatidylcholine
HSV tk	herpes simplex virus thymidine kinase
HV	herpes virus
HZ	hydrazide
I	ionic strength
ICAM	intercellular adhesion molecule
IFN	interferon
IFX	ifosamide
Ig	immunoglubulin
IGF	insulin-like growth factor
IL	interleukin
IM	intamuscular
IRES	internal ribosomal entry sites
IRIV	immunopotentiating reconstituted influenza virosomes
IU	international unit
IV	intravenous
kD	kilo Dalton
KS	Kaposi's sarcoma
L	liposomal
LD50	50%-lethal dose
LDL	low density lipoprotein
LEH	liposome encapsulated hemoglobin
LOV	large oligolamellar vesicle
LPC	lyso phosphatidyl choline
LPDI	lipid polymer complex I
LPS	lipopolysaccharide
LS	lung surfactant
LTI	Liposome technology, Inc.
LTP	lipid transfer protein
LTS	longe time survivor
LUV	large unilamellar vesicle
LVEF	left ventricular ejection fraction
mAb	monoclonal antibody
MAC	*mycobacterium avium* complex
MALT	mucosa associated lymphoid tissue
MC	maleimidecaproyilo
MCAF	monocyte chemotactic and activating factor
MDP	multidrug resistance
MDP	muramyl dipeptide
MFC	minimum fungicidal concentration
MHC	major histocompatibility complex
MIC	minimum inhibitory concentration
MLV	multilamellar vesicle
MMAD	mass median aerodynamic diameter
MMC	maleimido methylcyclohexyl carboxylic acid
MMC	maleimidomethyl cyclohexane
MP	methylphosphonates
MP	maleimido propionic acid
MPB	maleimidophenyl

MPEG	methoxy(polyethylene) glycol
MPLA	monophosphoryl lipid A
MPS	mononuclear phagocytic system or RES
MR	magnetic resonance
MRI	magnetic resonance imaging
MRP	multidrug resitance protein
MST	median survival time
MTA	monocyte cytotoxicity assay
MTD	maximal tolerated dose
MTP	muramyl tripeptide
MTX	methatrexate
NA	neuraminidase
NDA	new drug application
NDDP	diaminocyclohexane platinum analog
NG	n-glutaryl
NHL	non-Hodgkins lymphoma
NIH	National Institute of Health
NLS	nuclear localization sequence
NMR	nuclear magnetic resonance
NT	neutralization test
ODN	oligodeoxylnucleotide
ori	origin of replication
OVA	ovalbumin
PA	phosphatidic acid
PAC	perturbed angular correlation
PB	Poisson Boltzmann
PBS	phosphate buffered saline
PC	phosphatidylcholine
PCR	polymerase chain reaction
PDP	pyrydyldithiopropionyl
PE	phosphatidylethanolamine
PEG	(polyethylene) glycol
PEO	polyethylene oxide
PG	phosphatidylglycerol
PGE1	prostaglandin E1
PGK	phosphoglycerate kinase
PGP	P-glycoprotein
PHC	palmitoyl homocysteine
PI	phosphatidylinositol
PK	pharmacokinetics
PKC	phosphokinase C
PLL	poly-L-lysine
PO	oral (per)
POPC	palmitoyl oleoyl phosphatidylcholine
PR	partial responder
PS	phosphatidylserine
PS	phosphorothioates
RBC	red blood cell
RCR	reconstituted chylomicron remnants
RDS	respiratory distress syndrome
RES	reticuloendothelial system (also MPS)
RIA	radioactive immunoassay
RLF	recombinant protein antigen
RNA	ribinucleic acid
S	phosphorothioate
SA	stearylamine
SAR	structure-activity relationships
SATP	succinimidyl acetylthio propionate
SAXS	small angle x-ray scattering
SC	subcutaneous

SCID	severe combined immunodefiency
SIL	sterically stabilized immunoliposome
SM	sphingomyelin
SMPB	succinimidyl maleimidophenyl butyrate
SOD	superoxide dismutase
SPDP	succinimidyl pyrydyldithio propionate
Sph	sphingomyelin
SPM	sphingomyelin
SSL	sterically stabilized liposome
SUV	small unilamellar vesicle
SV	simian virus
TAP	transport and antigen processing
TAP	triamcinolone acetonide phosphate
TCR	T-cell receptor
TEAP	tetraethylammoniumperchlorate
THU	tetrahydrouridine
TLC	The Liposome Company
TNF	tumor necrosis factor
TTC	triphenyl tetrazolium chloride
UFCT	ultrafast compoturized tomography
USP	United States Pharmacopeia
V	volume
VEGF	vascular endothelial growth factor
VL	visceral leishmaniasis
VLDL	very low density lipoprotein
WBC	white blood cells
WHO	world health organization
YAC	yeast artificial chromosome

Lasic and Papahadjopoulos (eds.), Medical Applications of Liposomes

Appendix 2. Primary amino acids: abbreviations and polarity

Amino acid	Long abbreviation	Short abbreviation	Polarity/charge
alanine	ala	A	nonpolar
arginine	arg	R	cationic
asparagine	asn	N	polar
aspartic acid	asp	D	anionic
cysteine	cys	C	polar
glutamine	gln	Q	polar
glutamic acid	glu	E	anionic
glycine	gly	G	polar
histidine	his	H	cationic
isoleucine	ile	I	nonpolar
leucine	leu	L	nonpolar
lysine	lys	K	cationic
methionine	met	M	nonpolar
phenylalanine	phe	F	nonpolar
proline	pro	P	nonpolar
serine	ser	S	polar
threonine	thr	T	polar
tryptophan	trp	W	nonpolar
tyrosine	tyr	Y	polar
valine	val	V	nonpolar

Lasic and Papahadjopoulos (eds.), Medical Applications of Liposomes

Contributors

J.P. Adler-Moore
California State Polytechnic University, Pomona, 3801 West Temple Ave., Pomona, CA 91768-4032, U.S.A.

T. Allen
Department of Pharmacology, 9–31 MSB, University of Alberta, Edmonton, Alberta, Canada T6G 2H7. E-mail: terry.allen@ualberta.ca

C. Alving
Department of Membrane Biochemistry, Walter Reed Army Institute of Research, Washington, D.C. 20307–5100, U.S.A.
E-mail: ALVING@WRAIR-EMHI.ARMY.MIL

A. Arkema
Department of Physiological Chemistry, Groningen–Utrecht Institute for Drug Exploration (GUIDE), University of Groningen, Ant. Deusinglaan 1, 9713 AV Groningen, The Netherlands.

I.A.J.M. Bakker-Woudenberg
Department of Clinical Microbiology, Erasmus University, P.O. Box 1738, 3000 DR Rotterdam, The Netherlands.

M.B. Bally
Division of Medical Oncology, British Columbia Cancer Agency, 600 West 10th Avenue, Vancouver, BC, Canada V5Z 4E6. E-mail: mbally@bccancer.bc.ca

A. Bangham
17 High Green, The Cottages, Great Shelford, Cambridge, CB2 5EG, England.
E-mail: Alec.bangham@wwr.co.uk

Y. Barenholz
Department of Membrane Biochemistry, Hebrew University – Hadassah Medical School, P.O. Box 12272, Jerusalem 91120, Israel. E-mail: yb@cc.huji.ac.il

N. Barois
Centre d'Immunologie INSERM-CNRS de Marseille-Luminy, Parc Scientifique et Technologique de Luminy, Case 906, 13288 Marseille Cedex 9, France.

R. Beissinger
Department of Chemical Engineering, Illinois Institute of Technology, 140 Perlstein Building, 10 West 33rd Street, Chicago, IL 60616, U.S.A.

C.C. Benz
Department of Medicine, Division of Hematology/Oncology, University of California, San Francisco, CA 94143–1270, U.S.A.

S. Bhattacharya
Department of Pharmacology, University of Pittsburgh, School of Medicine, W1351 Biomedical Science Tower, Pittsburgh, PA 15261, U.S.A.

G. Colbern
SEQUUS Pharmaceuticals, Inc., 960 Hamilton Court, Menlo Park, CA 94025, U.S.A.

D.J.A. Crommelin
Department of Pharmaceutics, Faculty of Pharmacy, University of Utrecht, P.O. Box 80.082, NL - 3508 TB Utrecht, The Netherlands.
E-mail: D.J.A.Crommelin@far.ruu.nl

P.R. Cullis
Inex Pharmaceuticals Corp., 100-8900 Glenlyon Parkway, Burnaby, BC Canada V5J 5J8. E-mail: pcullis@inexpharm.com

T. Daemen
Department of Physiological Chemistry, University of Groningen, Ant. Deusinglaan 1, 9713 AV Groningen, The Netherlands.

A. De Haan
Department of Physiological Chemistry, University of Groningen, Ant. Deusinglaan 1, 9713 AV Groningen, The Netherlands.

D. Donovan
Department of Molecular Immunology, Roswell Park Cancer Institute, Buffalo, NY 14263, U.S.A.

N. Düzgüneş
School of Dentistry, Department of Microbiology, University of the Pacific, 2155 Webster Street, San Francisco, CA 94115, U.S.A.
E-mail: nduzgune@uop.edu

J. Fareed
Department of Pathology, Loyola Medical Center, Maywood, IL, U.S.A.

I.J. Fidler
Department of Cell Biology - 173, M.D. Anderson Cancer Center, 1515 Holcombe Boulevard, Houston, TX 77030, U.S.A.
E-mail: ifidler@notes.mdacc.tmc.edu

E.A. Forssen
NeXstar Pharmaceuticals, Inc., 2860 Wilderness Place, Boulder, CO 80301, U.S.A.

J. Freitag
The Liposome Company, Inc., 1 Research Way, Princeton Forrestal Center, Princeton, NJ 08540–6619, U.S.A.

A. Gabizon
Sharett Institute of Oncology, Hadassah University Hospital, Jerusalem 91120, Israel. E-mail: Gabizon@Hadassah.org.il

R. Glück
Swiss Serum and Vaccine Institute, P.O. Box CH 3001, Berne, Switzerland.

D. Goren
Department of Oncology, Hadassah Medical Center, Jerusalem 91120, Israel.

G. Gregoriadis
Centre for Drug Delivery Research, The School of Pharmacy, University of London, 29–39 Brunswick Square, London WC1N 1AX, England.
E-mail: Gregoriadis@cua.ulsop.ac.uk

C.B. Hansen
Department of Pharmacology, University of Alberta, Edmonton, Canada AB T6G 2H7.

K. Hong
California Pacific Medical Center Research Institute, Liposome Research Laboratory, 2200 Webster St., 2nd Floor, San Francisco, CA 94115, U.S.A.
E-mail: keelung@cooper.cpmc.org

L. Huang
Department of Pharmacology, University of Pittsburgh, School of Medicine, W1351 Biomedical Science Tower, Pittsburgh, PA 15261, U.S.A.
E-mail: Leaf@Prophet.pharm.pitt.edu

A.S. Janoff
The Liposome Company, Inc., 1 Research Way, Princeton Forrestal Center, Princeton, NJ 08540–6619, U.S.A. E-mail: ajanoff@lipo.com

J. Jato
Department of Chemical Engineering, Illinois Institute of Technology, 140 Perlstein Building, 10 West 33rd Street, Chicago, IL 60616, U.S.A.

D. Kirpotin
California Pacific Medical Center Research Institute, Liposome Research Laboratory, 2200 Webster St., 2nd Floor, San Francisco, CA 94115, U.S.A.
E-mail: kizpo@cooper.cpmc.org

E.S. Kleinerman
Department of Cell Biology, UTMD Anderson Cancer Center, 1515 Holcombe Boulevard, HMB-173, Houston, TX 77030, U.S.A.
E-mail: ekleinerman@odin.mdacc.tmc.edu

D.D. Lasic
Liposome Consultations, 7512 Birkdale Drive, Newark, CA 94560, U.S.A.
E-mail: DAN.LASIC@roche.com

L. Leserman
Centre d'Immunologie INSERM-CNRS de Marseille-Luminy, Parc Scientifique et Technologique de Luminy, Case 906, 13288 Marseille Cedex 9, France.
E-mail: leserman@ciml.univ-mrs.fr

L. Lieb
University of Utah Health Sciences Center, 50 North Medical Drive, Salt Lake City, UT 84132, U.S.A.

G. Lopez-Berestein
M.D. Anderson Cancer Center, 1515 Holcombe, Box 60, Houston, TX 77030, U.S.A. E-mail: GLOPEZ@UTMDACC.MDACC.TMC.EDU

B. McCormack
Centre for Drug Delivery Research, The School of Pharmacy, University of London, 29–39 Brunswick Square, London WC1N 1AX, England.

D. McCormick
Life Sciences, IIT Research Institute, Chicago, IL 60616, U.S.A.

F.J. Martin
Sequus Pharmaceuticals, Inc., 1050 Hamilton Court, Menlo Park, CA 94025, U.S.A.

L.D. Mayer
British Columbia Cancer Agency, 600 West 10th Avenue, Vancouver Cancer Centre, Div. of Medical Oncology, Vancouver, BC, Canada V5Z 4E6.
E-mail: lmayer@bccancer.bc.ca

O. Meyer
California Pacific Medical Center Research Institute, Liposome Research Laboratory, 2200 Webster St., 2nd Floor, San Francisco, CA 94115, U.S.A.
E-mail: omeyer@itsa.ucsf.edu

C. Nicolau
Blood Research and Development Laboratory, Harvard Medical School, 1256 Soldiers Field Road, Boston, MA 02135, U.S.A.
E-mail: nicolau@cbr.med.harvard.edu

D. Papahadjopoulos
California Pacific Medical Center Research Institute, Liposome Research Laboratory, 2200 Webster St., 2nd Floor, San Francisco, CA 94115, U.S.A.
E-mail: papah@cooper.cpmc.org

J. Park
Department of Medicine, Division of Hematology/Oncology, University of California, San Francisco, CA 94143–1270, U.S.A.
E-mail: john_park@quickmail.ucsf.edu

R. Perez-Soler
Section of Experimental Therapy, Department of Thoracic/Head, and Neck Medical Oncology, M.D. Anderson Cancer Center, 1515 Holcombe Blvd., Box 80, Houston, TX 77030, U.S.A. E-mail: rperezso@notes.mdacc.tmc.edu

Y. Perríe
Centre for Drug Delivery Research, The School of Pharmacy, University of London, 29–39 Brunswick Square, London WC1N 1AX, England.

R.T. Proffitt
NeXstar Pharmaceuticals, Inc., 2860 Wilderness Place, Boulder, CO 80301, U.S.A.

M. Rao
Department of Membrane Biochemistry, Walter Reed Army Institute of Research, Washington, DC 20307–5100, U.S.A.
E-mail: Dr._Mangala_Rao@wrsmtp.ccmail.army.mil

D. Ruff
Perkin Elmer, Applied Biosystem Division, Foster City, California, U.S.A.

R. Saffie
Centre for Drug Delivery Research, The School of Pharmacy, University of London, 29–39 Brunswick Square, London WC1N 1AX, England.

P.G. Schmidt
Drug Delivery Research, NeXstar Pharmaceuticals, Inc., 2860 Wilderness Place, Boulder, CO 80301, U.S.A. E-mail: gmead-brill@nexstar.com

H. Schreier
Advanced Therapy, 371 Bel Marin Keys, #210, Novato, CA 94949, U.S.A.
E-mail: mlsb86a@prodigy.com

J. Senior
DepoTech Corporation, 10450 Science Center Drive, San Diego, CA 92121, U.S.A. E-mail: Judy_Senior@Depotech.com

V. Shankey
Department of Pathology, Loyola Medical Center, Maywood, IL, U.S.A.

Y. Shao
California Pacific Medical Center Research Institute, Liposome Research Laboratory, 2200 Webster St., 2nd Floor, San Francisco, CA 94115, U.S.A.

R. Sherwood
Life Sciences, IIT Research Institute, Chicago, IL 60616, U.S.A.

B. Sternberg
California Pacific Medical Center Research Institute, Liposome Research Laboratory, 2200 Webster St., 2nd Floor, San Francisco, CA 94115, U.S.A.
E-mail: brigitt@itsa.ucsf.edu

D.D. Stuart
Department of Pharmacology, University of Alberta, Edmonton, Canada AB t6F 2H7.

C.E. Swenson
The Liposome Company, Inc., 1 Research Way, Princeton Forrestal Center, Princeton, NJ 08540–6619, U.S.A.

V.P. Torchilin
Massachusetts General Hospital, Center for Imaging and Pharmaceutical Research, Department of Radiology, 149 13th Street, Charlestown, MA 02129–2060, U.S.A. E-mail: vladimir@cipr.mgh.harvard.edu

P. Uster
SEQUUS Pharmaceuticals, Inc., 960 Hamilton Court, Menlo Park, CA 94025–1430, U.S.A.

J. Vaage
Cancer Research Science, Department of Experimental Pathology, Roswell Park Cancer Institute, Elm and Carlton Streets, Buffalo, NY 14263–0001, U.S.A. E-mail: vaage@sc3102.med.buffalo.edu

E.W.M. van Etten
Department of Clinical Microbiology, Erasmus University, P.O. Box 1738, 3000 DR Rotterdam, The Netherlands.

E.C.A. van Winden
Department of Pharmaceutics, Utrecht Institute for Pharmaceutical Sciences (UIPS), Utrecht University, Utrecht, The Netherlands.

K.M. Wasan
Division of Pharmaceutics and Biopharmaceutilcs, Faculty of Pharmaceuticals Sciences, The University of British Columbia, 2146 East Mall Avenue, Vancouver, BC, Canada V6T 1Z3.

A. Wegmann
Swiss Serum & Vaccine Institute Berne, P.O. Box, CH-3001, Berne, Switzerland.

N. Weiner
College of Pharmacy, University of Michigan, Ann Arbor, MI 48109, U.S.A.
E-mail: nweiner@umich.edu

J. Wilschut
Laboratory of Physiological Chemistry, University of Groningen, Ant. Deusinglaan 1, 9713 AV Groningen, The Netherlands.
E-mail: j.c.wilschut@med.rug.nl

M. Woodle
Genetics Therapy, Inc., 938 Clopper Road, Gaithersburg, MD 20878, U.S.A.
E-mail: martin.woodle@pharma.novartis.com

L.L. Worth
Department of Pediatrics, Box 87, M.D. Anderson Cancer Center, 1515 Holcombe Boulevard, Houston, TX 77030, U.S.A.

P. Working
Sequus Pharmaceuticals, Inc., 960 Hamilton Court, Menlo Park, CA 94025, U.S.A.

S. Zheng
Department of Chemical Engineering, Illinois Institute of Technology, 140 Perlstein Building, 10 West 33rd Street, Chicago, IL 60616, U.S.A.

W. Zheng
California Pacific Medical Center Research Institute, Liposome Research Laboratory, 2200 Webster St., 2nd Floor, San Francisco, CA 94115, U.S.A.
E-mail: win10@itsa.ucsf.edu

Y. Zou
Section of Experimental Therapy, Department of Thoracic/Head, and Neck Medical Oncology, M.D. Anderson Cancer Center, 1515 Holcombe Blvd., Box 80, Houston, TX 77030, U.S.A.

N.J. Zuidam
Department of Pharmaceutics, Utrecht Institute for Pharmaceutical Sciences (UIPS), Utrecht University, Utrecht, The Netherlands.

Lasic and Papahadjopoulos (eds.), Medical Applications of Liposomes

Subject Index